Learning MASTERCAM MILL Step by Step

James Valentino
Joseph Goldenberg

Industrial Press

Library of Congress Cataloging-in-Publication Data

Valentino, James.
Learning Mastercam Mill step by step/ James V. Valentino, Joseph Goldenberg.
p. cm
ISBN 0-8311-3177-2
1. CAD/CAM systems. I. Goldenberg, Joseph. II. Title.

TS155.6.V346 2003
670'.285--dc22

2003056637

Learning Mastercam Mill Step by Step

Cover Design: Janet Romano
Managing Editor: John Carleo

Cover Illustrations: Courtesy of CNC Software Inc.

Industrial Press Inc.
200 Madison Avenue
New York, New York 10016

10 9 8 7 6 5 4 3 2

DEDICATION

To my wife, Barbara and my children, Sarah and Andrew.
--James Valentino

To my lovely wife, Erica and our children, Janet, Simon and Vicky, for their love, support and inspiration.
-- Joseph Goldenberg

CONTENTS

CHAPTER 5 GENERATING HOLE OPERATIONS IN 2D SPACE 5-1

CHAPTER 6 PROFILING AND POCKETING IN 2D SPACE 6-1

CHAPTER 7 EDITING MACHINING OPERATIONS VIA THE OPERATIONS MANAGER 7-1

CHAPTER 8 USING TRANSFORM TO TRANSLATE, ROTATE OR MIRROR EXISTING TOOLPATHS 8-1

CHAPTER 9 USING A LIBRARY TO SAVE OR IMPORT MACHINING OPERATIONS 9-1

PREFACE

The CNC programmer now has a powerful tool to assist in the job of creating and verifying part programs. *Mastercam* CNC software provides the programmer with a full array of easy to use features. The benefits of using *Mastercam* include: automatic calculation of toolpath coordinates, determination of speeds and feeds, animation of the machining process off-line without tying up the CNC machine, and posprocessing the part program.

Mastercam is a robust PC based package. Its many cababilities must be presented in a clear and logical sequence. This text was written to provide a thorough introduction to *Mastercam's* MILL package for students with little or no prior experience.

Several learning aids have been designed throughout .

- Good graphical displays rather than long text and definitions are emphasised
- An overview of the process of generating a word address program is presented
- Key definitions are boxed in
- Examples provide step by step instructions with excellent graphical displays
- Needless cross-referencing has been eliminated. Each example is presented with all explanations appearing on the same page.
- Exercises are presented at the ends of chapters
- A process plan is provided for many machining exercises to indicate the machining operations to be performed and the tools to be used
- A CD provided with the text contains:
    - ***Mastercam*'s demo version of the software so students can practice interactively on their own PC's**
    - files keyed to selected examples so students can follow interactively when learning the procedure with the concepts presented.
    - files containing CAD parts for machining exercises

LEARNING Mastercam MILL can be used for many different types of training applications these include:

- Undergraduate one-semester or two semester CNC programming courses
- Computer assisted component of a CNC programming course
- Industry training courses
- Trade school courses on computer assisted CNC programming
- Seminar on computer assisted CNC programming
- Adult education courses
- Reference text for self-study

This text is designed to be used in many types of educational institutions such as:

- Four-year engineering schools
- Four-year technology schools
- Community colleges
- Trade schools
- Industrial training centers

CHAPTER - 1

INTRODUCTION TO *Mastercam*

1-1 Chapter Objectives

After completing this chapter you will be able to:

1. State the system requirements for installing *Mastercam*
2. Describe the general process of generating a word address program via *Mastercam*
3. Know the types of files created by *Mastercam*
4. Understand how to start *Mastercam*
5. Describe the screen menus and short-cut keys for entering commands
6. State how to set the system's working parameters
7. Know how to use the **Help**, **Save**, and **Exit** commands

1-2 *Mastercam* CNC Software

One of the most popular CNC software packages available today is *Mastercam* from CNC Software, Inc located at 671 Old Post Road, Tolland, CT 06084. CNC Software can also be reached at 860-875-5006 or at their web site *www.mastercam.com*.

This software has a short learning curve. It presents the user with an easy to follow menu system that works fully with the Windows 98/00, NT4.0 or higher, ME or XP operating systems. Part geometry can be easily created with *Mastercam*'s CAD package. The CAM package enables the operator to quickly select machining operations and cutting tools. The software allows the operator to identify the CAD geometry to be selected for a machining operation then quickly generates the required tool path. The operator simply selects the appropriate postprocessor from the system's library and directs *Mastercam* to generate the corresponding word address part program. A very powerful feature of the software is its ability to verify the part program by animating the entire machining process.

1-3 System Requirements for Version 9

Mastercam V9 is a 32-bit CNC software package.
The following minimum system hardware and software must be installed.

- WINDOWS 98 , 2000, 2000XP, 2000ME, NT4.0 or higher
  (NT, 2000 or XP recommemnded)
- A Pentium or Pentium-Pro processor
- CD-ROM drive
- 64 Mbytes of RAM(mininum), 128Mbytes(recommended)
- 350 Mbytes of free disk space in addition to windows requirements

1-4 Conventions Used Throughout the Text

The following conventions are used throughout this text.

DISPLAY	*MEANING*	*PICTORIAL*
	Directs the operator to *press* the Enter key on the keyboard.	
Click (N)	Means to move the mouse cursor to position (N) and press the ***LEFT*** mouse button	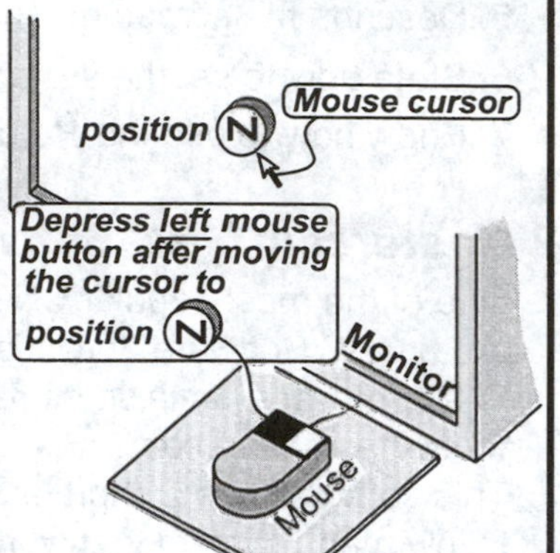
Bold	Commands to be ***typed at the keyboard*** appear in **bold**	
	When placed next to examples and exercises, this icon indicates a ***file by the same name is on the enclosed CD***. The student can ***get the file and follow the work interactively***. Note: the files on the CD are contained in chapter folders: CHAPTER2 CHAPTER3 ⋮	

1-5 Installation of *Mastercam* V.9 Demo CD Software for Student Use

From the Windows desktop:

- Click ① the **Start** button
- Click ② **Run**
- Click ③ in the Open box and enter **D: SETUP** (USE THE LETTER OF YOUR CD DRIVE HERE)
- Click ④ the OK button

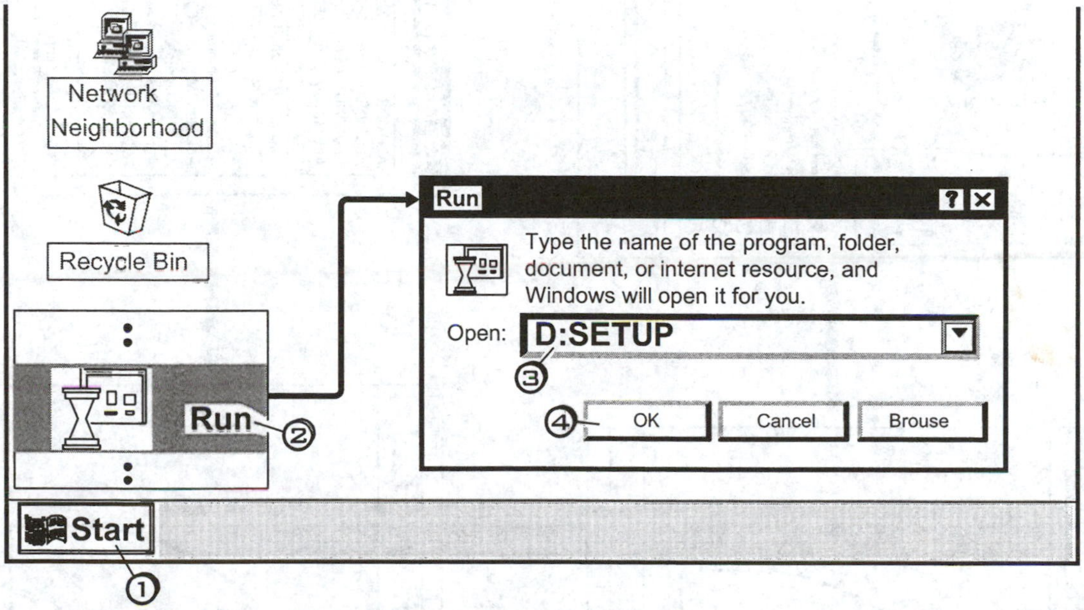

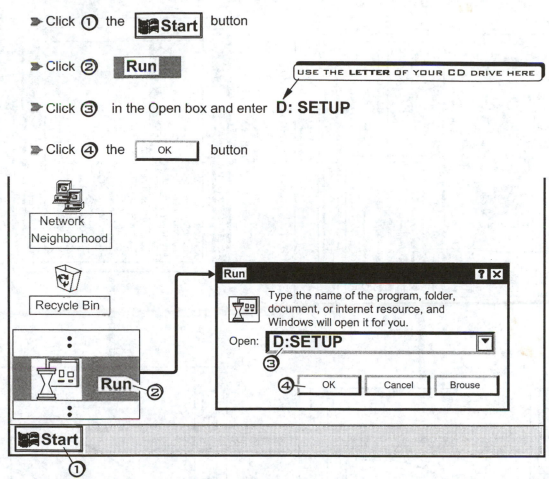

Follow the prompts automatically triggered by the software to complete the installation.

1-6 An Overview of the Process of Generating a Word Address Program Via *Mastercam*

Any part to be machined using *Mastercam* software must first be drawn using either the *Mastercam* computer aided drafting (CAD) package or imported from another CAD package such as AutoCAD or CADKEY. The part geometry created by the CAD package is used directly by the computer aided machining (CAM) package in specifying the location of machining cycles and in the determination of the corresponding tool paths. Ultimately, the CAM package produces a complete word address program for machining a specified part on a particular CNC machine tool (to learn more about word address programming refer to Introduction to Computer Numerical Control 3rd Ed by J Valentino and J Goldenberg, published by Prentice Hall and CNC Programming Handbook, 2nd Edition by Peter Smid, published by Industrial Press).

The sequence of steps to be followed to direct *Mastercam* to generate a word address program for a milling CNC machine tool are shown in Figure 1-1.

Step-1

CAD OPERATION

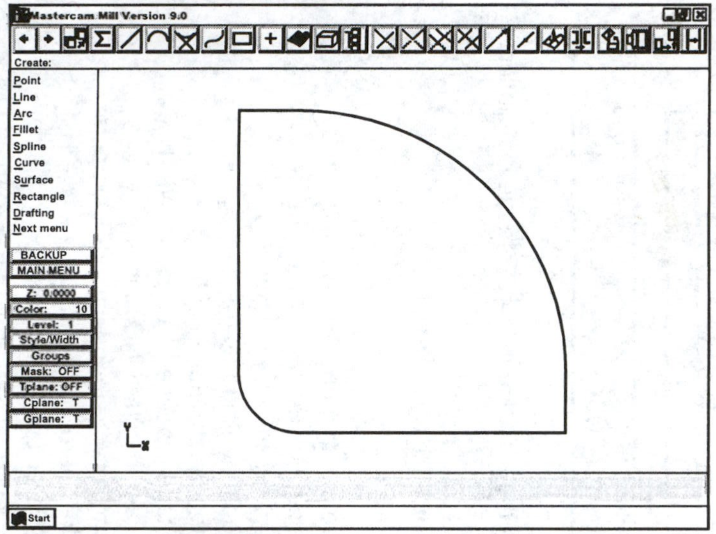

A CAD model of the part is created.

Step-2

CAM OPERATIONS

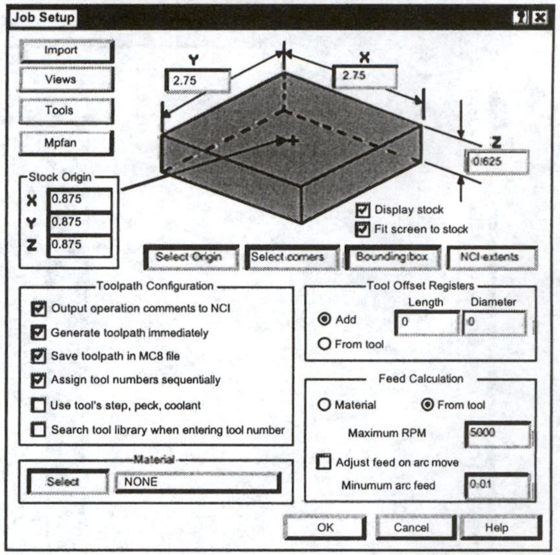

The stock to be machined is set up

Step-3

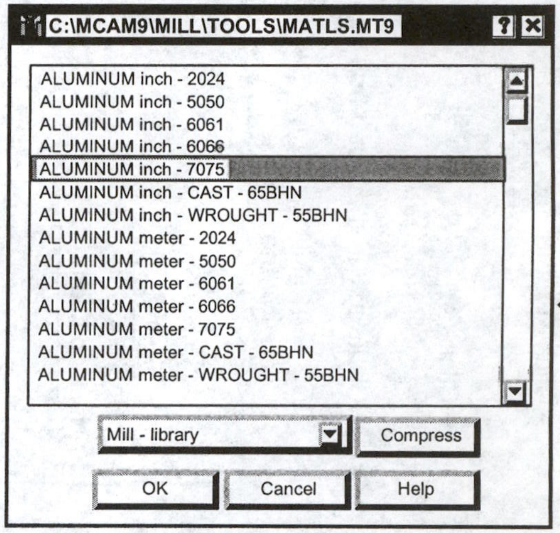

The material to be machined is selected.

Step-4

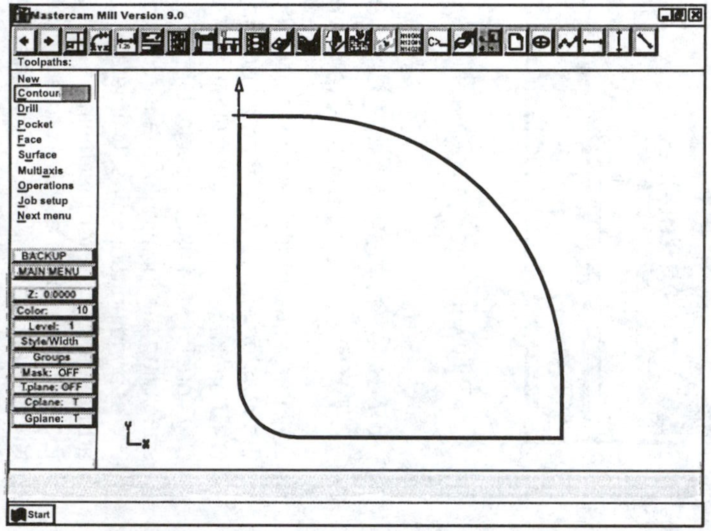

The operator selects the operation(contour) and clicks geometry to be milled.

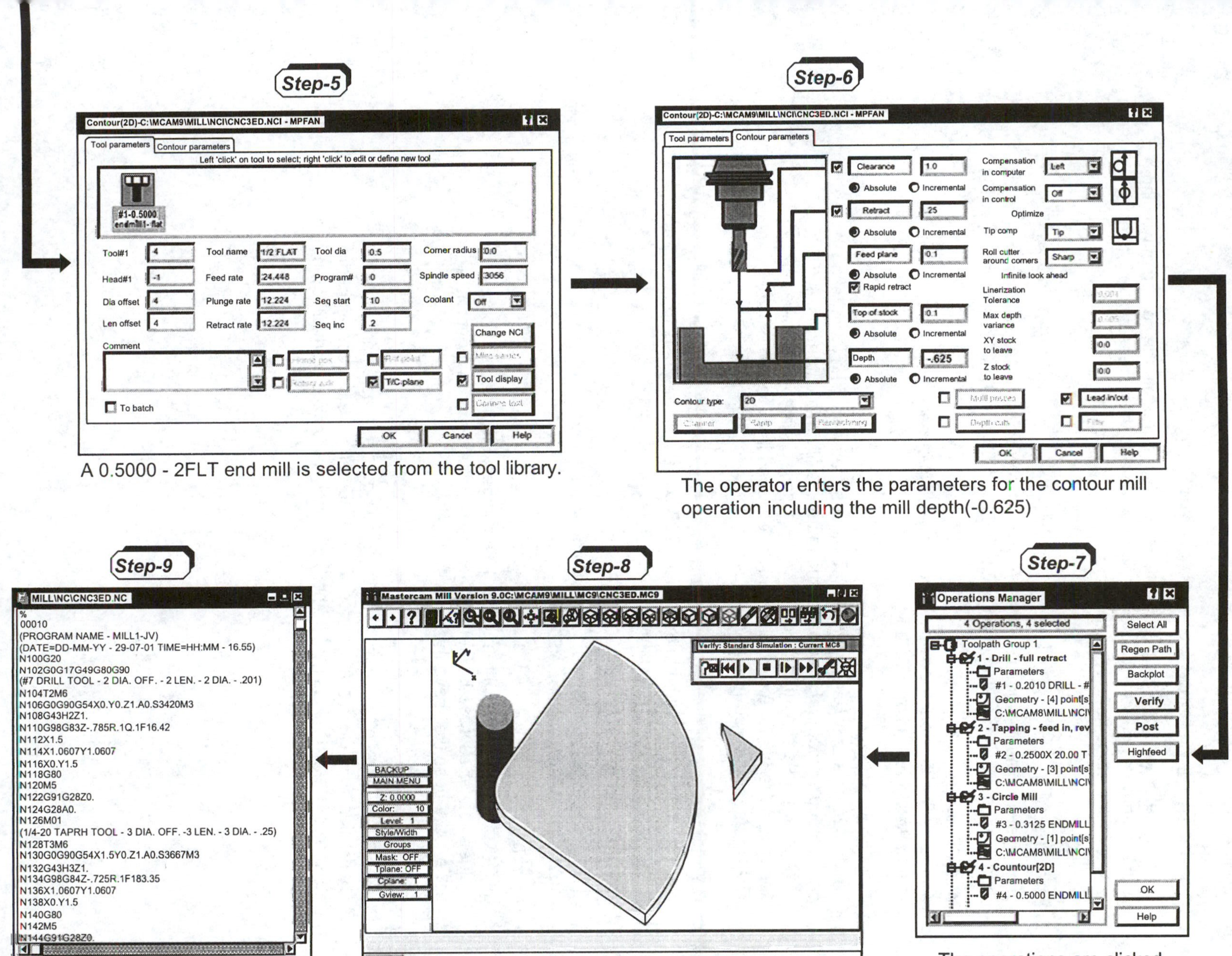

A 0.5000 - 2FLT end mill is selected from the tool library.

The operator enters the parameters for the contour mill operation including the mill depth(-0.625)

The operations are clicked and Verify is selected

Mastercam animates all the machining operations

Post is selected to direct *Mastercam* to generate the word address program

Figure 1-1 The sequence of steps for generating a part program with *Mastercam's* mill package

1-7 Types of Files Created by *Mastercam*

The operator will encounter various file extension names in the course of working with *Mastercam* software. The extension names and their meanings are listed below in one place for quick reference.

a) **CFG File**

The system default values such as *allocations, tolerances, NC settings* and *CAD settings* are stored in the configuration file that has the *extension* **.CFG**

b) **MC9 Files**

All the CAD mode *part geometry* and *tool path information* is stored in files that have the *extension* **.MC9**

c) **MT9 Files**

Mastercam has an extensive library of various materials. The information on a particular *material* is stored in the materials file with extension **.MT9** *Mastercam* uses this information to *automatically set the recommended speeds and feeds* for a particular cutting tool used in a machining operation. The operator can also *manually* enter desired speeds and feeds.

d) **TL9 Files**

The *tool library* files with the extension **.TL9** contain all the *tools* the operator needs to select in order to execute the machining of a part.

e) **PST Files**

The *post processors* for various CNC control units are stored in files that have the *extension* **.PST**

f) **NCI Files**

NCI(Numerical Control Intermediate) files contain the tool path coordinate values as well as speeds, feeds and other important machining information for a job. These files have the extension **. NCI**

f) **NC Files**

These files contain the *word address part programs*. *Mastercam* uses a particular **.PST** file and a selected **.NCI** file to generate the corresponding **.NC** file. The **.NC** file is sent to the CNC machine tool for producing the part.

1-8 Starting the *Mastercam* Mill Package

The *Mastercam* milling package is started by double clicking (1) on the *Mastercam* Mill icon appearing on the Windows desktop. See Figure 1-2.

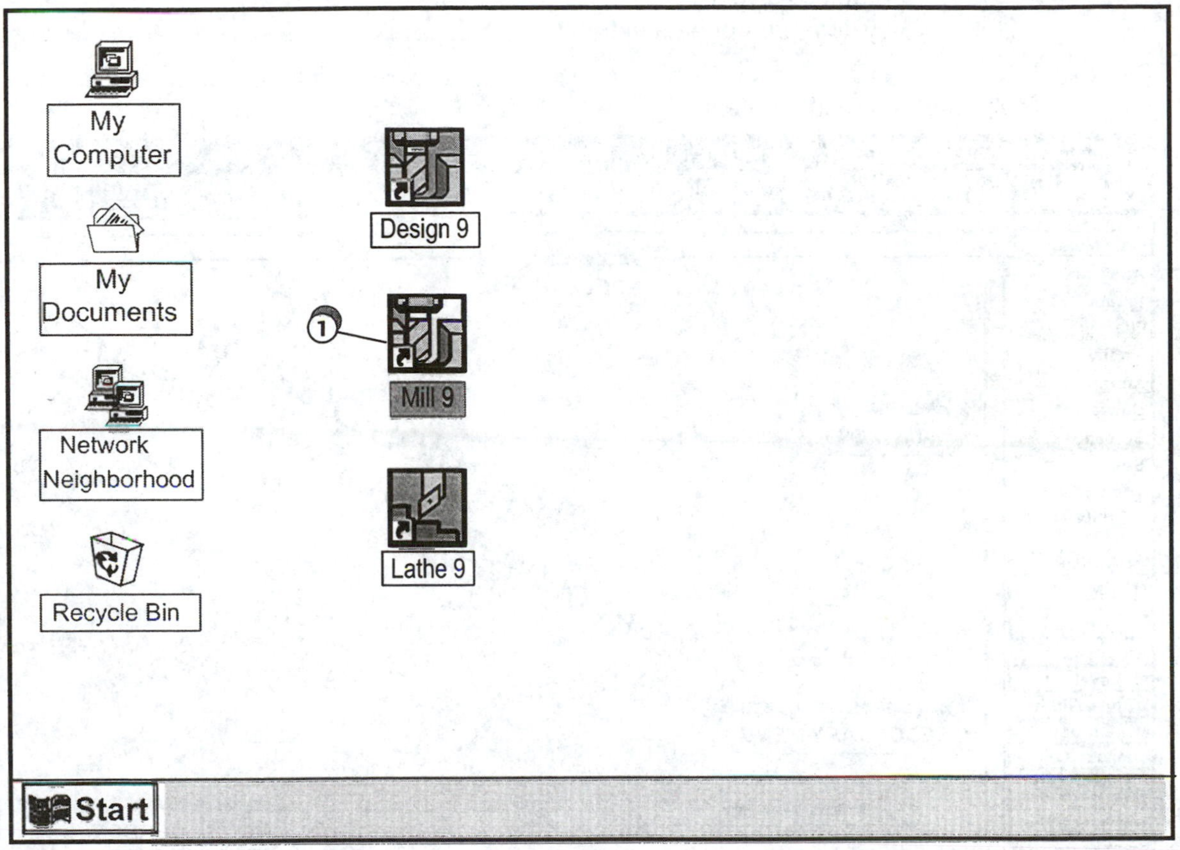

Figure 1-2 Starting the *Mastercam* Milling Package from the Windows Desktop

1-9 A Description of *Mastercam's* Design/Mill Main Window

Mastercam's main window is similar to any other window that is used in the Windows operating system. The main window consists of four general areas:

- toolbar menu
- screen menu(main and secondary)
- graphics display
- system prompt/data entry

These are shown in Figure 1-3

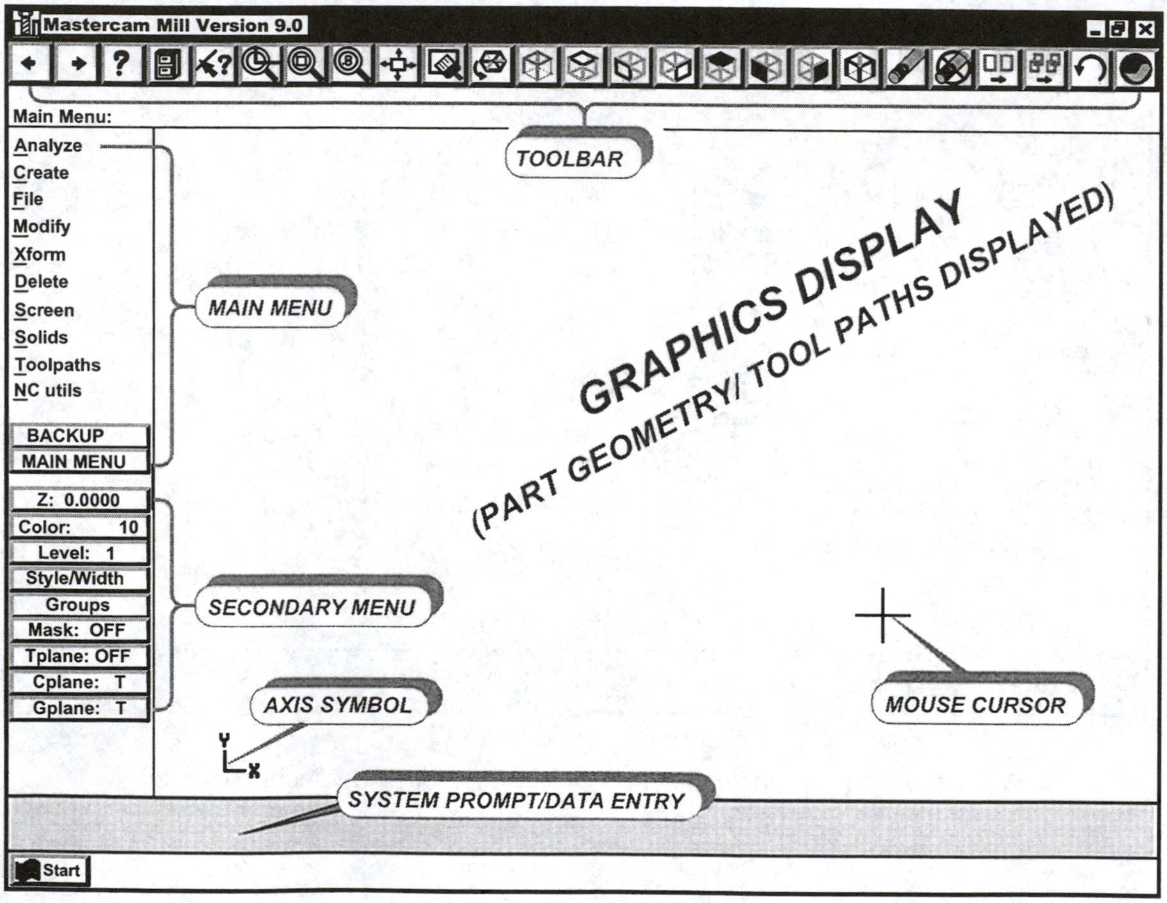

Figure 1-3 *Mastercam*'s design/mill main window

TOOLBAR MENU

This area consists of a set of icon buttons. ***Moving*** the mouse cursor to an icon causes the system to display its command function. Depressing the ***left*** mouse button on the icon ***executes*** the icon's command.

LEFT CLICK TO DISPLAY THE **PREVIOUS** TOOLBAR MENU

LEFT CLICK TO DISPLAY THE **NEXT** TOOLBAR MENU

Screen-Change Colors

MOUSE CURSOR
- **MOVE** OVER ICON TO **DISPLAY** ITS FUNCTION
- **LEFT** CLICK OVER THE ICON TO **EXECUTE** ITS FUNCTION

SCREEN MENU

MAIN MENU

The main screen menu consists of a set of primary commands that are ***activated*** by moving the mouse cursor ***over*** the command and pressing the ***left*** mouse button. The system may display other submenu menu options that branch from a primary command selected. This menu is used to execute such main operations as geometry creation and editing, tool and material selection, and the setting and verification of toolpaths.

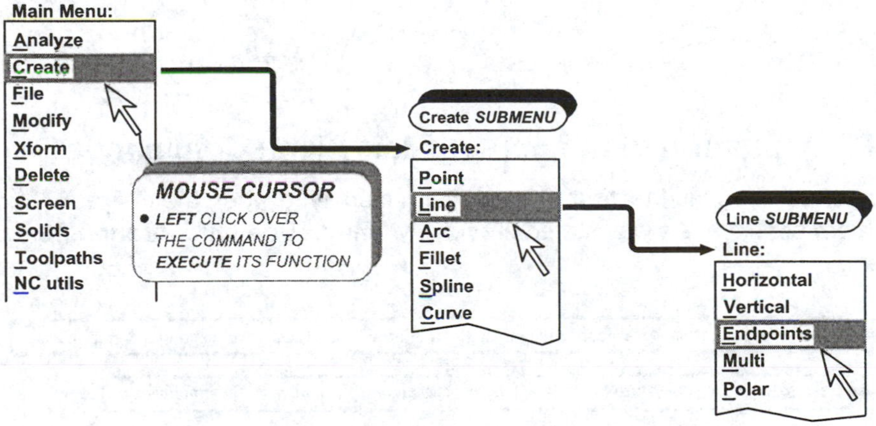

SECONDARY MENU

The setting of important parameters for a particular job is accomplished via the secondary menu. These include line widths, colors for geometry, and Z depth.

SYSTEM PROMPT/DATA ENTRY

System responses to inputted commands are displayed in this area. Some prompts require the operator to enter data values. These values are entered in this area.

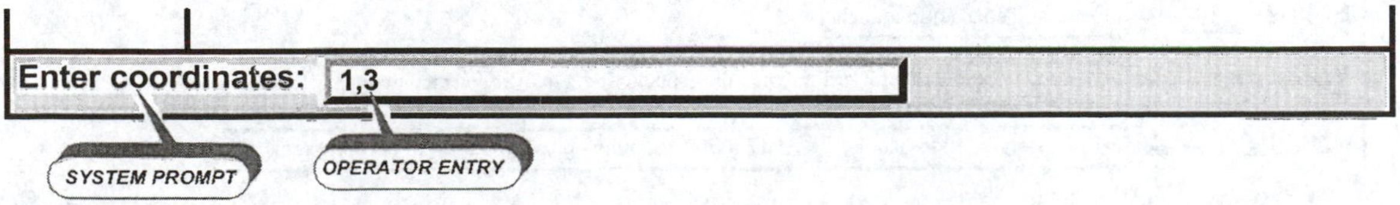

BACKUP and MAIN MENU

Clicking BACKUP or depressing the Esc key causes the system to ***return to the previous menu*** in the submenu tree structure.

Clicking MAIN MENU or ***continuously*** pressing the Esc key causes the system to ***return to the main menu*** from any currently displayed submenu

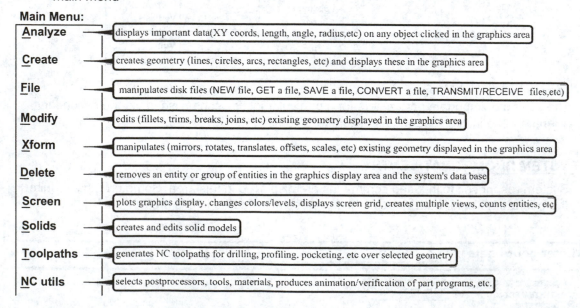

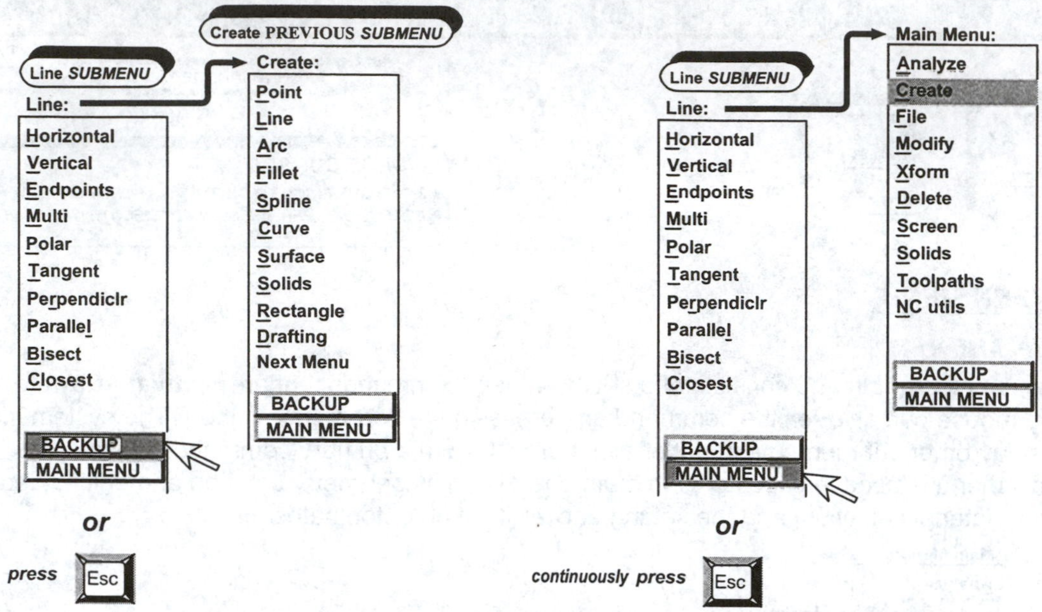

1-10 A Brief Explanation of the Screen Main Menu Commands

The main screen menu is the ***first*** or ***root*** menu displayed when the operator ***begins*** a *Mastercam* session . As was stated previously many ***submenus branch*** from the main menu

Main Menu:

Analyze	displays important data(XY coords, length, angle, radius,etc) on any object clicked in the graphics area
Create	creates geometry (lines, circles, arcs, rectangles, etc) and displays these in the graphics area
File	manipulates disk files (NEW file, GET a file, SAVE a file, CONVERT a file, TRANSMIT/RECEIVE files,etc)
Modify	edits (fillets, trims, breaks, joins, etc) existing geometry displayed in the graphics area
Xform	manipulates (mirrors, rotates, translates, offsets, scales, etc) existing geometry displayed in the graphics area
Delete	removes an entity or group of entities in the graphics display area and the system's data base
Screen	plots graphics display, changes colors/levels, displays screen grid, creates multiple views, counts entities, etc
Solids	creates and edits solid models
Toolpaths	generates NC toolpaths for drilling, profiling, pocketing, etc over selected geometry
NC utils	selects postprocessors, tools, materials, produces animation/verification of part programs, etc.

1-11 A Brief Explanation of the Screen Secondary Menu Commands

The secondary screen menu is ***continuously*** displayed as the main menu branches back and forth to its submenus

Z: 0.0000	displays/changes the current working depth
Color: 10	displays/changes the current working color for new geometry creation
Level: 1	is a layer level on which new geometry is placed, selected levels can be made visible or invisible
Style/Width	sets the current working line style and line width
Groups	creates a name for a group of selected entities enabling group manipulation by using the name
Mask: OFF	sets the layer levels that are to be visible but locked against *Mastercam* commands
Tplane: OFF	sets the tool plane or plane on which the CNC machine tool coordinate system is currently set
Cplane: T	sets the construction plane or plane on which geometry is created and manipulated
Gplane: T	changes the orientation of the geometry displayed in the graphics area

1-12 Mastercam's Short-Cut Keys for Entering Commands

In addition to using toolbar and screen menus, the operator can also use short-cut keys for quickly entering commands. Key use reduces the need for clicking into sub-menus thereby dramatically lowering the time it takes to complete operations in *Mastercam.* Keys can provide additional flexibility. For example, the keys that cause pan and zoom can be used *while* a line command is being executed.

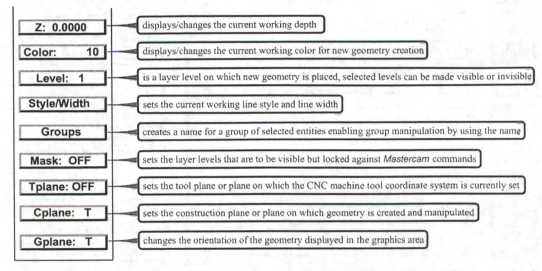

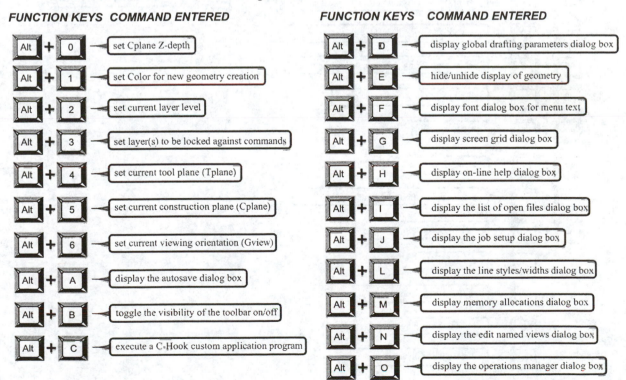

FUNCTION KEYS	***COMMAND ENTERED***
Alt + 0	set Cplane Z-depth
Alt + 1	set Color for new geometry creation
Alt + 2	set current layer level
Alt + 3	set layer(s) to be locked against commands
Alt + 4	set current tool plane (Tplane)
Alt + 5	set current construction plane (Cplane)
Alt + 6	set current viewing orientation (Gview)
Alt + A	display the autosave dialog box
Alt + B	toggle the visibility of the toolbar on/off
Alt + C	execute a C-Hook custom application program

FUNCTION KEYS	***COMMAND ENTERED***
Alt + D	display global drafting parameters dialog box
Alt + E	hide/unhide display of geometry
Alt + F	display font dialog box for menu text
Alt + G	display screen grid dialog box
Alt + H	display on-line help dialog box
Alt + I	display the list of open files dialog box
Alt + J	display the job setup dialog box
Alt + L	display the line styles/widths dialog box
Alt + M	display memory allocations dialog box
Alt + N	display the edit named views dialog box
Alt + O	display the operations manager dialog box

FUNCTION KEYS	COMMAND ENTERED
Alt + P	toggle display of prompt area on/off
Alt + Q	undo the last operation
Alt + R	edit the last operation
Alt + S	set display of full time shading on/off
Alt + T	turn tool path display on/off
Alt + U	undo the last operation
Alt + V	display the *Mastercam* version and SIM numbers
Alt + W	display the multiple viewports dialog box
Alt + X	set color, width, level based on an existing object
Alt + Z	display memory allocations dialog box
Alt + '	create a circle by specifying two points
Alt + Tab	display switch between applications dialog box
Alt + F1	zoom graphics display to fit screen extents
Alt + F2	zoom down current graphics display by .8
Alt + F3	set cursor tracking (XY coords display) on/off
Alt + F4	display exit *Mastercam* dialog box
Alt + F5	delete window
Alt + F7	blank
Alt + F8	display the configuration manager dialog box
Alt + F9	display the coordinate axis
F1	zoom window
F2	unzoom current display by .5
F3	repaint the graphics area
F4	display the **analyze** submenu

FUNCTION KEYS	COMMAND ENTERED
F5	display the **delete** submenu
F6	display the **file** submenu
F7	display the **modify** submenu
F8	display the **create** submenu
F9	display file name/size. CAD construction and tool origins (X0Y0)
F10	display the execute function dialog box
Esc	*cancel* current command
Pg Up	zoom graphics display *up*
Pg Dn	zoom graphics display *down*
→	pan graphics display *right*
←	pan graphics display *left*
↑	pan graphics display *up*
↓	pan graphics display *down*

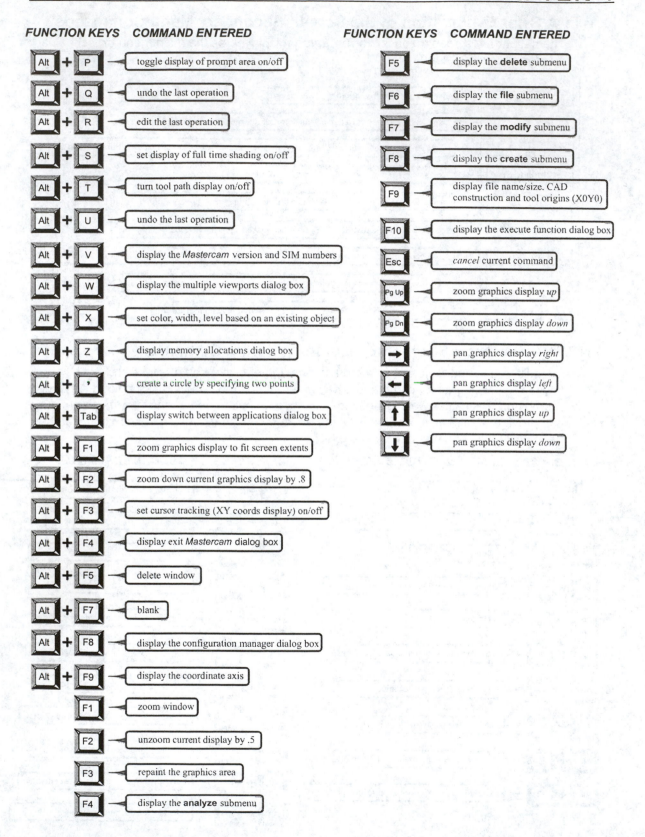

1-13 Setting Working Parameters Via the System Configuration Dialog Box

Important working parameters in *Mastercam* are specified in the System Configuration dialog box. These include tolerance setting, file subdirectory creation, screen settings, NC settings, CAD settings, etc. The configuration file is saved as **MILL9.CFG** for English units or **MILL9M.CFG** for metric units.

- Click ① MAIN MENU and ② Screen
- Click ③ Configure

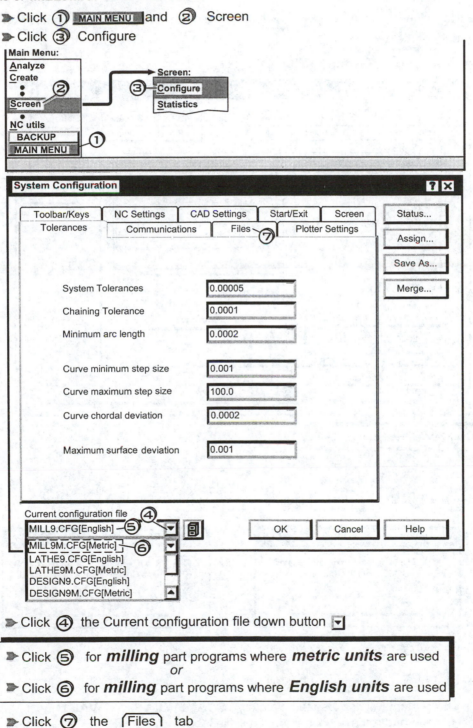

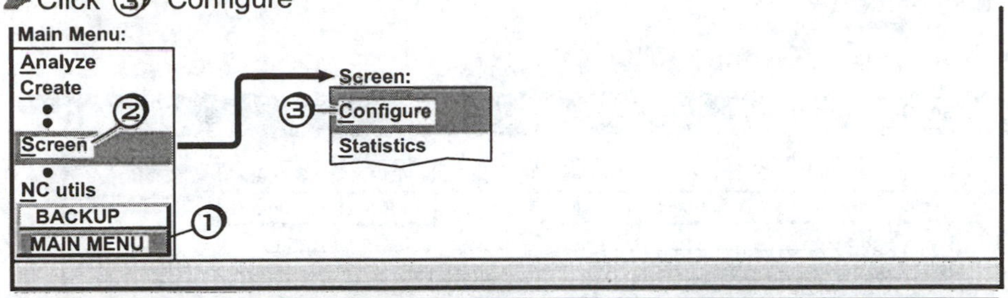

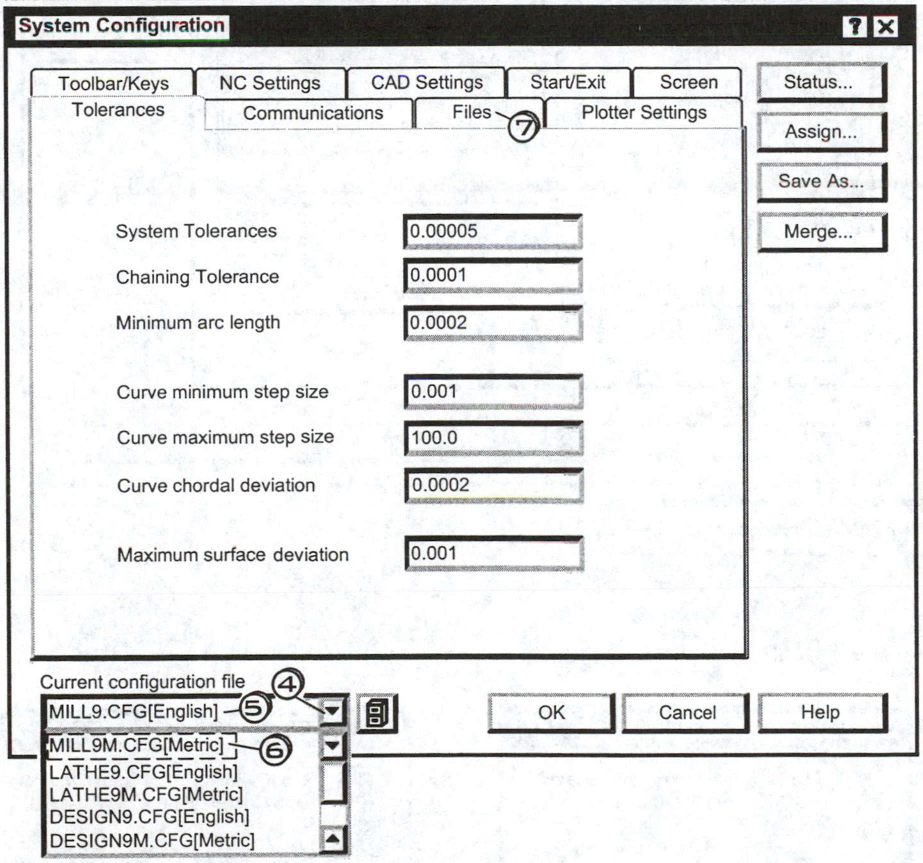

- Click ④ the Current configuration file down button

> - Click ⑤ for ***milling*** part programs where ***metric units*** are used
>
> *or*
>
> - Click ⑥ for ***milling*** part programs where ***English units*** are used

- Click ⑦ the Files tab

Create Your Own Directory for Storing and Getting Your .MC9 Student Exercise Files

Make sure the Mastercam[MC9] file is ***highlighted*** in the Data Paths area

- Click ⑧ in the Selected item's Data path
- Keep the direction *right* key → depressed until the cursor | is positioned to the right of the text
- Type (YOUR INITIALS)**-MILL**

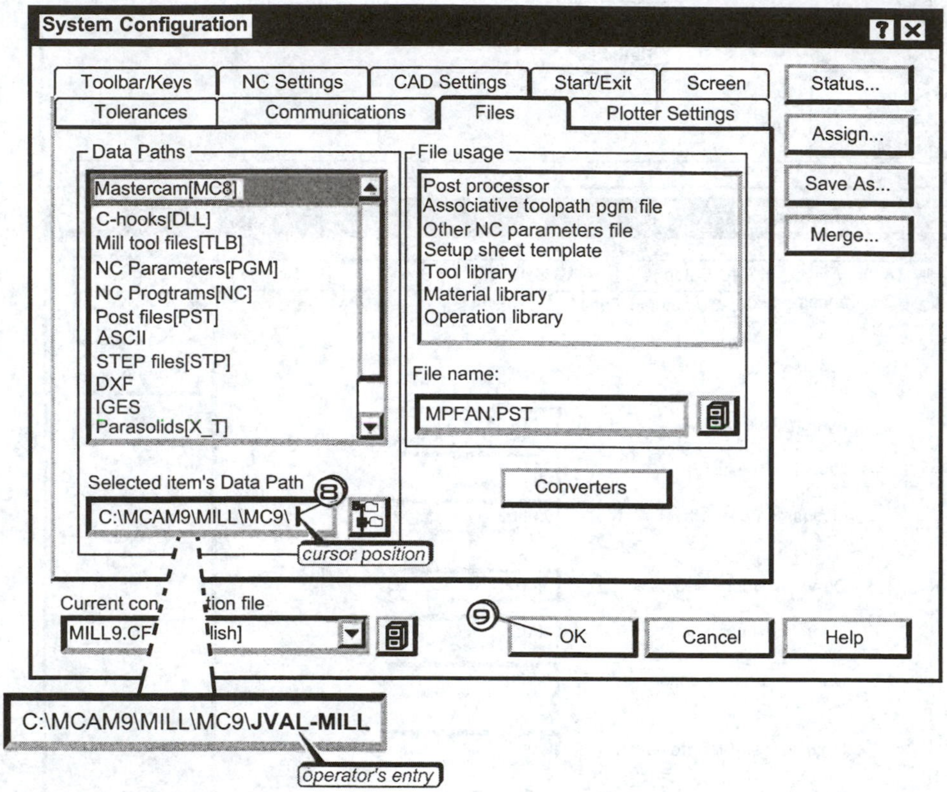

- Click ⑨ the OK button

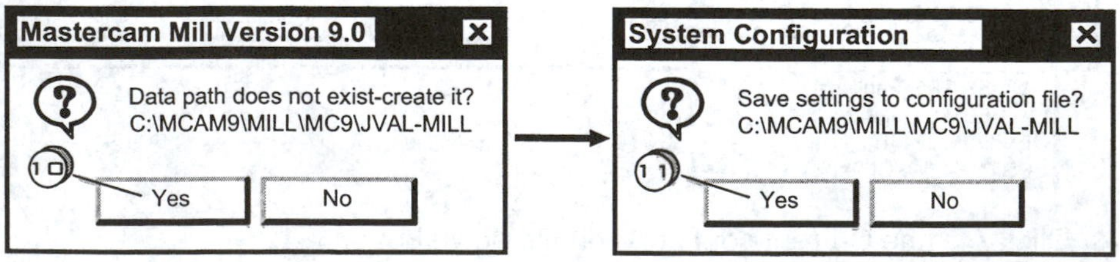

- Click ⑩ the Yes button
- Click ⑪ the Yes button

The system will now use the directory **MCAM9\MILL\MC9\JVAL-MILL** to save and retrieve your student mill file jobs

Specify the default settings for *point* and *line* styles

- Click (11) the CAD Settings tab
- Click (12) the down arrow
- Click (13) the desired point style
- Click (14) the down arrow
- Click (15) the desired line thickness

System Configuration

Tolerances | Communications | Files | Plotter Settings | Toolbar/Keys | NC Settings | CAD Settings | Start/Exit | Screen

Status... | Assign... | Save As... | Merge...

Spline/Surface creation type

NURBS

Draw highlight on back of surfaces

Stay in Create-Drafting-Note

Delete duplicate join lines

Drafting Settings

Dimension Factors | Note Factors | Globals | Delete

Group Colors | Attributes | Center Lines[arcs] | Solids Settings | Rectangular Shapes

Attributes

Color 12

Level 1 Select

Line Style: solid, solid, hidden, center, phantom, Zbreak

Line Width

Point Style: 3D star, 3D star, dot, cross, x, circle, square

Same as

OK | Cancel | Help

Current configuration file

MILL9.CFG[English]

OK | Cancel | Help

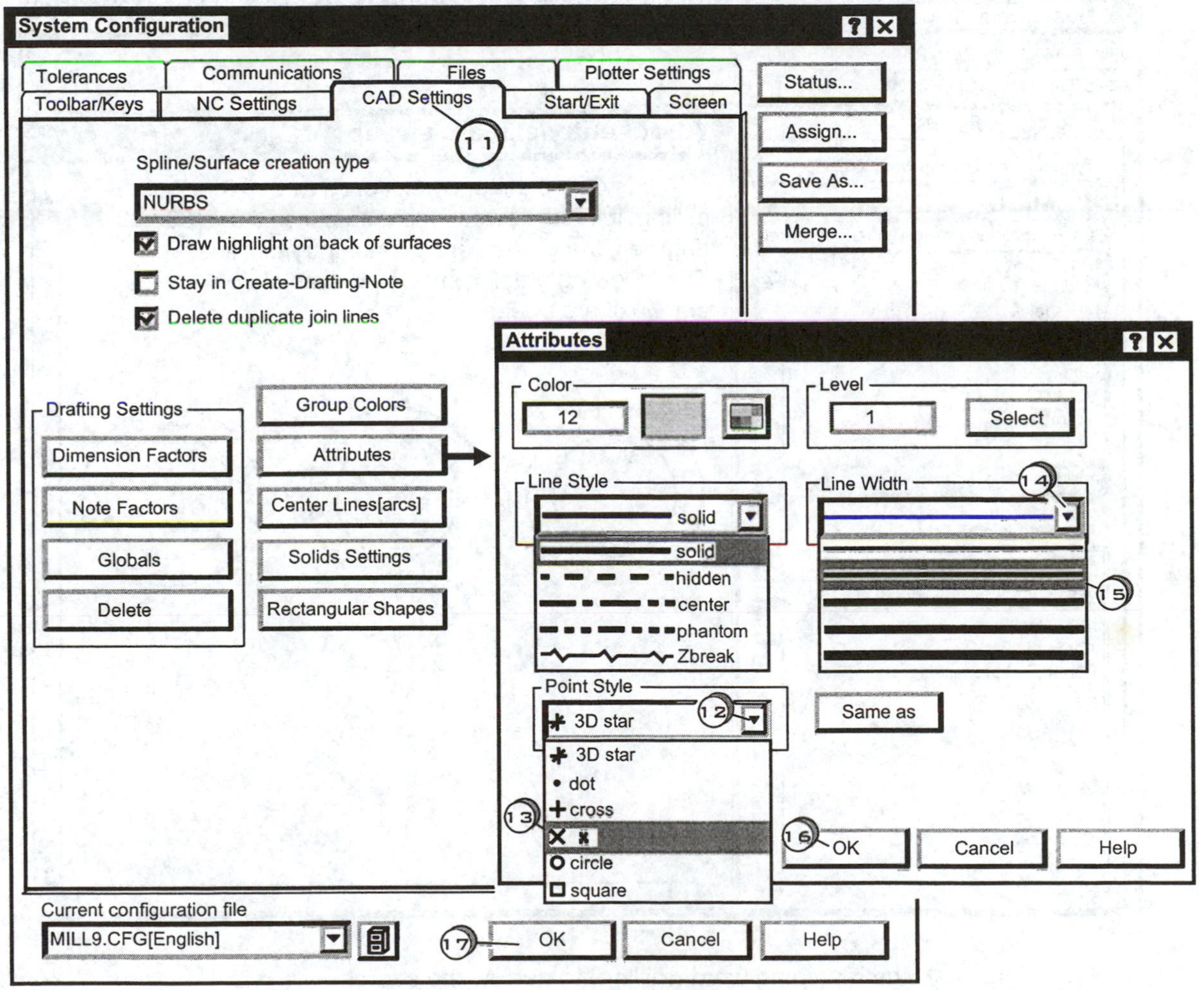

- Click (16) and (17) OK

Note: it is advisable to save your customized configuration file(s) with extension (**.CFG**) to a floppy or ZIP diskette so that they are readily avalable if restoration is needed.

1-14 Using On Line Help

Mastercam provides quick access to information regarding basic concepts, commands, tools,and information about the latest release of the software.

General Use Of Help

- Press the Alt + H keys
- Click ① the Search tab
- Click ② in the key word area and enter the key word you want to search: **MILL**

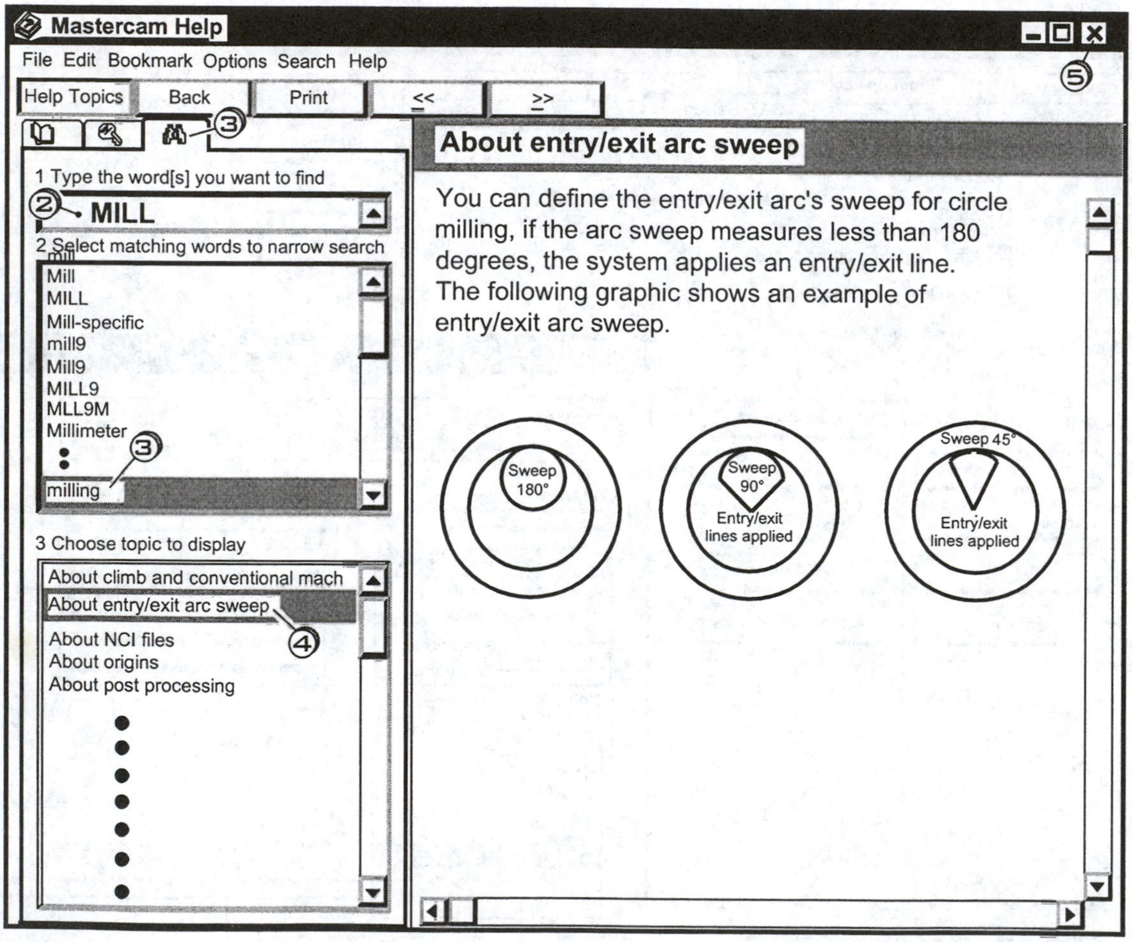

- Click ③ the matching word (milling) to narrow the search
- Click ④ the topic for which detailed information is to be displayed
- Click ⑤ the close button ☒ to *end* the help session

1-15 Saving a File

To save a file:

- Click ① MAIN MENU and ② File
- Click ③ Save

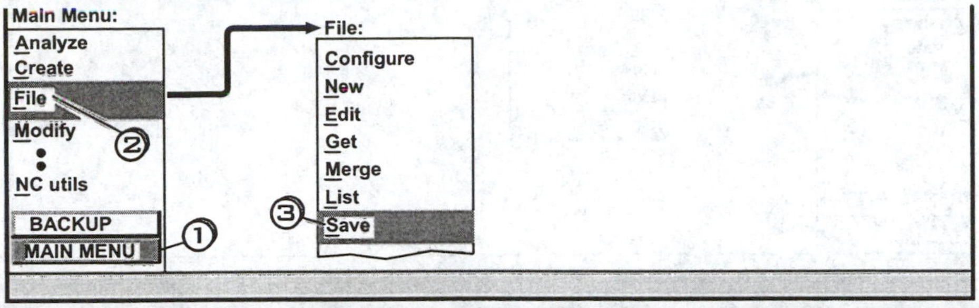

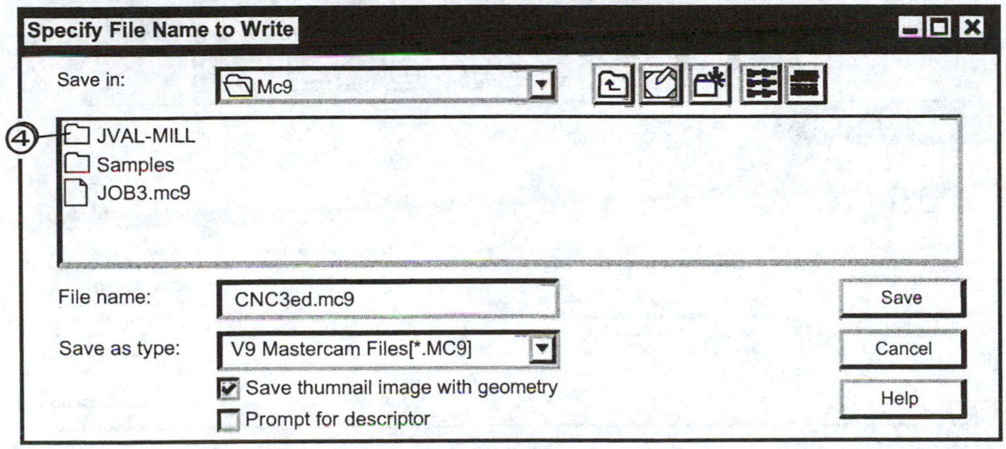

- *double* Click ④ on the **JVAL-MILL** file folder to open this directory

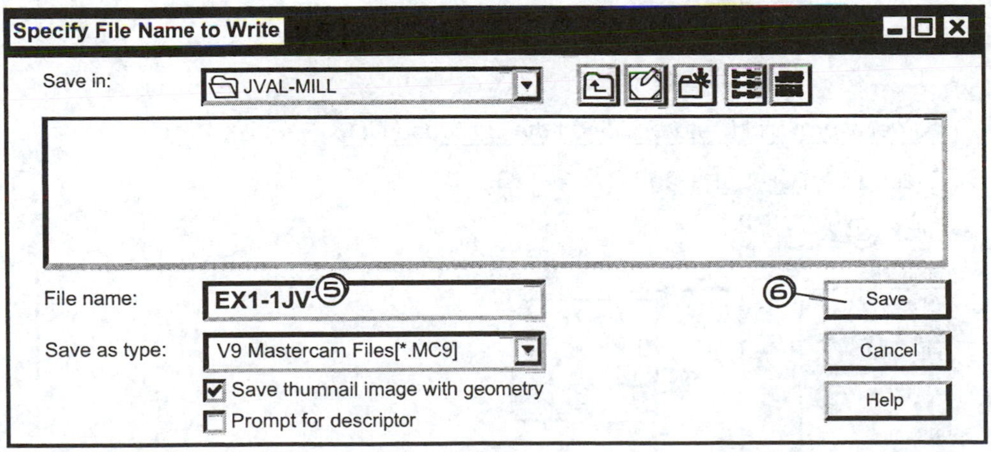

- Click ⑤ in the file name box and enter the name of the **.MC9** file to be saved, for example file, **EX1-1JV**
- Click ⑥ Save

1-16 Getting a File

a) *To get a previously stored file:*

- Click ① **MAIN MENU** and ② **File**
- Click ③ **Get**

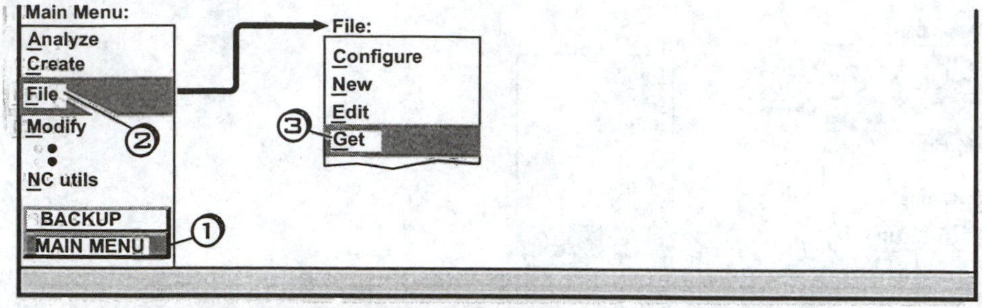

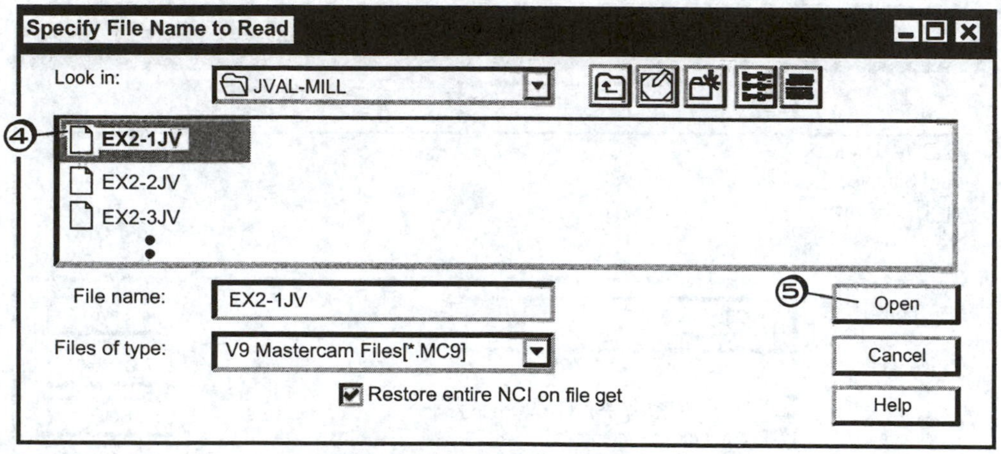

- Click ④ on the desired file to open **EX2-1JV**
- Click ⑤ **Open**

b) *To get a previously stored file on the Enclosed CD :*

- Click ① **MAIN MENU** and ② **File**
- Click ③ **Get**

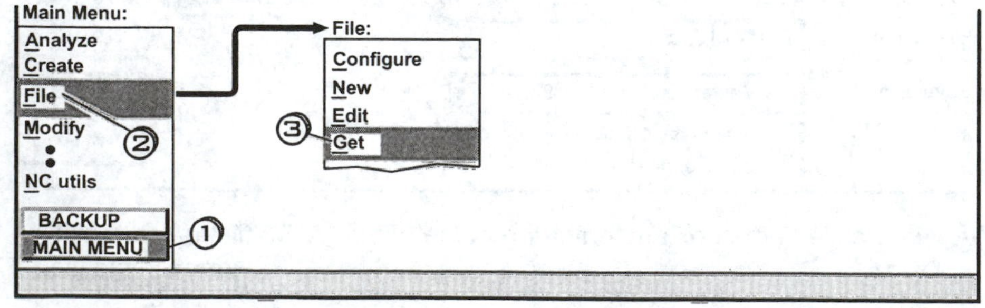

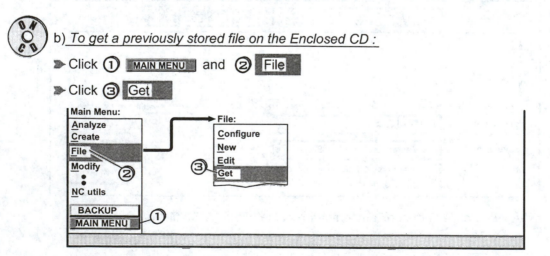

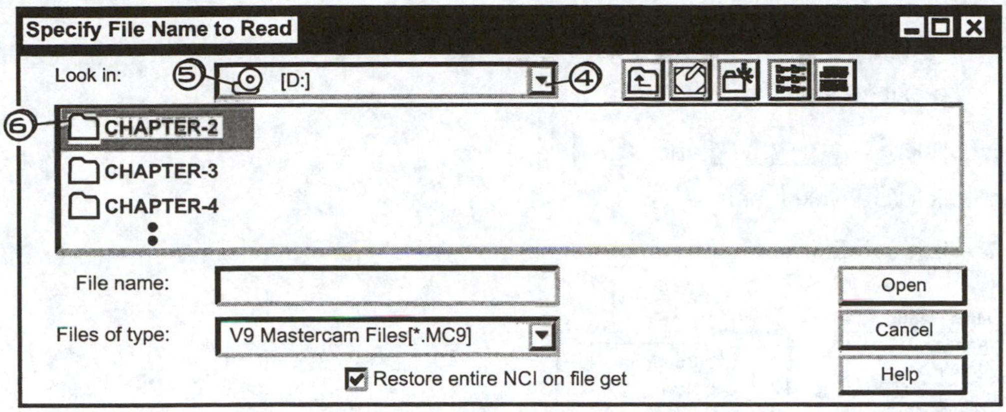

- Click ④ on the down button and ⑤ on the CD drive icon
- ***Double*** Click ⑥ on the file folder of the desired chapter

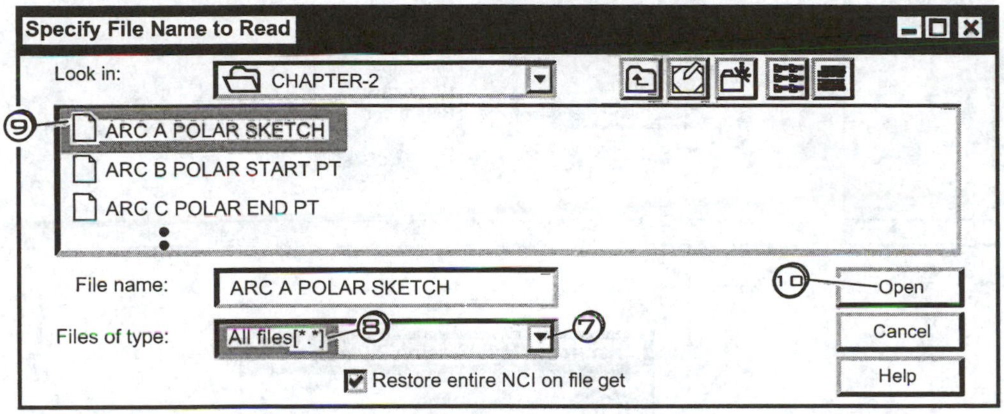

- Click ⑦ on the down button and ⑧ on **All Files[*.*]**
- Click ⑨ on the file to get
- Click ⑩ Open

1-17 Creating a New File

To create a new file while in *Mastercam*

- Click ① MAIN MENU and ② File
- Click ③ New

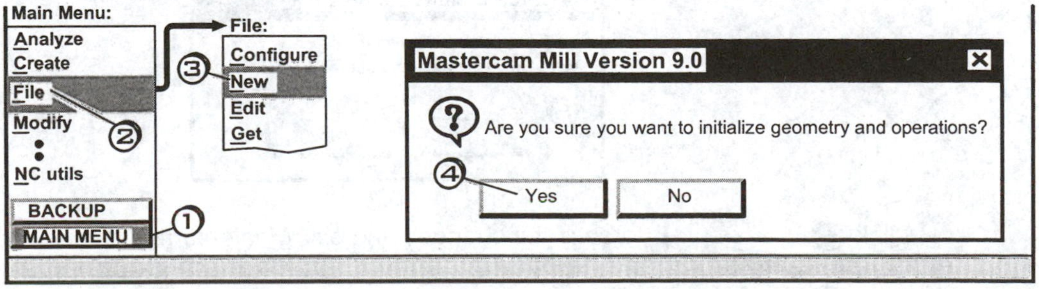

- Click ④ Yes

1-18 Exiting the *Mastercam* Mill package

To exit *Mastercam*:

- Click ① MAIN MENU and ② File
- Click ③ Next menu
- Click ④ Exit

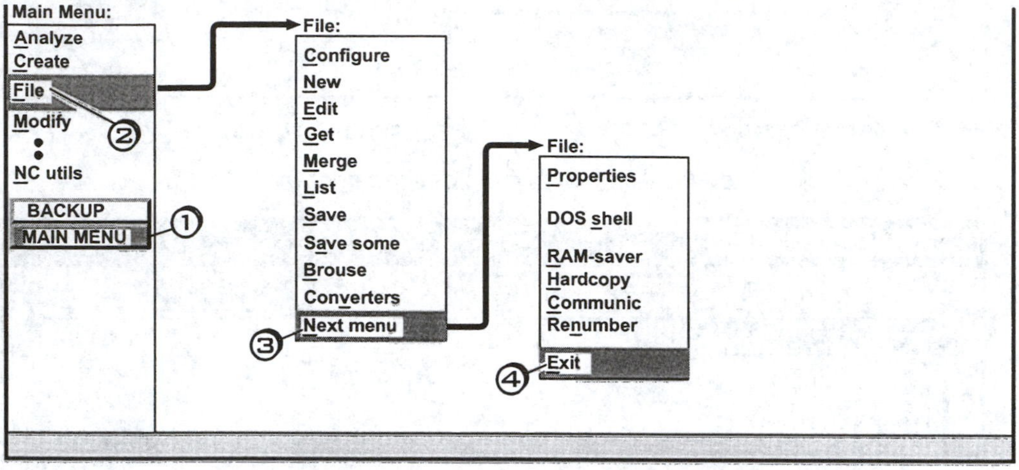

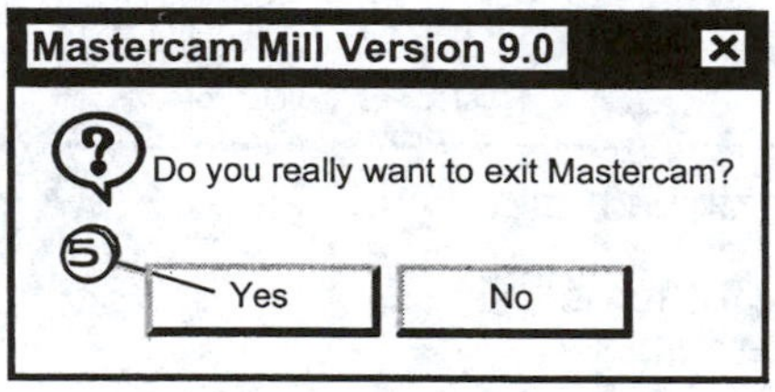

- Click ⑤ Yes

If any *changes* to the current **.MC9** file were *not saved* the system will display the dialog box below

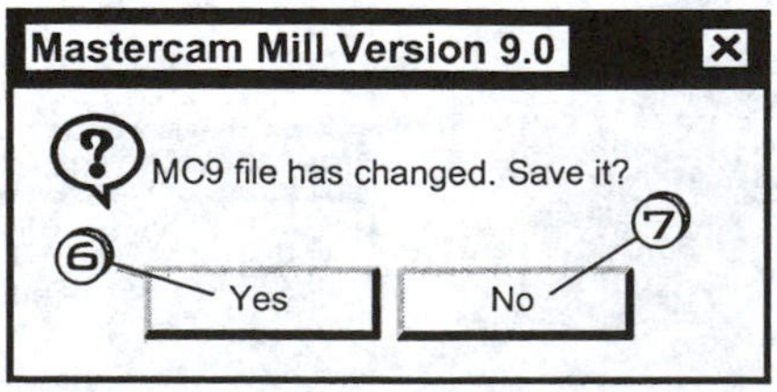

- Click ⑥ Yes to *save* the changes and exit *Mastercam*
- Click ⑦ No to *discard* the changes and exit *Mastercam*

EXERCISES

1-1) Execute the steps necessary to turn on the computer and start the *Mastercam* Mill package

1-2) What information is contained in the following files: a) **CFG** b) **MC9**

1-3) Identify and describe the use of each of the areas *A, B, C, D, E, F, G* of *Mastercam*'s main screen window shown in Figure 1p-1

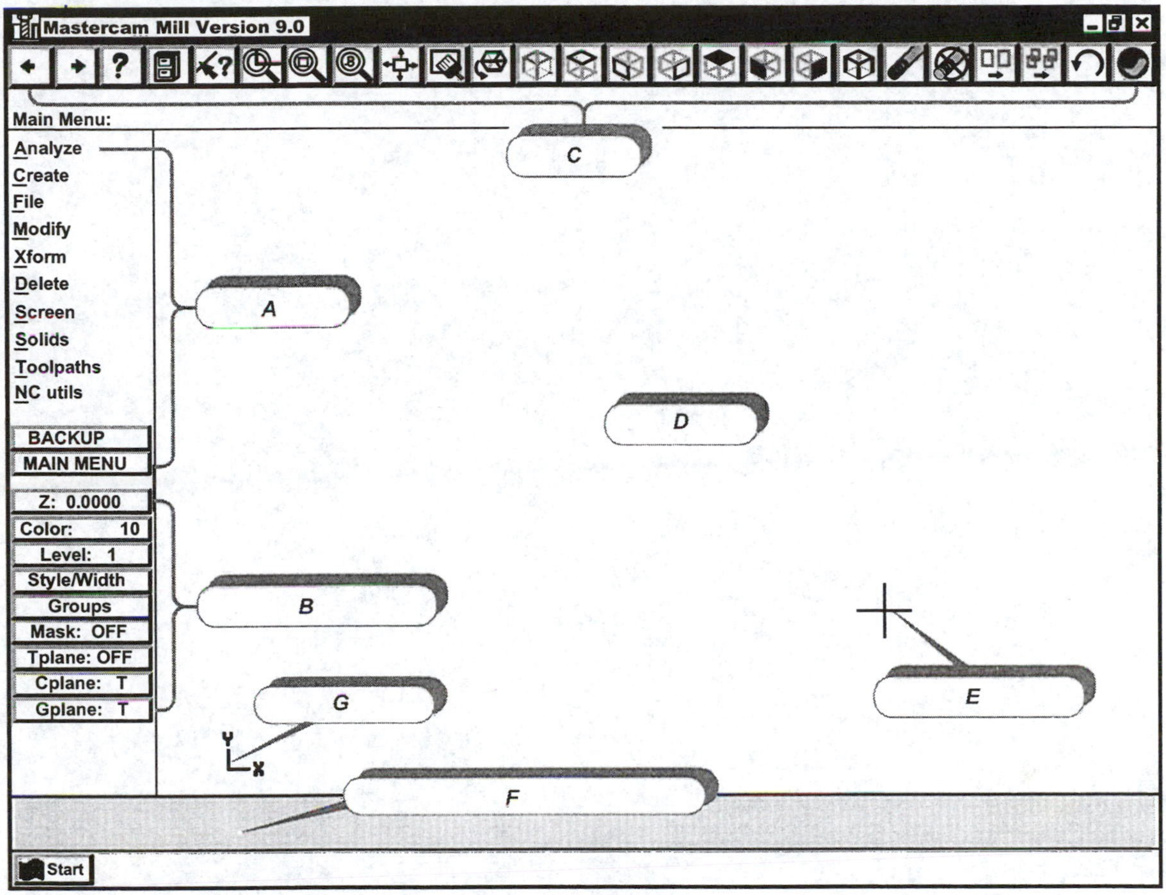

Figure 1-p1

1-4) What is the significance of using the Esc key

1-5) What are some advantages of using the short-cut keys

1-6) What functions are executed by the following short-cut keys

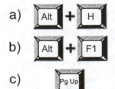

a) Alt + H

b) Alt + F1

c) Pg Up

d) →

1-7) Describe the steps required to set the system to *metric* part programming.

1-8) Describe the steps required to create the subdirectory **SPEEDY-MILL** for saving and getting *Mastercam* **.MC9** files.

1-9) Use the search word *mastercam* in on line help to find and print information about the topic *configuration file*.

1-10) Explain the steps for saving the file **EX1-2JV** in the sub-directory **JVAL-MILL** and exiting *Mastercam*.

CHAPTER - 2

CREATING 2D GEOMETRY

2-1 Chapter Objectives

After completing this chapter you will be able to:

1. Know how to construct basic 2D geometric entities such as points, lines, arcs, fillets and splines.
2. Execute rectangle, chamfer and letter constructions in 2D space.
3. Know how to construct 2D ellipse and polygon geometric entities.
4. Understand how to use *Mastercam*'s point positioning commands to quickly and accurately locate points on existing geometric entities.
5. Explain how to pan and zoom screen displays.
6. State the methods used for repainting and regenerating the screen displays.

2-2 Generating the CAD Model of a Part

The first step in generating toolpaths and the part program in *Mastercam* is to create the part's geometry using software's CAD package.

2-3 Point Constructions in 2D Space

A point is displayed on the screen as the symbol (**✕**) . Points can be used as references to specify locations for subsequent geometric constructions involving lines, circles, arcs, splines, etc

Creating Points-1

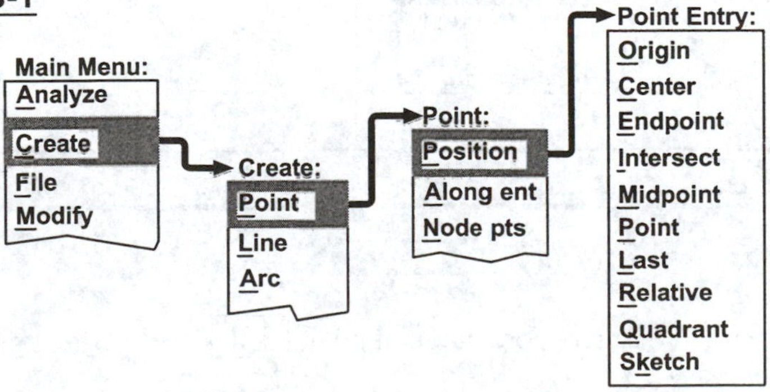

Note: For rectangular coordinates, *Mastercam* employs the *Cartesian coordinate system* for specifying the location of a point in space. This system consists of three directional lines, called axes, that mutually intersect at an angle of 90°. The point of intersection is called the origin. The XY coordinate plane is broken up into *four* quadrants. The location of a point is determined by *first* entering its x-distance, *then* y-distance and *finally* z-distance from the origin.

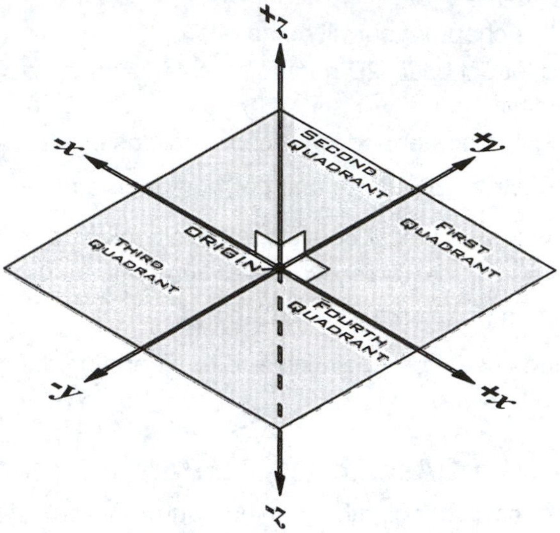

Origin

A *point* is created at the CAD *screen origin*(X0Y0) and at *any other* coordinate entered.

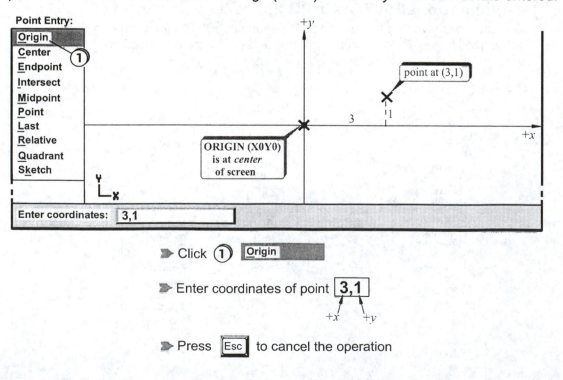

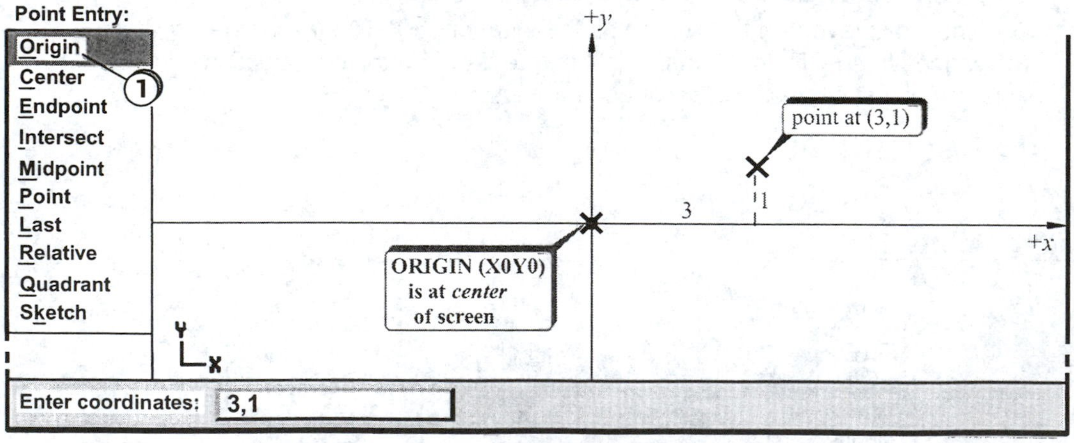

➤ Click (1) Origin

➤ Enter coordinates of point 3,1 (+x, +y)

➤ Press Esc to cancel the operation

Center

A *point* is created at the *center* of an existing *circle* or *arc* entity

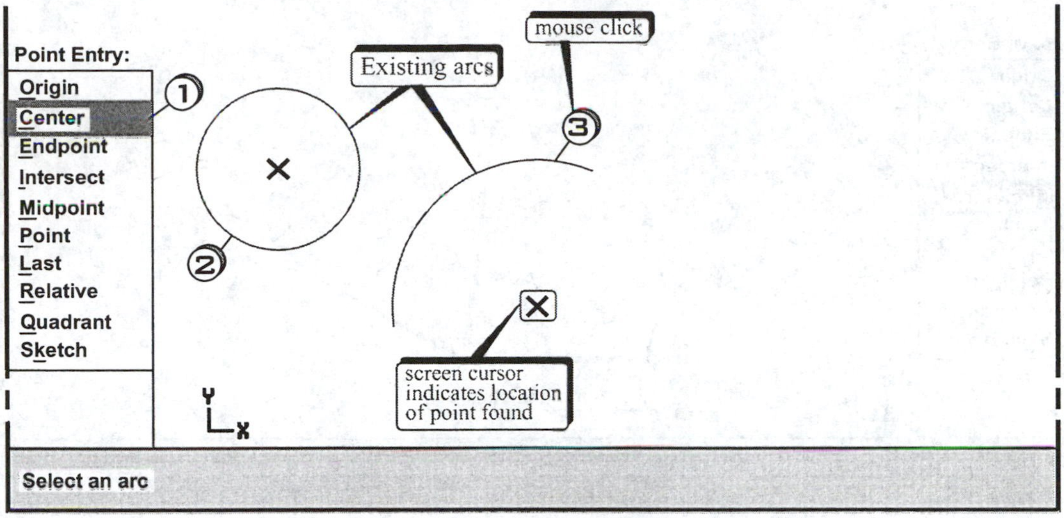

- Click ① Center
- Click ② ③
- Press Esc to cancel the operation

Endpoint

A *point* is created at the *ends* of an existing *line, arc or spline* entity.

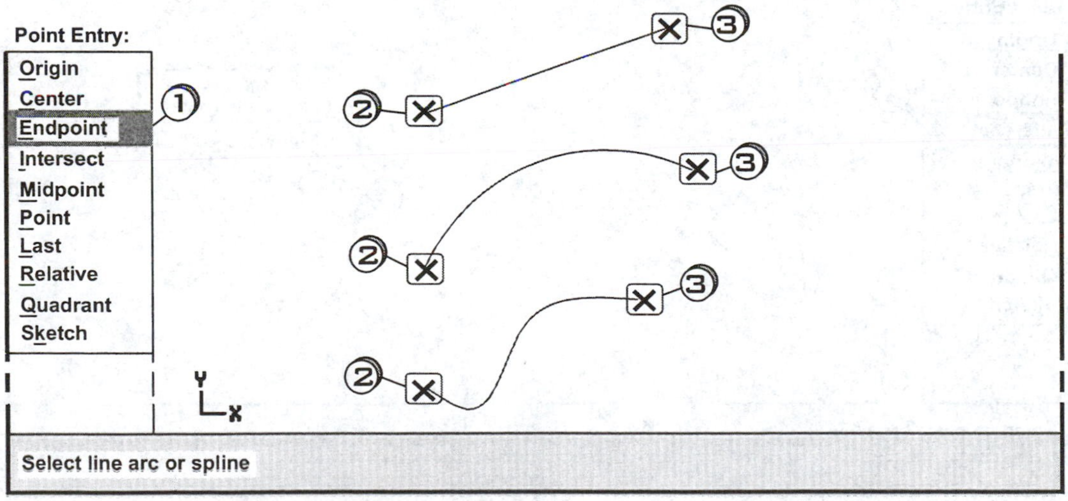

- Click ① Endpoint
- Click ② ③
- Press Esc to cancel the operation

Midpoint

A *point* is created at the *midpoint* of an existing *line, arc or spline* entity

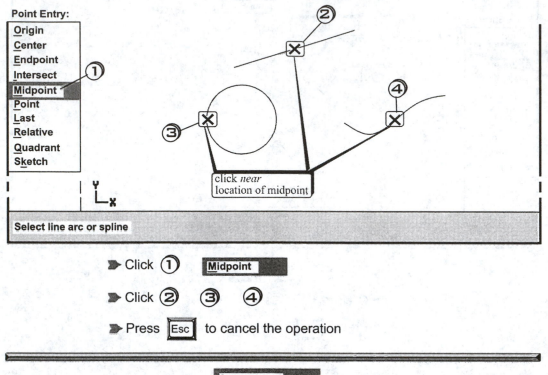

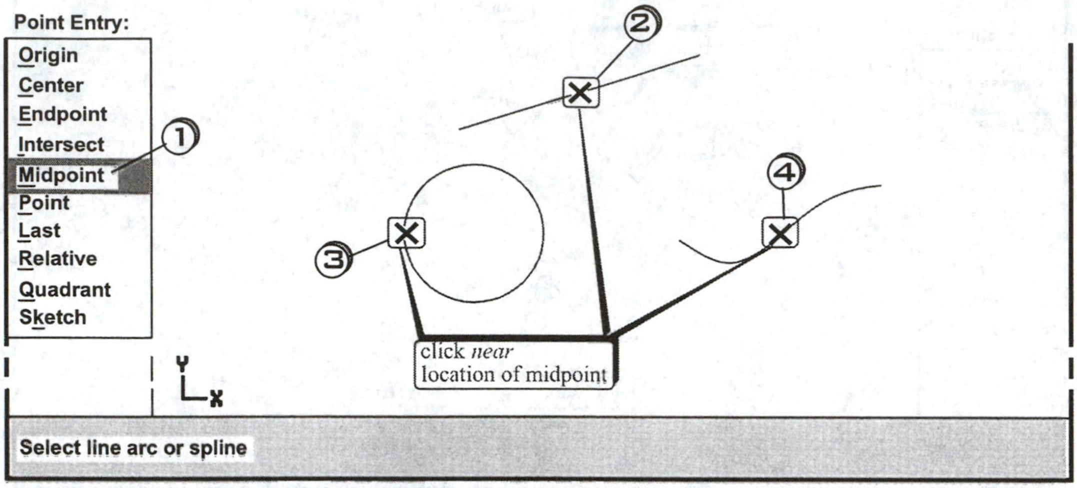

- Click (1) Midpoint
- Click (2) (3) (4)
- Press Esc to cancel the operation

Intersect

A *point* is created at the *intersection* of an existing *line, arc and spline* entities

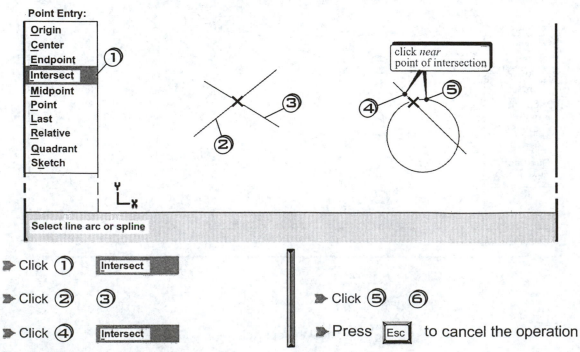

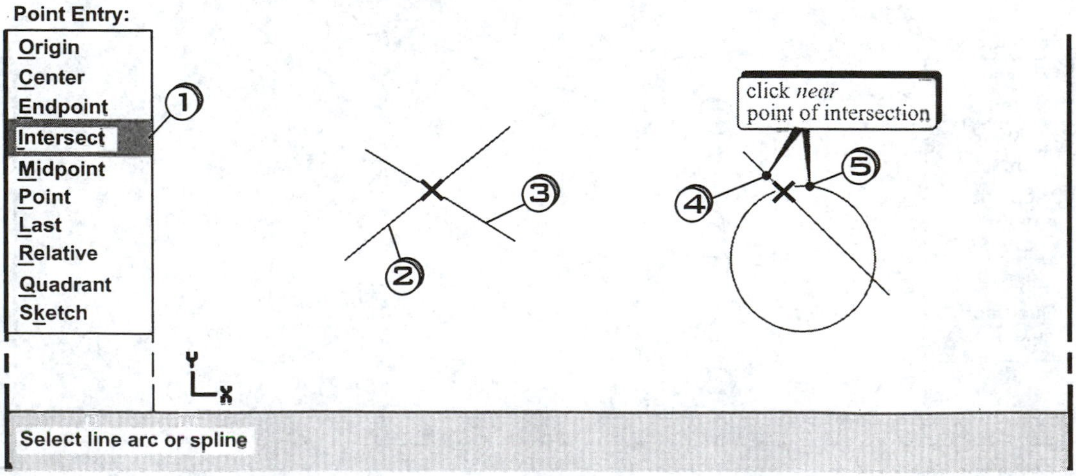

- Click (1) Intersect
- Click (2) (3)
- Click (4) Intersect
- Click (5) (6)
- Press Esc to cancel the operation

Point

A *point* is created at the location of an *existing point*

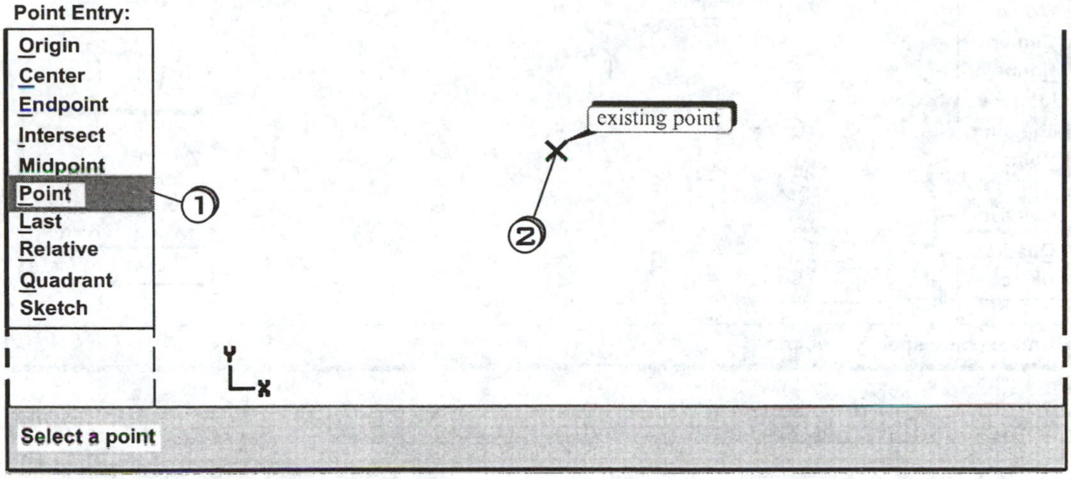

- Click ① **Point**
- Click ②
- Press **Esc** to cancel the operation

Last

A *point* is *automatically* created at the location of the *last point entered*

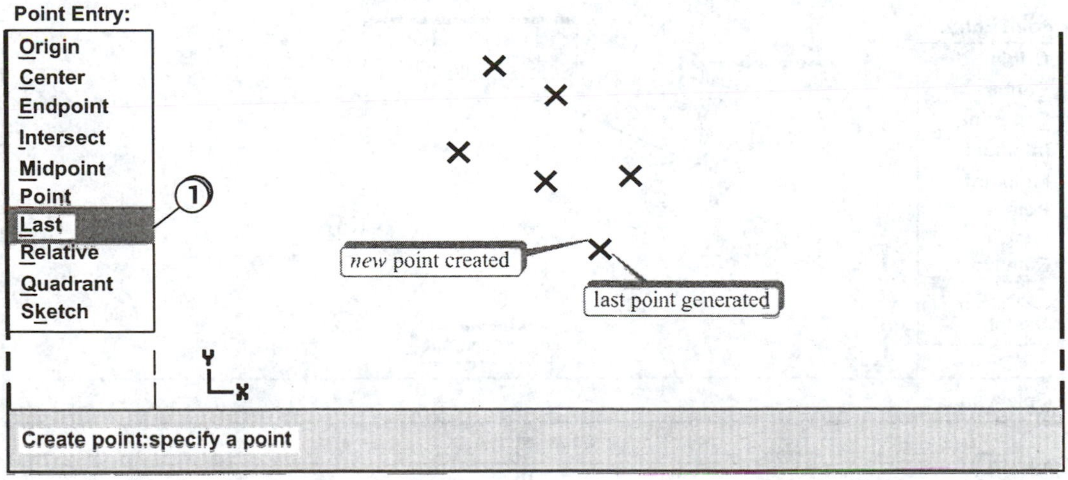

- Click ① **Last**
- Press **Esc** to cancel the operation

Relative

A *point* is created at a *relative* distance(*X,Y* or *Rad/Ang*) from an existing point

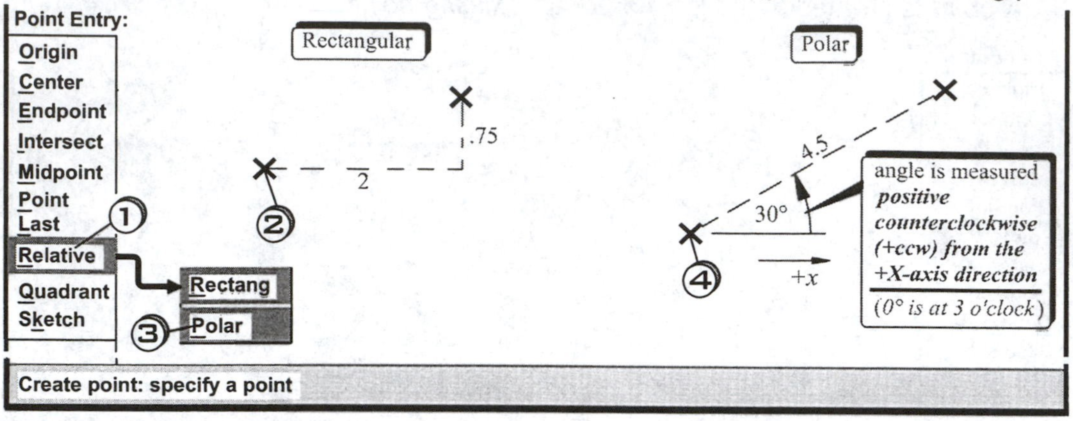

- Click ① Relative
- Click ②
- Enter relative rectang coords 2,.75 Enter
- Click ① Relative ; Click ③ Polar

- Click ④
- Enter the relative rad 4.5 Enter
- enter the relative angle: 30
- Press Esc to cancel the operation

Quadrant

A *point* is created at the *0°, 90°, 180° or 270°* angle on an existing *arc*

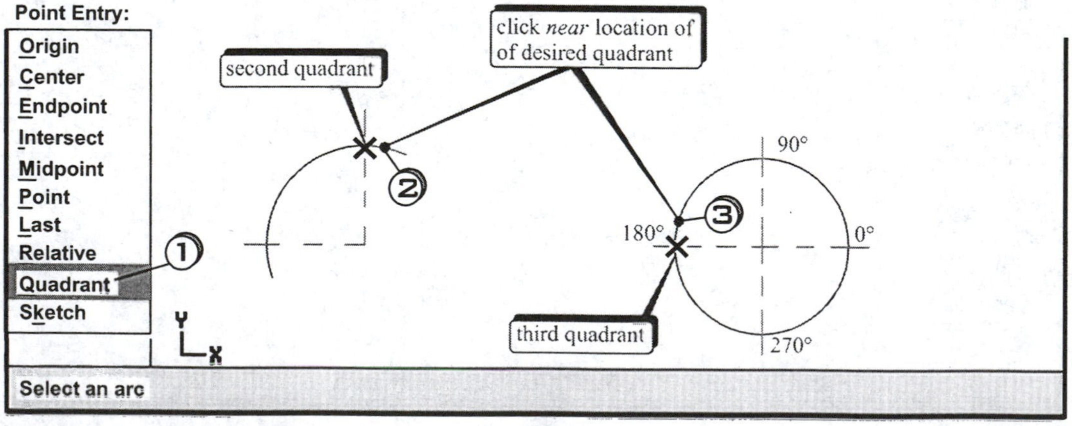

- Click ① Quadrant
- Click ② near the desired quadrant
- Click ① Quadrant

- Click ③ near the desired quadrant
- Press Esc to cancel the operation

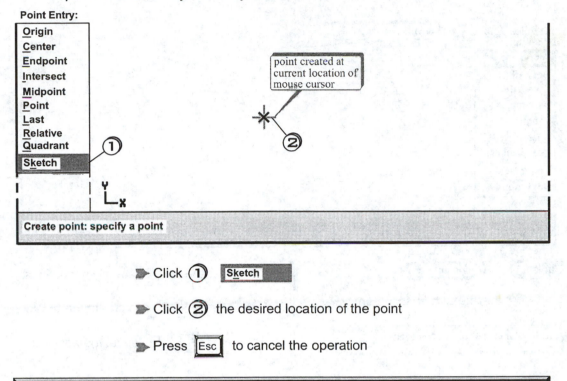

Sketch

A *point* is created *dynamically* at the *mouse click* location

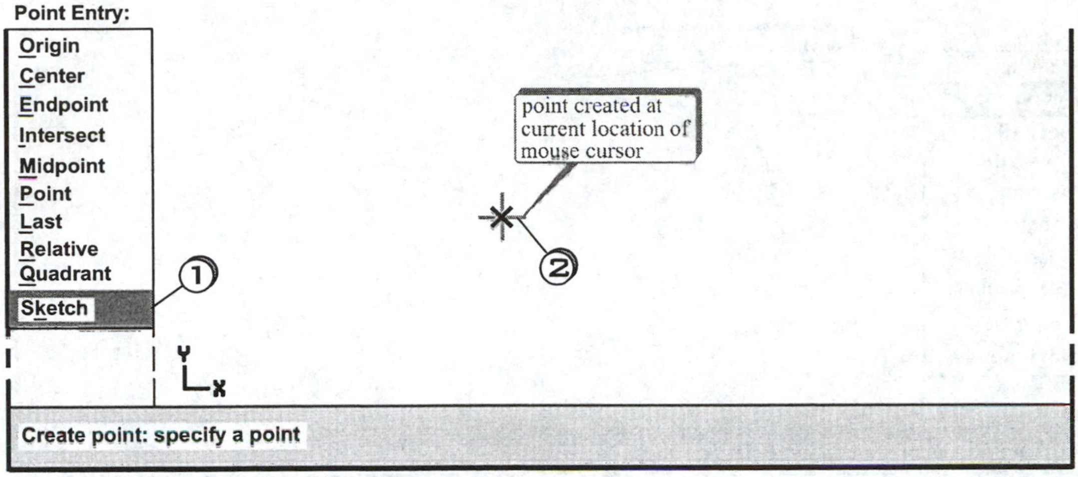

- Click ① **Sketch**
- Click ② the desired location of the point
- Press **Esc** to cancel the operation

Creating Points-2

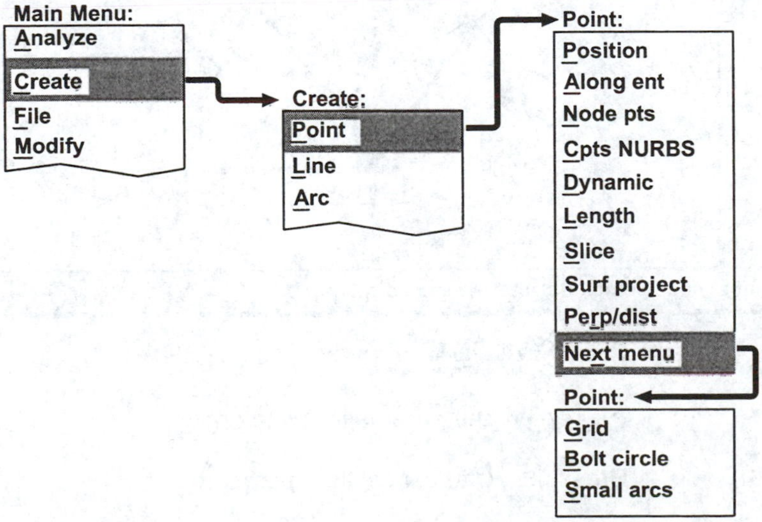

Along ent

Equally spaced points are created *along* an existing line, arc or spline entity

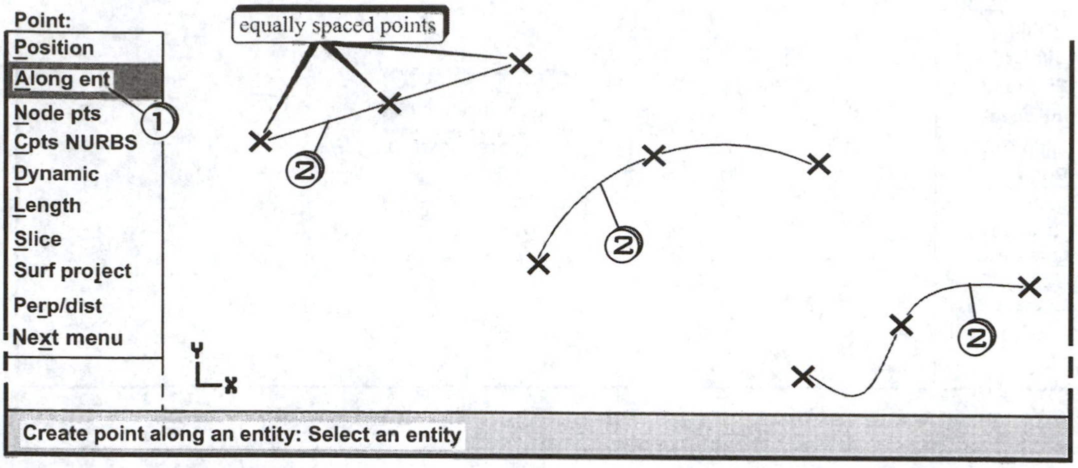

- Click ① **Along ent**
- Click ② the line, arc or spline entity

Enter number of points to create 3

- Enter the num of equally spaced pts 3
- Press Enter to create the points
- Press Esc to cancel the operation

Node pts

Points are created *at the node locations of a parametric(P) spline*

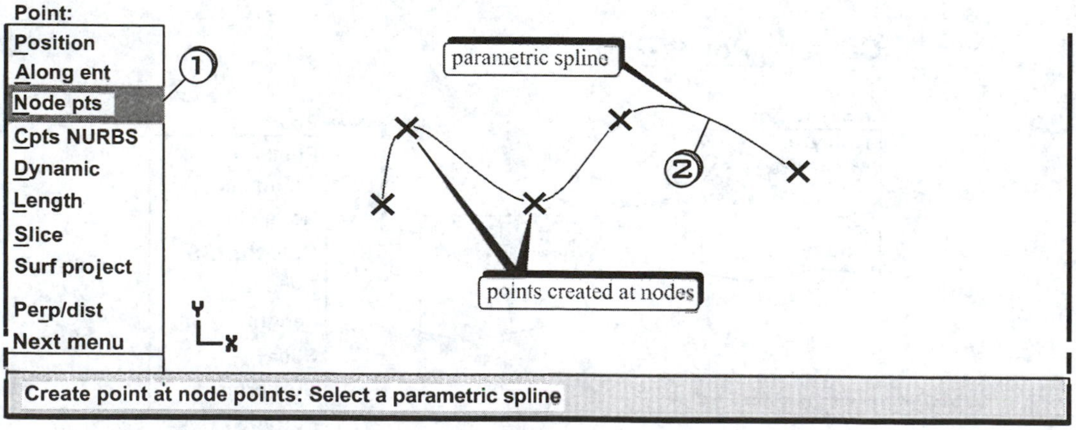

- Click ① **Node pts**
- Click ② the parametic spline entity
- Press Esc to cancel the operation

Cpts NURBS

Points are created *at the control locations* of an existing NURBS spline

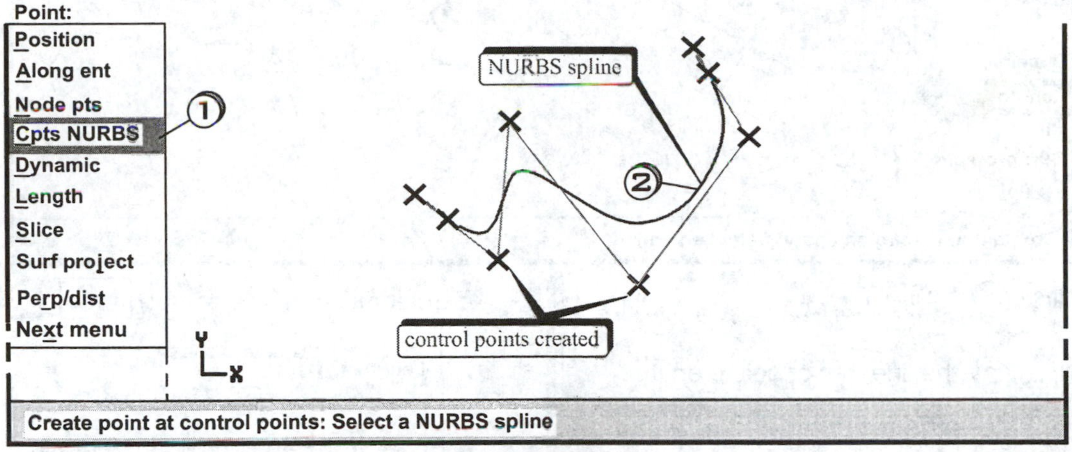

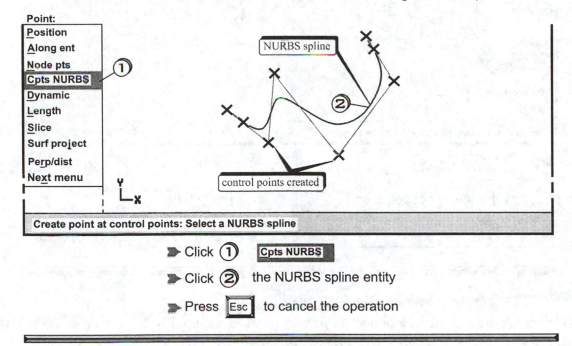

- Click ① Cpts NURBS
- Click ② the NURBS spline entity
- Press Esc to cancel the operation

Dynamic

Points are created *at any mouse click along* a line, arc or spline entity.

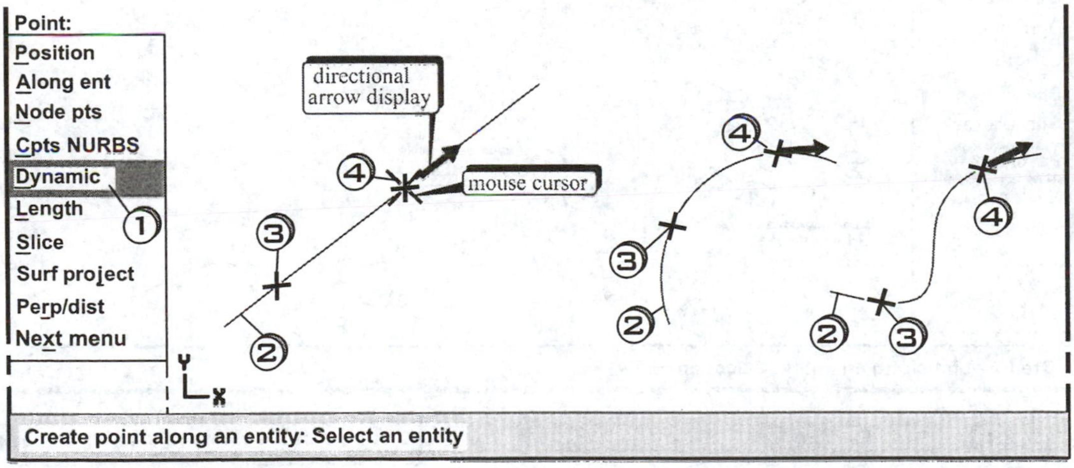

- Click ① Dynamic
- Click ② the line, arc or spline entity

Select points. Press<Esc> when done
Snapping is OFF - type<S> to turn off snapping

- Click the pt location(s) along the entitiy ③④
- Press Esc to cancel the operation

Length

Points are created at an *inputted length along* a line, arc or spline entity

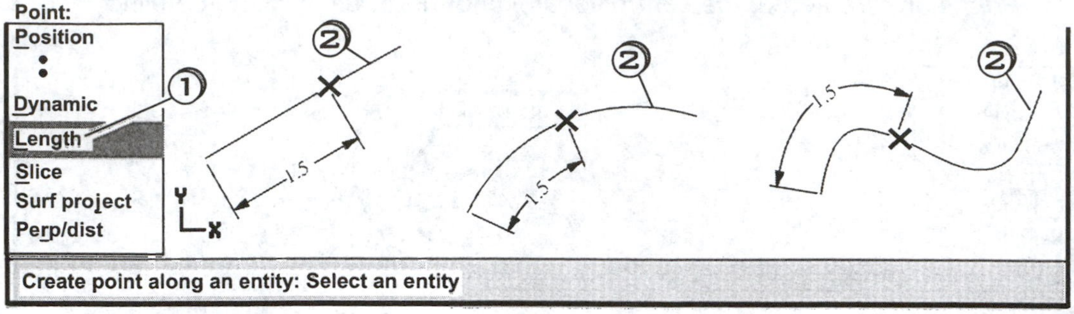

- Click ① Length
- Click ② the line, arc or spline entity

Length = 1

- Enter dist from the end pt 1.5 Enter

Length = 1.5

- Press Esc to cancel the operation

Perp/dist

Points are created at an *inputted perpendicular distance from* a line, arc or spline entity

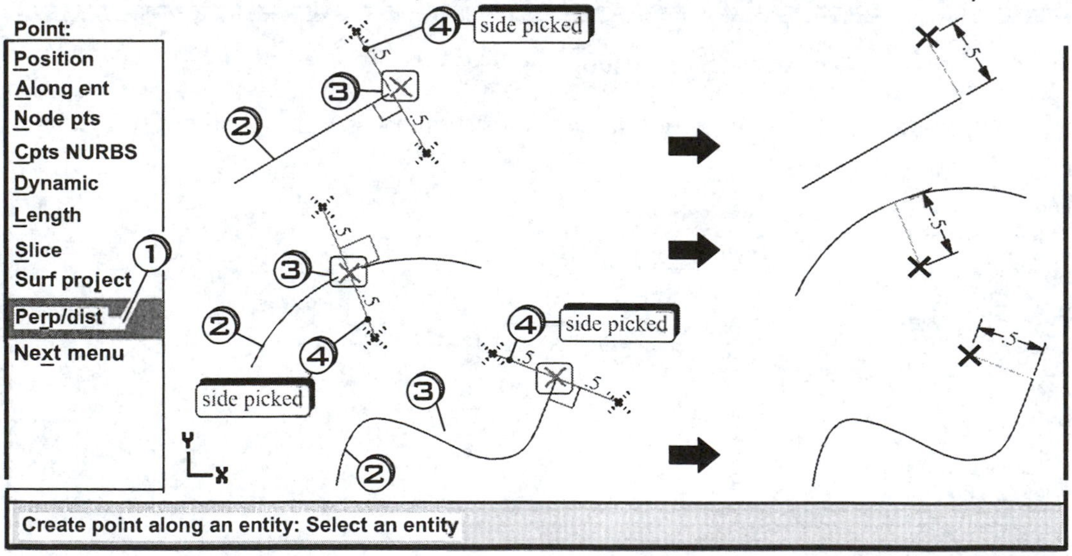

- Click ① Perp/dist
- Click ② the line, arc or spline entity

Select a point on or near the selected curve

- Click ③ the position along the entity, endpoint, center, etc

Dist from curve where pt is to be created

- Enter dist from the curve .5 Enter

Select a perp line to indicate side for new pt

- Click ④ the line indicating the side
- Press Esc to cancel the operation

Grid

Points are created in a rectangular grid pattern

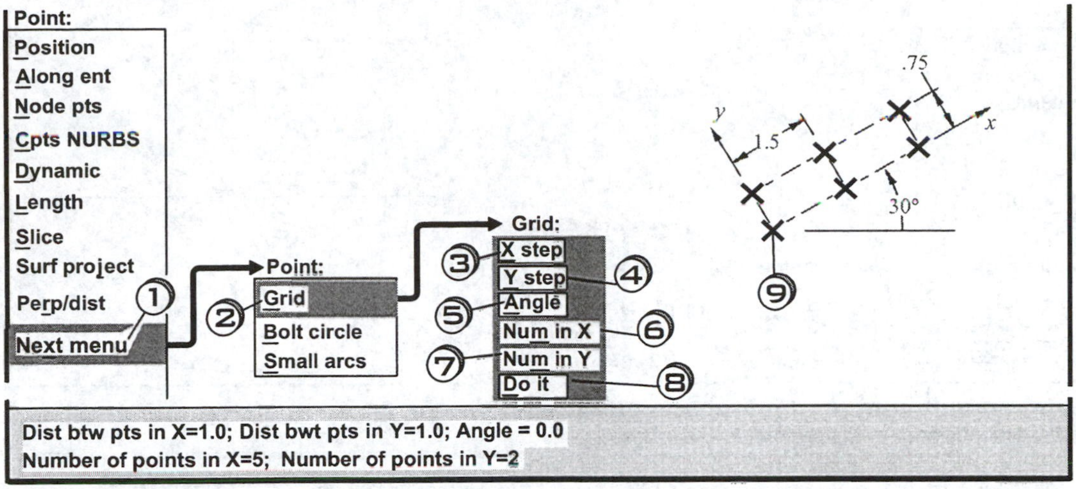

- Click ① **Next menu**
- Click ② **Grid**
- Click ③ **X step**

Distance between points in X=

- Enter the X distance **1.5** Enter
- Click ④ **Y step**

Distance between points in Y=

- Enter the Y distance **.75** Enter
- Click ⑤ **Angle**

Angle=

- Enter the angle **30** Enter
- Click ⑥ **Num in X**

Number of points in X=

- Enter the **num of pts in X: 3** Enter
- Click ⑦ **Num in Y**

Number of points in Y=

- Enter the **num of pts in Y: 2** Enter
- Click ⑧ **Do it**
- Click ⑨ the location of the grid
- Press Esc to cancel the operation

Bolt circle

Points are created in a *bolt circle* pattern

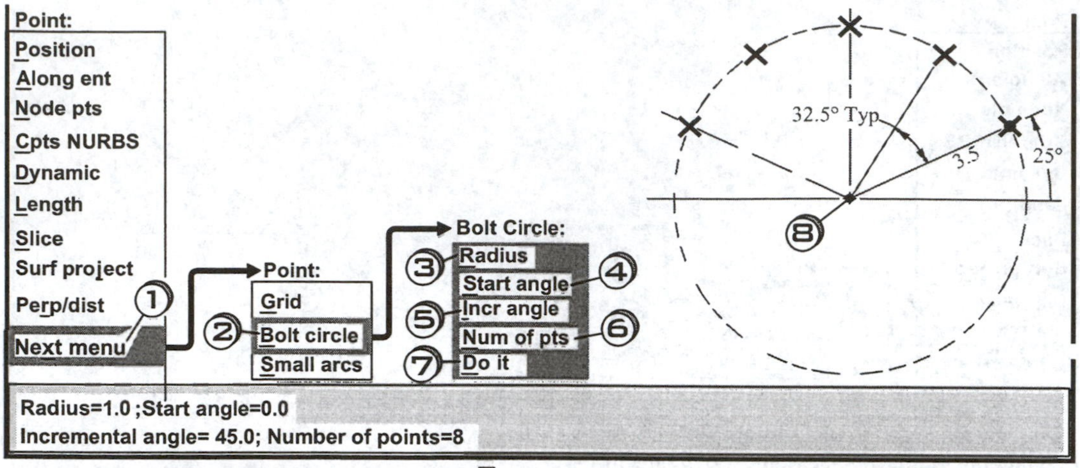

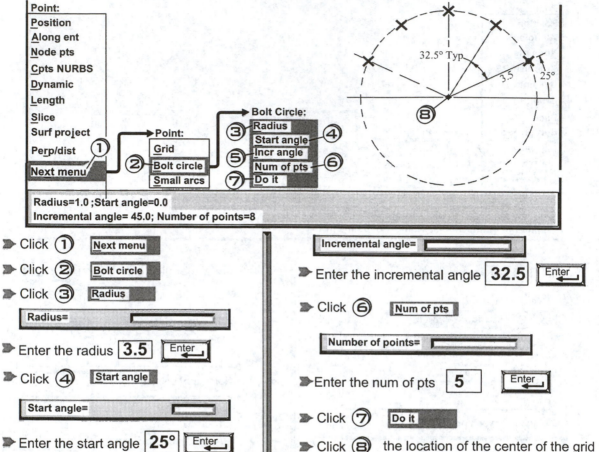

- Click ① **Next menu**
- Click ② **Bolt circle**
- Click ③ **Radius**

  Radius=

- Enter the radius **3.5** Enter
- Click ④ **Start angle**

  Start angle=

- Enter the start angle **25°** Enter
- Click ⑤ **Incr angle**

  Incremental angle=

- Enter the incremental angle **32.5** Enter
- Click ⑥ **Num of pts**

  Number of points=

- Enter the num of pts **5** Enter
- Click ⑦ **Do it**
- Click ⑧ the location of the center of the grid
- Press Esc to cancel the operation

Using the Toolbar Menus

Toolbar Menu Displayed when *Mastercam* is **Started**

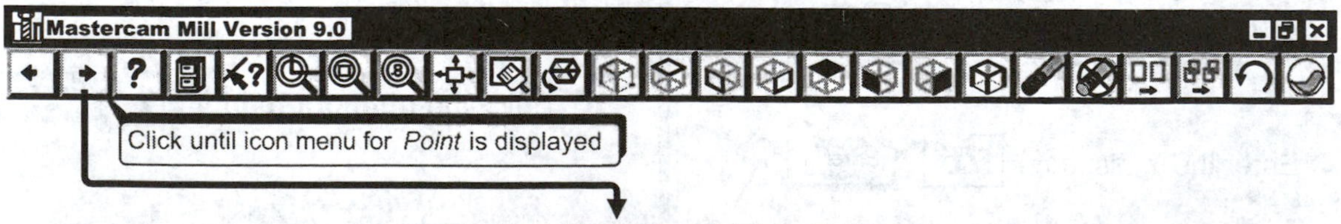

Toolbar Menu Containing ***Point*** Icon

Click Esc to cancel the points command

2-4 Line Constructions in 2D Space

The various commands for constructing lines via *Mastercam* are presented in this section.

Creating Lines

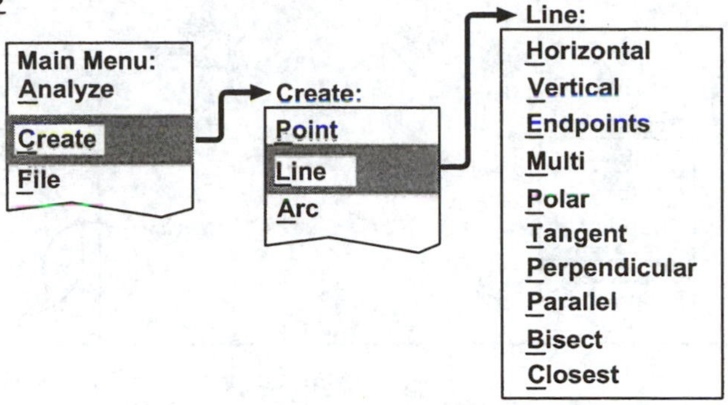

Horizontal

A *horizontal line* is created between inputted *start* and *end points.* The line is also *offset* a specified *distance* from the *x*-axis

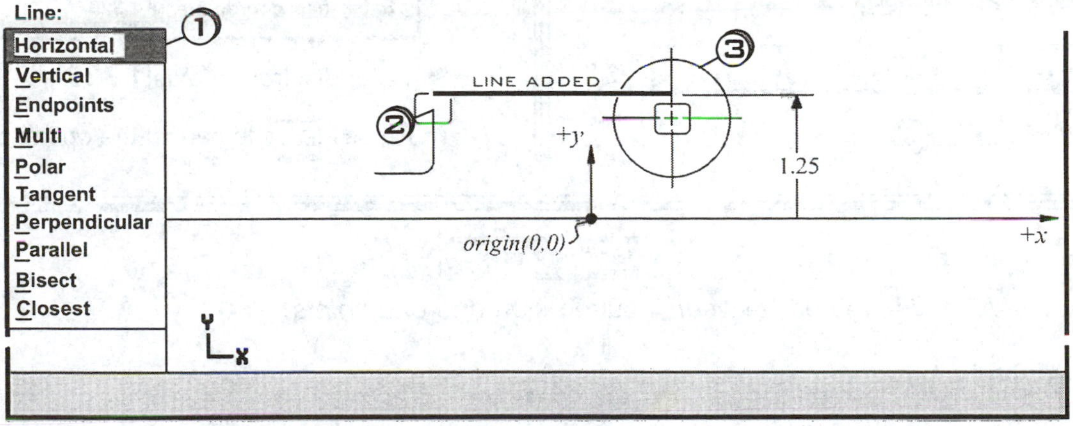

- Click ① **Horizontal**

Create line horizontal: enter the first endpoint

- Click ② near the endpoint of the line

Create line horizontal: enter the second endpoint

- Click ③ on the the circle's arc

Enter the y coordinate

- Enter dist from *x*-axis **1.25** Enter
- Press Esc to cancel the operation

Vertical

A *vertical line* is created between inputted *start* and *end points.* The line is also *offset* a specified *distance* from the y-axis

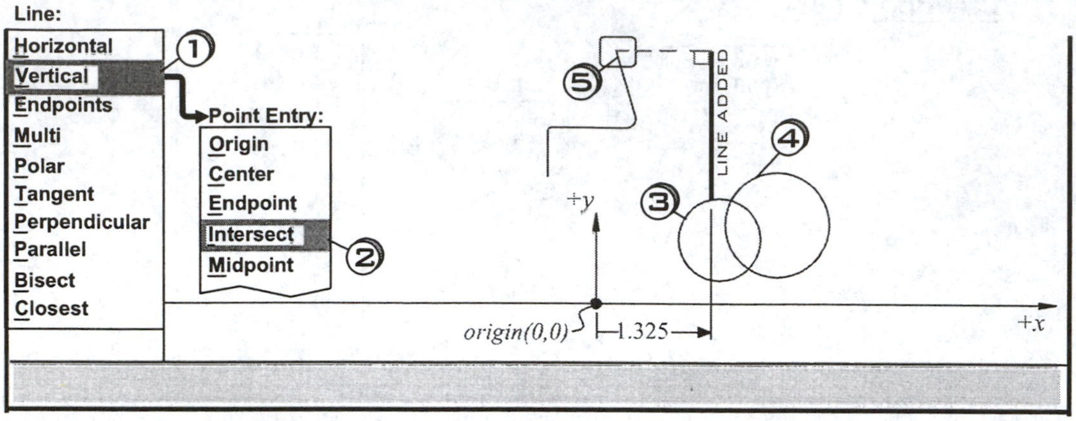

- Click ① **Vertical**

Create line vertical: enter the first endpoint

- Click ② **Intersect**

Select line arc or spline

- Click ③ ④

Create line vertical: enter the second endpoint

- Click ⑤ near the endpoint of the line

Enter the x coordinate **1.325**

- Enter dist from y-axis **1.325** Enter
- Press Esc to cancel the operation

Endpoints

A *line* is created *between* inputted *start* and *end points*

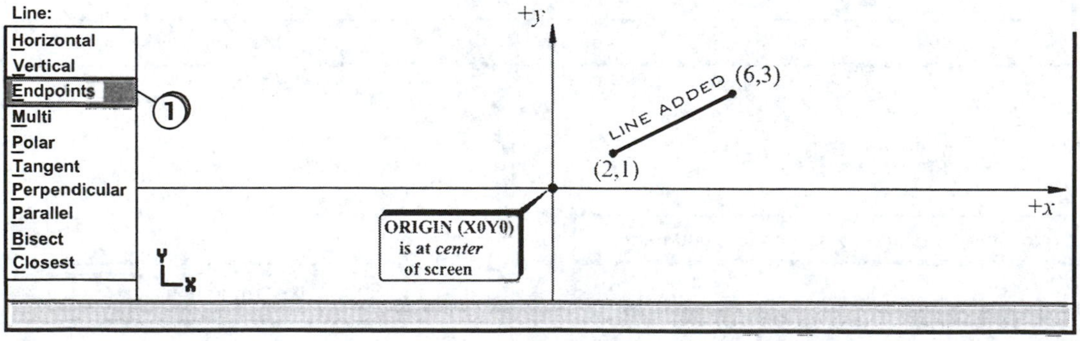

- Click ① **Endpoints**

Create line endpoints: Specify the first endpoint

- Enter the coordinates **2,1** Enter

Create line endpoints: Specify the second endpoint

- Enter the coordinates **6,3** Enter
- Press Esc to cancel the operation

Multi

Multiple lines are created between inputted *start* and *end points.*
Absolute coordinates are used: each new *X,Y is measured from the **origin***

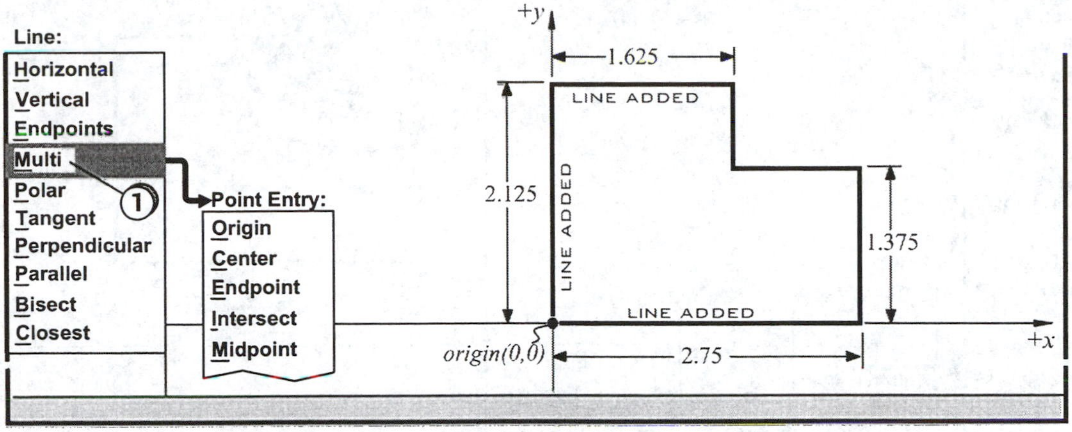

➤ Click ① Multi

Create line, multi: Specify endpoint 1

➤ Enter coords **0,0** Enter

➤ Enter coords **2.75,0** Enter

➤ Enter coords **2.75,1.375** Enter

➤ Enter coords **1.625,1.375** Enter

➤ Enter coords **1.625,2.125** Enter

➤ Enter coords **0,2.125** Enter

➤ Enter coords **0,0** Enter

➤ Press Esc to cancel the operation

Multi **Relative**

Multiple lines are created between inputted *start* and *end points.*

Relative coordinates are used: each new *X,Y is measured from the* ***last point entered***

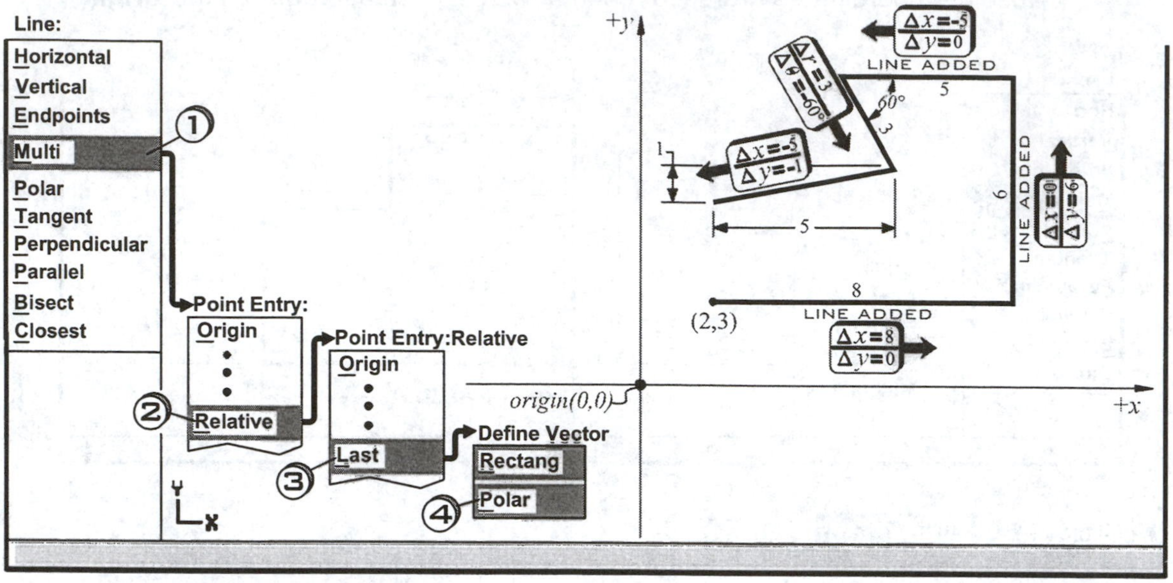

- Click ① **Multi**

Create line, multi: Specify endpoint 1

- Enter coords **2,3** Enter
- Click ② **Relative**
- Click ③ **Last**
- Enter relative rect coords **8,0** Enter
- Click ② **Relative**
- Click ③ **Last**
- Enter relative rect coords **0,6** Enter
- Click ② **Relative**
- Click ③ **Last**
- Enter relative rect coords **-5,0** Enter
- Click ② **Relative**
- Click ③ **Last**
- Click ④ **Polar**

Enter relative distance

- Enter the relative length **3** Enter

Enter relative angle

- Enter the relative angle **-60** Enter
- Click ② **Relative**
- Click ③ **Last**
- Enter relative rect coords **-5,-1** Enter
- Press Esc to cancel the operation

Polar

A *line* is created using *Angle*, *Length* *polar* coordinates

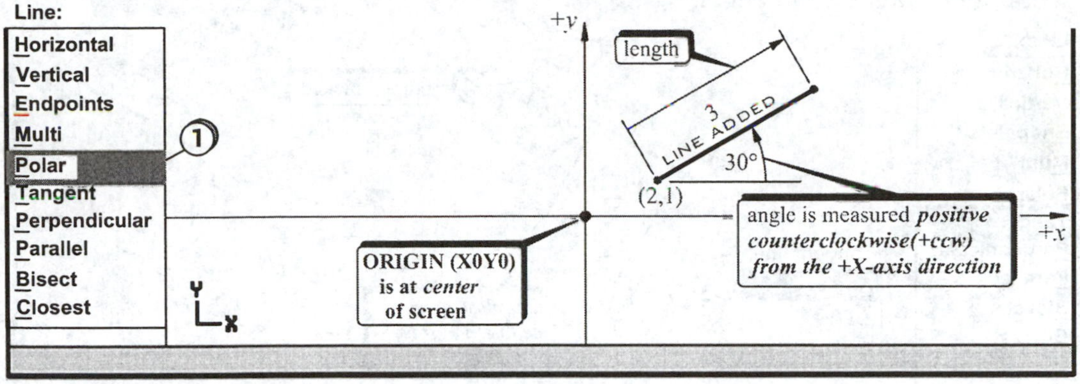

➤ Click ① **Polar**

Create line polar: Specify an endpoint

➤ Enter coords **2,1** Enter

Enter the angle in degrees

➤ Enter angle in deg **30** Enter

Enter the line length

➤ Enter line length **3** Enter

➤ Press Esc to cancel the operation

Tangent **Angle**

A *line* is created starting at a *tangency* to an *arc or spline* and having an inputted *Angle* and *Length.*

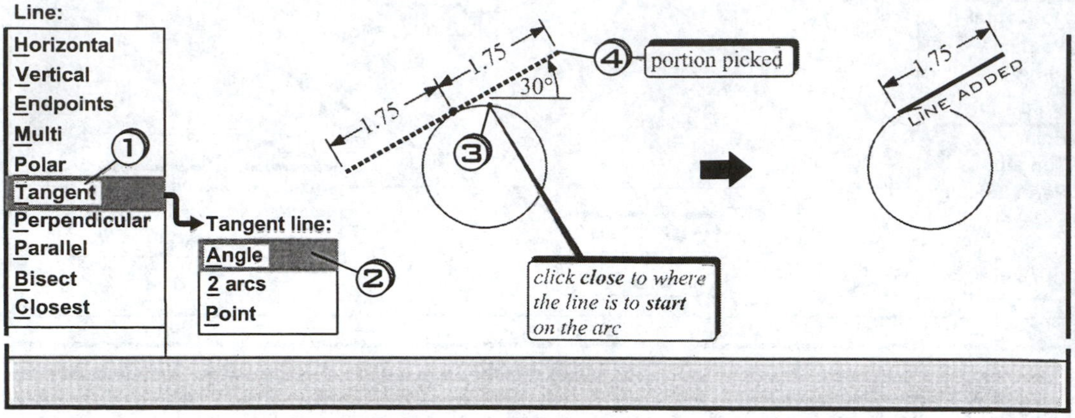

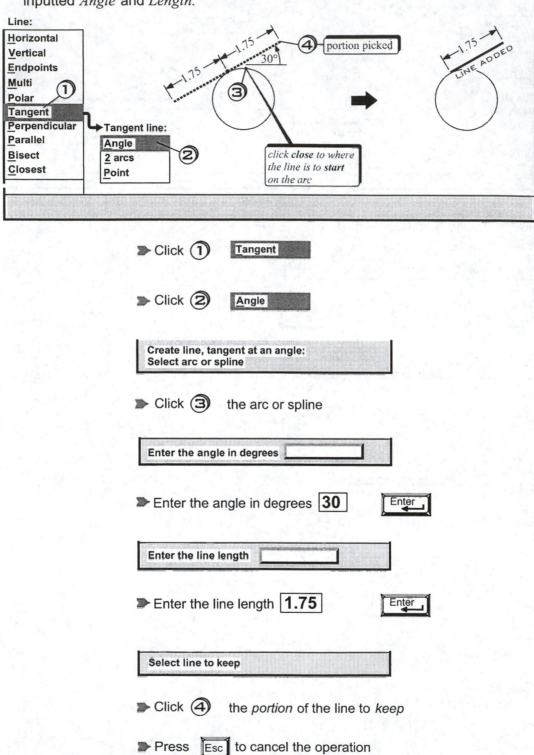

- Click ① **Tangent**
- Click ② **Angle**

Create line, tangent at an angle:
Select arc or spline

- Click ③ the arc or spline

Enter the angle in degrees

- Enter the angle in degrees **30** Enter

Enter the line length

- Enter the line length **1.75** Enter

Select line to keep

- Click ④ the *portion* of the line to *keep*
- Press Esc to cancel the operation

Tangent **2 arcs**

A line is created *tangent* to *two arcs*

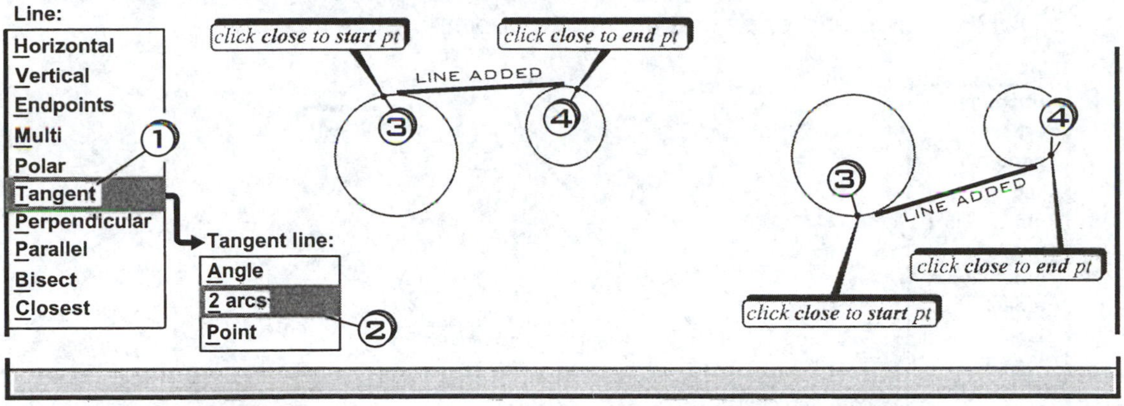

- Click ① **Tangent**
- Click ② **2 arcs**

Create line, tangent to two arcs: Select an arc

- Click ③ the arc for the line's tan start pt

Create line, tangent to two arcs: Select another arc

- Click ④ the arc for the line's tan end pt
- Press **Esc** to cancel the operation

Tangent **Point**

A *line* is created starting at a *tangency* to an *arc* or *spline* and *ending* at an inputted *point* location.

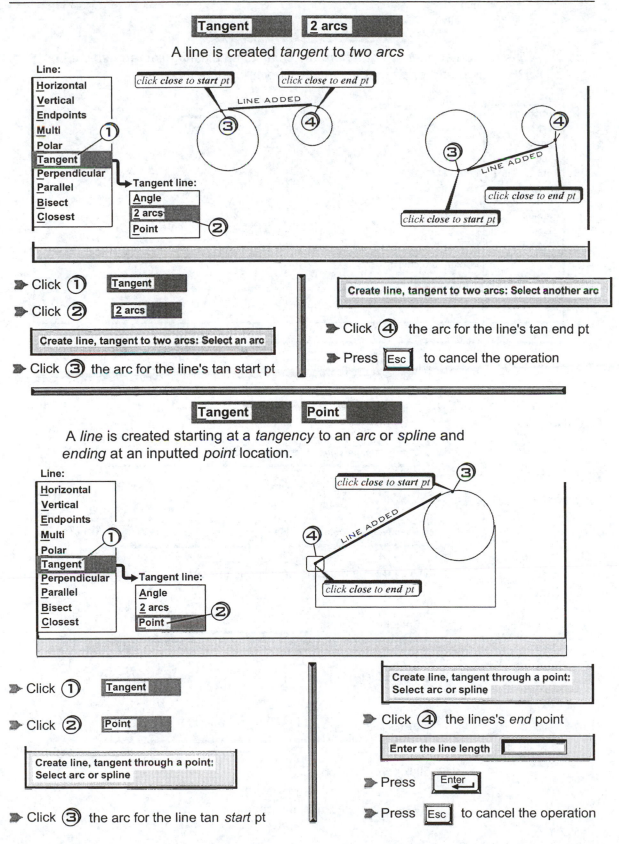

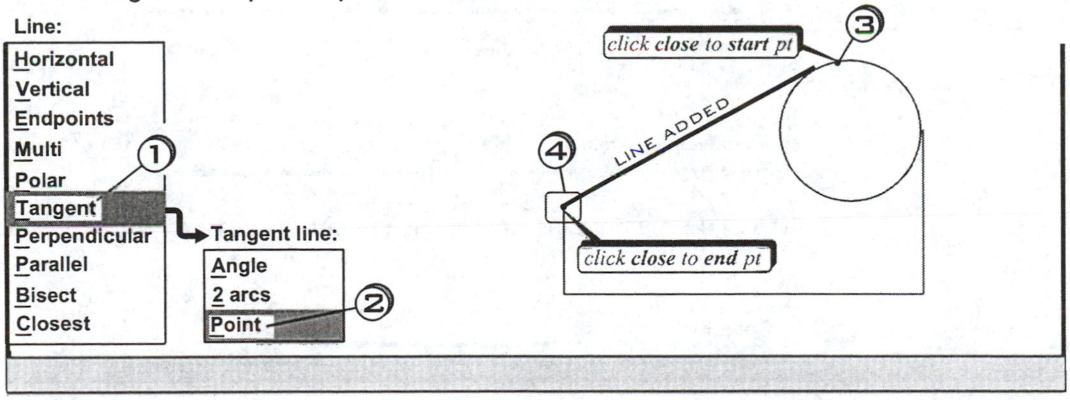

- Click ① **Tangent**
- Click ② **Point**

Create line, tangent through a point: Select arc or spline

- Click ③ the arc for the line tan *start* pt

Create line, tangent through a point: Select arc or spline

- Click ④ the lines's *end* point

Enter the line length

- Press **Enter**
- Press **Esc** to cancel the operation

Perpendicular | **Point**

A *line* is created *starting* at a *perpendicular* to an existing *line arc, spline* or NURBS curve and *ending* at an inputted *point* location.

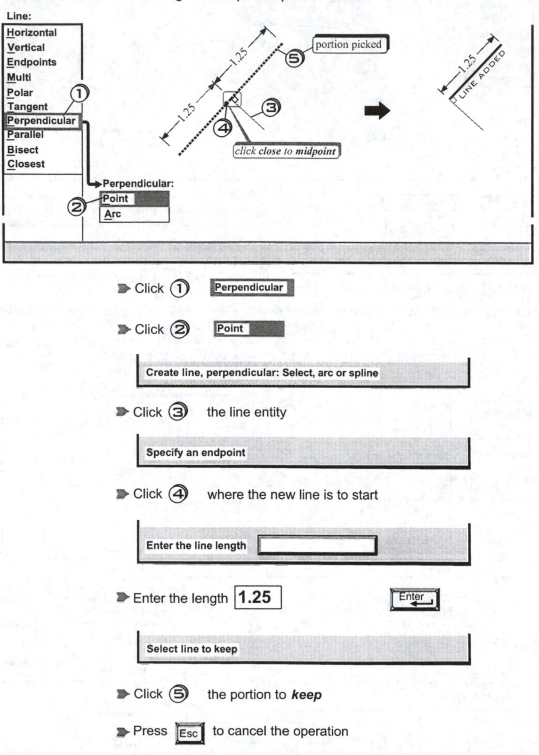

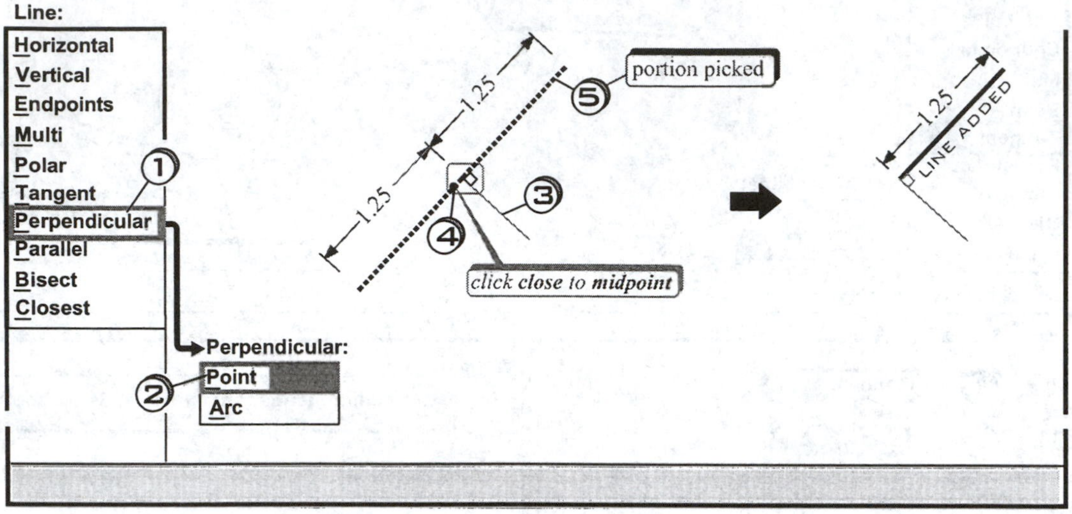

➤ Click ① **Perpendicular**

➤ Click ② **Point**

Create line, perpendicular: Select, arc or spline

➤ Click ③ the line entity

Specify an endpoint

➤ Click ④ where the new line is to start

Enter the line length

➤ Enter the length **1.25** Enter

Select line to keep

➤ Click ⑤ the portion to ***keep***

➤ Press Esc to cancel the operation

Perpendicular **Arc**

A *line* is created *starting* at a *perpendicular* to an existing *line* and *ending tangent* to an existing *arc.*

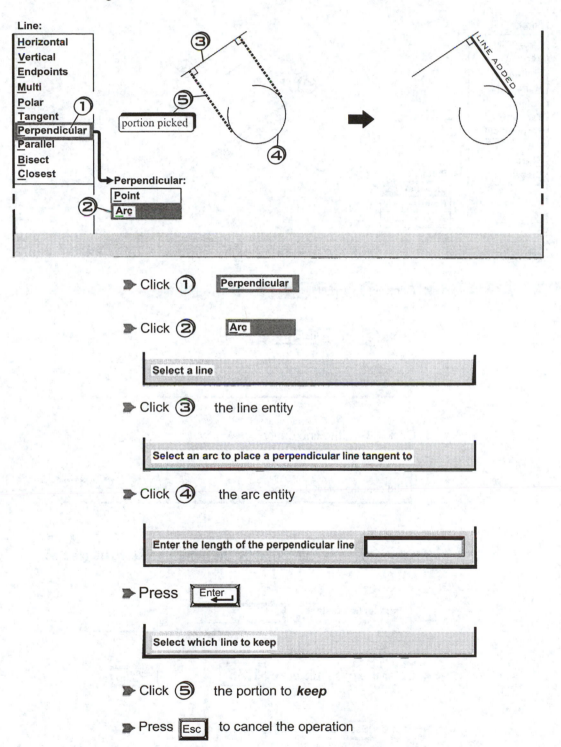

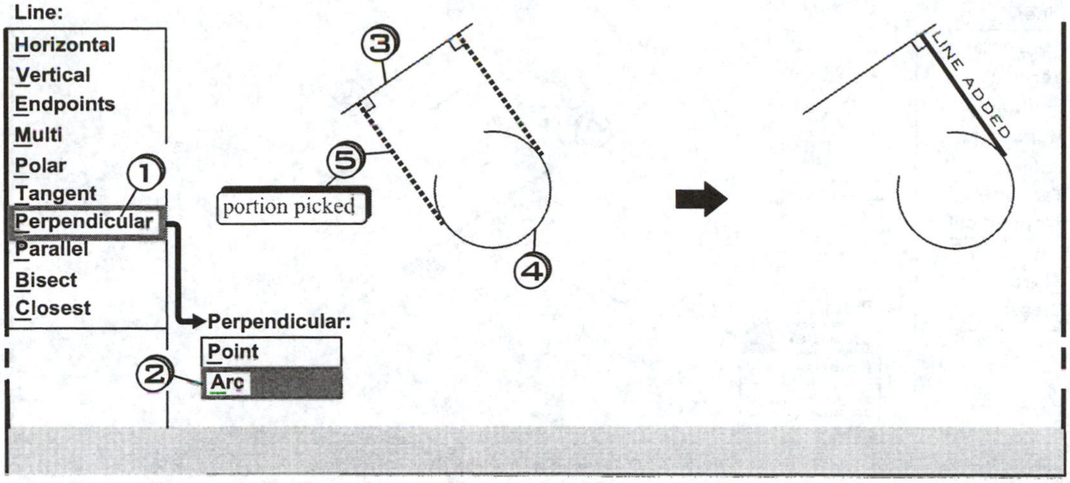

- Click ① **Perpendicular**
- Click ② **Arc**

Select a line

- Click ③ the line entity

Select an arc to place a perpendicular line tangent to

- Click ④ the arc entity

Enter the length of the perpendicular line

- Press Enter

Select which line to keep

- Click ⑤ the portion to ***keep***
- Press Esc to cancel the operation

Parallel **Side/dist**

A *line* is created *parallel* to an existing *line* at an *offset distance* inputted by the operator.

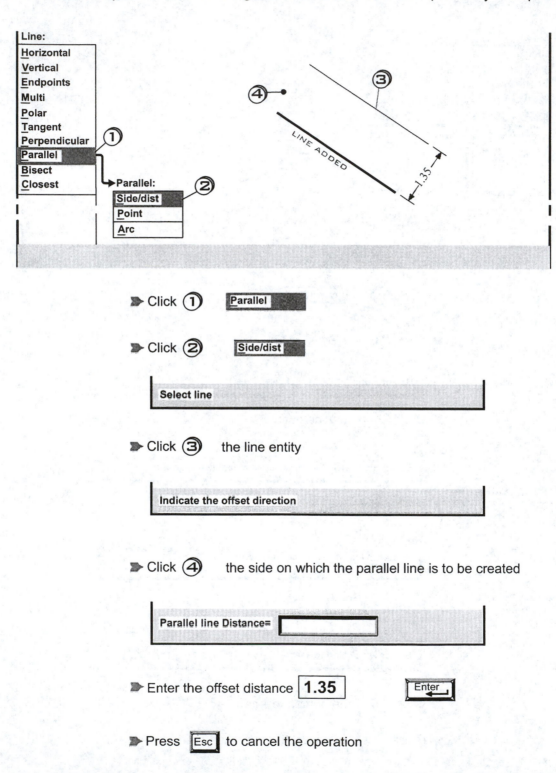

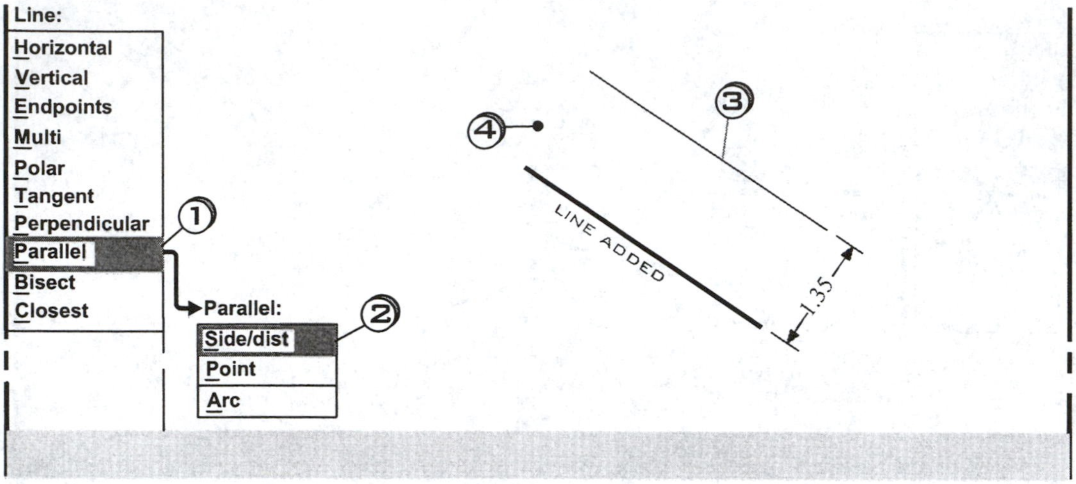

➢ Click ① **Parallel**

➢ Click ② **Side/dist**

Select line

➢ Click ③ the line entity

Indicate the offset direction

➢ Click ④ the side on which the parallel line is to be created

Parallel line Distance=

➢ Enter the offset distance **1.35** Enter

➢ Press Esc to cancel the operation

Parallel **Point**

A *line* is created *parallel* to an existing *line* and *passing through* a specified *point* location

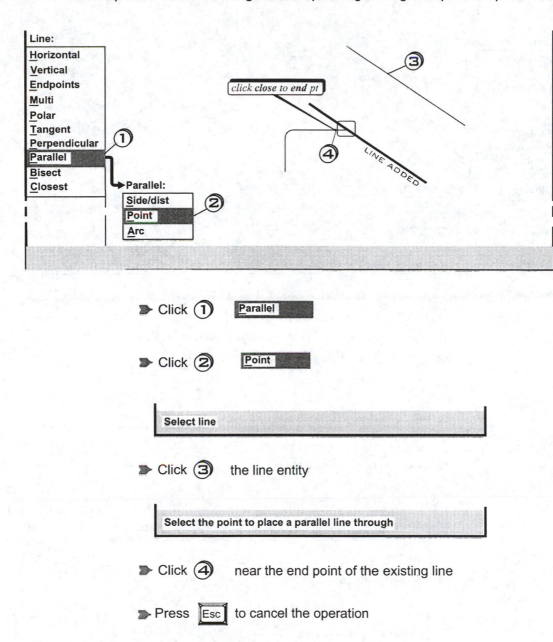

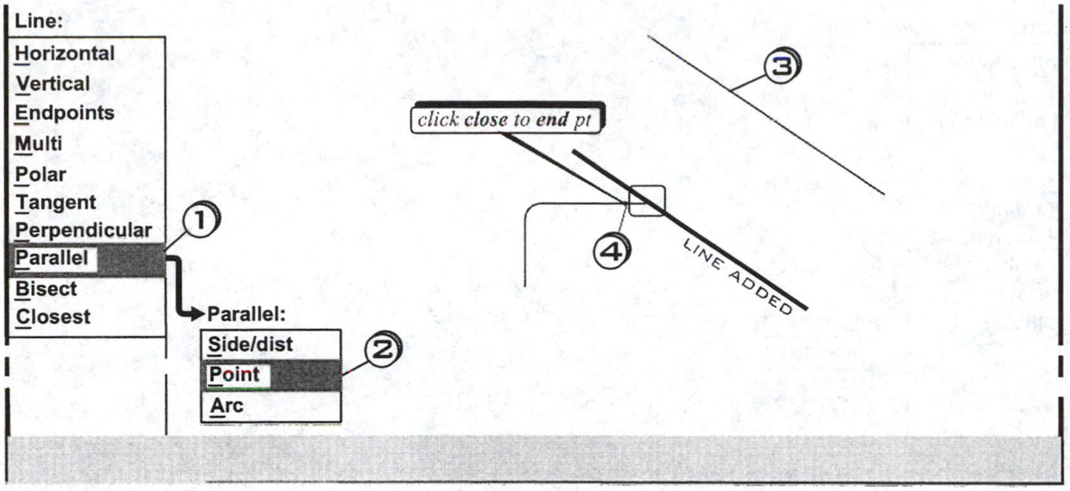

- Click ① **Parallel**
- Click ② **Point**

Select line

- Click ③ the line entity

Select the point to place a parallel line through

- Click ④ near the end point of the existing line
- Press **Esc** to cancel the operation

Parallel **Arc**

A *line* is created *parallel* to an existing *line* and *tangent* to an existing *arc*

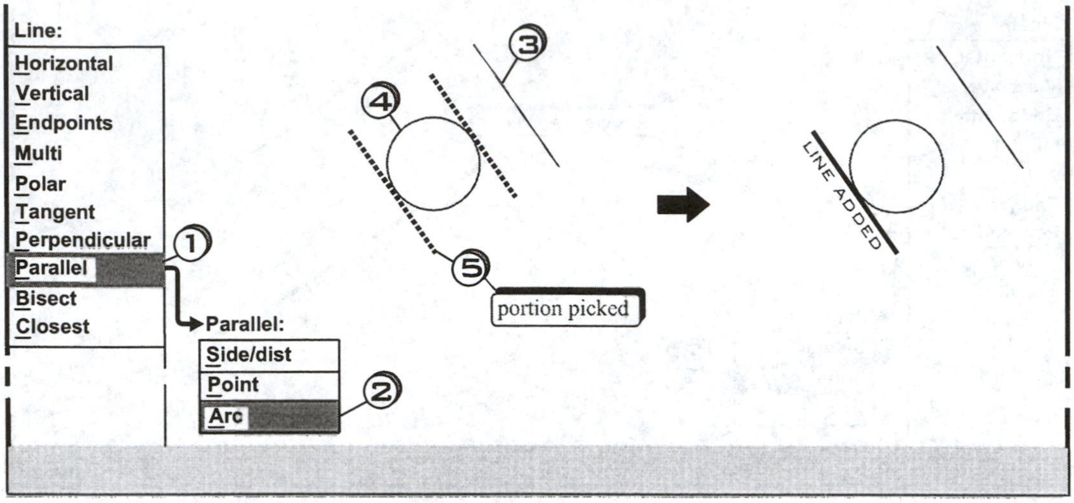

- Click ① **Parallel**
- Click ② **Arc**

Select line

- Click ③ the line entity

Select an arc to place a parallel line tangent to.

- Click ④ the arc entity

Select which line to keep

- Click ⑤ the portion to ***keep***
- Press **Esc** to cancel the operation

Bisect

A *line* of specified *length* is created that *bisects the angle between* two *intersecting lines*

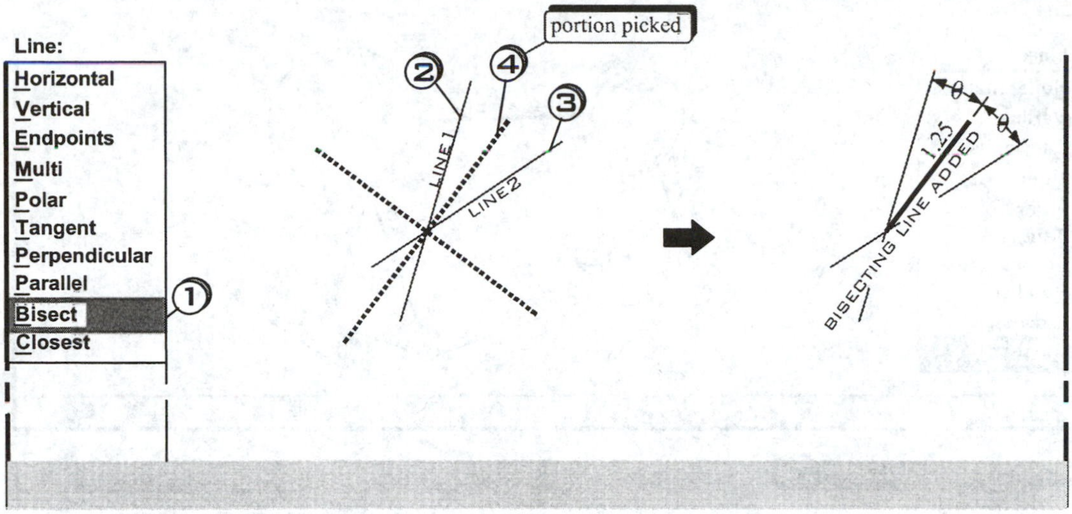

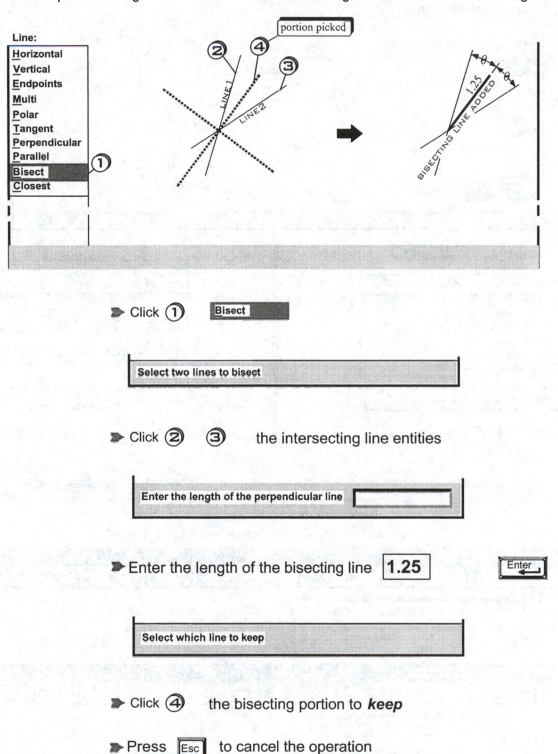

➤ Click ① **Bisect**

Select two lines to bisect

➤ Click ② ③ the intersecting line entities

Enter the length of the perpendicular line

➤ Enter the length of the bisecting line **1.25** Enter

Select which line to keep

➤ Click ④ the bisecting portion to ***keep***

➤ Press Esc to cancel the operation

Closest

A *line* of specified *length* is created *between* the *closest points* of *adjacent* lines, arcs or splines

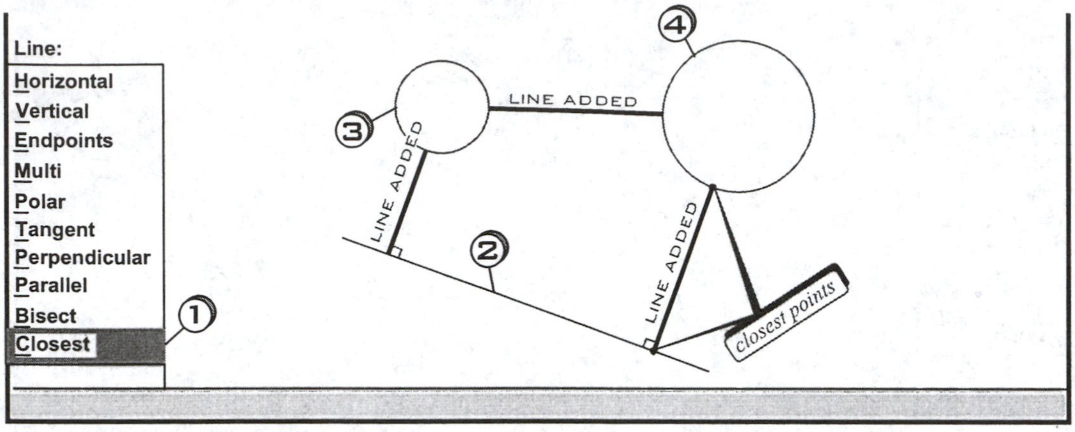

➤ Click ① **Closest**

Select line, arc, or spline

➤ Click ② ③ the line and circle

Select line, arc, or spline
2D Length: 1.12908

Select line, arc, or spline

➤ Click ③ ④ the circles

Select line, arc, or spline
2D Length: 1.12908

Select line, arc, or spline

➤ Click ② ④ the line and circle

Select line, arc, or spline
2D Length: 1.12908

➤ Press Esc to cancel the operation

Using the Toolbar Menus

Toolbar Menu Displayed when *Mastercam* is **Started**

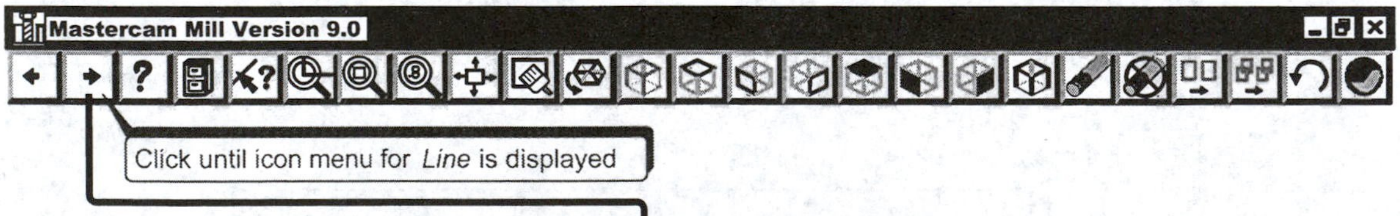

Toolbar Menu Containing ***Line*** Icon

Click Esc to cancel the line commands

2-5 Arc Constructions in 2D Space

The commands for constructing arcs(including circles) via *Mastercam* are considered in this section

Creating Arcs

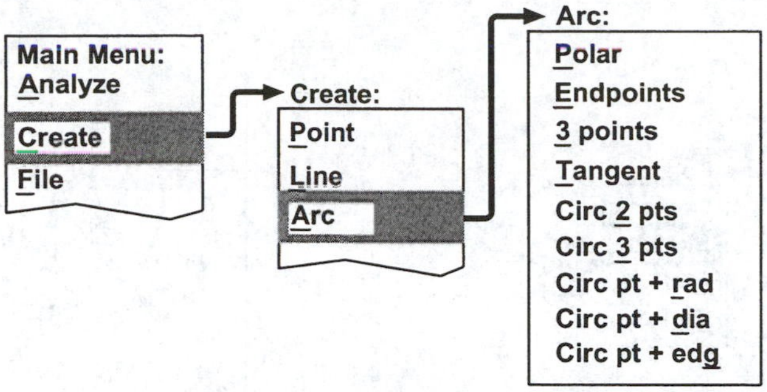

Polar **Center pt**

An *arc* is created at inputted *center point, radius, starting angle* and *ending angle* values.

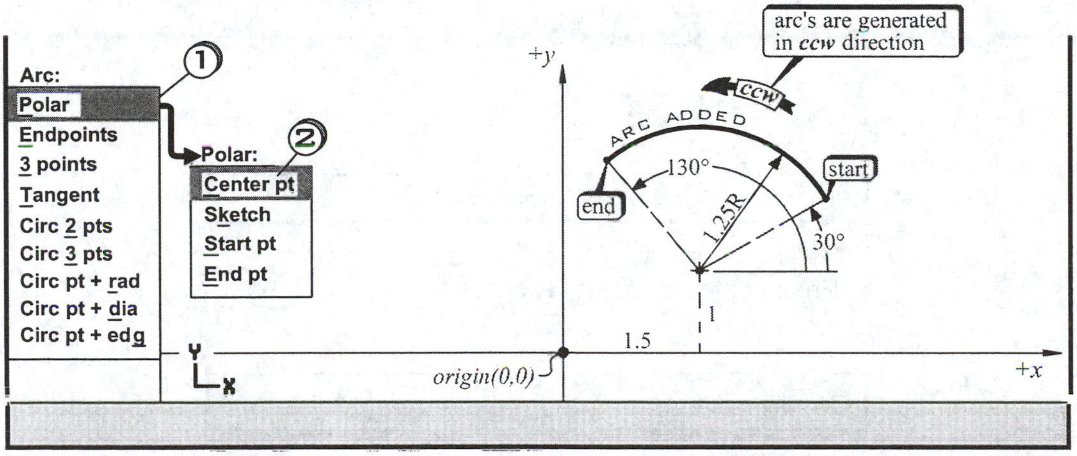

- Click ① **Polar**
- Click ② **Center pt**

Create arc, polar: Enter the center point

- Enter center pt coords **1.5,1** Enter
- Enter arc's radius **1.25** Enter

Enter the initial angle

- Enter arc's starting angle **30** Enter

Enter the final angle

- Enter arc's end angle **130** Enter
- Press Esc to cancel the operation

Polar **Sketch**

An *arc* is created at inputted *center point and radius values. Mouse clicks* specify the *start* and *end angles*.

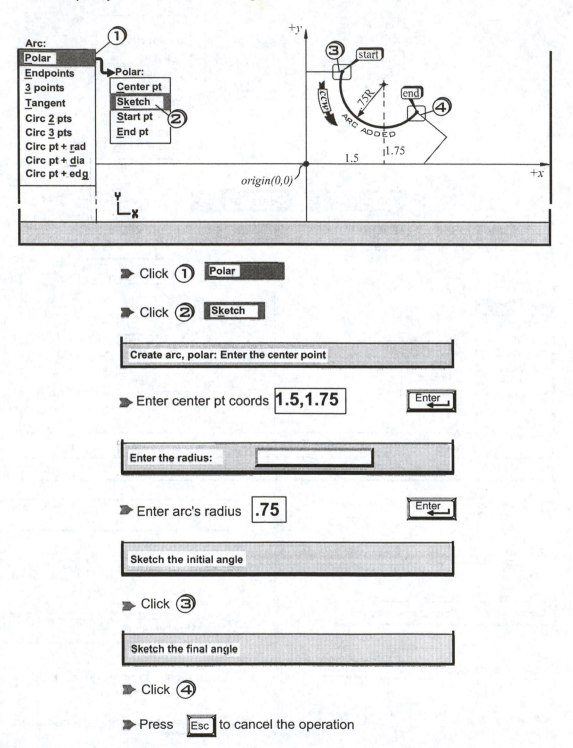

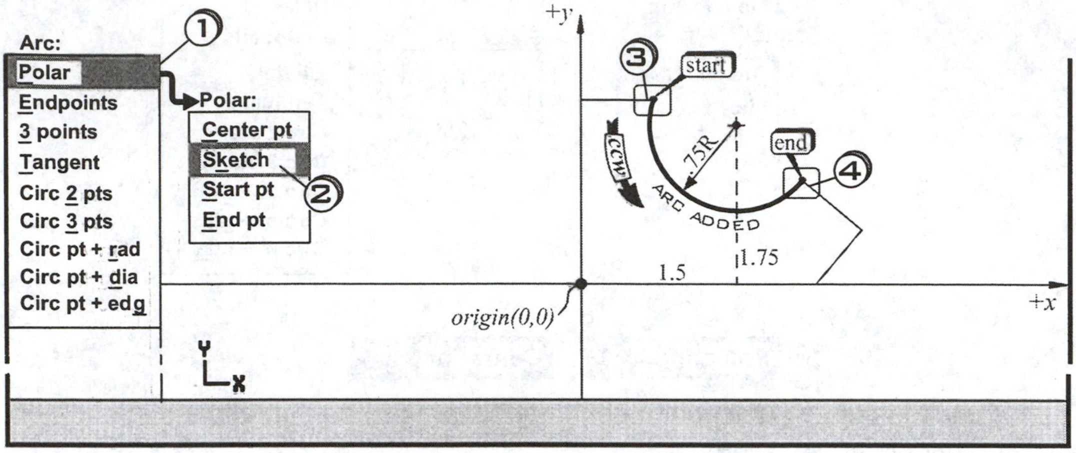

- Click ① **Polar**
- Click ② **Sketch**

Create arc, polar: Enter the center point

- Enter center pt coords **1.5,1.75** Enter

Enter the radius:

- Enter arc's radius **.75** Enter

Sketch the initial angle

- Click ③

Sketch the final angle

- Click ④
- Press Esc to cancel the operation

Polar **Start pt**

An *arc* is created *starting* at an inputted *start* point and having specified *values* for its *radius* and *start/end angles*.

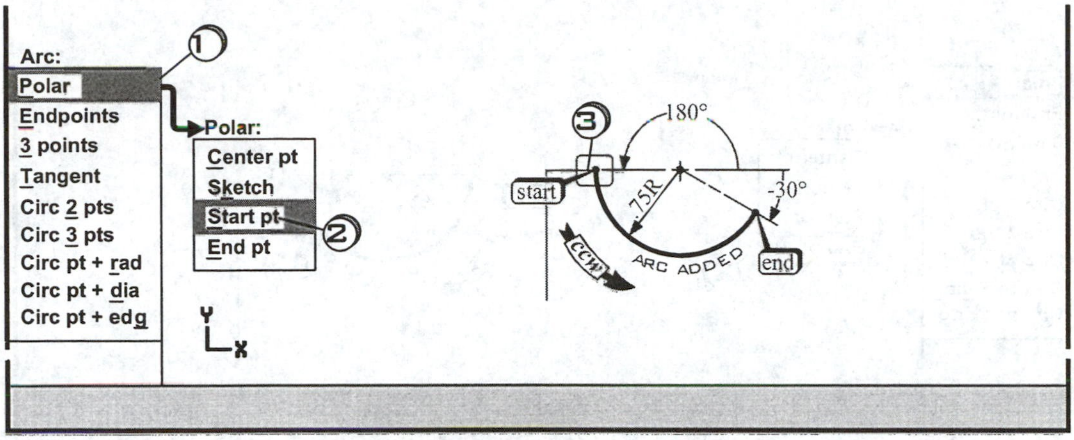

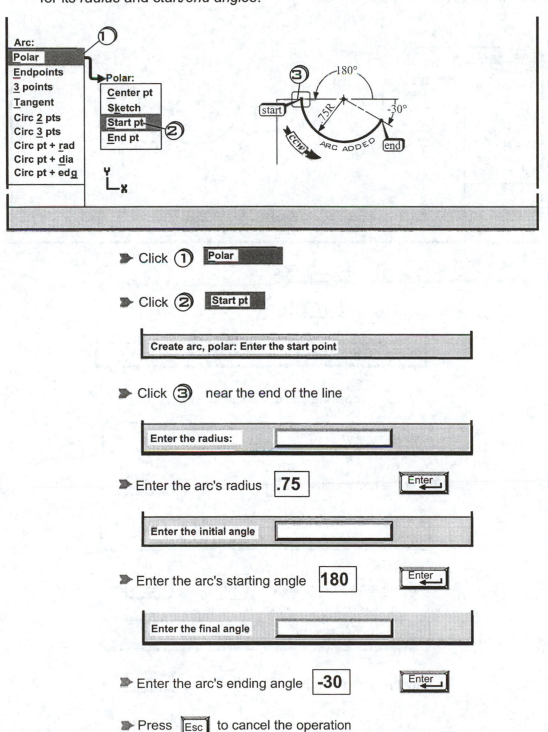

- Click ① **Polar**
- Click ② **Start pt**

Create arc, polar: Enter the start point

- Click ③ near the end of the line

Enter the radius:

- Enter the arc's radius **.75** Enter

Enter the initial angle

- Enter the arc's starting angle **180** Enter

Enter the final angle

- Enter the arc's ending angle **-30** Enter
- Press Esc to cancel the operation

Polar **End pt**

An *arc* is created *ending* at an inputted *end* point and having specified *values* for its *radius* and *start/end angles*.

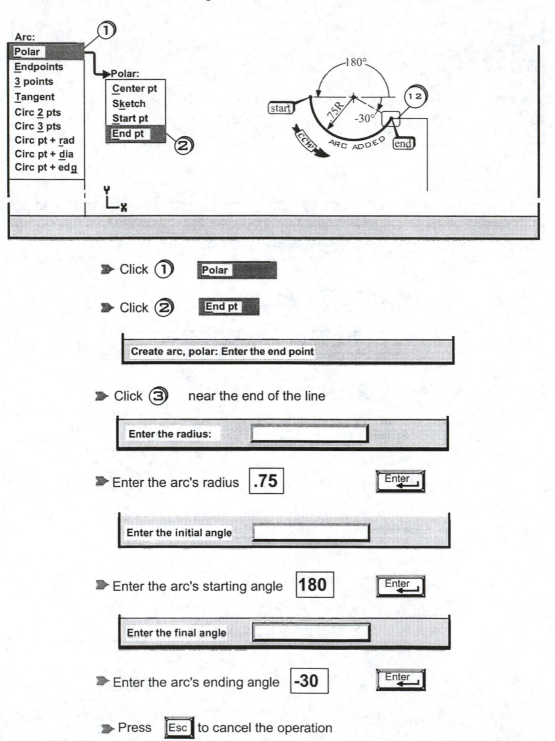

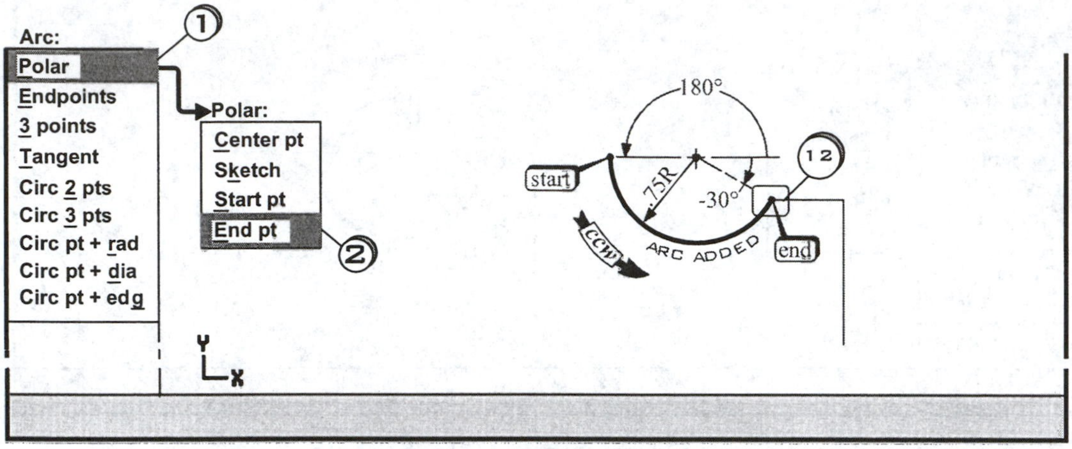

- Click ① **Polar**
- Click ② **End pt**

Create arc, polar: Enter the end point

- Click ③ near the end of the line

Enter the radius:

- Enter the arc's radius **.75** Enter

Enter the initial angle

- Enter the arc's starting angle **180** Enter

Enter the final angle

- Enter the arc's ending angle **-30** Enter
- Press Esc to cancel the operation

Endpoints

Arcs are created *starting/ending* at inputted *start/end* points and having a specified *radius value.* The operator *clicks* the *portion* of the arc to ***keep***.

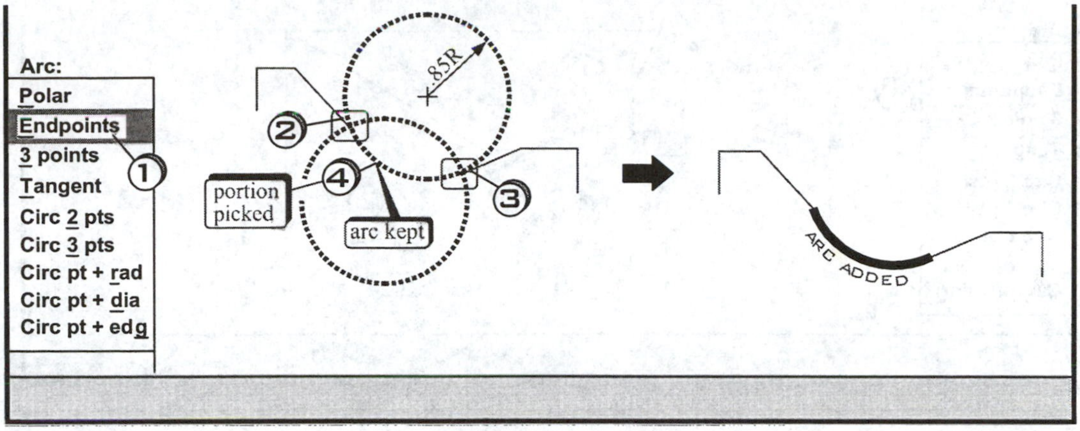

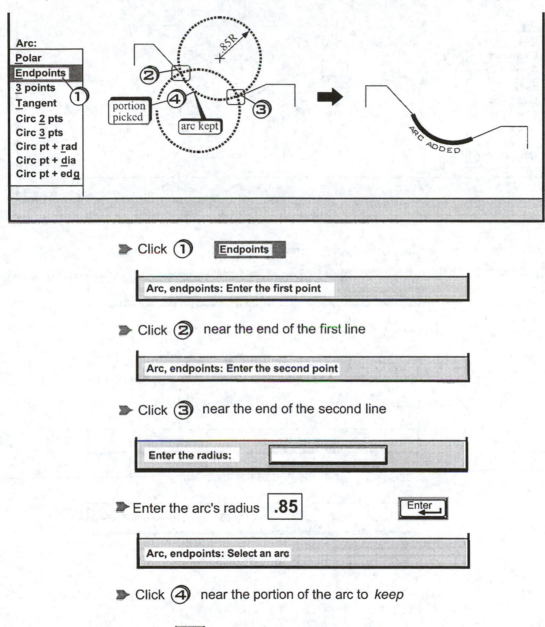

- Click ① **Endpoints**

Arc, endpoints: Enter the first point

- Click ② near the end of the first line

Arc, endpoints: Enter the second point

- Click ③ near the end of the second line

Enter the radius:

- Enter the arc's radius **.85** Enter

Arc, endpoints: Select an arc

- Click ④ near the portion of the arc to *keep*

- Press Esc to cancel the operation

3 points

An *arc* is created passing through *three* inputted points. The *first* point specifies the arc's *start* location and the *third* point the arc's *end* location.

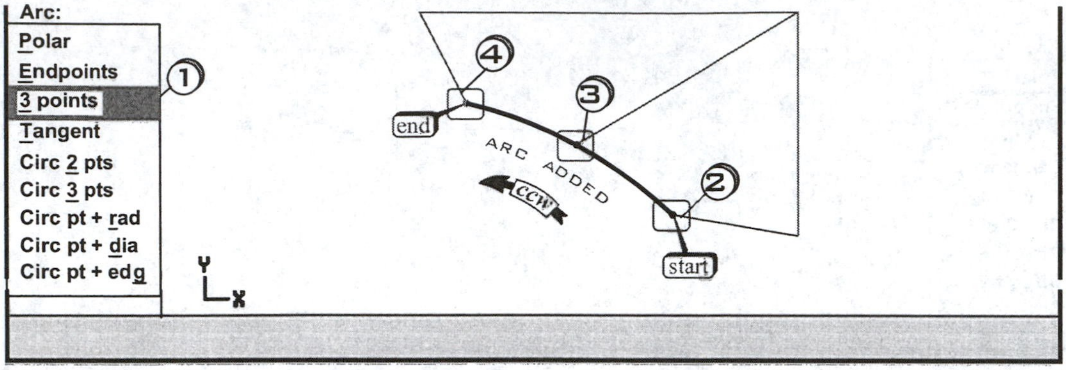

- Click ① **3 points**

  Arc, 3 points: Enter the first point

- Click ② near the end of the *first* line

  Arc, 3 points: Enter the second point

- Click ③ near the end of the *second* line

  Arc, 3 points: Enter the third point

- Click ④ near the end of the *third* line

  Radius=3.69234; Start Angle = 47.80050; Sweep Angle =28.55946

- Press Esc to cancel the operation

Tangent **1 entity**

180-degree arcs are created *tangent* to an existing line or arc entity at a specified *point along* the entity. The operator inputs the *radius* and *clicks* the arc *portion to keep*.

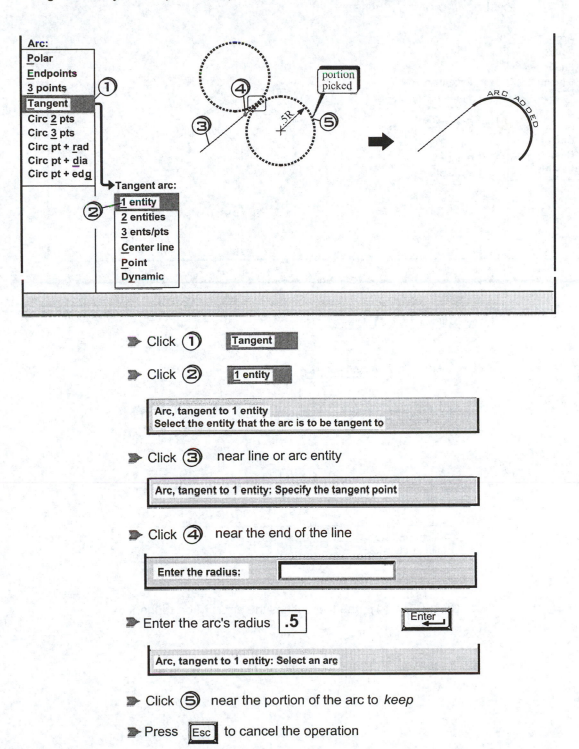

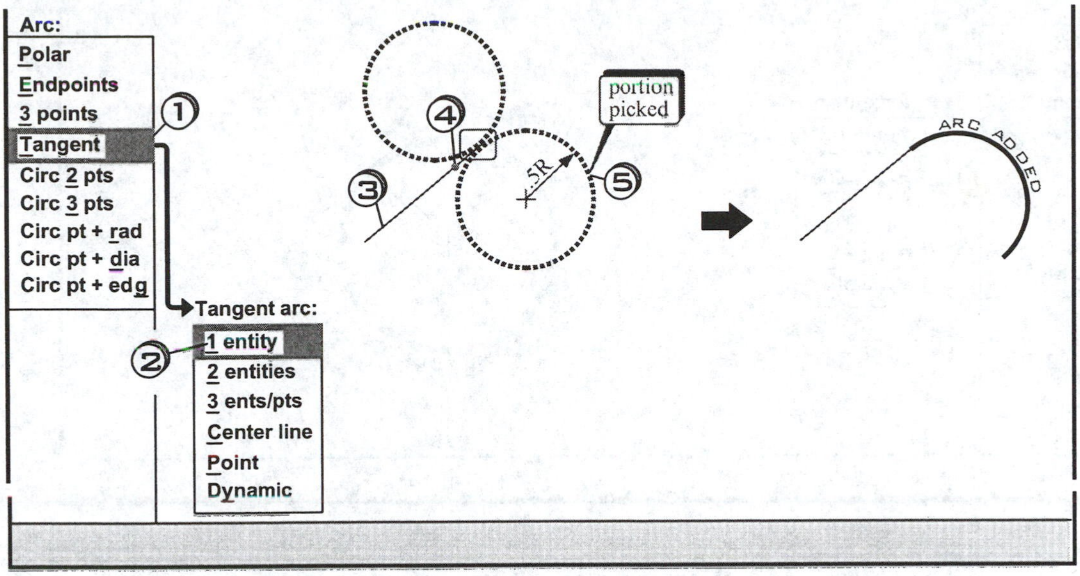

- Click ① **Tangent**
- Click ② **1 entity**

Arc, tangent to 1 entity
Select the entity that the arc is to be tangent to

- Click ③ near line or arc entity

Arc, tangent to 1 entity: Specify the tangent point

- Click ④ near the end of the line

Enter the radius:

- Enter the arc's radius **.5** Enter

Arc, tangent to 1 entity: Select an arc

- Click ⑤ near the portion of the arc to *keep*
- Press Esc to cancel the operation

Tangent **2 entities**

A *360-degree arc* of specified *radius* is created *tangent* to *two* other existing *line* or *arc* entities.

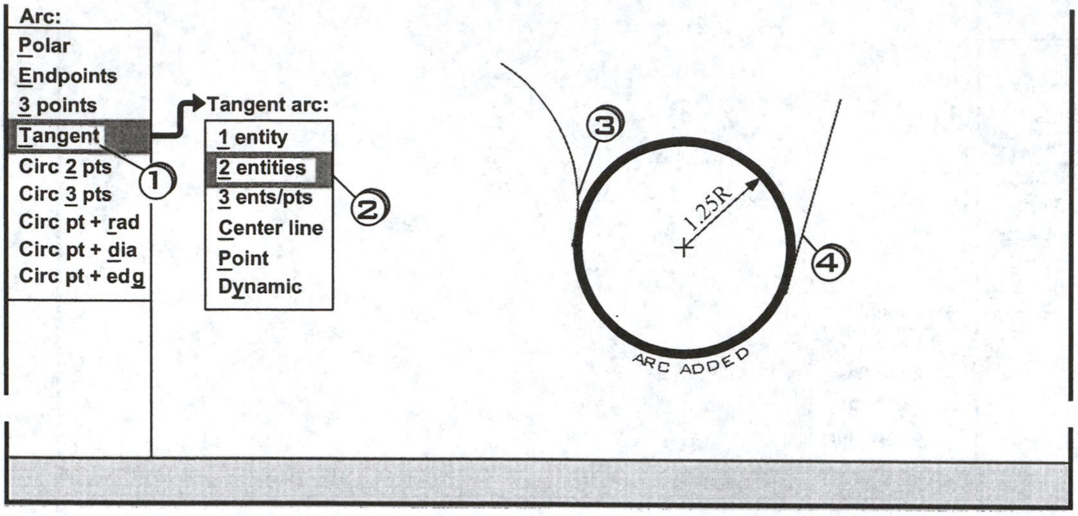

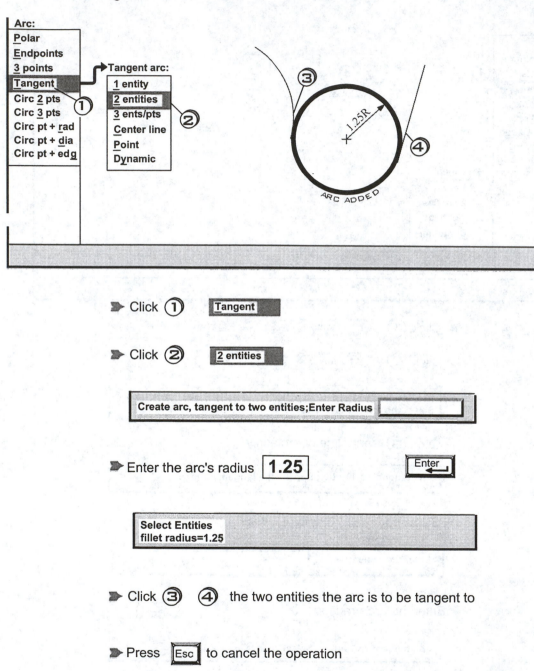

➤ Click ① **Tangent**

➤ Click ② **2 entities**

Create arc, tangent to two entities;Enter Radius

➤ Enter the arc's radius **1.25** Enter

Select Entities
fillet radius=1.25

➤ Click ③ ④ the two entities the arc is to be tangent to

➤ Press Esc to cancel the operation

Tangent **3 ents/pts**

A *360-degree arc* is created *tangent to three* existing *line* or *arc* entities

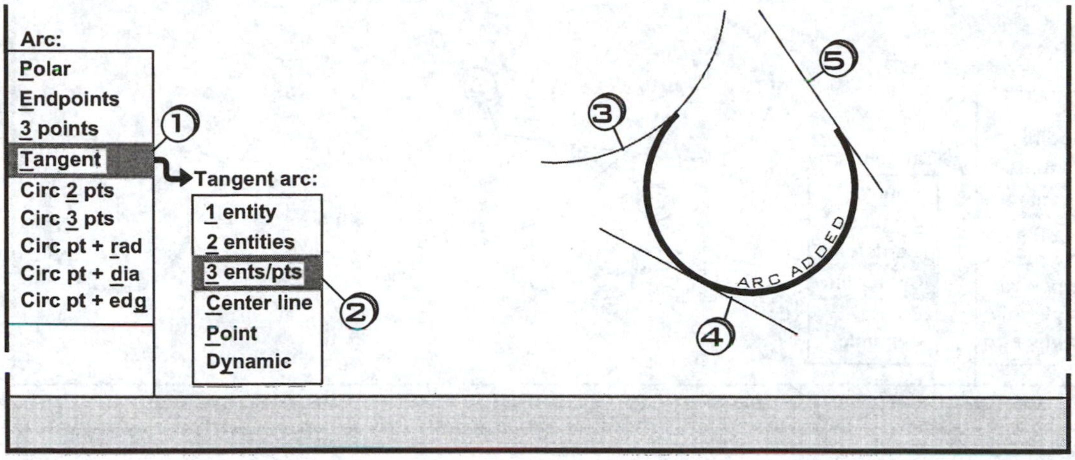

- Click ① **Tangent**

- Click ② **3 ents/pts**

Arc, tangent to 3: Select an entity or:

- Click ③ ④ ⑤ the line, arc entities

- Press **Esc** to cancel the operation

Tangent **Centerline**

360-degree arcs are created *tangent* to an existing *line* clicked and positioned such that a *second line* clicked passes through their *centers.* The radius is inputted and the operator clicks the *arc to **keep**.*

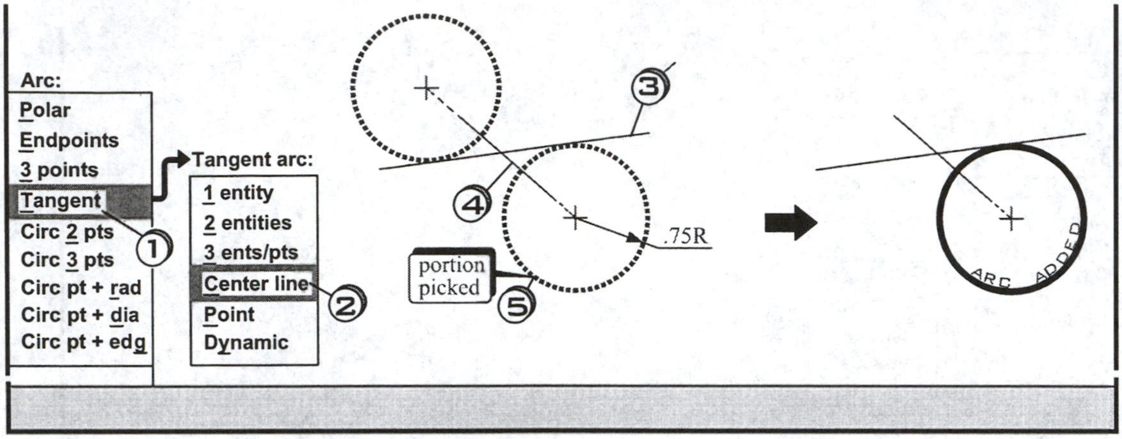

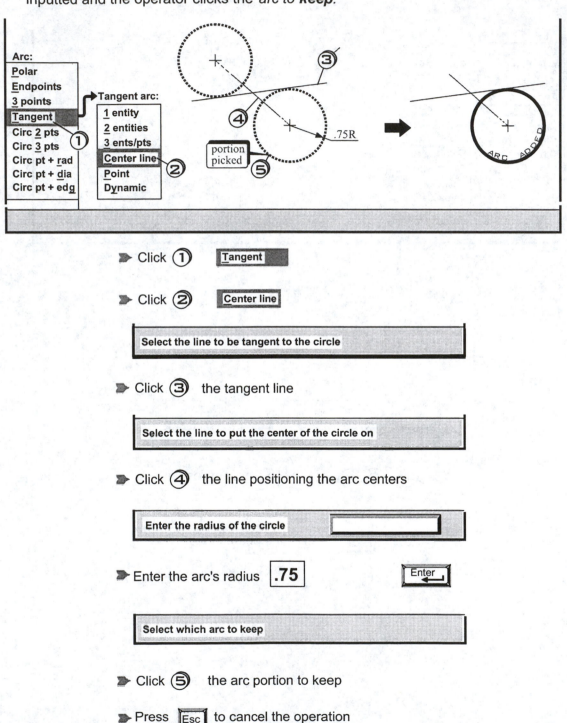

- Click ① **Tangent**
- Click ② **Center line**

Select the line to be tangent to the circle

- Click ③ the tangent line

Select the line to put the center of the circle on

- Click ④ the line positioning the arc centers

Enter the radius of the circle

- Enter the arc's radius **.75** Enter

Select which arc to keep

- Click ⑤ the arc portion to keep
- Press Esc to cancel the operation

Tangent **Point**

360-degree arcs are created *tangent* to an existing *line* clicked and positioned such that they *pass through a point* inputted. The radius is inputted and the operator clicks the *arc to* ***keep***.

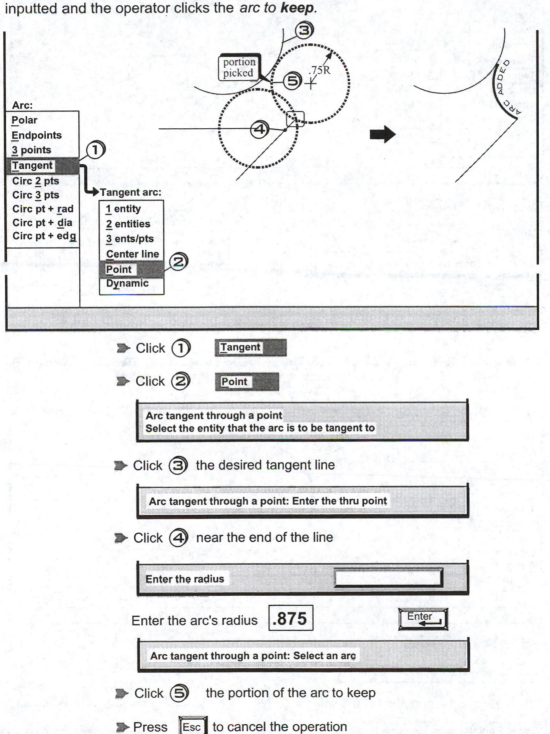

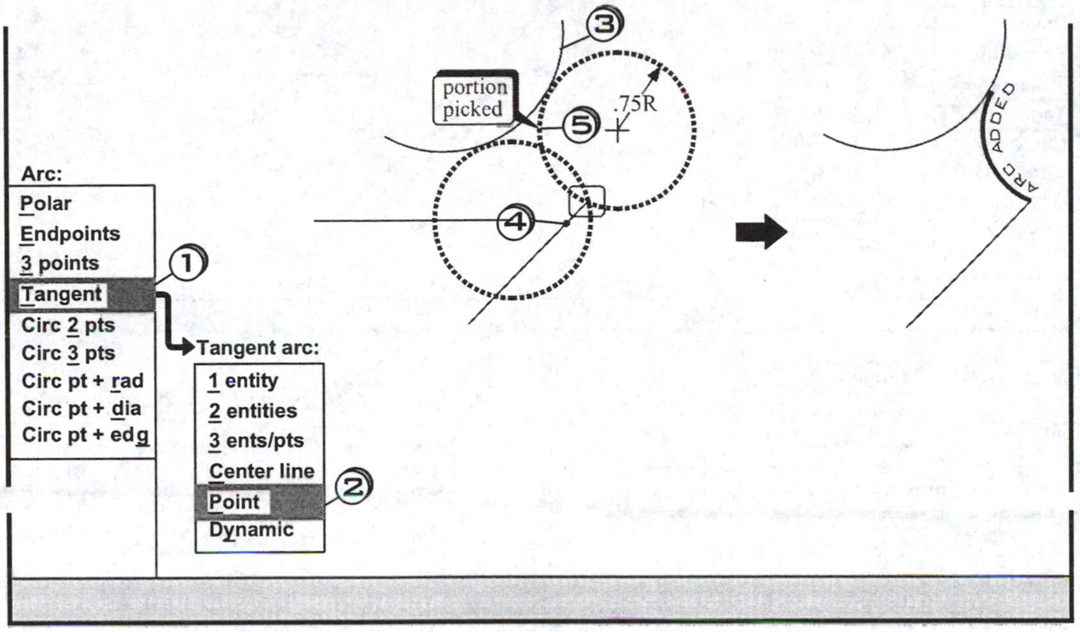

- Click ① **Tangent**
- Click ② **Point**

Arc tangent through a point
Select the entity that the arc is to be tangent to

- Click ③ the desired tangent line

Arc tangent through a point: Enter the thru point

- Click ④ near the end of the line

Enter the radius

Enter the arc's radius **.875** Enter

Arc tangent through a point: Select an arc

- Click ⑤ the portion of the arc to keep
- Press Esc to cancel the operation

Tangent | Dynamic

An *arc* is created starting at a point *dynamically* clicked on an existing *tangent line* or *arc* and *ending at a point* inputted.

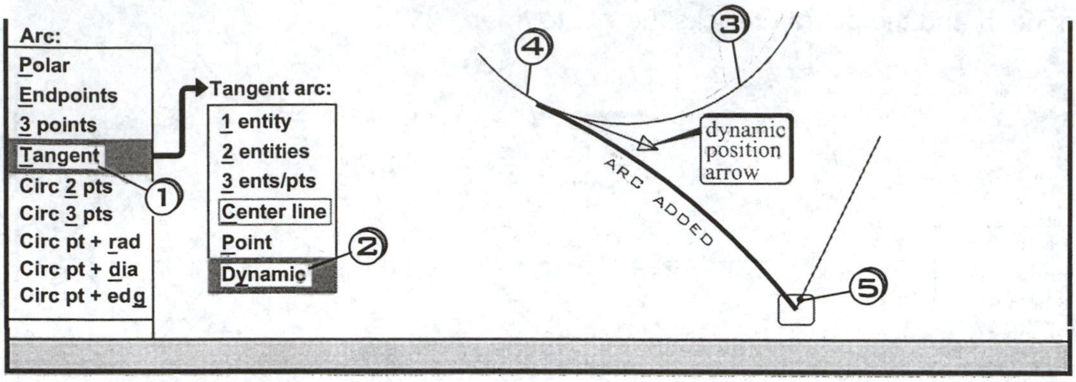

- Click ① **Tangent**
- Click ② **Dynamic**

Select the entity the arc is to be tangent to

- Click ③ the desired arc

Slide Arrow to position to be tangent to

- Click ④ the arc's tangency start point

Rad= 3.235; Start Ang=82.659; Sweep Ang=-21.071

- Click ⑤ near the end point of the line
- Press **Esc** to cancel the operation

Circ 2pts

A *circle* is created whose *diameter* is defined by two inputted *end point* locations

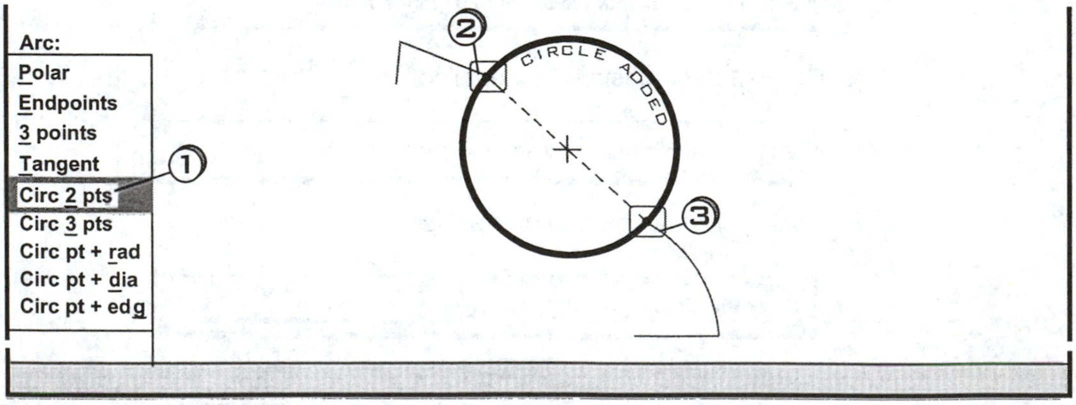

- Click ① **Circ 2 pts**

Circle, 2 points: Enter the first point

- Click ② near the end point of the line

Circle, 2 points: Enter the second point

- Click ③ near the end pt of the arc

Radius = 0.75

- Press **Esc** to cancel the operation

Circ 3pts

A *circle* is created whose *circumference* is defined by *three* inputted *point* locations

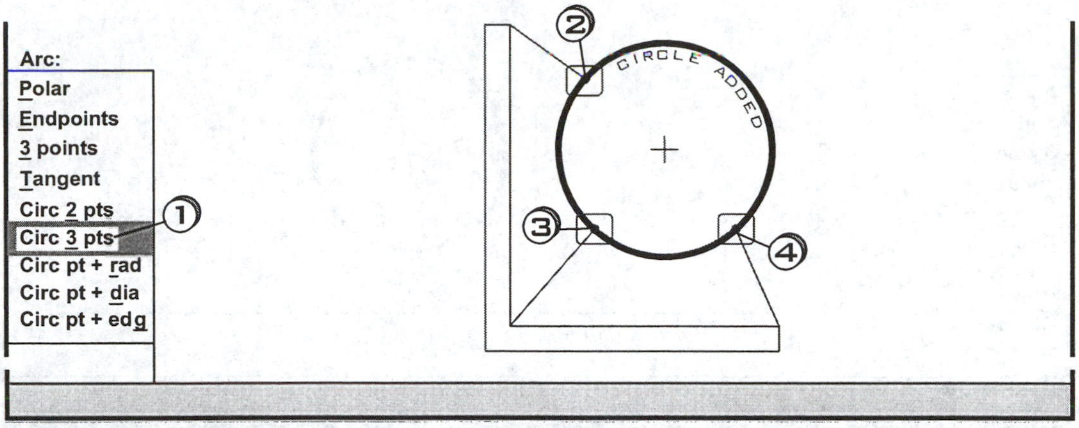

- Click ① **Circ 3 pts**

Circle, 3 points: Enter the first point

- Click ② near the end pt of the line

Circle, 3 points: Enter the second point

- Click ③ near the end pt of the line

Circle, 3 points: Enter the third point

- Click ④ near the end pt of the line

Radius = 0.75

- Press **Esc** to cancel the operation

Circ pt + rad

A *circle* is created by entering its *radius* and *center point* values

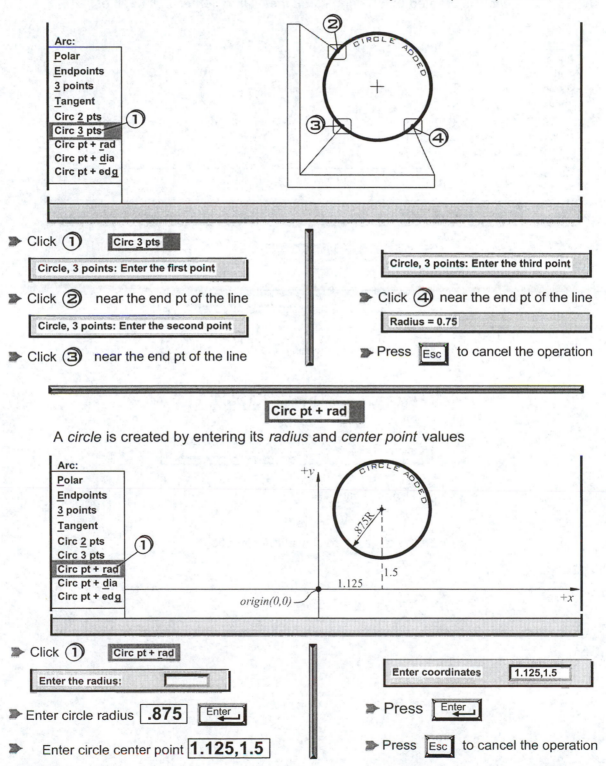

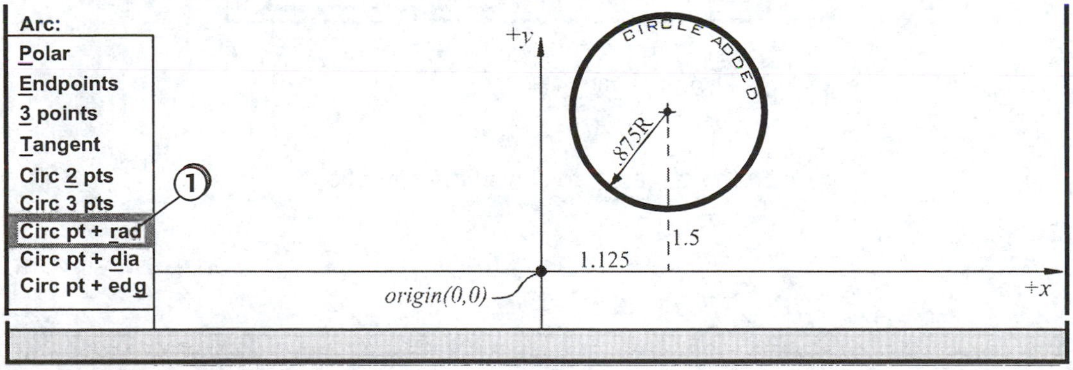

- Click ① **Circ pt + rad**

Enter the radius:

- Enter circle radius **.875** **Enter**
- Enter circle center point **1.125,1.5**

Enter coordinates **1.125,1.5**

- Press **Enter**
- Press **Esc** to cancel the operation

Circ pt + dia

A *circle* is created by entering its *diameter* and *center* point values.

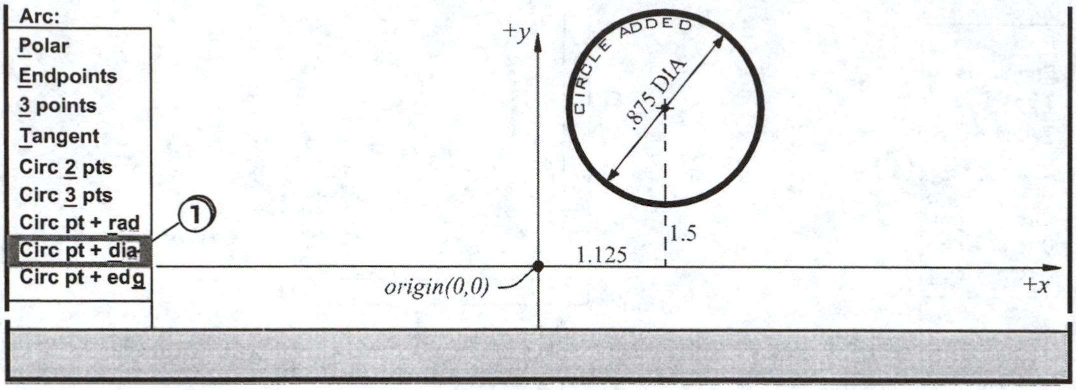

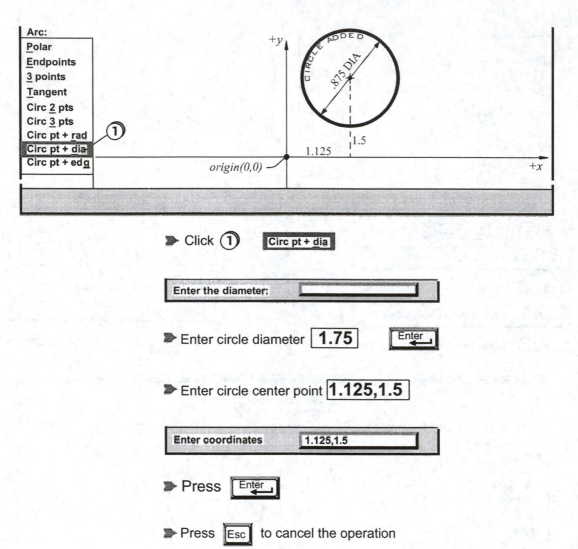

➤ Click ① **Circ pt + dia**

Enter the diameter:

➤ Enter circle diameter **1.75** Enter

➤ Enter circle center point **1.125,1.5**

Enter coordinates **1.125,1.5**

➤ Press Enter

➤ Press Esc to cancel the operation

Circ pt + dia | Relative | Polar

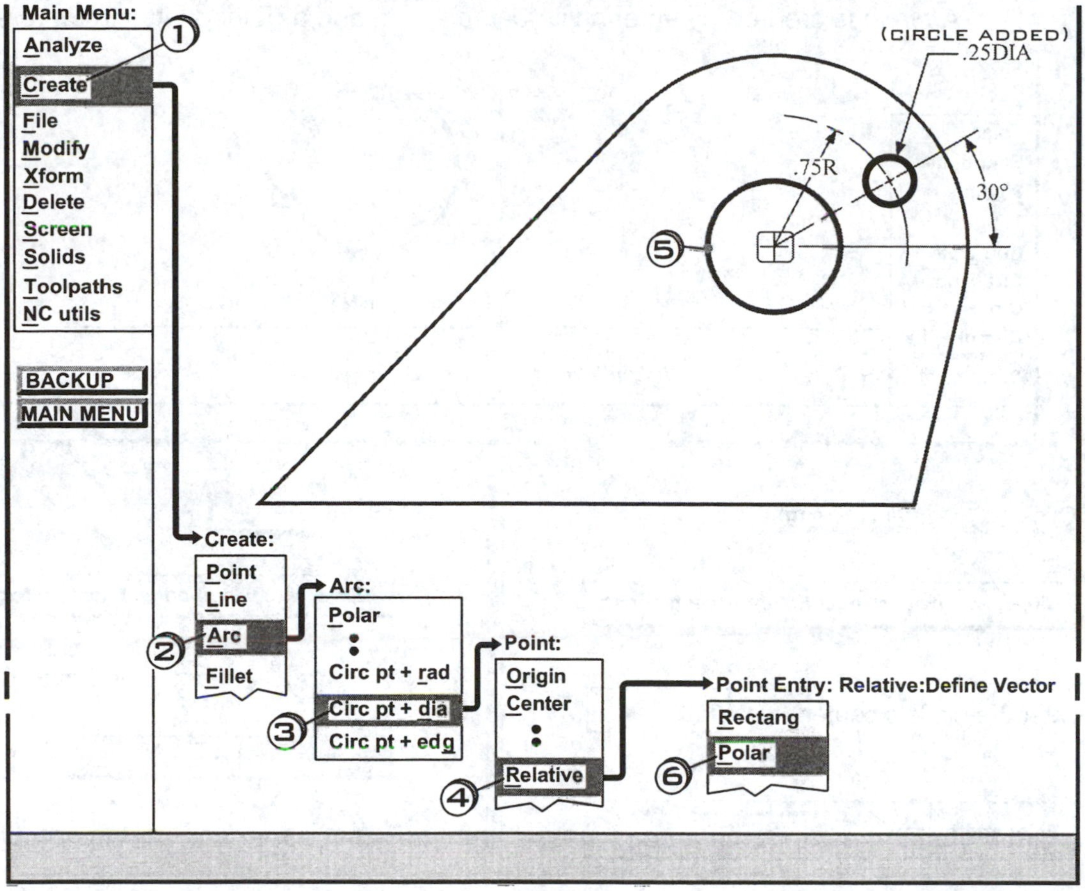

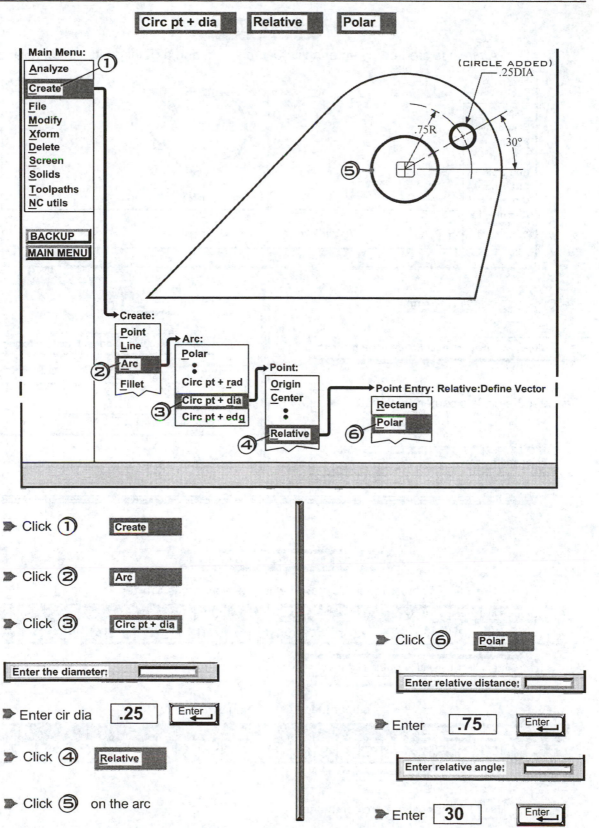

- Click ① **Create**
- Click ② **Arc**
- Click ③ **Circ pt + dia**

Enter the diameter:

- Enter cir dia **.25** Enter
- Click ④ **Relative**
- Click ⑤ on the arc
- Click ⑥ **Polar**

Enter relative distance:

- Enter **.75** Enter

Enter relative angle:

- Enter **30** Enter

Circ pt + edg

A *circle* is created by entering its *center* point and a point on its *circumference*.

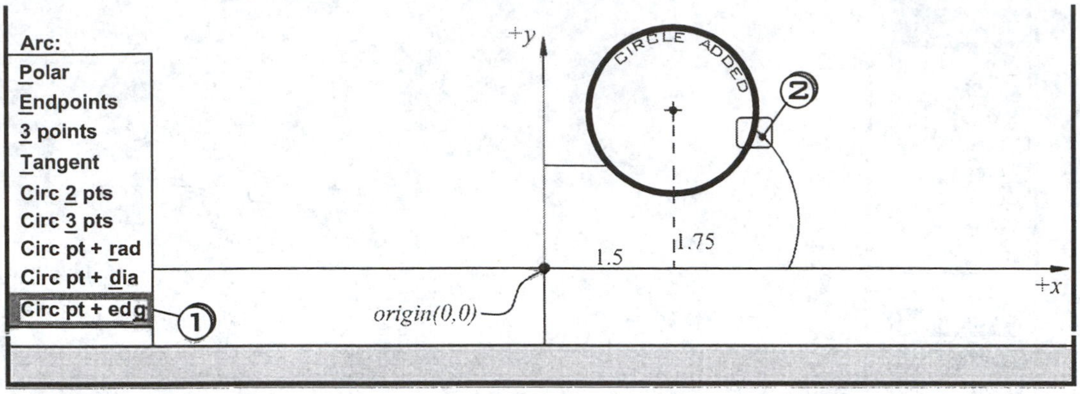

- Click ① **Circ pt + edg**

Circle, with center/edge: Enter the center point

- Enter the circle's center point **1.5,1.75**

Enter coordinates **1.5,1.75**

- Press **Enter**

Circle, with center/edge: Enter the edge pt

- Click ② near the end of the arc

Radius= 0.758696

- Press **Esc** to cancel the operation

Using the Toolbar Menus

Toolbar Menu Displayed when *Mastercam* is **Started**

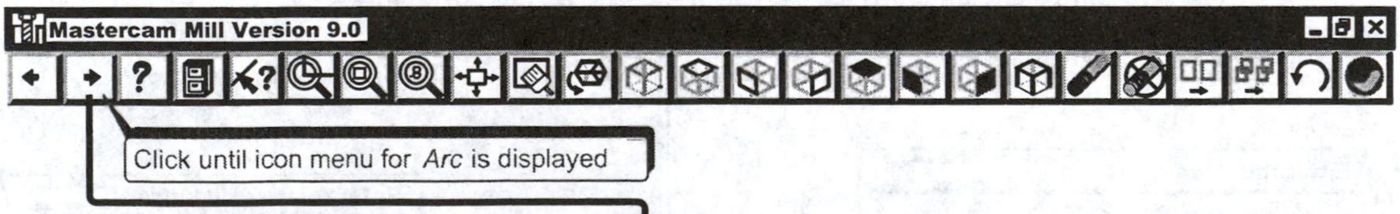

Toolbar Menu Containing ***Arc*** Icon

Click **Esc** to cancel the arc commands

2-6 Fillet Constructions in 2D Space

A fillet is an arc that is fitted *tangent* to two potentially intersecting line, arc or spline entities. *Mastercam's* filleting commands are presented in this section

Creating Fillets

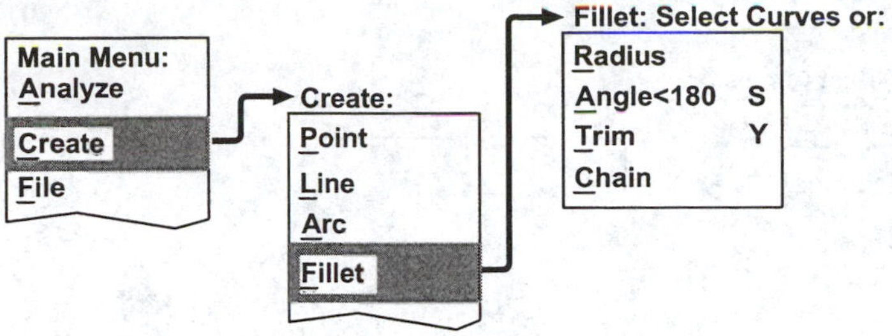

Radius | Angle<180 S | Trim Y

The operator *sets* the fillet *radius* for *all subsequent* fillets created. The filet *arc angle* is set *less than*(**S**) *180°*. The two entities fillited *will be*(**Y**) *trimmed* to their *tangency* point with the fillet.

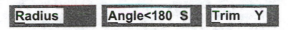

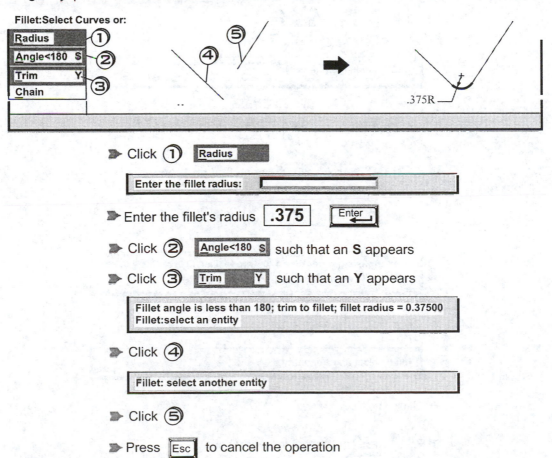

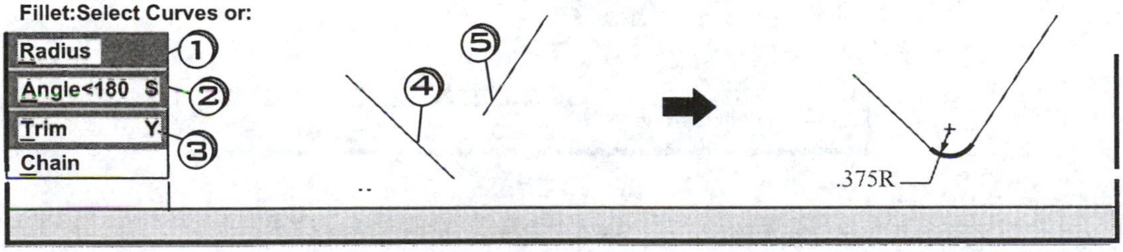

- Click ① Radius

Enter the fillet radius:

- Enter the fillet's radius .375 Enter
- Click ② Angle<180 S such that an **S** appears
- Click ③ Trim Y such that an **Y** appears

Fillet angle is less than 180; trim to fillet; fillet radius = 0.37500
Fillet:select an entity

- Click ④

Fillet: select another entity

- Click ⑤
- Press Esc to cancel the operation

Radius | **Angle<180 L** | **Trim Y**

The operator *sets* the fillet *radius* for *all subsequent* fillets created. The filet *arc angle* is set *greater than*(**L**) *180°*. The two entities fillited *will be*(**Y**) *trimmed* to their *tangency* point with the fillet.

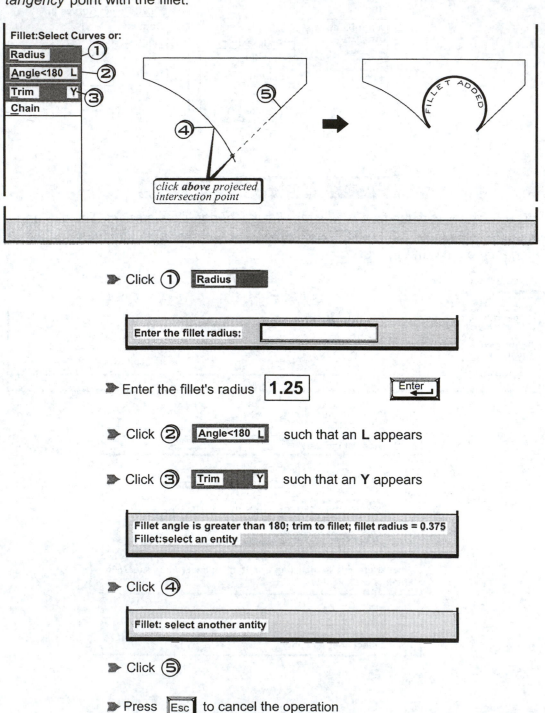

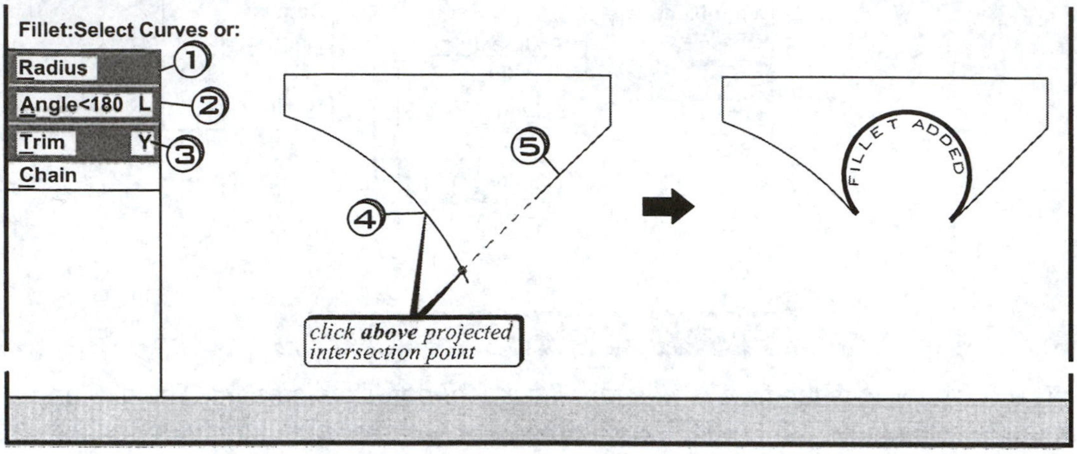

- Click (1) **Radius**

Enter the fillet radius:

- Enter the fillet's radius **1.25** Enter

- Click (2) **Angle<180 L** such that an **L** appears

- Click (3) **Trim Y** such that an **Y** appears

Fillet angle is greater than 180; trim to fillet; fillet radius = 0.375
Fillet:select an entity

- Click (4)

Fillet: select another antity

- Click (5)

- Press Esc to cancel the operation

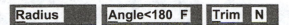

Radius | **Angle<180 F** | **Trim N**

The operator *sets* the fillet *radius* for *all subsequent* fillets created. The filet *arc angle* is set to a *full*(**F**) *360°*. The two entities fillited *will not be*(**N**) *trimmed* to their *tangency* point with the fillet.

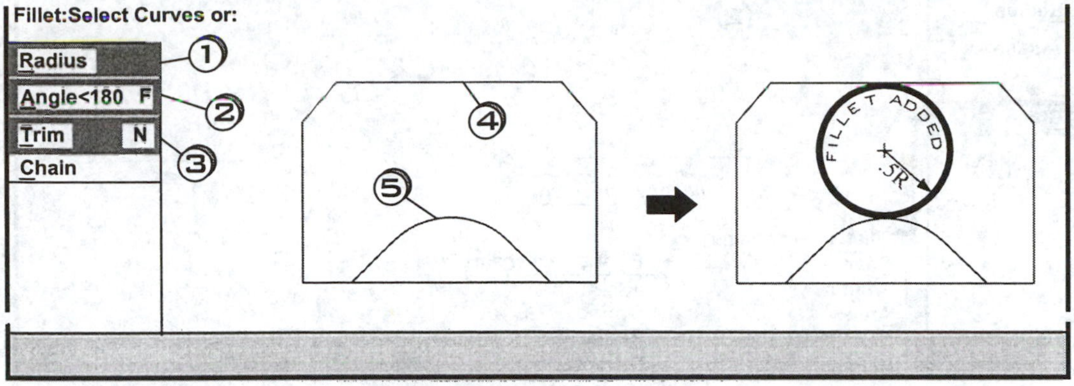

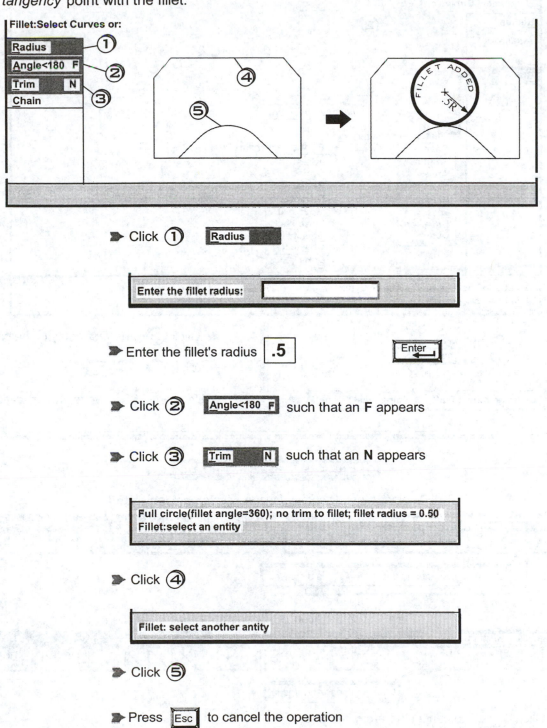

➤ Click ① **Radius**

Enter the fillet radius:

➤ Enter the fillet's radius **.5** [Enter]

➤ Click ② **Angle<180 F** such that an **F** appears

➤ Click ③ **Trim N** such that an **N** appears

Full circle(fillet angle=360); no trim to fillet; fillet radius = 0.50
Fillet:select an entity

➤ Click ④

Fillet: select another antity

➤ Click ⑤

➤ Press [Esc] to cancel the operation

Chain

Fillets are created at the corners of *several* entities that are *connected* end to end in a *chain*

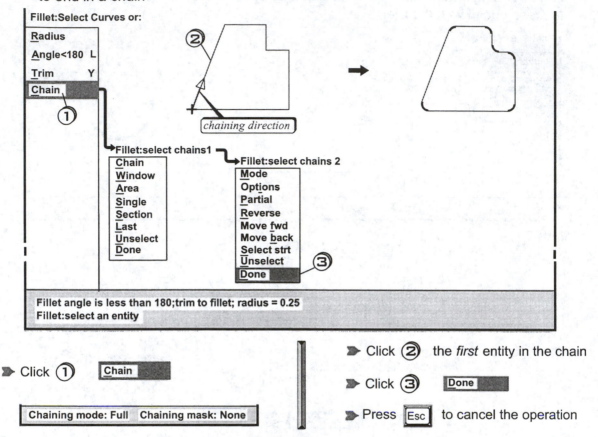

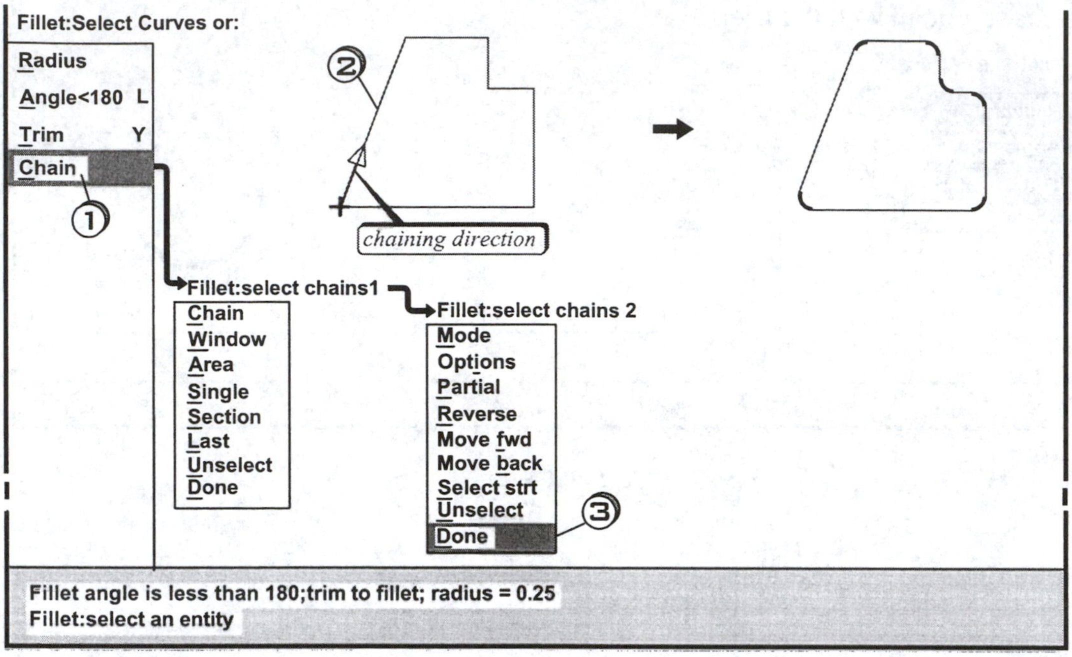

- Click ① **Chain**

Chaining mode: Full | Chaining mask: None

- Click ② the *first* entity in the chain
- Click ③ **Done**
- Press **Esc** to cancel the operation

Using the Toolbar Menus

Toolbar Menu Displayed when *Mastercam* is **Started**

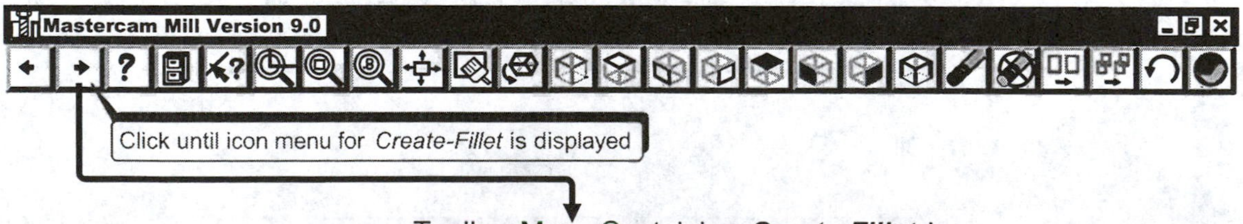

Toolbar Menu Containing ***Create-Fillet*** Icon

Click **Esc** to cancel the arc commands

2-7 Spline Constructions in 2D Space

A spline is a smooth freeform curve that passes through a set of operator generated node points

Mastercam can create *two* types of spline curves as follows:

Parametric spline(**P**) - is a curve that is made to pass through a desired set of points. If any of the control points are changed the operator must recreate the spline such that it passes through the new node point locations.

P - spline curve

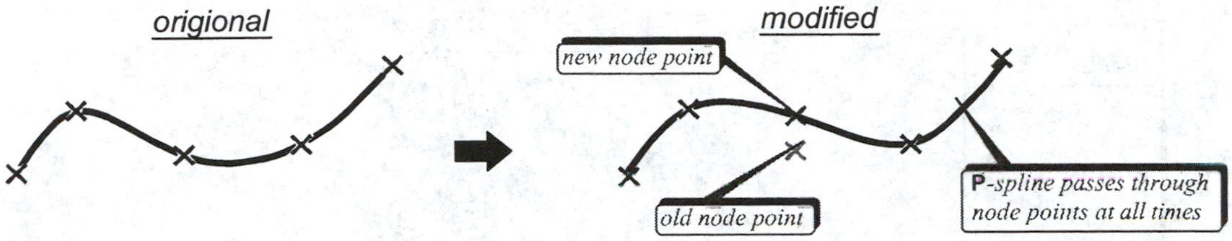

Non-Uniform Rational B-Spline/NURBS(**N**) - is a curve that is made to pass through a desired set of points. Its shape can be modified by moving its control points. In general it is a smoother curve than the P-spline curve.

N - spline curve

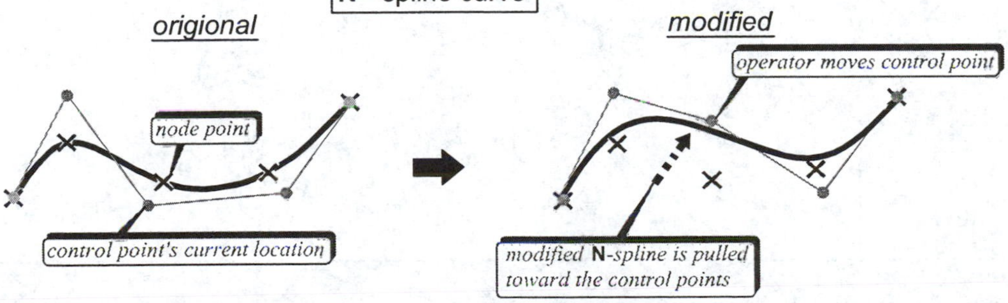

Splines are especially useful in the aircraft, automobile and shipbuilding industries.

Creating Splines

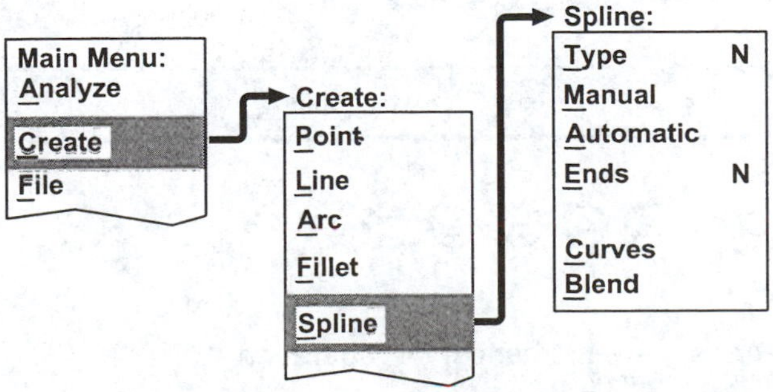

Type P **Manual**

A *Parametric(***P***) spline* curve is created. The operator must click *each node* point the spline passes through.

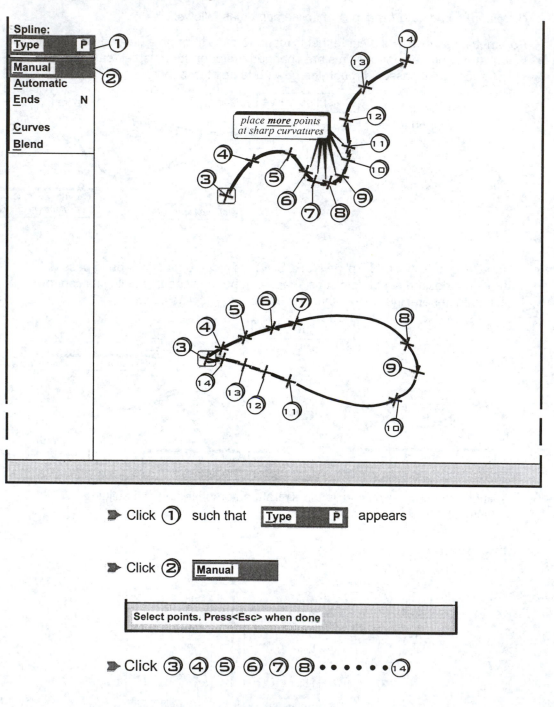

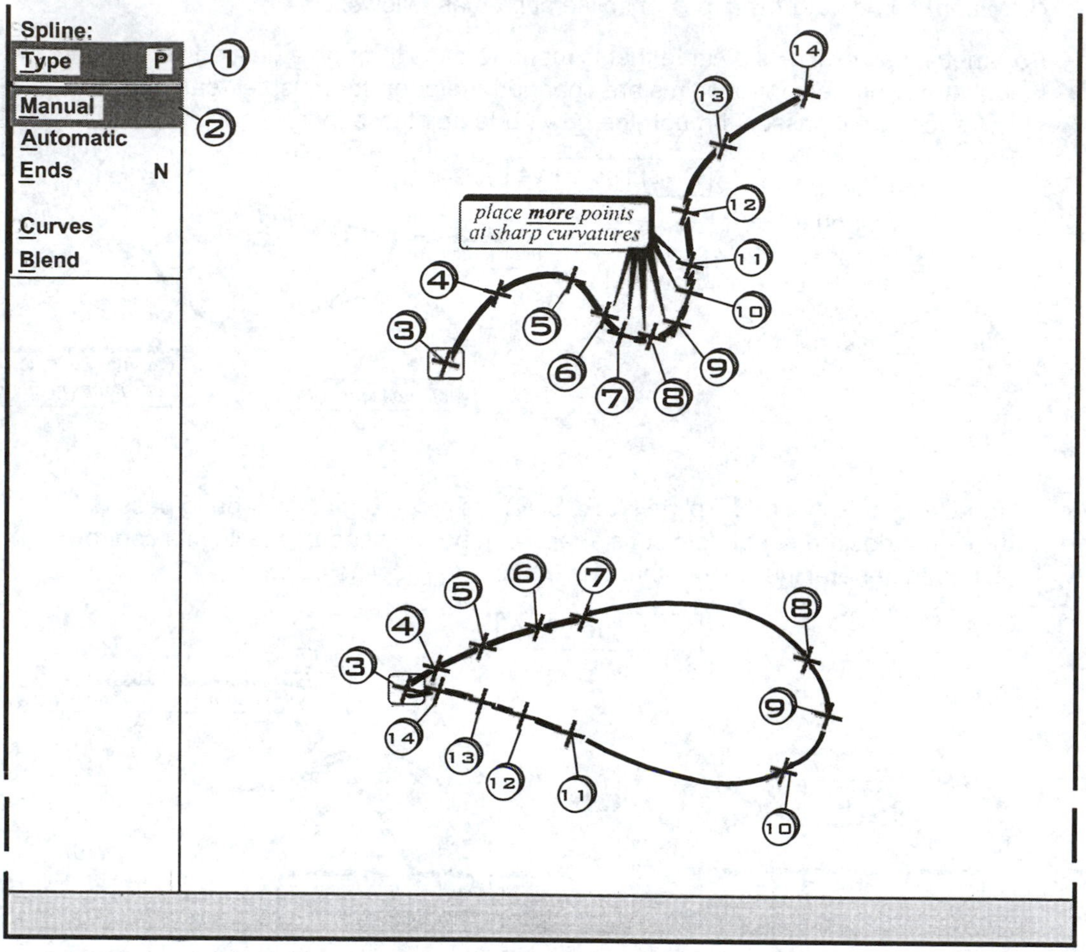

- Click (1) such that **Type P** appears
- Click (2) **Manual**

Select points. Press<Esc> when done

- Click (3) (4) (5) (6) (7) (8) • • • • • • (14)
- Press **Esc** to cancel the operation

Type N **Automatic**

A *NURBS(***N***) spline* curve is created. The operator must click the *first, second* and *last* points the curve is to *pass through.* The system will then *select any other* existing points to pass the curve through such that it *remains within the system's curve tolerance* value.

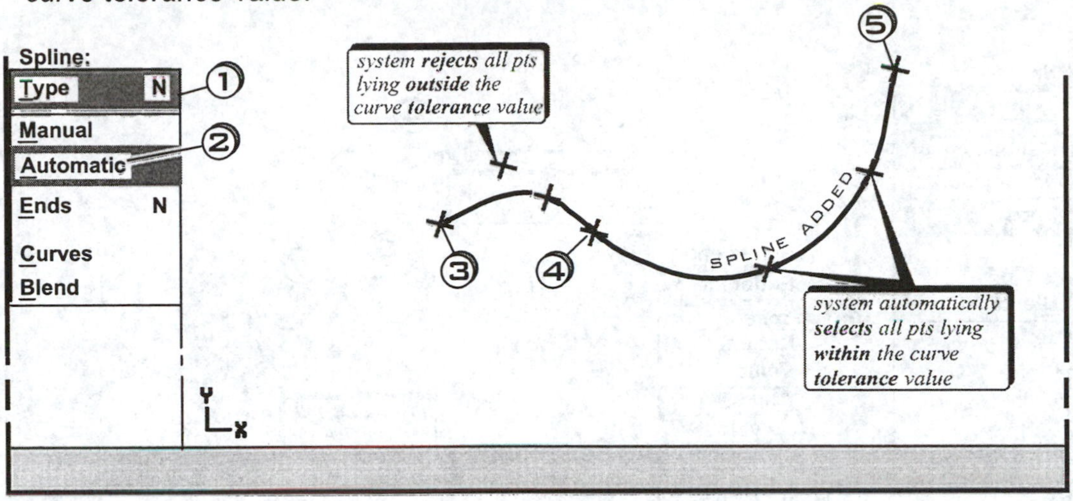

- Click ① such that **Type N** appears
- Click ② **Automatic**

Select the first point

- Click ③

Select the second point

- Click ④

Select the last point

- Click ⑤
- Press **Esc** to cancel the operation

Type N | **Manual** | **Ends**

A *NURBS(***N***) spline* curve is created. The operator must click *each node point* the curve is to pass through. The operator specifies the *first*(**F**) endpoint is *tangent* to a *0° virtual line* and the *last*(**L**) endpoint is *tangent* to an existing *arc*.

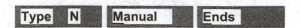

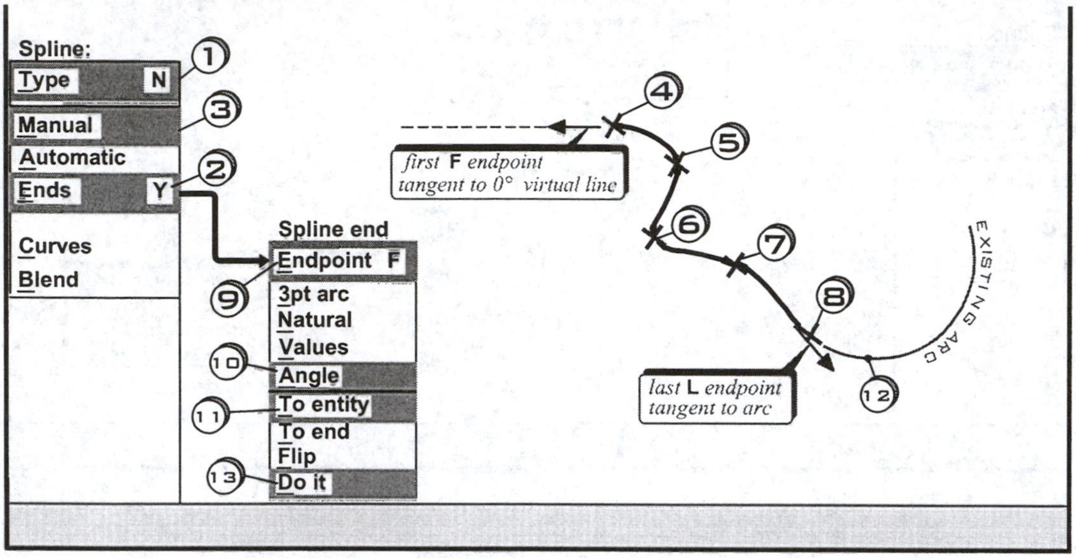

- Click ① such that **Type N** appears
- Click ② such that **Ends Y** appears
- Click ③ **Manual**

Select points. Press<Esc> when done

- Click ④ ⑤ ⑥ ⑦ ⑧
- Press the **Esc** key
- Click ⑨ such that **Endpoint F** appears
- Click ⑩ **Angle**

Enter the angle in degrees [ ]

- Enter the angle for the tangent line **0**
- Click ⑨ such that **Endpoint L** appears
- Click ⑪ **To entity**
- Click ⑫ the existing arc entity
- Click ⑬ **Do it**
- Press **Esc** to cancel the operation

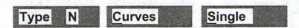

Type N | **Curves** | **Single**

*NURBS(***N***) spline* curves are *fitted to* and *replace* existing *line, arc, ellipse* or *spline* entities clicked by the operator.

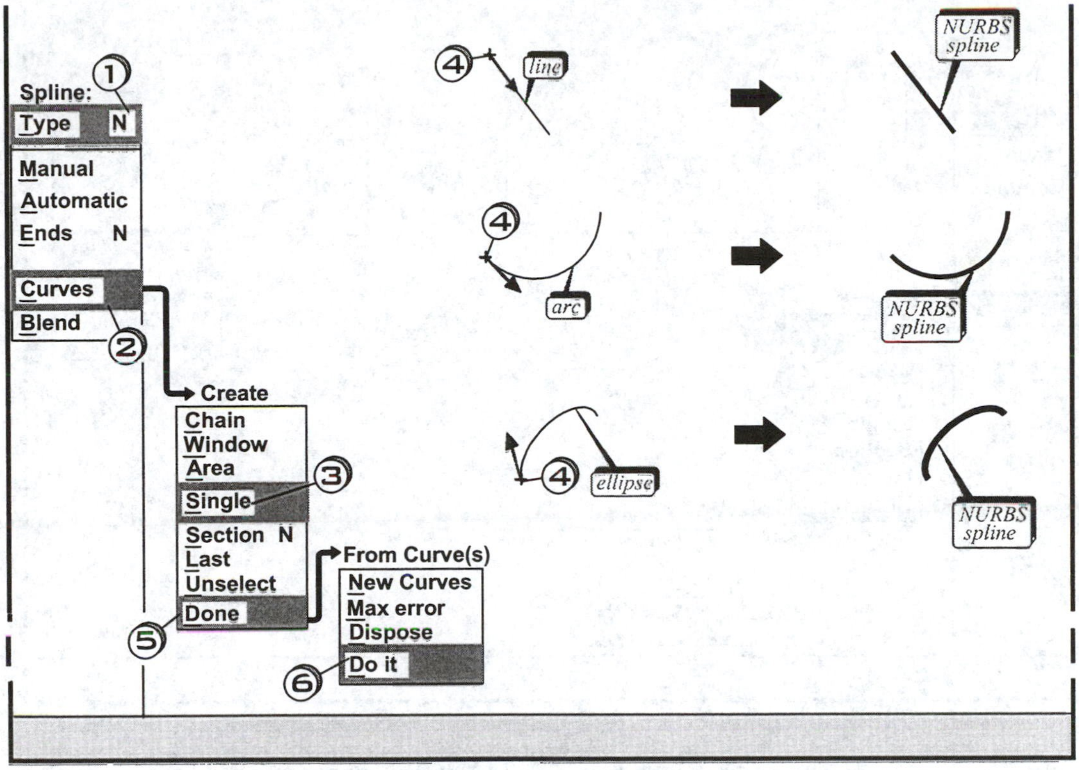

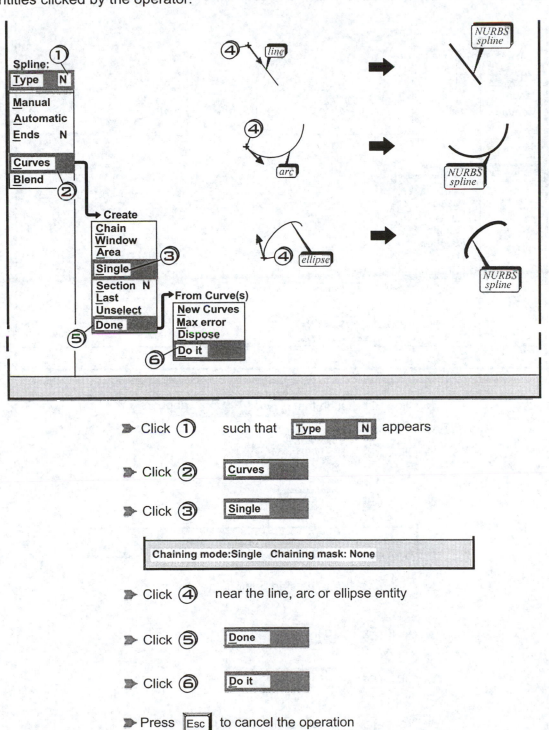

- Click ① such that **Type N** appears
- Click ② **Curves**
- Click ③ **Single**

Chaining mode:Single Chaining mask: None

- Click ④ near the line, arc or ellipse entity
- Click ⑤ **Done**
- Click ⑥ **Do it**
- Press **Esc** to cancel the operation

Type N **Curves** **Chain**

*NURBS(***N***) spline* curves are *fitted to* and *replace* existing *line, arc, ellipse* or *spline* entities connected end to end in a *chain*. The operator clicks the *first* entity in the chain.

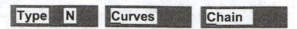

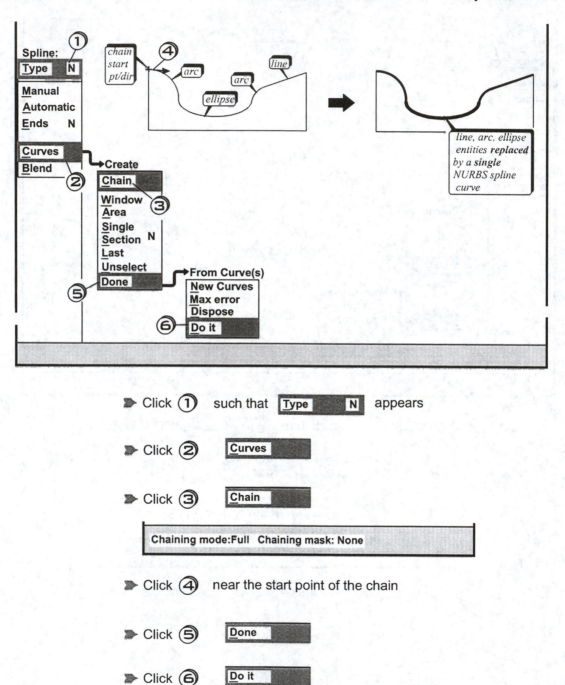

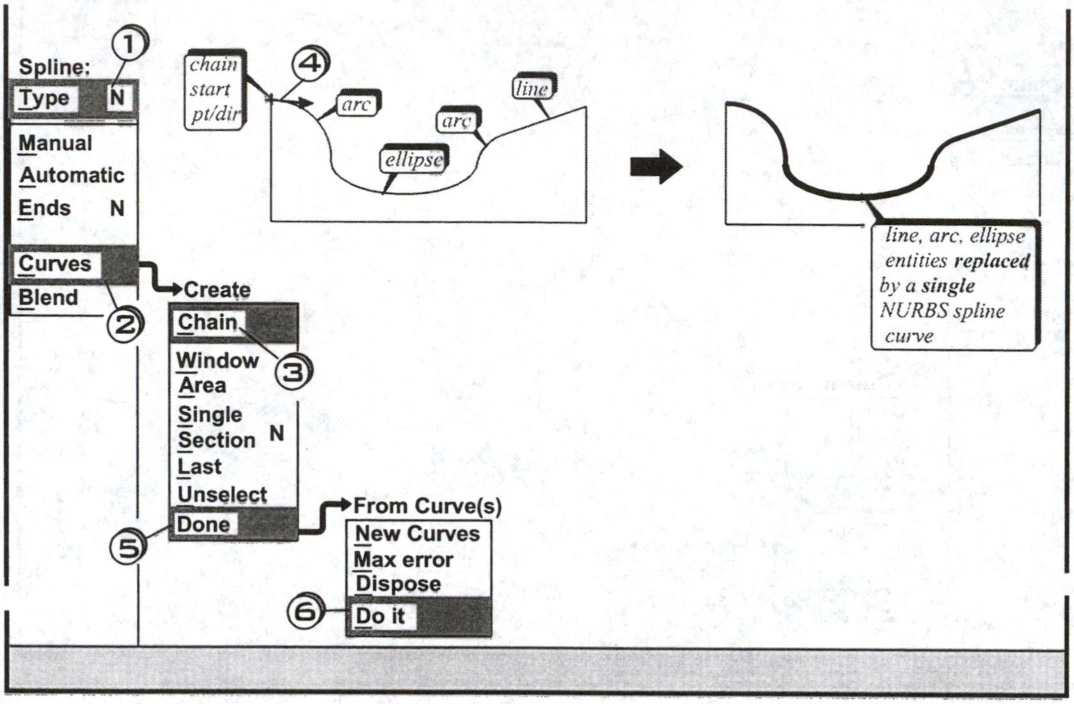

- Click ① such that **Type N** appears
- Click ② **Curves**
- Click ③ **Chain**

Chaining mode:Full Chaining mask: None

- Click ④ near the start point of the chain
- Click ⑤ **Done**
- Click ⑥ **Do it**
- Press **Esc** to cancel the operation

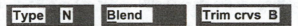

Type N **Blend** **Trim crvs B**

A *NURBS*(**N**) *spline* curve is *fitted tangent to two* existing line, arc, ellipse or spline entities clicked by the operator. Both(**B**) entities will be trimmed at their point of tangency.

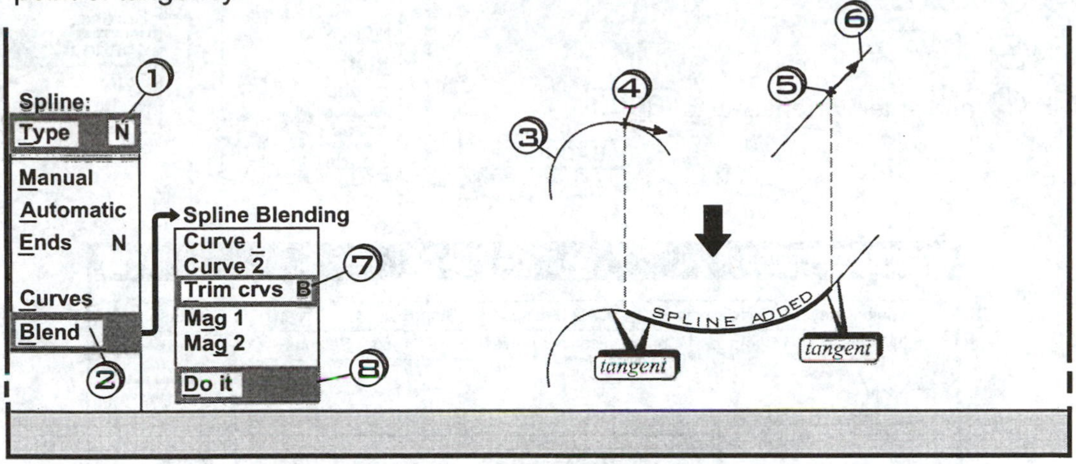

➤ Click ① such that **Type N** appears

➤ Click ② **Blend**

Select curve 1

➤ Click ③ the first existing entity

Slide arrow to position to blend onto
Snapping is OFF-type<S> to turn snapping ON

➤ Click ④ the first blending point

Select curve 2

➤ Click ⑤ the second existing entity

Slide arrow to position to blend onto
Snapping is OFF-type<S> to turn snapping ON

➤ Click ⑥ the second blending point

➤ Click ⑦ such that **Trim crvs B** appears

➤ Click ⑧ **Do it**

➤ Press **Esc** to cancel the operation

Adjusting the System Setting for the Maximum Number of Node Points for Spline Creation

- Click ① MAIN MENU
- Click ② Screen
- Click ③ Configure
- Click ④ Allocations
- Click ⑤ and enter current setting **3000**

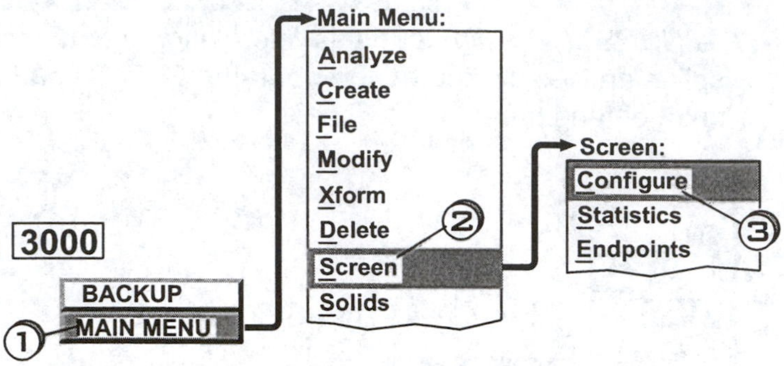

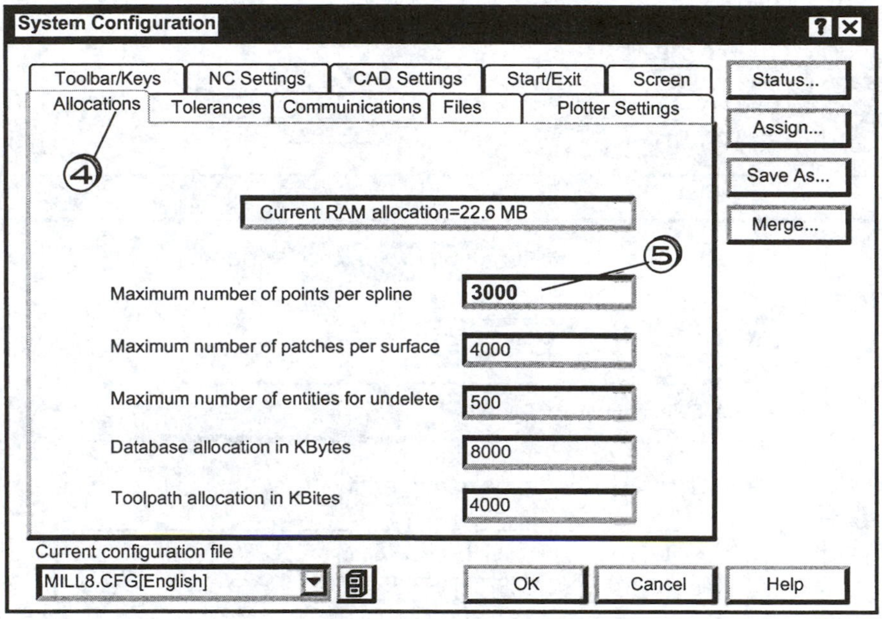

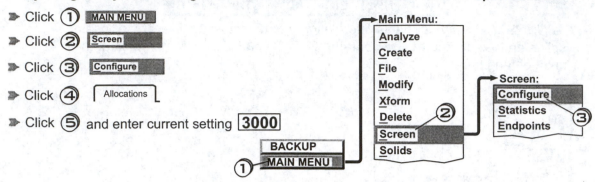

Using the Toolbar Menus

Toolbar Menu Displayed when *Mastercam* is **Started**

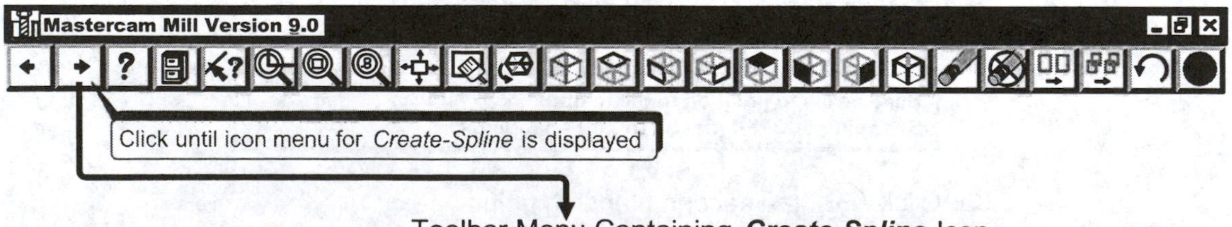

Toolbar Menu Containing ***Create-Spline*** Icon

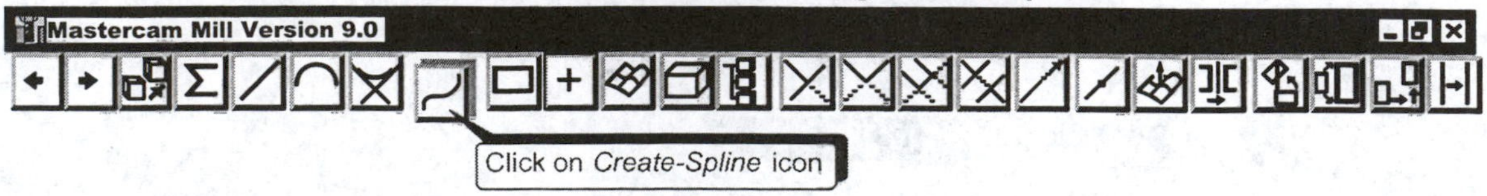

Click Esc to cancel the Spline commands

2-8 Rectangle Constructions in 2D Space

Creating Rectangles

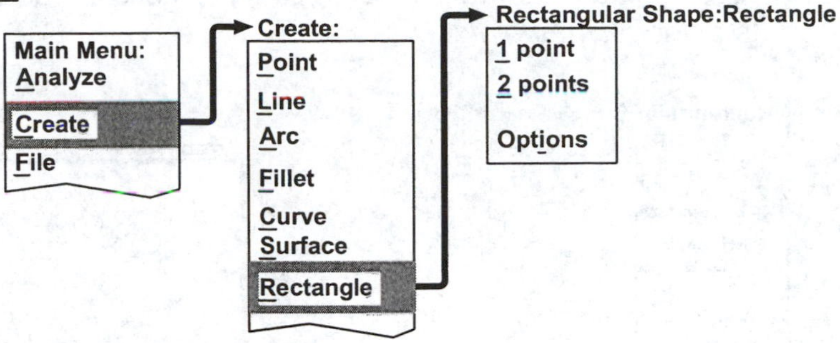

1 point

A *rectangle* is created. The operator specifies the *Width*, *Height* and *placement position*.

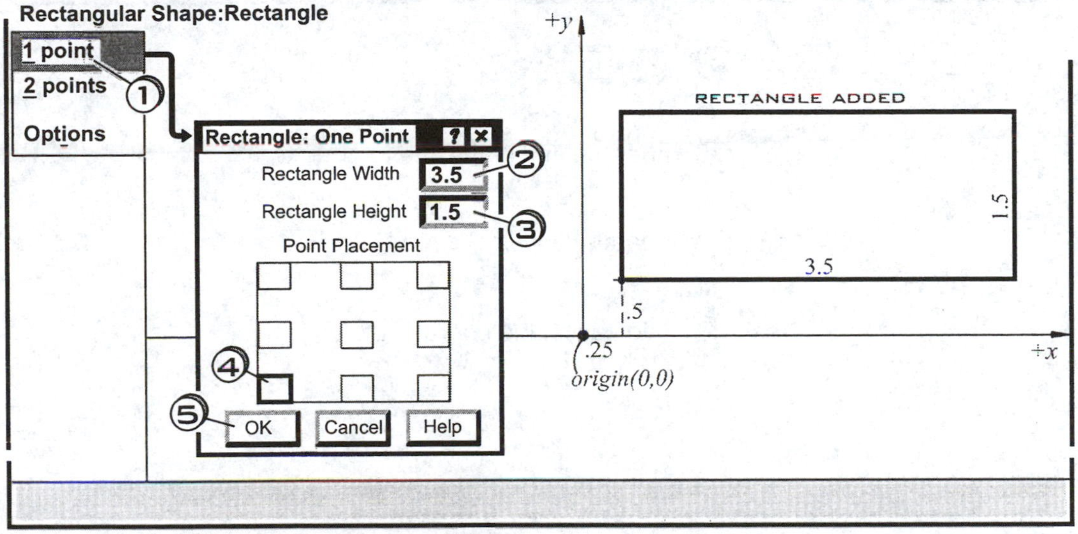

- Click ① **1 point**
- Click ② in the Width box and enter **3.5**
- Click ③ in the Height box and enter **1.5**
- Click ④ the left lower corner placement icon
- Click ⑤ the **OK** button
- Enter the XY placement coordinates **.25,.5**

- Press **Esc** to cancel the operation

2 points

A *rectangle* is created whose *size* and *location* is determined by clicking its *diagonal endpoints*.

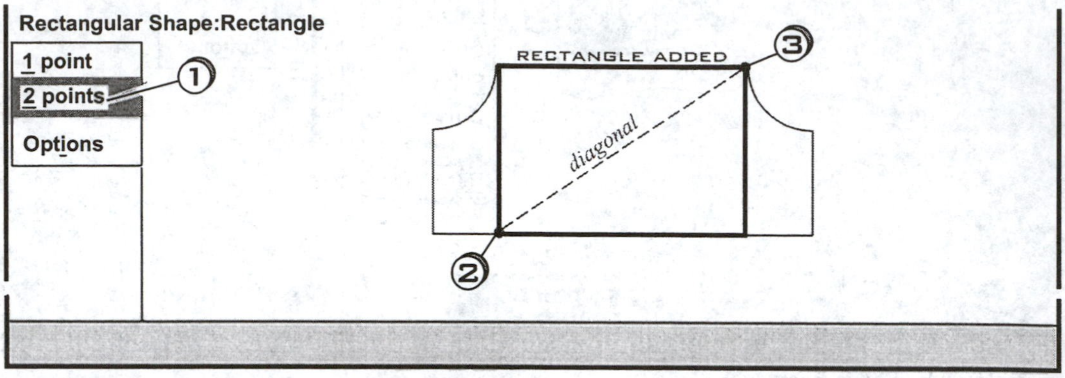

- Click ① **2 points**
- Click ② near the end of the line
- Click ③ near the end of the arc
- Press Esc to cancel the operation

Options **2 points**

Optional parameters such as the *shape* created(*Rectangular,Obround, Single D, Double D* or *Ellipse*)shape as well as *rotation angle* and *corner radii* can be specified. The *size and location* of the shape is determined by clicking its *diagonal endpoints.*

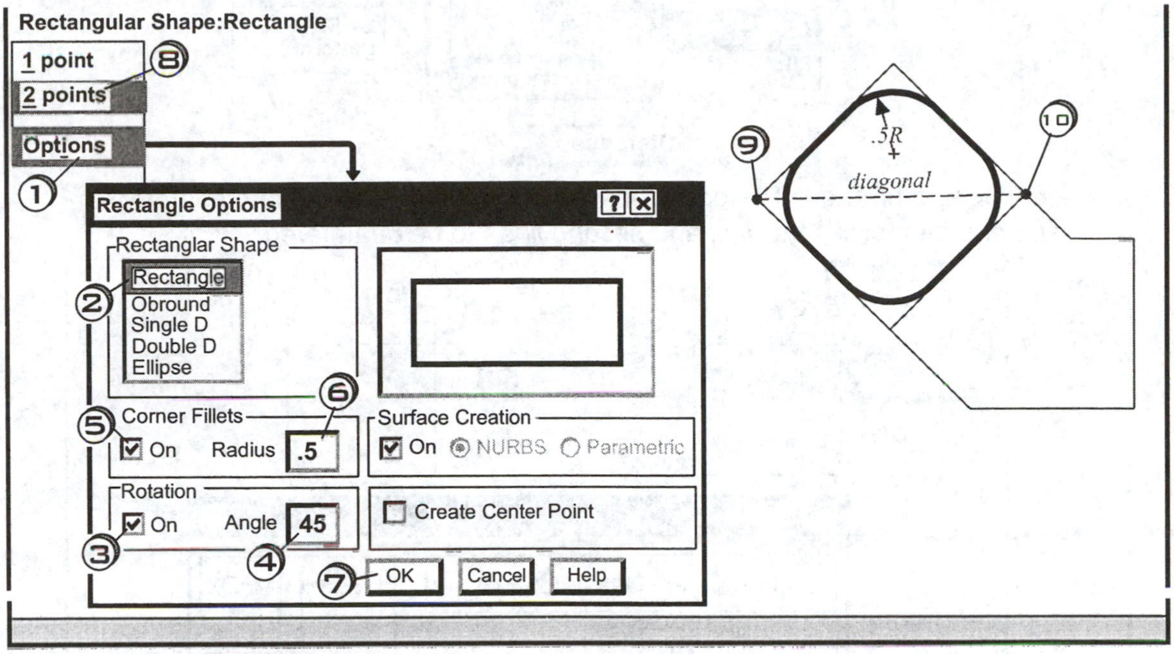

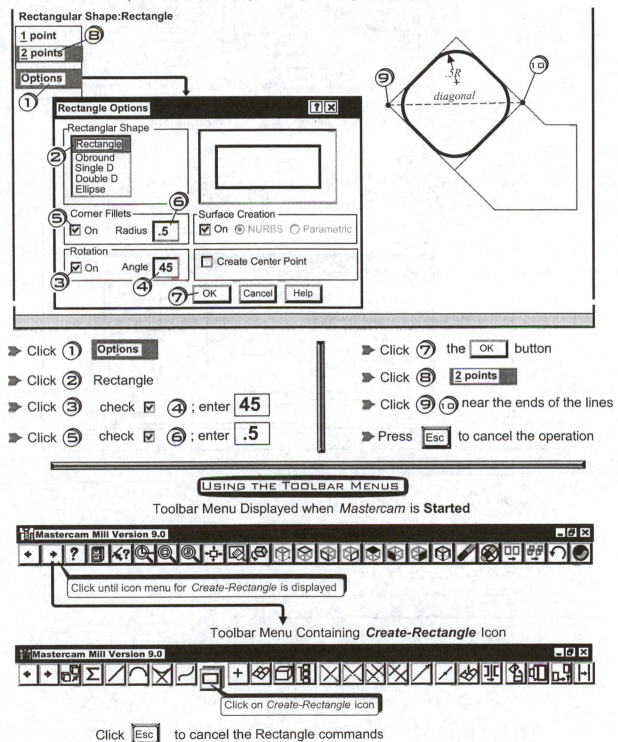

- Click ① **Options**
- Click ② Rectangle
- Click ③ check ☑ ④ ; enter **45**
- Click ⑤ check ☑ ⑥ ; enter **.5**
- Click ⑦ the OK button
- Click ⑧ **2 points**
- Click ⑨ ⑩ near the ends of the lines
- Press Esc to cancel the operation

USING THE TOOLBAR MENUS

Toolbar Menu Displayed when *Mastercam* is **Started**

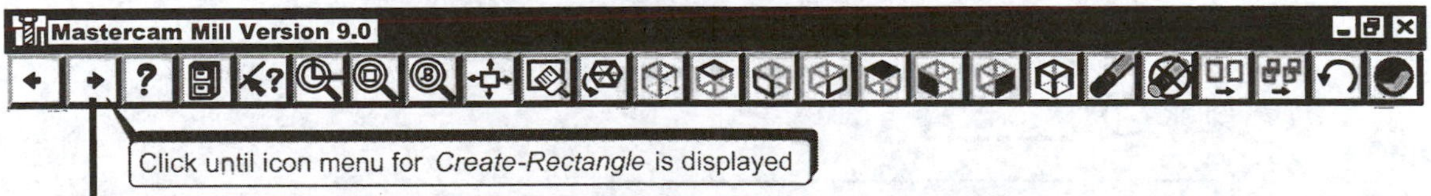

Toolbar Menu Containing ***Create-Rectangle*** Icon

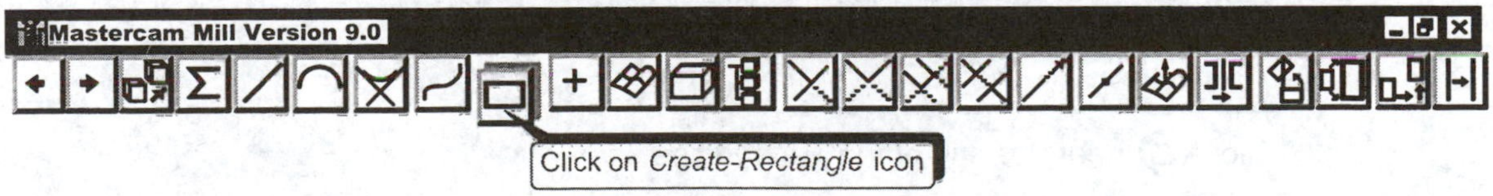

Click Esc to cancel the Rectangle commands

2-9 Chamfer Constructions in 2D Space

Creating Chamfers

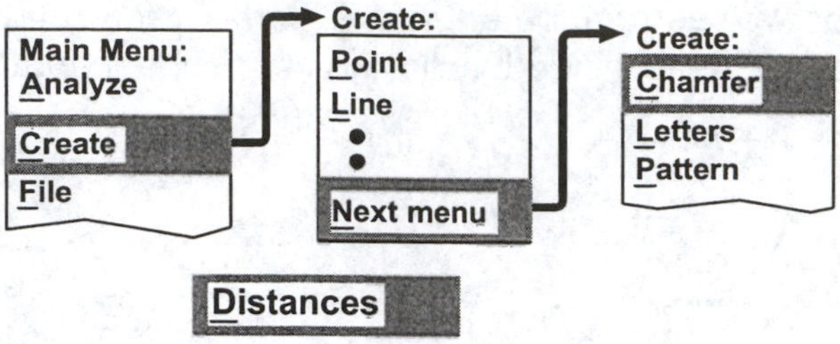

A *chamfer* is created. The operator specifies the *first* and *second* chamfer *distances* then clicks the *first* and *second lines* to be chamfered.

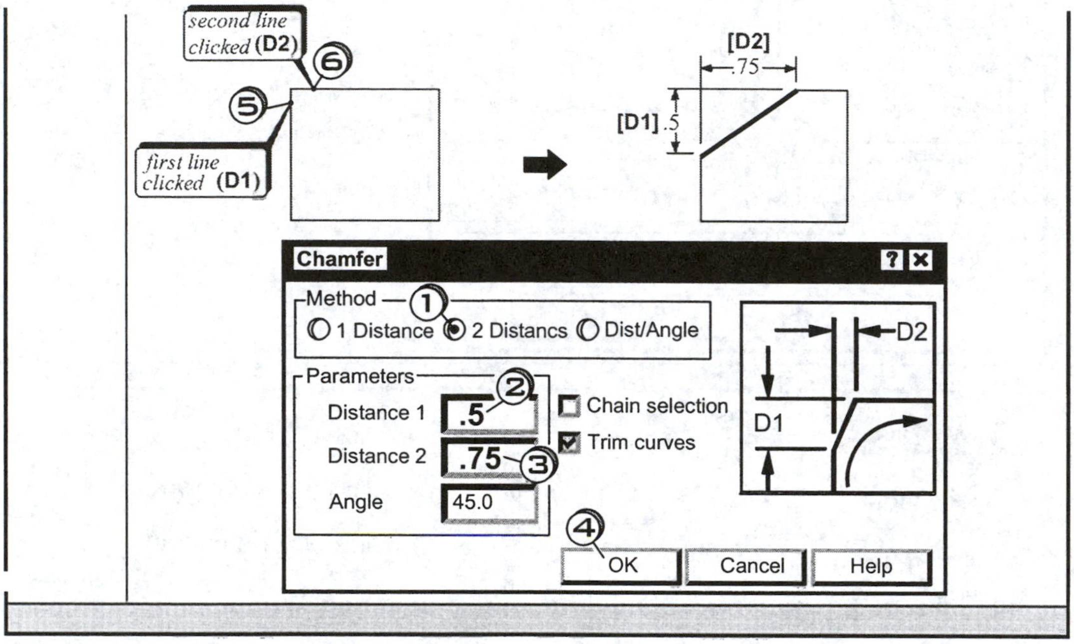

- Click ① the 2 Distances radio button ◉
- Click ② ; enter Distance 1 **.5**
- Click ③ ; enter Distance 2 **.75**
- Click ④ OK

Chamfer: select first line or arc

- Click ⑤ near the end of the *first* line

Chamfer: select second line

- Click ⑥ near the end of the *second* line
- Press Esc to cancel the operation

2-10 Geometric Letter Constructions in 2D Space

Geometric letters are used for creating text to be engraved(machined) into stock material. Geometric letters are composed of separate line and arc entities that are connected end to end. This is different from regular or non-geometric text which is taken as a single entity and cannot be machined.

Creating Letters

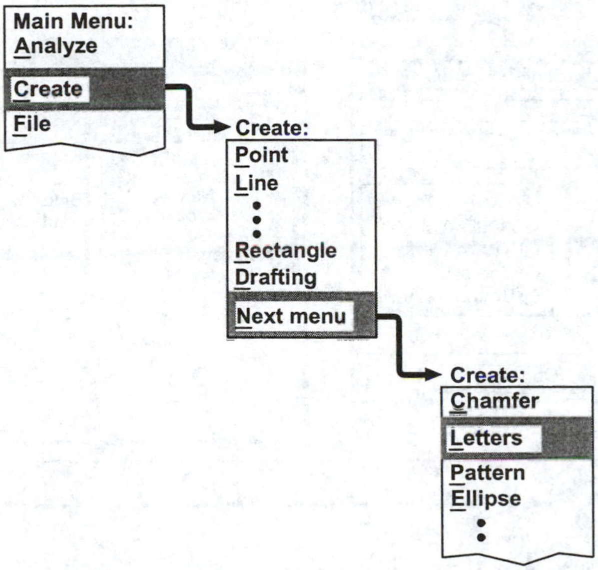

True Type[R] ◉ Horizontal

Geometric letters are created from *True Type* fonts. The operator clicks the desired *font, font style, size, and alignment* (Horizontal). Then enters the text *height and spacing* parameters. Finally, the operator clicks the starting *location* of the text.

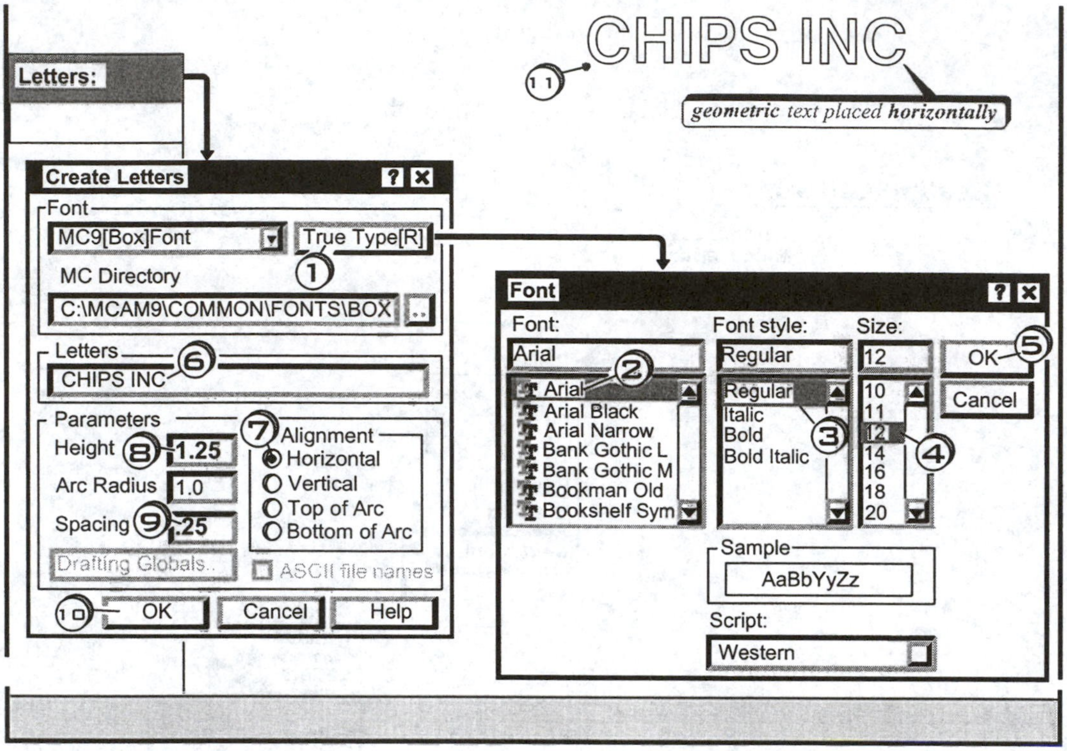

- Click ① the True Type[R] button
- Click ② Arial
- Click ③ Regular
- Click ④ 12
- Click ⑤ OK
- Click ⑥ inside Letters; enter the txt
  Letters: **CHIPS INC**
- Click ⑦ the Horizontal radio button *on*
  ◉ Horizontal
- Click ⑧ in the Height box; enter the text height **1.25**
- Click ⑨ in the Spacing box; enter the spacing between each letter **.25**
- Click ⑩ OK
  Enter starting location of letters
- Click ⑪
- Press Esc to cancel the operation

[True Type[R]] ◉ Top of Arc

Geometric letters are created from *True Type* fonts. The operator clicks the desired *font, font style, size, and alignment* (Top of Arc) .Then enters the text *height, arc radius and spacing* parameters. Finally, the operator clicks the starting *location* of the text.

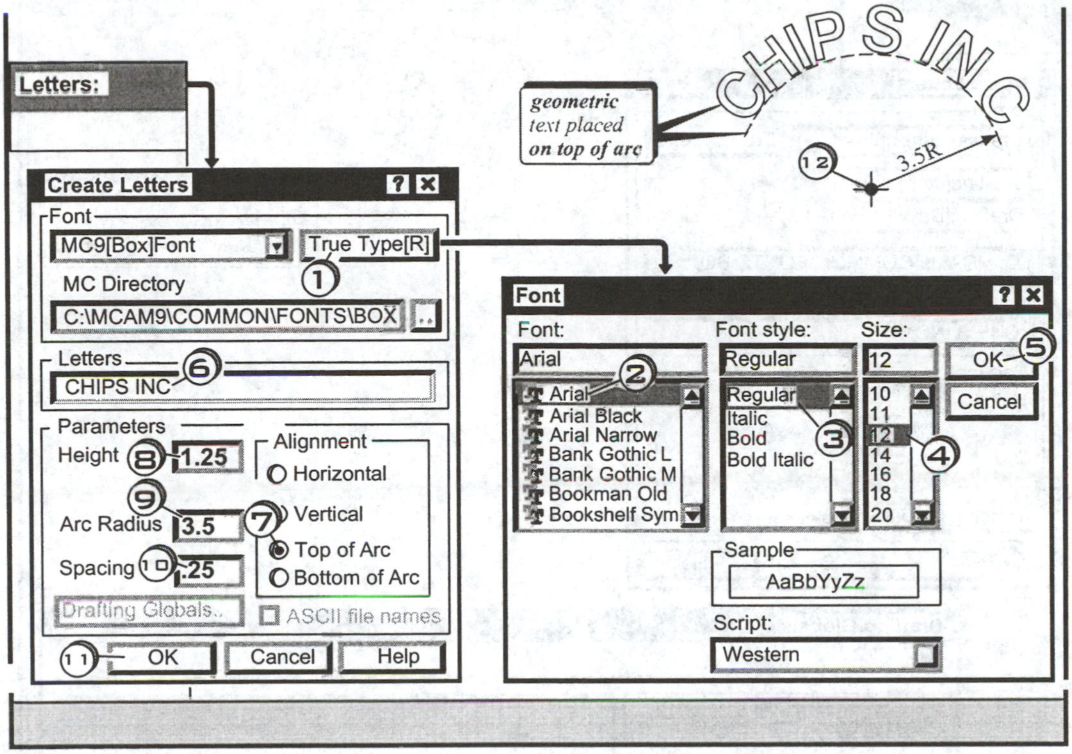

- Click ① the [True Type[R]] button
- Click ② [Arial]
- Click ③ [Regular]
- Click ④ [12]
- Click ⑤ [OK]
- Click ⑥ inside Letters; enter the txt

  Letters: CHIPS INC

- Click ⑦ the Top of Arc radio button *on*

  ◉ Top of Arc

- Click ⑧ in the Height box;
  enter the text height **1.25**
- Click ⑨ in the Arc Radius box;
  enter the text arc radius **3.25**
- Click ⑩ in the Spacing box;
  enter the spacing
  between each letter **.25**
- Click ⑪ [OK]

Enter coordinates of arc center

- Click ⑫
- Press [Esc] to cancel the operation

CONVERTING NON-MACHINABLE DRAFTING TEXT INTO MACHINABLE LETTERS

Drafting font text is, by default, *non-geometric* and therefore cannot be machined directly. A *new* feature in the **Create Letters** dialog box enables the operator to select a *drafting font* text style, enter parameters and then direct *Mastercam* to *automatically convert it to machinable geometric letters.*

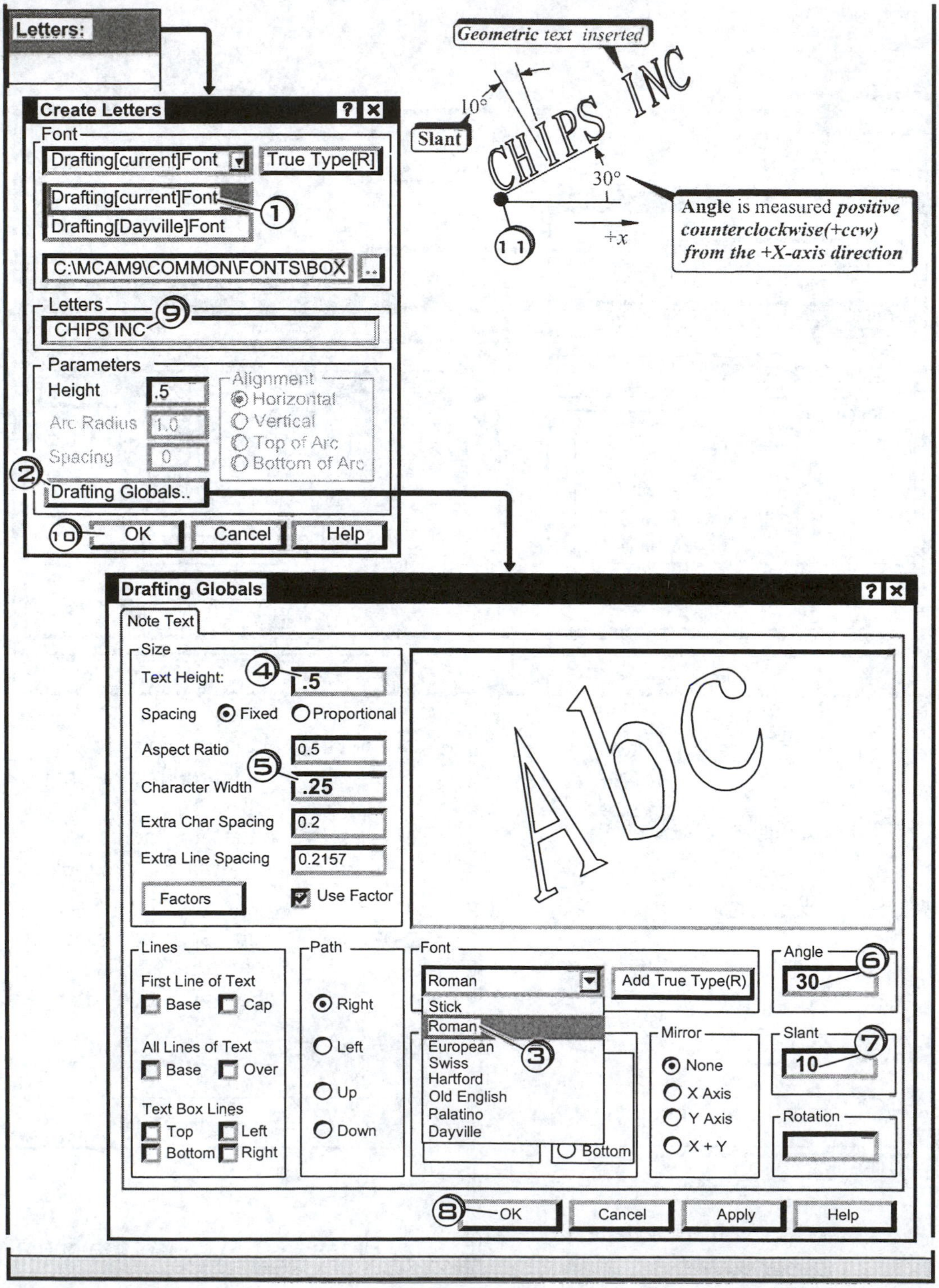

➤ Click ① the font style Drawing[Current]Font

➤ Click ② the Drafting Globals.. button

➤ Click ③ select the font style Roman

➤ Click ④ in the Height box; enter the text height .5

➤ Click ⑤ in the Character Width box; enter the text width .25

➤ Click ⑥ in the Angle box; enter the text angle 30

➤ Click ⑦ in the Slant box; enter the text slant angle 10

➤ Click ⑧ OK

➤ Click ⑨ inside Letters and enter the text

Letters
CHIPS INC

➤ Click ⑩ OK

Enter starting location of letters

➤ Click ⑪ *Mastercam* will insert the text as a set of geometric lines and arcs that can be used for machining

➤ Press Esc to cancel the operation

2-11 Ellipse Constructions in 2D Space

Creating Ellipses

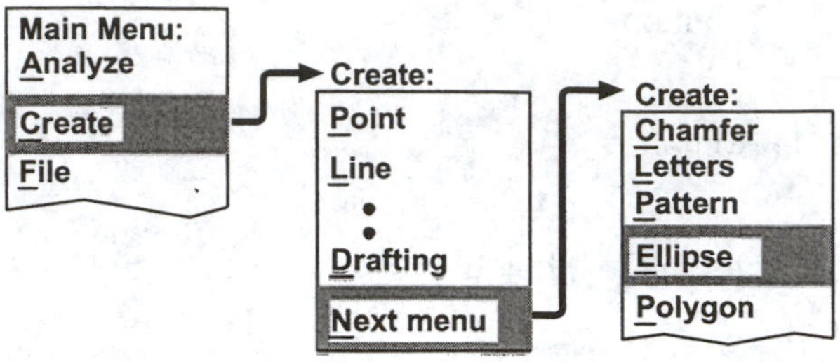

Ellipse

An *ellipse* is created. The operator specifies the **X Axis Radius**, **Y Axis Radius**, *arc* **Start angle**, **End angle**, **Rotation** angle *and the center point* location.

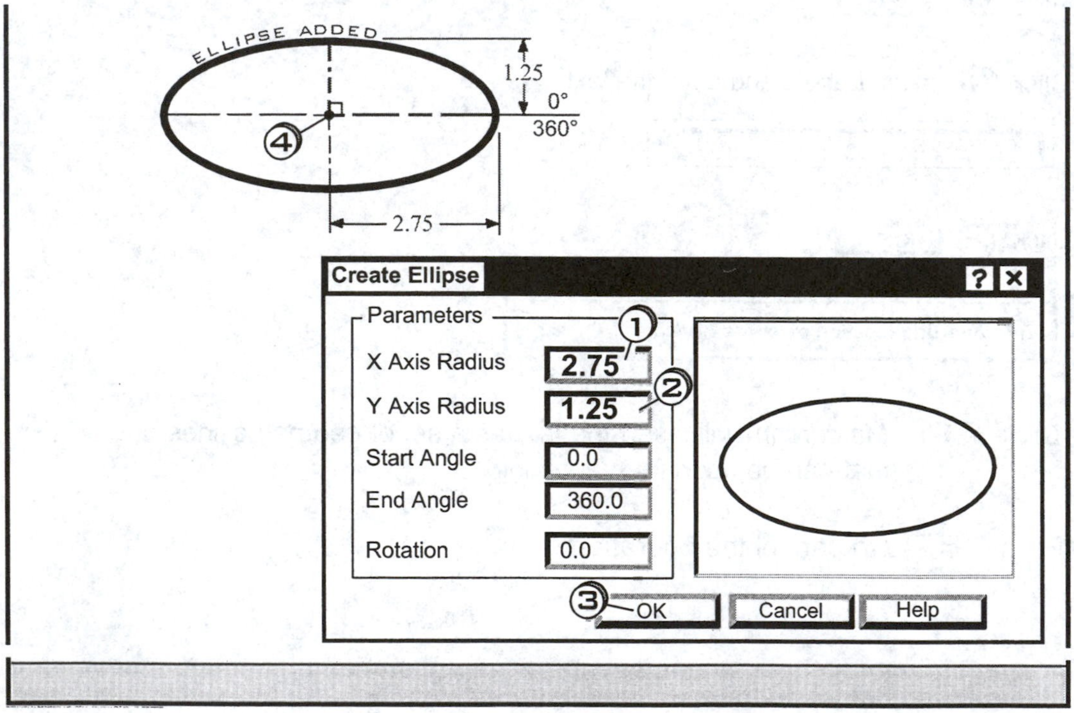

- Click ① in the X Axis Radius box; enter **2.75**
- Click ② in the Y Axis Radius box; enter **1.25**
- Click ③ the OK button

Enter the center point

- Click ④ the location of the center pt
- Press Esc to cancel the operation

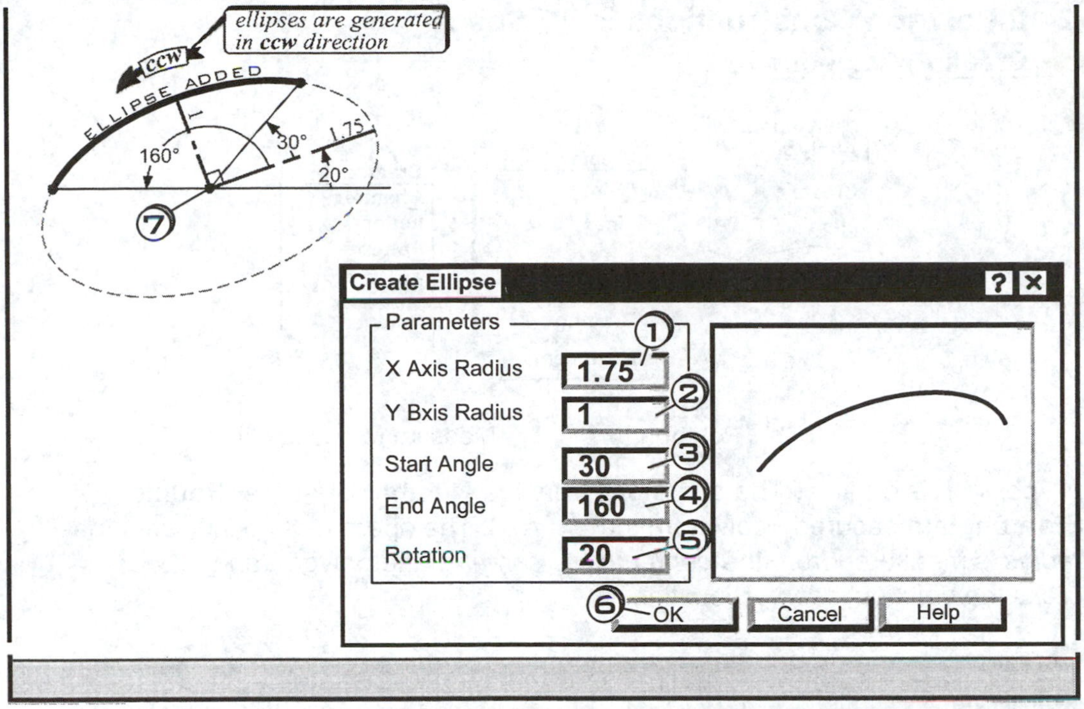

- Click ① in the X Axis box; enter **1.75**
- Click ② in the Y Axis box; enter **1**
- Click ③ in the Start Angle box; enter **30**
- Click ④ in the End Angle box; enter **160**
- Click ⑤ in the Rotation box; enter **20**
- Click ⑥ the OK button

Enter the center point

- Click ⑦ the location of the center point
- Press Esc to cancel the operation

Using the Toolbar Menus

Toolbar Menu Displayed when *Mastercam* is **Started**

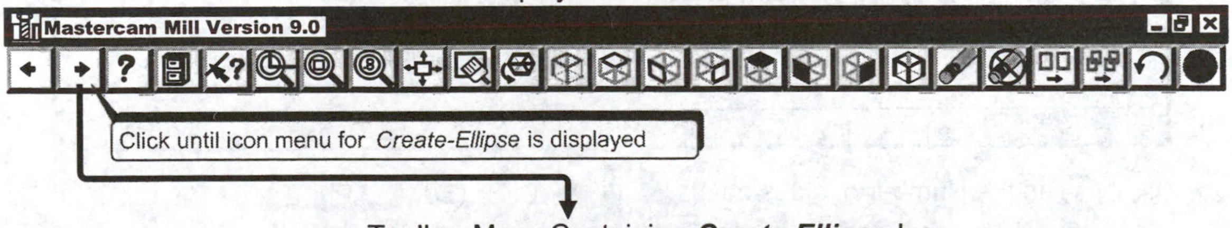

Toolbar Menu Containing ***Create-Ellipse*** Icon

Click Esc to cancel the Ellipse commands

2-12 Polygon Constructions in 2D Space

Creating Polygons

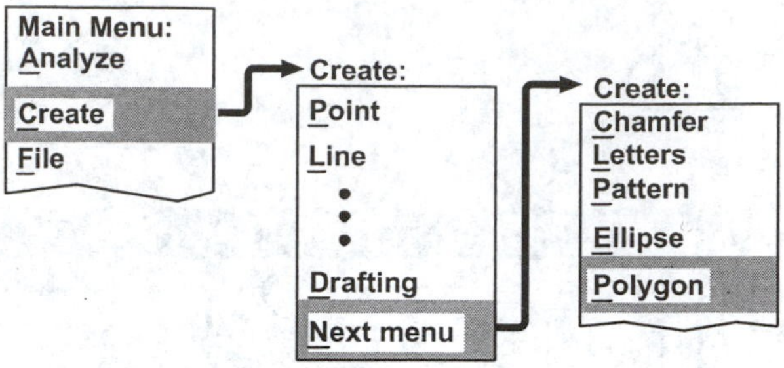

Polygon ☑ Measure radius to corner

A *polygon* is created. The operator enters the **Number of sides**, **Radius**, **Start angle**(measured *+ccw* from the *+x-axis*). The operator also indicates the *radius* is measured from the *center to the corner* of the polygon and clicks the polygon's *center* location.

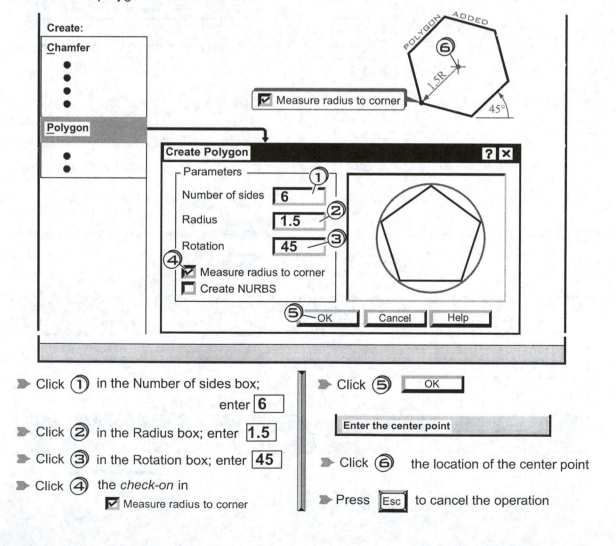

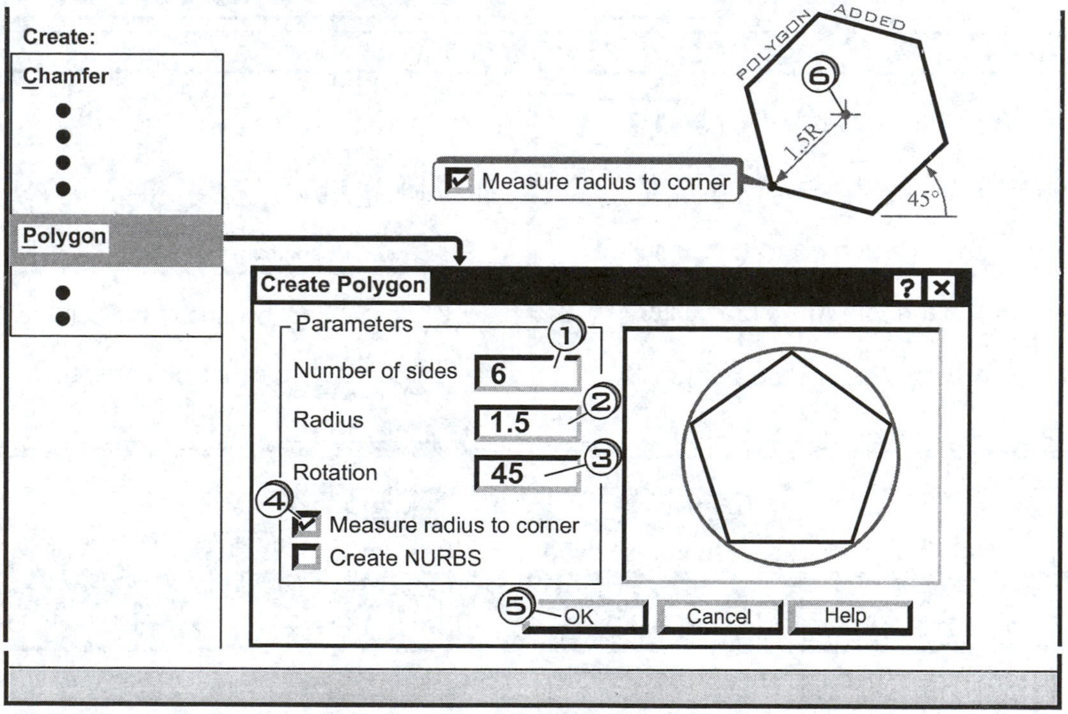

- Click ① in the Number of sides box; enter **6**
- Click ② in the Radius box; enter **1.5**
- Click ③ in the Rotation box; enter **45**
- Click ④ the *check-on* in ☑ Measure radius to corner
- Click ⑤ **OK**

Enter the center point

- Click ⑥ the location of the center point
- Press **Esc** to cancel the operation

Polygon ☐ Measure radius to corner

A *polygon* is created. The operator enters the **Number of sides**, **Radius**, **Start angle**(measured +*ccw* from the +*x-axis*). The operator also indicates the *radius* is measured from the *center to the midpoint of the flat side* of the polygon and clicks the polygon's *center* location.

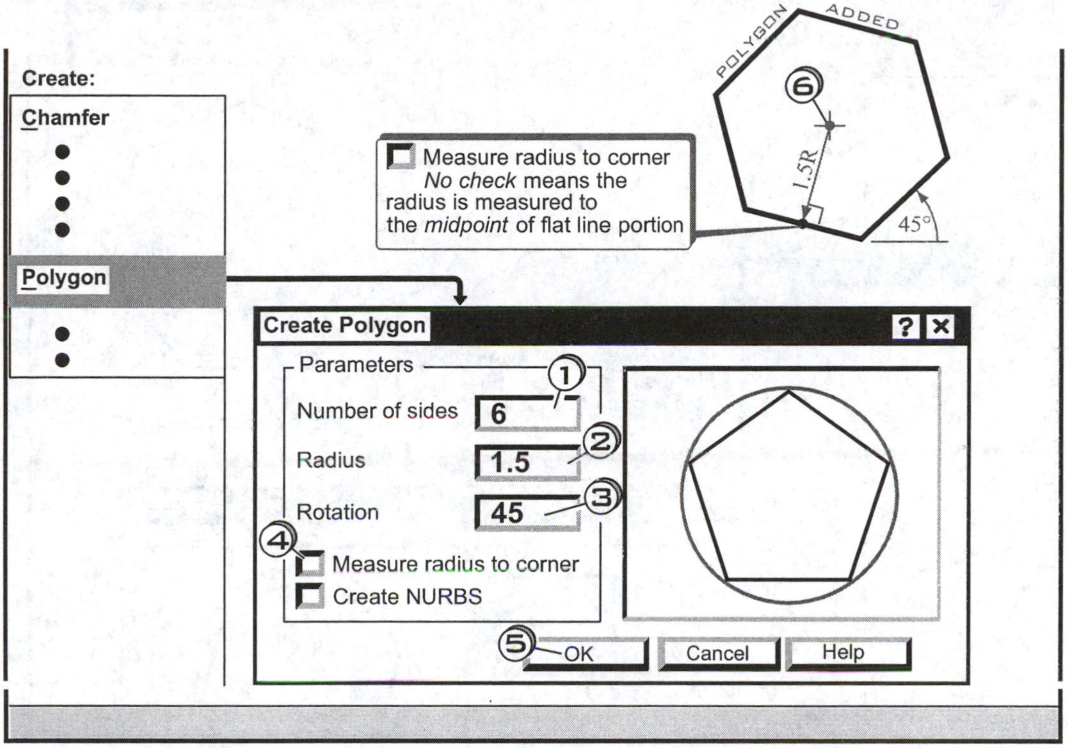

- Click ① in the Number of sides box; enter **6**
- Click ② in the Radius box; enter **1.5**
- Click ③ in the Rotation box; enter **45**
- Click ④ the *check-off* in ☐ Measure radius to corner
- Click ⑤ OK

Enter the center point

- Click ⑥ the location of the center point
- Press Esc to cancel the operation

Polygon ☐ Measure radius to corner ☑ NURBS

A *polygon* is created as a single NURBS spline. The operator enters the **Number of sides**, **Radius**,**Start angle**(measured *+ccw* from the *+x-axis*). The operator also indicates the *radius* is measured from the *center to the midpoint of the flat side* of the polygon and clicks the polygon's *center* location.

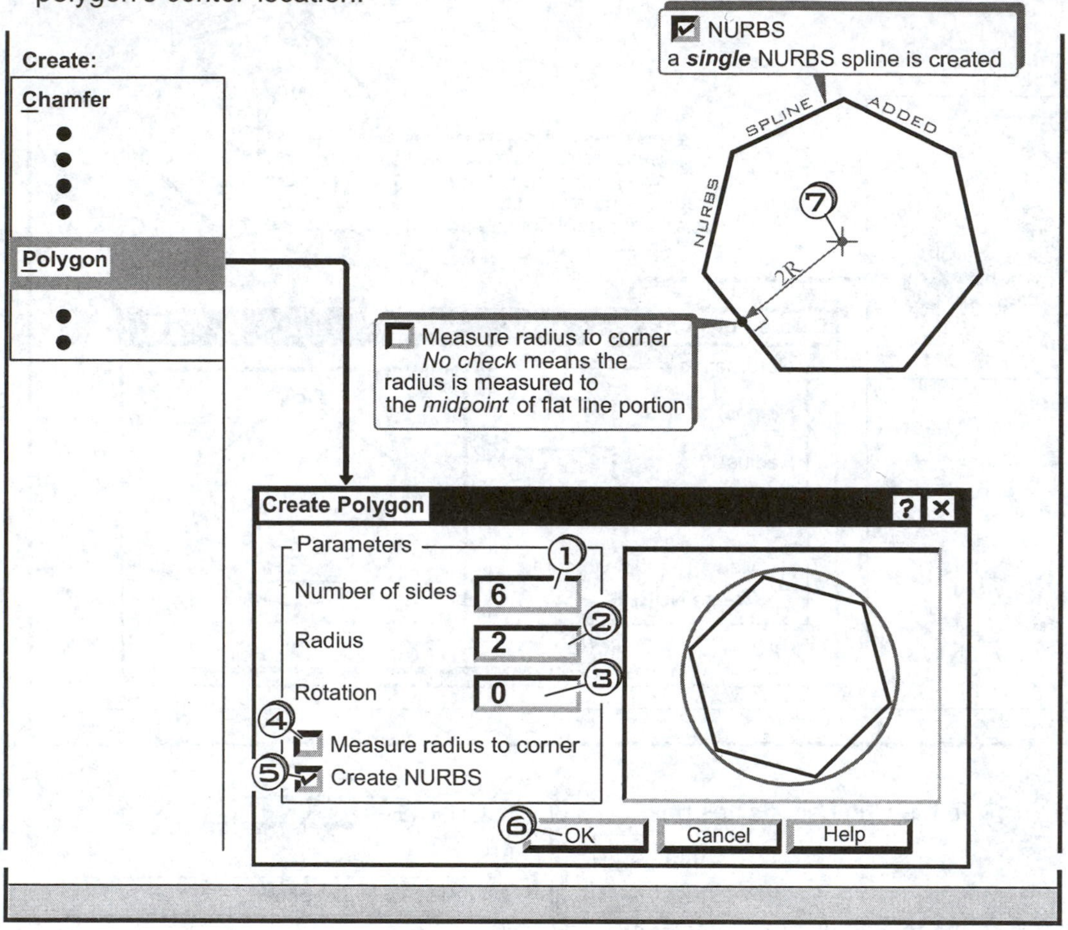

- Click ① in the Number of sides box; enter **6**
- Click ② in the Radius box; enter **2**
- Click ③ in the Rotation box; enter **0**
- Click ④ the *check-off* in ☐ Measure radius to corner
- Click ⑤ the *check-on* in ☑ Create NURBS
- Click ⑥ OK

Enter the center point

- Click ⑦ the location of the center point
- Press Esc to cancel the operation

2-13 Zooming In and Out on Screen Displays

Mastercam provides the user with the ability to control the size of the screen geometry displayed. Zooming in reveals needed details while zooming out or zooming the full view gives a complete picture of the overall part.

SCREEN ZOOM

Magnifies the geometry contained within a *rectangular* window clicked by the operator.

Toolbar Menu Displayed when *Mastercam* is **Started**

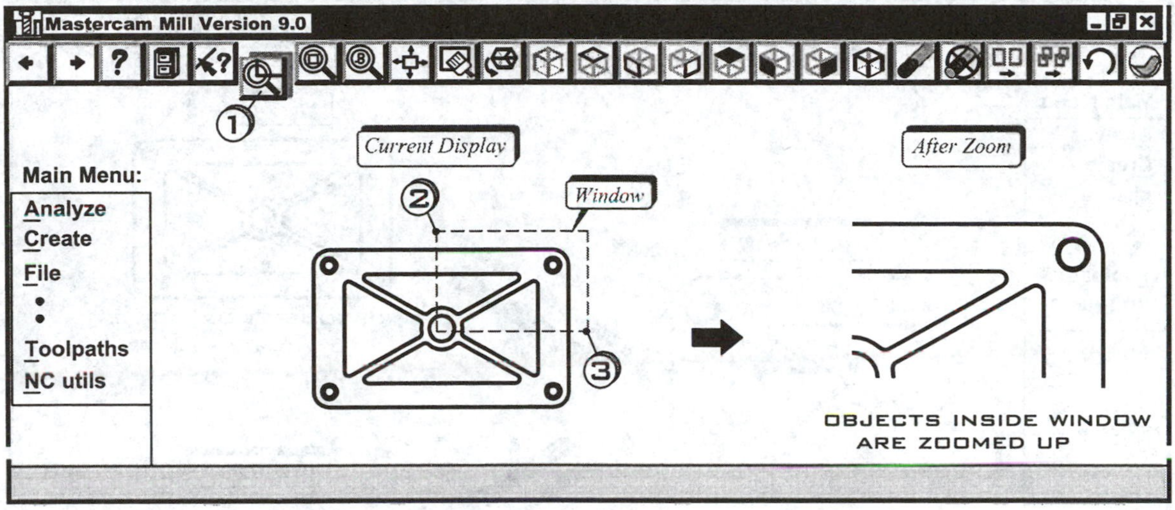

- Click ① the Zoom icon OR *press* F1
- Click ② ③ corners of the zoom window
- Press Esc to cancel the operation

SCREEN UNZOOM BY .5

Displays the geometry at the *scale of the previous zoom*. If *no* previous zoom *exists* the system will *reduce* the geometric display *by half its original size*. Unzoom can be used a *maximum* of *eight times* to reduce the current geometric display.

Toolbar Menu Displayed when *Mastercam* is **Started**

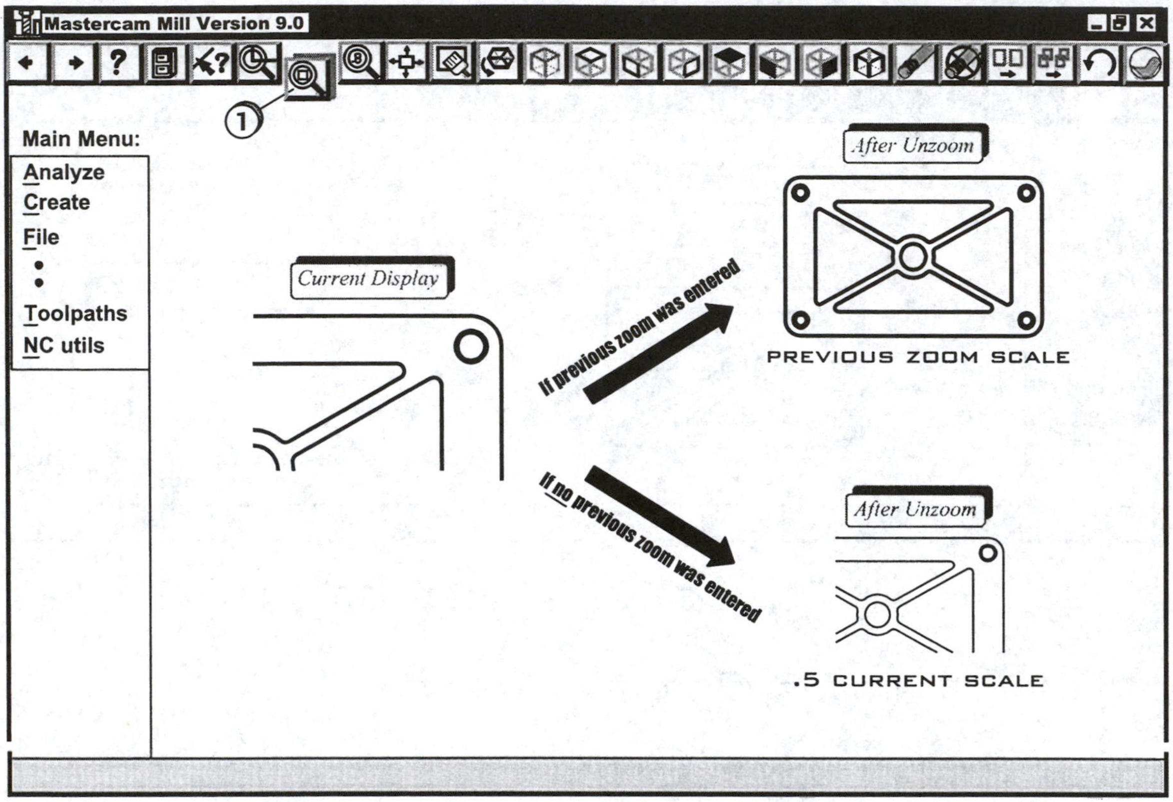

➡ Click ① the Unzoom icon OR *press* F2

SCREEN UNZOOM BY .8

Displays the geometry at the *scale of the previous zoom*. If *no* previous zoom *exists* the system will *reduce* the geometric display *by .8 its original size*. Unzoom can be used a *maximum* of *eight times* to reduce the current geometric display.

Toolbar Menu Displayed when *Mastercam* is **Started**

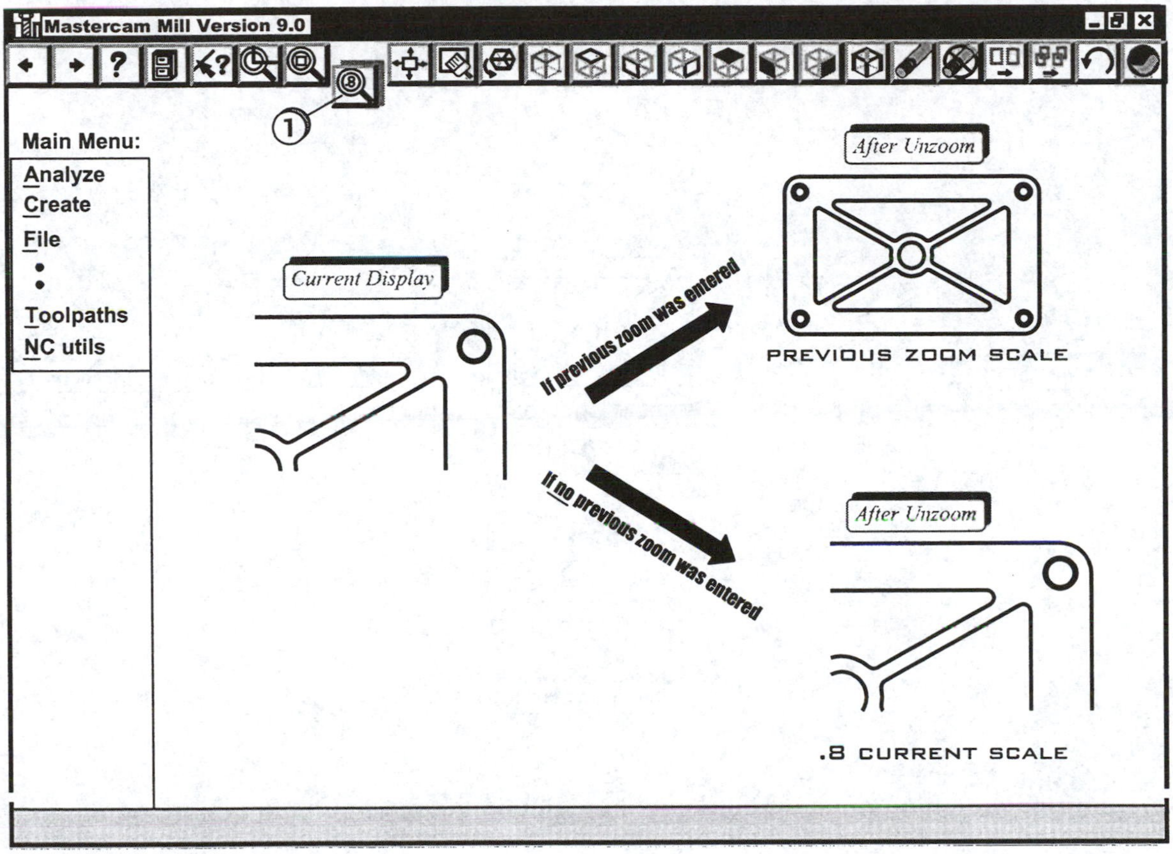

➤ Click ① the Unzoom .8 icon **OR** *press* Alt + F2

2-14 Panning Screen displays

The operator can *pan* the geometric screen display in any of *four directions*: *up, down, left* or *right*. This is accomplished by tapping the *directional arrow key* on the numeric keypad corresponding to the panning direction desired.

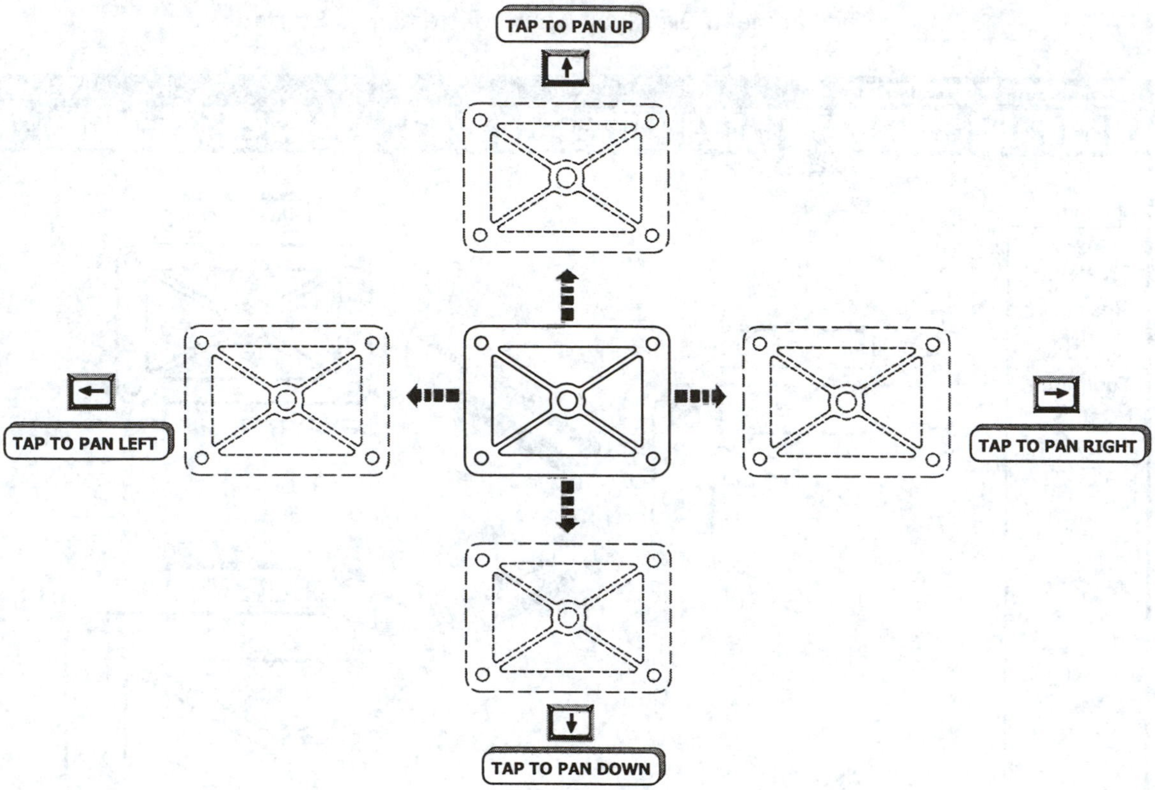

2-15 Fitting the Existing Geometry to the Screen Display Area

SCREEN FIT

Scales ***all*** the geometry such that it ***fits*** the *screen*

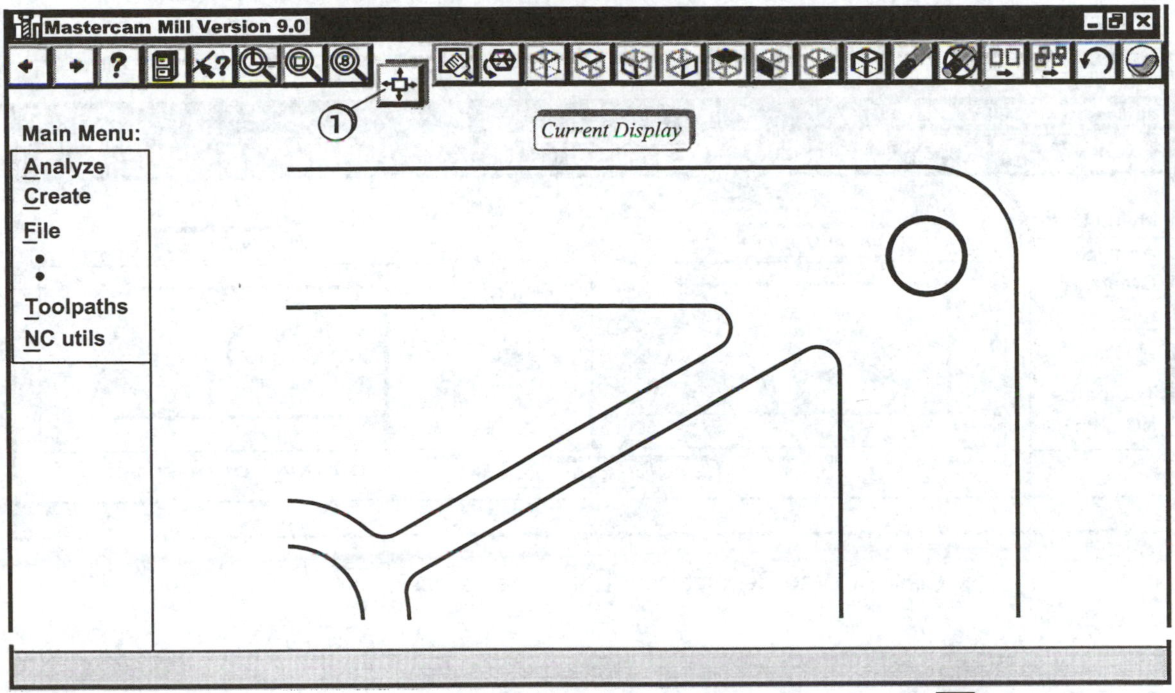

➤ Click ① the Screen-Fit icon OR *press* Alt + F1

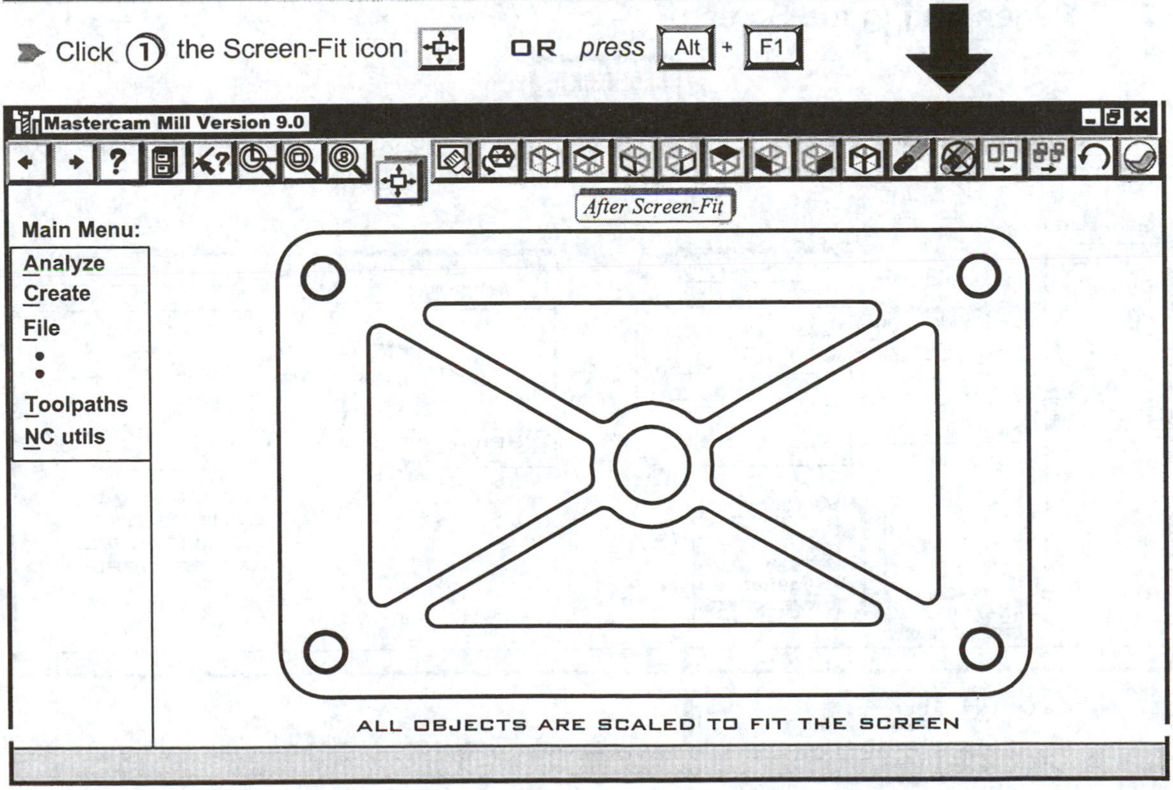

2-16 Repainting the Screen

SCREEN REPAINT

Ocassionally, *Mastercam* displays remanents of objects that have been removed via trim/erase etc, temporarily displays a different screen backround/geometry color or displays the ghost of another application when the operator switches between that application and Mastercam. Repaint is used to clear up these *ghost display* problems.

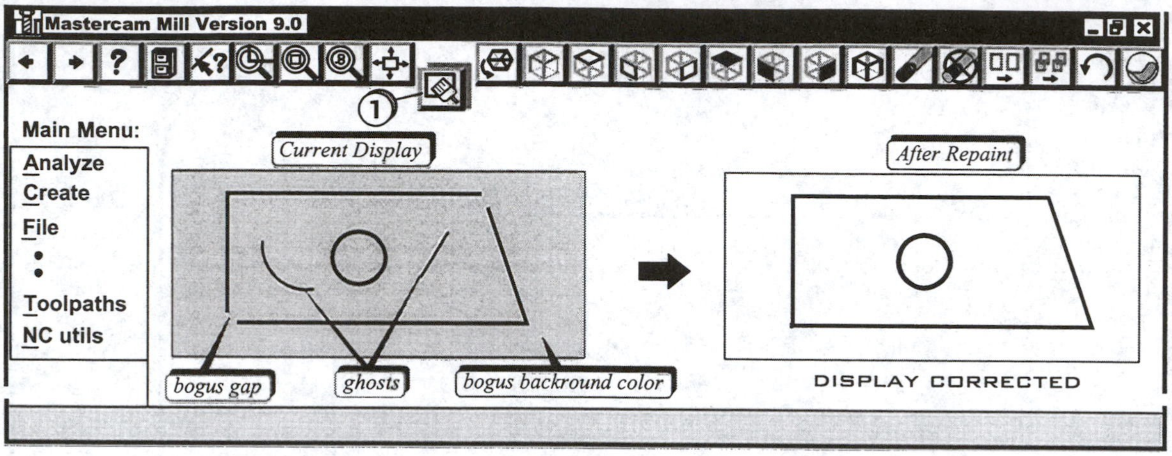

➤ Click ① the Repaint icon OR *press* F3

2-17 Regenerating the Screen

REGENERATE

Regenerates all the graphic entities and *displays all* the *current existing geometry. This* command is used if Repaint *fails* to restore all the existing graphics.

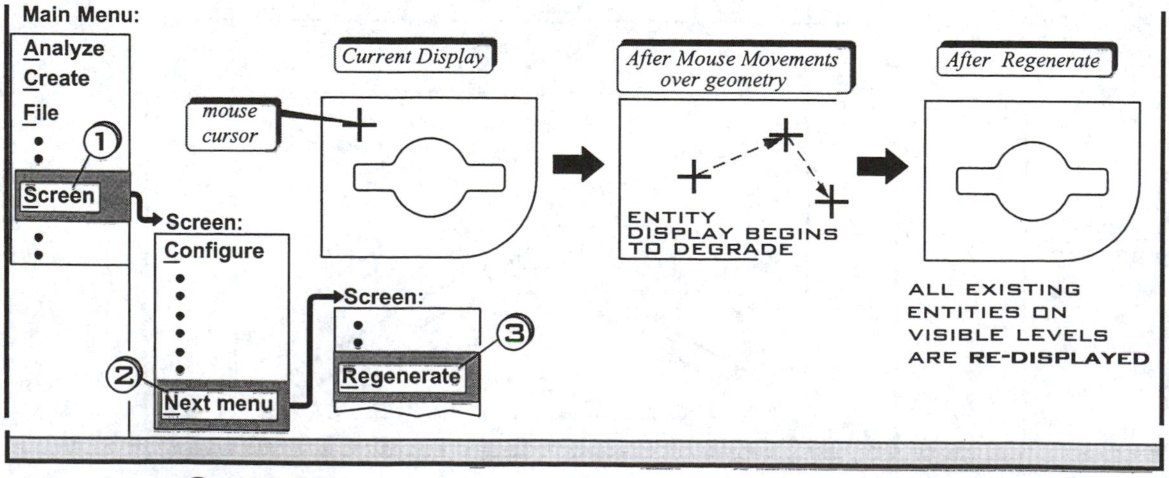

➤ Click ① Screen

➤ Click ② Next menu

➤ Click ③ Regenerate

Using the Toolbar Menus

Toolbar Menu Displayed when *Mastercam* is **Started**

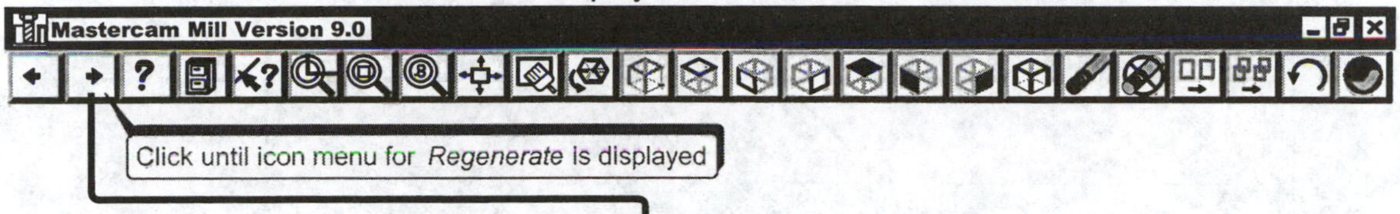

Toolbar Menu Containing ***Regenerate*** Icon

Click [Esc] to cancel the Regenerate commands

2-18 Undoing

UNDO

Undoes the *last* action performed. ***The last action cannot be undone if the operator exits the menu* of the *operation to be undone.***

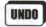

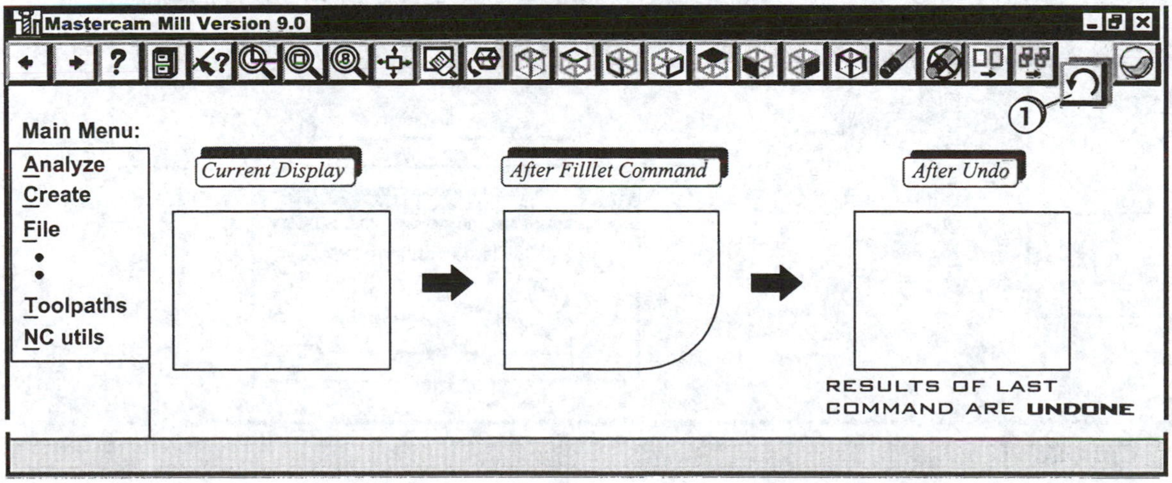

➤ Click ① the Undo icon OR *press* [Alt] + [U]

EXERCISES

In each exercise create the part geometry using *Mastercam's* CAD package.

2-1) File Name: **EX2-1JV** ← YOUR INITIALS

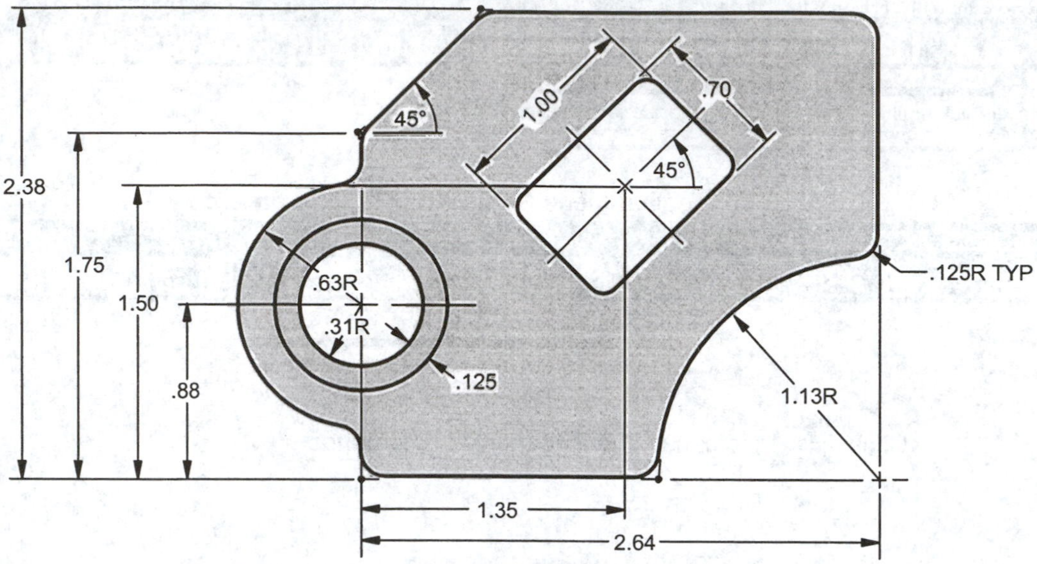

Figure 2-p1

Enter *Mastercam* mill and open a new **.MC9** file.

➤ Click ① the *Mastercam* **MILL** icon 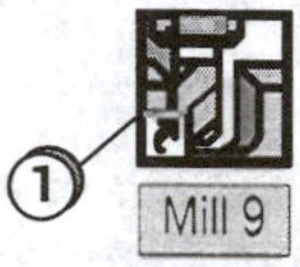 to enter the Mill package

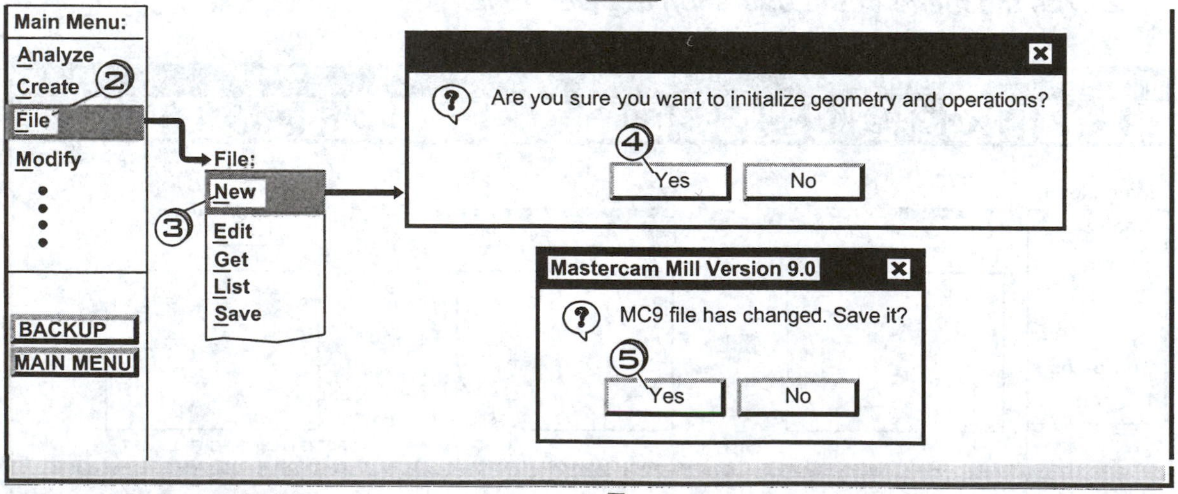

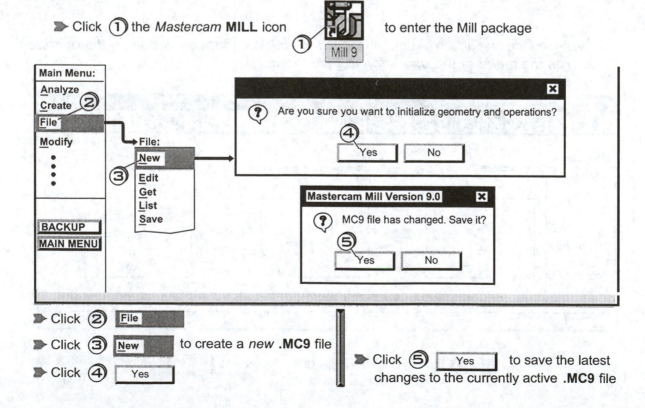

➤ Click ② File

➤ Click ③ New to create a *new* **.MC9** file

➤ Click ④ Yes

➤ Click ⑤ Yes to save the latest changes to the currently active **.MC9** file

Create the CAD model.

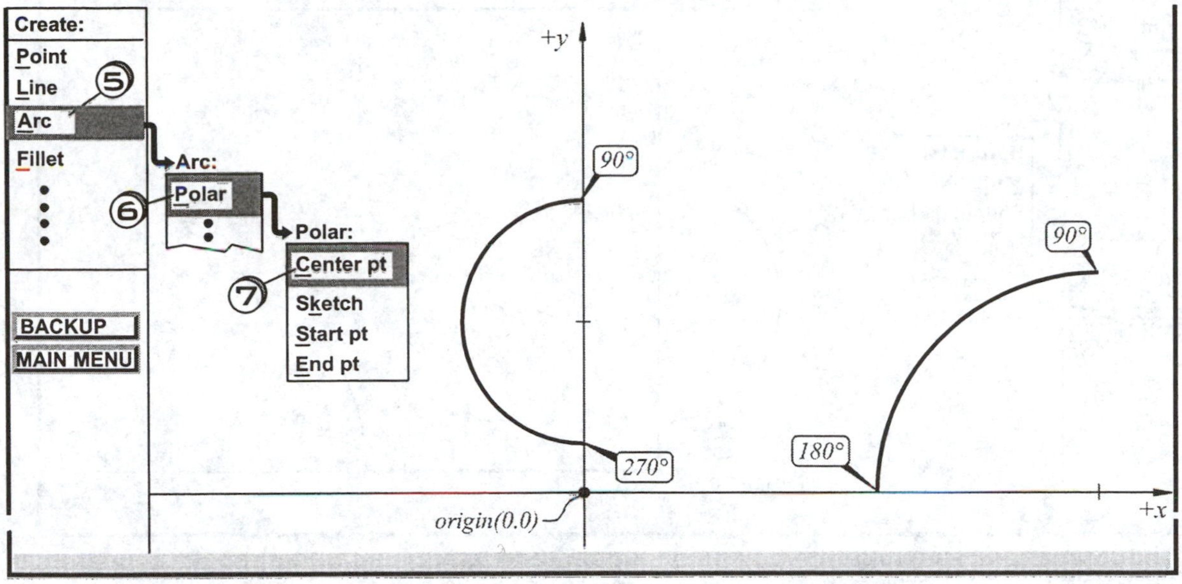

➤ MAIN MENU → Create:

NOTE: FOR THIS EXERCISE, THE ABOVE IS TO SIGNIFY THE OPERATOR IS TO FIRST Click THE MAIN MENU BUTTON THEN Click Create:

➤ Click ⑤ Arc

➤ Click ⑥ Polar

➤ Click ⑦ Center pt

Create arc, polar: Enter the center point

➤ Enter center pt coords **0,.88** Enter

Enter the radius .25

➤ Enter arc's radius **.63** Enter

Enter the initial angle 0

➤ Enter arc's starting angle **90** Enter

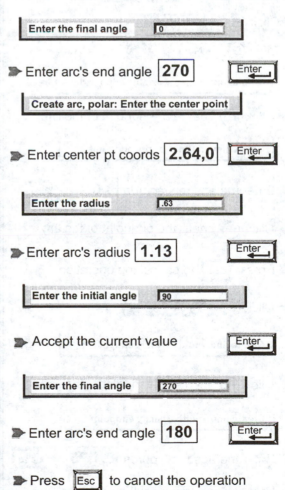

Enter the final angle 0

➤ Enter arc's end angle **270** Enter

Create arc, polar: Enter the center point

➤ Enter center pt coords **2.64,0** Enter

Enter the radius .63

➤ Enter arc's radius **1.13** Enter

Enter the initial angle 90

➤ Accept the current value Enter

Enter the final angle 270

➤ Enter arc's end angle **180** Enter

➤ Press Esc to cancel the operation

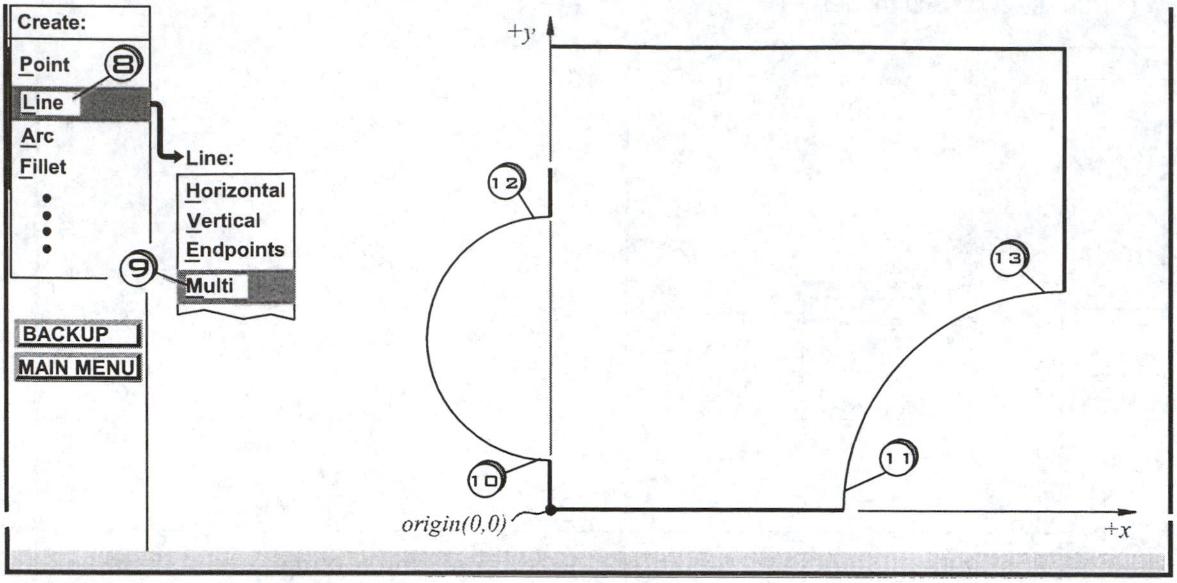

- MAIN MENU → Create:
- Click (8) Line
- Click (9) Multi

Create line, multi: Specify endpoint 1

- Click (10) near the endpoint of the line

Create line, multi: Specify endpoint 2

- Enter the abs rect coords **0,0** Enter
- Click (11) near the endpoint of the arc
- Press Esc to cancel the operation
- Click (9) Multi

Create line, multi: Specify endpoint 1

- Click (12) near the endpoint of the line

Create line, multi: Specify endpoint 2

- Enter the abs rect coords **0,1.75** Enter
- Press Esc to cancel the operation

- Click (9) Multi
- Enter the abs rect coords **0,2.38** Enter

Create line, multi: Specify endpoint 2

- Enter the absolute rectang coords **2.64,2.38** Enter

Create line, multi: Specify endpoint 3

- Click (13) near the endpoint of the arc
- Press Esc to cancel the operation

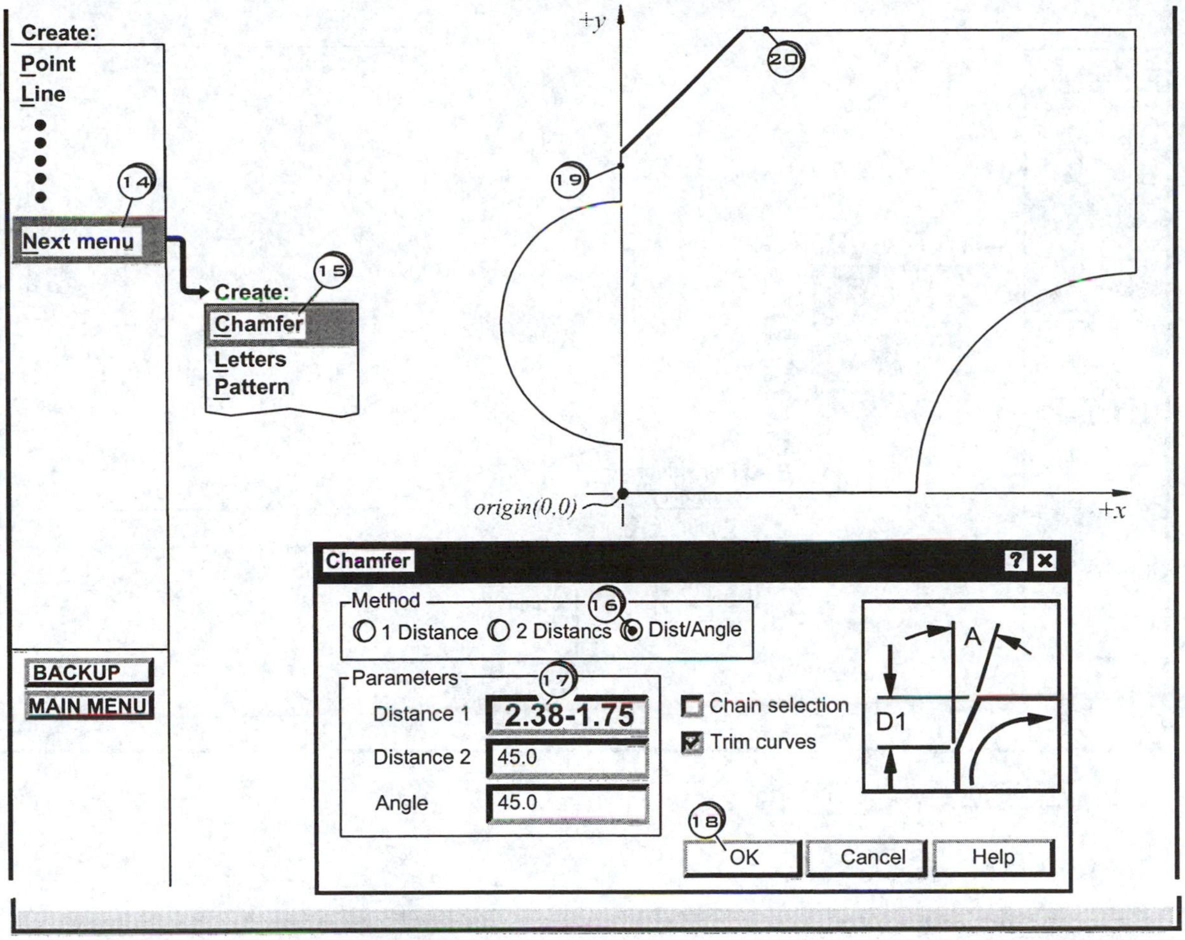

- MAIN MENU → Create:
- Click (14) Next menu
- Click (15) Chamfer
- Click (16) the Dist/Angle radio button
- Click (17) ; enter Distance 1 **2.28-1.75**
- Click (18) OK

  Chamfer: select first line or arc

- Click (19) near the end of the *first* line

  Chamfer: select second line

- Click (20) near the end of the *second* line
- Press Esc to cancel the operation

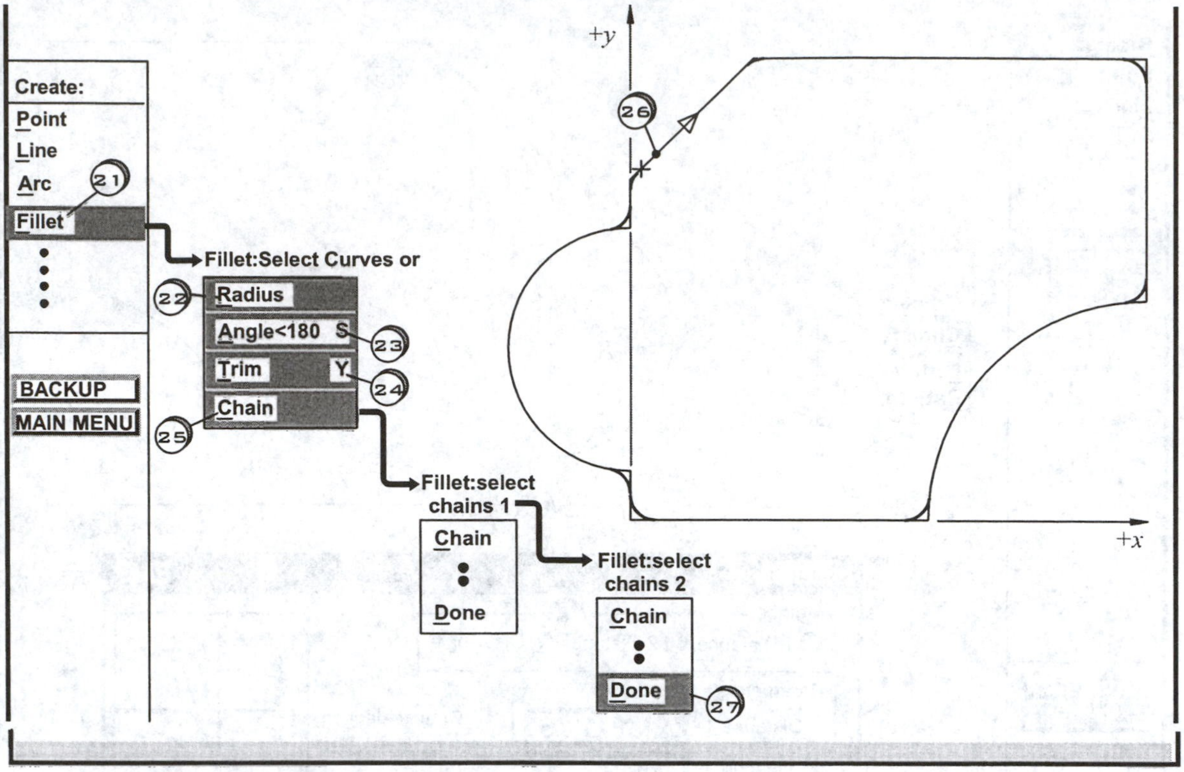

➤ MAIN MENU → Create:

➤ Click (21) Fillet

➤ Click (22) Radius

Enter the fillet radius:

➤ Enter the fillet's radius .125 Enter

➤ Click (23) Angle<180 S such that an **S** appears

➤ Click (24) Trim Y such that an **Y** appears

➤ Click (24) Trim Y

➤ Click (25) Chain

Chaining mode: Full Chaining mask: None

➤ Click (26) near the end of the chamfer line

➤ Click (27) Done

➤ Press Esc to cancel the operation

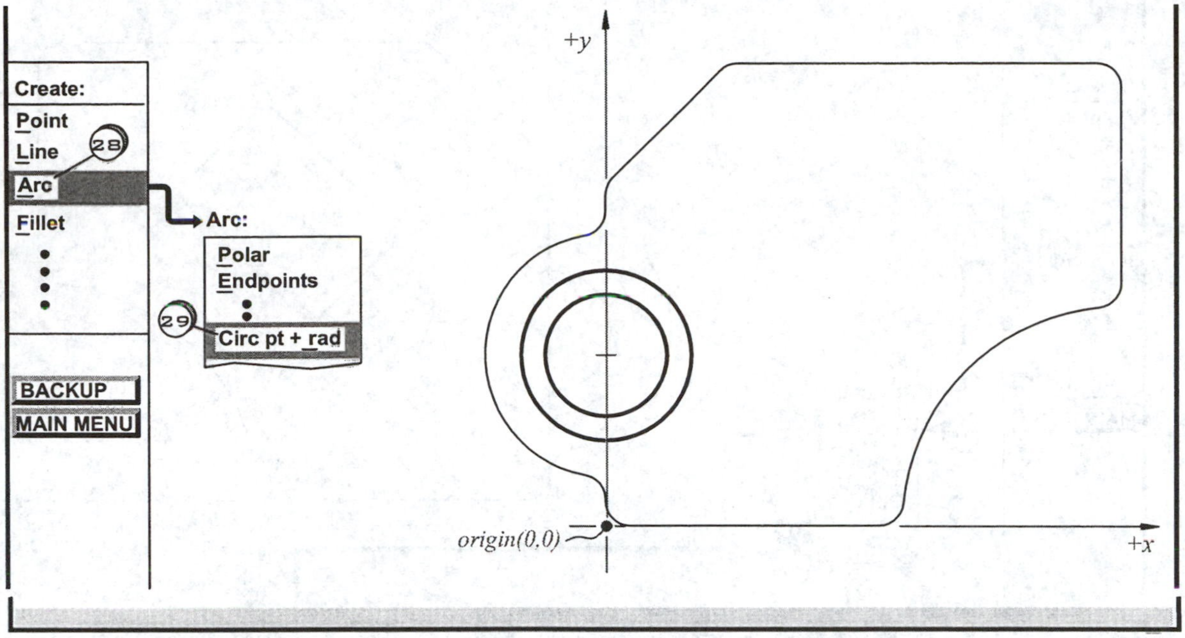

- MAIN MENU → Create:
- Click 28 Arc
- Click 29 Circ pt + rad

Enter the radius:

- Enter the circle's radius **.31** Enter
- Enter the rectang coords of the circle's center point **0,.88** Enter
- Press Esc to cancel the operation
- Click 29 Circ pt + rad

Enter the radius: .31

- Enter the circle's radius **.31+.125** Enter
- Enter the rectang coords of the circle's center point **0,.88** Enter
- Press Esc to cancel the operation

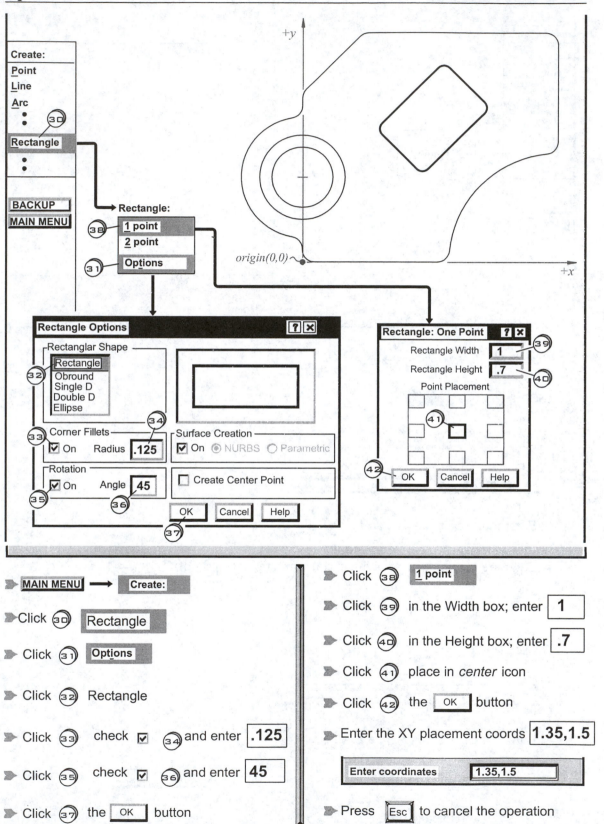

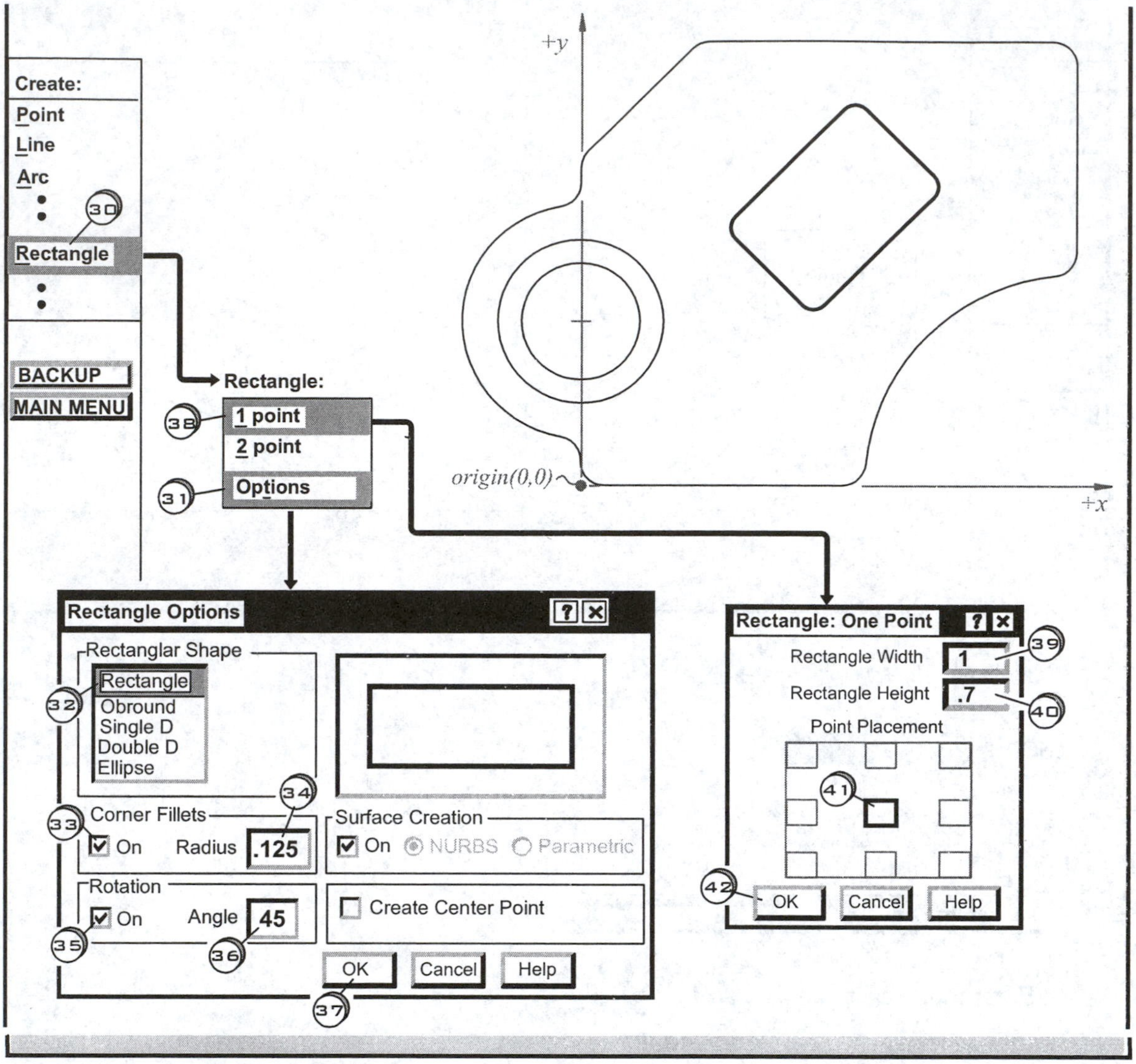

- MAIN MENU → Create:
- Click (30) Rectangle
- Click (31) Options
- Click (32) Rectangle
- Click (33) check ☑ (34) and enter .125
- Click (35) check ☑ (36) and enter 45
- Click (37) the OK button
- Click (38) 1 point
- Click (39) in the Width box; enter 1
- Click (40) in the Height box; enter .7
- Click (41) place in *center* icon
- Click (42) the OK button
- Enter the XY placement coords 1.35,1.5

  Enter coordinates 1.35,1.5

- Press Esc to cancel the operation

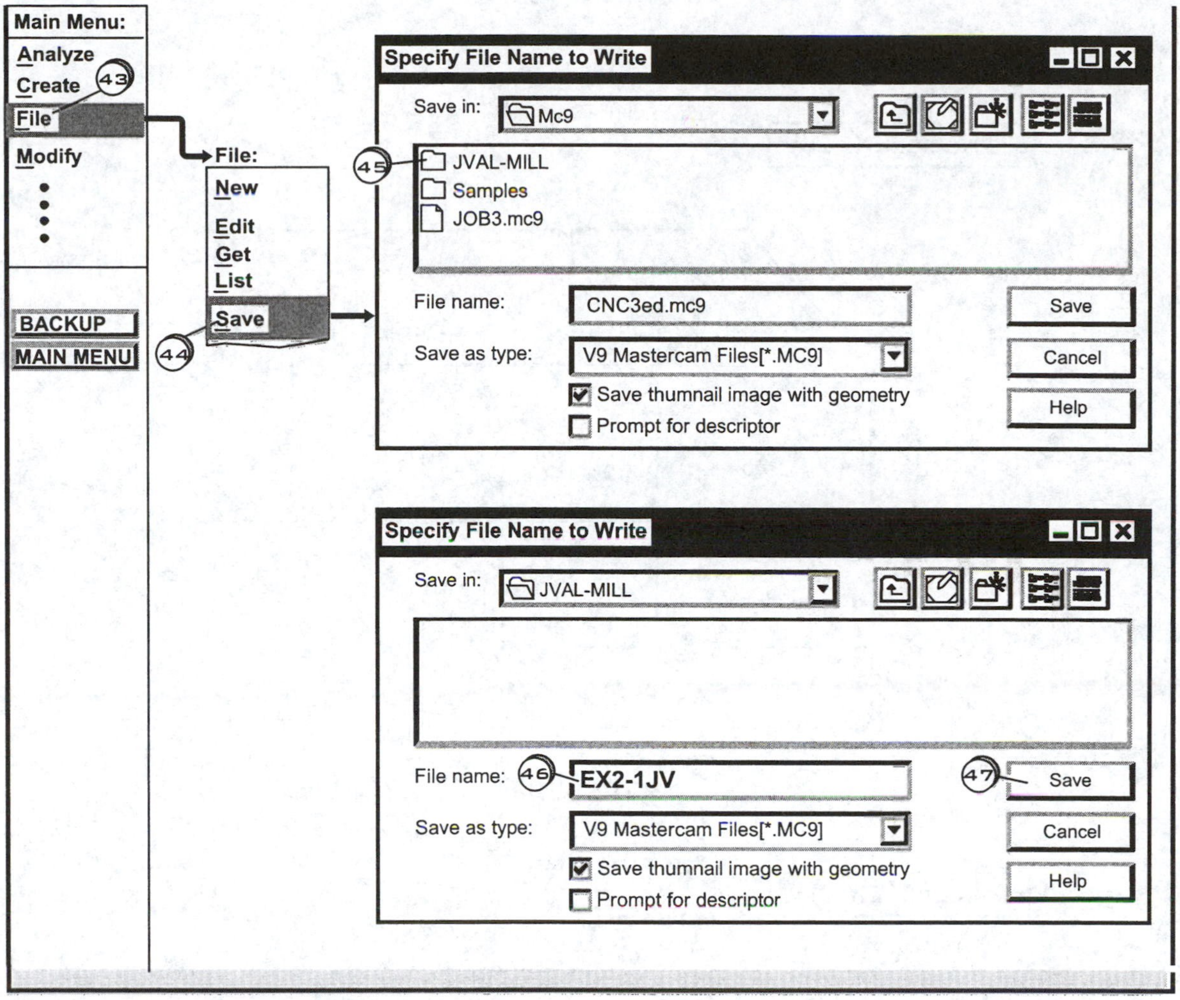

- Click (43) **File**
- Click (44) **Save**
- *double* Click (45) on the **JVAL-MILL** file folder to open this directory
- Click (46) in the file name box and enter the name of the **.MC9** file to be saved: **EX2-1JV**
- Click (47) Save

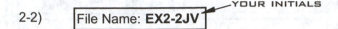

2-2) File Name: **EX2-2JV** ← YOUR INITIALS

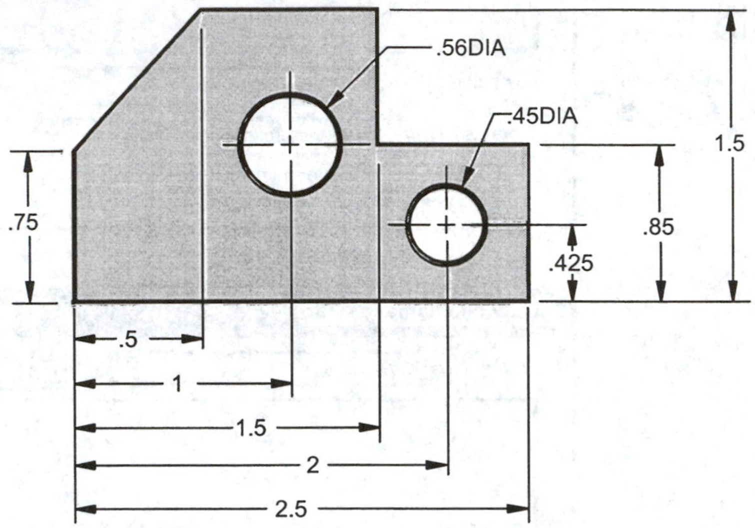

Figure 2-p2

2-3) File Name: **EX2-3JV** ← YOUR INITIALS

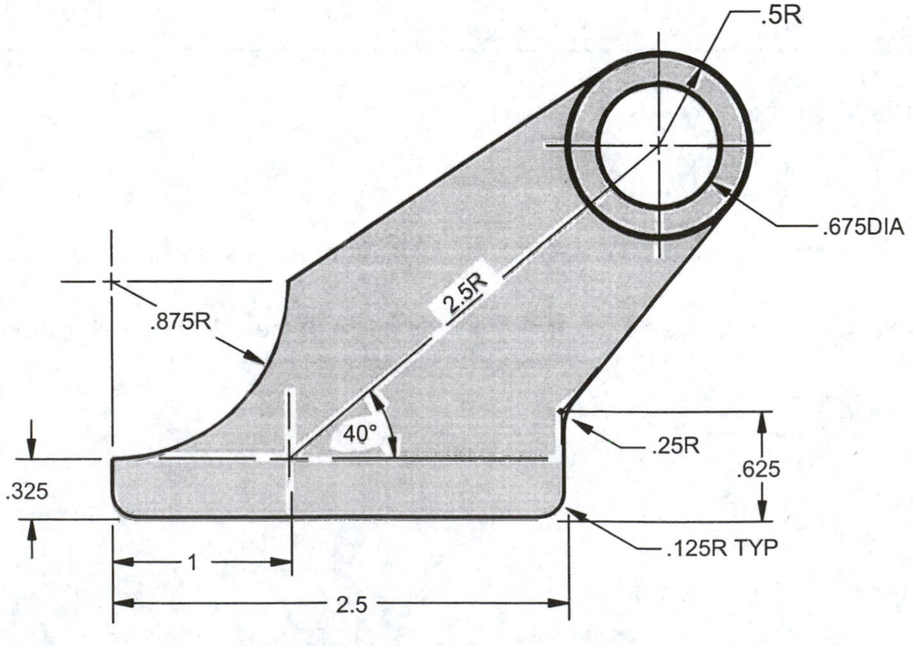

Figure 2-p3

2-4) File Name: **EX2-4JV** ← YOUR INITIALS

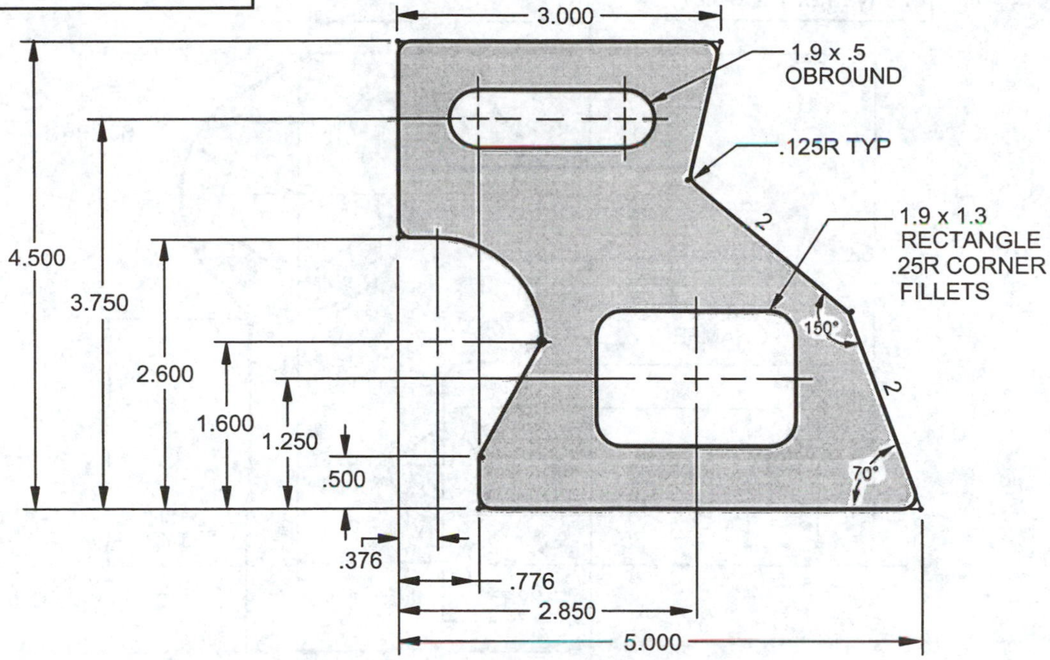

Figure 2-p4

2-5) File Name: **EX2-5JV** ← YOUR INITIALS

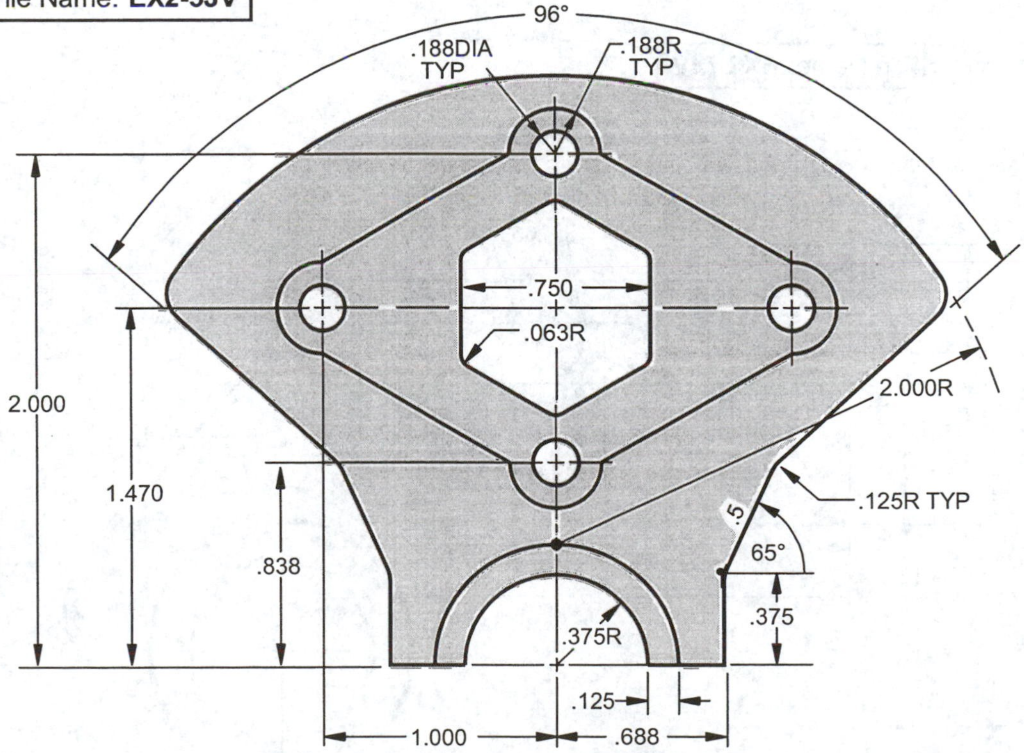

Figure 2-p5

2-6) File Name: **EX2-6JV** ← YOUR INITIALS

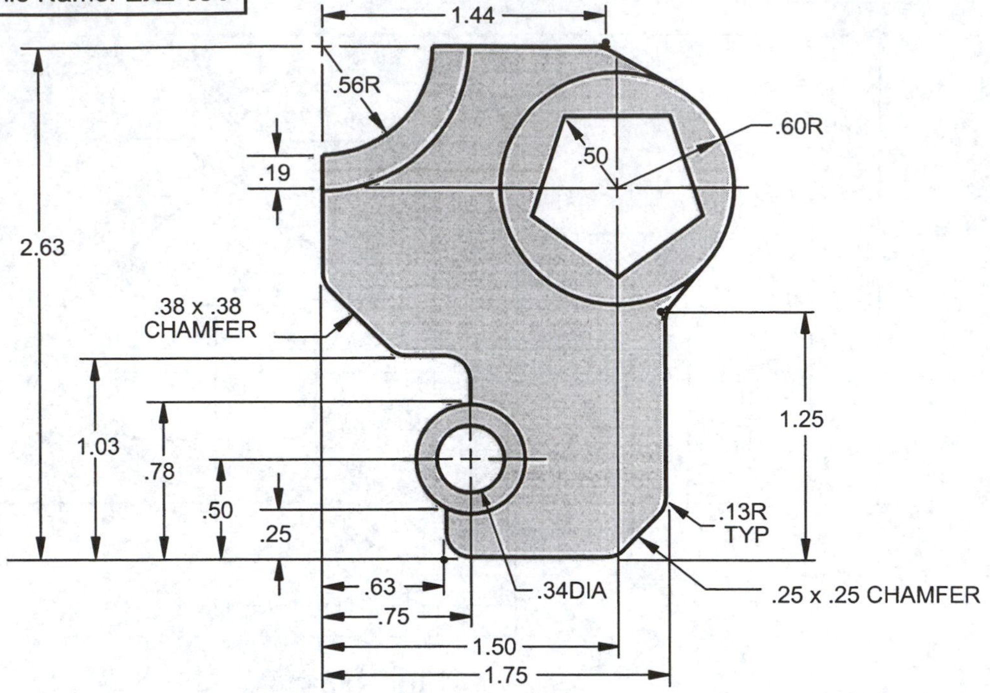

Figure 2-p6

2-7) File Name: **EX2-7JV** ← YOUR INITIALS

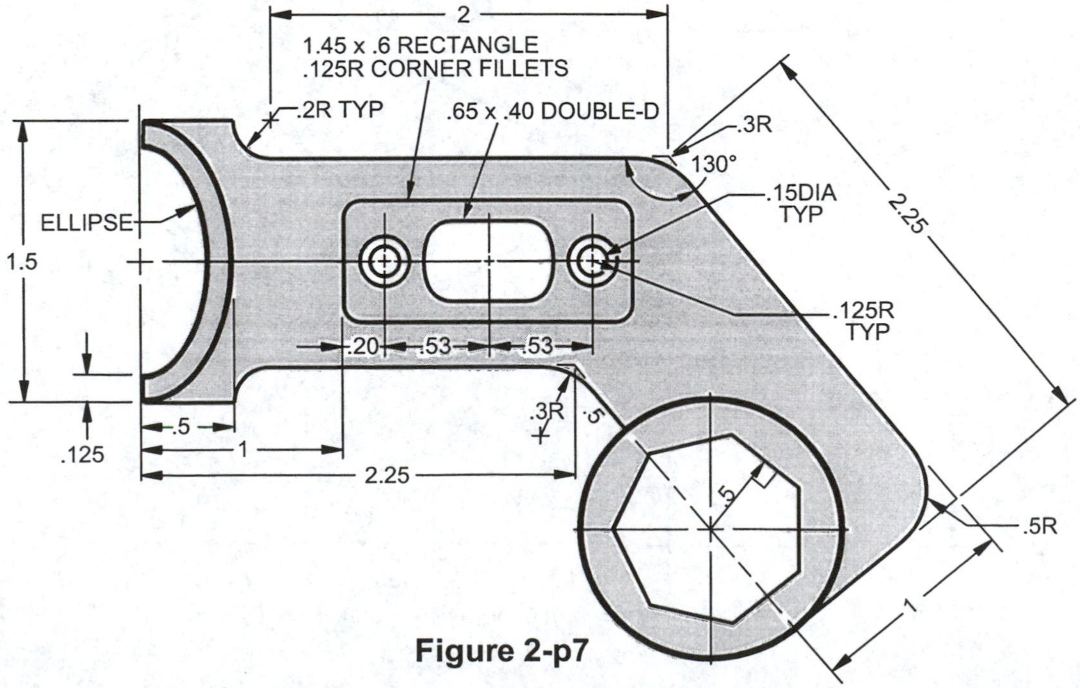

Figure 2-p7

2-8) File Name: EX2-8JV ← YOUR INITIALS

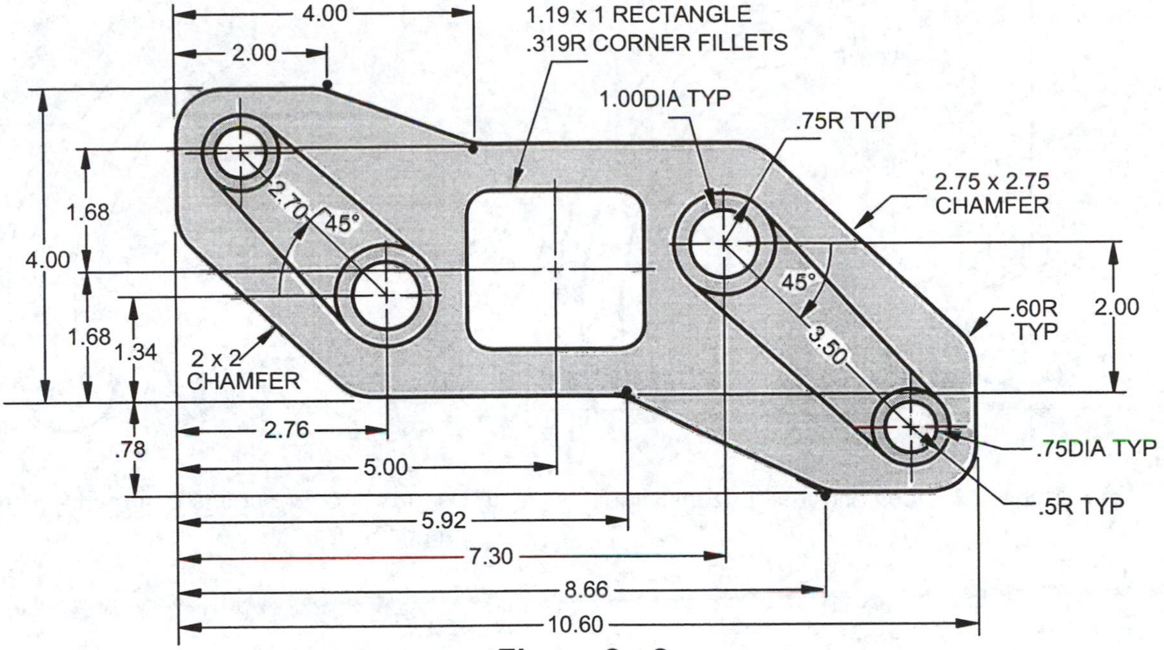

Figure 2-p8

2-9) File Name: EX2-9JV ← YOUR INITIALS

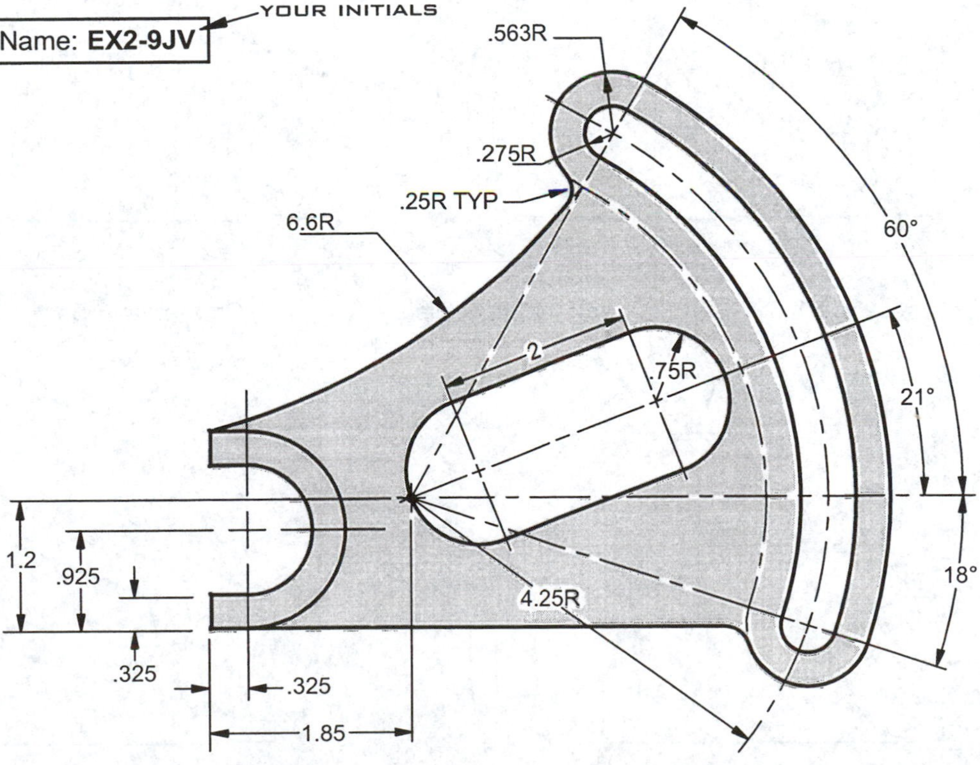

Figure 2-p9

a) Use the *Arc*, *Polar* command to create the .6R, .925R, 3.687R, 3.975R, 4.525R and 4.813R arcs

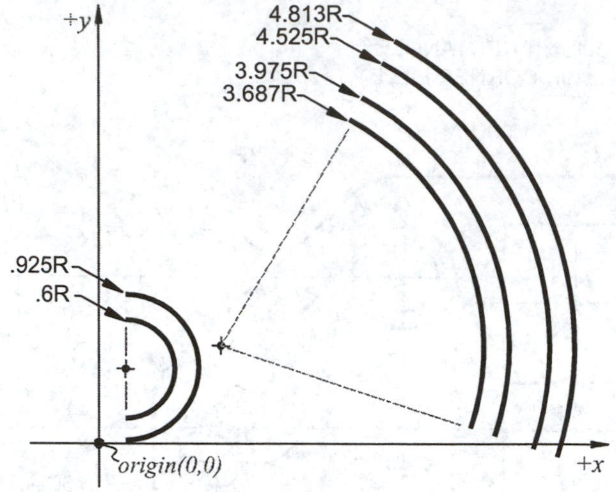

b) Use the *Arc*, *Endpoints* command to create the .275R and .563R arcs

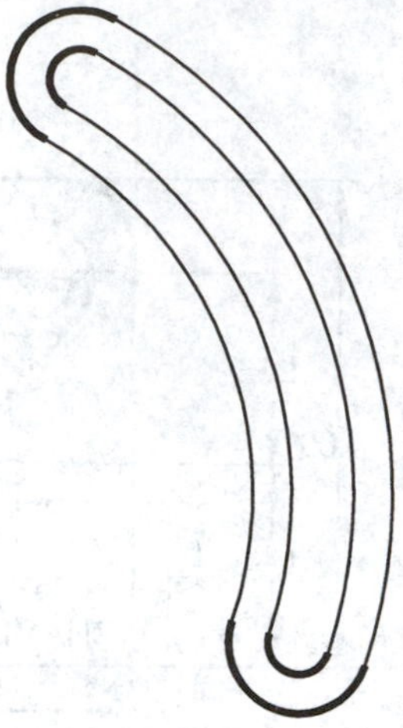

c) Add the .325 lines and the horizontal line

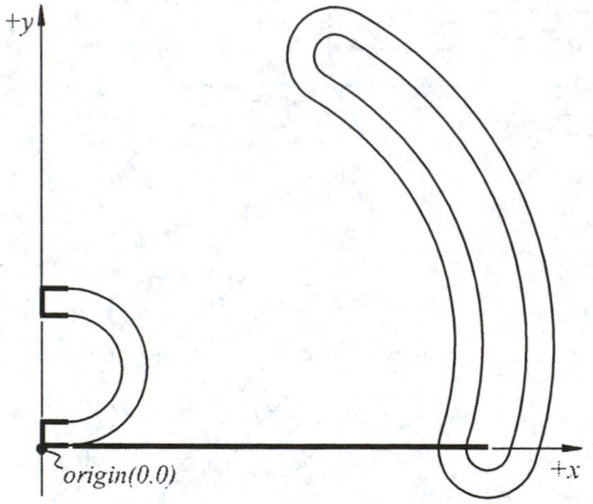

d) Use the *Arc, Endpoints* command to create the 6.6R arc

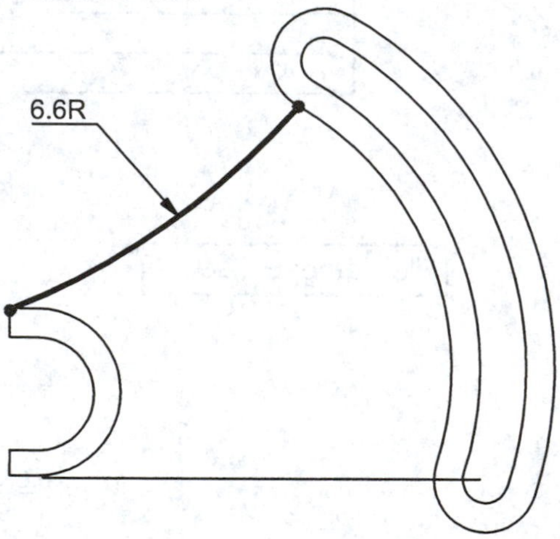

e) Erase the 3.687R arc and add the .25R fillets

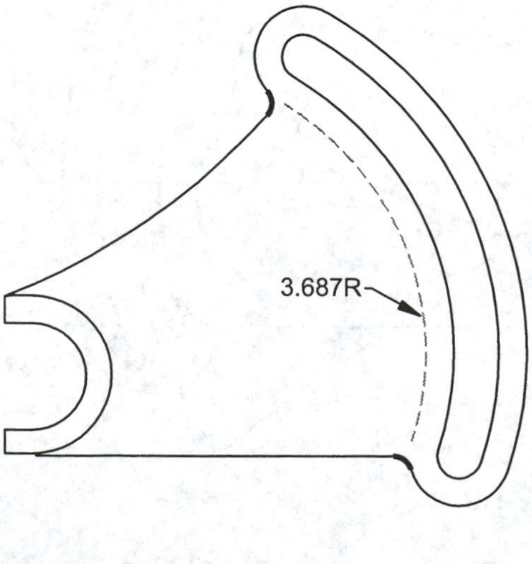

f) Create the 3.5 x 1.5 Obround at 21°

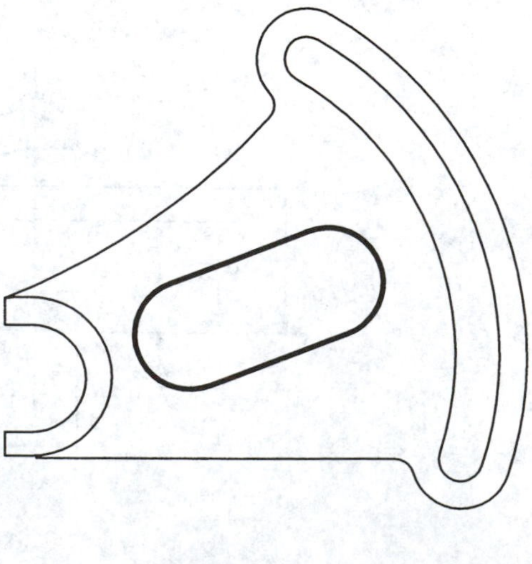

2-10) File Name: **EX2-10JV** ← YOUR INITIALS

Generate the CAD entities and the geometric letters as shown in Figure 2p-10

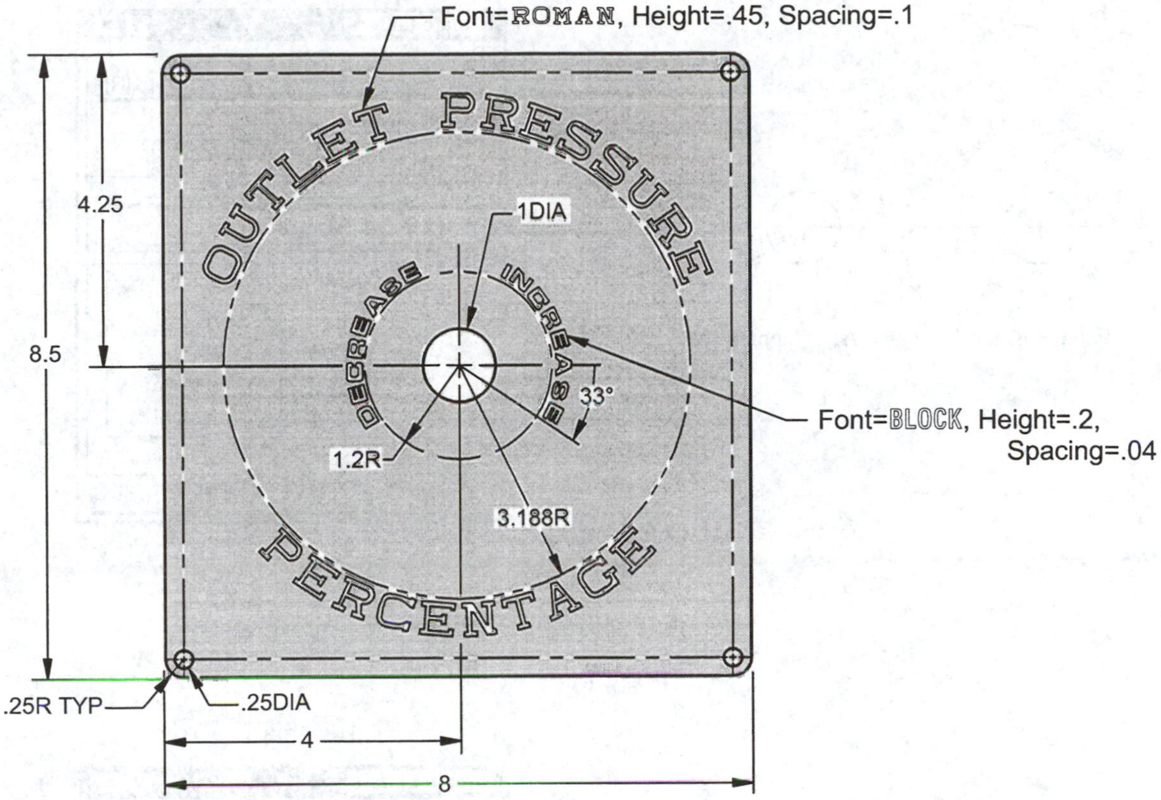

Figure 2-p10

a) Create the 8.5 x 8 plate, .25, 1DIA circles and ROMAN font lettering

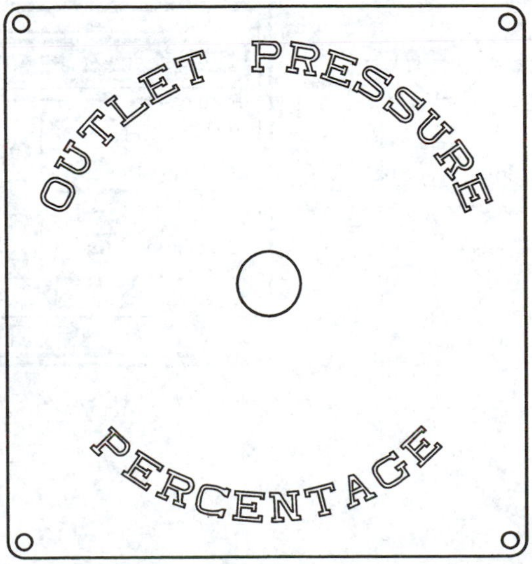

b) Create the BLOCK font lettering

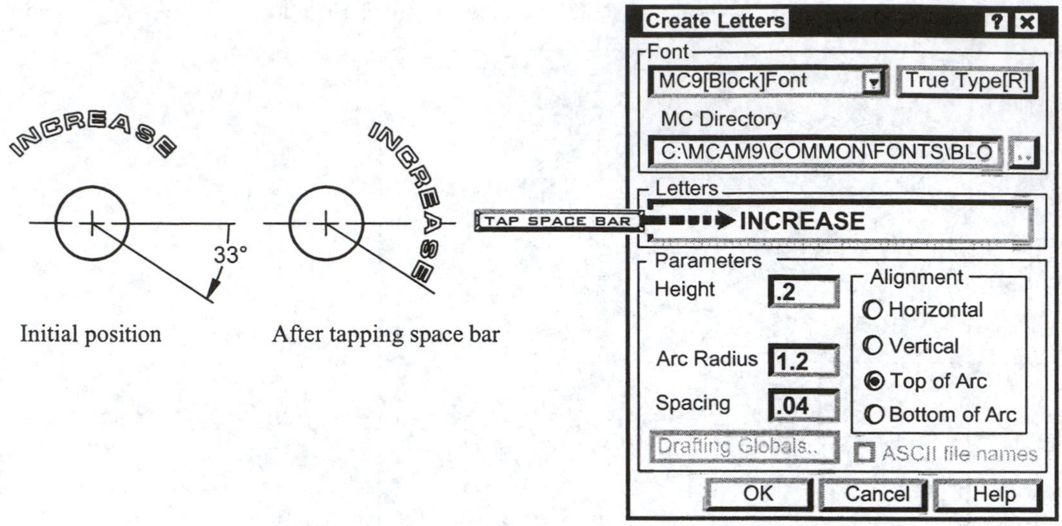

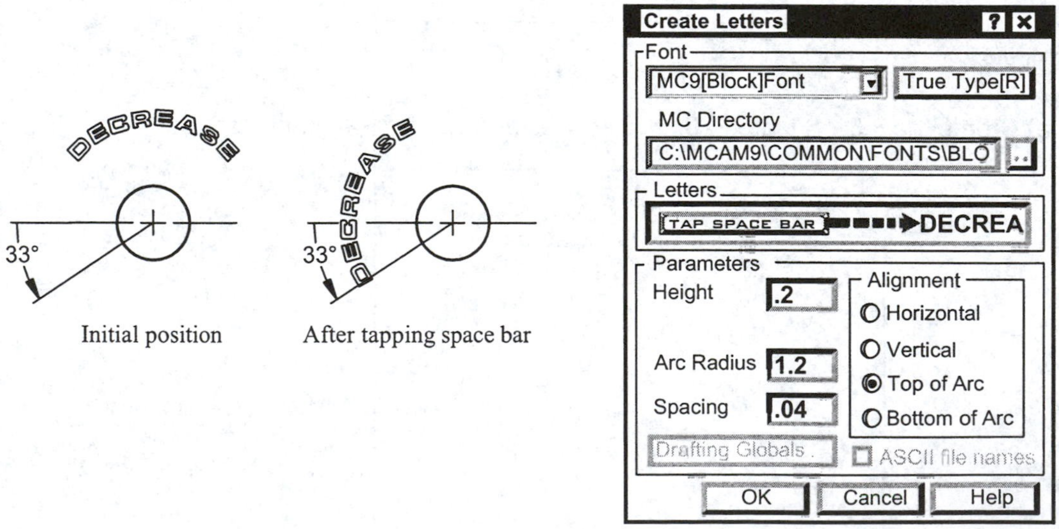

2-11) File Name: **EX2-11JV** ← YOUR INITIALS

Create the CAD entities and the geometric letters for the plaque shown in Figure 2p-11

+y

Font=HARTFORD , Height=1.5, Width=.75
Angle= 45°, Slant= 25°

.25R TYP

PLEASENTVILLE
ENGRAVING
INDUSTRIAL
3.5R
.5R TYP
30° 2.25R 30°
+x
6
.25
COMMERCIAL
ELLIPSE
X Axis Radius: .75
Y Axis Radius: .5
Rotion: 90°
Center:(1.25,1.25)
SERVICES
INCORPORATED
18x18RECTANGLE
45°

Font=ARIAL, Height=1.125,
Spacing=.13
on 4R arc

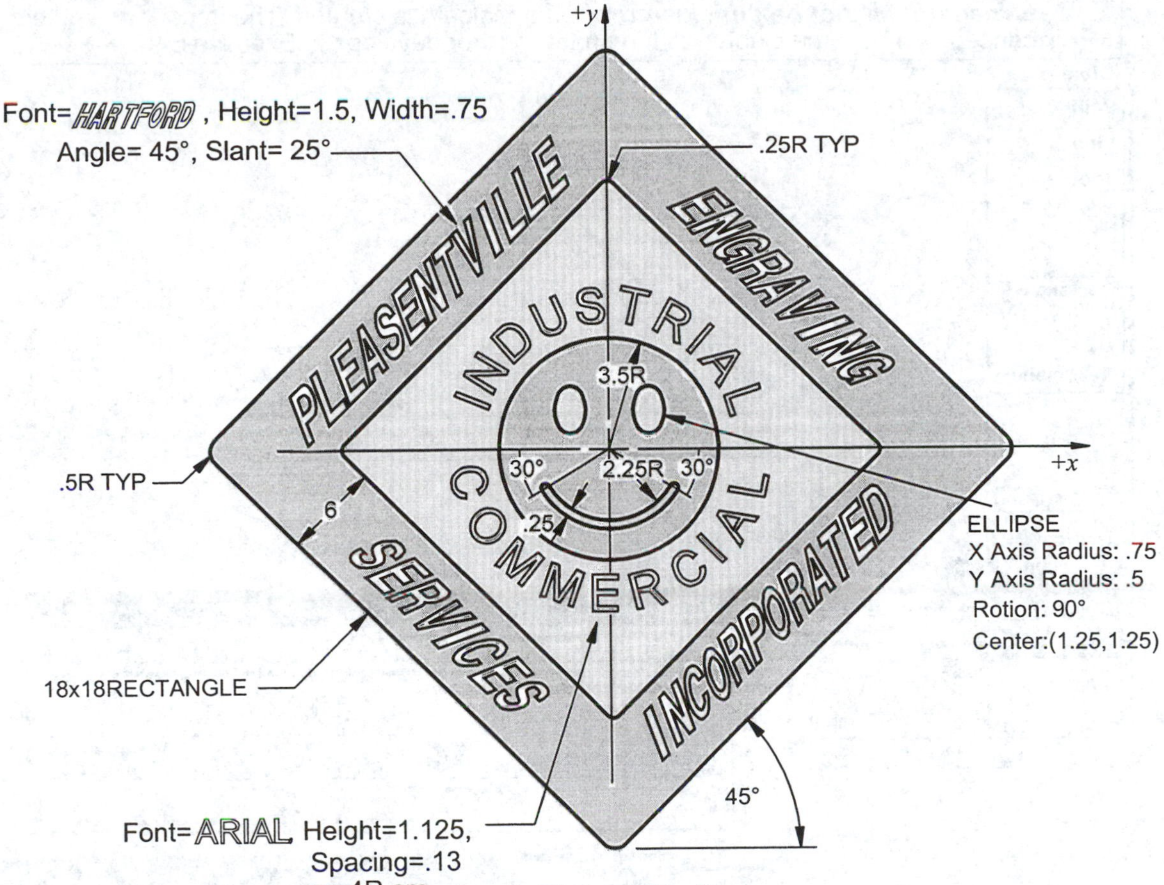

Figure 2-p11

a) Create the 18 x 18 and 12 x 12 rectangles on ***level 1*** *(default level)* in ***color red(12)***

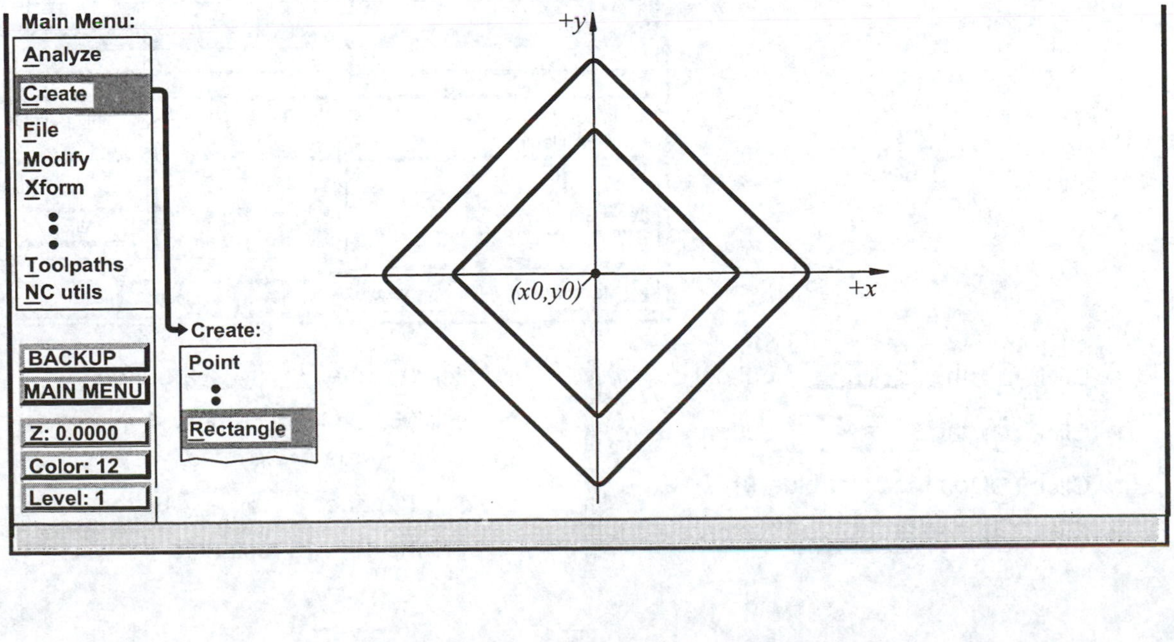

b) Create a 15 x 15 rectangle on ***level 4*** in ***color blue(9)***

Note: the primary reason for placing geometry on a level is to control its ***visibility*** when later using it to specify a toolpath for machining. Geometry on a level whose visibility is ***turned off*** will ***not be considered*** when a toolpath is created. The importance of this concept will become clearer as the project further develops in Exercise 6-9.

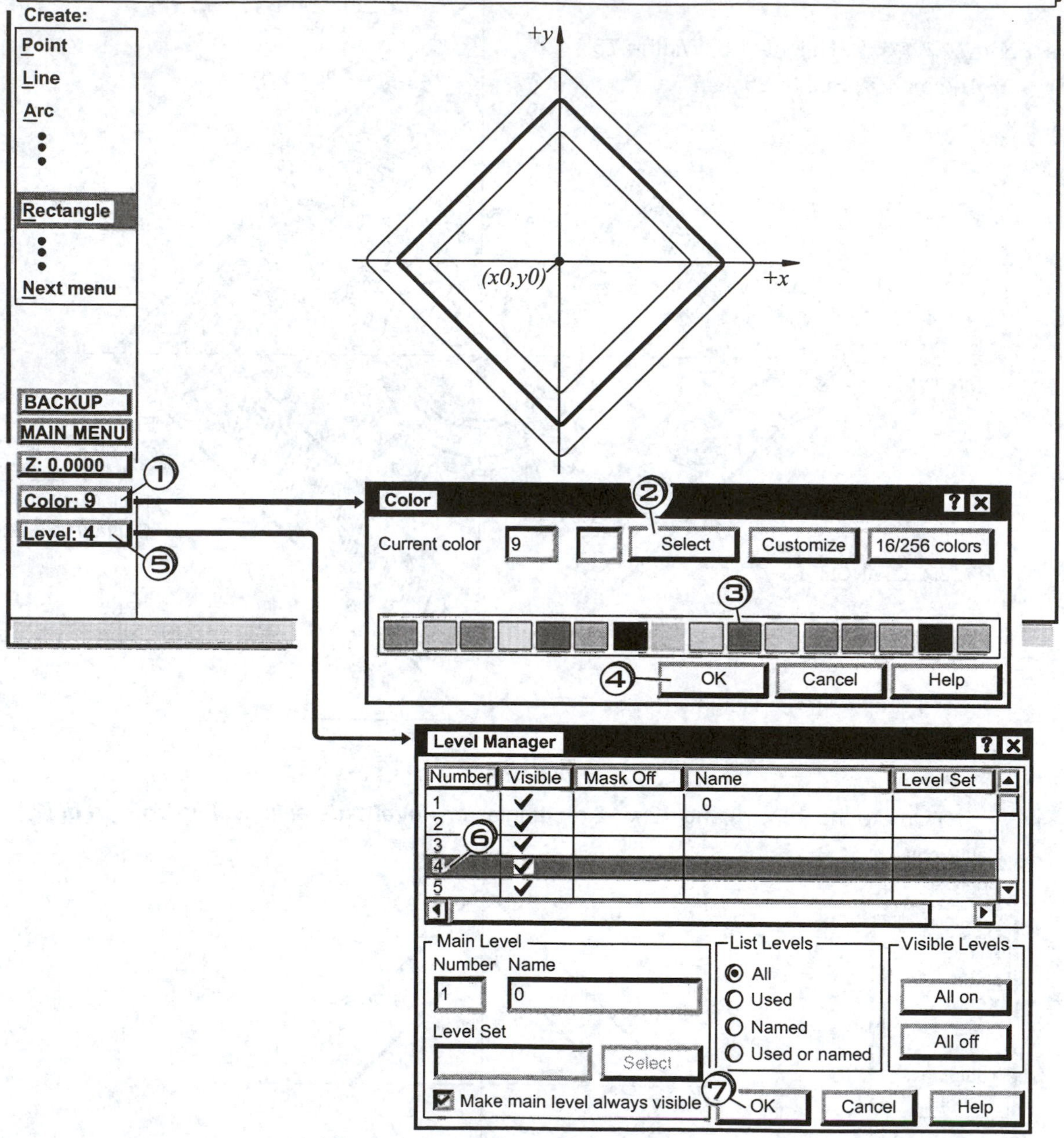

- Click ① the **Color:** button
- Click ② the Select button
- Click ③ on the color blue(**9**)
- Click ④ the OK button
- Click ⑤ the **Level:** button
- Click ⑥ on level number **4** such that it is *highlighted*
- Click ⑦ the OK button

Create the 15 x 15 rectangle

c) Insert the *HARTFORD* font geometric lettering on ***level 2*** in ***color green(10)***
Place each string at the *midpoint* of of the sides of the 15 x 15 rectangle

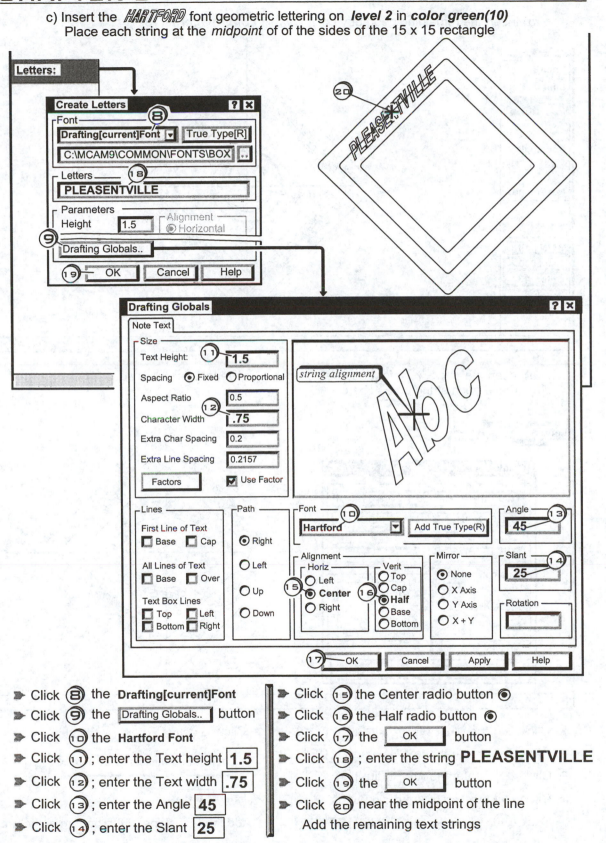

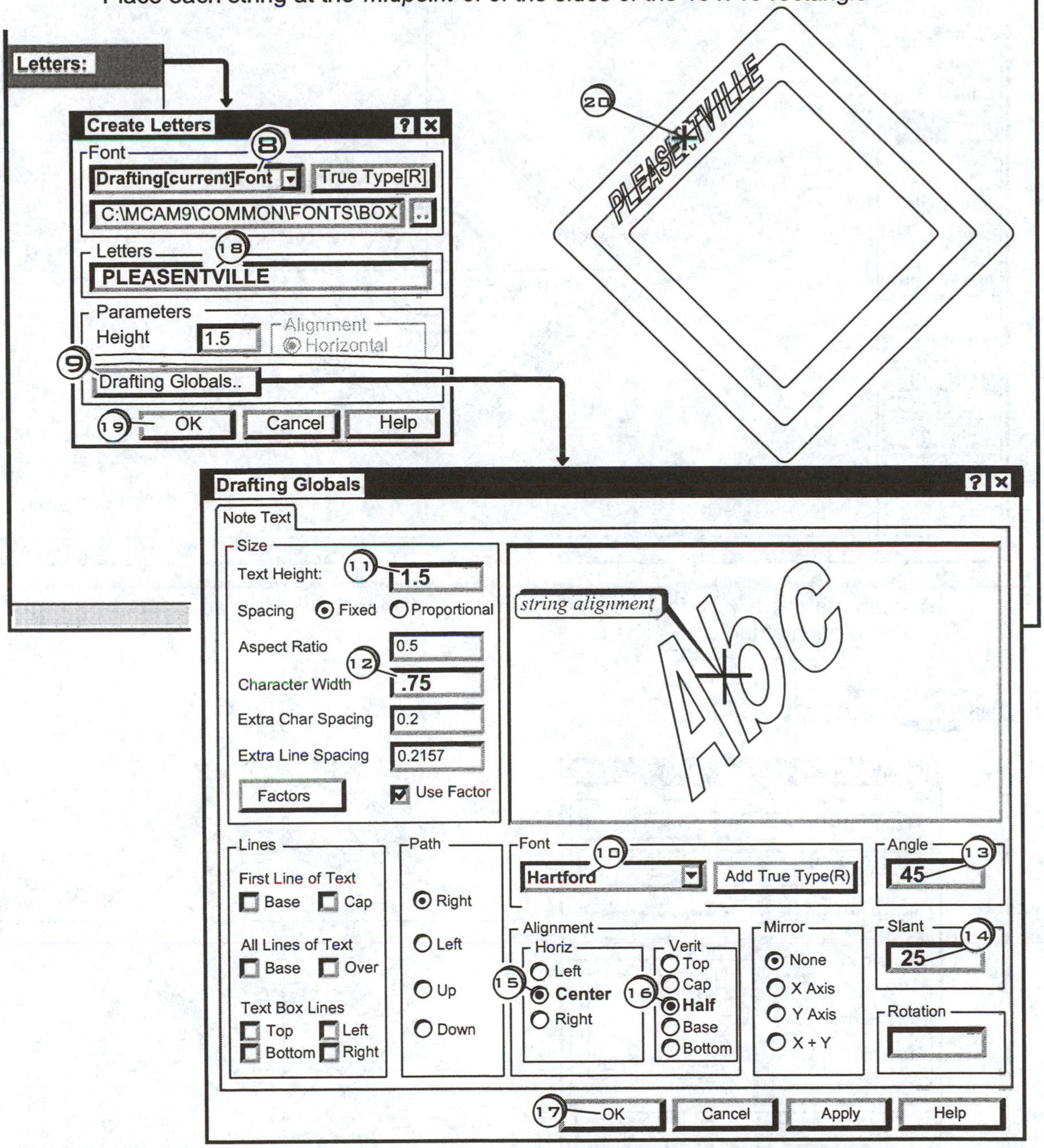

- Click (8) the **Drafting[current]Font**
- Click (9) the [Drafting Globals..] button
- Click (10) the **Hartford Font**
- Click (11); enter the Text height **1.5**
- Click (12); enter the Text width **.75**
- Click (13); enter the Angle **45**
- Click (14); enter the Slant **25**
- Click (15) the Center radio button ◉
- Click (16) the Half radio button ◉
- Click (17) the [OK] button
- Click (18); enter the string **PLEASENTVILLE**
- Click (19) the [OK] button
- Click (20) near the midpoint of the line

Add the remaining text strings

d) Insert the remaining geometry on ***level 3*** in ***color magenta(13)***

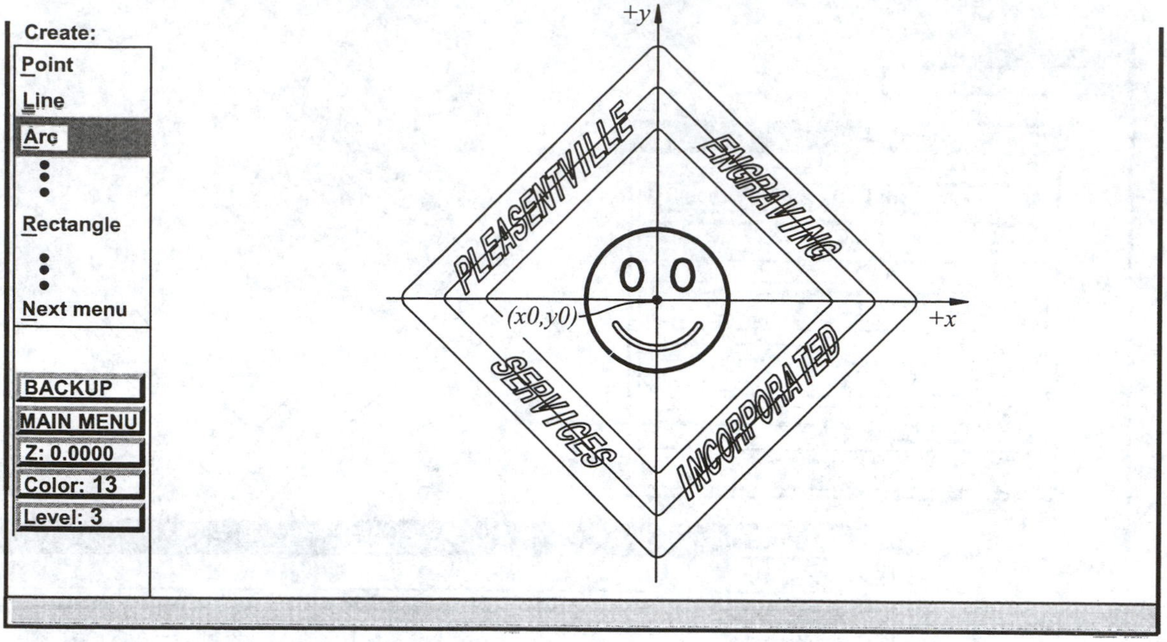

e) Insert the ARIAL font geometric lettering on ***level 3*** in ***color aqua(11)***
Place each string on a 4R arc.

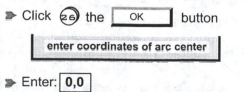

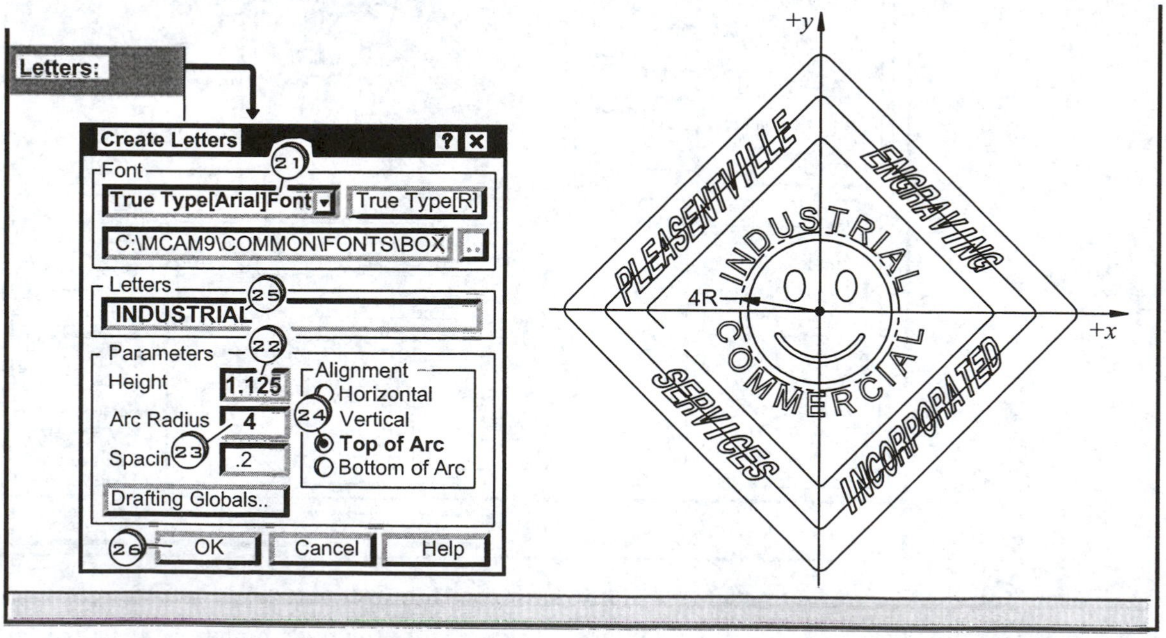

- Click (21) the **True Type[Arial]Font**
- Click (22) ; enter the text Height **1.125**
- Click (23) ; enter the Arc Radius **4**
- Click (24) the Top of Arc radio button ◉
- Click (25) ; enter the string **INDUSTRIAL**

- Click (26) the OK button

enter coordinates of arc center

- Enter: **0,0**

Add the text string at the bottom of the arc

CHAPTER - 3

EDITING 2D GEOMETRY

3-1 Chapter Objectives

After completing this chapter you will be able to:

1. Explain how to delete and undelete existing geometrc entities.
2. Know how to shorten or extend an existing geometric entity to its intersection (s) with another entity or entities via the **Trim** command
3. Explain how to break an existing geometric entity into sparate entities via **Break**
4. Understand how to shorten or extend a line or arc entity by a specific ammount via the **Extend** command.
5. State the uses of the **Xform** command for translating, mirroring and rotating existing geometric entities.
6. State the uses of the **Xform** command for creating copies of existing geometric entities by translating, mirroring and rotating.

3-2 Deleting Geometry

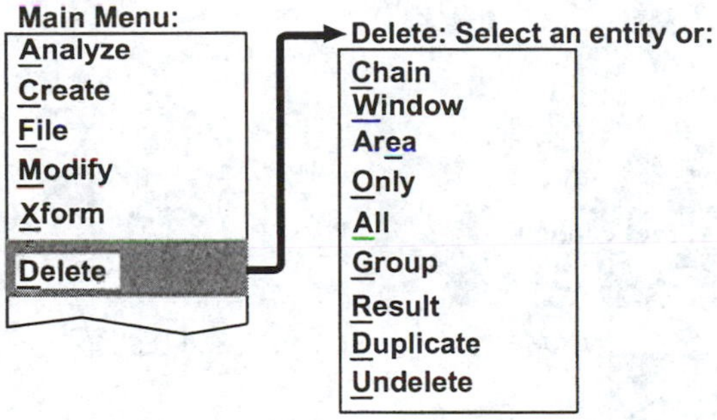

Chain

Deletes all entities connected end to end in a *continuous chain.*

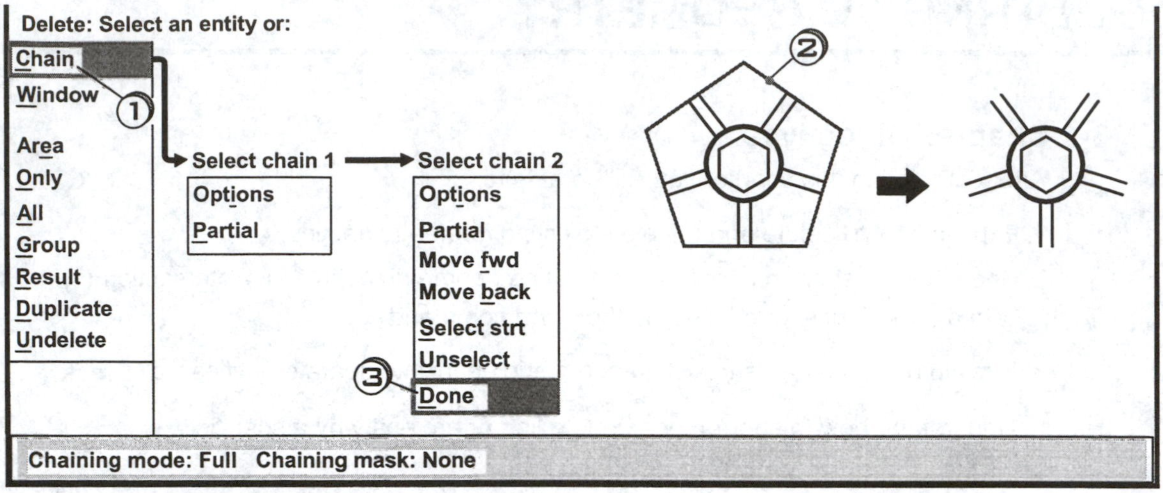

- Click ① **Chain**
- Click ② an entity in the desired chain
- Click ③ **Done**
- Press **Esc** to cancel the operation

Rectangle **Inside +**

All entities contained *inside* a *rectangular window* clicked are deleted

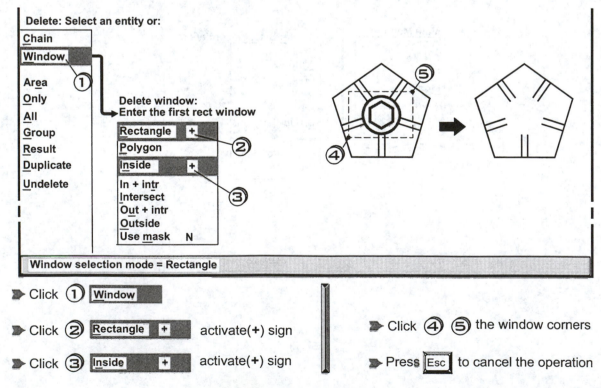

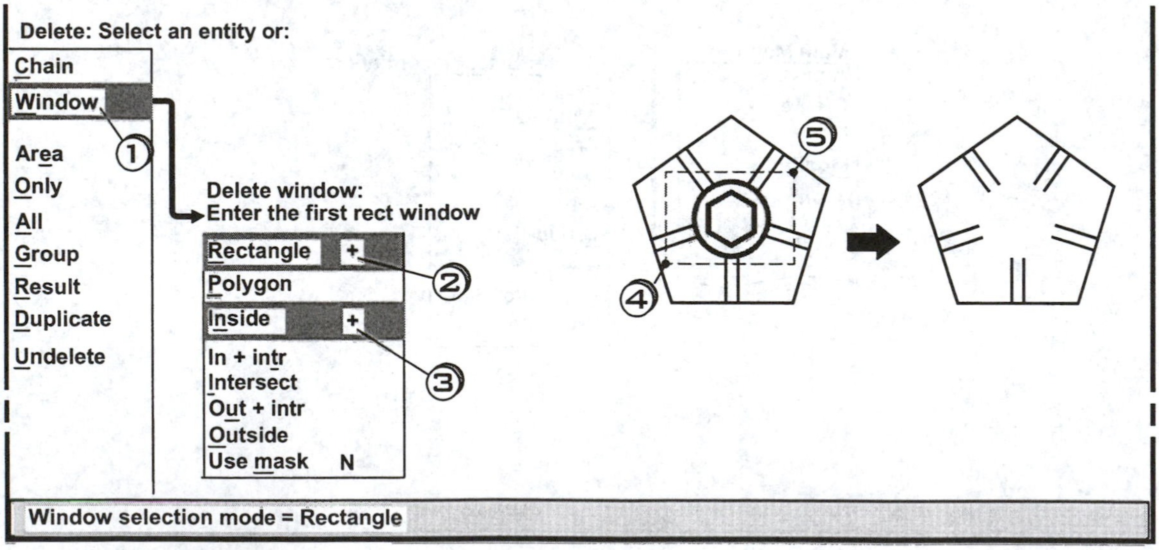

- Click ① **Window**
- Click ② **Rectangle +** activate(+) sign
- Click ③ **Inside +** activate(+) sign
- Click ④ ⑤ the window corners
- Press **Esc** to cancel the operation

Window | **Rectangle +** | **In + intr +**

All entities contained *inside* and crossing a *rectangular window* clicked are deleted

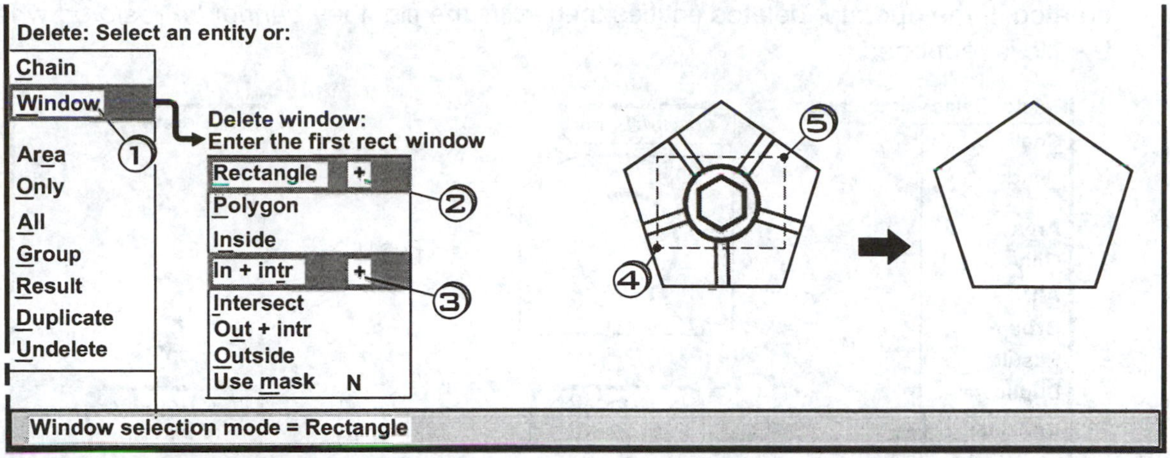

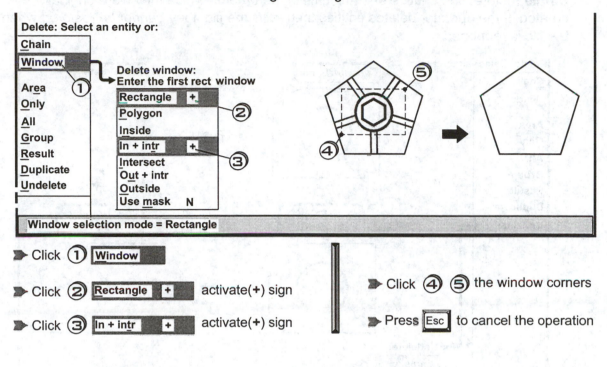

- Click ① **Window**
- Click ② **Rectangle +** activate(+) sign
- Click ③ **In + intr +** activate(+) sign
- Click ④ ⑤ the window corners
- Press **Esc** to cancel the operation

Window | **Polygon +** | **Inside +**

All entities contained *inside* a *polygon window clicked* are *deleted*

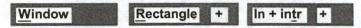

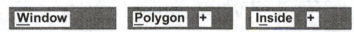

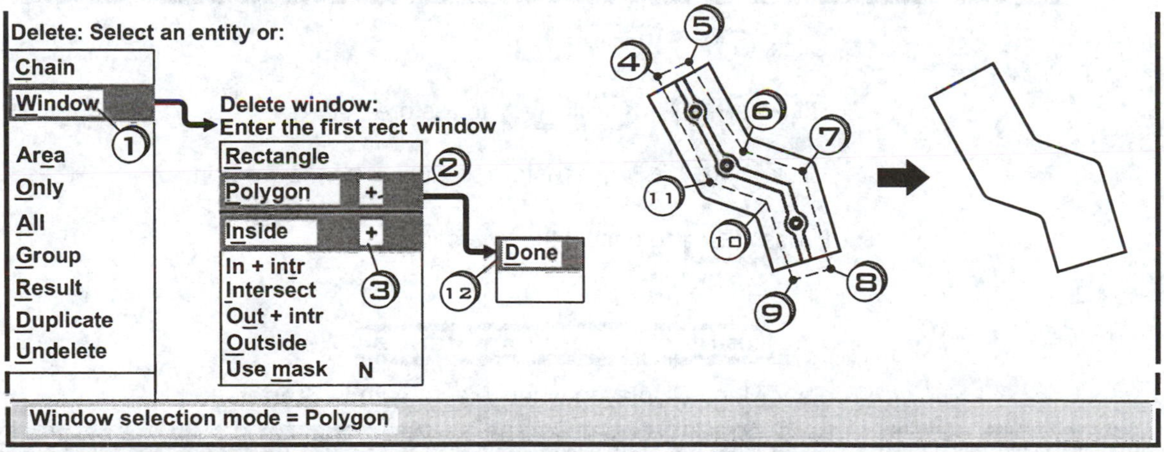

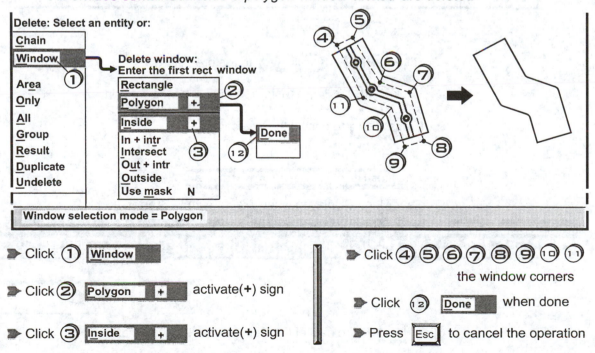

- Click ① **Window**
- Click ② **Polygon +** activate(+) sign
- Click ③ **Inside +** activate(+) sign
- Click ④ ⑤ ⑥ ⑦ ⑧ ⑨ ⑩ ⑪ the window corners
- Click ⑫ **Done** when done
- Press **Esc** to cancel the operation

Undelete **Single**

Restores deleted entities within the ***currently active drawing***. Repeatedly clicking **Single** restores the entities *one* at a time in the *order* in which they were originally created. If the operator deletes entities then *exits* the file, they *cannot be restored* when the file is reopened

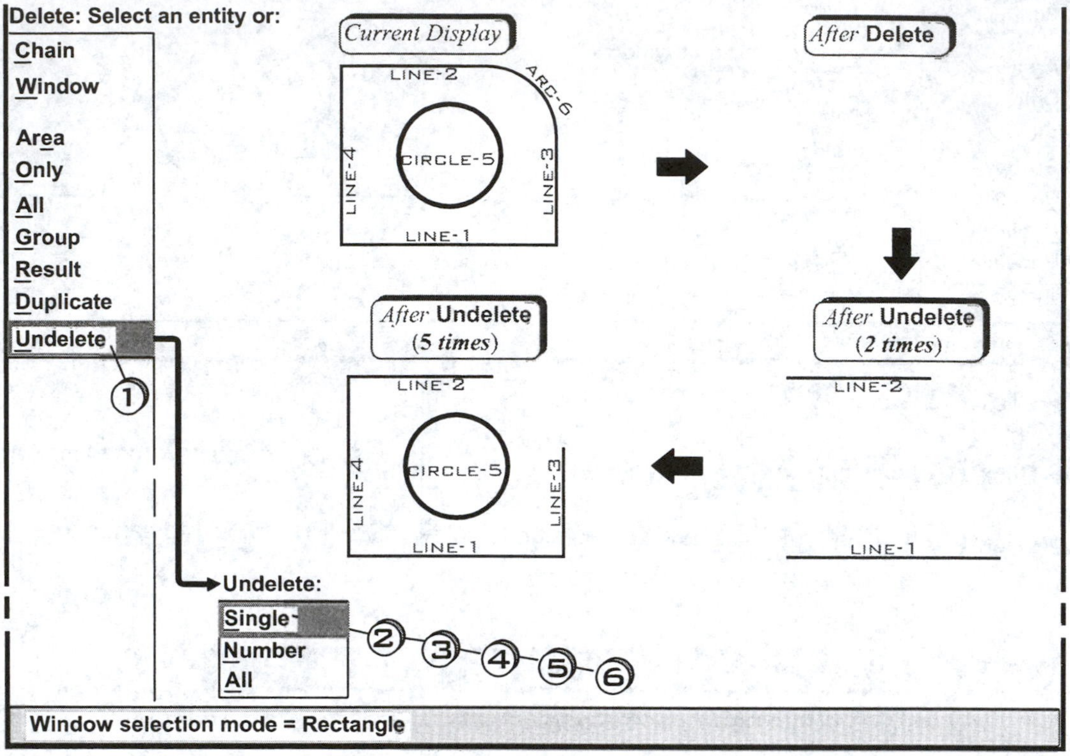

- Click ① **Undelete**
- Click ② ③ **Single** to restore LINES-1,2
- Click ④ ⑤ ⑥ to restore LINES-3,4;CIRCLE-1
- Press Esc to cancel the operation

Using the Toolbar Menus

Toolbar Menu Displayed when *Mastercam* is **Started**

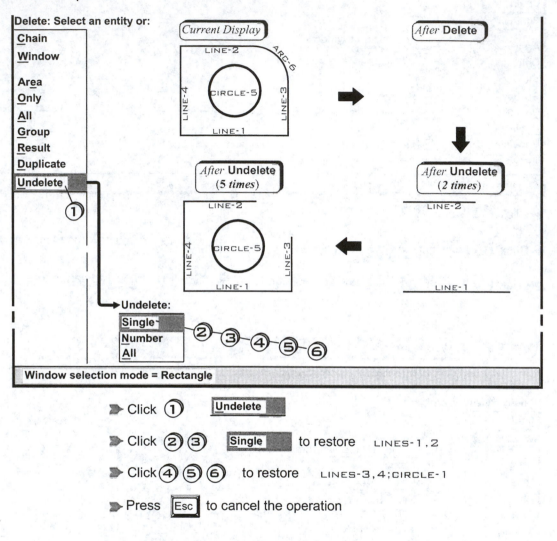

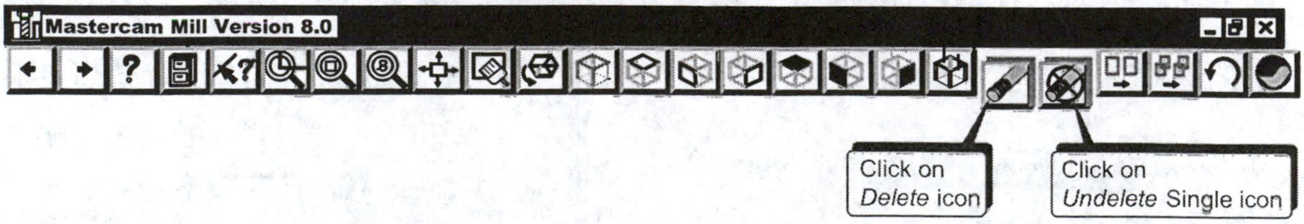

Click Esc to cancel the Delete, Undelete commands

3-3 Modifying Existing 2D Geometry

The various commands for modifying existing entities via *Mastercam* are presented in this section.

Trimming Entities

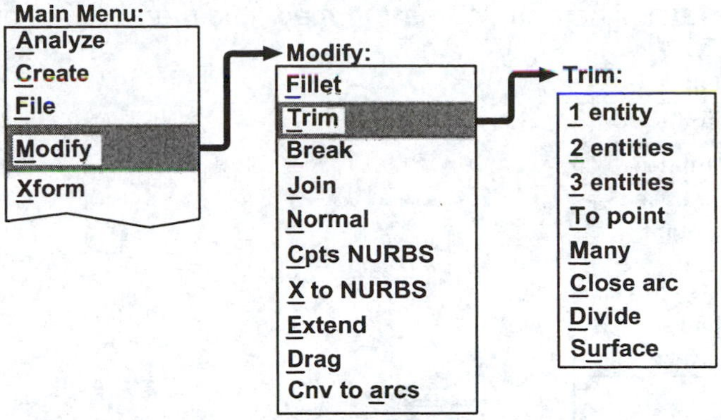

Trim **1 entity**

Trims(shortens or extends) an existing line, arc, fillet, ellipse or spline entity to its *intersection* with another entity.The operator *first* clicks the entity *to be trimmed* on the **portion to remain.** The entity to be *trimmed* **to** is *then* clicked.

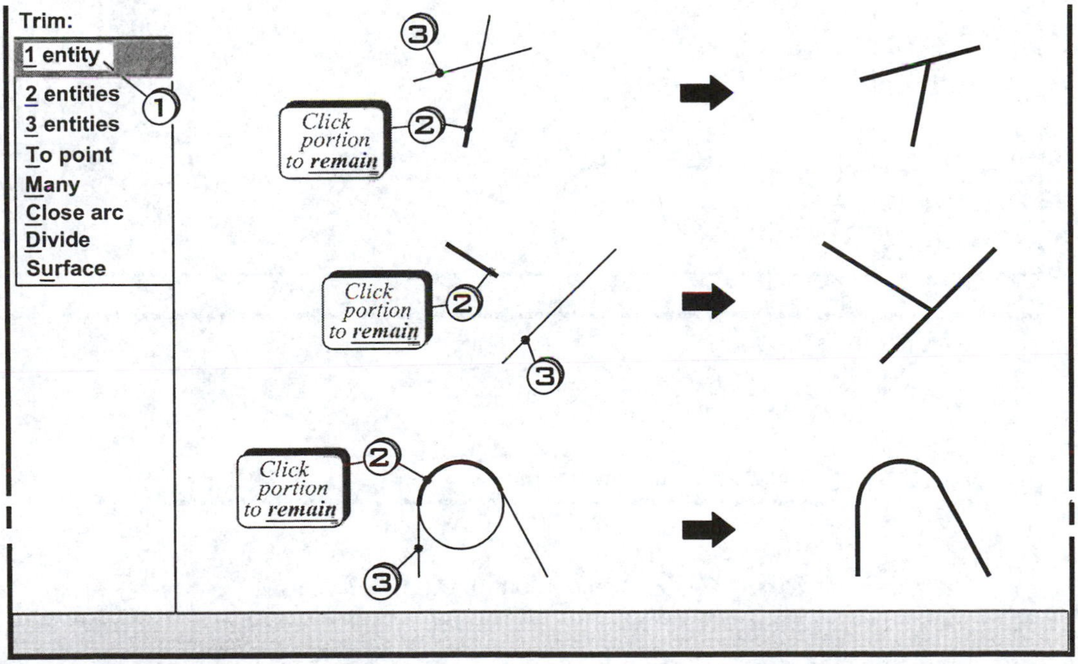

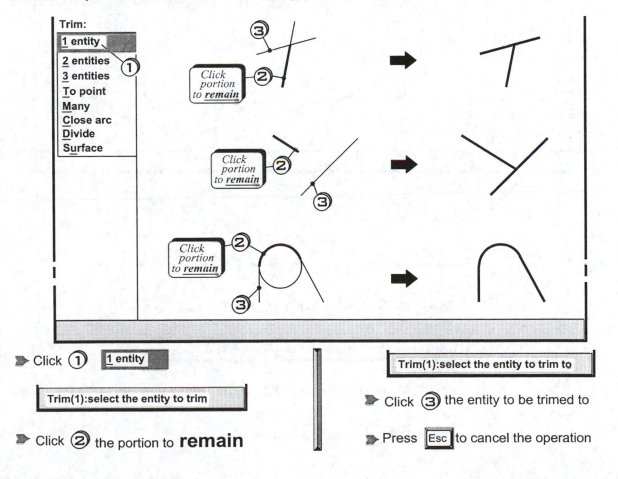

➤ Click ① **1 entity**

Trim(1):select the entity to trim

➤ Click ② the portion to **remain**

Trim(1):select the entity to trim to

➤ Click ③ the entity to be trimed to

➤ Press Esc to cancel the operation

Trim **2 entities**

Trims(shortens or extends) existing line, arc, fillet, ellipse and spline entities to their *intersection point*.The operator *first* clicks the entity *to be trimmed* on the **portion to remain.** The entity to be *trimmed to is then* clicked on the **portion to remain**.

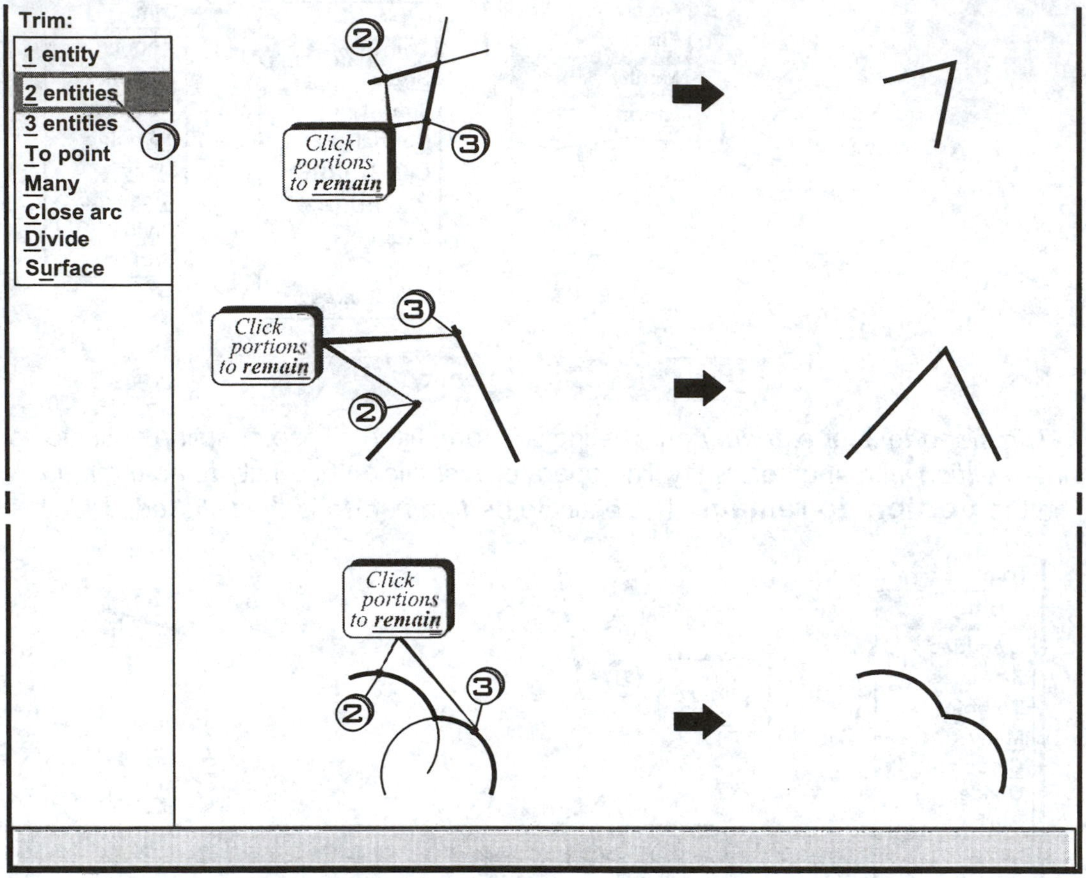

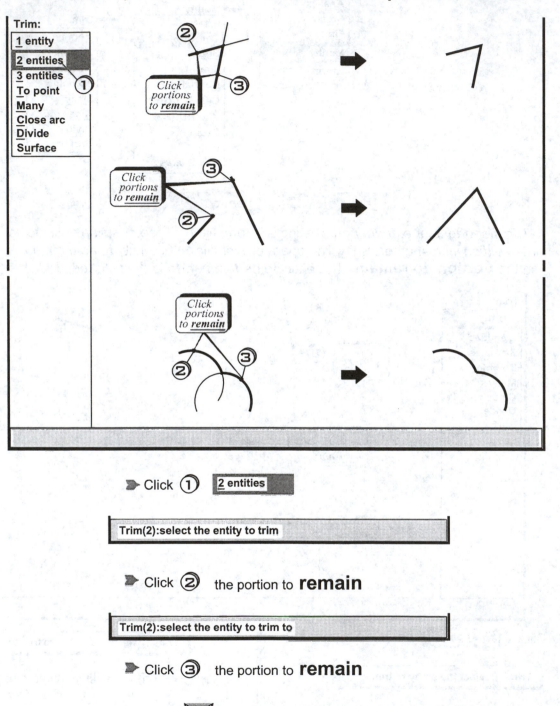

Click ① **2 entities**

Trim(2):select the entity to trim

Click ② the portion to **remain**

Trim(2):select the entity to trim to

Click ③ the portion to **remain**

Press **Esc** to cancel the operation

Trim **3 entities**

Trims(shortens or extends) an existing line, arc, fillet, ellipse and spline entitiy to its *intersection* with *two other* entities. The operator *first* clicks the *two trimming* entities on the **portions to remain.** The entity *to be trimmed* on the **portion to remain.** is then clicked.

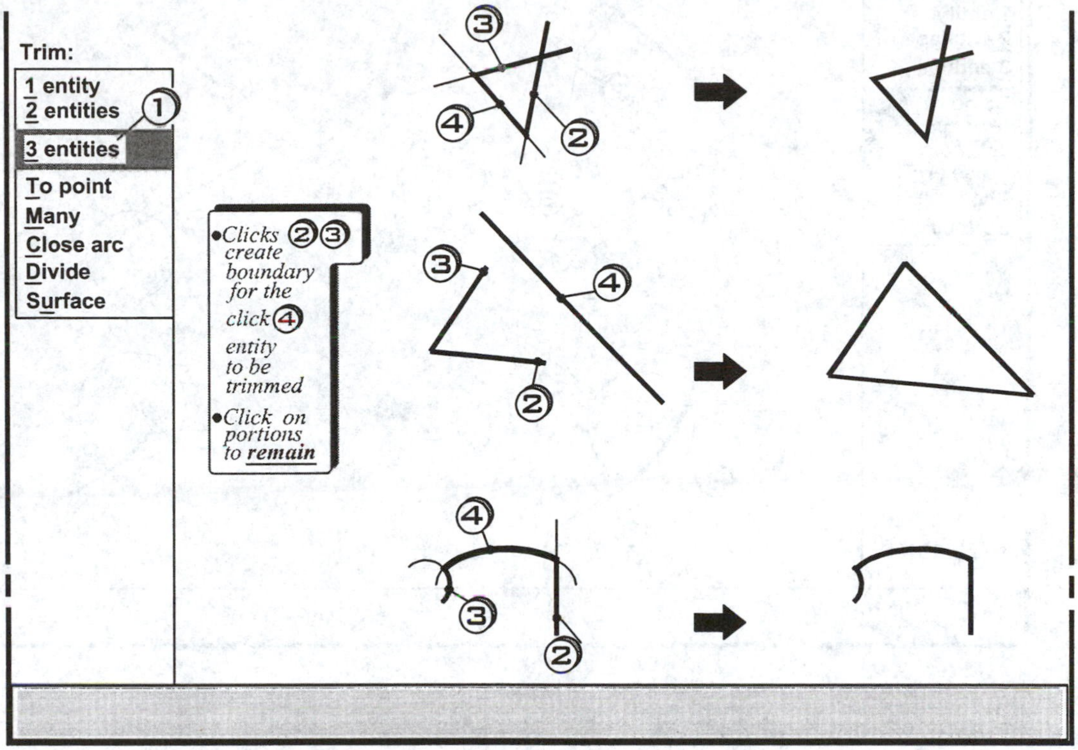

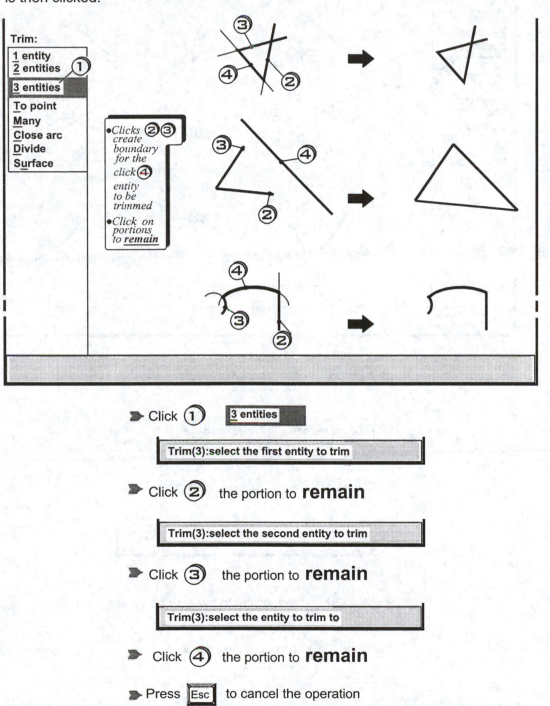

➤ Click ① **3 entities**

Trim(3):select the first entity to trim

➤ Click ② the portion to **remain**

Trim(3):select the second entity to trim

➤ Click ③ the portion to **remain**

Trim(3):select the entity to trim to

➤ Click ④ the portion to **remain**

➤ Press **Esc** to cancel the operation

Trim **To point**

Trims/Extends an existing line, arc, fillet, ellipse and spline entity to an apparent *intersection* with a *perpendicular* to it from a *point* clicked.

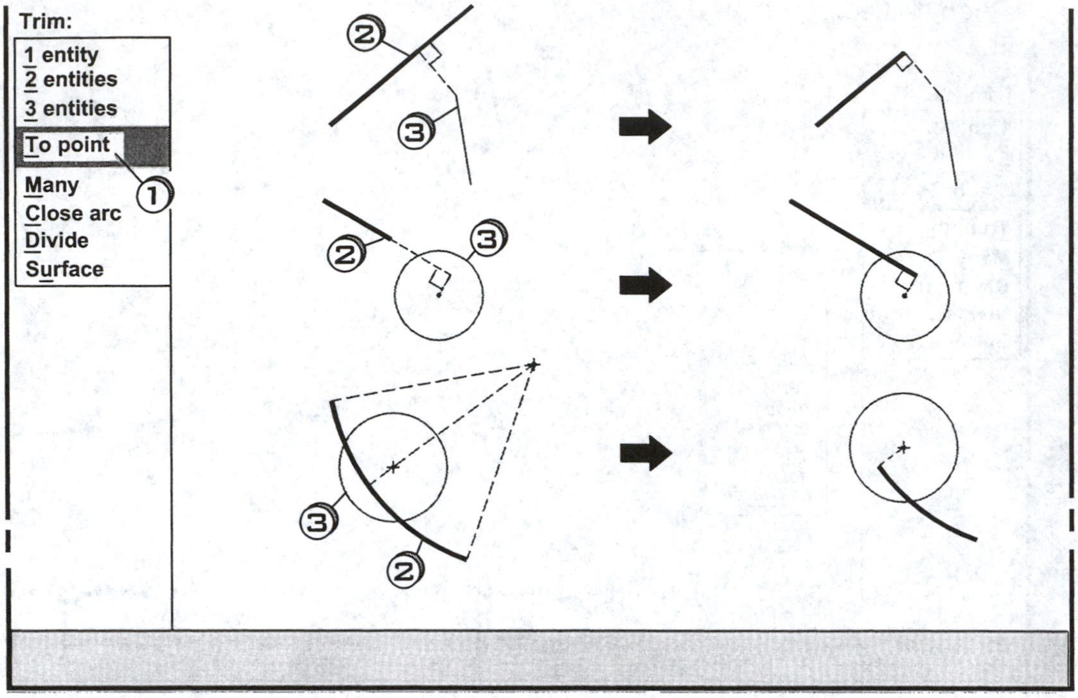

Click ① **To point**

Select the entity to Trim/Extend

Click ② the portion to **remain**

Indicate Trim/Extend location

Click ③ the location to trim/extend to

Press Esc to cancel the operation

Trim **Many**

Trims many existing line, arc, fillet, ellipse and spline entities to an intersection entity. The operator first clicks the entities *to be trimmed* then the entity to be trimmed to and finally clicks the **side on which portions of the entities are to remain**.

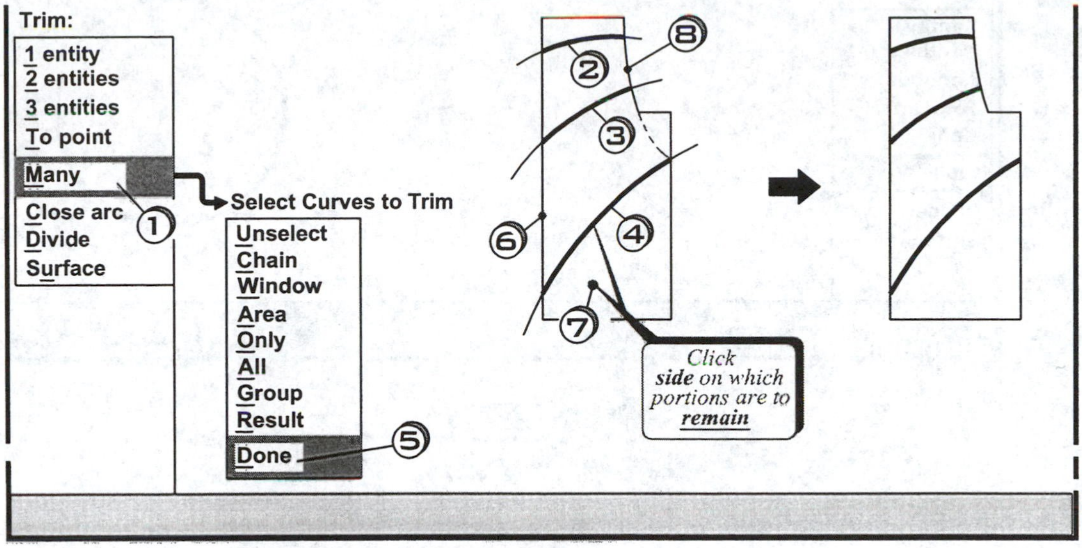

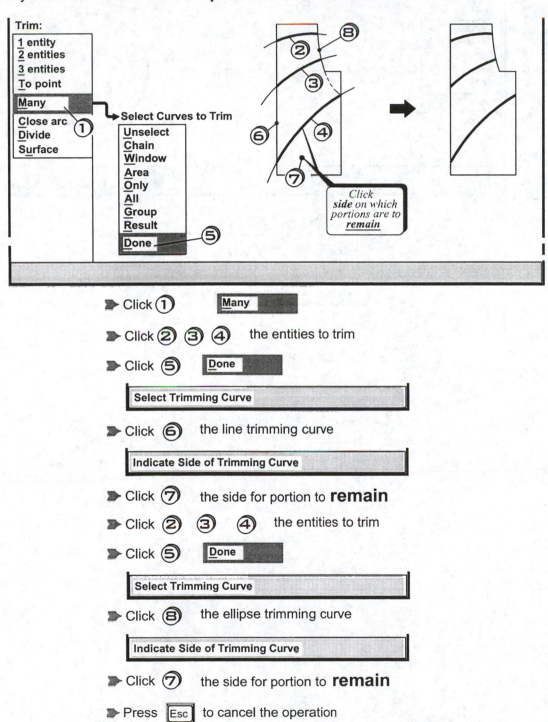

- Click ① **Many**
- Click ② ③ ④ the entities to trim
- Click ⑤ **Done**

Select Trimming Curve

- Click ⑥ the line trimming curve

Indicate Side of Trimming Curve

- Click ⑦ the side for portion to **remain**
- Click ② ③ ④ the entities to trim
- Click ⑤ **Done**

Select Trimming Curve

- Click ⑧ the ellipse trimming curve

Indicate Side of Trimming Curve

- Click ⑦ the side for portion to **remain**
- Press Esc to cancel the operation

Trim **Close arc**

Trims an arc clicked such that it *closes to a circle.*

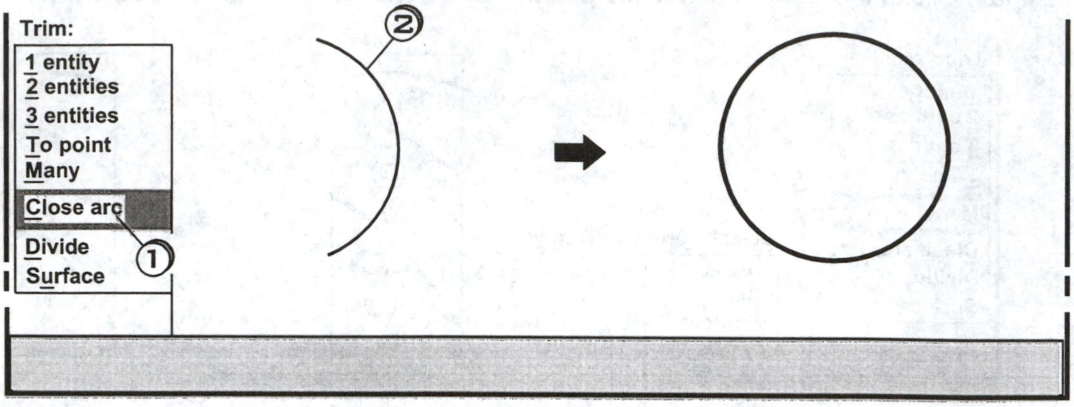

➤ Click ① **Close arc**

Select an Arc to convert to a full circle

➤ Click ② the arc to close to a circle

➤ Press Esc to cancel the operation

Trim **Divide**

Trims an existing line, arc, fillet, ellipse or spline entitiy by *dividing it at its intersection* with *two* other entities. The operator first clicks the **portion of the entitiy to be removed** then clicks the two other intersecting entities

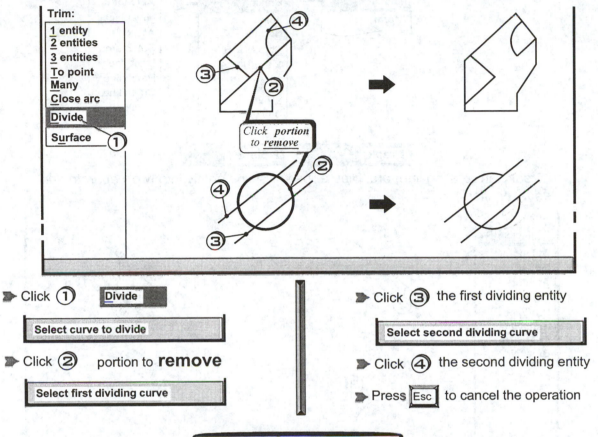

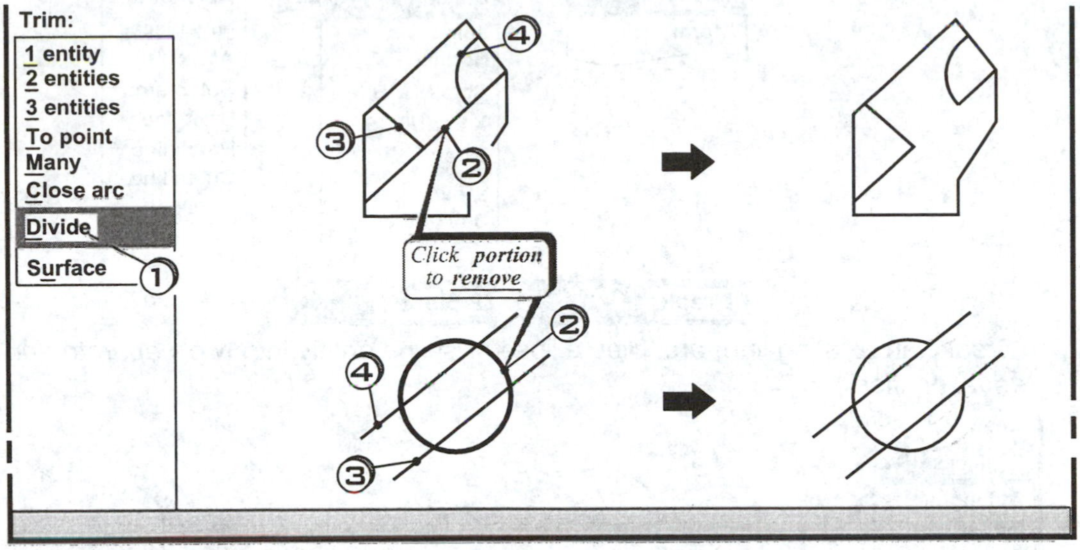

- Click ① **Divide**

Select curve to divide

- Click ② portion to **remove**

Select first dividing curve

- Click ③ the first dividing entity

Select second dividing curve

- Click ④ the second dividing entity
- Press Esc to cancel the operation

Using the Toolbar Menus

Toolbar Menu Displayed when *Mastercam* is **Started**

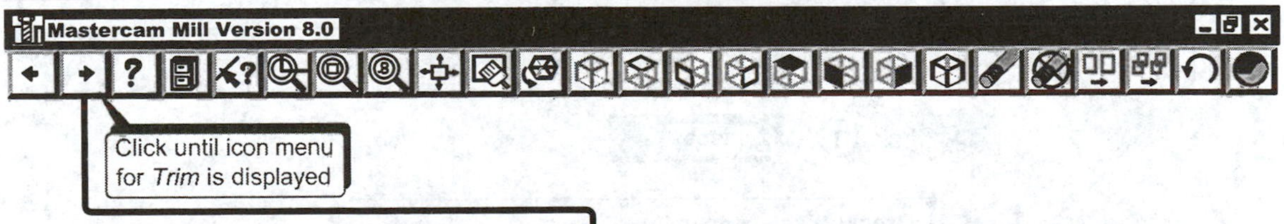

Toolbar Menu Containing Trim Icons

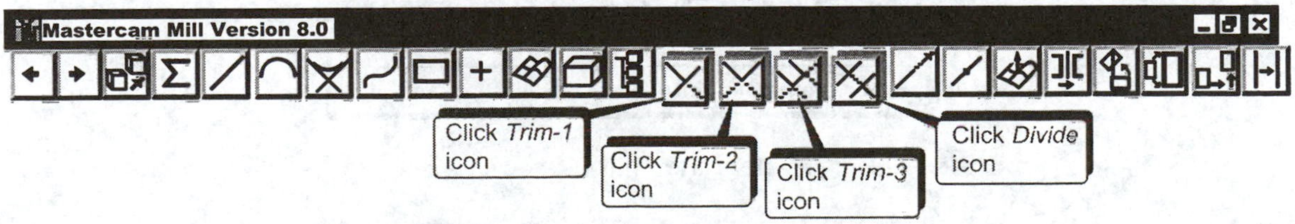

Click Esc to cancel the Trim commands

Breaking Entities

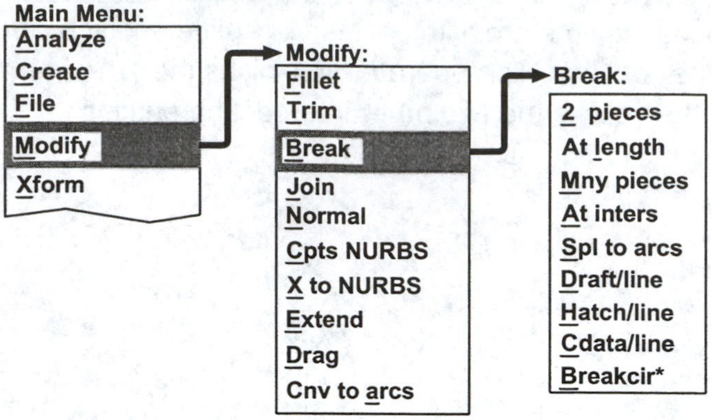

Break **2 pieces**

Breaks an existing line, arc, fillet, ellipse or spline entitiy into two *separate and equal* entities

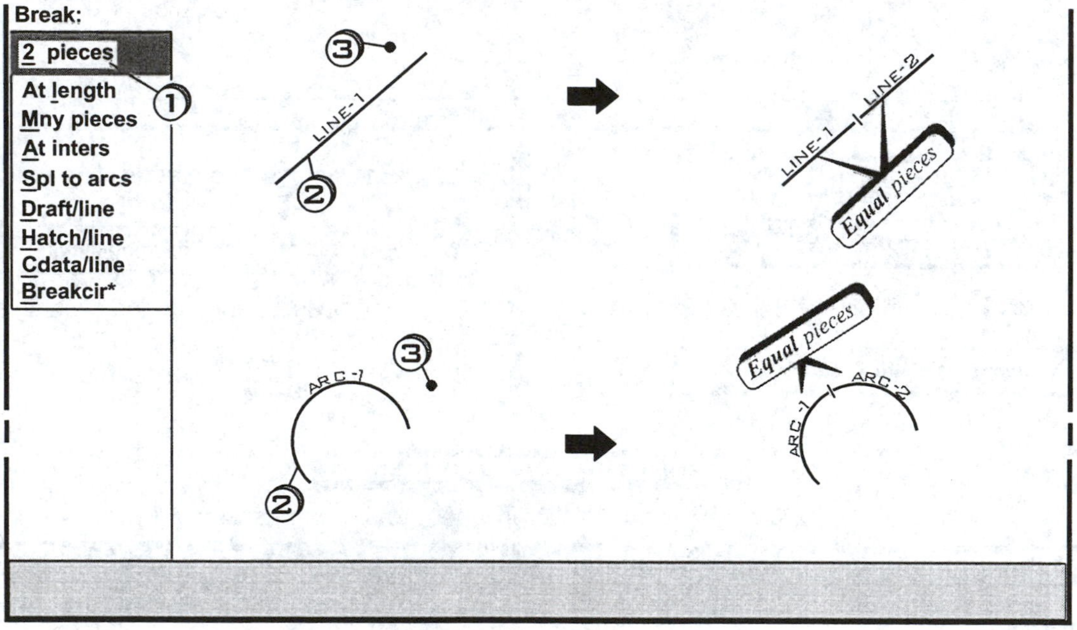

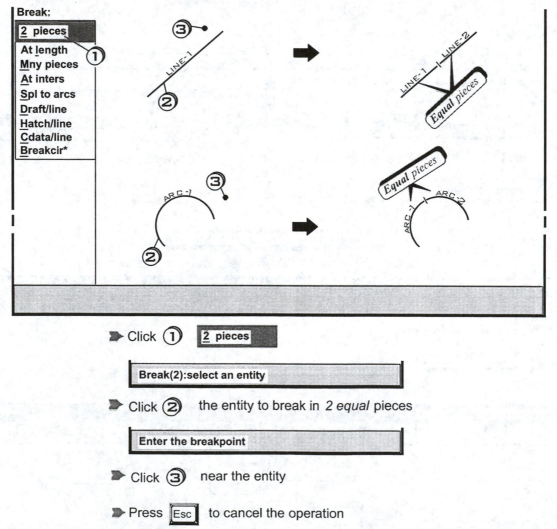

➤ Click ① **2 pieces**

Break(2):select an entity

➤ Click ② the entity to break in *2 equal* pieces

Enter the breakpoint

➤ Click ③ near the entity

➤ Press **Esc** to cancel the operation

Break **At length**

Breaks an existing line, arc, fillet, ellipse or spline entitiy into two *separate length* entities. The operator clicks near the *end* of the entity to have the *desired length* and enters the length value.

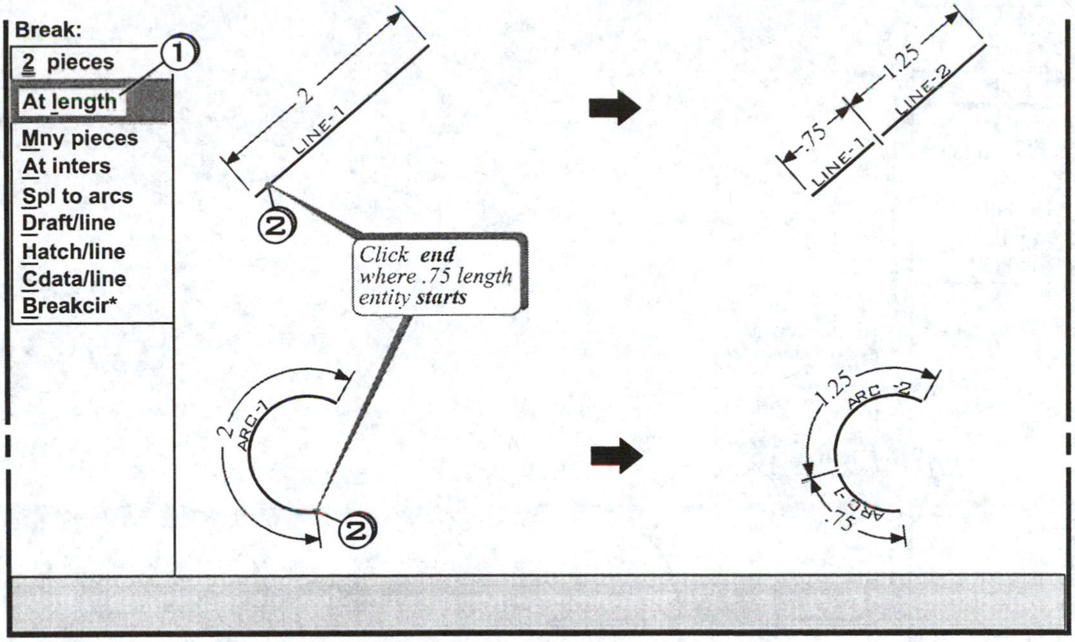

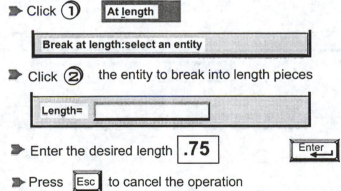

- Click ① **At length**

Break at length:select an entity

- Click ② the entity to break into length pieces

Length=

- Enter the desired length **.75** Enter
- Press Esc to cancel the operation

Break **Mny pieces**

Breaks an existing line, arc, fillet, ellipse or spline entitiy into a specified *number of equal and separate* entities.

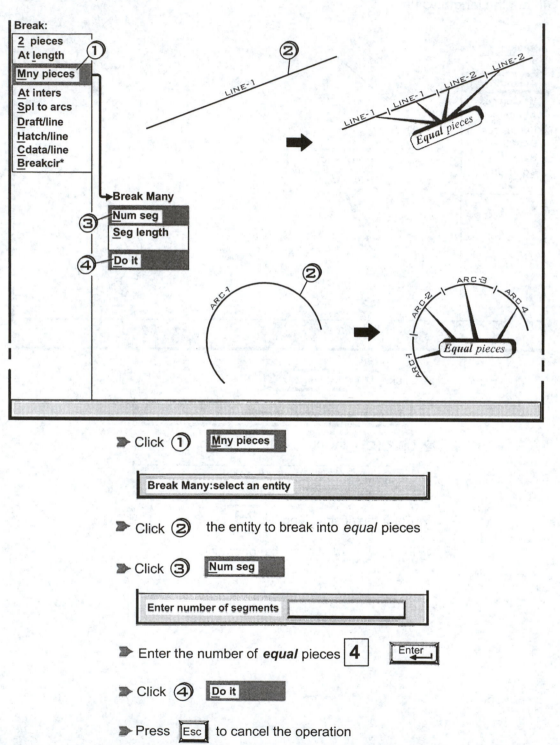

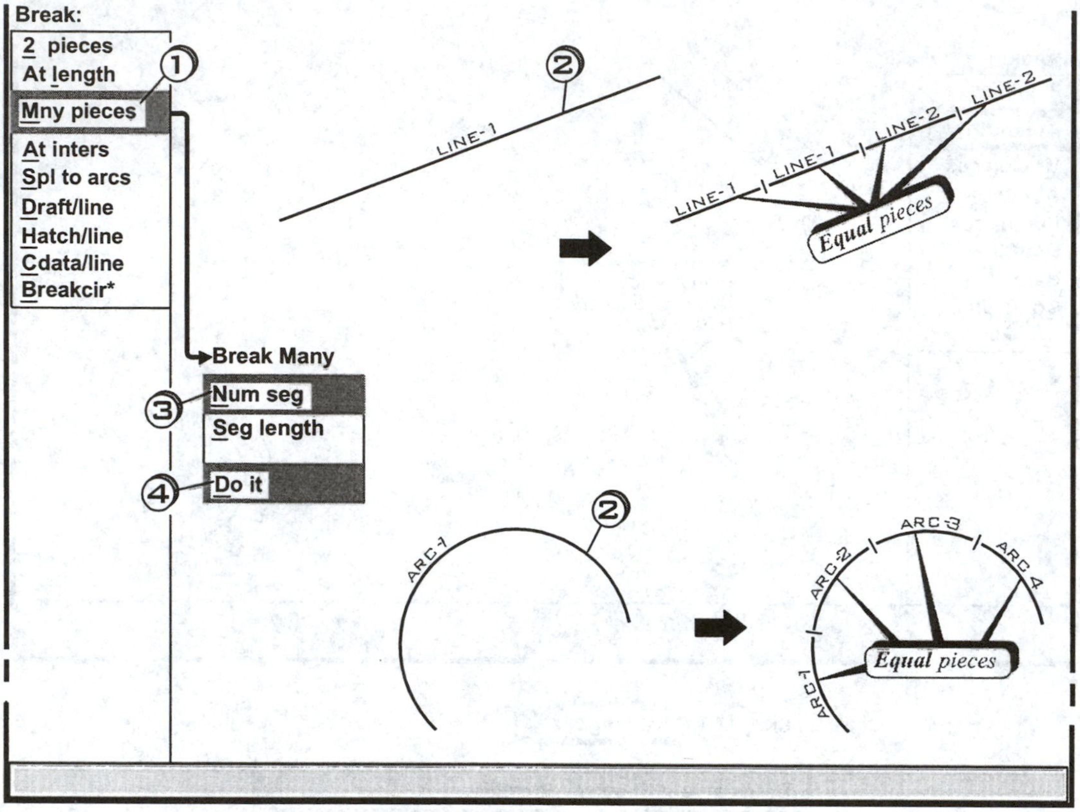

- Click ① **Mny pieces**

Break Many:select an entity

- Click ② the entity to break into *equal* pieces
- Click ③ **Num seg**

Enter number of segments

- Enter the number of ***equal*** pieces **4** Enter
- Click ④ **Do it**
- Press Esc to cancel the operation

Break **At inters**

Breaks existing line, arc, fillet, ellipse or spline entities into a *separate length entities* at their *points of intersection*

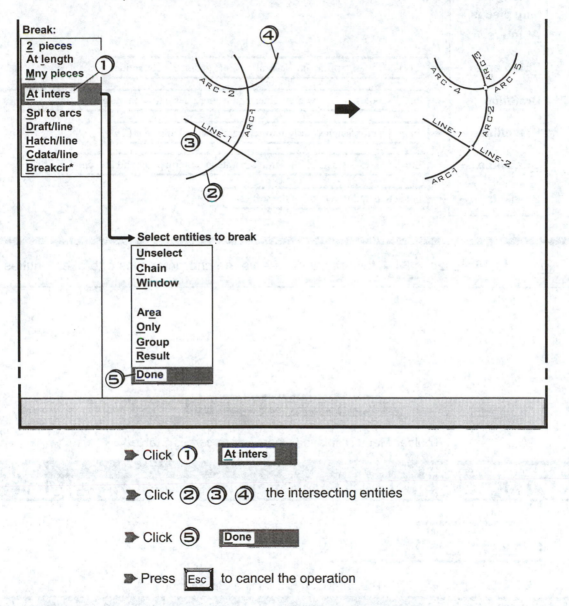

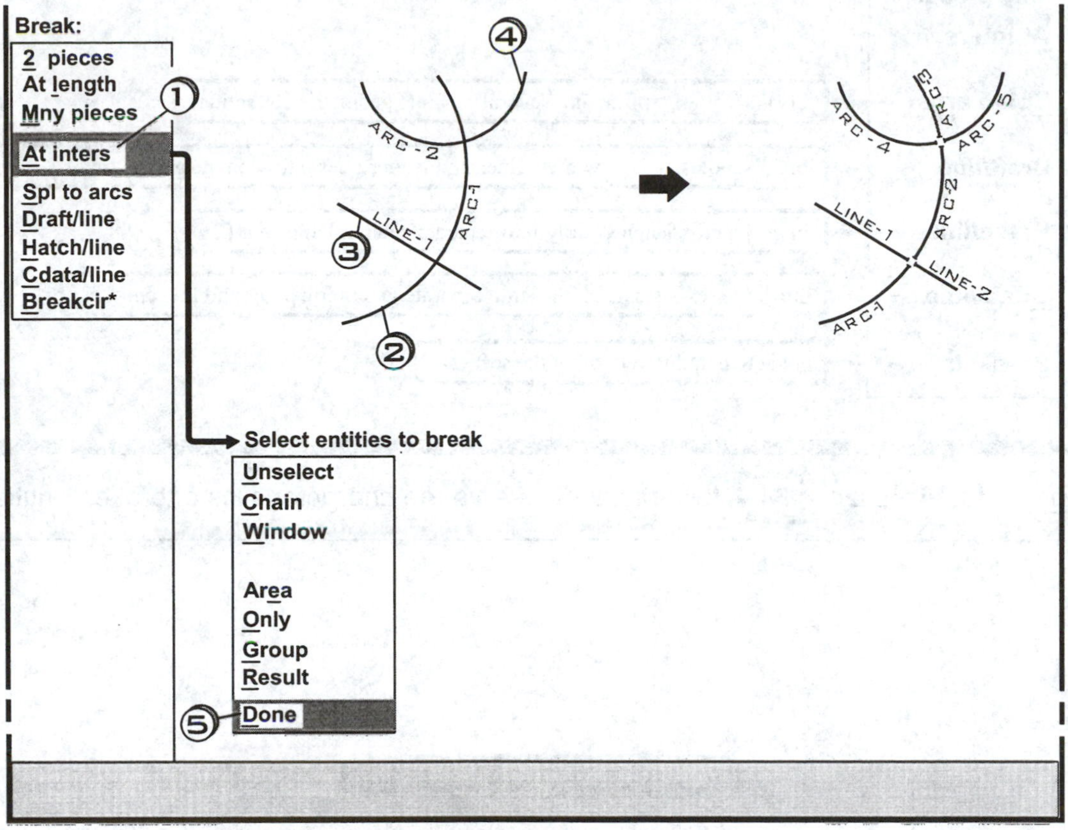

- Click ① **At inters**
- Click ② ③ ④ the intersecting entities
- Click ⑤ **Done**
- Press **Esc** to cancel the operation

Additional Break Commands

Break:

- **2 pieces**
- **At length**
- **Mny pieces**
- **At inters**
- **Spl to arcs** — breaks a single spline into several separate geometric arc entities
- **Draft/line** — breaks notes, labels, witness lines and leader lines into separate geometric line entities
- **Hatch/line** — breaks a crosshatch entitity into separate geometric line entities
- **Cdata/line** — breaks a copious data entities into separate geometric point and line entitites
- **Breakcir*** — not active in this version of the software

Note: Use **Analyze** [F4] to verify and observe the characteristics of broken entities

Using the Toolbar Menus

Toolbar Menu Displayed when *Mastercam* is **Started**

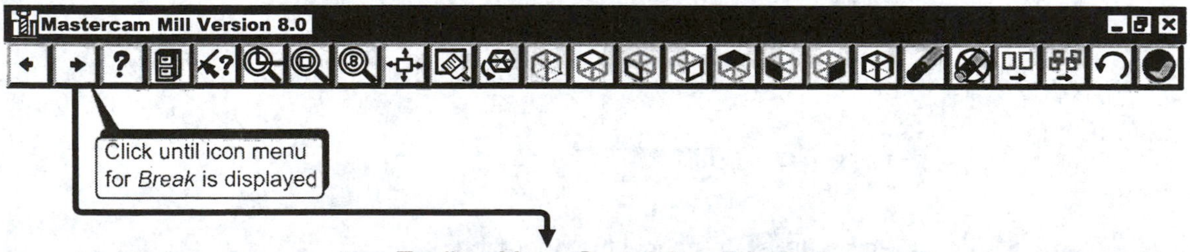
Mastercam Mill Version 8.0
Click until icon menu for *Break* is displayed

Toolbar Menu Containing ***Break*** Icon

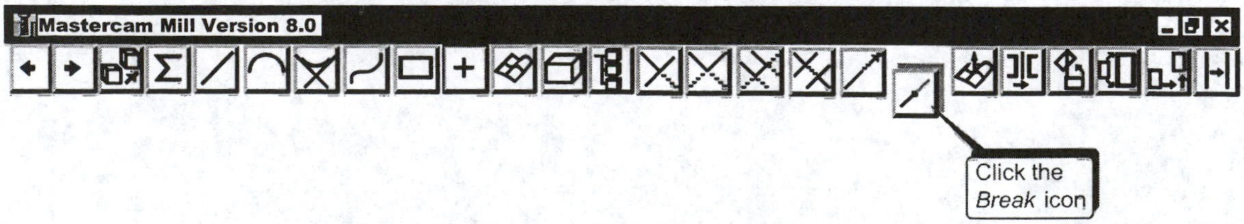
Mastercam Mill Version 8.0
Click the *Break* icon

Click [Esc] to cancel the Break command

Joining, Modifying Control Points, Chg to NURBS Splines and Extending Entities

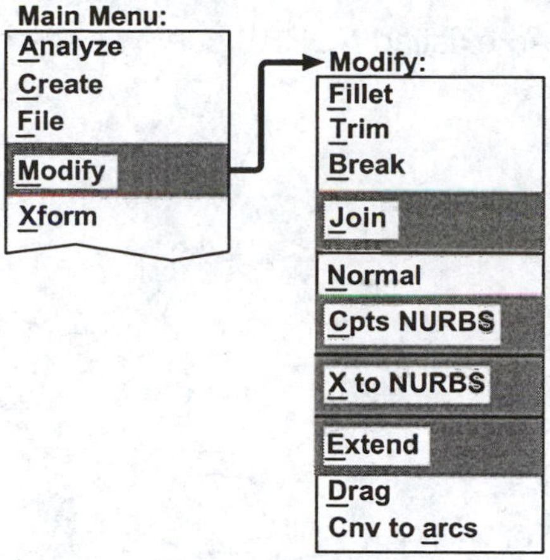

Join

Joins existing *separate colinear lines* or *concentric arcs* having the same radii into a *single* line or arc entity.

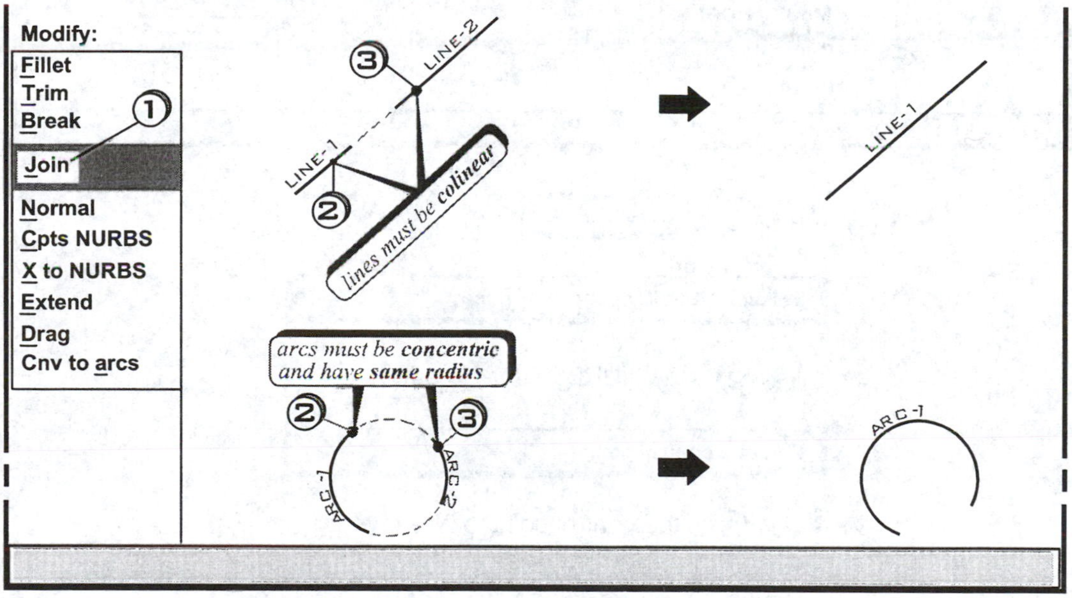

- Click ① **Join**

Join: select an entity

- Click ② the first entity join

Join: select another line

OR

Join: select another arc

- Click ③ second entity to join, etc
- Press **Esc** to cancel the operation

Cpts NURBS

Modifies the shape of an existing NURBS curve as the operator *moves* the curve's *control points.*

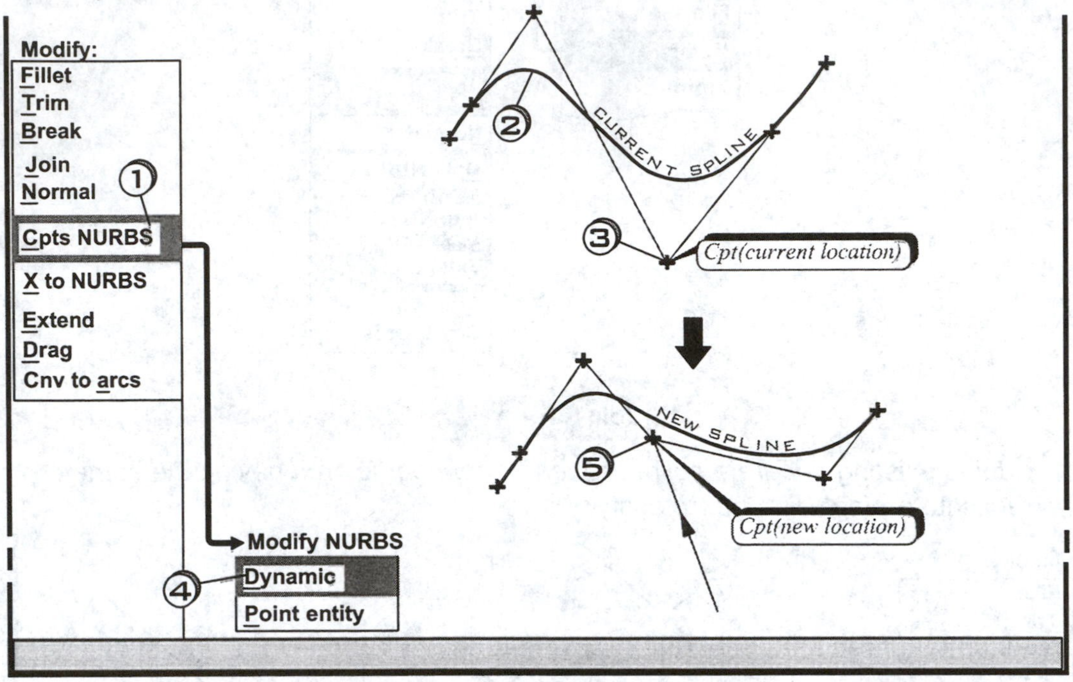

- Click ① **Cpts NURBS**

Select a NURBS spline or surface

- Click ② the NURBS curve to edit

Select a control point

- Click ③ the control point to be moved

- Click ④ **Dynamic**

Drag: enter the starting point

- Click ③ and move the cursor to the control point's *new* location and Click ⑤

- Press Esc to cancel the operation

X to NURBS

Changes line, arc, ellipse, P-spline entities to *NURBS spline* curves.

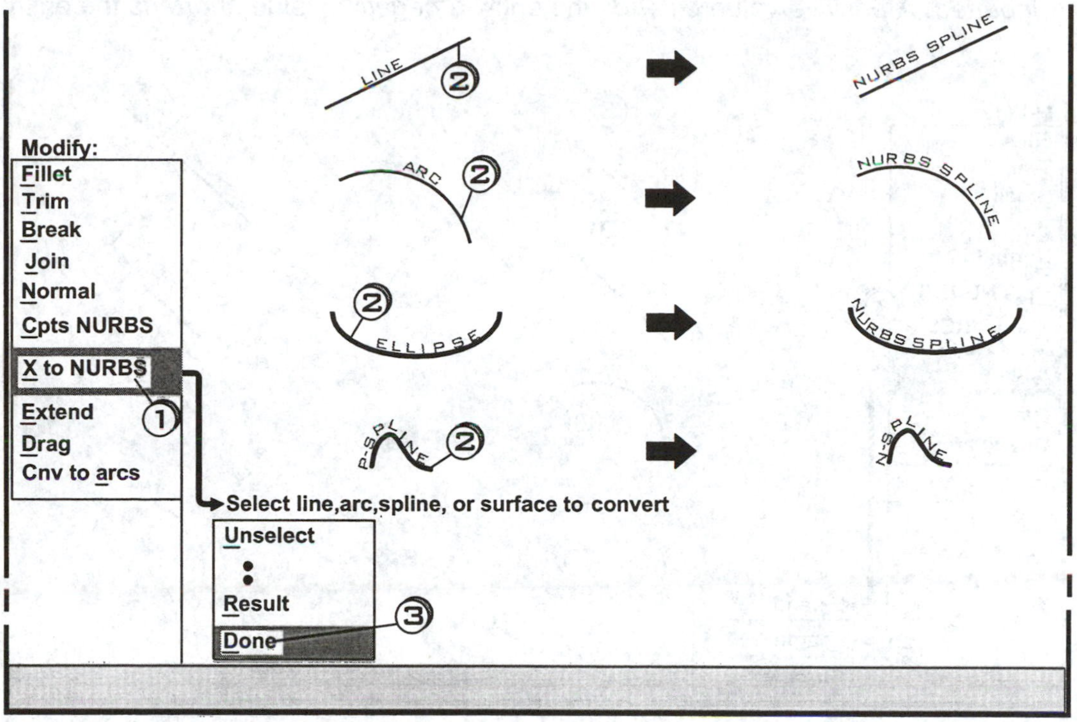

- Click ① **X to NURBS**
- Click ② the entity to convert to a N-spline
- Click ③ **Done**
- Press **Esc** to cancel the operation

Extend

Extends or shortens a line, arc, ellipse, or pline entities by a *length value inputted*. A *positive* value *extends* the entity a *negative* value *shortens* the entity.

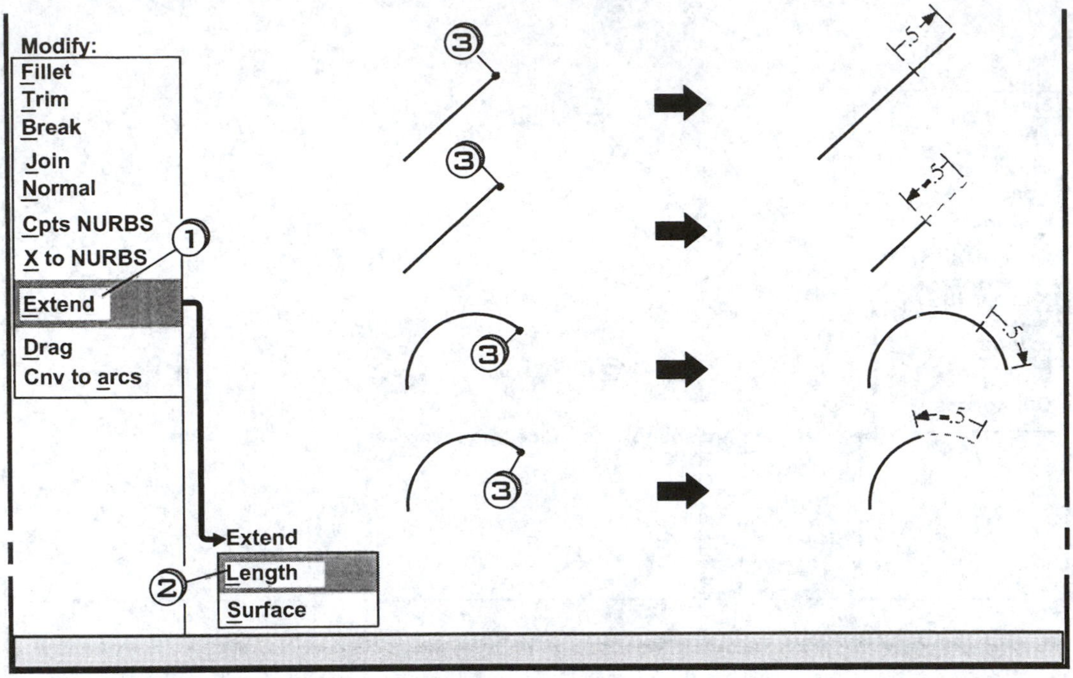

- Click ① **Extend**
- Click ② **Length**

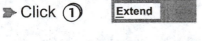

Enter length to extend [ ]

- Enter the length to ***add*** **.5** Enter

OR

- Enter the length to ***subtract*** **-.5** Enter

- Click ③ near the *end* to be *extended*

OR

shortened

- Press Esc to cancel the operation

Dragging and Converting Circular Splines or NURBS Spline Entities To Arcs

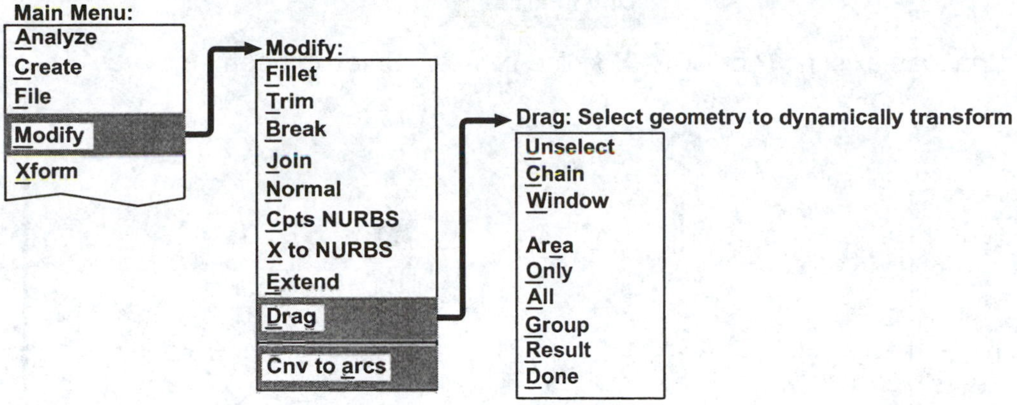

Drag ⦿ Move ⦿ Translate

Dynamically moves or *copies* selected entities as the operator moves the mouse cursor

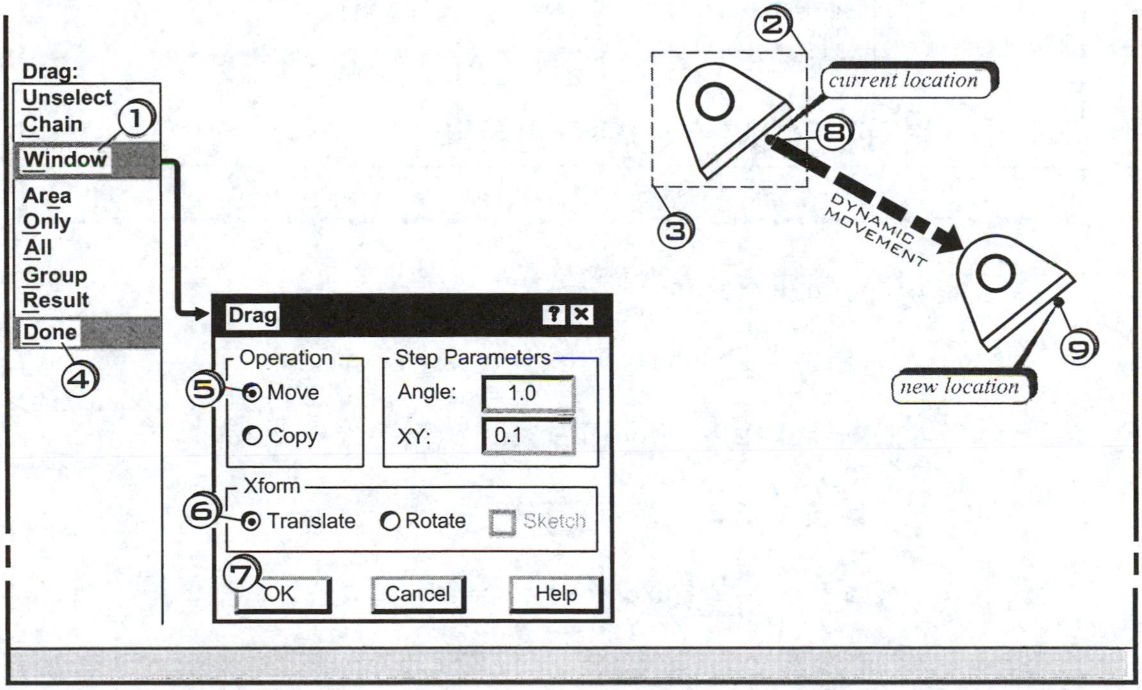

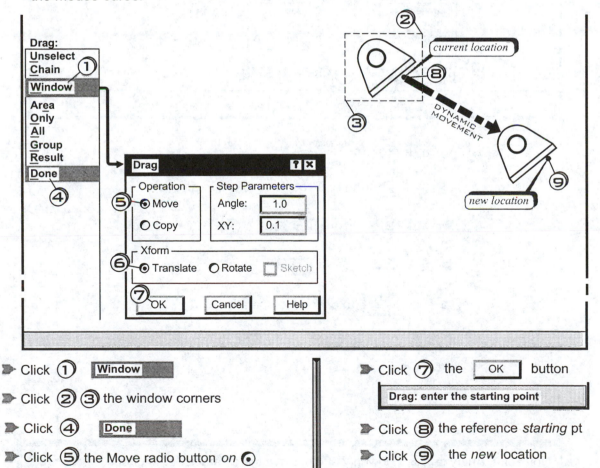

- Click ① **Window**
- Click ② ③ the window corners
- Click ④ **Done**
- Click ⑤ the Move radio button *on* ⦿
- Click ⑥ the Translate radio button *on* ⦿
- Click ⑦ the OK button

Drag: enter the starting point

- Click ⑧ the reference *starting* pt
- Click ⑨ the *new* location
- Press Esc to cancel the operation

Cnv to arcs

Changes existing P or NURBS spline *circular* curves to geometric *arcs*

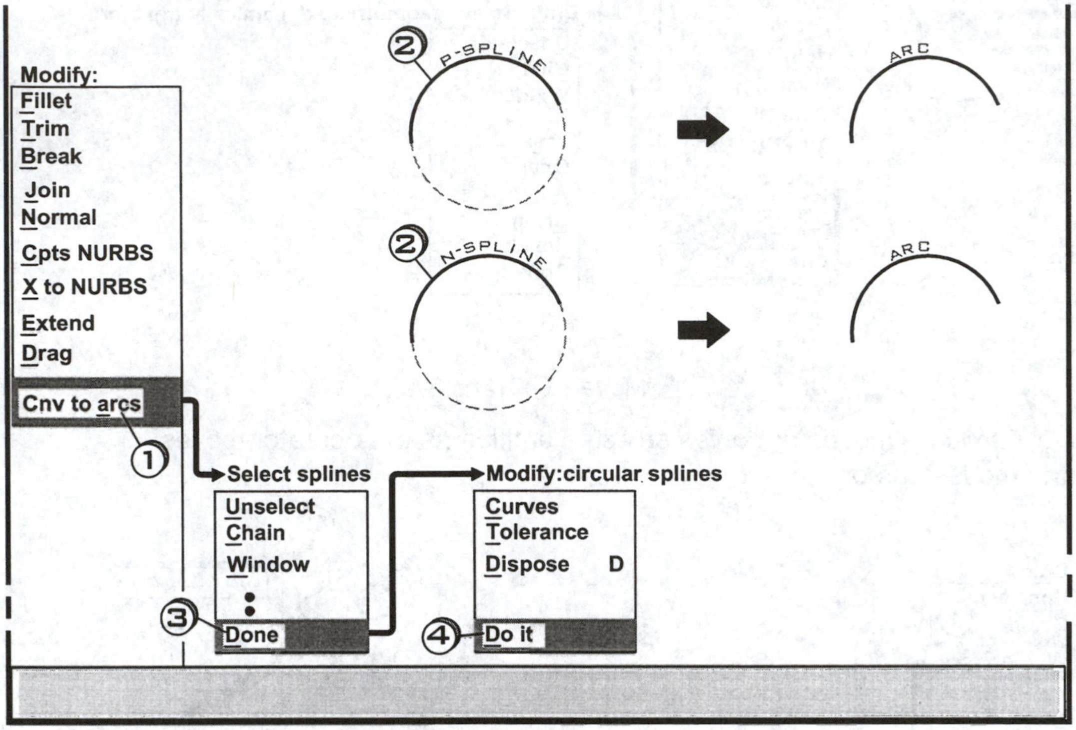

- Click ① **Cnv to arcs**
- Click ② the P or N-splines to chg to arcs
- Click ③ **Done**
- Click ④ **Do it**
- Press **Esc** to cancel the operation

Note: Use **Analyze** F4 to verify and observe the chracteristics of entities changed

Using Xform Functions to Edit Entities

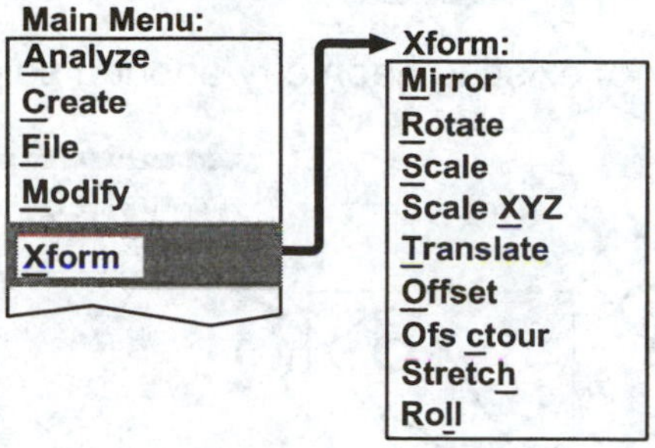

Mirror ⊙ Move

Moves existing geometry into a *mirror image* orientation about a selected *mirror line*

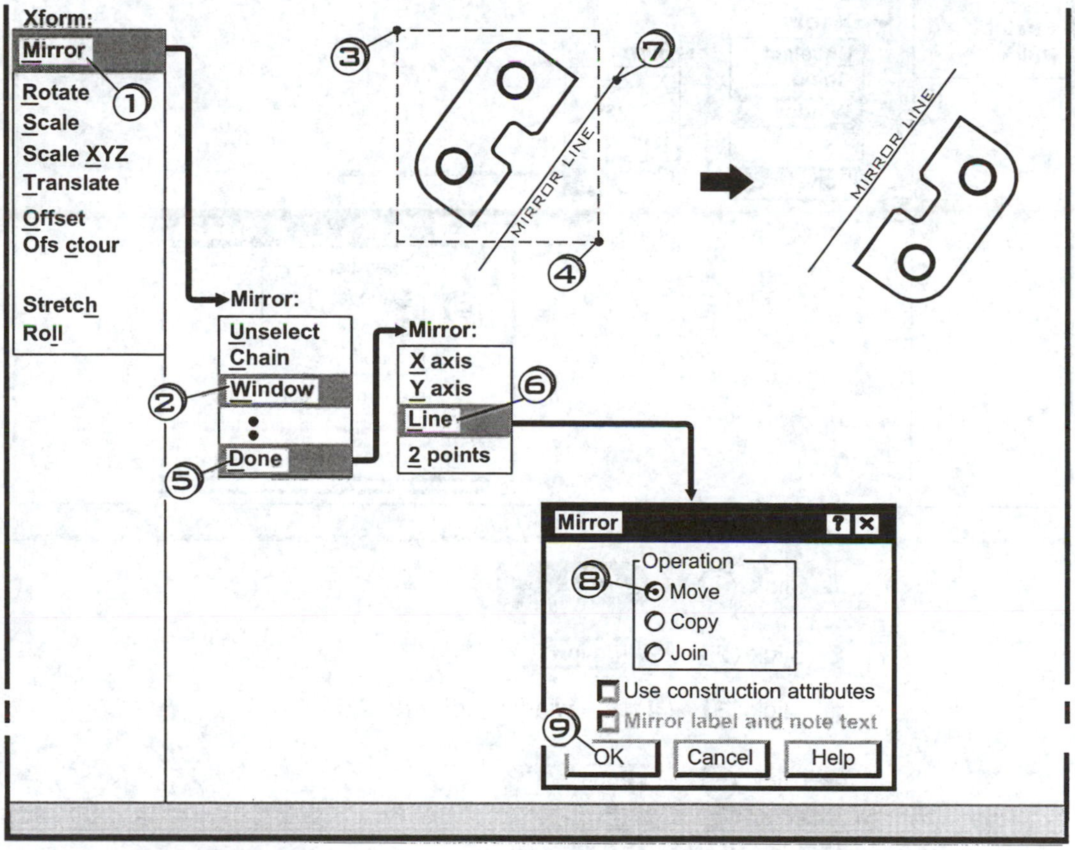

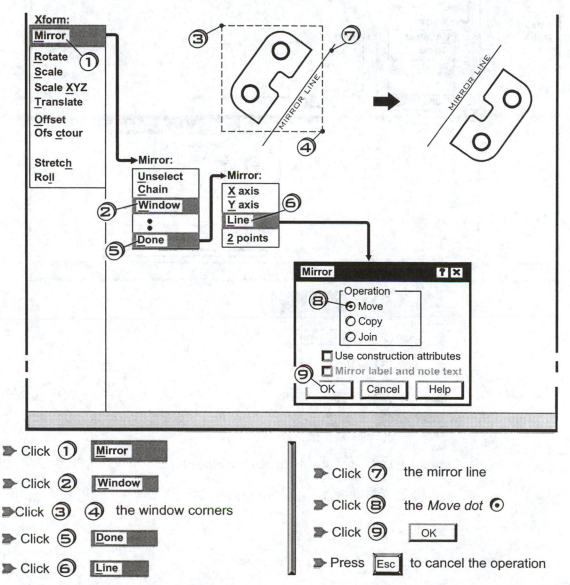

- Click ① **Mirror**
- Click ② **Window**
- Click ③ ④ the window corners
- Click ⑤ **Done**
- Click ⑥ **Line**
- Click ⑦ the mirror line
- Click ⑧ the *Move dot* ⊙
- Click ⑨ OK
- Press Esc to cancel the operation

Mirror ⊙ Copy

Creates a *mirror copy* of existing geometry about a selected *mirror* line

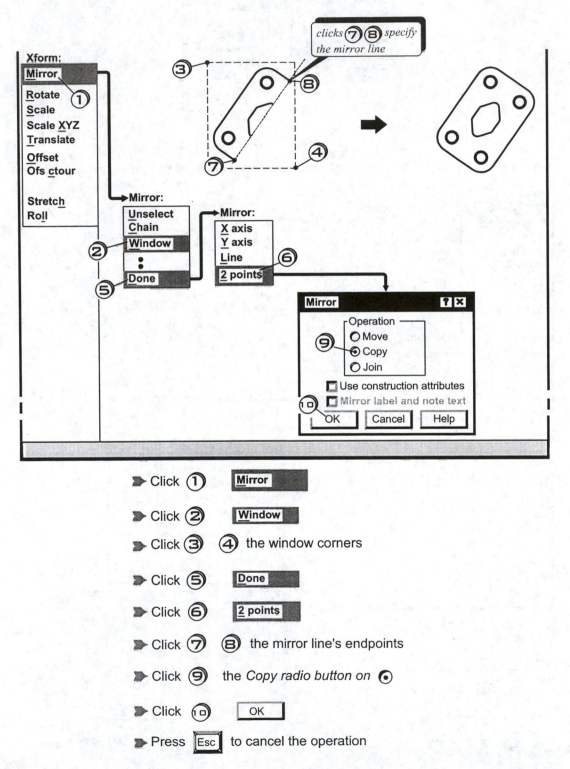

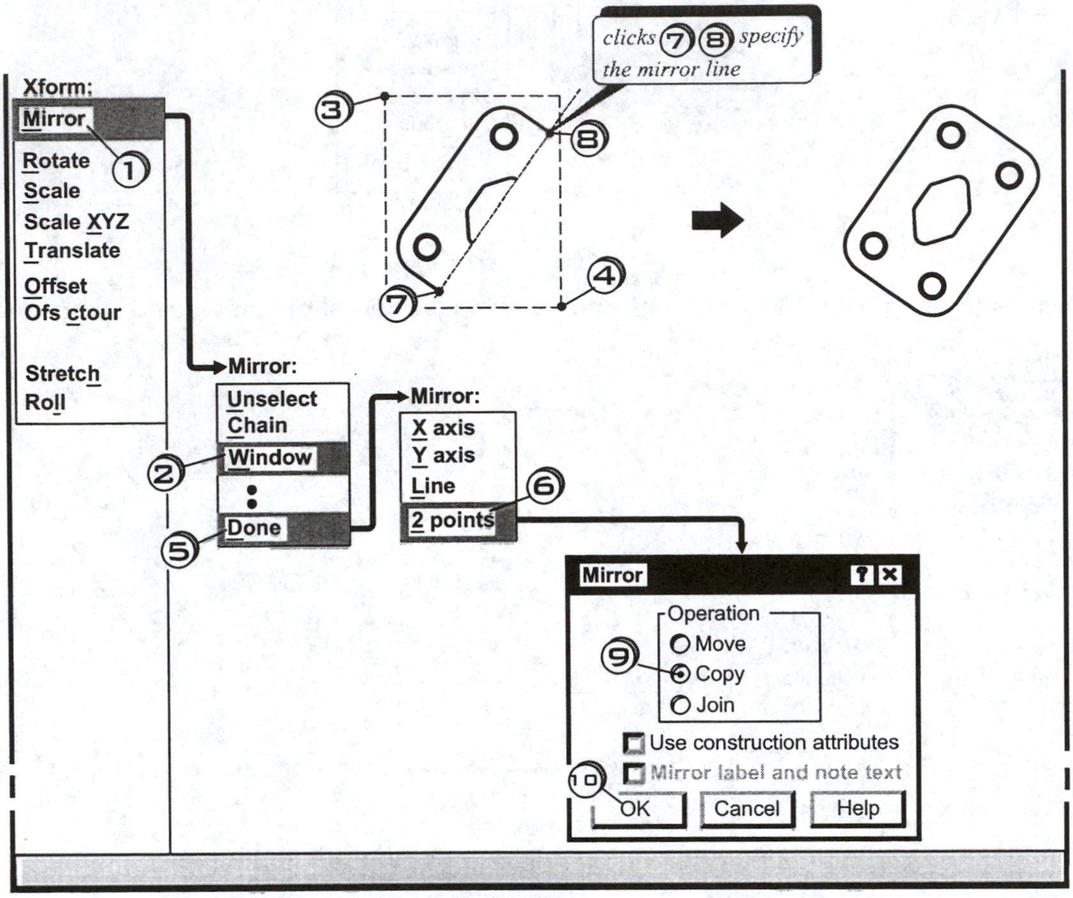

- Click ① **Mirror**
- Click ② **Window**
- Click ③ ④ the window corners
- Click ⑤ **Done**
- Click ⑥ **2 points**
- Click ⑦ ⑧ the mirror line's endpoints
- Click ⑨ the *Copy radio button on* ⊙
- Click ⑩ OK
- Press Esc to cancel the operation

Rotate ⊙ Move

Rotates existing geometry about a *specified center* and through an *inputted angle*.

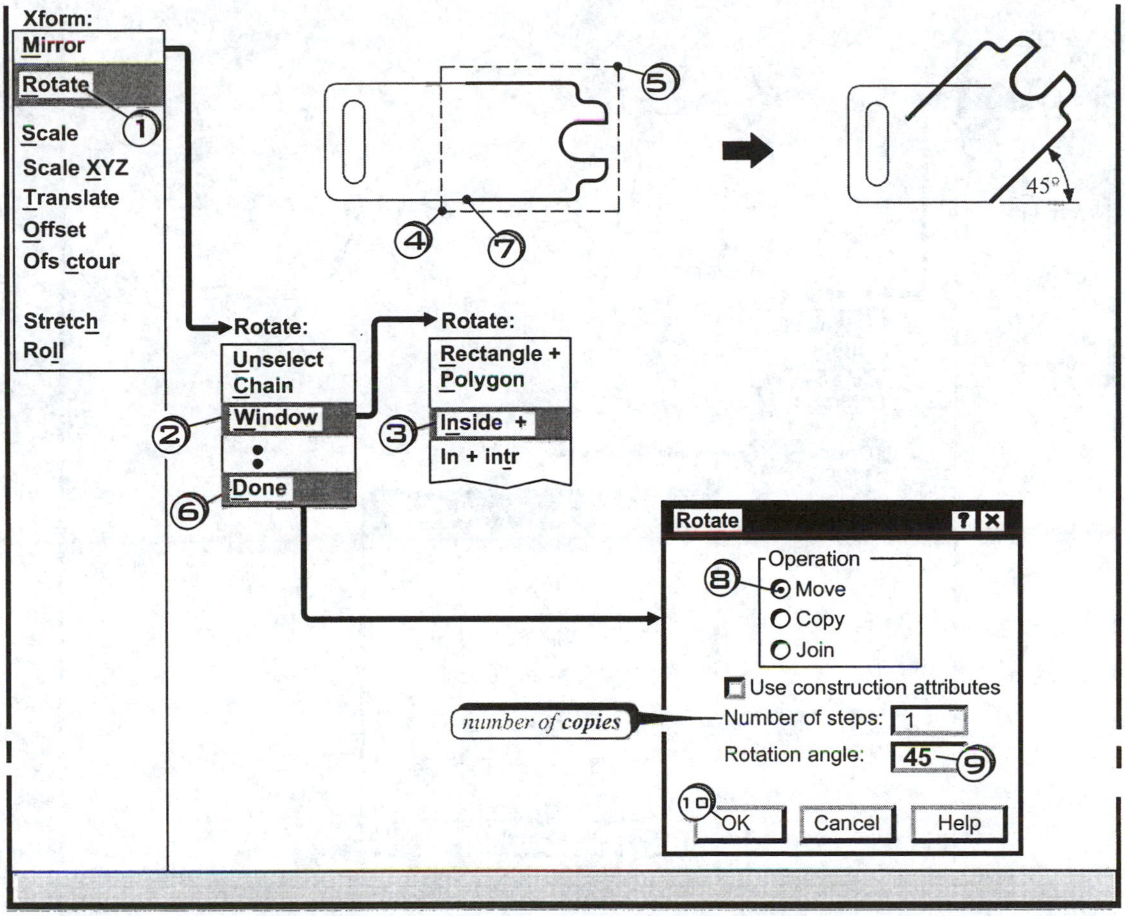

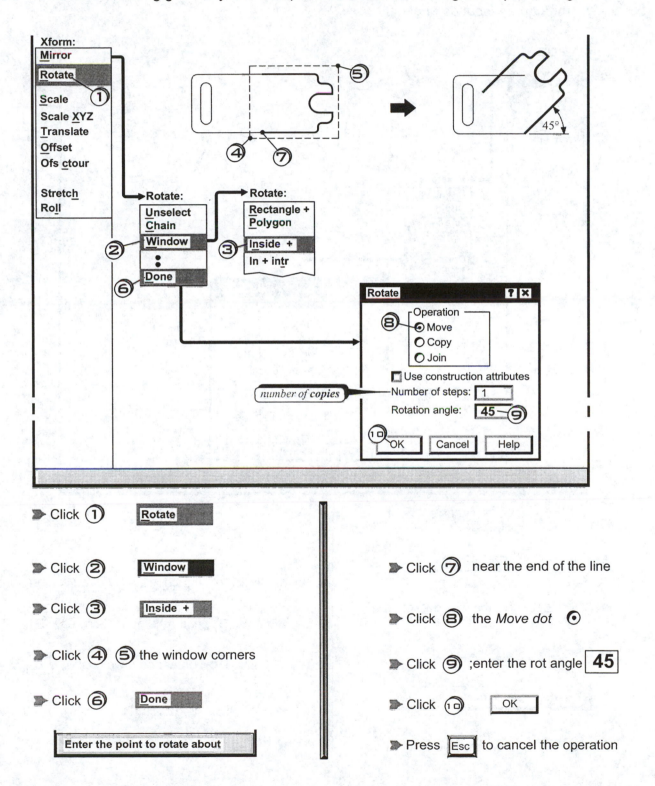

- Click ① **Rotate**
- Click ② **Window**
- Click ③ **Inside +**
- Click ④ ⑤ the window corners
- Click ⑥ **Done**

Enter the point to rotate about

- Click ⑦ near the end of the line
- Click ⑧ the *Move dot* ⊙
- Click ⑨ ;enter the rot angle **45**
- Click ⑩ OK
- Press Esc to cancel the operation

Rotate ⊙ Copy

Creates rotated copies of existing geometry about a *specified center* and through an *inputted angle*.

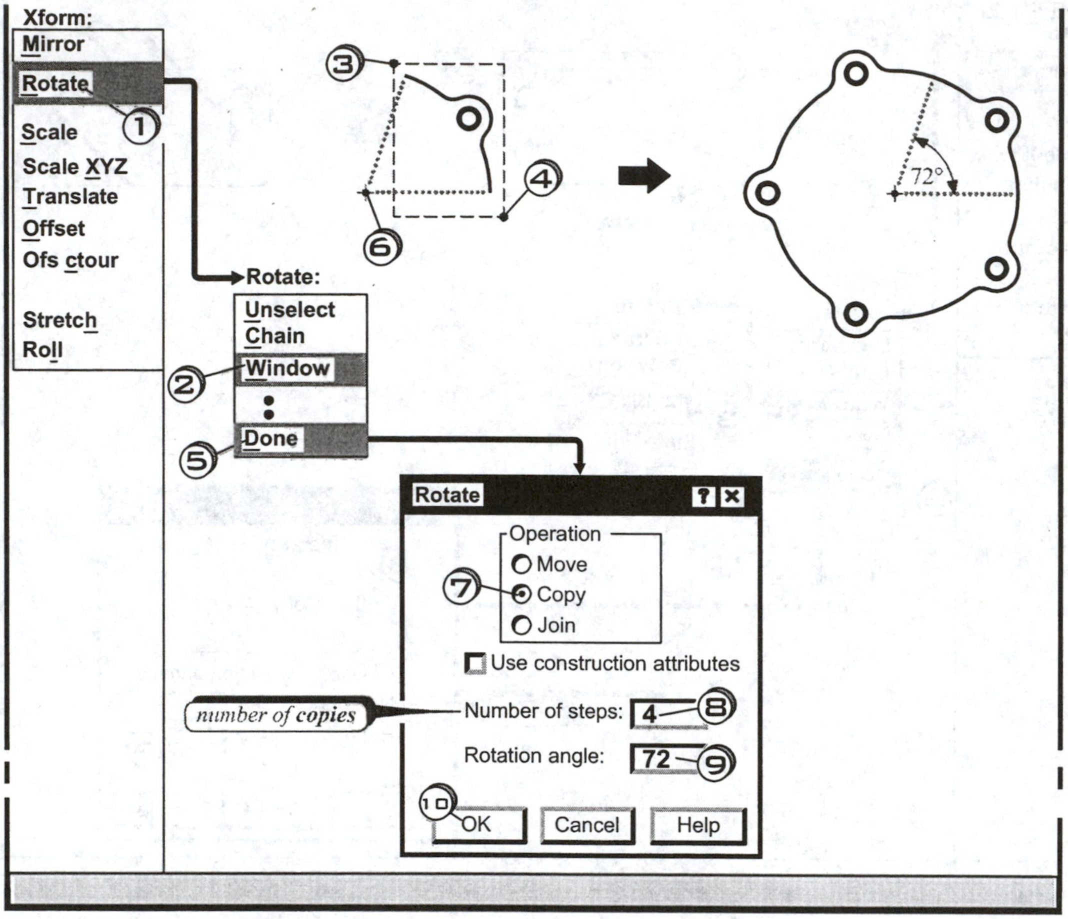

- Click ① **Rotate**
- Click ② **Window**
- Click ③ ④ the window corners
- Click ⑤ **Done**

Enter the point to rotate about

- Click ⑥ near the end of the line
- Click ⑦ the *copy dot* ⊙
- Click ⑧ ; enter the num of copies **4**
- Click ⑨ ; enter ang btwn copies **72**
- Click ⑩ OK
- Press Esc to cancel the operation

Translate | Between pts | ⊙ Move

Moves existing geometry from a *present* reference point to a *new* reference point

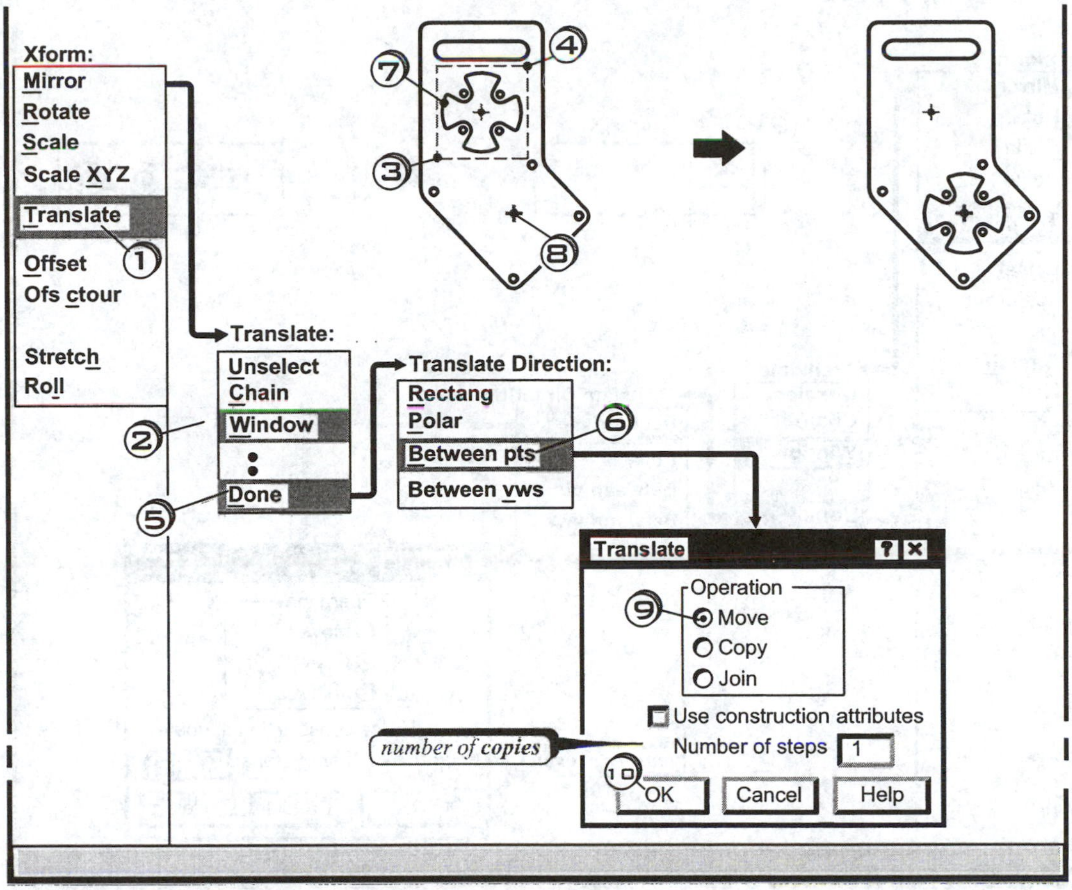

- Click ① Translate
- Click ② Window
- Click ③ ④ the window corners
- Click ⑤ Done
- Click ⑥ Between pts

Enter the point to translate from

- Click ⑦ near the arc

Enter the point to translate to

- Click ⑧ near the point
- Click ⑨ the *move dot* ⊙
- Click ⑩ OK
- Press Esc to cancel the operation

Translate **Rectang** ⊙ Copy

Creates translated copies of existing geometry. The *X,Y distances* between *each copy* are entered

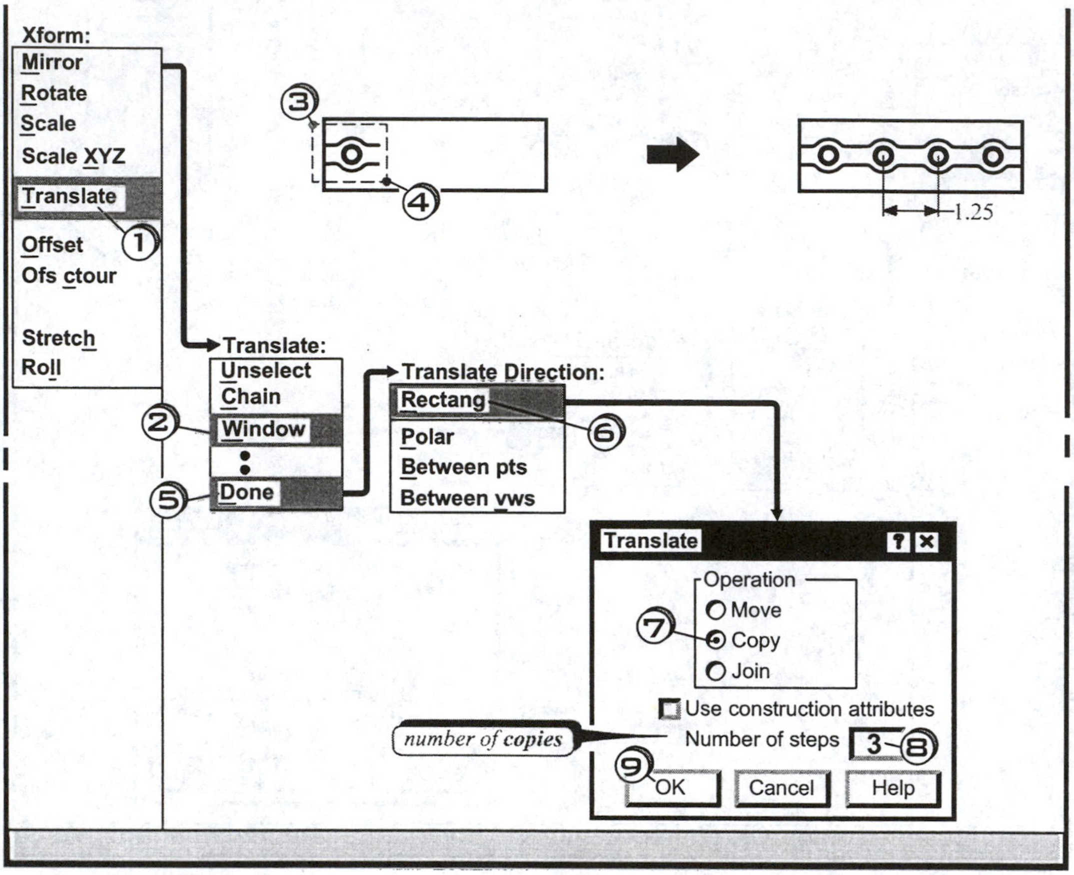

- Click ① **Translate**
- Click ② **Window**
- Click ③ ④ the window corners
- Click ⑤ **Done**
- Click ⑥ **Rectang**

Enter the translation vector [ ]

- Enter the X,Y distances between each copy **1.25,0** Enter
- Click ⑦ the *copy dot* ⊙
- Click ⑧ ;enter the num of copies **3**
- Click ⑨ OK
- Press Esc to cancel the operation

Translate **Polar** ⊙ Copy

Creates translated copies of existing geometry. The *polar distance between and the angle* for *each copy* are entered

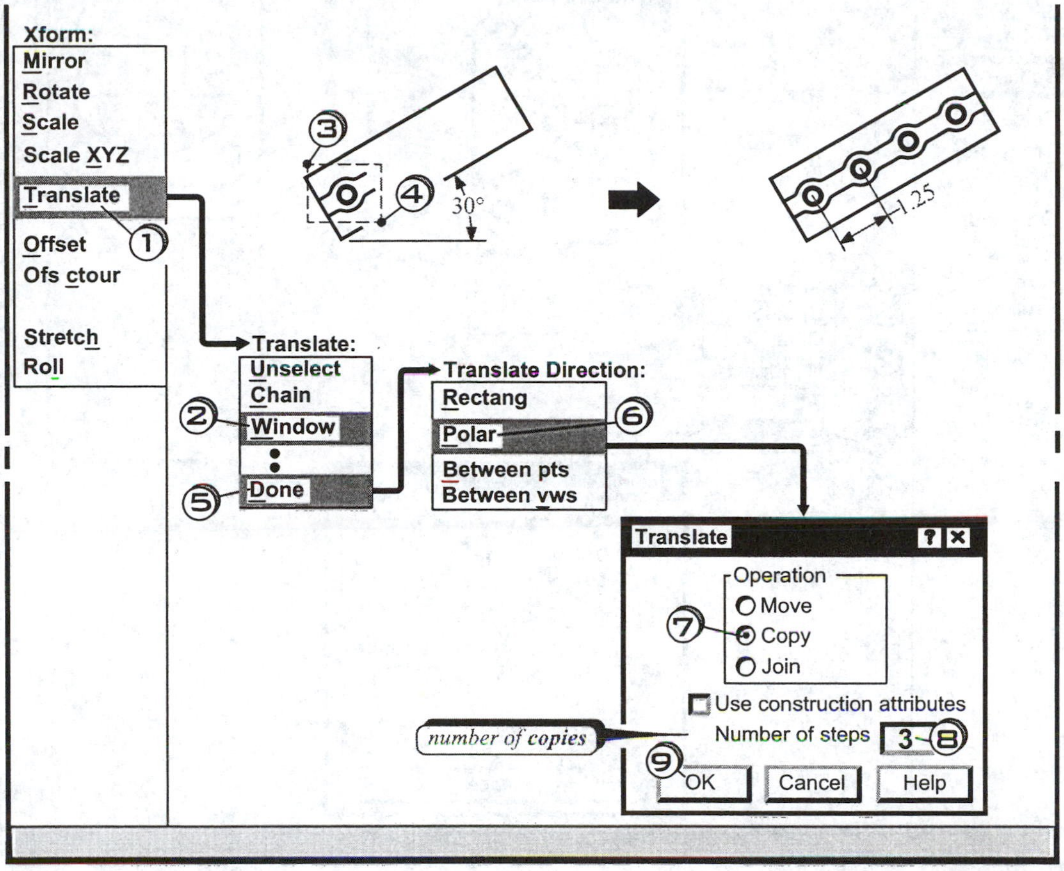

- Click ① **Translate**
- Click ② **Window**
- Click ③ ④ the window corners
- Click ⑤ **Done**
- Click ⑥ **Polar**

Enter the translation distance

- Enter the *distance* between each copy **1.25** Enter

Enter the translation angle

- Enter the *angle* for copies **30** Enter
- Click ⑦ the a *copy dot* ⊙
- Click ⑧ ; enter the num of copies **3**
- Click ⑨ OK
- Press Esc to cancel the operation

Scale ⊙ Uniform

Uniformly scales the *size* of existing geometry *up or down*

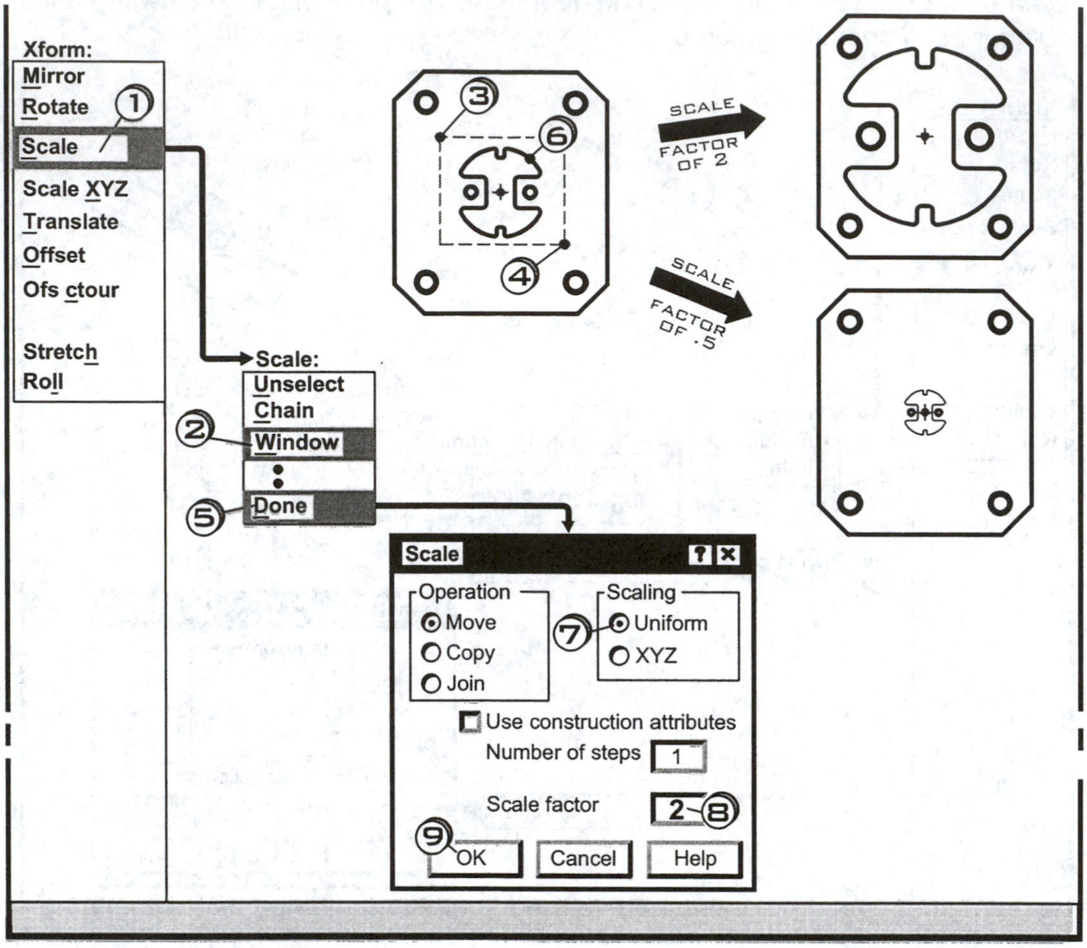

➤ Click ① **Scale**

➤ Click ② **Window**

➤ Click ③ ④ the window corners

➤ Click ⑤ **Done**

Enter the point to scale about

➤ Click ⑥ on the arc

➤ Click ⑦ the Uniform dot *on* ⊙

➤ Click ⑧ in the scale factor box:

enter **2**

OR

enter **.5**

➤ Click ⑨ OK

➤ Press Esc to cancel the operation

Scale ⊙ XYZ

Selectively scales the *size* of existing geometry in the *X,Y* and *Z* directions

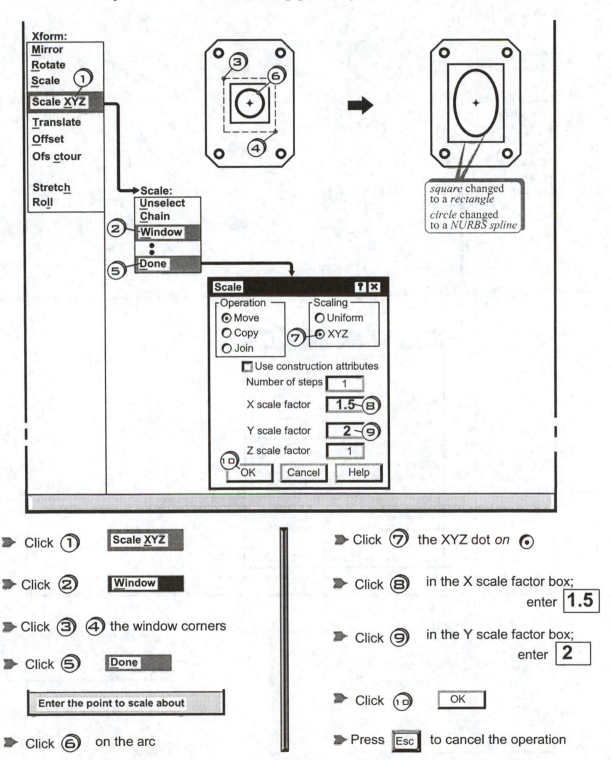

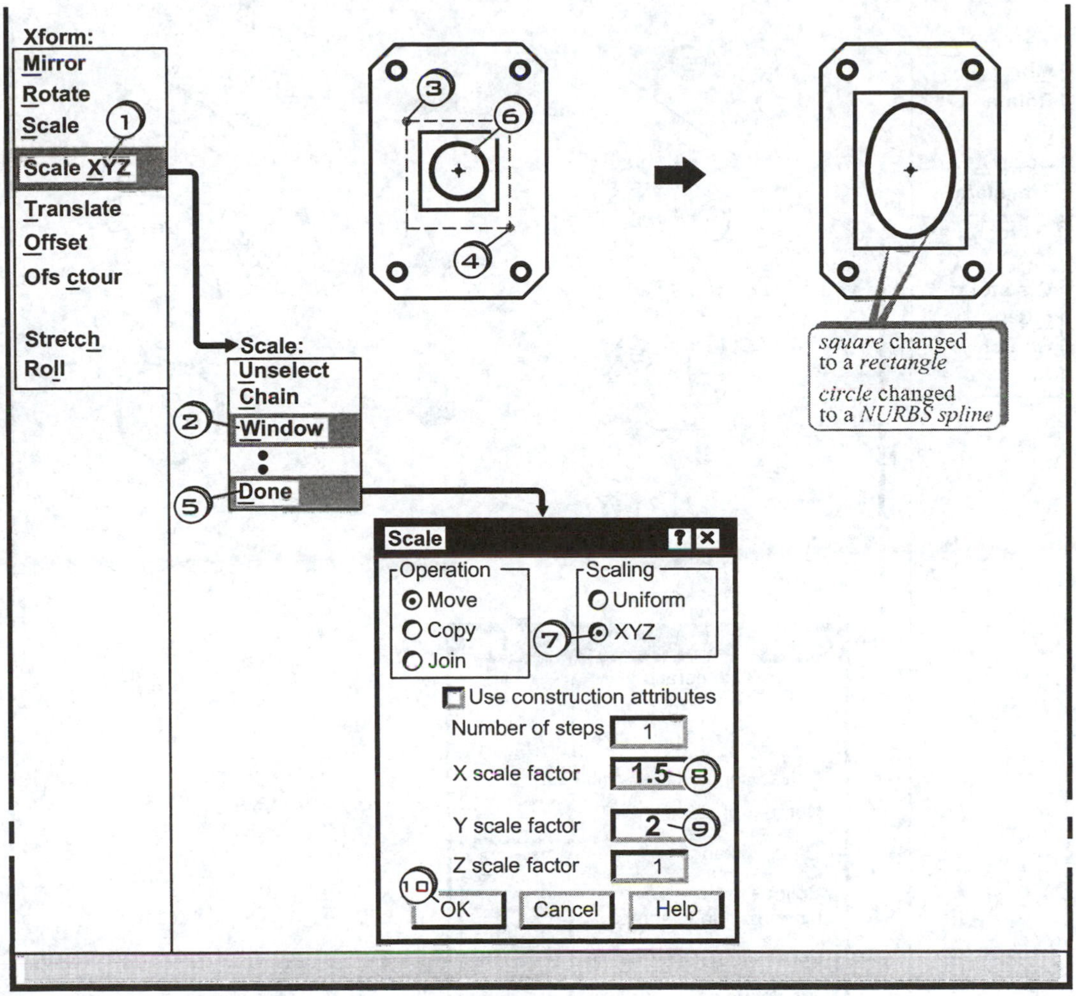

- Click ① **Scale XYZ**
- Click ② **Window**
- Click ③ ④ the window corners
- Click ⑤ **Done**

Enter the point to scale about

- Click ⑥ on the arc
- Click ⑦ the XYZ dot *on* ⊙
- Click ⑧ in the X scale factor box; enter **1.5**
- Click ⑨ in the Y scale factor box; enter **2**
- Click ⑩ OK
- Press Esc to cancel the operation

Offset ⊙ Copy

Creates an offset copy in a *perpendicular* direction to an existing line, arc, spline or curve entity.

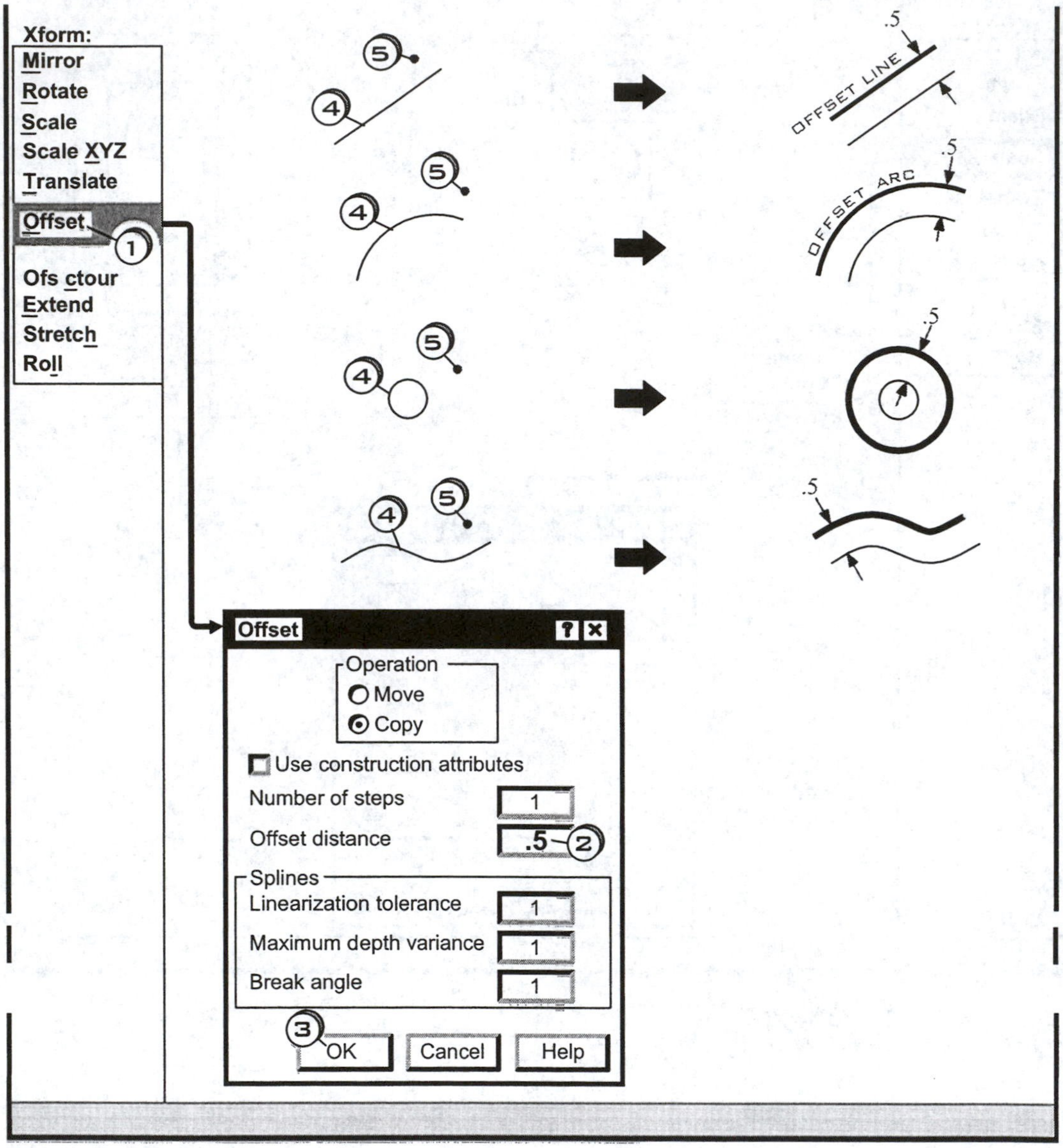

- Click ① **Offset**
- Click ② in the offset distance box ; enter **.5**
- Click ③ **OK**

Select the line, arc, spline or curve to offset

- Click ④ on the entity to offset

Indicate the offset direction

- Click ⑤ on the side to offset
- Press **Esc** to cancel the operation

Ofs ctour ⦿ Copy

Creates an *offset copy* in a *perpendicular* direction to the existing contour. The contour may consist of connected line, arc, spline or curve entities.

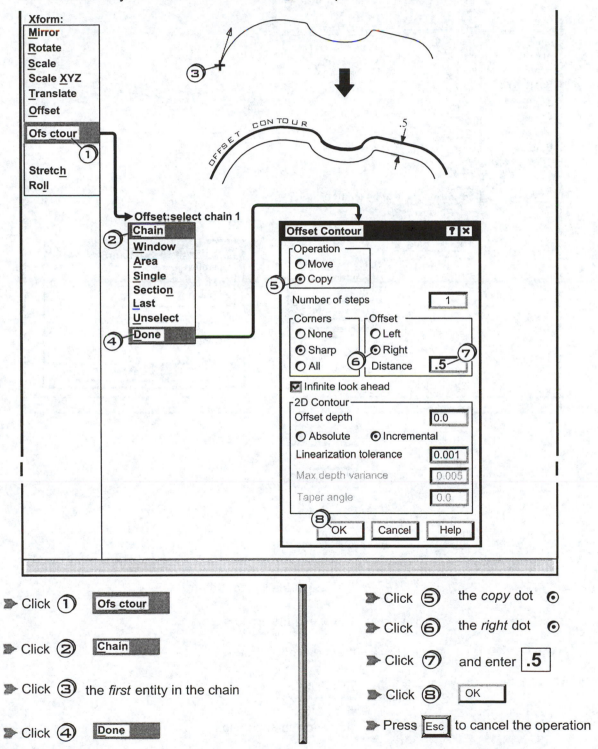

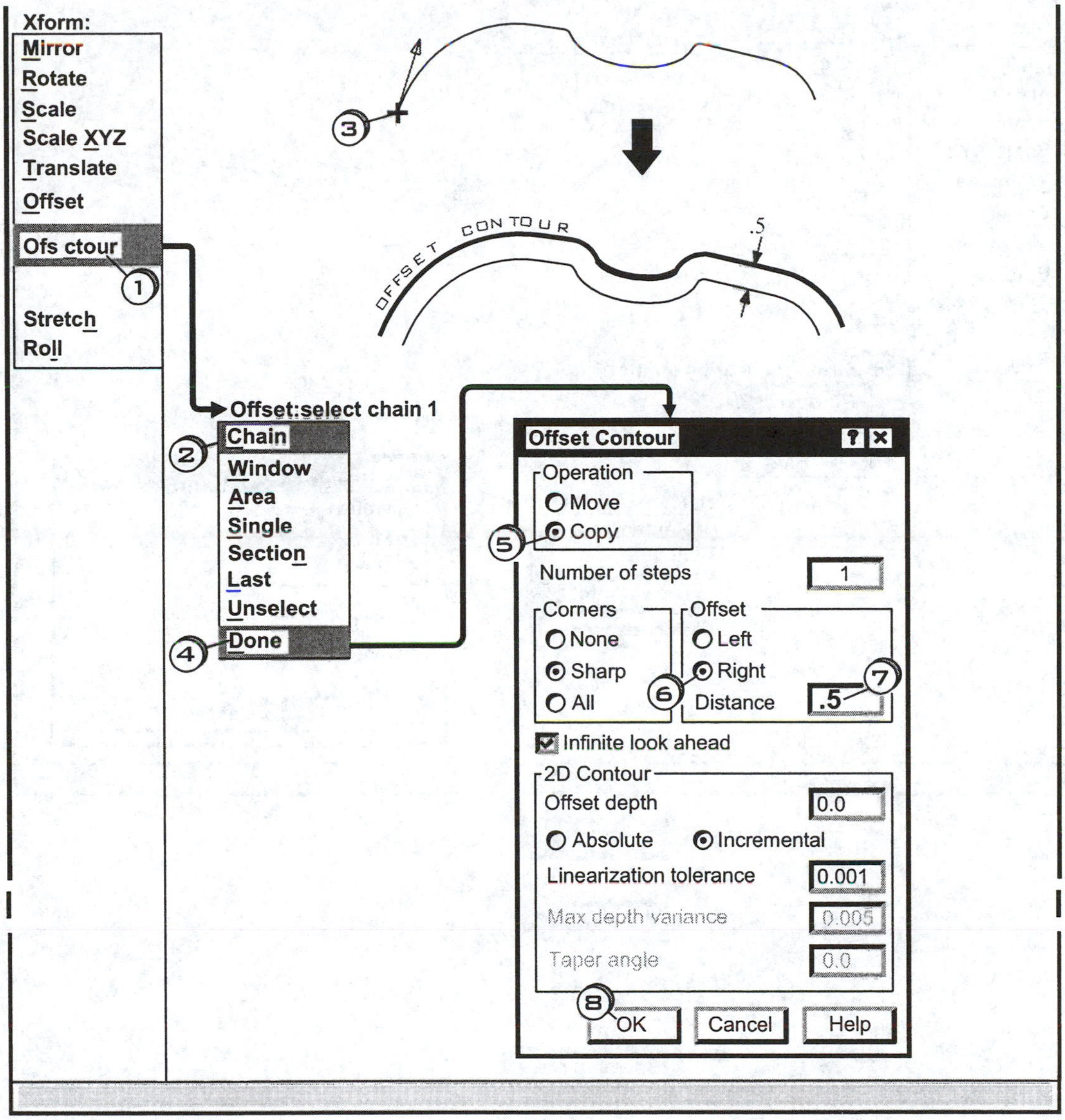

- Click ① **Ofs ctour**
- Click ② **Chain**
- Click ③ the *first* entity in the chain
- Click ④ **Done**
- Click ⑤ the *copy* dot ⦿
- Click ⑥ the *right* dot ⦿
- Click ⑦ and enter **.5**
- Click ⑧ OK
- Press Esc to cancel the operation

Stretch ⊙ Move

Stretches the *end* of an *existing line* entity between *two* inputted *points*

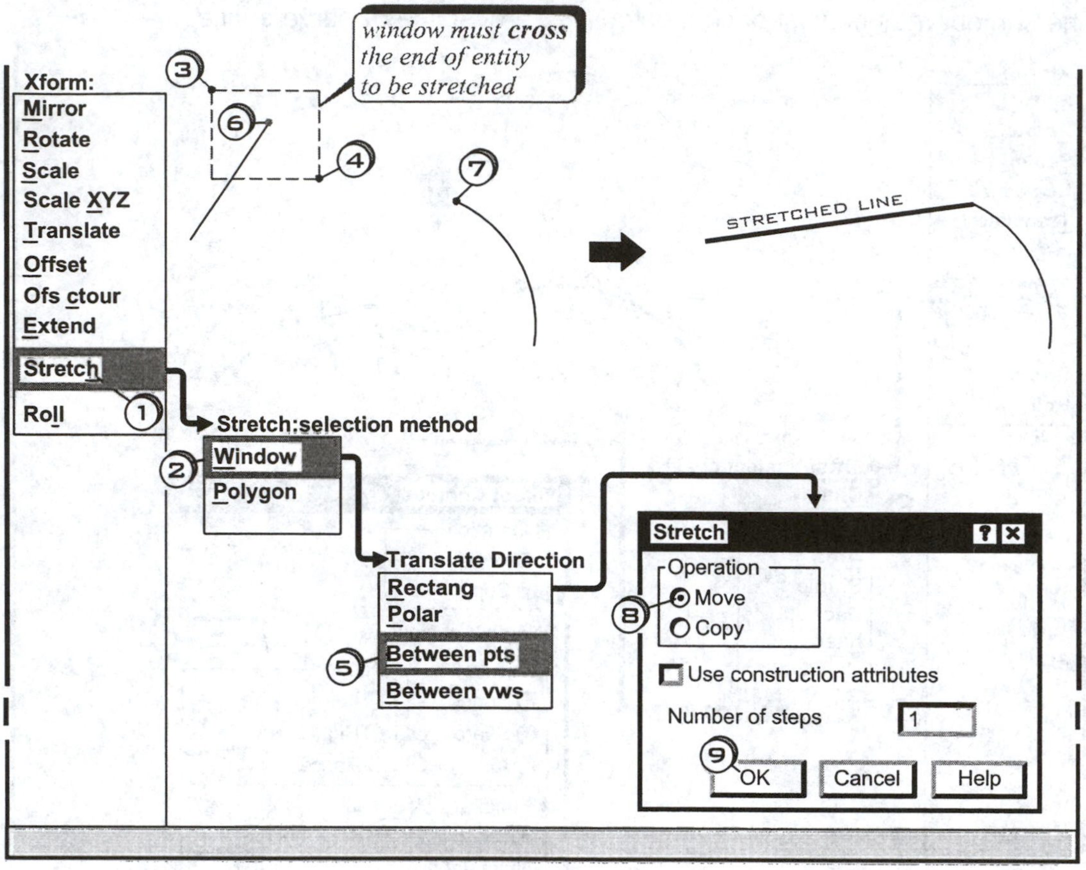

- Click ① **Stretch**
- Click ② **Window**
- Click ③ ④ corners of the window
- Click ⑤ **Between pts**
- Click ⑥ the end of the line to stretch
- Click ⑦ the end of the arc to stretch to
- Click ⑧ the *move* dot ⊙ Move
- Click ⑨ OK
- Press Esc to cancel the operation

Roll ⦿ Y axis ⦿ Nurbs

Changes an existing line to a NURBS spline that is *rolled around a cylinder*.
The operator specifies the cylinder orientation and diameter

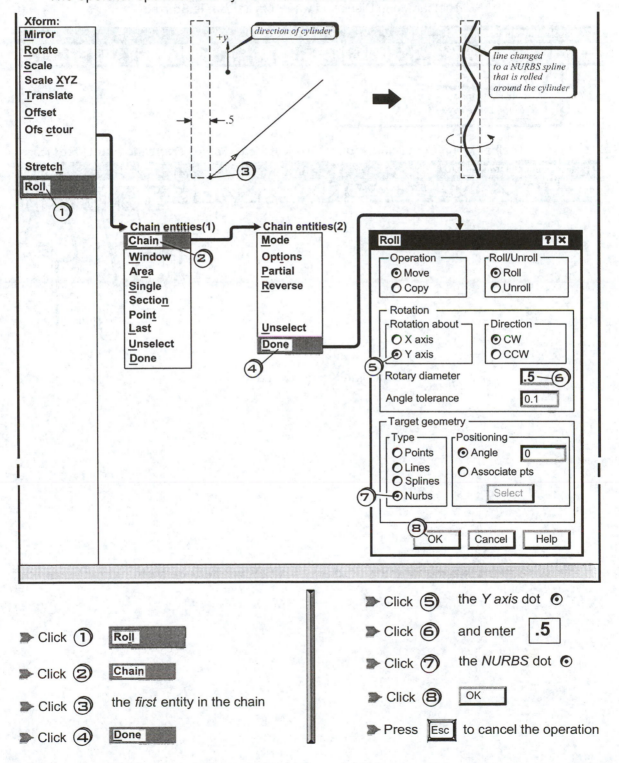

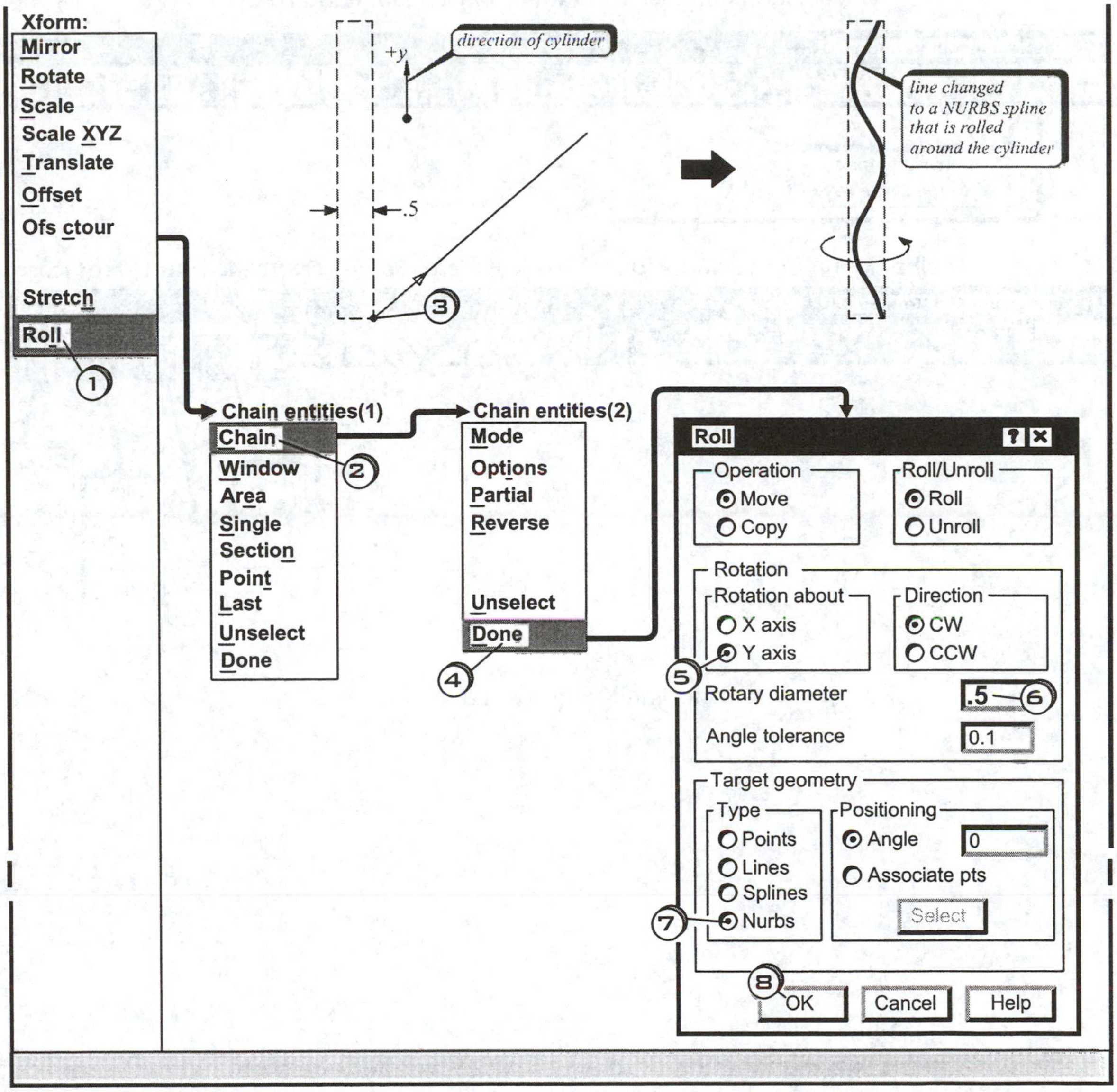

- Click ① **Roll**
- Click ② **Chain**
- Click ③ the *first* entity in the chain
- Click ④ **Done**
- Click ⑤ the *Y axis* dot ⦿
- Click ⑥ and enter **.5**
- Click ⑦ the *NURBS* dot ⦿
- Click ⑧ OK
- Press Esc to cancel the operation

Using the Toolbar Menus

Toolbar Menu Displayed when *Mastercam* is **Started**

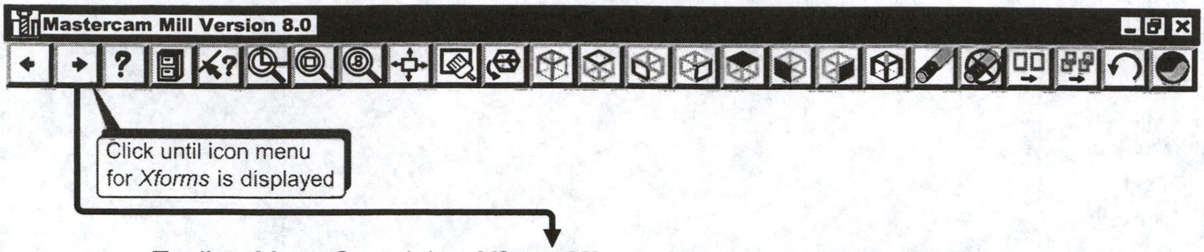
Mastercam Mill Version 8.0
Click until icon menu for *Xforms* is displayed

Toolbar Menu Containing ***Xform-Mirror, Rotate, Scale, Translate and Offset*** icons

Mastercam Mill Version 8.0
Click the *Mirror* icon
Click the *Rotate* icon
Click the *Scale* icon
Click the *Trans-late* icon
Click the *Offset* icon

Click Esc to cancel an Xform command

EXERCISES

3-1) Use *Mastercam's* CAD package to create the part shown in Figure 3-p1.

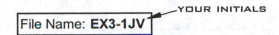

File Name: **EX3-1JV** ← YOUR INITIALS

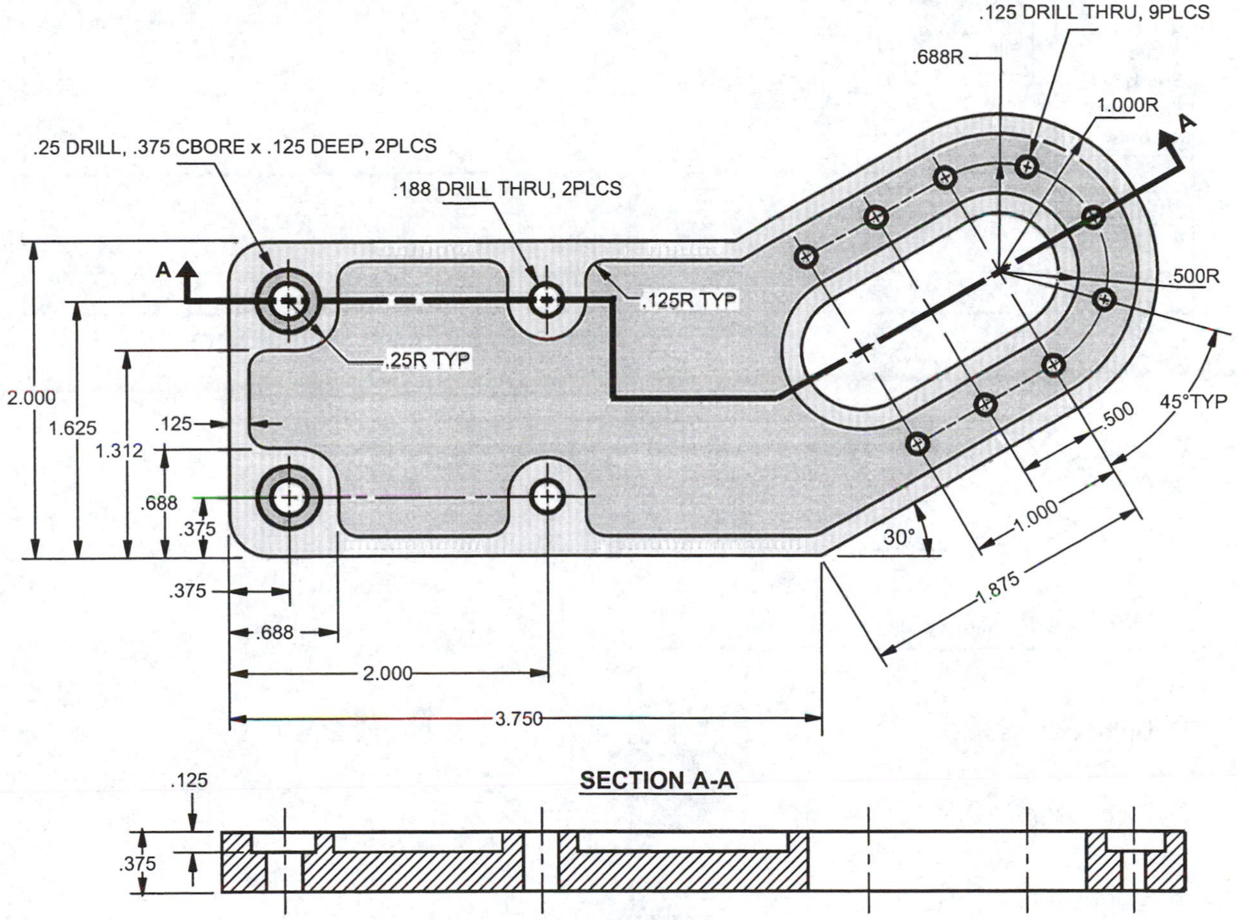

Figure 3-p1

➤ Click ① the *Mastercam* **MILL** icon 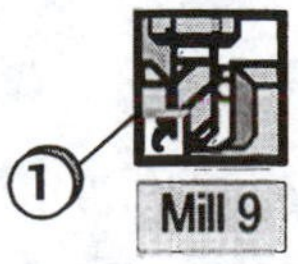 to enter the Mill package

➤ Direct *Mastercam* to create a new file. Refer to page 2p-1 to review this step.

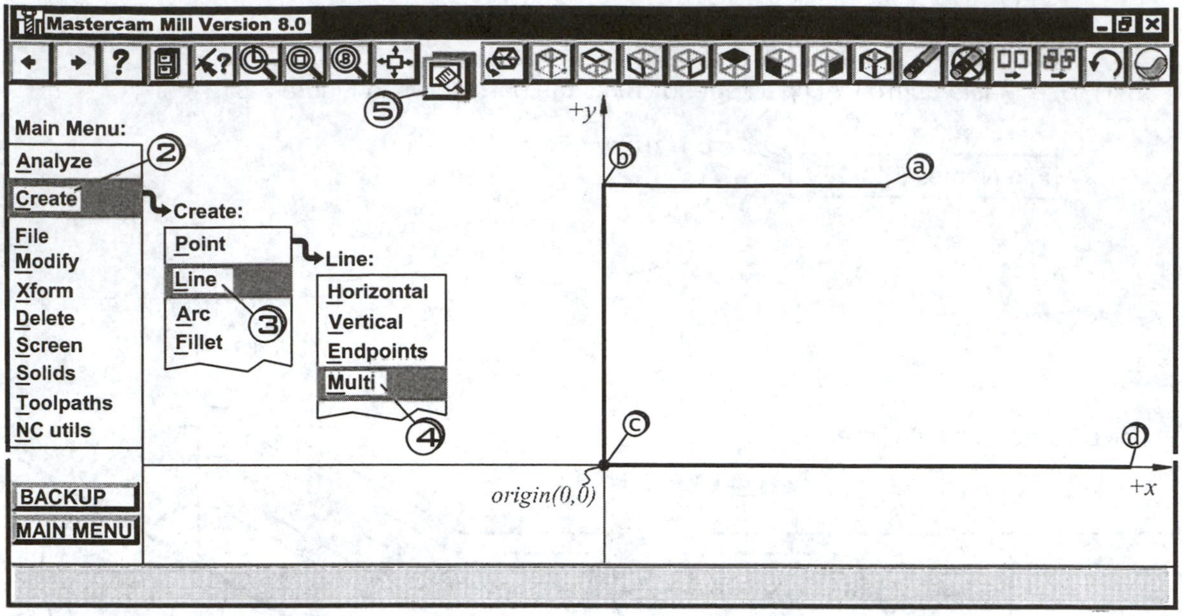

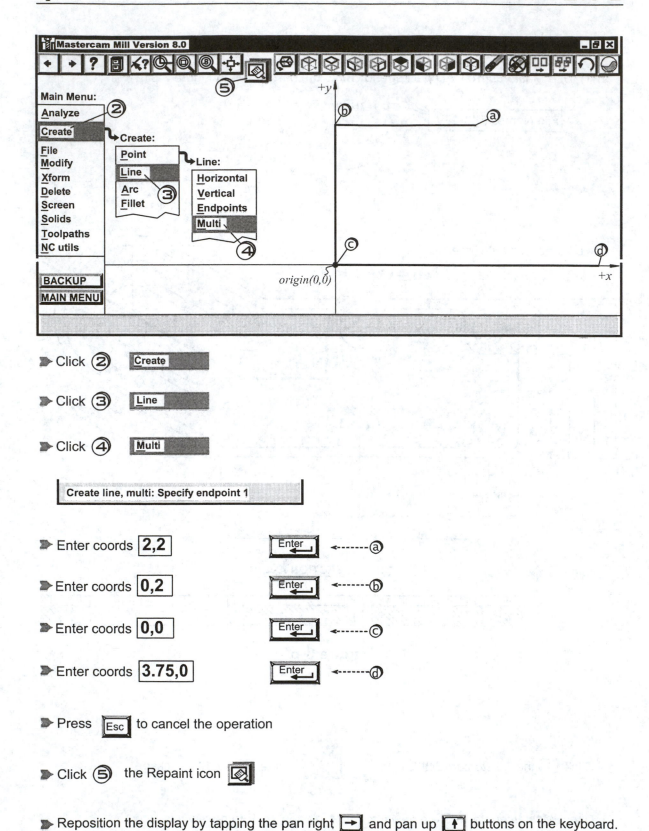

- Click ② **Create**
- Click ③ **Line**
- Click ④ **Multi**

Create line, multi: Specify endpoint 1

- Enter coords **2,2** Enter ←-------ⓐ
- Enter coords **0,2** Enter ←-------ⓑ
- Enter coords **0,0** Enter ←-------ⓒ
- Enter coords **3.75,0** Enter ←-------ⓓ
- Press **Esc** to cancel the operation
- Click ⑤ the Repaint icon
- Reposition the display by tapping the pan right **→** and pan up **↑** buttons on the keyboard.

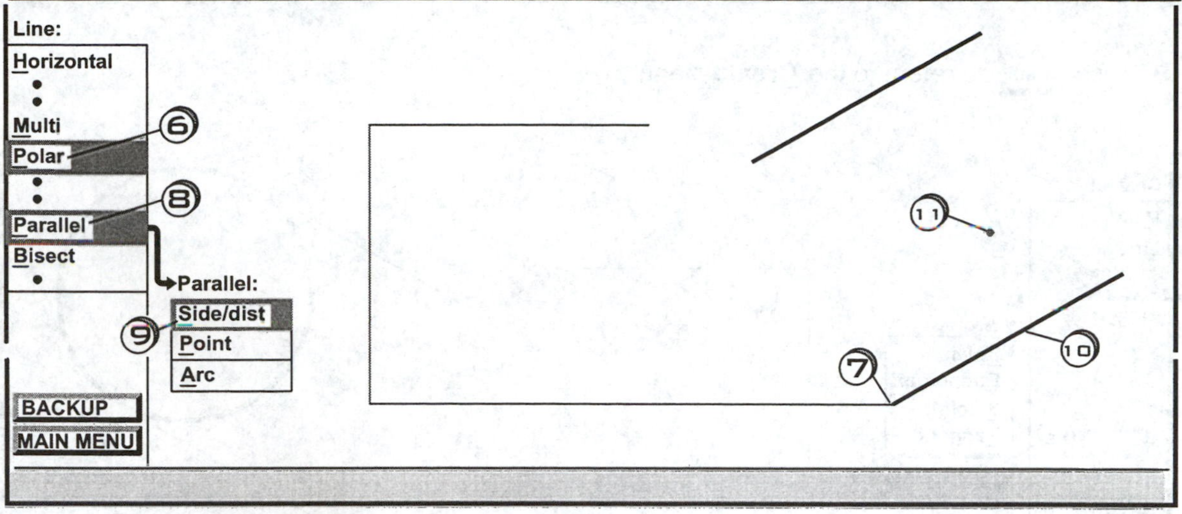

- Click ⑥ **Polar**

Create line polar: Specify an endpoint

- Click ⑦ near the endpoint of the line

Enter the angle in degrees

- Enter angle in deg **30** Enter

Enter the line length

- Enter line length **1.875** Enter
- Press Esc to cancel the operation
- Click ⑧ **Parallel**
- Click ⑨ **Side/dist**

Select line

- Click ⑩ the line entity

Indicate the offset direction

- Click ⑪ the side on which the parallel line is to be created

Parallel line Distance=

- Enter the offset distance **2** Enter
- Press Esc to cancel the operation

Press Esc to return to the **Create** menu

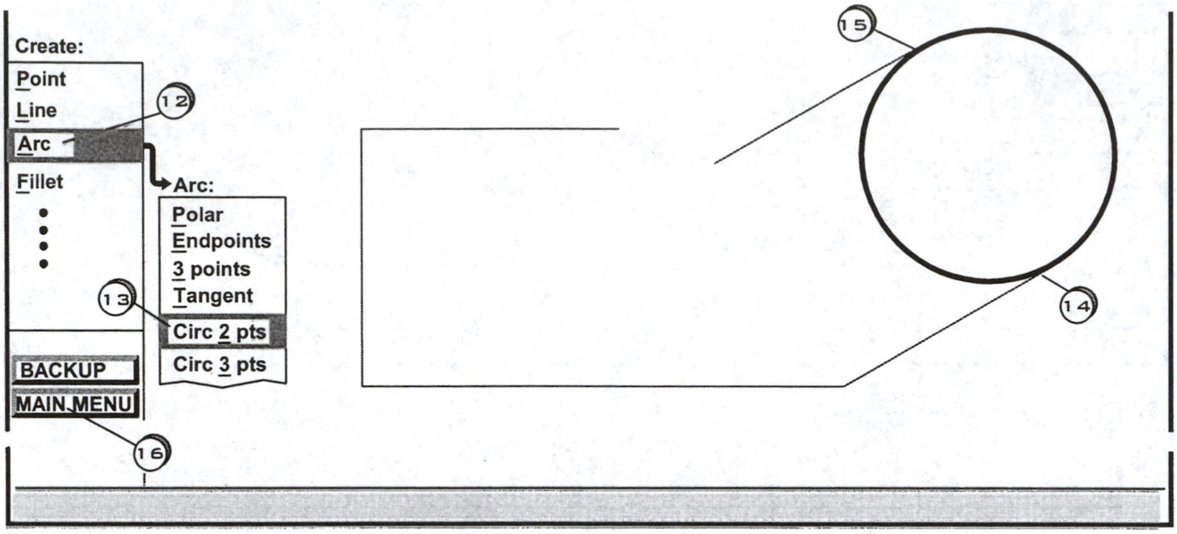

- Click (12) **Arc**

- Click (13) **Circ 2 pts**

Circle, 2 points: Enter the first point

- Click (14) near the end point of the line

Circle, 2 points: Enter the second point

- Click (15) near the end point of the arc

Radius = 1

- Press Esc to cancel the operation

- Click (16) **MAIN MENU**

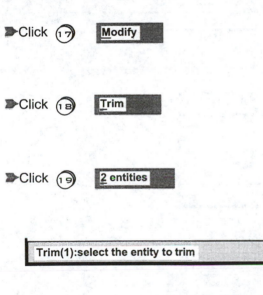

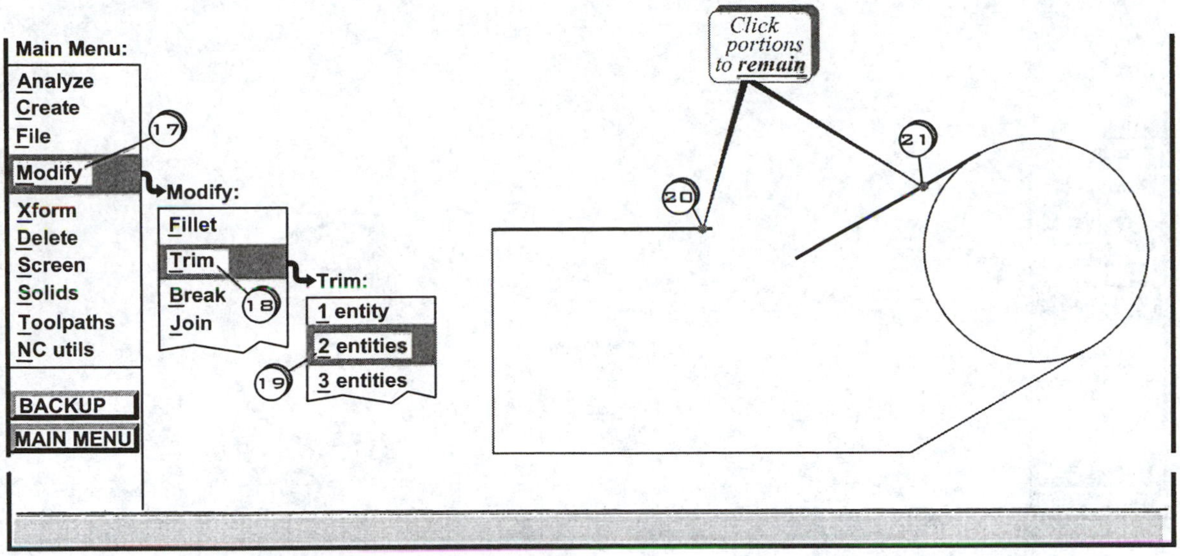

➤Click (17) **Modify**

➤Click (18) **Trim**

➤Click (19) **2 entities**

Trim(1):select the entity to trim

➤Click (20) the portion to **remain**

Trim(1):select the entity to trim to

➤Click (21) the entity to be trimed to

➤Press [Esc] to cancel the operation

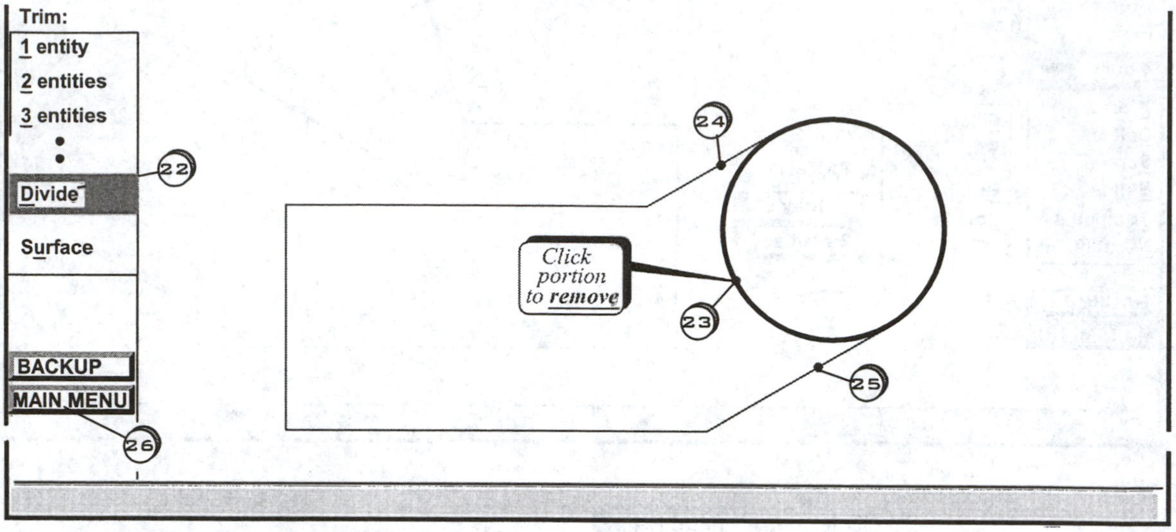

➤Click (22) **Divide**

Select curve to divide

➤Click (23) portion to **remove**

Select first dividing curve

➤Click (24) the first dividing entity

Select second dividing curve

➤Click (25) the second dividing entity

➤Press **Esc** to cancel the operation

➤Click (26) **MAIN MENU**

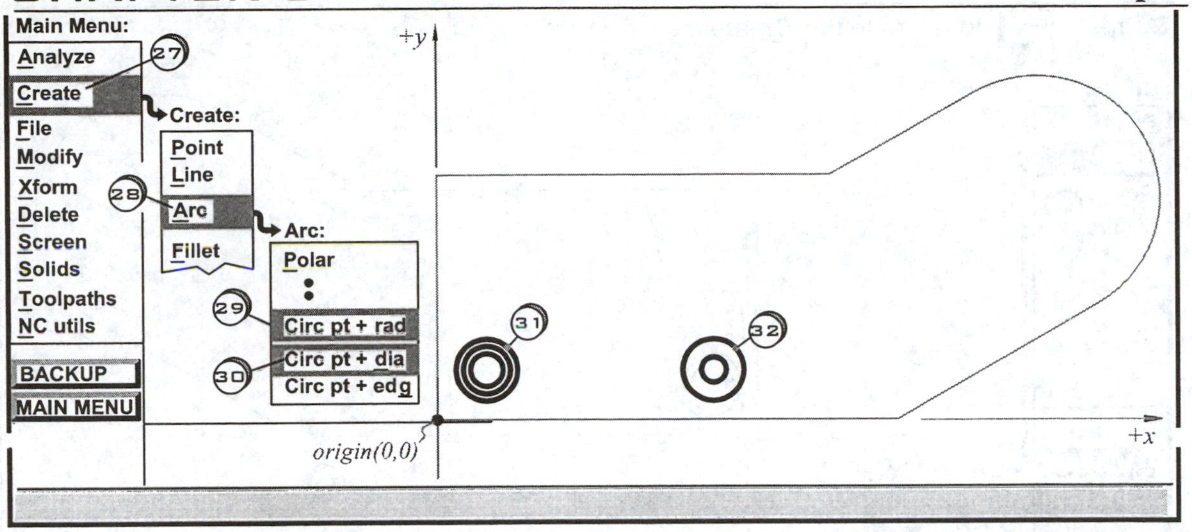

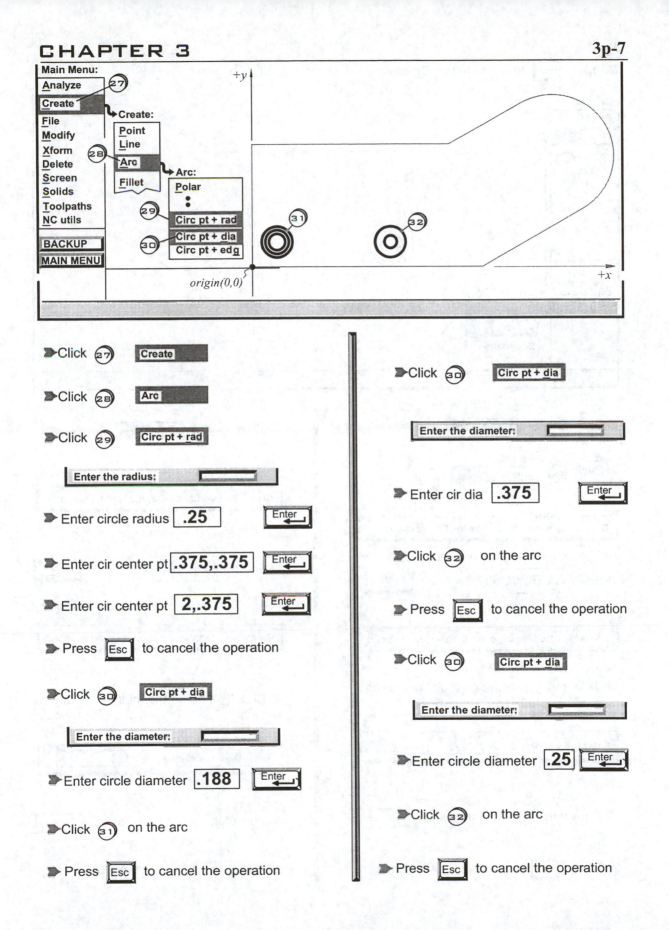

- Click (27) **Create**
- Click (28) **Arc**
- Click (29) **Circ pt + rad**

  Enter the radius:

- Enter circle radius **.25** Enter
- Enter cir center pt **.375,.375** Enter
- Enter cir center pt **2,.375** Enter
- Press Esc to cancel the operation
- Click (30) **Circ pt + dia**

  Enter the diameter:

- Enter circle diameter **.188** Enter
- Click (31) on the arc
- Press Esc to cancel the operation
- Click (30) **Circ pt + dia**

  Enter the diameter:

- Enter cir dia **.375** Enter
- Click (32) on the arc
- Press Esc to cancel the operation
- Click (30) **Circ pt + dia**

  Enter the diameter:

- Enter circle diameter **.25** Enter
- Click (32) on the arc
- Press Esc to cancel the operation

➢Press **Esc** to return to the **Create** menu

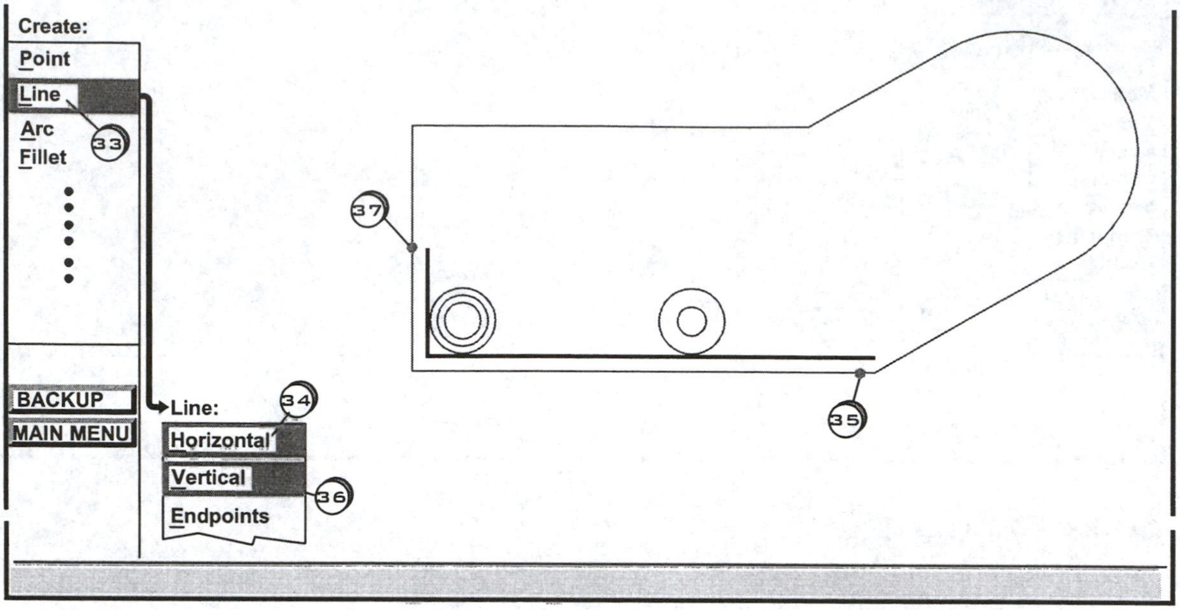

➢Click 33 **Line**

➢Click 34 **Horizontal**

Create line horizontal: enter the first endpoint

➢Enter **.125,.125** Enter

Create line horizontal: enter the second endpoint

➢Click 35 near the endpoint of the line

Enter the y coordinate **.125**

➢Accept the value by pressing Enter

➢Press Esc to cancel the operation

➢Click 36 **Vertical**

Create line vertical: enter the first endpoint

➢Enter **.125,.125** Enter

Create line vertical: enter the second endpoint

➢Click 37 near the midpoint of the line

Enter the x coordinate **.125**

➢Accept the value by pressing Enter

➢Press Esc to cancel the operation

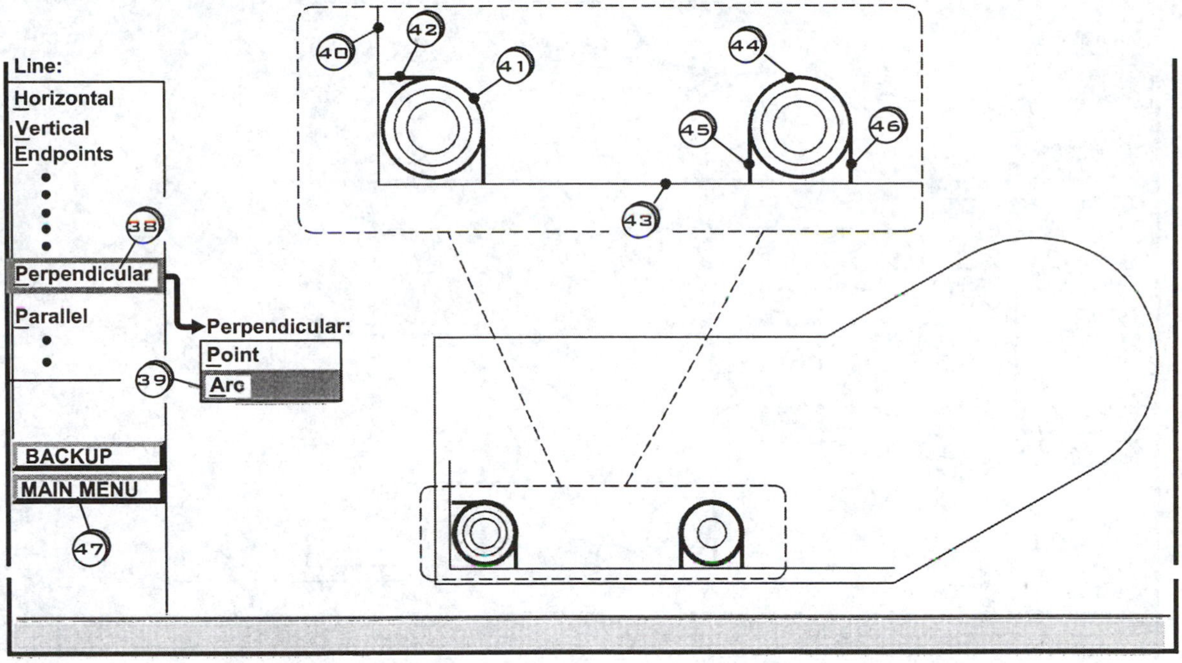

➤Click (38) **Perpendicular**

➤Click (39) **Arc**

Select a line

➤Click (40) the line entity

Select an arc to place a perp line tangent to

➤Click (41) the arc entity

Enter the length of the perp

➤Press **Enter**

Select which line to keep

➤Click (42) the portion to keep

Select a line

➤Click (43) the line entity

Select an arc to place a perp line tangent to

➤Click (44) the arc entity

Enter the length of the perp

➤Press **Enter**

Select which line to keep

➤Click (45) the portion to keep

Select a line

➤Click (43) the line entity

Select an arc to place a perp line tangent to

➤Click (44) the arc entity

Enter the length of the perp

➤Press **Enter**

Select which line to keep

➤Click (46) the portion to keep

➤Click (47) **MAIN MENU**

 3p-10

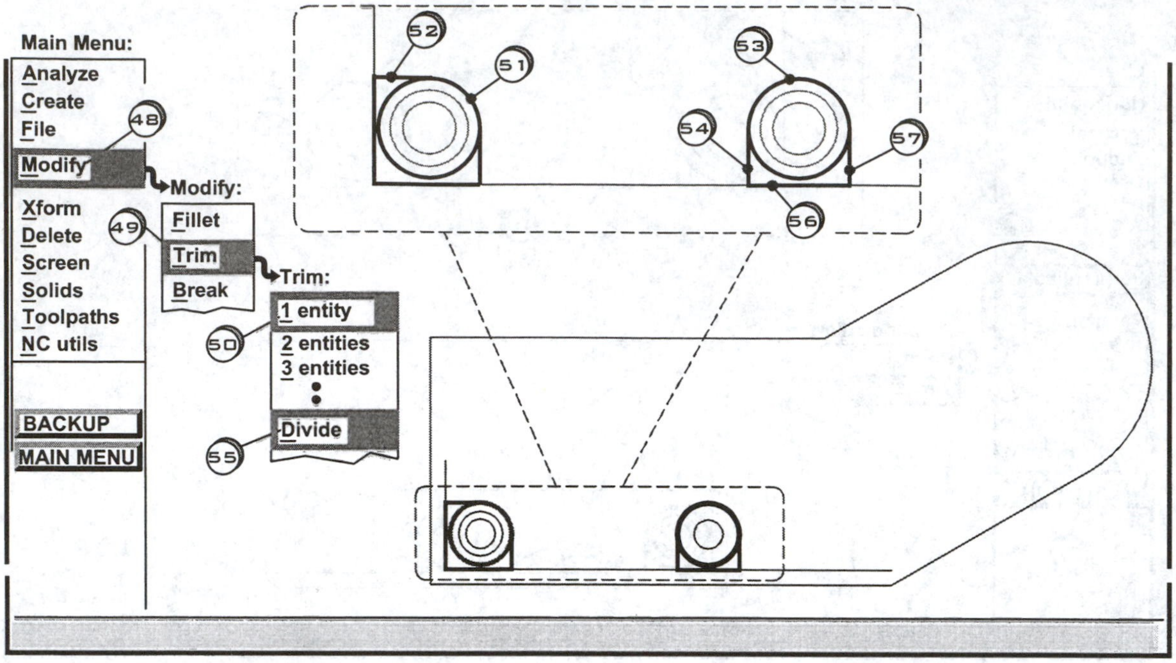

➤Click (48) **Modify**

➤Click (49) **Trim**

➤Click (50) **1 entitity**

Trim(1):select the entity to trim

➤Click (51) the portion to **remain**

Trim(1):select the entity to trim to

➤Click (52) the entity to be trimed to

Trim(1):select the entity to trim

➤Click (53) the portion to **remain**

Trim(1):select the entity to trim to

➤Click (54) the entity to be trimed to

➤ Press Esc to cancel the operation

➤Click (55) **Divide**

Select curve to divide

➤Click (56) portion to **remove**

Select first dividing curve

➤Click (54) the first dividing entity

Select second dividing curve

➤Click (57) the second dividing entity

➤ Press Esc to cancel the operation

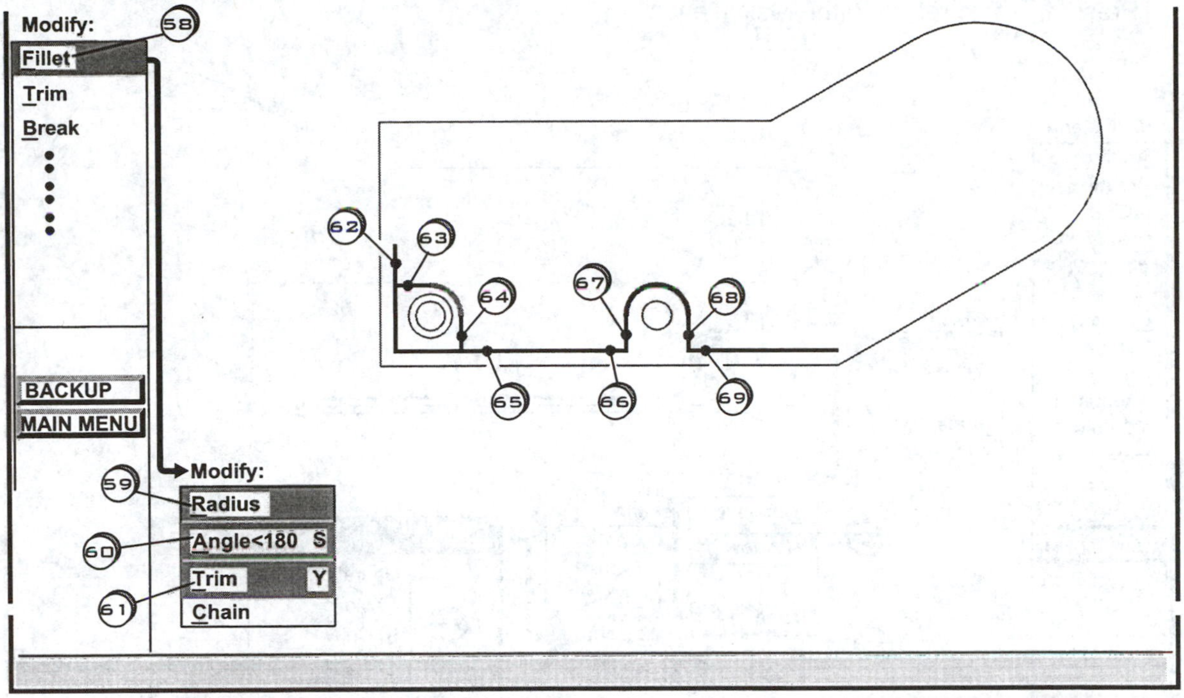

- Click (58) Fillet
- Click (59) Radius

Enter the fillet radius:

- Enter the fillet's radius .125 Enter
- Click (60) the S in Angle<180 S
- Click (61) the Y in Trim Y

Fillet radius = 0.12500
Fillet:select an entity

- Click (62) the line entity

Fillet: select another entity

- Click (63) the line entity

Fillet radius = 0.12500
Fillet:select an entity

- Click (64) the line entity

Fillet: select another entity

- Click (65) the line entity

Fillet radius = 0.12500
Fillet:select an entity

- Click (66) the line entity

Fillet: select another entity

- Click (67) the line entity

Fillet radius = 0.12500
Fillet:select an entity

- Click (68) the line entity

Fillet: select another entity

- Click (69) the line entity
- Press Esc to cancel the operation

➢ Press Esc to return to the Main menu

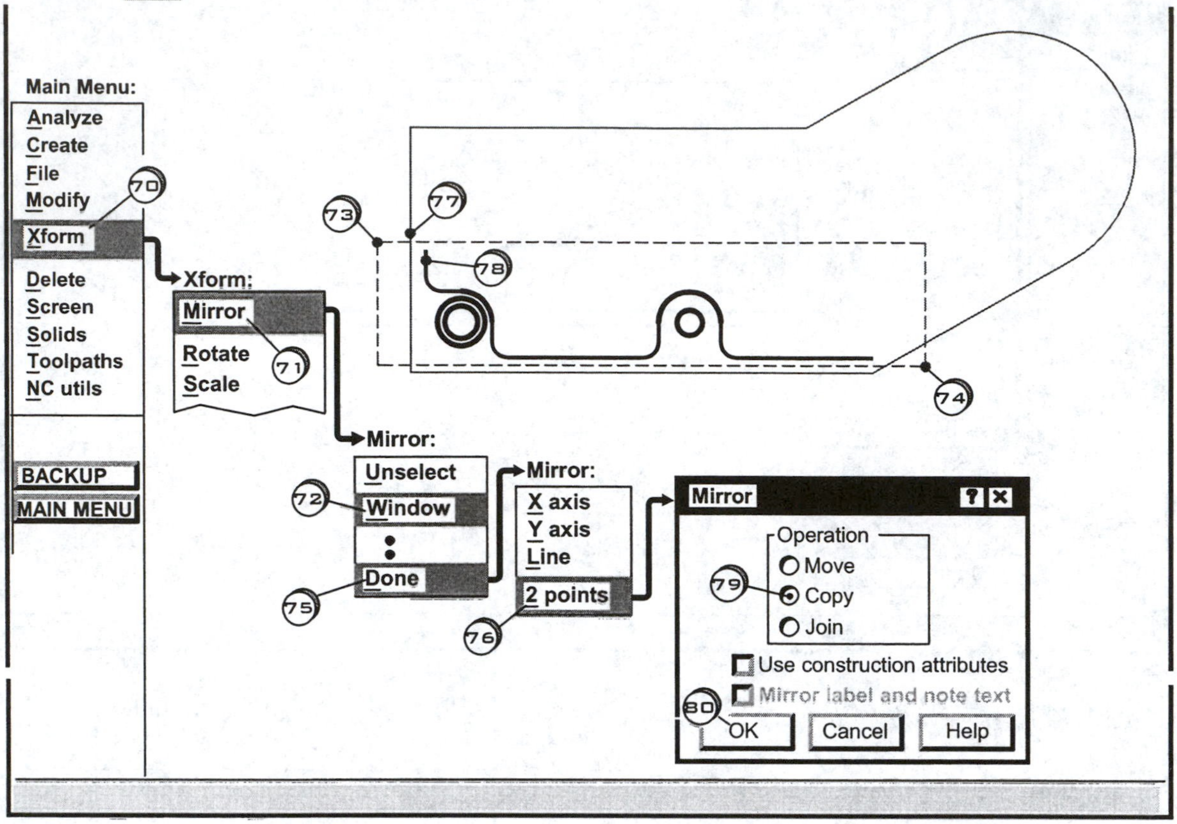

➢Click 70 **Xform**

➢Click 71 **Mirror**

➢Click 72 **Window**

➢Click 73 74 the window corners

➢Click 75 **Done**

➢Click 76 **2 points**

➢Click 77 78 the mirror line's endpoints

➢Click 79 the *copy dot* ⦿

➢Click 80 OK

➢ Press Esc to cancel the operation

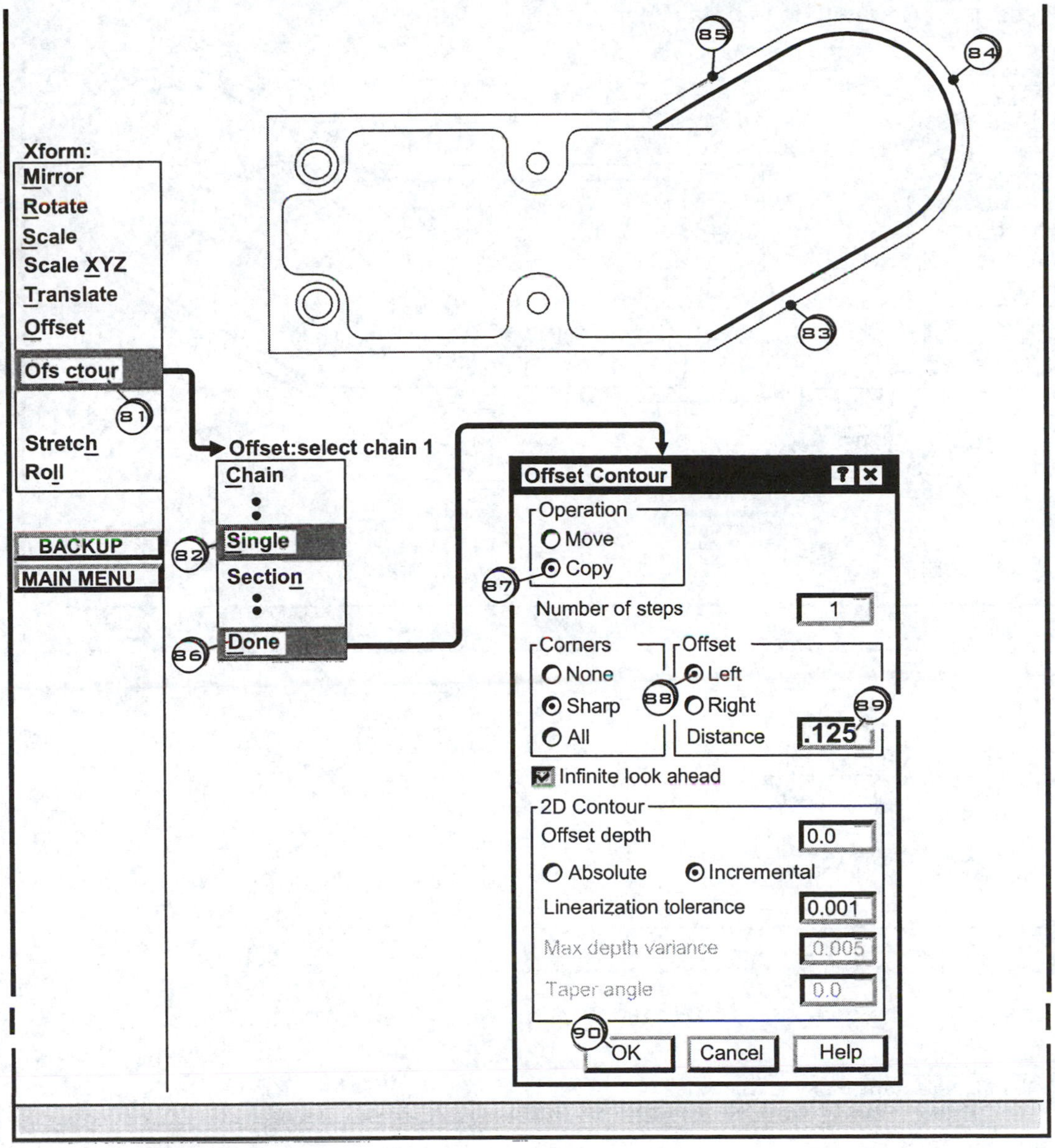

➢Click (81) **Ofs ctour**

➢Click (82) **Single**

➢Click (83) (84) (85)

➢Click (86) **Done**

➢Click (87) the *copy* dot ⊙

➢Click (88) the *left* dot ⊙

➢Click (89) and enter **.125**

➢Click (90) OK

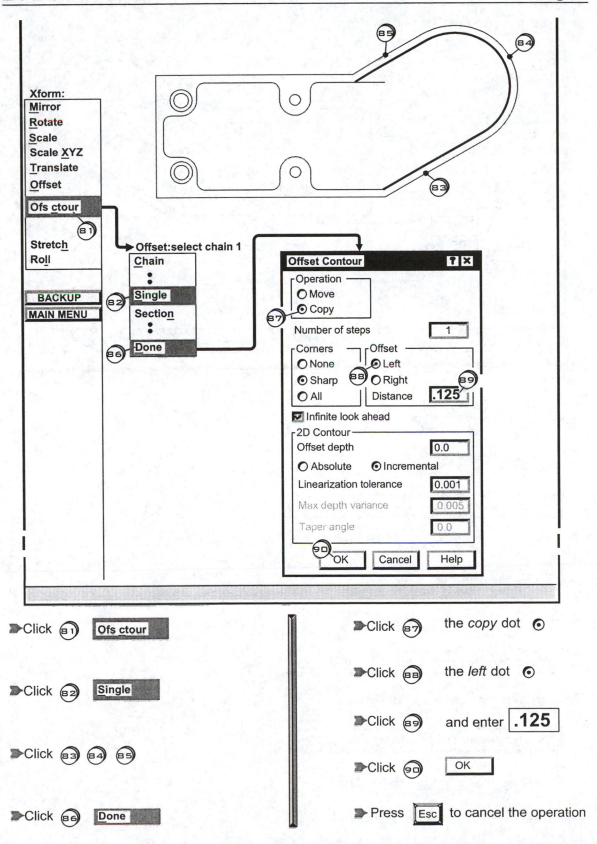

➢Press Esc to cancel the operation

Press Esc to return to the **Main Menu**

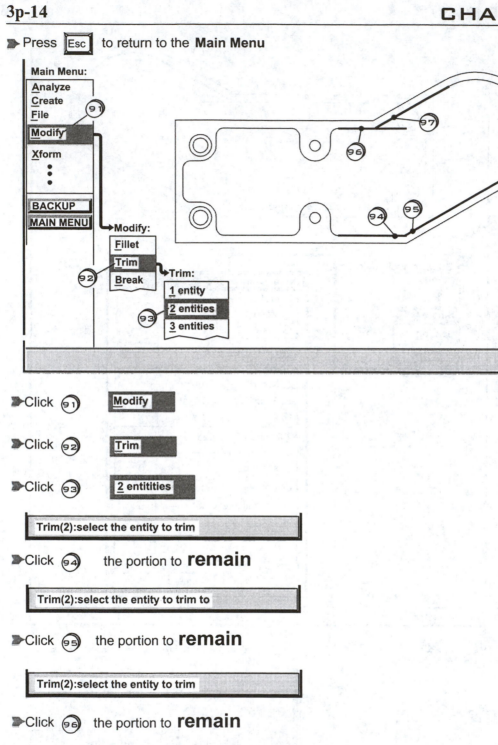

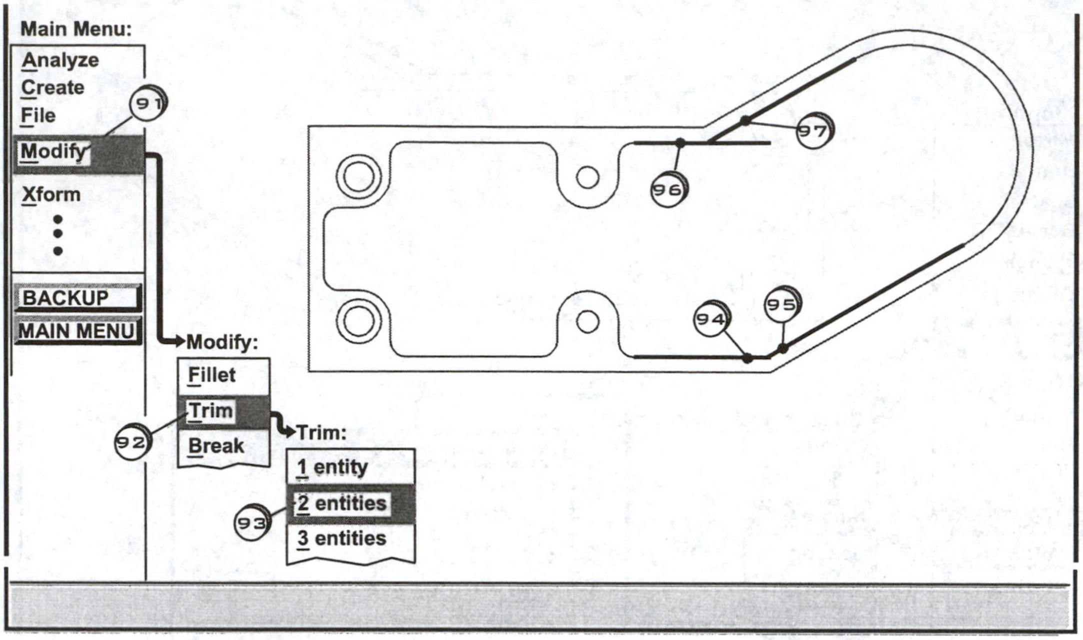

Click 91 Modify

Click 92 Trim

Click 93 2 entitities

Trim(2):select the entity to trim

Click 94 the portion to **remain**

Trim(2):select the entity to trim to

Click 95 the portion to **remain**

Trim(2):select the entity to trim

Click 96 the portion to **remain**

Trim(2):select the entity to trim to

Click 97 the portion to **remain**

Press Esc to cancel the operation

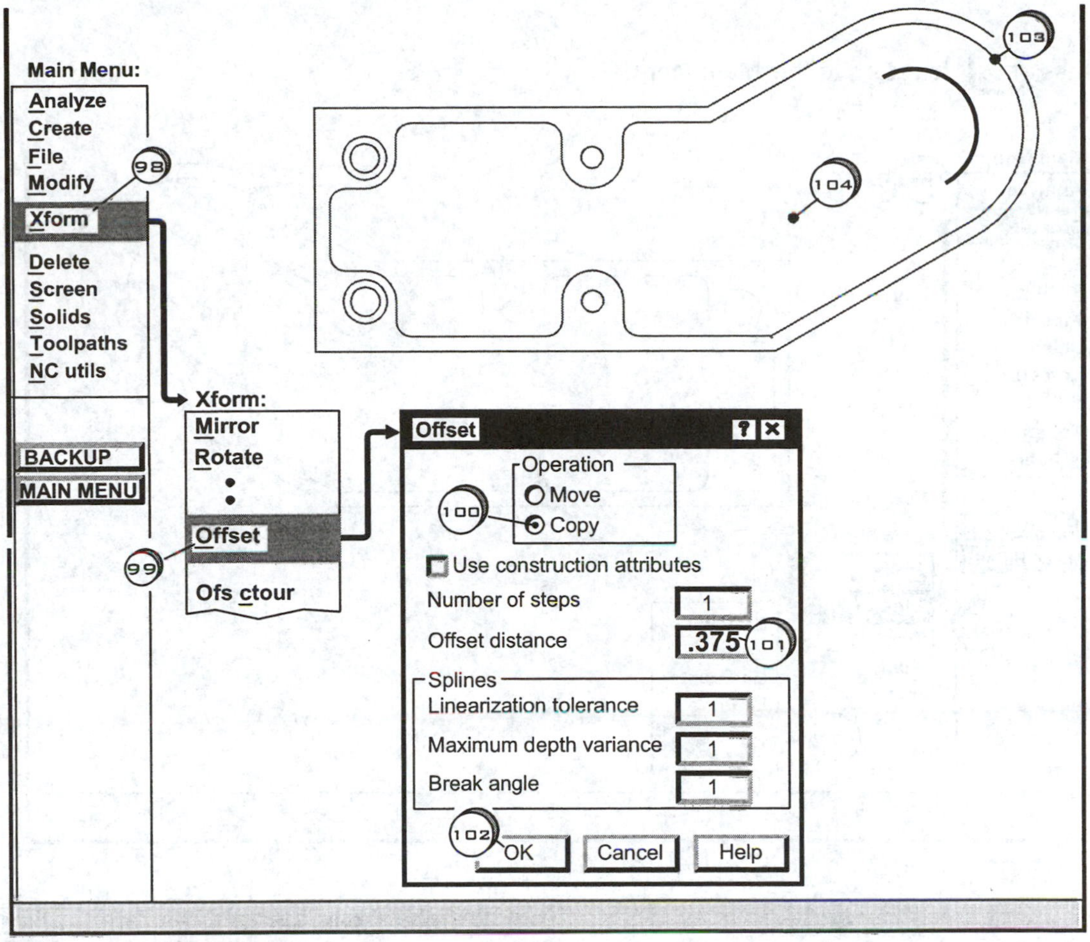

➤Click (98) Xform

➤Click (99) Offset

➤Click (100) the copy button

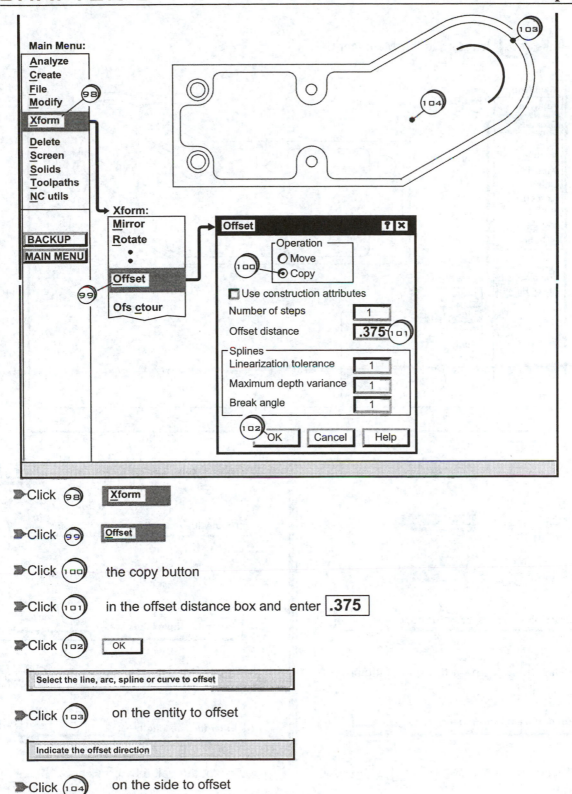

➤Click (101) in the offset distance box and enter .375

➤Click (102) OK

Select the line, arc, spline or curve to offset

➤Click (103) on the entity to offset

Indicate the offset direction

➤Click (104) on the side to offset

➤Press Esc to cancel the operation

Press [Esc] to return to the **Main Menu**

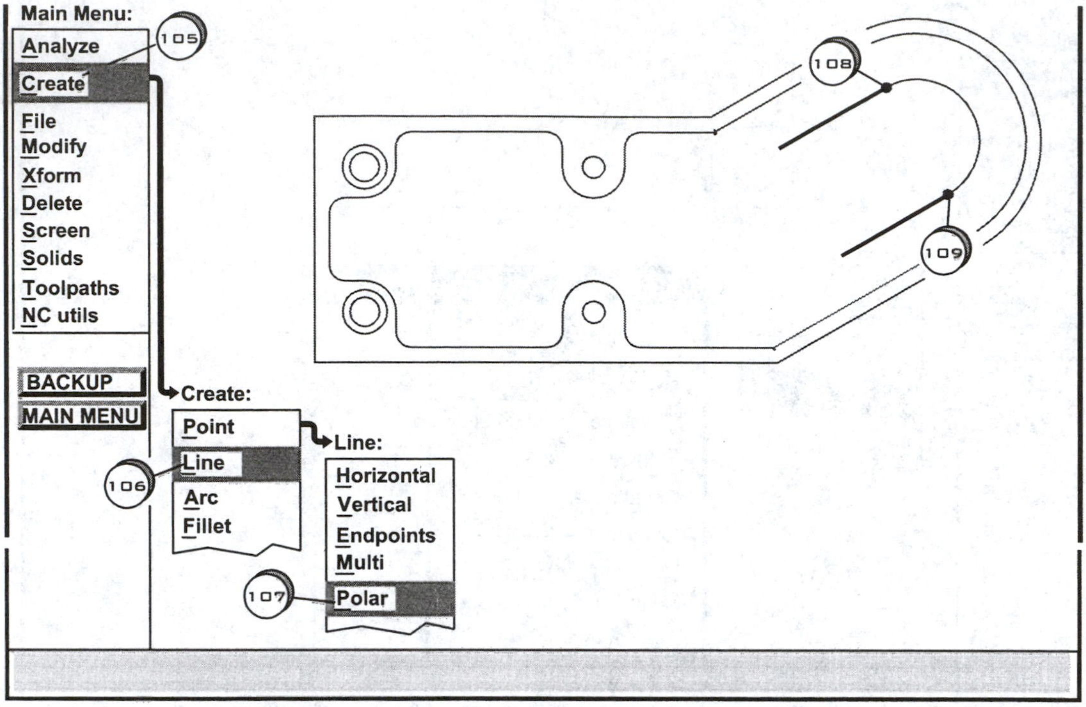

Click (105) **Create**

Click (106) **Line**

Click (107) **Polar**

Create line polar: Specify an endpoint

Click (108) near the *end* of the arc

Enter the angle in degrees

Enter the ang in deg **-150** [Enter]

Enter the line length .5

Enter the line length **1** [Enter]

Create line polar: Specify an endpoint

Click (109) near the *end* of the arc

Enter the angle in degrees -150

Accept the value by pressing [Enter]

Enter the line length .5

Accept the value by pressing [Enter]

Press [Esc] to cancel the operation

Press [Esc] to return to the **Main Menu**

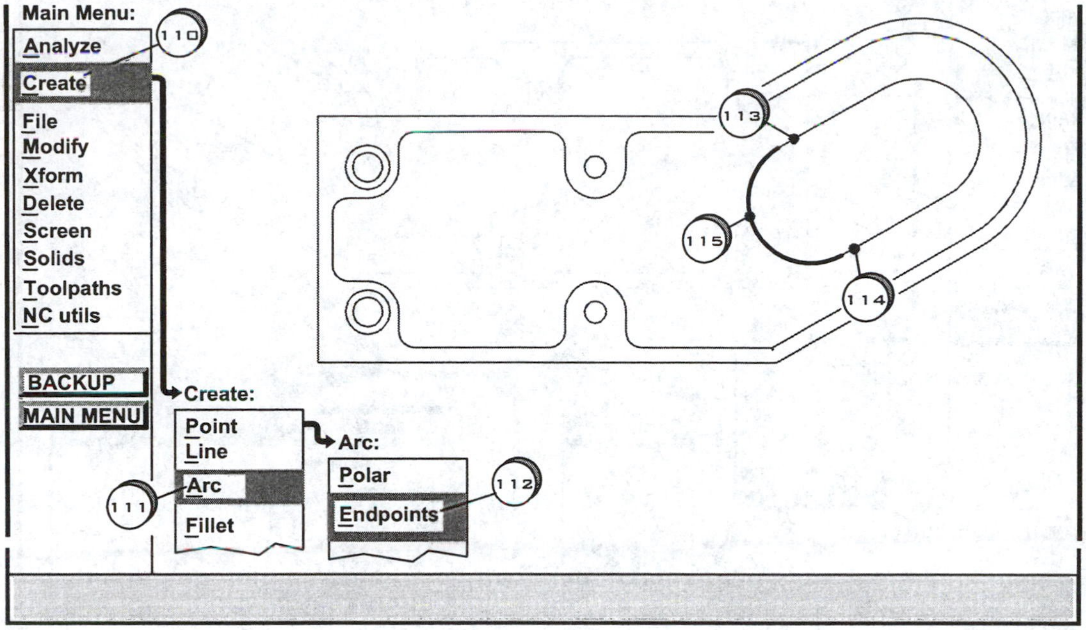

Click (110) **Create**

Click (111) **Arc**

Click (112) **Endpoints**

Arc, endpoints: Enter the first point

Click (113) near the end of the line

Arc, endpoints: Enter the second point

Click (114) near the end of the line

Enter the radius:

Enter the arc's radius **.5** [Enter]

Arc, endpoints: Select an arc

Click (115) near the portion to ***keep***

Press [Esc] to cancel the operation

➤ Press Esc to return to the **Main Menu**

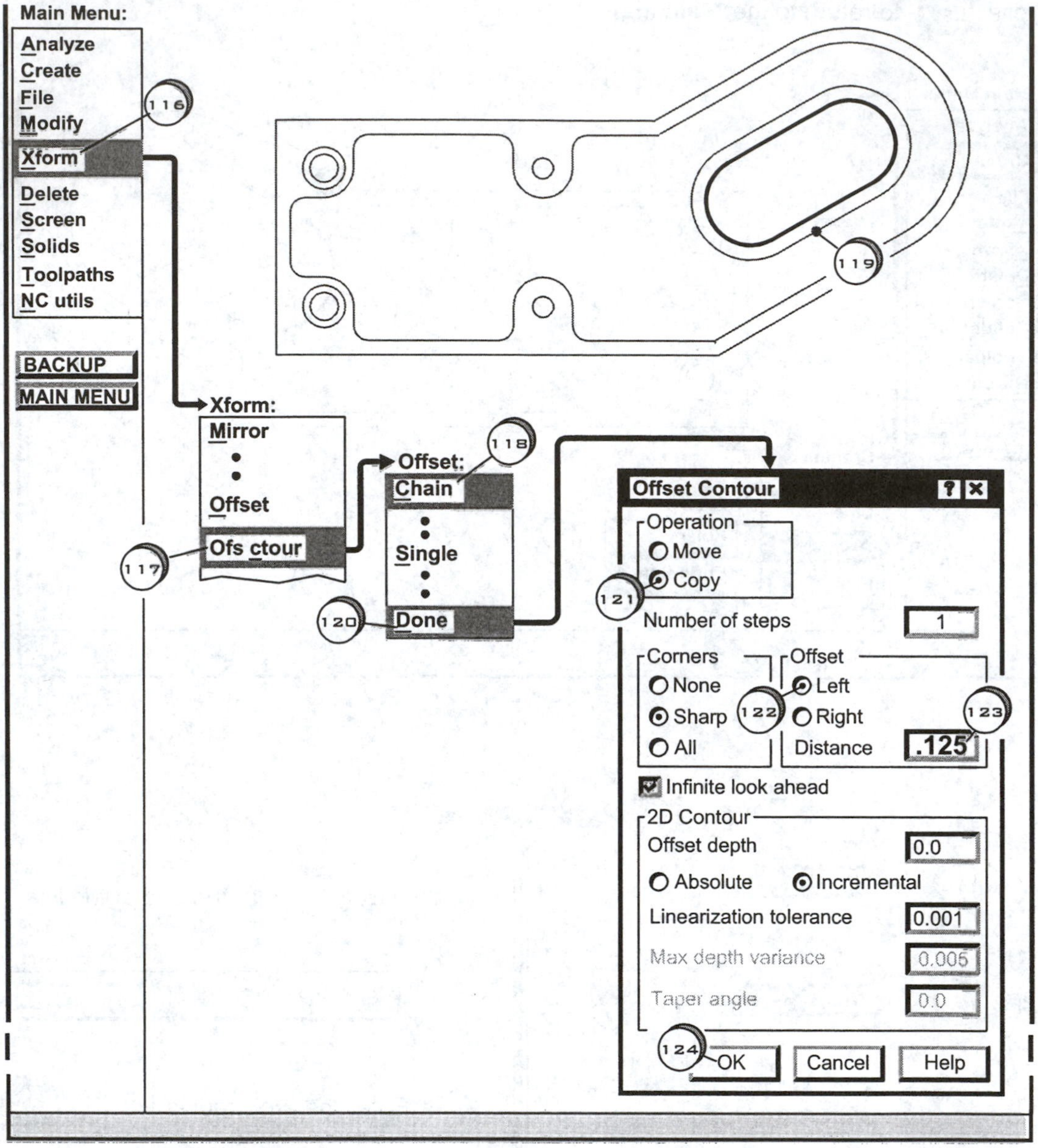

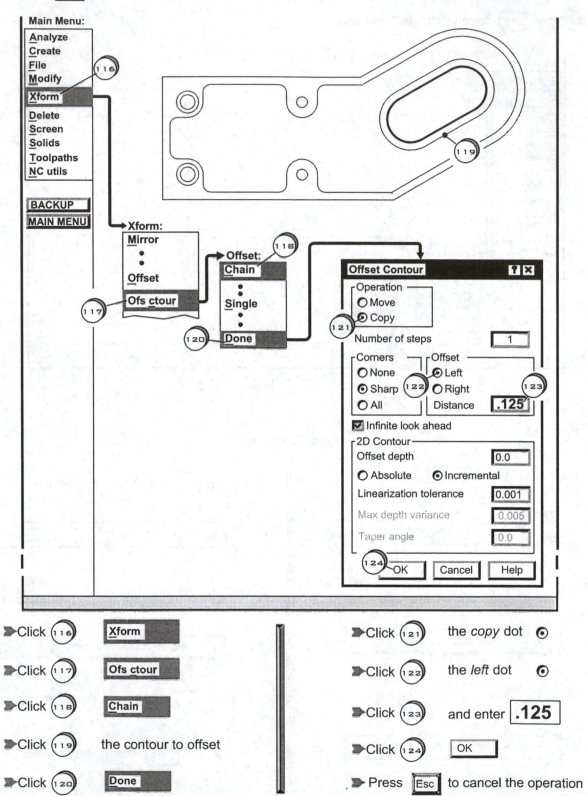

➤ Click (116) **Xform**

➤ Click (117) **Ofs ctour**

➤ Click (118) **Chain**

➤ Click (119) the contour to offset

➤ Click (120) **Done**

➤ Click (121) the *copy* dot ⦿

➤ Click (122) the *left* dot ⦿

➤ Click (123) and enter **.125**

➤ Click (124) OK

➤ Press Esc to cancel the operation

➤ Press Esc to return to the **Main Menu**

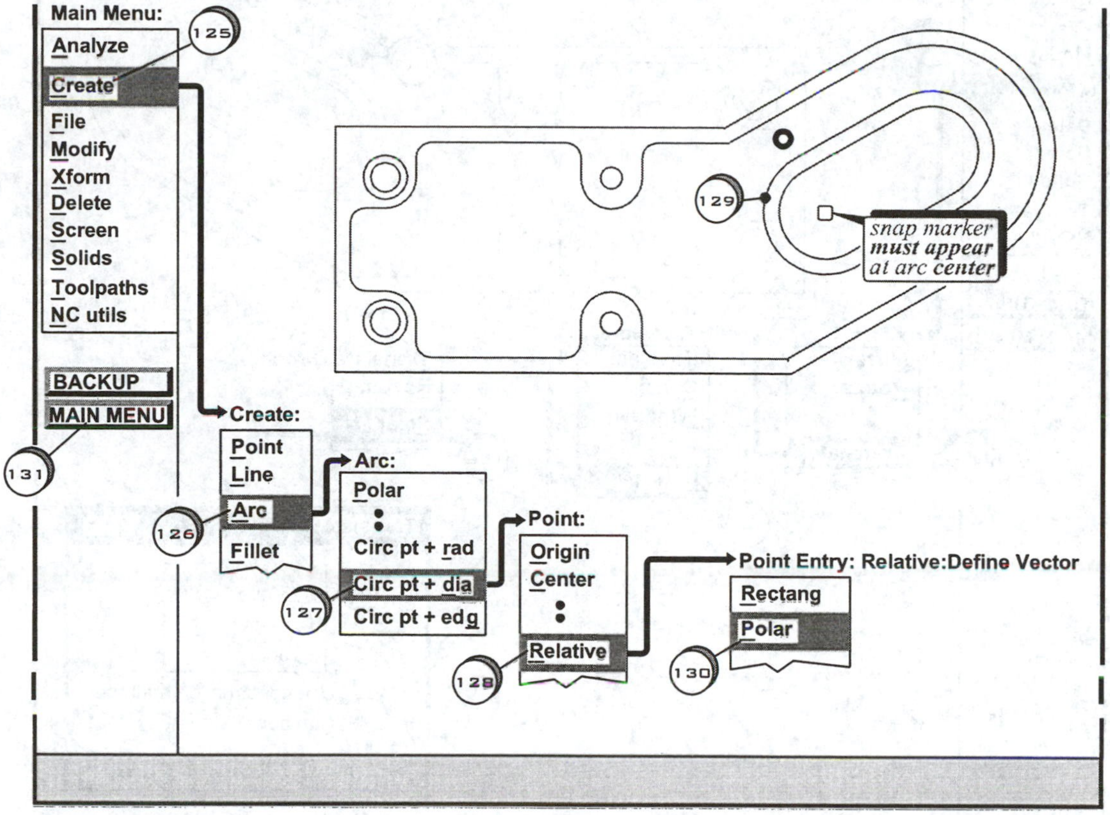

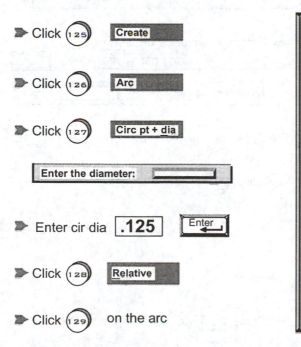

➤ Click 125 Create

➤ Click 126 Arc

➤ Click 127 Circ pt + dia

Enter the diameter:

➤ Enter cir dia .125 Enter

➤ Click 128 Relative

➤ Click 129 on the arc

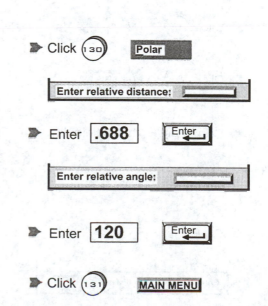

➤ Click 130 Polar

Enter relative distance:

➤ Enter .688 Enter

Enter relative angle:

➤ Enter 120 Enter

➤ Click 131 MAIN MENU

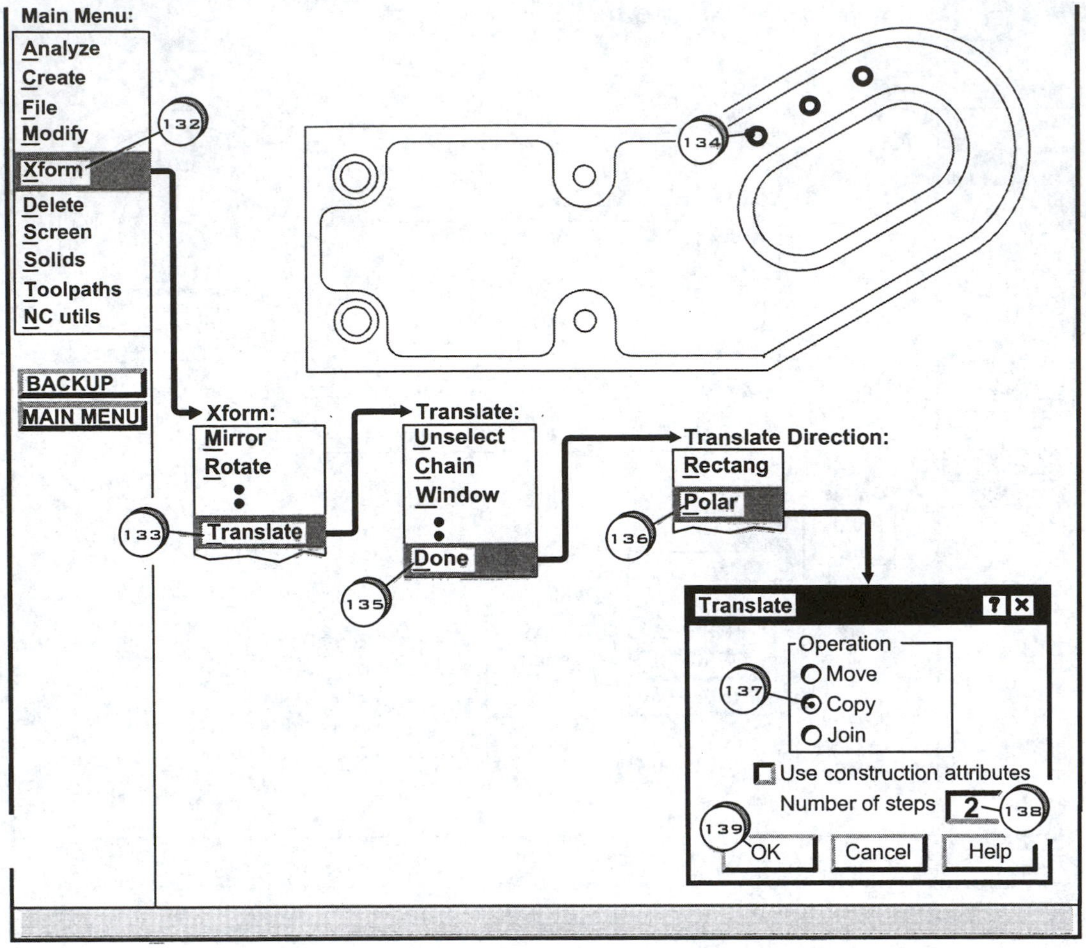

- Click (132) Xform
- Click (133) Translate
- Click (134) the circle entity
- Click (135) Done
- Click (136) Polar

Enter the translation distance

- Enter the *distance* between each copy .5 Enter

Enter the translation angle

- Enter the *angle* for the copies 30 Enter
- Click (137) the a *copy dot* ⊙
- Click (138) and enter the num of copies 2
- Click (139) OK
- Press Esc to cancel the operation

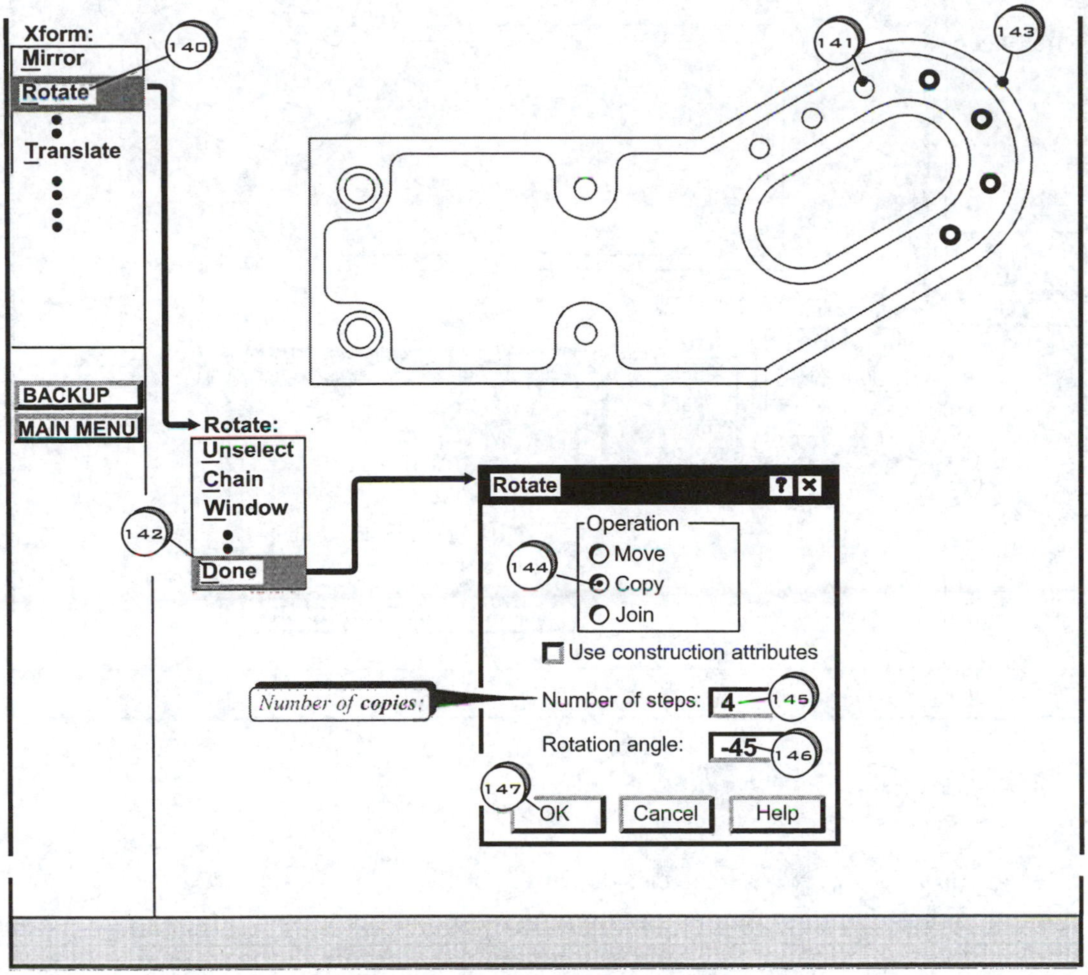

- Click (140) Rotate
- Click (141) the circle to copy
- Click (142) Done

Enter the point to rotate about

- Click (143) near the arc entity
- Click (144) the *copy dot* ⊙
- Click (145); enter the num of copies **4**
- Click (146); enter the angle btwn copies **-45**
- Click (147) OK
- Press Esc to cancel the operation

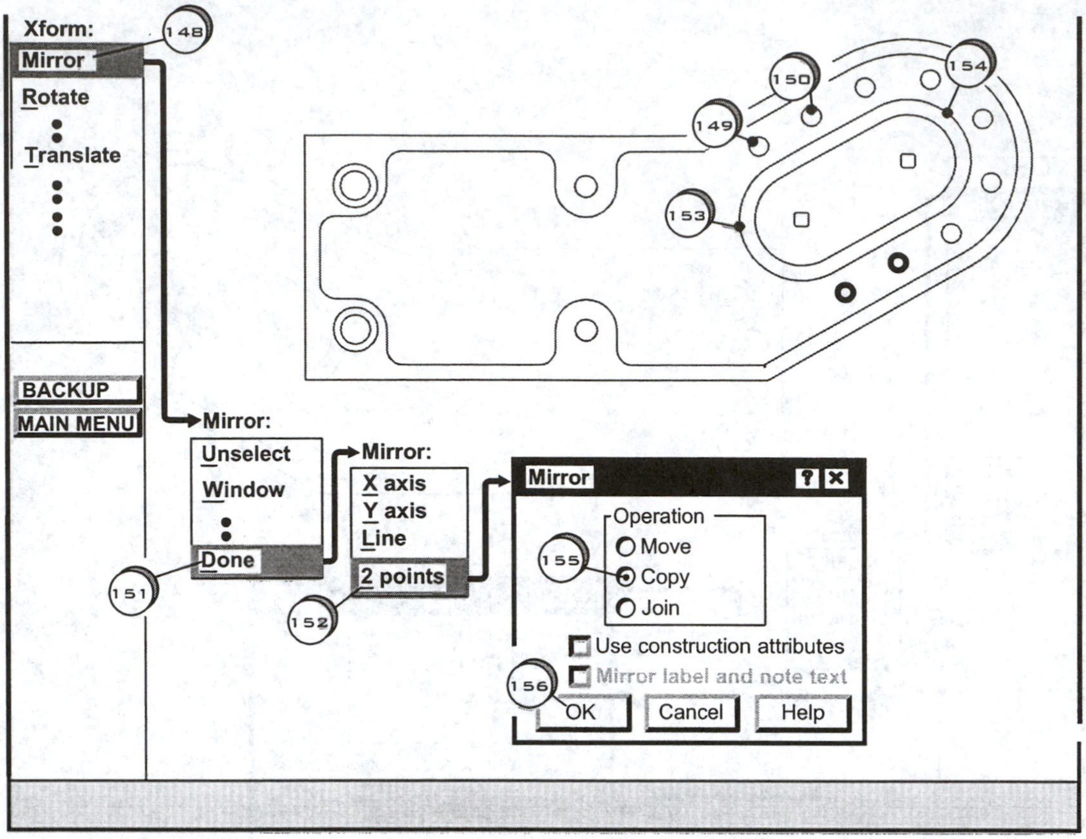

- Click (148) **Mirror**
- Click (149) (150) the circles
- Click (151) **Done**
- Click (152) **2 points**
- Click (153) (154) on the arcs
- Click (155) the *copy dot* ⊙
- Click (156) OK
- Press Esc to cancel the operation

➤ Press **Esc** to return to the **Main** menu

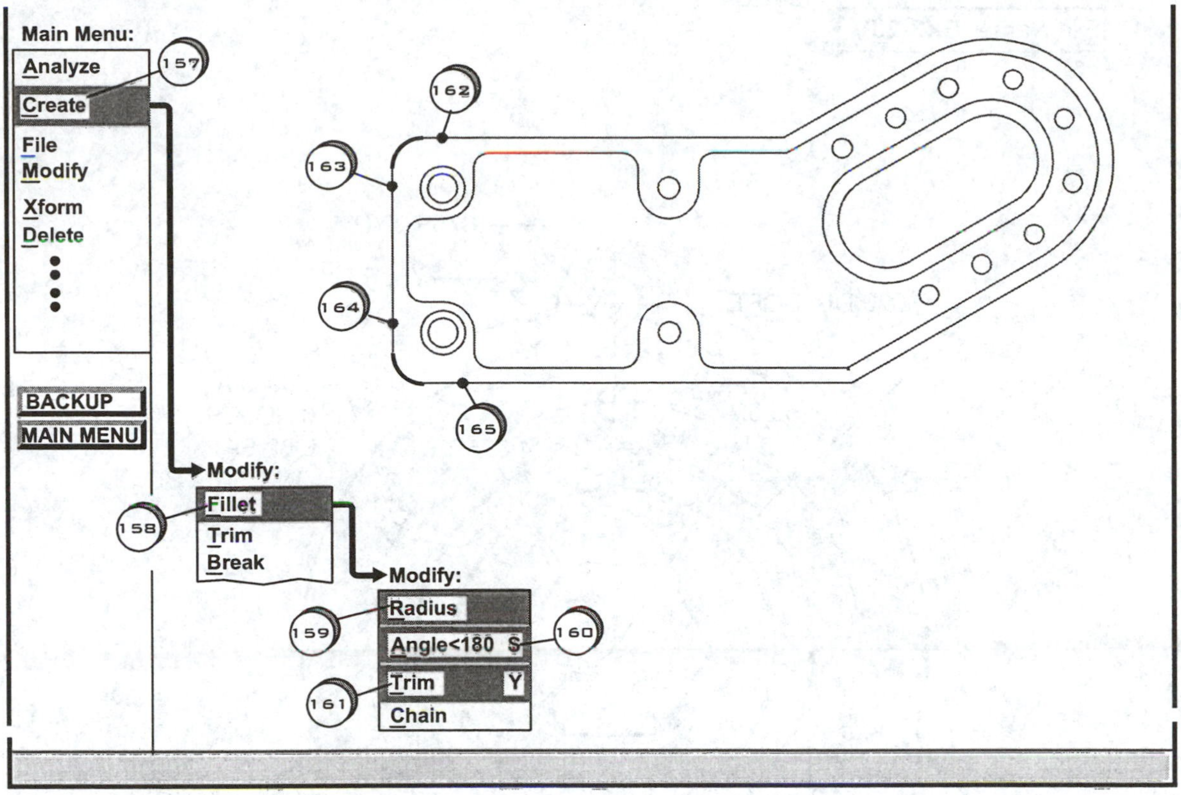

➤ Click (157) **Create**

➤ Click (158) **Fillet**

➤ Click (159) **Radius**

Enter the fillet radius:

➤ Enter the fillet's radius **.25** **Enter**

➤ Click (160) the **S** in **Angle<180 S**

➤ Click (161) the **Y** in **Trim Y**

Fillet radius = 0.2500
Fillet:select an entity

➤ Click (162) near the end of the line entity

Fillet: select another entity

➤ Click (163) near the end of the line entity

Fillet radius = 0.2500
Fillet:select an entity

➤ Click (164) near the end of the line entity

Fillet: select another entity

➤ Click (165) near the end of the line entity

➤ Press **Esc** to cancel the operation

3-2) Create the part shown in Figure 3-p2 using *Mastercam's* CAD package

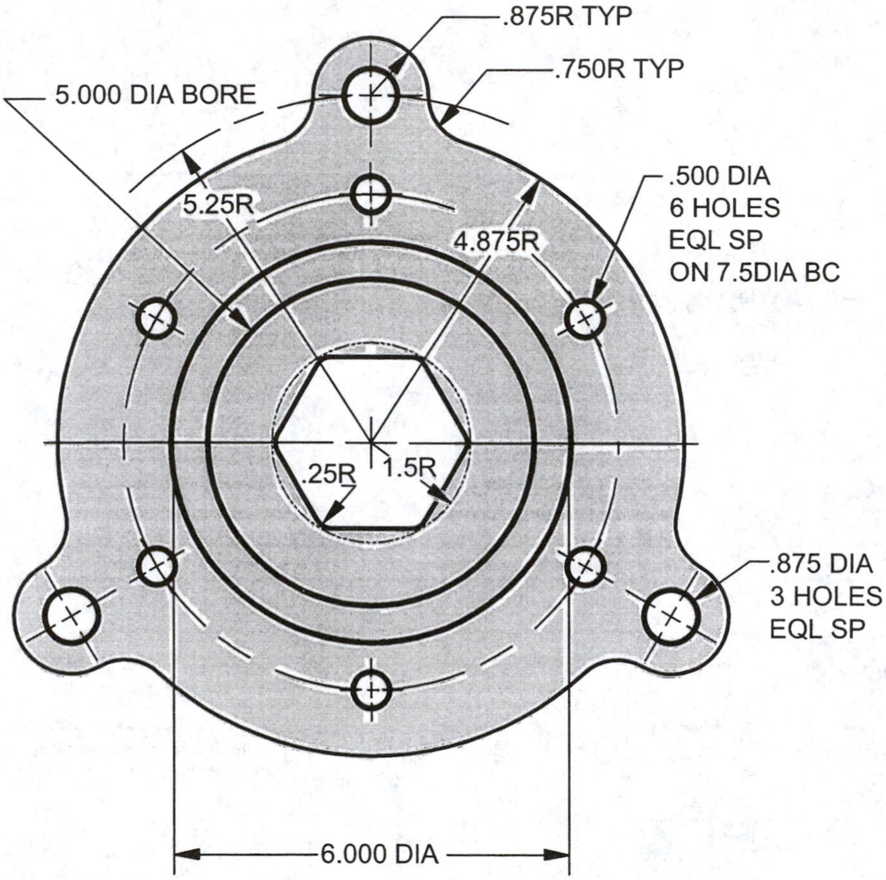

Figure 3-p2

a) Use the *Arc*, *Polar* command with Radius set to 4.875, initial angle set to 30 and final angle set to 150

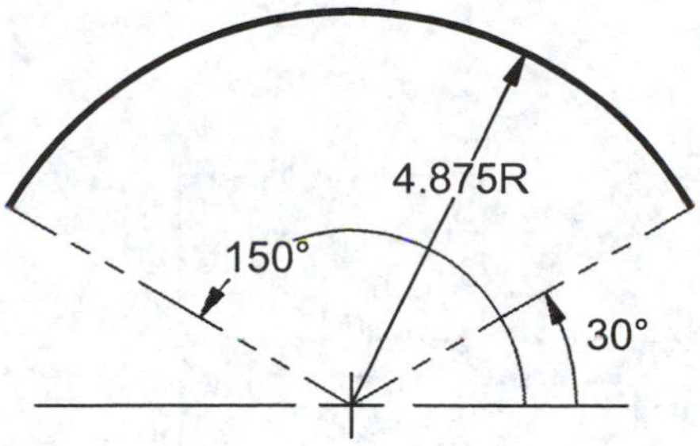

b) Add the .875R circle and *trim*

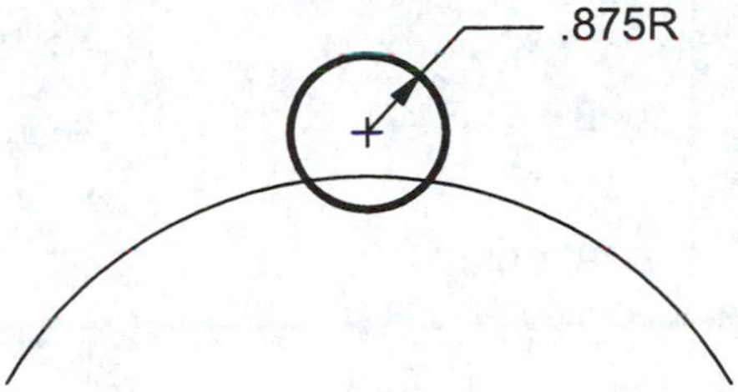

c) Add the .75R *fillets* and the .875DIA circle

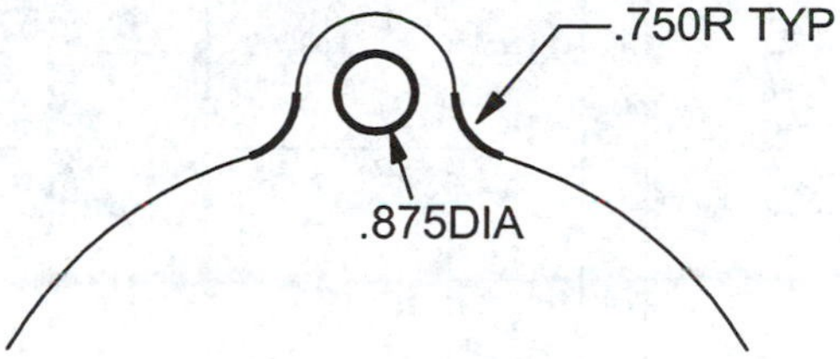

d) Use the *Xform, Rotate* command to make *two* copies. Add the .5 DIA, 5 DIA and 6DIA circles

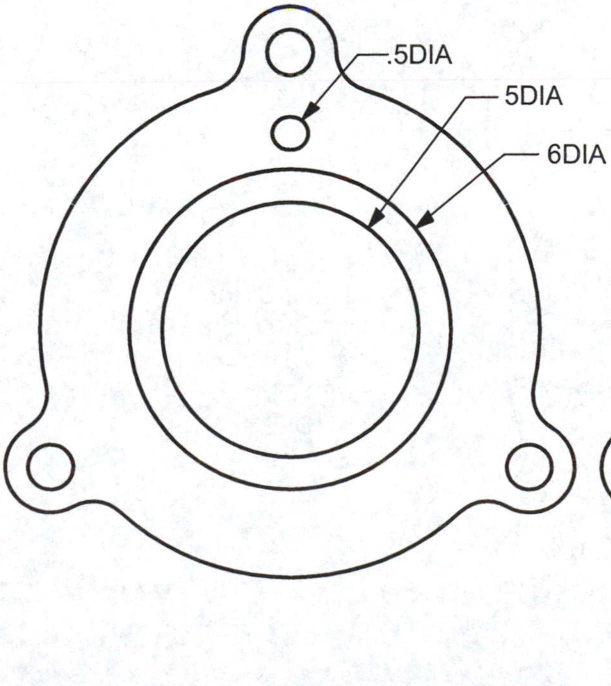

e) Use the *Xform, Rotate* command to make *five* copies of the .5Dia circle. . Use the *Polygon* command to create the 1.5R inscribed hexagon

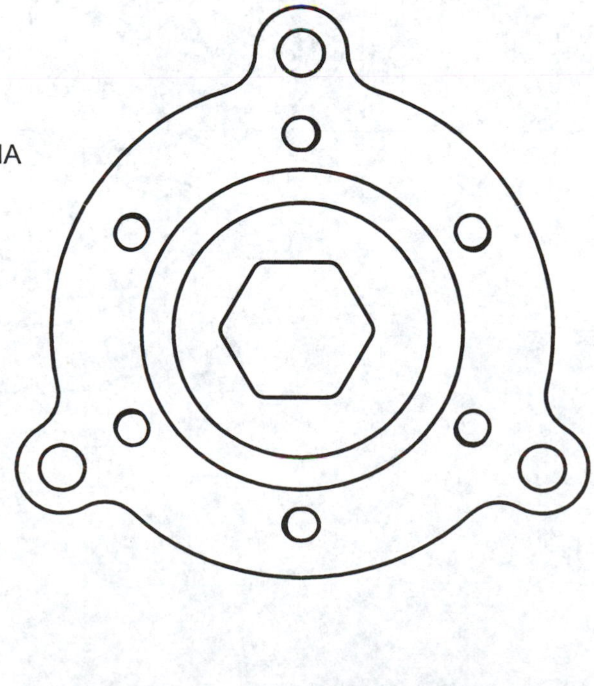

3-3) Create the part shown in Figure 3-p3 using *Mastercam's* CAD package

File Name: **EX3-3JV** ← YOUR INITIALS

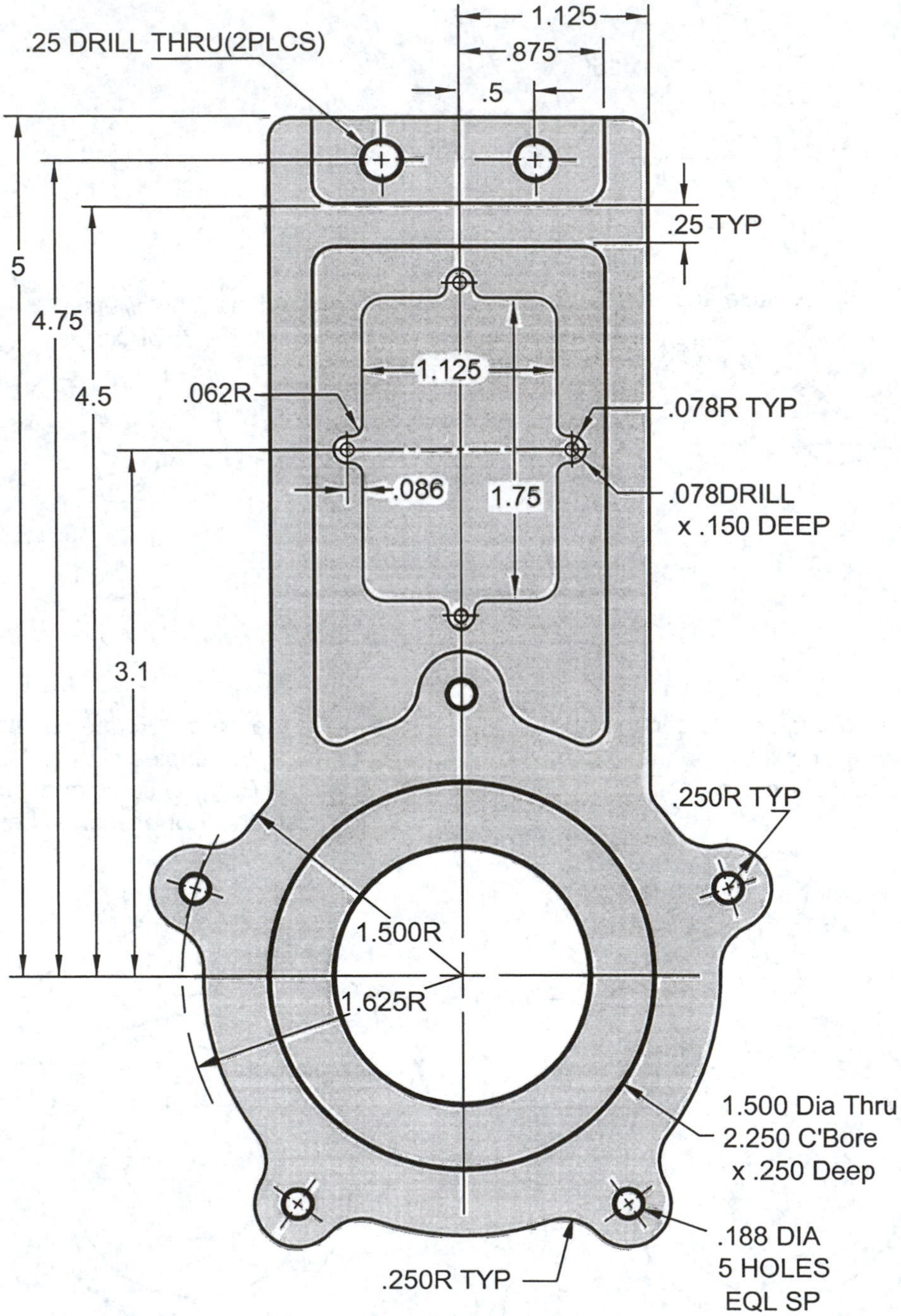

Figure 3-p3

3-4) Create the part shown in Figure 3-p4 using *Mastercam's* CAD package

File Name: **EX3-4JV** ← YOUR INITIALS

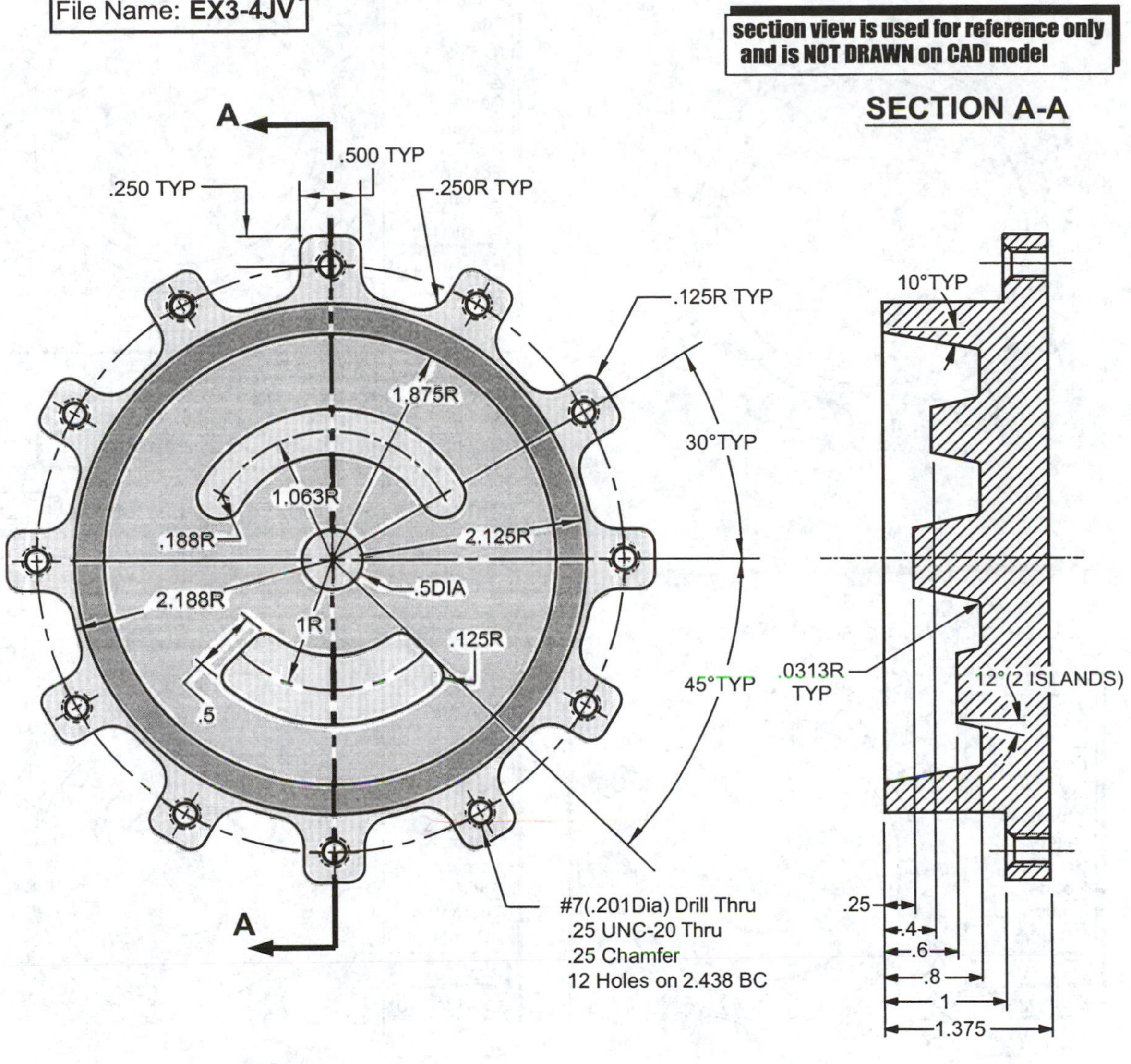

Figure 3-p4

a) Use the *Arc*, *Polar* command with Radius set to 2.188, initial angle set to 75 and final angle set to 105

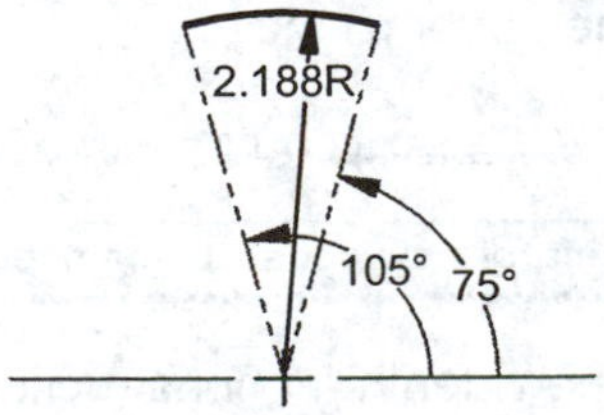

b) Complete the flange template as shown below

c) Use the *Xform, Rotate* command to make *11* copies. Add the 1.875R, and 2.125R circles.

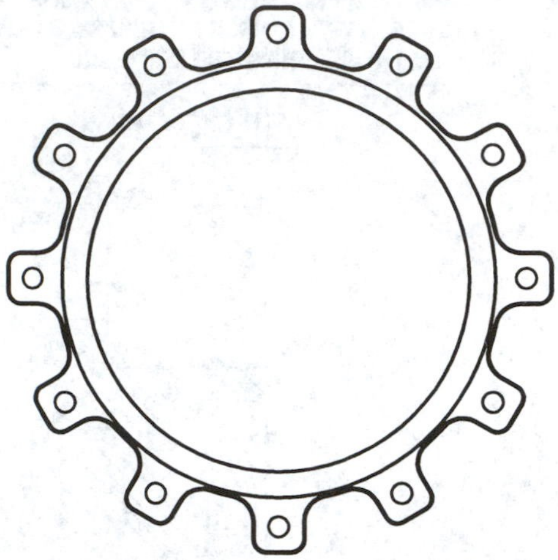

d) Create the *circle* island at a depth of -.25.

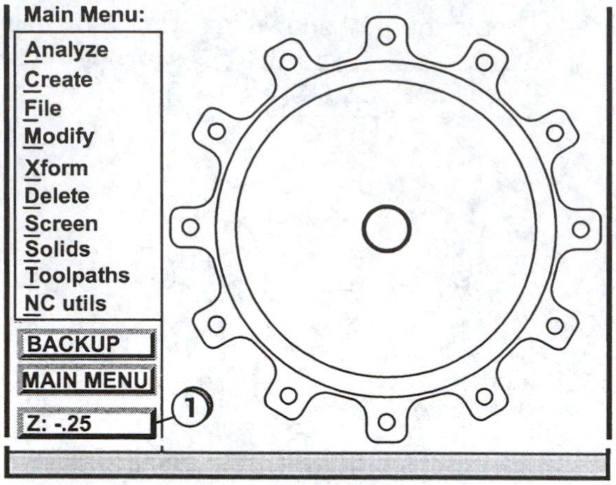

➤ Click ① Z: button

Select point for new construction depth

➤ Enter the new depth for constructions **-.25**

Create the .5Dia circle island

e) Create the *lower* island at a depth of -.4.

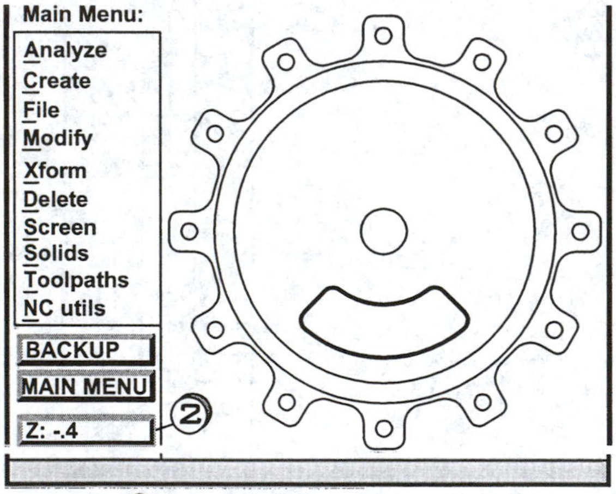

➤ Click ② Z: button

Select point for new construction depth

➤ Enter the new depth for constructions **-.4**

Create the lower island

f) Create the *upper* island at a depth of -.6.

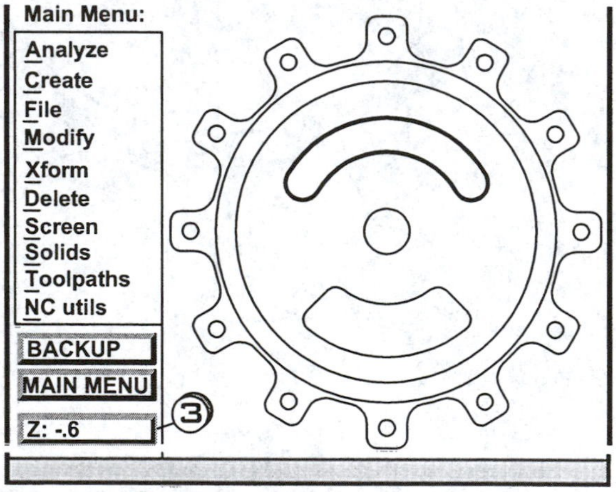

➤ Click ③ Z: button

Select point for new construction depth

➤ Enter the new depth for constructions **-.6**

Create the upper island

➤ Click ③ Z: button

Select point for new construction depth

➤ Enter the new depth for constructions **0**

3-5) Create the part shown in Figure 3-p5 using *Mastercam's* CAD package

File Name: **EX3-5JV** ← YOUR INITIALS

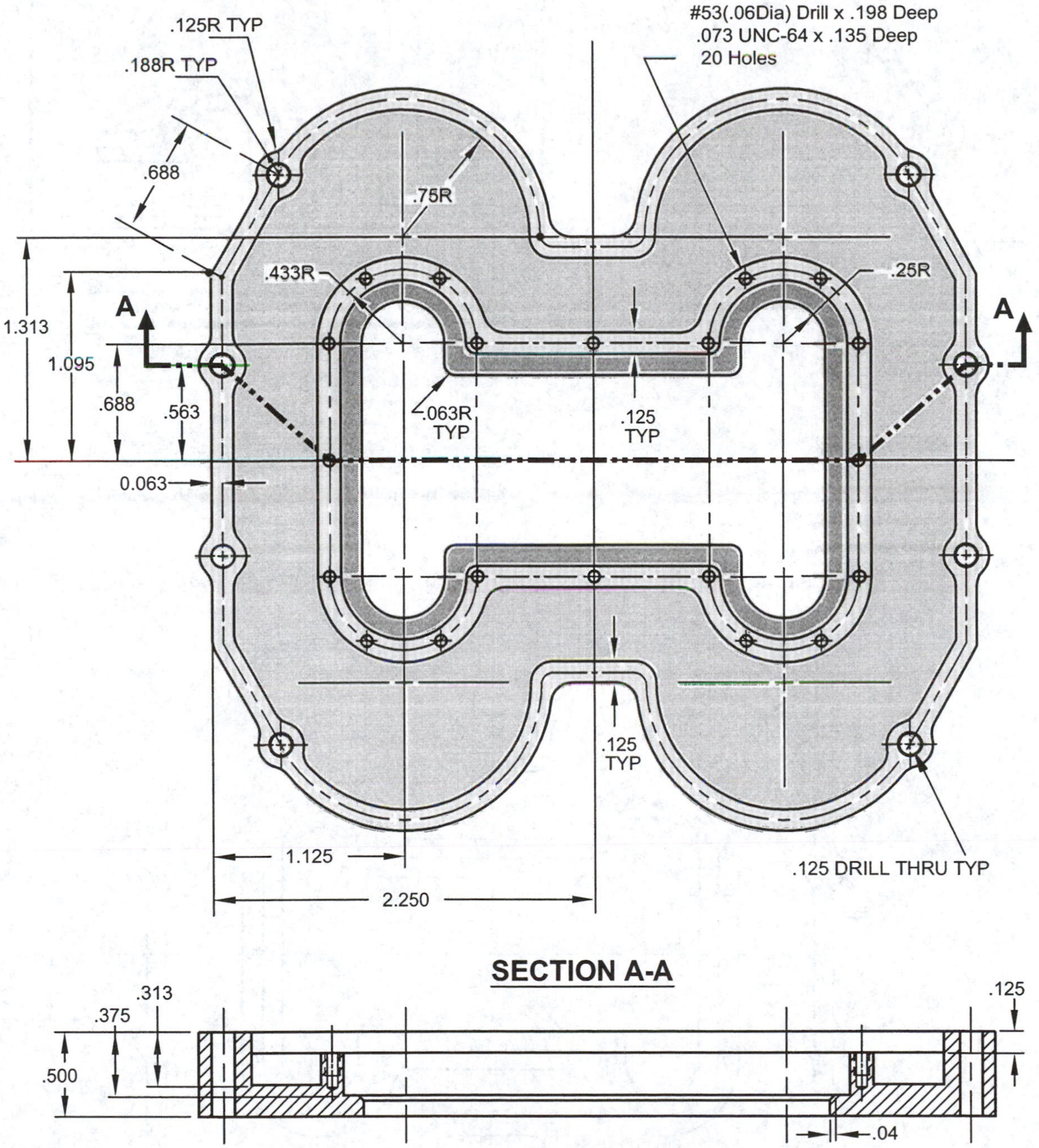

Figure 3-p5

section view is used for reference only and is NOT DRAWN on CAD model

a) Create the upper left quarter of the outside part geometry.

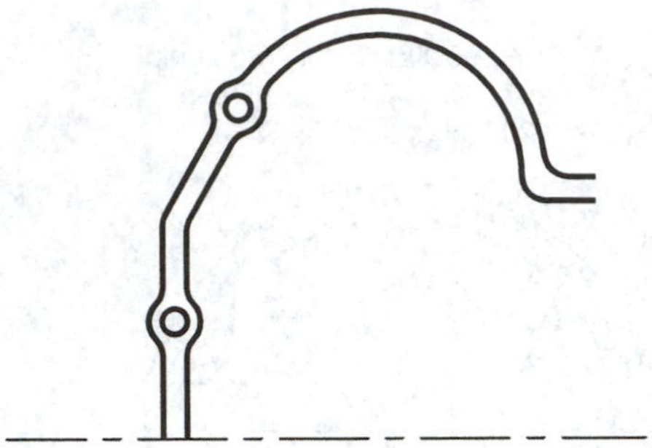

b) Create the inside geometry at a depth of -.125

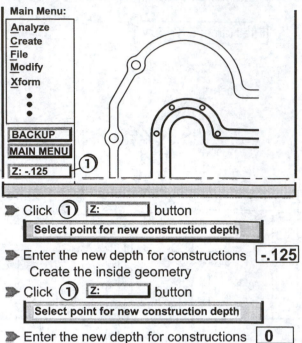

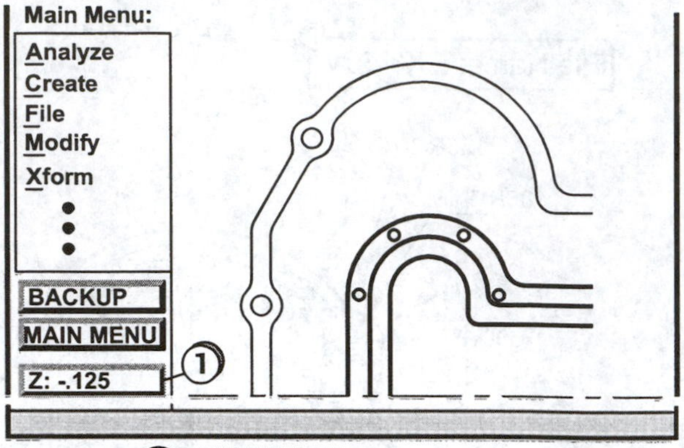

- Click (1) Z: button

Select point for new construction depth

- Enter the new depth for constructions **-.125**

  Create the inside geometry

- Click (1) Z: button

Select point for new construction depth

- Enter the new depth for constructions **0**

c) Use the *Xform, Mirror* command to complete the part geometry

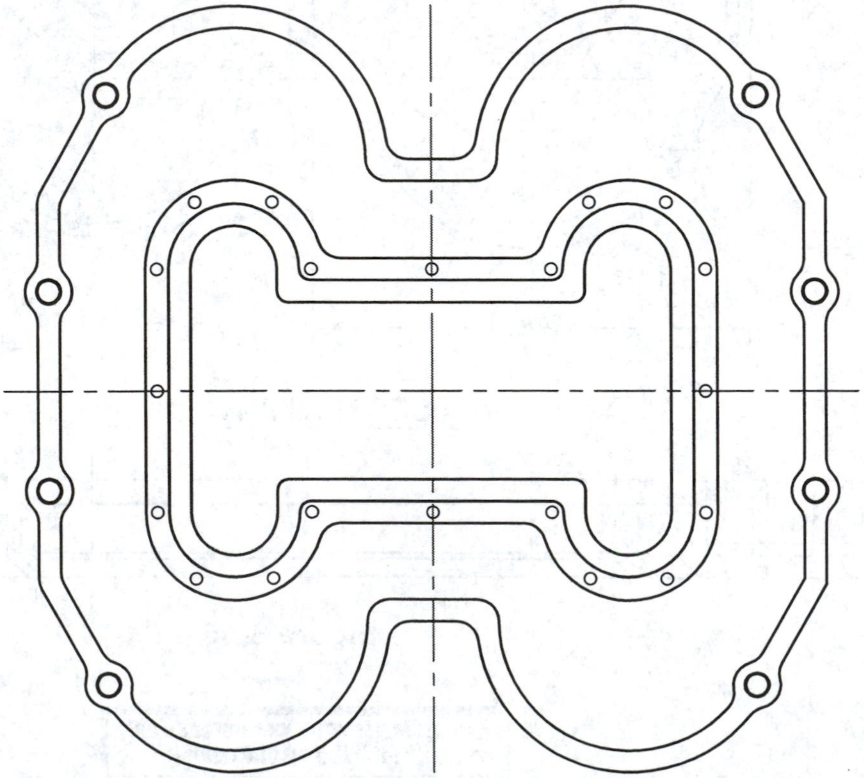

3-6) Create the part shown in Figure 3-p6 using *Mastercam's* CAD package

File Name: **EX3-6JV** ← YOUR INITIALS

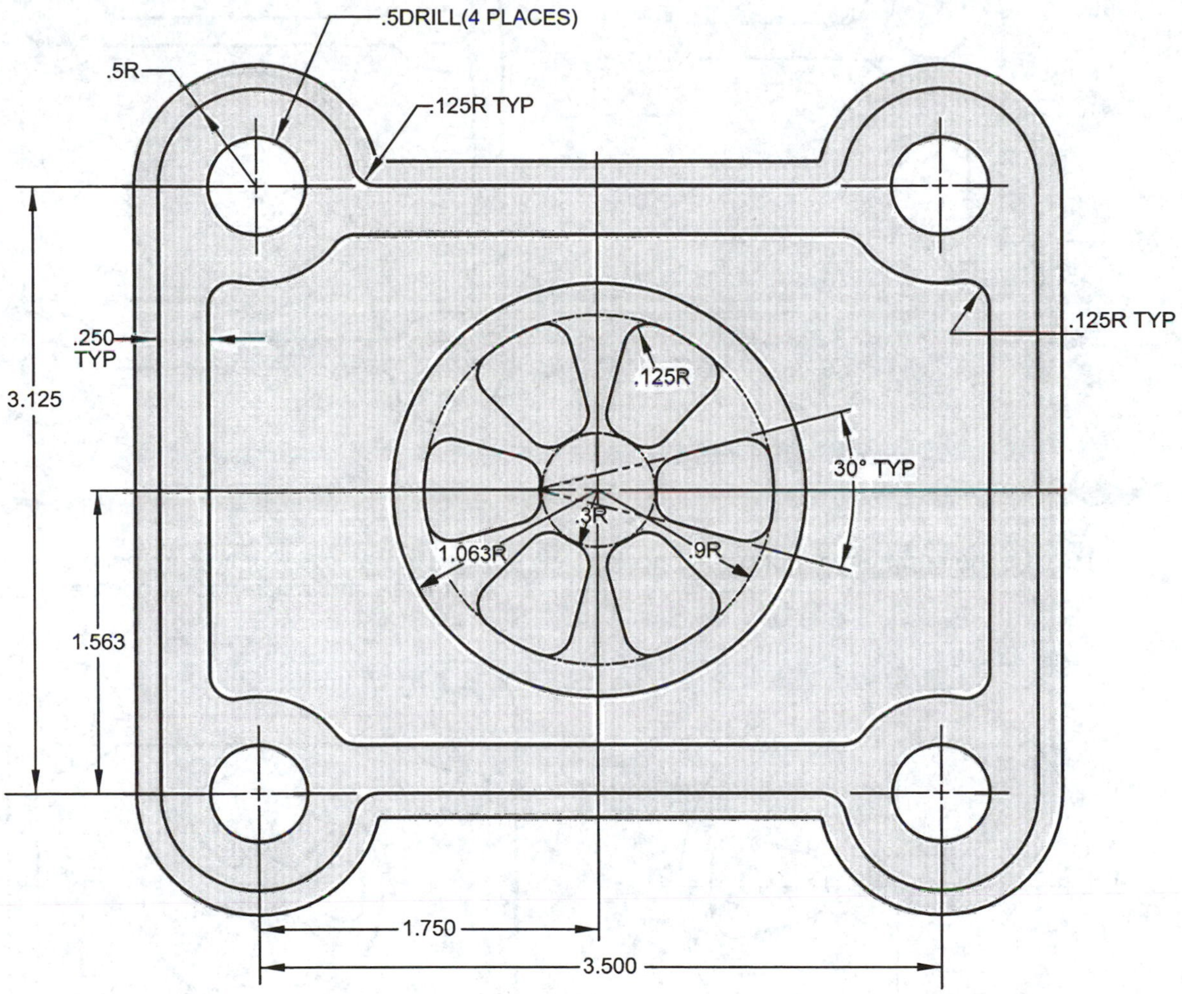

Figure 3-p6

a) Create the upper left quarter of the outside part geometry.

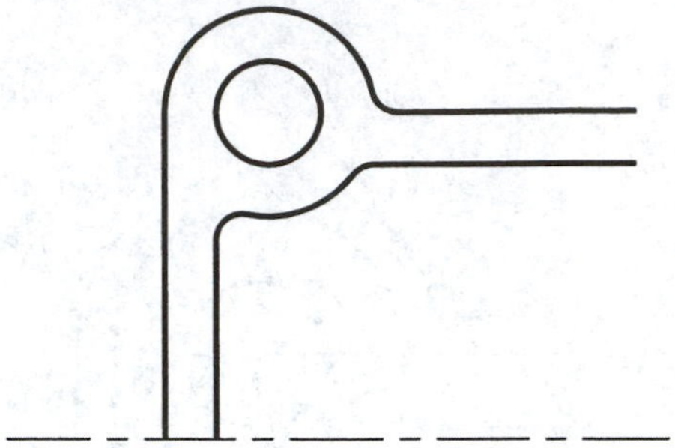

b) Use the Xform, *Mirror* command to complete the outside part geometry. add the 1.063R inner circle

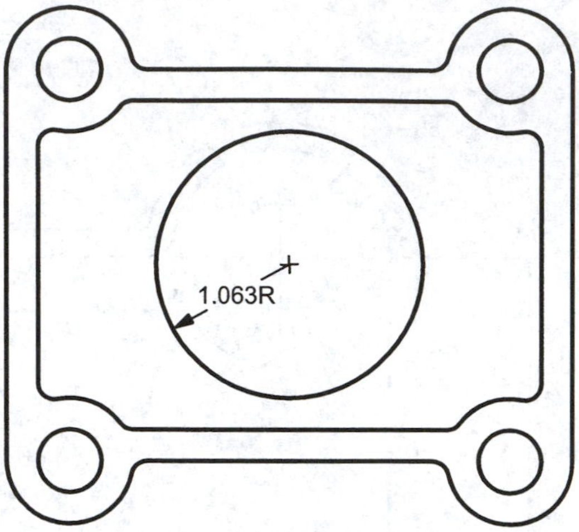

c) Create the geometry for one inner pocket

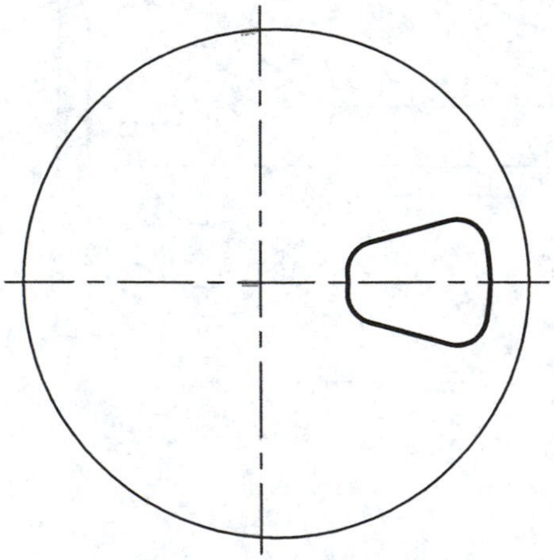

d) Use the Xform, *Rotate* command to complete the inside pocket geometry.

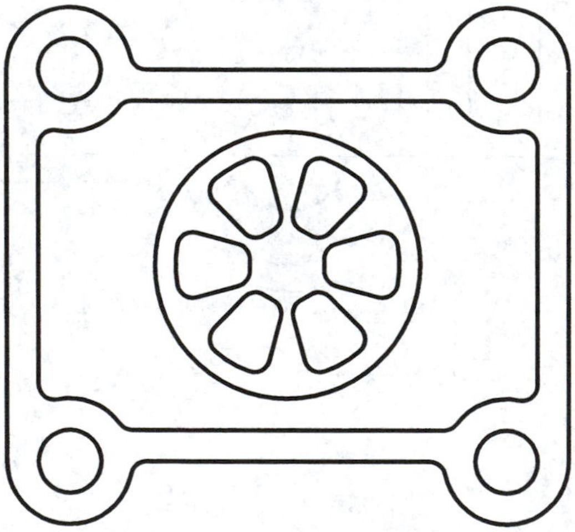

3-7) Create the part shown in Figure 3-p7 using *Mastercam's* CAD package

section view is used for reference only and is NOT DRAWN on CAD model

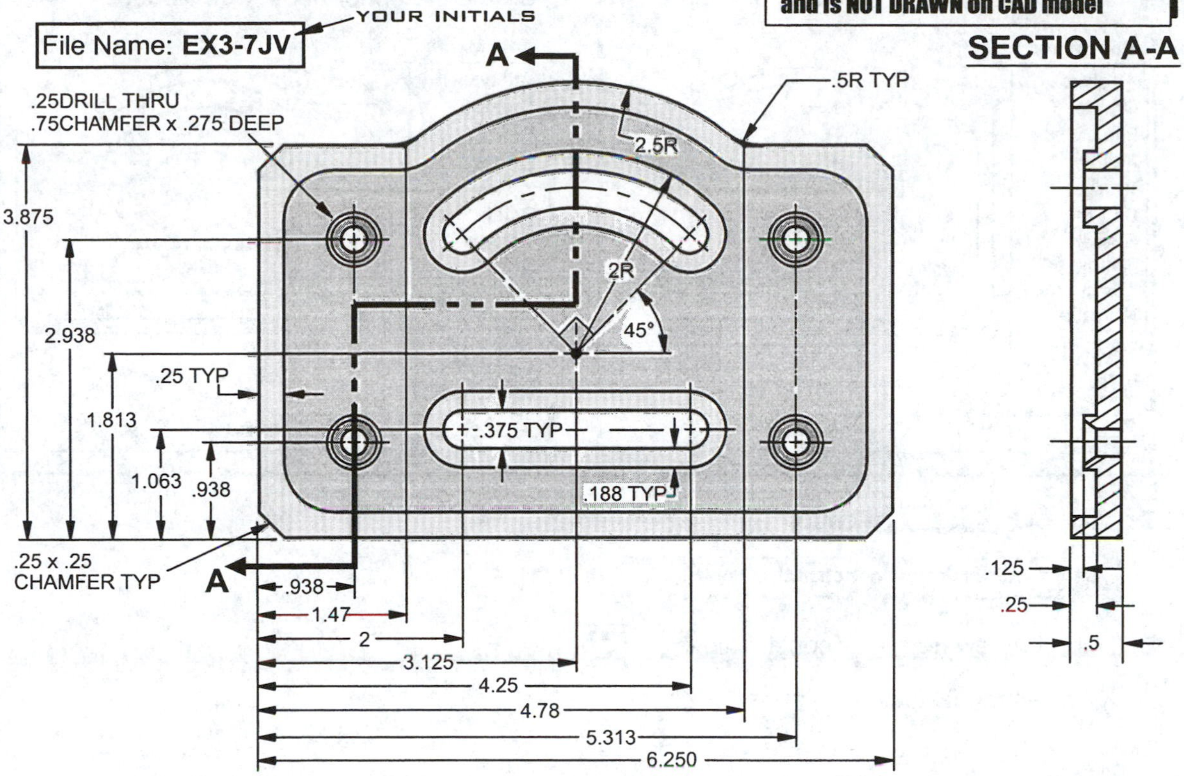

Figure 3-p7

a) Begin the creation of the outside contour

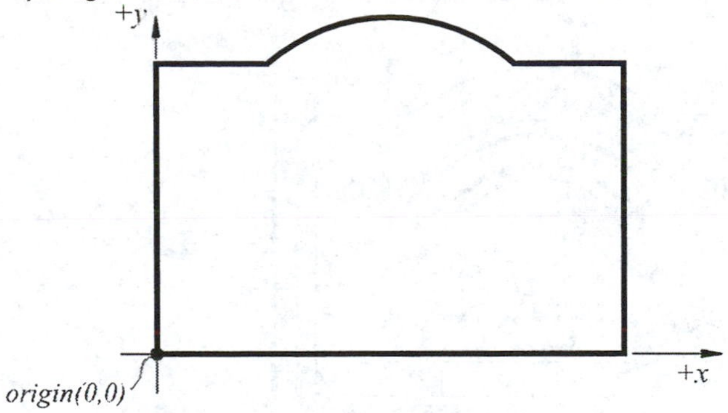

b) Add the .5R Fillets and .25x.25 Chamfers

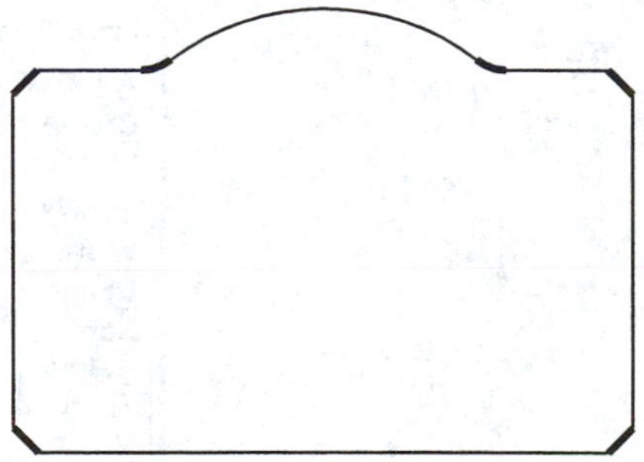

c) Use *Xform, Offset* to layout the inside contour

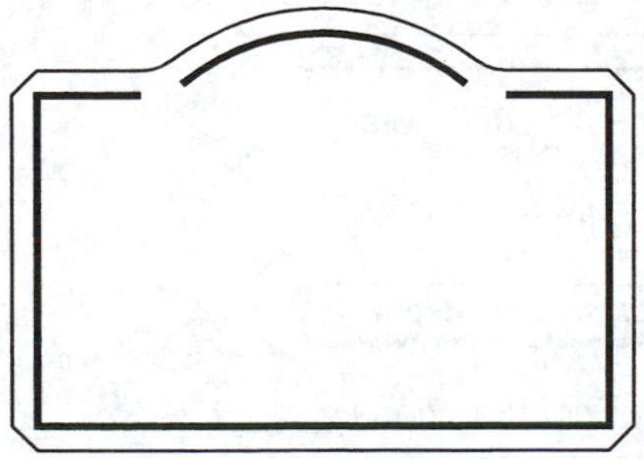

d) Add the .5R Fillets to the inside geometry

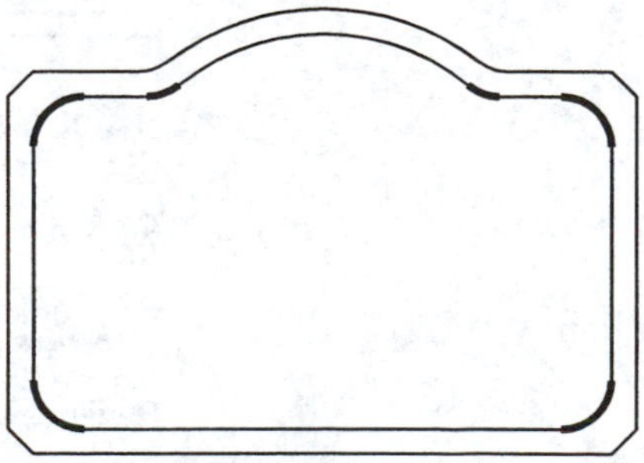

e) Create the islands at a depth of -.125.

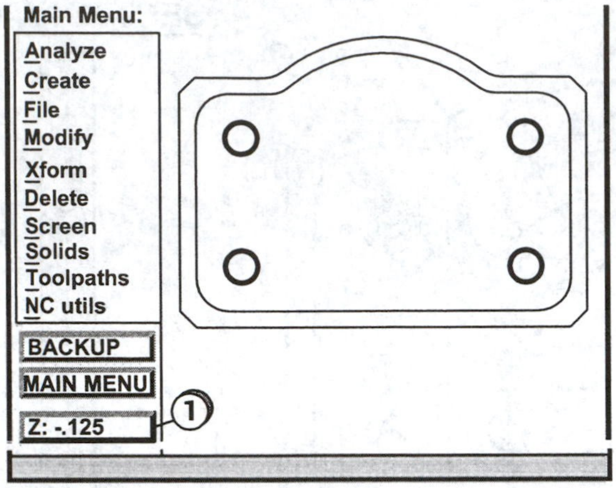

f) Create the upper and lower islands

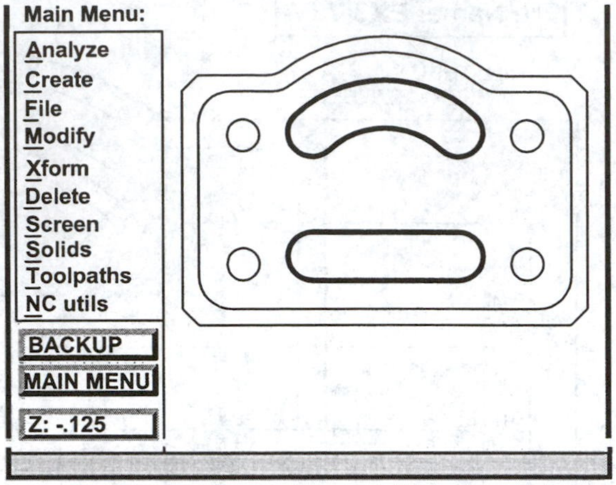

➤ Click ① [Z:] button

Select point for new construction depth

➤ Enter the new depth for constructions **-.125**

Create the .5Dia circle islands

g) Use *Xform, Offset* to create the slot cutouts and add the drill holes

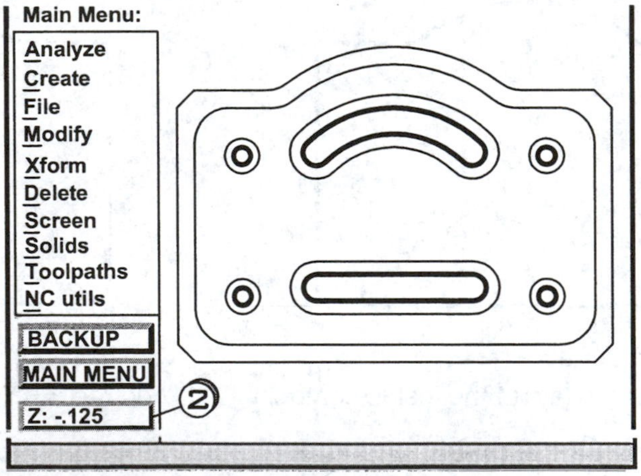

After creating slot cutouts and drill holes

➤ Click ② [Z:] button

Select point for new construction depth

➤ Enter the new depth for constructions **0**

3-8) File Name: **EX3-8JV** ← YOUR INITIALS

Get the file EX2-10JV completed in exercise 2-10. Add the Increase/Decrease arrows, the 2.25R circle, the line graduations and the Hartford Font text. See Figure 3p-8.
Save the file as **EX3-8JV.**

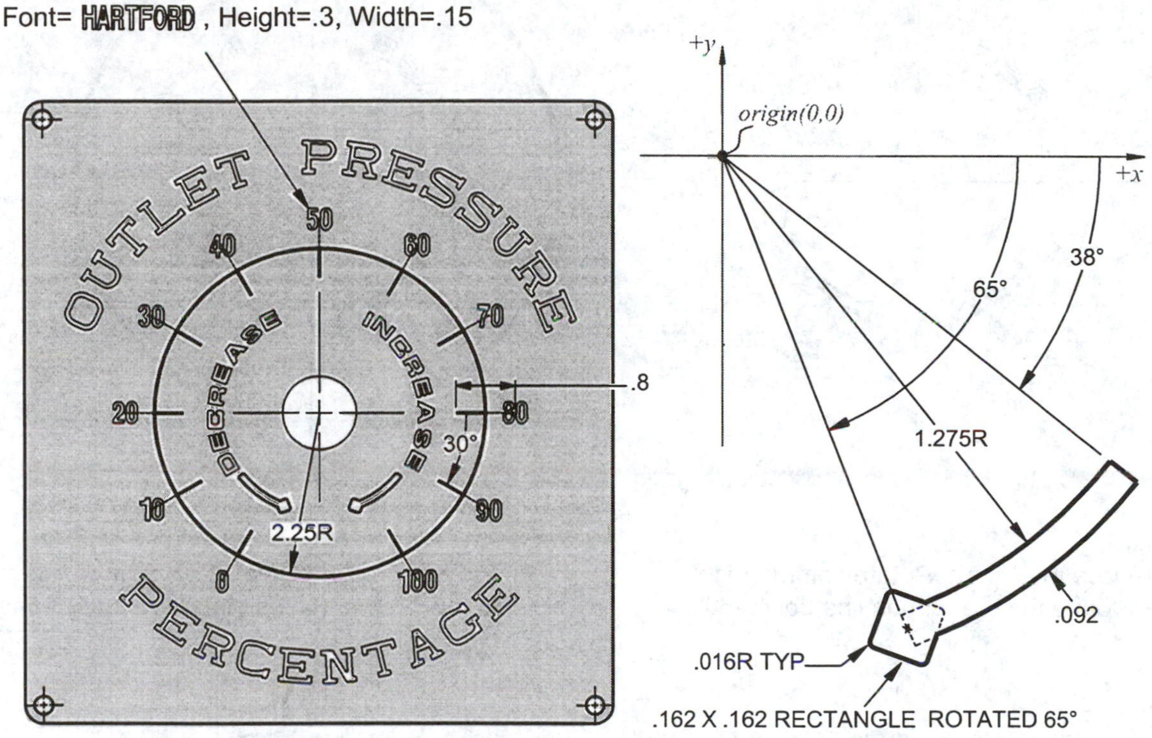

Figure 3-p8

a) Use *Create Arc* and *Rectangle* commands to begin the construction of the directional arrow.

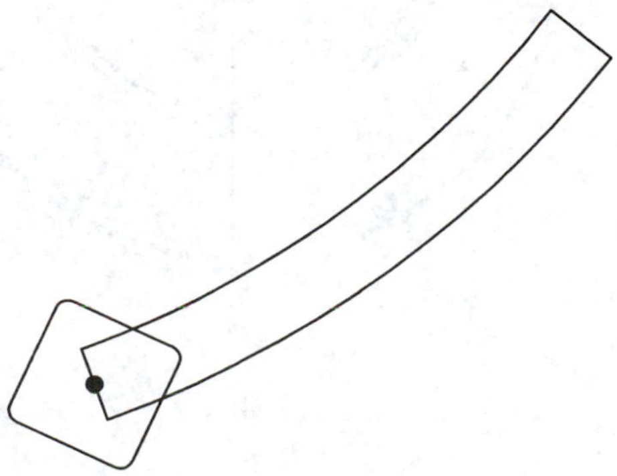

b) Use *Delete* and *Trim* commands to complete the drectional arrow design.

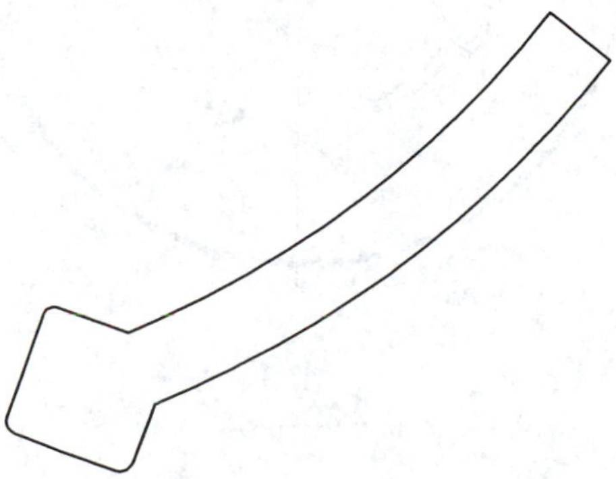

c) Use *Xform, Mirror, Copy* commands to generate the left copy of the directional arrow

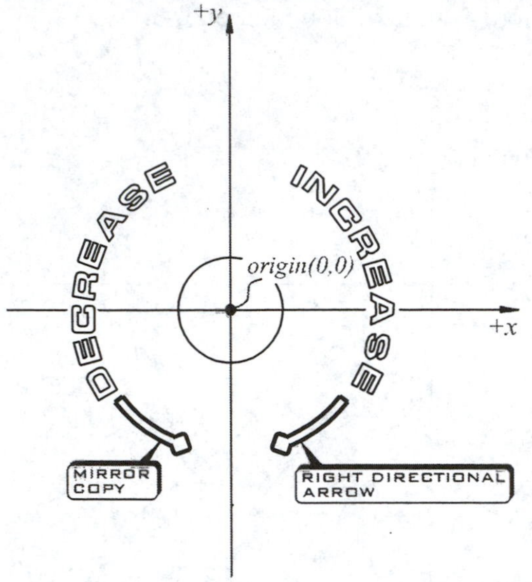

d) Create the 2.25R Circle and .8 graduation line

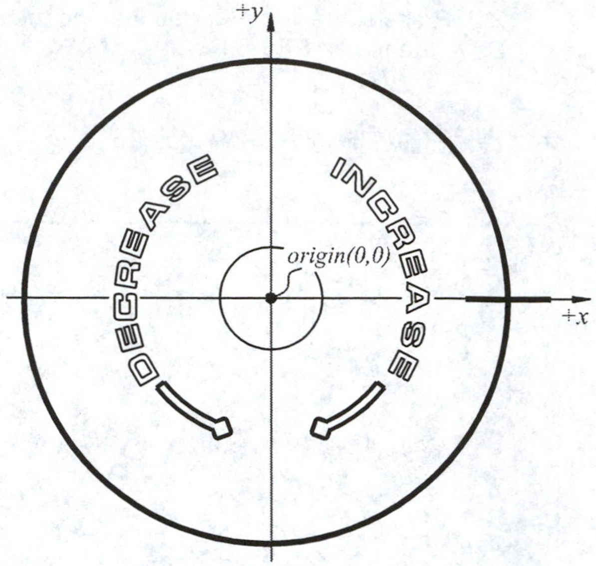

e) Use the Xform, Rotate command to rotate the graduation line down -60°

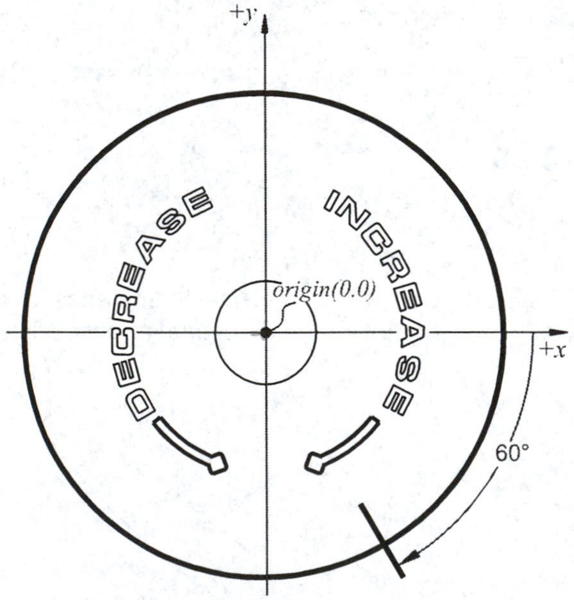

f) Use the *Xform, Rotate, Copy* command to make *nine* copies of the graduation line

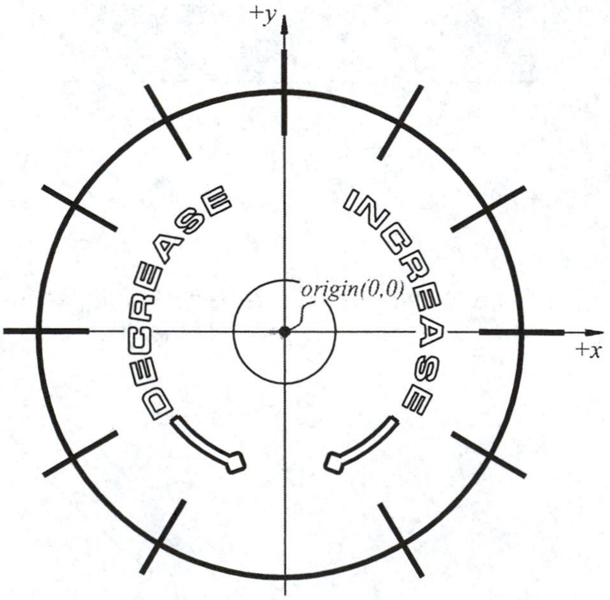

g) Insert the HARTFORD font geometric lettering at the ***endpoint*** of each graduation line.

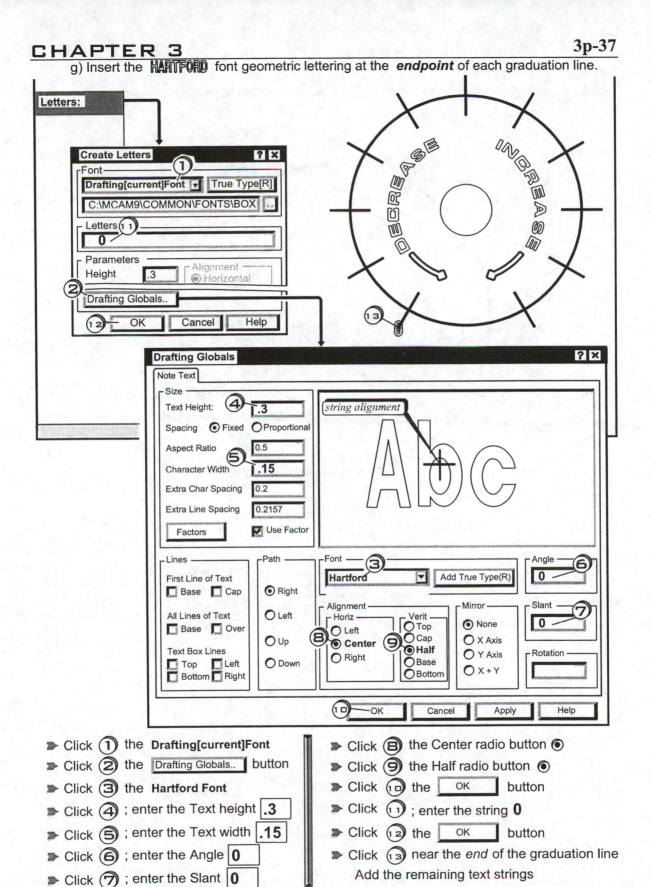

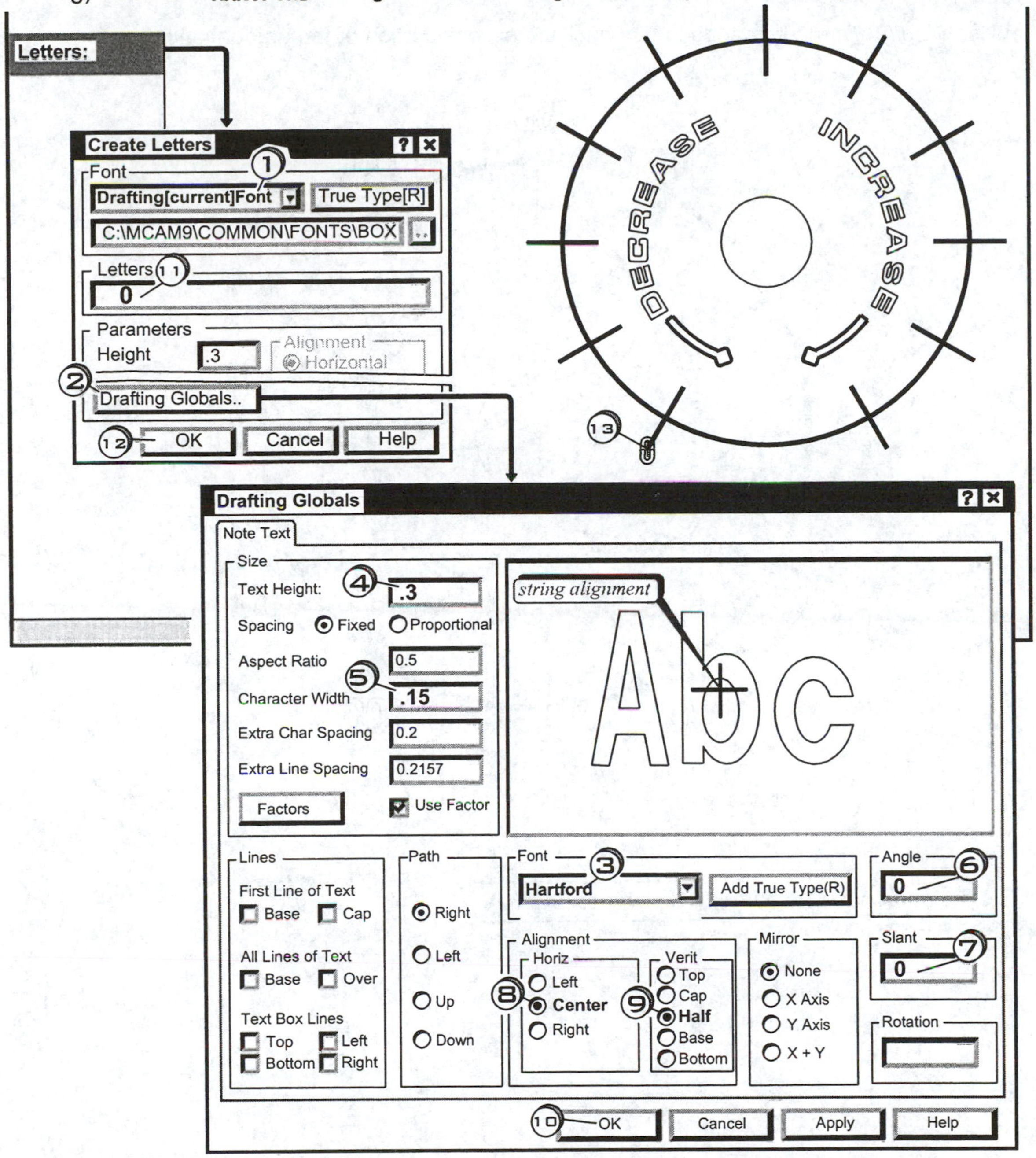

- Click ① the **Drafting[current]Font**
- Click ② the Drafting Globals.. button
- Click ③ the **Hartford Font**
- Click ④ ; enter the Text height **.3**
- Click ⑤ ; enter the Text width **.15**
- Click ⑥ ; enter the Angle **0**
- Click ⑦ ; enter the Slant **0**
- Click ⑧ the Center radio button ◉
- Click ⑨ the Half radio button ◉
- Click ⑩ the OK button
- Click ⑪ ; enter the string **0**
- Click ⑫ the OK button
- Click ⑬ near the *end* of the graduation line

Add the remaining text strings

e) Use *Modify, Trim* commands to trim back the excess portion of the graduation lines

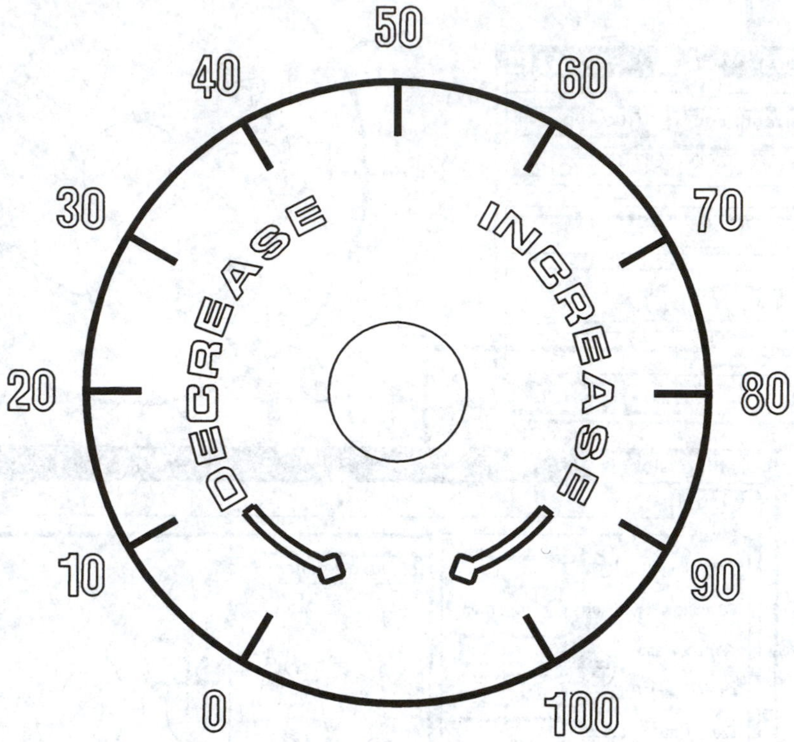

File the part under the name File Name: **EX3-8JV** ← YOUR INITIALS

CHAPTER - 4

ADDITIONAL TOOLS FOR CAD

4-1 Chapter Objectives

After completing this chapter you will be able to:

1. Use the Dimension command to check the dimensional correctness of CAD models.
2. Explain how to execute import/export operations between *Mastercam* and other CAD packages such as AutoCAD.

4-2 Checking the CAD Model for Dimensional Correctness

It is highly recommended to ***check*** the CAD model for correctness of size and the *location* of geometric features such as holes. If any dimensional errors are found, the operator needs to quickly make adjustments in size and location ***before*** applying machining toolpaths. *Mastercam* features the **Dimension** command for this purpose.

EXAMPLE 4-1

Verify that the existing CAD model has the dimensions as shown in Figure 4-1

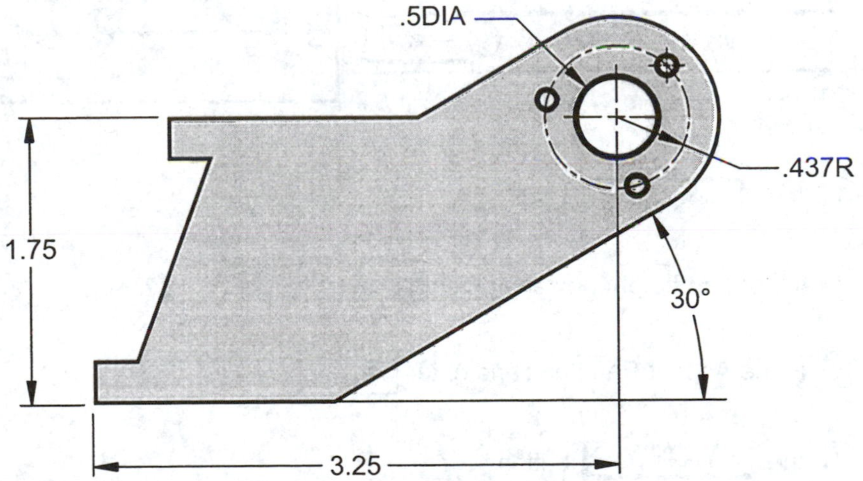

The existing CAD model to be checked

Figure 4-1

A) Set the desired global dimensioning parameters

➤ Press Alt + D

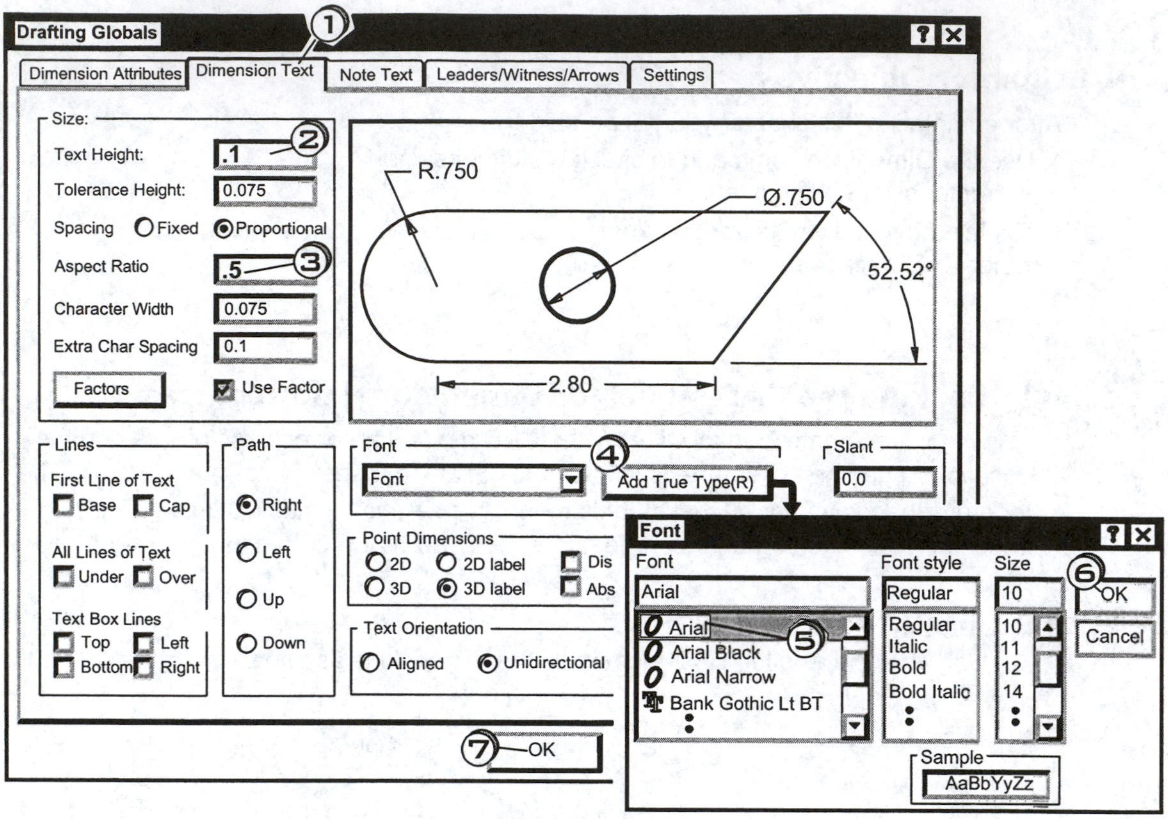

➤ Click ① the Dimension Text tab

➤ Click ② in the text height box and enter .1

➤ Click ③ in the Aspect Ratio box and enter .5

➤ Click ④ the Add True Type(R) button

➤ Click ⑤ the Arial font

➤ Click ⑥ the OK button

➤ Click ⑦ the OK button

B) CHECK THE APPROPRIATE DIMENSIONS

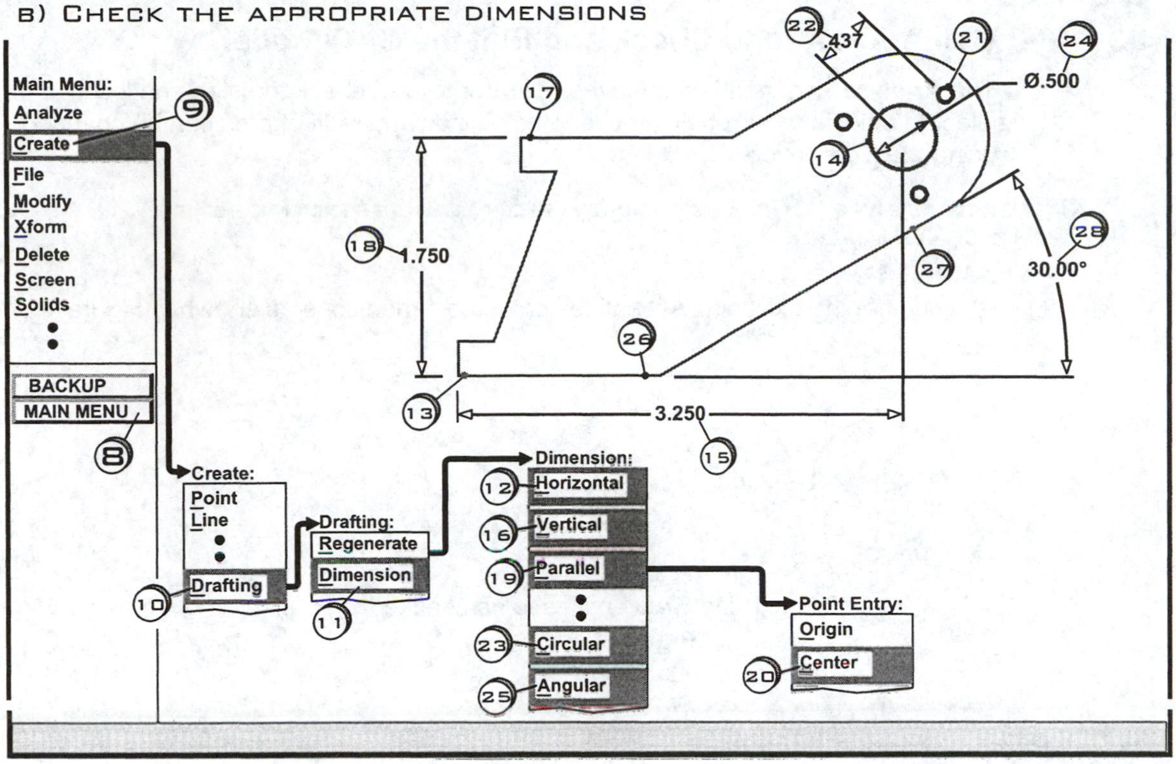

▶ Click (8) the MAIN MENU button

▶ Click (9) Create

▶ Click (10) Drafting

▶ Click (11) Dimension

▶ Click (12) Horizontal

▶ Click (13) near the *end* of the *line*

(14) *on* the *arc* (15) the text loc

▶ Press Esc to cancel the operation

▶ Click (16) Vertical

▶ Click (13) (17) *near* the *ends* of the *lines*

(18) the text location

▶ Press Esc to cancel the operation

▶ Click (19) Parallel

▶ Click (14) *on* the *arc*

▶ Click (20) Center

▶ Click (21) *on* the *arc* (22) the text loc

▶ Press Esc to cancel the operation

▶ Click (23) Circular

▶ Click (14) *on* the *arc* (24) the text loc

▶ Press Esc to cancel the operation

▶ Click (25) Angular

▶ Click (26) (27) *near* the *ends* of the *lines*

(28) the text location

▶ Press Esc to cancel the operation

4-3 Using Analyze to Check and Edit the CAD Model

The **Analyze** command enables the operator to spot check the size and location of selected objects. Additionally, the lengths of lines and radii of arcs can be changed using the **Edit** option.

A) Using Analyze to Check Dimensional Correctness of the CAD Model

EXAMPLE 4-2

Verify that the existing CAD entities have the dimensions as shown in Figure 4-2

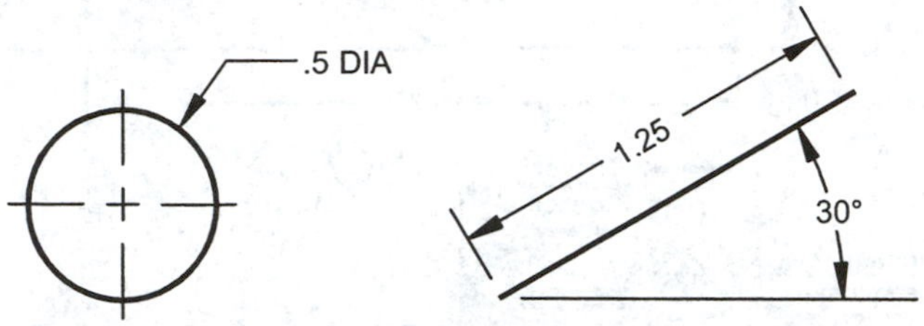

The existing CAD entities to be checked

Figure 4-2

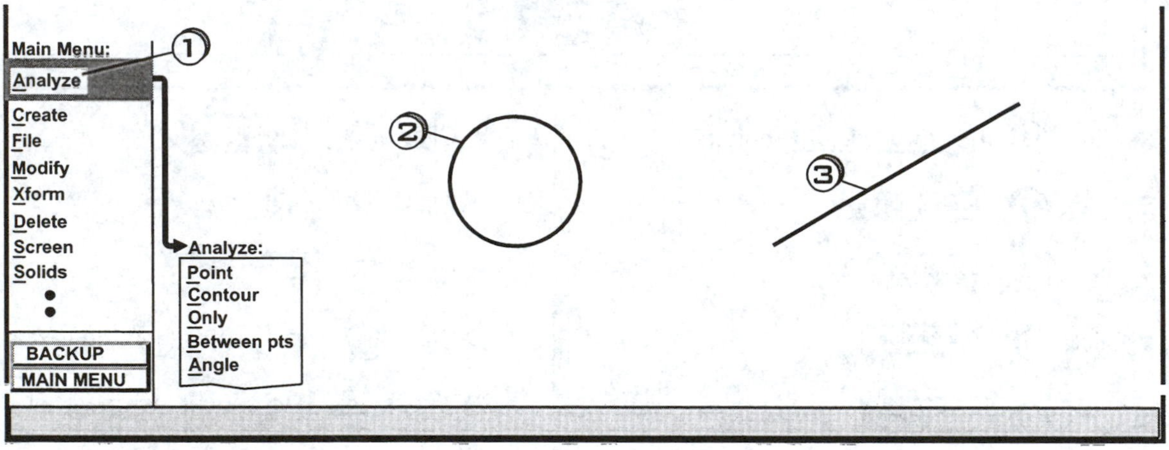

- Click ① Analyze
- Click ② the arc entity

ARC: View 1 Level:1 Color:0 # Geometry Refs: 0
Style: solid Width: 1 Entity#8 Creation date: Tue Dec25 01:03:39 AM 2001 # Toolpath Refs:0
Radius: 0.25000 Diameter: 0.50000 Z:0.00000
Center: X.75000 Y:1.25000 Z: 0.00000 Angle:0.00000 Sweep:360.00000 3D Length:

- Click ③ the line entity

LINE: Level 1 Color:0 # Geometry Refs: 0
Style: solid Width: 1 Entity#9 Creation date: Tue Dec25 01:03:39 AM 2001 # Toolpath Refs:0
3D Length: 1.25000 2D Length: 1.25000 Angle: 30.00000
Startpoint: X:2.50000 Y:2.25000 Z: 0.00000 Endpoint: X3.58253 Y:2.87500 Z:0.00000

B) USING Analyze TO EDIT THE DIMENSIONS OF THE CAD MODEL

EXAMPLE 4-3

Change the .5DIA circle to .25DIA. Change the 1.25 long line to a length of .75 .

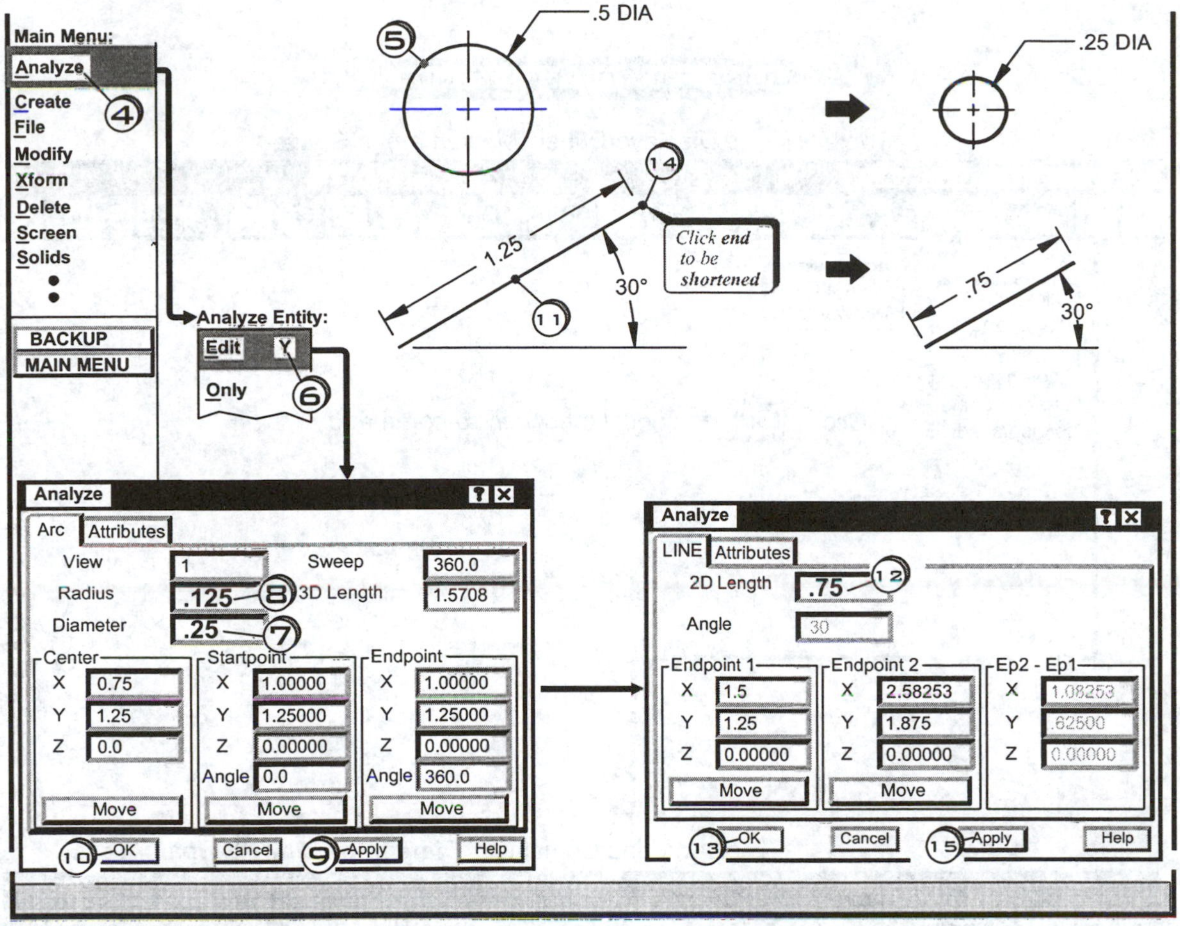

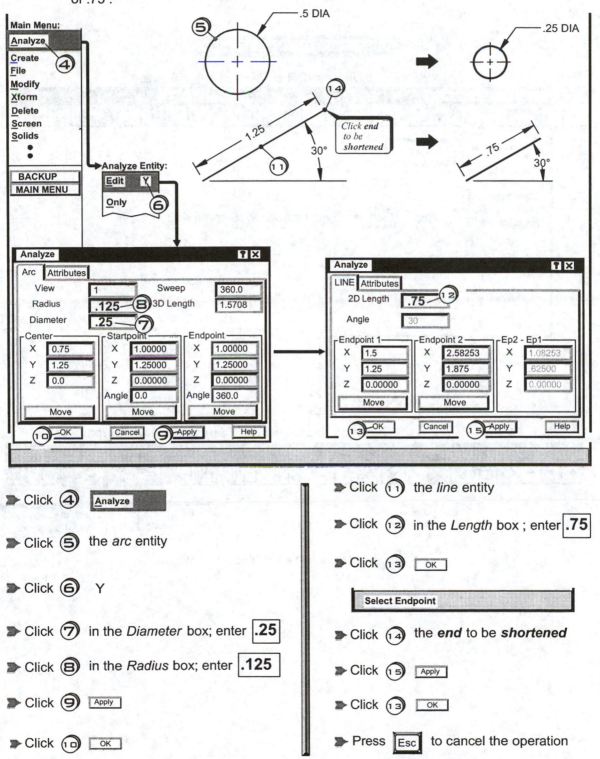

- Click ④ **Analyze**
- Click ⑤ the *arc* entity
- Click ⑥ Y
- Click ⑦ in the *Diameter* box; enter **.25**
- Click ⑧ in the *Radius* box; enter **.125**
- Click ⑨ Apply
- Click ⑩ OK
- Click ⑪ the *line* entity
- Click ⑫ in the *Length* box ; enter **.75**
- Click ⑬ OK

Select Endpoint

- Click ⑭ the ***end*** to be ***shortened***
- Click ⑮ Apply
- Click ⑬ OK
- Press Esc to cancel the operation

Using the Toolbar Menus

Toolbar Menu Displayed when *Mastercam* is **Started**

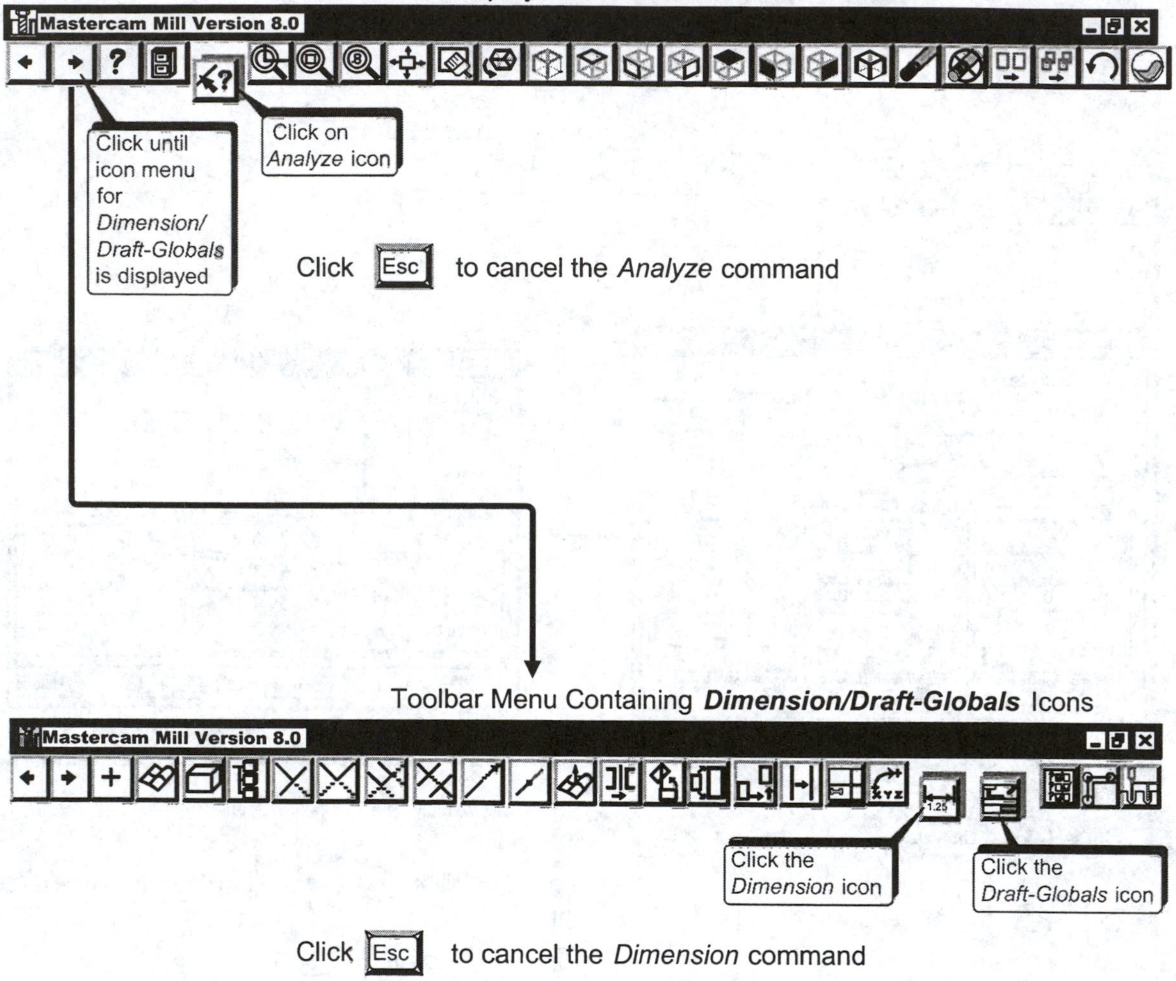
Mastercam Mill Version 8.0
Click until icon menu for *Dimension/ Draft-Globals* is displayed
Click on *Analyze* icon
Click Esc to cancel the *Analyze* command
Toolbar Menu Containing ***Dimension/Draft-Globals*** Icons
Mastercam Mill Version 8.0
1.25
Click the *Dimension* icon
Click the *Draft-Globals* icon
Click Esc to cancel the *Dimension* command

4-4 Import/Export Operations With Other CAD Packages

The operator has the full flexibility of importing a CAD model from other CAD packages such as AutoCAD into *Mastercam* or from *Mastercam* to AutoCAD. The commands and procedures for executing these important operations are considered in this section.

A) IMPORTING CAD MODELS FROM AutoCAD TO *Mastercam*

- The location of the part with respect to ***AutoCAD's World Coordinate System(WCS)*** will be the ***same*** as that for ***Mastercam's XY coordinate system***

EXAMPLE 4-4

Import the AutoCAD drawing file, **EXAMPLE4-4.DWG**, shown below to *Mastercam*

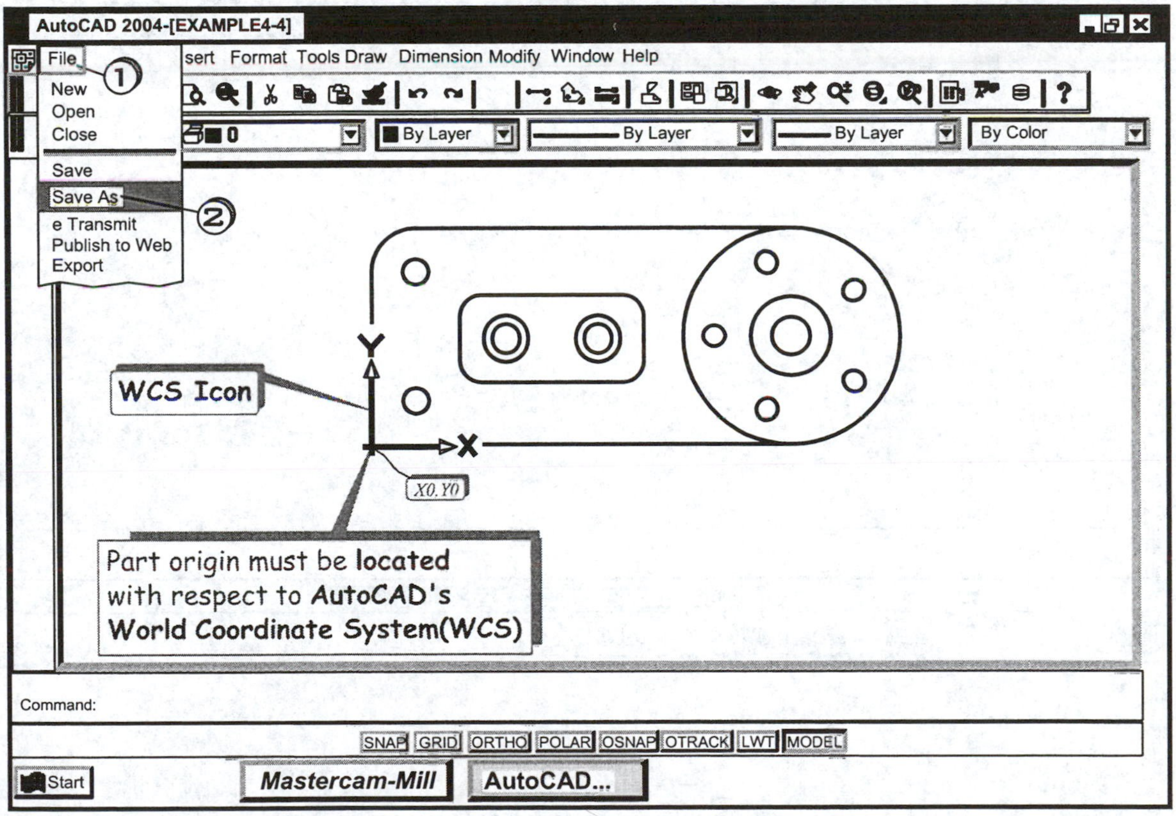

➤ Click ① File

➤ Click ② Save As

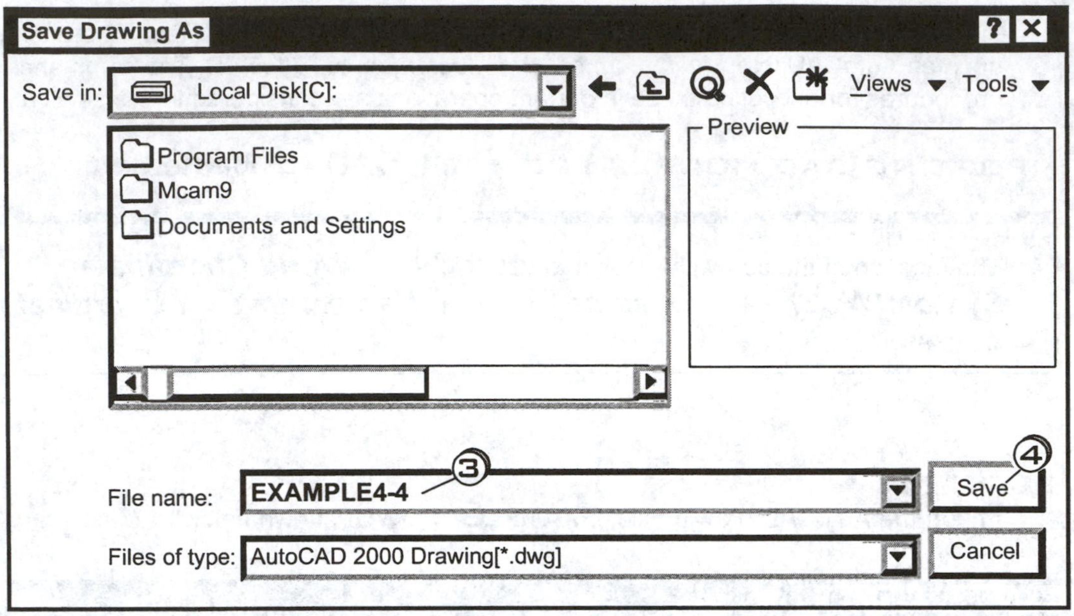

➤ Click ③ in the File name box and enter: **EXAMPLE4-4**

➤ Click ④ Save

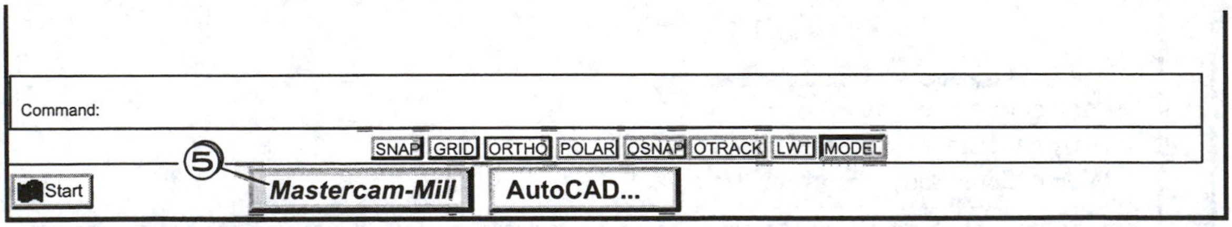

➤ Click ⑤ the ***Mastercam-Mill*** button or enter the *Mastercam* Mill package from the Windows desktop

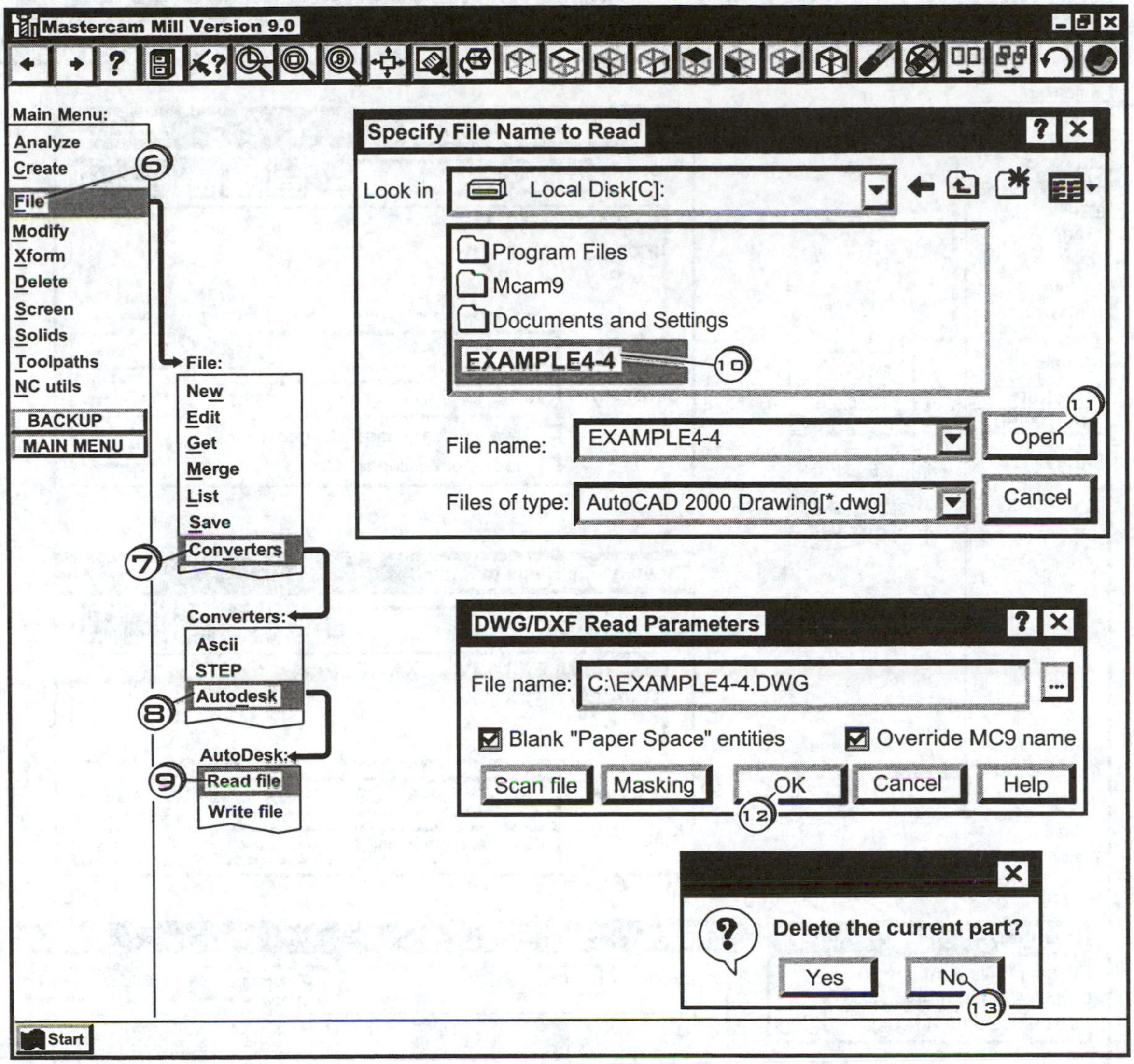

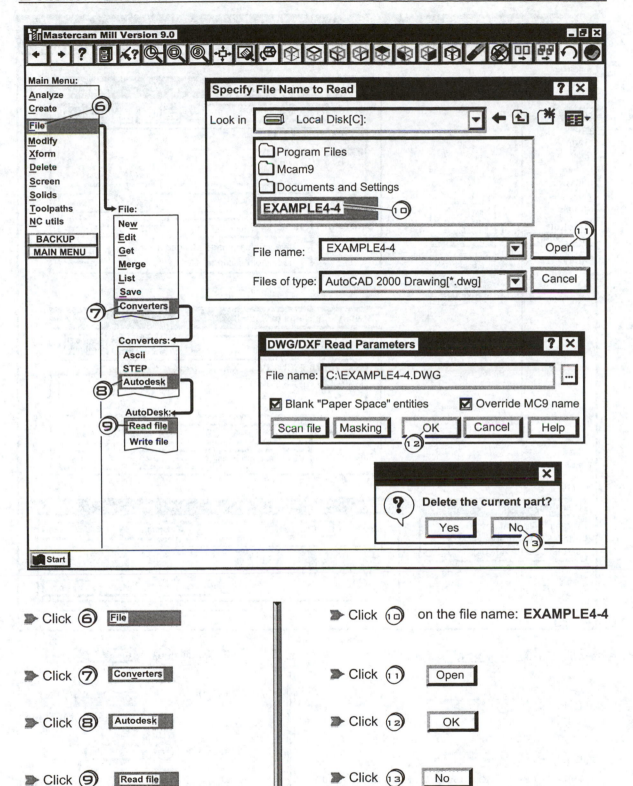

- Click (6) File
- Click (7) Converters
- Click (8) Autodesk
- Click (9) Read file
- Click (10) on the file name: **EXAMPLE4-4**
- Click (11) Open
- Click (12) OK
- Click (13) No

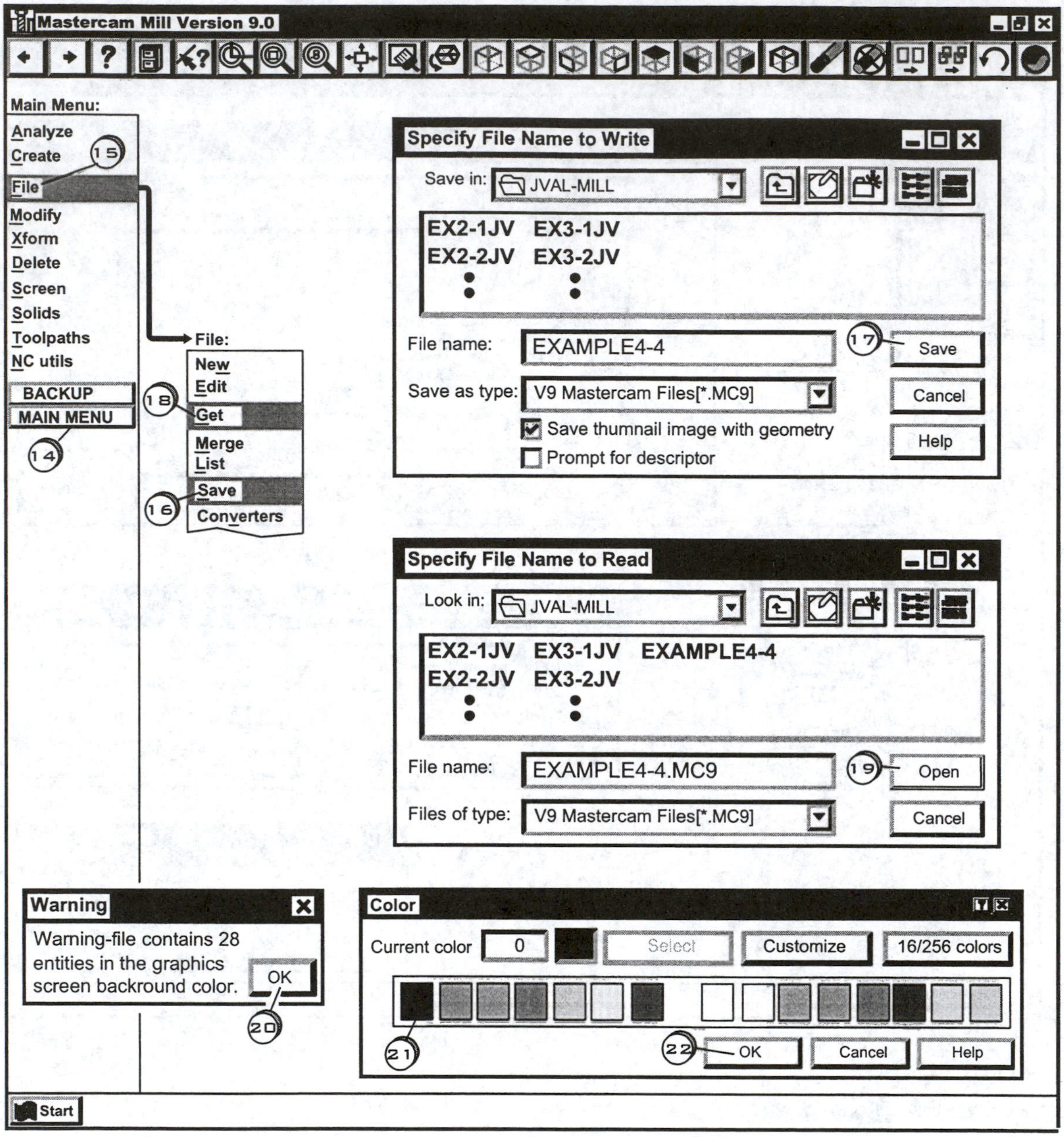

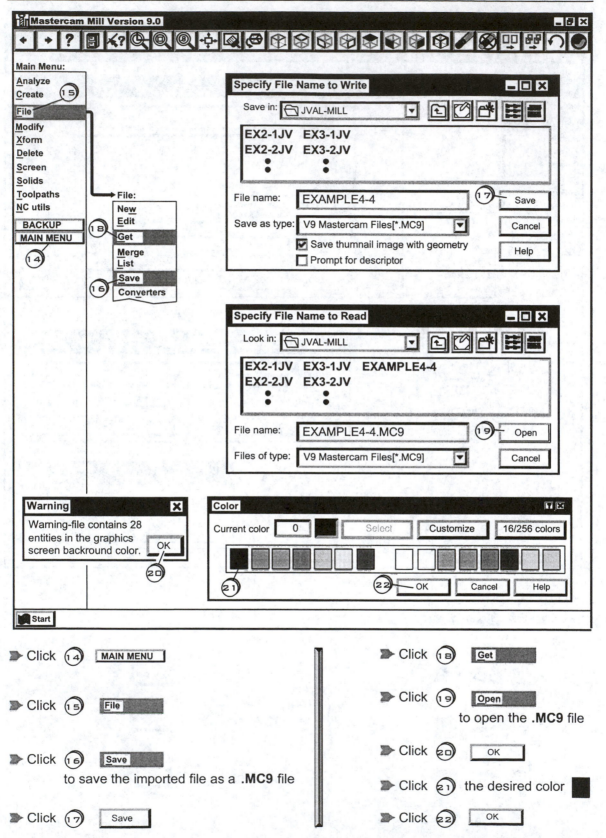

- Click (14) MAIN MENU
- Click (15) File
- Click (16) Save
  to save the imported file as a **.MC9** file
- Click (17) Save
- Click (18) Get
- Click (19) Open
  to open the **.MC9** file
- Click (20) OK
- Click (21) the desired color ■
- Click (22) OK

Mastercam will then open the file **EXAMPLE4-4.MC9**. *Mastercam* commands can be applied as needed.

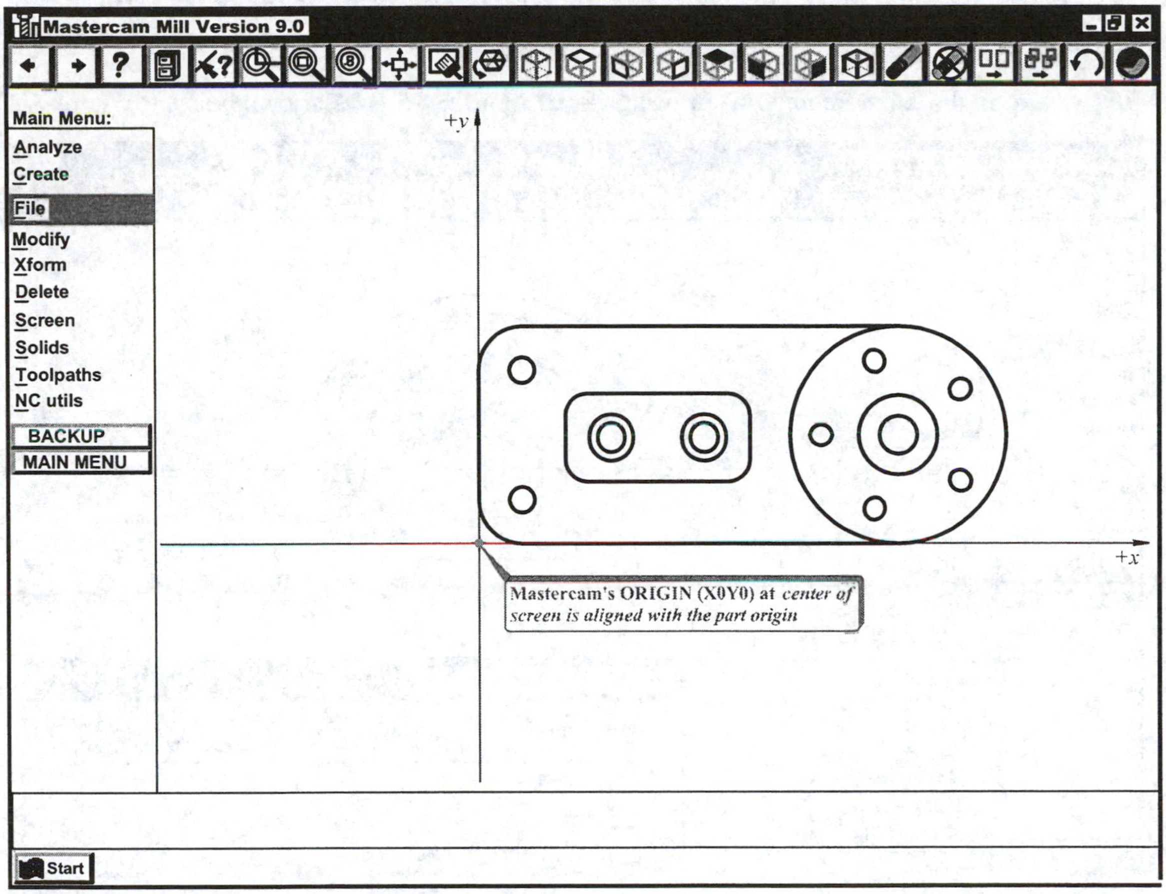

B) EXPORTING CAD MODELS FROM *Mastercam* TO AutoCAD

- The location of the part with respect to ***Mastercam's XY coordinate system*** will be the same with respect to ***AutoCAD's World Coordinate System(WCS)***

EXAMPLE 4-5

Export the *Mastercam* CAD drawing file **EXAMPLE4-5.MC9** shown below to AutoCAD

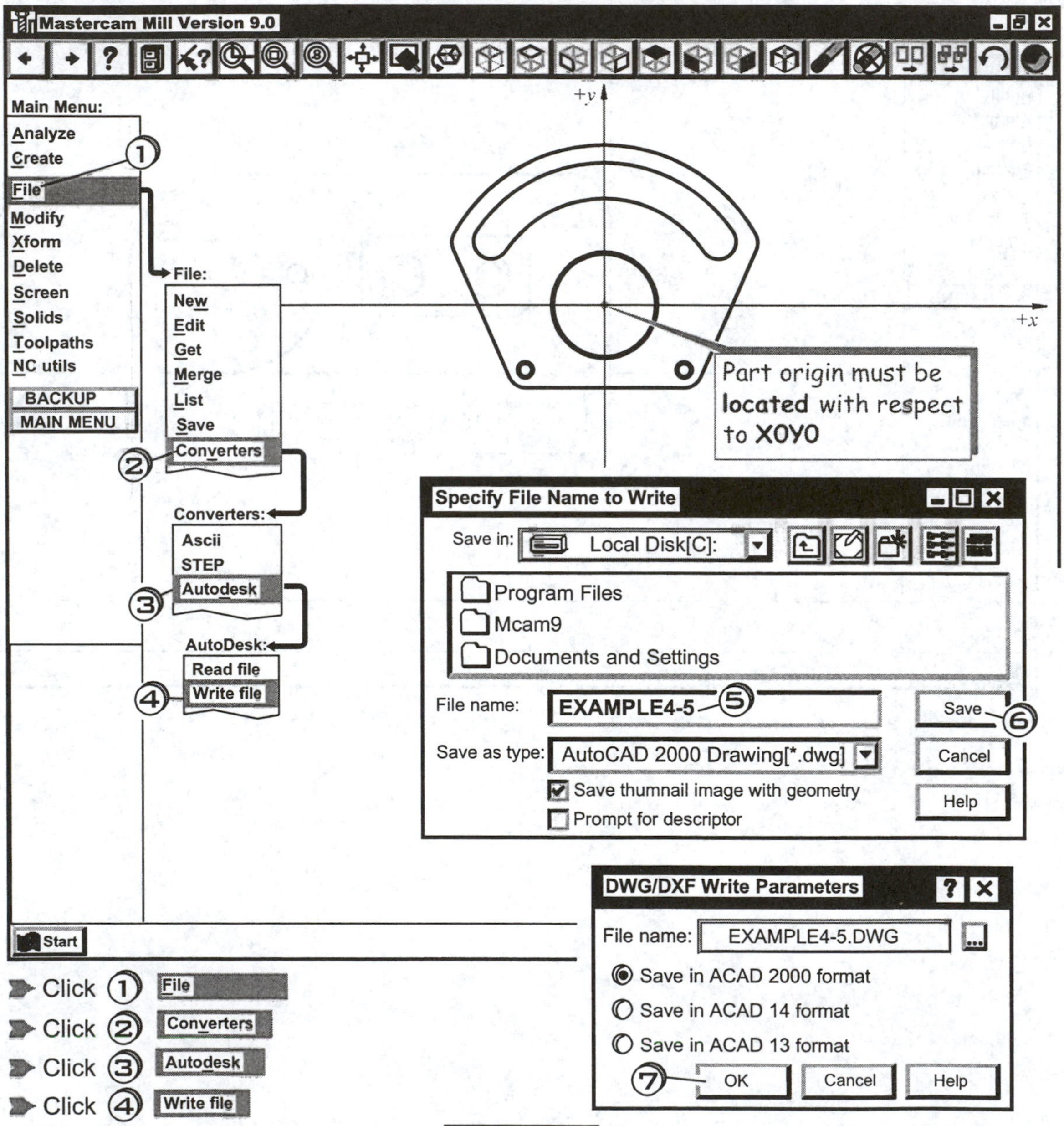

- Click ① File
- Click ② Converters
- Click ③ Autodesk
- Click ④ Write file
- Click ⑤ in the file name box; enter EXAMPLE4-5
- Click ⑥ Save
- Click ⑦ OK

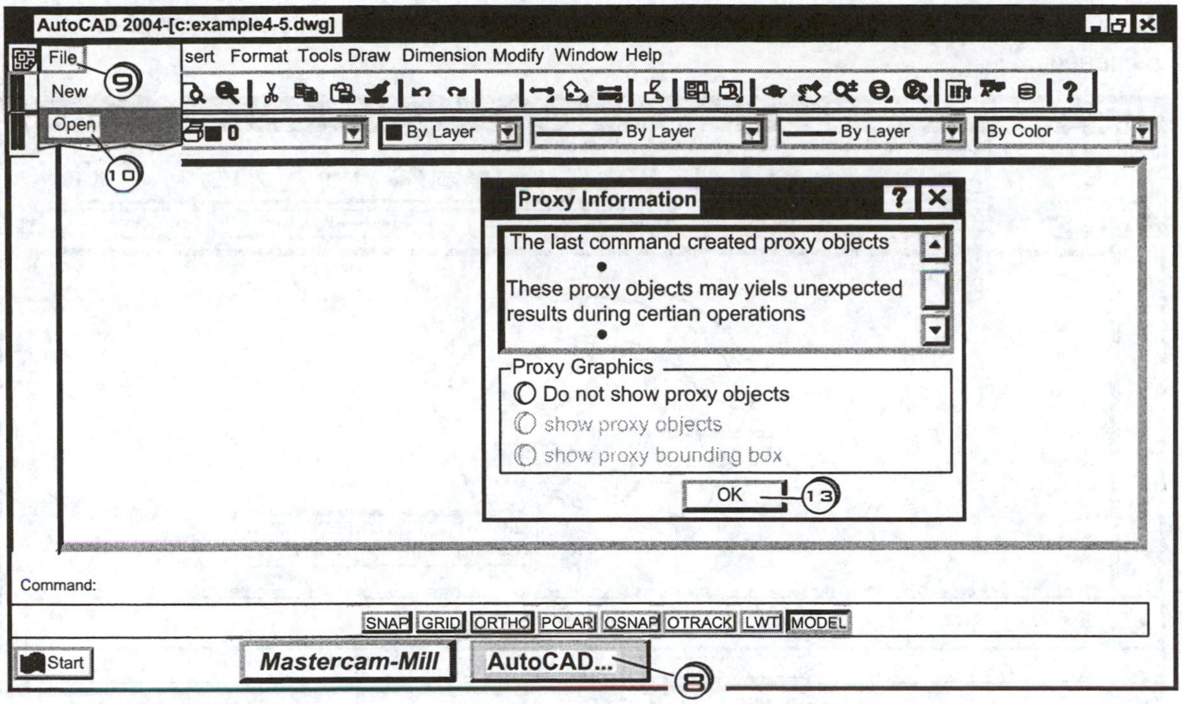

➤ Click ⑧ **AutoCAD...** or enter the AutoCAD package from the Windows desktop

➤ Click ⑨ File

➤ Click ⑩ **Open**

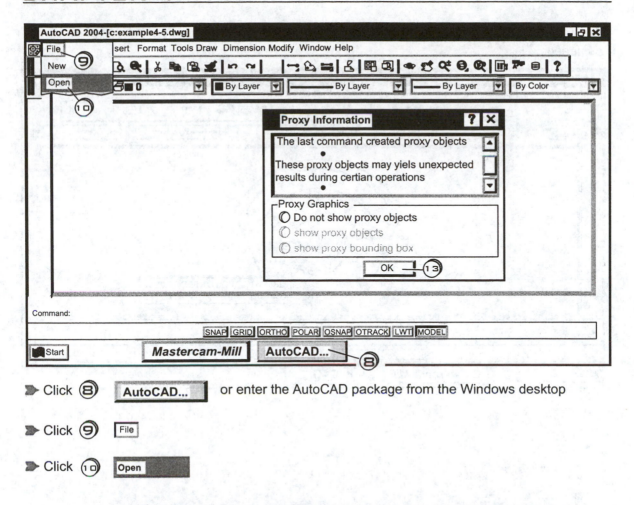

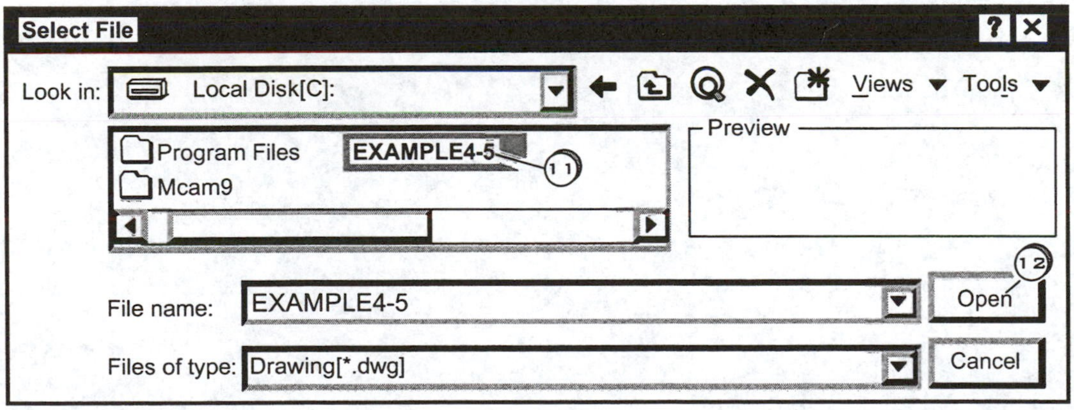

➤ Click ⑪ on the file **EXAMPLE4-5**

➤ Click ⑫ Open

➤ Click ⑬ OK

AutoCad will then open the file **EXAMPLE4-5.DWG**. AutoCad commands can be applied as neeed.

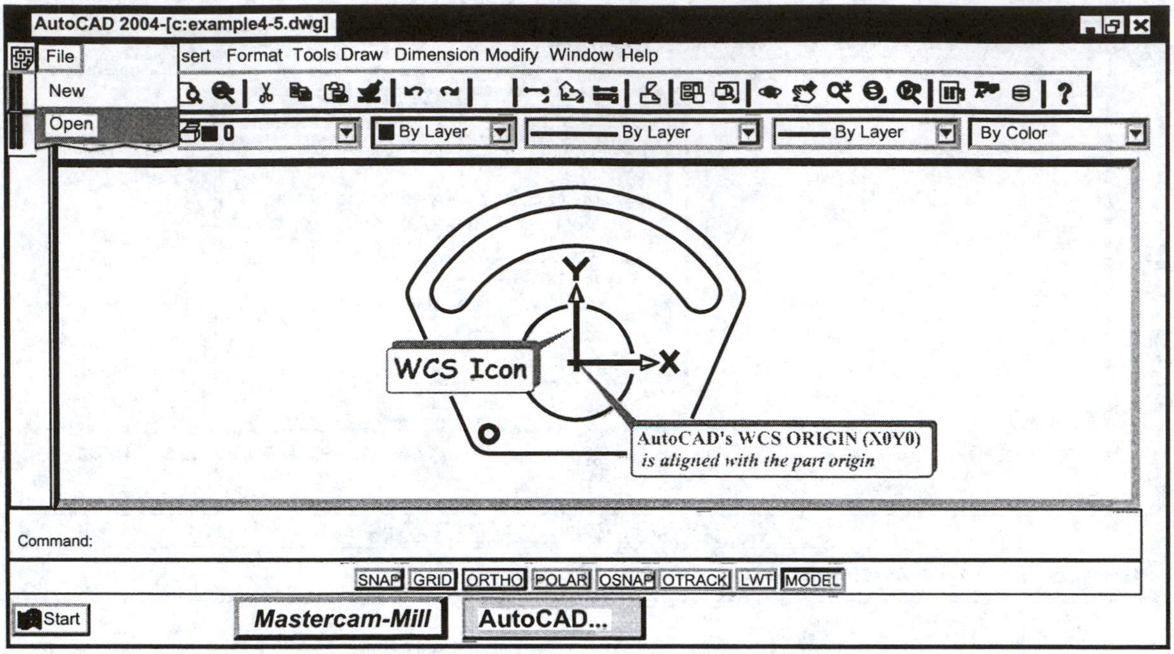

EXERCISES

4-1) The CAD model shown in Figure 4p-1 is stored in the file folder CHAPTER4 as file **EX4-1**. Get this file in *Mastercam* and use *Mastercam*'s **Drafting, Dimension** commands to verify the model has the dimensions as indicated. Edit the CAD model as needed to assure its dimensional correctness.

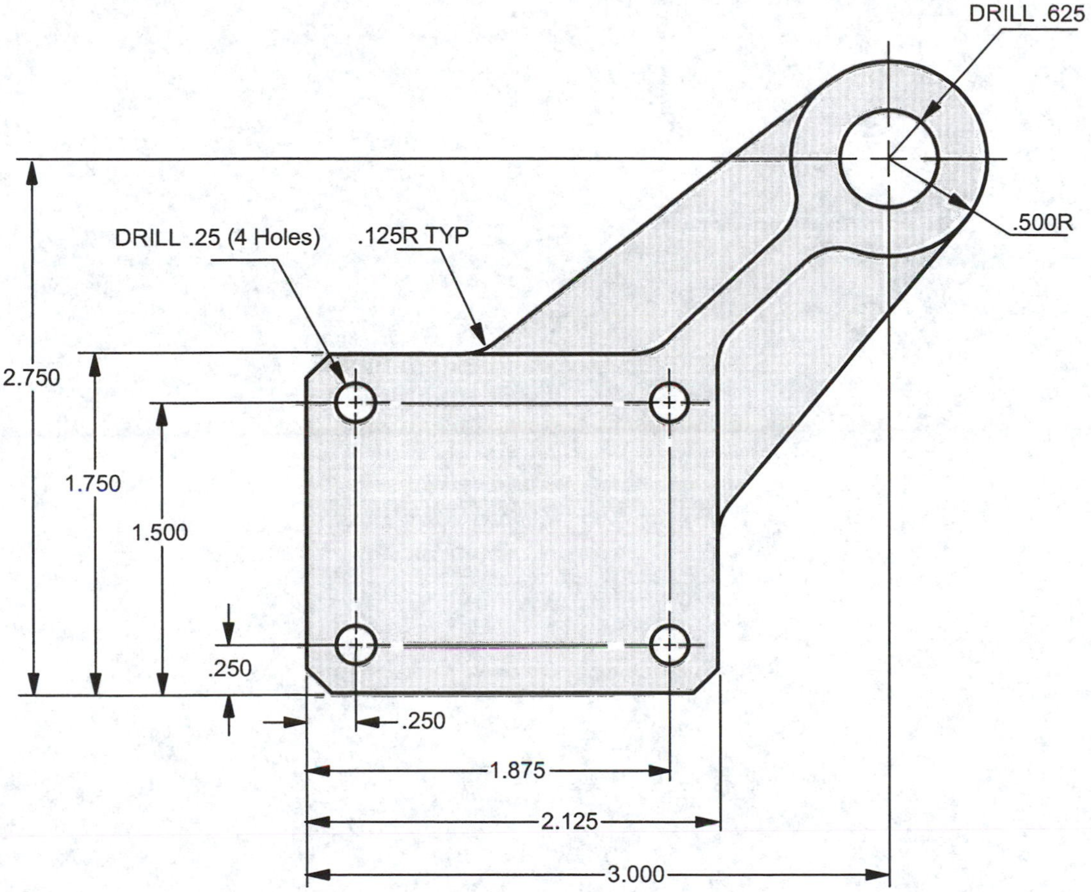

Figure 4-p1

Note: To move entities to proper locations use the **Xform, Translate, Rectang** commands. Refer to Chapter 3, p 3-28 for a review.

CHAPTER - 5

GENERATING HOLE OPERATIONS IN 2D SPACE

5-1 Chapter Objectives

After completing this chapter you will be able to:

1. Create tools and tool libraries.
2. Execute commands for specifying drilling locations on a CAD model.
3. Understand how to use the drill module to execute drilling in a part.
4. Explain how to check toolpaths via the backplotter.
5. Know how to use the verifier to view real time solid model animation of the entire machining process.
5. Explain how to pan and zoom and section the machined part within the maching verifier.
6. State how to machine a circle in stock using the circle mill module.
7. Understand the process of machining holes at different depths and retract heights.
8. Know how to edit existing drill tool paths.
9. Explain the steps taken in generating a word address part program to run on a CNC machine tool(postprocessing).

5-2 Defining Tools and Tool Libraries

The first step that must be taken in order to execute machining operations on the CAD model of a part is to define all the required cutting tools. Often, it is convenient to select tools from Mastercam's large tool library to create a *smaller, customized* tool library. This approach provides the operator greater access to the types of tools frequently used in jobs. The customized library can contain the specific dimensions of the tools at hand as well as parameters associated with the tool such as speed and feed set for the types of materials most often machined.

As was mentioned in Chapter 1 information about a cutting tool currently in the tool library database is stored in the tool file with extension **.TL9.**

use ***your*** *initials instead of* **JVAL**

EXAMPLE 5-1

Create a customized tool library called **JVAL** and place the tools listed in Table 5-1 into this library.

TABLE 5-1

Tooling	Dimensions
1/8 Spot Drill	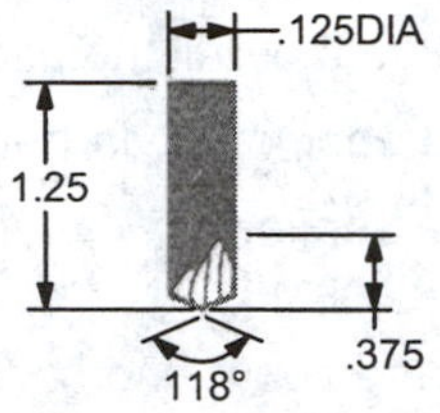
1/4 Drill	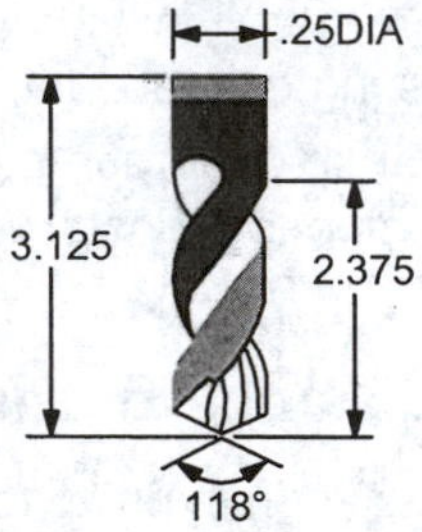
1/8-40UNC Tap	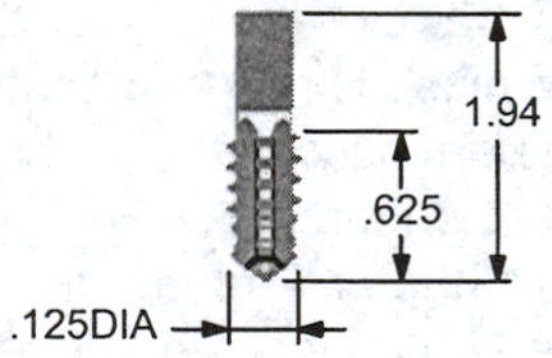

A) ENTER ***Mastercam's*** MAIN TOOL LIBRARY. SEARCH AND RETRIEVE A .125DIA SPOT DRILL, .25DIA DRILL AND .25DIA FLAT END MILL.

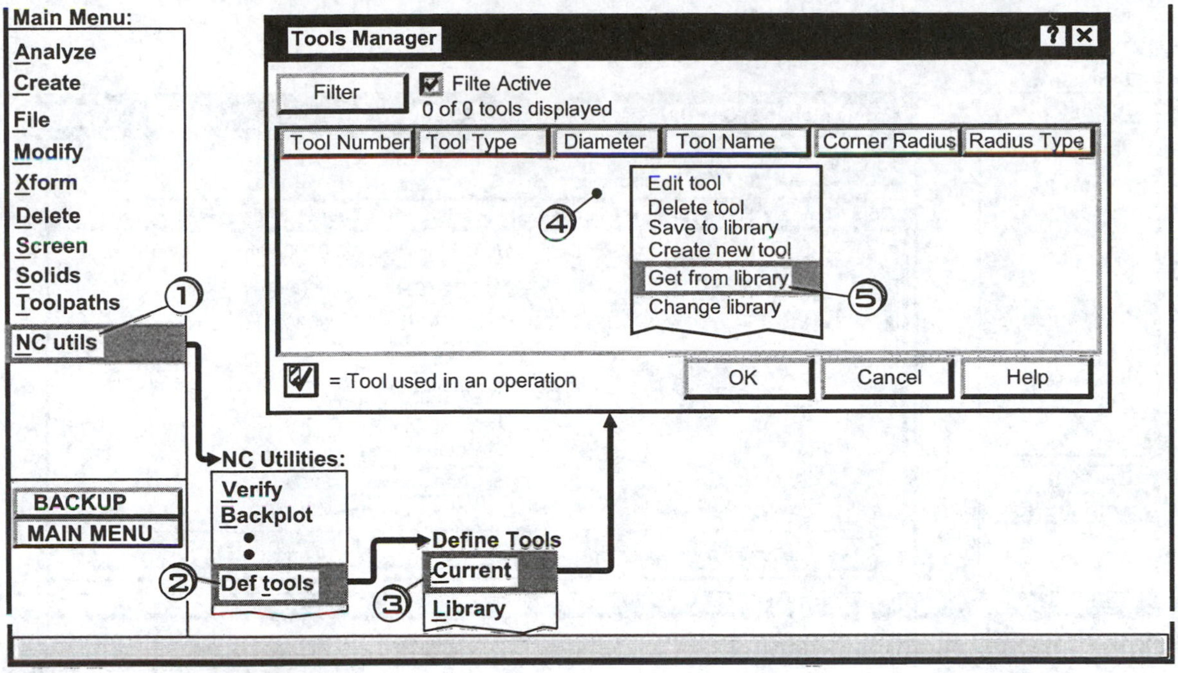

- Click ① NC utils
- Click ② Def tools
- Click ③ Current
- ***Right click*** ④ ***in the white area***, move the cursor down and Click ⑤ Get from library

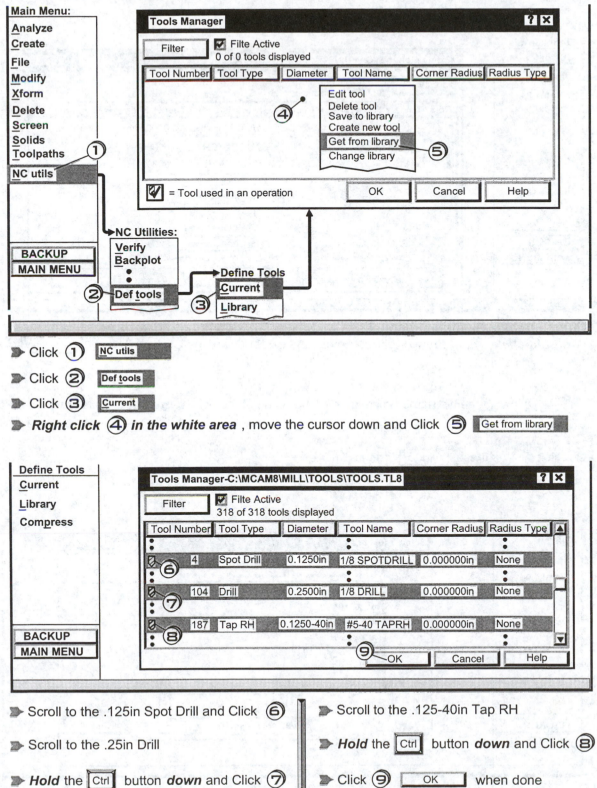

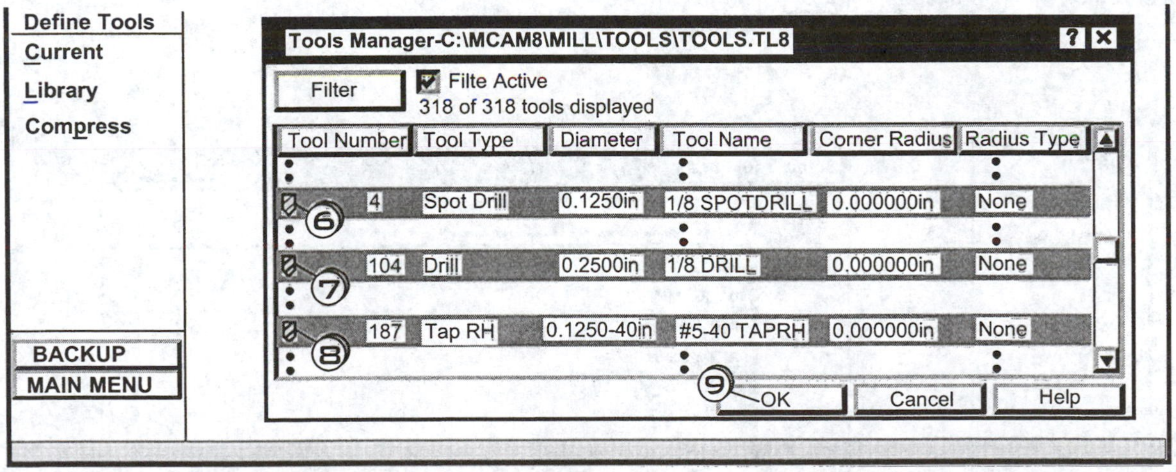

- Scroll to the .125in Spot Drill and Click ⑥
- Scroll to the .25in Drill
- ***Hold*** the Ctrl button ***down*** and Click ⑦
- Scroll to the .125-40in Tap RH
- ***Hold*** the Ctrl button ***down*** and Click ⑧
- Click ⑨ OK when done

B) Create your own library containing the required tools.

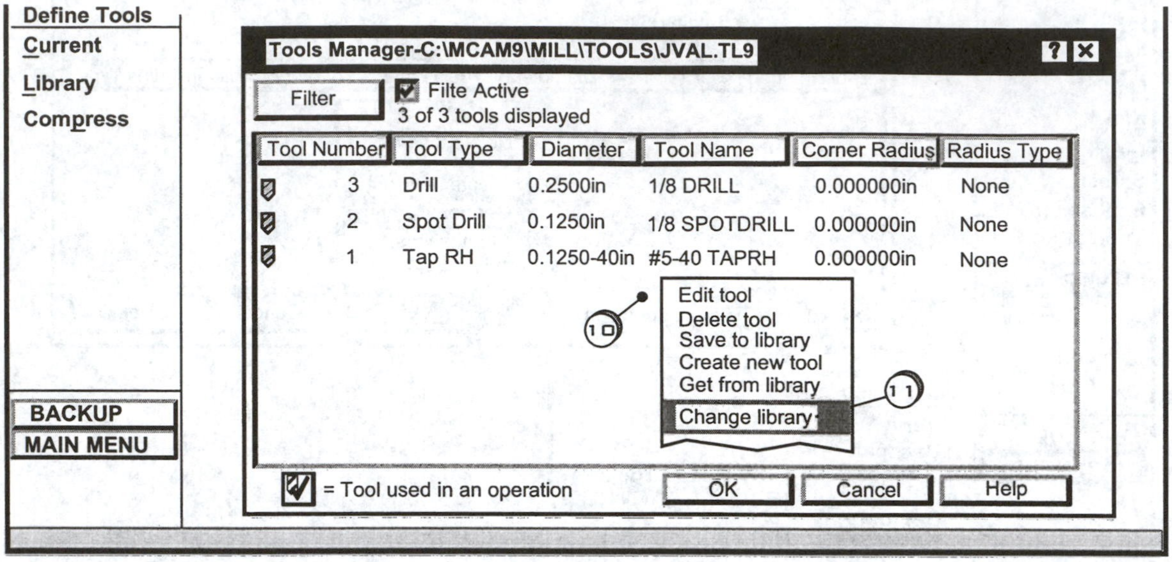

➤ ***Right click*** (10) ***in the white area*** , move the cursor down and Click (11) [Change library]

➤Click (12) in the file name box and enter the name of the new *customized* tool library: **JVAL**

➤Click (13) the [Save] button

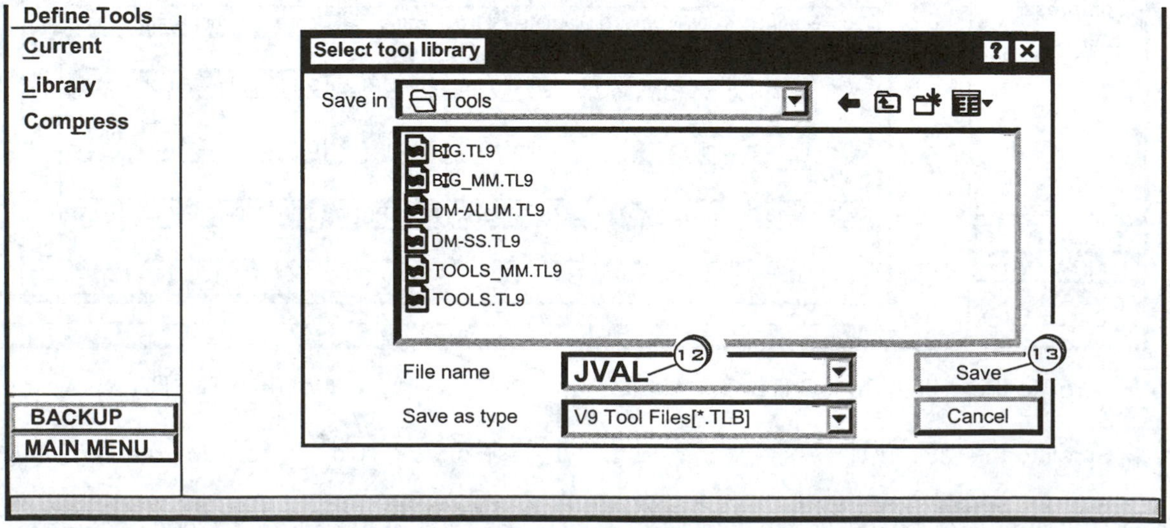

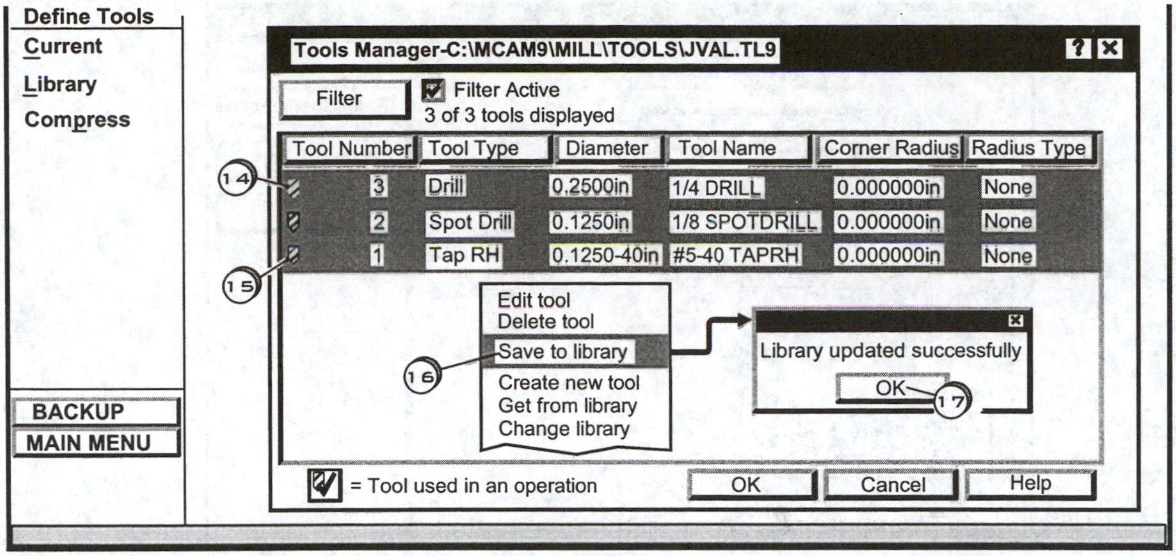

➤ Click (14) on the Drill

➤ Hold the Shift key down and Click (15) on the Tap RH tool.

➤ Release the Shift key and Click (16) Save to library

➤ Click (17) OK

C) Edit any tool dimensions to reflect those given in Table 5-1

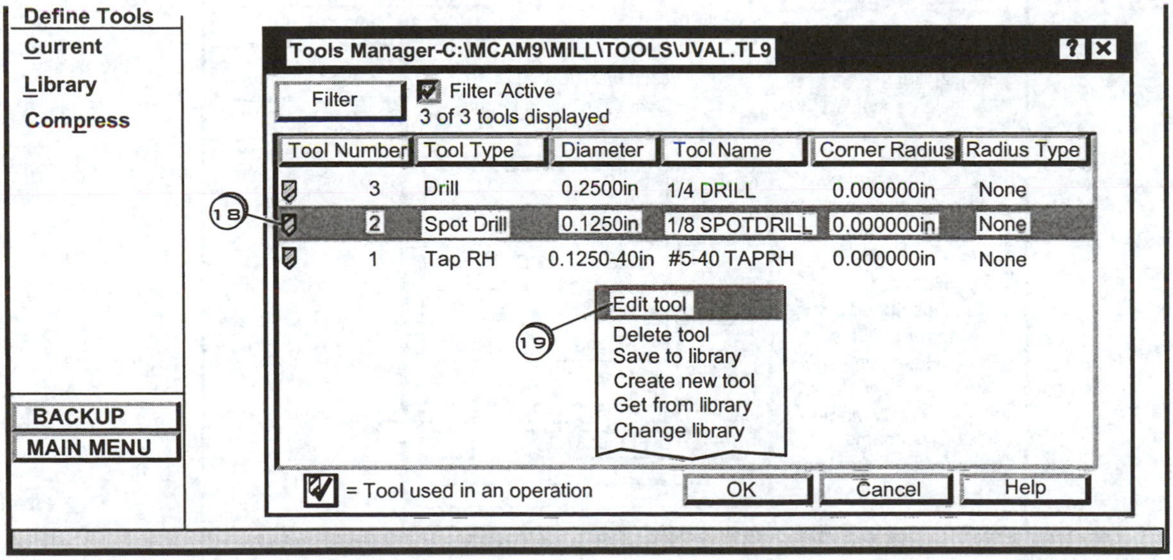

➤ Click (18) on the Spot Drill

➤ Click (19) Edit tool

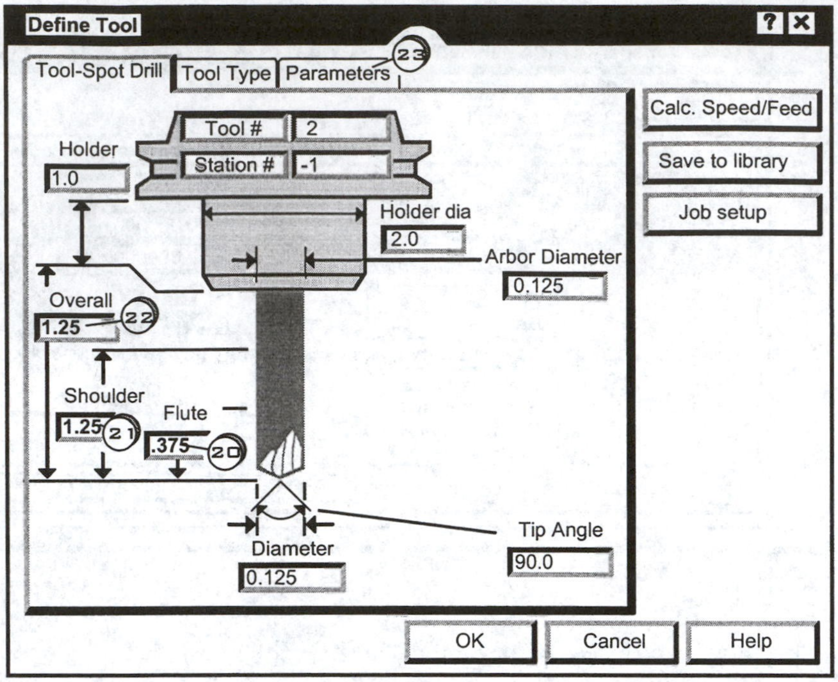

- Click 20 in the Flute box; enter **.375**
- Click 21 in the Shoulder box; enter **1.25**
- Click 22 in the Overall box; enter **1.25**
- Click 23 the Parameters tab

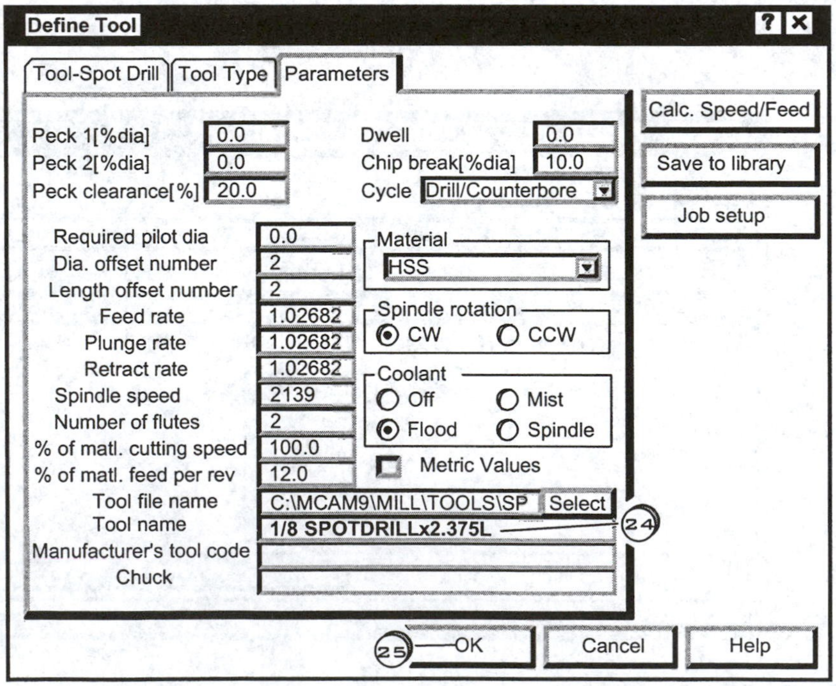

- Click 24 in the Tool name box; enter **1/8 SPOTDRILLx2.375L**
- Click 25 OK

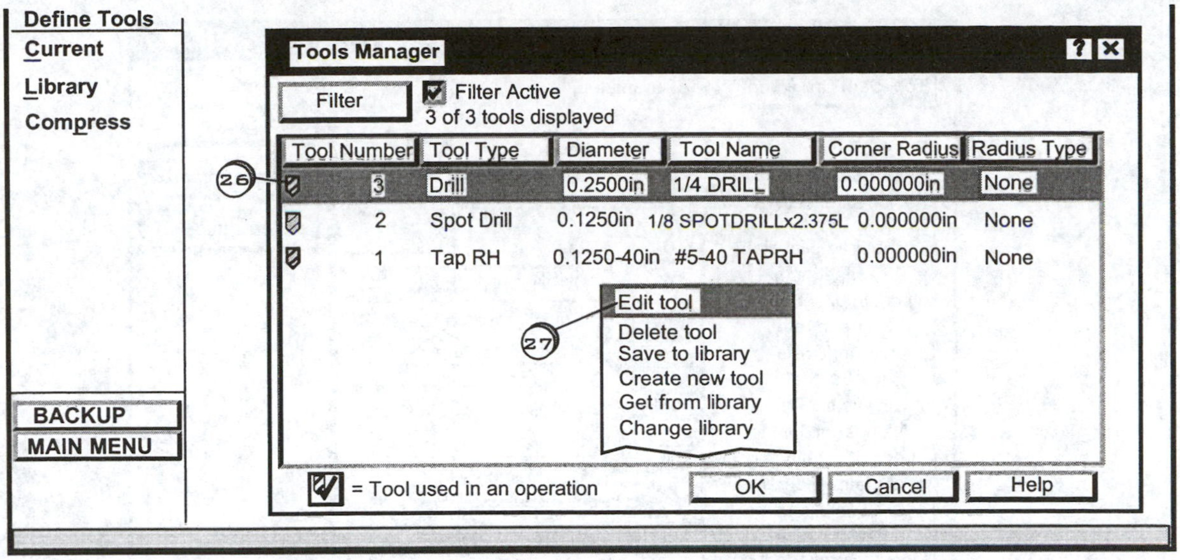

- Click 26 on the Drill
- Click 27 Edit tool

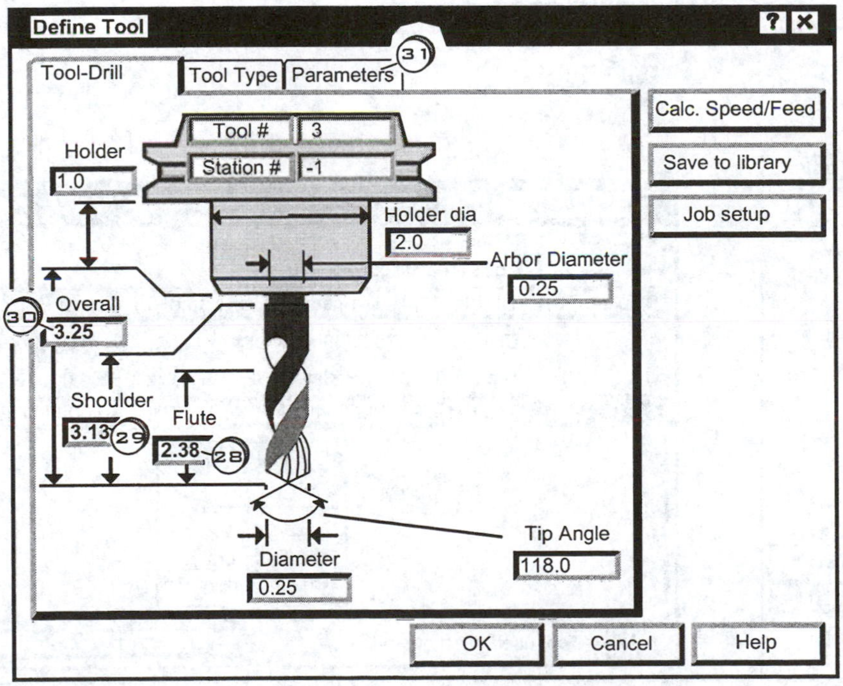

- Click 28 in the Flute box; enter **2.38**
- Click 29 in the Shoulder box; enter **3.13**
- Click 30 in the Overall box; enter **3.25**
- Click 31 the Parameters tab

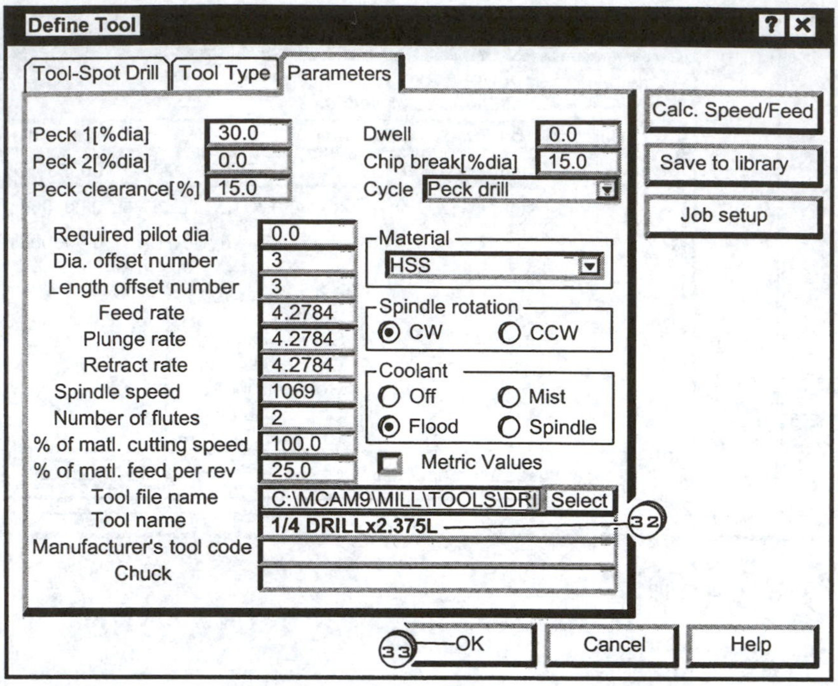

➤ Click (32) in the Tool name box; enter

1/4 DRILLx 2.375L

➤ Click (33) OK

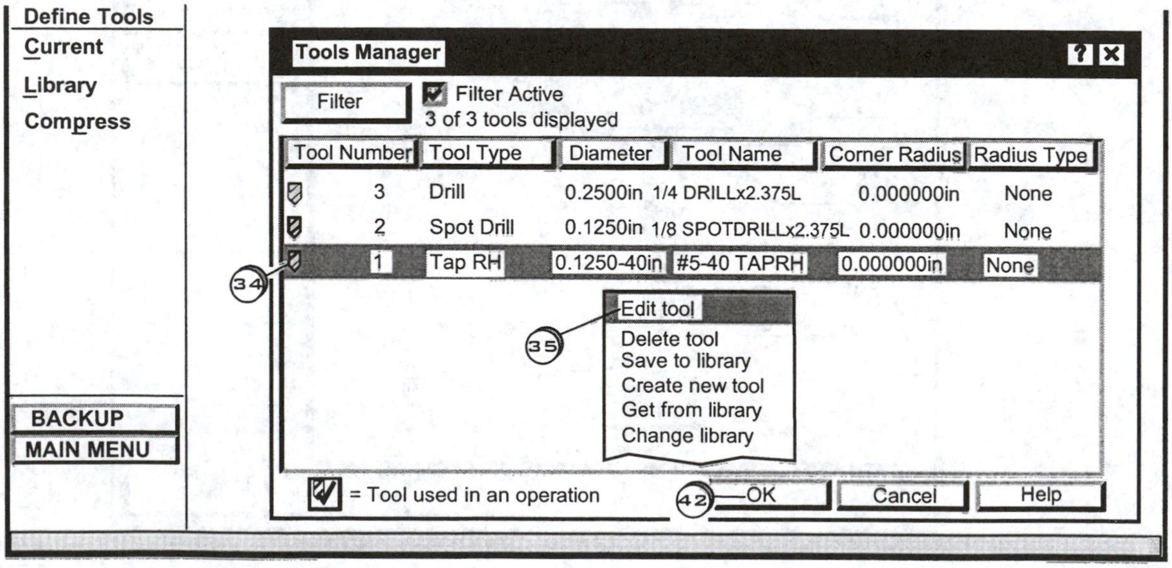

➤ Click (34) on the Tap RH

➤ Click (35) Edit tool

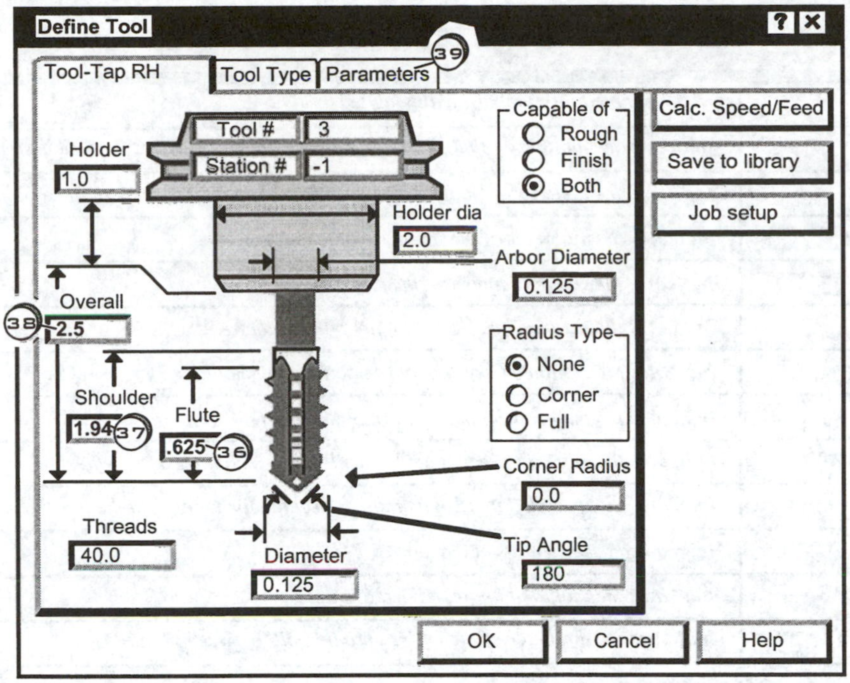

➤ Click (36) in the Flute box; enter **.625**

➤ Click (37) in the Shoulder box; enter **1.94**

➤ Click (38) in the Overall box; enter **2.5**

➤ Click (39) the Parameters tab

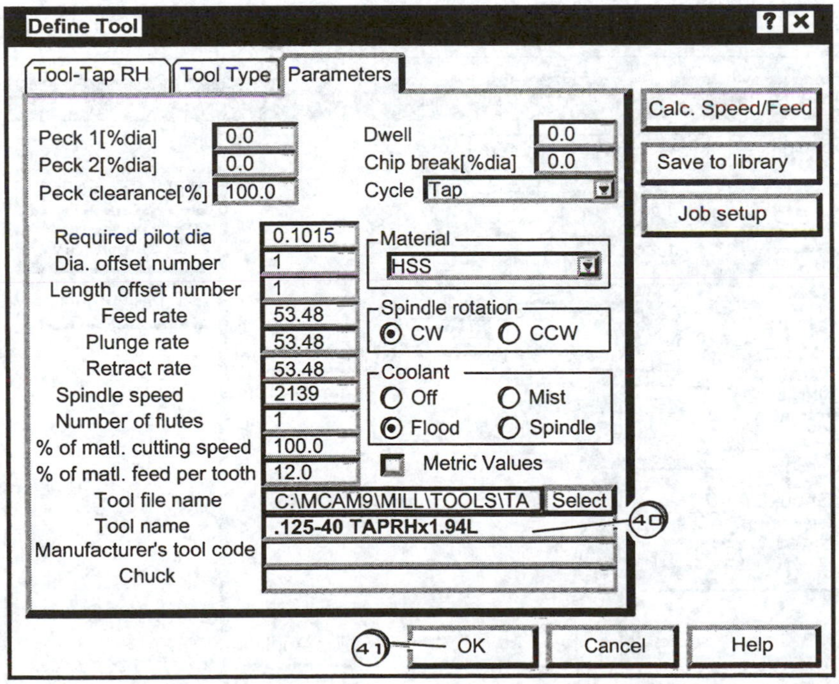

➤ Click (40) in the Tool name box; enter

1/4 FLATx3.125L

➤ Click (41) OK

➤ Click (42) OK . See p5-8

TERMINOLOGY- Define Tool

Tool #	*Indicates the tool number identifying the tool*
Station#	*Indicates the number of the station at the CNC machine where the tool is stored*
Holder	*Indicates the distance from the bottom surface to the top surface of the tool holder*
Holder Diameter	*Indicates the diameter of the top piece of the holder*
Arbor Diameter	*Indicates the shank diameter of the tool*
Overall	*Indicates the distance from the tool tip to the holder end*
Shoulder	*Indicates the distance from the tool tip to the tool shoulder*
Flute	*Indicates the length from the tool tip to the top of the tool flutes*
Shoulder Angle	*Indicates the angle between a center drill's tool's shoulder and centerline*
Threads	*Indicates the threads/inch(English units) or pitch for tapping tools*
Corner Radius	*Indicates corner radius at the tool tip*
Tip Angle	*Indicates the included angle of the tool's tip*
Outside Diameter	*Indicates the diameter of the entire chamfer mill or face mill tool*
Diameter	*Indicates the tool diameter*
Taper Angle	*Indicates the angle between the tool's centerline and outer cylindrical surface*
◉	*Indicates the type of machining operation:* ○rough ○finish ○both *for the tool*
Radius Type ◉	*Indicates the type of radius:* ○none ○corner ○full *for the tool*
Calc. Speed/Feed	Mastercam ***automatically assigns*** *the tool's* ***speed, feed, plunge and retract rates*** *(the stock* ***material must first be selected before*** *this function can be used)*

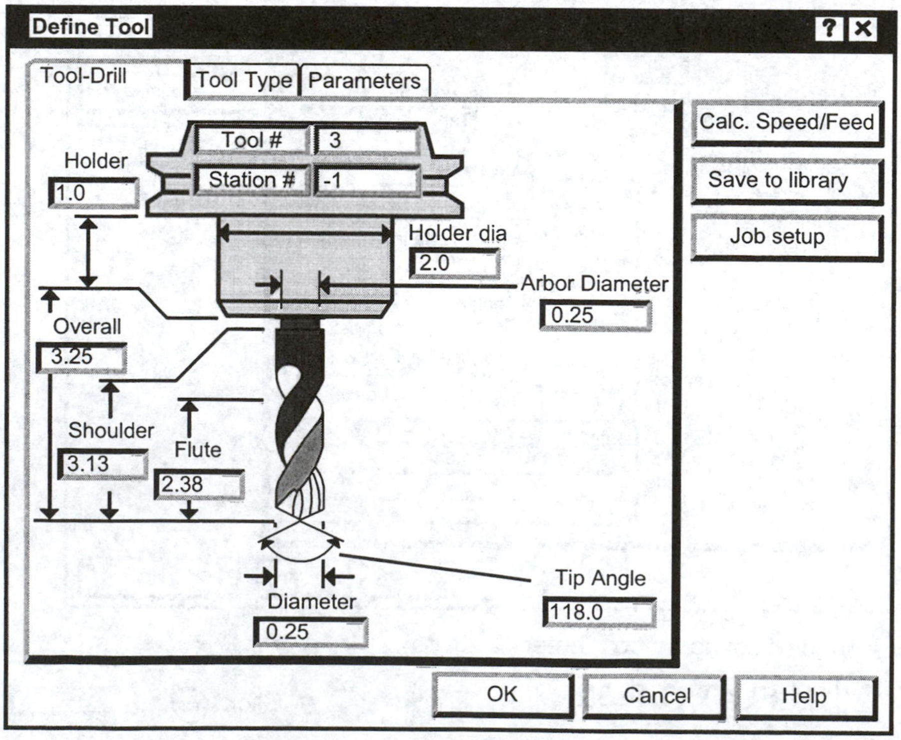

TERMINOLOGY- Parameters

1st Peck	*Indicates the drill depth of the first peck as a % of tool diameter(used for peck drill and chip break cycles only).*
Subsequent Peck	*Indicates the drill depth of pecks following the first peck as a % of tool diameter (used for peck drill and chip break cycles only).*
Peck Clear	*Indicates the depth the tool rapids down to between each peck*
Shift Value	*Indicates the distance a boring bar moves away from the wall of the toolpath before retracting to prevent gouging(used for a bore bar tool only)*
Dwell	*Indicates the time(in seconds) to remain at the bottom of a hole during drilling*
Retract	*Indicates the height the tool rapid retracts to just before it makes a peck move (used for chip break cycles only).*
Cycle	*Sets one of eight different types of drillling cycles:* Drill/Counterbore Peck Drill Chip Break, Tapping, Bore#1, Bore#2, Misc.#1, Misc.#2
Thread Root Diameter	*Indicates the root diameter of the threads*
Required Pilot Diameter	*Indicates the minimum diameter of the tool necessary to enter the toolpath (this is used by the system for reference only)* 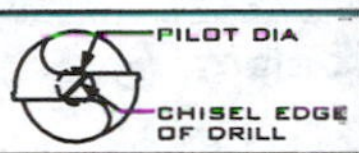
Diameter Offset Number	*Indicates the register at the CNC machine that stores the cutter radius offset value*
Length Offset Number	*Indicates the register at the CNC machine that stores the tool length offset value*
Feed Rate	*Indicates speed of the feed movement of the cutter (IPM or MMPM)*
Plunge Rate	*Indicates speed of the tool's plunge movement into the stock(-Z direction only)*
Retract Rate	*Indicates speed at rapid of the tool's movement out of the stock(+Z direction only)*
Spindle Speed	*Indicates spindle speed(RPM) for the tool*
Number of Flutes	*Indicates number of flutes on the tool*

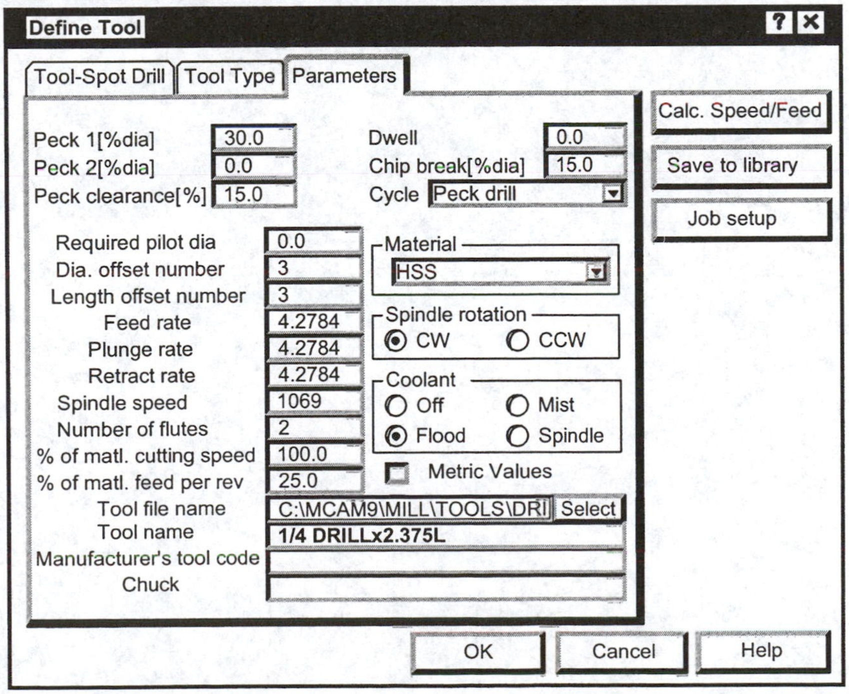

TERMINOLOGY- Parameters

Tool Name	*Indicates the name that has been created by the operator to identify the tool*
Comment	*Indicates any special comment the operator enters to further describe the tool*
Manufacturers Tool Code	*Indicates company from which the tool was purchased*
Chuck	*Indicates information concerning the chuck at the CNC machine*
Material	*Sets the tool material which Mastercam uses to automatically determine tool speed, feed and plunge rate. Six material types are available:* HSS-high speed steel Carbide C Carbide Ceramic Borzon Unknown
Spindle Rotation	*Indicates the direction of spin for the tool at the CNC machine: (CW)-clockwise or (CCW)-counterclockwise direction*
Coolant	*Indicates the coolant type to be used with the tool:* Off *- coolant is off* Flood *- coolant on(flood mode)* Mist *- coolant on(mist mode)* Tool *- coolant on(thru tool)*
Metric Values/Inch Values	*Indicates the current tool uses metric or inch values*

5-3 Specifying Drilling Locations on the CAD Model of a Part

The several methods of specifying the drilling points on a CAD model for a part are presented in this section.

Specifying Drilling Points

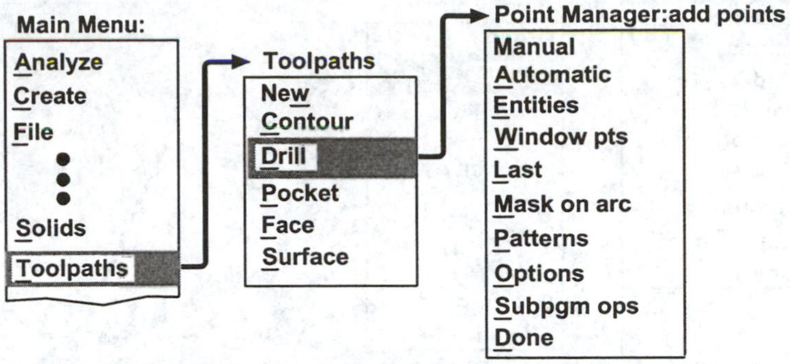

Manual **Origin**

The operator specifies the ***X and Y coordinates*** of each drilling point

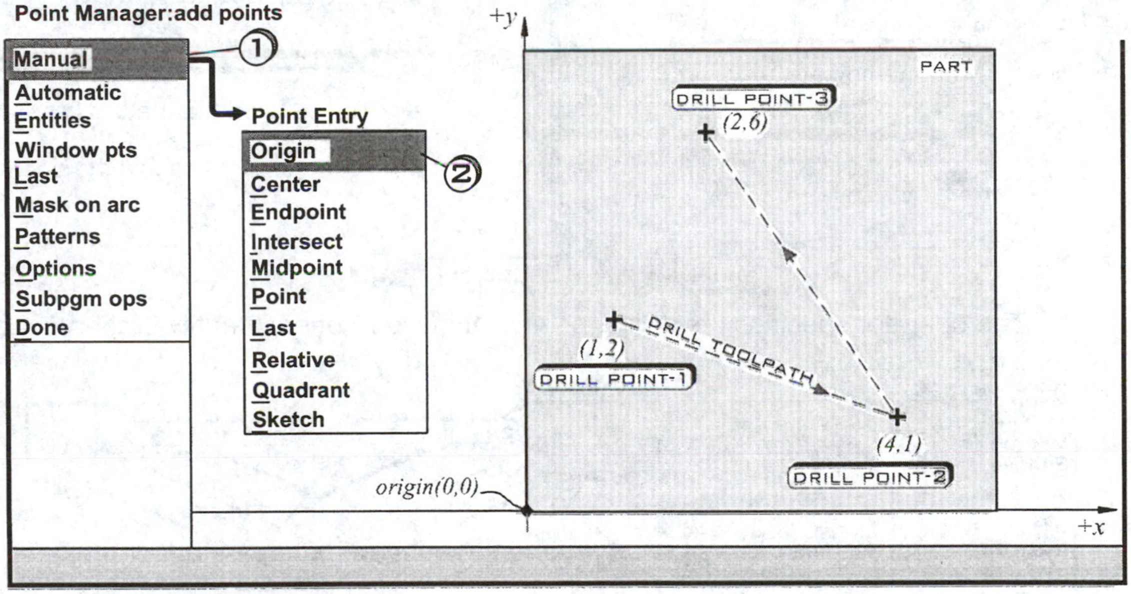

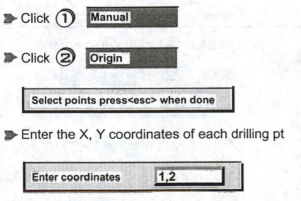

- Click ① **Manual**
- Click ② **Origin**

Select points press<esc> when done

- Enter the X, Y coordinates of each drilling pt

Enter coordinates **1,2**

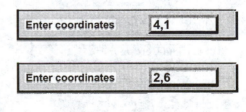

Enter coordinates **4,1**

Enter coordinates **2,6**

- Press **Esc** to cancel the operation

Manual **Center**

The operator specifies the drilling point is at the ***center*** of each arc clicked

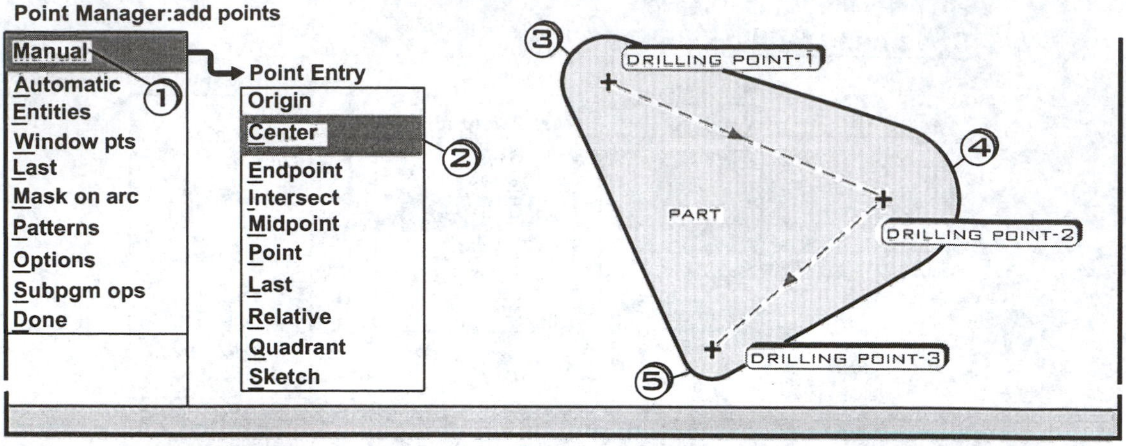

- Click ① **Manual**
- Click ② **Center**

Select an arc

- Click ③ on the first arc
- Click ② **Center**

Select an arc

- Click ④ on the second arc
- Click ② **Center**

Select an arc

- Click ⑤ on the the third arc
- Press **Esc** to cancel the operation

Manual **Endpoint**

The operator specifies the drilling point is at the ***end*** of each entity clicked

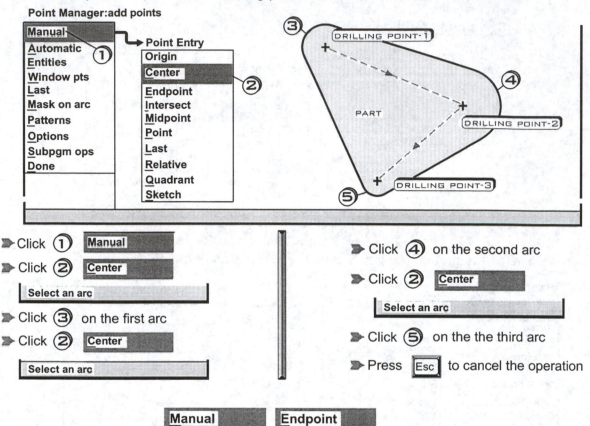

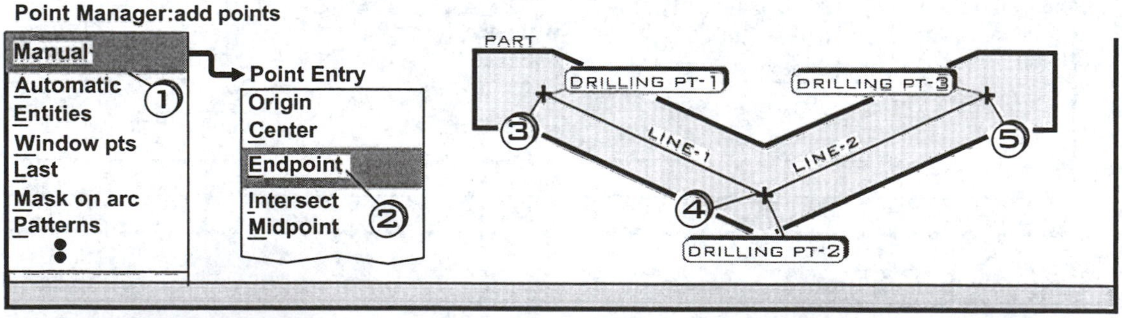

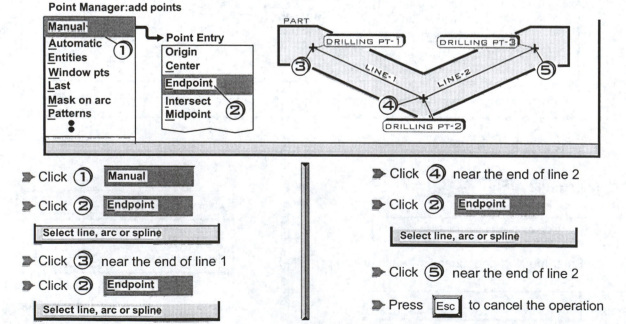

- Click ① **Manual**
- Click ② **Endpoint**

Select line, arc or spline

- Click ③ near the end of line 1
- Click ② **Endpoint**

Select line, arc or spline

- Click ④ near the end of line 2
- Click ② **Endpoint**

Select line, arc or spline

- Click ⑤ near the end of line 2
- Press **Esc** to cancel the operation

Manual **Intersect**

The operator specifies the drilling point is at the ***intersection*** of two entities clicked

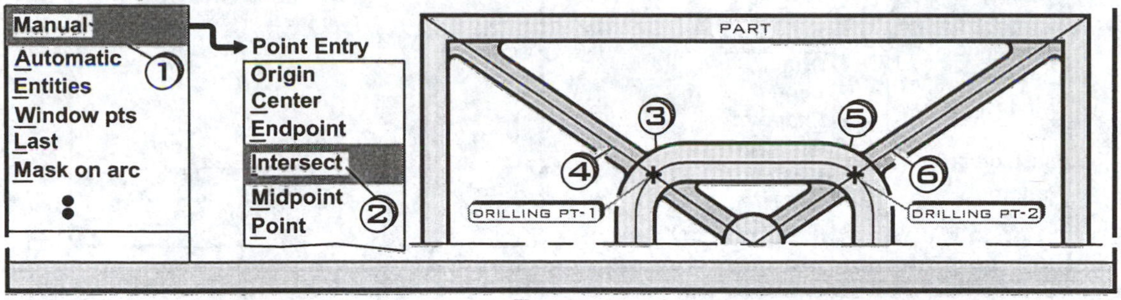

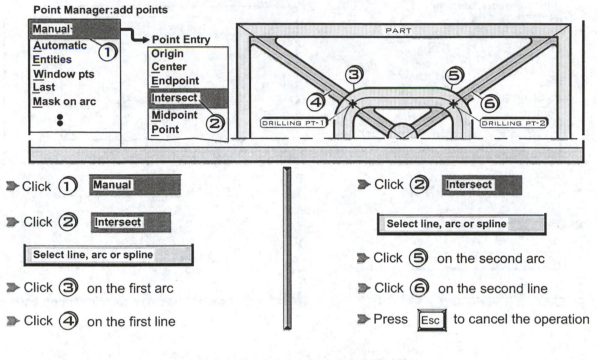

- Click ① **Manual**
- Click ② **Intersect**

Select line, arc or spline

- Click ③ on the first arc
- Click ④ on the first line
- Click ② **Intersect**

Select line, arc or spline

- Click ⑤ on the second arc
- Click ⑥ on the second line
- Press **Esc** to cancel the operation

Manual **Midpoint**

The operator specifies the drilling point is at the ***midpoint*** of each entity clicked

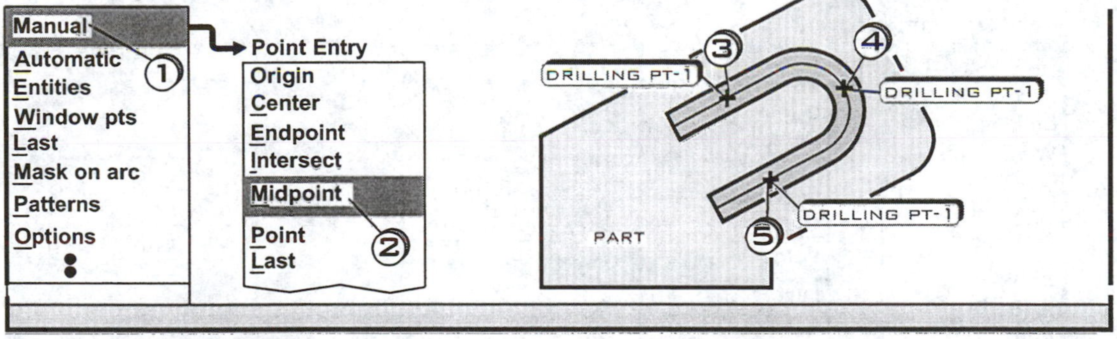

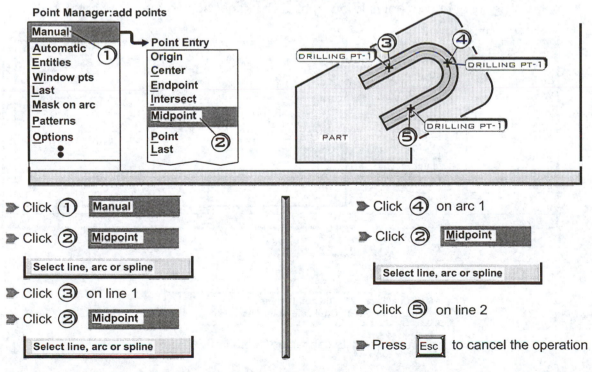

- Click ① **Manual**
- Click ② **Midpoint**

Select line, arc or spline

- Click ③ on line 1
- Click ② **Midpoint**

Select line, arc or spline

- Click ④ on arc 1
- Click ② **Midpoint**

Select line, arc or spline

- Click ⑤ on line 2
- Press **Esc** to cancel the operation

Manual **Point**

The operator specifies the drilling point is at the ***existing point*** clicked

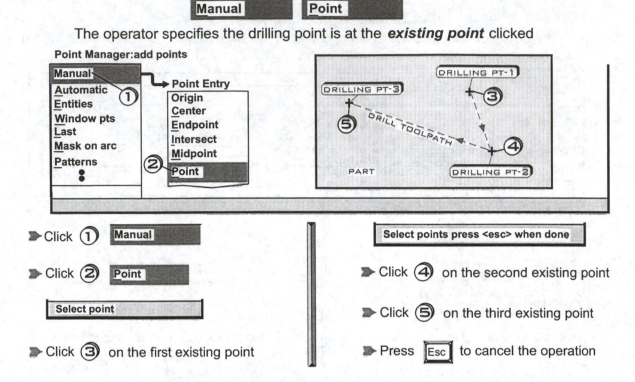

Point Manager:add points

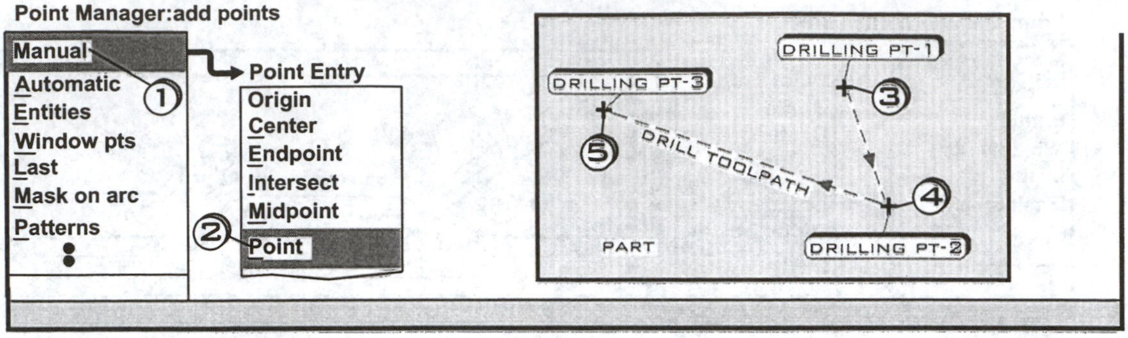

➤ Click ① **Manual**

➤ Click ② **Point**

Select point

➤ Click ③ on the first existing point

Select points press <esc> when done

➤ Click ④ on the second existing point

➤ Click ⑤ on the third existing point

➤ Press Esc to cancel the operation

Manual **Last**

The operator specifies that *Mastercam* is to ***copy the last set*** of drilling point locations and ***use these as the new drill point set*** locations. Useful for drilling a set of pilot holes followed by drilling and tapping the same set.

Point Manager:add points

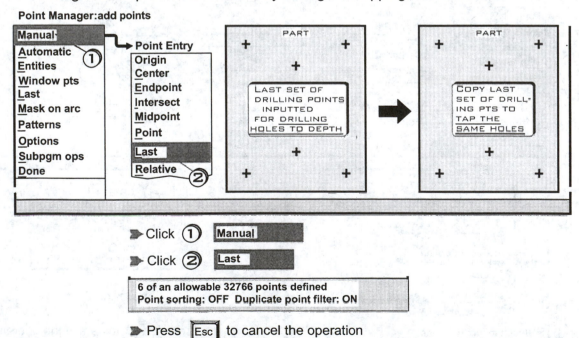

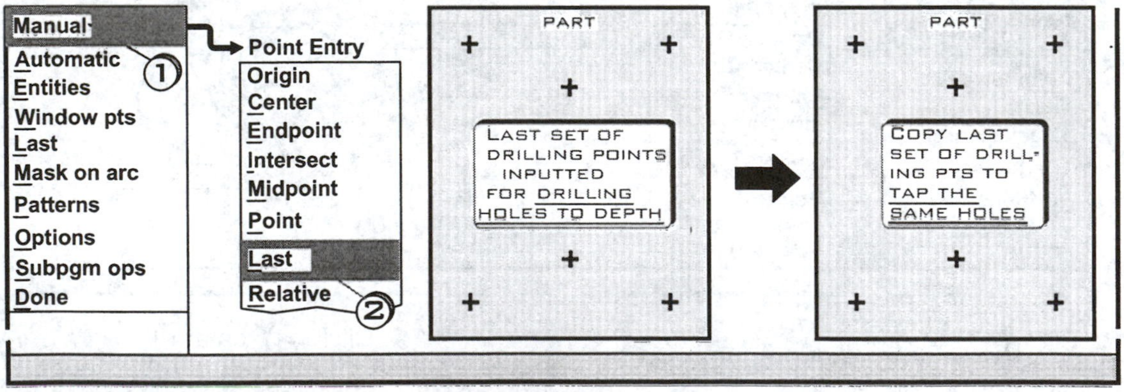

➤ Click ① **Manual**

➤ Click ② **Last**

6 of an allowable 32766 points defined
Point sorting: OFF Duplicate point filter: ON

➤ Press Esc to cancel the operation

Manual **Relative**

The operator specifies the drilling point is taken ***relative to*** an ***existing point*** clicked

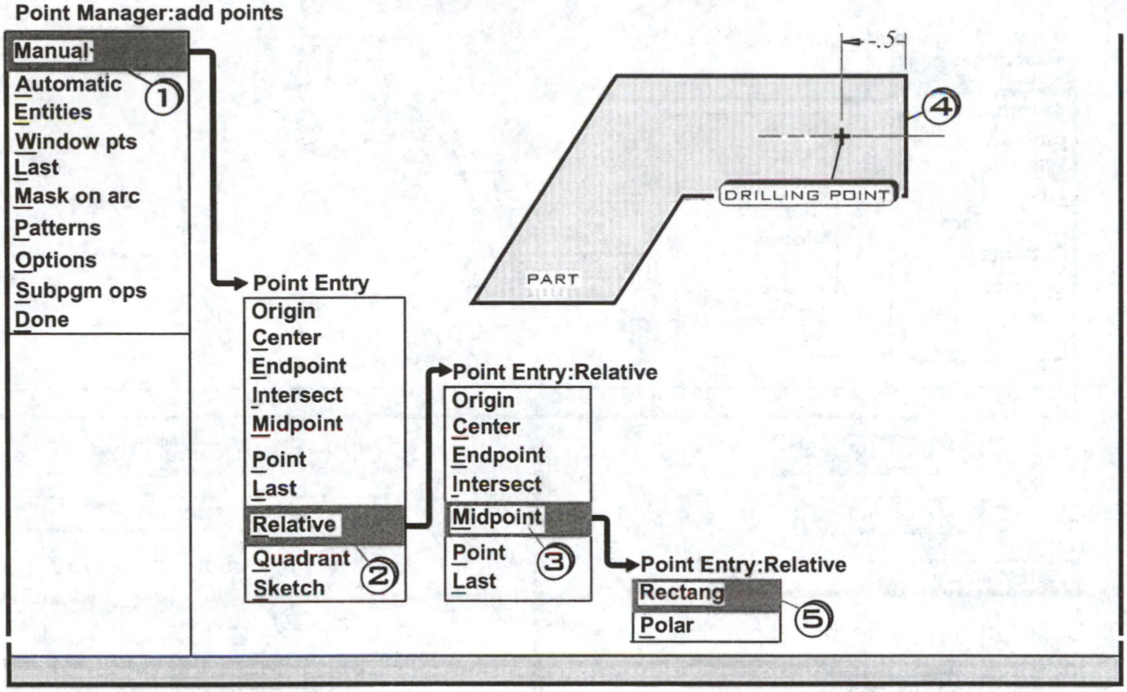

- Click ① **Manual**
- Click ② **Relative**
- Click ③ **Midpoint**

Select line, arc or spline

- Click ④ on the line from which relative distances will be measured

- Click ⑤ **Rectang**

Enter relative coordinates -.5,0

- Enter the relative X, Y values **-.5, 0**

- Press Esc to cancel the operation

Manual **Quadrant**

The operator specifies the drilling point is located at the *0°, 90°, 180°, 270°* ***quadrant position*** on the ***circumference of an arc or circle*** clicked

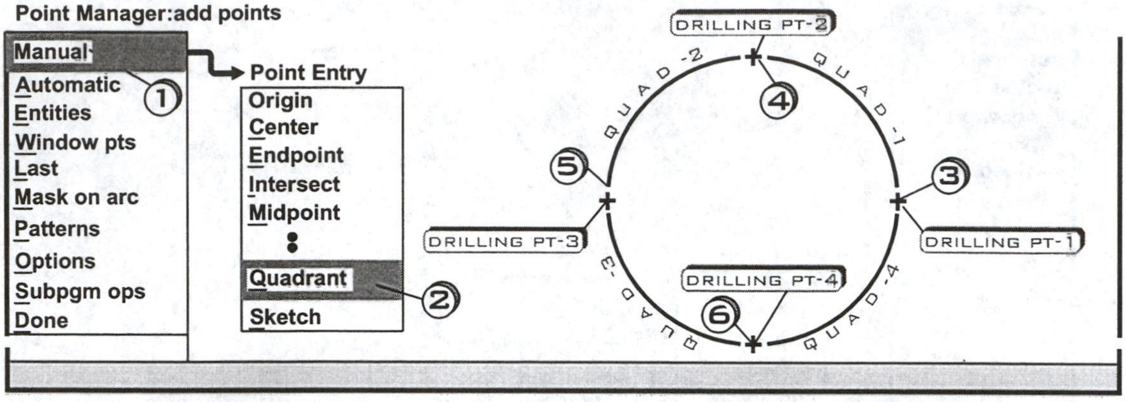

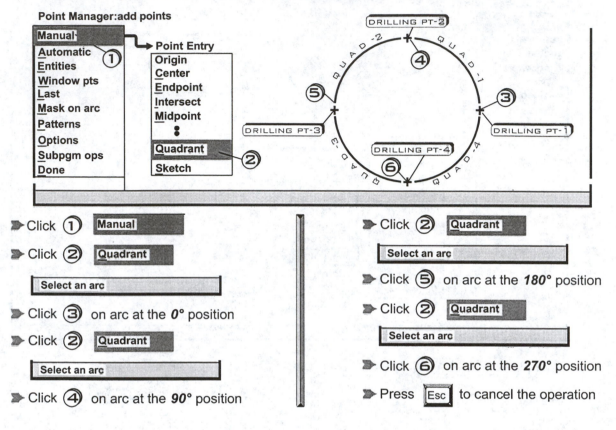

- Click ① **Manual**
- Click ② **Quadrant**

Select an arc

- Click ③ on arc at the ***0°*** position
- Click ② **Quadrant**

Select an arc

- Click ④ on arc at the ***90°*** position
- Click ② **Quadrant**

Select an arc

- Click ⑤ on arc at the ***180°*** position
- Click ② **Quadrant**

Select an arc

- Click ⑥ on arc at the ***270°*** position
- Press **Esc** to cancel the operation

Manual **Sketch**

The operator specifies the drilling point is located at ***selected mouse clicks***

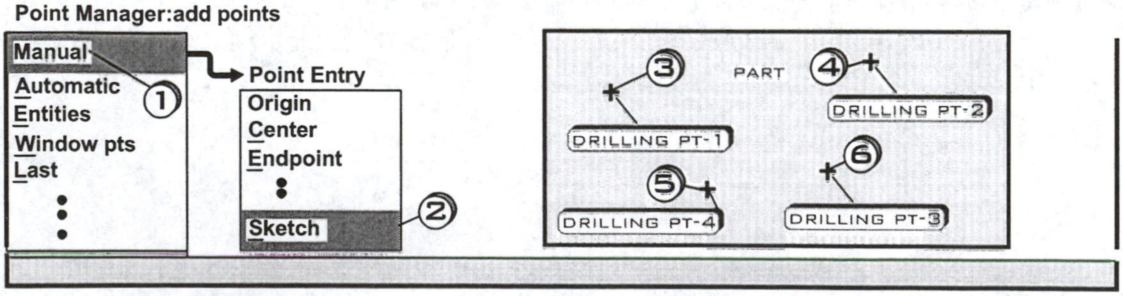

- Click ① **Manual**
- Click ② **Sketch**

Select points press <esc> when done

- Click ③ ④ ⑤ ⑥ the drill point locations
- Press **Esc** to cancel the operation

Automatic

The operator automatically specifies a ***set of drilling points from an existing array of points:*** ● ***first*** point clicked indicates the ***first hole to be drilled***,
● ***second*** point clicked indicates the ***tool path direction***
● ***third*** point clicked indicates the ***last hole to be drilled.***

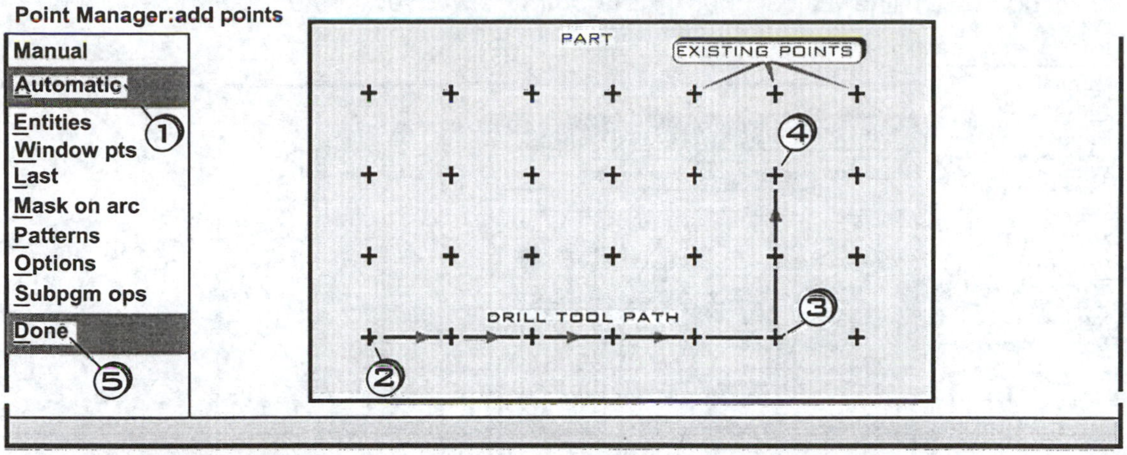

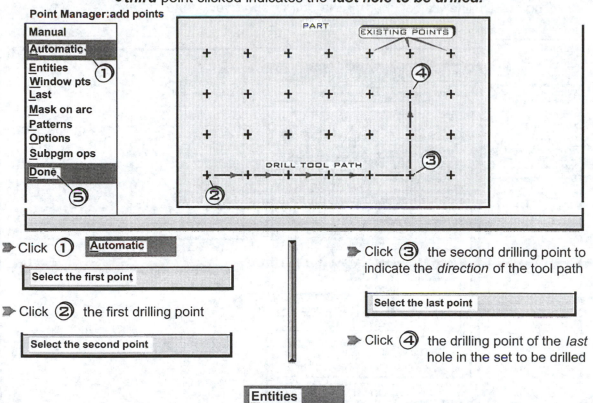

➤ Click ① **Automatic**

Select the first point

➤ Click ② the first drilling point

Select the second point

➤ Click ③ the second drilling point to indicate the *direction* of the tool path

Select the last point

➤ Click ④ the drilling point of the *last* hole in the set to be drilled

Entities

The operator specifies the drilling points are located at the ***ends*** of existing entities clicked

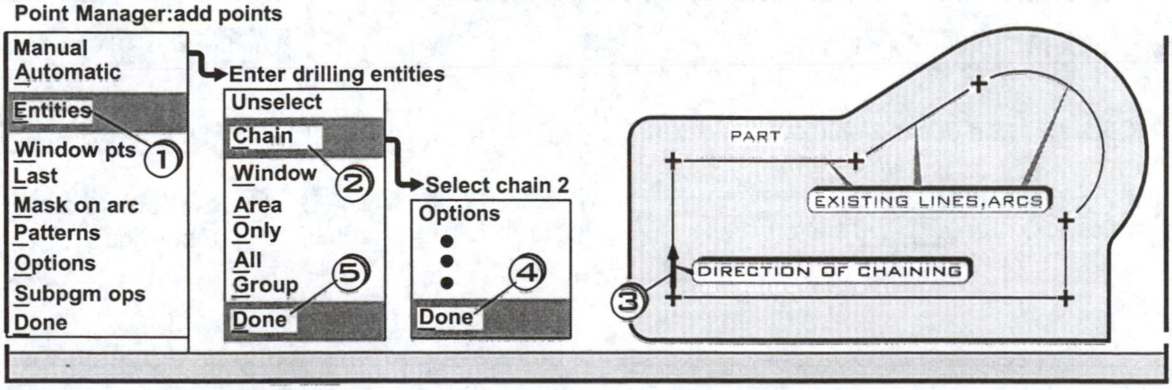

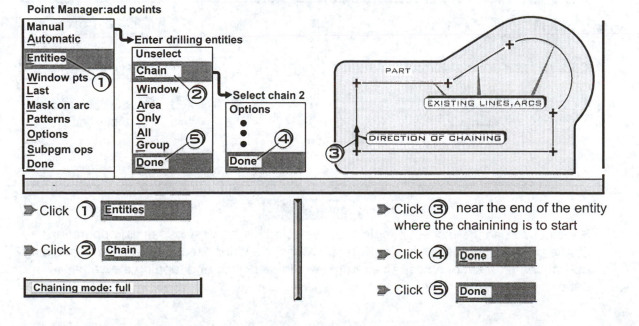

➤ Click ① **Entities**

➤ Click ② **Chain**

Chaining mode: full

➤ Click ③ near the end of the entity where the chainining is to start

➤ Click ④ **Done**

➤ Click ⑤ **Done**

Window pts

The operator ***first selects*** a point ***sorting pattern*** to indicate the ***type of tool path pattern*** the drill tool is to follow when machining at the point locations in the set. The operator then clicks a ***rectangular window*** to enclose the existing set of points. Drilling will occur at ***all the points enclosed*** in the rectangular window.

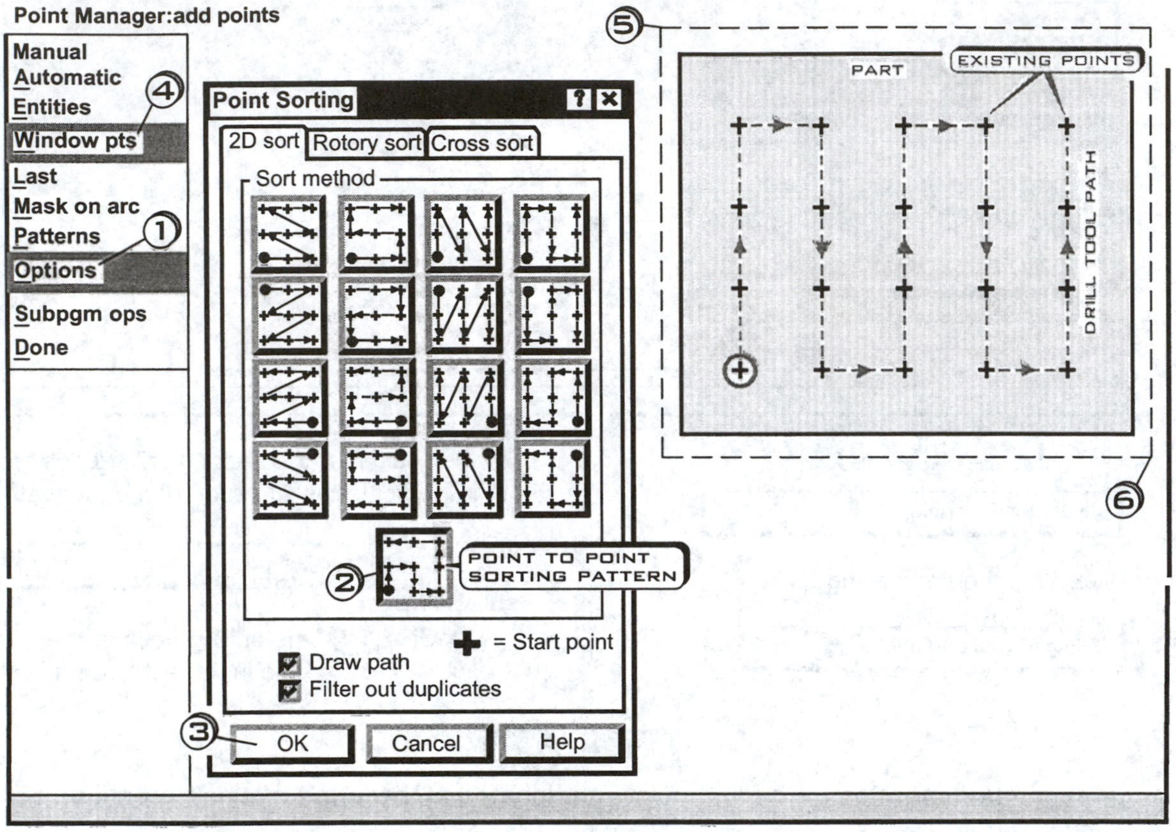

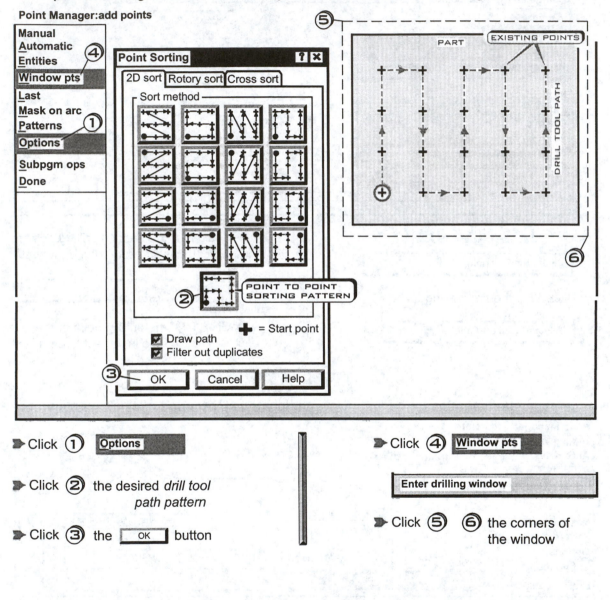

- Click ① **Options**
- Click ② the desired *drill tool path pattern*
- Click ③ the OK button
- Click ④ **Window pts**

Enter drilling window

- Click ⑤ ⑥ the corners of the window

Last

Specifies to use the *last* set of points just selected for the *previous* drilling operation as the point set for the *current* drilling operaton. Useful for executing *multiple machining* operations on a set of holes. An example would be drilling and tapping the same set of holes.

Mask on arc

Directs Mastercam to *automatically find and select* the *center* of all ***holes that match the diameter*** of an ***initial*** hole clicked.

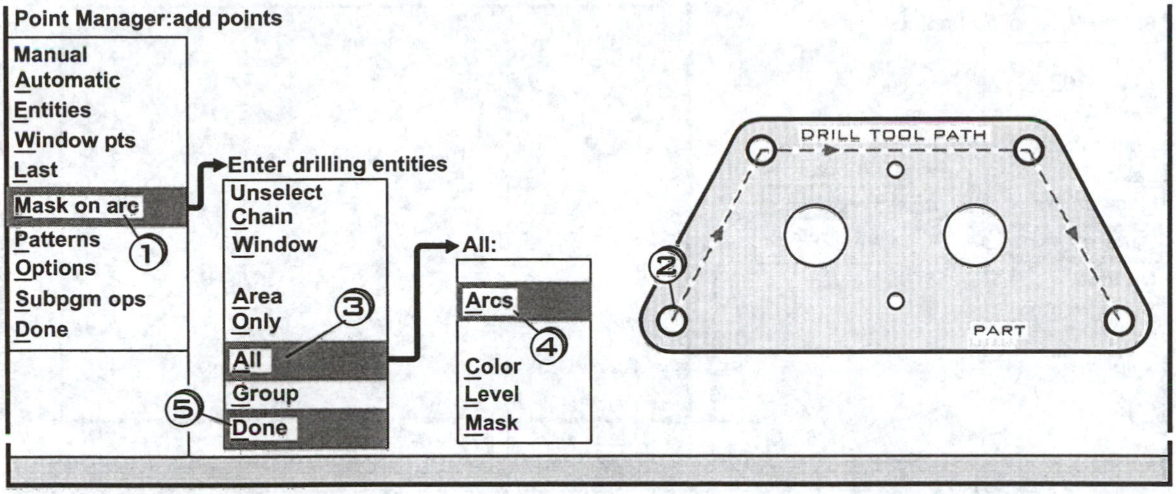

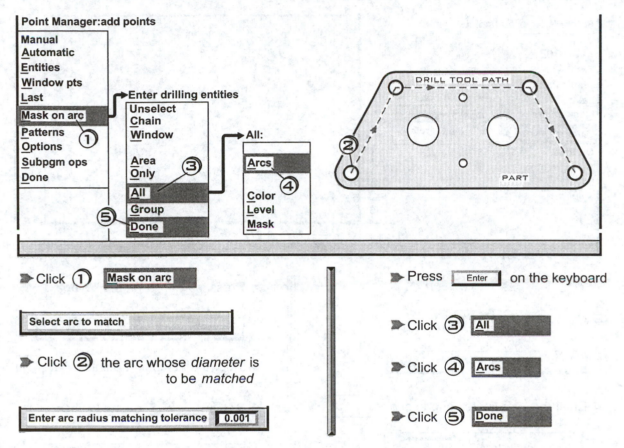

➤ Click ① **Mask on arc**

Select arc to match

➤ Click ② the arc whose *diameter* is to be *matched*

Enter arc radius matching tolerance **0.001**

➤ Press **Enter** on the keyboard

➤ Click ③ **All**

➤ Click ④ **Arcs**

➤ Click ⑤ **Done**

Patterns **Rectangular**

Directs Mastercam to ***automatically create*** a set of drill points according to a ***rectangular grid*** *pattern.*

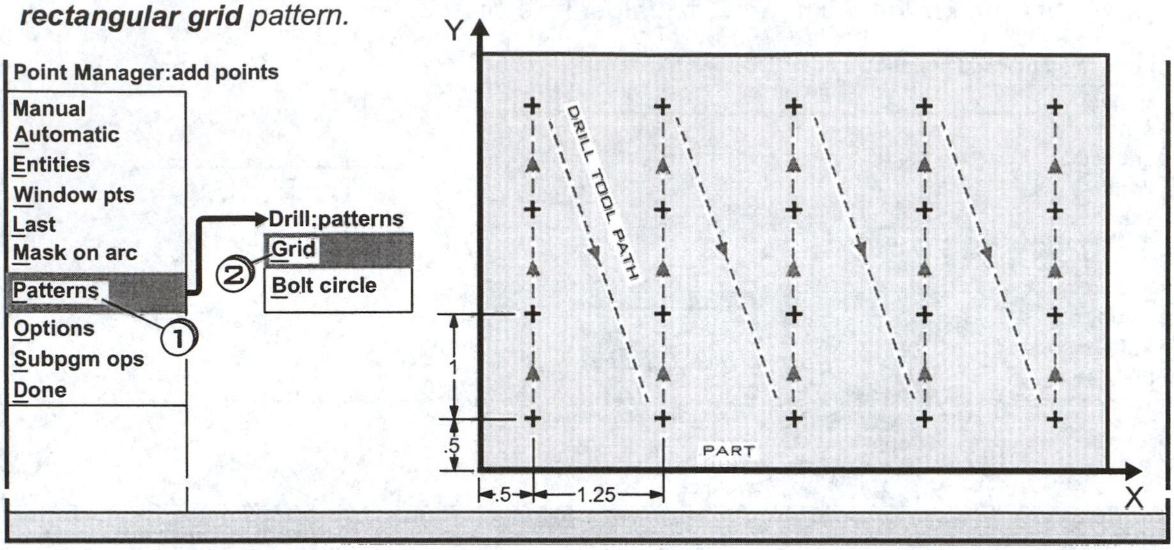

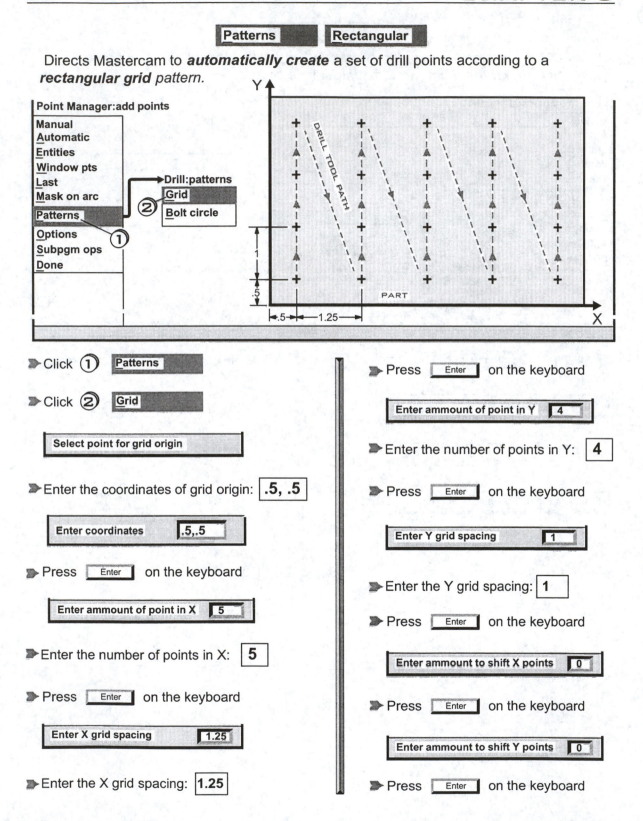

- Click ① **Patterns**
- Click ② **Grid**

Select point for grid origin

- Enter the coordinates of grid origin: **.5, .5**

Enter coordinates .5,.5

- Press Enter on the keyboard

Enter ammount of point in X 5

- Enter the number of points in X: **5**
- Press Enter on the keyboard

Enter X grid spacing 1.25

- Enter the X grid spacing: **1.25**
- Press Enter on the keyboard

Enter ammount of point in Y 4

- Enter the number of points in Y: **4**
- Press Enter on the keyboard

Enter Y grid spacing 1

- Enter the Y grid spacing: **1**
- Press Enter on the keyboard

Enter ammount to shift X points 0

- Press Enter on the keyboard

Enter ammount to shift Y points 0

- Press Enter on the keyboard

Patterns **Bolt circle**

Directs Mastercam to ***automatically create*** a set of drill points according to a ***bolt circle*** *pattern.*

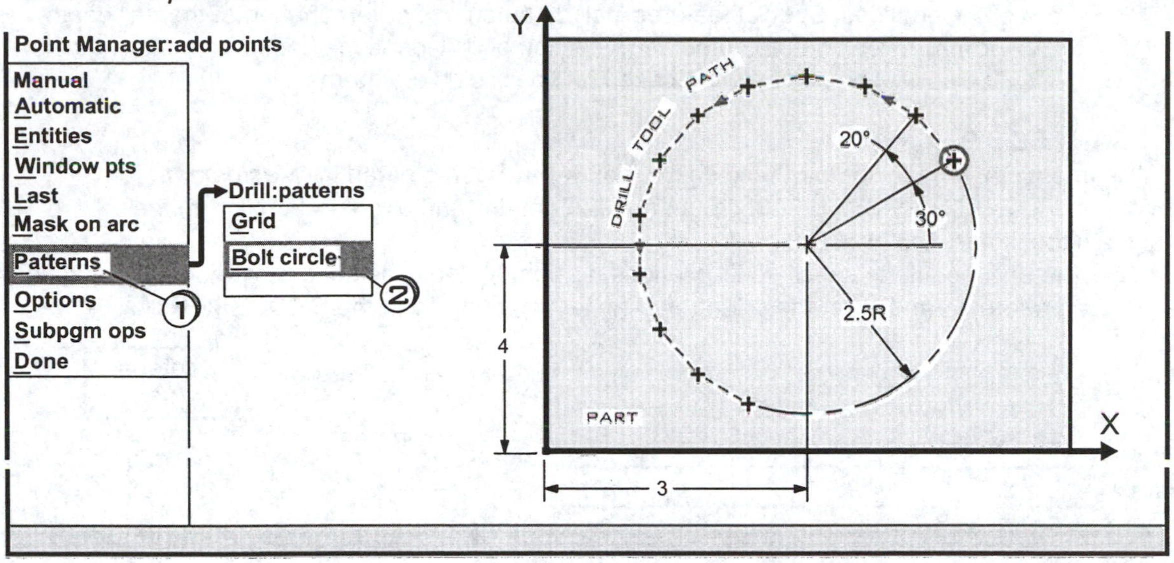

- Click ① **Patterns**
- Click ② **Bolt circle**

Select bolt circle origin

- Enter bolt circle origin: **3, 4**

Enter coordinates 3,4

- Press Enter on the keyboard

Enter bolt circle radius 2.5

- Enter bolt circle radius: **2.5**
- Press Enter on the keyboard

Enter start angle(0=3 o'clock) 30

- Enter the start angle **30**
- Press Enter

Enter angle between points 20

- Enter the angle between points **20**
- Press Enter

Enter number of points 12

- Enter number of points: **12**
- Press Enter on the keyboard

5-4 Using the Drill Module on A CAD Model

Drilling operations are carried out with the use of *Mastercam's* drill module. The module contains cycles for drilling, tapping and boring the CAD model at selected point locations. Special features include automatic determination of the drill depth for *thru* holes and the automatic selection of drill locations based on matching hole size as well as point sorting based on specific hole patterns.

EXAMPLE 5-2

Assume a CAD model the part face shown below has been created in *Mastercam*.

- Open the tool library **JVAL.TL9**: create a .032 Spot drill and add a #43 drill from the main tool library **TOOLS.TL9**
- Use the drill module to specify drill locations, assign the tools, select the type of hole operations to be preformed and enter the drilling parameters as outlined in the PROCESS PLAN 5-2

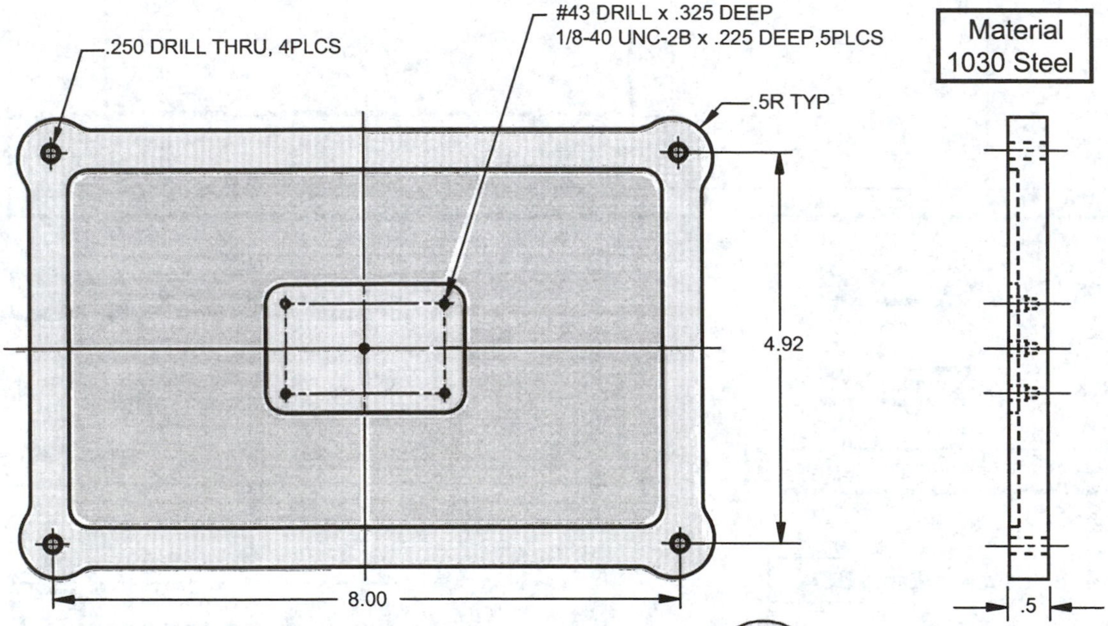

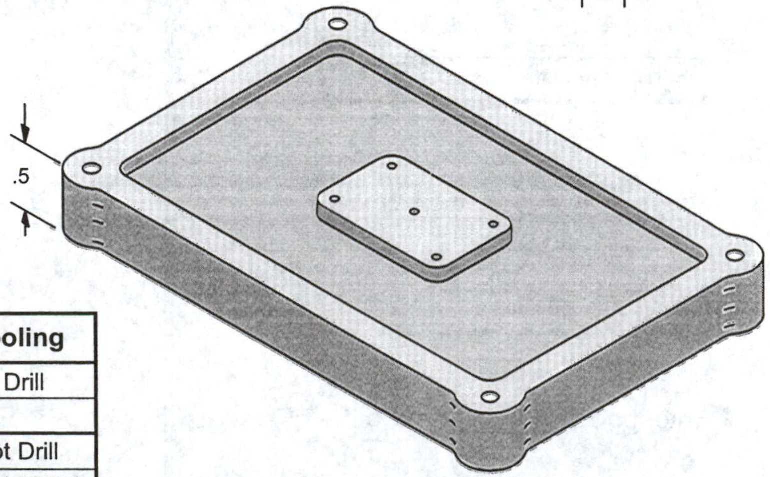

PROCESS PLAN 5-2

Operation	Tooling
Spot Drill x .166 Deep	1/8 Spot Drill
Peck Drill Through	1/4 Drill
Spot Drill x .125 Deep	1/32 Spot Drill
Peck Drill x .325 Deep	#43 Drill(.089in)
Tap x .225 Deep	1/8-40 UNC Tap

Figure 5-1

A) ENTER *Mastercam's* JOB SETUP DIALOG BOX

- SELECT SIZE OF THE STOCK

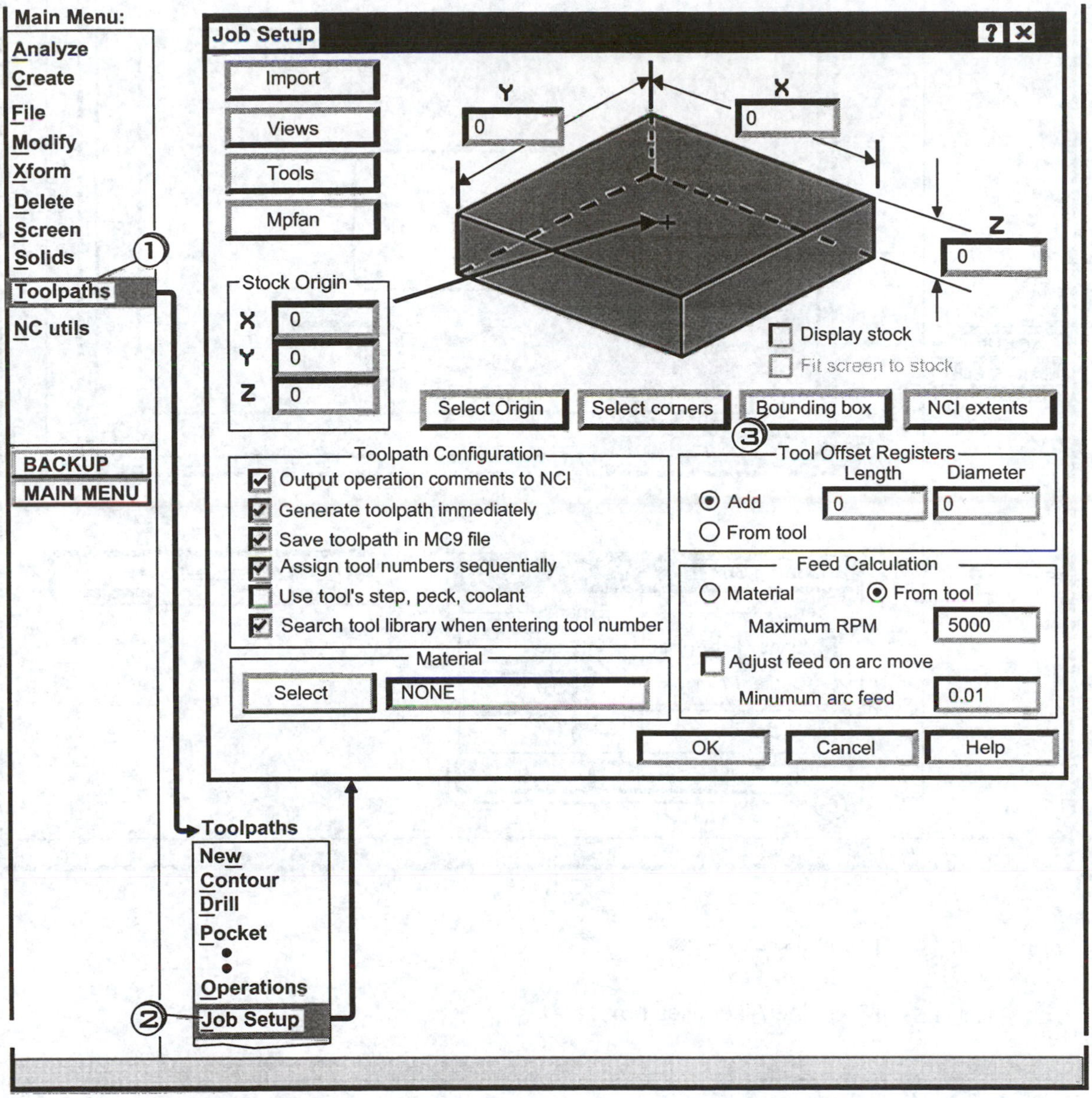

- Click ① Toolpaths
- Click ② Job Setup
- Click ③ Bounding box to direct *Mastercam* to automatically compute the stock size based on a box that just fits around the CAD model.

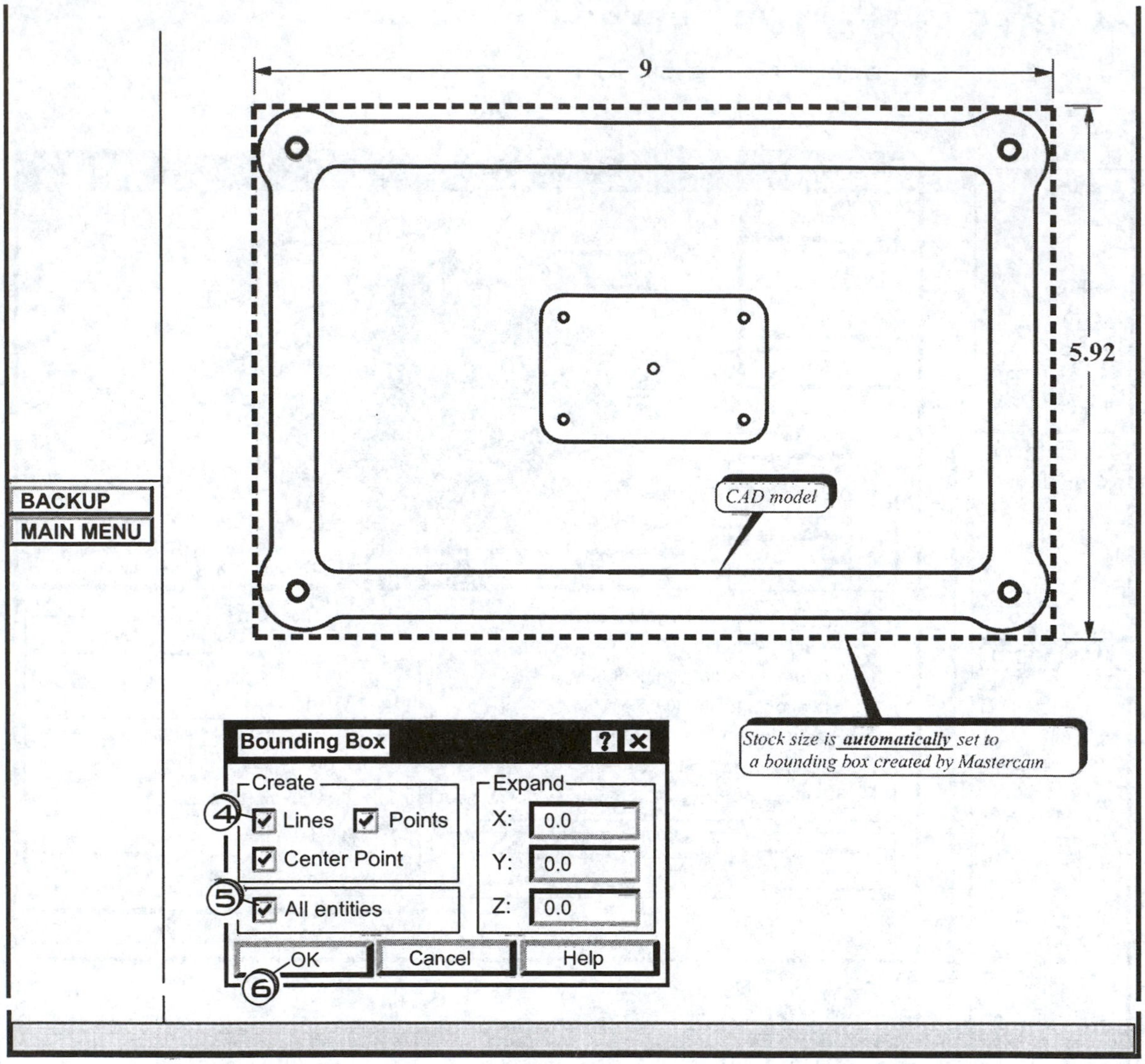

➤ Click ④ ☑ in the Lines box

➤ Click ⑤ ☑ in the All entities box

➤ Click ⑥ OK

◆ IDENTIFY THE STOCK MATERIAL TO BE MACHINED

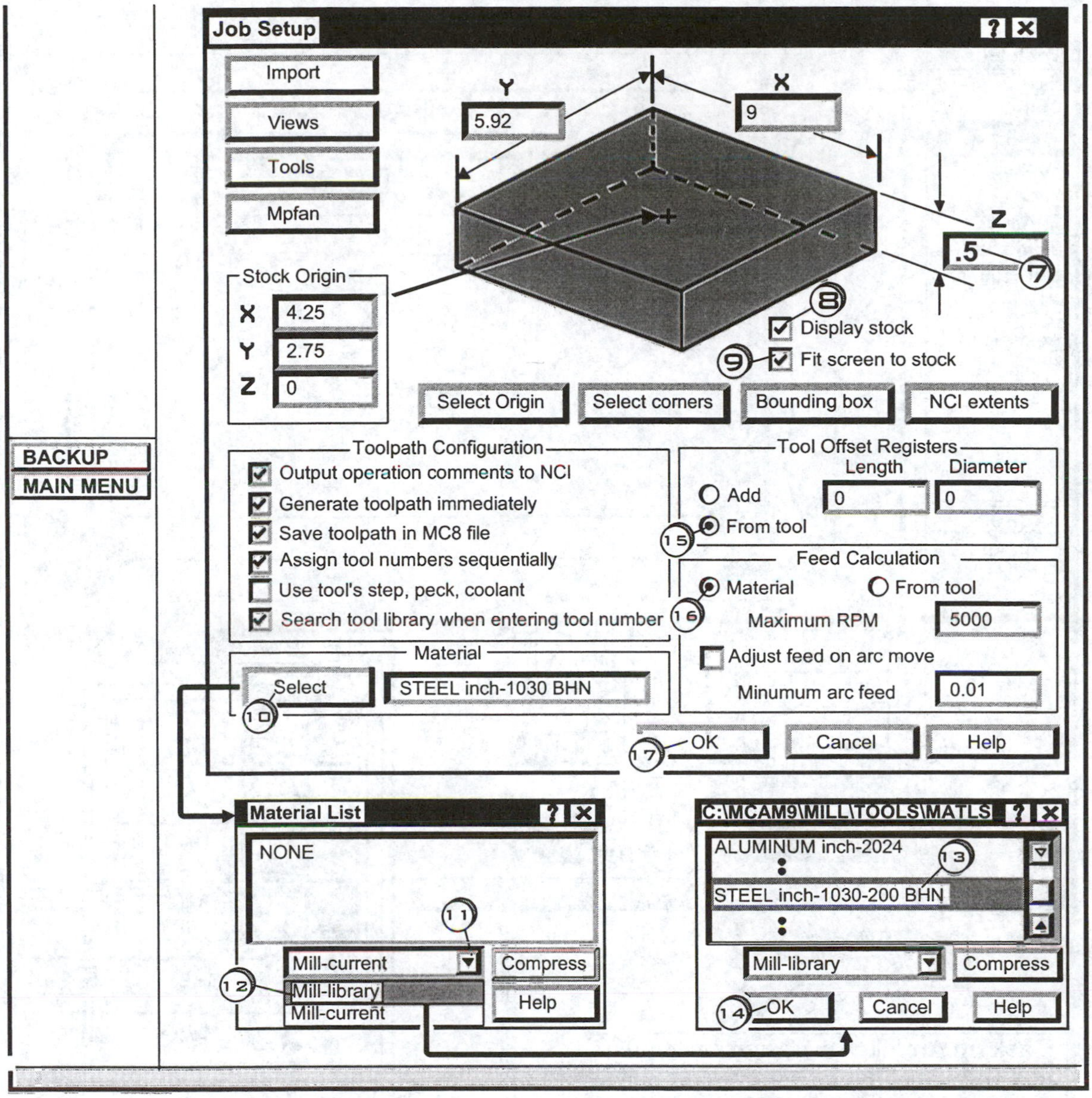

- Click ⑦ in the Z box ; enter **.5**
- Click ⑧ ⑨ the checks ☑ ☑ in Display stock and Fit screen to stock
- Click ⑩ Select
- Scroll down ⑪ ; Click ⑫ Mill-library
- Scroll down; Click ⑬ STEEL inch-1030-200 BHN
- Click ⑭ OK

- Click ⑮ the From Tool radio button

  Mastercam will use the length and diameter compensation values ***directly from the tool*** when computing the required tool path

- Click ⑯ the Material radio button

  Mastercam will input the tool feed from the value given in the material file

- Click ⑰ OK

B) DRILL THE .125DIA PILOT HOLES X .166 DEEP

◆ CREATE THE PILOT HOLE TOOLPATH

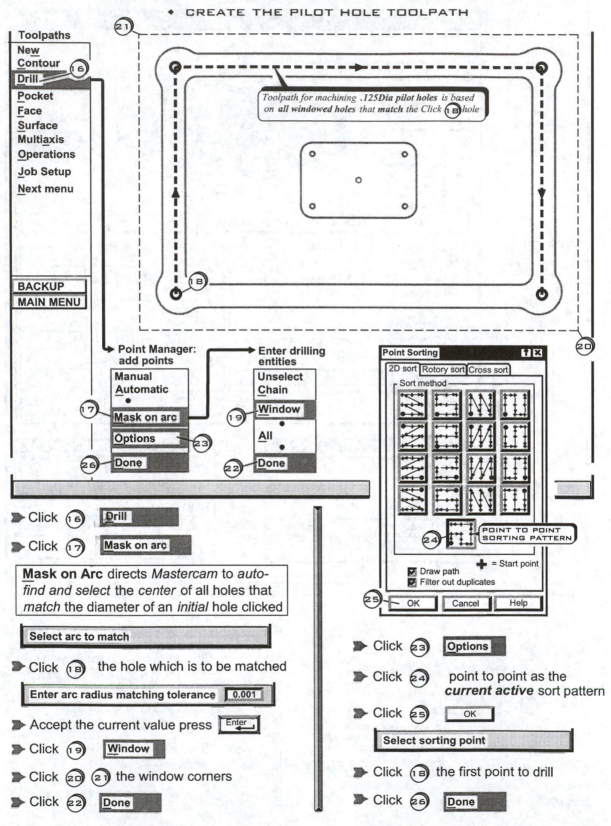

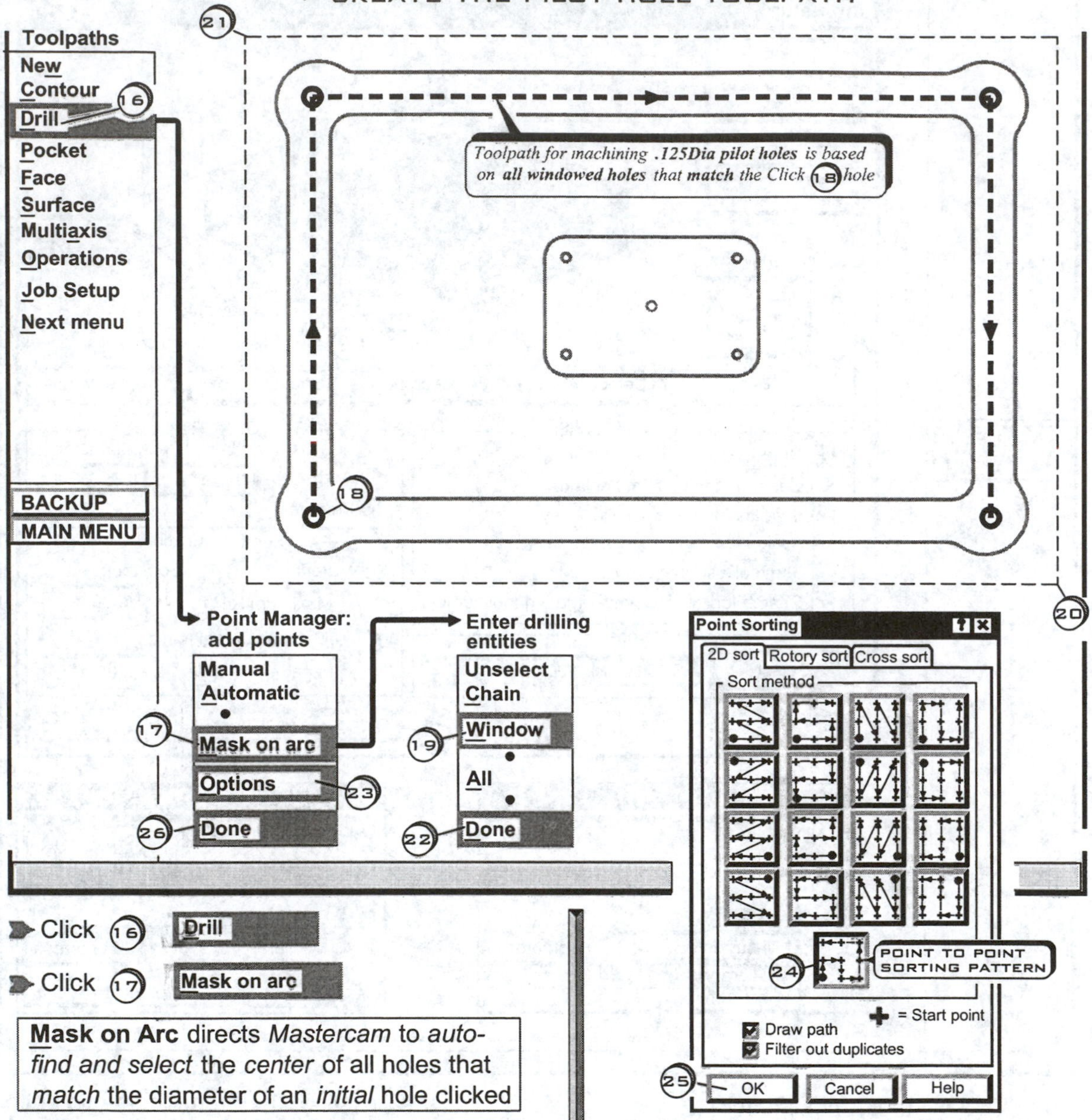

- Click (16) **Drill**
- Click (17) **Mask on arc**

Mask on Arc directs *Mastercam* to *auto-find and select* the *center* of all holes that *match* the diameter of an *initial* hole clicked

Select arc to match

- Click (18) the hole which is to be matched

Enter arc radius matching tolerance 0.001

- Accept the current value press Enter
- Click (19) **Window**
- Click (20) (21) the window corners
- Click (22) **Done**
- Click (23) **Options**
- Click (24) point to point as the ***current active*** sort pattern
- Click (25) **OK**

Select sorting point

- Click (18) the first point to drill
- Click (26) **Done**

◆ OBTAIN THE NEEDED .125 DIA SPOT DRILL TOOL

Mastercam will activate and display the Tool and Drill Parameters dialog box after the drill tool path has been created.

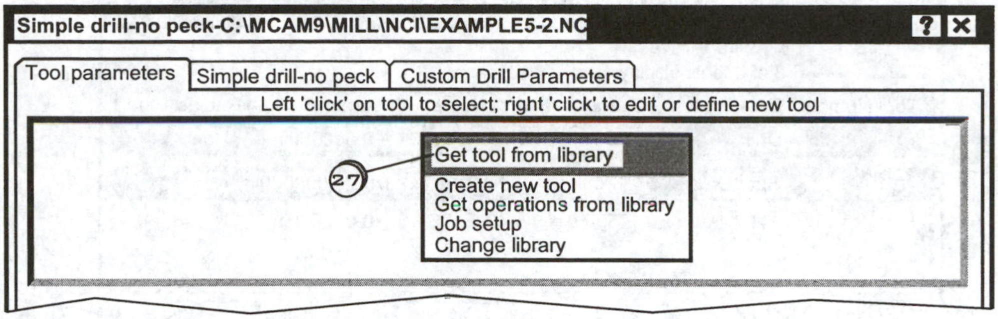

➤ Press the ***right*** mouse button in the ***white area***

➤ Click (27) Get tool from library

Mastercam will display the main Tool Library dialog box

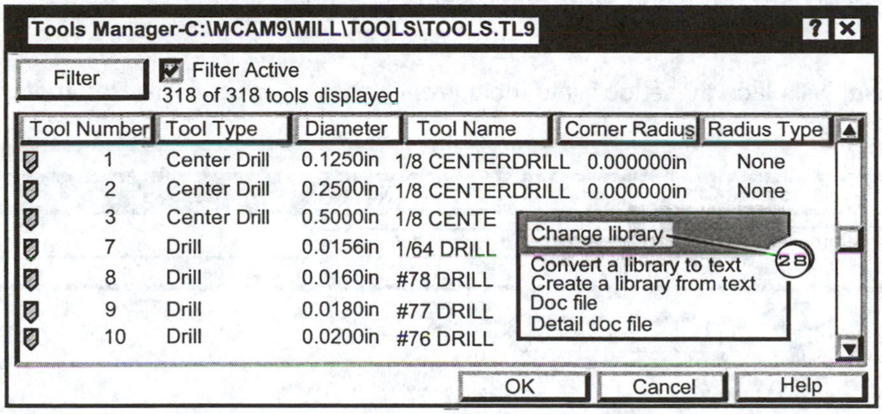

➤ Press the ***right*** mouse button in the ***white area***

➤ Click (28) Change library to change to the customized tool library **JVAL**

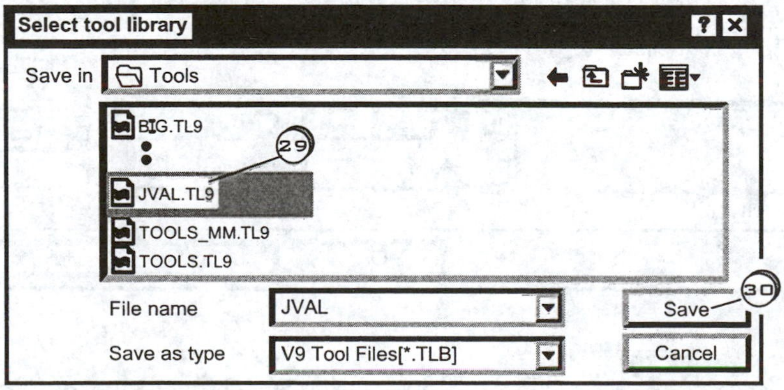

➤ Click (29) JVAL.TL9

➤ Click (30) Save

Select the .125in Spot Drill and the .25in Drill from the **JVAL** tool library

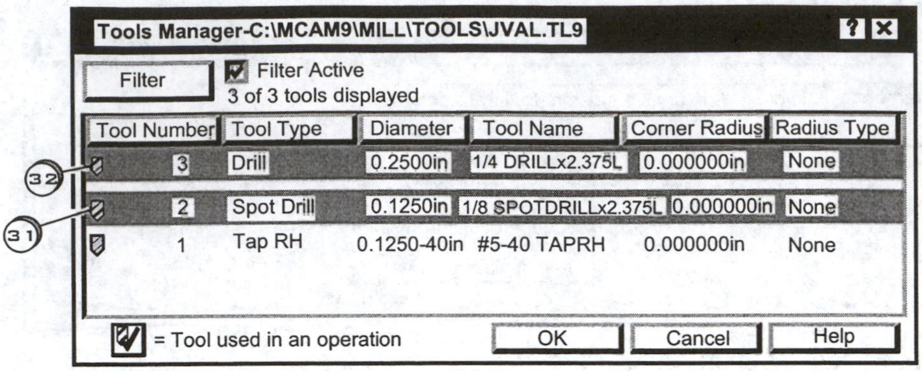

➤ Click (31) on the Spot Drill

➤ Hold the Shift key down and Click (32) on the Drill tool.

Mastercam will place these tools into the currently active Tool and Drill Parameters dialog box

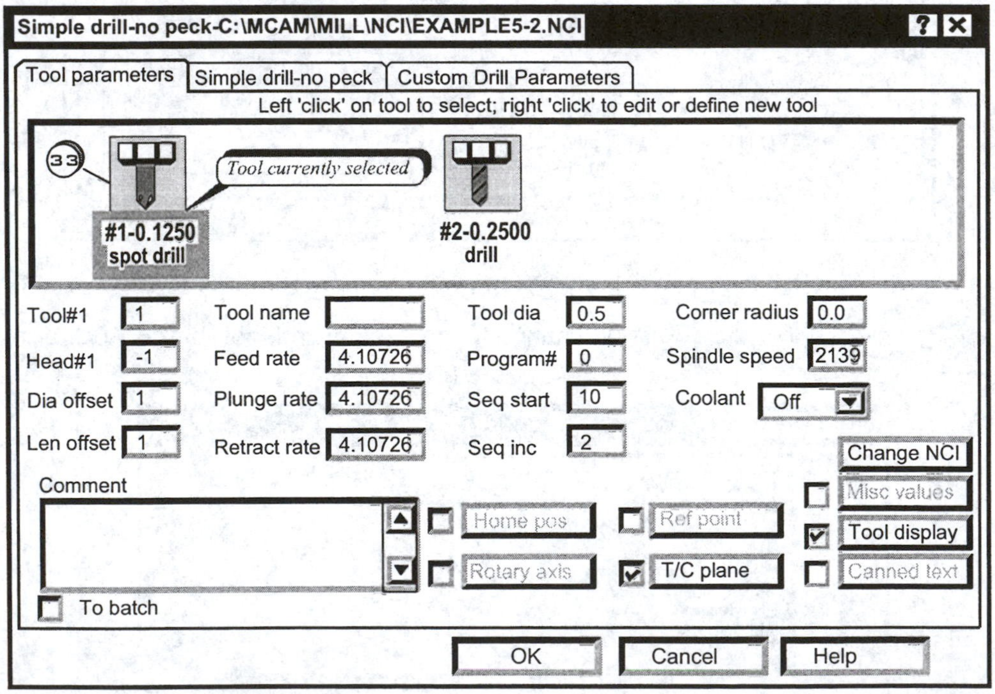

➤ ***Right*** click (33)

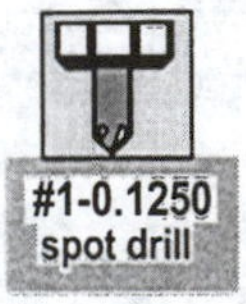

on the .125Spot Drill to select this as the ***active tool*** and direct *Mastercam* to display the Define Tool dialog box

➤ Click (34) Parameters

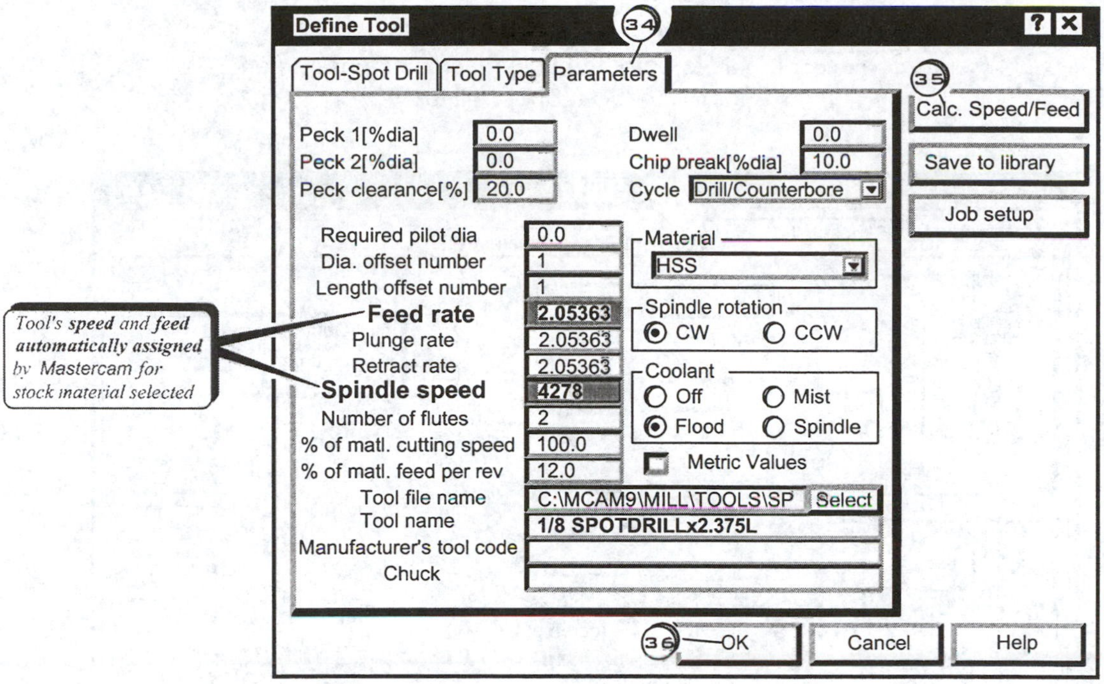

➤ Click (35) Calc Speed/Feed to direct Mastercam to ***automatically assign speeds and feeds*** to the Spotdrill tool for the ***stock material*** selected in STEP A

➤ Click (36) OK

Mastercam will redisplay the Tool and Drill Parameters dialog box with the tool's updated speed and feed

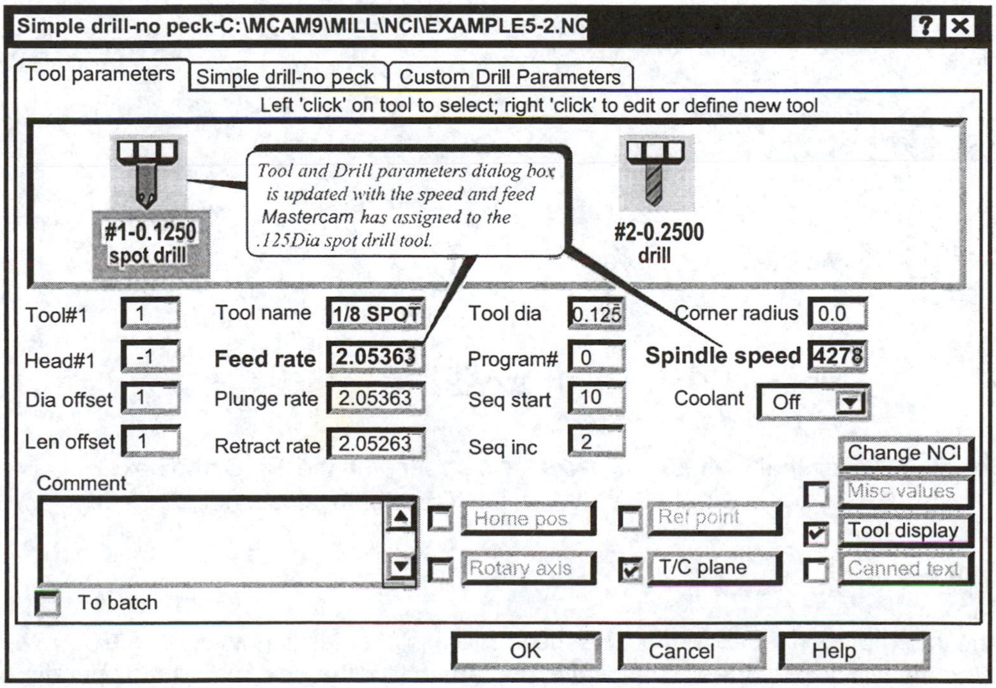

◆ ENTER THE REQUIRED .125 DIA PILOT HOLE MACHINING PARAMETERS

➤ Click (37) [Simple drill-no peck] tab to enter the hole machining parameters for the .125Dia Spot Drill tool

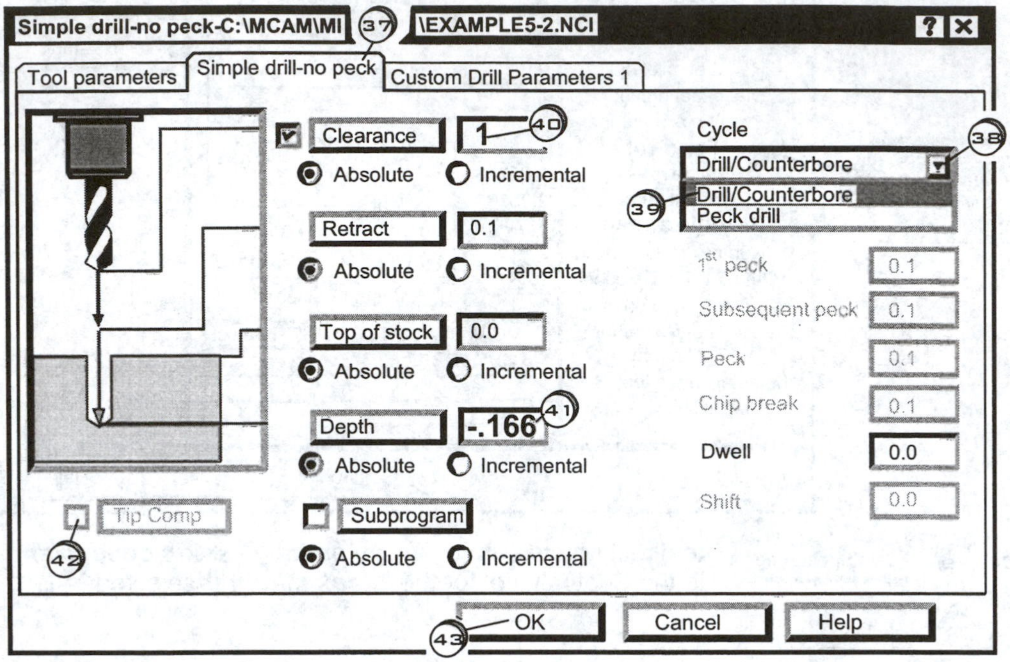

➤ Click (38) [▼] the toggle down button

➤ Click (39) [Drill/Counterbore]

➤ Click (40) in the Clearance box and enter [1]

➤ Click (41) in the Depth box and enter [-.166]

➤ Click (42) such that a check ☐ ***does not*** appear in the Tip Comp box (Tip Comp is used for *thru* holes)

➤ Click (43) [OK]

All the information needed to machine the .125Dia pilot holes is now complete. *Mastercam* will display the CAD model in preparation for the next machining operation.

C) PECK DRILL THE .25DIA HOLES THRU

◆ CREATE THE DRILL THRU TOOLPATH

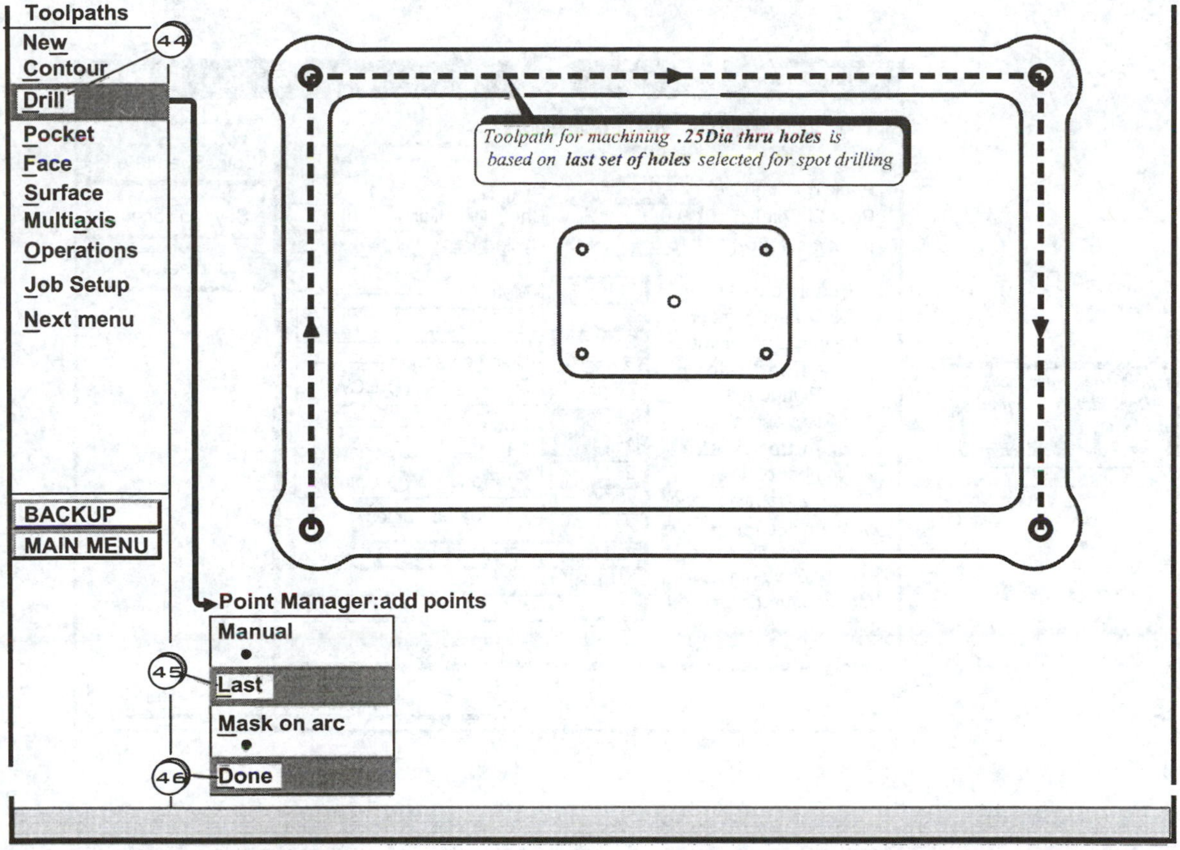

➤ Click (44) **Drill**

➤ Click (45) **Last**

Last directs *Mastercam* to *re-enter the last set* of holes selected

➤ Click (46) **Done**

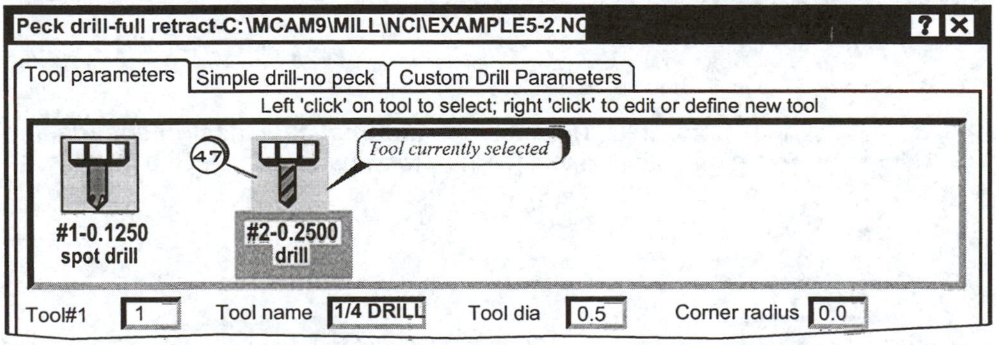

➤Right click (47)

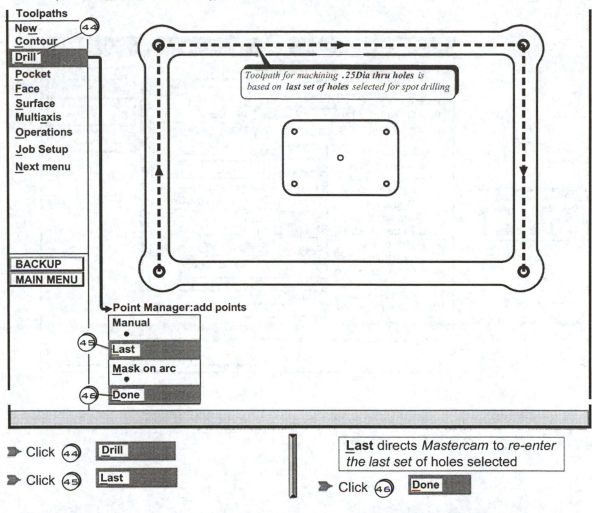

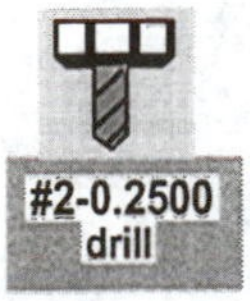

on the .25Drill to select this as the ***active tool*** and direct *Mastercam* to display the Define Tool dialog box

➤Click (48) Parameters

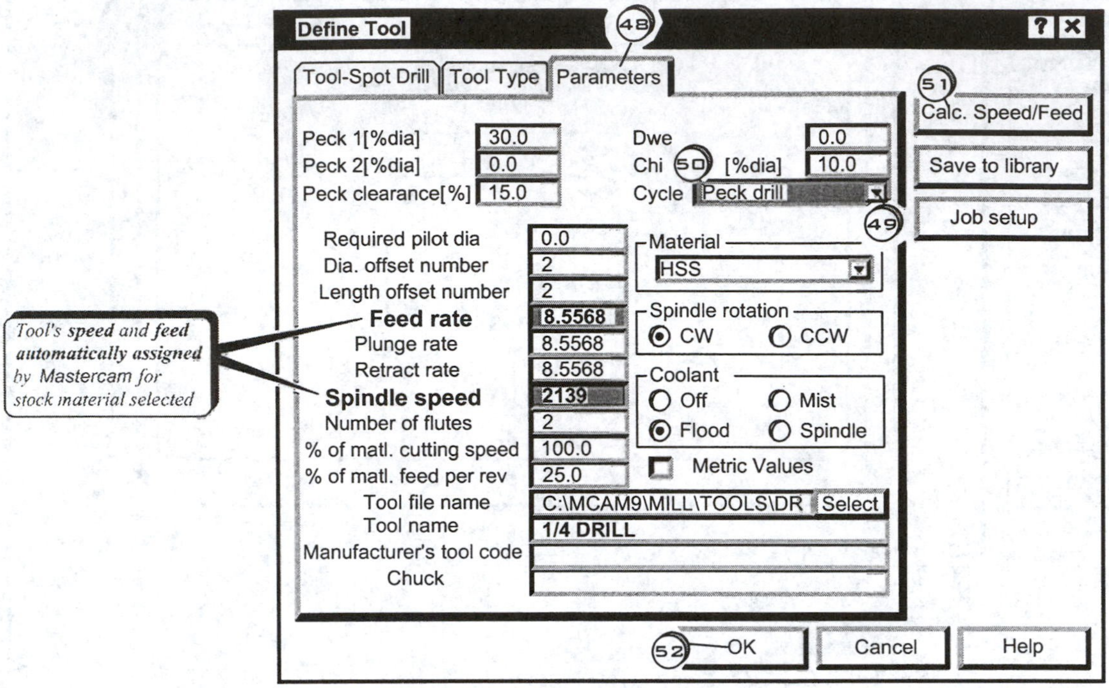

➤Click (49) the scroll down button ▼

➤Click (50) Peck drill

➤Click (51) Calc Speed/Feed to direct *Mastercam* to ***automatically assign speeds and feeds*** to the Drill tool for the ***stock material*** selected in **STEP A**

➤Click (52) OK

Mastercam will redisplay the Tool and Drill Parameters dialog box with the tool's updated speed and feed

- **ENTER THE REQUIRED .25 DIA THRU HOLE MACHINING PARAMETERS**

➤ Click (53) [Peck drill-full retract] tab to enter the hole machining parameters for the .125Dia Spot Drill tool

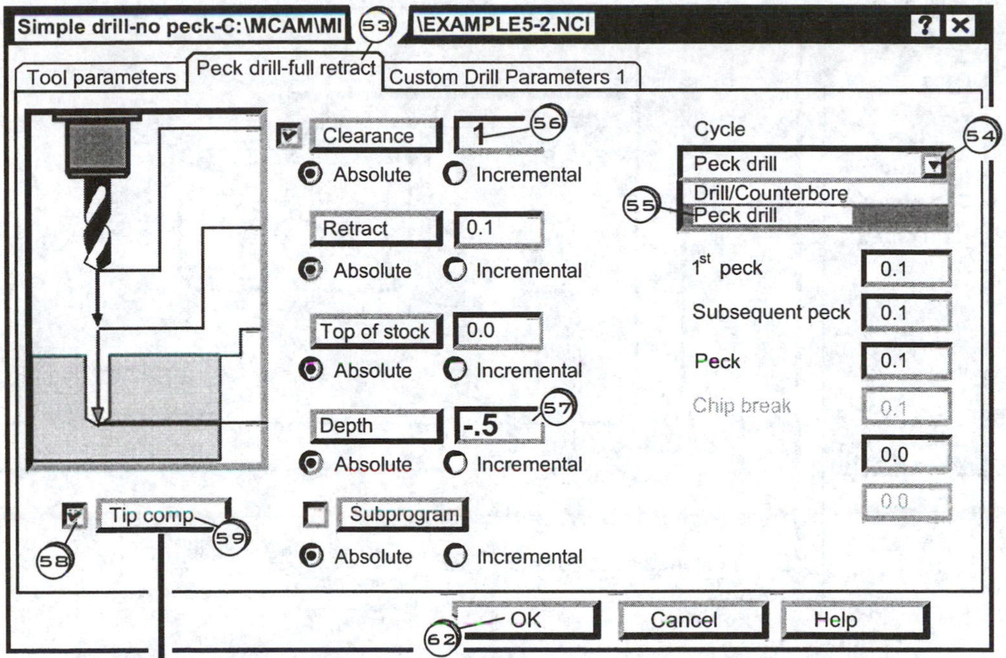

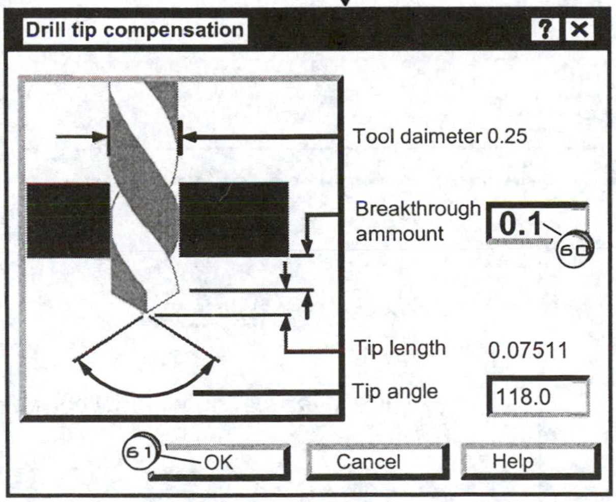

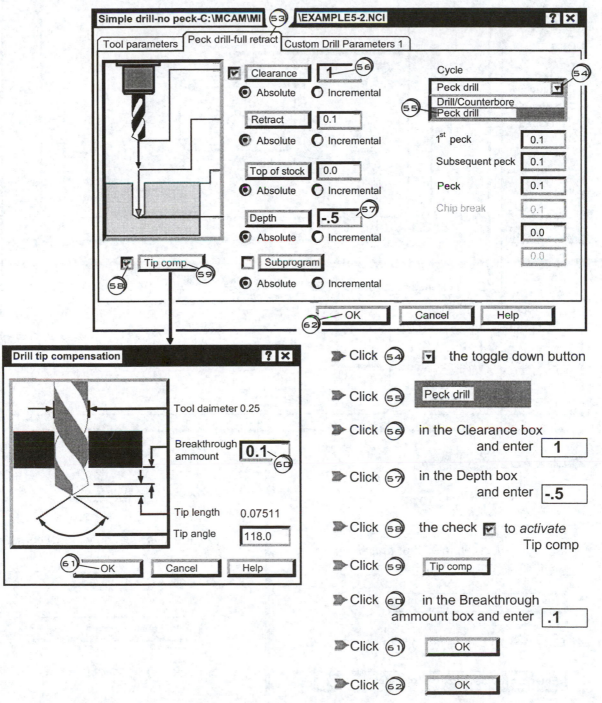

➤ Click (54) [▼] the toggle down button

➤ Click (55) [Peck drill]

➤ Click (56) in the Clearance box and enter [1]

➤ Click (57) in the Depth box and enter [-.5]

➤ Click (58) the check [☑] to *activate* Tip comp

➤ Click (59) [Tip comp]

➤ Click (60) in the Breakthrough ammount box and enter [.1]

➤ Click (61) [OK]

➤ Click (62) [OK]

All the information needed to machine the .25Dia thru holes is now complete.
Mastercam will display the CAD model in preparation for the next machining operation.

D) DRILL THE .031DIA PILOT HOLES X .125 DEEP

- ◆ CREATE THE PILOT HOLE TOOLPATH

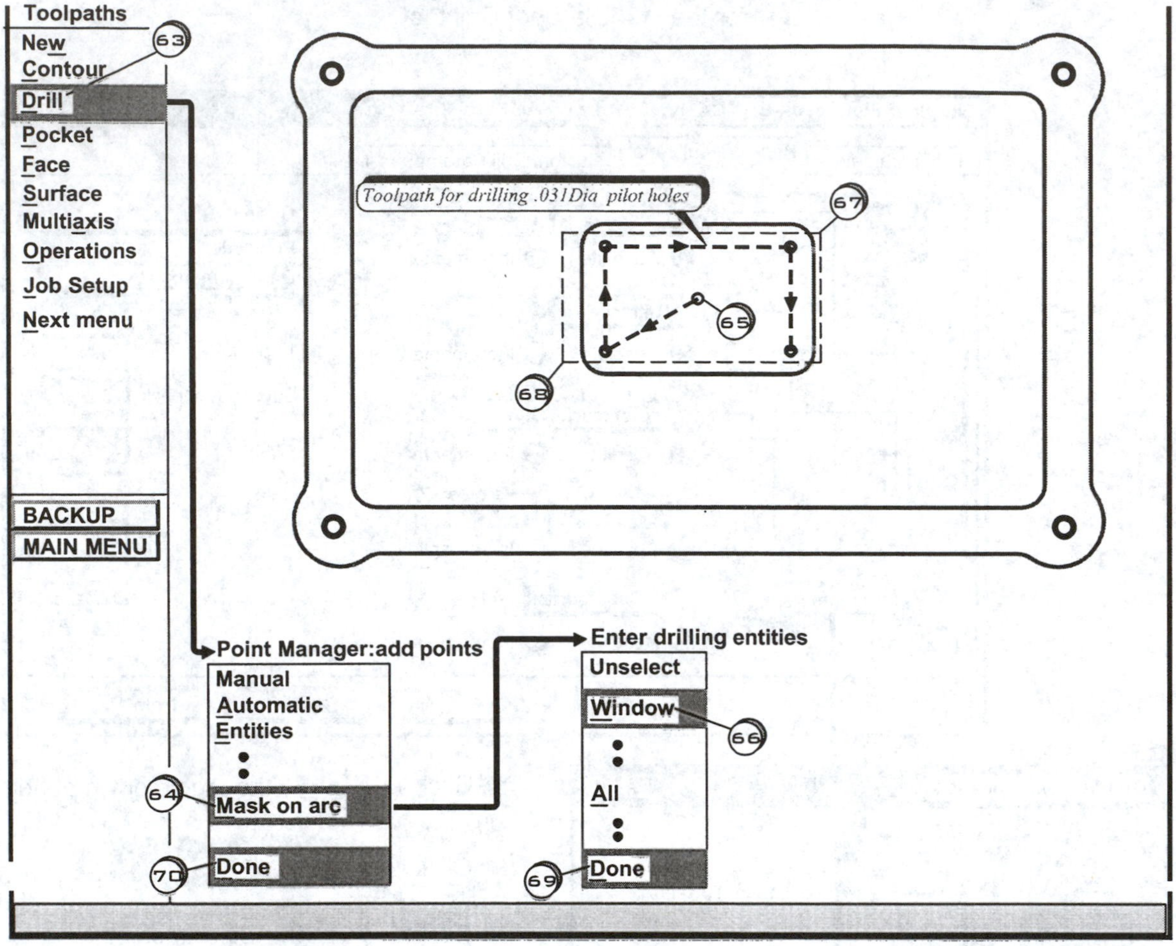

➤ Click (63) **Drill**

➤ Click (64) **Mask on arc**

to direct *Mastercam* to *automatically find and select* all holes that *match* the diameter of an initial hole clicked

Select arc to match

➤ Click (65) the hole which is to be matched

Enter arc radius matching tolerance **0.001**

➤ Accept the current value press Enter

➤ Click (66) **Window**

➤ Click (67) (68) the window corners

➤ Click (69) **Done**

Select sorting point

➤ Click (65) the first point to drill

➤ Click (70) **Done**

◆ CREATE THE NEEDED .031 DIA SPOT DRILL TOOL

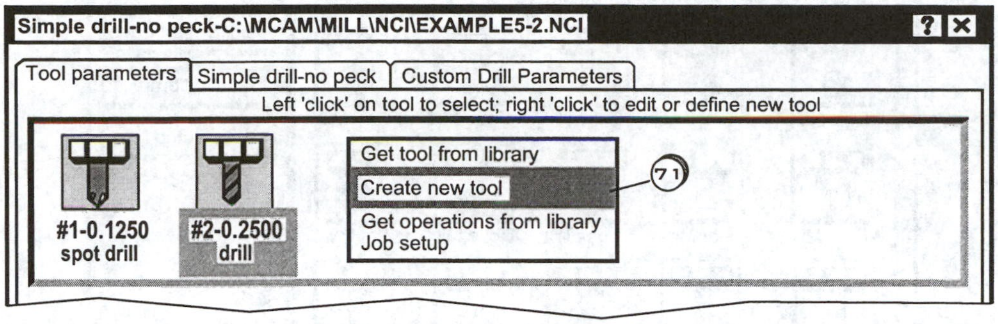

➤ ***Right*** Click move the mouse cursor *down* and Click (71) Create new tool

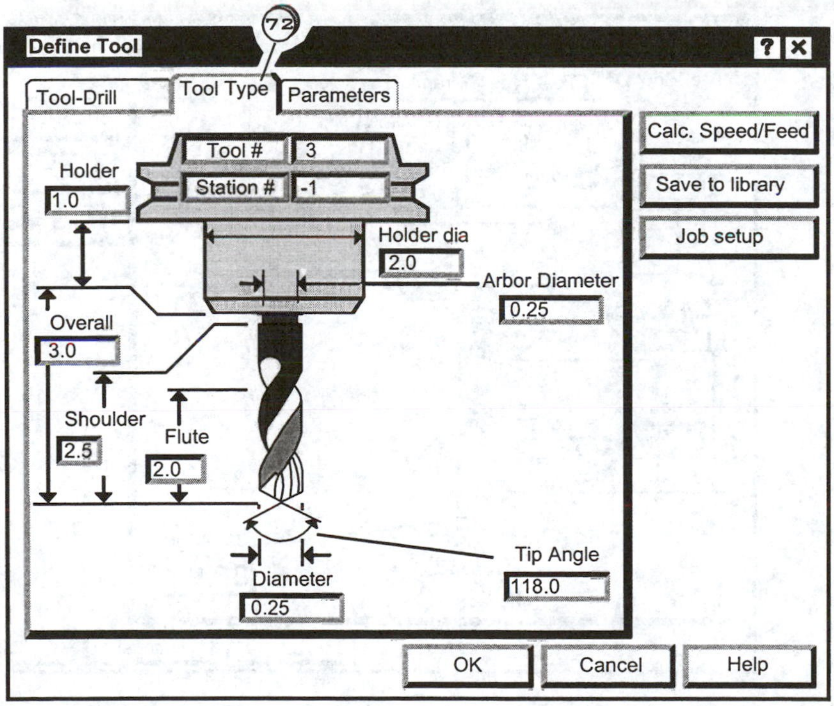

➤ Click (72) the Tool Type tab

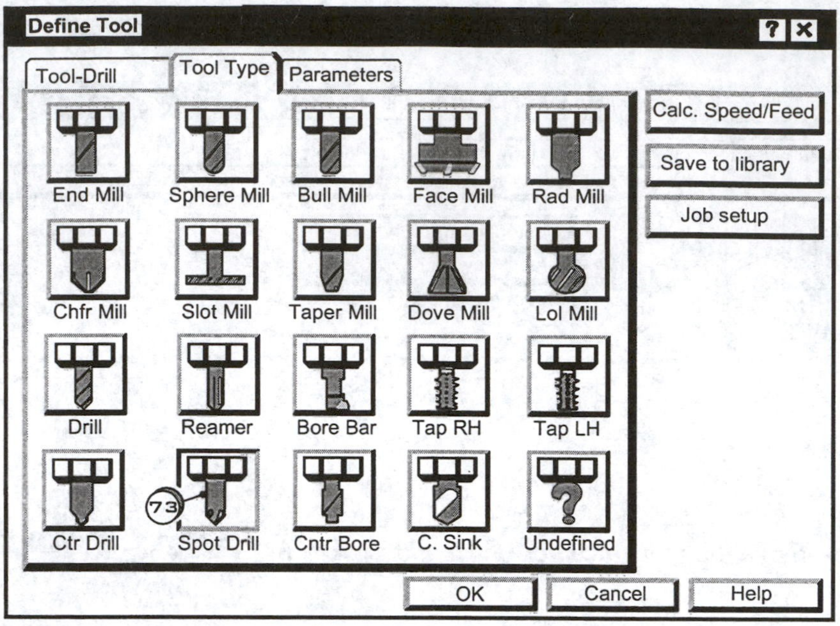

➤ Click (73) the Spot Drill icon

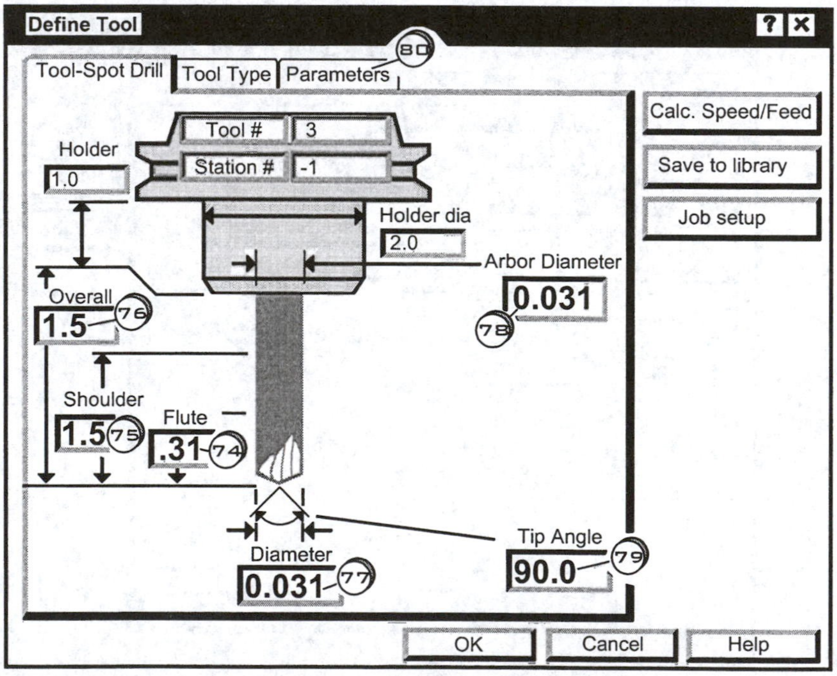

➤ Click (74) in the Flute box; enter .31

➤ Click (75) in the Shoulder box; enter 1.5

➤ Click (76) in the Overall box; enter 1.5

➤ Click (77) in the Diam box; enter .031

➤ Click (78) the Arbor Dia box; enter .031

➤ Click (79) in the Tip Angle box; enter 90

➤ Click (80) the Parameters tab

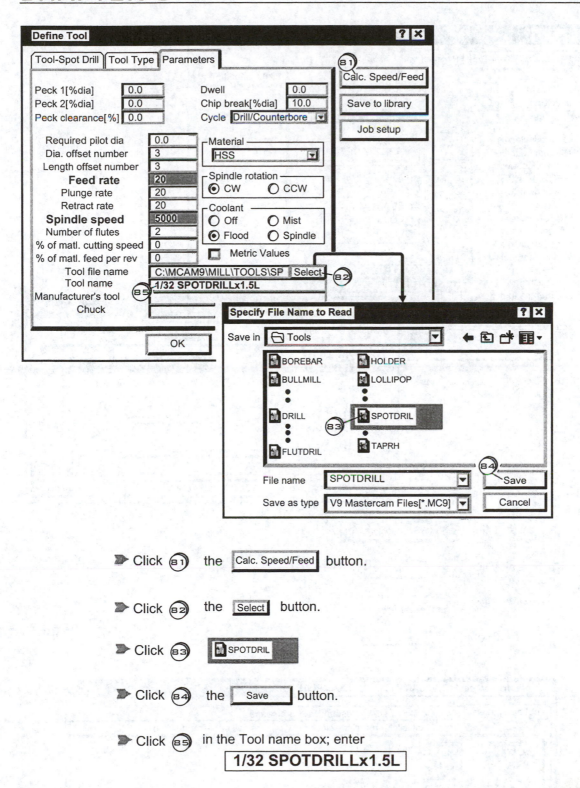

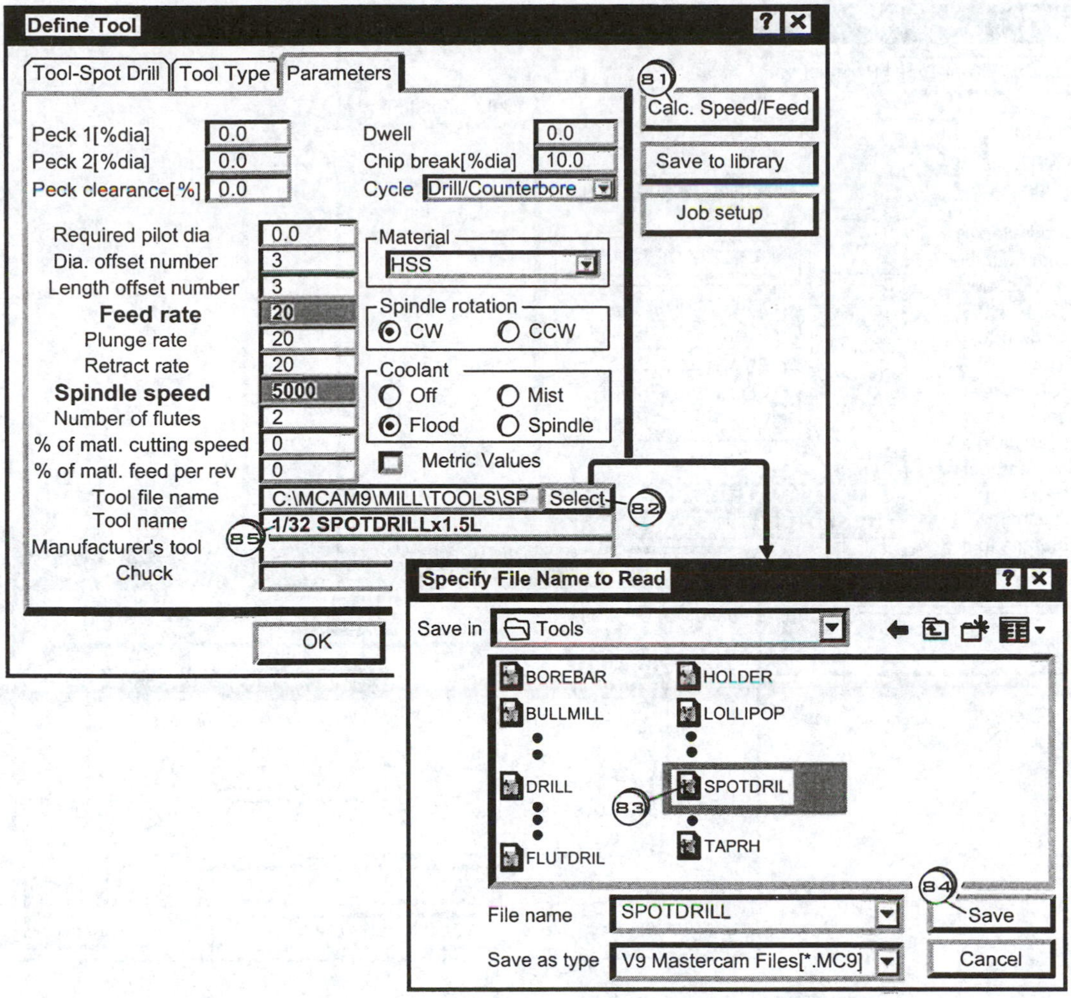

- Click B1 the Calc. Speed/Feed button.
- Click B2 the Select button.
- Click B3 SPOTDRIL
- Click B4 the Save button.
- Click B5 in the Tool name box; enter **1/32 SPOTDRILLx1.5L**

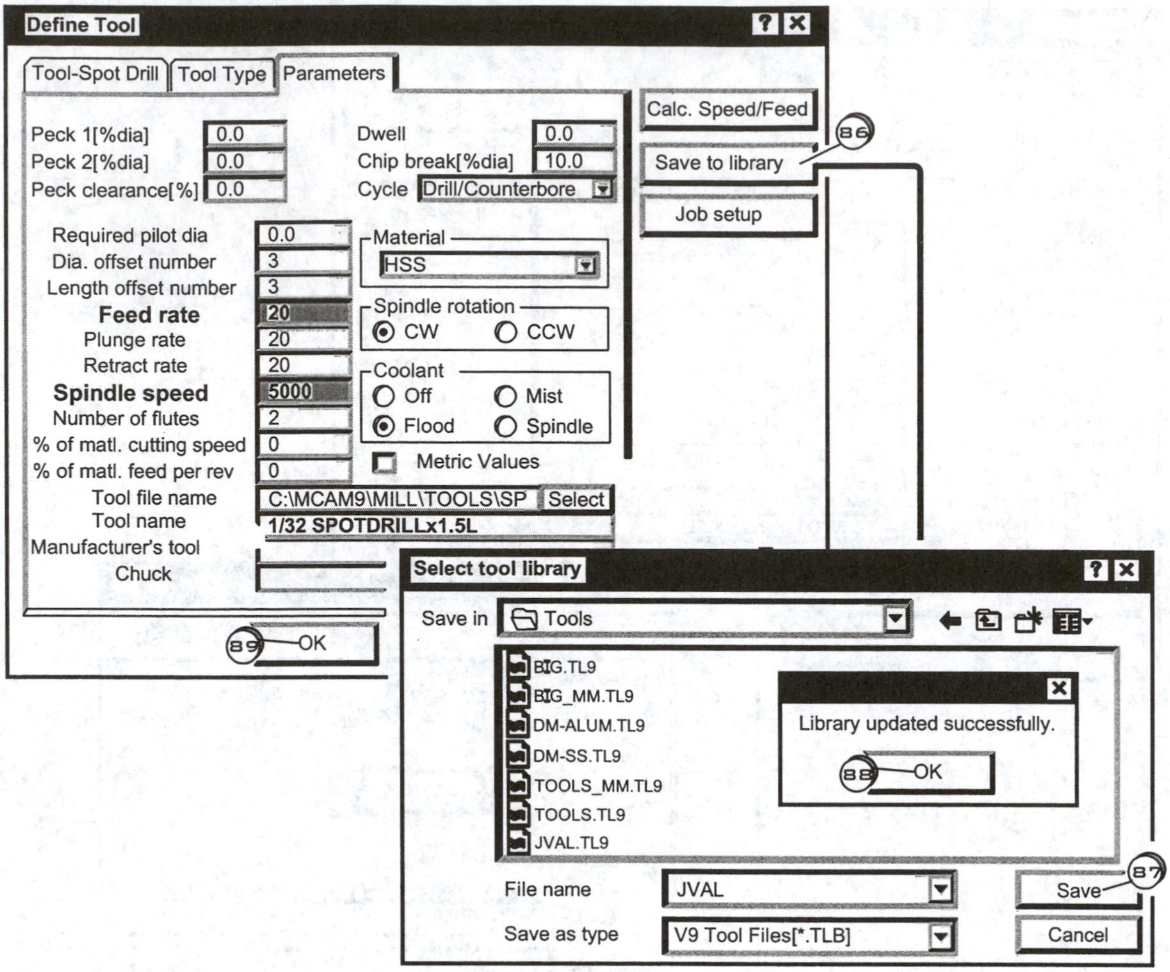
Define Tool
Tool-Spot Drill
Tool Type
Parameters
Peck 1[%dia] 0.0
Peck 2[%dia] 0.0
Peck clearance[%] 0.0
Dwell 0.0
Chip break[%dia] 10.0
Cycle Drill/Counterbore
Calc. Speed/Feed
Save to library
Job setup
86
Required pilot dia 0.0
Dia. offset number 3
Length offset number 3
Feed rate 20
Plunge rate 20
Retract rate 20
Spindle speed 5000
Number of flutes 2
% of matl. cutting speed 0
% of matl. feed per rev 0
Tool file name C:\MCAM9\MILL\TOOLS\SP
Select
Tool name 1/32 SPOTDRILLx1.5L
Manufacturer's tool
Chuck
Material
HSS
Spindle rotation
CW
CCW
Coolant
Off
Mist
Flood
Spindle
Metric Values
89
OK
Select tool library
Save in
Tools
BIG.TL9
BIG_MM.TL9
DM-ALUM.TL9
DM-SS.TL9
TOOLS_MM.TL9
TOOLS.TL9
JVAL.TL9
Library updated successfully.
88
OK
File name JVAL
Save as type V9 Tool Files[*.TLB]
Save
87
Cancel

Click 86 Save to library

Click 87 Save

Click 88 OK

Click 89 OK

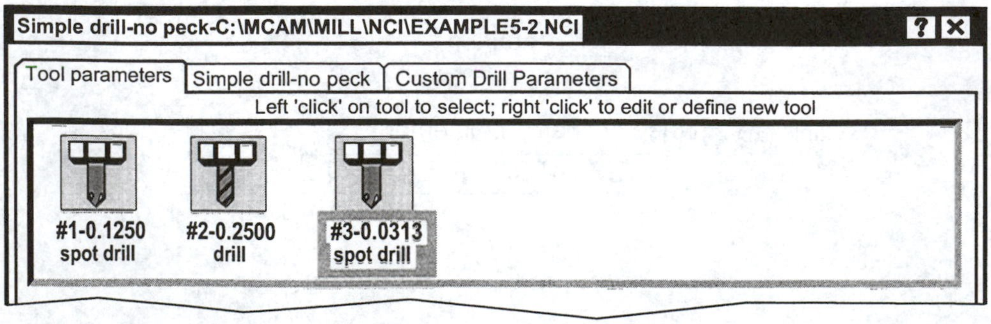
Simple drill-no peck-C:\MCAM\MILL\NCI\EXAMPLE5-2.NCI
Tool parameters
Simple drill-no peck
Custom Drill Parameters
Left 'click' on tool to select; right 'click' to edit or define new tool
#1-0.1250
spot drill
#2-0.2500
drill
#3-0.0313
spot drill

◆ **ENTER THE REQUIRED .031 DIA PILOT HOLE MACHINING PARAMETERS**

➤ Click (90) [Peck drill-full retract] tab to enter the hole machining parameters for the .125Dia Spot Drill tool

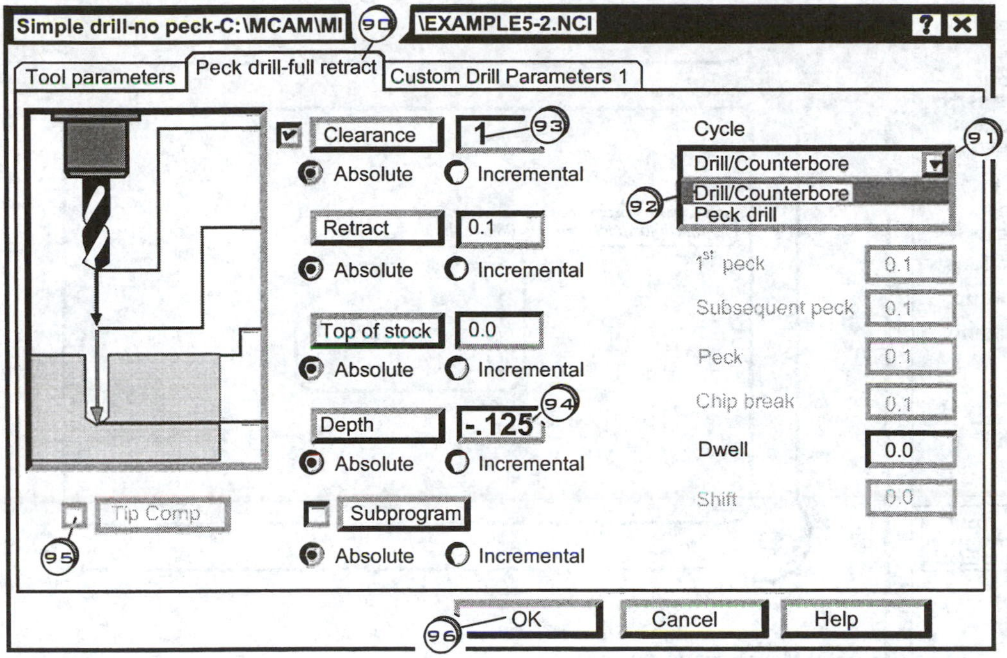

➤ Click (91) [▼] the toggle down button

➤ Click (92) [Drill/Counterbore]

➤ Click (93) in the Clearance box and enter [1]

➤ Click (94) in the Depth box and enter [-.125]

➤ Click (95) such that a check ☐ ***does not*** appear in the Tip Comp box (Tip Comp *off* for *blind* holes)

➤ Click (96) [OK]

All the information needed to machine the .125Dia pilot holes is now complete. *Mastercam* will display the CAD model in preparation for the next machining operation.

E) PECK DRILL THE .089DIA HOLES x .325 DEEP

- CREATE DRILL TOOLPATHS
- IDENTIFY THE NEEDED .089 DIA DRILL TOOL
- ENTER THE REQUIRED HOLE OPERATION PARAMETERS

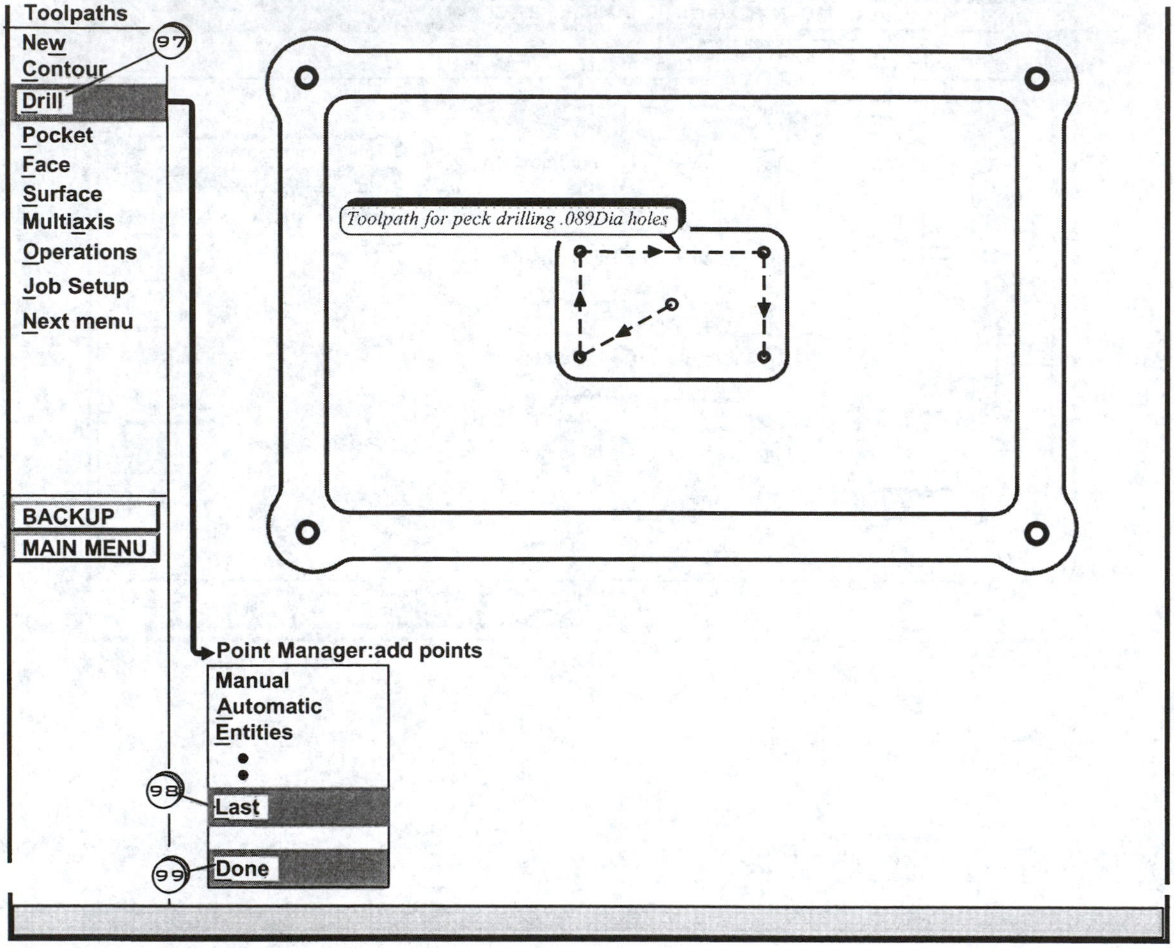

➤ Click (97) **Drill**

➤ Click (98) **Last**

to direct *Mastercam* to use the *last* set of drill points selected.

➤ Click (99) **Done**

◆ OBTAIN THE NEEDED #43 DRILL TOOL

Mastercam will activate and display the Tool and Drill Parameters dialog box after the .031 spot drill tool path has been created.

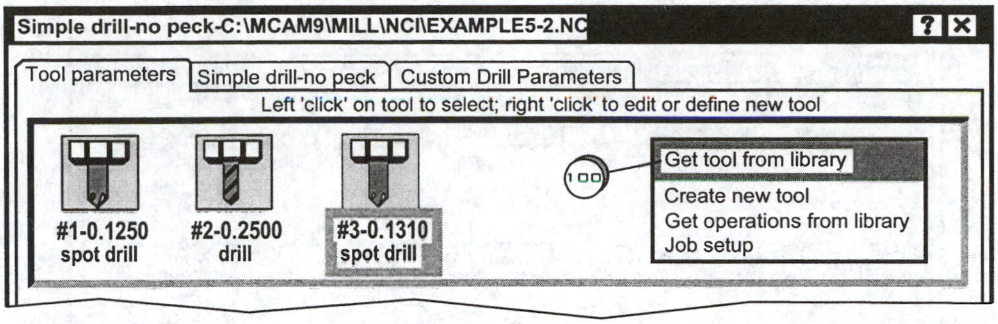

➤ ***Right*** Click move the mouse cursor *down* and Click (100) Get tool from library

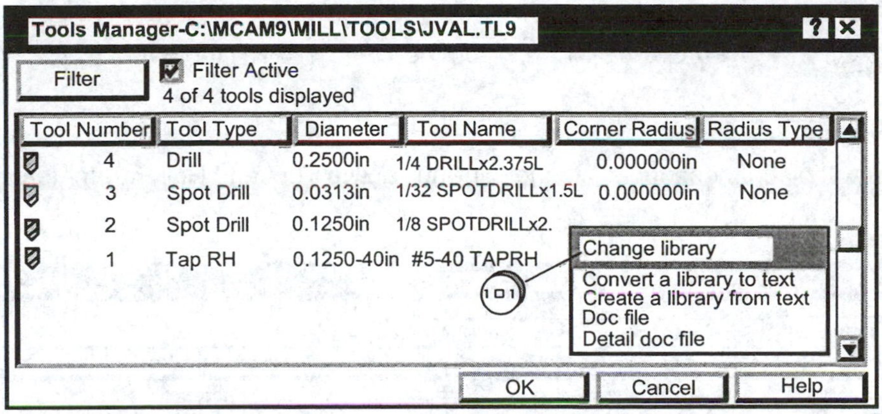

➤ Press the ***right*** mouse button in the ***white area***

➤ Click (101) Change library to change to the tool library **TOOLS.TLB**

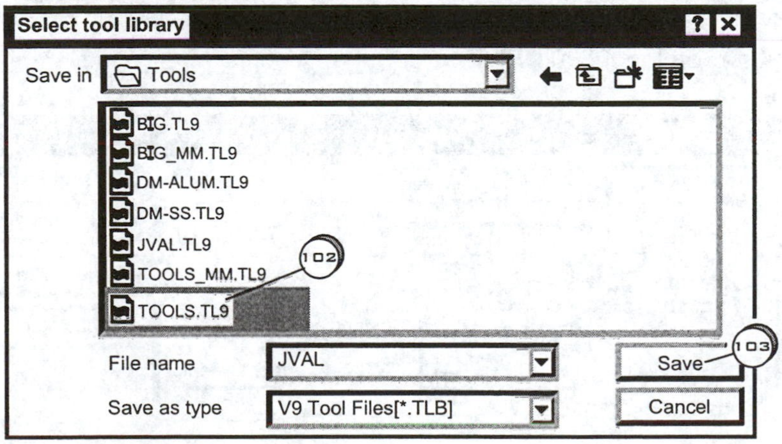

➤ Click (102) TOOLS.TL9

➤ Click (103) Save

Select the #43 Drill and the .125-40 Tap from the **TOOLS** tool library

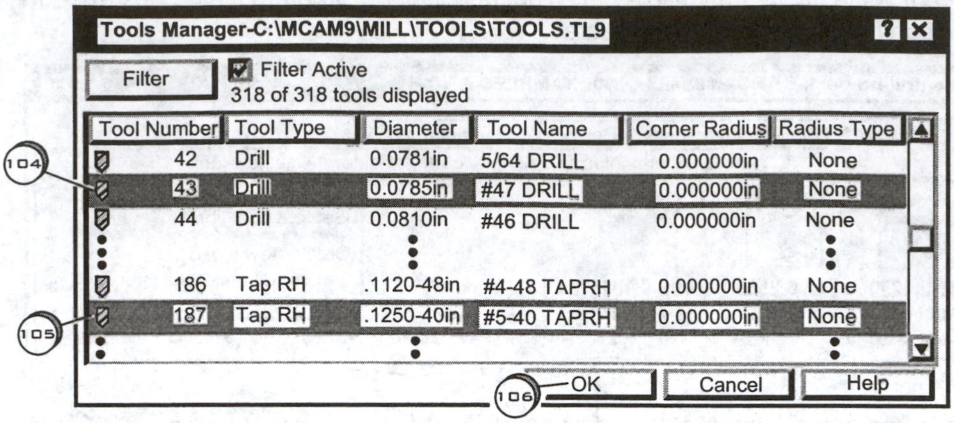

➤ Click (104) on the #43 DRILL

➤ Hold the Ctrl key down and Click (105) on the #5-40 TAPRH tool.

➤ Click (106) OK

Mastercam will place these tools into the currently active Tool and Drill Parameters dialog box

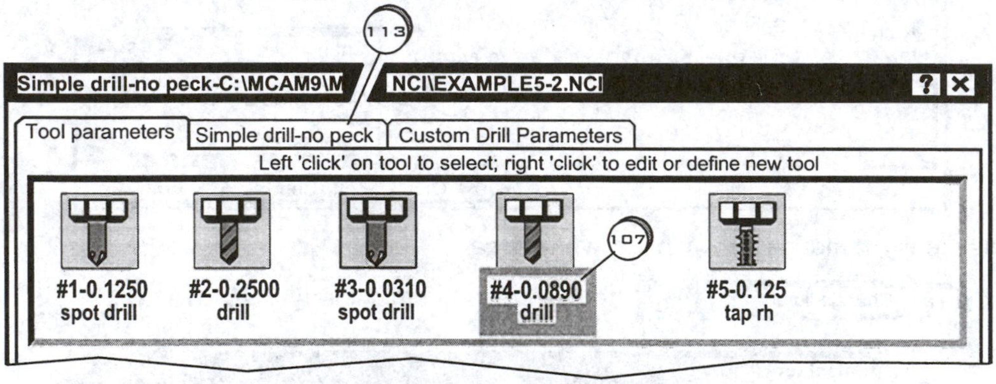

➤ ***Right*** Click (107) on the .0890 DRILL

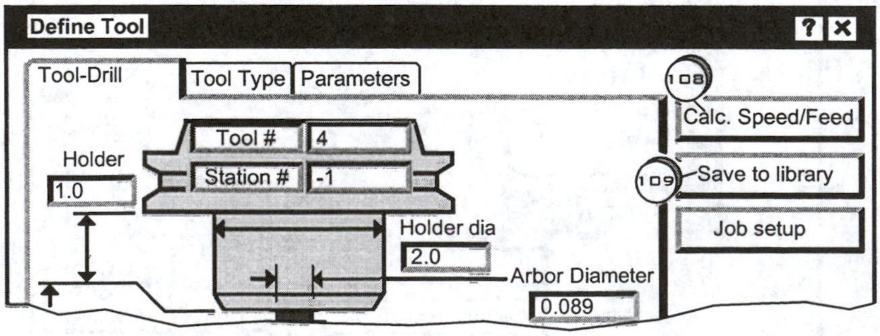

➤ Click (108) the Calc. Speed/Feed key

➤ Click (109) the Save to library key

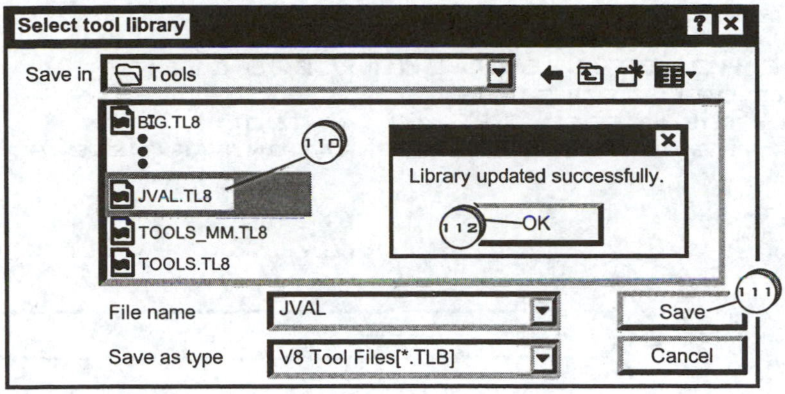

- Click (110) the **JVAL** tool library
- Click (111) the Save key
- Click (112) OK

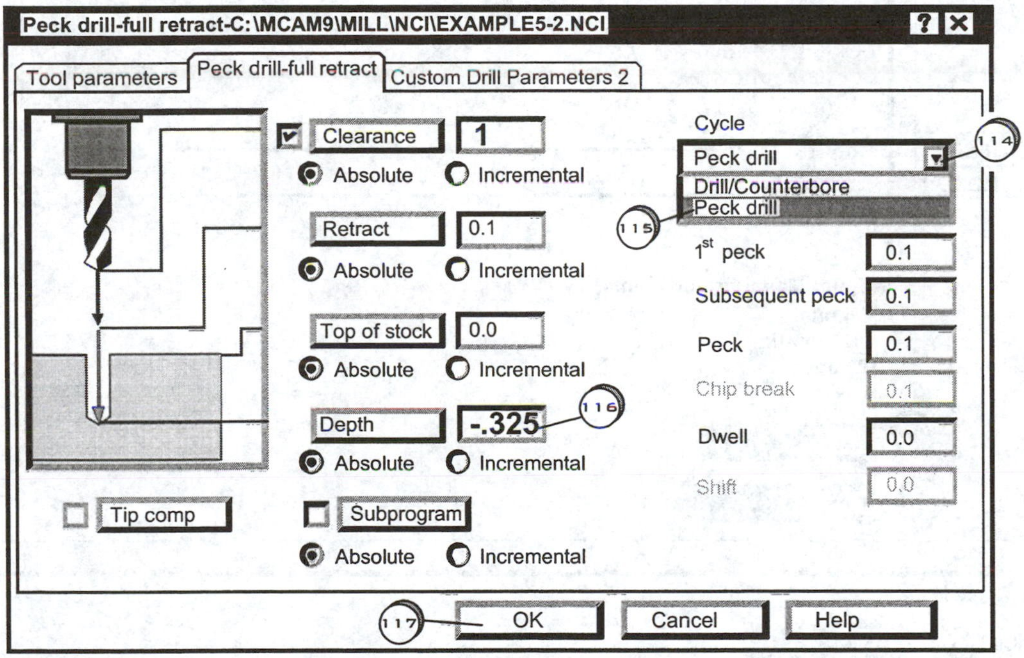

- Click (113) the Simple drill-no peck tab. See page 5-44.
- Click (114) the cycle down button
- Click (115) the Peck drill cycle
- Click (116) in the Depth box and enter **-.325**
- Click (117) the OK button

D) TAP THE HOLES .125-40-UNC x .225 DEEP

- CREATE DRILL TOOLPATHS
- OBTAIN THE NEEDED .125-40 TAP TOOL
- ENTER THE REQUIRED HOLE OPERATION PARAMETERS

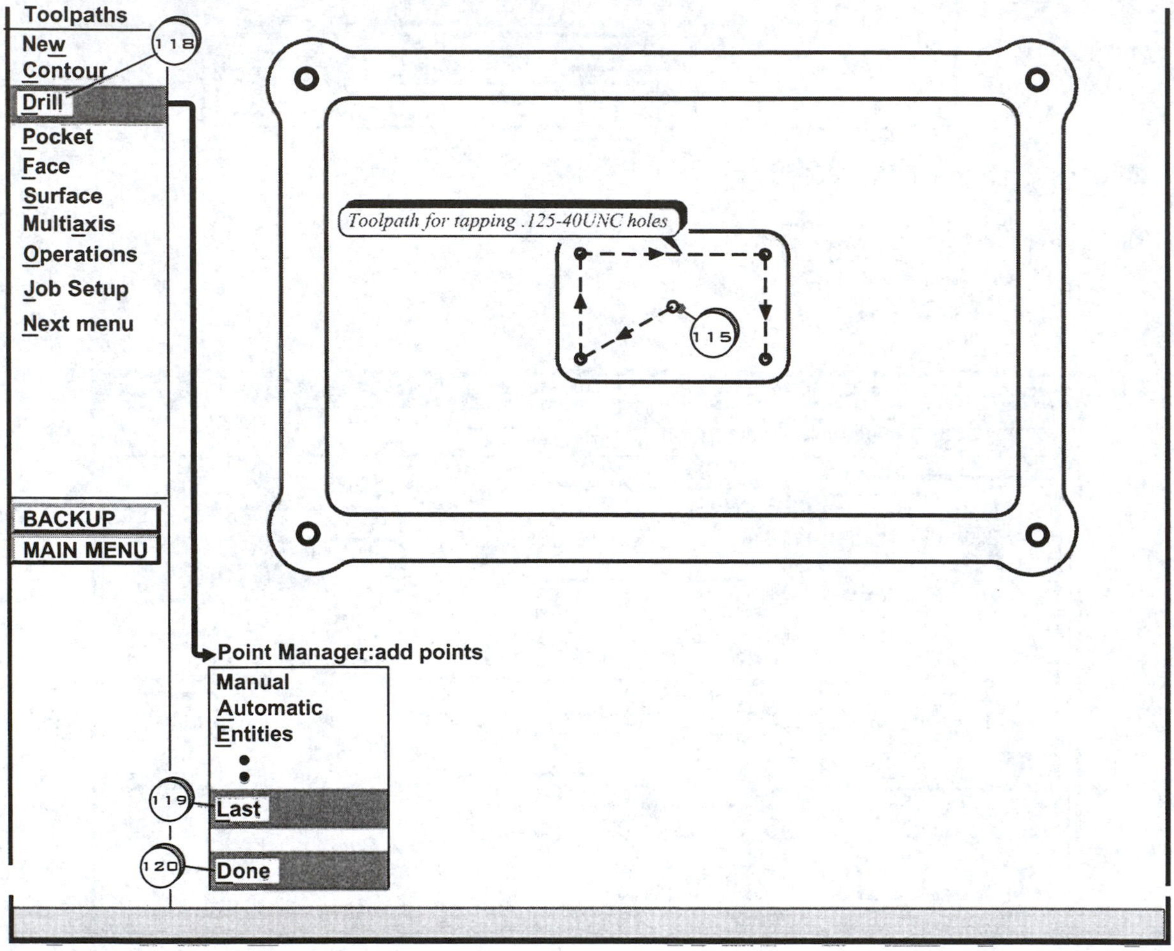

➤Click (118) Drill

➤Click (119) Last

to direct *Mastercam* to use the *last* set of drill points selected.

➤Click (120) Done

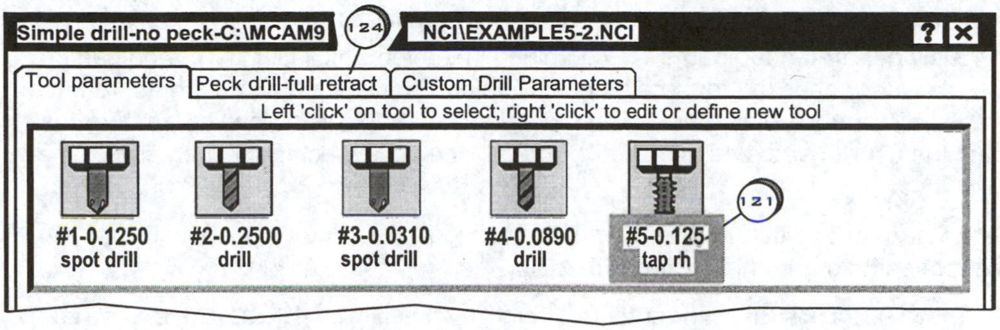

➤ ***Right*** Click (121) on the 0.125 TAP RH tool

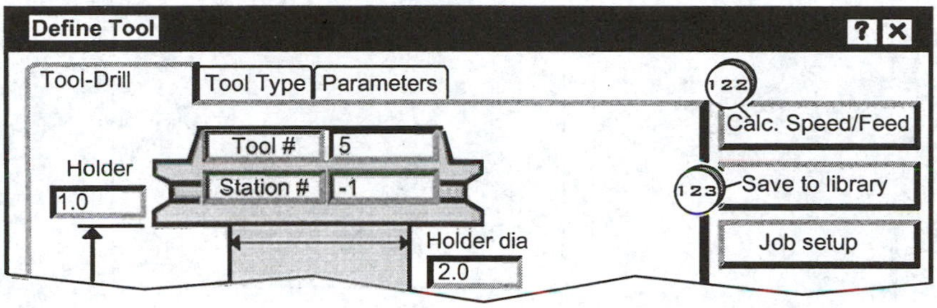

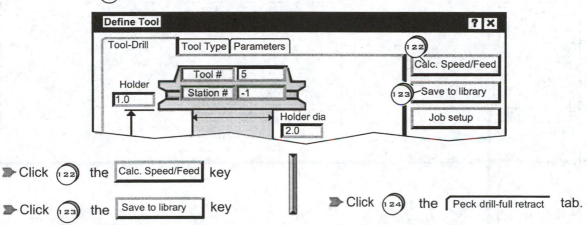

➤ Click (122) the Calc. Speed/Feed key

➤ Click (123) the Save to library key

➤ Click (124) the Peck drill-full retract tab.

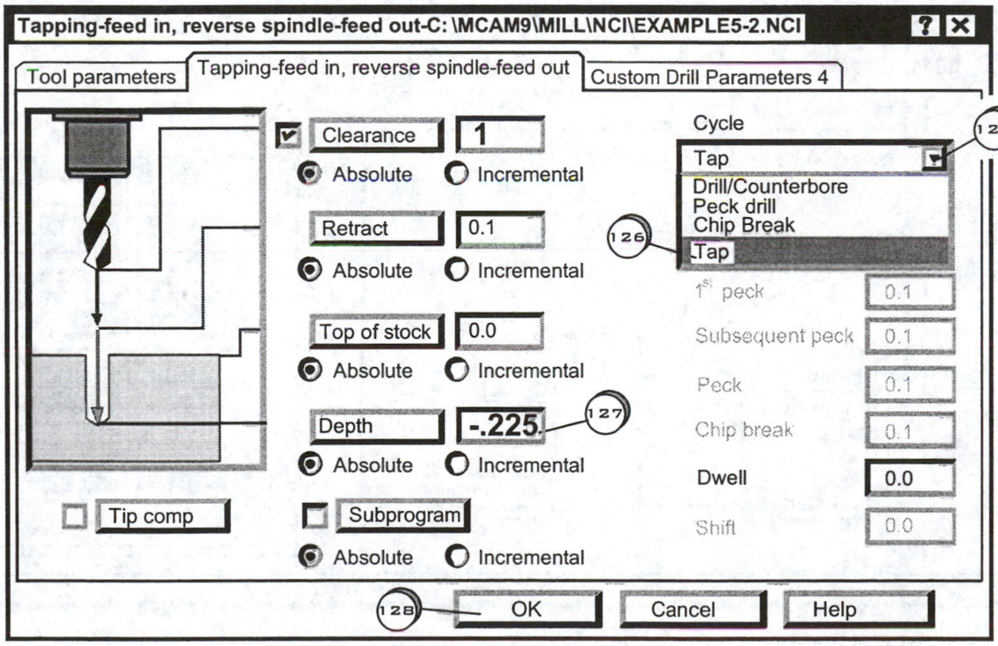

➤ Click (125) the ▾ cycle down button

➤ Click (126) the Tap cycle

➤ Click (127) in the Depth box; enter **-.225**

➤ Click (128) the OK button

5-5 Backplotting Hole Machining Operations

Previously generated toolpaths are contained in the job's NCI file(*.NCI) and can be played back and visually checked for correctness by selecting the **Backplot** module from the **Operations Manager** dialog box. The Operations manager is a powerful feature that enables the user to see, backplot, verify and edit the job's machining operations.

EXAMPLE 5-3

Use the Backplot function in the Operations Manager dialog box to display the drilling tool paths created in Example 5-2.

A) BACKPLOTTING THE TOOLPATHS FOR ALL OPERATIONS

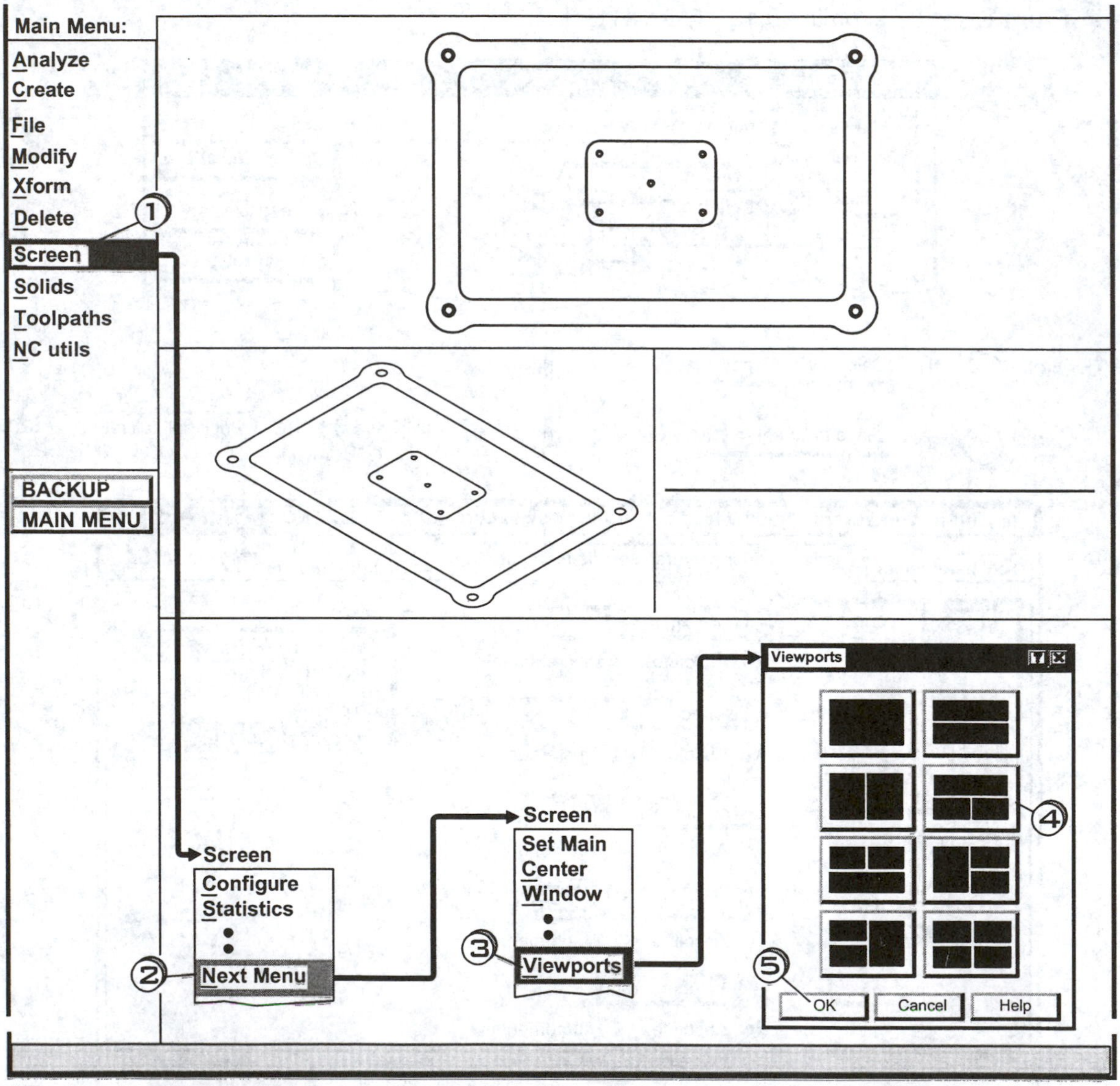

- Click ① Screen
- Click ② Next Menu
- Click ③ Viewports
- Click ④ the desired viewport configuration
- Click ⑤ the OK button

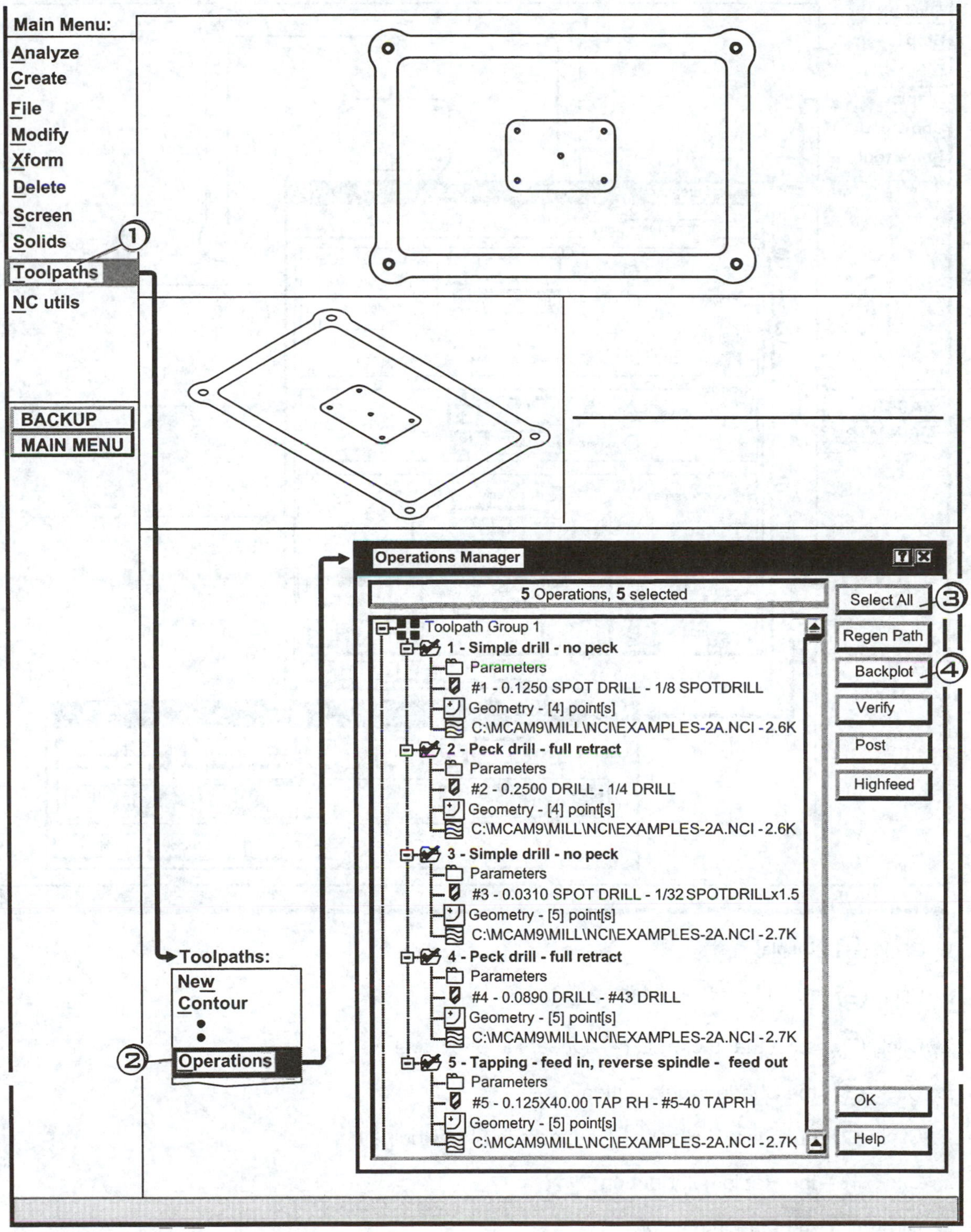
Main Menu:
Analyze
Create
File
Modify
Xform
Delete
Screen
Solids
Toolpaths
NC utils
BACKUP
MAIN MENU
Toolpaths:
New
Contour
Operations
Operations Manager
5 Operations, 5 selected
Toolpath Group 1
1 - Simple drill - no peck
Parameters
#1 - 0.1250 SPOT DRILL - 1/8 SPOTDRILL
Geometry - [4] point[s]
C:\MCAM9\MILL\NCI\EXAMPLES-2A.NCI - 2.6K
2 - Peck drill - full retract
Parameters
#2 - 0.2500 DRILL - 1/4 DRILL
Geometry - [4] point[s]
C:\MCAM9\MILL\NCI\EXAMPLES-2A.NCI - 2.6K
3 - Simple drill - no peck
Parameters
#3 - 0.0310 SPOT DRILL - 1/32 SPOTDRILLx1.5
Geometry - [5] point[s]
C:\MCAM9\MILL\NCI\EXAMPLES-2A.NCI - 2.7K
4 - Peck drill - full retract
Parameters
#4 - 0.0890 DRILL - #43 DRILL
Geometry - [5] point[s]
C:\MCAM9\MILL\NCI\EXAMPLES-2A.NCI - 2.7K
5 - Tapping - feed in, reverse spindle - feed out
Parameters
#5 - 0.125X40.00 TAP RH - #5-40 TAPRH
Geometry - [5] point[s]
C:\MCAM9\MILL\NCI\EXAMPLES-2A.NCI - 2.7K
Select All
Regen Path
Backplot
Verify
Post
Highfeed
OK
Help

- Click ① **Toolpaths**
- Click ② **Operations**
- Click ③ the **Select All** button
- Click ④ the **Backplot** button

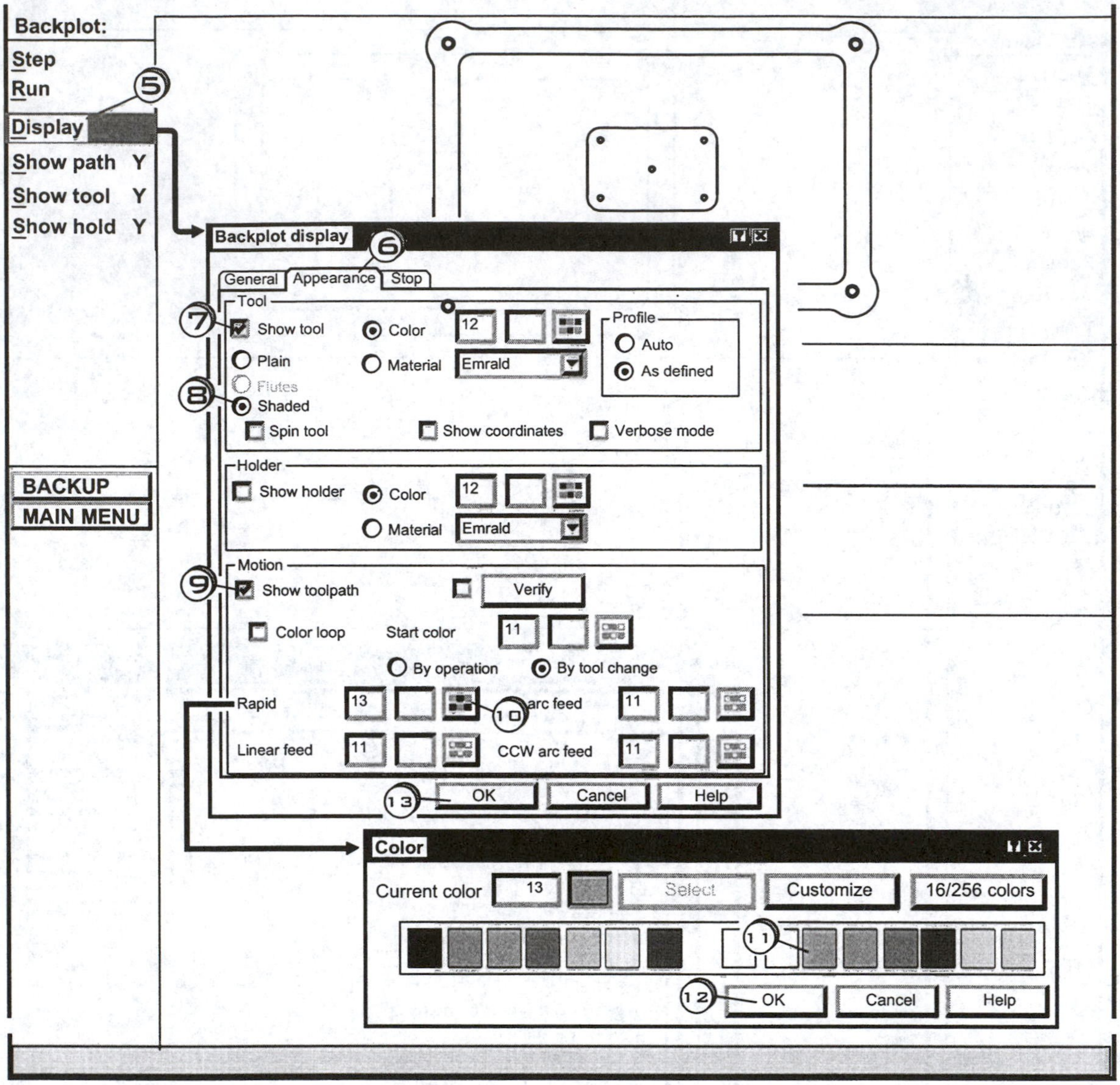

- Click ⑤ Display
- Click ⑥ the Appearance tab
- Click ⑦ such that a ckeck ☑ appears in Show tool
- Click ⑧ the Shaded radio button ◉
- Click ⑨ such that a ckeck ☑ appears in Show toolpath
- Click ⑩ the Rapid color button
- Click ⑪ the Magenta color
- Click ⑫ the OK button
- Click ⑬ the OK button

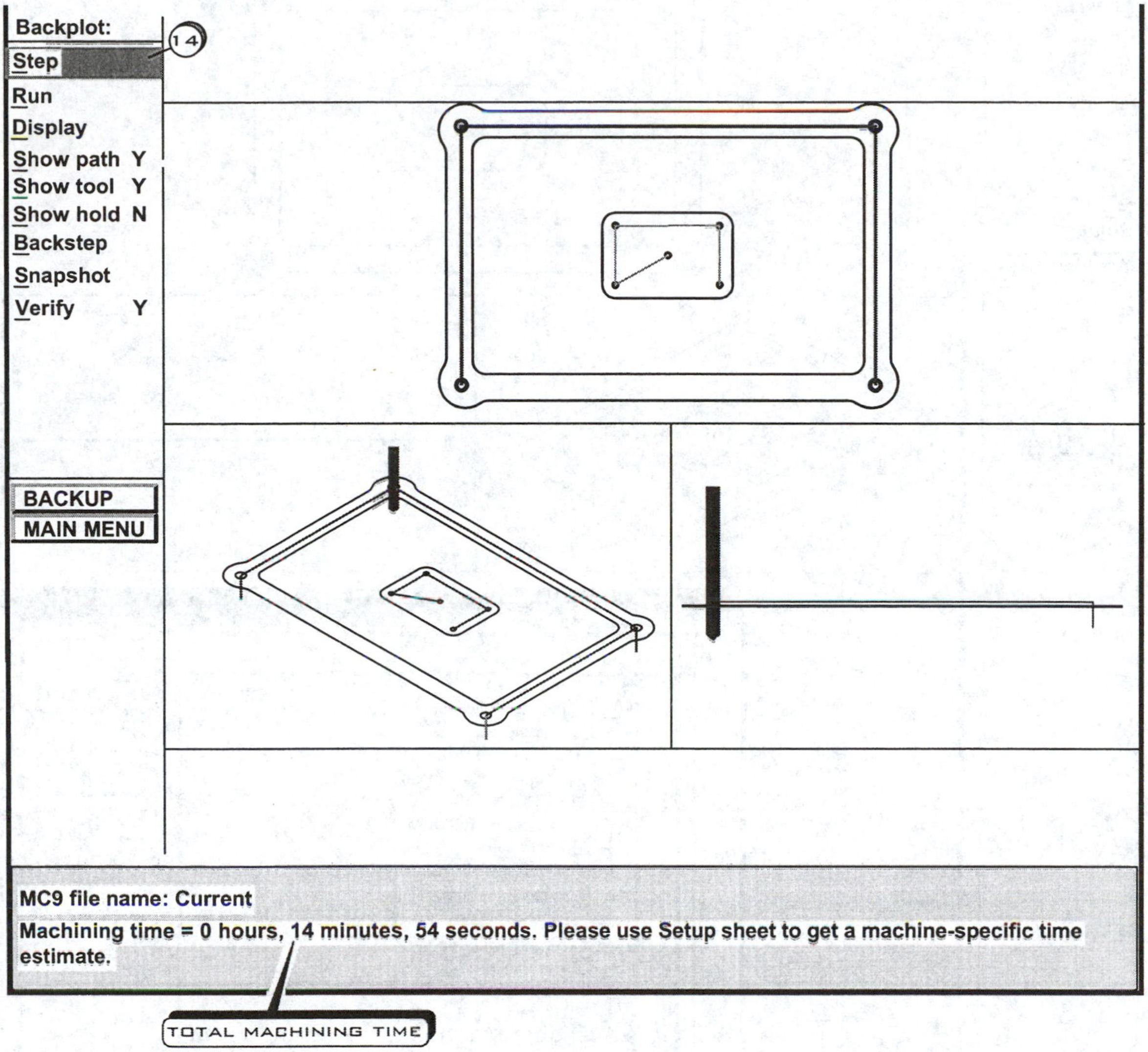

➤ ***Repeatedly*** Click (14) **Step** to see a step by step animation of tool movement

Mastercam will display the *total machining time for the operations selected*

B) Backplotting the Toolpaths for Selected Operations

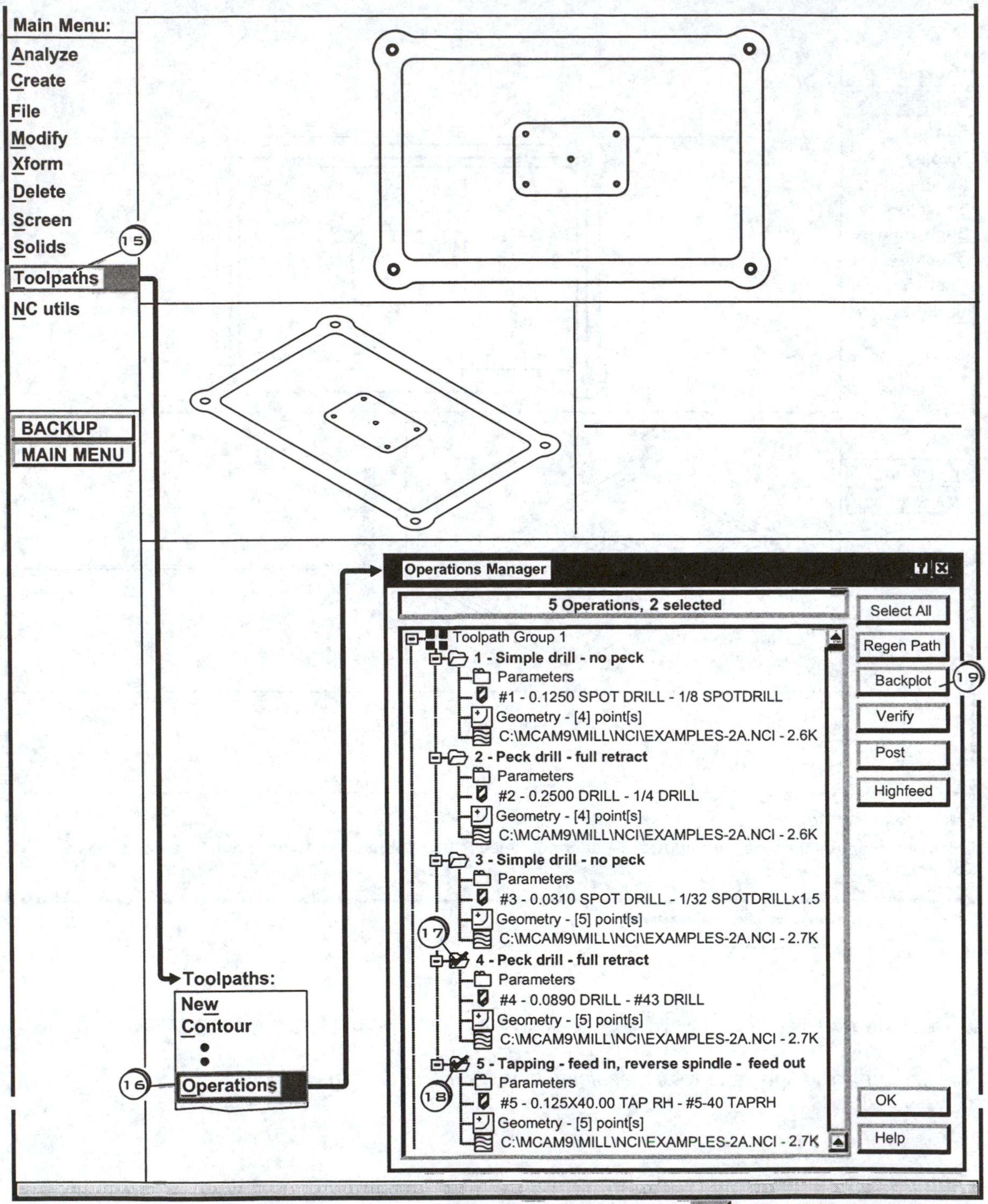

- Click (15) **Toolpaths**
- Click (16) **Operations**
- Click (17) on the ***first*** operation ✔
- Depress the Ctrl key on the keyboard and ***keeping it depressed***

  Click (18) on the ***next*** operation(s) ✔ etc
- Click (19) the Backplot button

5-6 Verifying Hole Machining Operations

Mastercam's verifier uses the NCI file to produce the most realistic animation of previously defined machining operations. The part is displayed as a solid model. The operator can see material being removed as the tool moves along a specified toolpath. Verification provides for *off line* checking or checking that *does not* cause a CNC machine to be tied up for the job proveout. This reduces possible tool and machine damage, increases the capability of machining a correct part on the first run and thus *increases the return on investment* for the job.

EXAMPLE 5-4

Use the Verify function in the Operations Manager dialog box to generate a proveout run of the drilling tool paths created in Example 5-2.

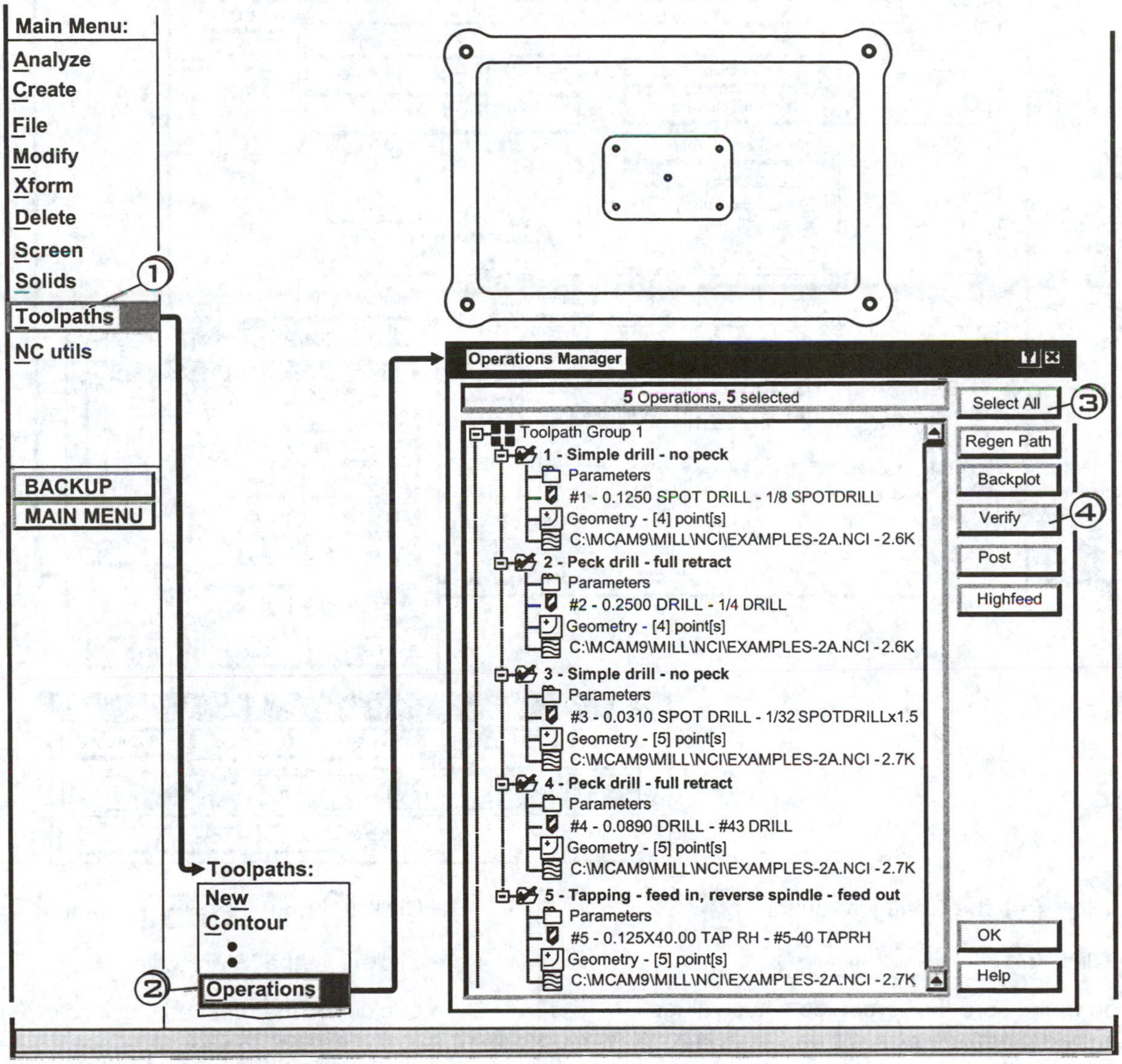

- Click ① Toolpaths
- Click ② Operations
- Click ③ the Select All button
- Click ④ the Verify button

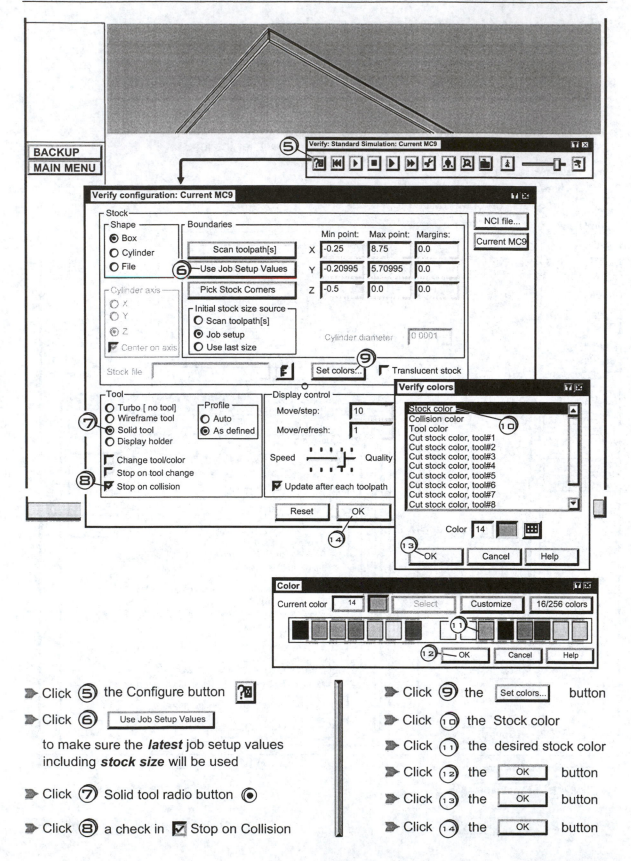

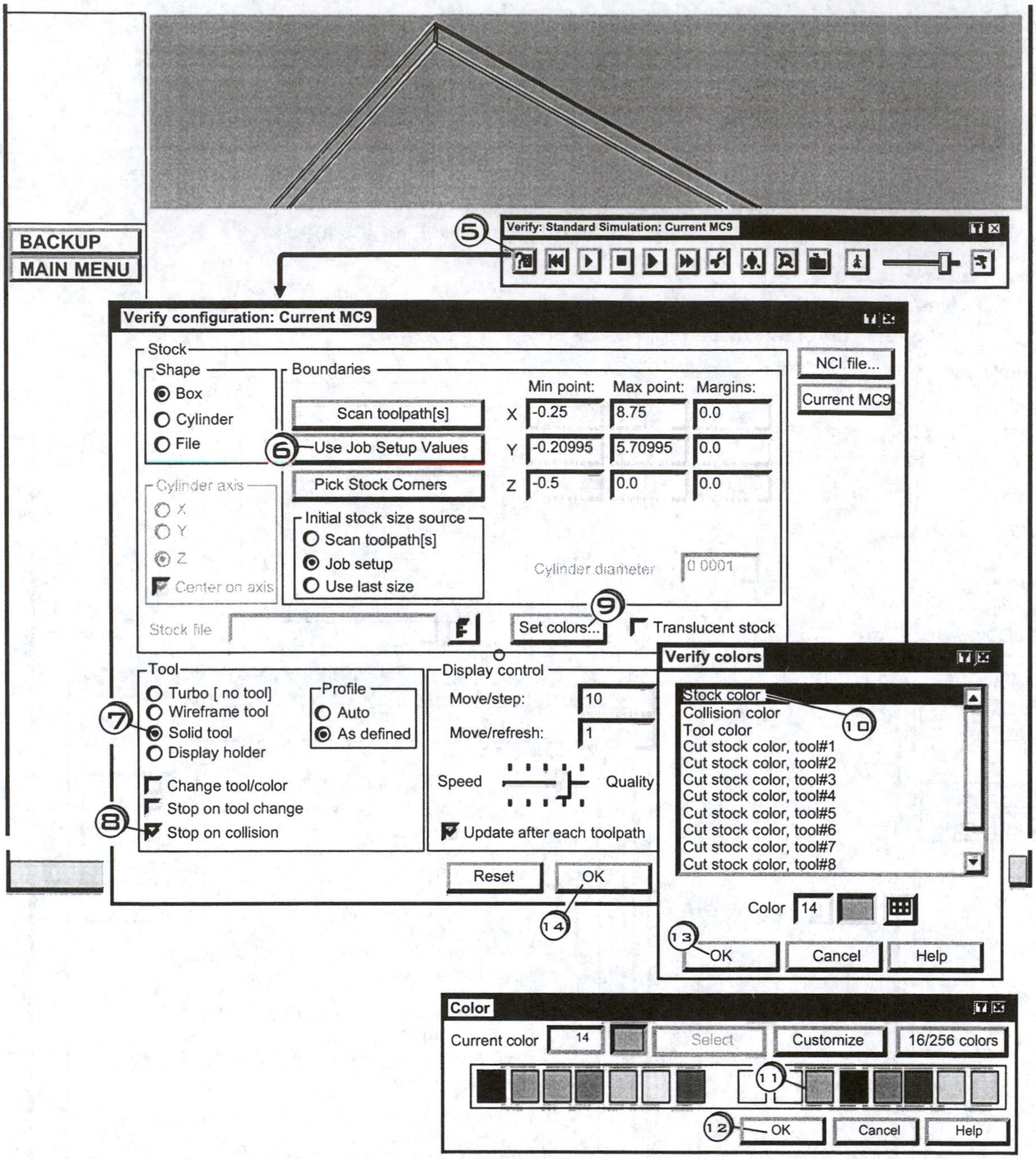

- Click ⑤ the Configure button
- Click ⑥ Use Job Setup Values

  to make sure the ***latest*** job setup values including ***stock size*** will be used

- Click ⑦ Solid tool radio button
- Click ⑧ a check in Stop on Collision
- Click ⑨ the Set colors... button
- Click ⑩ the Stock color
- Click ⑪ the desired stock color
- Click ⑫ the OK button
- Click ⑬ the OK button
- Click ⑭ the OK button

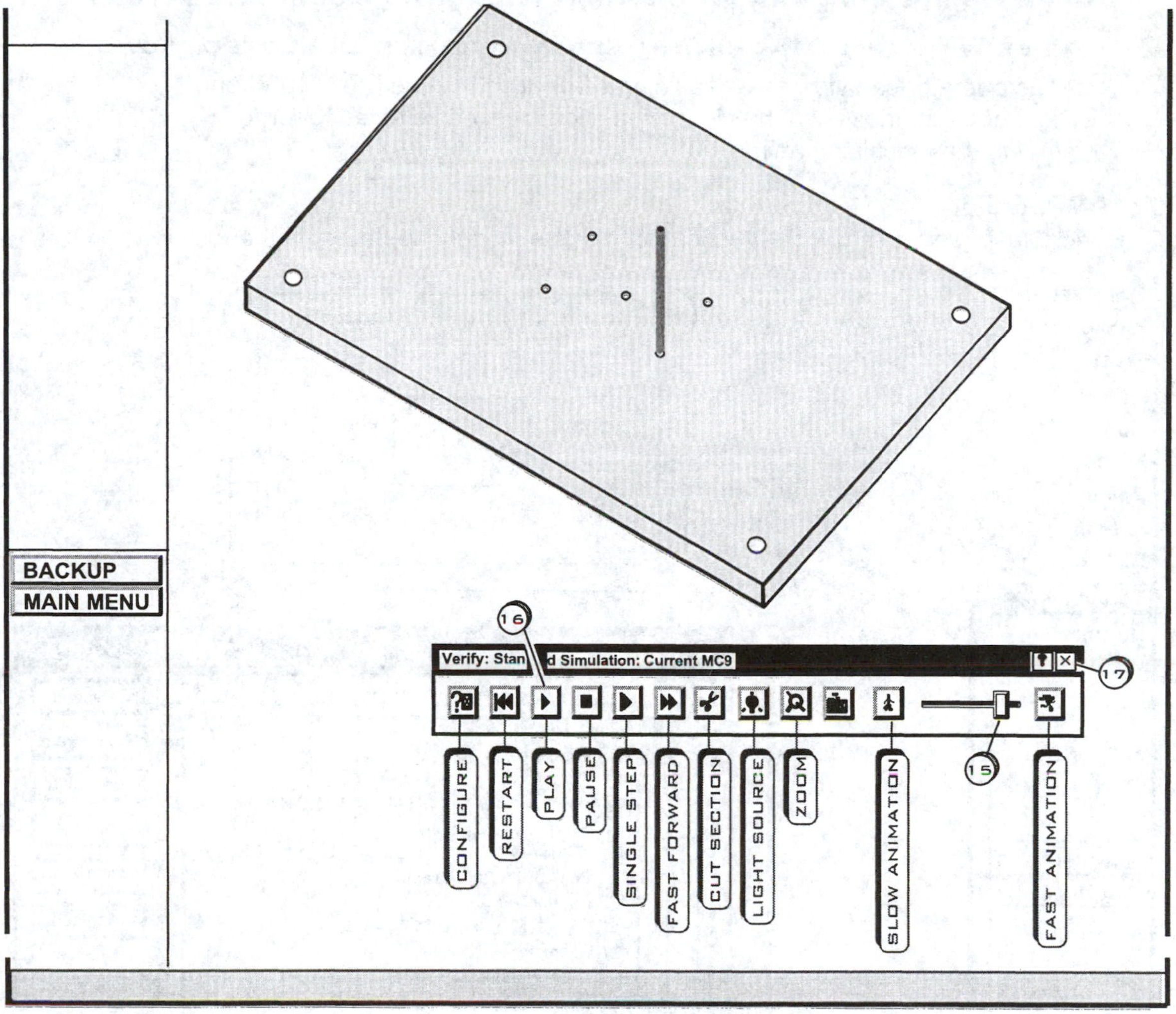

- Click (15) on the animation speed bar and drag it to a desired speed
- Click (16) the play button ▶
- Click (17) the exit button ☒ to exit the verifier

5-7 Views, Panning and Sectioning Within the Machining Verifier

A) Veiwing the Machining Verification in Selected Views

The operator can view the machining verification in either the Top, Front, Right or Isometric view. The view orientation must be selected *prior to starting* the machining animation.

EXAMPLE 5-5

Direct *Mastercam* to show the verification with the model orientated in the top view.

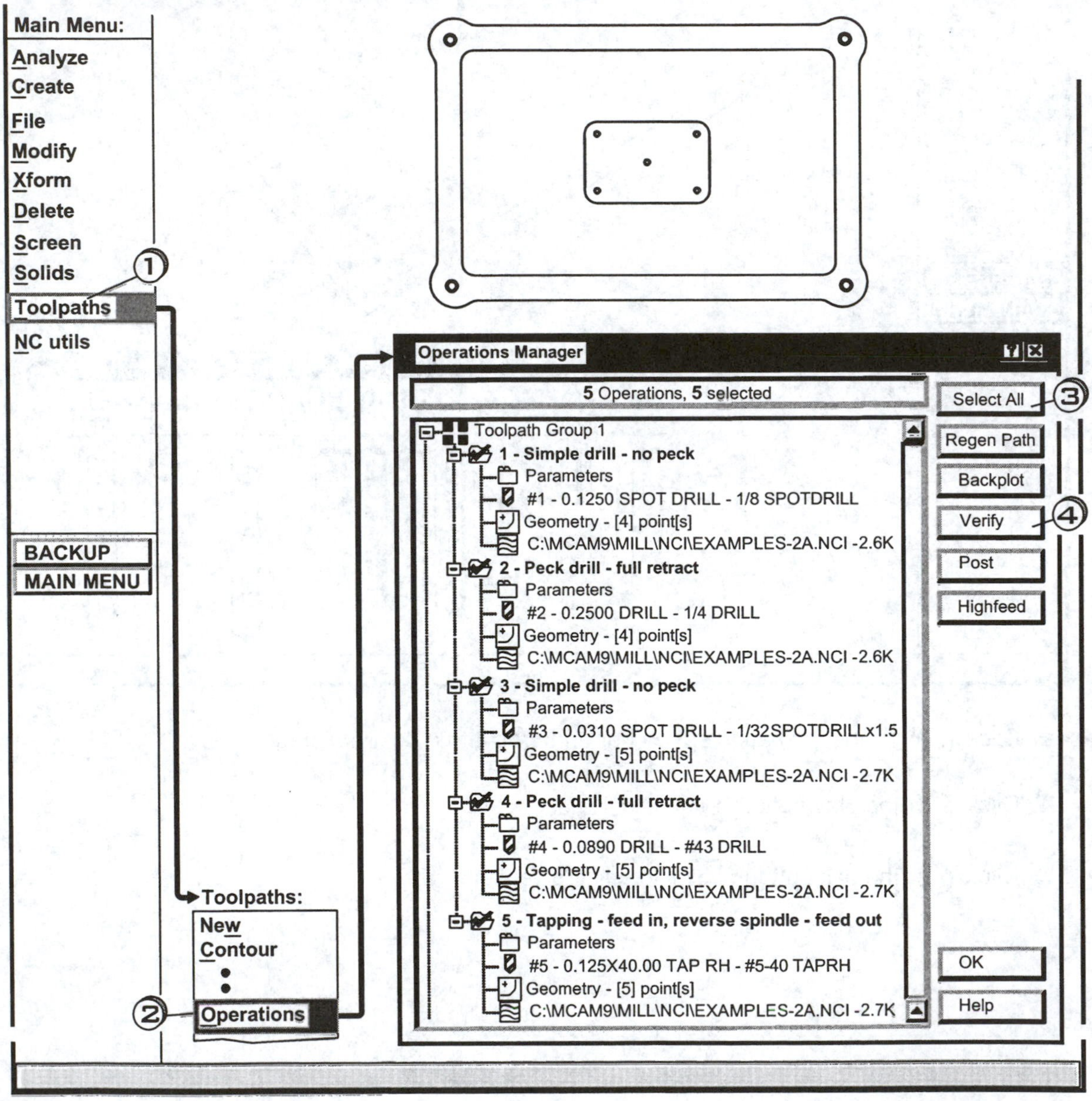

➤ Click ① Toolpaths

➤ Click ② Operations

➤ Click ③ the Select All button

➤ Click ④ the Verify button

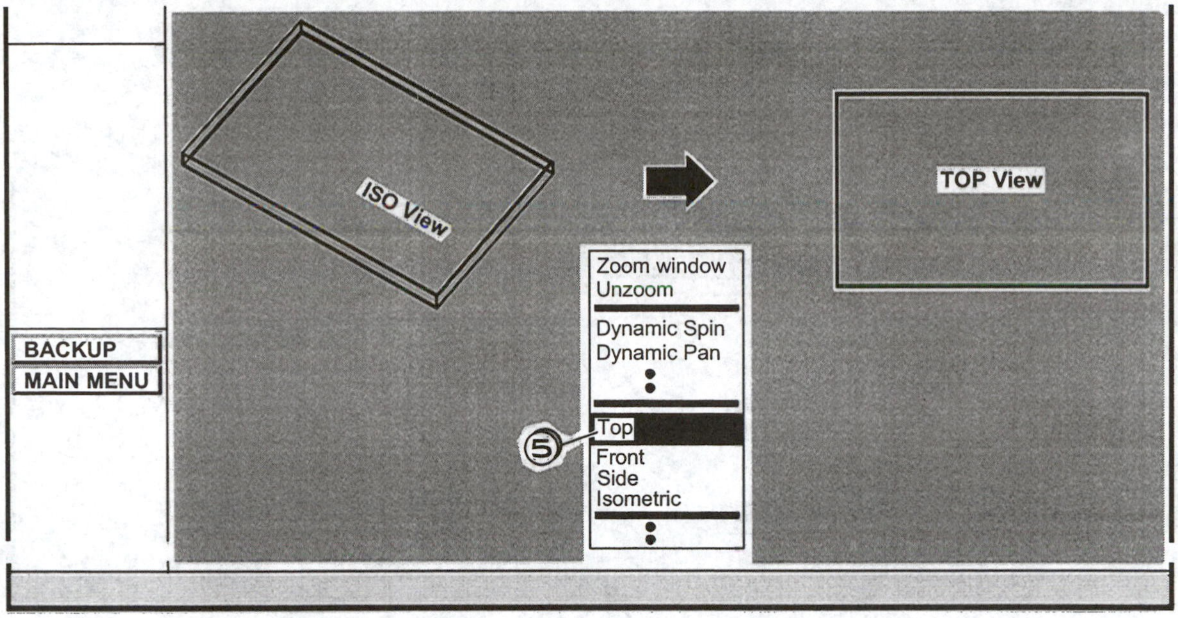

➤ ***Right*** Click; move the mouse cursor to Top and Click ⑤

the verification can then be observed in the Top view

B) Dynamically Panning the Machined Model

Use the Pan option to dynamically rotate the machined model within the verifier.

EXAMPLE 5-6

Dynamically pan machined model as shown below

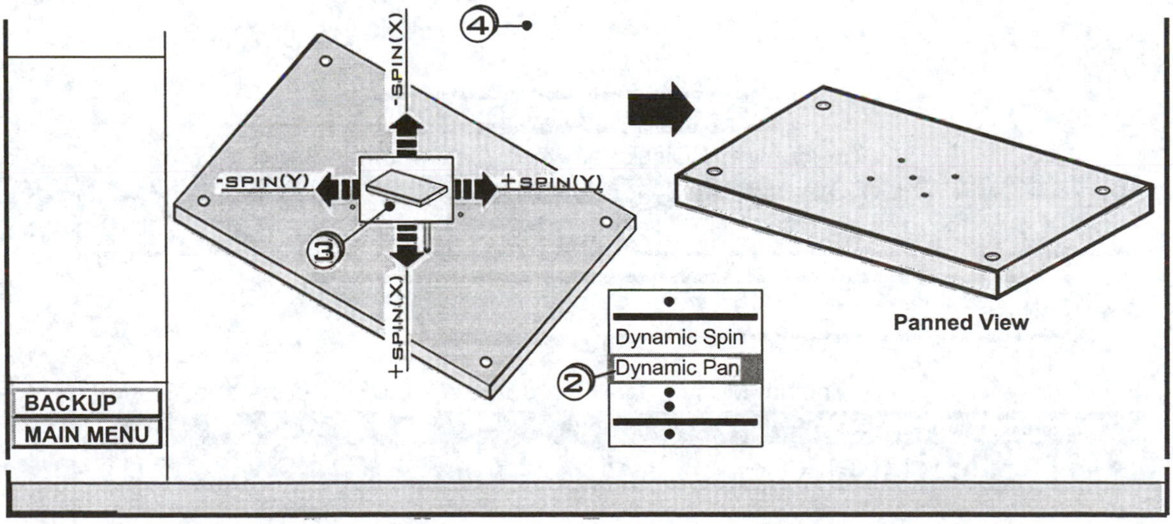

➤ ***Right*** Click ① ;move the mouse cursor to ② and click

Pick a point to begin dynamics

➤ Click ③ and ***drag*** the mouse cursor *right* and *left* *(Y-spinning)* or *up* and *down(X-spinning)*

➤ Click ④ to *set* the rotation and exit

C) Sectioning the Machined Model

The machined model can be sectioned within the verifier.

EXAMPLE 5-7

Section the machined model shown below

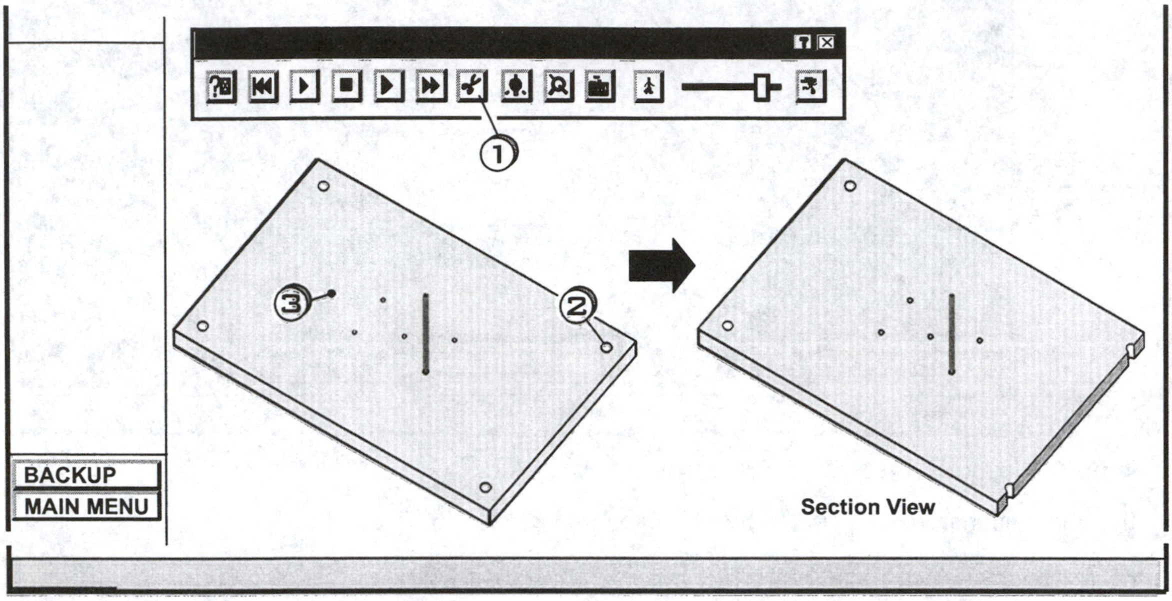

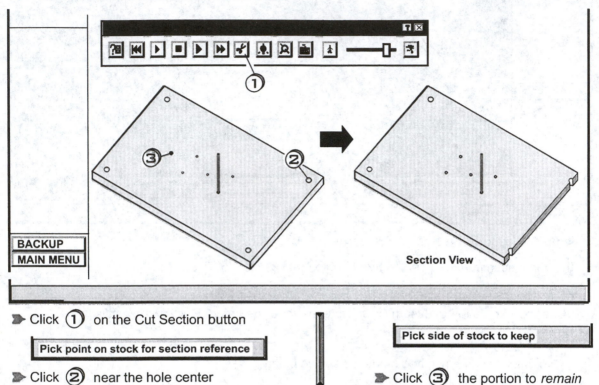

➤ Click ① on the Cut Section button

Pick point on stock for section reference

➤ Click ② near the hole center

Pick side of stock to keep

➤ Click ③ the portion to *remain*

Using the Toolbar Menus

Toolbar Menu Displayed when *Mastercam* is **Started**

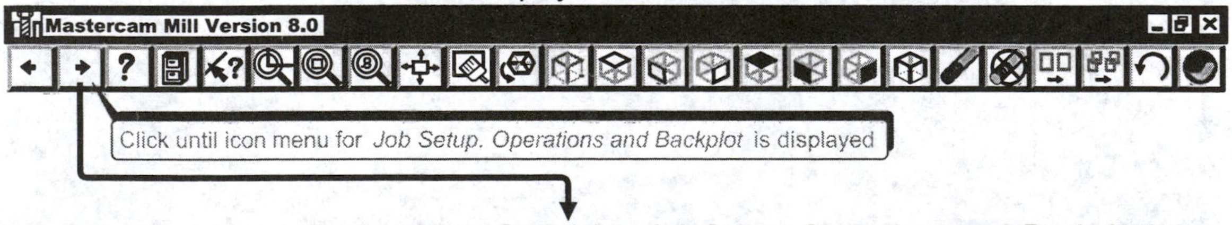

Toolbar Menu Containing ***Job Setup, Operations and Backplot*** icons

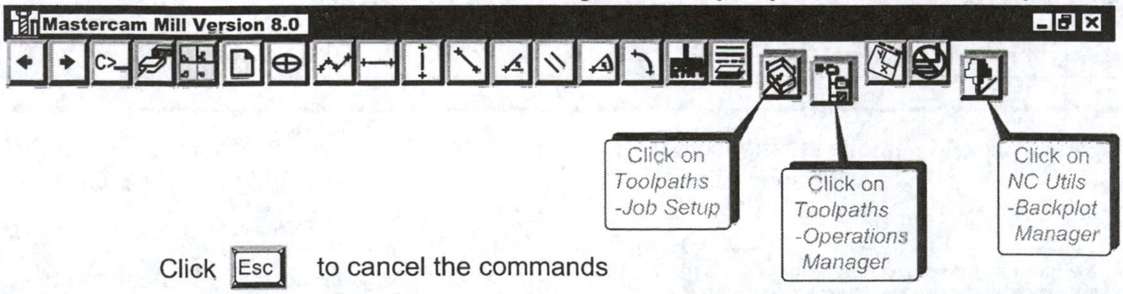

Click Esc to cancel the commands

5-8 Using the Circle Mill Module

The circle mill module automatically creates tool paths for machining a circle in stock. This is especially useful when using an endmill to produce a counterbore or a spotface.

EXAMPLE 5-6

Use a .75Dia Endmill to machine a 1.5Dia Counterbore .25 deep in the stock shown in Figure 5-2.

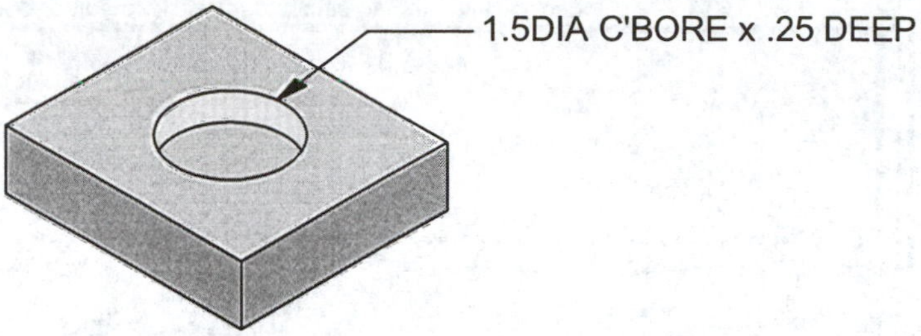

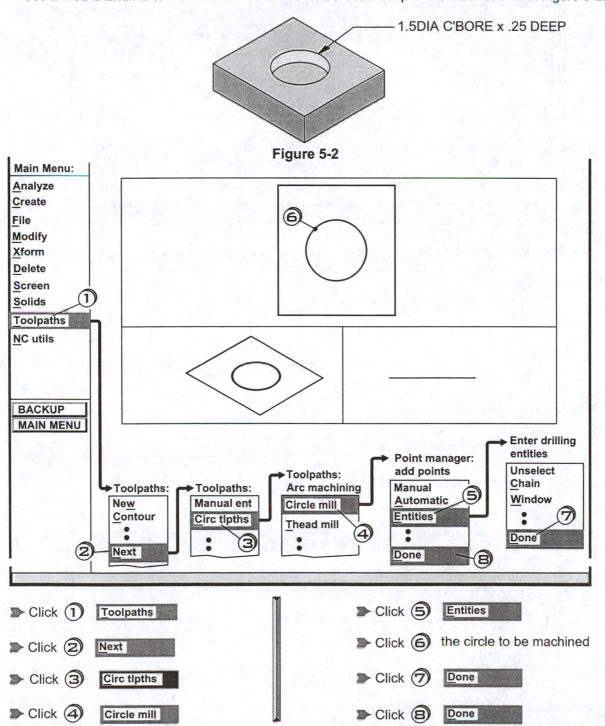

Figure 5-2

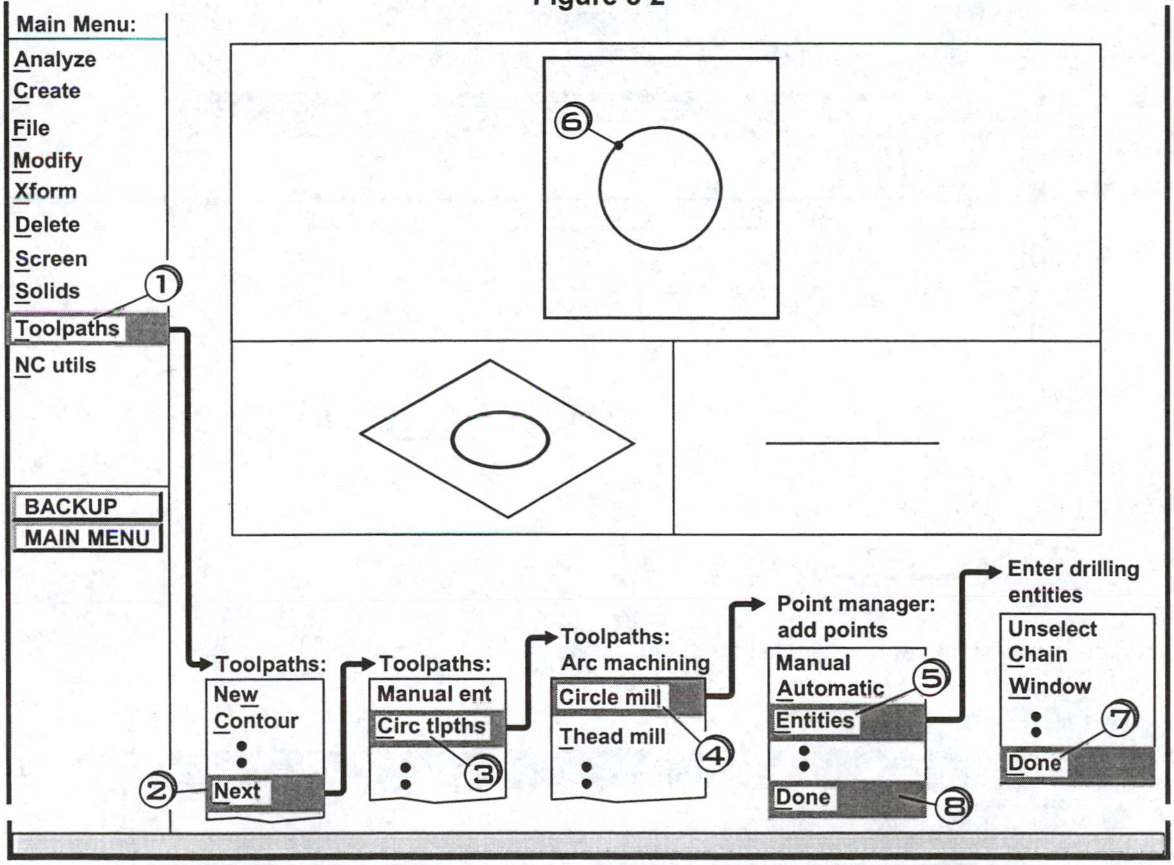

- Click ① Toolpaths
- Click ② Next
- Click ③ Circ tlpths
- Click ④ Circle mill
- Click ⑤ Entities
- Click ⑥ the circle to be machined
- Click ⑦ Done
- Click ⑧ Done

◆ OBTAIN THE NEEDED 3/4 END MILL TOOL

Mastercam will activate and display the Tool and Drill Parameters dialog box after the circle mill tool path has been created.

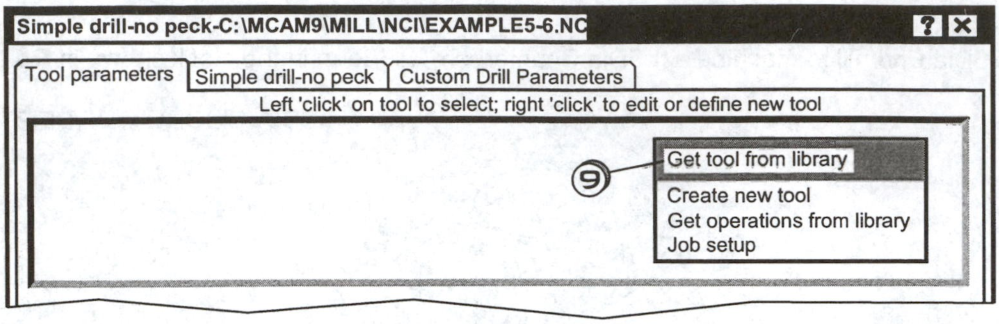

➤ ***Right*** Click move the mouse cursor *down* and Click ⑨ Get tool from library

Select the 3/4in Dia end mill tool from the **TOOLS** tool library

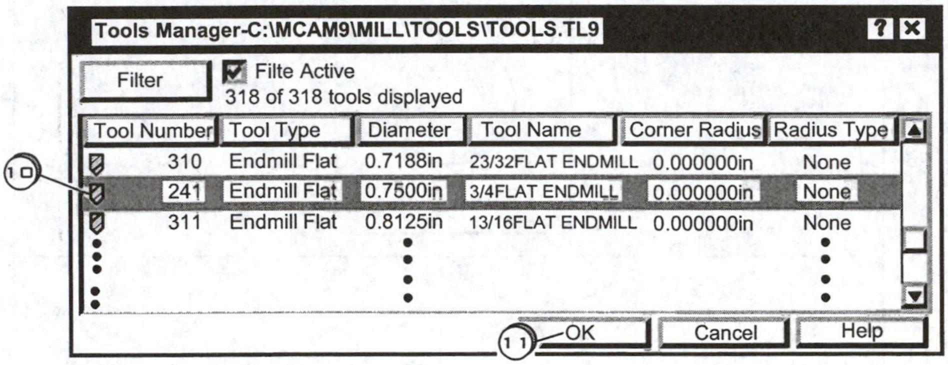

➤ Click ⑩ on the 3/4 FLAT ENDMILL

➤ Click ⑪ the OK button

Mastercam will place these tools into the currently active Tool and Drill Parameters dialog box

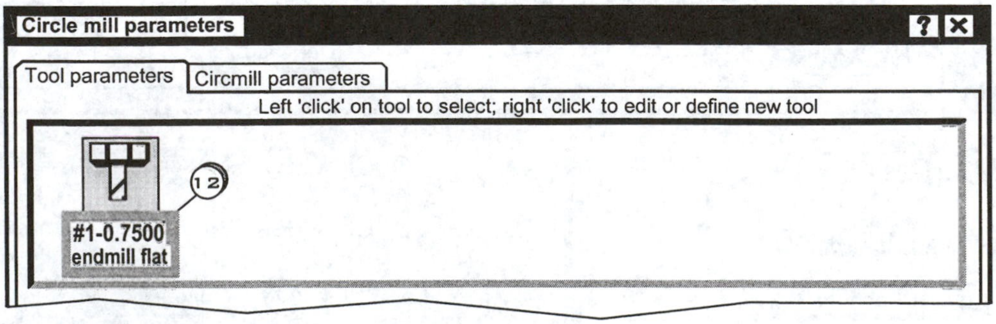

➤ ***Right*** Click ⑫ on the .7500 endmill flat tool

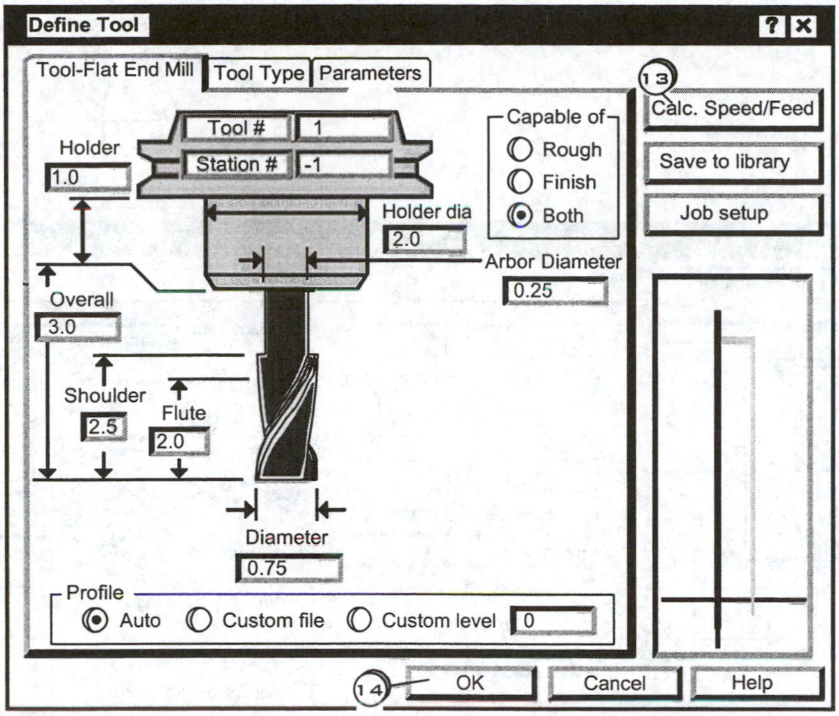

➤ Click (13) the [Calc. Speed/Feed] key

➤ Click (14) the [OK] key

◆ SPECIFY THE CIRCLE MILL TOOL PATH PARAMETERS

Mastercam executes the circle mill module by moving the tool as shown in Figure 5-3

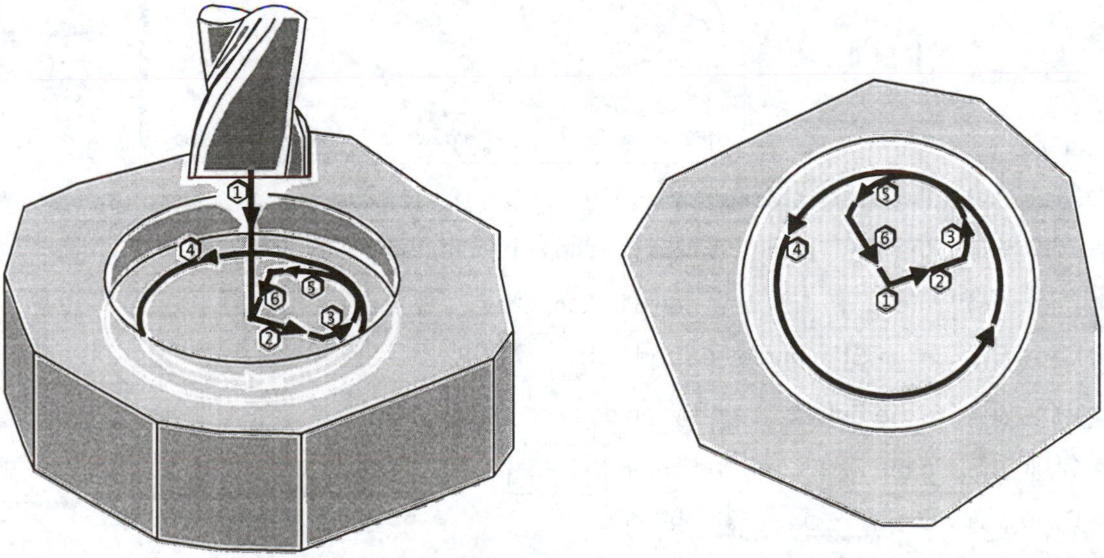

Circle mill module tool path generated by *Mastercam*

Figure 5-3

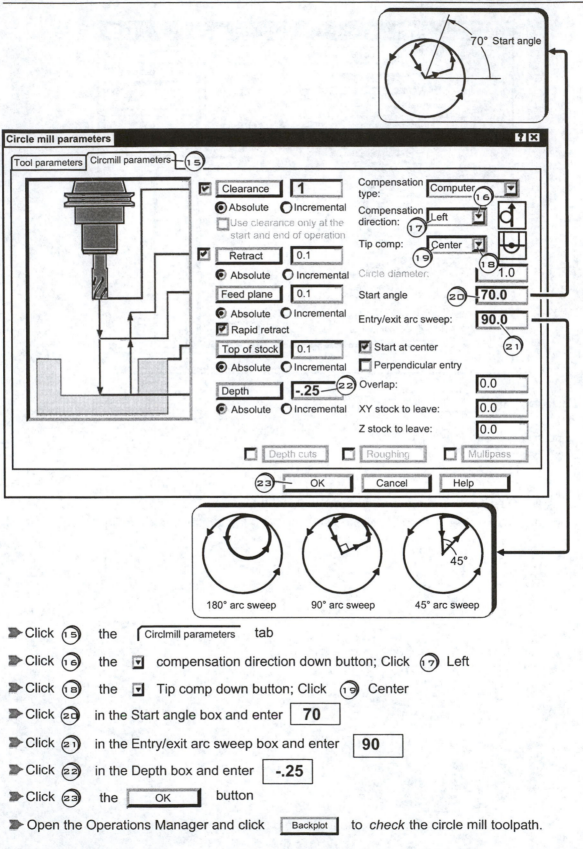

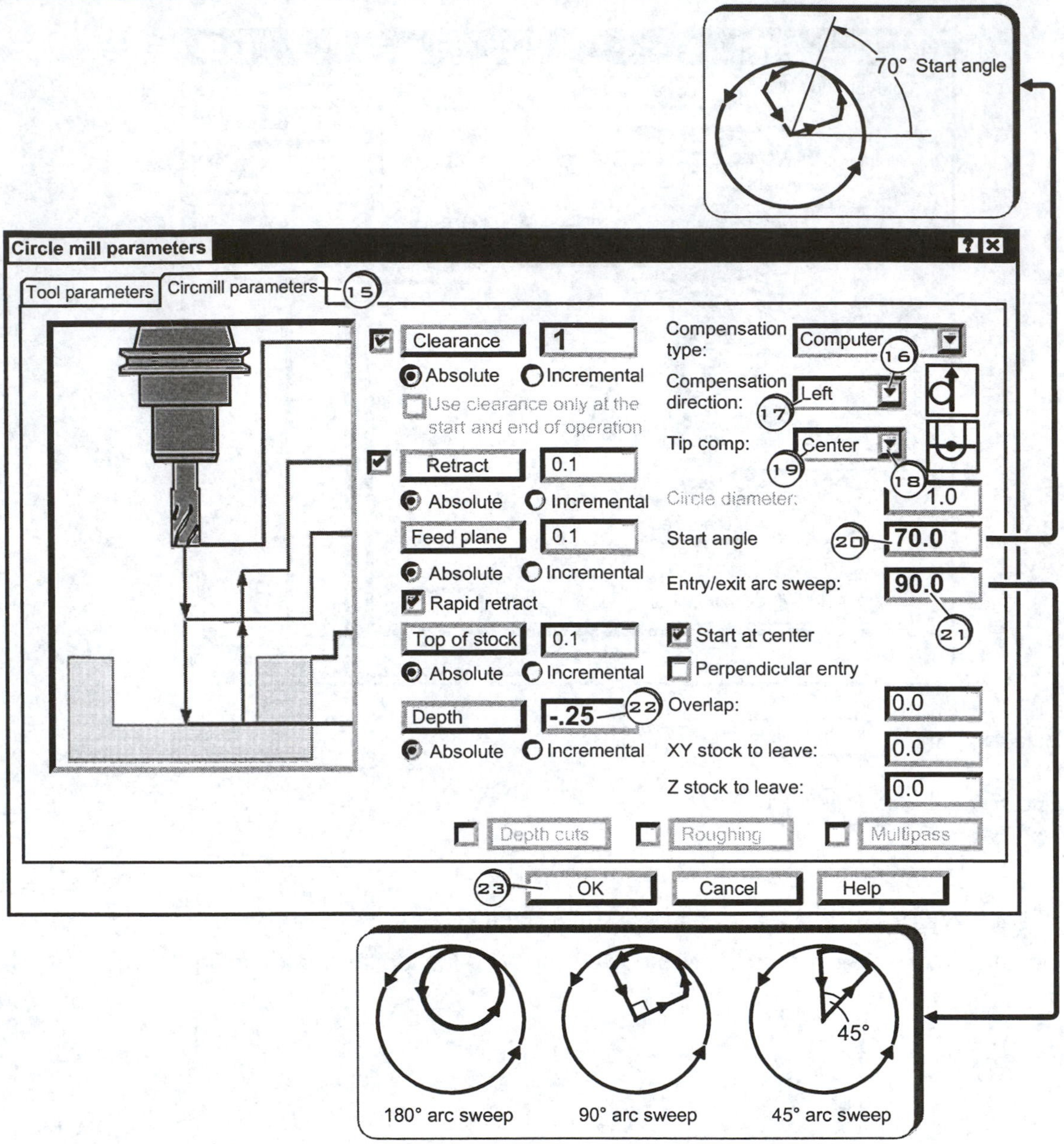

- Click (15) the [Circmill parameters] tab
- Click (16) the [▼] compensation direction down button; Click (17) Left
- Click (18) the [▼] Tip comp down button; Click (19) Center
- Click (20) in the Start angle box and enter [70]
- Click (21) in the Entry/exit arc sweep box and enter [90]
- Click (22) in the Depth box and enter [-.25]
- Click (23) the [OK] button
- Open the Operations Manager and click [Backplot] to *check* the circle mill toolpath.

5-9 Machining Holes at Different Depths and Retract Heights

Some drilling operations can be more efficiently carried out by specifying different depths and different retract heights. This is the case for parts whose thickness changes within the drill area. The operator may also need to change the retract height in order to avoid collisions with clamps or other workholding devices.

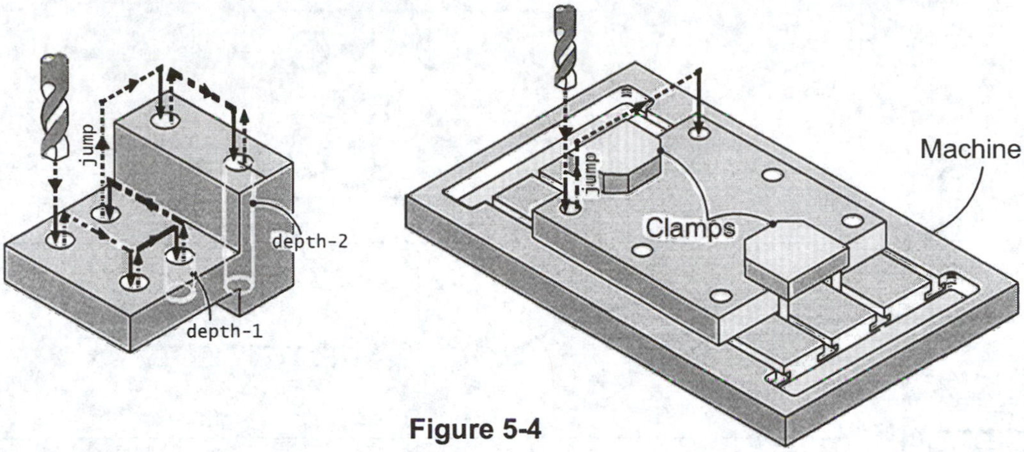

Figure 5-4

EXAMPLE 5-7

Use a .25Dia drill to machine the through holes in the part shown in Figure 5-5.

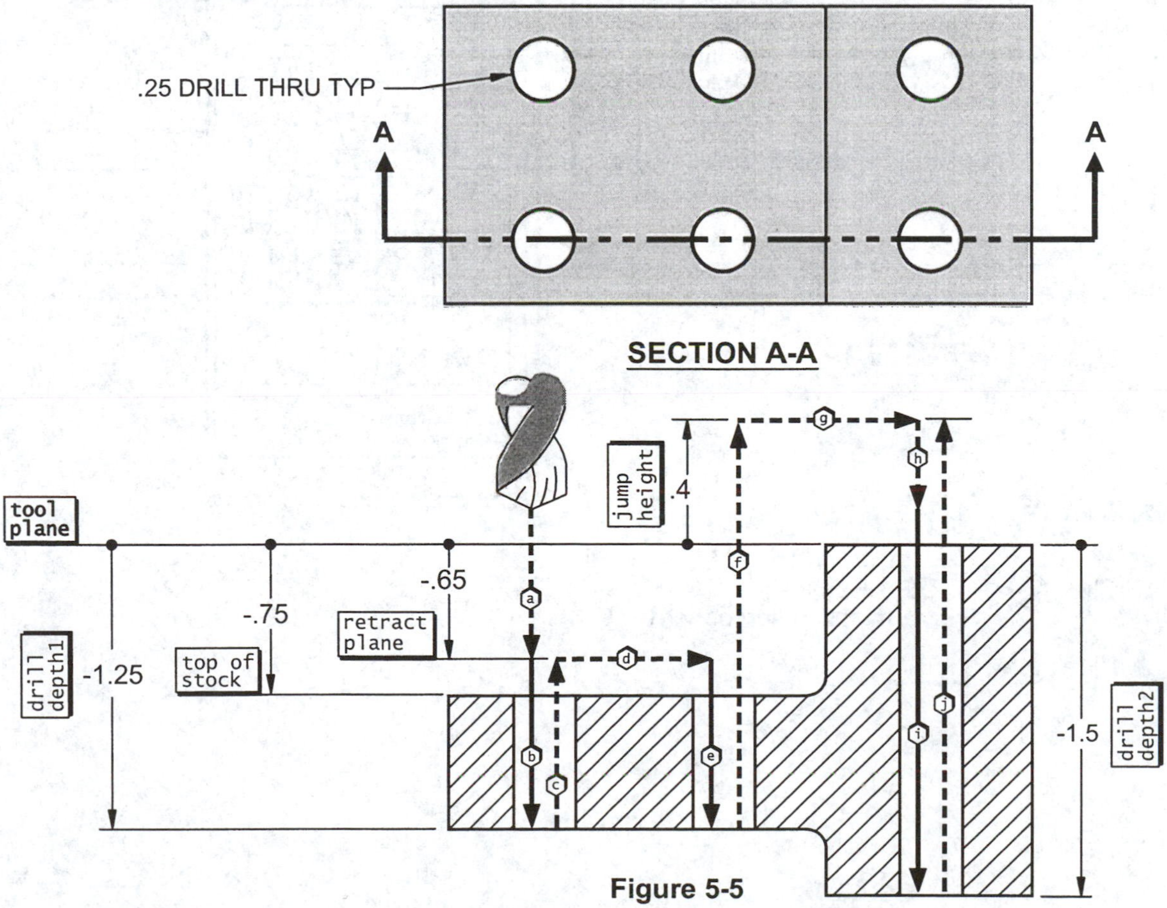

Figure 5-5

◆ SELECT THE Y ZIG X+ DRILL SORTING PATTERN

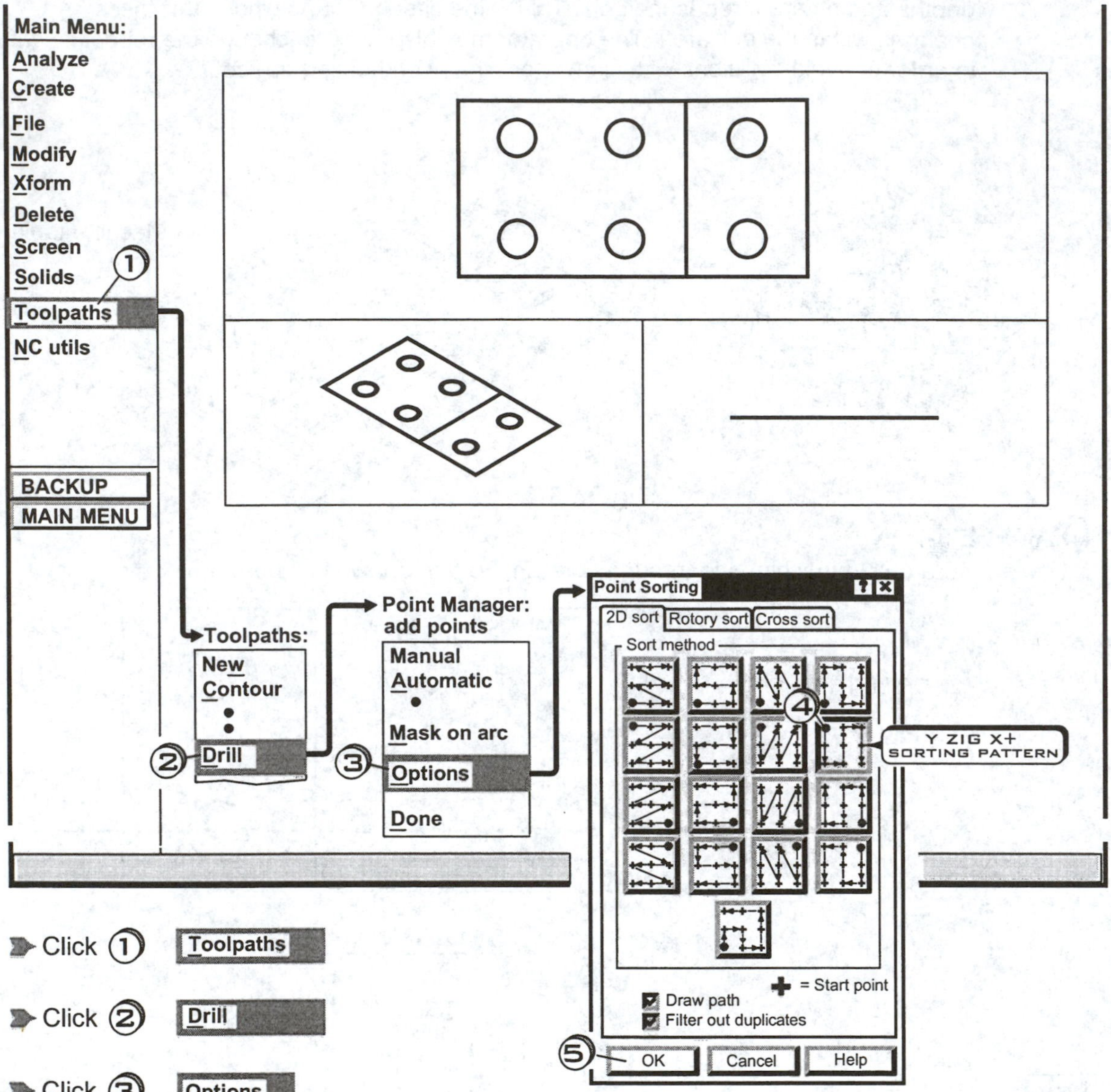

- Click ① **Toolpaths**
- Click ② **Drill**
- Click ③ **Options**
- Click ④ **Y ZIG X+** as the ***current active*** sort pattern
- Click ⑤ OK

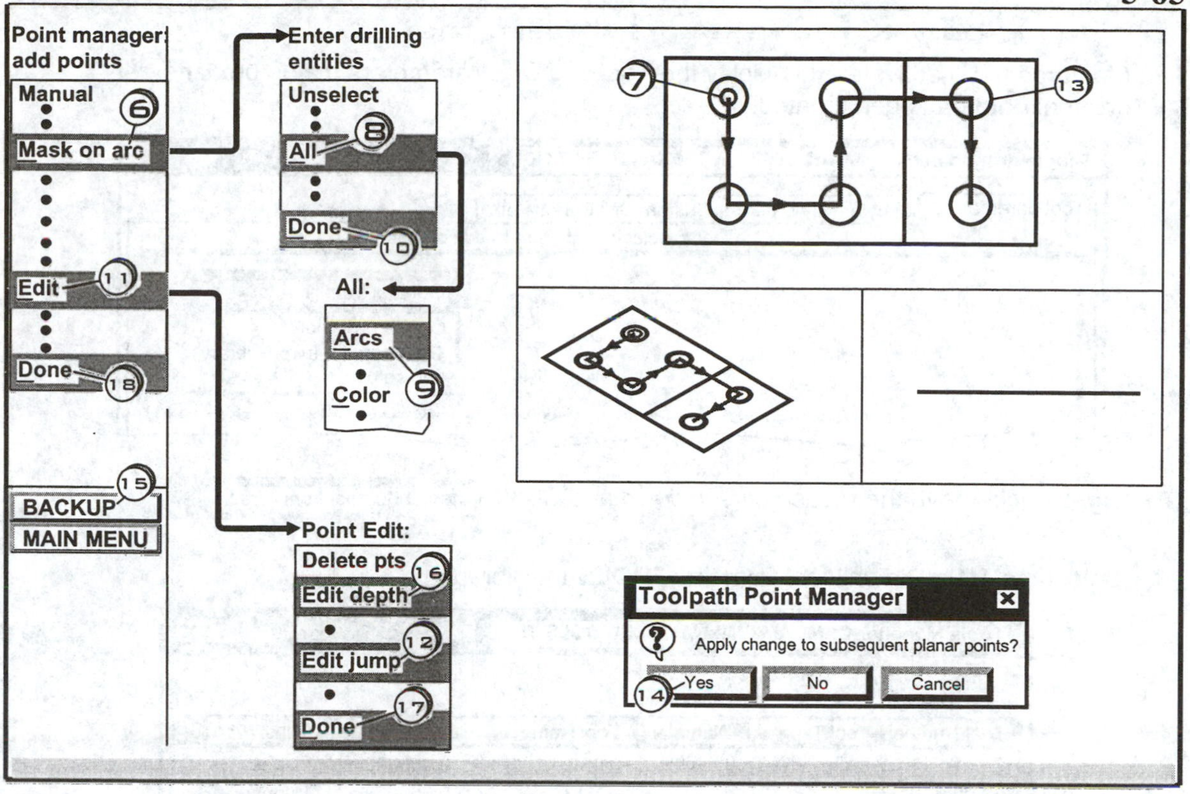

- Click ⑥ Mask on arc

Select arc to match

- Click ⑦ the arc to be matched

Enter arc rad match tol 0.001

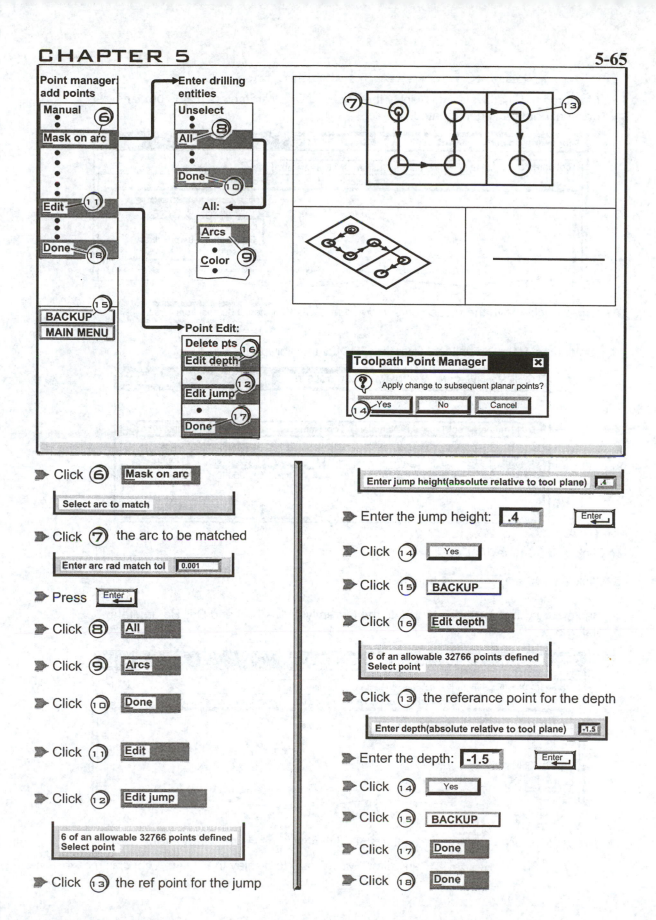

- Press Enter
- Click ⑧ All
- Click ⑨ Arcs
- Click ⑩ Done
- Click ⑪ Edit
- Click ⑫ Edit jump

6 of an allowable 32766 points defined
Select point

- Click ⑬ the ref point for the jump

Enter jump height(absolute relative to tool plane) .4

- Enter the jump height: .4 Enter
- Click ⑭ Yes
- Click ⑮ BACKUP
- Click ⑯ Edit depth

6 of an allowable 32766 points defined
Select point

- Click ⑬ the referance point for the depth

Enter depth(absolute relative to tool plane) -1.5

- Enter the depth: -1.5 Enter
- Click ⑭ Yes
- Click ⑮ BACKUP
- Click ⑰ Done
- Click ⑱ Done

◆ OBTAIN THE NEEDED 1/4 DRILL TOOL

Mastercam will activate and display the Tool and Drill Parameters dialog box after the drill tool path has been created.

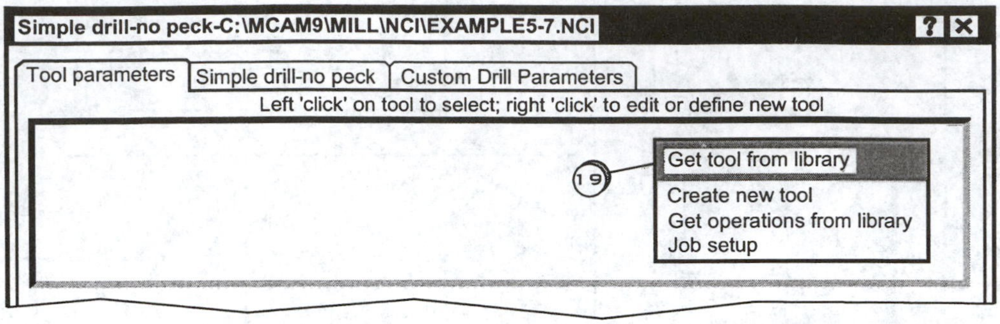

➤ ***Right*** Click move the mouse cursor *down* and Click (19) 

Select the 1/4in Dia drill tool from the **TOOLS** tool library

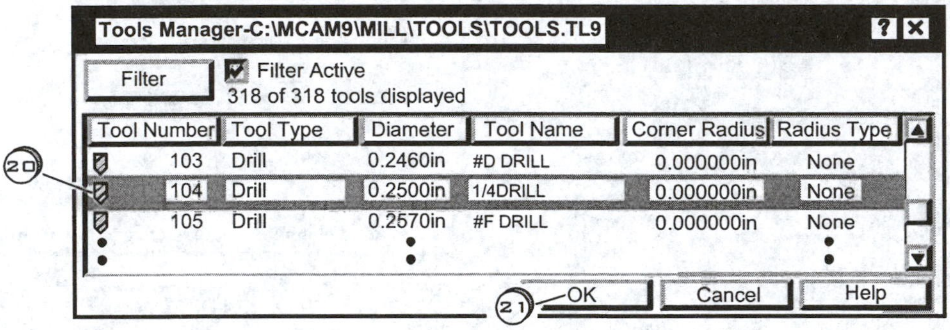

➤ Click (20) on the 1/4 DRILL

➤ Click (21) the [OK] button

Mastercam will place these tools into the currently active Tool and Drill Parameters dialog box

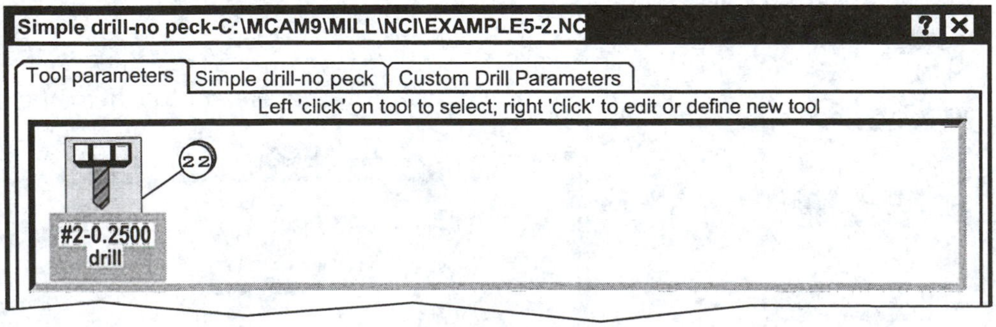

➤ ***Right*** click (22) 

on the .25Drill to select this as the ***active tool*** and direct *Mastercam* to display the Define Tool dialog box

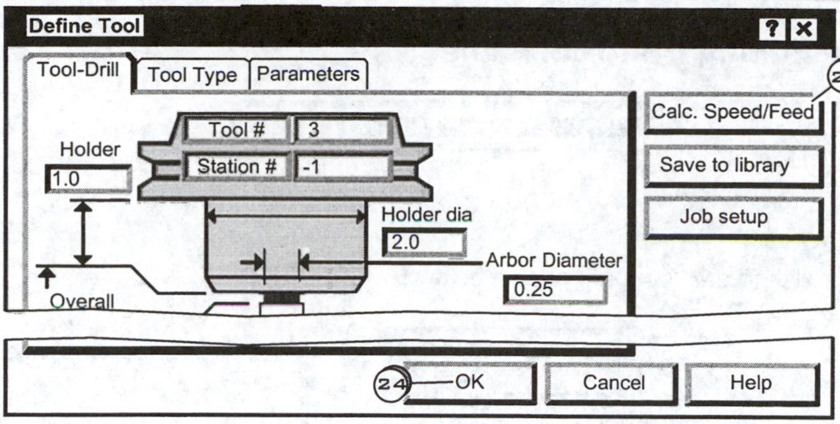

➤ Click (23) [Calc Speed/Feed] ➤ Click (24) [OK]

◆ ENTER THE REQUIRED .25 DIA THRU HOLE MACHINING PARAMETERS

➤ Click (25) [Peck drill-full retract] tab to enter the .25DIA hole machining parameters.

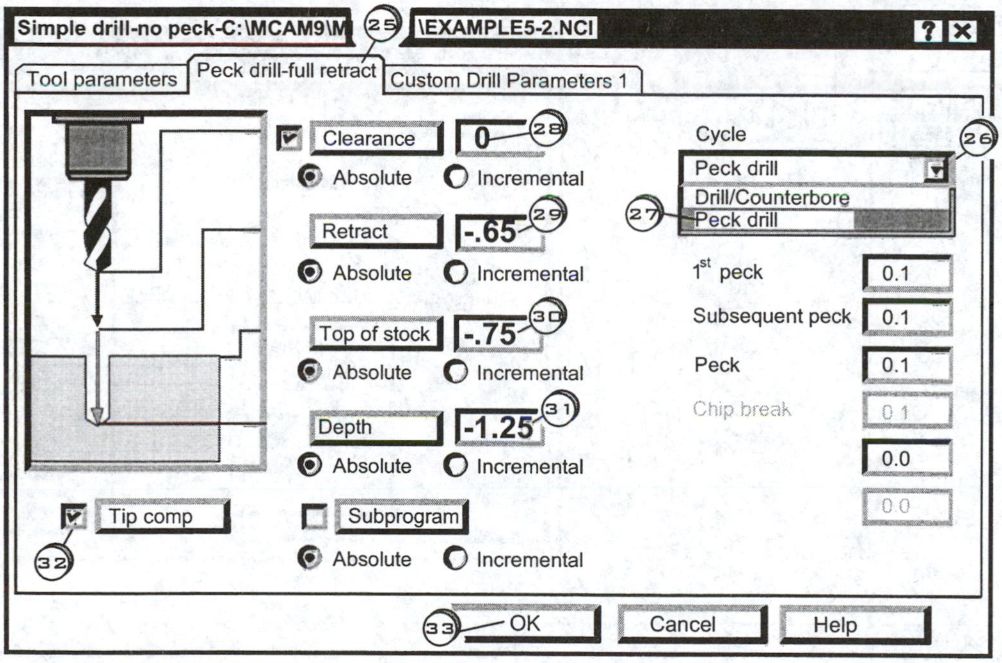

➤ Click (26) [▼] the toggle down button

➤ Click (27) [Peck drill]

➤ Click (28) in the Clearance box; enter [0]

➤ Click (29) in the Retract box; enter [-.65]

➤ Click (30) in the Top of stock box ; enter [-.75]

➤ Click (31) in the Depth box; enter [-1.25]

➤ Click (32) the tip comp check [☑]

➤ Click (33) [OK]

➤ Open the Operations Manager and click [Backplot] to ckeck the drill toolpaths be sure to *continuously* click the **Step** command in **Backplot:**

5-10 Editing Drill Tool Paths

DELETING AN EXISTING DRILL TOOL PATH

EXAMPLE 5-7

Delete the existing drill tool path shown below

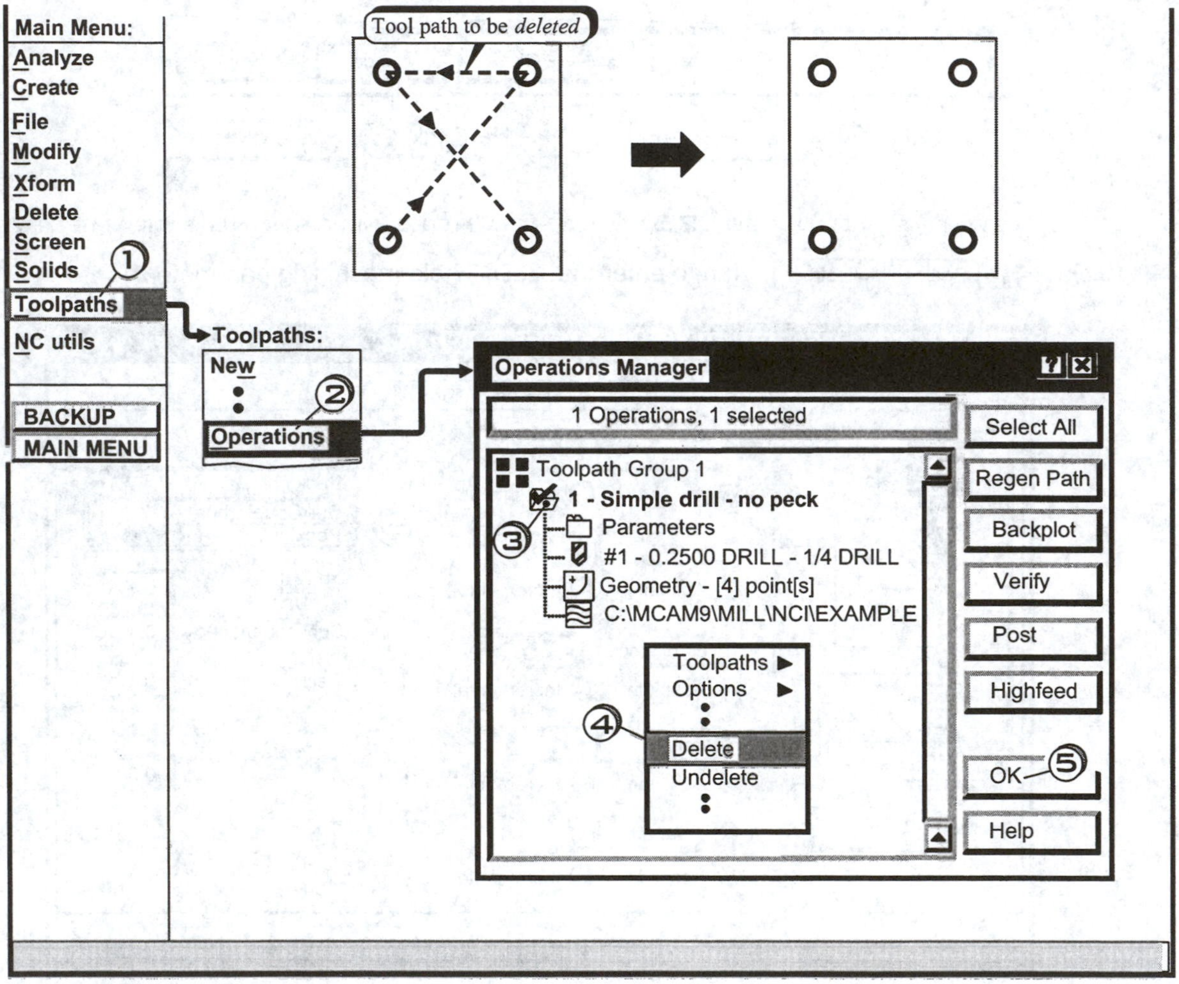

➤ Click ① **Toolpaths**

➤ Click ② **Operations**

➤ Click ③ on the file folder of the operation to delete such that a check appears

➤ ***Right*** Click move the cursor down to **Delete**; Click ④

➤ Click ⑤ the OK button

ADDING DRILL POINTS TO AN EXISTING DRILL TOOL PATH

EXAMPLE 5-8

Add the additional drill points to the existing drill tool path shown below

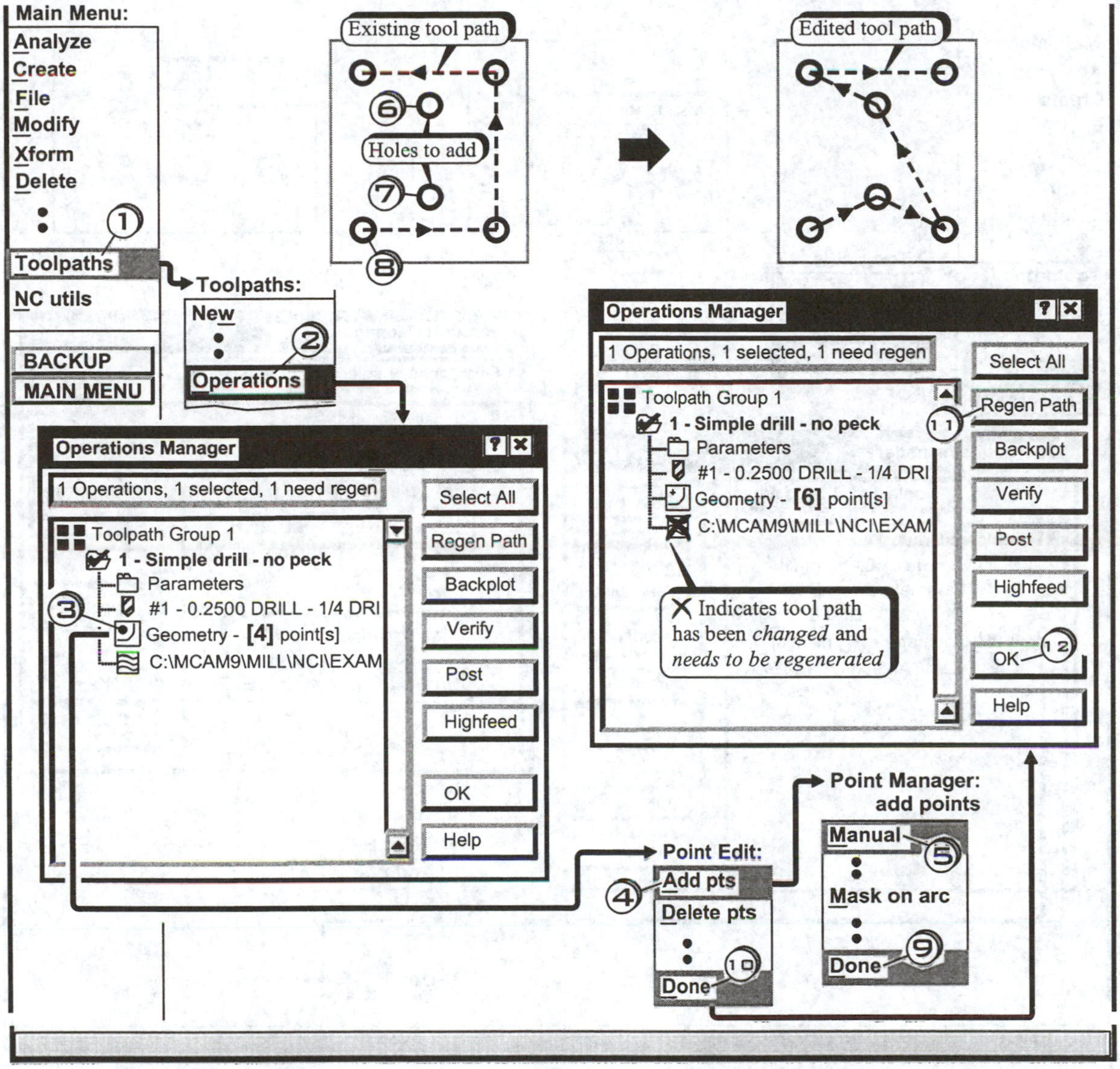

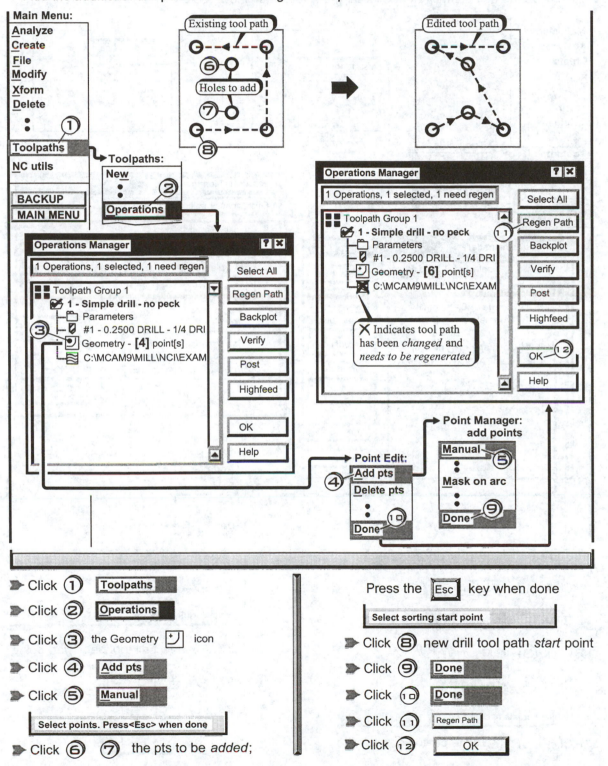

- Click ① Toolpaths
- Click ② Operations
- Click ③ the Geometry icon
- Click ④ Add pts
- Click ⑤ Manual

Select points. Press<Esc> when done

- Click ⑥ ⑦ the pts to be *added*;

Press the Esc key when done

Select sorting start point

- Click ⑧ new drill tool path *start* point
- Click ⑨ Done
- Click ⑩ Done
- Click ⑪ Regen Path
- Click ⑫ OK

DELETING POINTS FROM AN EXISTING DRILL TOOL PATH

EXAMPLE 5-9

Remove the drill points from the existing drill tool path shown below

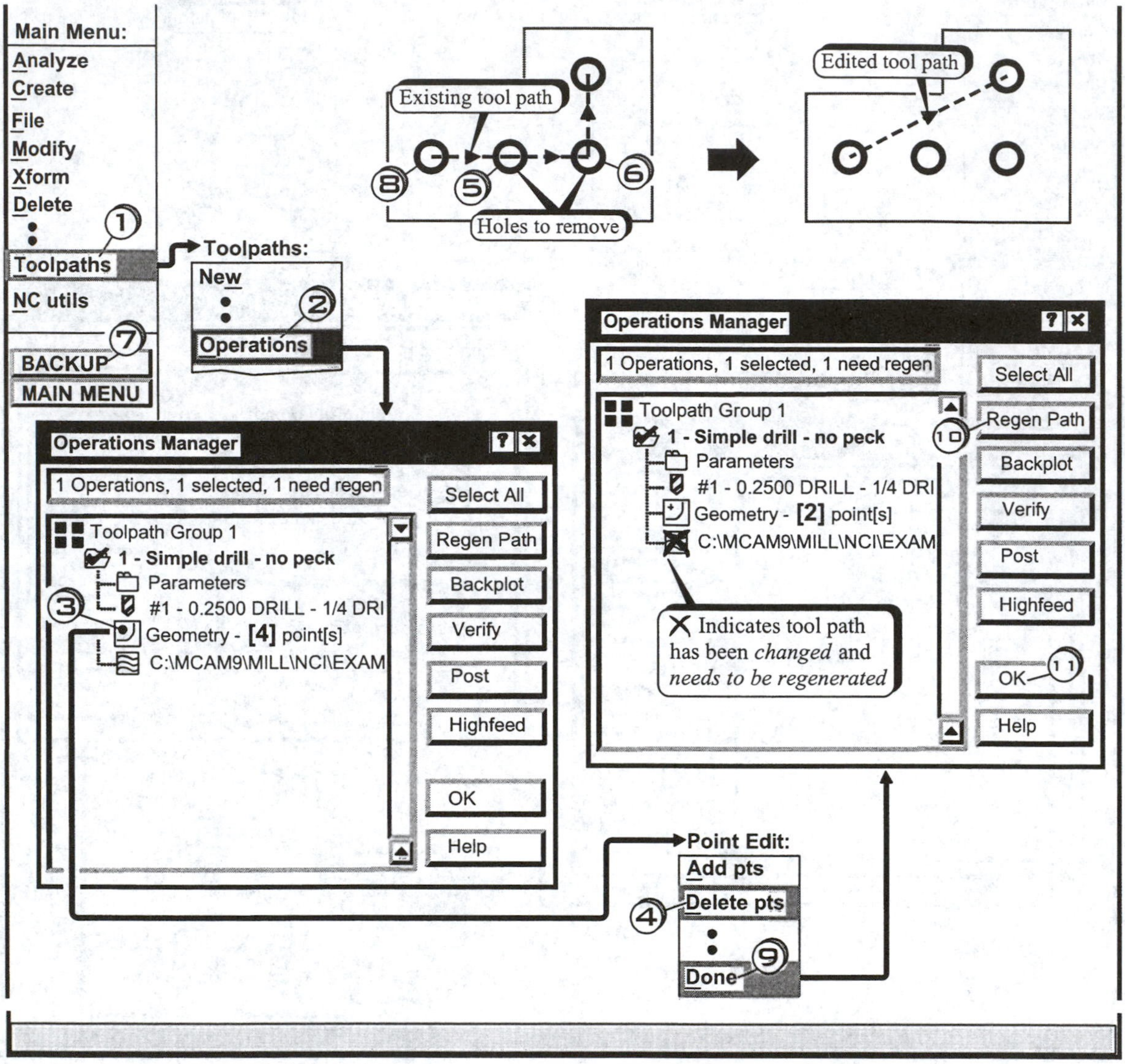

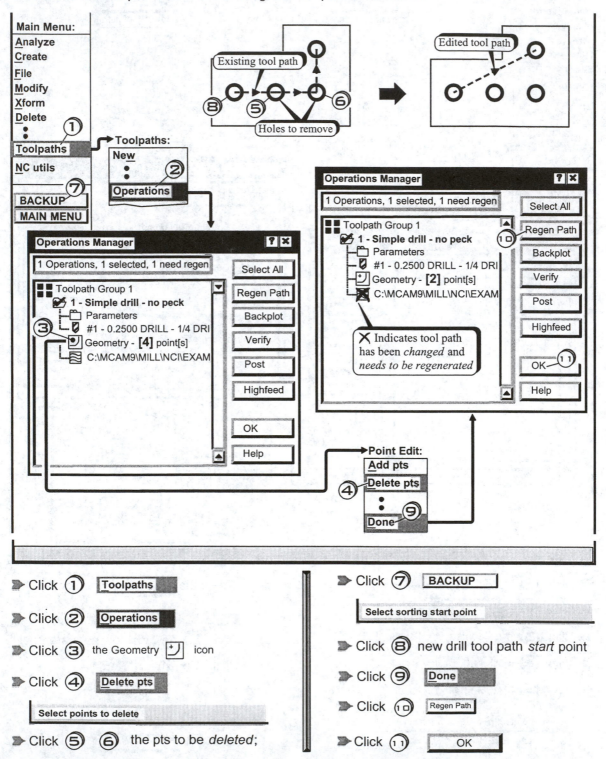

- Click ① **Toolpaths**
- Click ② **Operations**
- Click ③ the Geometry icon
- Click ④ **Delete pts**

  Select points to delete

- Click ⑤ ⑥ the pts to be *deleted*;
- Click ⑦ **BACKUP**

  Select sorting start point

- Click ⑧ new drill tool path *start* point
- Click ⑨ **Done**
- Click ⑩ **Regen Path**
- Click ⑪ **OK**

5-11 Postprocessing

Postprocessing involves the generation of a word address part program for producing a part on a *particular* CNC machine tool.

The operator must select the appropriate post (FANUC, FADAL, MAZAK,etc) for the CNC machine to be used. *Mastercam* provides a library of posts for many types of CNC machines. Customized posts can also be built to satisfy particular needs.

The **.NCI** file for a job must also be accessed. Recall the . **NCI** file contains important machining information for the job such as the tool path coordinates, tool speeds and feeds, etc. *Mastercam* uses the post file **.PST** and the **.NCI** file to *automatically* generate the word address part program or **.NC** file.

> Note: *Mastercam's* sample post file **MPFAN.PST** will be used as the default post for this text. Click [Change Post] to select *other* posts.

EXAMPLE 5-10

Direct Mastercam to generate a word address part program for the machining operations as listed below.

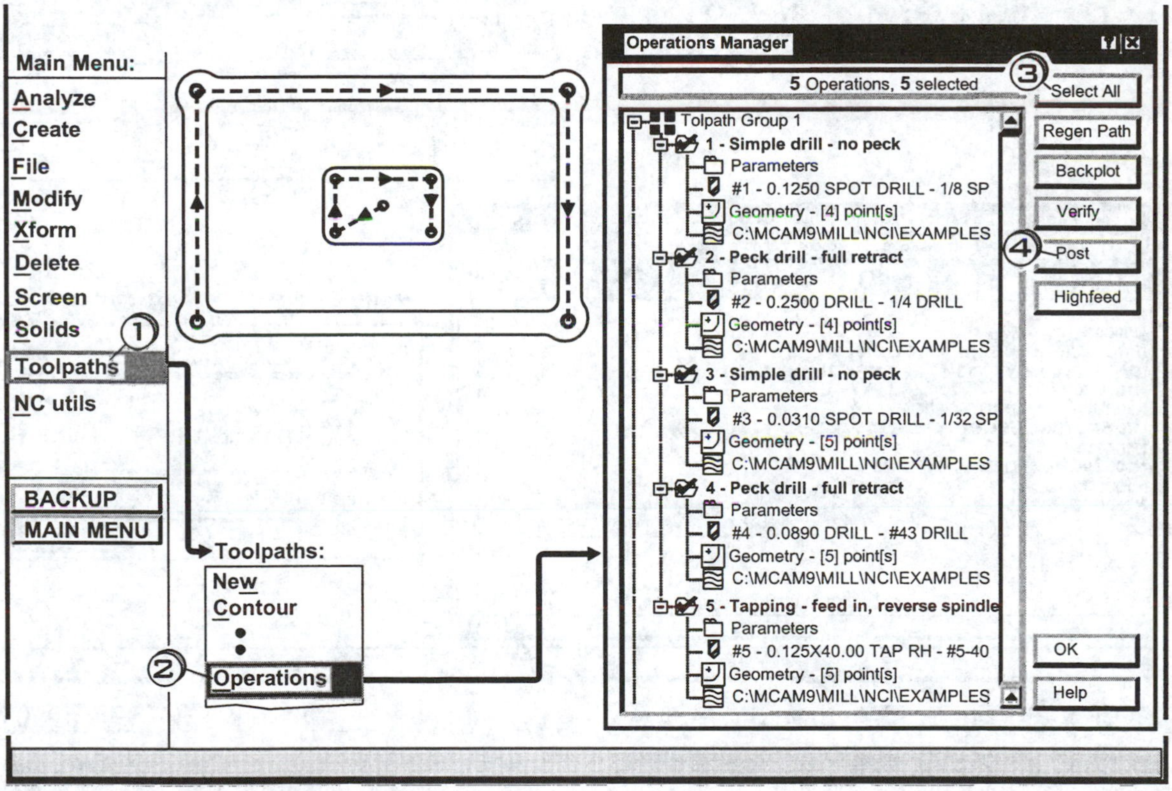

- Click ① [Toolpaths]
- Click ② [Operations]
- Click ③ the [Select All] button
- Click ④ the [Post] button

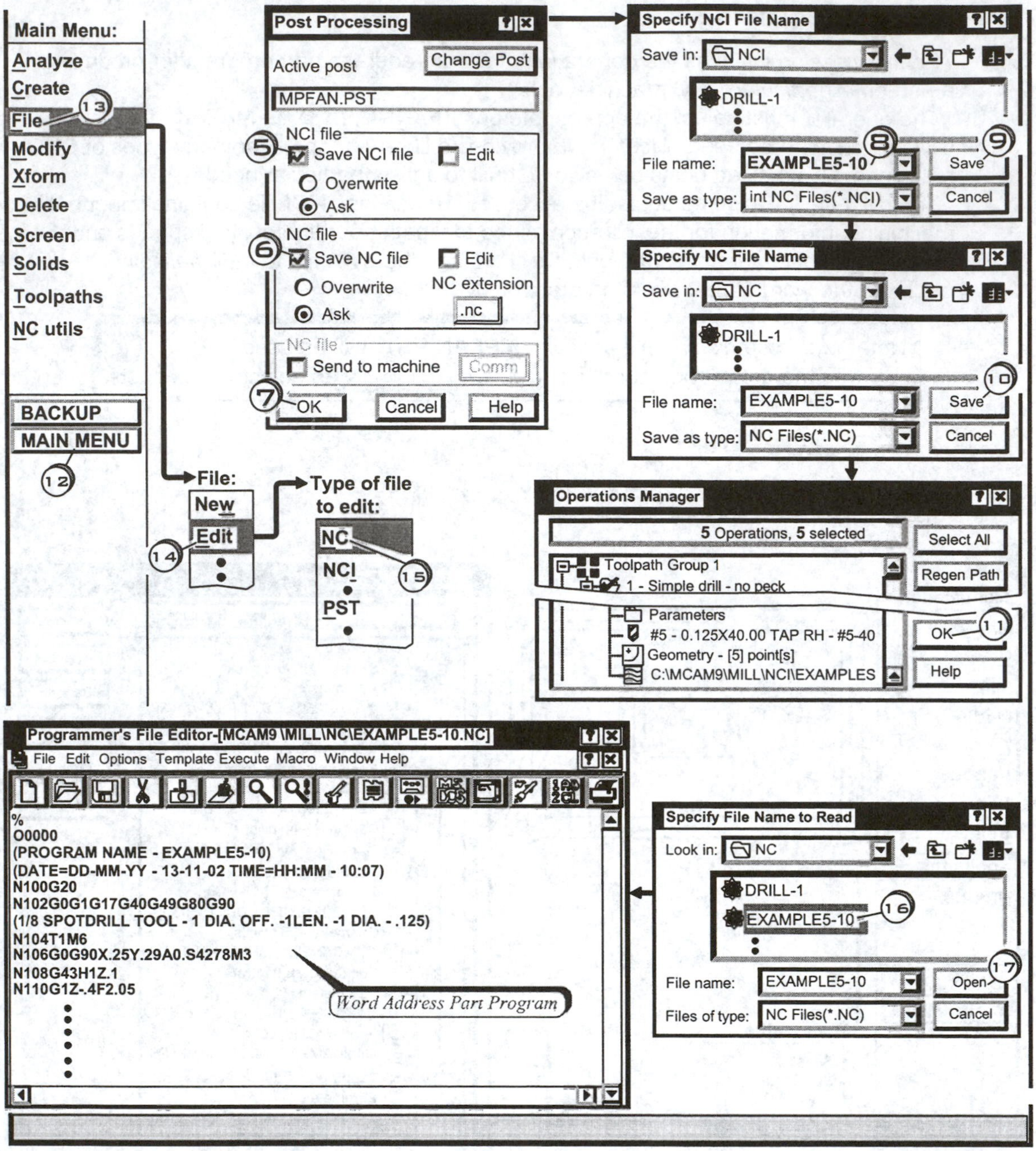

- Click ⑤ check ☑ in Save NCI file
- Click ⑥ check ☑ in Save NC file
- Click ⑦ OK
- Click ⑧ in the File name box; enter **EXAMPLE5-10**
- Click ⑨ Save
- Click ⑩ Save
- Click ⑪ OK
- Click ⑫ MAIN MENU
- Click ⑬ File
- Click ⑭ Edit
- Click ⑮ NC
- Click ⑯ EXAMPLE5-10
- Click ⑰ Open

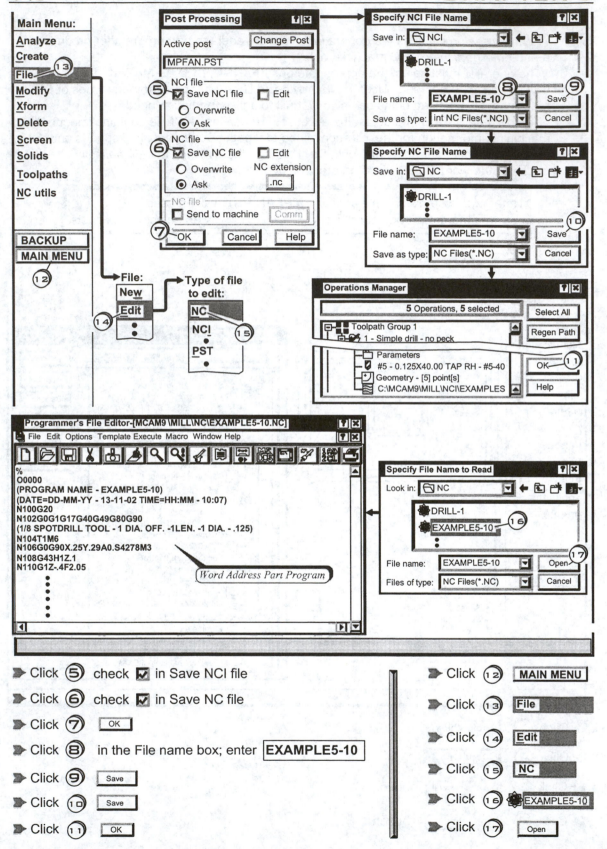

Using the Toolbar Menus

Toolbar Menu Displayed when *Mastercam* is **Started**

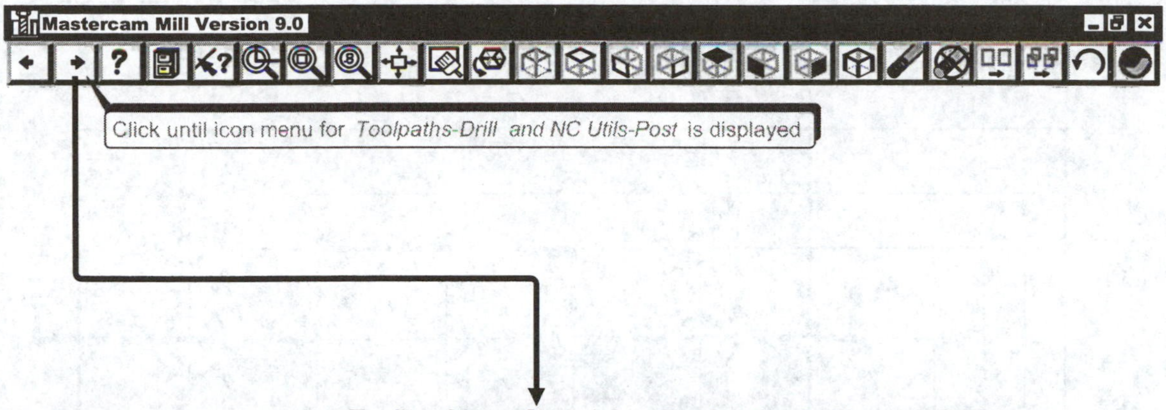
Mastercam Mill Version 9.0
Click until icon menu for *Toolpaths-Drill and NC Utils-Post* is displayed

Toolbar Menu Containing ***Toolpaths-Drill*** and ***NC Utils-Post*** Icons

Mastercam Mill Version 9.0
N10G00
N12G91
N14G28
Click on *Toolpaths-Drill* icon
Click on *NC Utils-Post*

Click Esc to cancel the commands

EXERCISES

5-1) Generate a part program for executing the operations in PROCESS PLAN 5P-1

File Name: **EX5-1JV** ← YOUR INITIALS

Material: 1030 Steel

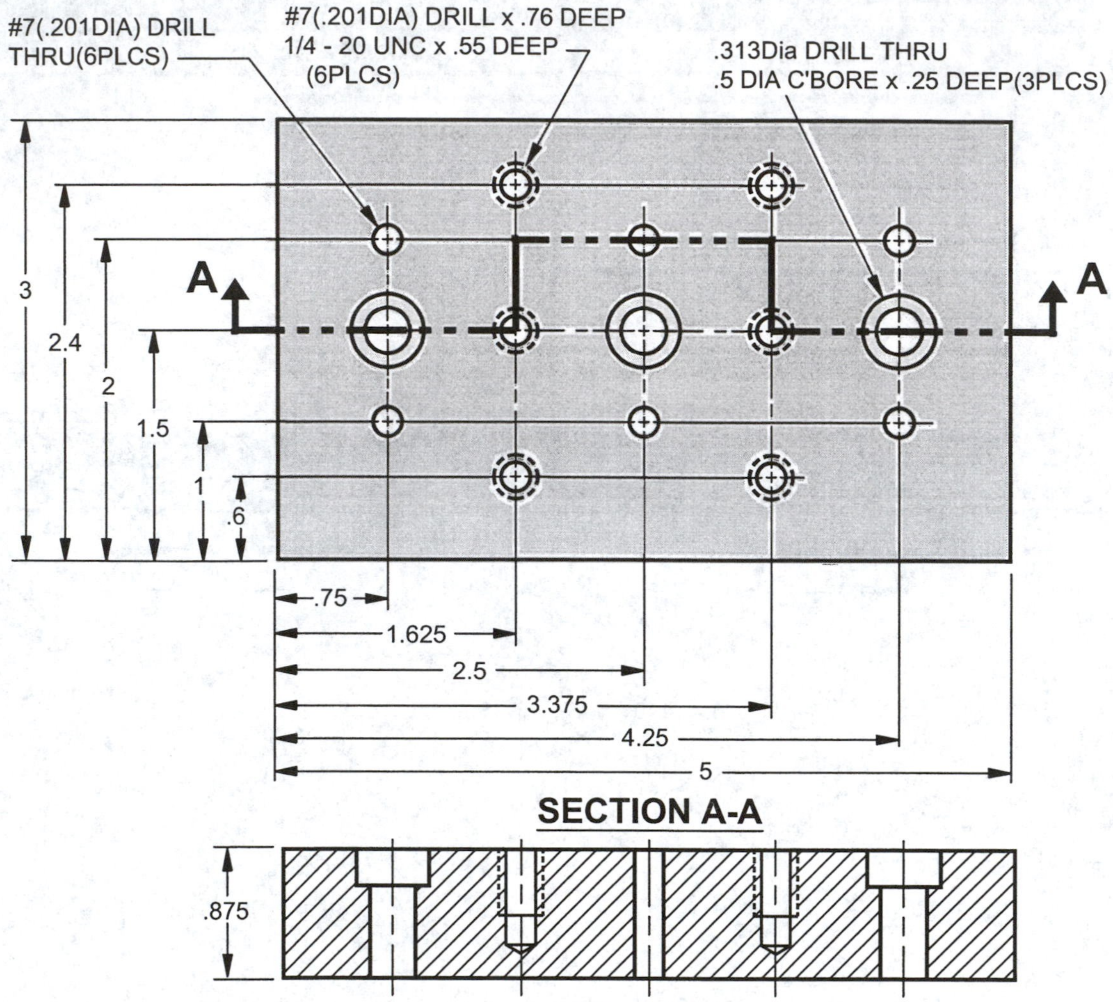

Figure 5p-1

PROCESS PLAN 5P-1

No.	Operation	Tooling
1	Center Drill x .2 deep(all holes)	1/8 Center drill
2	Peck drill thru(6PLCS)	#7(.201) Drill
3	Peck drill x .76 Deep(6PLCS)	#7(.201) Drill
4	Tap x .55 Deep(6PLCS)	1/4-20UNC Tap
5	Peck drill Thru (3PLCS)	5/16 Drill
6	Circle Mill x .25 Deep (3PLCS)	3/8 Flat Endmill

A) CREATE A CAD MODEL OF THE PART

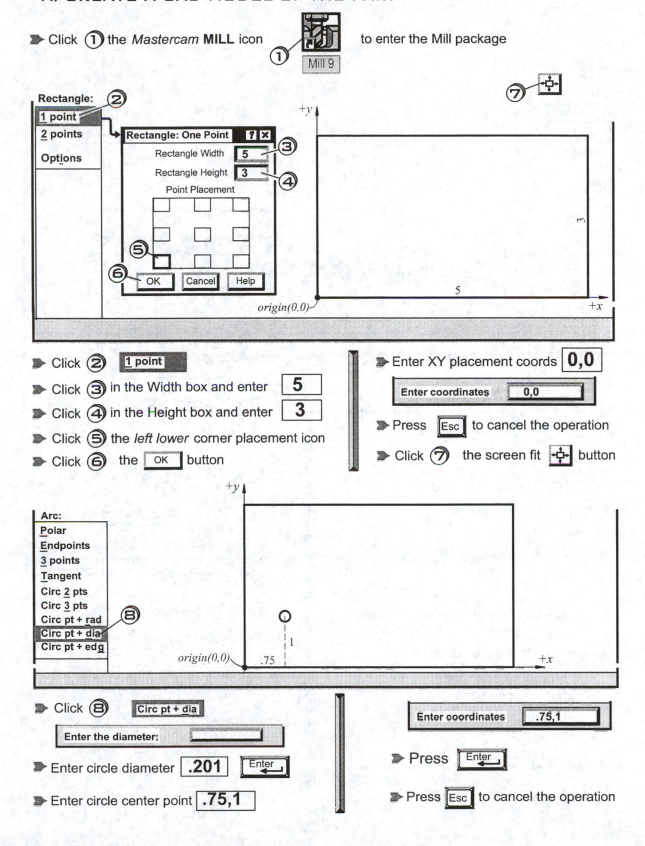

➤ Click ① the *Mastercam* **MILL** icon

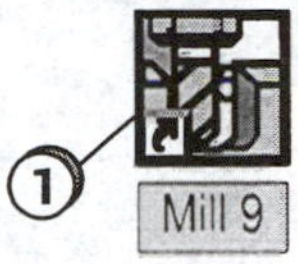

to enter the Mill package

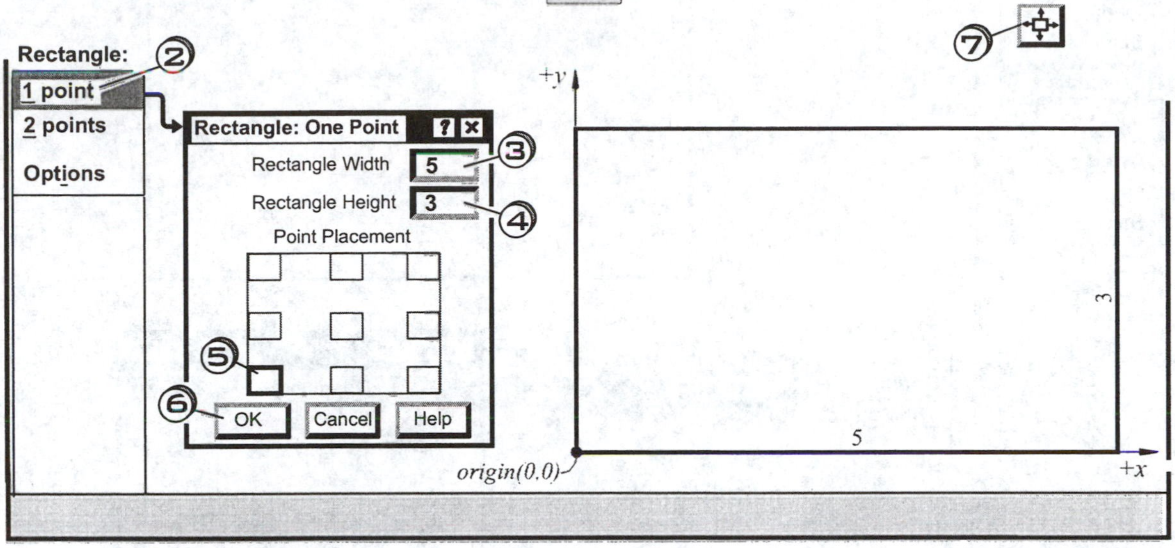

➤ Click ② **1 point**

➤ Click ③ in the Width box and enter **5**

➤ Click ④ in the Height box and enter **3**

➤ Click ⑤ the *left lower* corner placement icon

➤ Click ⑥ the OK button

➤ Enter XY placement coords **0,0**

Enter coordinates 0,0

➤ Press Esc to cancel the operation

➤ Click ⑦ the screen fit button

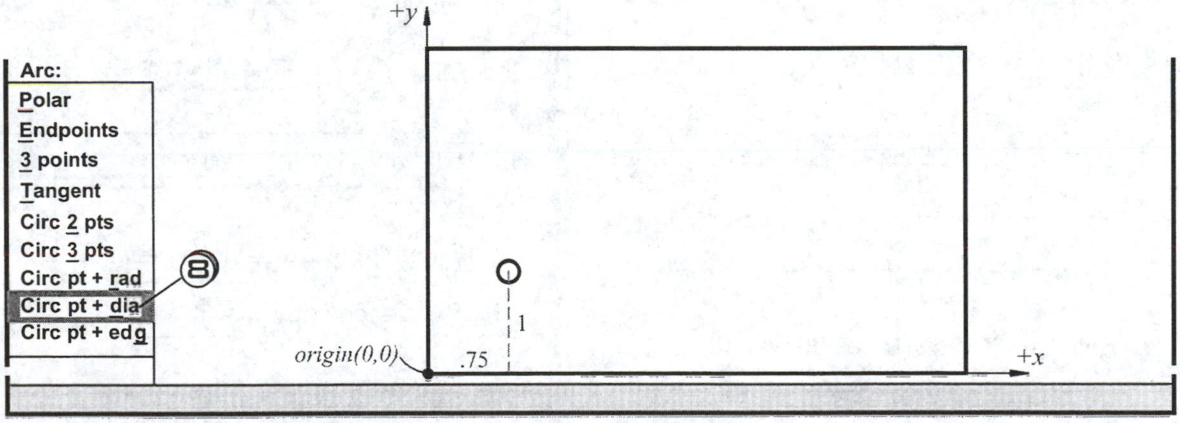

➤ Click ⑧ **Circ pt + dia**

Enter the diameter:

➤ Enter circle diameter **.201** Enter

➤ Enter circle center point **.75,1**

Enter coordinates .75,1

➤ Press Enter

➤ Press Esc to cancel the operation

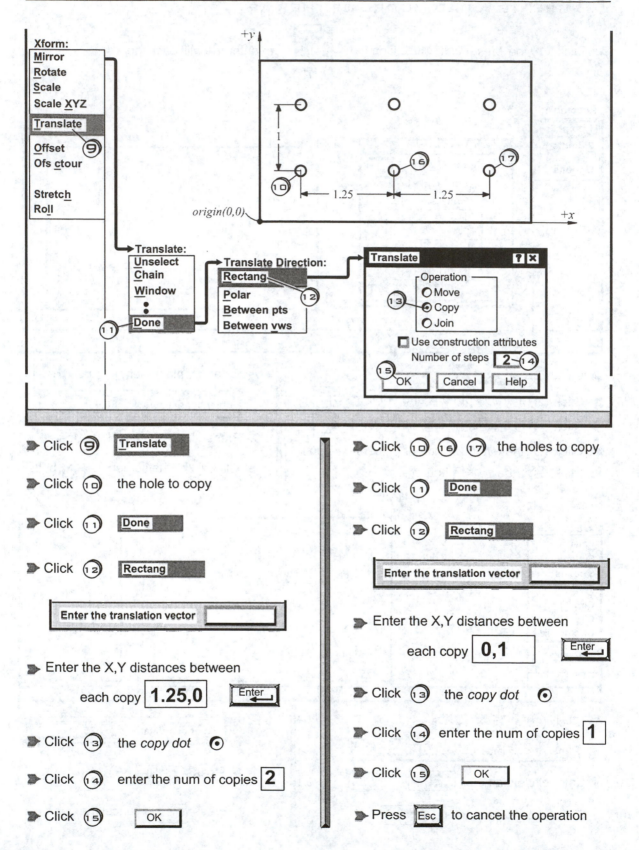

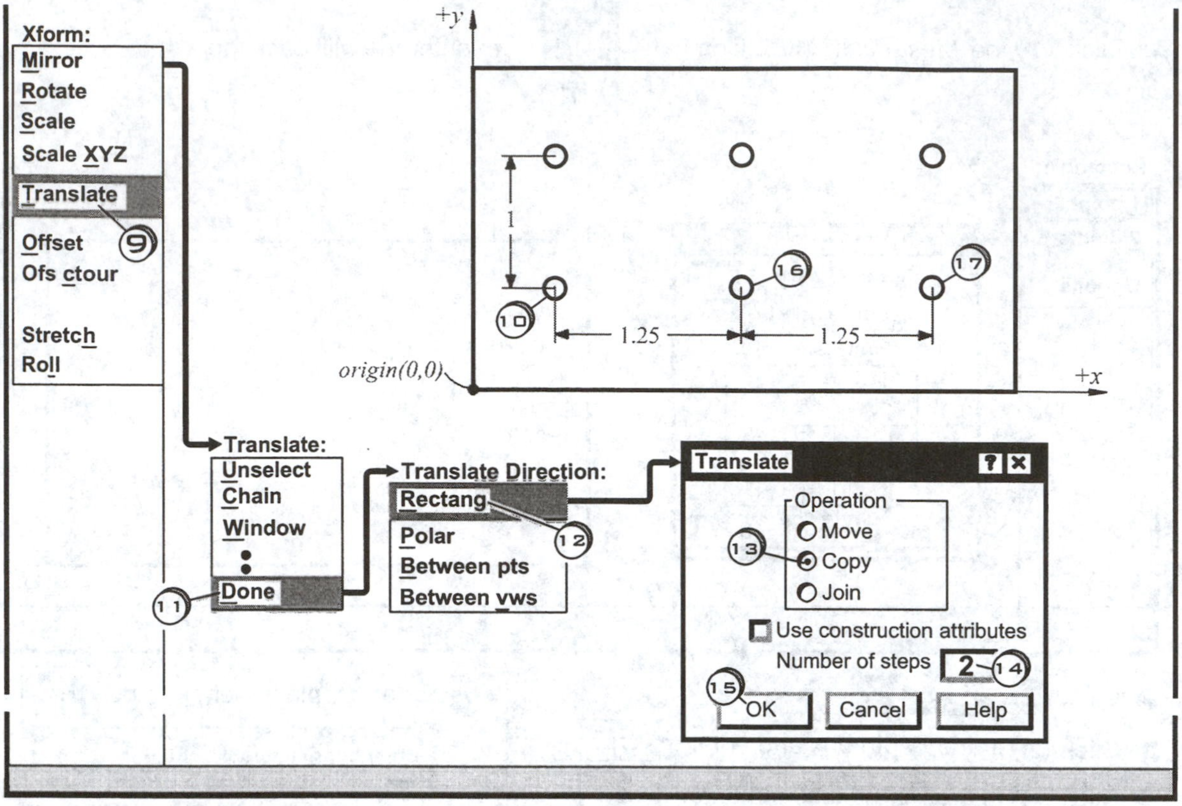

- Click (9) Translate
- Click (10) the hole to copy
- Click (11) Done
- Click (12) Rectang

Enter the translation vector

- Enter the X,Y distances between each copy **1.25,0** Enter
- Click (13) the *copy dot* ⊙
- Click (14) enter the num of copies **2**
- Click (15) OK

- Click (10) (16) (17) the holes to copy
- Click (11) Done
- Click (12) Rectang

Enter the translation vector

- Enter the X,Y distances between each copy **0,1** Enter
- Click (13) the *copy dot* ⊙
- Click (14) enter the num of copies **1**
- Click (15) OK
- Press Esc to cancel the operation

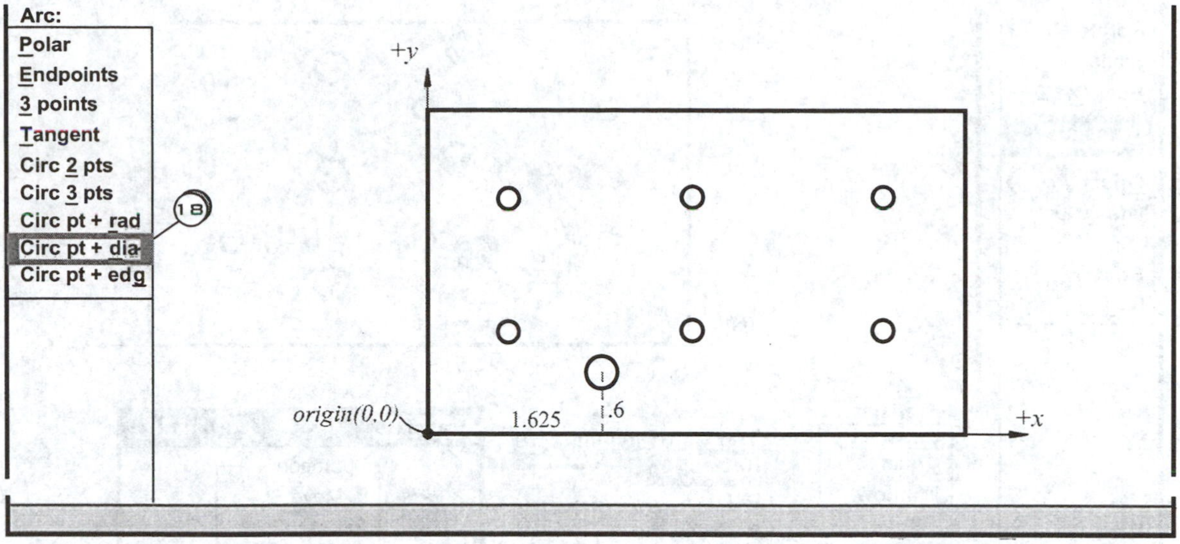

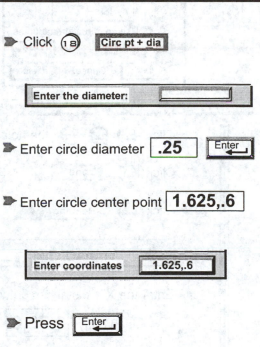

➤ Click (1 B) **Circ pt + dia**

Enter the diameter:

➤ Enter circle diameter **.25** Enter

➤ Enter circle center point **1.625,.6**

Enter coordinates 1.625,.6

➤ Press Enter

➤ Press Esc to cancel the operation

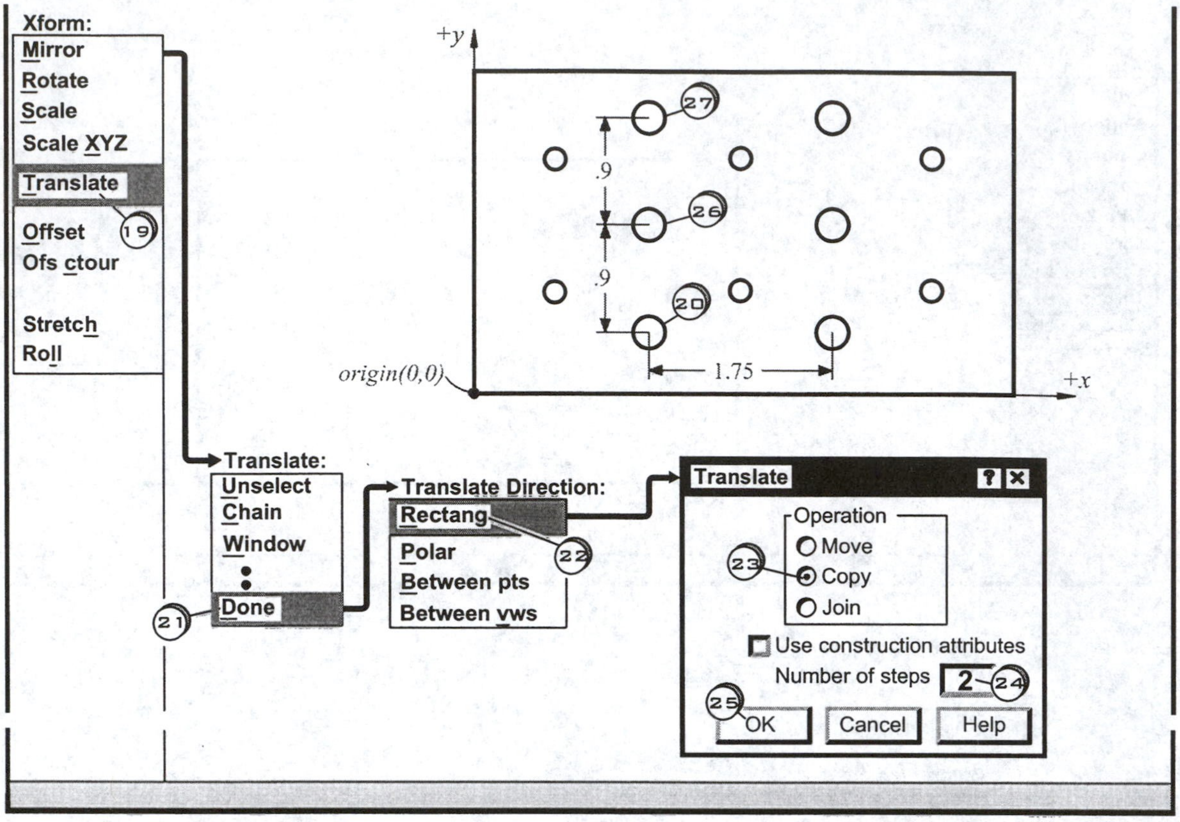

- Click (19) **Translate**
- Click (20) the hole to copy
- Click (21) **Done**
- Click (22) **Rectang**

Enter the translation vector [ ]

- Enter the X,Y distances between each copy **0,.9** Enter
- Click (23) the *copy dot* ⊙
- Click (24) enter the num of copies **2**
- Click (25) OK

- Click (20) (26) (27) the holes to copy
- Click (21) **Done**
- Click (22) **Rectang**

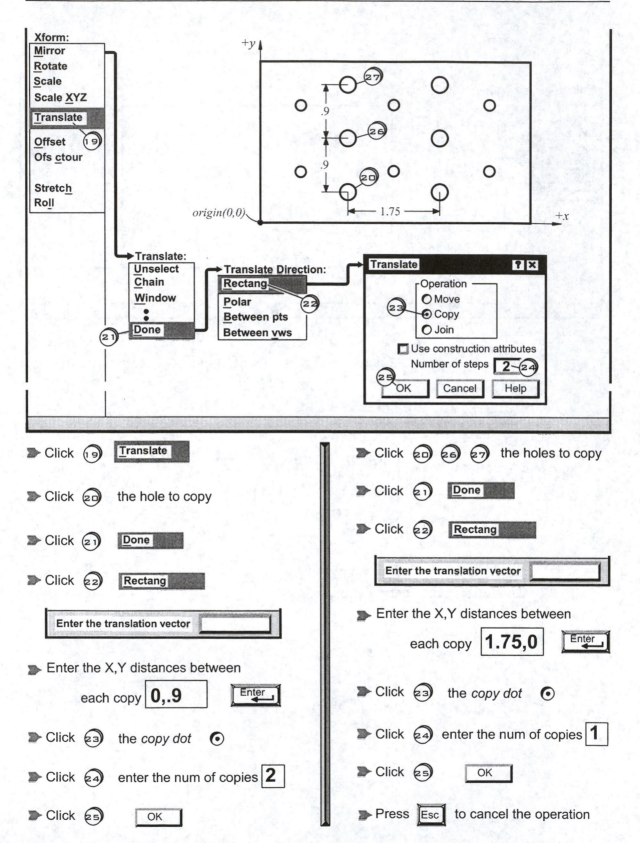

Enter the translation vector [ ]

- Enter the X,Y distances between each copy **1.75,0** Enter
- Click (23) the *copy dot* ⊙
- Click (24) enter the num of copies **1**
- Click (25) OK
- Press Esc to cancel the operation

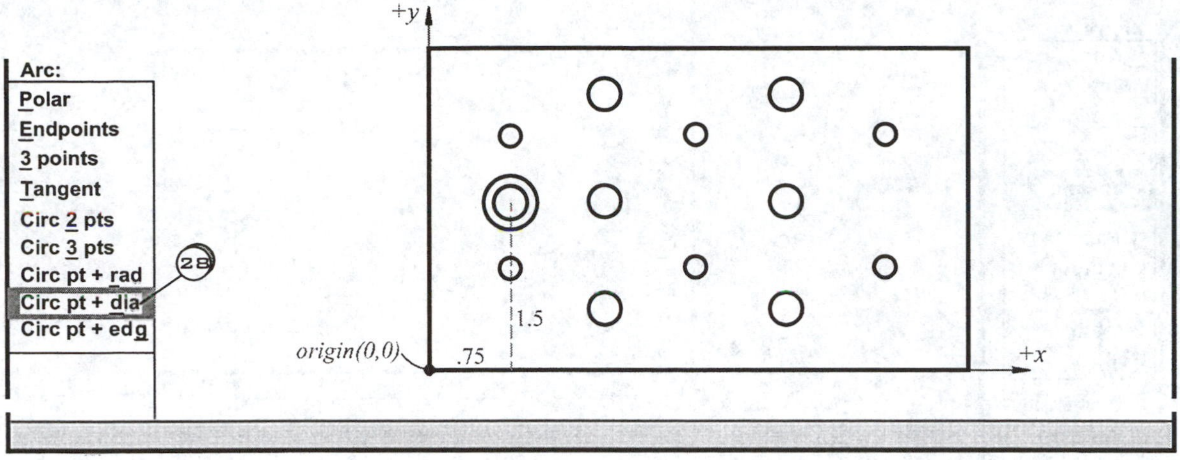

- Click 28 **Circ pt + dia**

Enter the diameter: [ ]

- Enter circle diameter **.313** [Enter]
- Enter circle center point **.75,1.5**

Enter coordinates [.75,1.5]

- Press [Enter]
- Press [Esc] to cancel the operation
- Click 28 **Circ pt + dia**

Enter the diameter: [ ]

- Enter circle diameter **.5** [Enter]
- Enter circle center point **.75,1.5**

Enter coordinates [ ]

- Press [Enter]
- Press [Esc] to cancel the operation

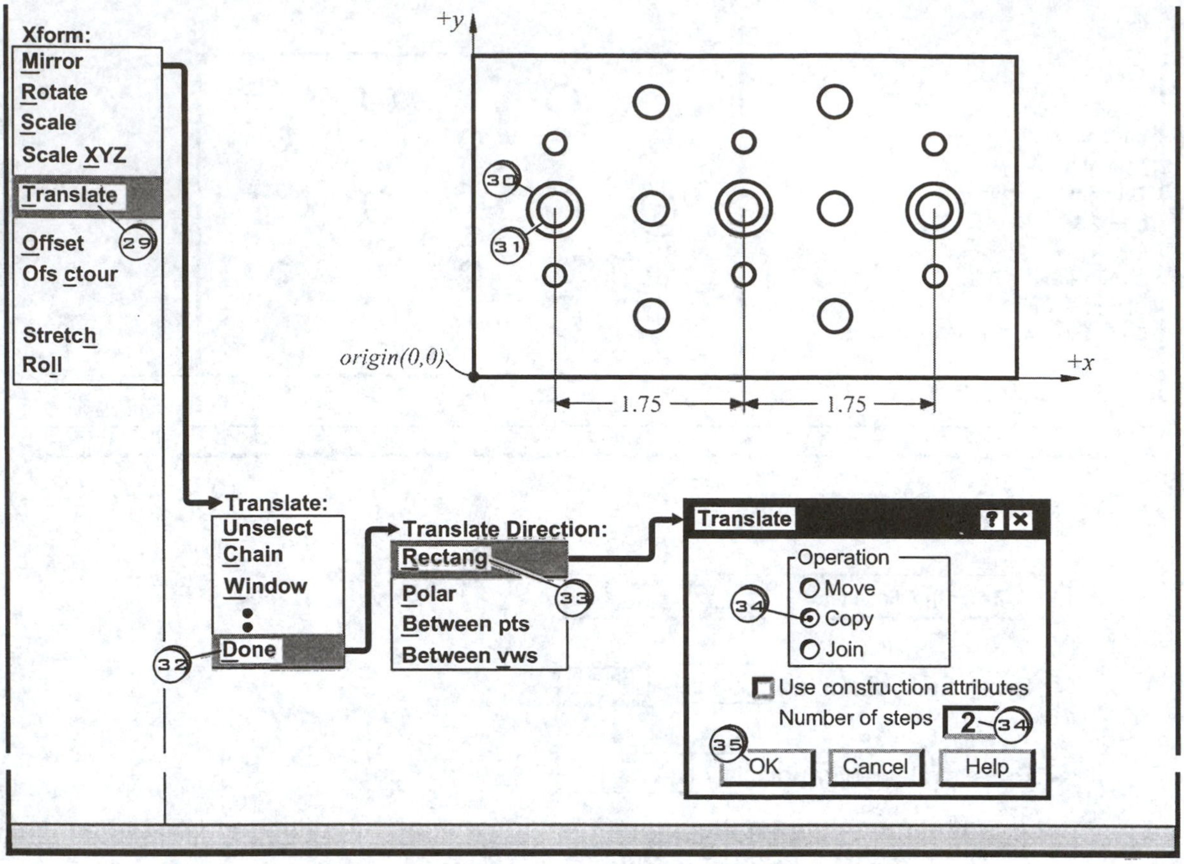

- Click (29) **Translate**
- Click (30) (31) the hole to copy
- Click (32) **Done**
- Click (33) **Rectangle**

Enter the translation vector [ ]

- Enter the X,Y distances between each copy **1.75,0** ; Enter
- Click (34) the *copy dot* ⊙
- Click (34) enter the num of copies **2**
- Click (35) OK
- Press Esc to cancel the operation

B) ENTER *Mastercam's* JOB SETUP DIALOG BOX

◆ SELECT SIZE OF THE STOCK

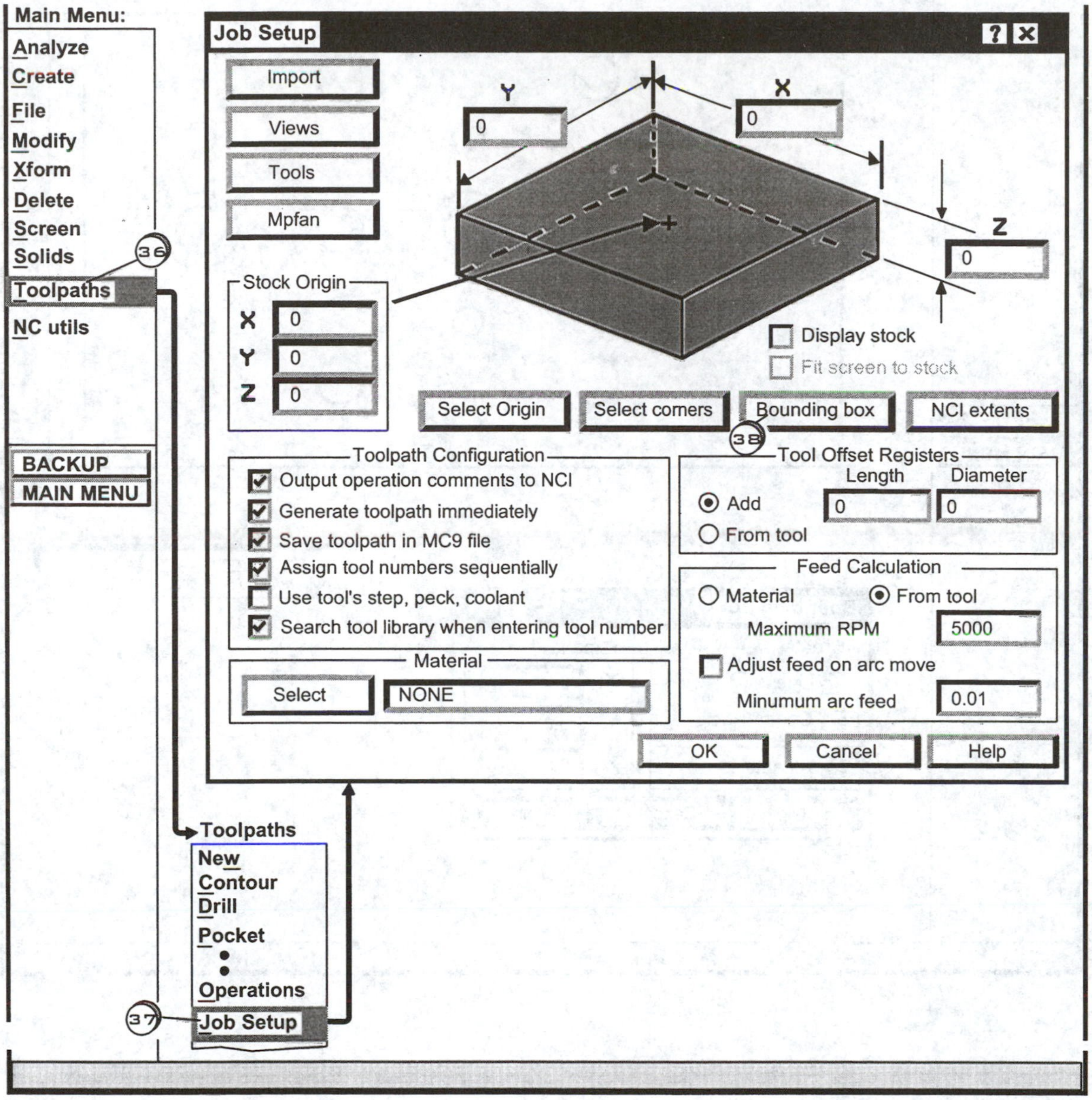

➤ Click (36) Toolpaths

➤ Click (37) Job Setup

➤ Click (38) Bounding box to direct *Mastercam* to automatically compute the stock size based on a box that just fits around the CAD model.

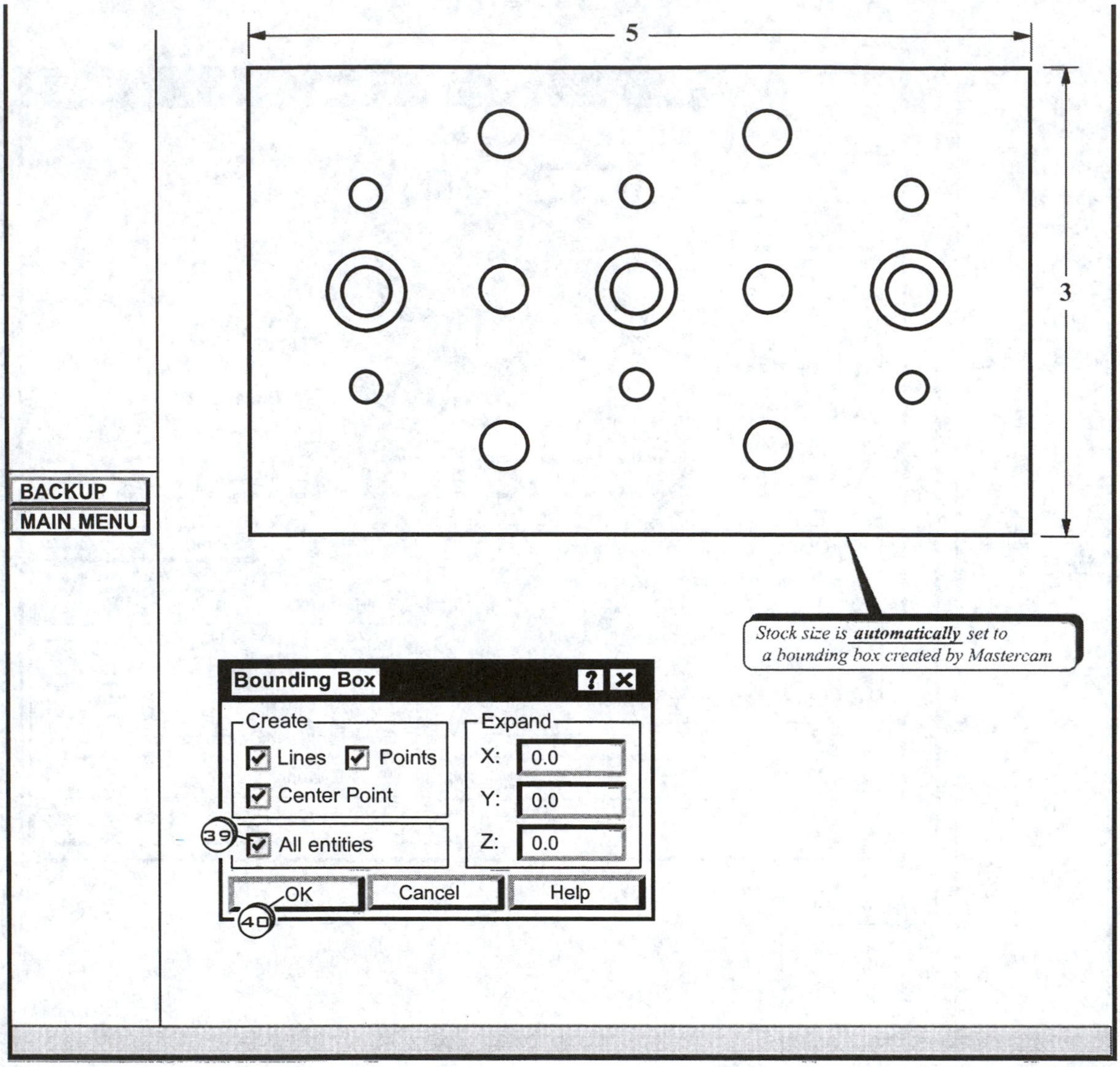

➤ Click (39) the All entities check ☑

➤ Click (40) OK

◆ IDENTIFY THE STOCK MATERIAL TO BE MACHINED

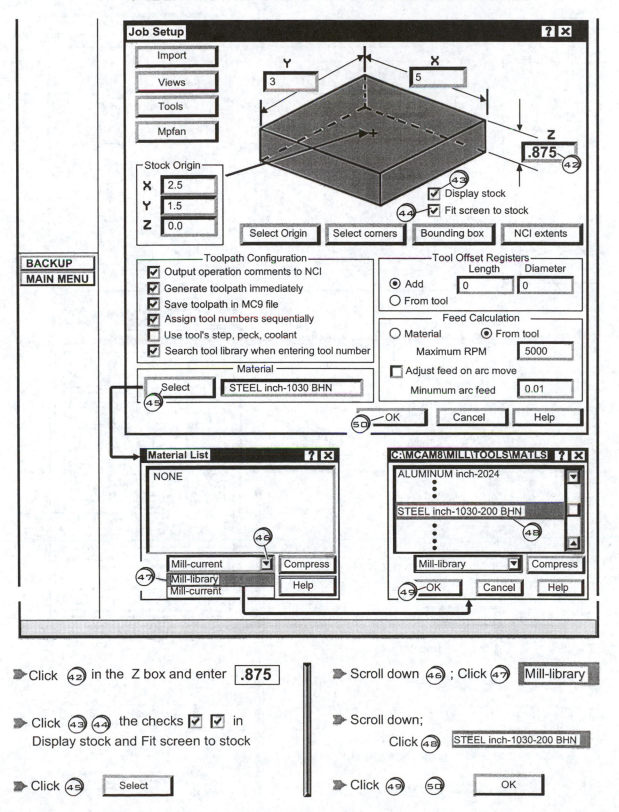

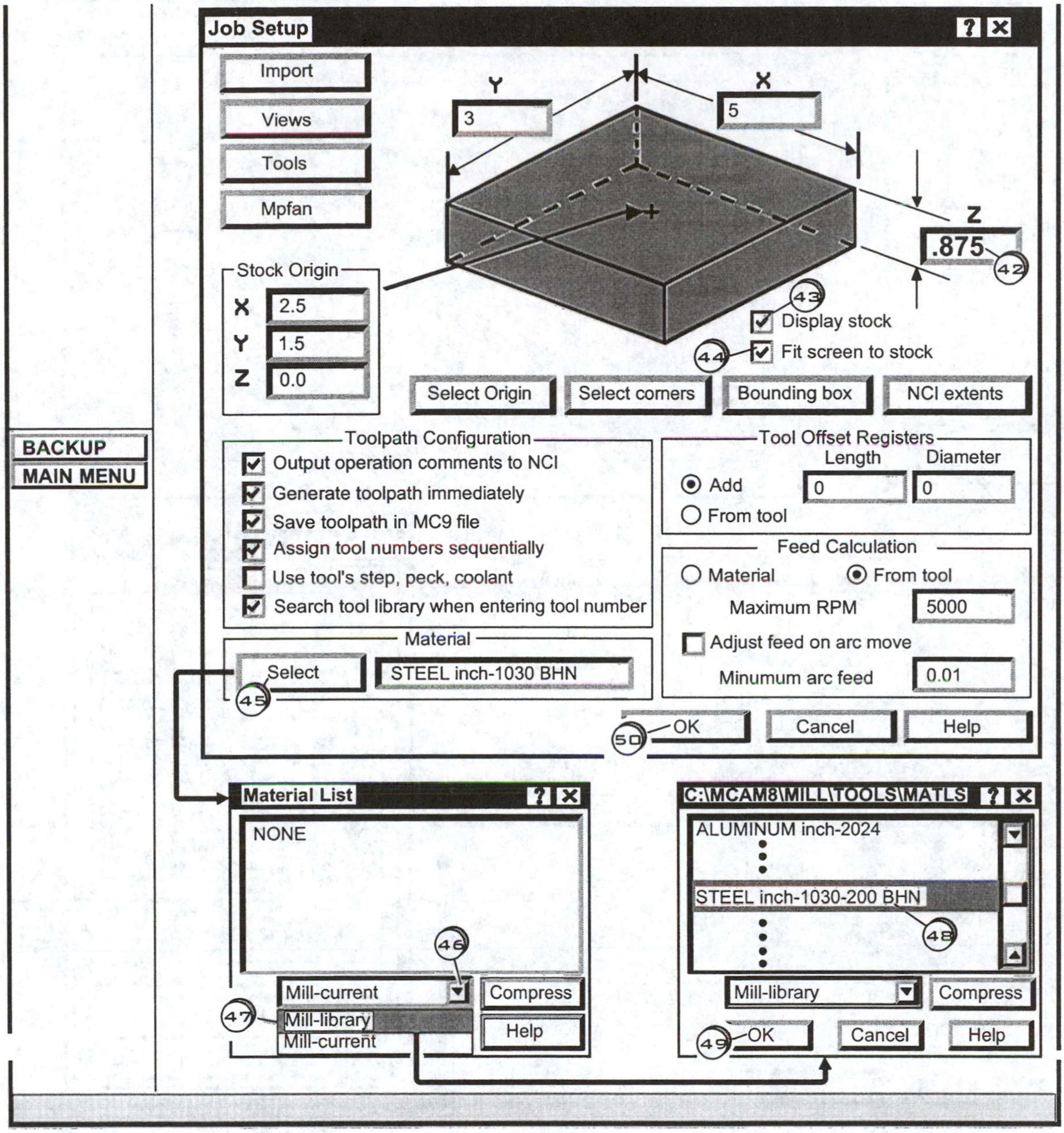

➤ Click (42) in the Z box and enter .875

➤ Click (43) (44) the checks ☑ ☑ in Display stock and Fit screen to stock

➤ Click (45) Select

➤ Scroll down (46) ; Click (47) Mill-library

➤ Scroll down;
Click (48) STEEL inch-1030-200 BHN

➤ Click (49) (50) OK

C) DRILL THE .125DIA CENTER DRILL HOLES X .166 DEEP

◆ CREATE THE CENTER DRILL TOOLPATH

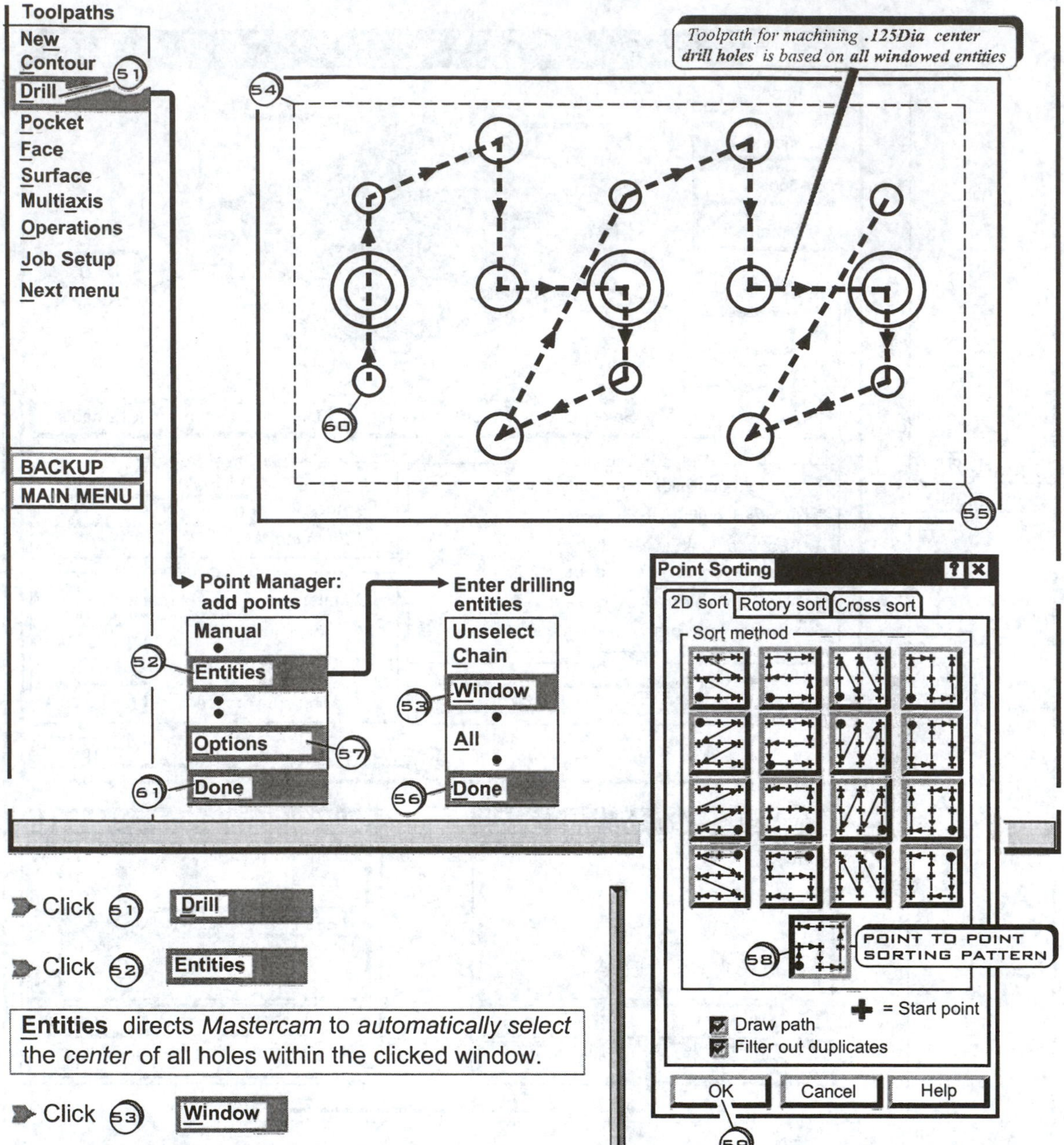

➤ Click 51 **Drill**

➤ Click 52 **Entities**

Entities directs *Mastercam* to *automatically select* the *center* of all holes within the clicked window.

➤ Click 53 **Window**

➤ Click 54 55 the window corners

➤ Click 56 **Done**

➤ Click 57 **Options**

➤ Click 58 point to point as the ***current active*** sort pattern

➤ Click 59 OK

Select sorting point

➤ Click 60 the first point to drill

➤ Click 61 **Done**

◆ OBTAIN THE NEEDED .125 DIA CENTER DRILL TOOL

Mastercam will activate and display the Tool and Drill Parameters dialog box after the drill tool path has been created.

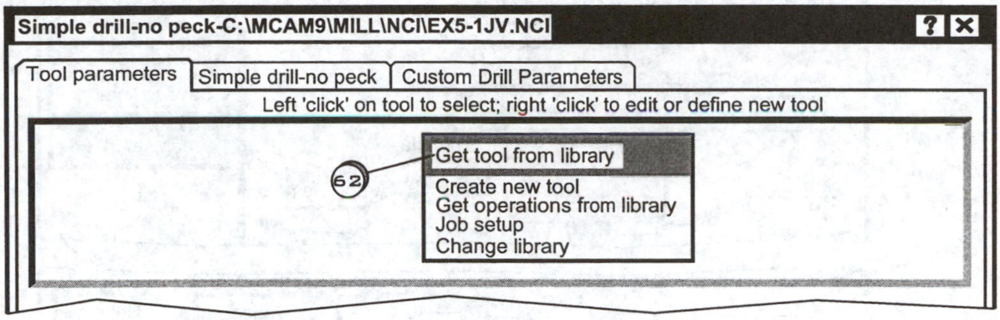

➤ Press the ***right*** mouse button in the ***white area***

➤ Click (62) Get tool from library

Mastercam will display the main Tool Library dialog box

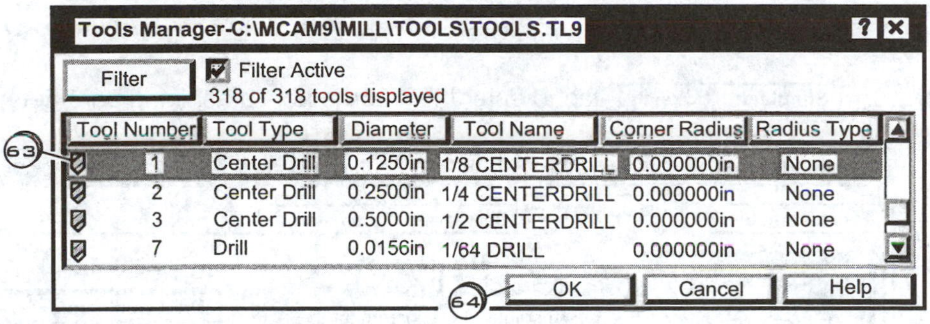

➤ Click (63) the .125Dia center Drill tool

➤ Click (64) OK

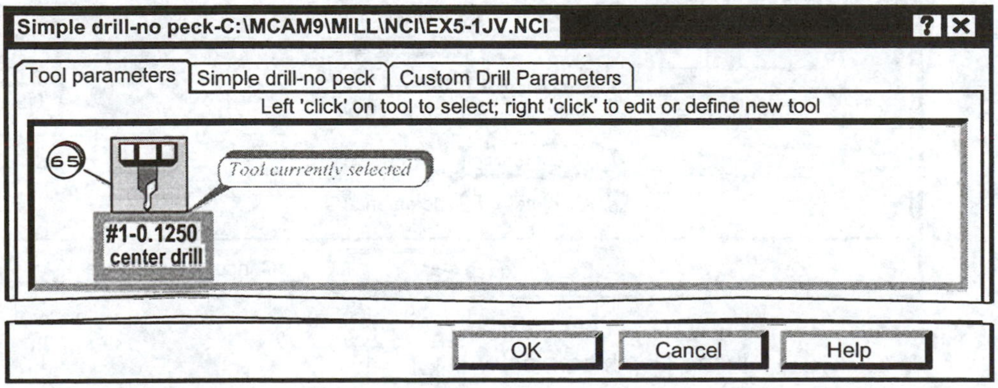

➤ ***Right*** click (65)

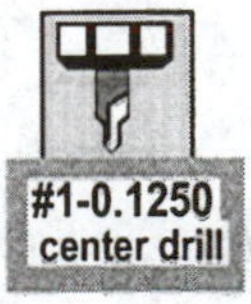

on the #1-0.1250 center drill to select this as the ***active tool*** and direct *Mastercam* to display the Define Tool dialog box

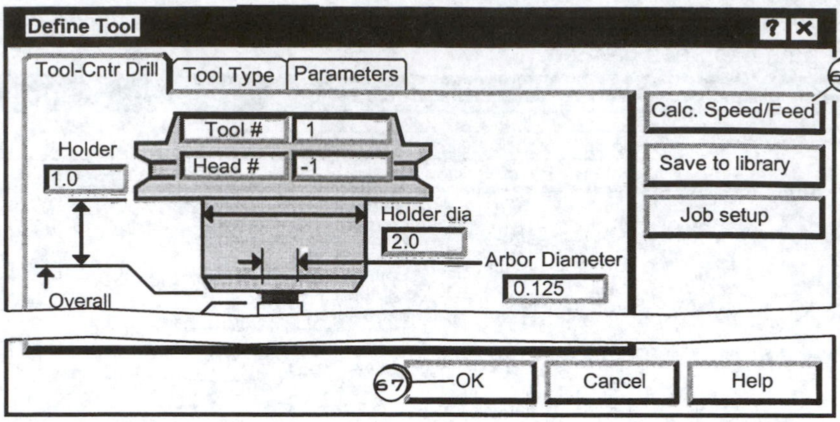

➤ Click 66 [Calc Speed/Feed] ➤ Click 67 [OK]

◆ ENTER THE .125 DIA CENTER DRILL HOLE MACHINING PARAMETERS

➤ Click 68 [Simple drill-no peck] tab to enter the .125Dia hole machining parameters.

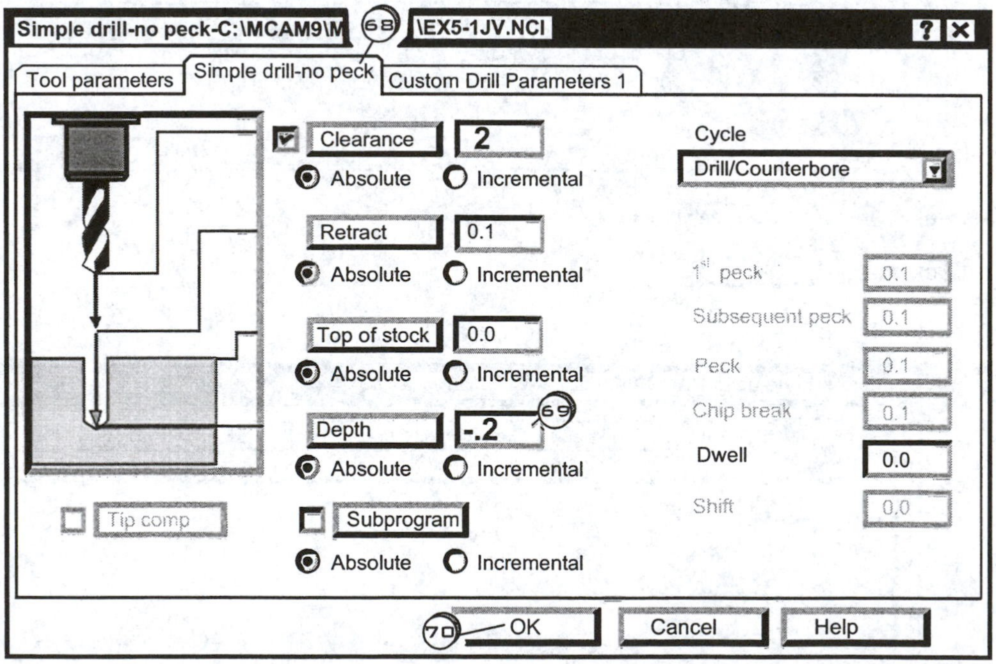

➤ Click 69 in the Depth box; enter [-.2]

➤ Click 70 [OK]

The .125Dia center drilling operation will be added to *Mastercam's* Operations Manager

D) PECK DRILL THE .201DIA HOLES THRU

◆ CREATE THE PECK DRILL THRU TOOLPATH

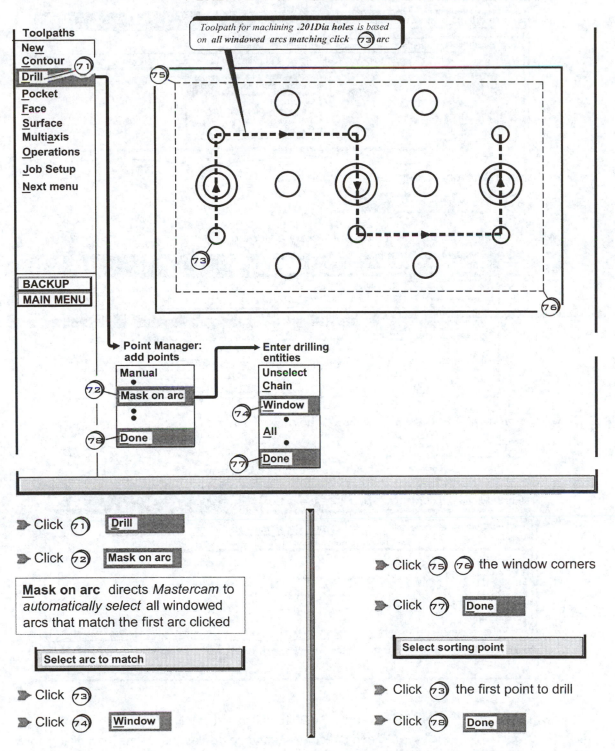

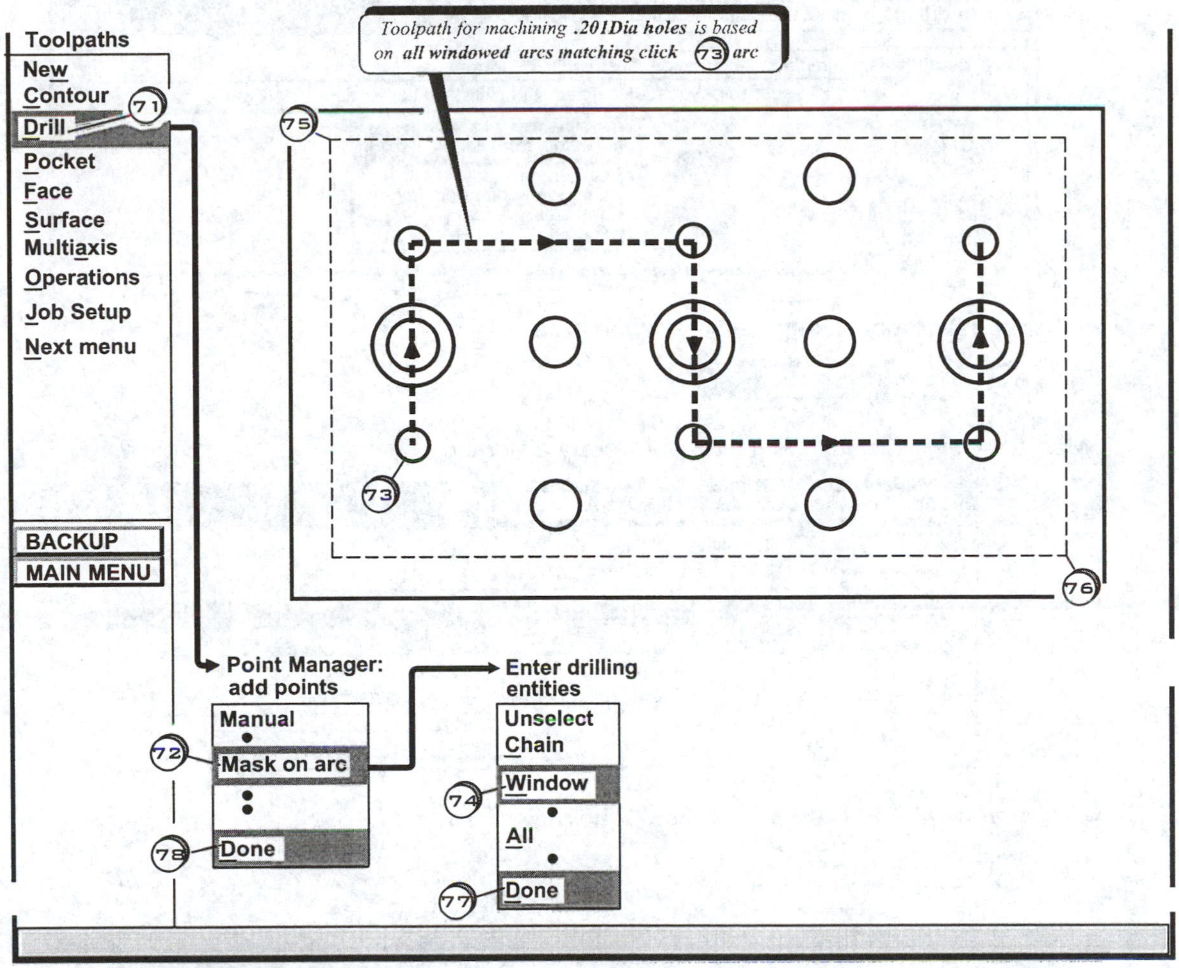

➤ Click 71 **Drill**

➤ Click 72 **Mask on arc**

Mask on arc directs *Mastercam* to *automatically select* all windowed arcs that match the first arc clicked

Select arc to match

➤ Click 73

➤ Click 74 **Window**

➤ Click 75 76 the window corners

➤ Click 77 **Done**

Select sorting point

➤ Click 73 the first point to drill

➤ Click 78 **Done**

◆ OBTAIN THE NEEDED .201DIA DRILL TOOL

Mastercam will activate and display the Tool and Drill Parameters dialog box after the drill tool path has been created.

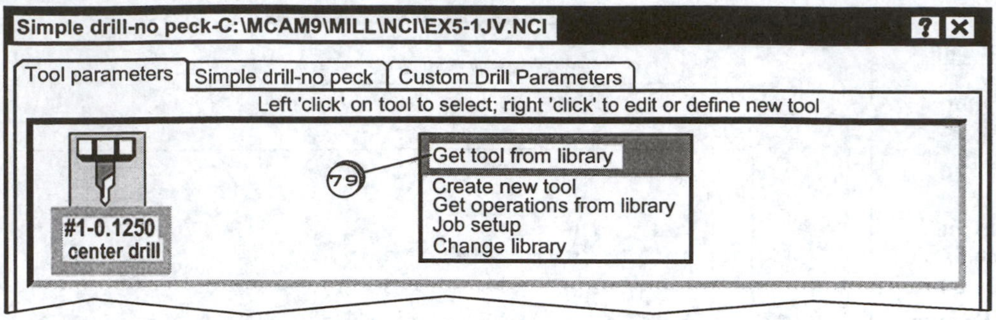

➤ Press the ***right*** mouse button in the ***white area***

➤ Click (79) Get tool from library

Mastercam will display the main Tool Library dialog box

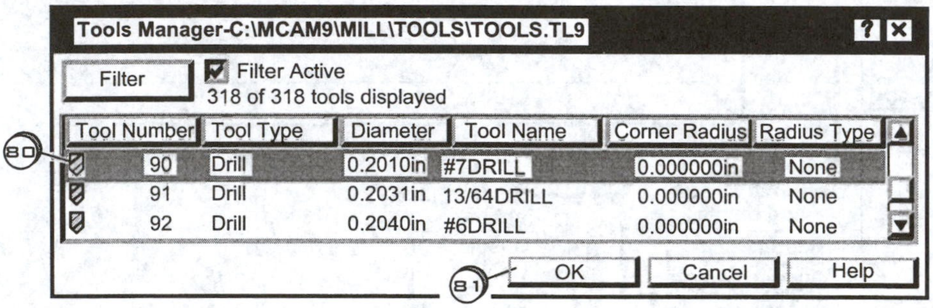

➤ Click (80) the .201Dia Drill tool

➤ Click (81) OK

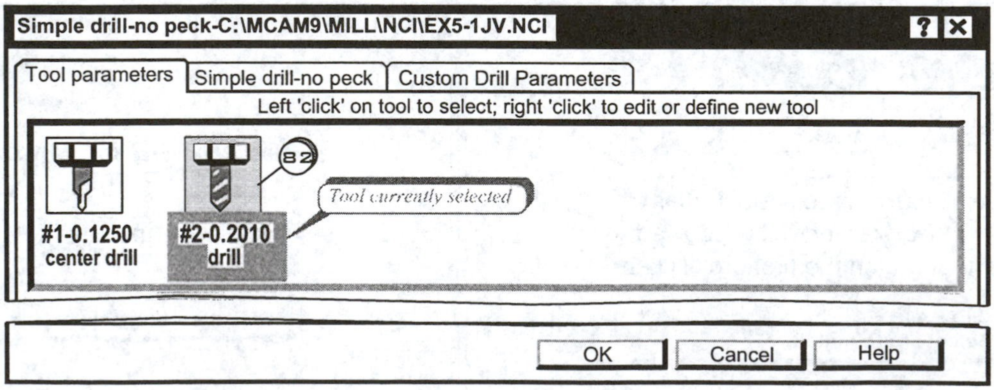

➤ ***Right*** click (82)

on the #2-0.2010 drill to select this as the ***active tool*** and direct *Mastercam* to display the Define Tool dialog box

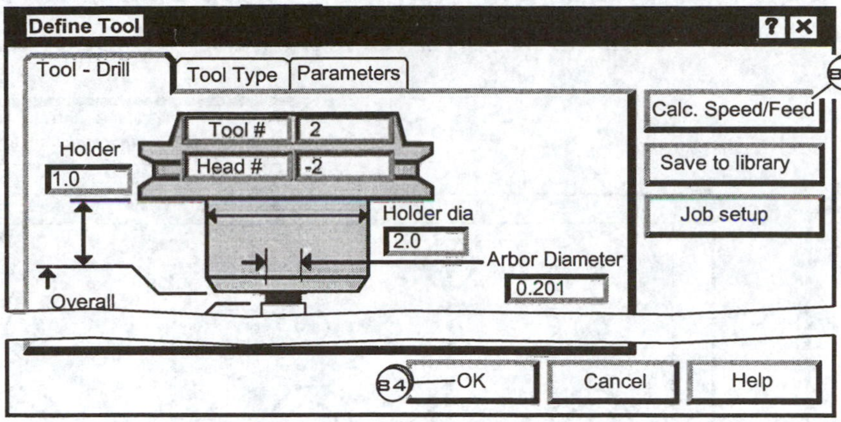

➤ Click 83 [Calc Speed/Feed] ➤ Click 84 [OK]

◆ ENTER THE .201DIA PECK DRILL THRU MACHINING PARAMETERS

➤ Click 85 [Peck drill-full retract] tab to enter the .201Dia peck drill machining parameters.

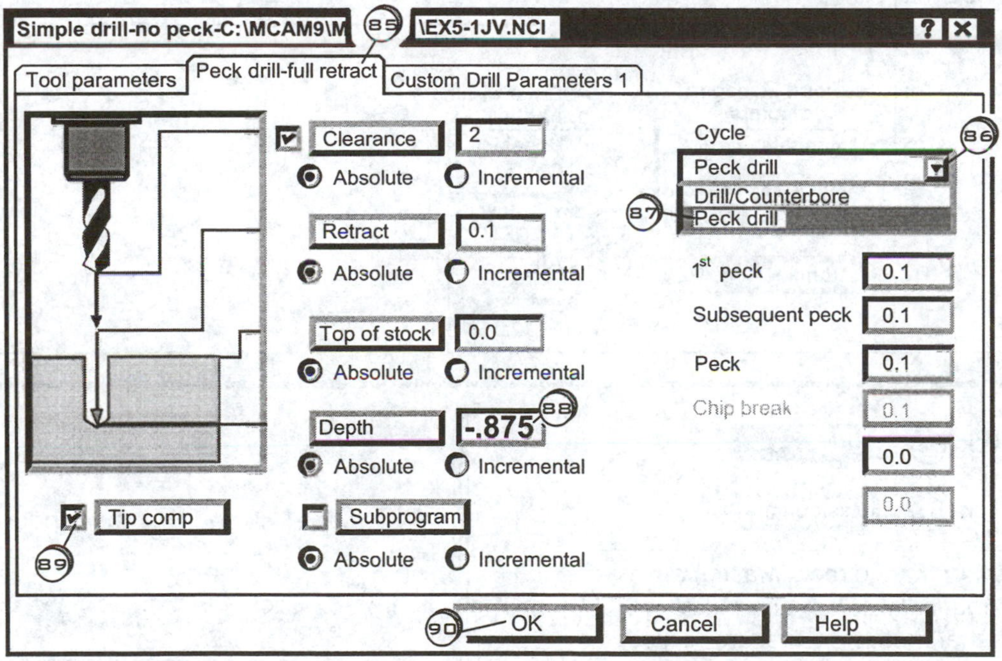

➤ Click 86 [▼] the toggle down button

➤ Click 87 [Peck drill]

➤ Click 88 in the Depth box; enter [-.875]

➤ Click 89 the tip comp check *on* [✔] for drilling *thru* holes

➤ Click 90 [OK]

The .201Dia peck drilling thru operation will be added to *Mastercam's* Operations Manager

E) PECK DRILL THE .201DIA HOLES x .76 DEEP

◆ CREATE THE PECK DRILL X .76 DEEP TOOLPATH

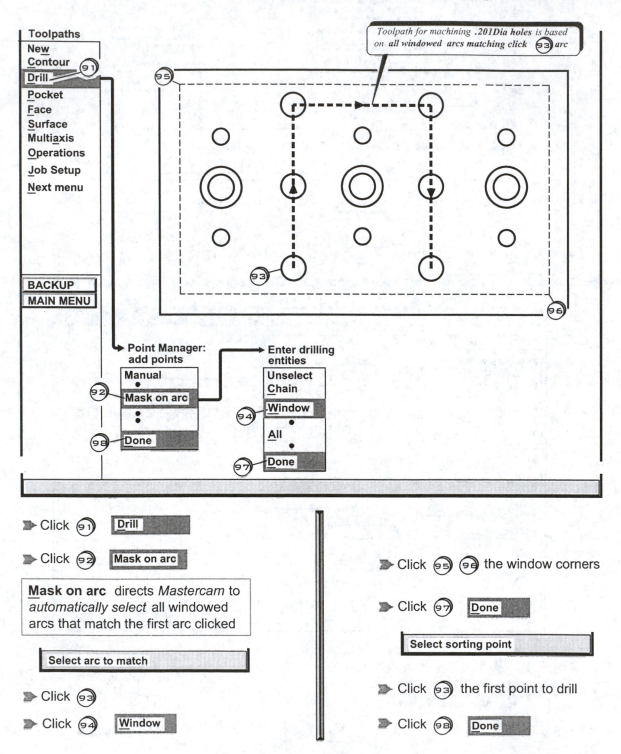

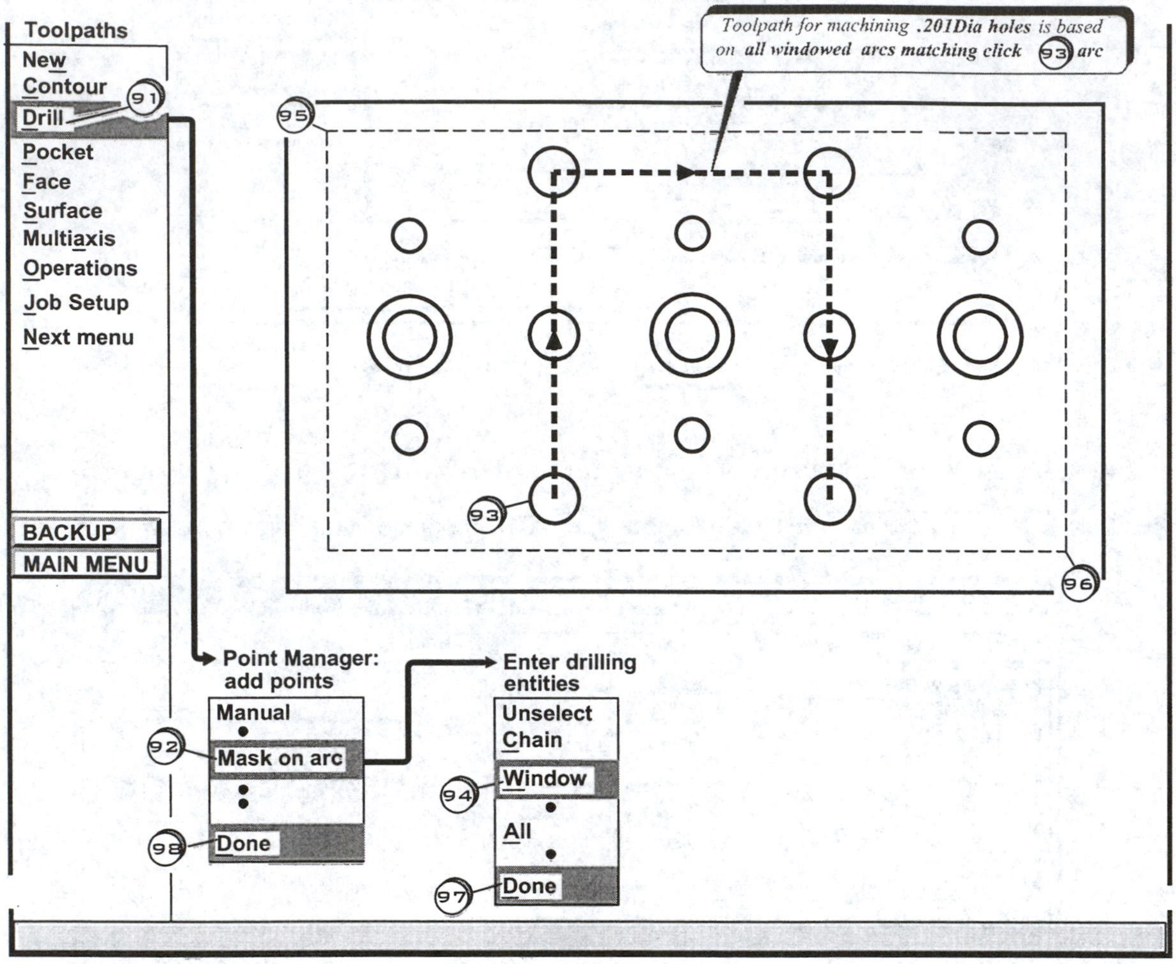

➤ Click 91 **Drill**

➤ Click 92 **Mask on arc**

Mask on arc directs *Mastercam* to *automatically select* all windowed arcs that match the first arc clicked

Select arc to match

➤ Click 93

➤ Click 94 **Window**

➤ Click 95 96 the window corners

➤ Click 97 **Done**

Select sorting point

➤ Click 93 the first point to drill

➤ Click 98 **Done**

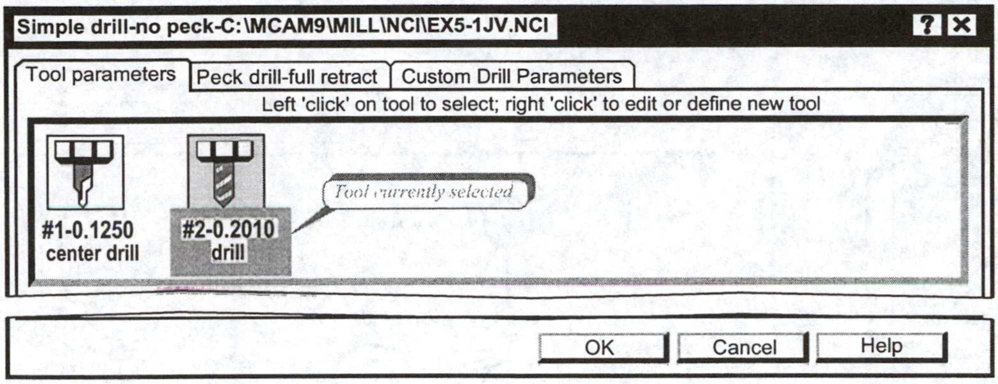

- ENTER THE .201DIA PECK DRILL X .76 DEEP MACHINING PARAMETERS

➤ Click (99) Peck drill-full retract tab to enter the .201Dia peck drill machining parameters.

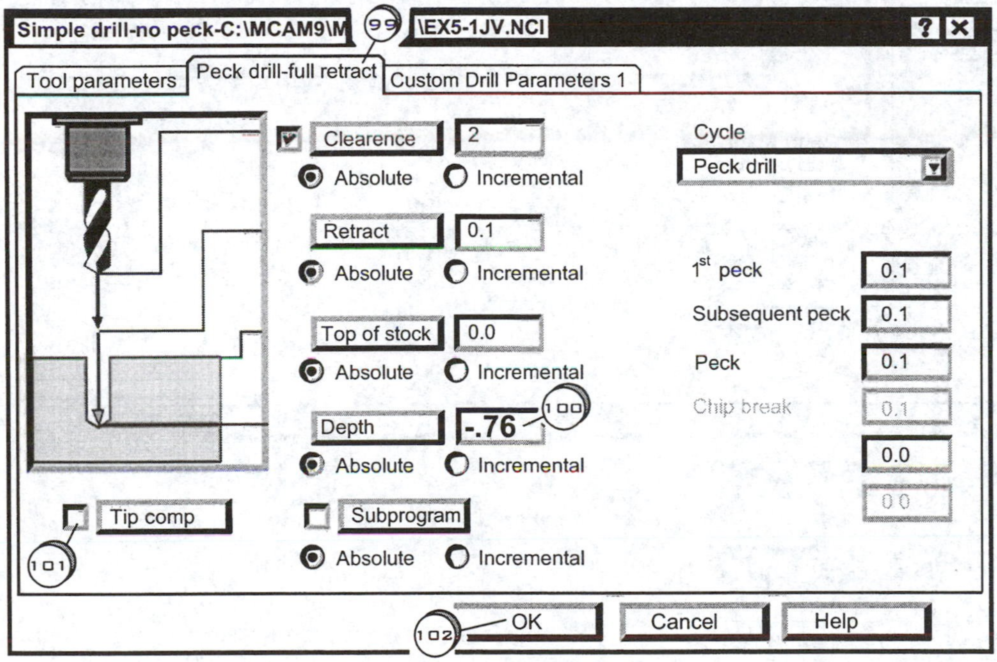

➤ Click (100) in the Depth box; enter -.76

➤ Click (101) the tip comp check *off* ☐ when drilling *blind* holes

➤ Click (102) OK

The .201Dia peck drilling x .76 deep operation will be added to *Mastercam's* Operations Manager

F) TAP THE HOLES 1/4-20 UNC X .55 DEEP

◆ CREATE THE TAP 1/4-20 UNC TOOLPATH

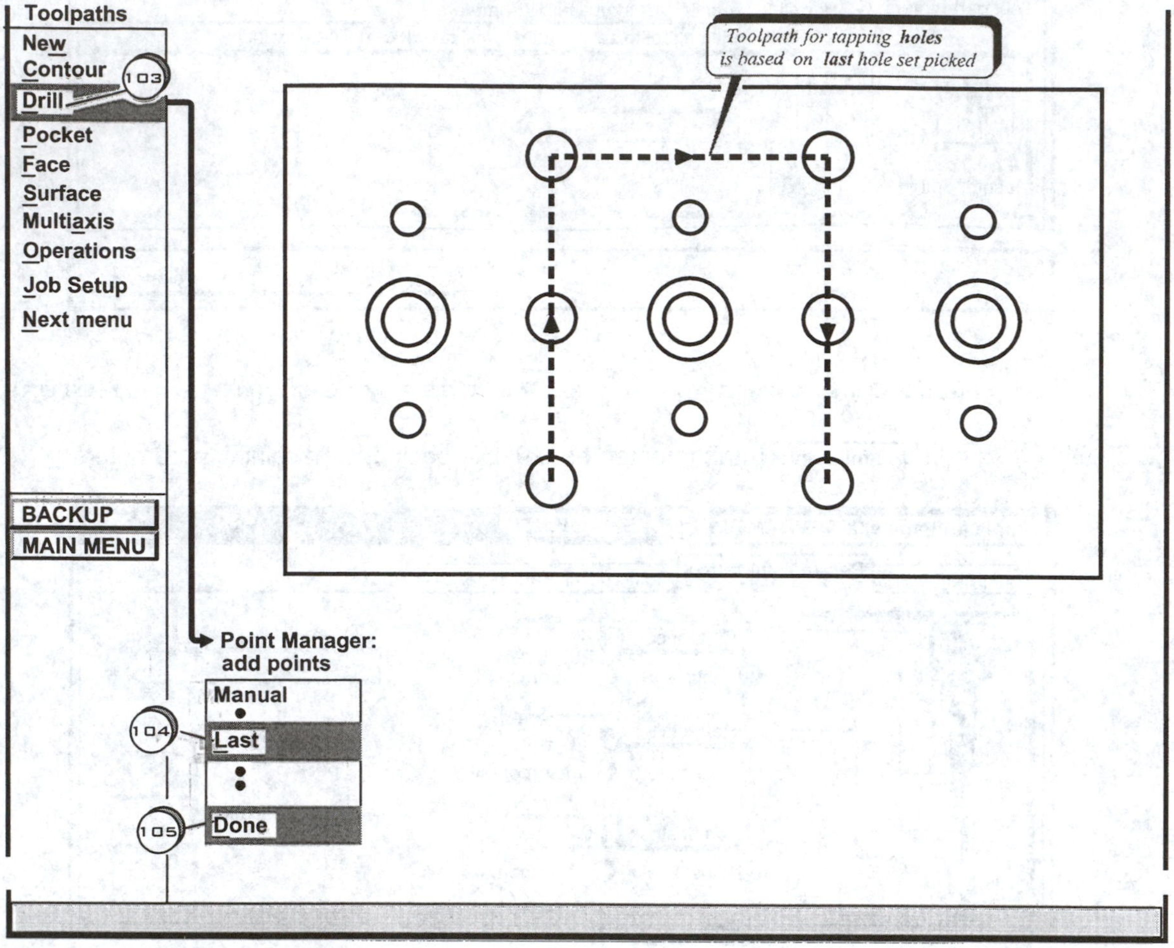

➢ Click 103 **Drill**

➢ Click 104 **Last**

Last directs *Mastercam* to *automatically select* the *last* set of drill points

➢ Click 105 **Done**

◆ OBTAIN THE NEEDED 1/4-20 UNC TAP TOOL

Mastercam will activate and display the Tool and Drill Parameters dialog box after the drill tool path has been created.

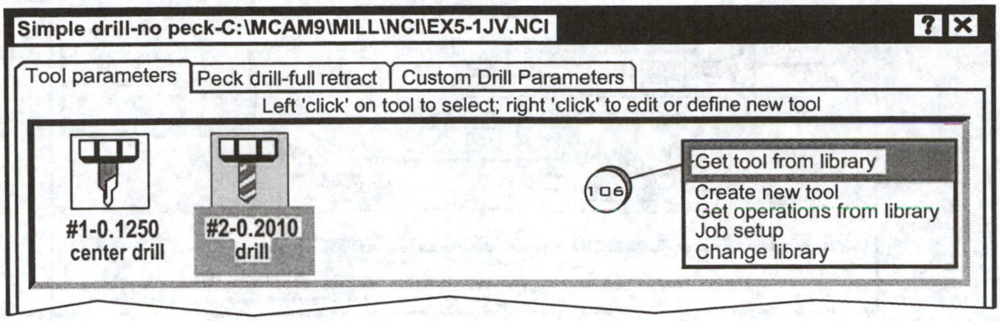

➤ Press the ***right*** mouse button in the ***white area***

➤ Click (106) Get tool from library

Mastercam will display the main Tool Library dialog box

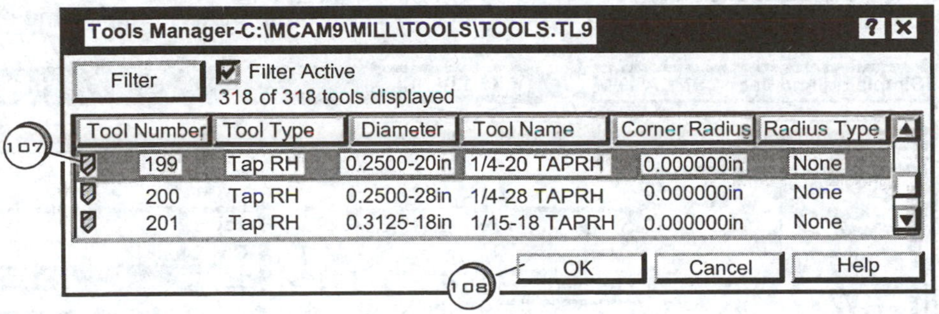

➤ Click (107) the 1/4-20 TAPRH tool

➤ Click (108) OK

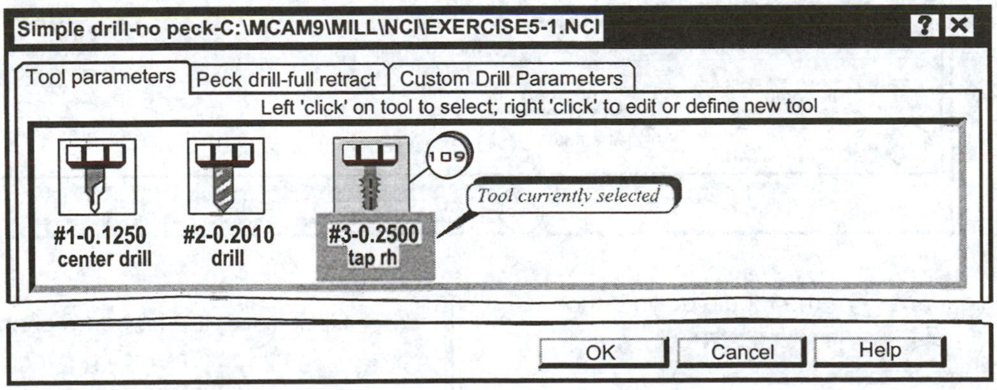

➤ ***Right*** click (109)

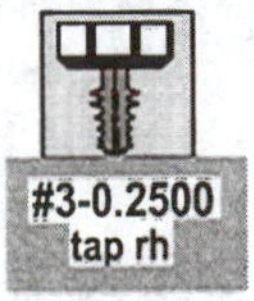

on the #3-0.2500 tap rh icon to select this as the ***active tool*** and direct *Mastercam* to display the Define Tool dialog box

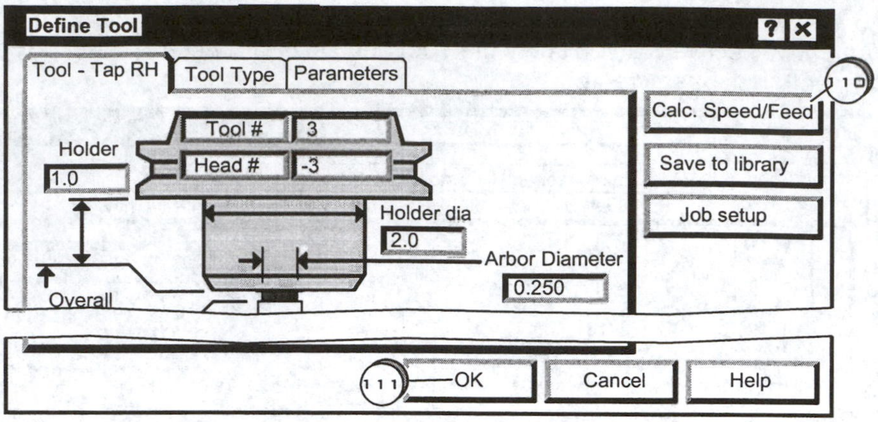

➤ Click (110) [Calc Speed/Feed] | ➤ Click (111) [OK]

- ENTER THE 1/4-20UNC TAP MACHINING PARAMETERS

➤ Click (112) [Peck drill-full retract] tab to enter the .25-20UNC tap machining parameters.

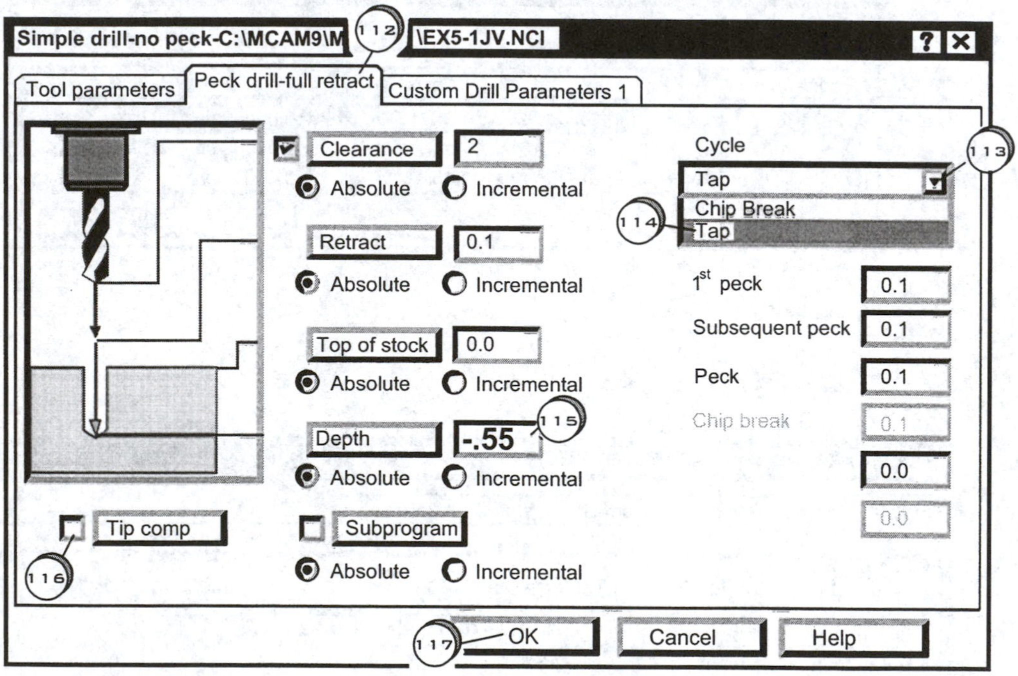

➤ Click (113) [▼] the toggle down button

➤ Click (114) [Tap]

➤ Click (115) in the Depth box; enter **-.55**

➤ Click (116) the tip comp check *off* ☐ for tapping *blind* holes

➤ Click (117) [OK]

The 1/4-20UNC tapping operation will be added to *Mastercam's* Operations Manager

G) PECK DRILL THE .313DIA HOLES THRU

◆ CREATE THE PECK DRILL THRU TOOLPATH

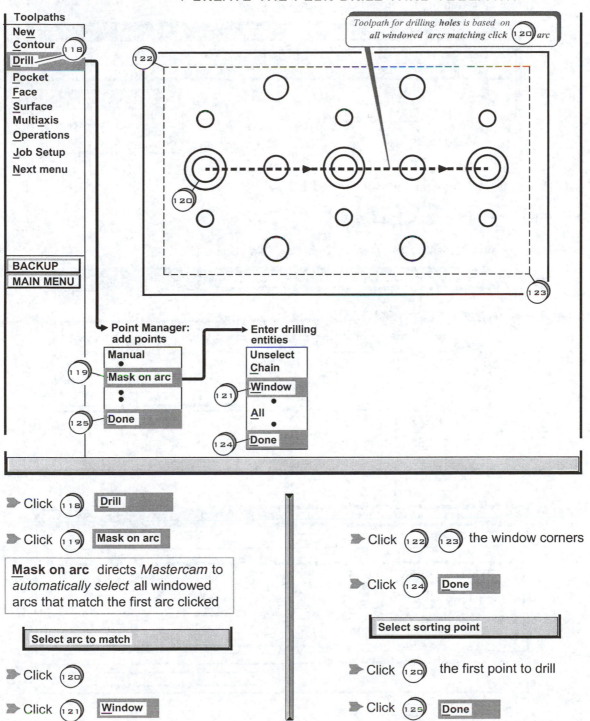

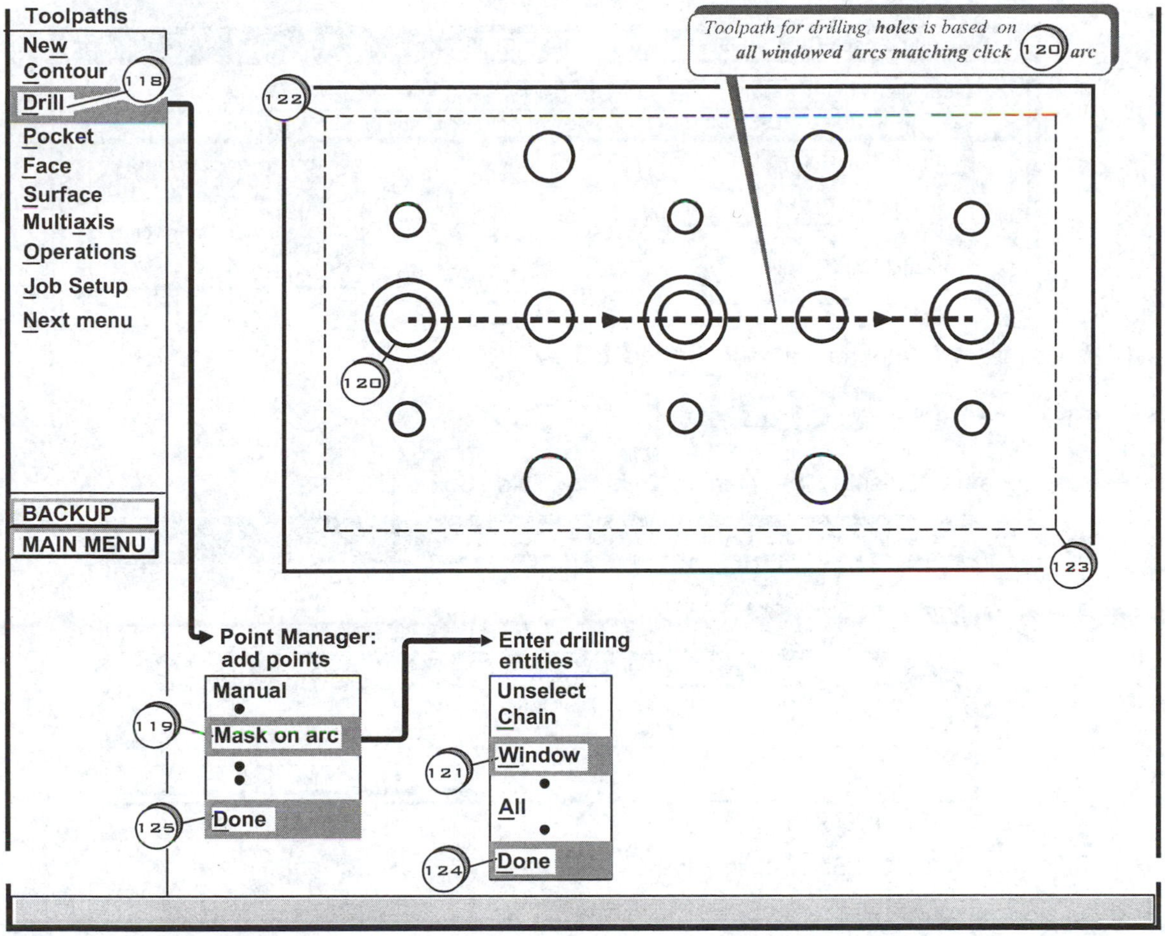

➤ Click 118 **Drill**

➤ Click 119 **Mask on arc**

Mask on arc directs *Mastercam* to *automatically select* all windowed arcs that match the first arc clicked

Select arc to match

➤ Click 120

➤ Click 121 **Window**

➤ Click 122 123 the window corners

➤ Click 124 **Done**

Select sorting point

➤ Click 120 the first point to drill

➤ Click 125 **Done**

◆ OBTAIN THE NEEDED .25UNC-20 TAP TOOL

Mastercam will activate and display the Tool and Drill Parameters dialog box after the drill tool path has been created.

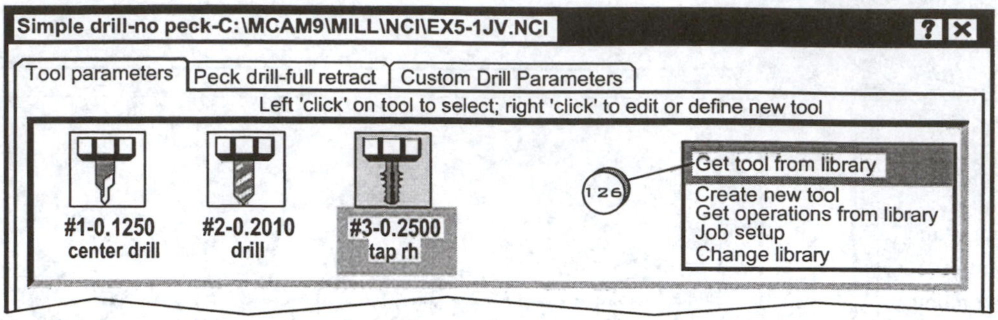

➤ Press the ***right*** mouse button in the ***white area***

➤ Click (126) Get tool from library

Mastercam will display the main Tool Library dialog box

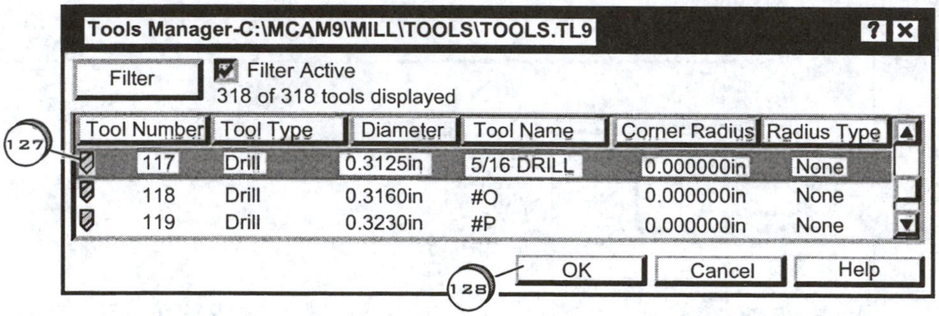

➤ Click (127) the .3125Dia Drill tool

➤ Click (128) OK

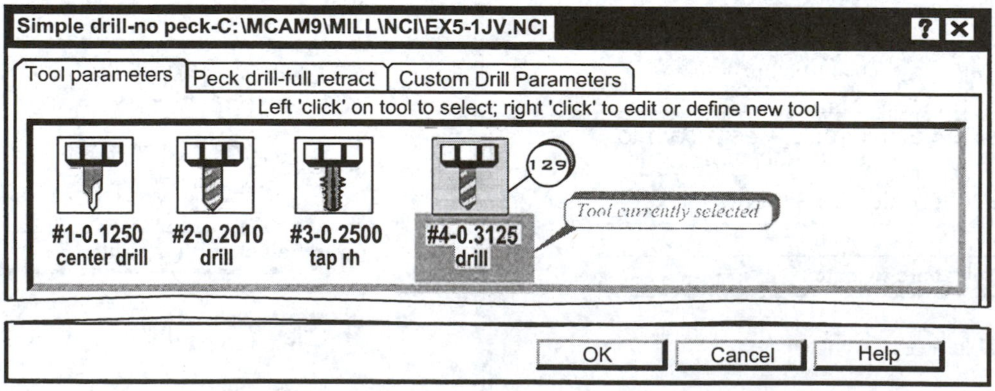

➤ ***Right*** click (129)

on the #4-0.3125 drill to select this as the ***active tool*** and direct *Mastercam* to display the Define Tool dialog box

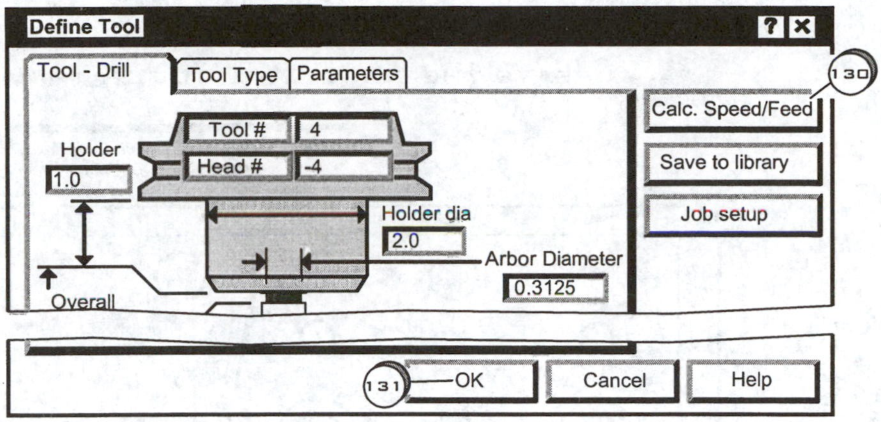

➤ Click (130) Calc Speed/Feed

➤ Click (131) OK

- ENTER THE .313DIA PECK DRILL THRU MACHINING PARAMETERS

➤ Click (132) Peck drill-full retract tab to enter the .201Dia peck drill machining parameters.

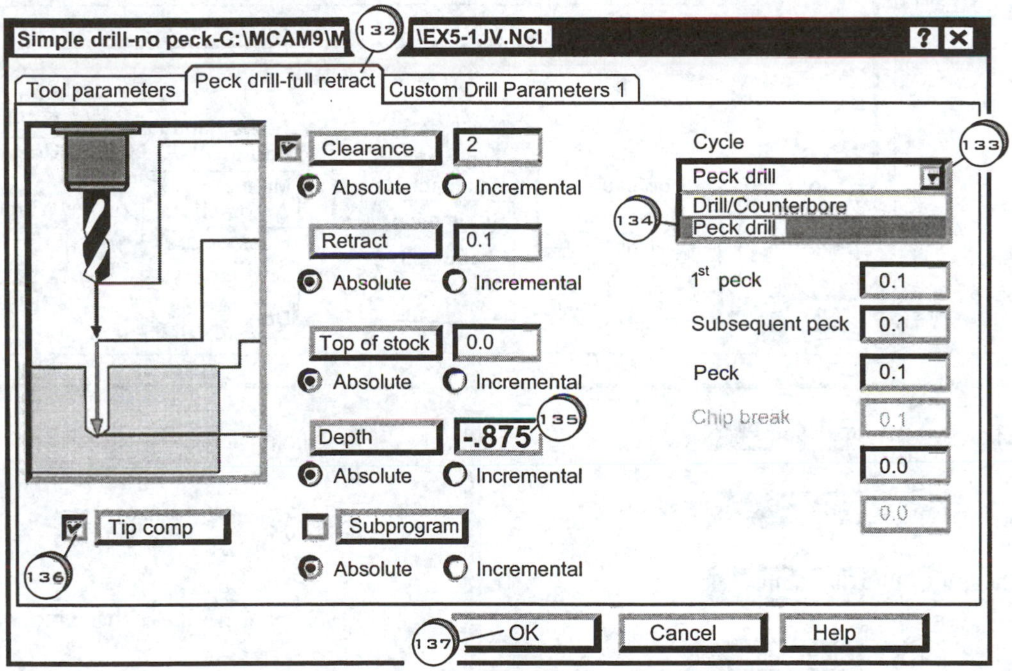

➤ Click (133) the toggle down button

➤ Click (134) Peck drill

➤ Click (135) in the Depth box; enter -.875

➤ Click (136) the tip comp check *on* when drilling *thru* holes

➤ Click (137) OK

The .313Dia peck drilling thru operation will be added to *Mastercam's* Operations Manager

H) CIRCLE MILL .5DIA HOLES X .25 DEEP

◆ CREATE THE CIRCLE MILL TOOLPATH

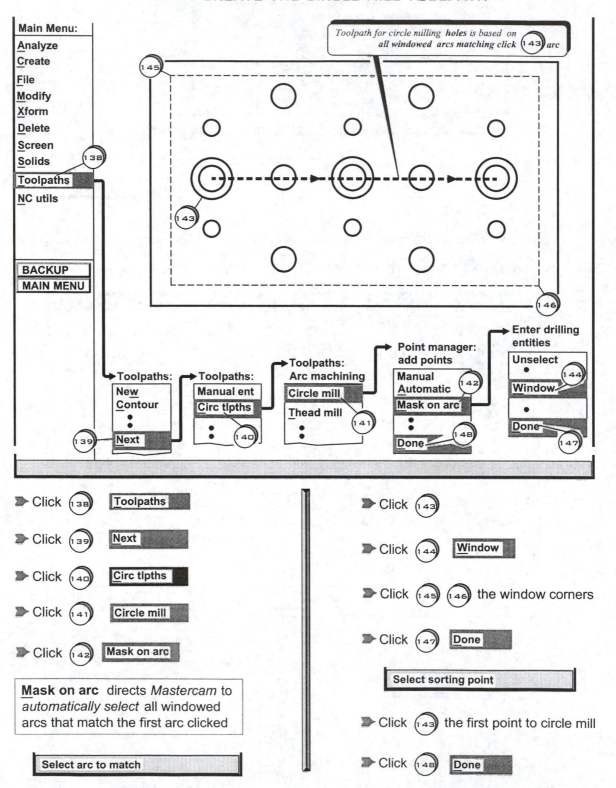

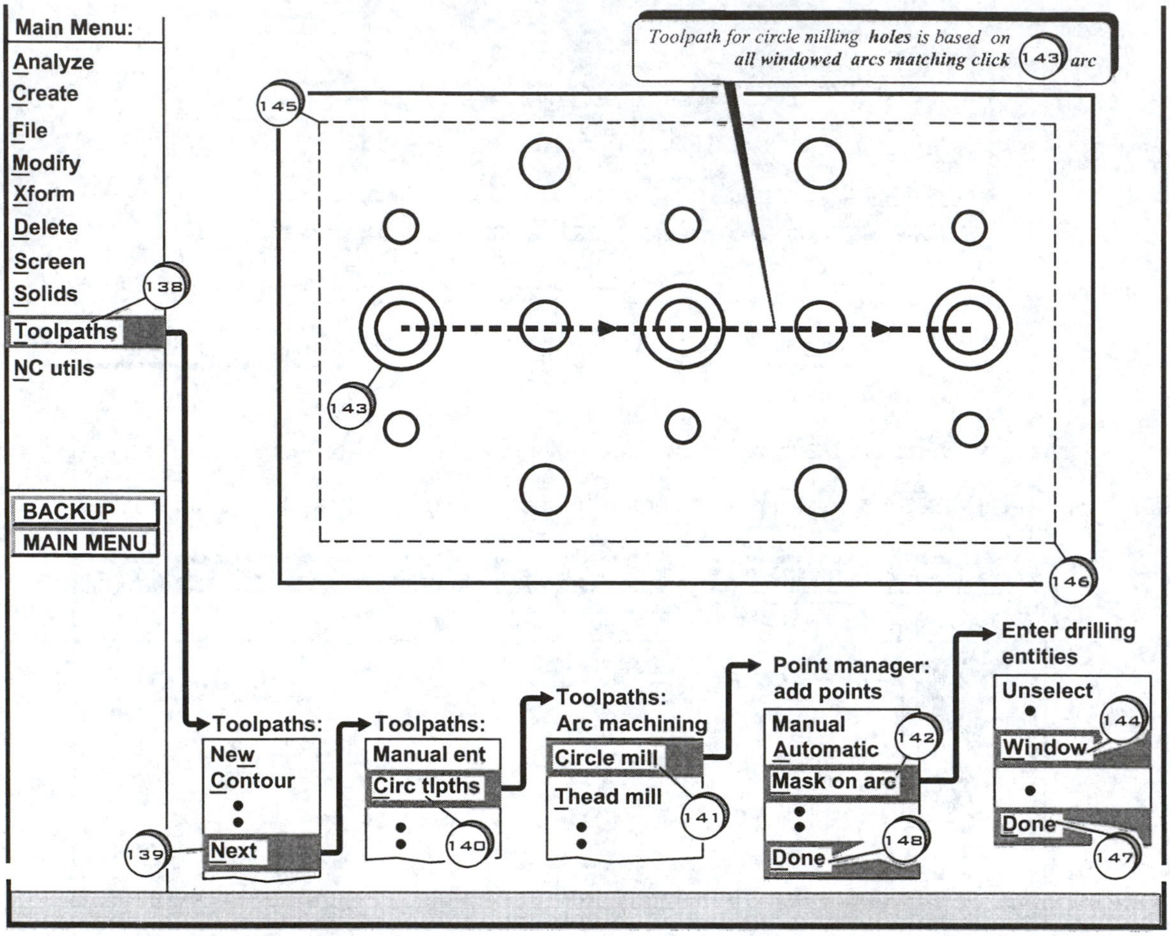

➤ Click 138 **Toolpaths**

➤ Click 139 **Next**

➤ Click 140 **Circ tlpths**

➤ Click 141 **Circle mill**

➤ Click 142 **Mask on arc**

Mask on arc directs *Mastercam* to *automatically select* all windowed arcs that match the first arc clicked

Select arc to match

➤ Click 143

➤ Click 144 **Window**

➤ Click 145 146 the window corners

➤ Click 147 **Done**

Select sorting point

➤ Click 143 the first point to circle mill

➤ Click 148 **Done**

◆ OBTAIN THE NEEDED 3/8 END MILL TOOL

Mastercam will activate and display the Tool and Drill Parameters dialog box after the circle mill tool path has been created.

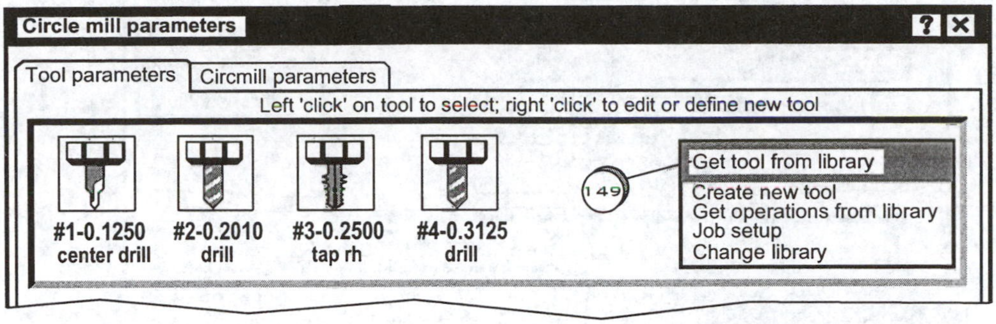

➤ ***Right*** Click move the mouse cursor *down* and Click (149) Get tool from library

Select the 3/8in Dia end mill tool from the **TOOLS** tool library

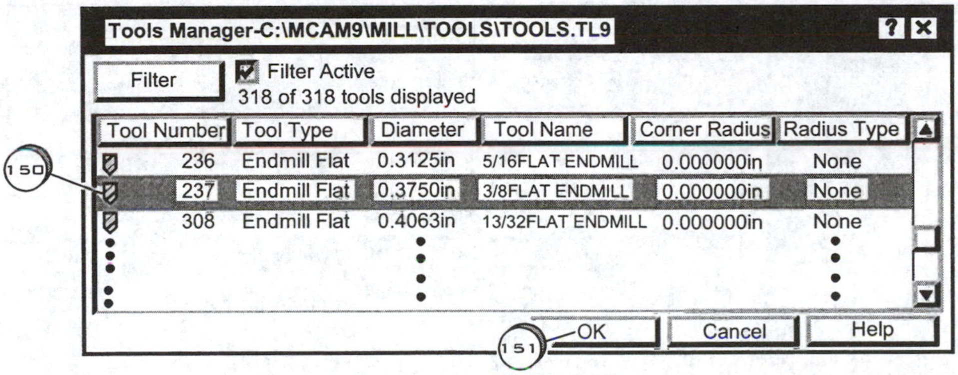

➤ Click (150) on the 3/8 FLAT ENDMILL tool

➤ Click (151) the OK button

Mastercam will place these tools into the currently active Tool and Drill Parameters dialog box

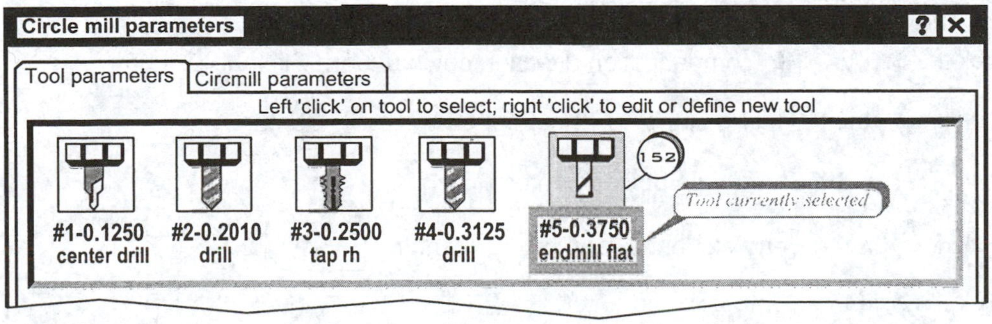

➤ ***Right*** Click (152) on the #5-0.3750 endmill flat tool icon

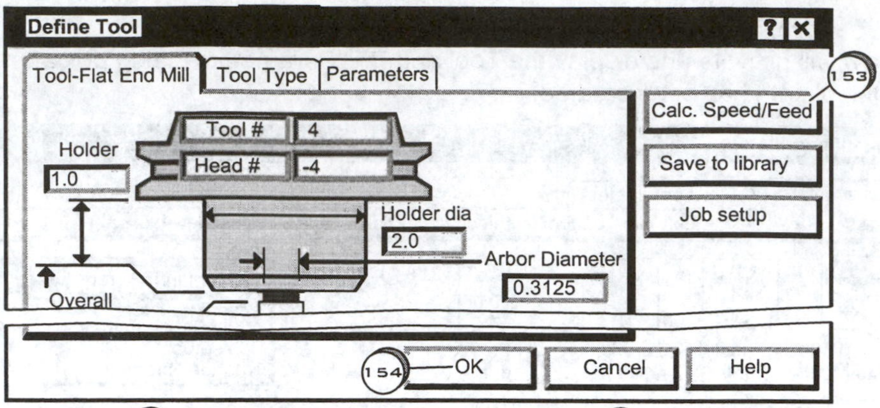

➤ Click 153 Calc Speed/Feed ➤ Click 154 OK

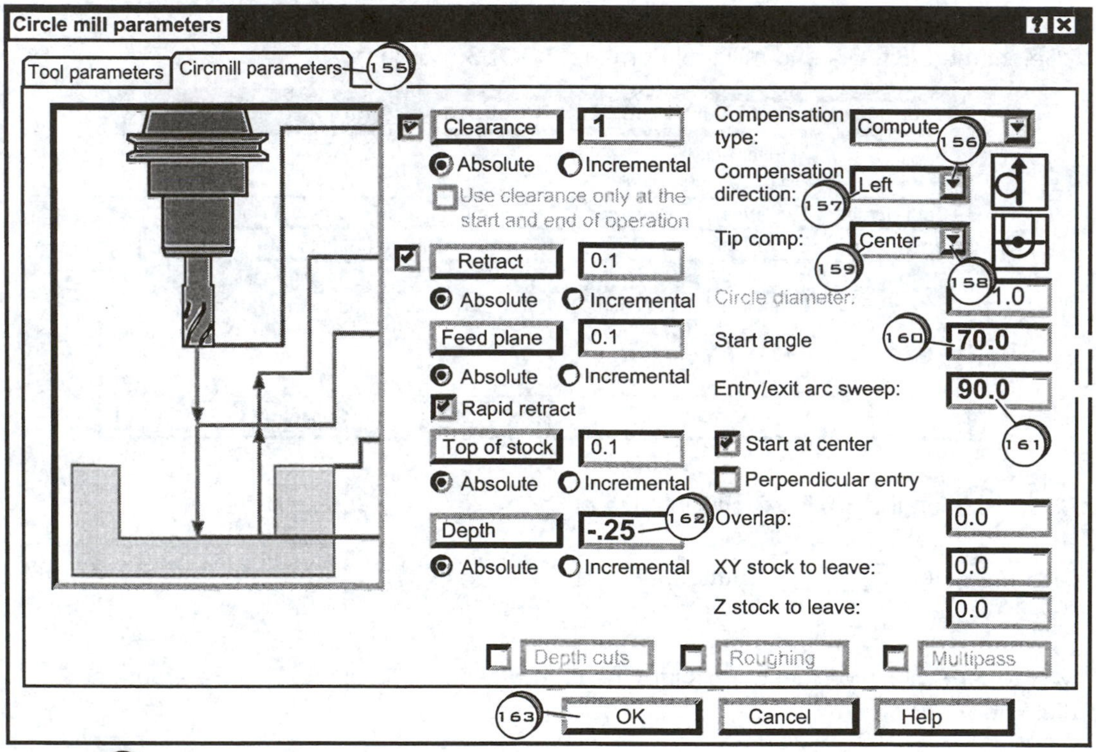

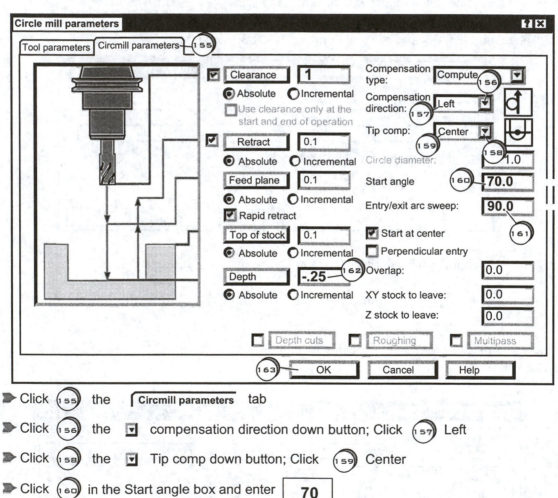

➤ Click 155 the **Circmill parameters** tab

➤ Click 156 the ▼ compensation direction down button; Click 157 Left

➤ Click 158 the ▼ Tip comp down button; Click 159 Center

➤ Click 160 in the Start angle box and enter **70**

➤ Click 161 in the Entry/exit arc sweep box and enter **90**

➤ Click 162 in the Depth box and enter **-.25**

➤ Click 163 the OK button

I) Verify all the machining operations

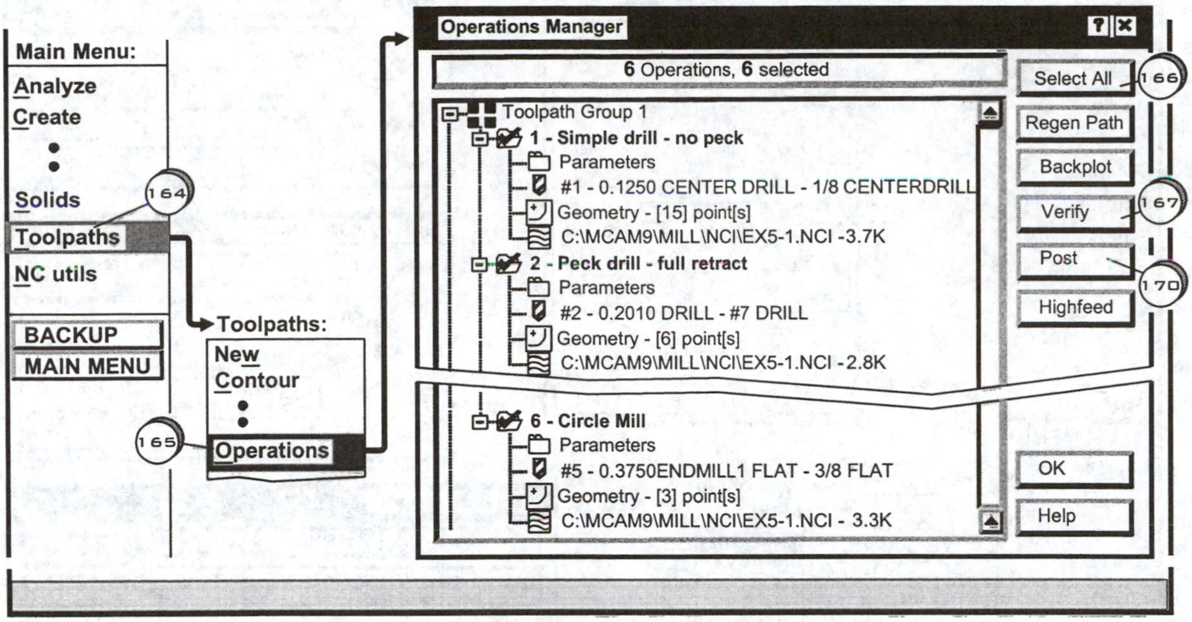

➤ Click (164) Toolpaths

➤ Click (165) Operations

➤ Click (166) the Select All button

➤ Click (167) the Verify button

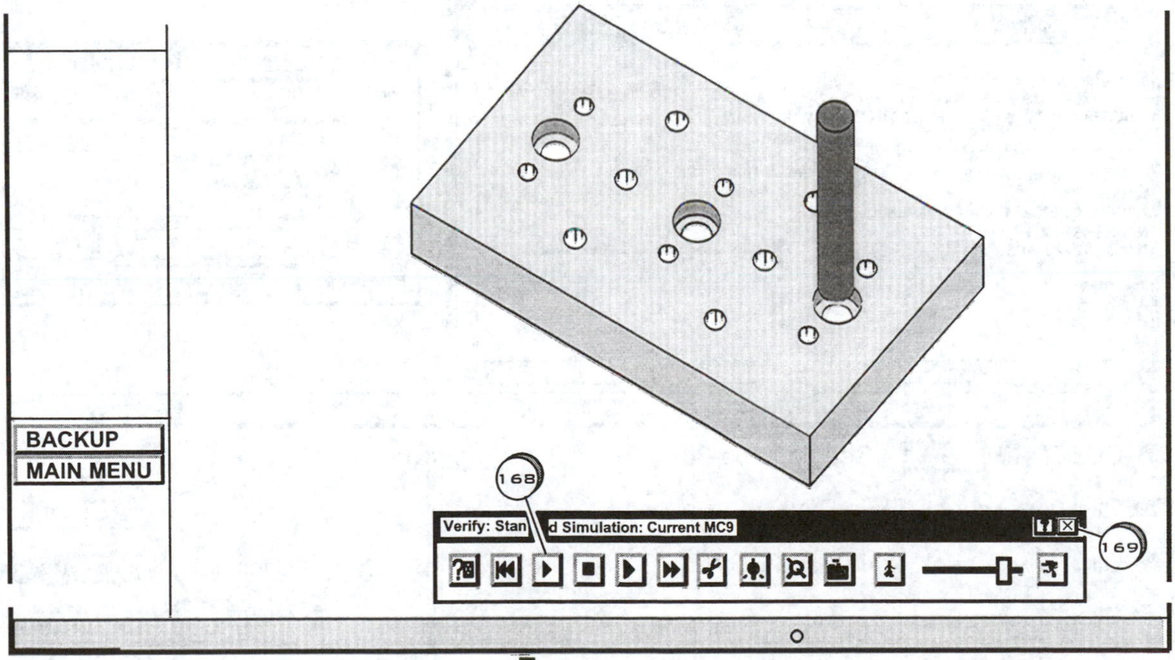

➤ Click (168) the play button ▶

➤ Click (169) the exit button ☒ to exit the verifier

J) Generate, save and print the word address part program

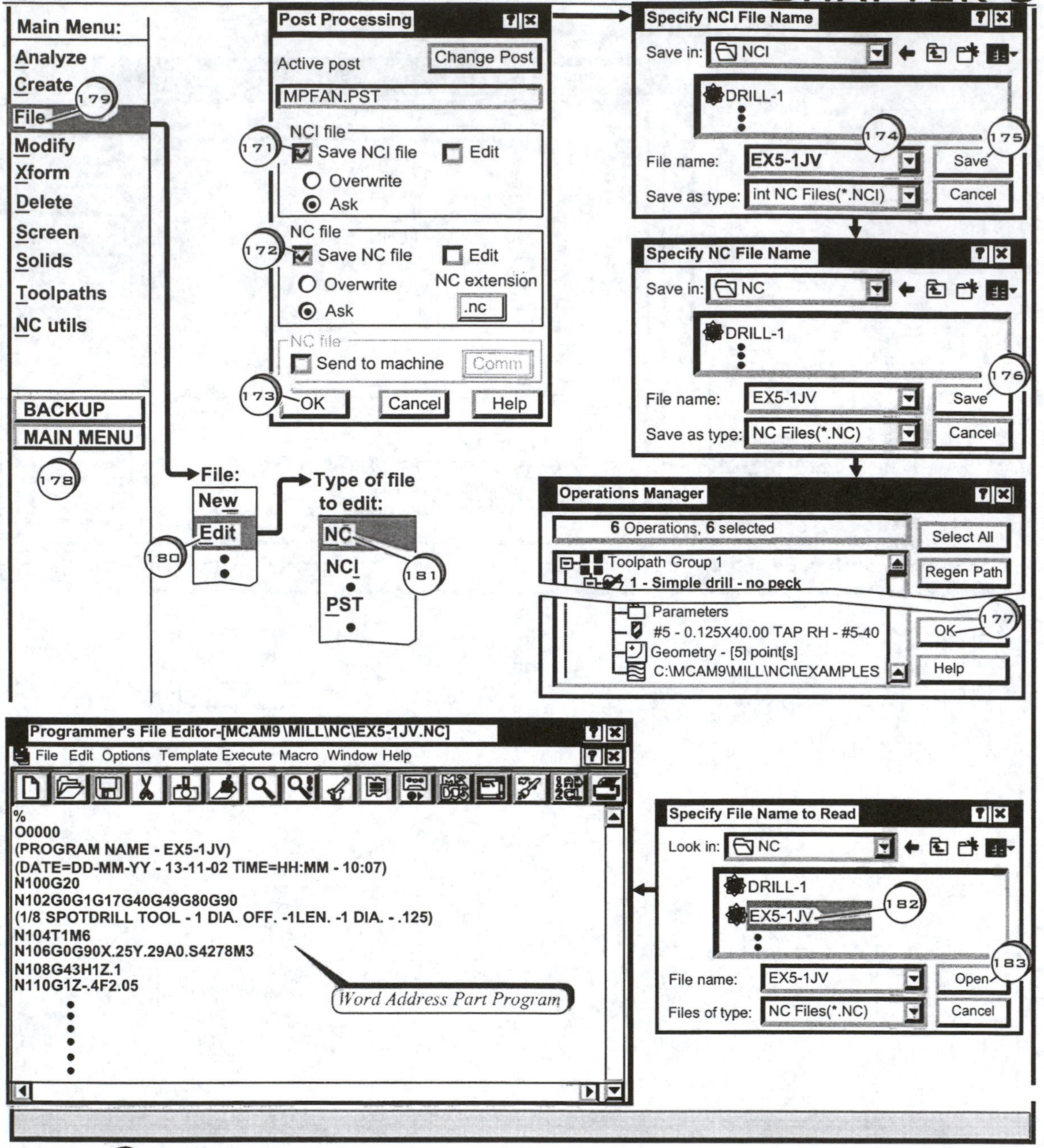
Main Menu:
Analyze
Create
File
Modify
Xform
Delete
Screen
Solids
Toolpaths
NC utils
BACKUP
MAIN MENU
Post Processing
Active post
Change Post
MPFAN.PST
NCI file
Save NCI file
Edit
Overwrite
Ask
NC file
Save NC file
Edit
Overwrite
NC extension
Ask
.nc
NC file
Send to machine
Comm
OK
Cancel
Help
Specify NCI File Name
Save in: NCI
DRILL-1
File name: EX5-1JV
Save
Save as type: int NC Files(*.NCI)
Cancel
Specify NC File Name
Save in: NC
DRILL-1
File name: EX5-1JV
Save
Save as type: NC Files(*.NC)
Cancel
File:
New
Edit
Type of file to edit:
NC
NCI
PST
Operations Manager
6 Operations, 6 selected
Select All
Toolpath Group 1
Regen Path
1 - Simple drill - no peck
Parameters
#5 - 0.125X40.00 TAP RH - #5-40
OK
Geometry - [5] point[s]
C:\MCAM9\MILL\NCI\EXAMPLES
Help
Programmer's File Editor-[MCAM9 \MILL\NC\EX5-1JV.NC]
File Edit Options Template Execute Macro Window Help
%
O0000
(PROGRAM NAME - EX5-1JV)
(DATE=DD-MM-YY - 13-11-02 TIME=HH:MM - 10:07)
N100G20
N102G0G1G17G40G49G80G90
(1/8 SPOTDRILL TOOL - 1 DIA. OFF. -1LEN. -1 DIA. - .125)
N104T1M6
N106G0G90X.25Y.29A0.S4278M3
N108G43H1Z.1
N110G1Z-.4F2.05
Word Address Part Program
Specify File Name to Read
Look in: NC
DRILL-1
EX5-1JV
File name: EX5-1JV
Open
Files of type: NC Files(*.NC)
Cancel
Click 170 Post .See page 5p-28.
Click 171 check in Save NCI file
Click 172 check in Save NC file
Click 173 OK
Click 174 in File name box ; enter EX5-1JV
Click 175 Save
Click 176 Save
Click 177 OK
Click 178 MAIN MENU
Click 179 File
Click 180 Edit
Click 181 NC
Click 182 EX5-1JV
Click 183 Open

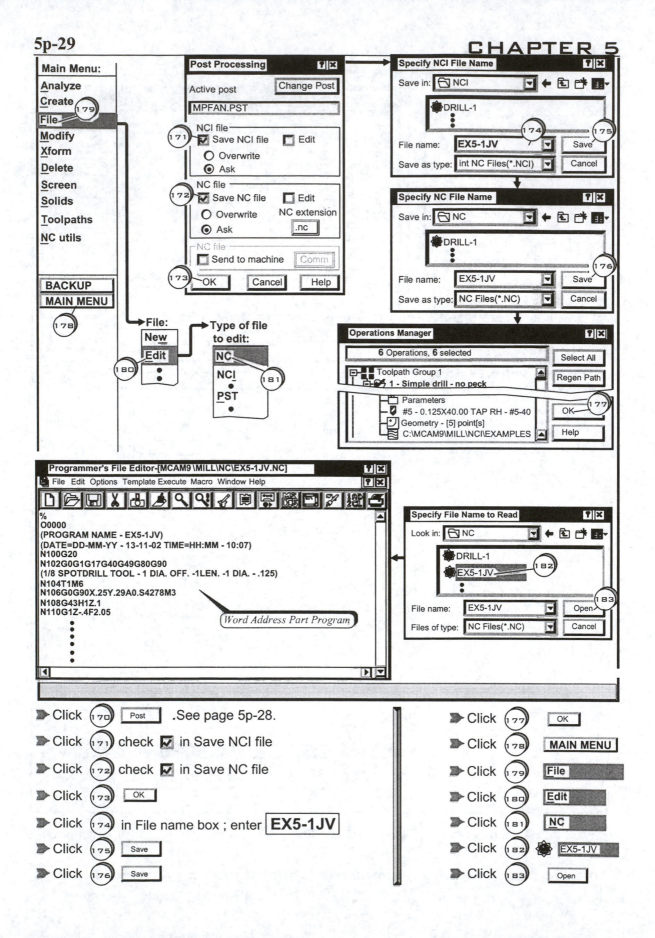

5-2) a) Generate the required drill tool paths for executing PROCESS PLAN 5P-2

b) Backplot and verify the tool paths

c) Direct Mastercam to generate the word address part program

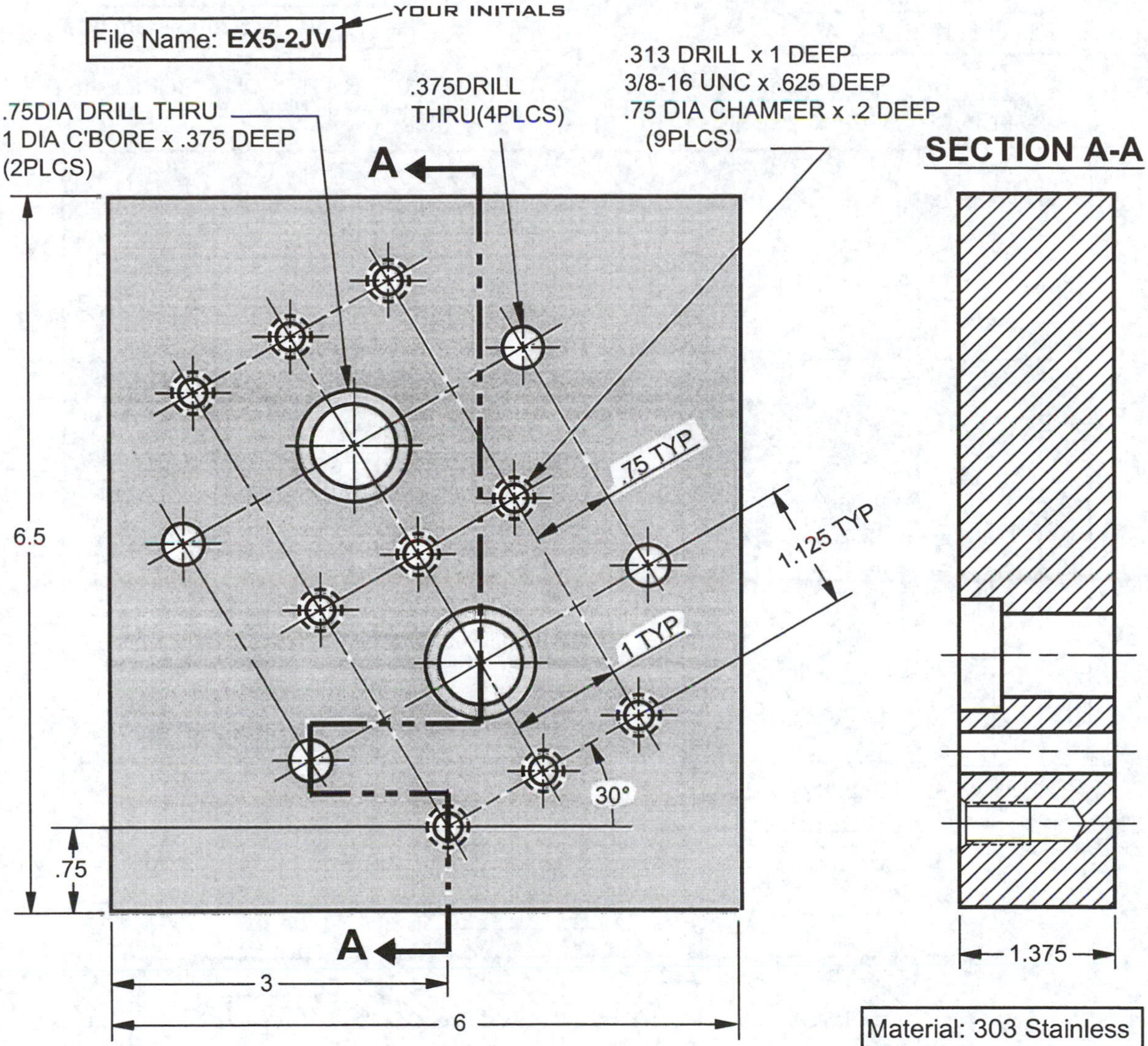

Figure 5p-2

PROCESS PLAN 5P-2

No.	Operation	Tooling
1	Center Drill x .2 deep(all holes)	1/8 Center drill
2	Peck drill x 1 Deep(9PLCS)	5/16 Drill
3	Tap x .625 Deep(9 PLCS)	.375-16UNC Tap
4	Chamfer x .2 Deep(9PLCS)	3/4 Chamfer Mill
5	Peck drill thru(4PLCS)	3/8 Drill
6	Peck drill thru(2PLCS)	3/4 Drill
7	Circle Mill x .375 Deep(2PLCS)	1 Flat Endmill

5-3) a) Generate the required drill tool paths for executing PROCESS PLAN 5P-3

b) Backplot and verify the tool paths

c) Direct Mastercam to generate the word address part program

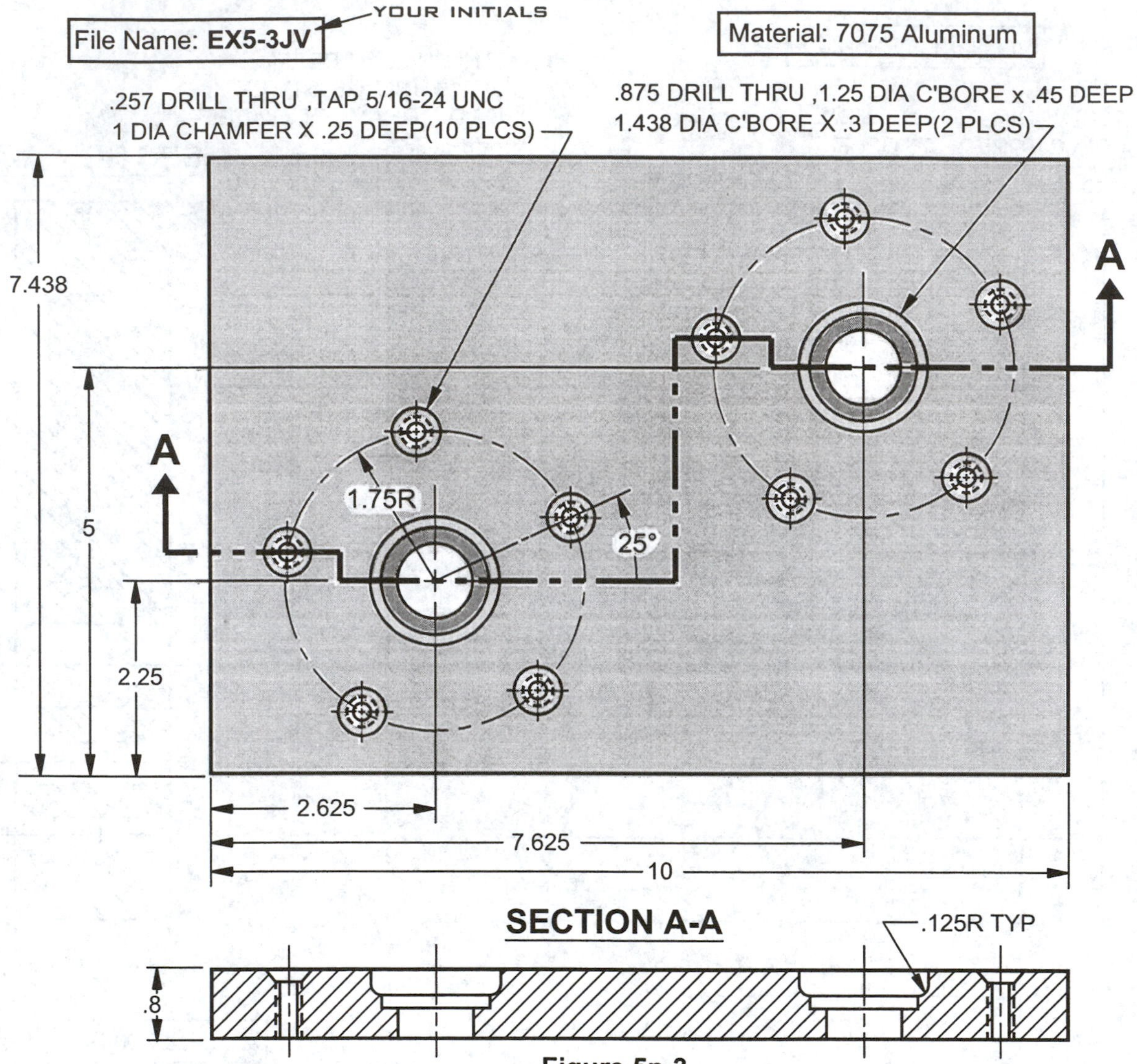

Figure 5p-3

PROCESS PLAN 5P-3

No.	Operation	Tooling
1	CENTER DRILL X .2 DEEP(ALL HOLES)	1/8 CENTER DRILL
2	PECK DRILL THRU(2PLCS)	7/8 DRILL
3	CIRCLE MILL X .45 DEEP (2PLCS)	3/4 FLAT ENDMILL
4	CIRCLE MILL X .3 DEEP (2PLCS)	3/4 BULL ENDMILL .125R
5	PECK DRILL THRU(10PLCS)	LTR 'F' DRILL
6	TAP THRU(10PLCS)	.313-24UNC TAP
7	CHAMFER X .25 DEEP(10PLCS)	1 CHAMFER MILL

5-4) a) Generate the required drill tool paths for executing PROCESS PLAN 5P-4

b) Backplot the tool paths

c) Direct Mastercam to generate the word address part program

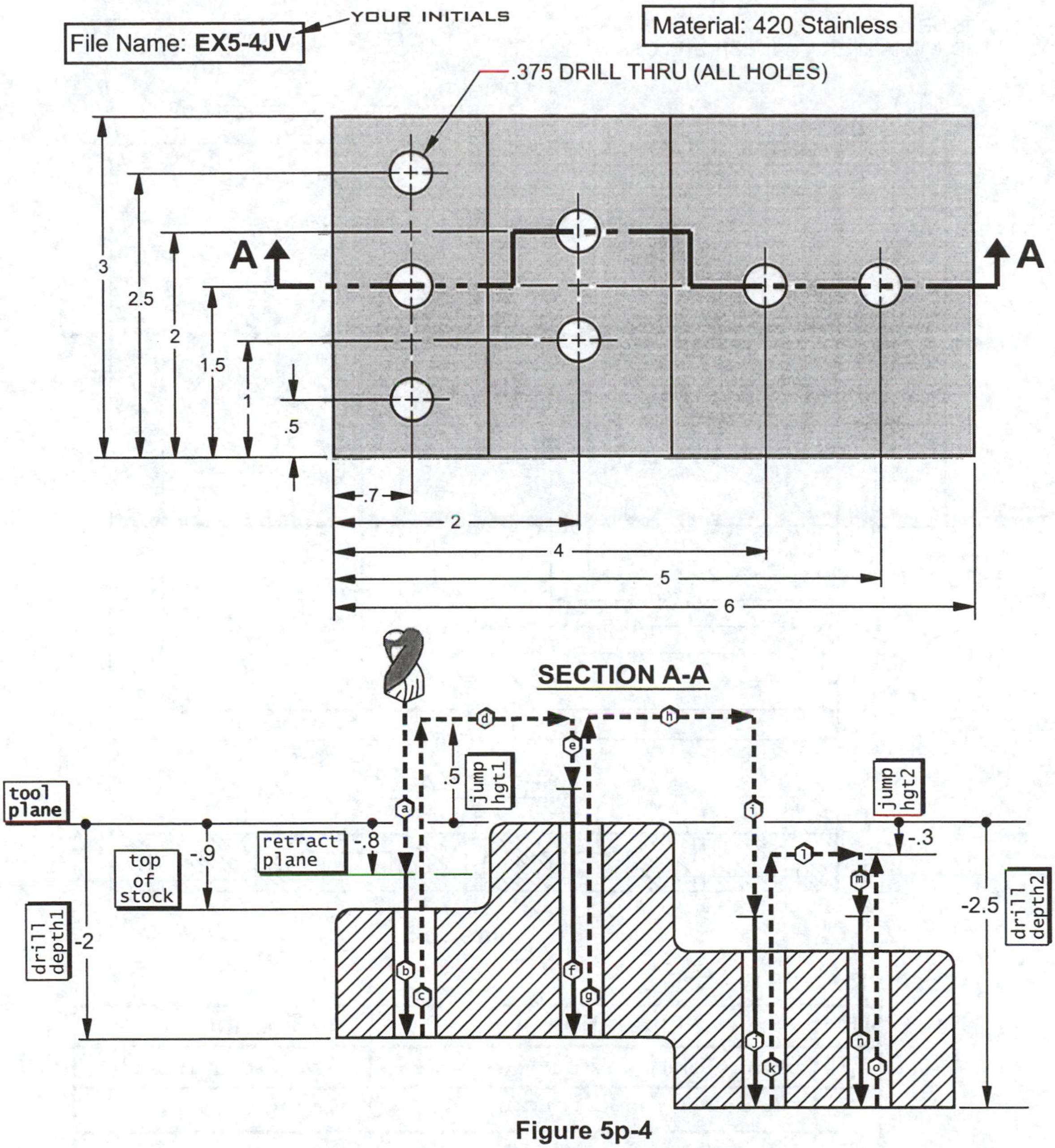

Figure 5p-4

PROCESS PLAN 5P-4

No.	Operation	Tooling
1	Center Drill x .2 deep(all holes)	1/8 Center drill
2	Peck drill thru(all holes)	3/8 Drill

5-5) a) Generate the required drill tool paths for executing PROCESS PLAN 5P-5
b) Verify the tool paths

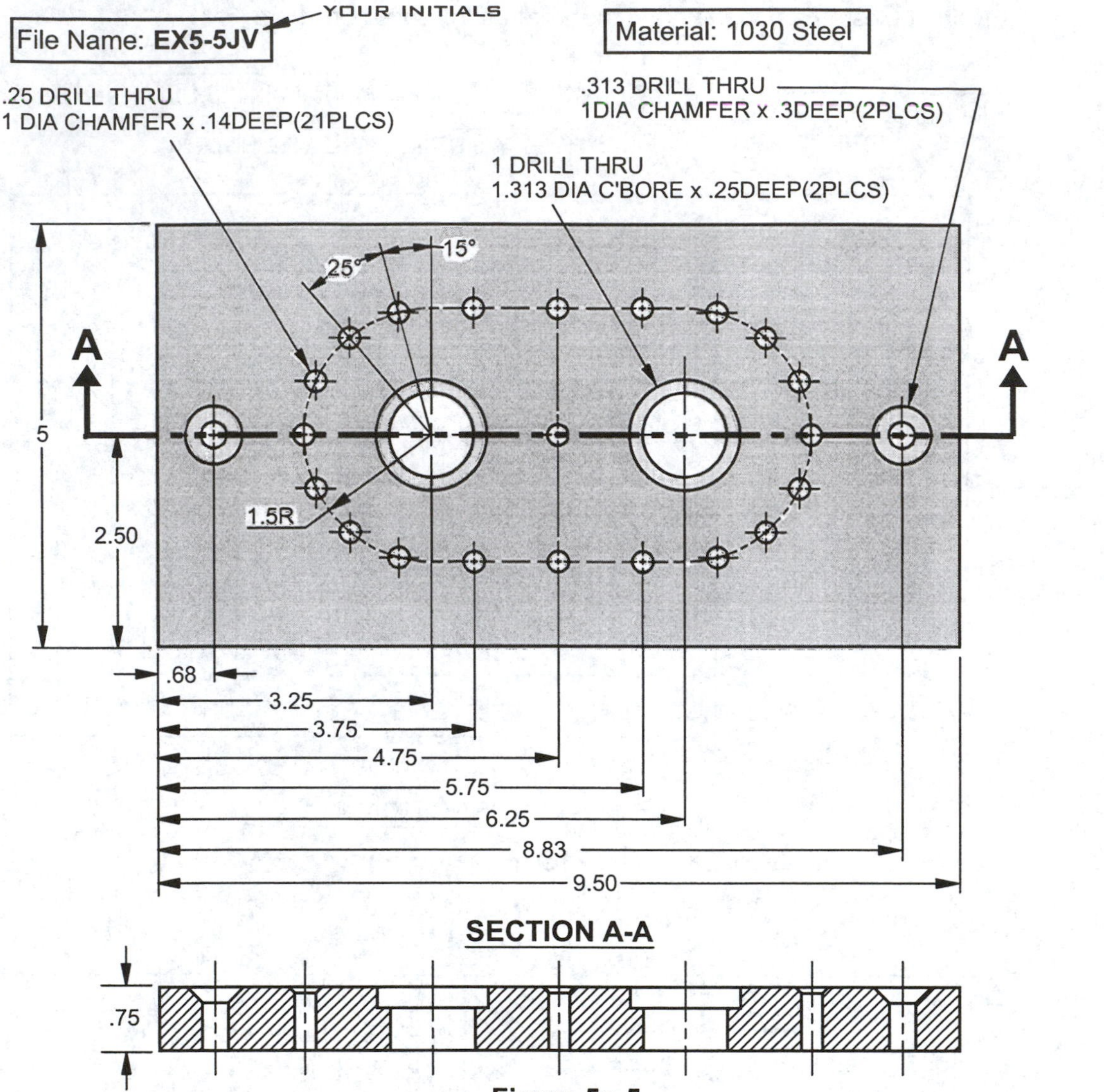

Figure 5p-5

PROCESS PLAN 5P-5

No.	Operation	Tooling
1	CENTER DRILL X .2 DEEP(ALL HOLES)	1/8 CENTER DRILL
2	PECK DRILL THRU(21PLCS)	1/4 DRILL
3	CHAMFER X .14 DEEP(21PLCS)	1 CHAMFER MILL
4	PECK DRILL THRU(2PLCS)	1 DRILL
5	CIRCLE MILL X .25 DEEP (2PLCS)	1 FLAT ENDMILL
6	PECK DRILL THRU(2PLCS)	5/16 DRILL
7	CHAMFER X .3 DEEP(2PLCS)	1 CHAMFER MILL

The hole design has been changed as shown in Figure 5p-6.

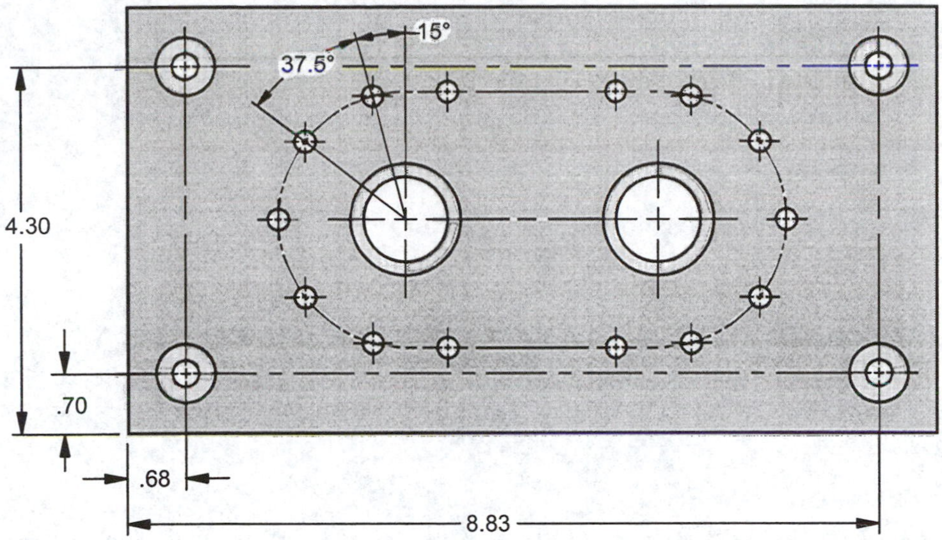

Figure 5p-6

c) Use the Geometry icon  in the Operations Manager to delete the appropriate drill points.

d) Erase the required hole geometry

e) Add the new hole geometry

f) Use the Geometry icon  in the Operations Manager to add the appropriate drill points.

g) Use **Regen Path** in the Operations Manager dialog box to regenerate the edited tool paths

h) Verify the edited tool paths

CHAPTER - 6

PROFILING AND POCKETING IN 2D SPACE

6-1 Chapter Objectives

After completeing this chapter you will be able to:

1. Know how to create 2D contours for profile machining
2. Prepare a part for 2D profile machining
3. Specify contouring parameters in the 2D contouring module
4. Know how to create 2D contours for pocketing
5. Specify pocketing parameters in the 2D pocketing module
6. Explain how to backplot contouring and pocketing toolpaths
6. State how to use *Mastercam's* verifier to produce solid model playback animations of contouring and pocketing machining.
4. Understand how to generate word address part programs via *Mastercam*'s post processing module

6-2 Creating a 2D Contour for Profiling

A set of lines, arcs and or splines connected *end to end* can be joined together into a *single* curve called a machining *contour*. When *Mastercam* is directed to contour a part it does so by moving the cutting tool along a prevlously defined contour. The various methods of creating a contour are described below.

Creating Contours

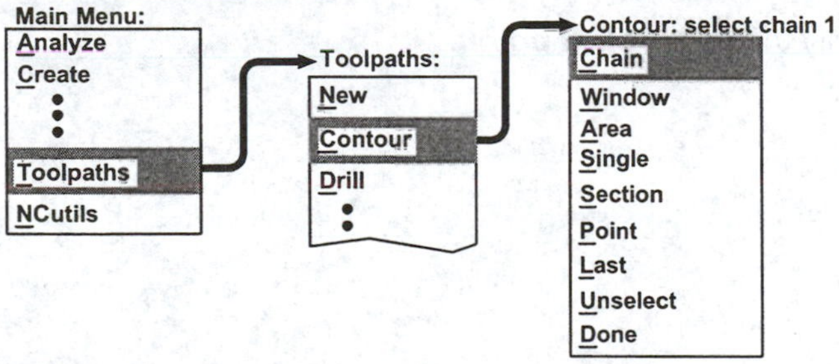

Chain

The operator clicks the chain start point on an entity. Chaining will proceed ***from the start*** *point* in the ***direction toward the midpoint*** of the entity. ***All*** entities connected ***end to end will be included*** in the chain

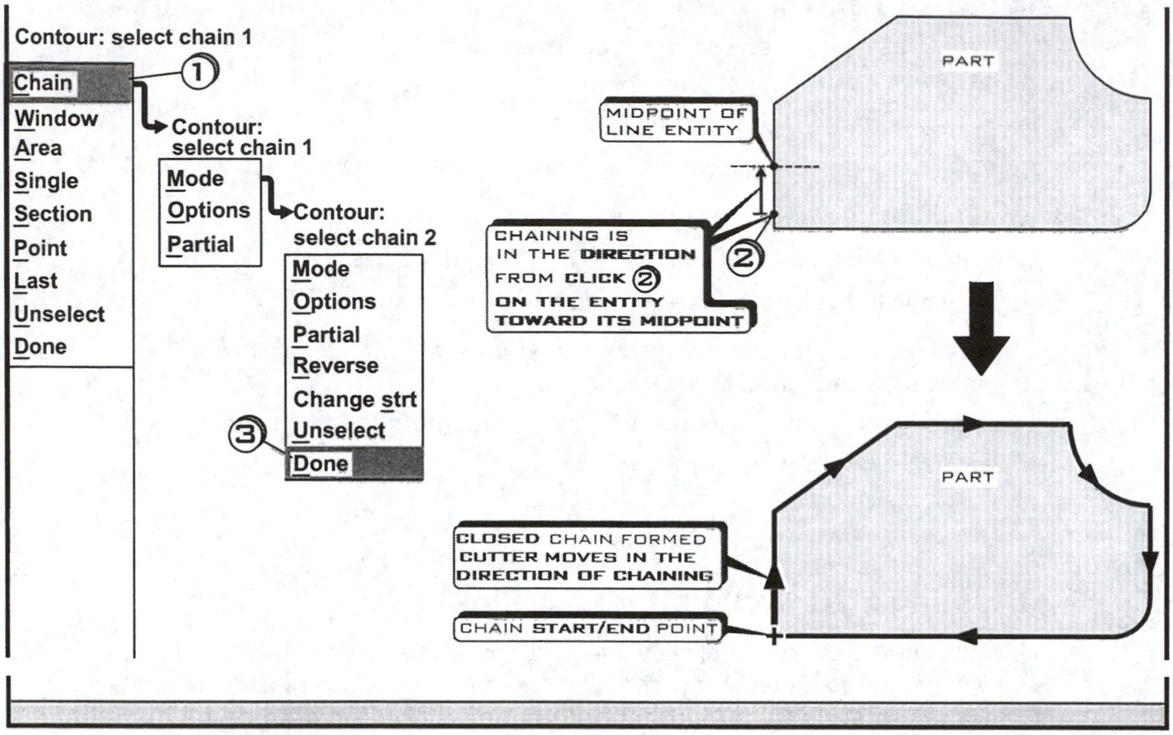

- Click ① **Chain**
- Click ② on the entity to specify the chain ***start point***
- Click ③ **Done**

Chain **Options**

Options gives the operator ***control*** over several important chaining parameters. ***Entity mask***, for example, allows the operator to selectively ***ignore or include*** point, line, arc and spline entities when executing chaining.

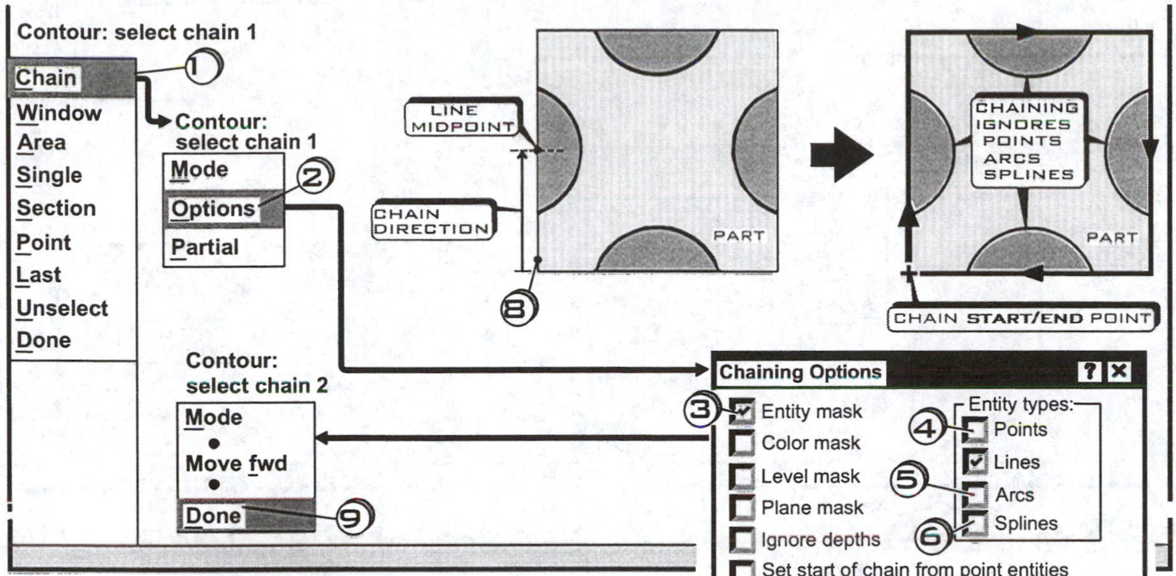

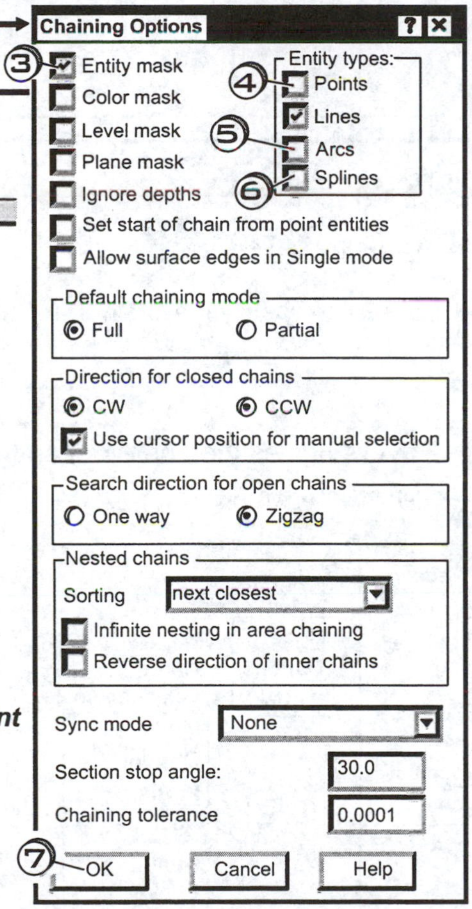

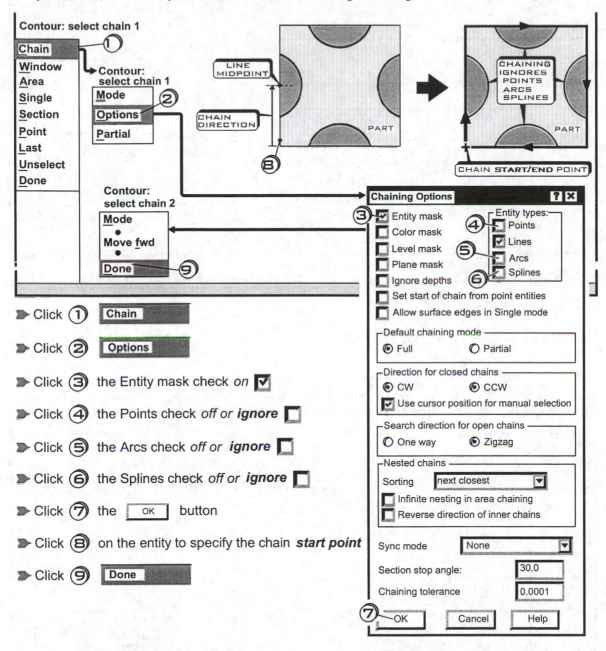

- Click ① **Chain**
- Click ② **Options**
- Click ③ the Entity mask check *on* ☑
- Click ④ the Points check *off or* ***ignore*** ☐
- Click ⑤ the Arcs check *off or* ***ignore*** ☐
- Click ⑥ the Splines check *off or* ***ignore*** ☐
- Click ⑦ the OK button
- Click ⑧ on the entity to specify the chain ***start point***
- Click ⑨ **Done**

Note: Open the **Chaining Options** dialog box and click ***on*** the checks for **Points, Lines, Arcs** and **Splines** *before* proceeding to the next set of chaining features.

Chain **Partial**

Partial gives the operator control over the ***start and end points*** of the chain

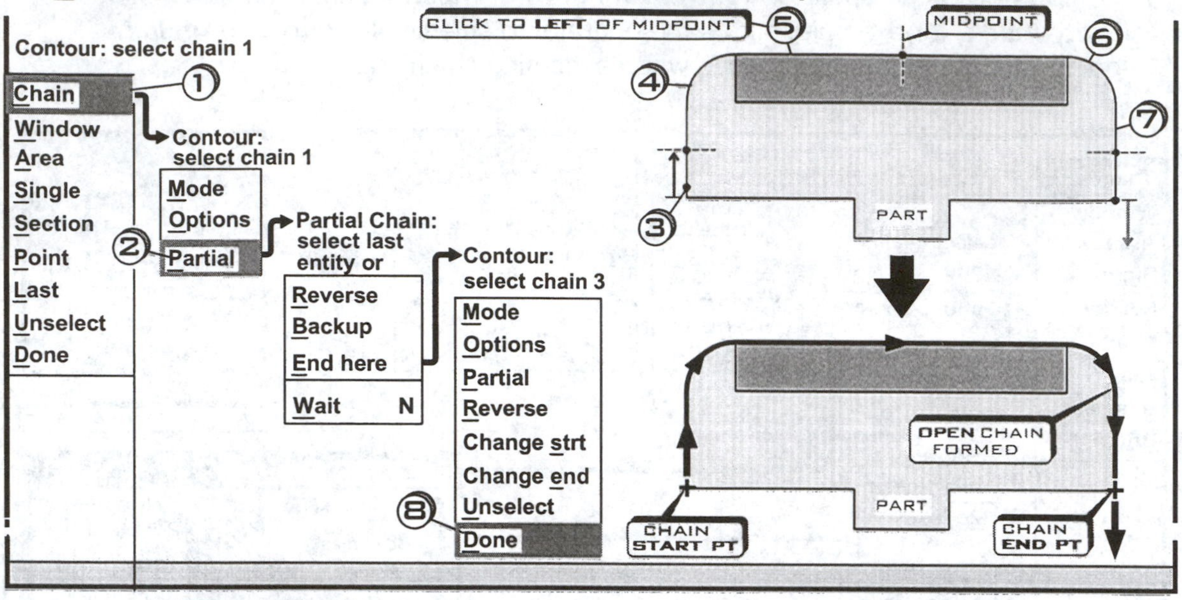

- Click ① **Chain**
- Click ② **Partial**
- Click ③ to specify the chain ***start point***
- Click ④ ⑤ ⑥ ⑦ the chain entities
- Click ⑧ **Done** to specify the chain ***end point***

Chain **Reverse**

Reverse gives the operator control over the ***chaining direction***

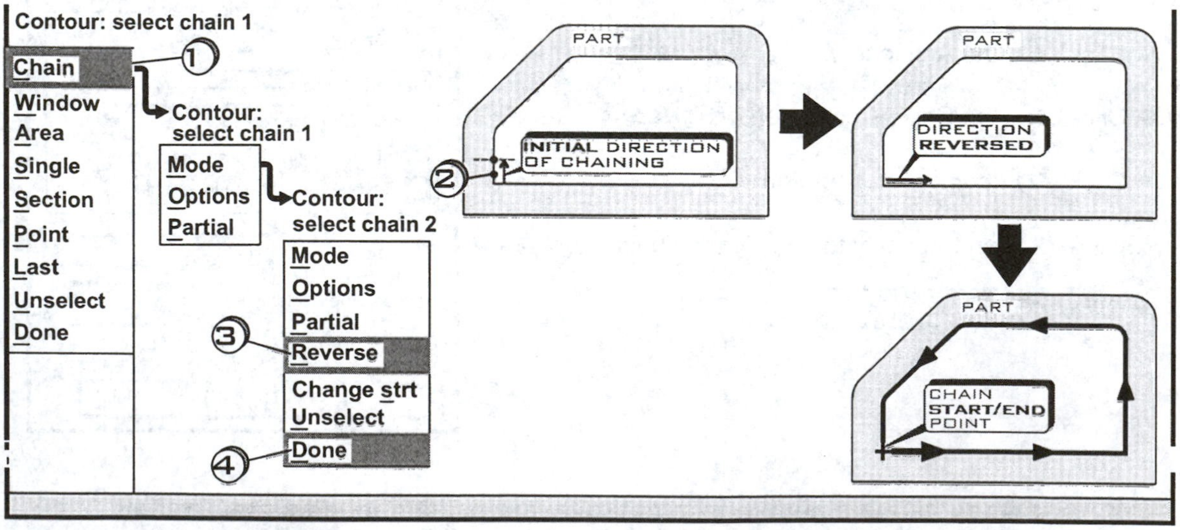

- Click ① **Chain**
- Click ② the chain ***start point***
- Click ③ **Reverse** to ***reverse*** the chaining direction
- Click ④ **Done**

Chain **Change strt**

Change strt gives the operator control over the ***location of the chain start point*** **Move fwd** moves the start point *foward one entity* in the chain. **Move back** moves the start point *back one entity* in the chain.

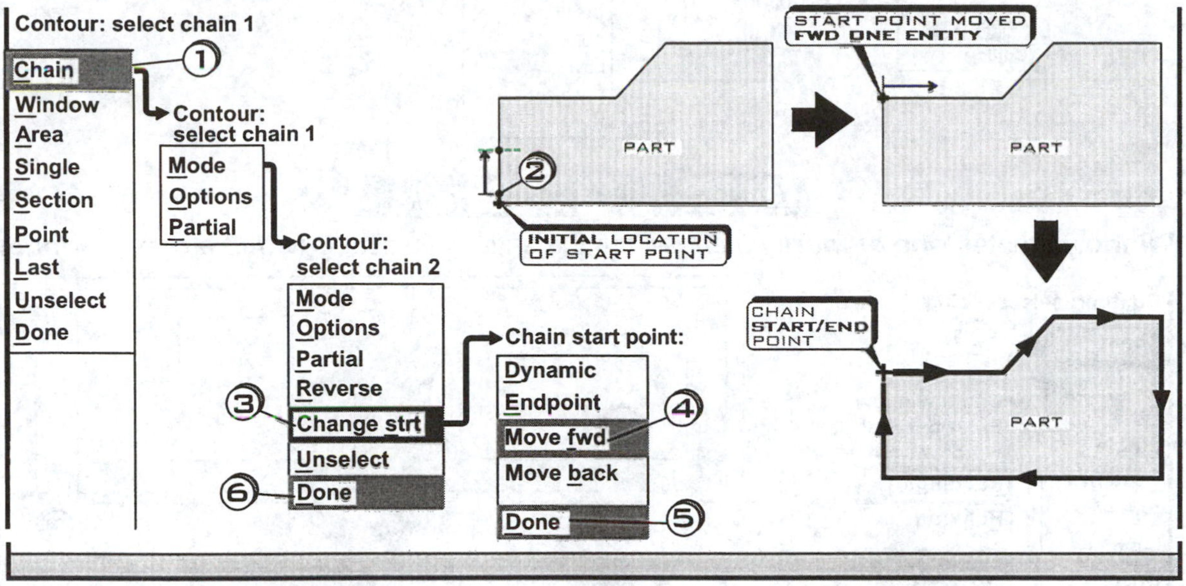

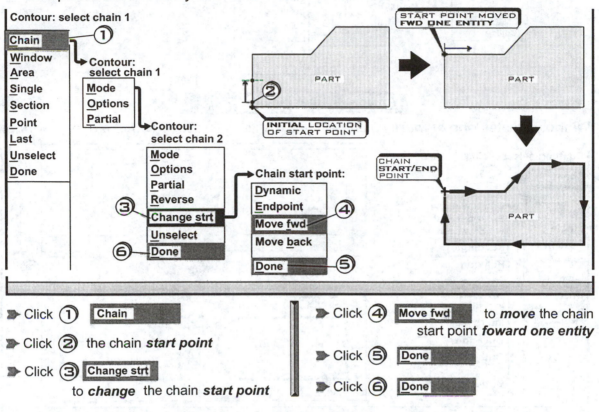

- Click ① **Chain**
- Click ② the chain ***start point***
- Click ③ **Change strt** to ***change*** the chain ***start point***
- Click ④ **Move fwd** to ***move*** the chain start point ***foward one entity***
- Click ⑤ **Done**
- Click ⑥ **Done**

Chain **Unselect**

Unselect cancels the last chain segment

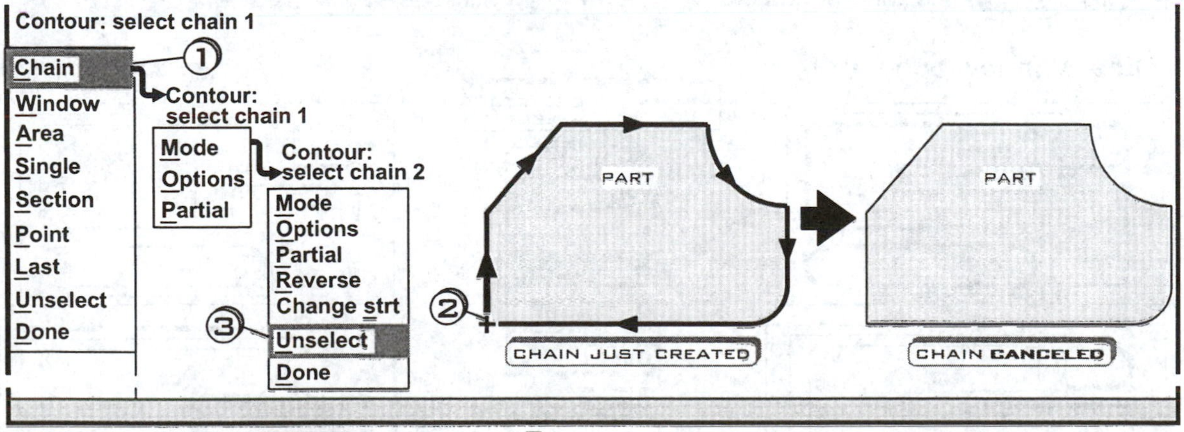

- Click ① **Chain**
- Click ② the chain ***start point***
- Click ③ **Unselect** to ***cancel the chain***

Creating Contours

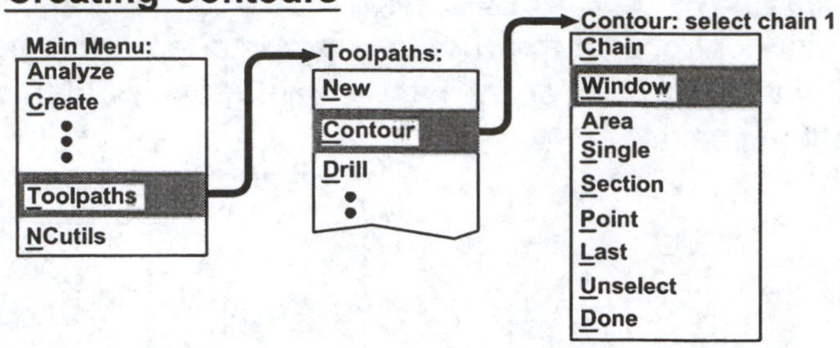

Window | Inside

Window creates ***one or more chaines*** from the entities enclosed ***within*** a clicked ***window.***

Contour: select chain 1
Chain ①
Window → Chain window: Enter the first window corner
Area
Single
Section
Point
Last
Unselect
Done

Rectangle +
Polygon
Inside +
In + intr
Intersect
Out + intr
Outside
Options

② ③ ④
PART
CHAIN 1
CHAIN 2

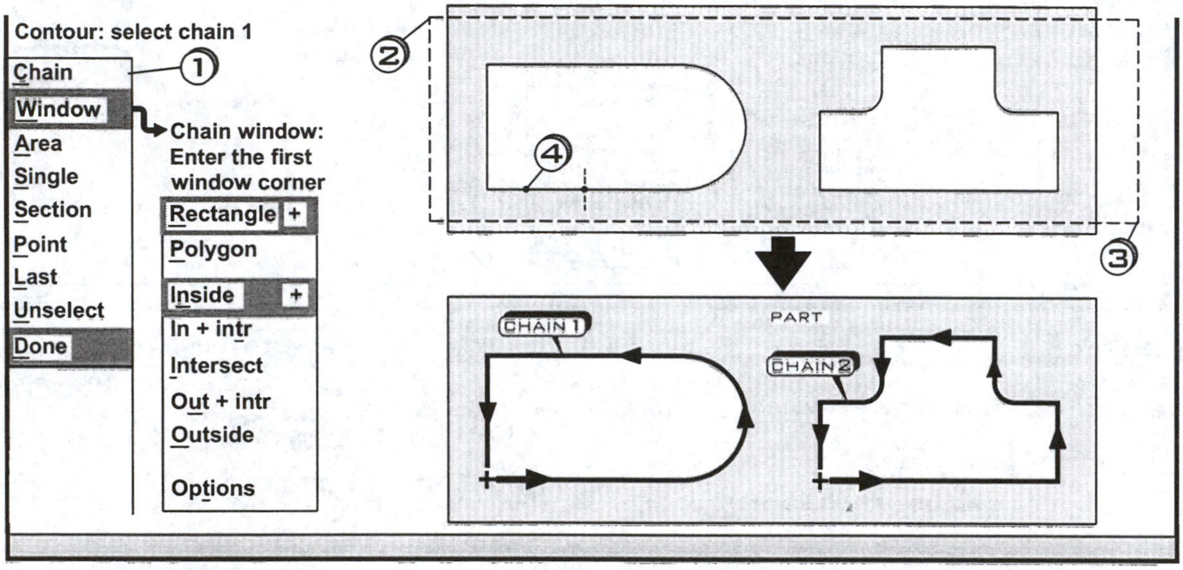

➤ Click ① Window

Note: **Rectangle +**; **Inside +** are set as *default*

➤ Click ② ③ the window corners

Enter search point

➤ Click ④ the point where chaining begins

➤ Click ⑤ Done

Other Window options:

= OBJECTS SELECTED

POLYGON WINDOW

Rectangle +
Polygon
Inside +
In + intr
Intersect
Out + intr
Outside
Options

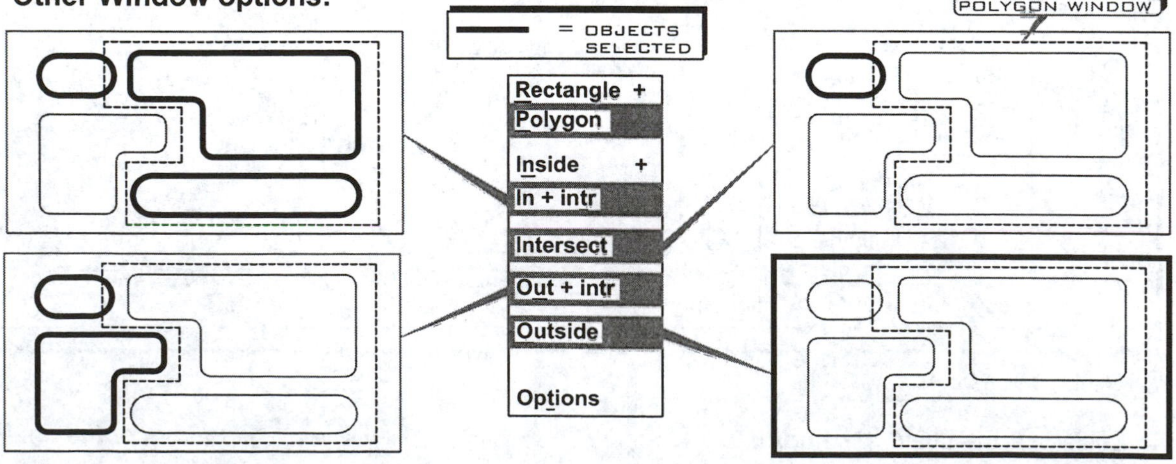

Creating Contours

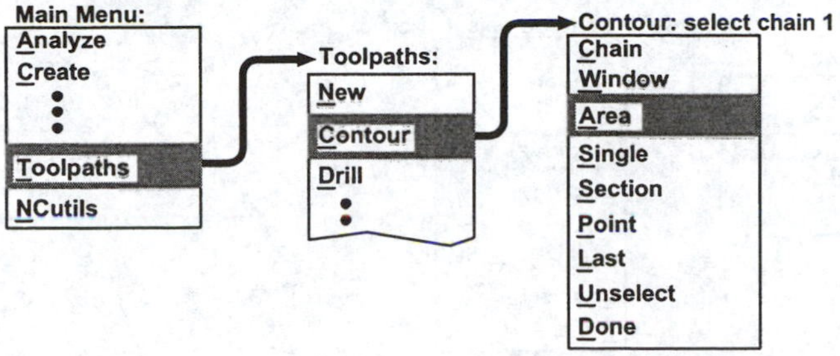

Area

Area creates ***one or more chains*** from the ***boundaries in the area*** that ***surround a point clicked.***

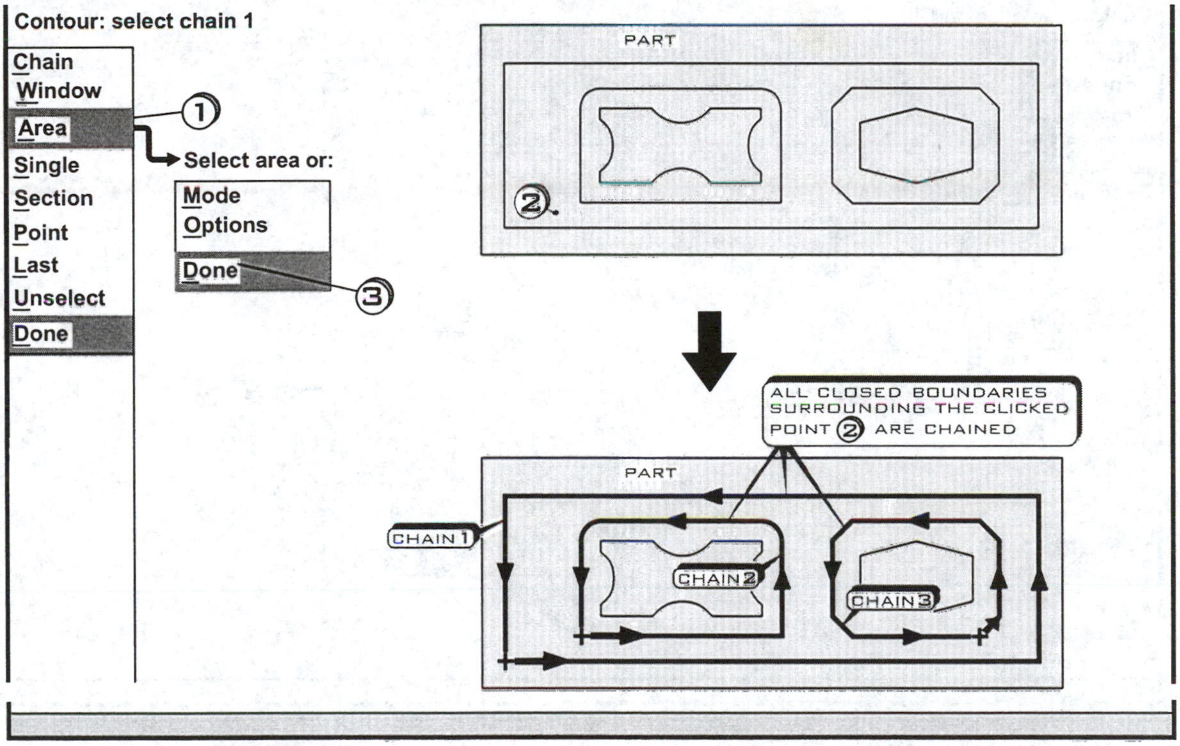

- Click ① **Area**
- Click ② the location of the point

  Mastercam will search for ***all the boundaries*** in the ***area surrounding the point*** and ***automatically*** create the ***corresponding chains***.
- Click ③ **Done**

Creating Contours

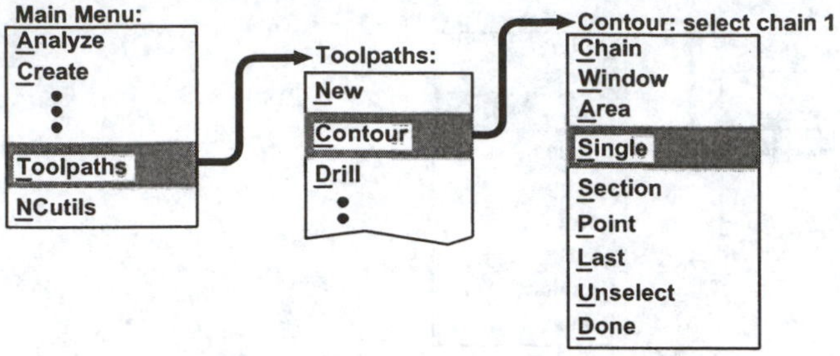

Single

Single creates a ***chain*** composed of a ***single entitiy only***

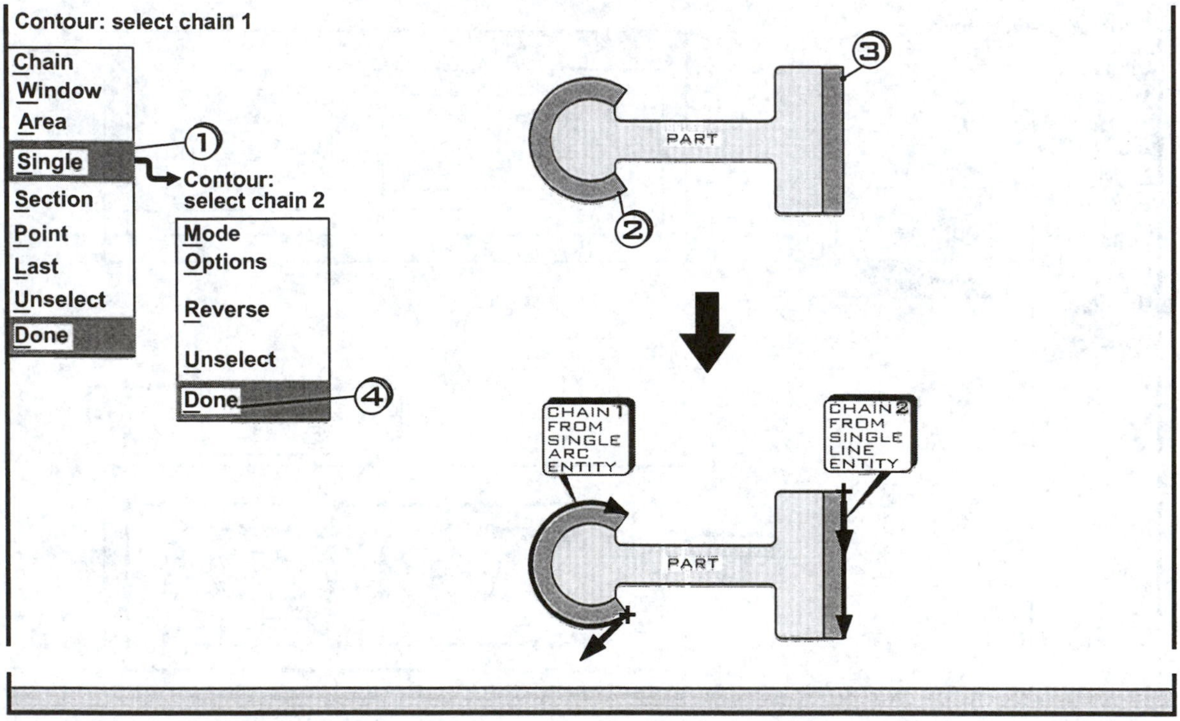

- Click ① **Single**
- Click ② ③ the location of the ***start point*** of each ***single entity*** chain
- Click ④ **Done**

Creating Contours

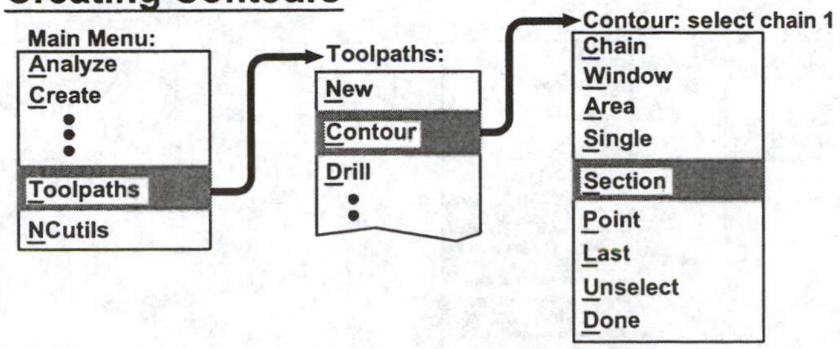

Section

Section creates a chain composed of a section of entities. *Mastercam* determines the section **begins and ends** at the ***intersection of two consecutive line entities.*** The default chaining stop angle for the line entities is 30°

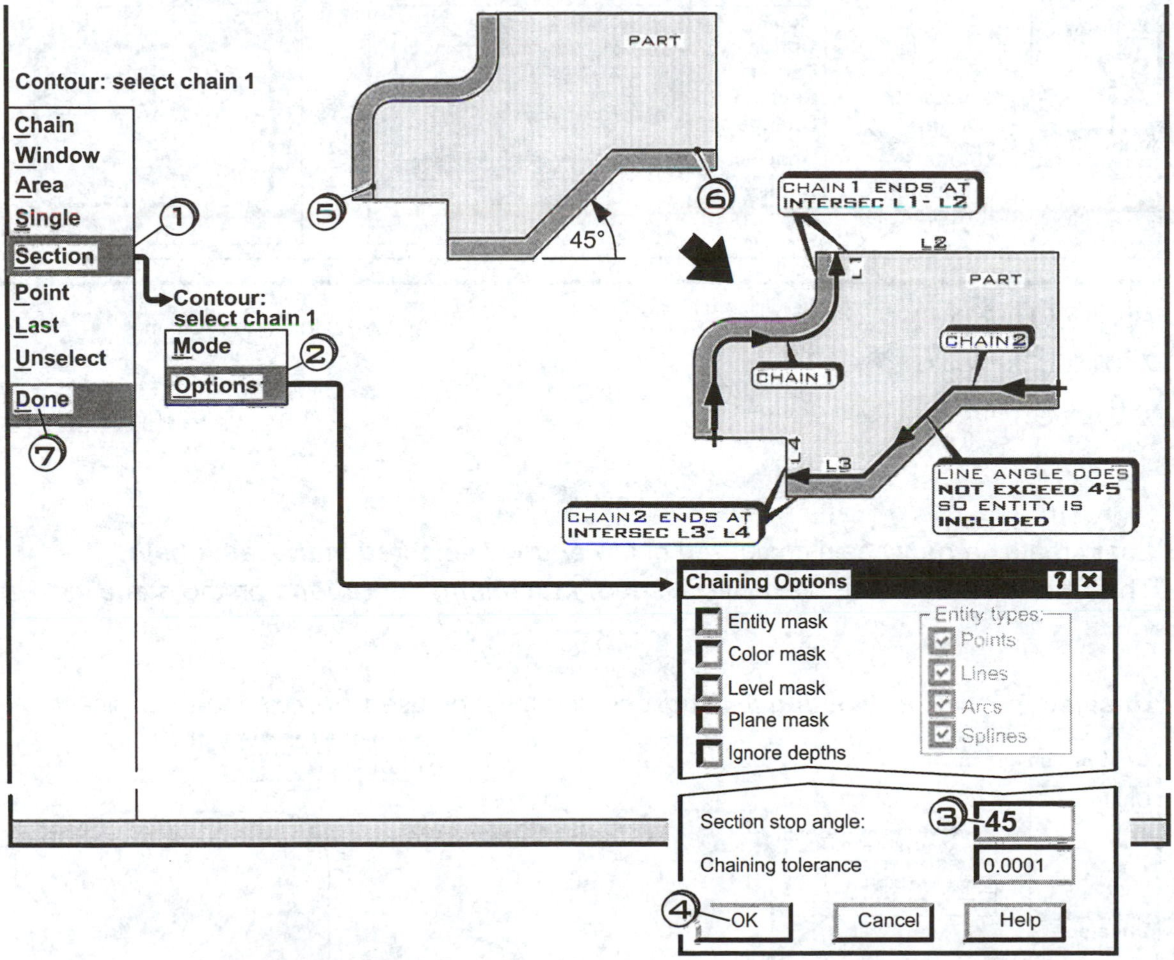

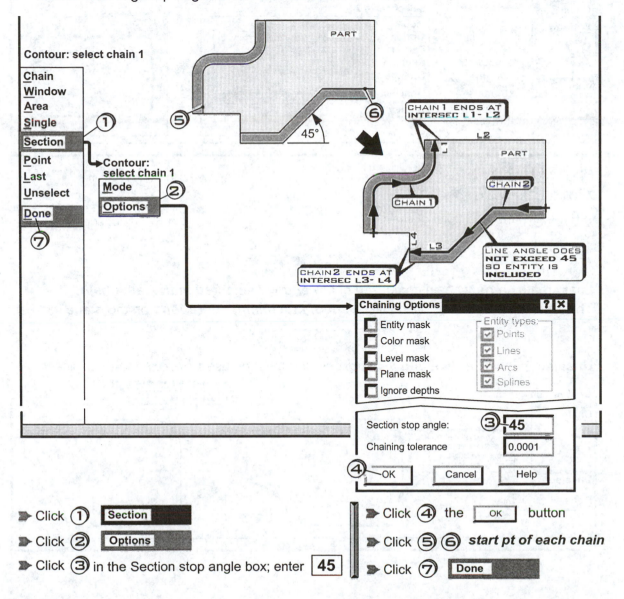

- Click ① **Section**
- Click ② **Options**
- Click ③ in the Section stop angle box; enter **45**
- Click ④ the **OK** button
- Click ⑤ ⑥ ***start pt of each chain***
- Click ⑦ **Done**

Creating Contours

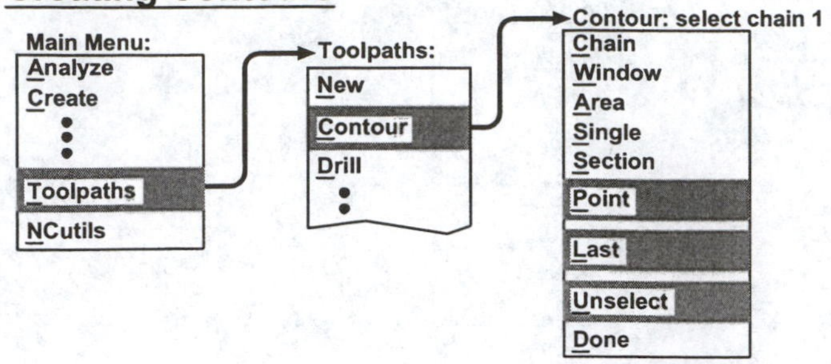

Point

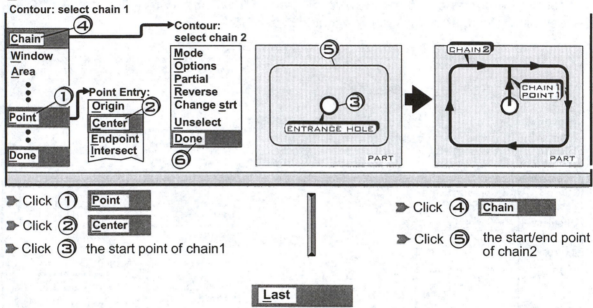

Point creates a ***chain*** consisting of a point location

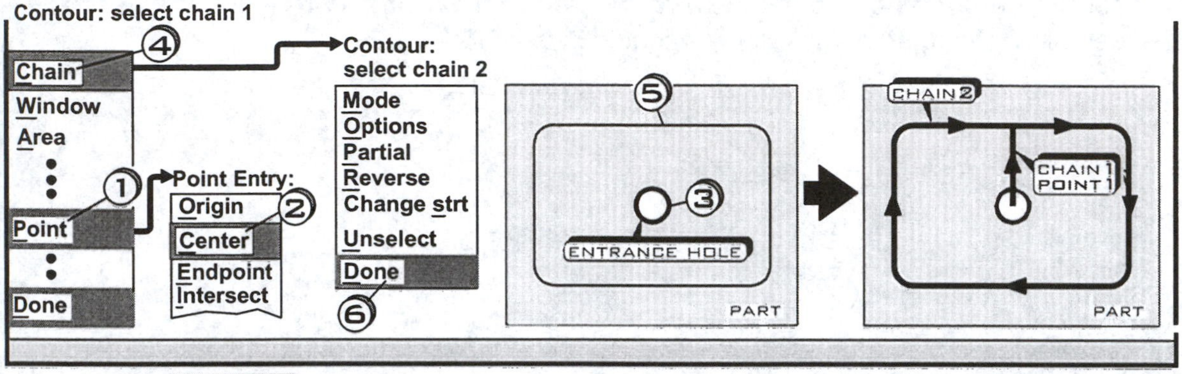

- Click ① **Point**
- Click ② **Center**
- Click ③ the start point of chain1
- Click ④ **Chain**
- Click ⑤ the start/end point of chain2

Last

Last creates a ***new*** chain composed of the ***entities selected in the last chain***
This allows the operator to execute a ***set of machining operations*** on the ***same*** tool path.

Unselect

Unselect cancels a ***chaining in progress*** and must be used ***before*** **Done** is clicked

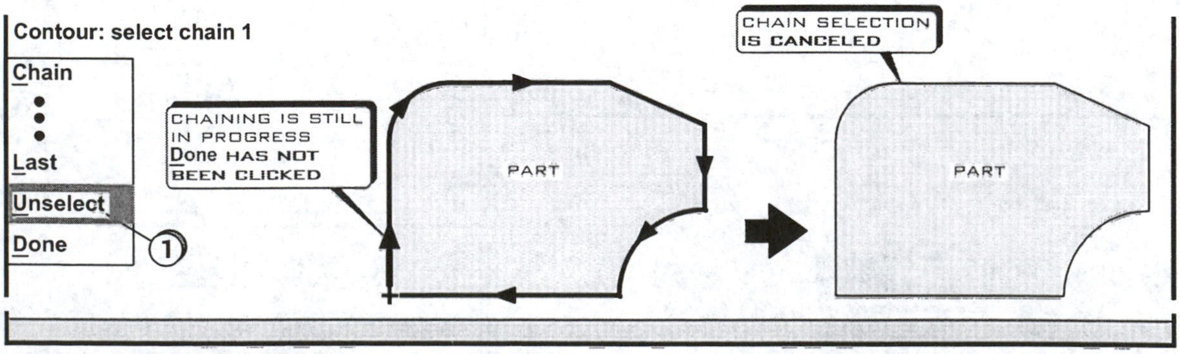

- Click ① **Unselect**

6-3 Common Problems Encountered in 2D Chaining

Below are listed the types of problems most commonly encounteted in attempting to chain 2D contours.

Problem	What's Wrong	Remedy	Result
Chaining stops at first entity of the boundary 	All entities in the boundary are ***not connected end to end***	Use the **Trim** function with radius set to **0** to trim the lines together	Entire boundary is chained 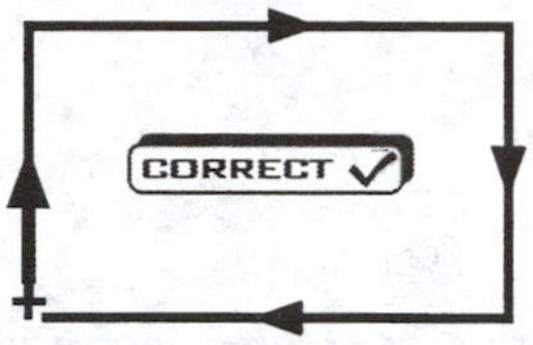
Chaining stops at first entity of the boundary 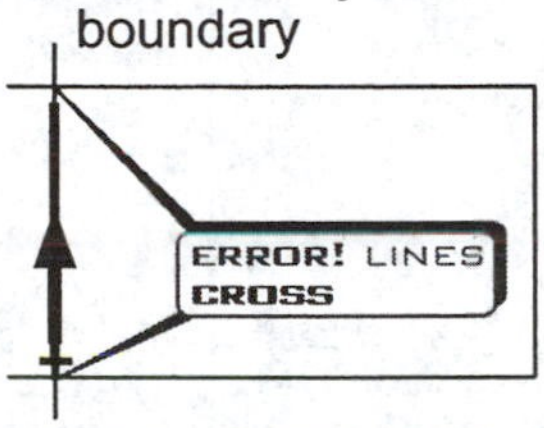	All entities in the boundary are ***not connected end to end. Entities that cross will cause chaining to stop***	Use the **Trim** function with radius set to **0** to trim the lines together	Entire boundary is chained 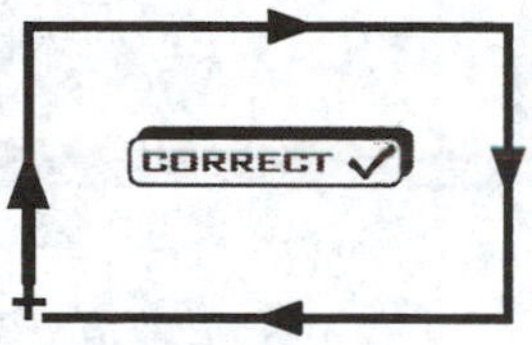
Chaining stops at first entity of the boundary 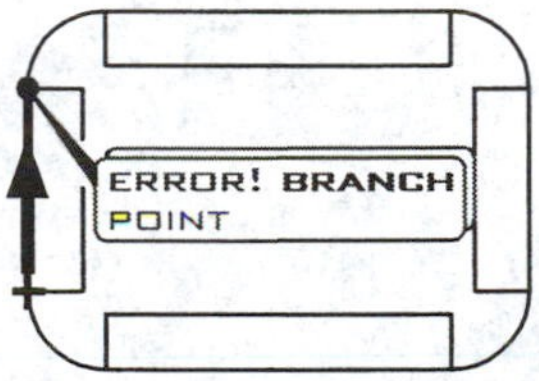	Branch points where ***more than two entities intersect will cause chaining to stop***	Use the **Chain** function and select **Partial** The operator must now click ***each entity that forms the desired boundary.***	Desired boundary is chained 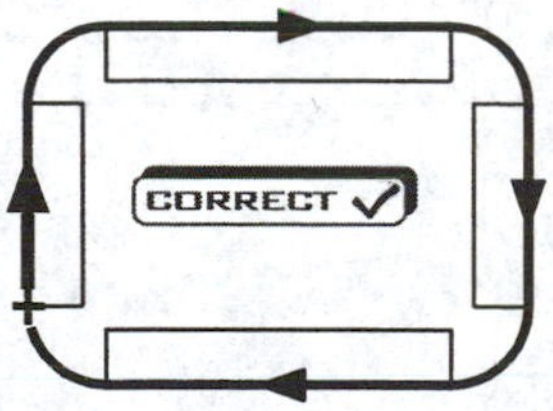
Chaining stops at a point on an entity 	***Duplicate*** entities ***on top of each other*** will cause ***erratic chaining***	Use the **Delete** function and select **Duplicate** and **Entities**	Desired boundary is chained 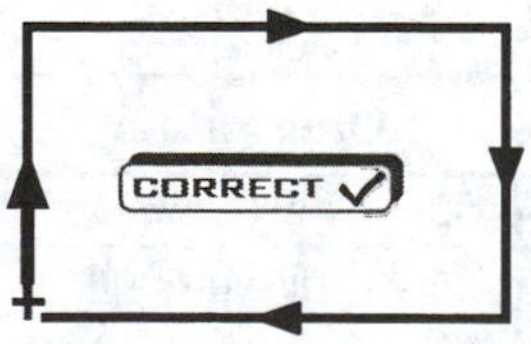

6-4 Preparing a Part for 2D Profile Machining

To prepare a part for profile machining the operator must first create the contour by an appropriate chaining method and then assign the necessary cutting tool.

EXAMPLE 6-1

Prepare the part given in Example 5-2. for outside boundary profile machining

- Create a machining contour by chaining
- Assign the contouring tool as listed in PROCESS PLAN 6-1

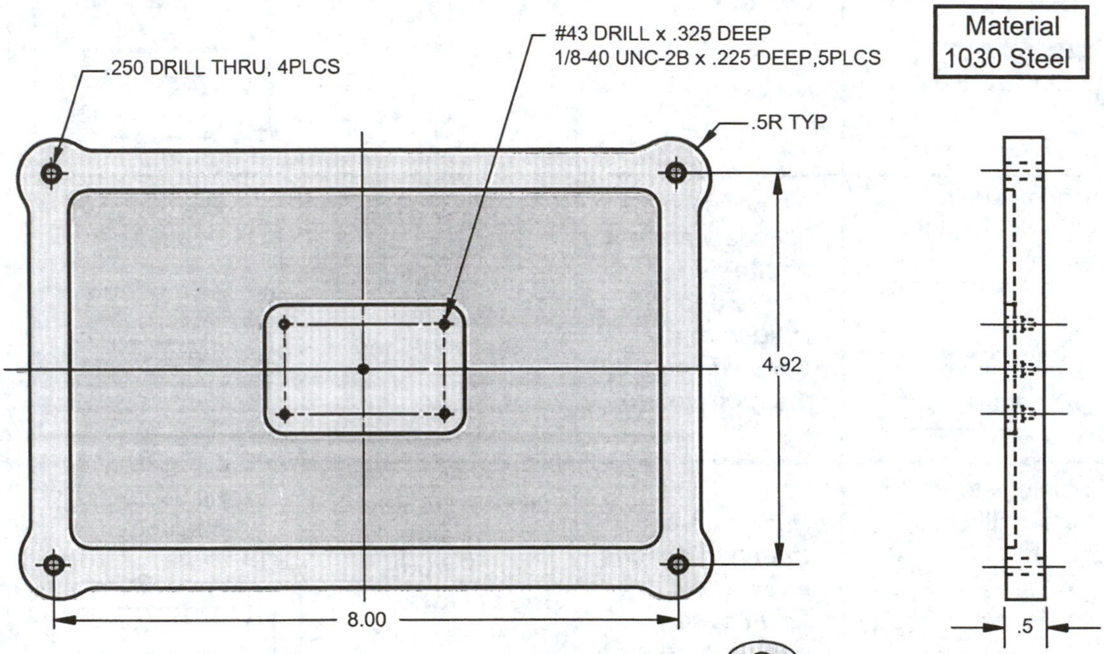

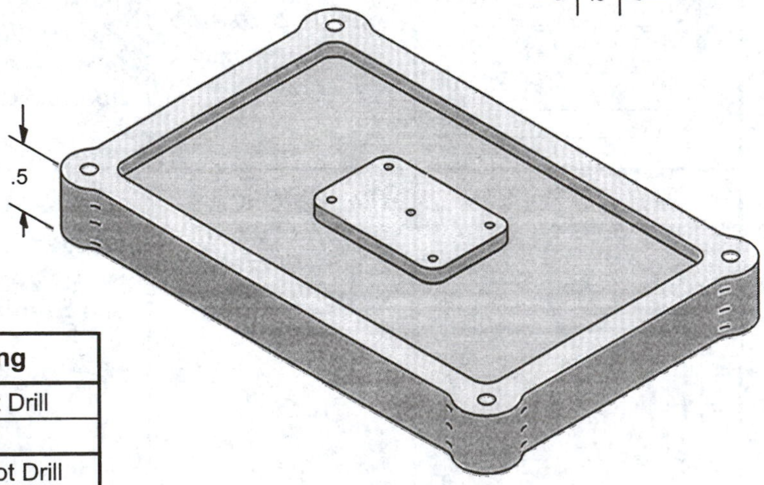

Figure 6-1

PROCESS PLAN 6-1

Operation	Tooling
Spot Drill x .166 Deep	1/8 Spot Drill
Peck Drill Through	1/4 Drill
Spot Drill x .125 Deep	1/32 Spot Drill
Peck Drill x .325 Deep	#43 Drill(.089in)
Tap x .225 Deep	1/8-40 UNC Tap
Profile x .5 Deep leave .01 for finish cuts in XY and Z	**1/4-End Mill**

A) Create a machining contour by chaining

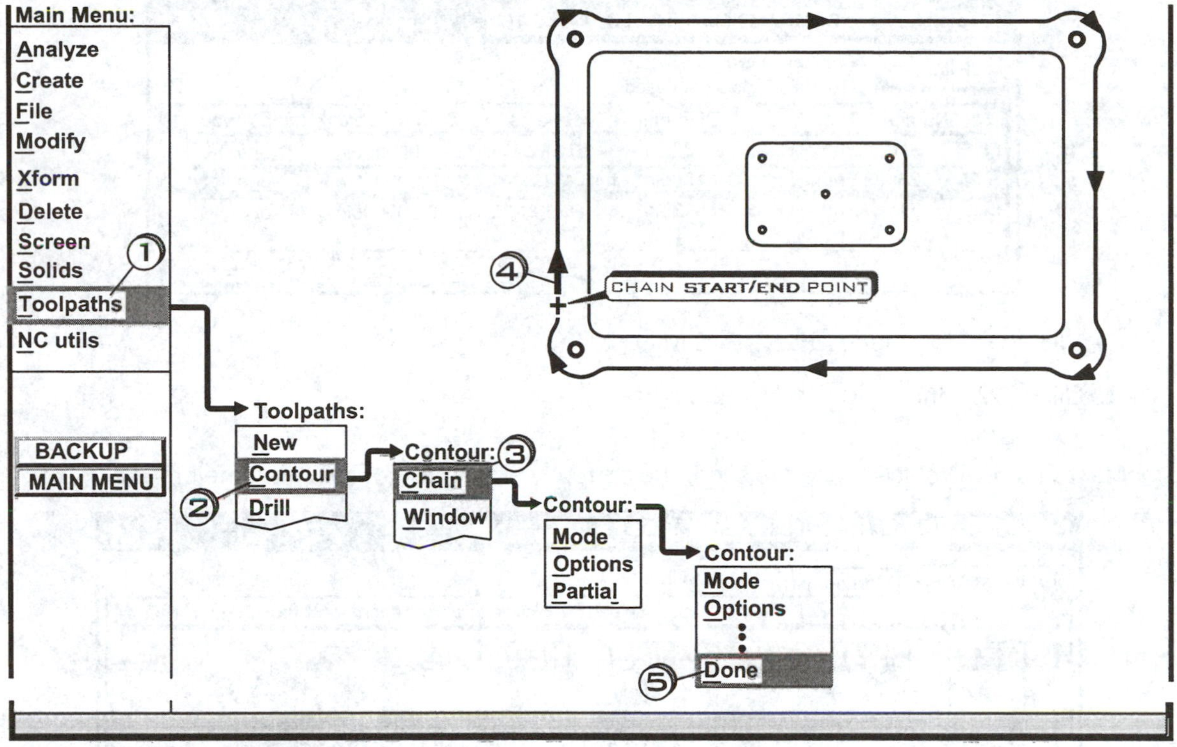

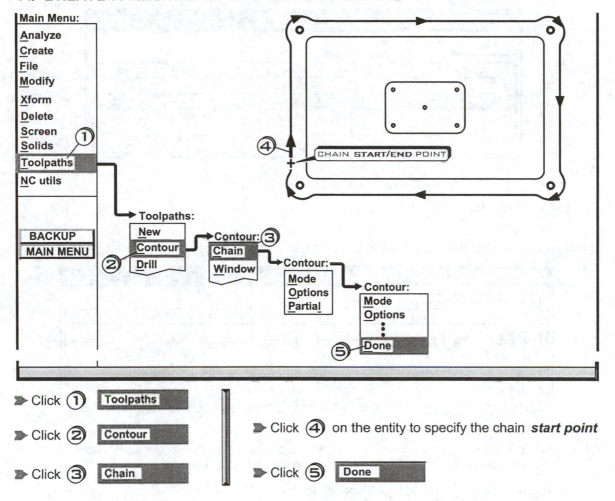

- Click ① Toolpaths
- Click ② Contour
- Click ③ Chain
- Click ④ on the entity to specify the chain ***start point***
- Click ⑤ Done

B) Obtain the needed 1/4 End Mill tool

- ***Right*** Click move the mouse cursor *down* and Click ⑥ Get tool from library

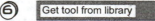

Select the 1/4in Dia End mill tool from the **TOOLS** tool library

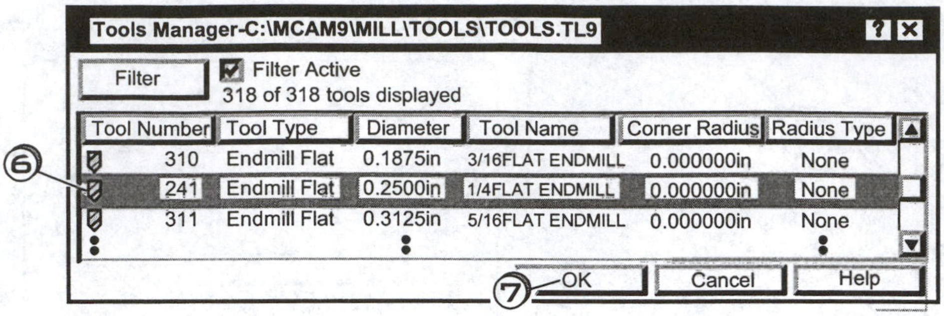

➤ Click ⑥ on the 1/4 FLAT ENDMILL

➤ Click ⑦ the [OK] button

Mastercam will place these tools into the currently active Contour 2D Parameters dialog box

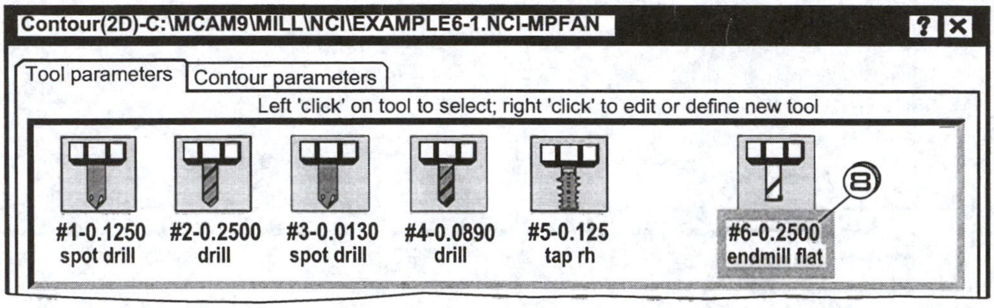

C) DIRECT *Mastercam* TO ASSIGN THE TOOL SPEED/FEED BASED ON THE STOCK MATERIAL SELECTED IN **Job Setup** IN **Example 5-2.**

➤ ***Right*** Click ⑧ on the .2500 endmill flat tool

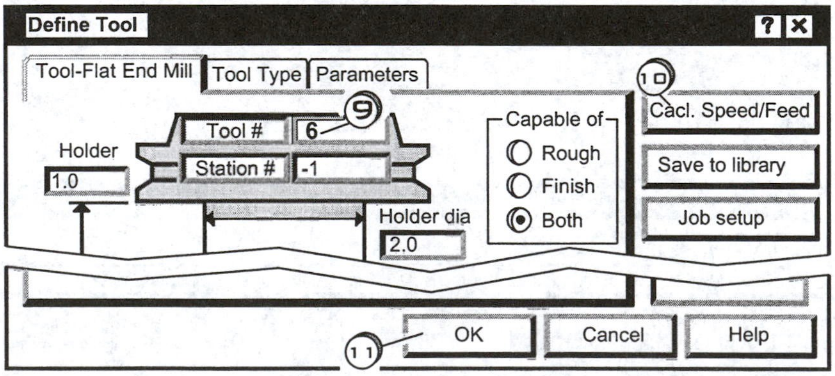

➤ Click ⑨ in the Tool# box and enter the tool number for the operation **6**

➤ Click ⑩ the [Cal. Speed/Feed] key

➤ Click ⑪ the [OK] key

6-5 Specifying Contouring Parameters in the Contour(2D) Module

The Contour(2D) module uses the *chained contour* and other key *contour machining parameters inputted by the operator* to generate the required contour tool paths.

A description of the *contouring parameters* and their effect on tool paths is considered in this section.

EXAMPLE 6-2

Direct *Mastercam* to display the **Contour parameters** dialog box in Contour 2D module for the profile machining described in **EXAMPLE 6-1**

➤ Click (1 2) the Contour parameters tab

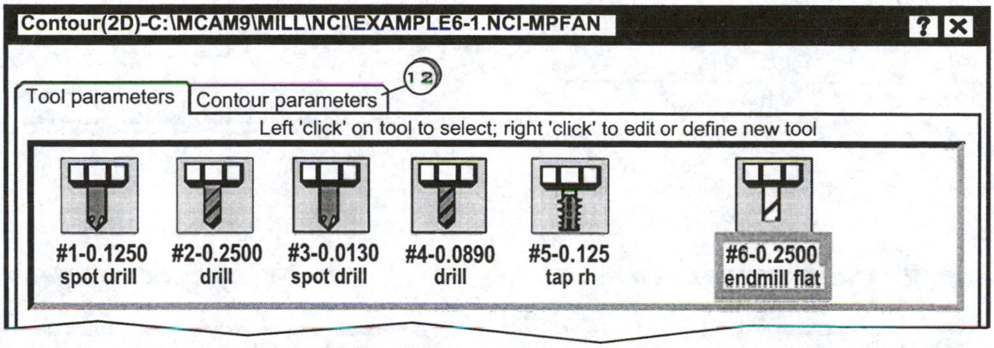

The system will display the *contouring parameters* dialog box as shown in p6-16

Z- DEPTH PARAMETERS

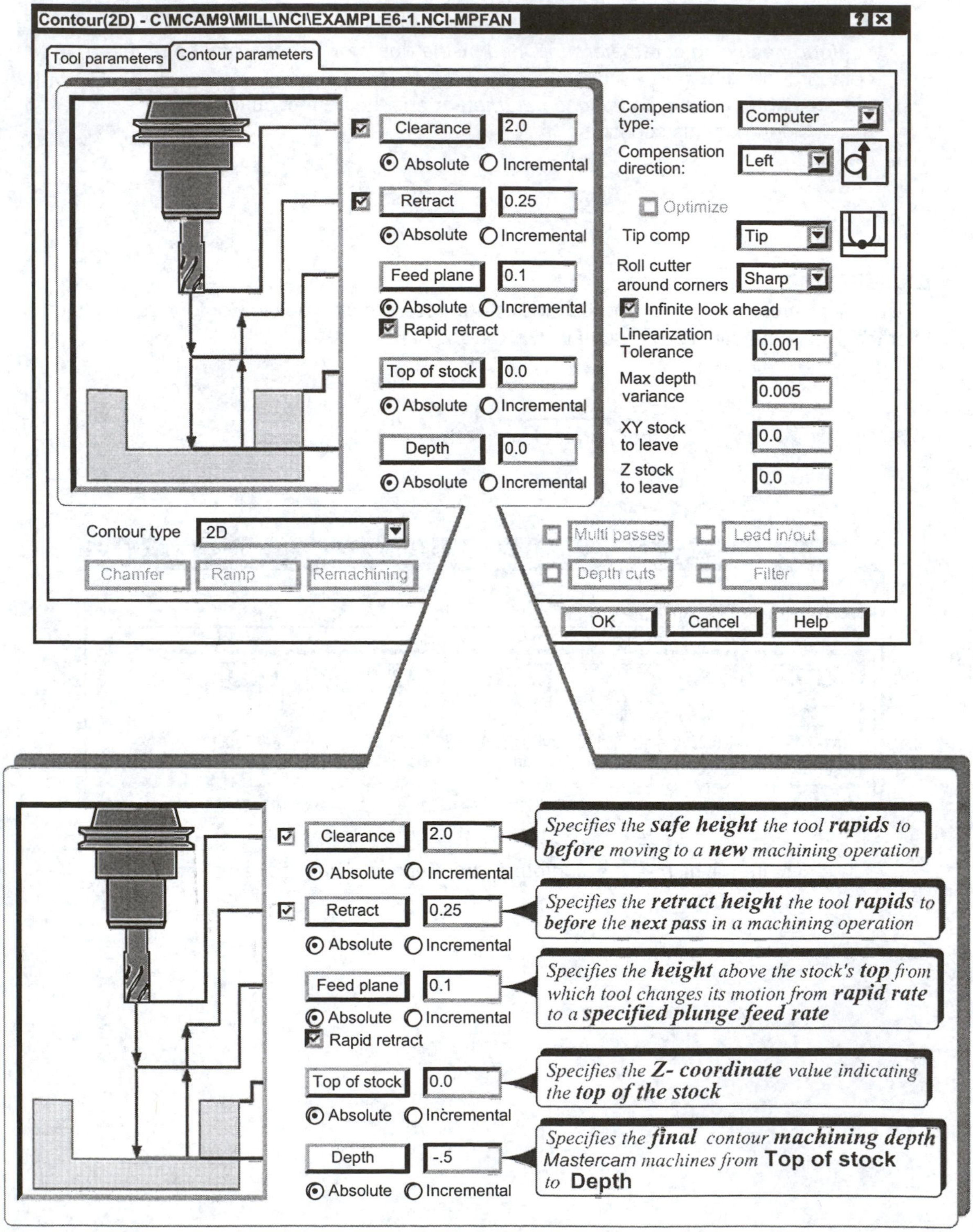
Contour(2D) - C\MCAM9\MILL\NCI\EXAMPLE6-1.NCI-MPFAN
Tool parameters
Contour parameters
Clearance
2.0
Absolute
Incremental
Retract
0.25
Feed plane
0.1
Rapid retract
Top of stock
0.0
Depth
Compensation type:
Computer
Compensation direction:
Left
Optimize
Tip comp
Tip
Roll cutter around corners
Sharp
Infinite look ahead
Linearization Tolerance
0.001
Max depth variance
0.005
XY stock to leave
0.0
Z stock to leave
0.0
Contour type
2D
Chamfer
Ramp
Remachining
Multi passes
Lead in/out
Depth cuts
Filter
OK
Cancel
Help
-.5
Specifies the safe height the tool rapids to before moving to a new machining operation
Specifies the retract height the tool rapids to before the next pass in a machining operation
Specifies the height above the stock's top from which tool changes its motion from rapid rate to a specified plunge feed rate
Specifies the Z- coordinate value indicating the top of the stock
Specifies the final contour machining depth Mastercam machines from Top of stock to Depth

DEPTH CUTS

The depth cuts parameters allow the operator to control the ***number*** and ***depth*** of ***roughing*** and ***finishing*** passes in the Z direction

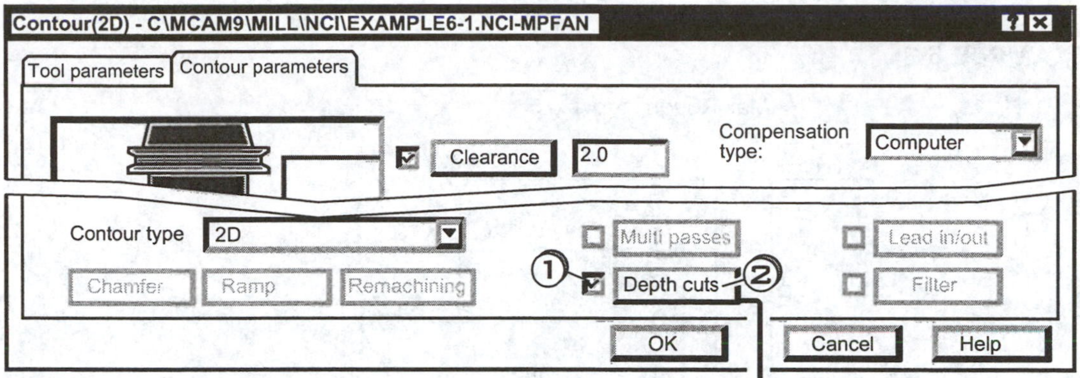

- Click ① the check ***on***
- Click ② the Depth cuts button

*Specifies the **maximum allowable** cutting depth per pass in the Z direction* ──▶ Max rough step: 0.5

*Specifies the **number of finishing passes** to be made in the **Z direction*** ──▶ # Finish cuts: 1

*Specifies the cutting **depth per pass** for finishing cuts in the Z direction* ──▶ Finish step: 0.05

☑ *Causes the tool to **plunge directly** from its current Z depth to the **next pass Z depth*** ──▶ Keep tool down

☐ *Causes the tool to **rapid to the feed plane** before plunging to the next pass Z depth* ── Subprogram

Depth cut order:
⦿ By contour ○ By depth

☐ Tapered walls
Taper angle 0.0

⦿ Absolute ○ Incremental

OK Cancel Help

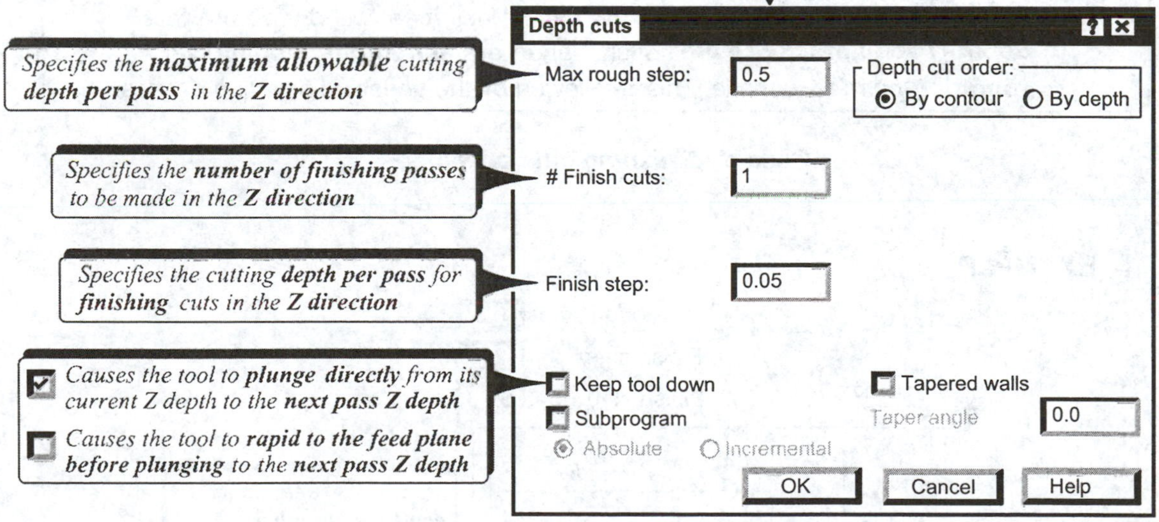

Case-a : ***No finishing cuts are specified***. In this case *Mastercam* will ***divide the total depth (Top of stock+ Depth)*** such that an ***equal number of roughing passes*** is made. Depth per pass ***will not exceed*** current Max rough step value.

Case-a : ***No finishing cuts specified***

EXAMPLE 6-3

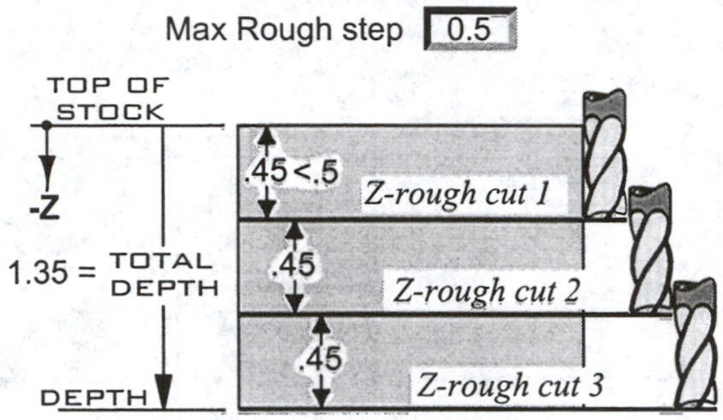

Figure 6-2

Case-b: ***Finishing cuts are specified.*** In this case *Mastercam* will ***divide the total depth (Top of stock+ Depth)*** such that an ***equal number*** of ***shallower roughing passes*** is made. This is followed by the specified finishing passes.

Case-b : ***Finishing cuts specified***

EXAMPLE 6-4

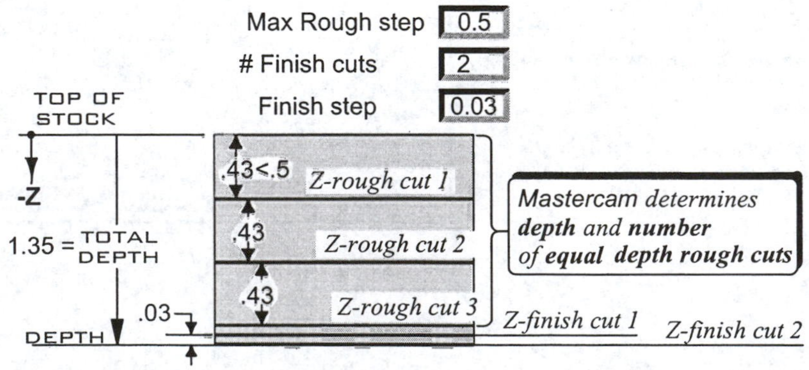

Figure 6-3

MULTI PASSES

The multi passes parameters allow the operator to control the ***number*** and ***stepover distances*** of ***roughing*** and ***finishing*** passes in the X and Y directions

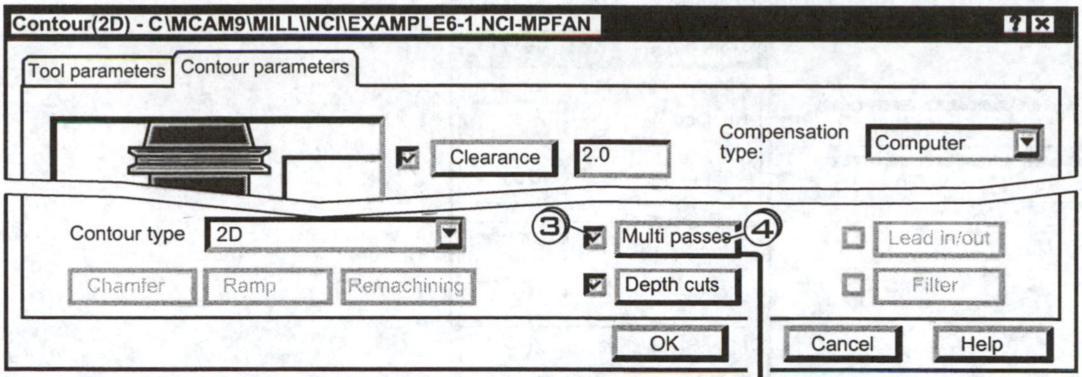

➤ Click ③ the check ***on*** ☑

➤ Click ④ the [Multi passes] button

Specifies the ***number of roughing passes*** *to be made in the* ***XY direction***

Specifies the ***stepover distance per pass*** *for* ***roughing cuts*** *in the* ***XY direction***

Specifies the ***number of finishing passes*** *to be made in the* ***XY direction***

Specifies the ***stepover distance per pass*** *for* ***finishing cuts*** *in the* ***XY direction***

◉ Final depth ***Stepover finish passes*** *are made at the* ***final Z-depth only***
◉ All depths ***Stepover finish passes*** *are be made at after* ***every Z-depth pass***

☑ *Tool* ***moves directly*** *from its current position to the* ***next pass XY stepover***
☐ *Tool* ***rapids to the feed plane before moving*** *to the* ***next pass XY stepover***

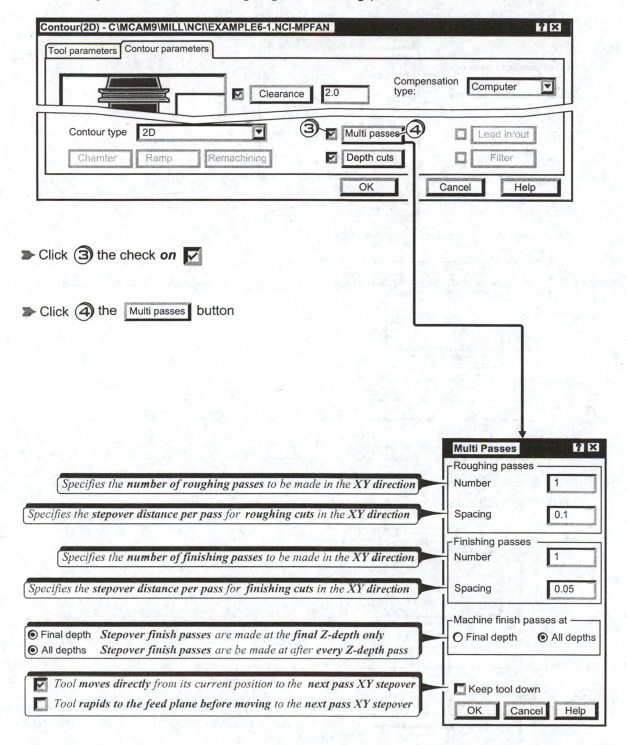

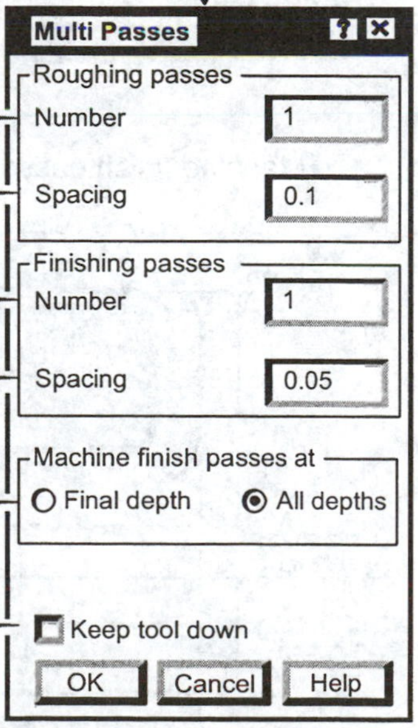

Determination of Spacing and Number of cuts for Roughing passes

Spacing between cuts ≈ (.60 to .75) x Tool Dia

Number of cuts ≈ XY Stock to remove/ Spacing

EXAMPLE 6-5 Tool Dia =.5in ; XY Stock to remove = .65in

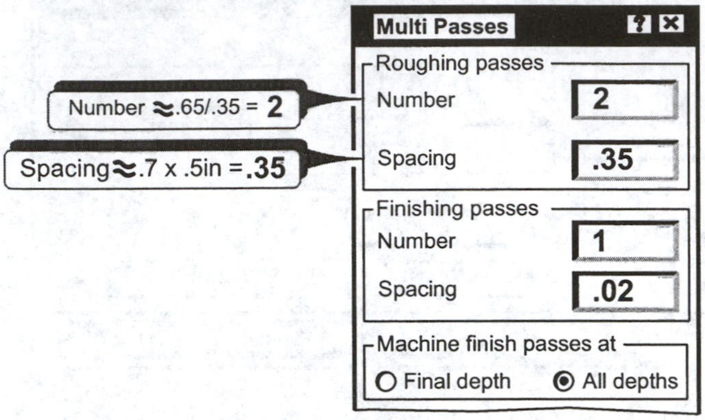

Machine finish passes at ⦿ **Final depth**

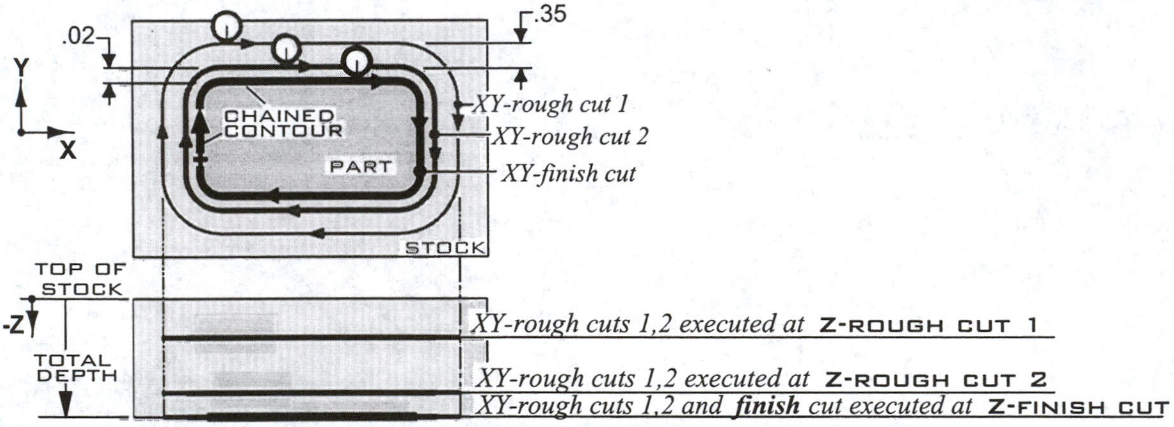

Machine finish passes at ⦿ **All depths**

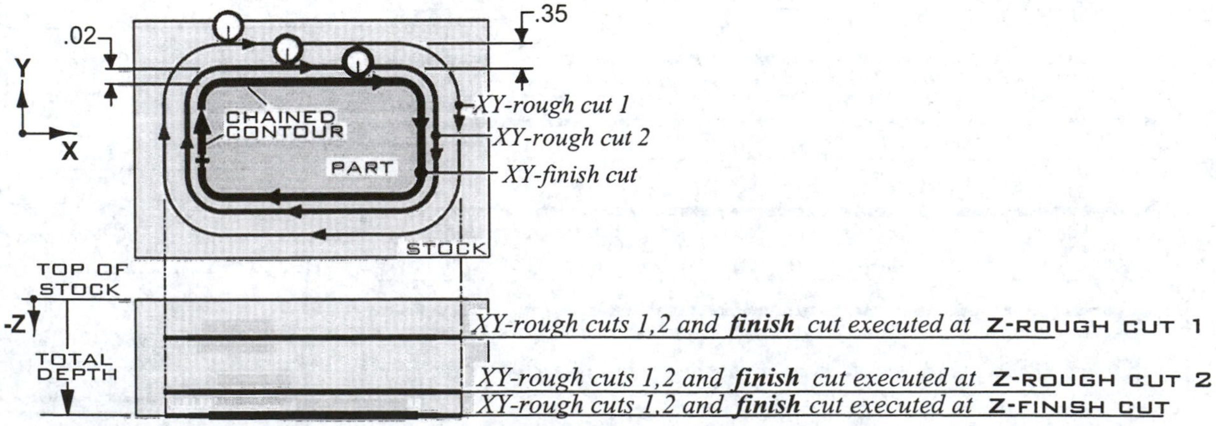

Figure 6-4

CUTTER COMPENSATION

The compensation parameters allow the operator to specify ***whether or not*** the cutter is positioned ***offset*** from the ***chained contour it follows***. When compensation is set to ***OFF*** the cutter ***center*** moves along the contour in the direction of chaining. When compensation is set to **ON** *Mastercam* ***automatically offsets*** the cutter by its ***radius*** as it moves along the contour in the direction of chaining.
The parameters also control the ***offset direction***(LEFT or RIGHT) with respect to the contour.

The tip compensation parameters specify whether the tool is to be moved by its ***tip*** or ***center*** point. This parameter only affects such tools as ball and bull end mills, etc.

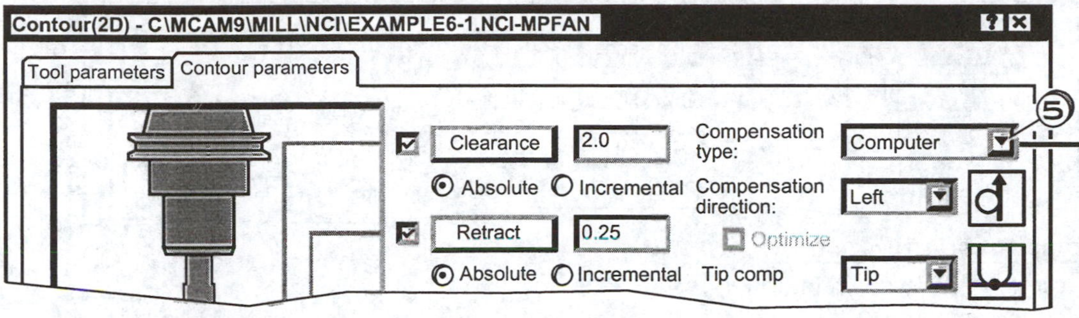

Compensation Type

➤ Click ⑤ the down button

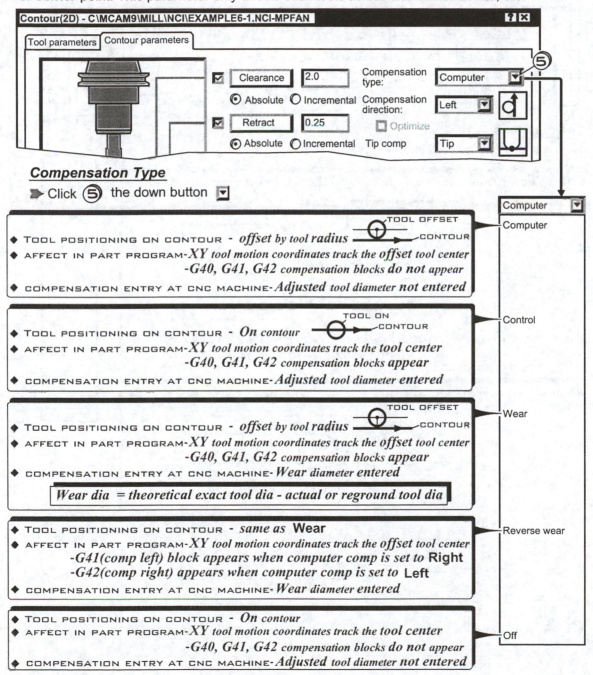

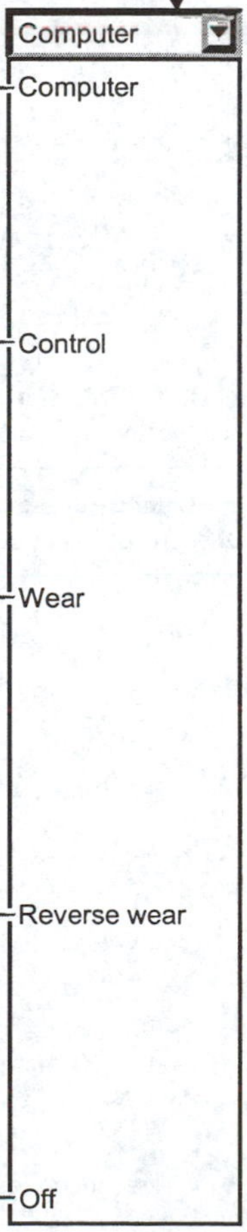

Computer

TOOL OFFSET — CONTOUR

- TOOL POSITIONING ON CONTOUR - *offset by tool radius*
- AFFECT IN PART PROGRAM-*XY tool motion coordinates track the offset tool center*
  -G40, G41, G42 compensation blocks do not appear
- COMPENSATION ENTRY AT CNC MACHINE-*Adjusted tool diameter not entered*

Control

TOOL ON CONTOUR

- TOOL POSITIONING ON CONTOUR - *On contour*
- AFFECT IN PART PROGRAM-*XY tool motion coordinates track the tool center*
  -G40, G41, G42 compensation blocks appear
- COMPENSATION ENTRY AT CNC MACHINE-*Adjusted tool diameter entered*

Wear

TOOL OFFSET — CONTOUR

- TOOL POSITIONING ON CONTOUR - *offset by tool radius*
- AFFECT IN PART PROGRAM-*XY tool motion coordinates track the offset tool center*
  -G40, G41, G42 compensation blocks appear
- COMPENSATION ENTRY AT CNC MACHINE- *Wear diameter entered*

Wear dia = theoretical exact tool dia - actual or reground tool dia

Reverse wear

- TOOL POSITIONING ON CONTOUR - *same as* **Wear**
- AFFECT IN PART PROGRAM-*XY tool motion coordinates track the offset tool center*
  -G41(comp left) block appears when computer comp is set to **Right**
  -G42(comp right) appears when computer comp is set to **Left**
- COMPENSATION ENTRY AT CNC MACHINE- *Wear diameter entered*

Off

- TOOL POSITIONING ON CONTOUR - *On contour*
- AFFECT IN PART PROGRAM-*XY tool motion coordinates track the tool center*
  -G40, G41, G42 compensation blocks do not appear
- COMPENSATION ENTRY AT CNC MACHINE-*Adjusted tool diameter not entered*

> Note: In many cases the best option is to choose **WEAR** . This allows the setup person to input the ***difference*** between the theoretical ***exact*** cutter ***diameter*** and the ***actual diameter*** into CNC machine control unit.

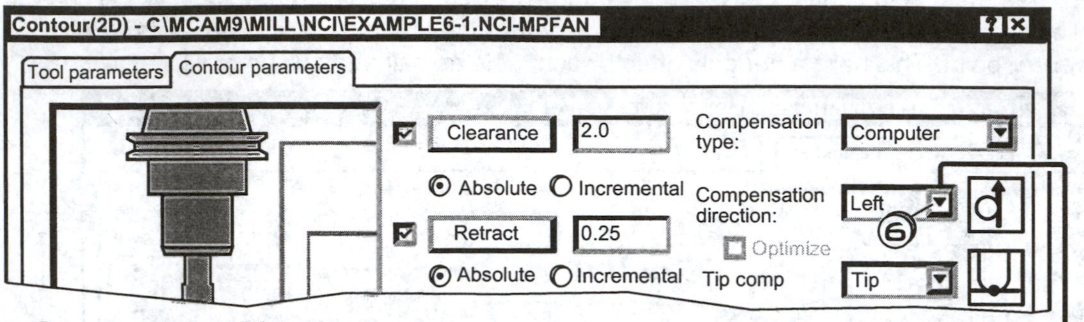

Compensation Direction

➡ Click ⑥ the down button

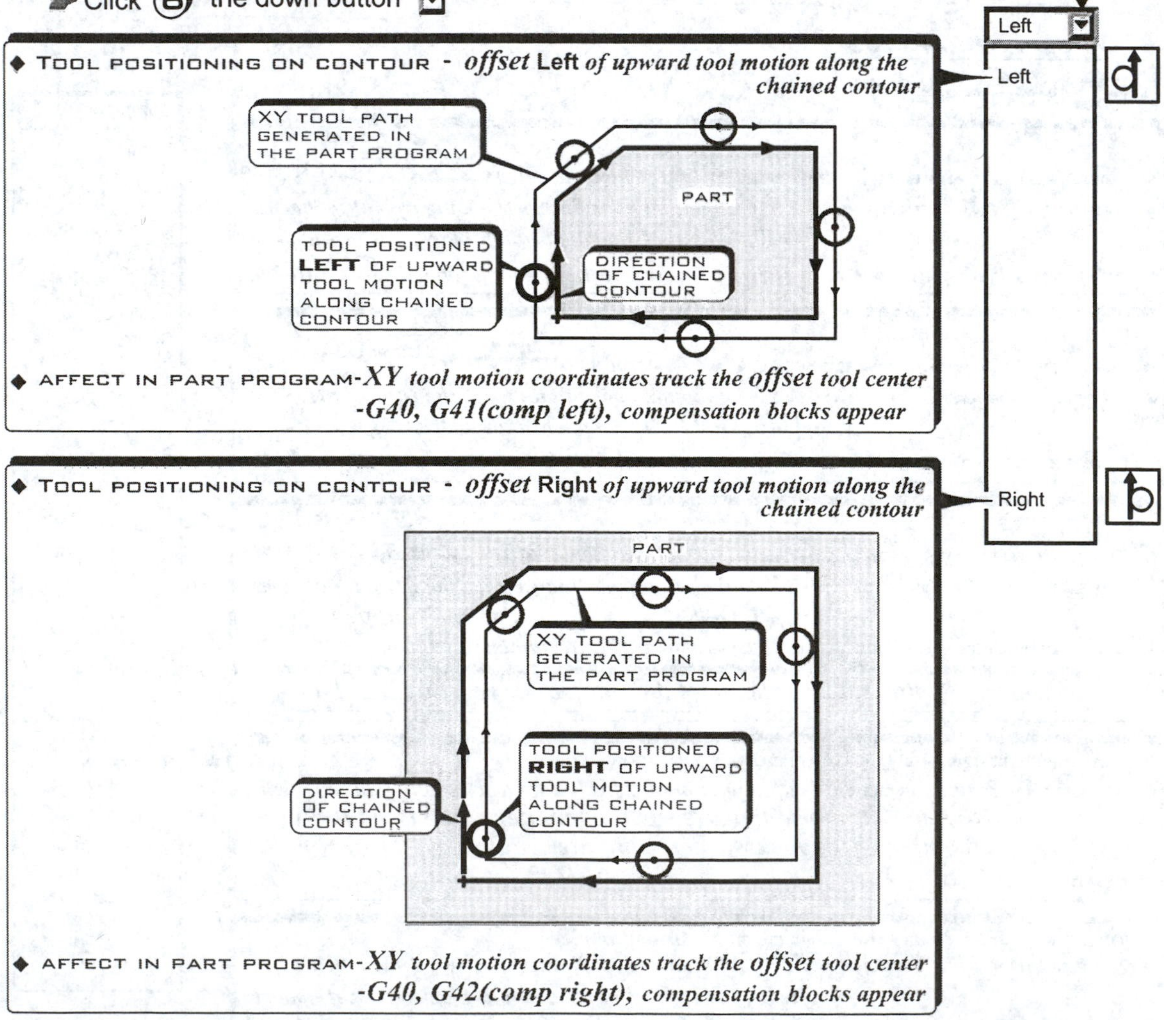

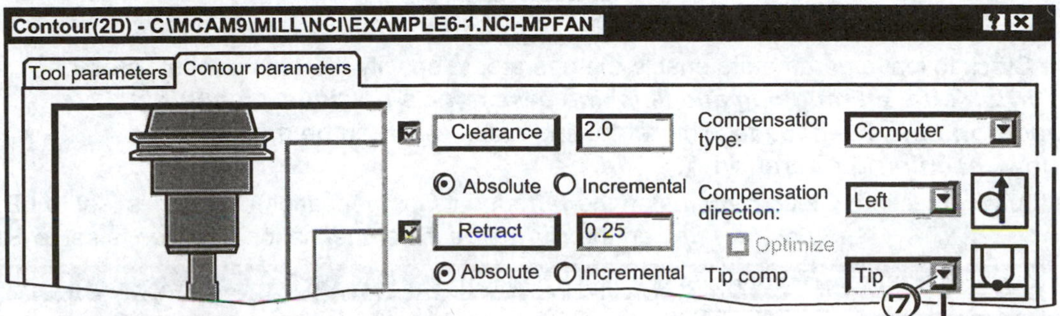

Tip Compensation

➤ Click (7) the down button

◆ AFFECT IN PART PROGRAM- *Tool machines to prescribed depth in part Z motion coordiantes track the tool tip*

TOP OF STOCK
TOTAL DEPTH=.5

PART

Z coordinate of tool **Mastercam** *enters in part program*

◆ AFFECT IN PART PROGRAM- *Tool machines to prescribed depth in part Z motion coordiantes track the tool center*

TOP OF STOCK
TOTAL DEPTH=.5

PART

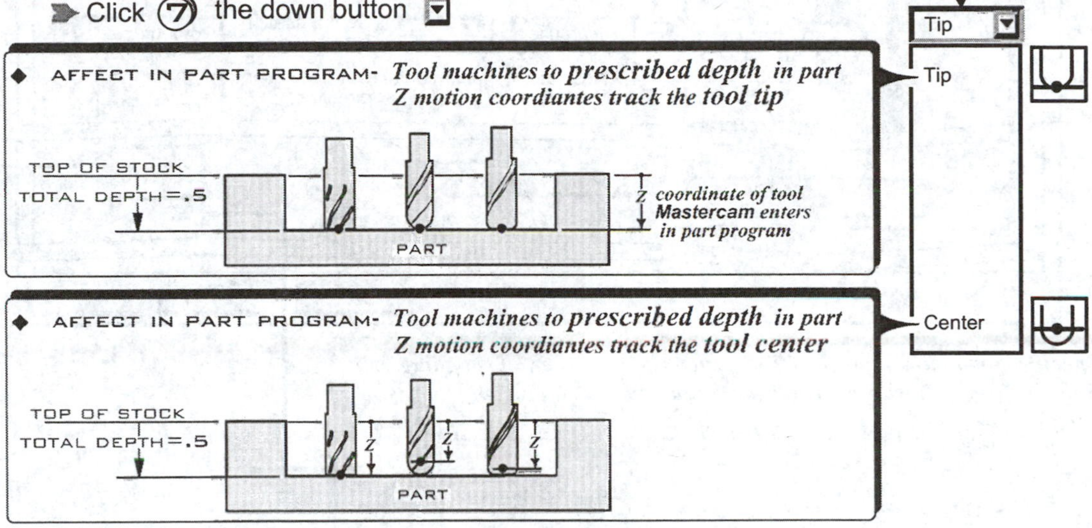

INFINITE LOOK AHEAD

When enabled, the Infinite look ahead parameter instructs Mastercam to search the ***entire chained contour*** for any ***self intersections*** that can cause ***tool gouging***. The determination is made based on the current ***offset distance*** and ***cutter compensation***. *Mastercam* ***automatically adjusts*** the ***tool path*** to ***prevent gouging.***

☑ *Mastercam adjusts tool path to avoid gouging*

STEP<TOOL RADIUS

☐ *Mastercam stops tool at point of gouging*

TOOL STOPPED AT GOUGING PT GOUGING

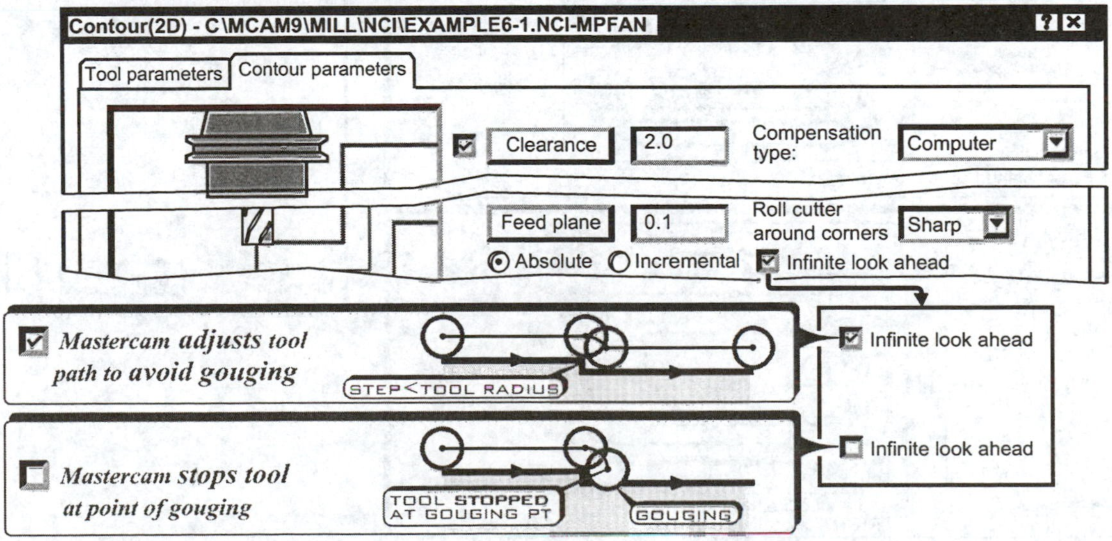

STOCK TO LEAVE

The Stock to leave parameters enable the operator to specify the ammount of ***stock to be left after all roughing and finishing pases for a particular machining operation have been made***. The remaining stock can then be machined via a ***new machining operation***.

In the XY direction: Stock ***remains*** on the ***left side*** if Compensation direction is set to **LEFT**
Stock ***remains*** on the ***right side*** if Compensation direction is set to **RIGHT**

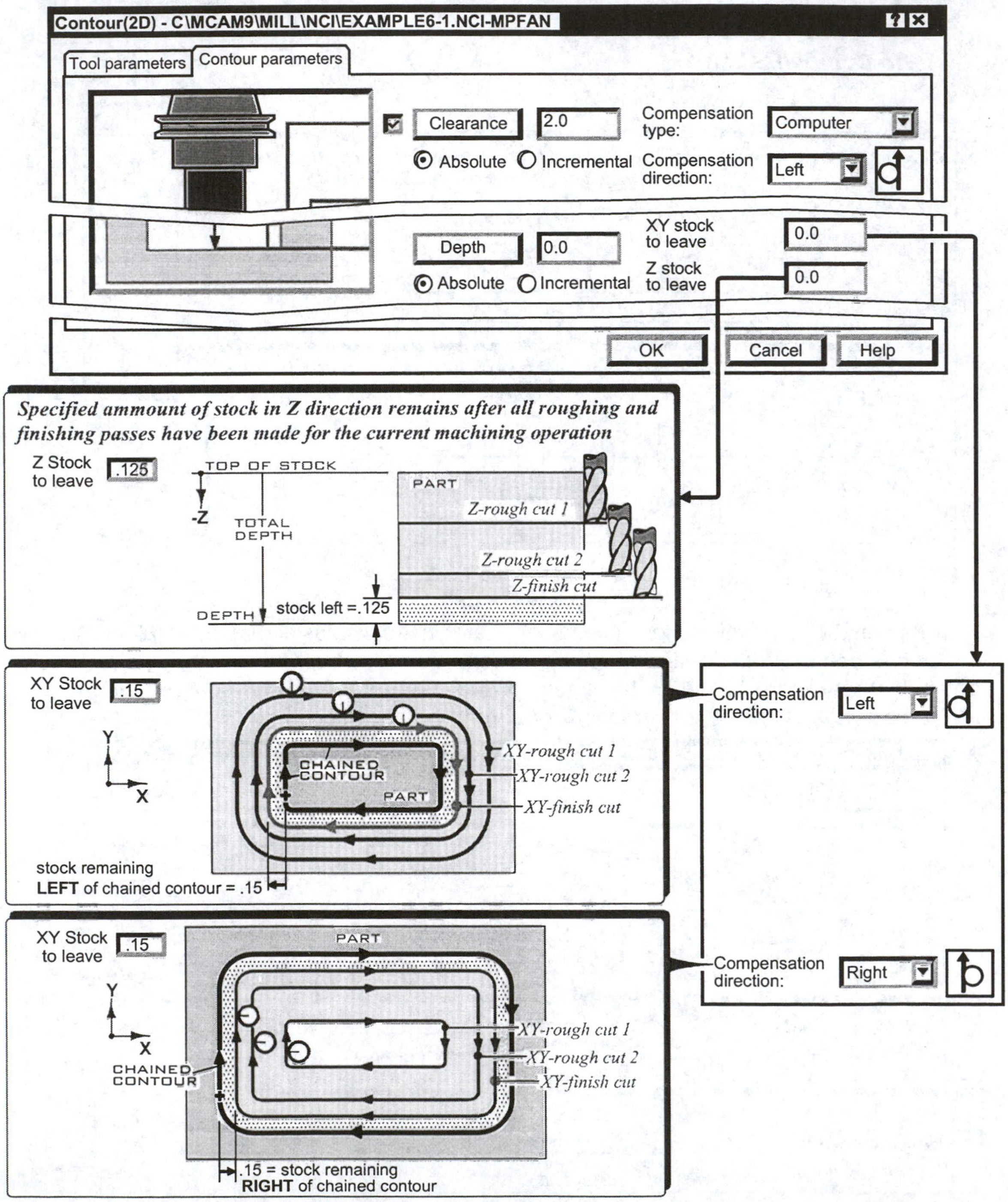

ROLL CUTTER AROUND CORNERS

The Roll cutter around corners parameters allows the operator to choose whether the cutter is to experience a ***sudden change*** in motion at a ***corner*** where two lines meet or have a ***smoother motion*** by following an ***arc path(rolling)*** around the corner.

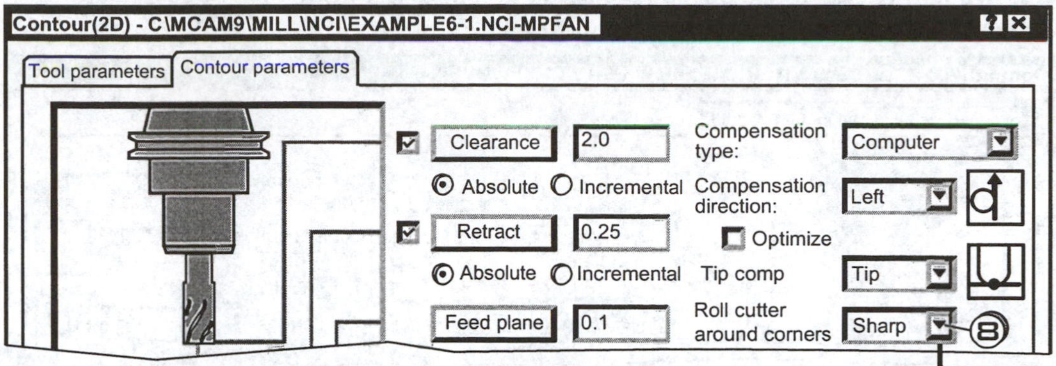

➤ Click ⑧ the down button

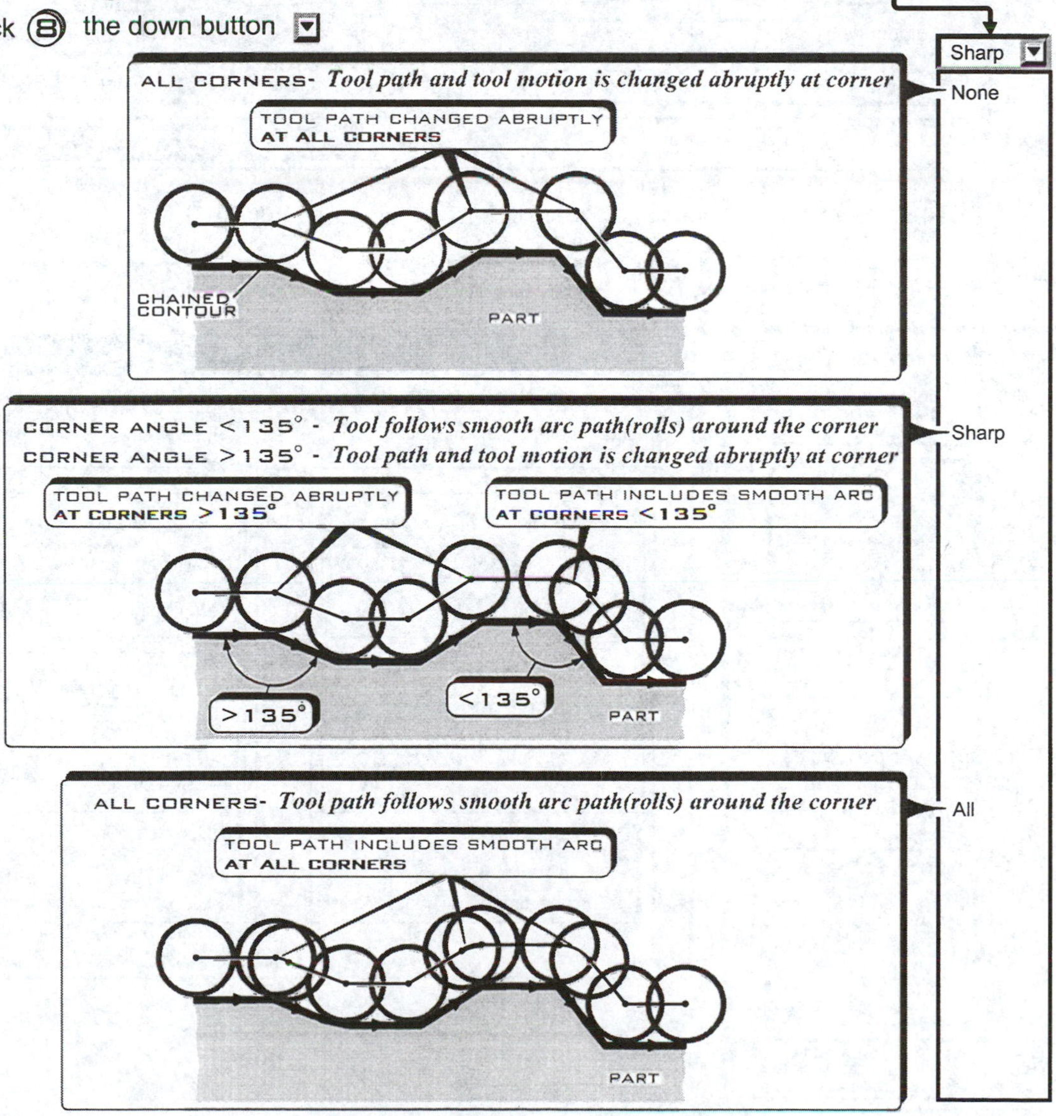

LEAD IN/OUT

The Lead in/out parameters enable the operator to control the ***tool's entry path into the part*** and after machining the specified contour its ***exit path out of part.*** *Mastercam* features several choices for in/out tool paths: ***straight lines, lines and arcs or arcs***. The lead in/out tool paths created by *Mastercam* are designed to insure the cutter moves ***smoothly into and out of the part***.

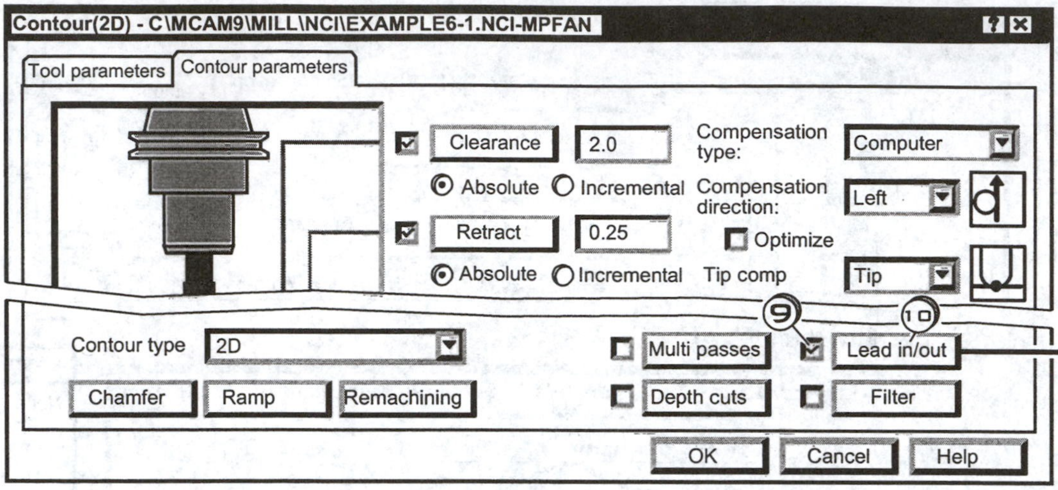

- Click (9) the such that a check appears ☑ to enable Lead in/Out
- Click (10) the Lead in/out button

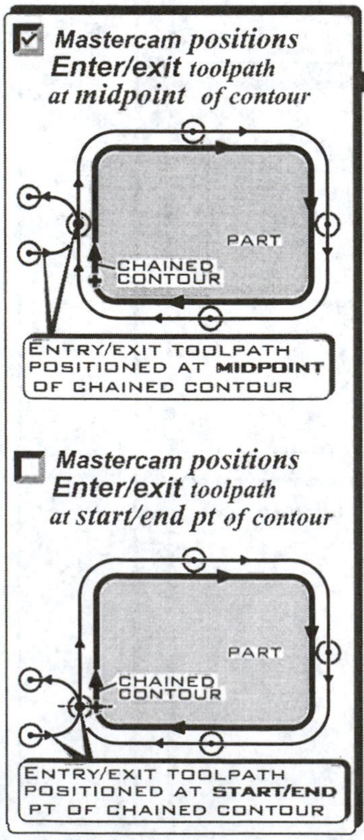

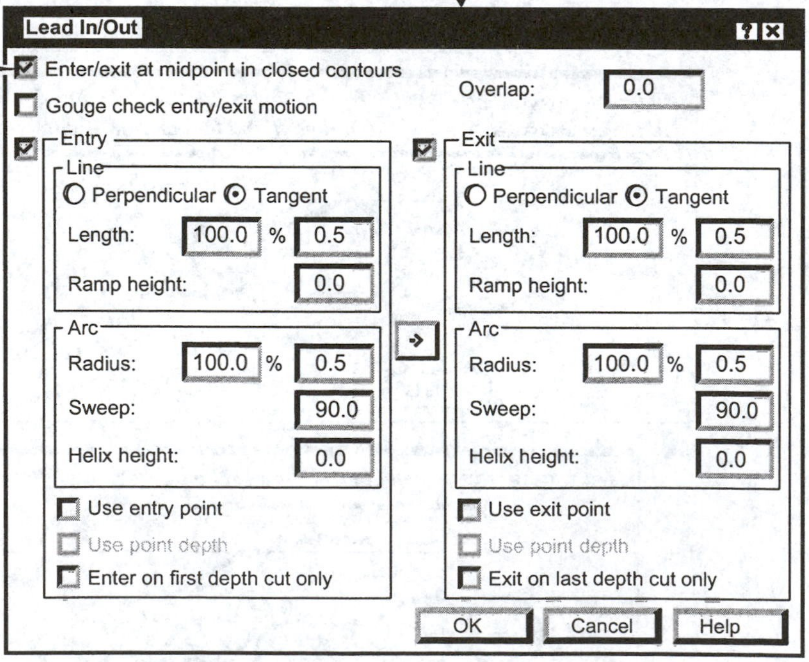

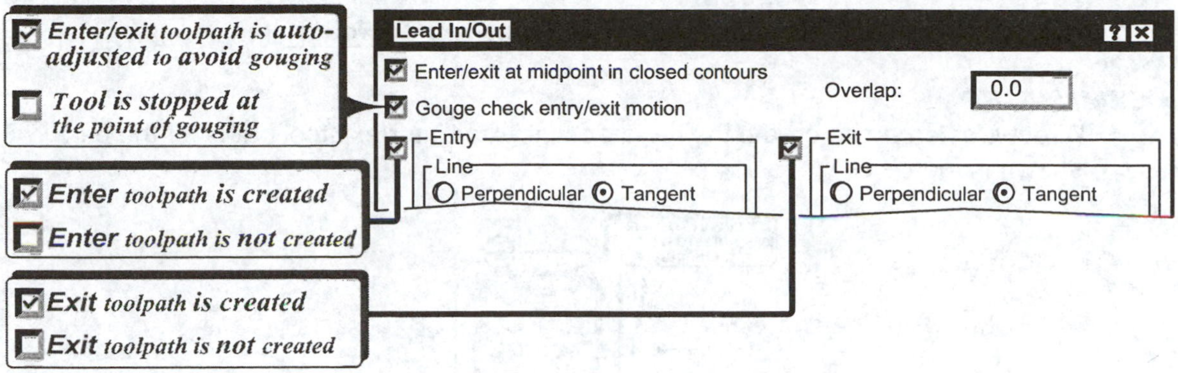

Enter: Line / Exit: Line Toolpaths

EXAMPLE 6-6

The contour shown in Figure 6-5 is to be machined using a 1/4 Flat end mill. Direct *Mastercam* to create the required Line-Line Enter/exit tool paths.

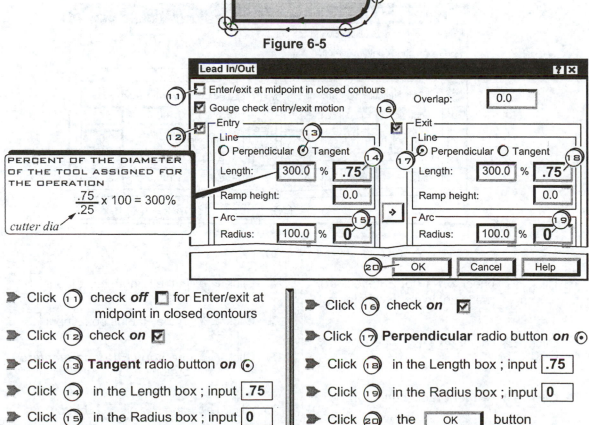

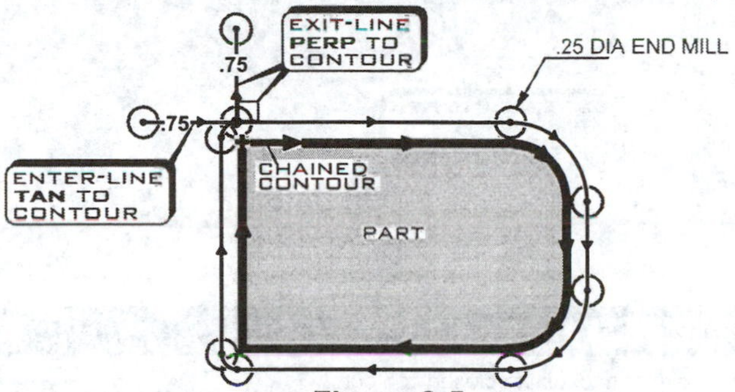

Figure 6-5

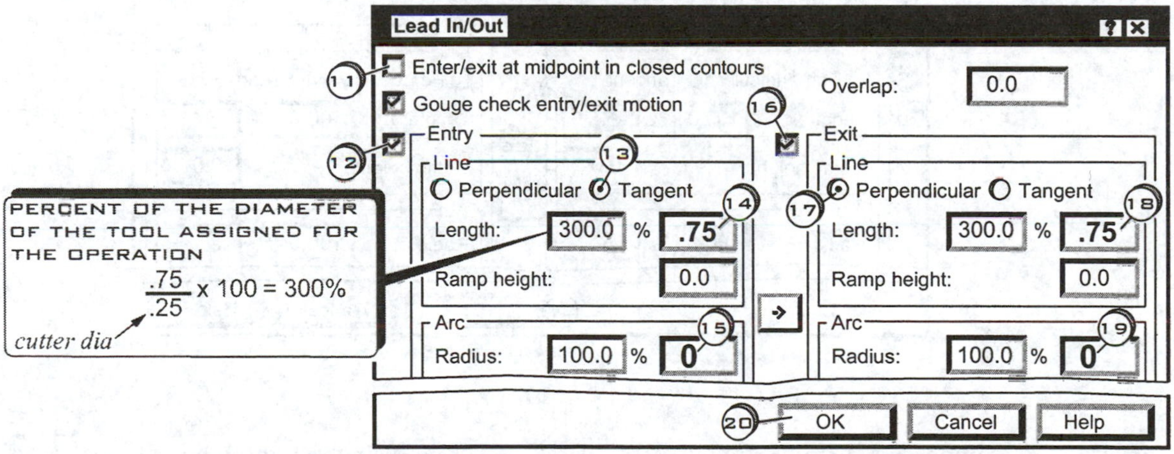

- Click (11) check ***off*** ☐ for Enter/exit at midpoint in closed contours
- Click (12) check ***on*** ☑
- Click (13) **Tangent** radio button ***on*** ⊙
- Click (14) in the Length box ; input **.75**
- Click (15) in the Radius box ; input **0**
- Click (16) check ***on*** ☑
- Click (17) **Perpendicular** radio button ***on*** ⊙
- Click (18) in the Length box ; input **.75**
- Click (19) in the Radius box ; input **0**
- Click (20) the OK button

Enter: Line-Arc / Exit: Line-Arc Toolpaths

EXAMPLE 6-7

Direct *Mastercam* to create the required Line-Arc Enter/exit tool paths for the part shown in Figure 6-6.

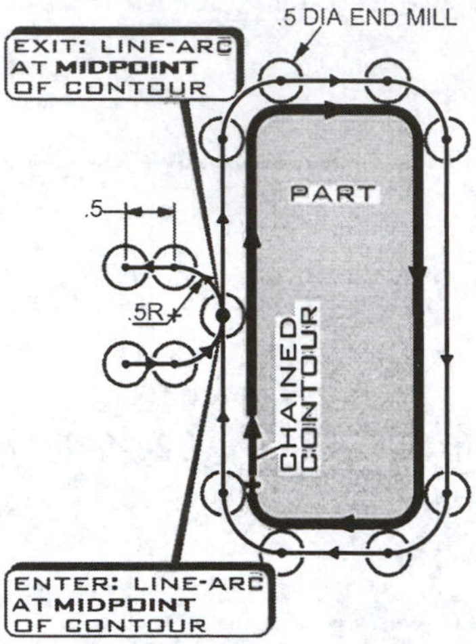

Figure 6-6

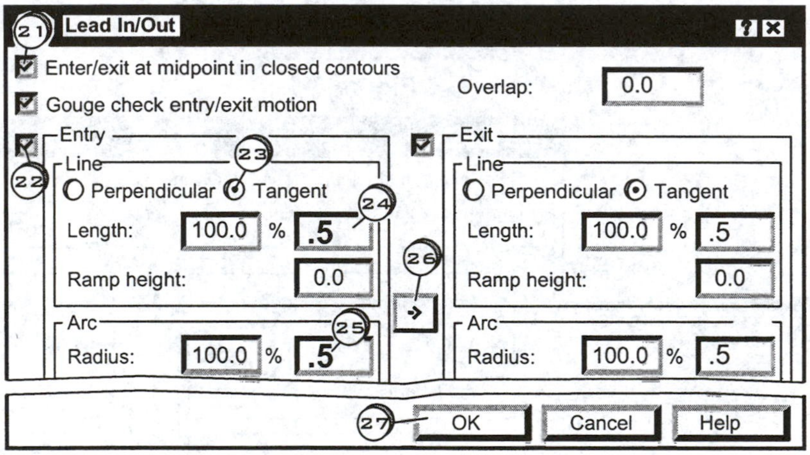

- Click (21) check ***on*** ☑ for Enter/exit at midpoint in closed contours
- Click (22) check ***on*** ☑
- Click (23) **Tangent** radio button ***on*** ⊙
- Click (24) in the Length box ; input **.5**
- Click (25) in the Radius box ; input **.5**
- Click (26) the ***copy Entry parameters*** button [→]
- Click (27) the [OK] button

Enter: Arc / Exit: Arc Toolpaths

EXAMPLE 6-8

Direct *Mastercam* to create the required Arc Enter/exit tool paths as indicated in Figure 6-7.

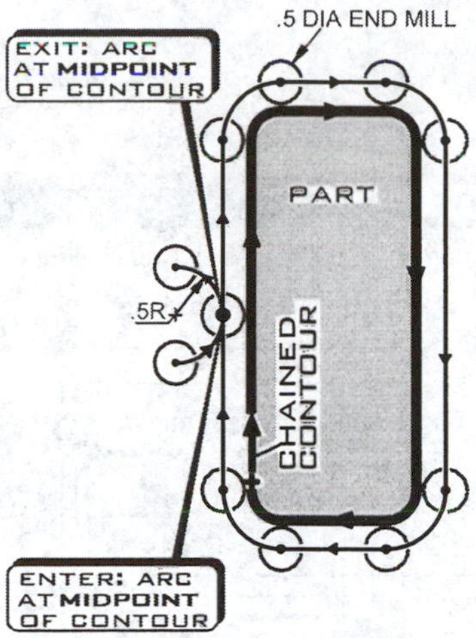

Figure 6-7

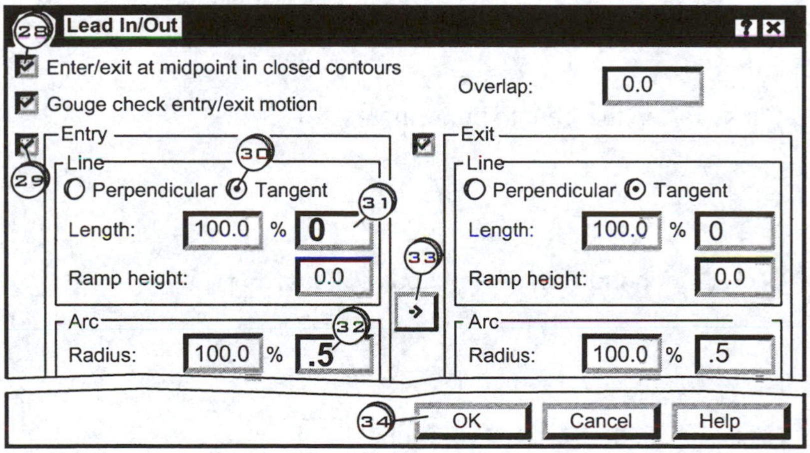

- Click (28) check ***on*** ☑ for Enter/exit at midpoint in closed contours
- Click (29) check ***on*** ☑
- Click (30) **Tangent** radio button ***on*** ⊙
- Click (31) in the Length box ; input 0
- Click (32) in the Radius box ; input .5
- Click (33) the ***copy Entry parameters*** button →
- Click (34) the OK button

Enter: Line Ramp Height / Exit: Line Ramp Height Toolpaths

EXAMPLE 6-9

Direct *Mastercam* to create the required Line Ramp Height Enter/exit tool paths.

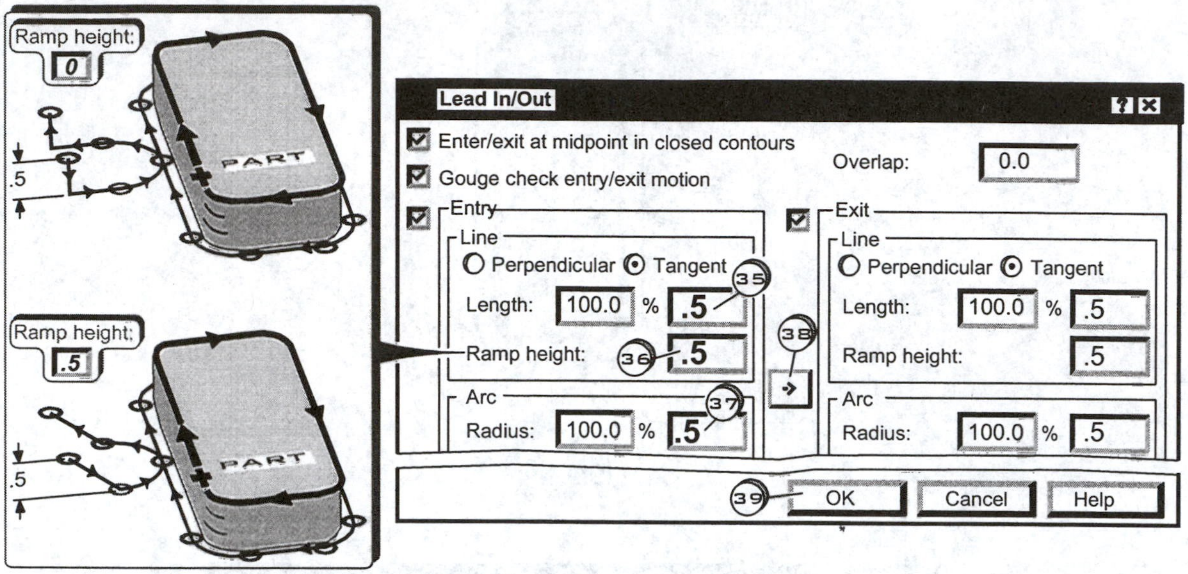

➤ Click (35) in the Length box ; input .5

➤ Click (36) in the Ramp height box ; input .5

➤ Click (37) in the Radius box ; input .5

➤ Click (38) the ***copy Entry parameters*** button →

➤ Click (39) the OK button

Enter: Arc-Helix Height / Exit: Arc-Helix Height Toolpaths

EXAMPLE 6-10

Direct *Mastercam* to create the required Arc-Helix Height Enter/exit tool paths.

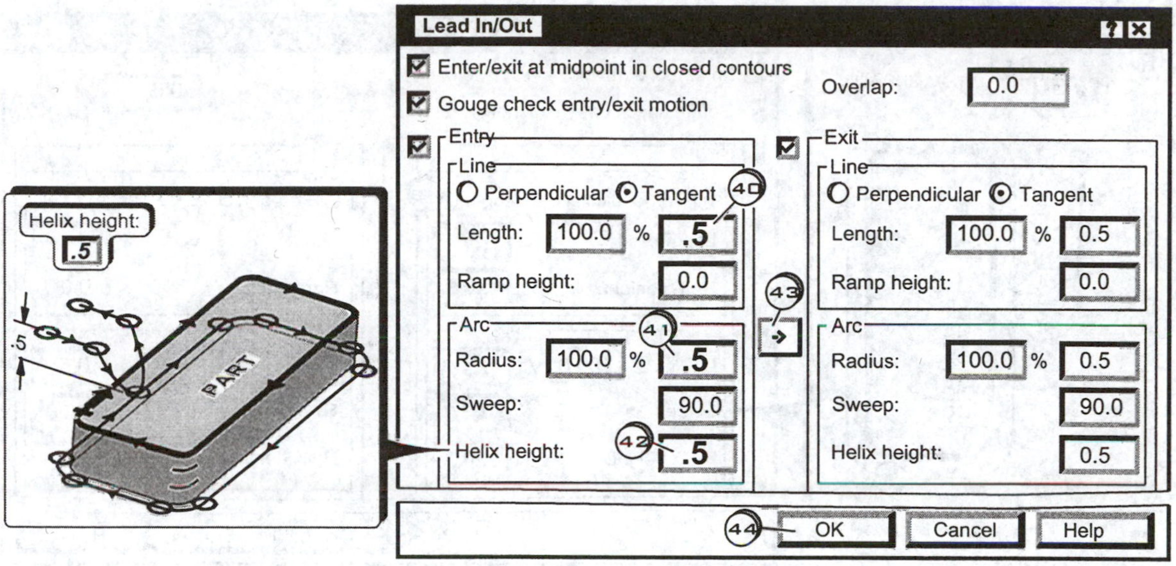

- Click (40) in the Length box ; input .5

- Click (41) in the Radius box ; input .5

- Click (42) in the Helix height box ; input .5

- Click (43) the ***copy Entry parameters*** button →

- Click (44) the OK button

Enter: Line-Arc Sweep / Exit: Line-Arc Sweep Toolpaths

EXAMPLE 6-11

Direct *Mastercam* to create the required Line-Arc Sweep Enter/exit tool paths.

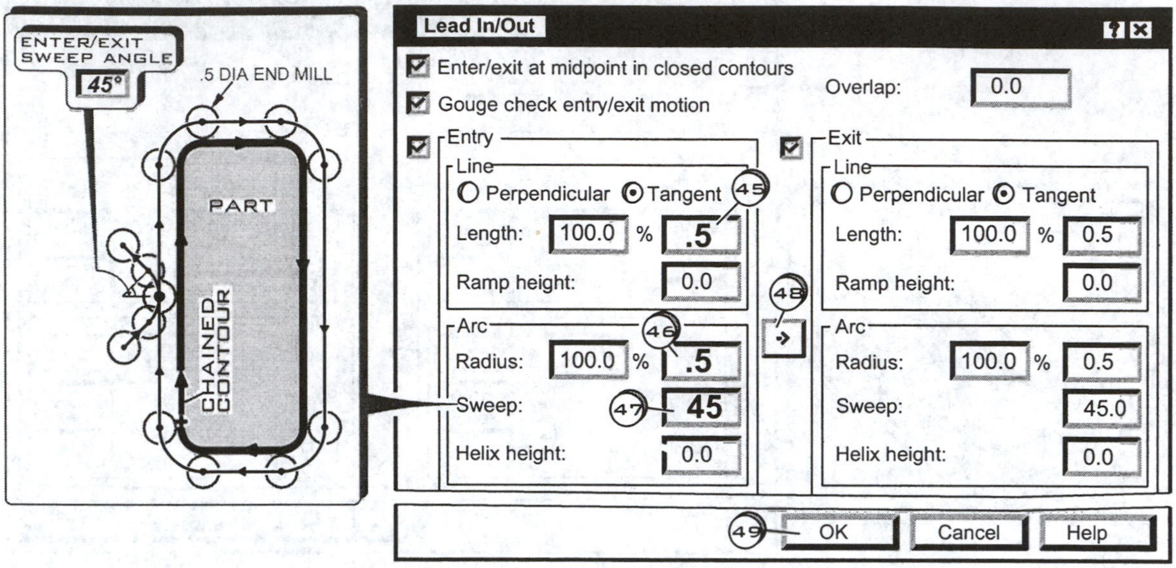

- Click (45) in the Length box ; input **.5**

- Click (46) in the Radius box ; input **.5**

- Click (47) in the Sweep box ; input **45**

- Click (48) the ***copy Entry parameters*** button [→]

- Click (49) the [OK] button

Enter: Line-Arc from Point/ Exit: Line-Arc to Point Toolpaths

EXAMPLE 6-12

Direct *Mastercam* to create the required Line-Arc from/to Point Enter/exit tool paths.

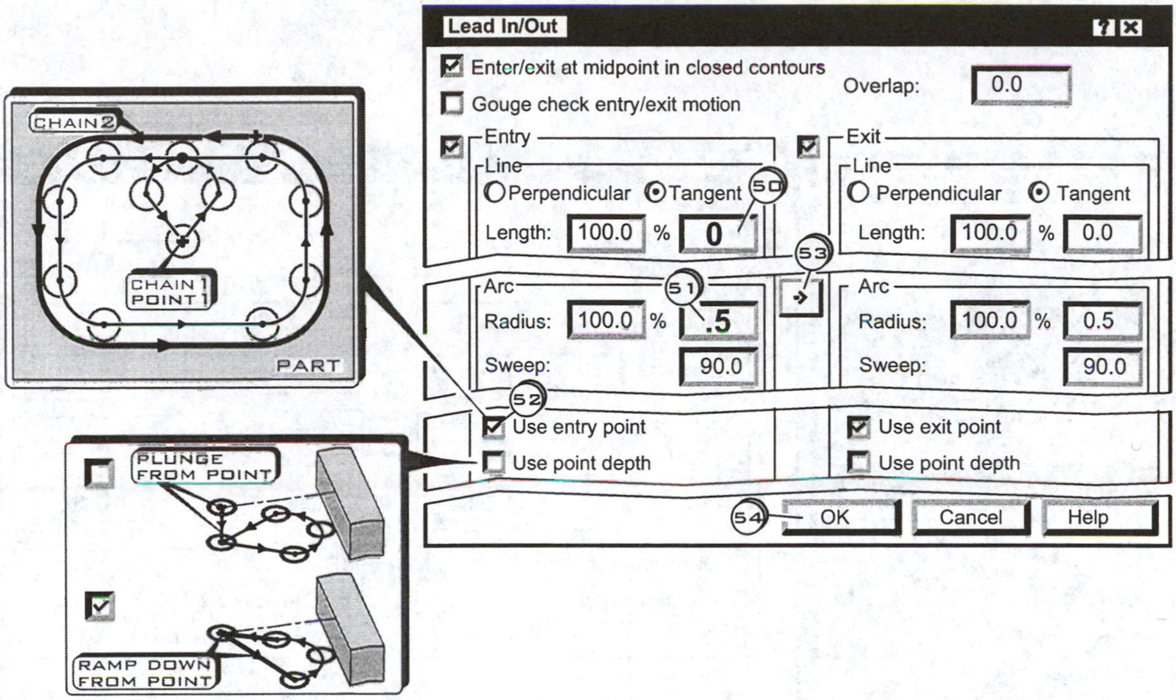

- Click (50) in the Length box ; input **0**
- Click (51) in the Radius box ; input **.5**
- Click (52) the Use entry point check ☑
- Click (53) the ***copy Entry parameters*** button [→]
- Click (54) the [OK] button

Enter: Line-Arc for First Depth Cut Only/ Exit: Line-Arc for Last Depth Cut Only

EXAMPLE 6-13

Direct *Mastercam* to create the required Enter/exit tool paths.

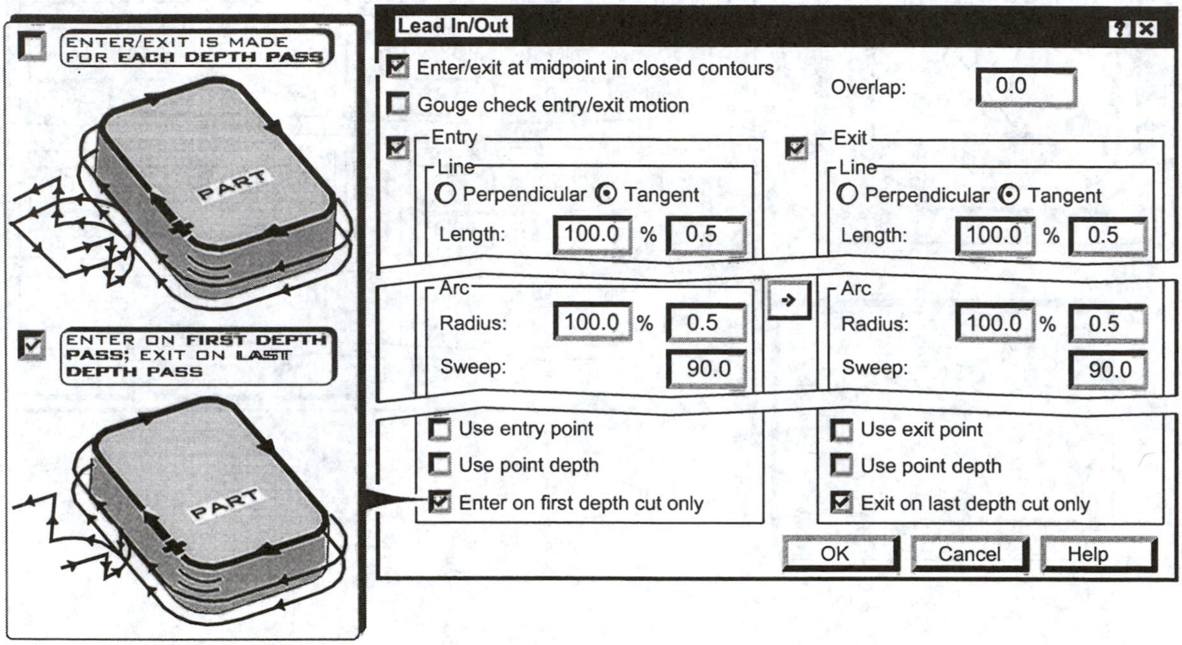

Enter: Line-Arc / Exit: Line-Arc Overlap Toolpaths

Note: The Overlap parameter can ***only be used for closed*** contours

EXAMPLE 6-14

Direct *Mastercam* to create the required Line-Arc Overlap Enter/exit tool paths.

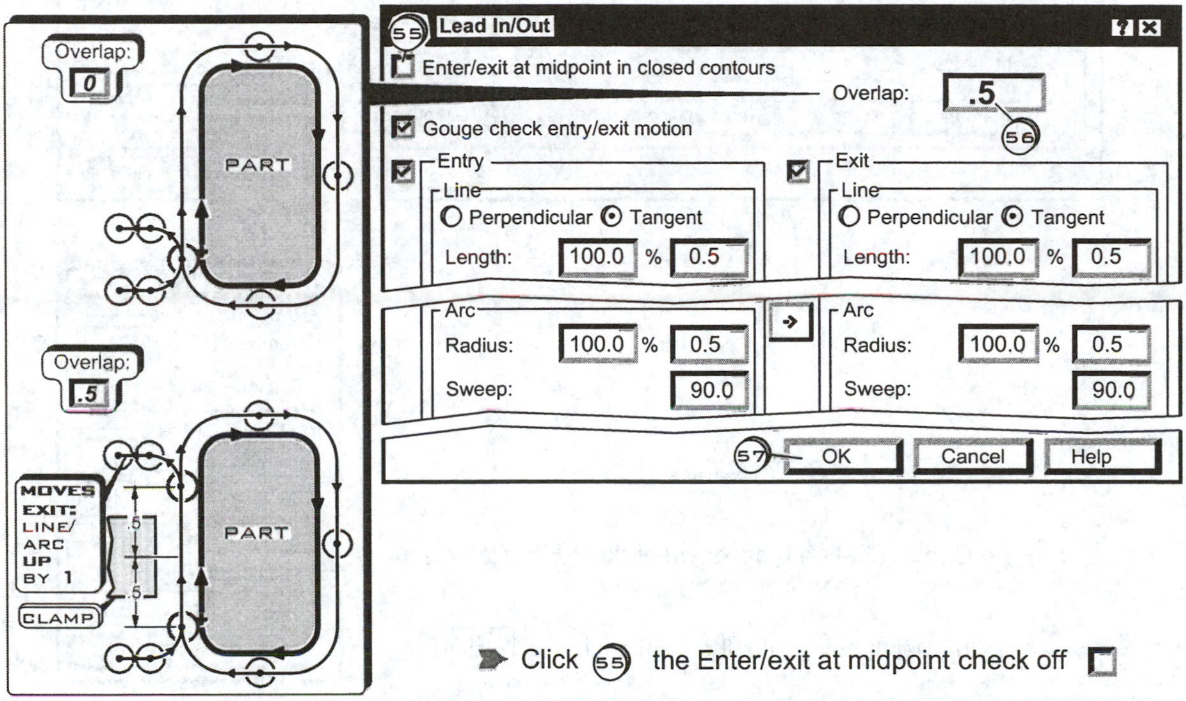

➤ Click (55) the Enter/exit at midpoint check off ☐

➤ Click (56) in the Overlap box ; input .5

➤ Click (57) the OK

EXAMPLE 6-15

Continue the steps began in **EXAMPLE 6-2.**
Enter the required data in the **Contour parameters** dialog box to execute the contour machining described in the PROCESS PLAN of **EXAMPLE 6-1**

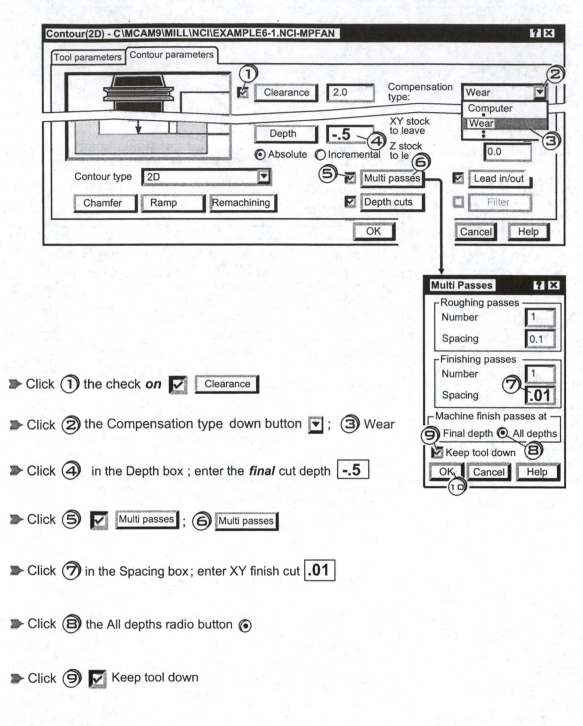

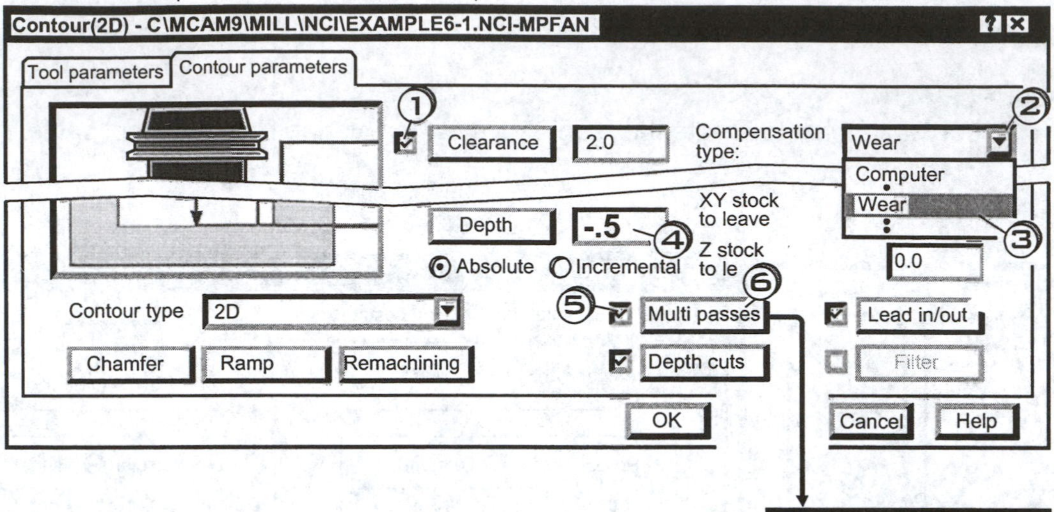

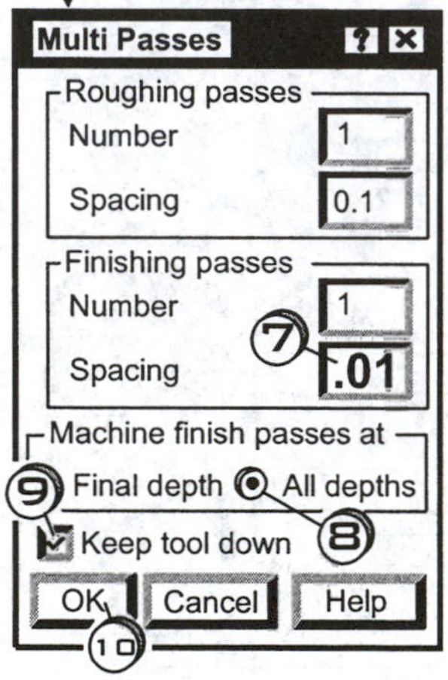

- Click ① the check ***on*** ☑ Clearance
- Click ② the Compensation type down button ▼ ; ③ Wear
- Click ④ in the Depth box ; enter the ***final*** cut depth **-.5**
- Click ⑤ ☑ Multi passes ; ⑥ Multi passes
- Click ⑦ in the Spacing box; enter XY finish cut **.01**
- Click ⑧ the All depths radio button ◉
- Click ⑨ ☑ Keep tool down
- Click ⑩ OK

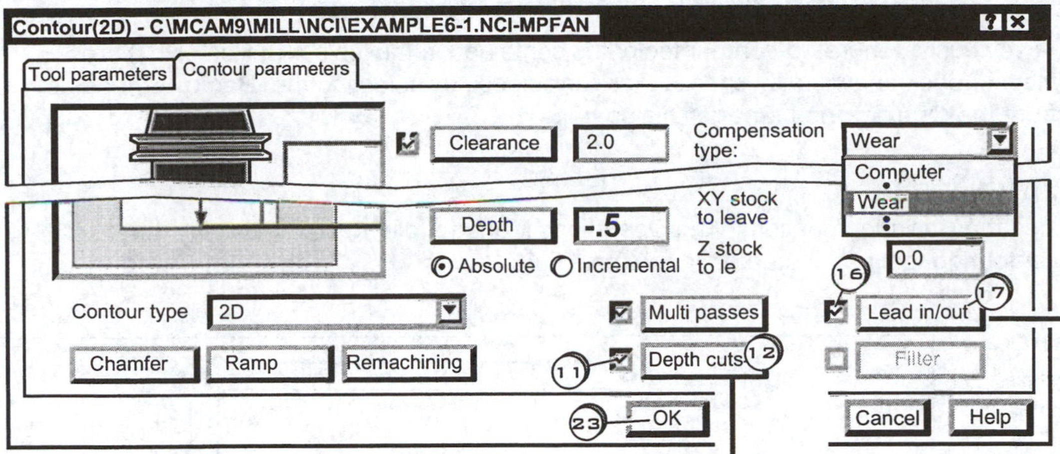

➤ Click (11) ☑ Depth cuts ; (12) Depth cuts

➤ Click (13) Finish step; enter Z finish cut .01

➤ Click (14) the check ***on*** ☑ Keep tool down

➤ Click (15) OK

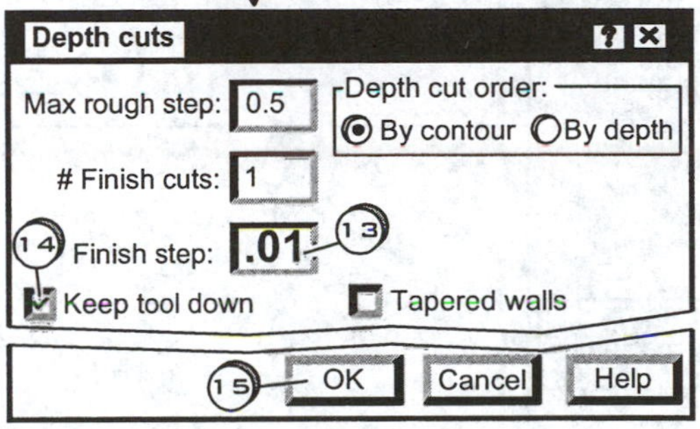

➤ Click (16) ☑ Lead in/out ; (17) Lead in/out

➤ Click (18) ☑ Enter/exit at midpoint

➤ Click (19) ☑ Gouge check

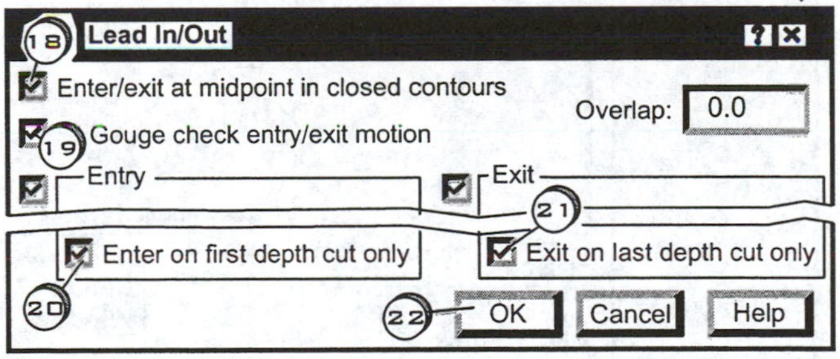

ACCEPT **DEFAULT** ENTRY/EXIT TOOL PATH CONSISTING OF **LINES AND ARCS.**
LINE LENGTH = TOOL DIA
ARC RADIUS = TOOL DIA

➤ Click (20) ☑ Enter on first depth cut

➤ Click (21) ☑ Exit on last depth cut

➤ Click (22) OK ; (23) OK

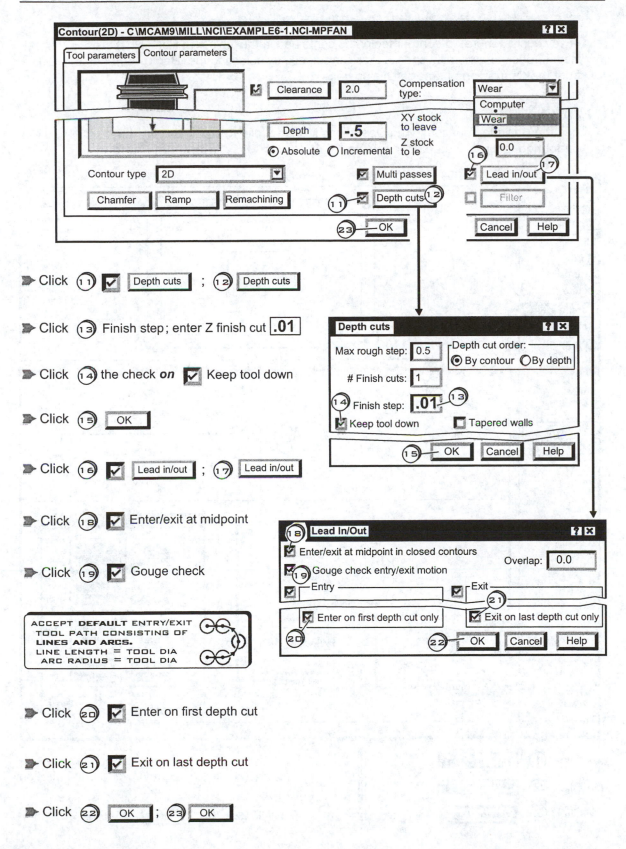

6-6 Backplotting Profile Machining Operations

Previously generated contour toolpaths contained in the job's NCI file(* **.NCI**) are played back and visually checked for correctness by selecting the **Backplot** module from the **Operations Manager** dialog box.

EXAMPLE 6-16

Use the Backplot function in the Operations Manager dialog box to display the contouring tool paths created in Example 6-15.

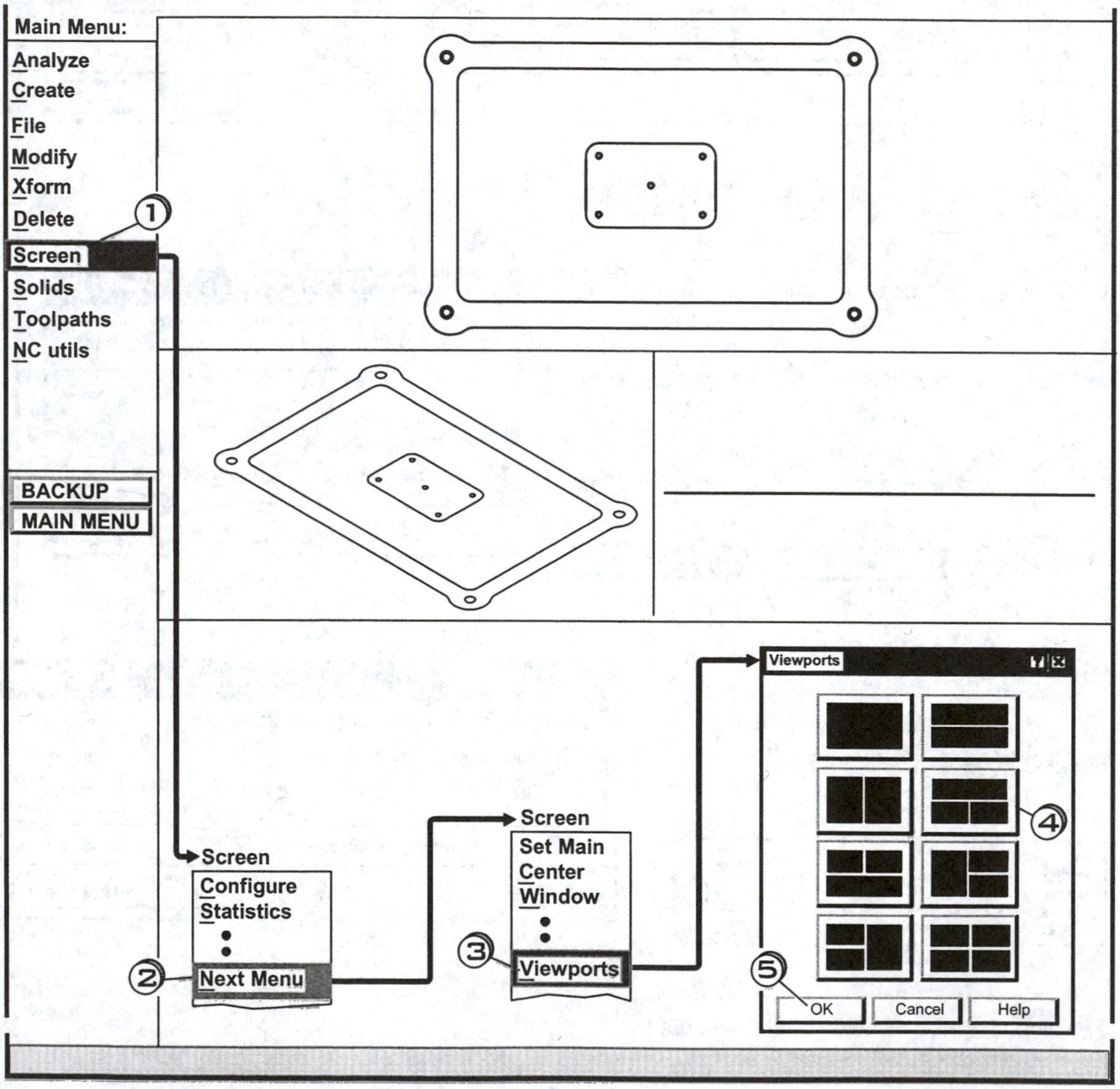

- Click ① Screen
- Click ② Next Menu
- Click ③ Viewports
- Click ④ the desired viewport configuration
- Click ⑤ the OK button

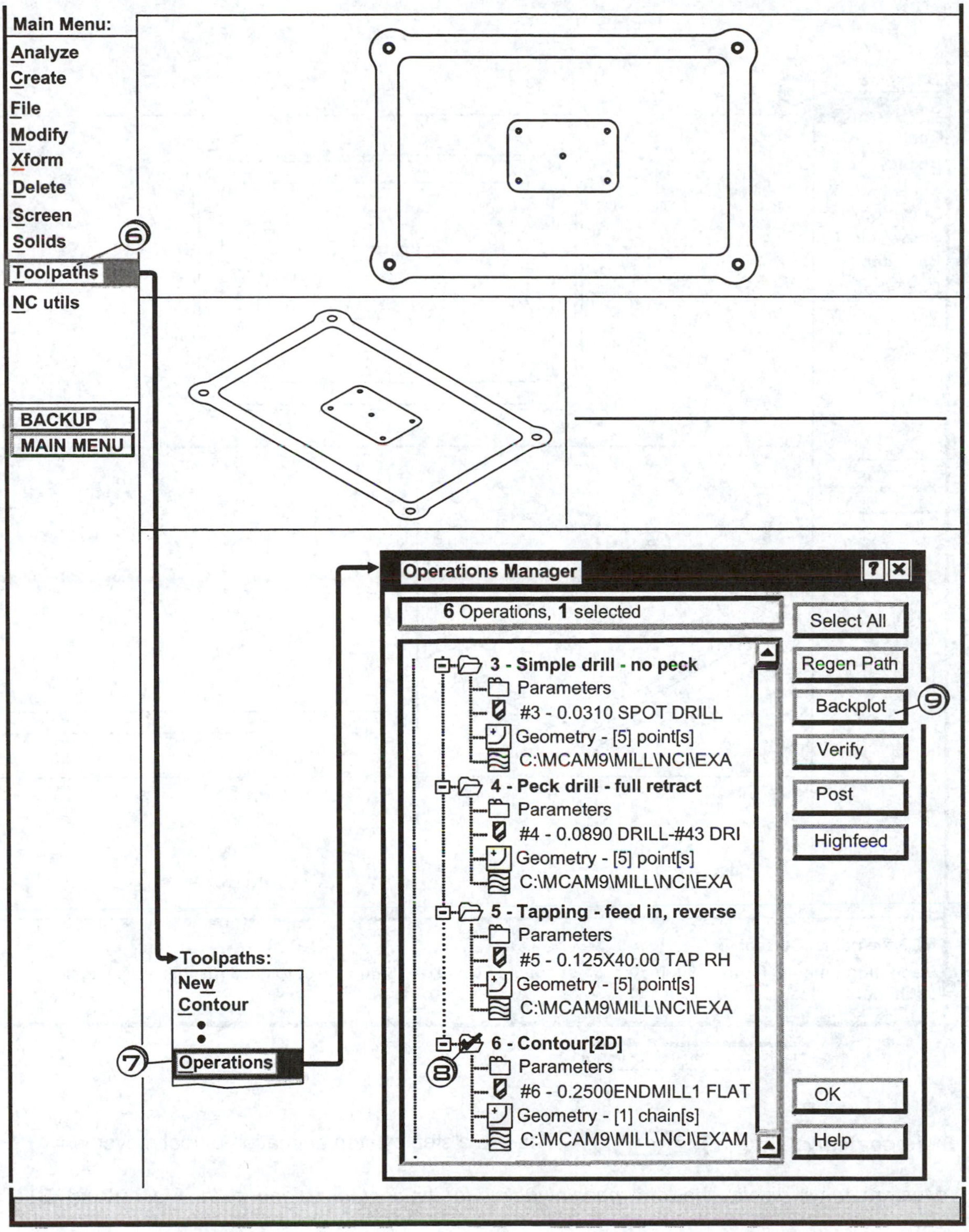

- Click ⑥ Toolpaths
- Click ⑦ Operations
- Click ⑧ on **6 - Contour[2D]**
- Click ⑨ the Backplot button

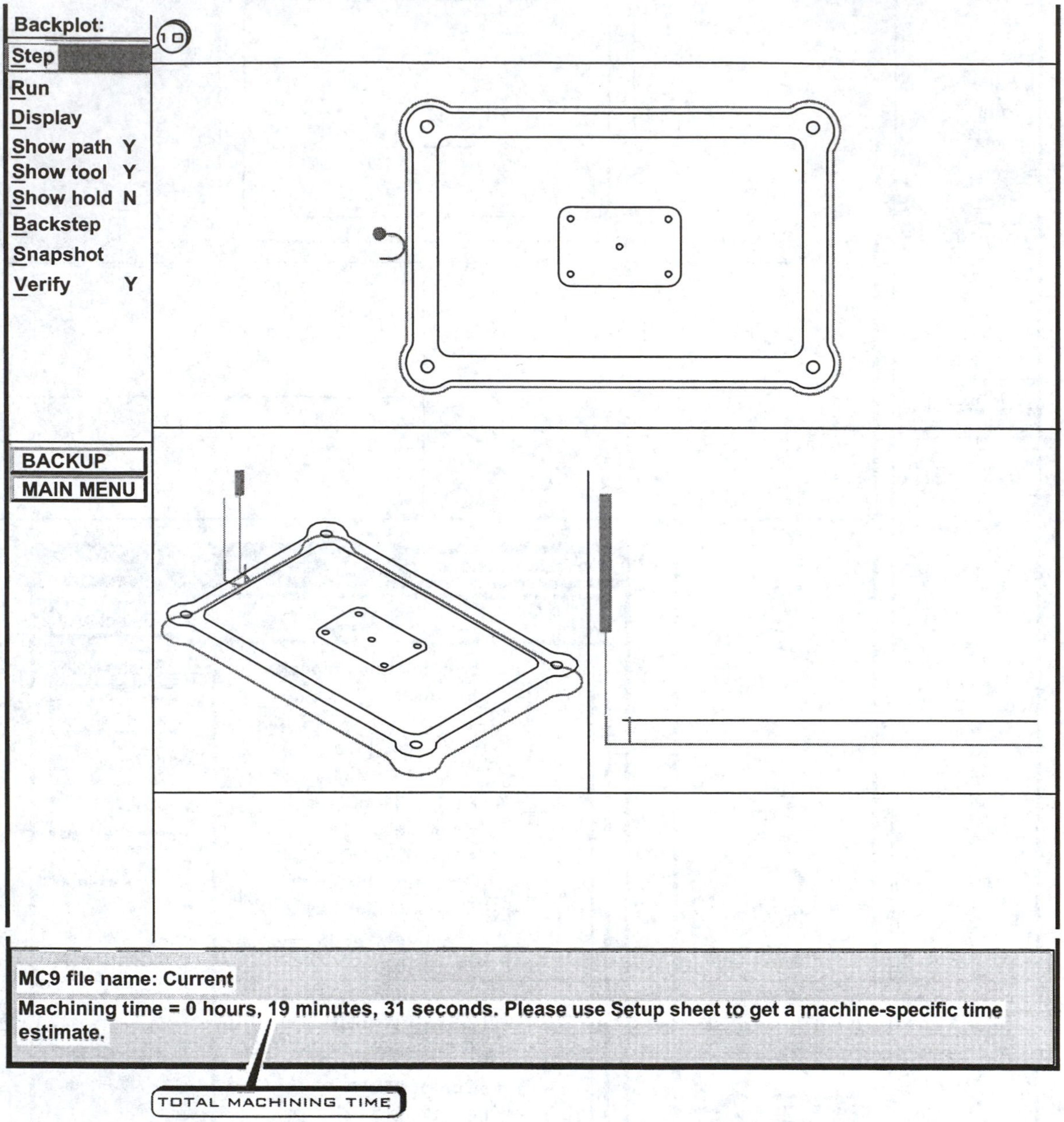

➡ ***Repeatedly*** Click (1 D) **Step** to see a step by step animation of tool movement

Mastercam will display the *total machining time for the contouring operation* **6- Contour[2D]** *selected*

6-7 Creating a Pocketing Contour

Mastercam's pocketing module generates tool paths to remove material from a ***closed*** contour chained or ***face mill*** the top of the part ***flat***.

EXAMPLE 6-17

The part shown in Figure 6-8 is to be prepared for pocket machining

- Create the pocket machining contour by chaining
- Assign the pocketing tool as listed in PROCESS PLAN 6-2

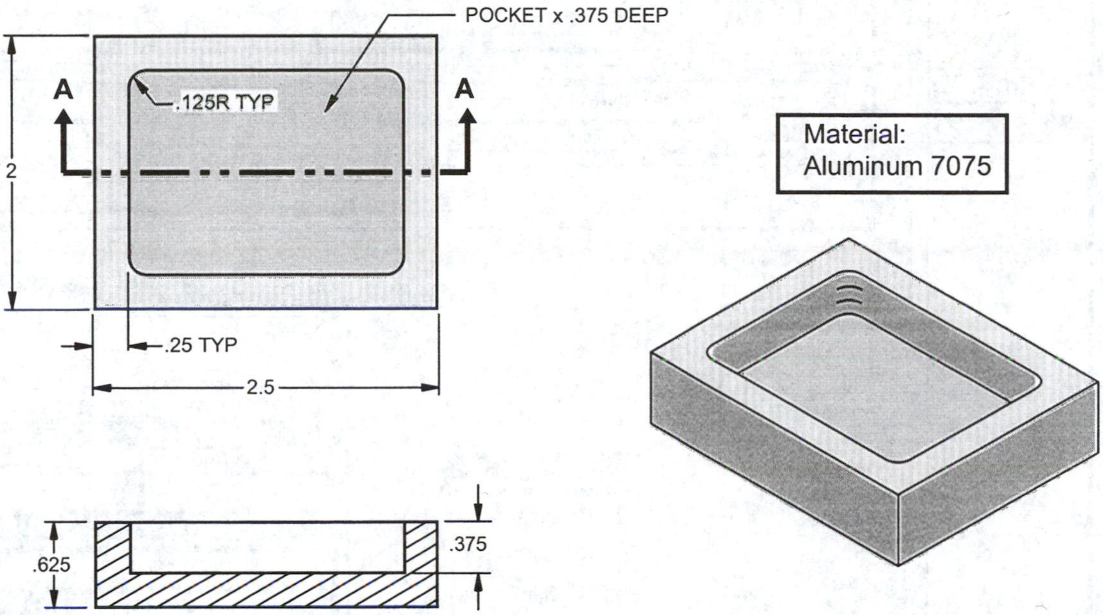

Figure 6-8

PROCESS PLAN 6-2

Operation	Tooling
Pocket x .375 Deep leave .03 for finish cuts in XY and Z	**1/4-End Mill**

A) OPEN JOB SETUP. SPECIFY THE STOCK SIZE AND MATERIAL

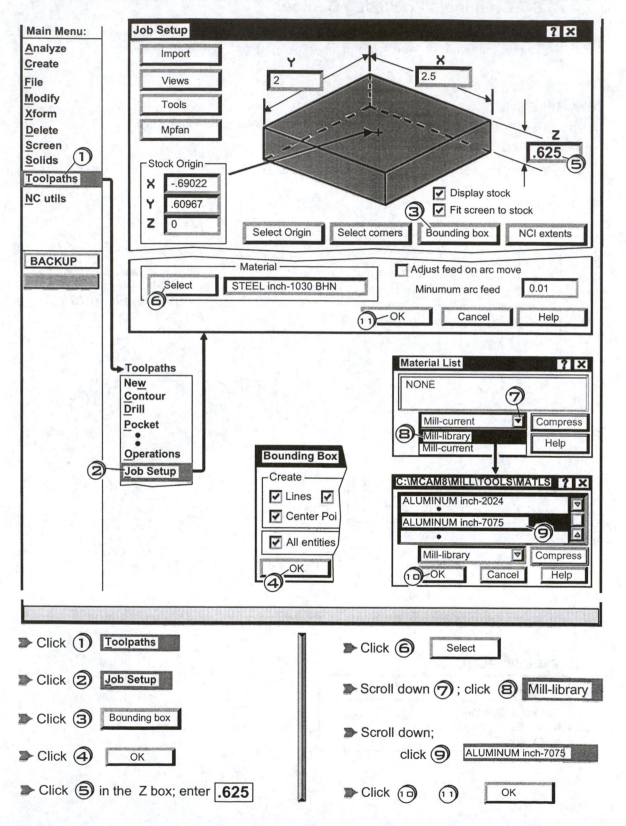

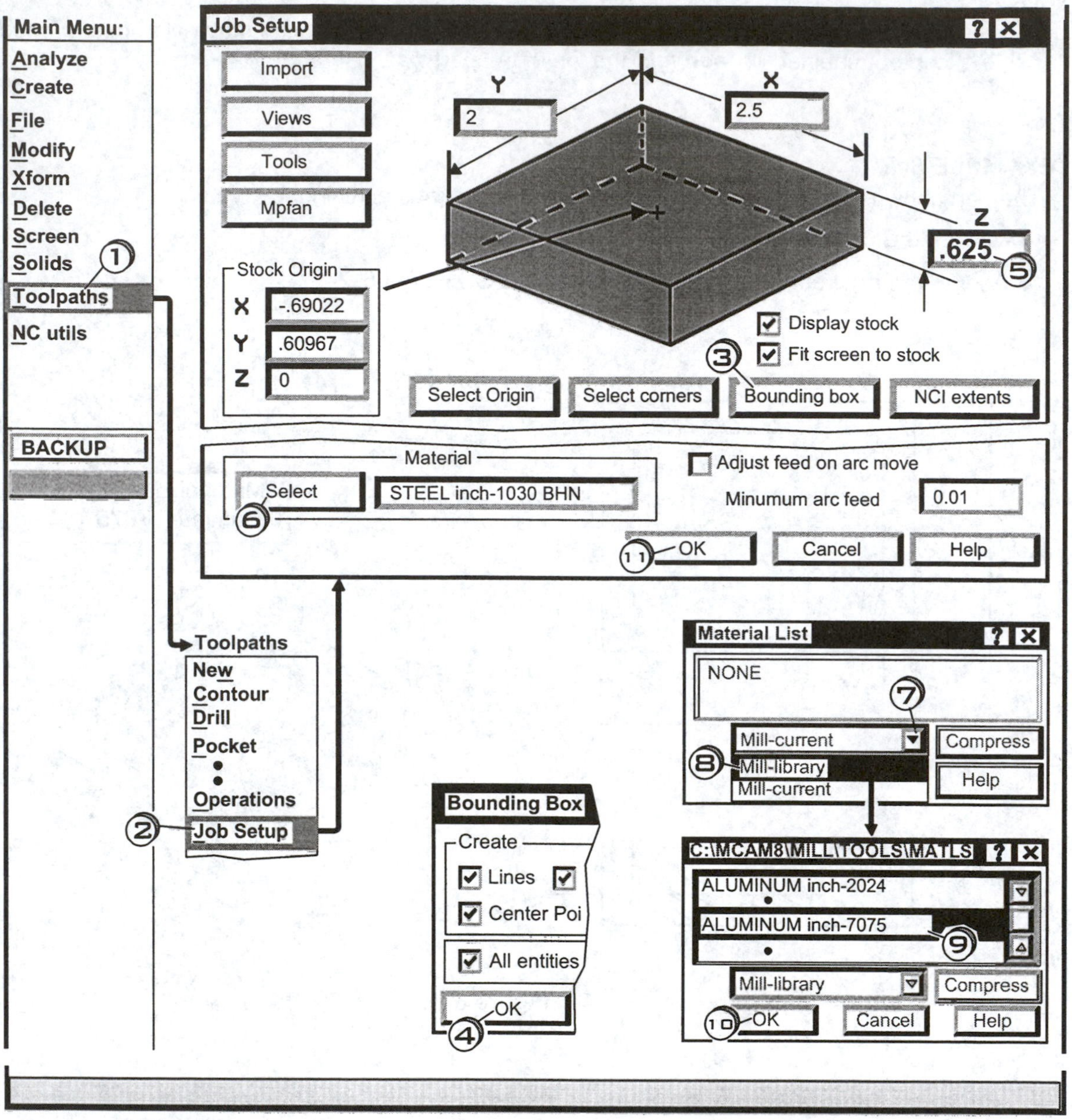

- Click ① Toolpaths
- Click ② Job Setup
- Click ③ Bounding box
- Click ④ OK
- Click ⑤ in the Z box; enter .625
- Click ⑥ Select
- Scroll down ⑦ ; click ⑧ Mill-library
- Scroll down; click ⑨ ALUMINUM inch-7075
- Click ⑩ ⑪ OK

B) CREATE A MACINING CONTOUR BY CHAINING

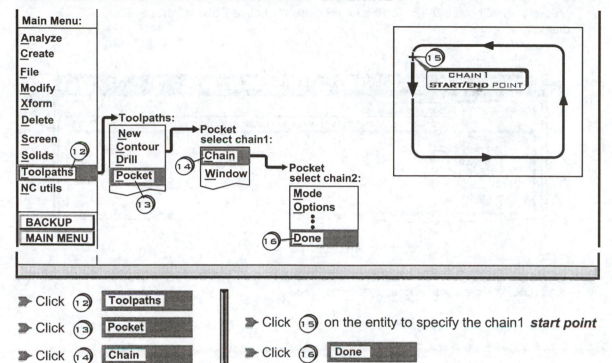

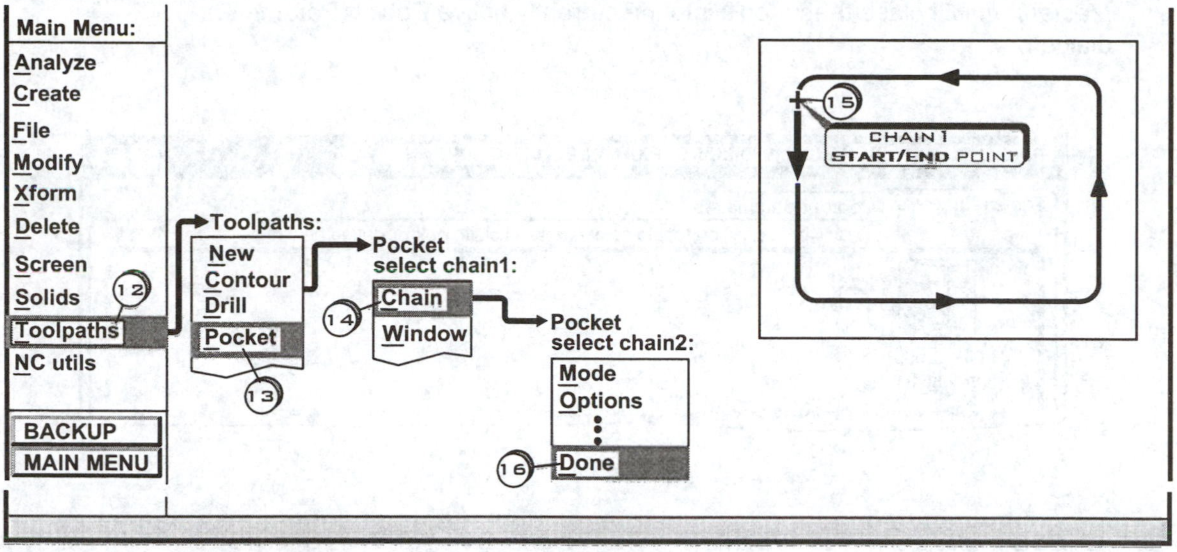

- Click (12) **Toolpaths**
- Click (13) **Pocket**
- Click (14) **Chain**
- Click (15) on the entity to specify the chain1 ***start point***
- Click (16) **Done**

C) OBTAIN THE NEEDED 1/4 END MILL TOOL

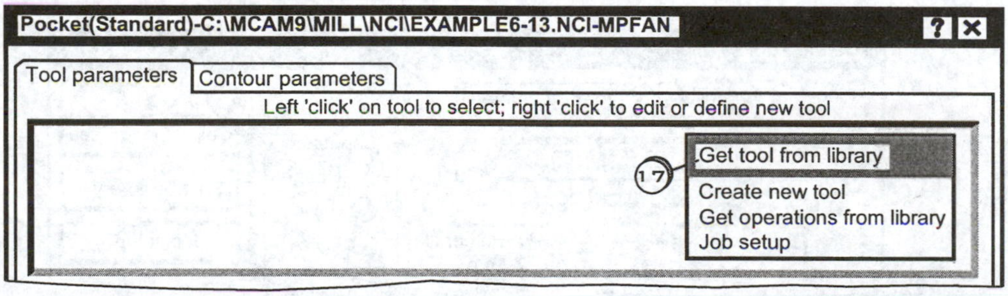

- ***Right*** Click move the mouse cursor *down* and Click (17) Get tool from library

Select the 1/4in Dia End mill tool from the **TOOLS** tool library

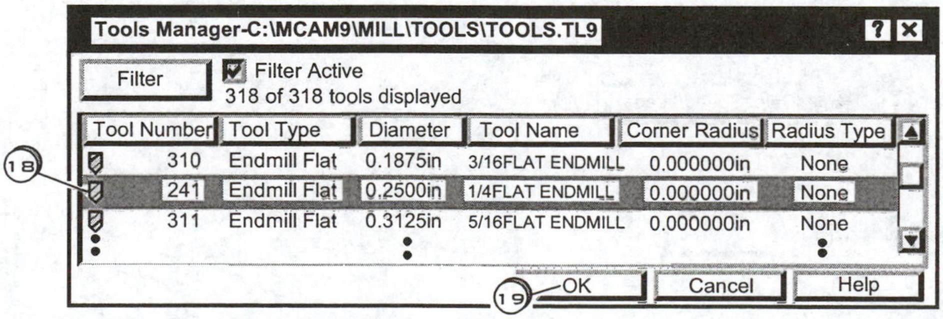

- Click (18) on the 1/4 FLAT ENDMILL
- Click (19) the OK button

Mastercam will place these tools into the currently active Pocket Parameters dialog box

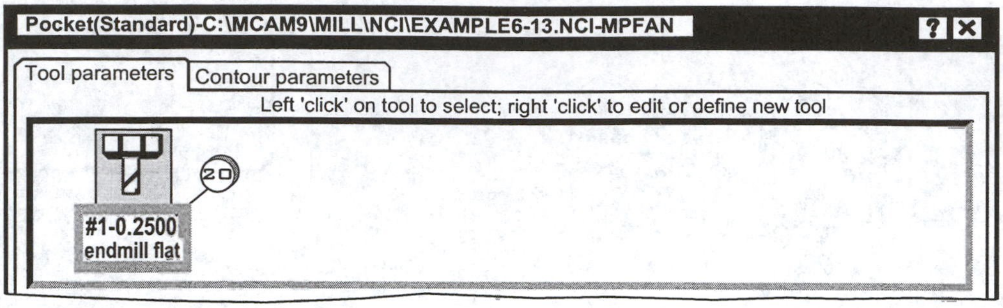

D) DIRECT *Mastercam* TO ASSIGN THE TOOL SPEED/FEED BASED ON THE STOCK MATERIAL SELECTED IN **Job Setup**

➤ ***Right*** Click (20) on the .2500 endmill flat tool

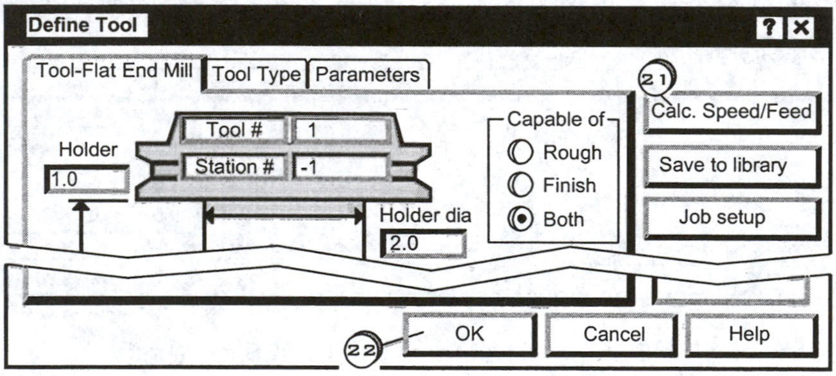

➤ Click (21) the [Calc. Speed/Feed] key

➤ Click (22) the [OK] key

6-8 Specifying Pocketing Parameters in the Pocketing Module

The Pocketing module uses the *chained contour* and other key pocketing *machining parameters inputted by the operator* to generate the required pocketing tool paths.

A description of the *pocketing parameters* and their effect on tool paths is considered in this section.

EXAMPLE 6-18

Direct *Mastercam* to display the **Pocketing parameters** dialog box in Pocketing module for the pocket machining described in **EXAMPLE 6-17**

➤ Click (23) the **Pocketing parameters** tab

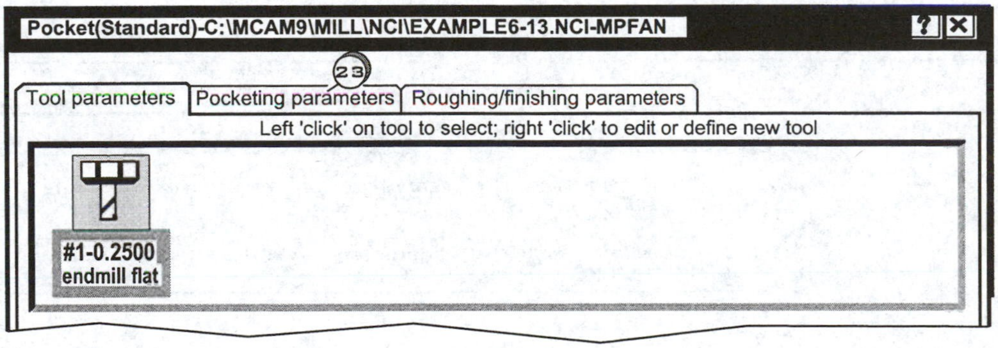

The system will display the *pocketing parameters* dialog box as shown on p6-46

Z- DEPTH PARAMETERS

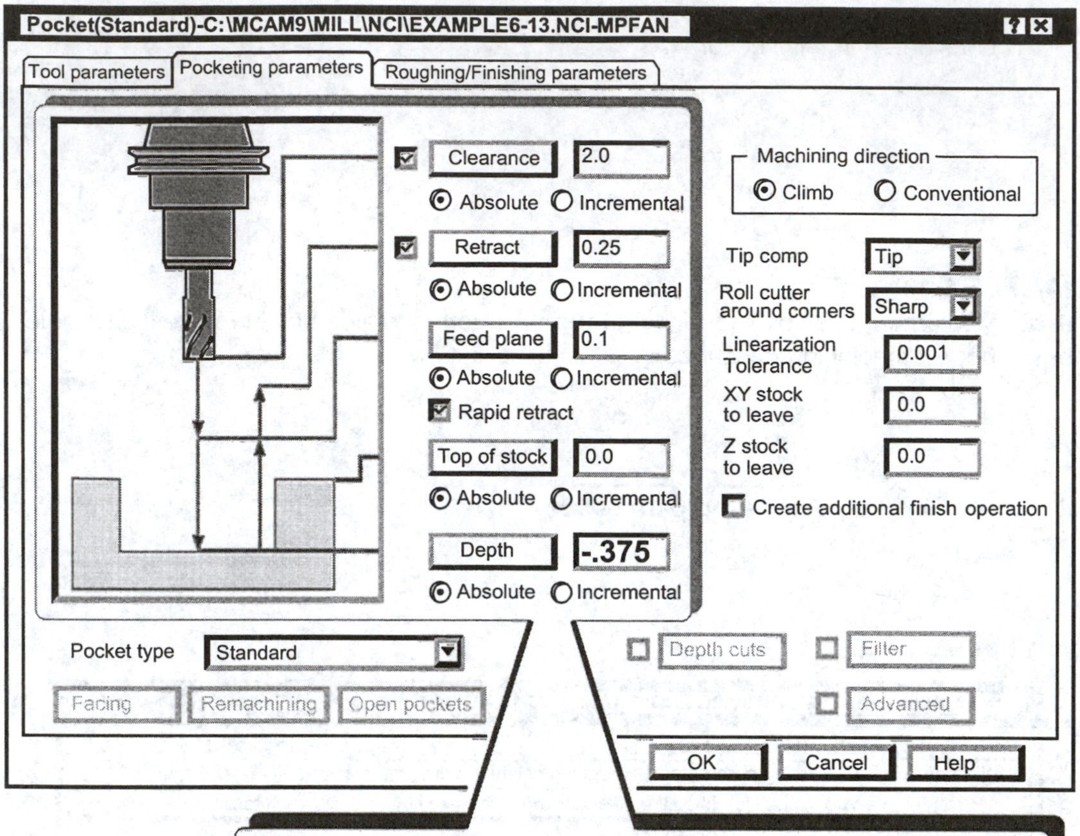

Z depth pocketing parameters are ***identical*** *to and carry the* ***same meaning*** *as those specified in* ***2D Contouring.***

NOTE: Depth -.375 SPECIFIES THE **FINAL POCKET DEPTH**

MACHINING DIRECTION

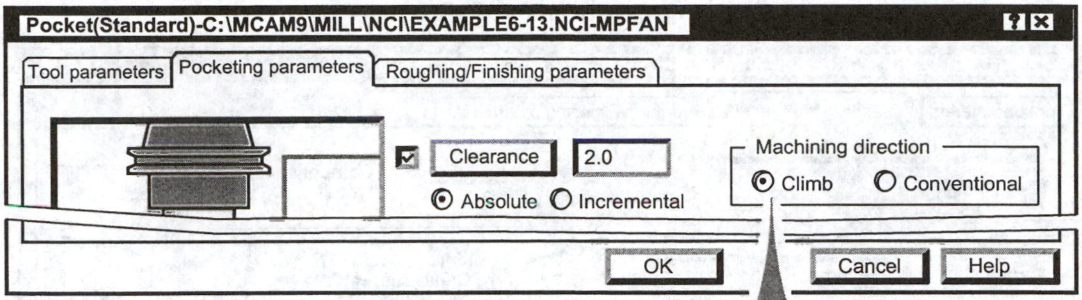

Climb

Mastercam machines the pocket by ***moving the tool in the direction of spindle rotation.*** *The cutting forces work into the machine not into the clamps. The cutter deflection is directed outside the contour and , therefore, the part ends up full rather than being violated. Climb milling is* ***preferred for use with CNC machine tools.*** *It requires less horespower to produce a part.*

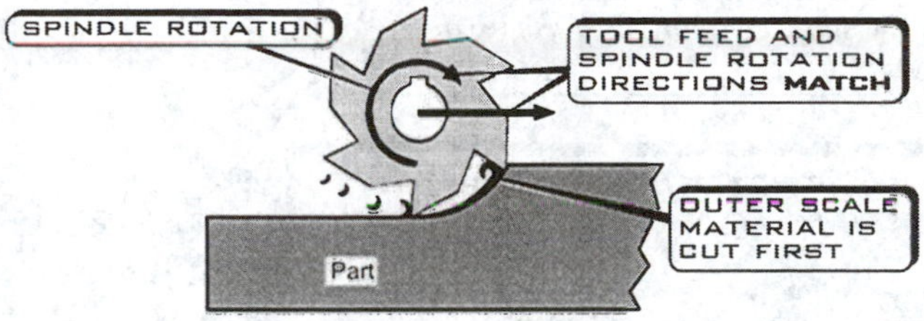

Conventional

Mastercam machines the pocket by ***moving the tool in the direction opposite of spindle rotation.*** *Conventional milling cuts the softer inner material first, results in a better finish and improves tool life. It is recommended* ***for light finish cuts or when rough machining castings*** *(tool cuts the soft inner material first and then breaks the hard outer scale)*

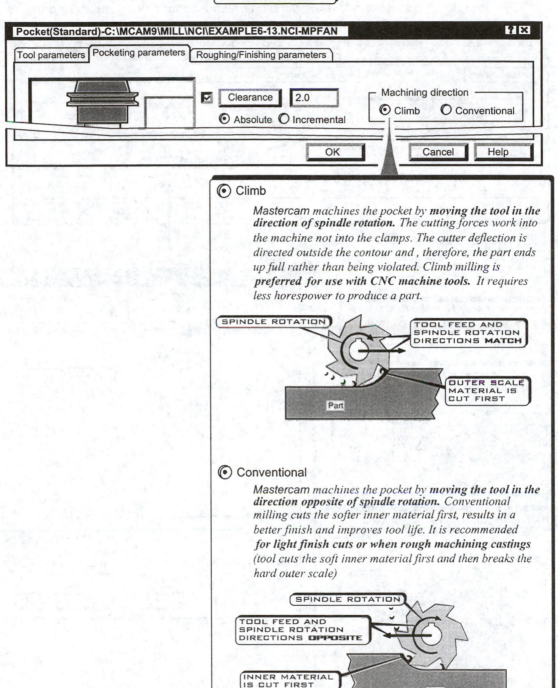

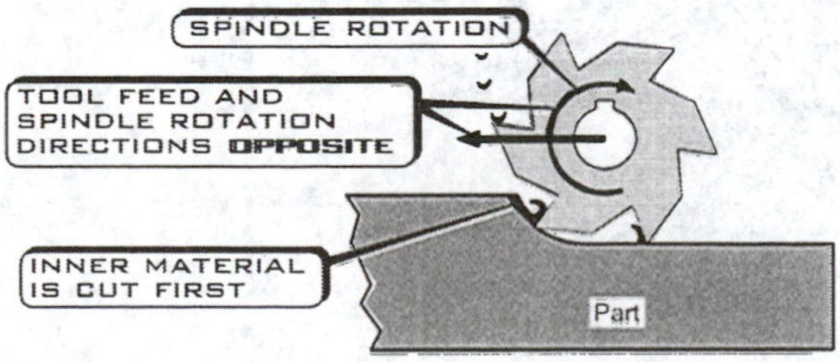

DEPTH CUTS

The depth cuts parameters allow the operator to control the ***number*** and ***depth*** of ***roughing*** and ***finishing*** passes in the Z direction

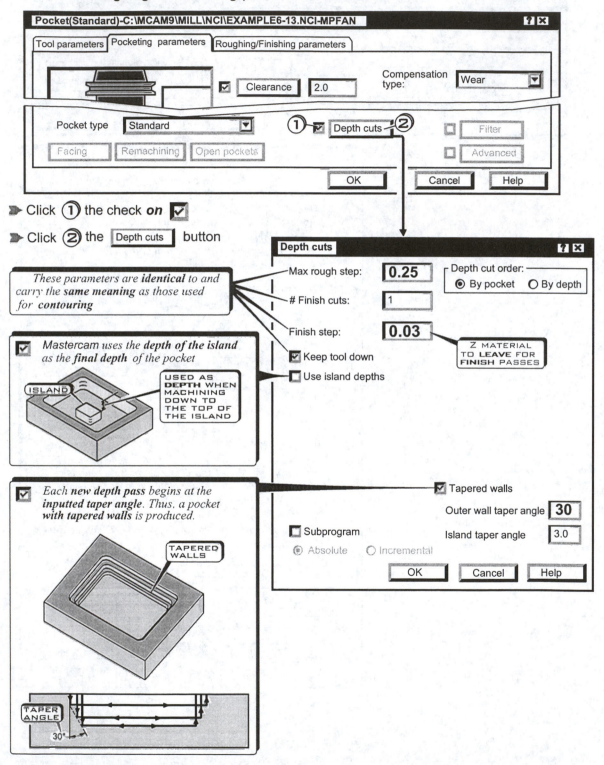

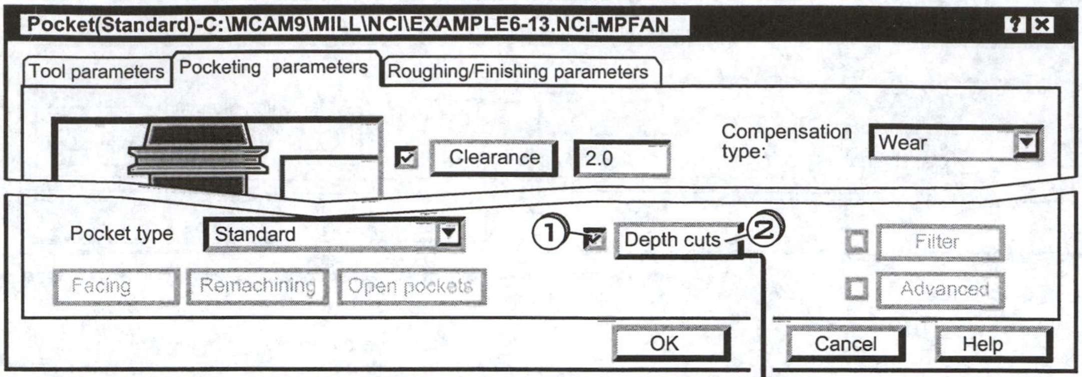

- Click ① the check ***on*** ☑
- Click ② the [Depth cuts] button

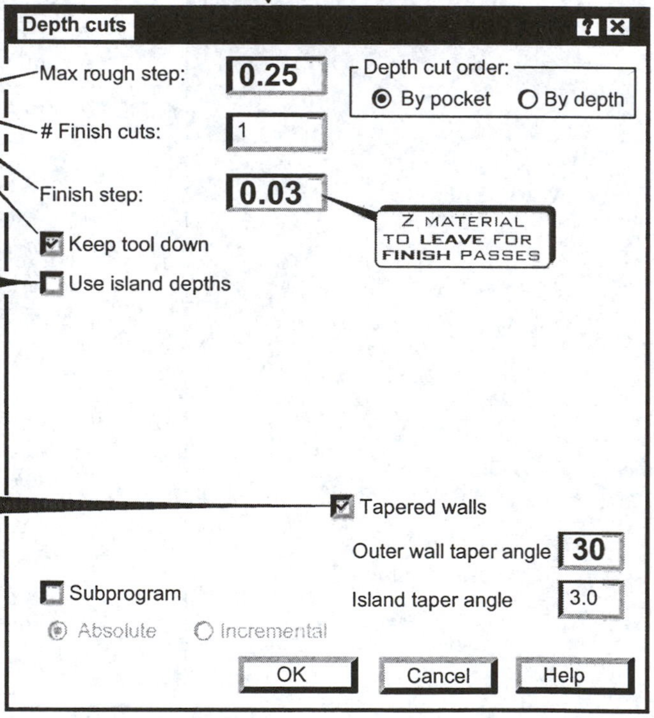

*These parameters are **identical** to and carry the **same meaning** as those used for **contouring***

☑ *Mastercam uses the **depth of the island** as the **final depth** of the pocket*

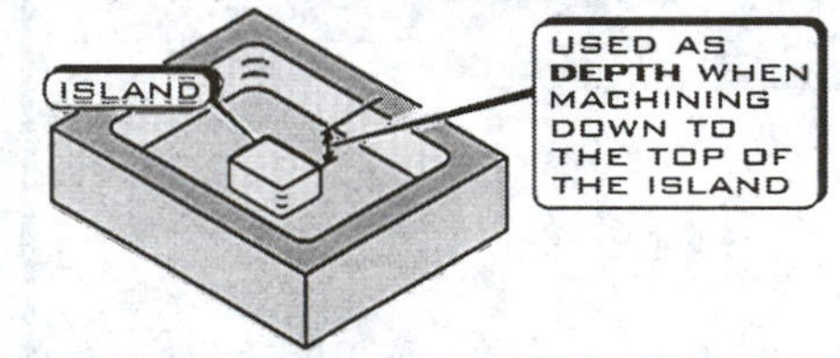

☑ *Each **new depth pass** begins at the **inputted taper angle**. Thus, a pocket **with tapered walls** is produced.*

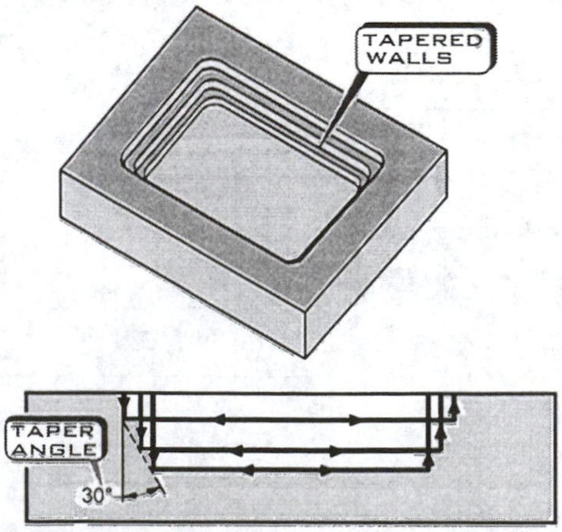

POCKET TYPE

The Pocket type parameters specify whether a ***Standard, Facing***, ***Island facing***, ***Remachining*** or ***Open*** pocket is to be machined.

- Click ① the Pocket type down button ▼
- Click ② the pocket type to machine

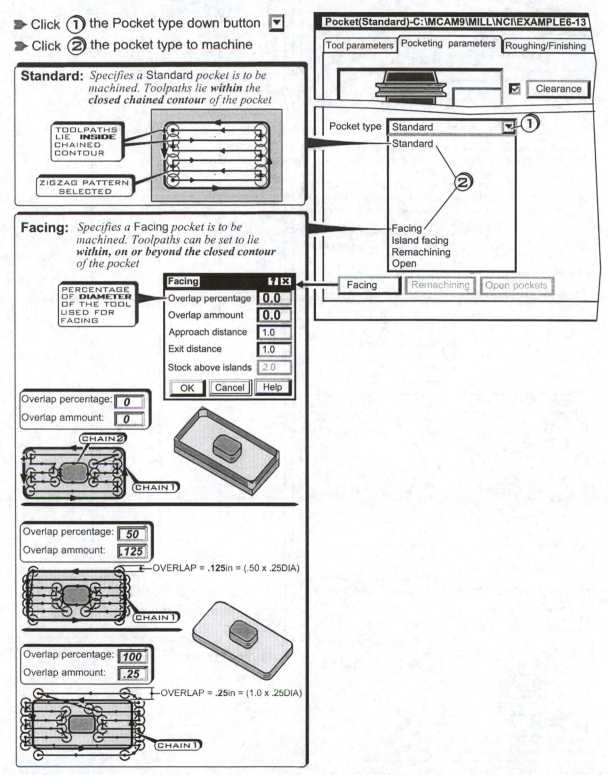

Standard: *Specifies a* Standard *pocket is to be machined. Toolpaths lie* ***within*** *the* ***closed chained contour*** *of the pocket*

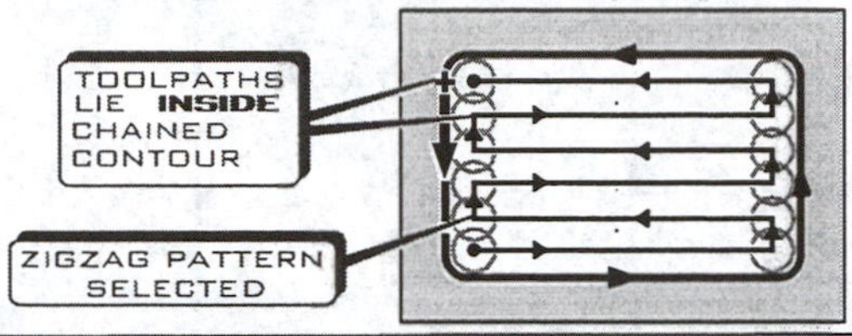

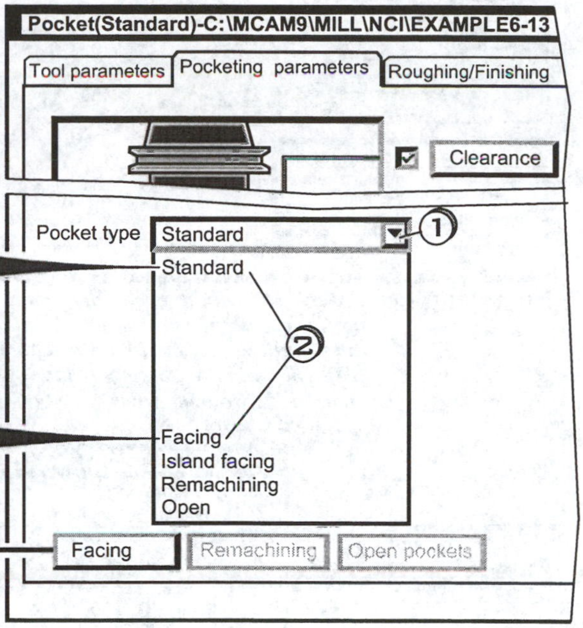

Facing: *Specifies a* Facing *pocket is to be machined. Toolpaths can be set to lie* ***within, on or beyond the closed contour*** *of the pocket*

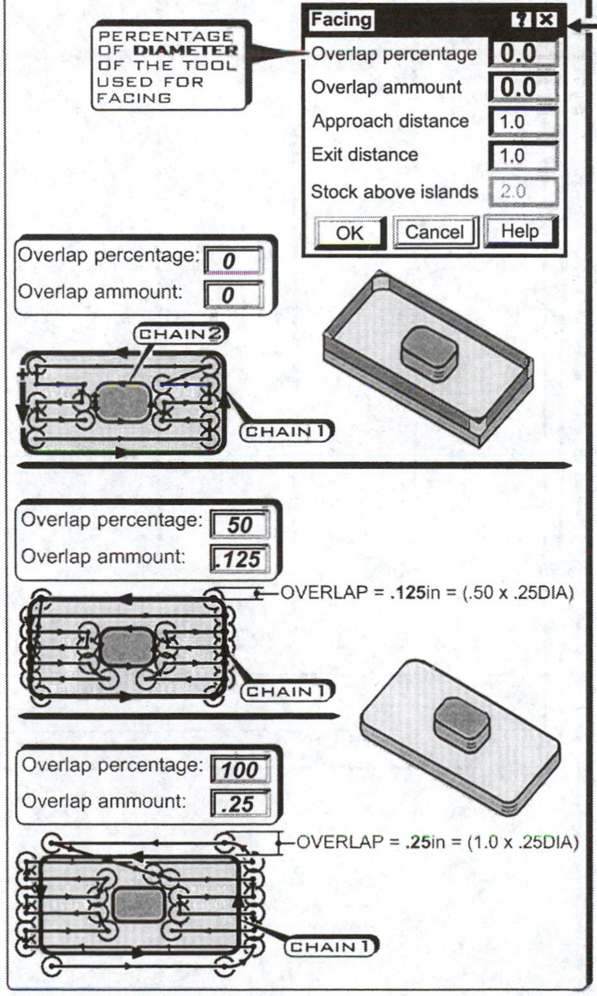

Island facing: *Specifies* ***only islands*** *in the part are to be faced. Toolpaths lie* ***within*** *the* ***closed contours*** *of the islands*

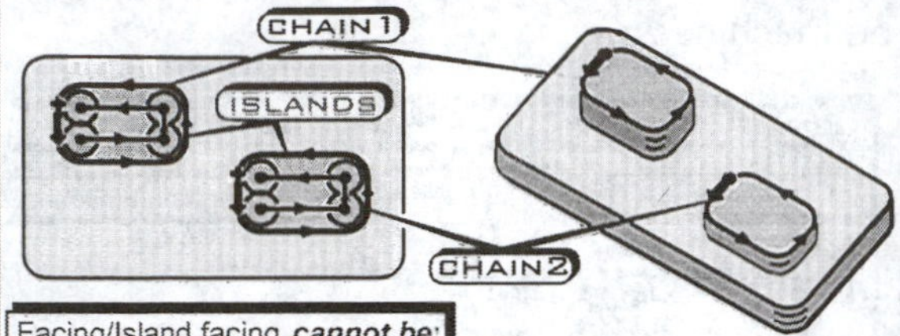

Facing/Island facing ***cannot be***:
- executed on Open pockets
- used with Remachinig
- used with Morph Spirling

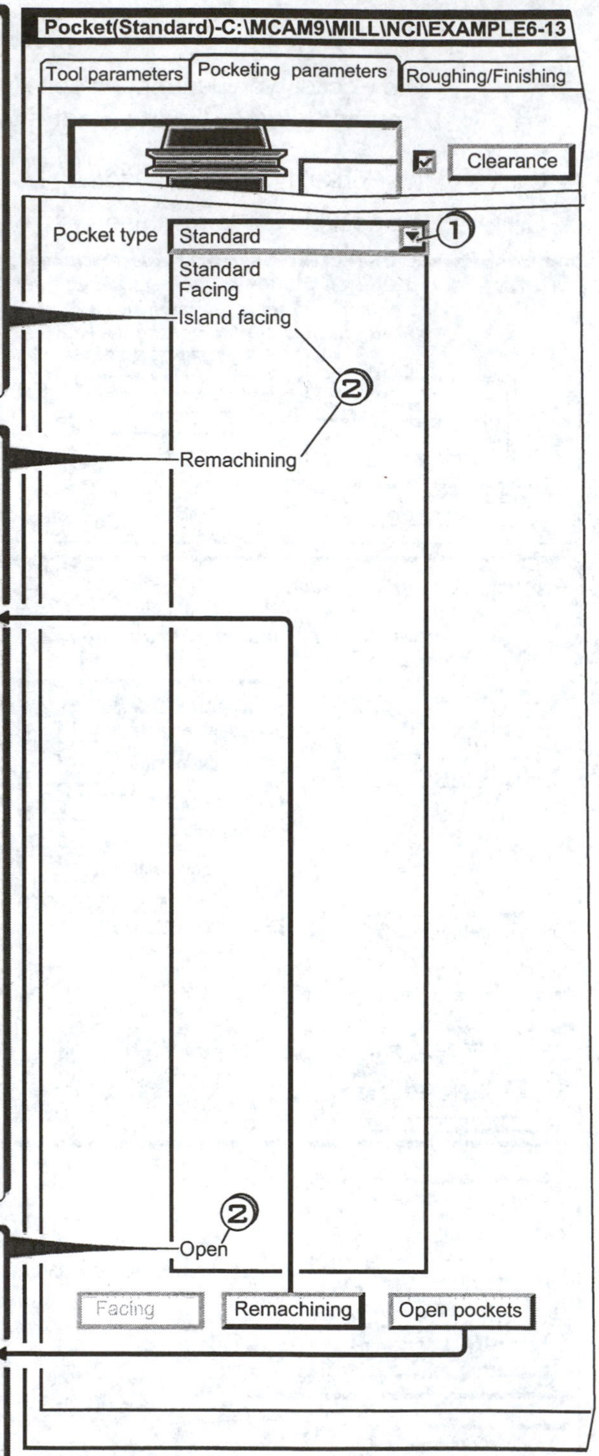

Remachining: *Specifies Mastercam is to detect any* ***tight areas*** *of the chained contour that* ***the roughing tool could not machine.*** *A* ***finish toolpath*** *is computed to* ***clear*** *these* ***problem areas*** *to within a* ***specified*** **Clearance** ***value***

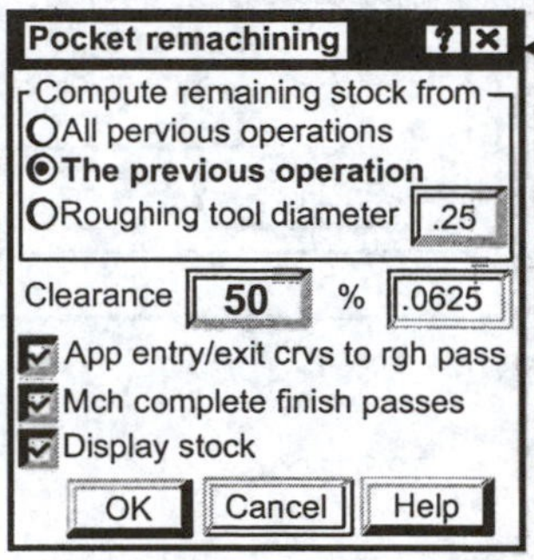

Remachining ***cannot be***:
- executed on Open pockets
- used with Facing

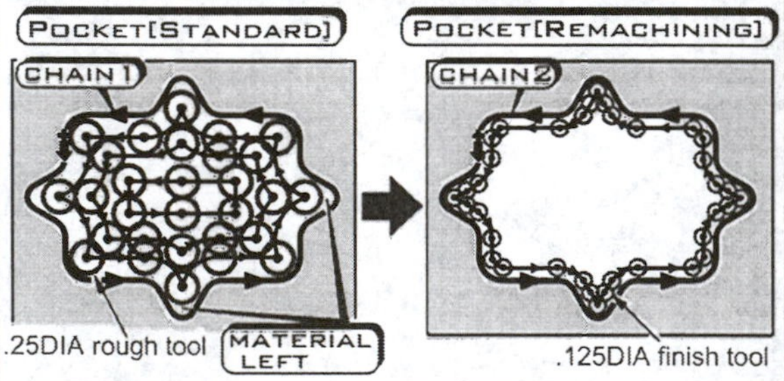

Open: *Specifies a pocket is to be machined within an* ***open chain contour***

Open ***cannot be***:
- used with Facing
- used with Remachining
- used for machining islands

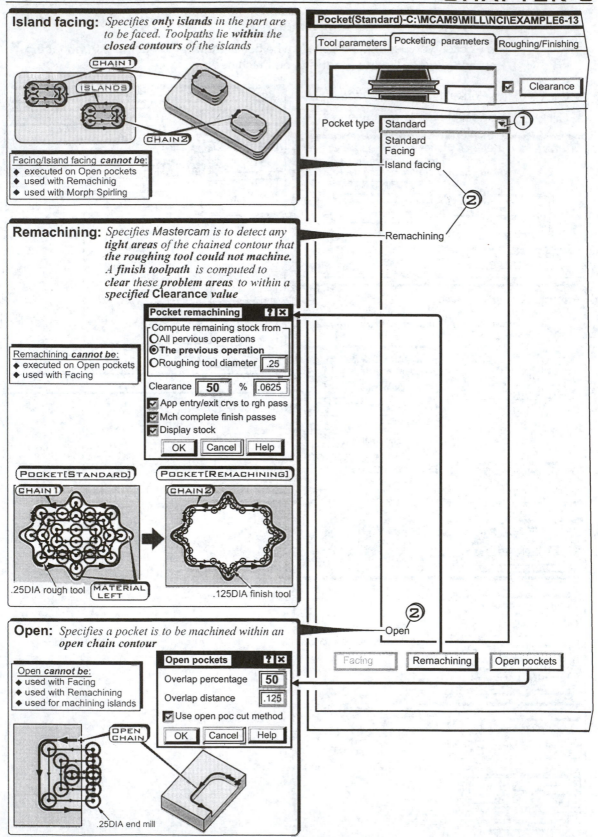

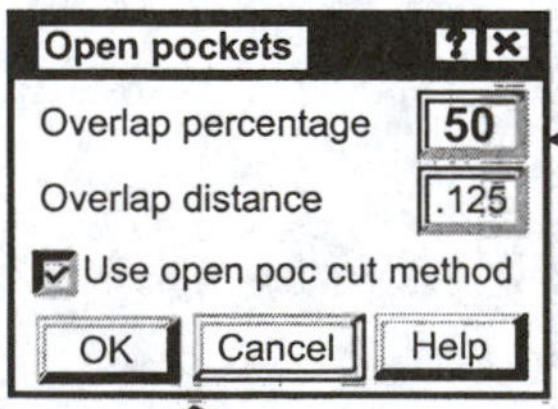

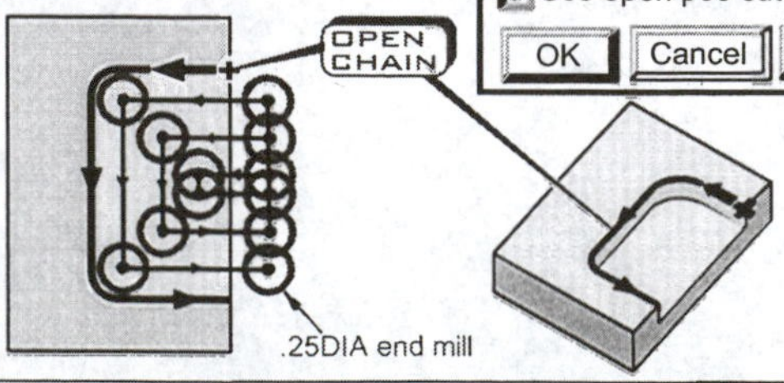

ADVANCED PARAMETERS

The advanced parameters control the ***tolerance*** for ***remachining*** as well as the the ***constant overlap spiral*** pocket machining pattern.

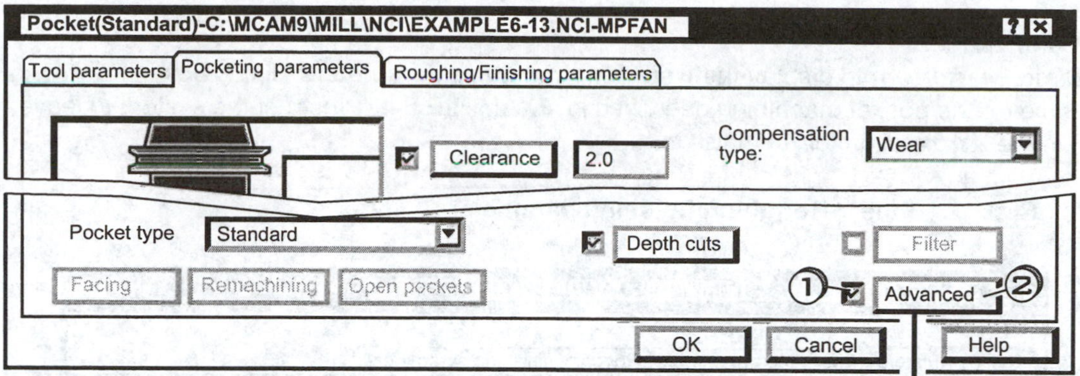

➤ Click ① the check ***on*** ☑

➤ Click ② the [Advanced] button

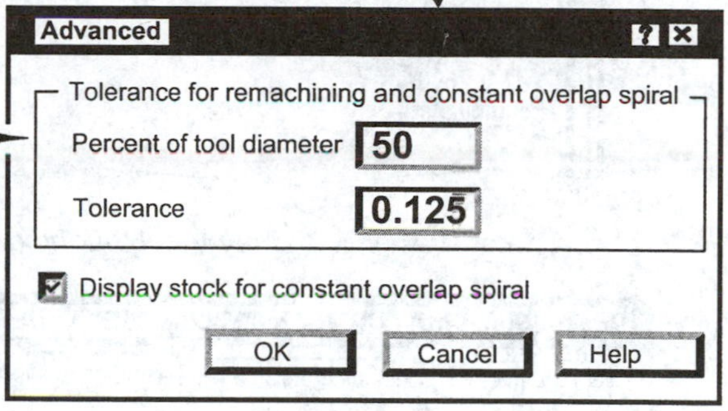

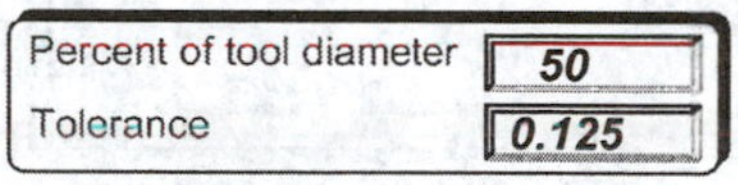

Medium *accuracy tool paths created for remachining and constant overlap spiral tool paths*
Less *processing time is required.*

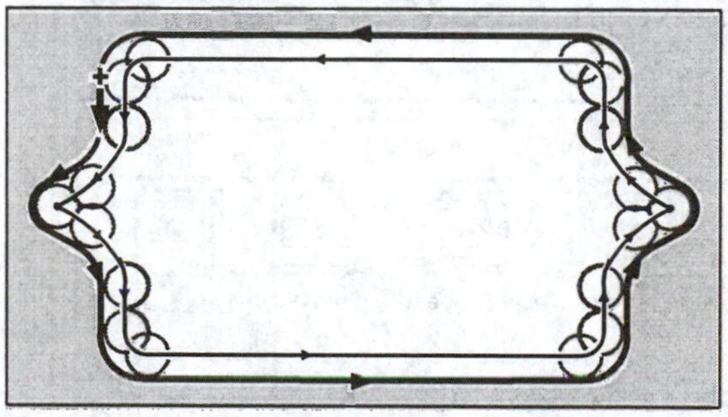

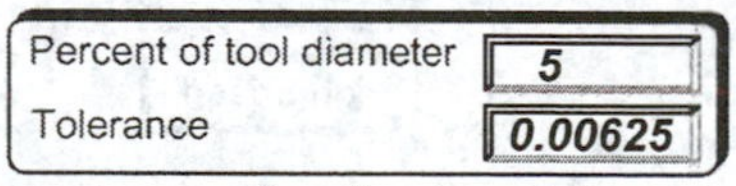

High *accuracy tool paths created for remachining and constant overlap spiral tool paths*
More *processing time is required.*

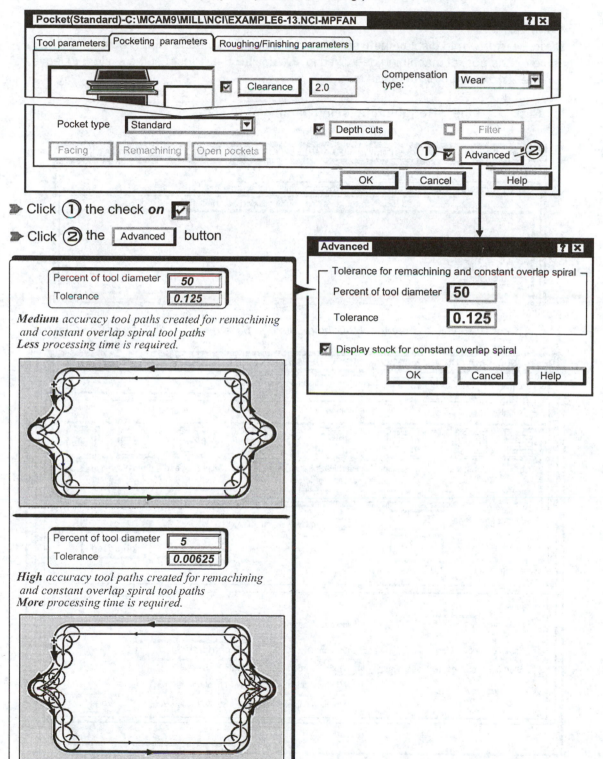

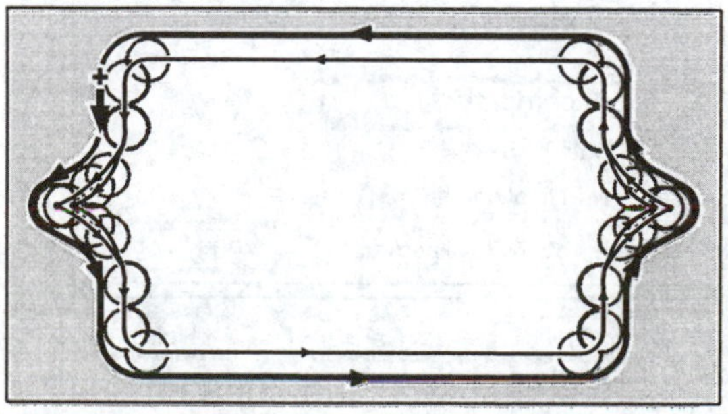

6-9 Specifying Roughing/Finishing Parameters in the Pocketing Module

A description of the *Roughing/finishing parameters* and their effect on tool paths is considered in this section.

EXAMPLE 6-19

Direct *Mastercam* to display the **Roughing/finishing parameters** dialog box in Pocketing module for the pocket machining described in **EXAMPLE 6-17.** Instruct the system to leave .03 in the XY direction for the finish cut.

➤ Click (24) the **Roughing/finishing parameters** tab

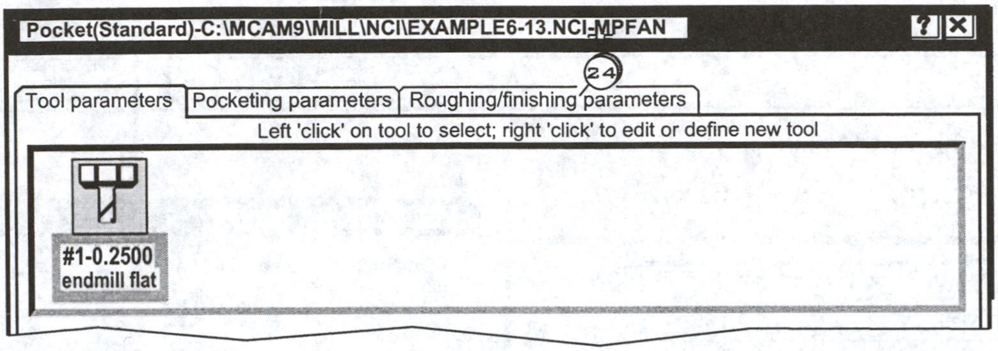

The system will display the *Roughing/finishing parameters* dialog box shown below

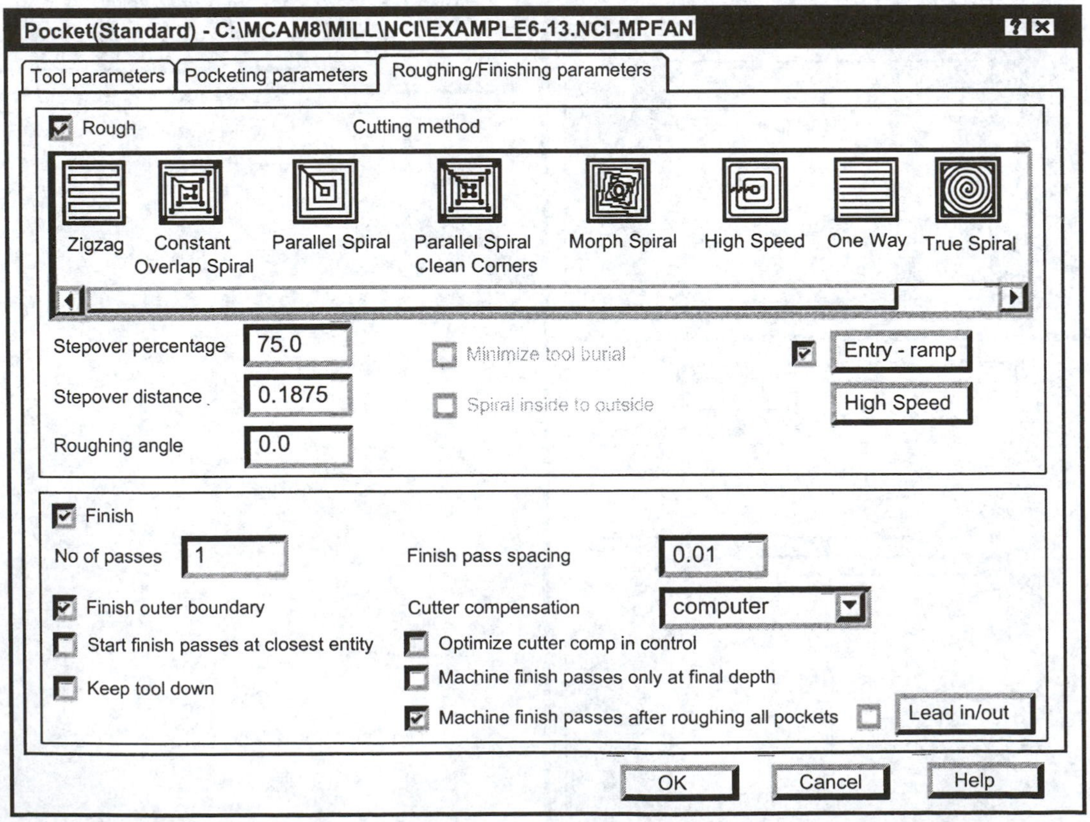

☑ Rough **CUTTING METHOD**

Mastercam features ***seven*** types of toolpaths for machining pockets, ***Zigzag, One Way***, ***Constant Overlap Spiral, Parallel Spiral, Parallel Spiral Clean Corners, Morph Spiral, High Speed Spiral*** and ***True Spiral***

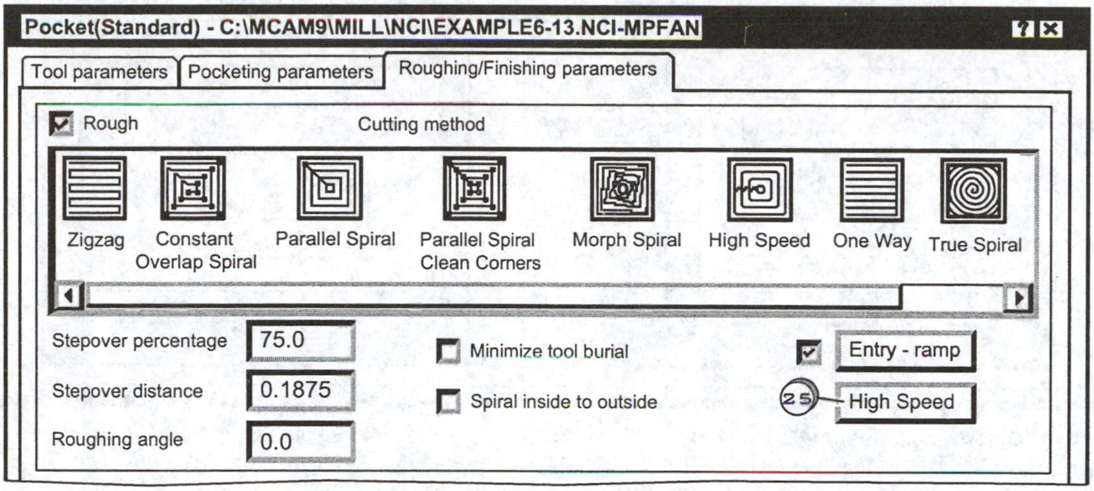

Zigzag: ***Rough*** *cuts the pocket by moving the tool back and forth along* ***straight lines****. The* ***orientation*** *of the pattern is set by the (* ***0°, 90°, 180°,*** *or* ***270°****) value of the* ***roughing angle***

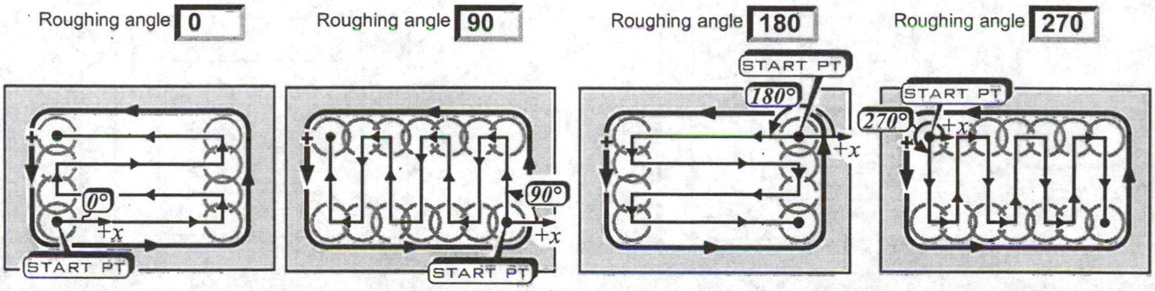

Minimize tool burial: *Useful when using the* ***Zigzag*** *pattern to machine* ***around islands*** *with* ***small tools*** *instead of across areas where tool burial damage typically occurs.*
This option may increase machining time but decrease chance of tool damage.

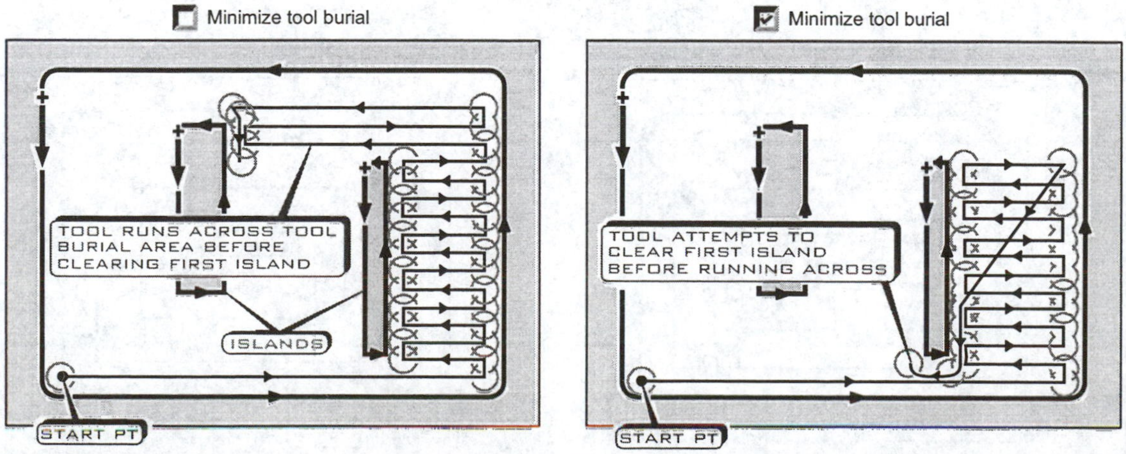

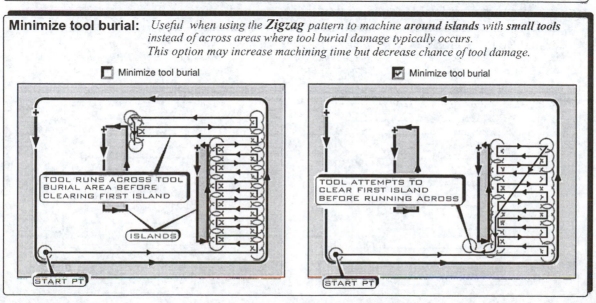

Constant Overlap Spiral: *Executes one roughing pass then determines the next pass based on the remaining area to be cut. This process is repeated until the pocket is cleared. The constant overlap spiral pattern uses more small linear moves to clear more material than the parallel spiral pattern.*

☑ Spiral inside to outside

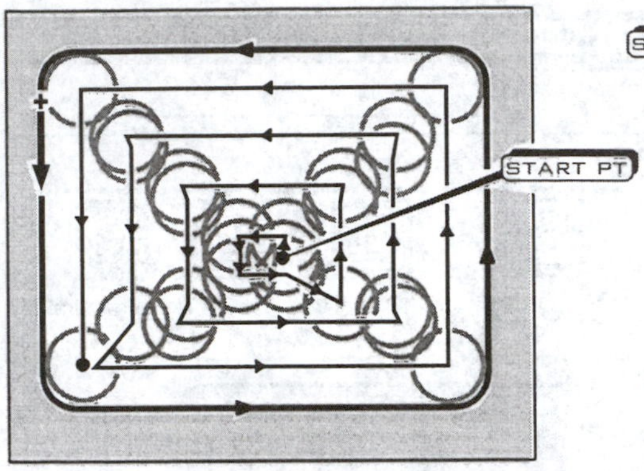

☐ Spiral inside to outside

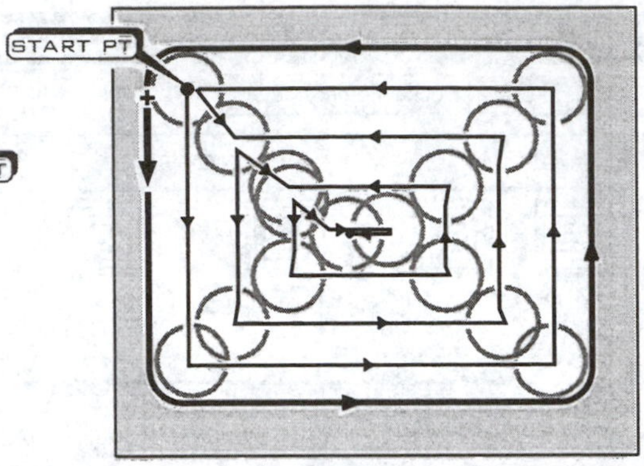

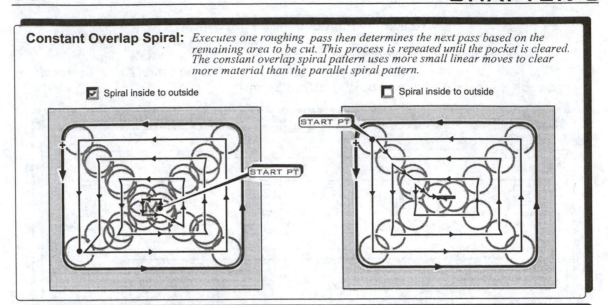

Parallel Spiral: ***Roughing** toolpaths follow a spiral pattern from inside to outside or outside to inside. Each new pass is **offset from the pervious pass by the stepover value entered**. This pattern does **not guarantee cleanout** of the pocket*

☑ Spiral inside to outside

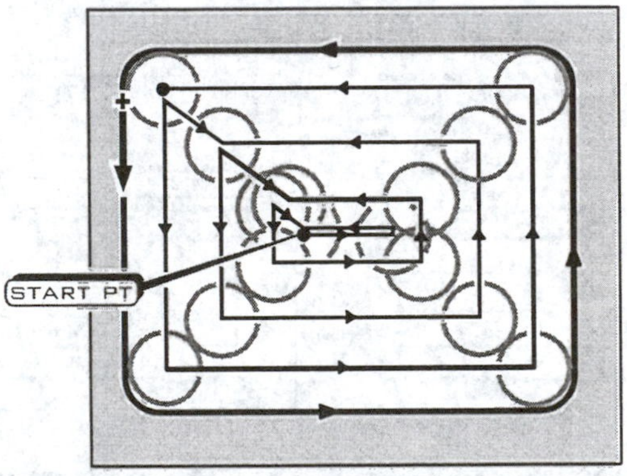

☐ Spiral inside to outside

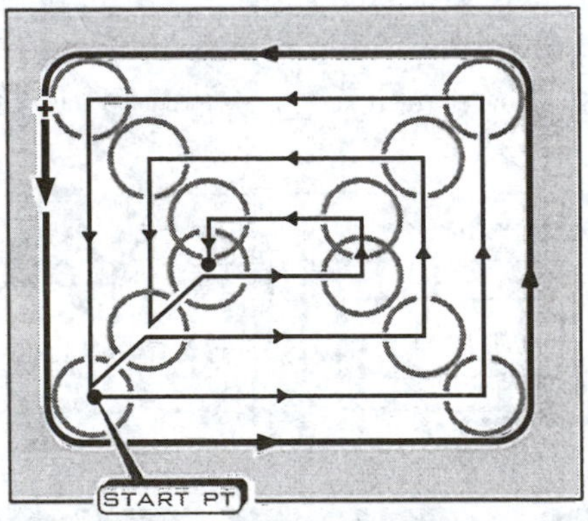

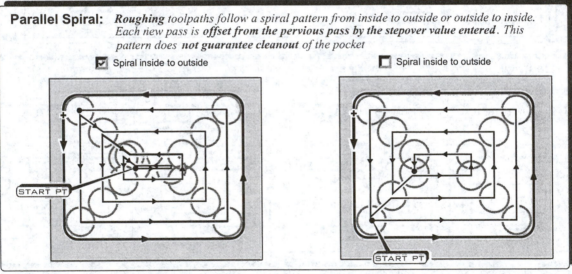

Parallel Spiral, Clean Corners: *Follows the same type of tool paths as the parallel spiral pattern except that **corner clean out moves are added**. This pattern increases the possibility but does not guarantee total cleanout of the pocket*

☑ Spiral inside to outside

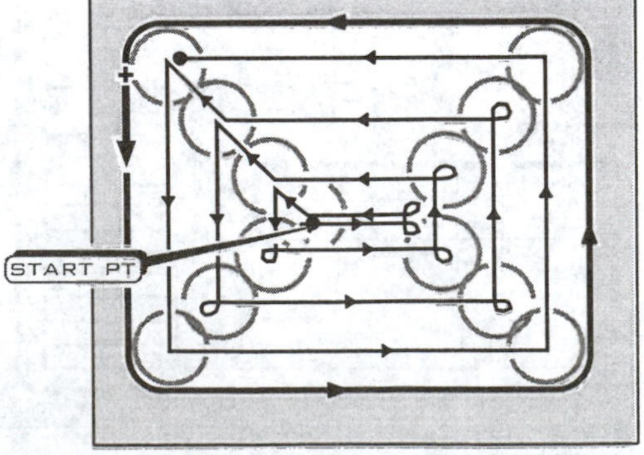

☐ Spiral inside to outside

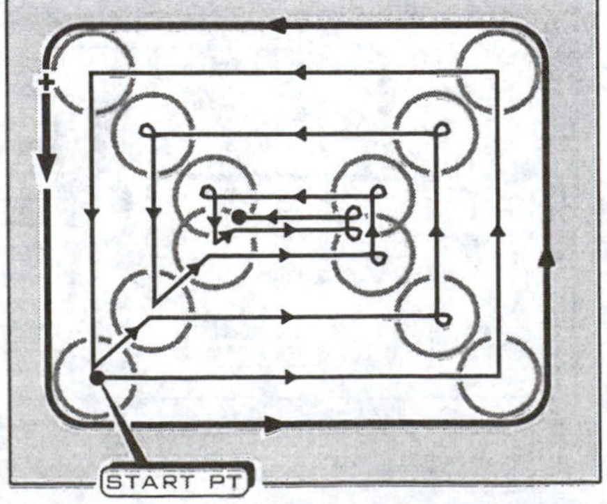

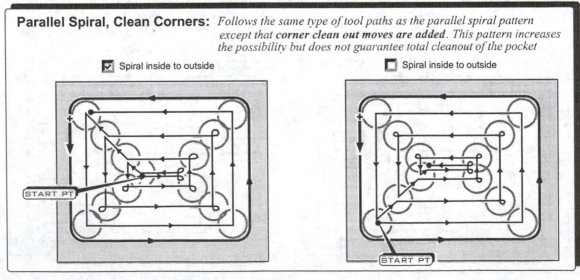

Morph Spiral: ***Roughing*** *Toolpaths are generated bu gradually interpolating between the outer bpundary of the pocket and the islands.*

☑ Spiral inside to outside

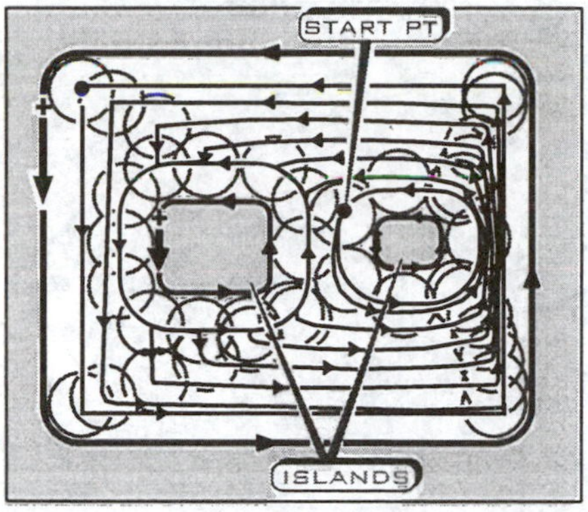

☐ Spiral inside to outside

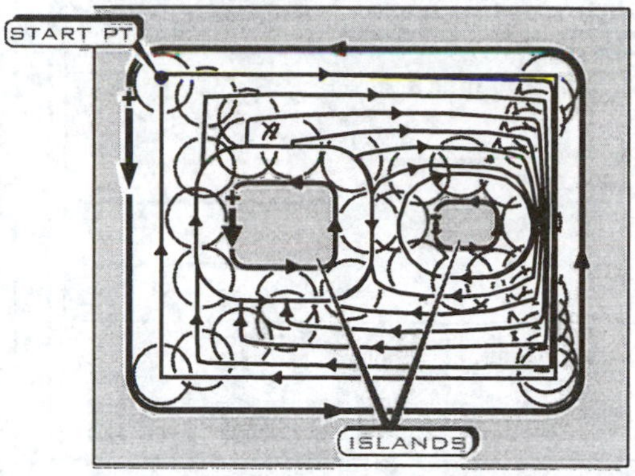

True Spiral: ***Roughing*** *toolpaths follow a spiral pattern with* ***tangent arc motion*** *from inside to outside or outside to inside. Tool moves* ***smoothly*** *with* ***minimal NC code generated*** *and* ***good pocket cleanout*** *resulting.*

☑ Spiral inside to outside

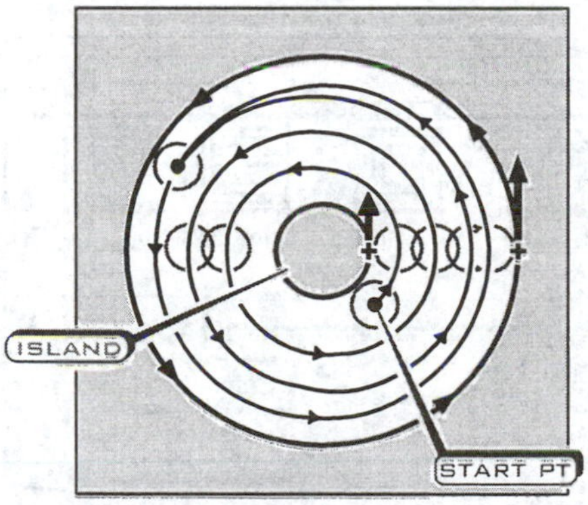

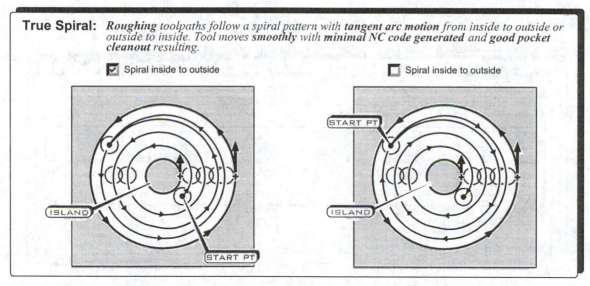

☐ Spiral inside to outside

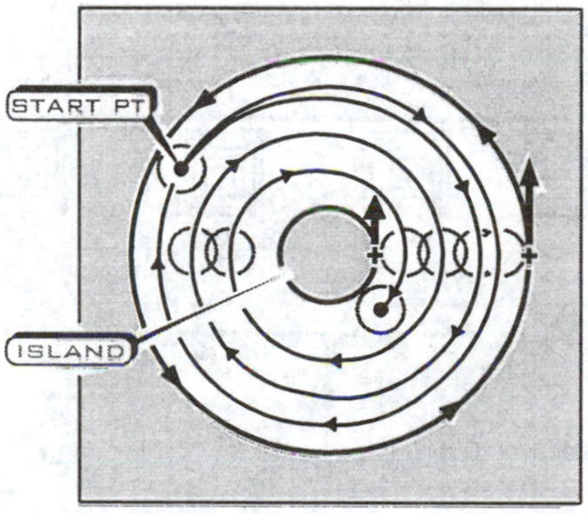

One Way: ***Roughing*** *tool paths are* ***linear*** *and cause tool to* ***cut*** *the pocket in* ***one direction.***

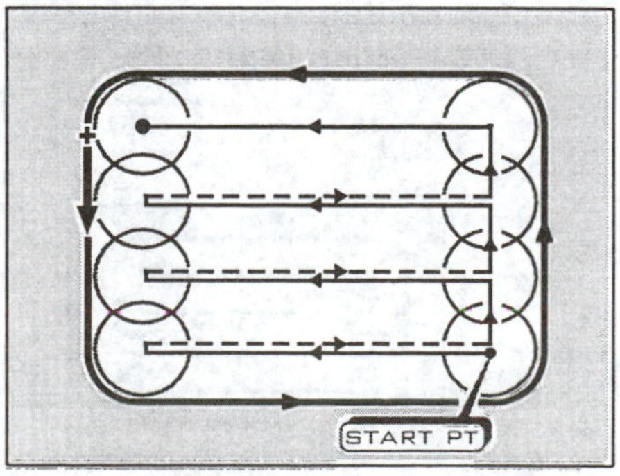

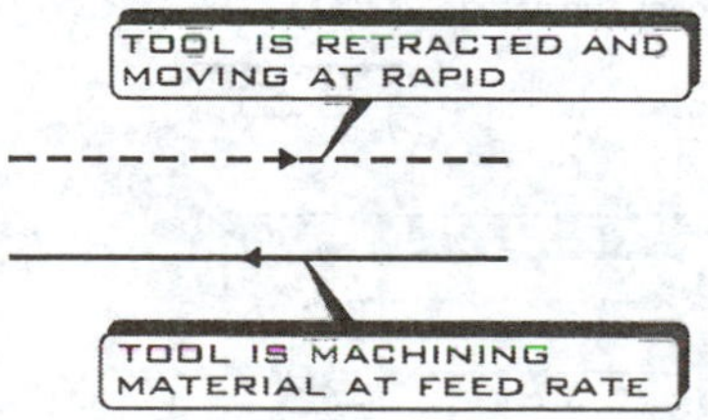

High Speed: ***Roughing*** *toolpaths follow smooth arc patterns that optionally* ***eliminates tool burial****.*

Click (25) *the* High Speed *button(See page 6-53), to select the machining parameters.*

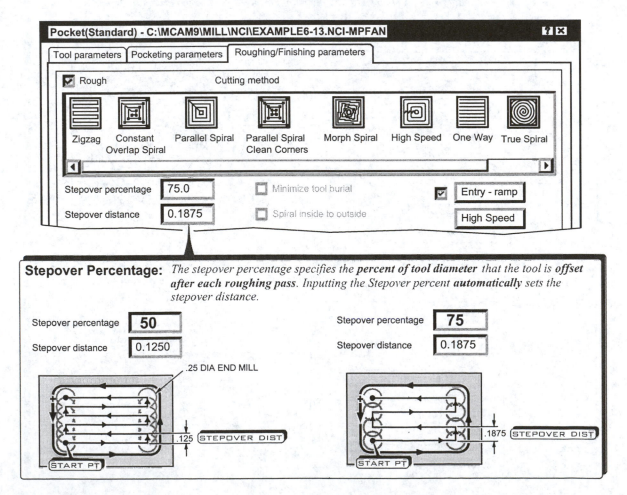

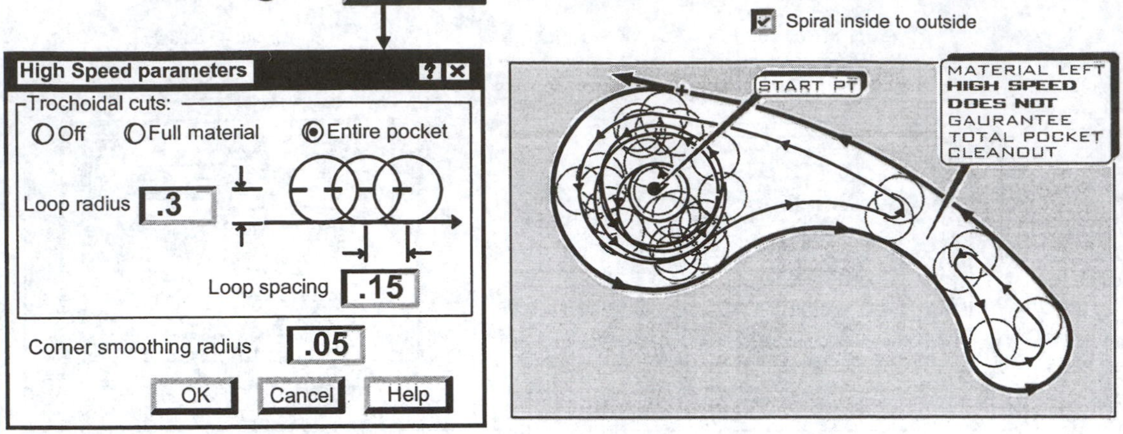

Rough **STEPOVER PARAMETERS**

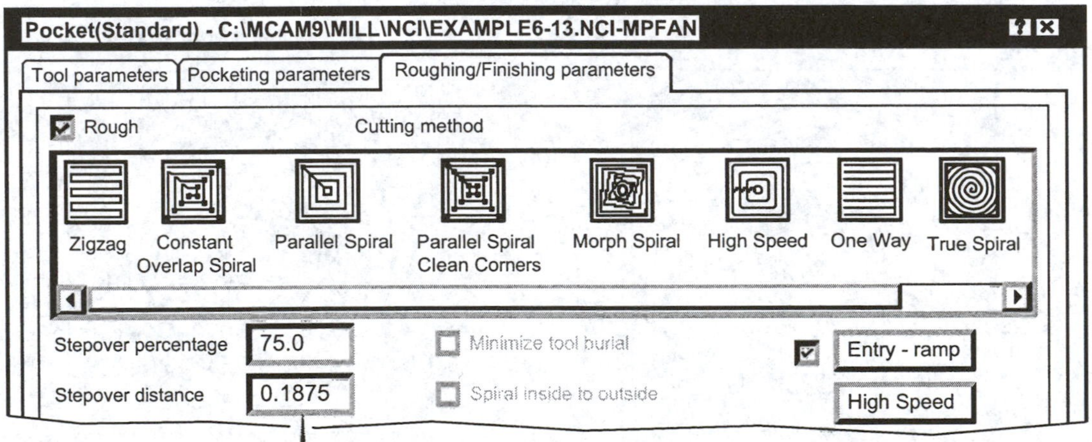

Stepover Percentage: *The stepover percentage specifies the* ***percent of tool diameter*** *that the tool is* ***offset after each roughing pass****. Inputting the Stepover percent* ***automatically*** *sets the stepover distance.*

Stepover percentage	**50**
Stepover distance	0.1250

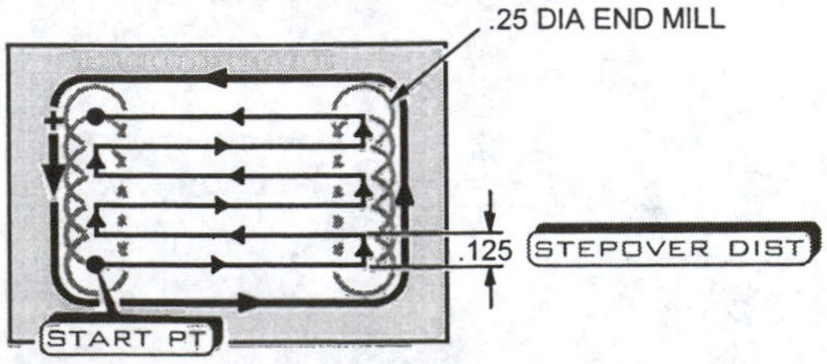

Stepover percentage	**75**
Stepover distance	0.1875

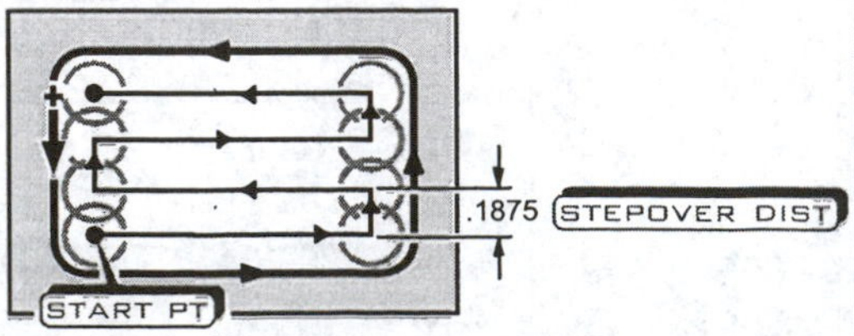

☑ Rough **ENTRY-HELIX/RAMP PARAMETERS**

The operator can select ***one*** of ***Three*** types of ***entry*** tool paths for ***roughing a pocket : Off(straight plunge entry), Ramp(enter on zig and zag angles), Helix(enter by following a downward spiral curve).***

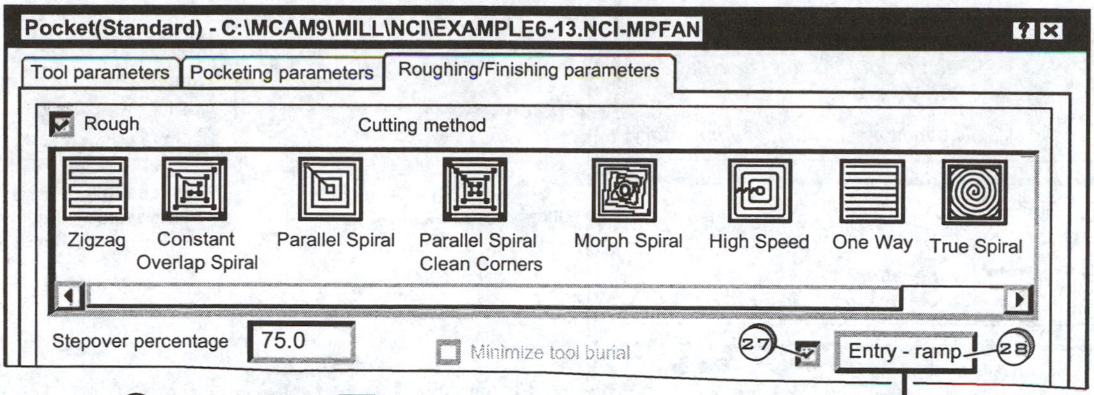

➤ Click (27) the check ***on*** ☑

➤ Click (28) the [Entry-ramp] button

Ramp: *Entry tool path is a series of **straight zig/zag lines inclined at** user specified **zig/zag angles***

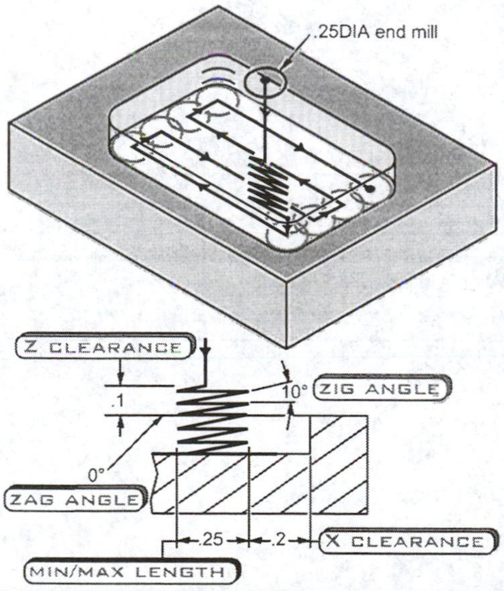

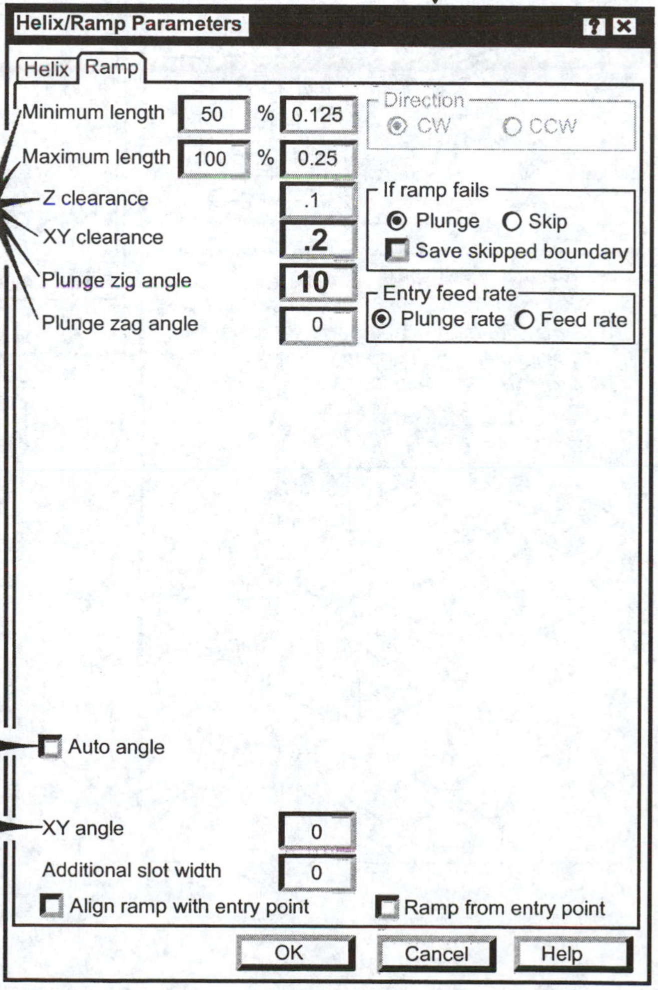

☑ Auto angle *Directs Mastercam to automatically assign the XY angle*

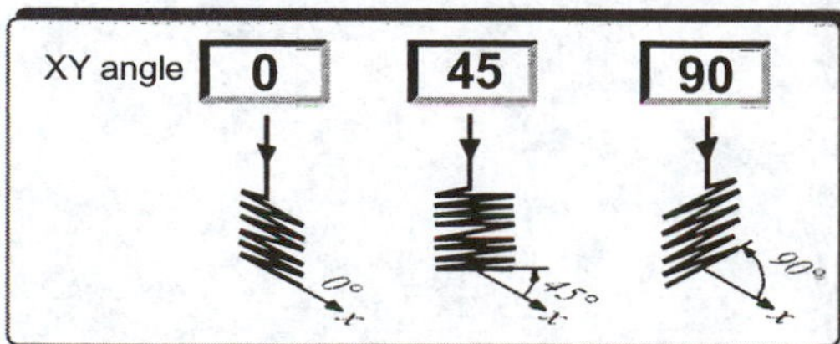

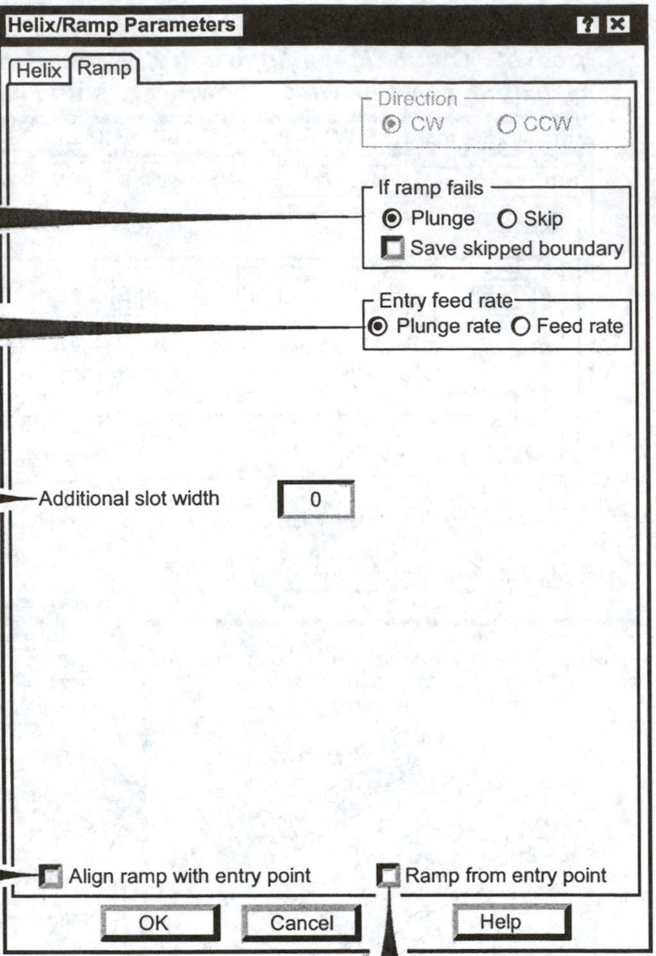

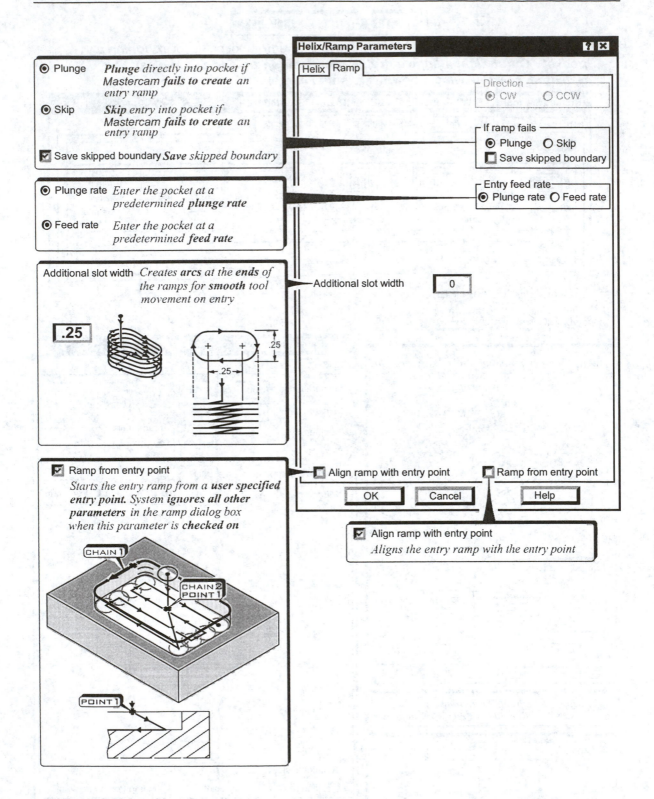

Plunge — ***Plunge** directly into pocket if Mastercam **fails to create** an entry ramp*

Skip — ***Skip** entry into pocket if Mastercam **fails to create** an entry ramp*

Save skipped boundary — ***Save** skipped boundary*

Plunge rate — *Enter the pocket at a predetermined **plunge rate***

Feed rate — *Enter the pocket at a predetermined **feed rate***

Additional slot width — *Creates **arcs** at the **ends** of the ramps for **smooth** tool movement on entry*

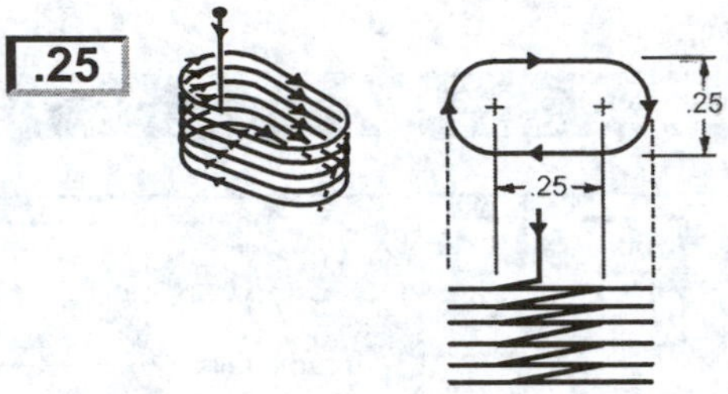

Ramp from entry point

*Starts the entry ramp from a **user specified entry point.** System **ignores all other parameters** in the ramp dialog box when this parameter is **checked on***

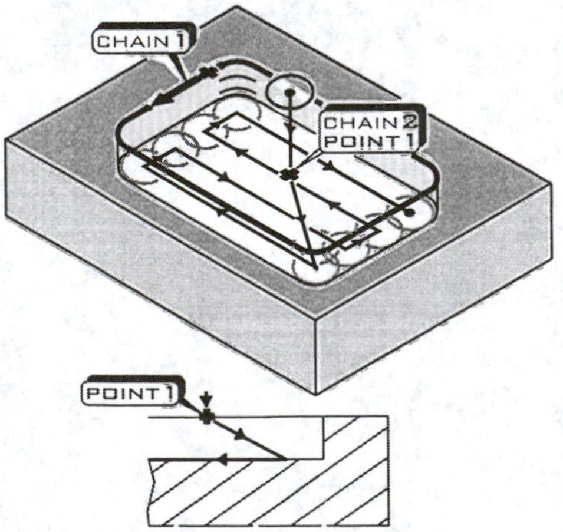

Align ramp with entry point

Aligns the entry ramp with the entry point

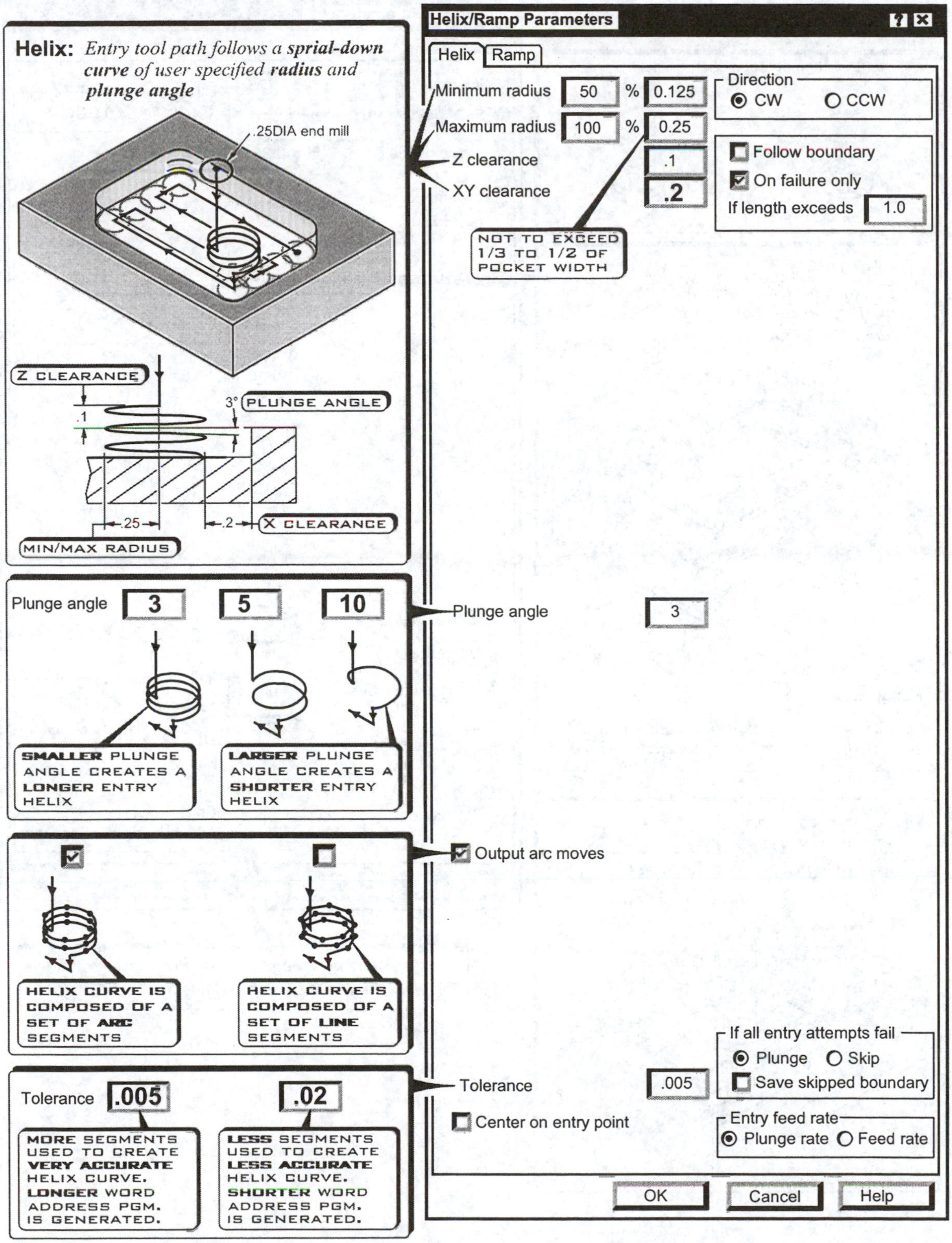
Helix: Entry tool path follows a sprial-down curve of user specified radius and plunge angle
.25DIA end mill
Z CLEARANCE
.1
3°
PLUNGE ANGLE
.25
.2
X CLEARANCE
MIN/MAX RADIUS
Plunge angle
3
5
10
SMALLER PLUNGE ANGLE CREATES A LONGER ENTRY HELIX
LARGER PLUNGE ANGLE CREATES A SHORTER ENTRY HELIX
HELIX CURVE IS COMPOSED OF A SET OF ARC SEGMENTS
HELIX CURVE IS COMPOSED OF A SET OF LINE SEGMENTS
Tolerance
.005
.02
MORE SEGMENTS USED TO CREATE VERY ACCURATE HELIX CURVE. LONGER WORD ADDRESS PGM. IS GENERATED.
LESS SEGMENTS USED TO CREATE LESS ACCURATE HELIX CURVE. SHORTER WORD ADDRESS PGM. IS GENERATED.
Helix/Ramp Parameters
Helix
Ramp
Minimum radius
50
%
0.125
Maximum radius
100
%
0.25
Z clearance
.1
XY clearance
.2
NOT TO EXCEED 1/3 TO 1/2 OF POCKET WIDTH
Direction
CW
CCW
Follow boundary
On failure only
If length exceeds
1.0
Plunge angle
3
Output arc moves
If all entry attempts fail
Plunge
Skip
Save skipped boundary
Tolerance
.005
Center on entry point
Entry feed rate
Plunge rate
Feed rate
OK
Cancel
Help

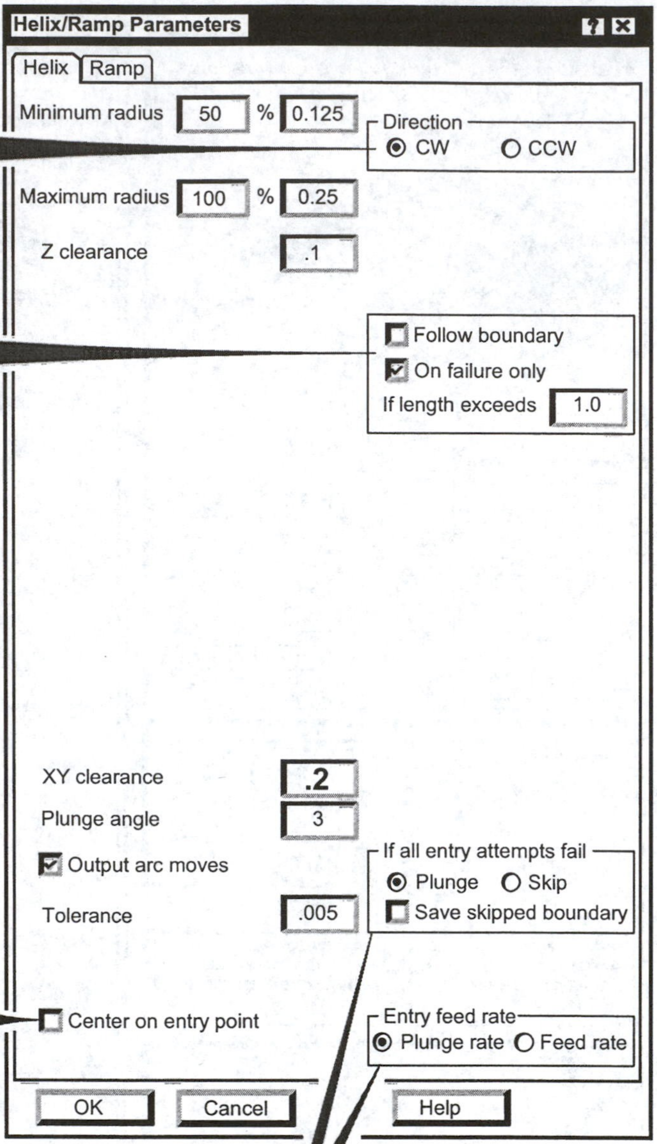

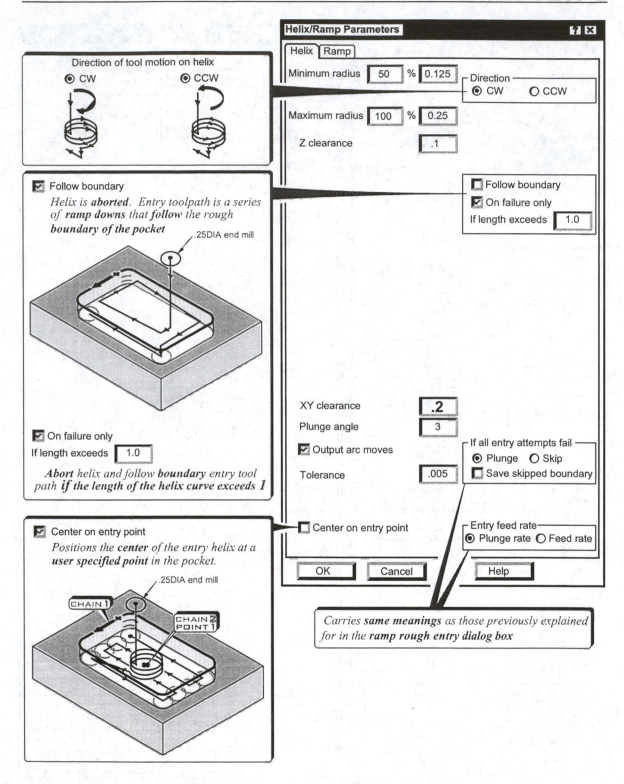

Direction of tool motion on helix

◉ CW ◉ CCW

☑ Follow boundary

Helix is ***aborted****. Entry toolpath is a series of* ***ramp downs*** *that* ***follow*** *the rough* ***boundary of the pocket***

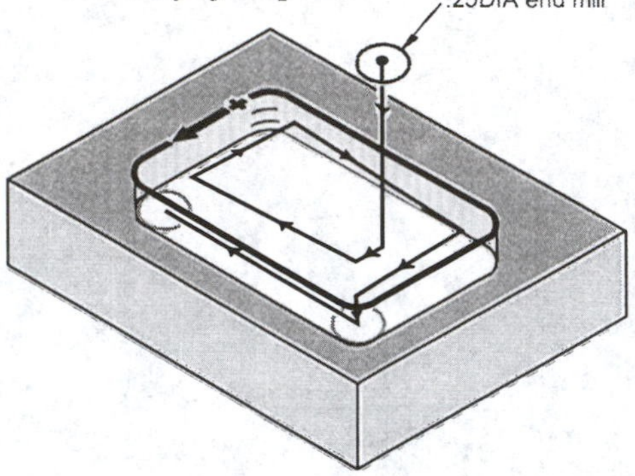

☑ On failure only

If length exceeds 1.0

Abort *helix and follow* ***boundary*** *entry tool path* ***if the length of the helix curve exceeds 1***

☑ Center on entry point

Positions the ***center*** *of the entry helix at a* ***user specified point*** *in the pocket.*

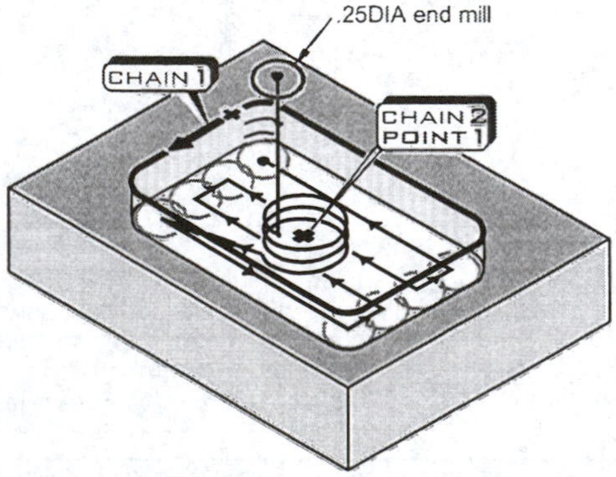

Carries ***same meanings*** *as those previously explained for in the* ***ramp rough entry dialog box***

☑ Finish **LEAD IN/OUT PARAMETERS**

It is recommended to add a lead in/out to the finish pass to insure that a ***dwell mark does not occur*** on the part at the end of the tool path.

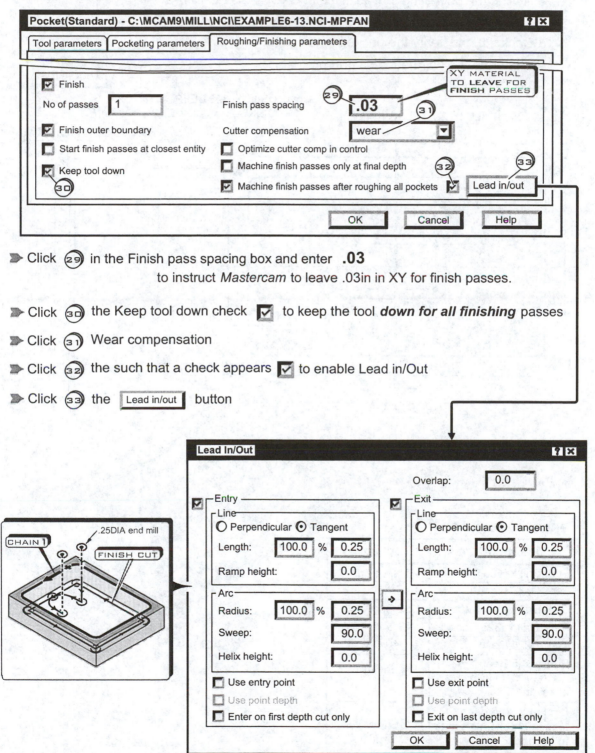

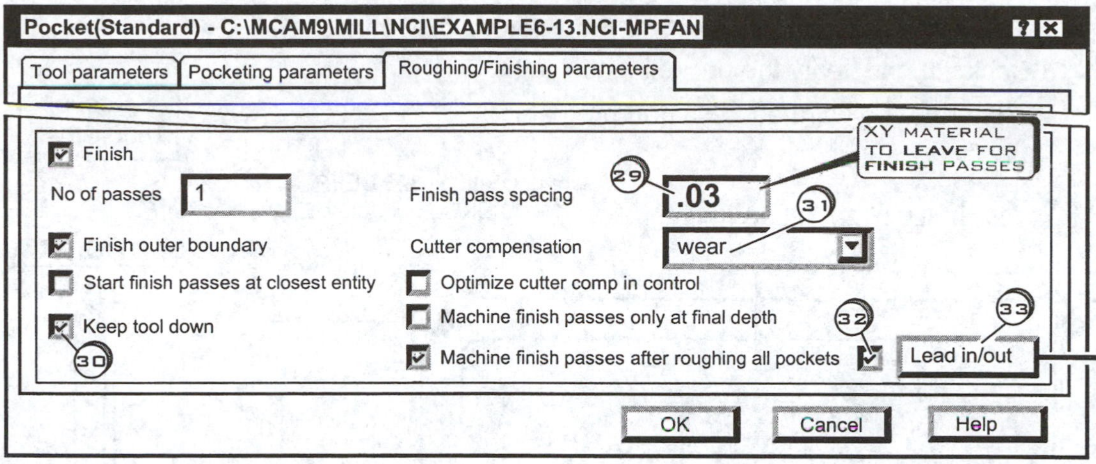

➤ Click (29) in the Finish pass spacing box and enter **.03** to instruct *Mastercam* to leave .03in in XY for finish passes.

➤ Click (30) the Keep tool down check ☑ to keep the tool ***down for all finishing*** passes

➤ Click (31) Wear compensation

➤ Click (32) the such that a check appears ☑ to enable Lead in/Out

➤ Click (33) the [Lead in/out] button

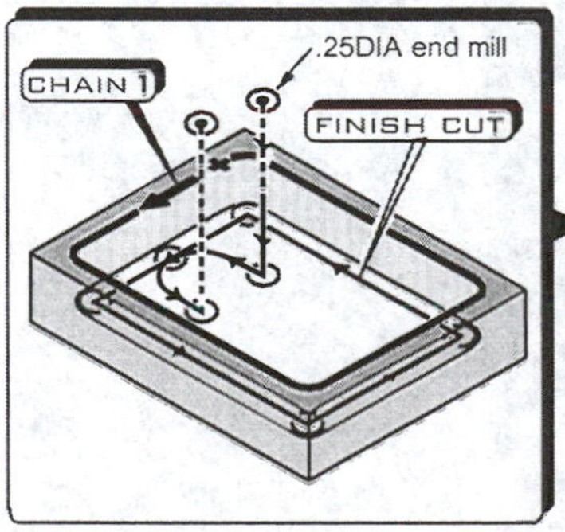

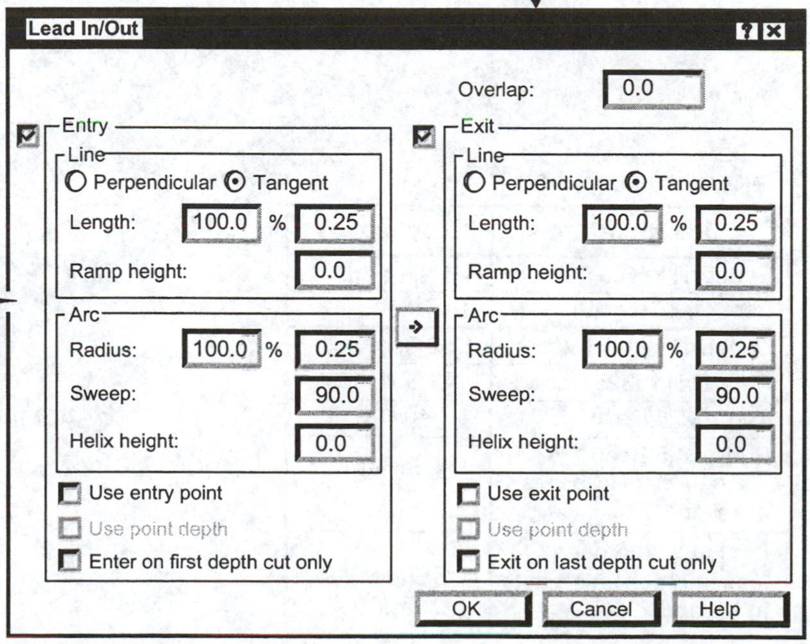

EXAMPLE 6-20

Direct *Mastercam* to execute the operations listed below on the part shown in Figure 6-9

- Create the required pocket machining contours by chaining
- Assign the pocketing tool as listed in PROCESS PLAN 6-3
- Enter the data in the **Pocketing parameters** dialog box
- Backplot and verify the pocketing toolpaths
- Generate the word address part program

Material
1030 Steel

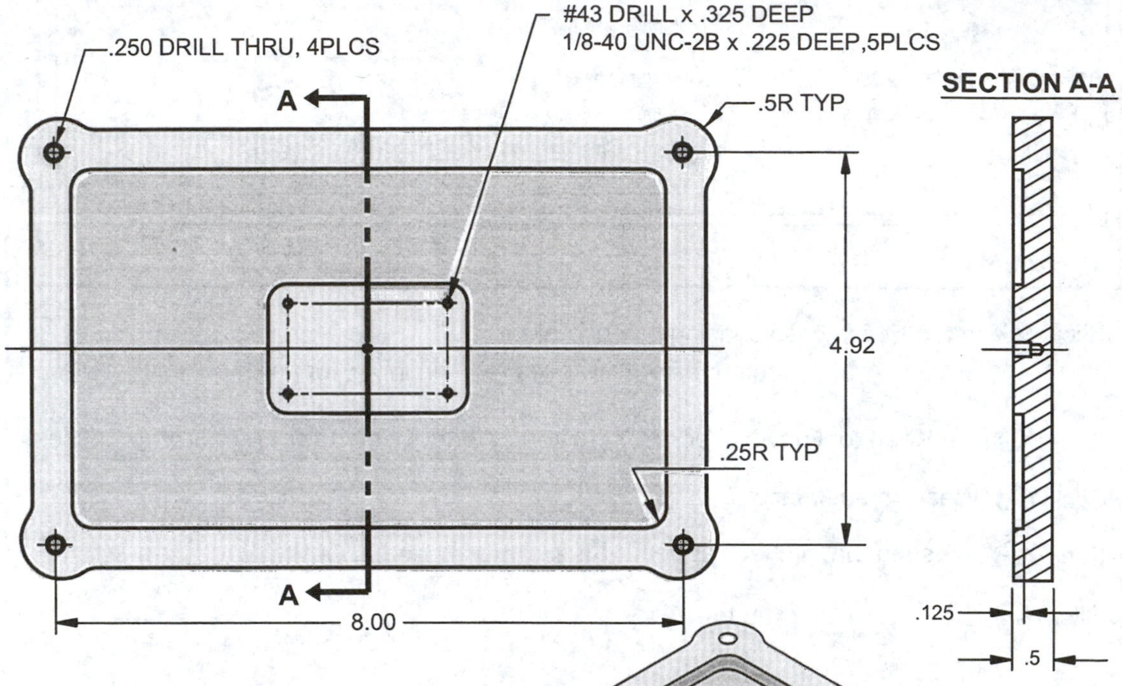

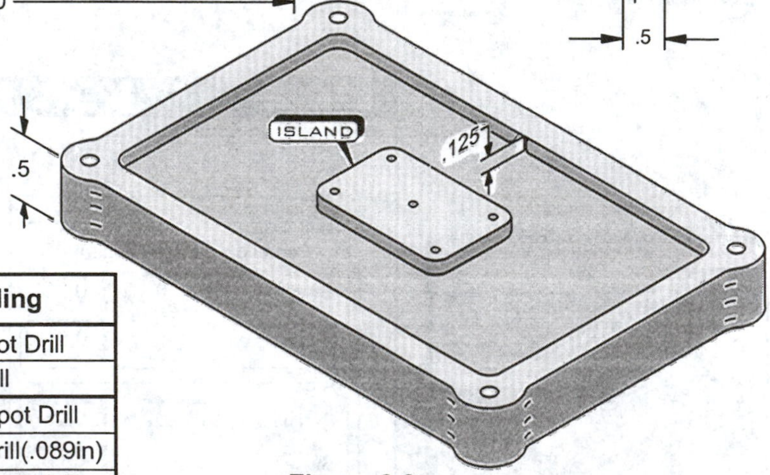

PROCESS PLAN 6-3

Operation	Tooling
Spot Drill x .166 Deep	1/8 Spot Drill
Peck Drill Through	1/4 Drill
Spot Drill x .125 Deep	1/32 Spot Drill
Peck Drill x .325 Deep	#43 Drill(.089in)
Tap x .225 Deep	1/8-40 UNC Tap
Profile x .5 Deep leave .01 for finish cuts in XY and Z	1/4-End Mill
Pocket x .125 deep leave .02 for finish cuts in XY and Z	1/2-End Mill

Figure 6-9

A) Create the Pocket and Island Macining Contours by Chaining

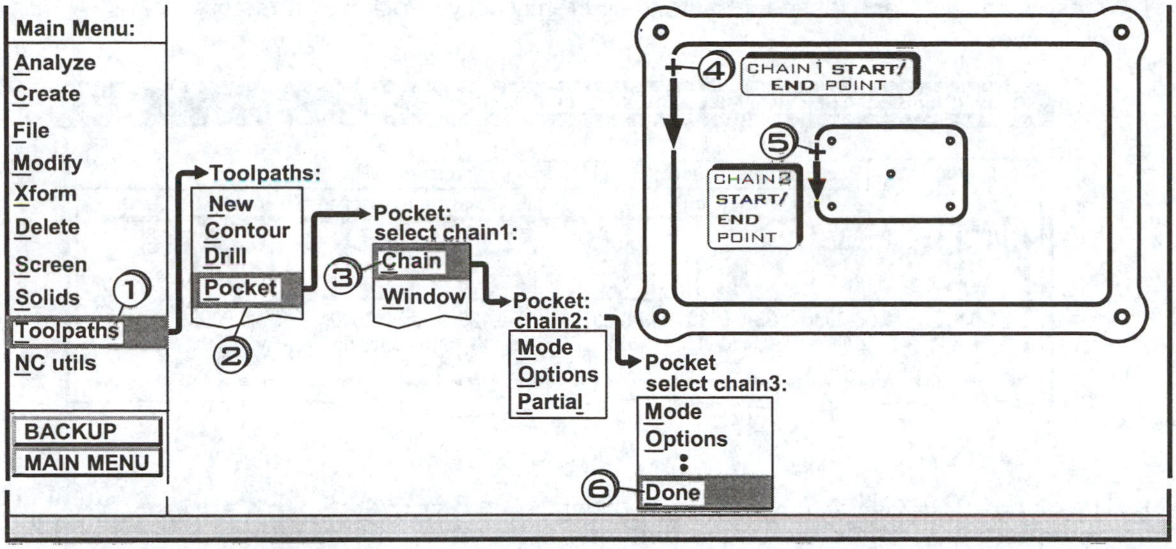

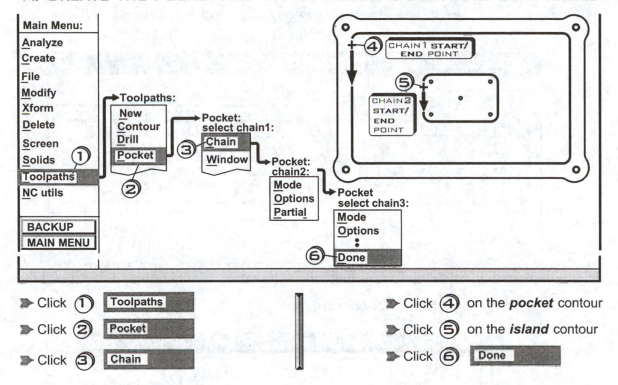

- Click ① **Toolpaths**
- Click ② **Pocket**
- Click ③ **Chain**
- Click ④ on the ***pocket*** contour
- Click ⑤ on the ***island*** contour
- Click ⑥ **Done**

B) Obtain the Needed 1/2 End Mill Tool

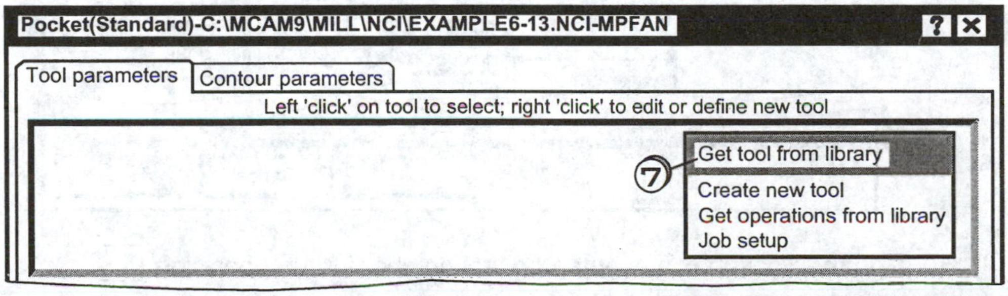

- ***Right*** Click move the mouse cursor *down* and Click ⑦ Get tool from library

Select the 1/2in Dia End mill tool from the **TOOLS** tool library

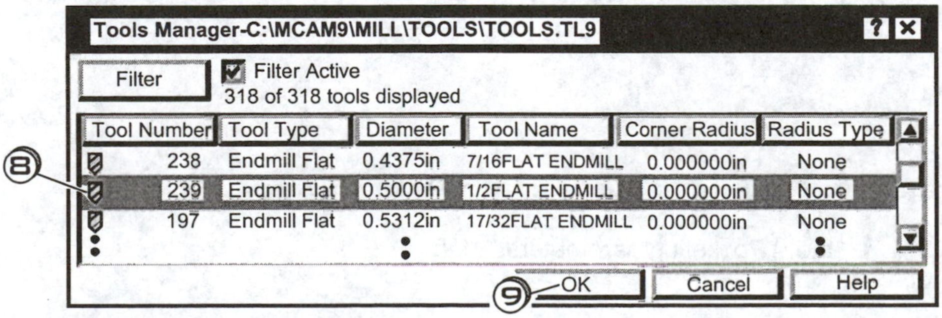

- Click ⑧ on the 1/2 FLAT ENDMILL
- Click ⑨ the OK button

Mastercam will place these tools into the currently active Pocket Parameters dialog box

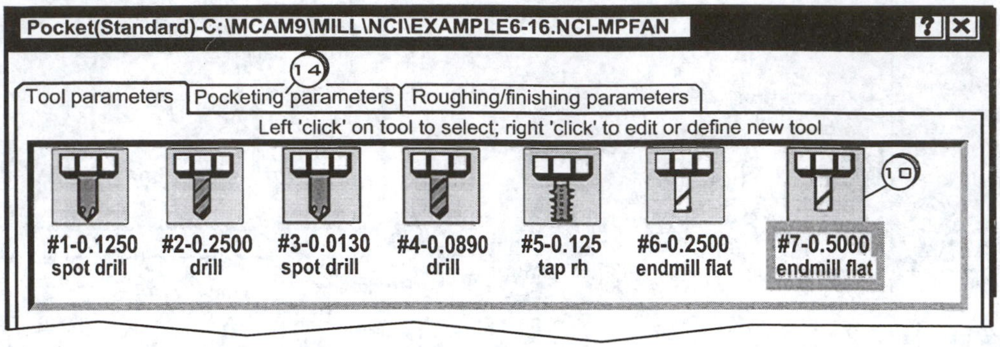

C) DIRECT *Mastercam* TO ASSIGN THE TOOL SPEED/FEED BASED ON THE STOCK MATERIAL SELECTED IN **Job Setup**

➤ ***Right*** Click (10) on the .5000 endmill flat tool

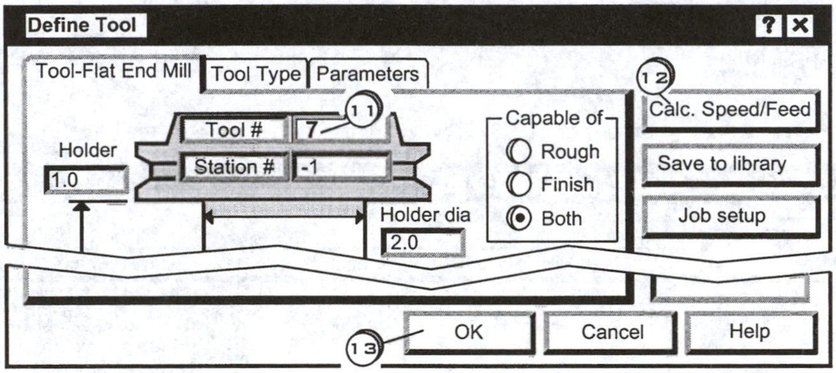

➤ Click (11) in the Tool# box and enter the tool number for the operation [7]

➤ Click (12) the [Calc. Speed/Feed] key

➤ Click (13) the [OK] key

D) ENTER THE REQUIRED POCKETING PARAMETERS

➤ Click (14) the [Pocketing parameters] tab

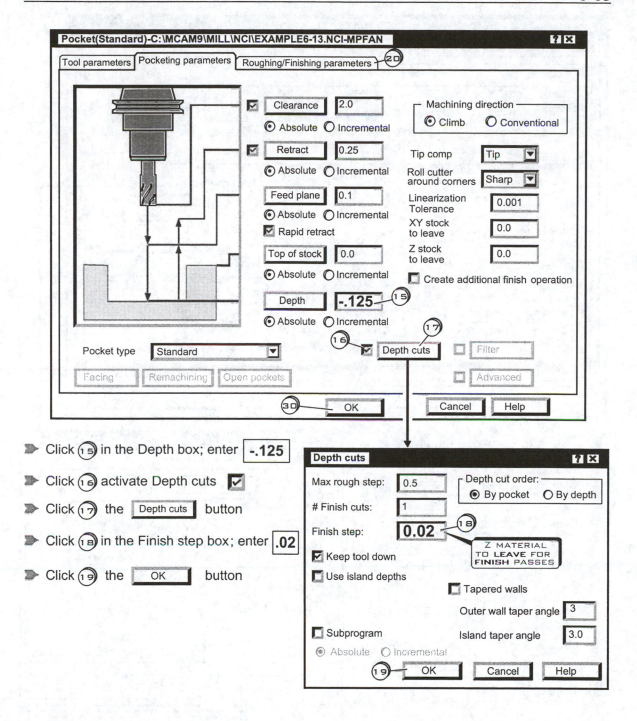

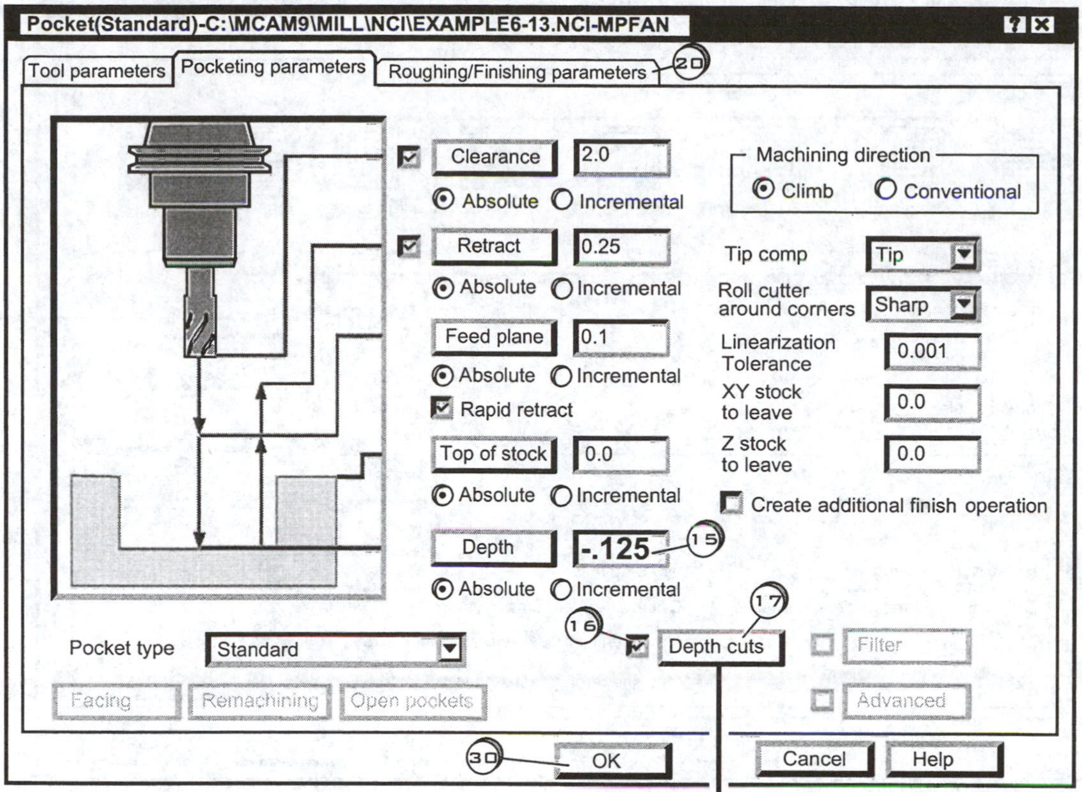

➤ Click 15 in the Depth box; enter -.125

➤ Click 16 activate Depth cuts ☑

➤ Click 17 the Depth cuts button

➤ Click 18 in the Finish step box; enter .02

➤ Click 19 the OK button

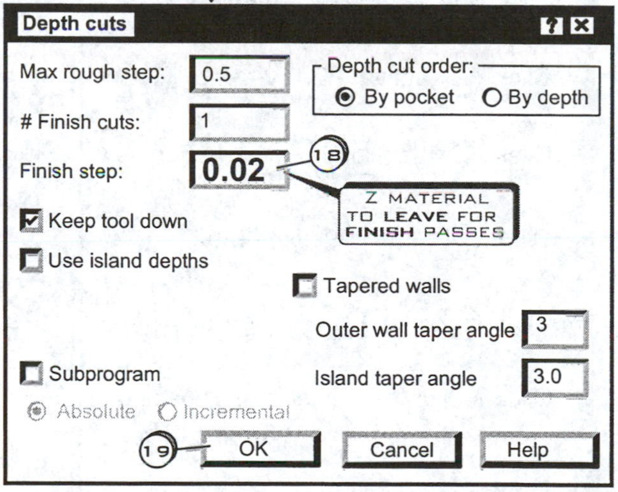

➤ Click 20 the Roughing/finishing parameters tab

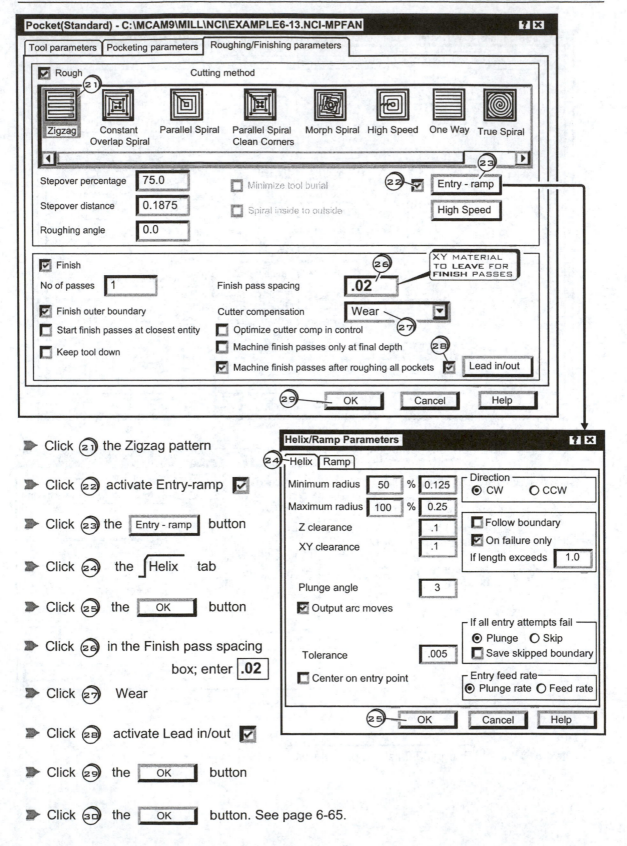

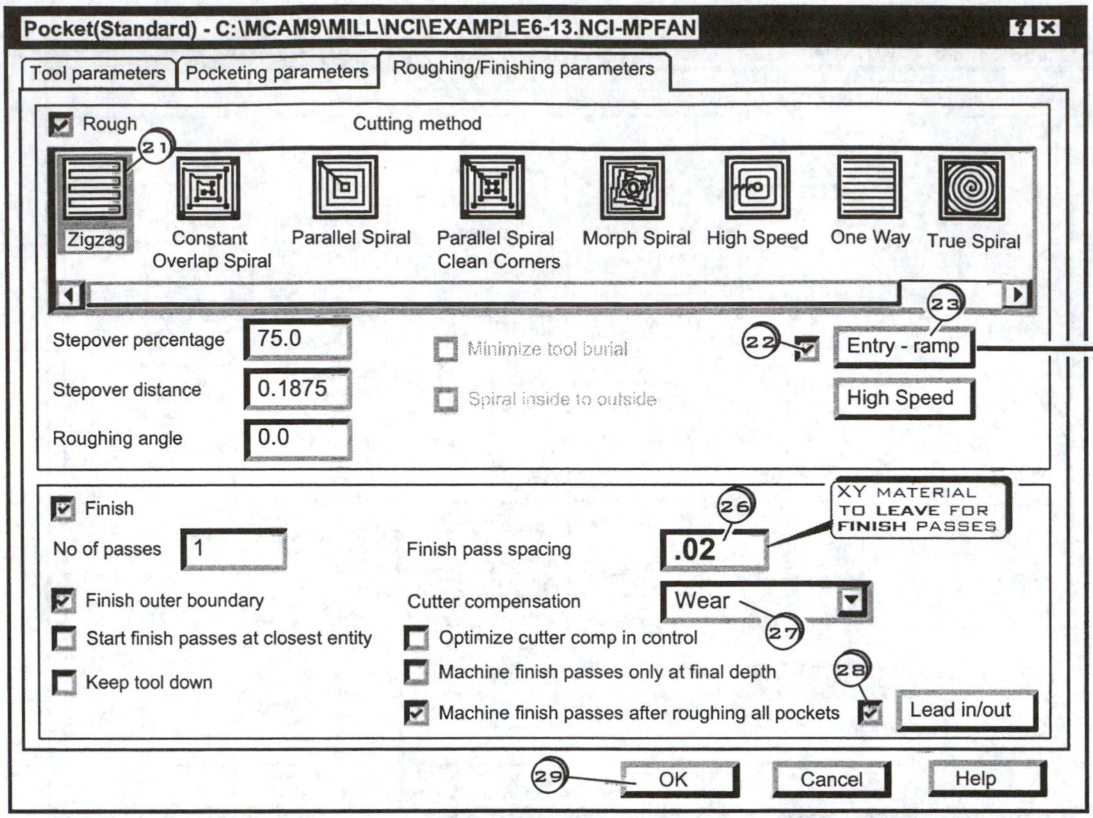

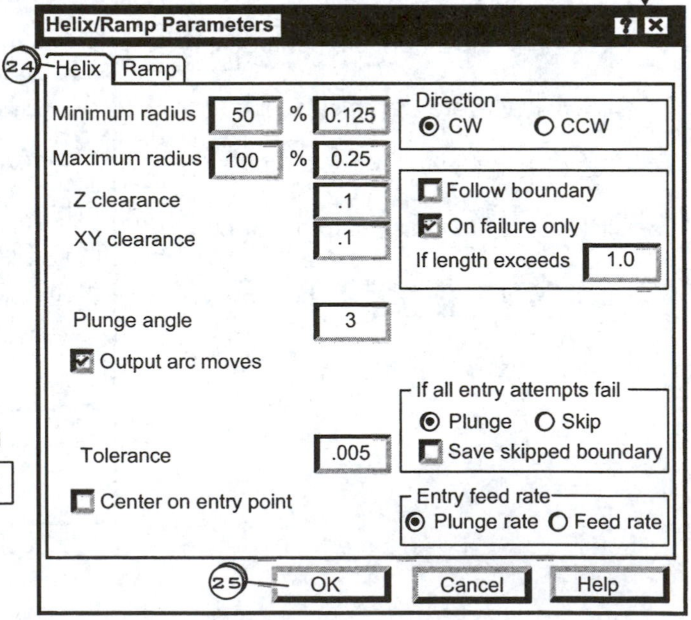

- Click (21) the Zigzag pattern
- Click (22) activate Entry-ramp ☑
- Click (23) the [Entry - ramp] button
- Click (24) the Helix tab
- Click (25) the [OK] button
- Click (26) in the Finish pass spacing box; enter **.02**
- Click (27) Wear
- Click (28) activate Lead in/out ☑
- Click (29) the [OK] button
- Click (30) the [OK] button. See page 6-65.

E) Backplot the Pocketing Machining Operation

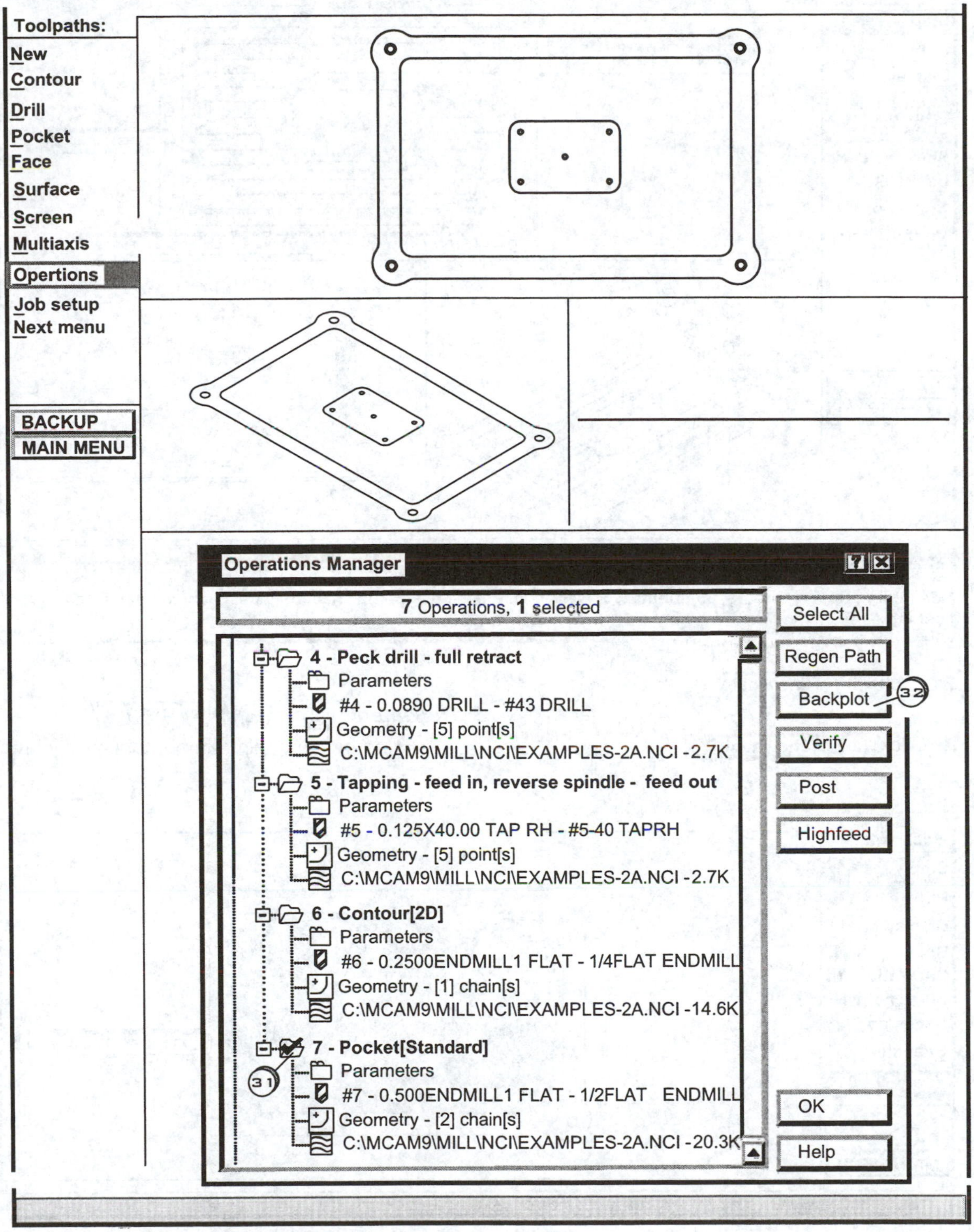

- Click (31) on **7 - Pocket[Standard]**
- Click (32) the Backplot button

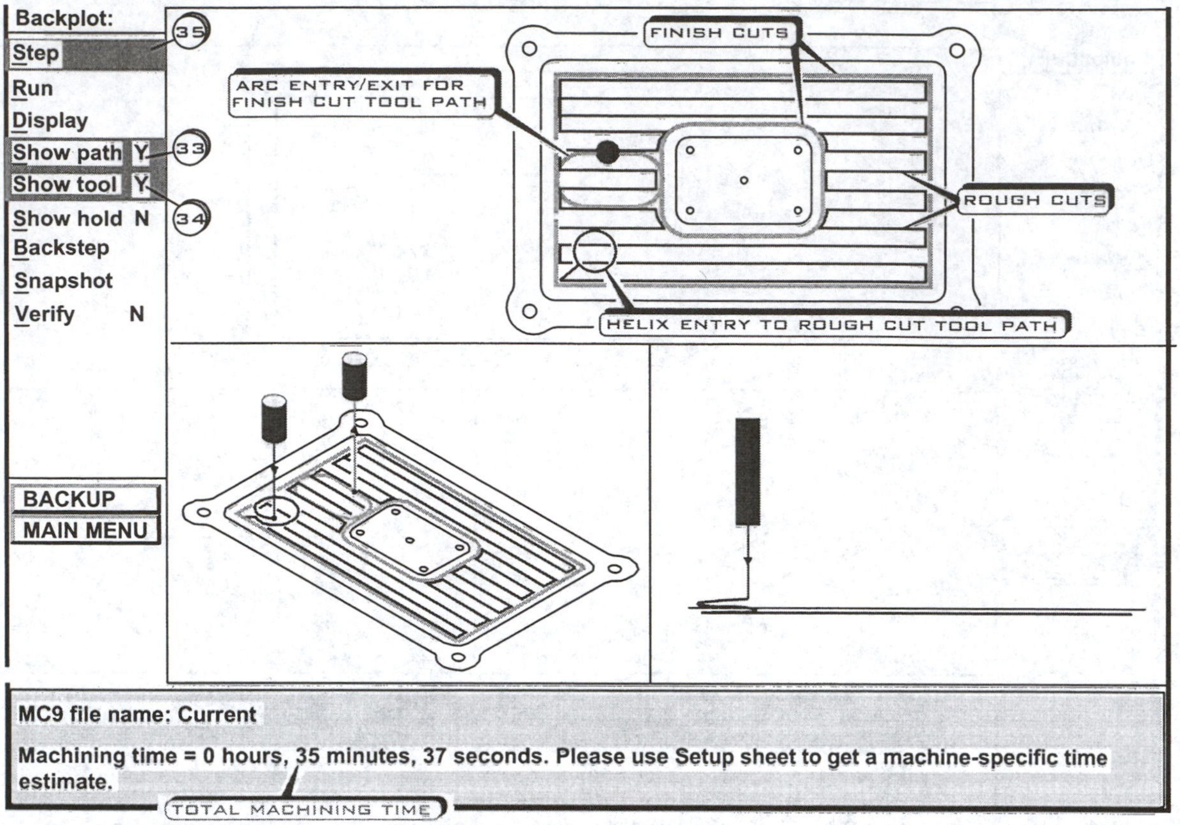

- Click (33) **Show path Y**
- Click (34) **Show tool Y**
- ***Repeatedly*** Click (35) **Step** to see a *step by step* animation of tool movement

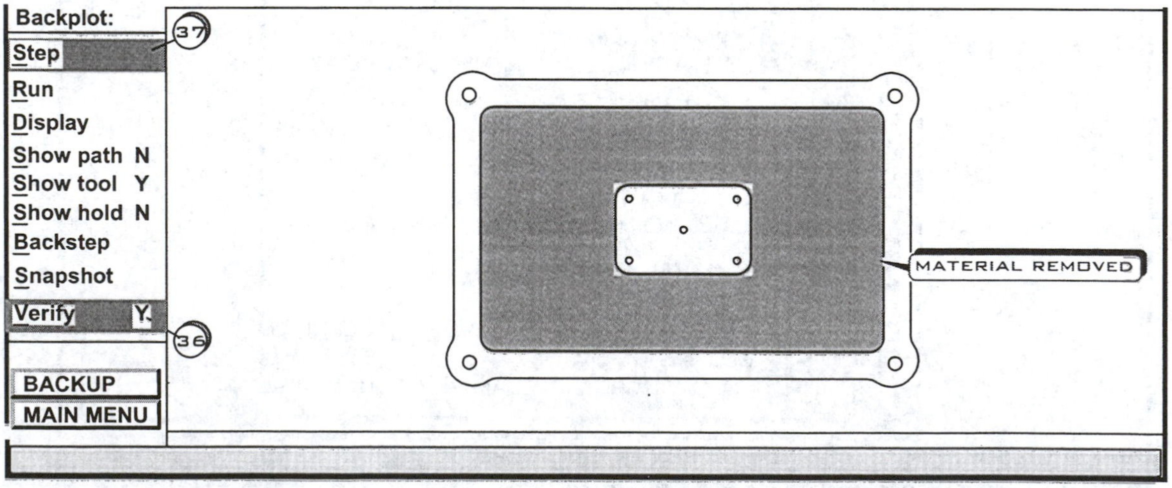

- Click (36) **Verify Y**
- ***Repeatedly*** Click (37) **Step** to see a *step by step* animation of ***pocket material removed***

F) VERIFY THE TOOLPATHS FOR ALL THE MACHINING OPERATIONS

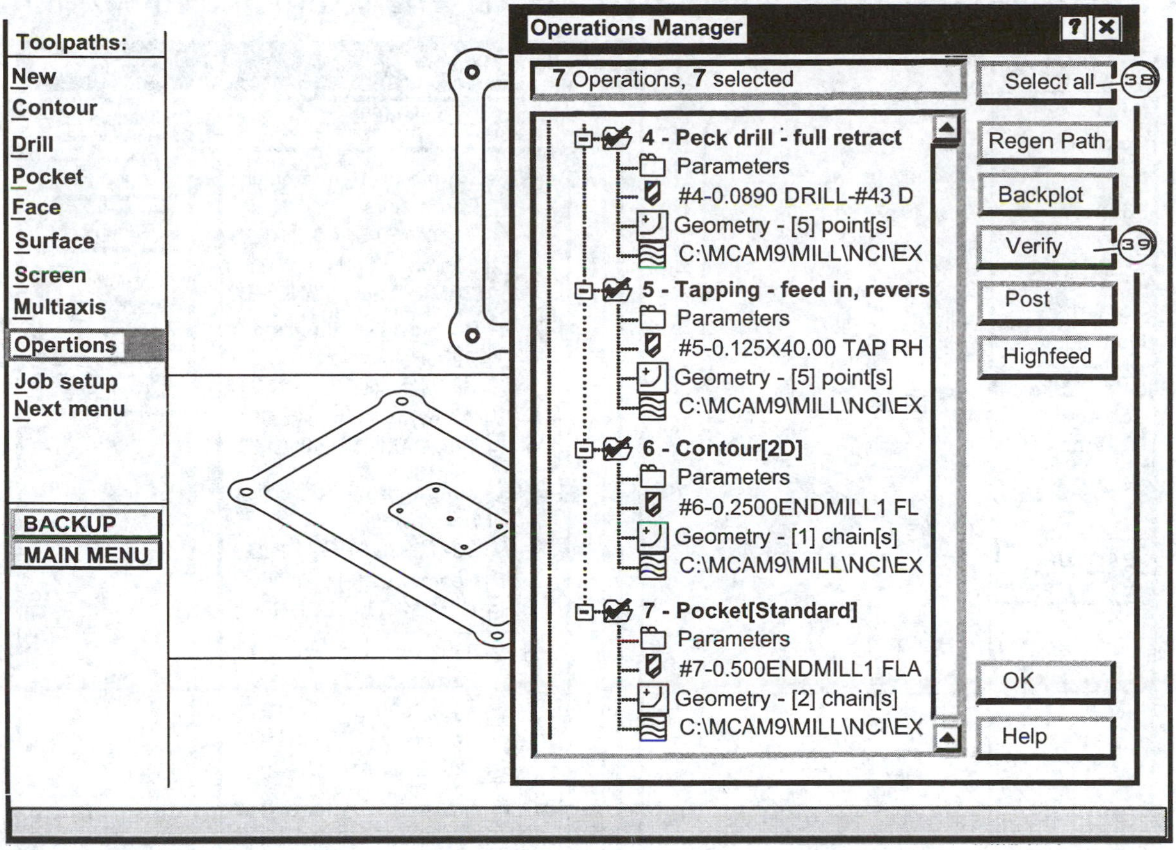

➤ Click (38) the Select All button

➤ Click (39) the Verify button

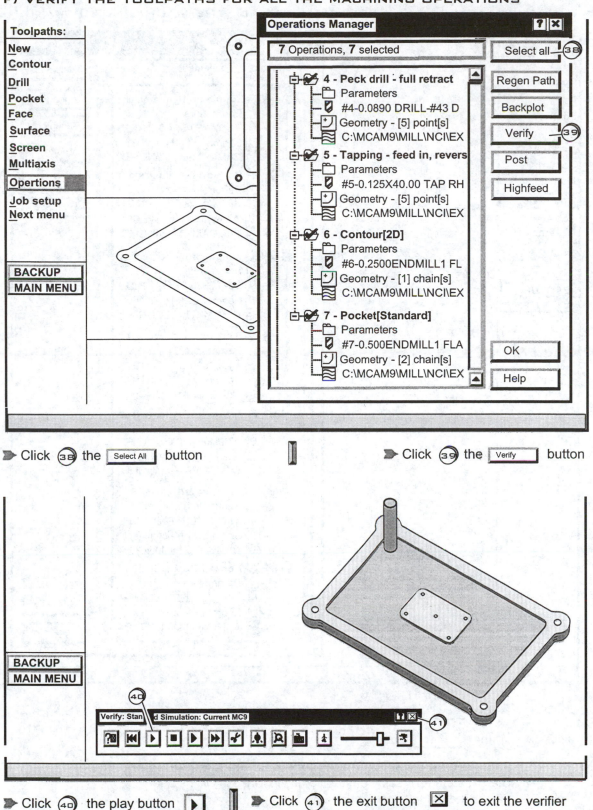

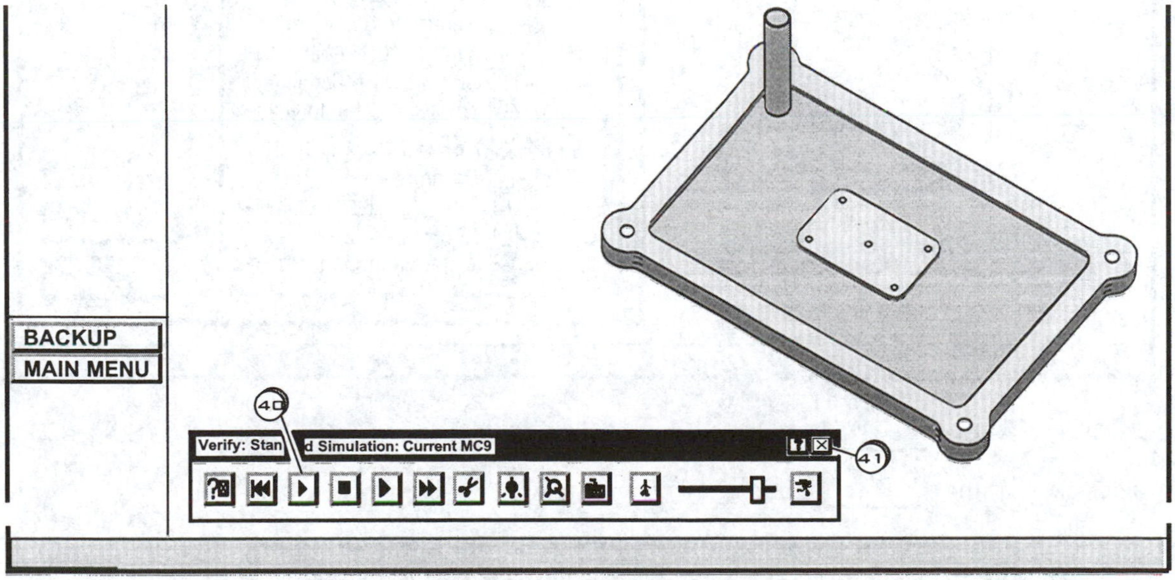

➤ Click (40) the play button

➤ Click (41) the exit button to exit the verifier

G) Generate the part program for all the machining operations

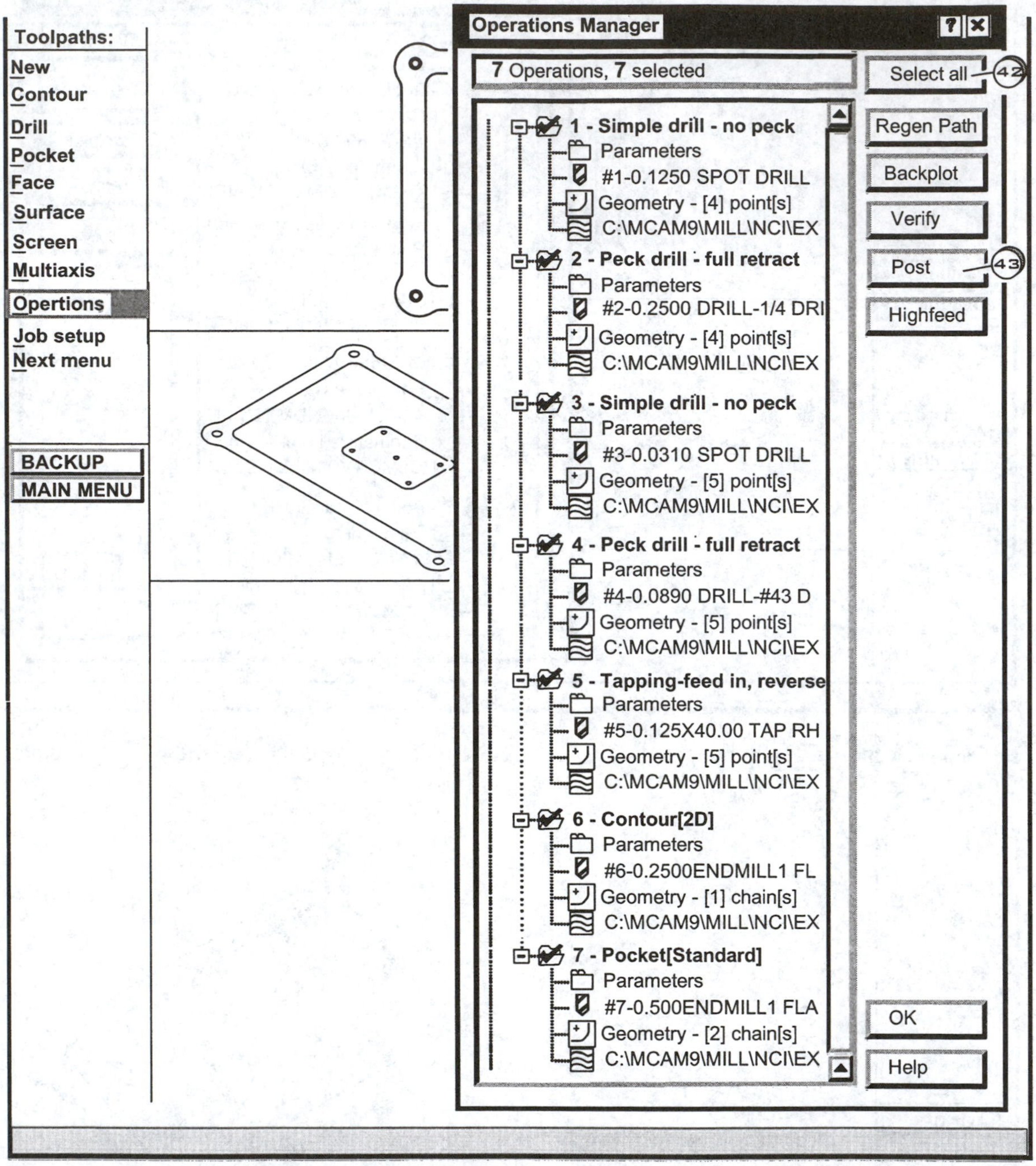

- Click (42) the [Select All] button
- Click (43) the [Post] button

> Note: *Mastercam's* sample post file **MPFAN.PST** will be used as the default post for this text.

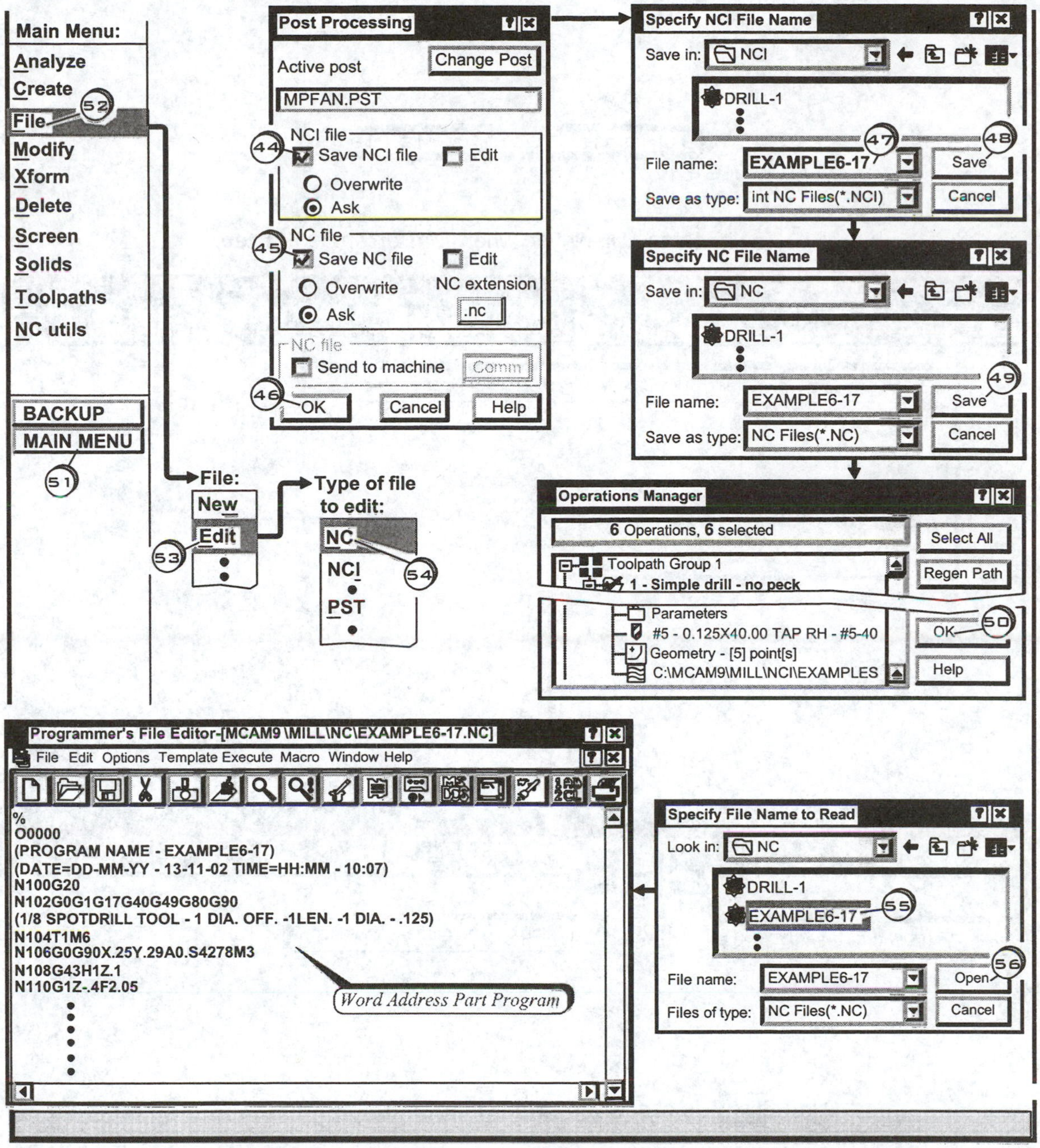

- Click 44 check ☑ in Save NCI file
- Click 45 check ☑ in Save NC file
- Click 46 OK
- Click 47 ; enter **EXAMPLE6-17**
- Click 48 Save
- Click 49 Save
- Click 50 OK
- Click 51 **MAIN MENU**
- Click 52 **File**
- Click 53 **Edit**
- Click 54 **NC**
- Click 55 NC file EXAMPLE6-13
- Click 56 Open

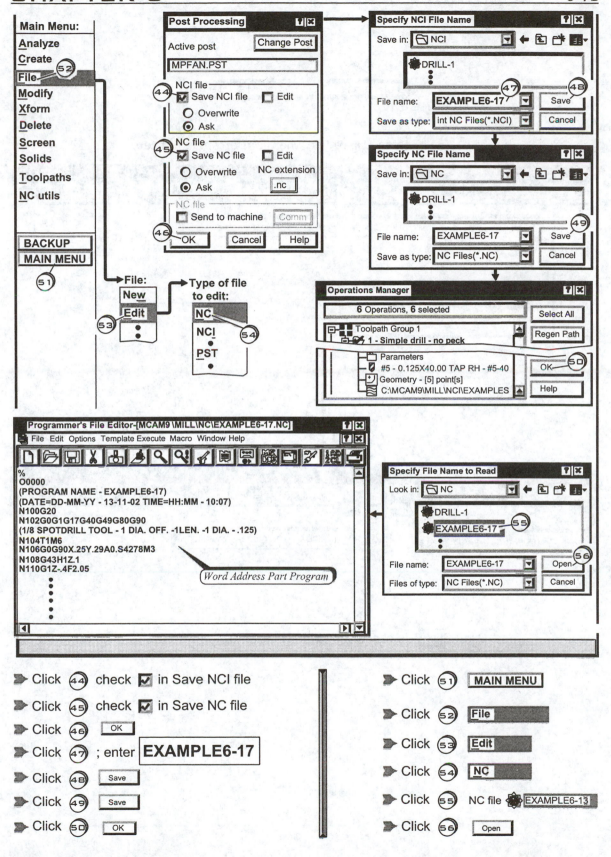

Using the Toolbar Menus

Toolbar Menu Displayed when *Mastercam* is **Started**

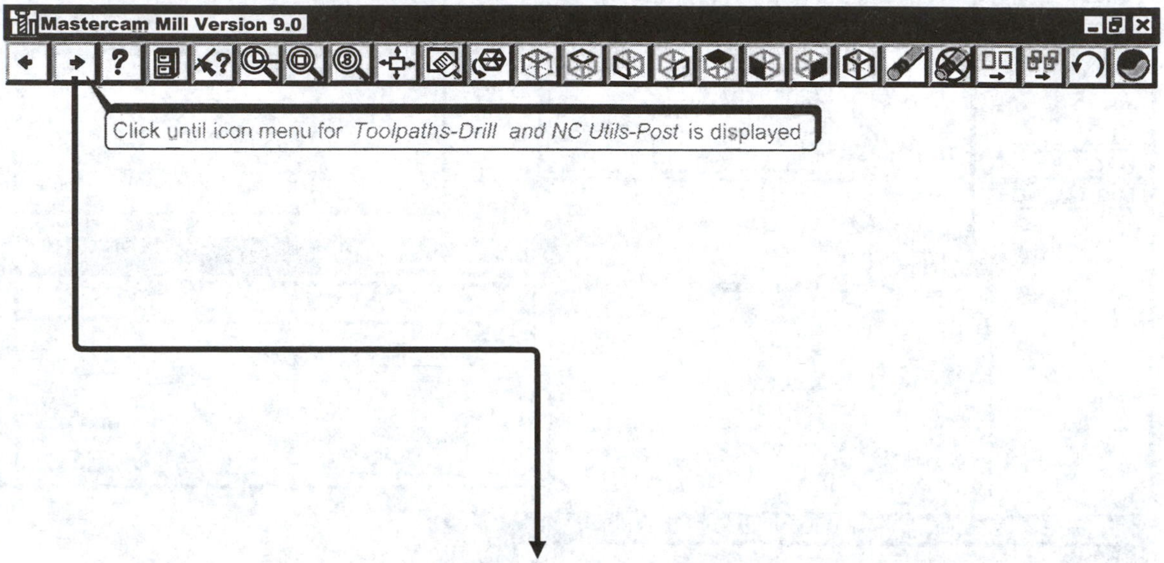
Mastercam Mill Version 9.0
Click until icon menu for *Toolpaths-Drill and NC Utils-Post* is displayed

Toolbar Menu Containing ***Toolpaths-Contour*** and ***Pocket*** Icons

Mastercam Mill Version 9.0
Click on *Toolpaths-Contour* icon
Click on *Toolpaths-Pocket* icon

Click Esc to cancel the commands

EXERCISES

6-1) Get the CAD model file **EX3-1JV** :

a) from the file generated in exercise 3-1

or

b) from the CD provided at the back of this text(file is located in the folder CHAPTER6).

Direct *Mastercam* to create a part program for executing the PROCESS PLAN 6P-1 on the part shown in Figure 6p-1.

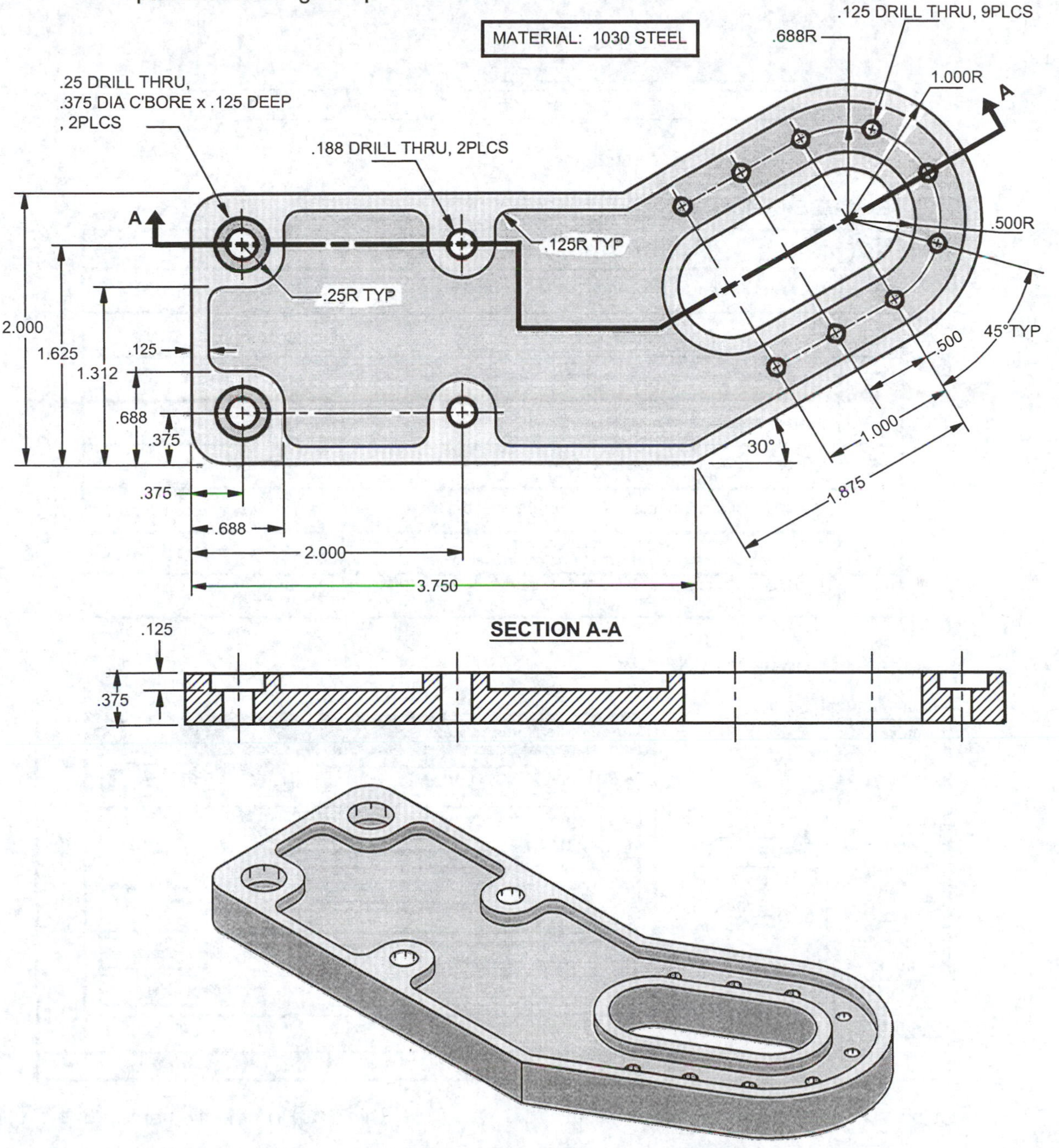

Figure 6-p1

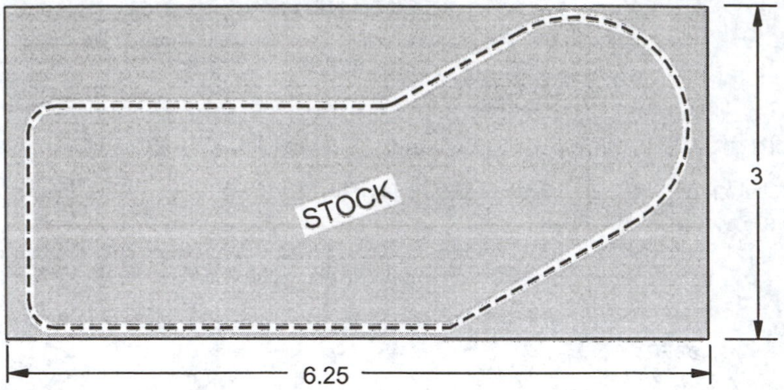

Figure 6-p2

PROCESS PLAN 6P-1

No.	Operation	Tooling
1	CENTER DRILL X .166 DEEP(ALL HOLES)	1/8 CENTER DRILL
2	PECK DRILL THRU(9PLCS)	1/8 DRILL
3	PECK DRILL TRHU(2PLCS)	3/16 DRILL
4	PECK DRILL TRHU(3PLCS)	1/4 DRILL
5	CIRCLE MILL X .125 DEEP	1/4 FLAT END MILL
6	ROUGH AND FINISH SLOT X .375 DEEP	1/4 FLAT END MILL
7	ROUGH AND FINISH POCKET X .125 DEEP LEAVE .01 FOR FINISH CUT IN XY AND Z.	1/4 FLAT END MILL
8	ROUGH OUTSIDE X .375 DEEP LEAVE .01 FOR FINISH CUT IN XY .	1/2 FLAT END MILL

Get the CAD model file **EX3-1JV**

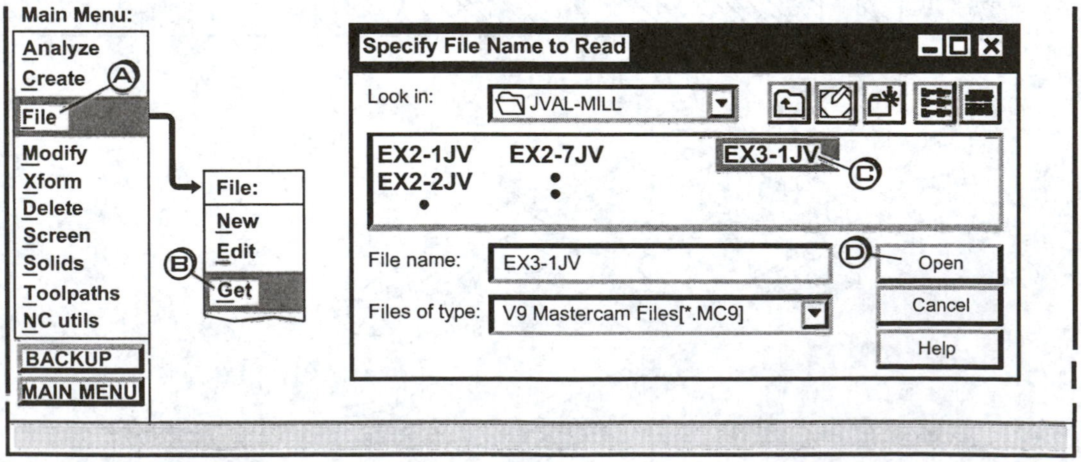

- Click Ⓐ **File**
- Click Ⓑ **Get**
- Click Ⓒ on the file name **EX3-1JV**
- Click Ⓓ Open

A) Create the .25 Dia Entrance Hole On the CAD Model

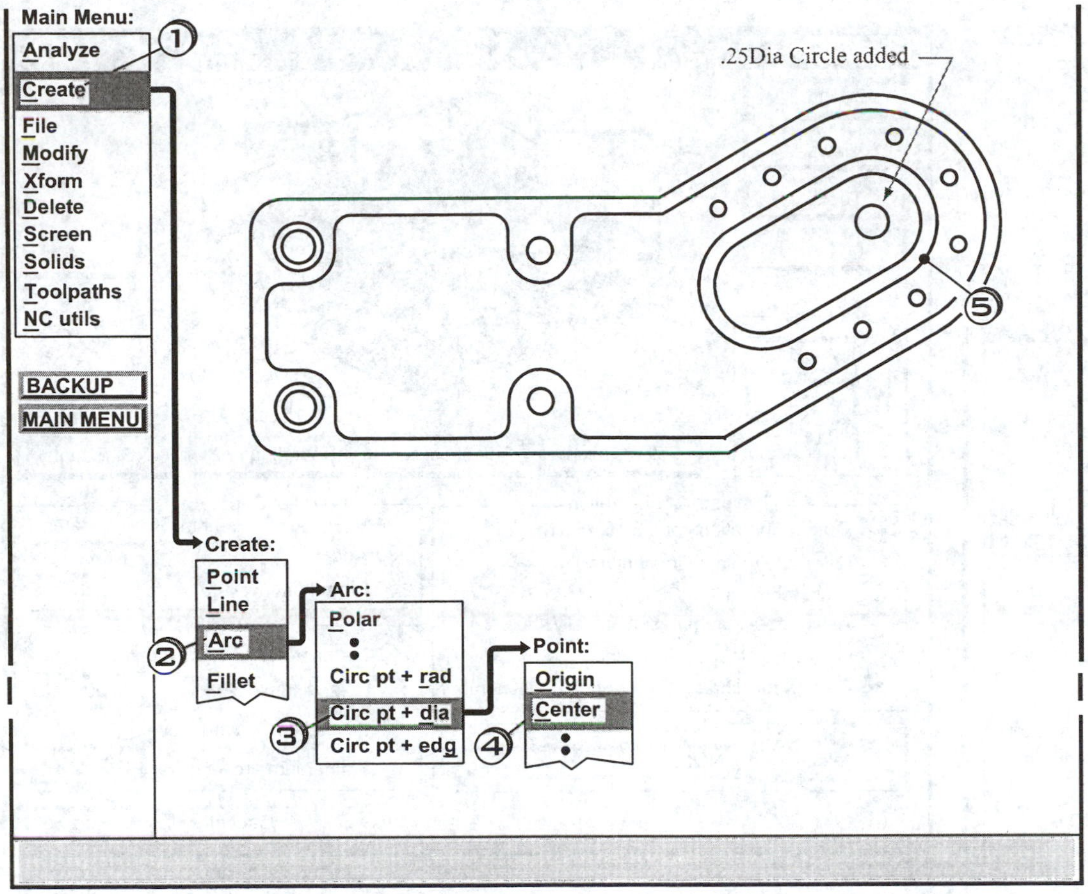

- Click ① Create
- Click ② Arc
- Click ③ Circ pt + dia

Enter the diameter:

- Enter the circle diameter .25 ; Press Enter
- Click ④ Center
- Click ⑤ on the arc
- Press Esc to cancel the operation

B) ENTER *Mastercam's* JOB SETUP DIALOG BOX

◆ SELECT SIZE OF THE STOCK

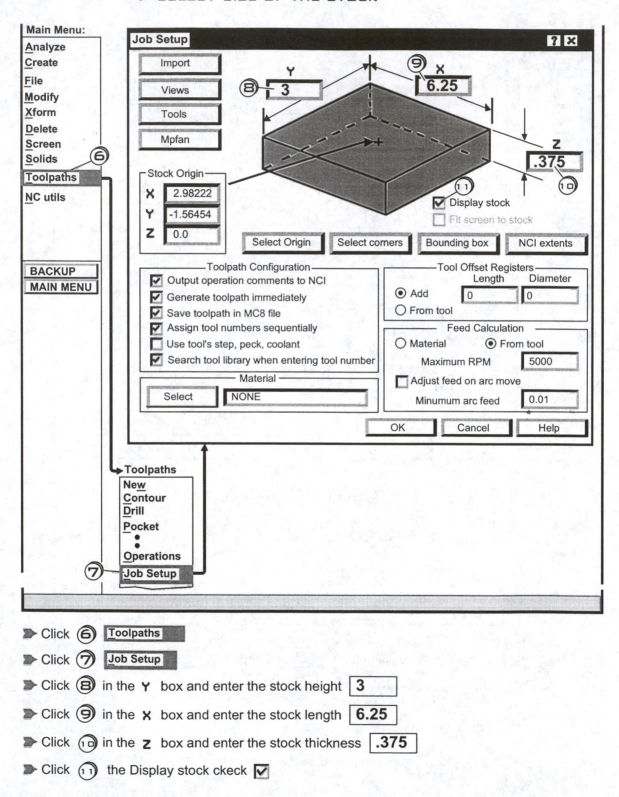

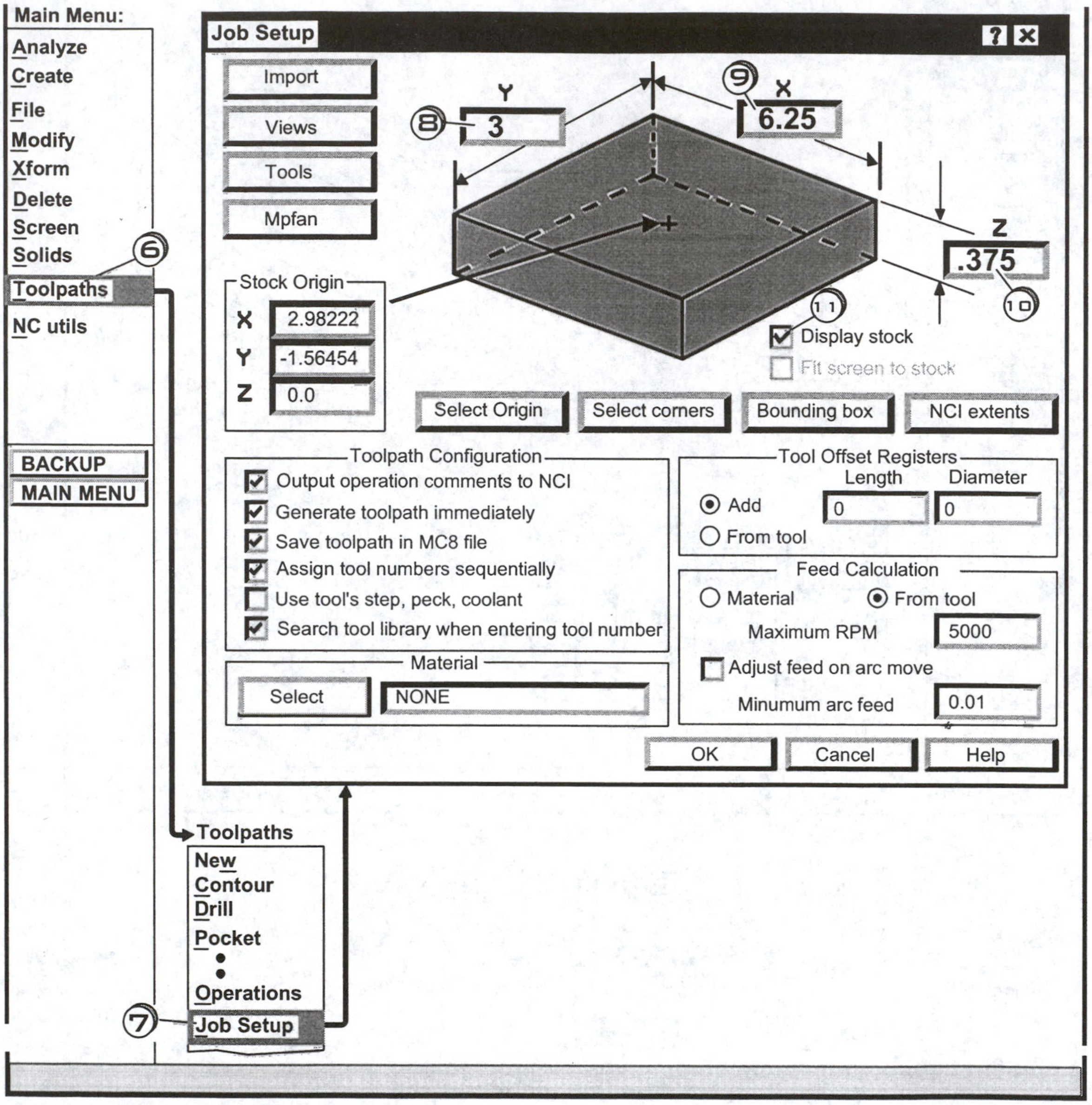

- Click ⑥ Toolpaths
- Click ⑦ Job Setup
- Click ⑧ in the Y box and enter the stock height 3
- Click ⑨ in the X box and enter the stock length 6.25
- Click ⑩ in the Z box and enter the stock thickness .375
- Click ⑪ the Display stock ckeck ☑

◆ IDENTIFY THE STOCK MATERIAL TO BE MACHINED

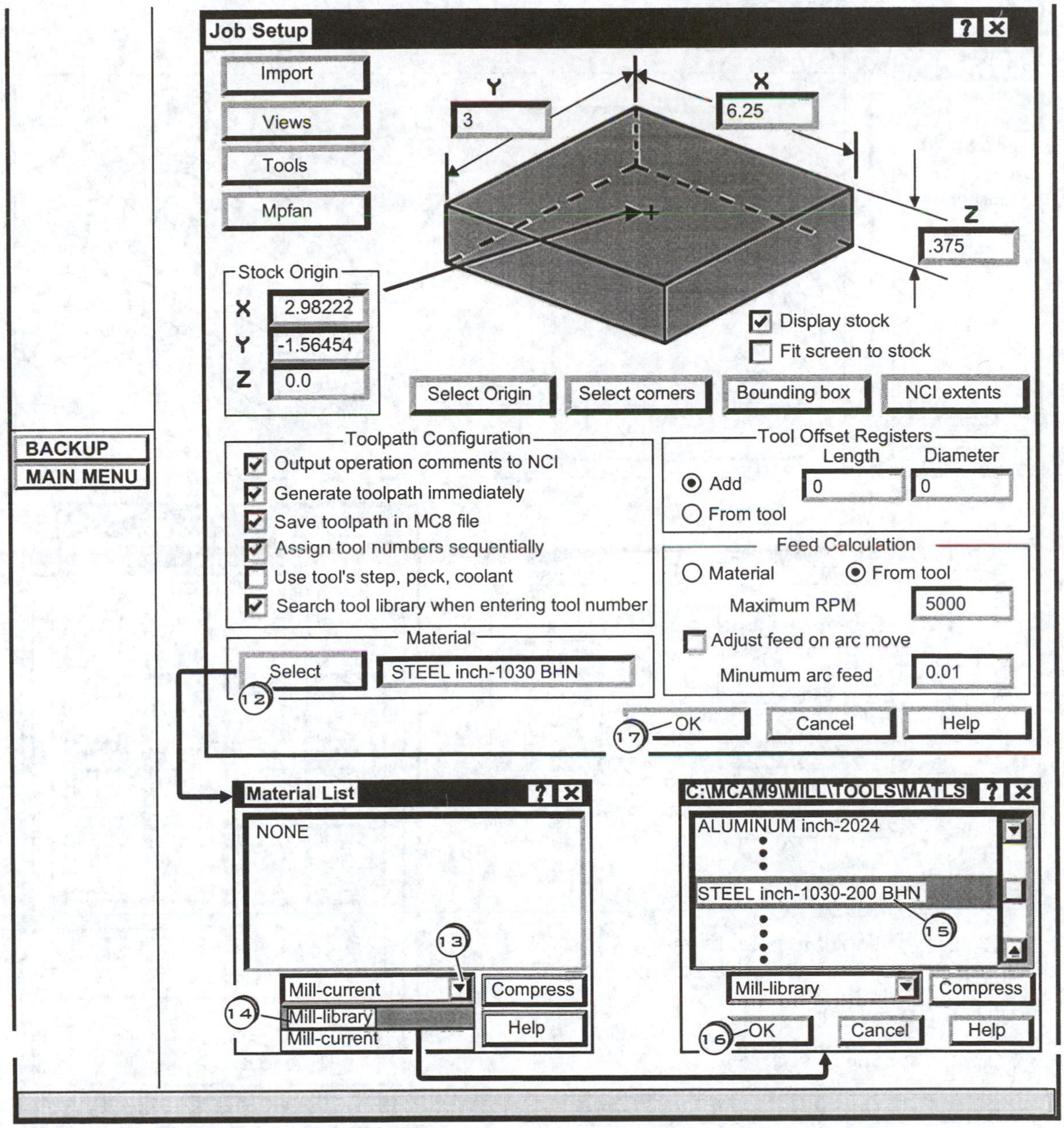

➤ Click (12) Select

➤ Scroll down (13) and click (14) Mill-library

➤ Scroll down and click (15) STEEL inch-1030-200 BHN

➤ Click (16) (17) OK

C) OPERATION#1- CNTR DRL .125 DIA ALL HOLES X .125 DEEP

◆ CREATE THE .125 DIA CENTER DRILL TOOLPATH

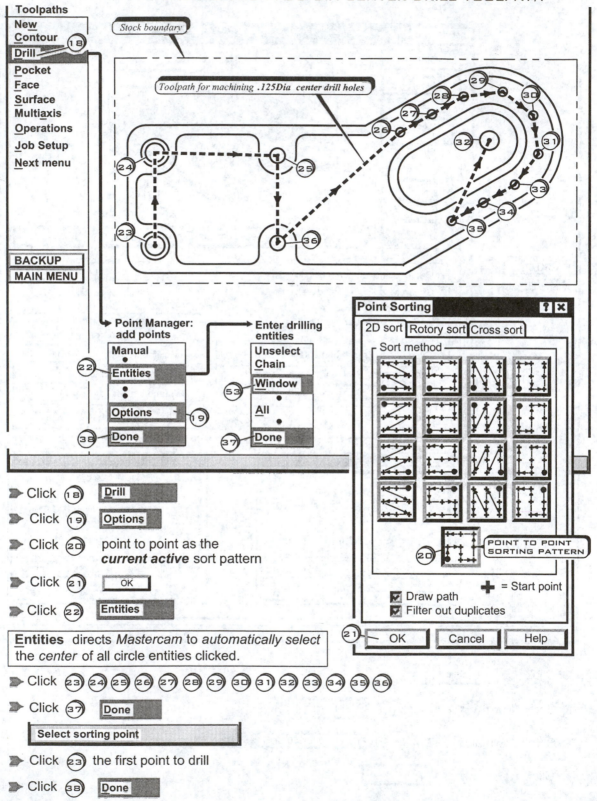

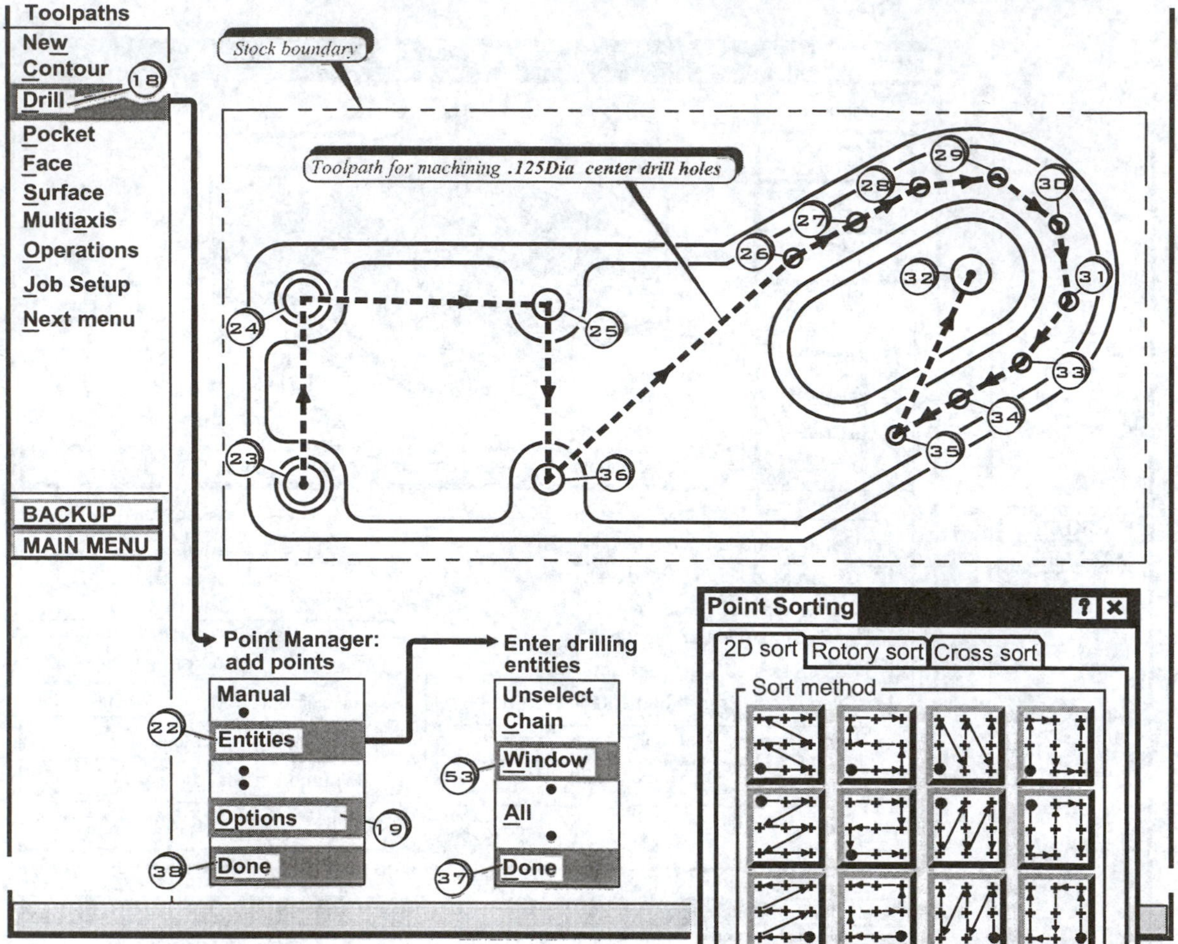

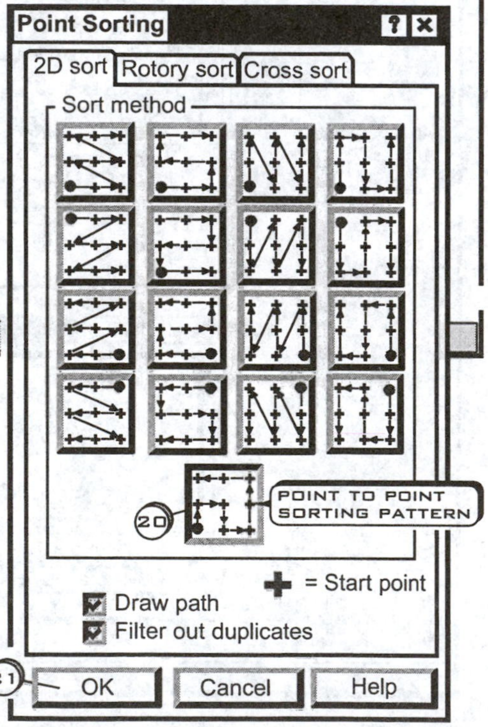

- Click (18) **Drill**
- Click (19) **Options**
- Click (20) point to point as the ***current active*** sort pattern
- Click (21) **OK**
- Click (22) **Entities**

Entities directs *Mastercam* to *automatically select* the *center* of all circle entities clicked.

- Click (23) (24) (25) (26) (27) (28) (29) (30) (31) (32) (33) (34) (35) (36)
- Click (37) **Done**

Select sorting point

- Click (23) the first point to drill
- Click (38) **Done**

◆ OBTAIN THE NEEDED .125 DIA CENTER DRILL TOOL

Mastercam will activate and display the Tool and Drill Parameters dialog box after the drill tool path has been created.

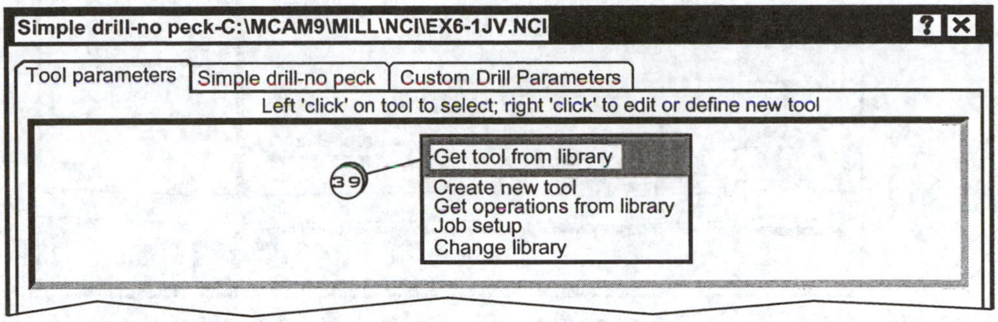

➤ Press the ***right*** mouse button in the ***white area***

➤ Click (39) Get tool from library

Mastercam will display the main Tool Library dialog box

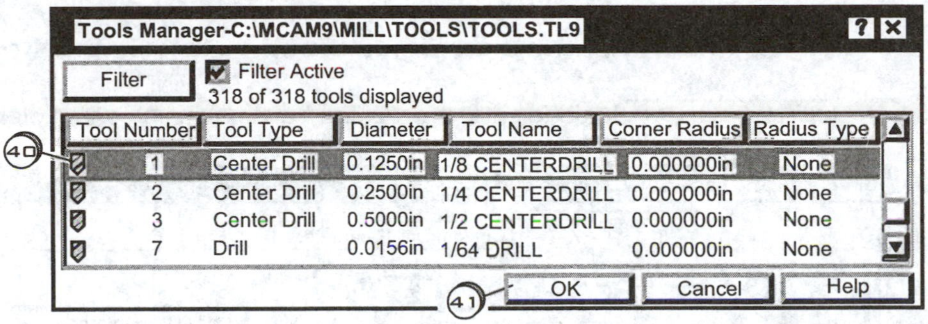

➤ Click (40) the .125Dia center Drill tool

➤ Click (41) OK

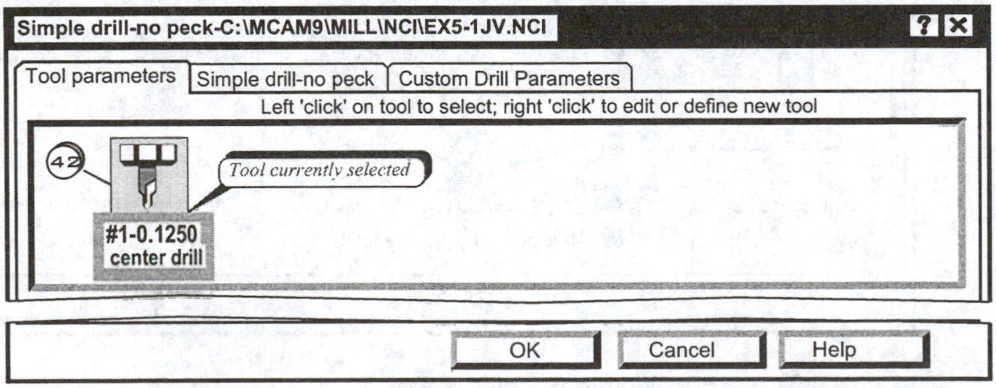

➤ ***Right*** click (42)

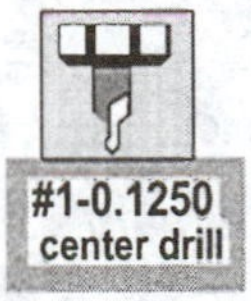

on the #1-0.1250 center drill to select this as the ***active tool*** and direct *Mastercam* to display the Define Tool dialog box

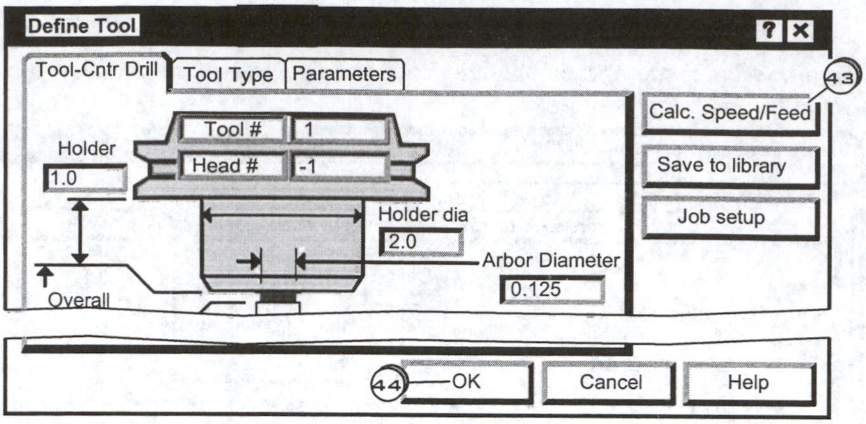

➤ Click 43 Calc Speed/Feed

➤ Click 44 OK

◆ ENTER THE .125 DIA CENTER DRILL HOLE MACHINING PARAMETERS

➤ Click 45 **Simple drill-no peck** tab to enter the .125Dia hole machining parameters.

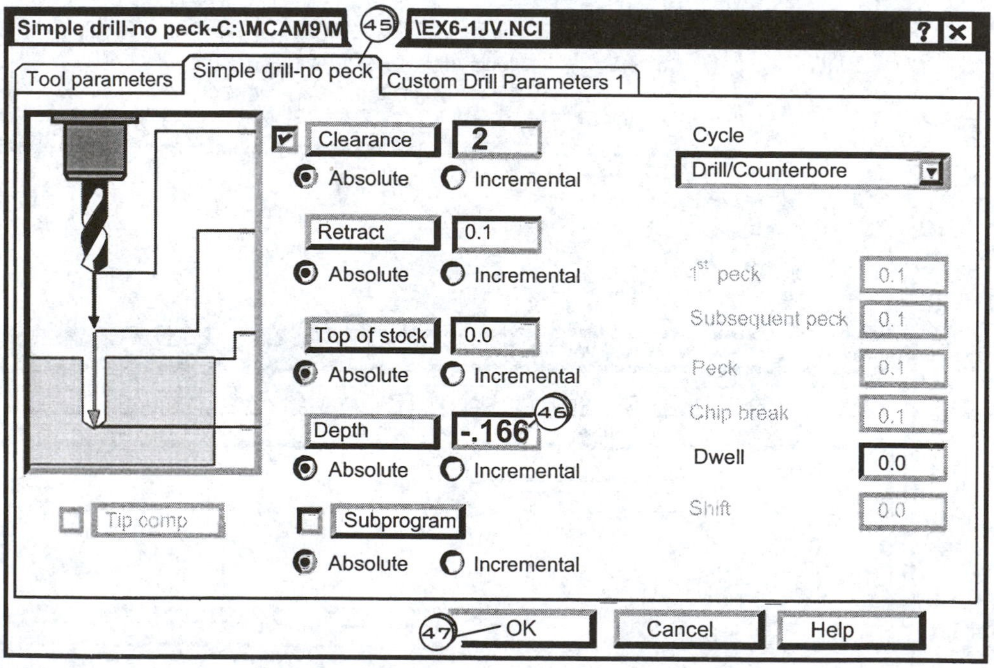

➤ Click 46 in the Depth box; enter **-.166**

➤ Click 47 OK

The .125Dia center drilling operation will be added to *Mastercam's* Operations Manager

D) OPERATION#2- PECK DRILL THE .125 DIA HOLES THRU

◆ CREATE THE .125 DIA PECK DRILL THRU TOOLPATH

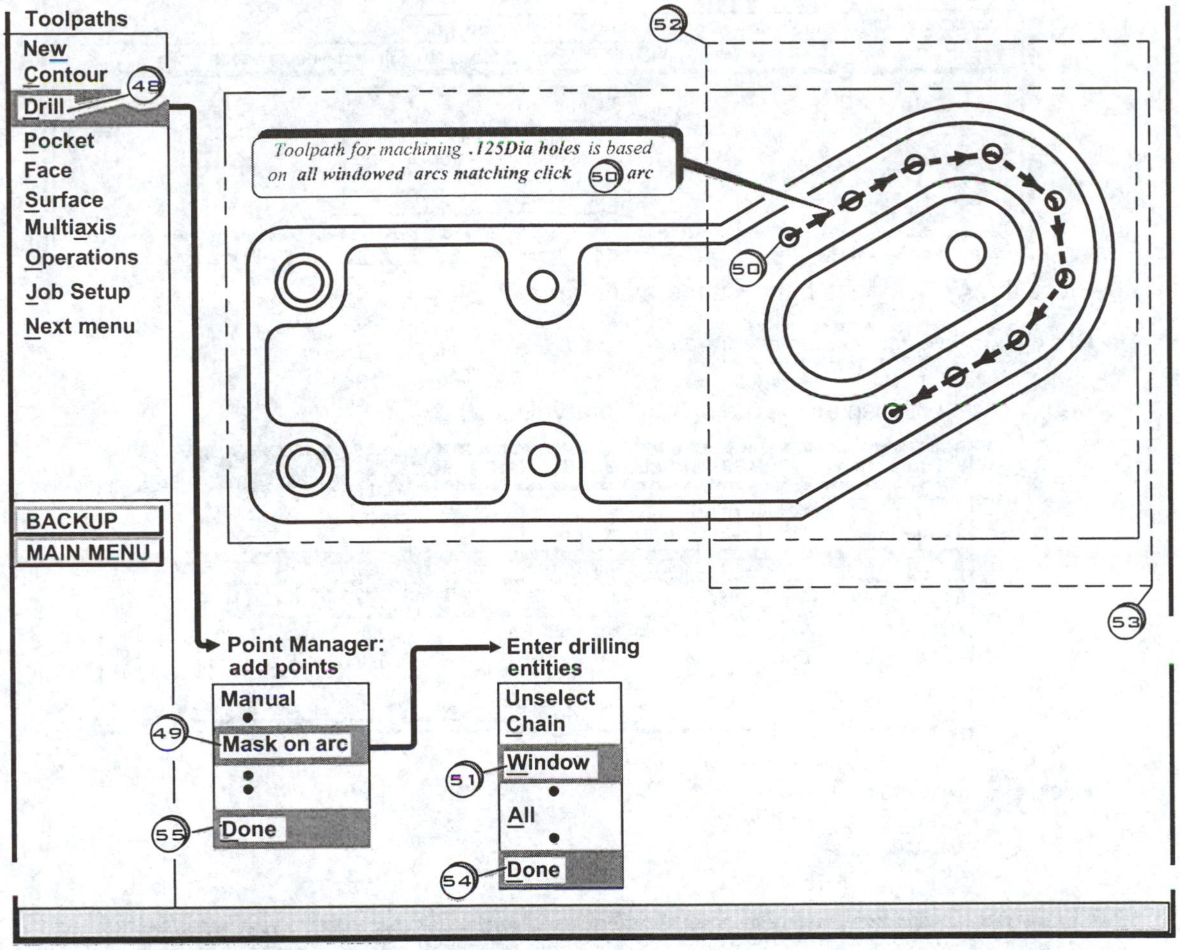

- Click 48 **Drill**
- Click 49 **Mask on arc**

Mask on arc directs *Mastercam* to *automatically select* all windowed arcs that match the first arc clicked

Select arc to match

- Click 50
- Click 51 **Window**
- Click 52 53 the window corners
- Click 54 **Done**

Select sorting point

- Click 50 the first point to drill
- Click 55 **Done**

◆ OBTAIN THE NEEDED .125DIA DRILL TOOL

Mastercam will activate and display the Tool and Drill Parameters dialog box after the drill tool path has been created.

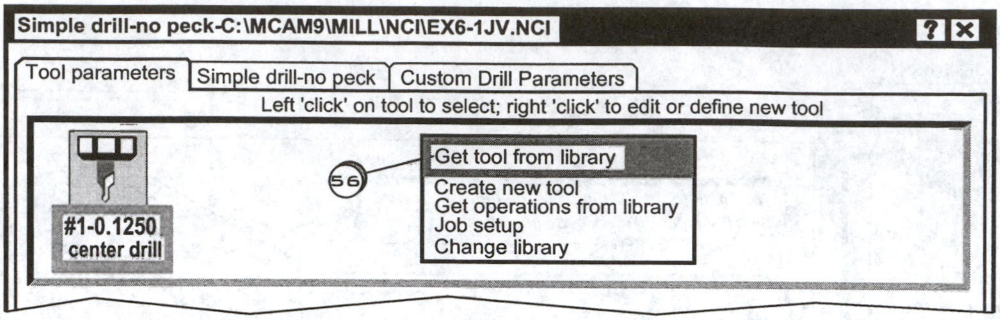

➤ Press the ***right*** mouse button in the ***white area***

➤ Click (56) Get tool from library

Mastercam will display the main Tool Library dialog box

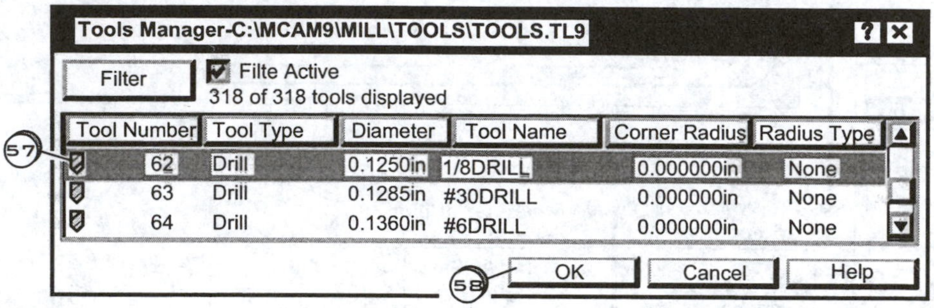

➤ Click (57) the .125Dia Drill tool

➤ Click (58) OK

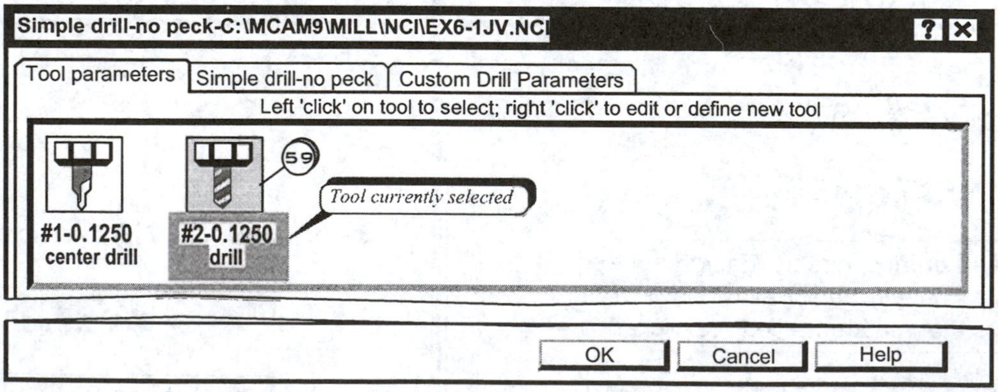

➤ ***Right*** click (59)

on the #2-0.1250 drill to select this as the ***active tool*** and direct *Mastercam* to display the Define Tool dialog box

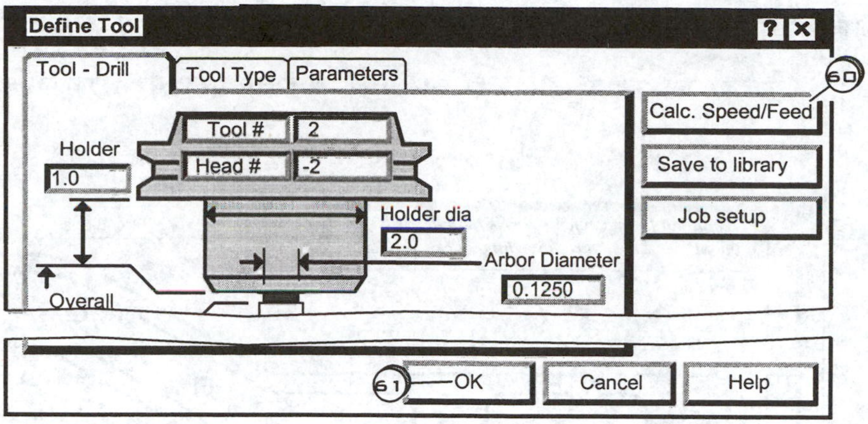

➤ Click 60 [Calc Speed/Feed] ➤ Click 61 [OK]

- ENTER THE .125DIA PECK DRILL THRU MACHINING PARAMETERS

➤ Click 62 [Peck drill-full retract] tab to enter the .125Dia peck drill machining parameters.

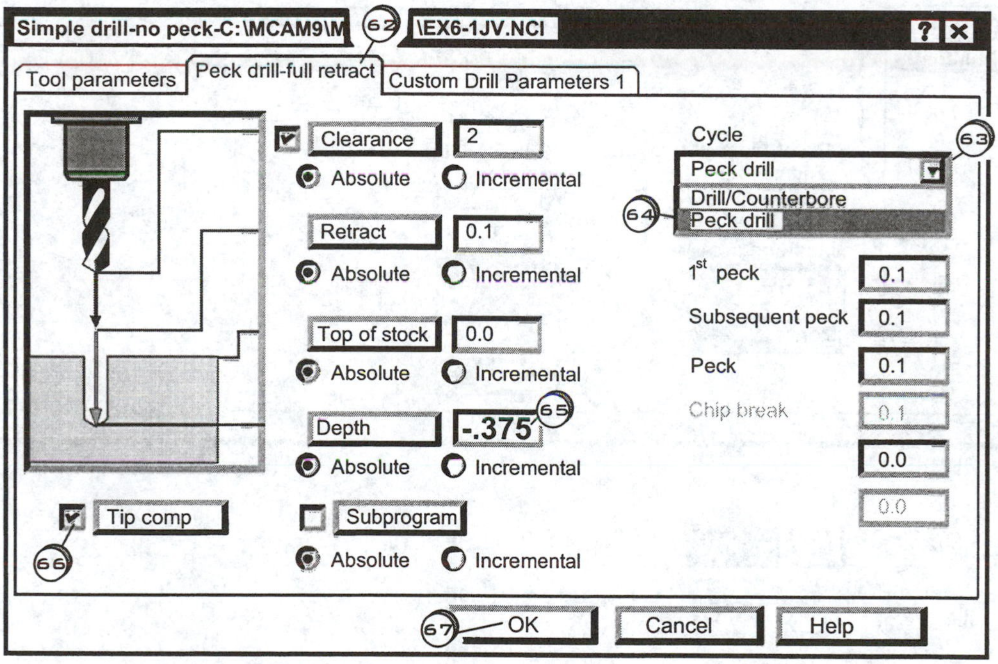

➤ Click 63 [▼] the toggle down button

➤ Click 64 [Peck drill]

➤ Click 65 in the Depth box; enter [-.375]

➤ Click 66 the tip comp for drill **thru**(***check appears***) ☑

➤ Click 67 [OK]

The .125Dia peck drilling thru operation will be added to *Mastercam's* Operations Manager

E) OPERATION#3- PECK DRILL THE .1875 DIA HOLES THRU

◆ CREATE THE .1875 DIA PECK DRILL THRU TOOLPATH

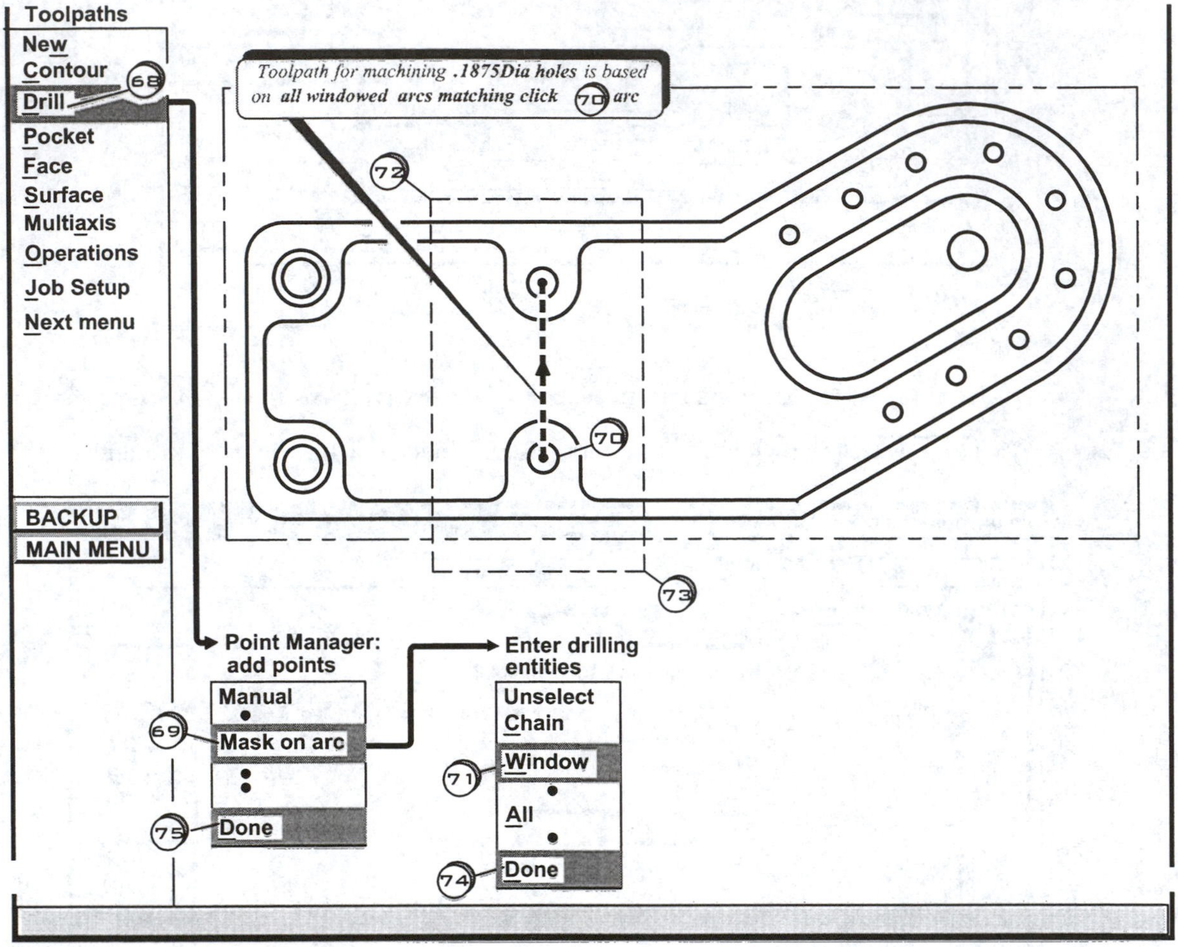

➤ Click (68) **Drill**

➤ Click (69) **Mask on arc**

Mask on arc directs *Mastercam* to *automatically select* all windowed arcs that match the first arc clicked

Select arc to match

➤ Click (70)

➤ Click (71) **Window**

➤ Click (72) (73) the window corners

➤ Click (74) **Done**

Select sorting point

➤ Click (70) the first point to drill

➤ Click (75) **Done**

◆ OBTAIN THE NEEDED .1875DIA DRILL TOOL

Mastercam will activate and display the Tool and Drill Parameters dialog box after the drill tool path has been created.

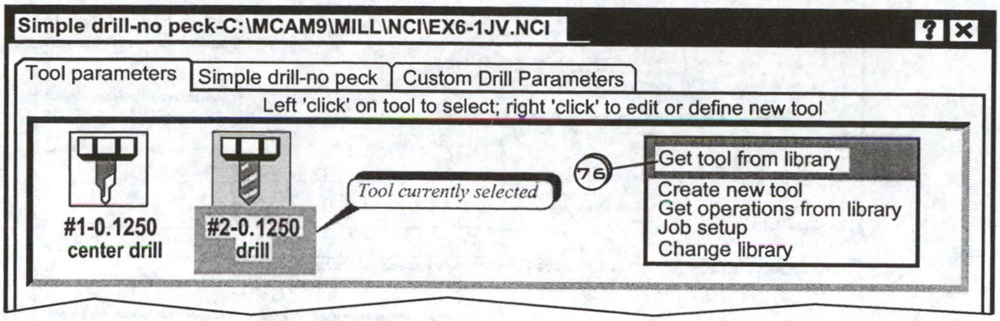

➤ Press the ***right*** mouse button in the ***white area***

➤ Click (76) Get tool from library

Mastercam will display the main Tool Library dialog box

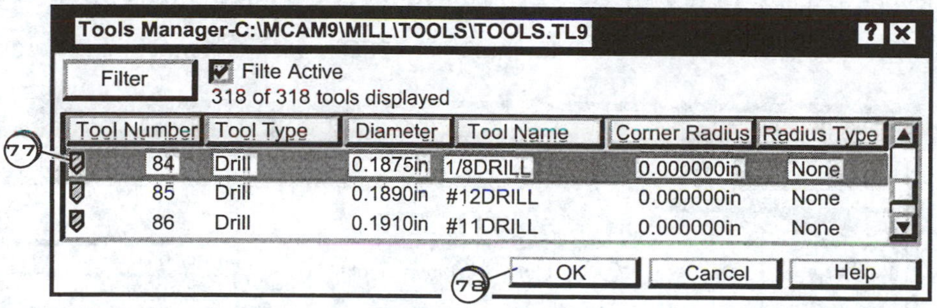

➤ Click (77) the .1875Dia Drill tool

➤ Click (78) OK

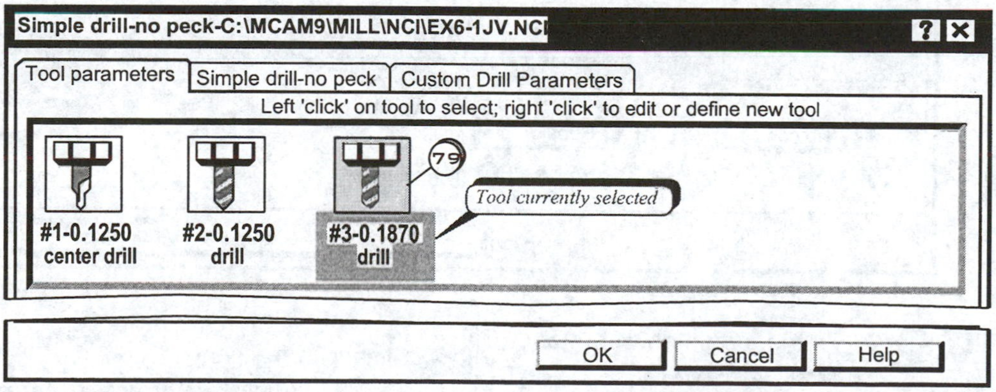

➤ ***Right*** click (79)

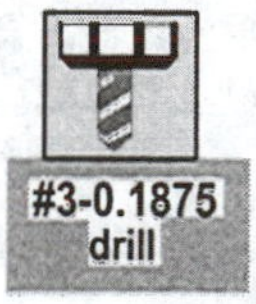

on the #3-0.1875 drill to select this as the ***active tool*** and direct *Mastercam* to display the Define Tool dialog box

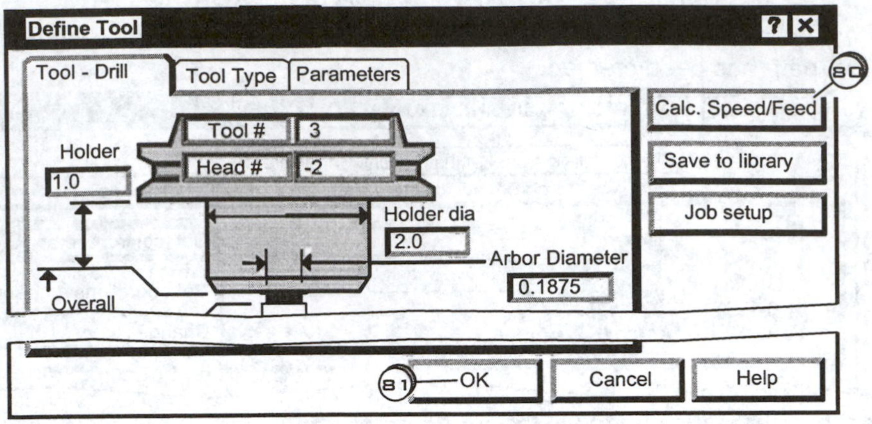

➤ Click 80 Calc Speed/Feed ➤ Click 81 OK

- ENTER THE .1875DIA PECK DRILL THRU MACHINING PARAMETERS

➤ Click 82 Peck drill-full retract tab to enter the .125Dia peck drill machining parameters.

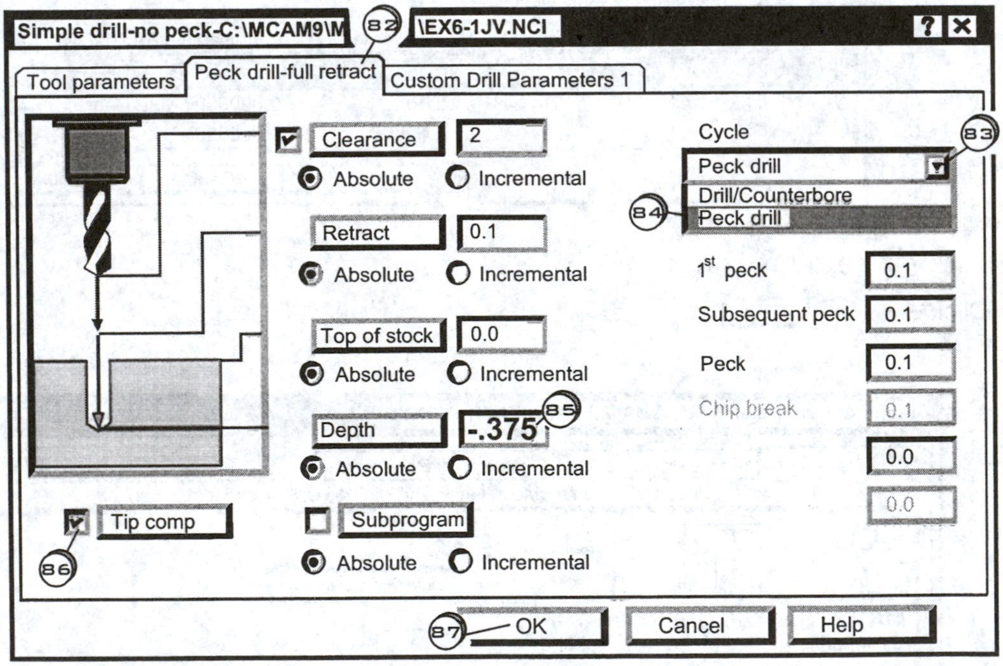

➤ Click 83 the toggle down button

➤ Click 84 Peck drill

➤ Click 85 in the Depth box; enter -.375

➤ Click 86 the tip comp for drill **thru**(***check appears)***

➤ Click 87 OK

The .1875Dia peck drilling thru operation will be added to *Mastercam's* Operations Manager

F) OPERATION#4- PECK DRILL THE .25 DIA HOLES THRU

◆ CREATE THE .25 DIA PECK DRILL THRU TOOLPATH

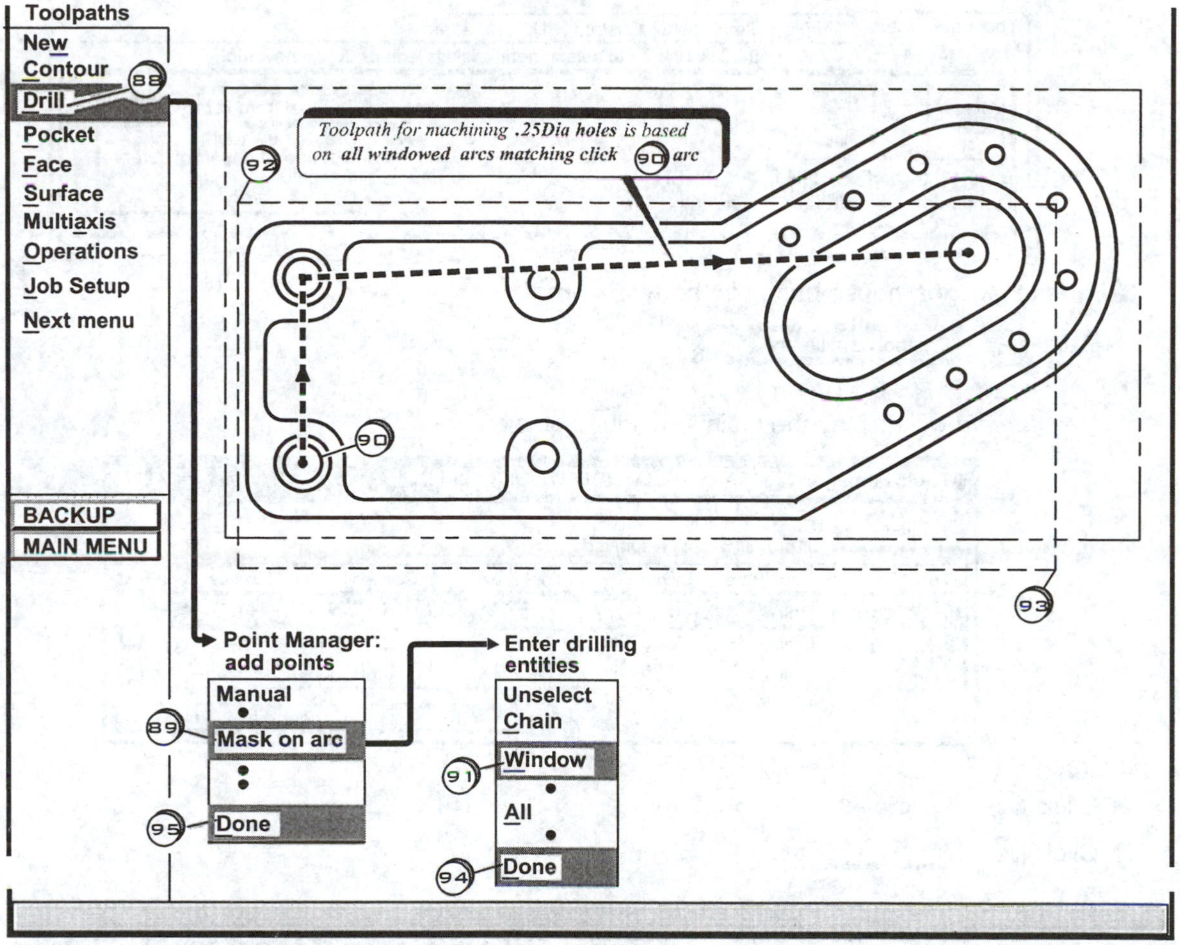

- Click 88 **Drill**
- Click 89 **Mask on arc**

Mask on arc directs *Mastercam* to *automatically select* all windowed arcs that match the first arc clicked

Select arc to match

- Click 90
- Click 91 **Window**
- Click 92 93 the window corners
- Click 94 **Done**

Select sorting point

- Click 90 the first point to drill
- Click 95 **Done**

◆ OBTAIN THE NEEDED .25DIA DRILL TOOL

Mastercam will activate and display the Tool and Drill Parameters dialog box after the drill tool path has been created.

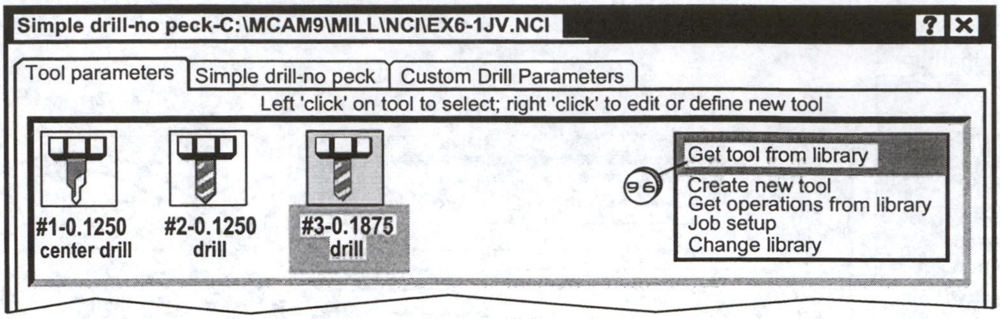

➤ Press the ***right*** mouse button in the ***white area***

➤ Click (96) 

Mastercam will display the main Tool Library dialog box

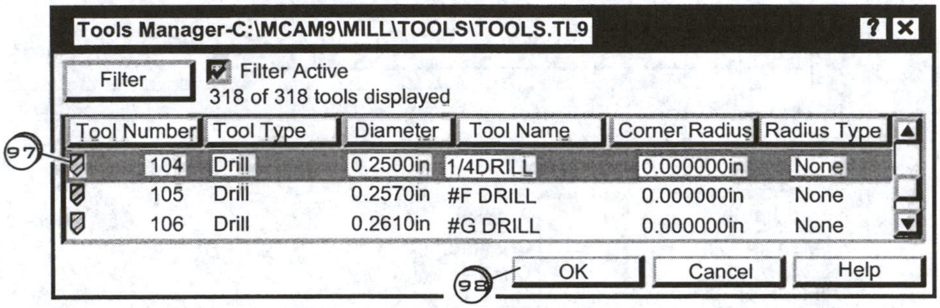

➤ Click (97) the .2500Dia Drill tool

➤ Click (98) 

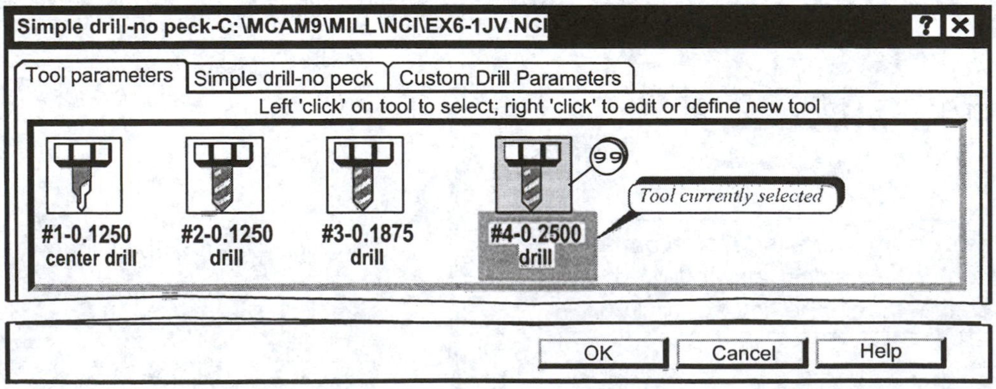

➤ ***Right*** click (99) 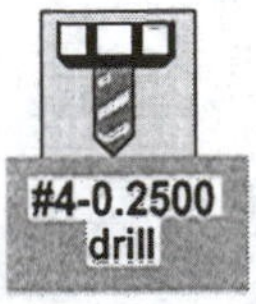

on the #4-0.2500 drill to select this as the ***active tool*** and direct *Mastercam* to display the Define Tool dialog box

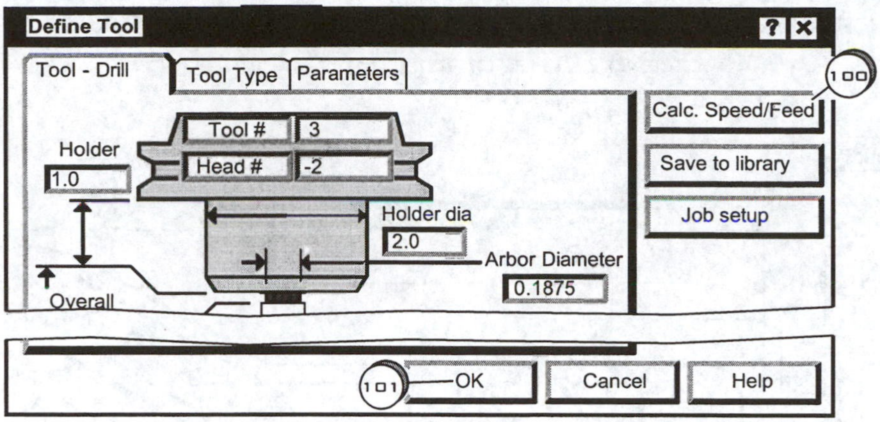

➤ Click (100) Calc Speed/Feed

➤ Click (101) OK

◆ ENTER THE .25DIA PECK DRILL THRU MACHINING PARAMETERS

➤ Click (102) Peck drill-full retract tab to enter the .125Dia peck drill machining parameters.

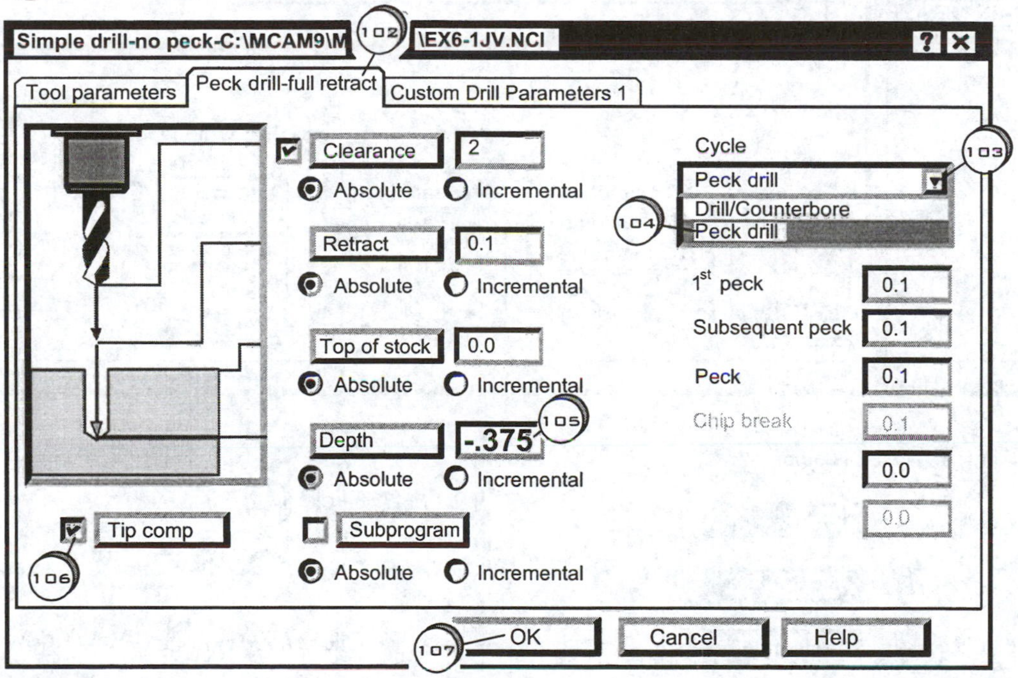

➤ Click (103) the toggle down button

➤ Click (104) Peck drill

➤ Click (105) in the Depth box; enter **-.375**

➤ Click (106) the tip comp check *on* when drilling *thru* holes

➤ Click (107) OK

The .25Dia peck drilling thru operation will be added to *Mastercam's* Operations Manager

G) OPERATION#5- CIRCLE MILL .375 DIA HOLES X .125 DEEP

◆ CREATE THE CIRCLE MILL TOOLPATH

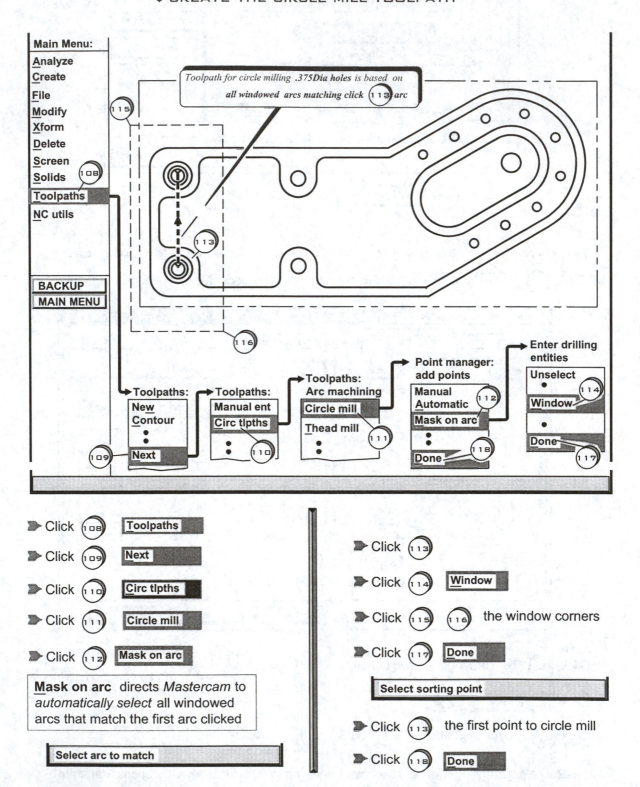

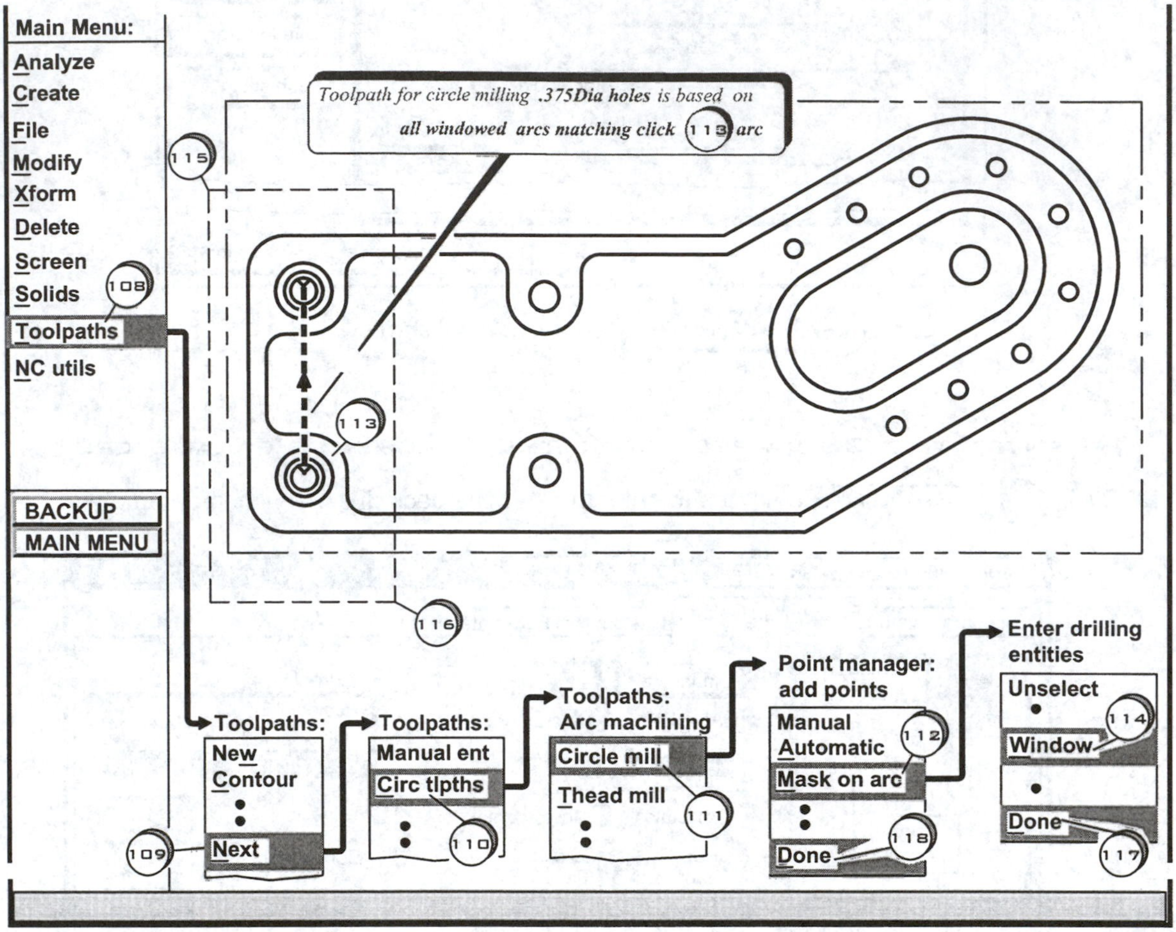

- Click 108 **Toolpaths**
- Click 109 **Next**
- Click 110 **Circ tlpths**
- Click 111 **Circle mill**
- Click 112 **Mask on arc**

Mask on arc directs *Mastercam* to *automatically select* all windowed arcs that match the first arc clicked

Select arc to match

- Click 113
- Click 114 **Window**
- Click 115 116 the window corners
- Click 117 **Done**

Select sorting point

- Click 113 the first point to circle mill
- Click 118 **Done**

◆ OBTAIN THE NEEDED 1/4 END MILL TOOL

Mastercam will activate and display the Tool and Drill Parameters dialog box after the circle mill tool path has been created.

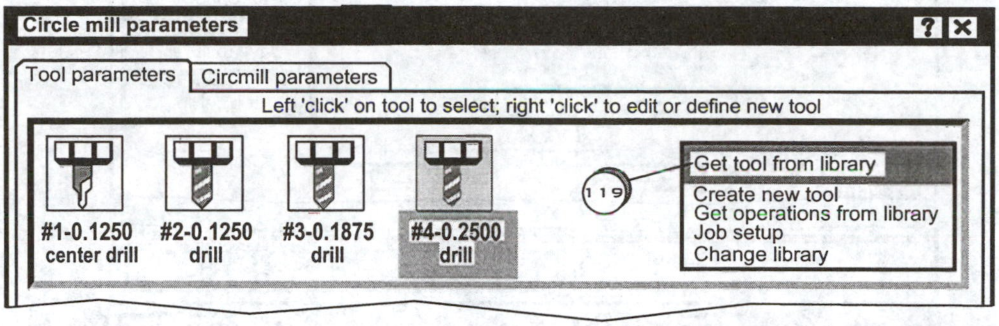

➤ ***Right*** Click move the mouse cursor *down* and Click (119) Get tool from library

Select the 1/4in Dia end mill tool from the **TOOLS** tool library

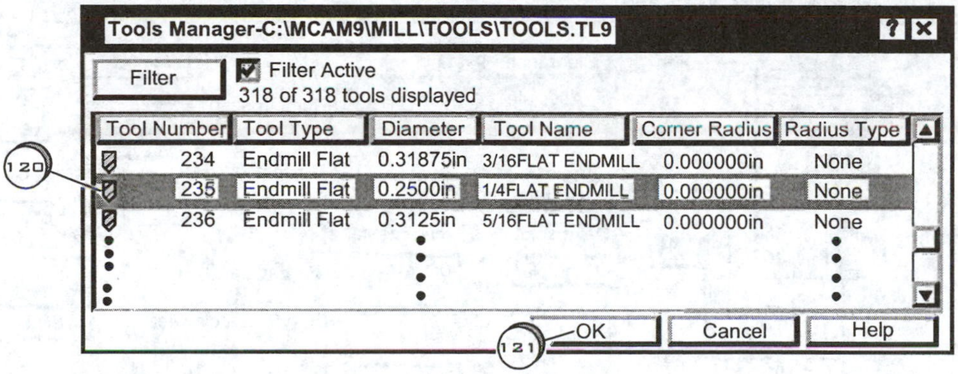

➤ Click (120) on the 1/4 FLAT ENDMILL tool

➤ Click (121) the OK button

Mastercam will place these tools into the currently active Tool and Drill Parameters dialog box

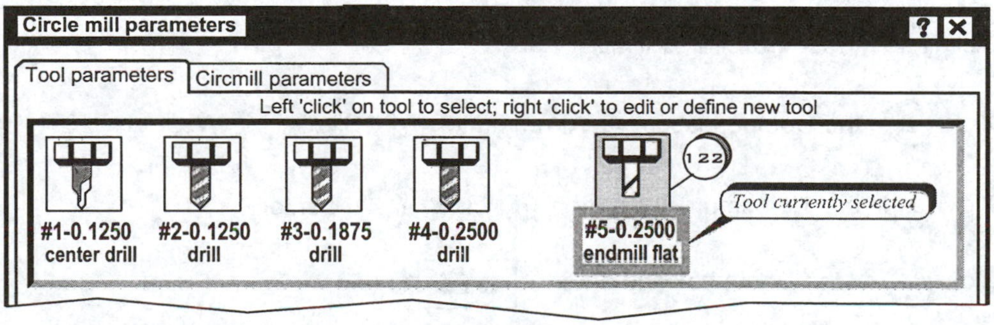

➤ ***Right*** Click (122) on the #5-0.2500 endmill flat tool

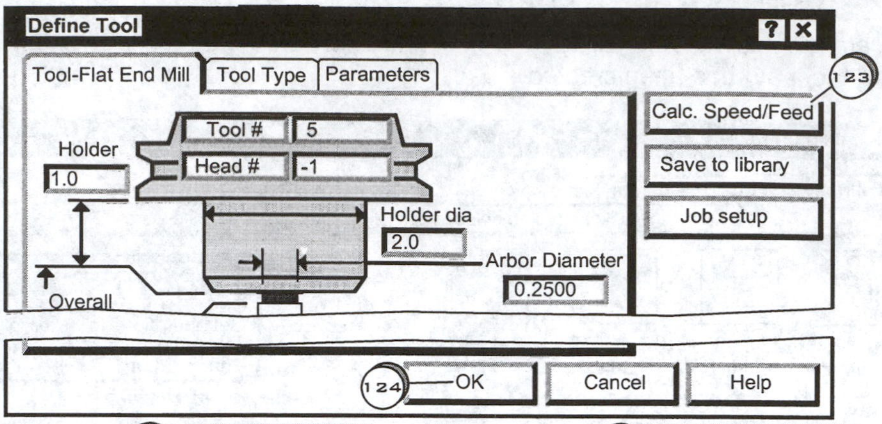

➤ Click (123) Calc Speed/Feed ║ ➤ Click (124) OK

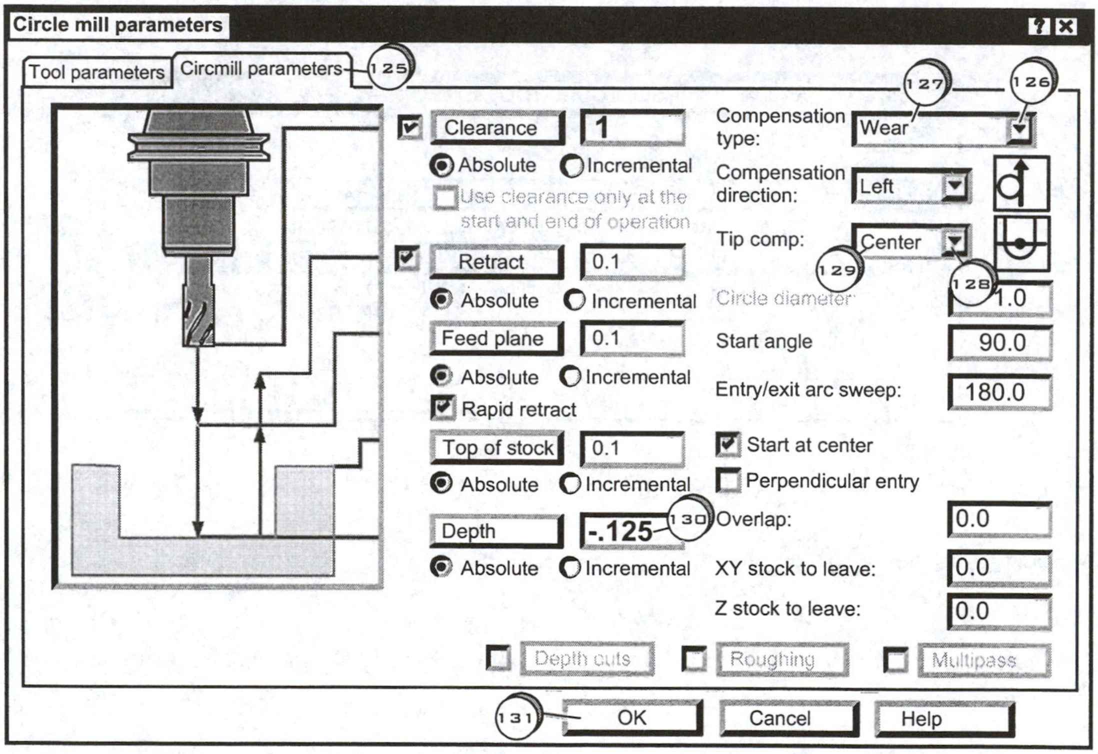

➤ Click (125) the **Circmill parameters** tab

➤ Click (126) the Compensation type down button ▼ ; Click (127) Wear

➤ Click (128) the Tip comp down button ▼ ; Click (129) Center

➤ Click (130) in the Depth box and enter **-.125**

➤ Click (131) the OK button

H) OPERATION#6- ROUGH AND FINISH SLOT X .375 DEEP

- THE SLOT CONTOUR IS TO BEGIN AT POINT1 IN CHAIN1 PROCEED TO THE START POINT IN CHAIN2 AND FOLLOW CHAIN2.

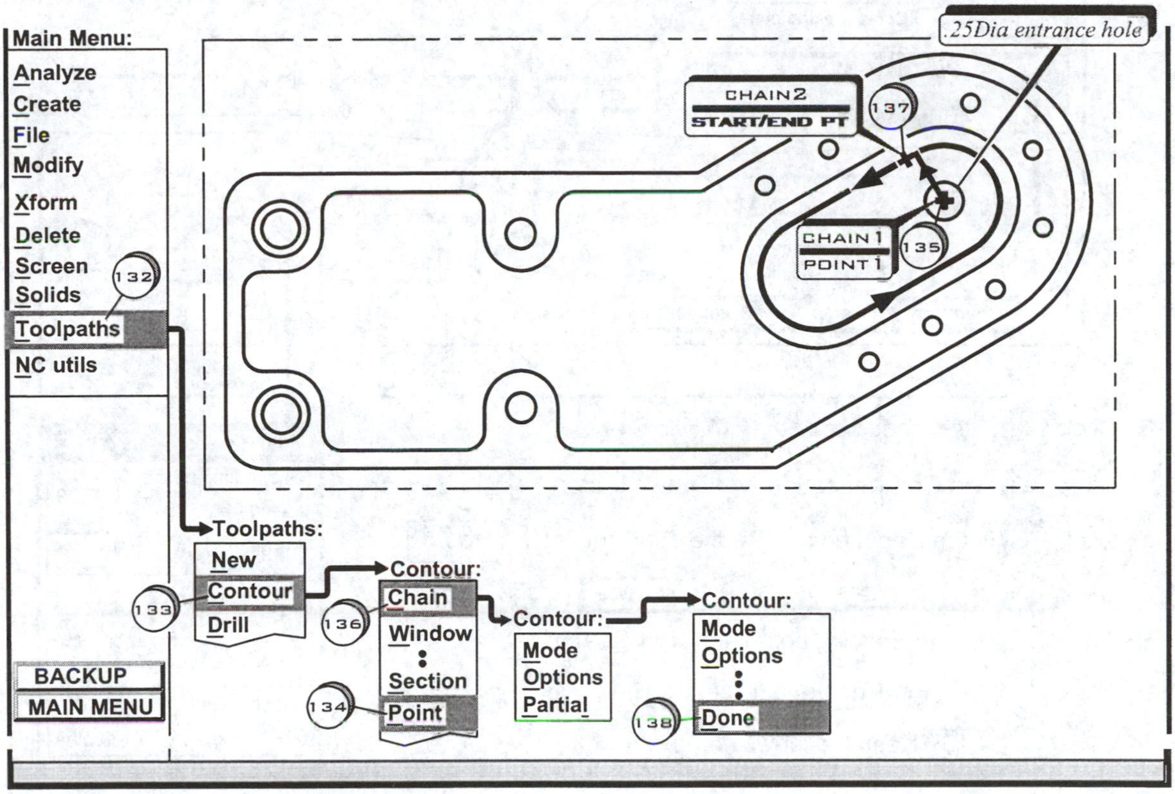

- Click 132 Toolpaths
- Click 133 Contour
- Click 134 Point
- Click 135 be sure the ***cursor*** is in the ***center*** of the .25 Dia entrance hole
- Click 136 Chain
- Click 137 near the ***end*** of the ***line*** element
- Click 138 Done

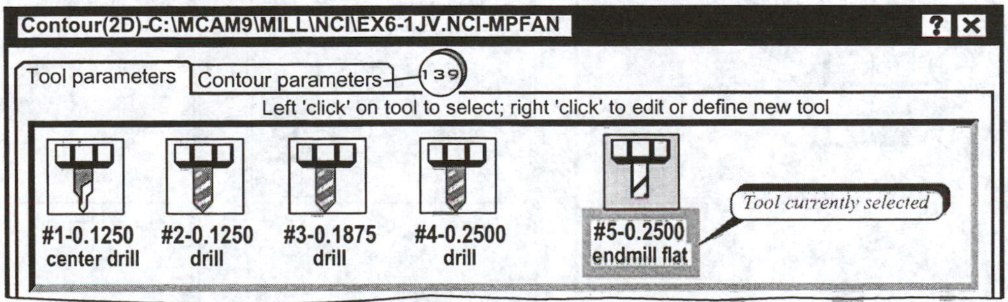

- Click 139 the Contour parameters tab

◆ ENTER THE REQUIRED CONTOUR PARAMETERS

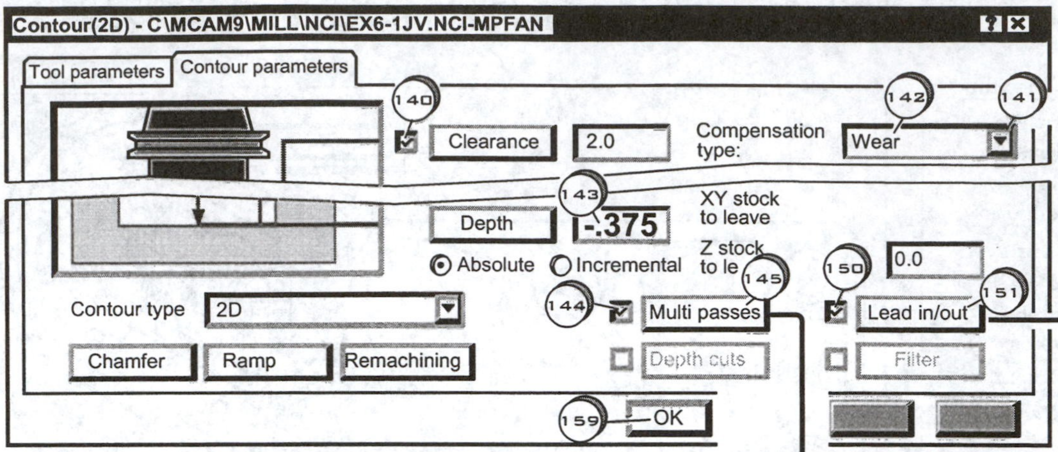

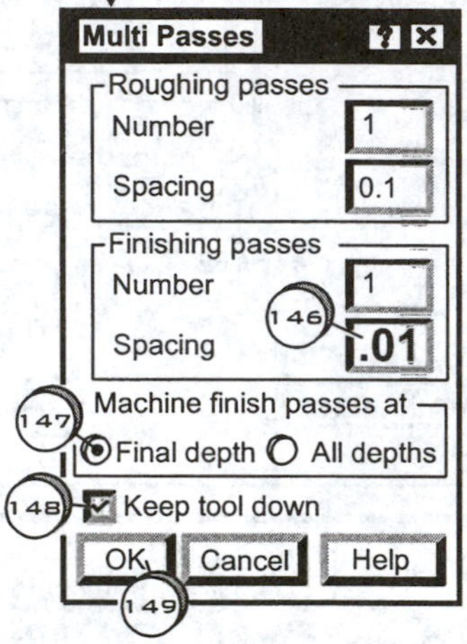

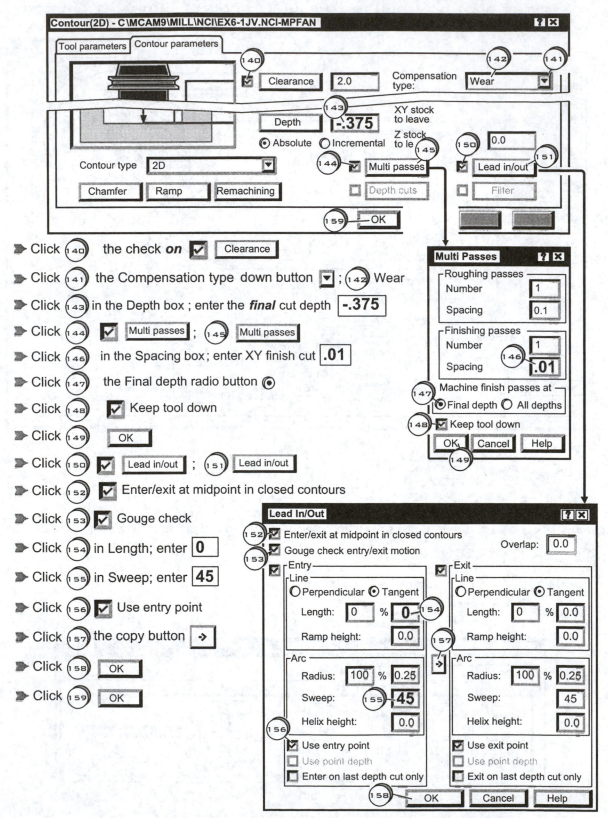

- ➤ Click (140) the check ***on*** ☑ [Clearance]
- ➤ Click (141) the Compensation type down button [▼] ; (142) Wear
- ➤ Click (143) in the Depth box ; enter the ***final*** cut depth [-.375]
- ➤ Click (144) ☑ [Multi passes] ; (145) [Multi passes]
- ➤ Click (146) in the Spacing box; enter XY finish cut [.01]
- ➤ Click (147) the Final depth radio button ◉
- ➤ Click (148) ☑ Keep tool down
- ➤ Click (149) [OK]
- ➤ Click (150) ☑ [Lead in/out] ; (151) [Lead in/out]
- ➤ Click (152) ☑ Enter/exit at midpoint in closed contours
- ➤ Click (153) ☑ Gouge check
- ➤ Click (154) in Length; enter [0]
- ➤ Click (155) in Sweep; enter [45]
- ➤ Click (156) ☑ Use entry point
- ➤ Click (157) the copy button [->]
- ➤ Click (158) [OK]
- ➤ Click (159) [OK]

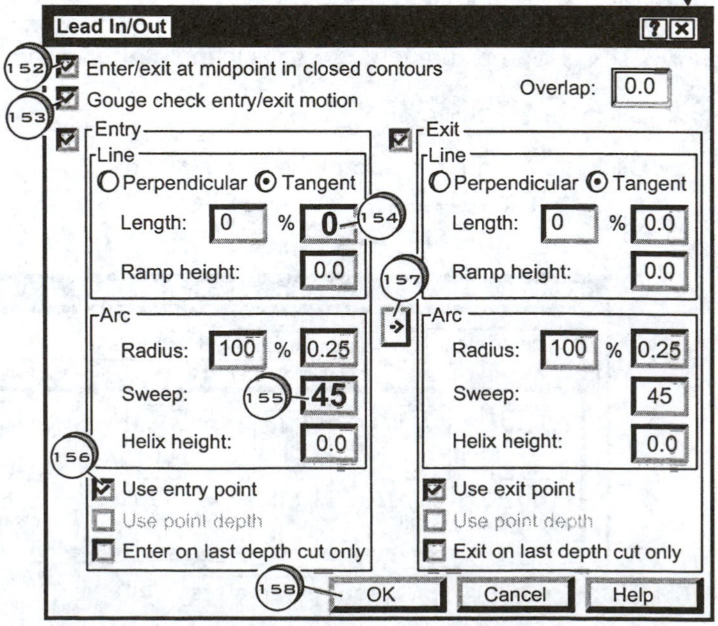

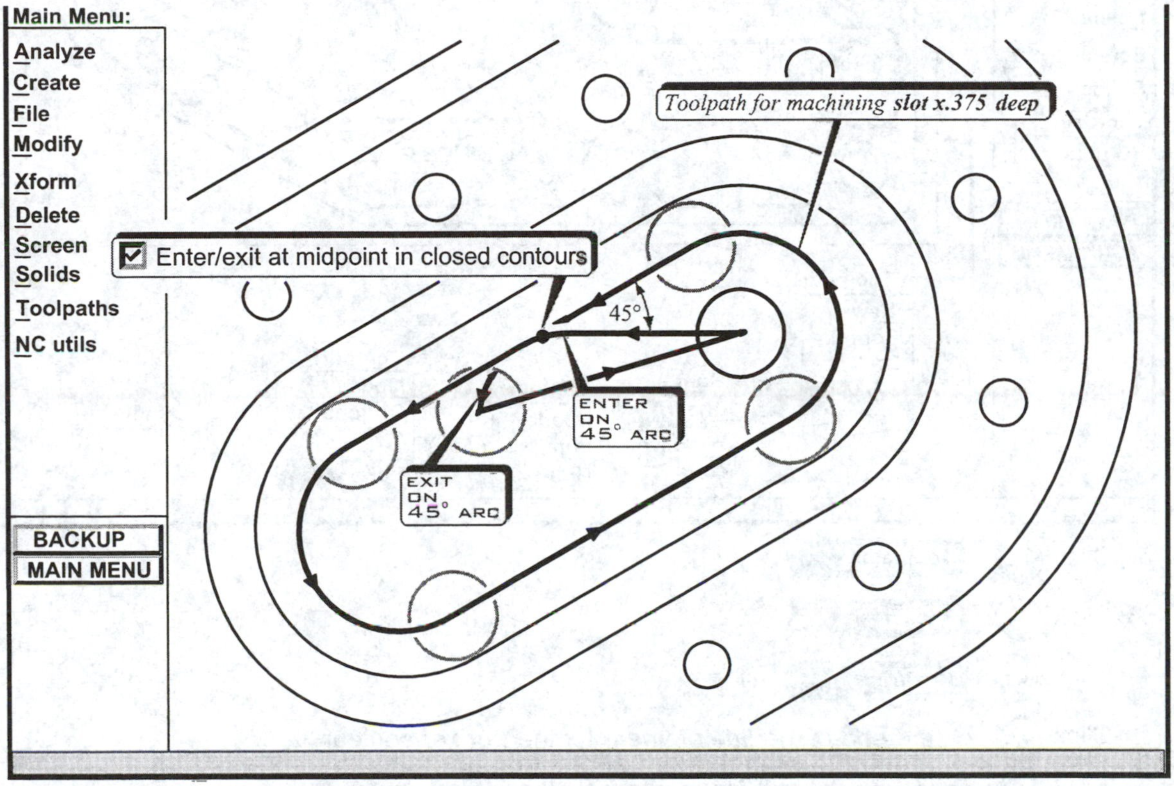
Main Menu:
Analyze
Create
File
Modify
Xform
Delete
Screen
Solids
Toolpaths
NC utils
BACKUP
MAIN MENU
Toolpath for machining slot x.375 deep
Enter/exit at midpoint in closed contours
45°
ENTER ON 45° ARC
EXIT ON 45° ARC

I) OPERATION#7- ROUGH AND FINISH POCKET X .125 DEEP

◆ CREATE THE POCKETING CONTOURS

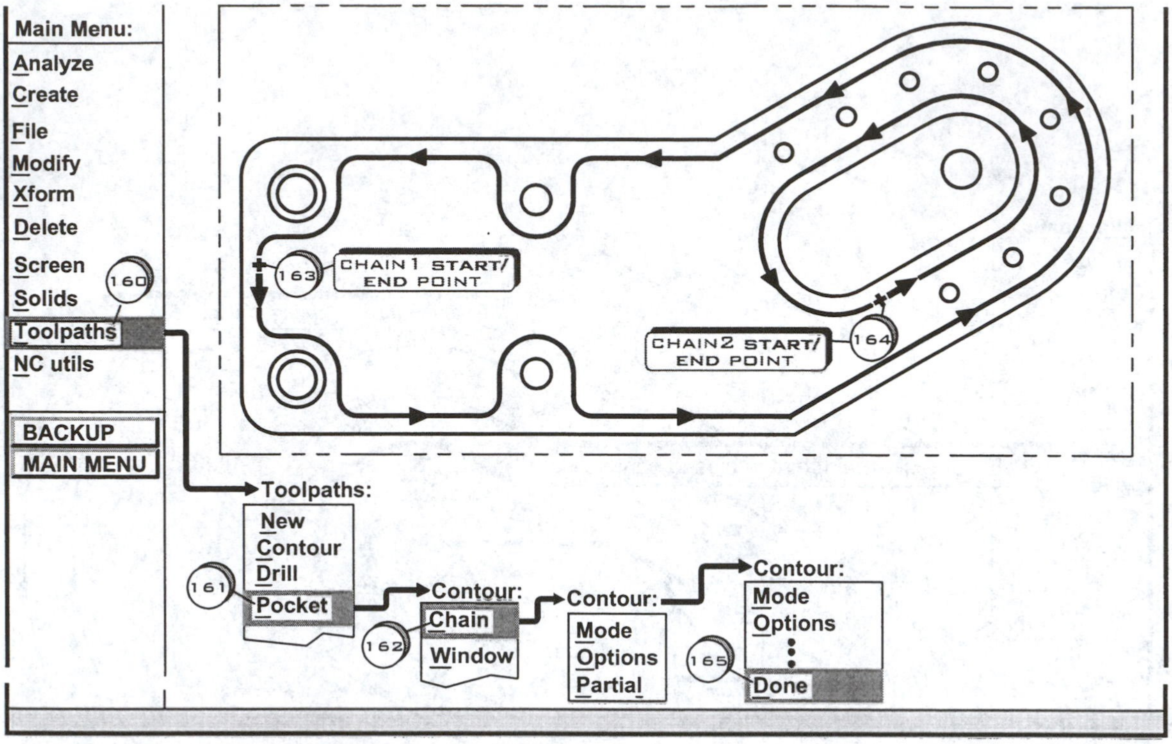

➢ Click (160) **Toolpaths**

➢ Click (161) **Pocket**

➢ Click (162) **Chain**

➢ Click (163) on the ***end*** of the ***line element*** for the ***pocket*** contour

➢ Click (164) on the ***end*** of the ***line element*** for the ***island*** contour

➢ Click (165) **Done**

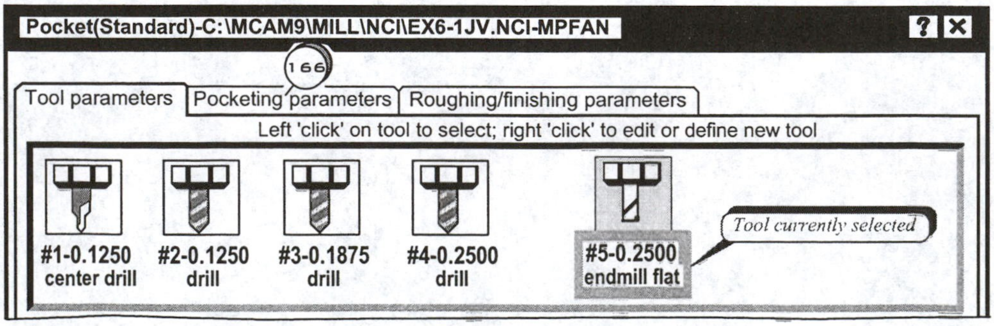

➢ Click (166) the **Pocketing parameters** tab

◆ ENTER THE REQUIRED POCKETING PARAMETERS

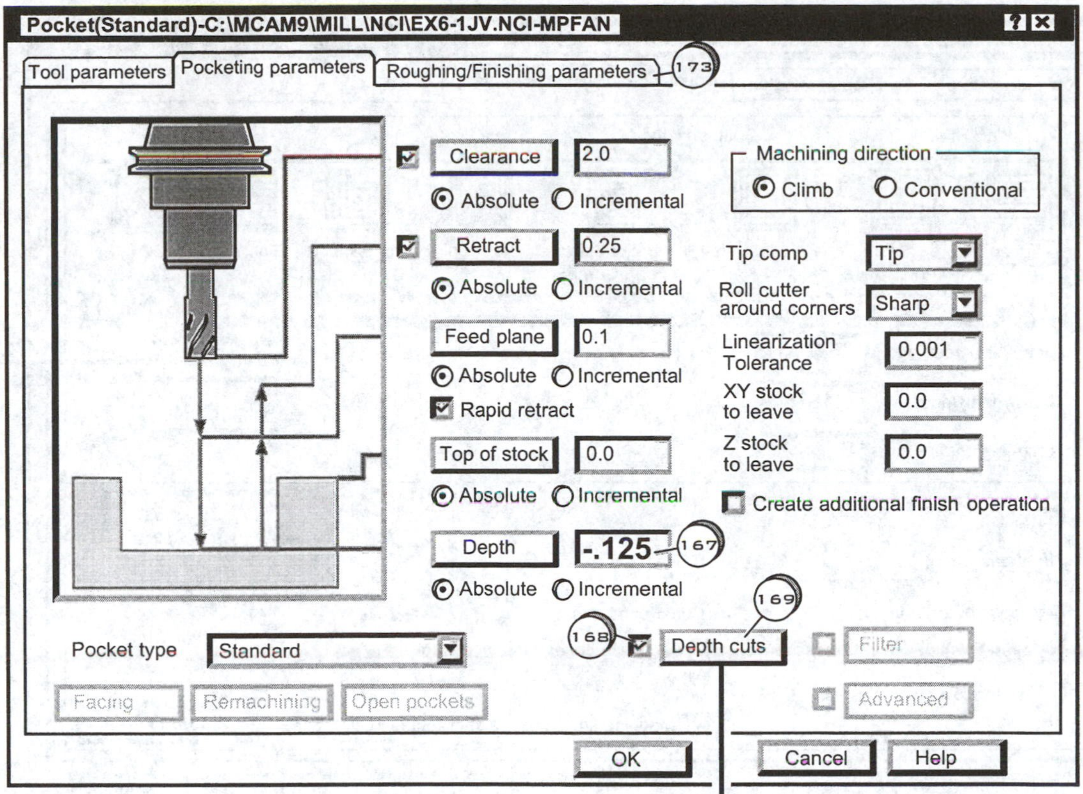

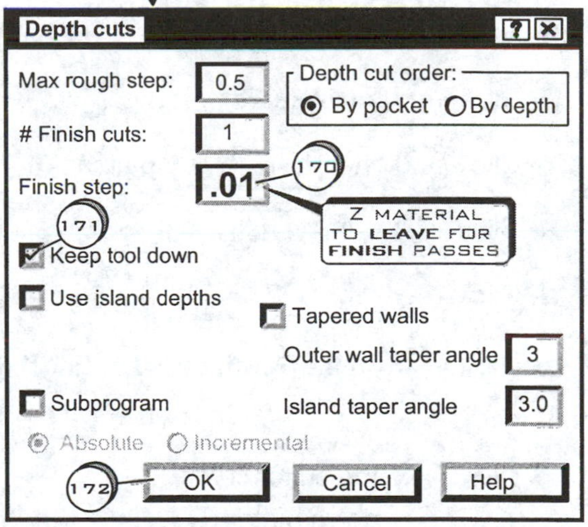

- Click (167) in the Depth box; enter **-.125**
- Click (168) activate Depth cuts ☑
- Click (169) the Depth cuts button
- Click (170) in the Finish step box; enter **.01**
- Click (171) ☑ Keep tool down
- Click (172) the OK button
- Click (173) the **Roughing/finishing parameters** tab

◆ ENTER THE REQUIRED ROUGHING/FINISHING PARAMETERS

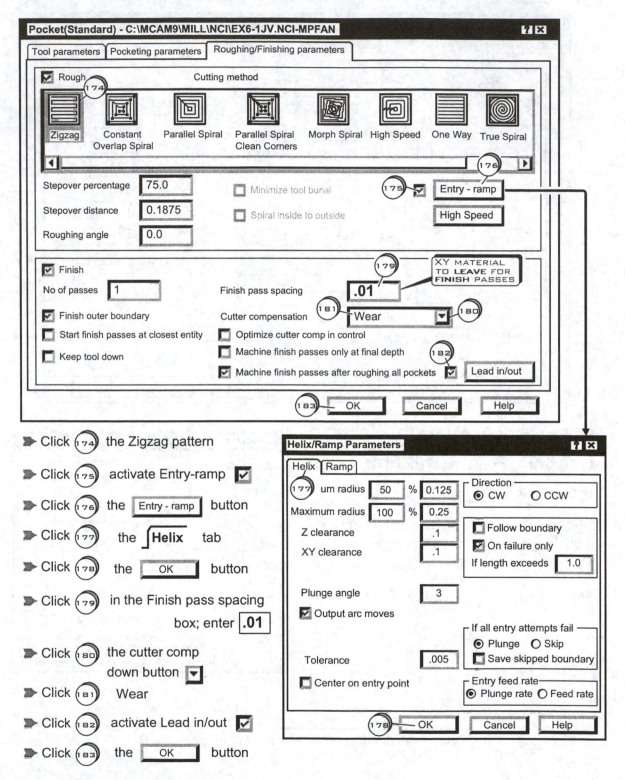

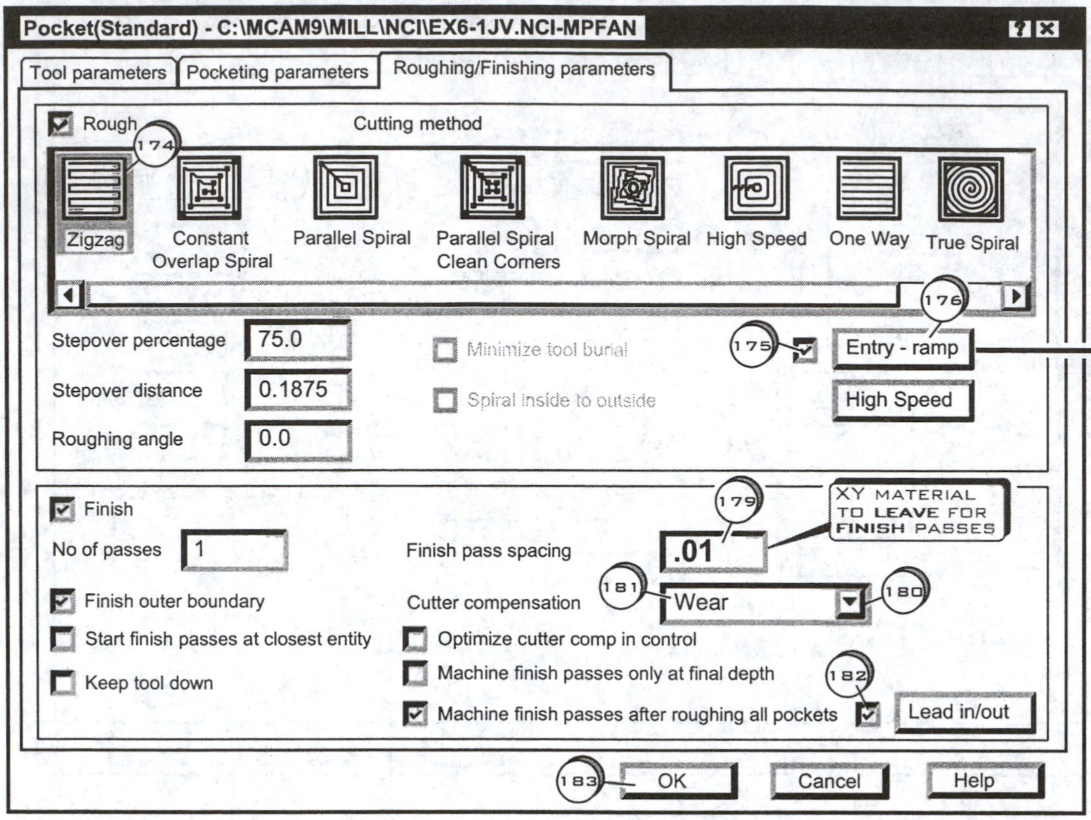

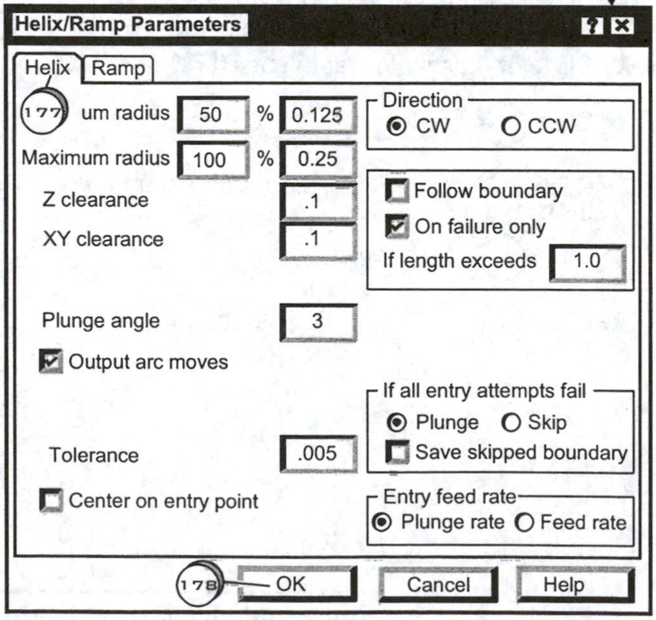

- Click (174) the Zigzag pattern
- Click (175) activate Entry-ramp ☑
- Click (176) the [Entry - ramp] button
- Click (177) the [Helix] tab
- Click (178) the [OK] button
- Click (179) in the Finish pass spacing box; enter .01
- Click (180) the cutter comp down button ▼
- Click (181) Wear
- Click (182) activate Lead in/out ☑
- Click (183) the [OK] button

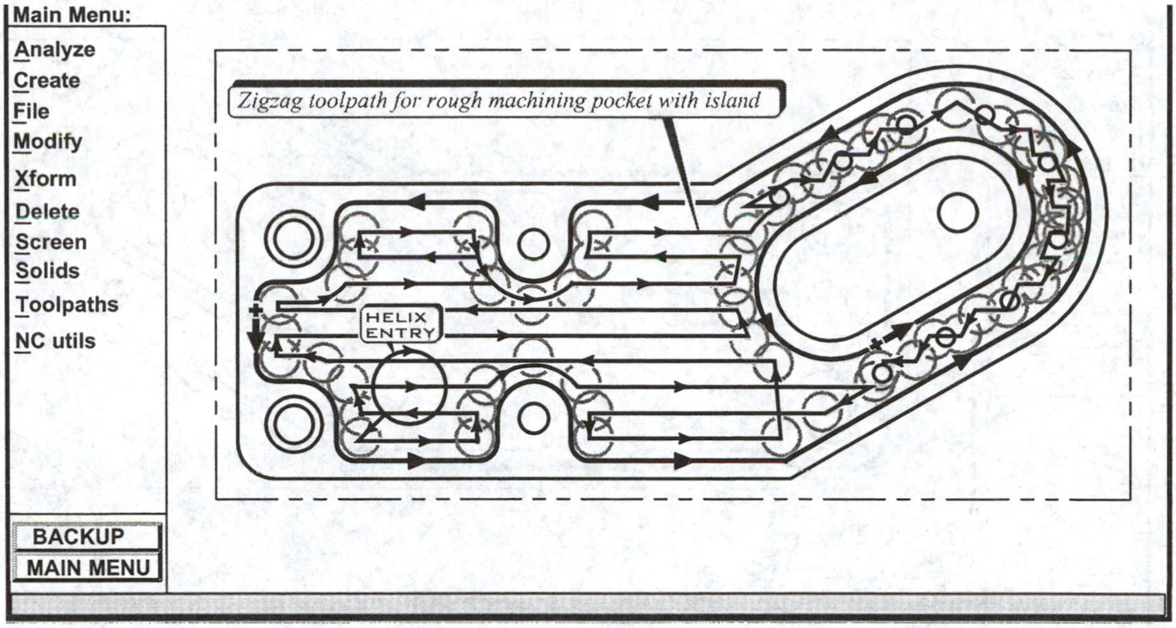
Main Menu:
Analyze
Create
File
Modify
Xform
Delete
Screen
Solids
Toolpaths
NC utils
BACKUP
MAIN MENU
Zigzag toolpath for rough machining pocket with island
HELIX ENTRY

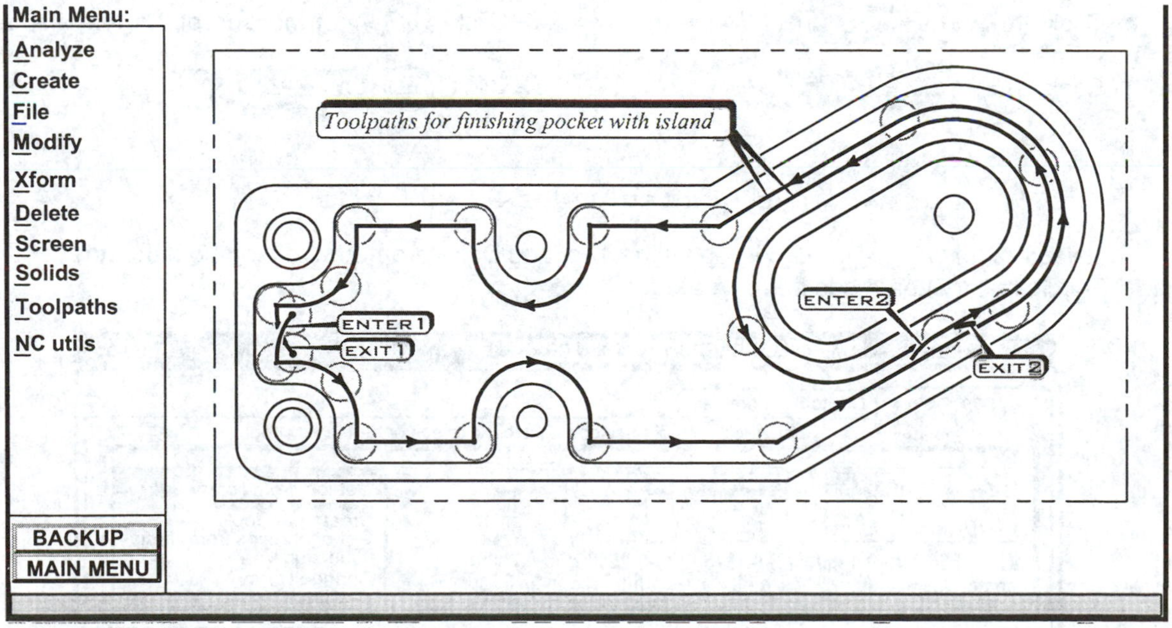
Main Menu:
Analyze
Create
File
Modify
Xform
Delete
Screen
Solids
Toolpaths
NC utils
BACKUP
MAIN MENU
Toolpaths for finishing pocket with island
ENTER1
EXIT1
ENTER2
EXIT2

J) OPERATION#8- ROUGH AND FINISH SUTSIDE X .375 DEEP

- CREATE THE CONTOUR FOR MACHINING THE OUTSIDE

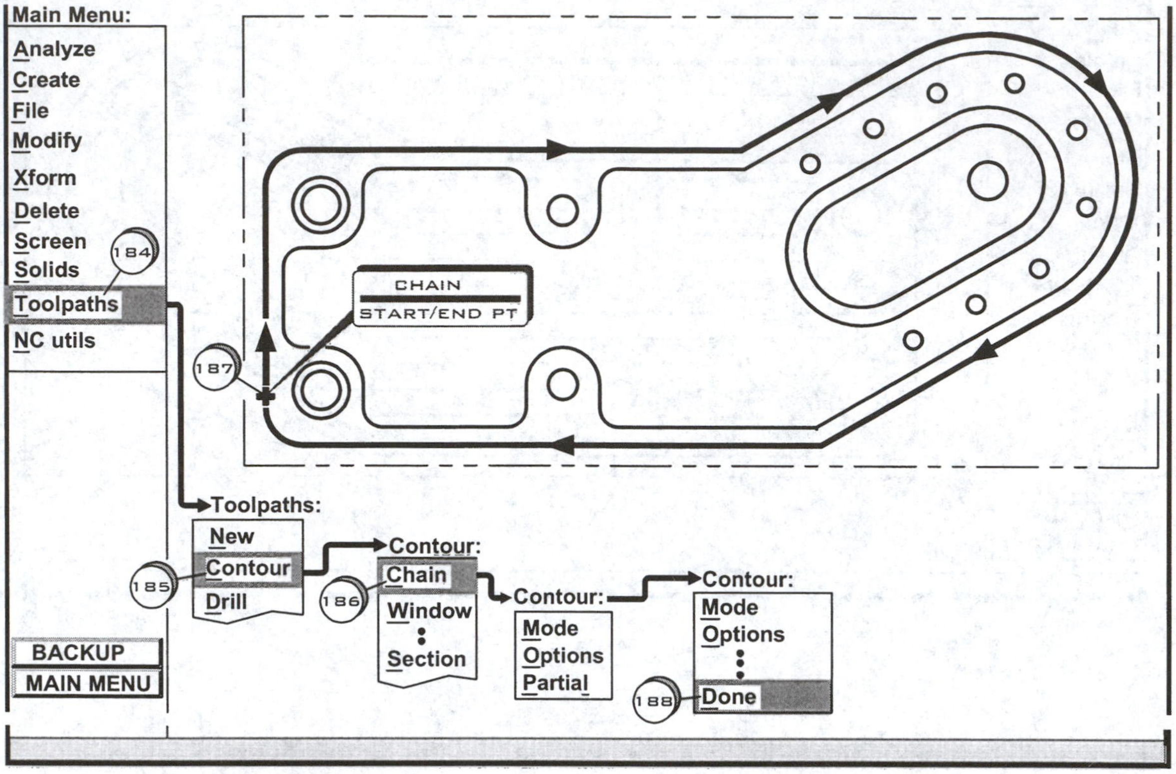

- Click 184 **Toolpaths**
- Click 185 **Contour**
- Click 186 **Chain**
- Click 187 near the ***end*** of the ***line*** element
- Click 188 **Done**

- OBTAIN THE NEEDED 1/4 END MILL TOOL

Mastercam will activate and display the Tool and Drill Parameters dialog box after the circle mill tool path has been created.

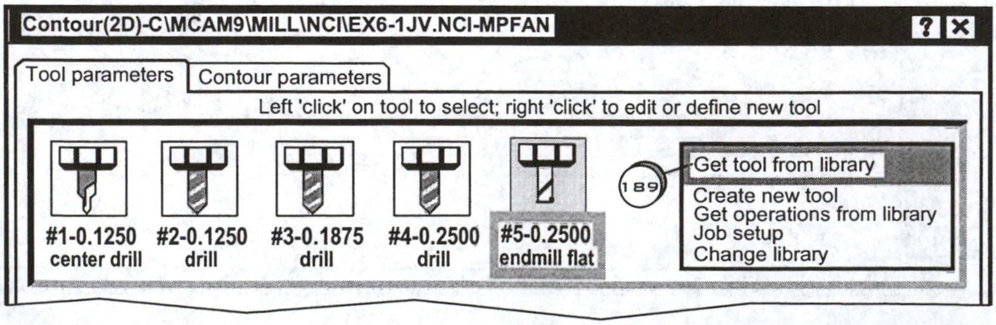

- ***Right*** Click move the mouse cursor *down* and Click 189 Get tool from library

Select the 1/2in Dia end mill tool from the **TOOLS** tool library

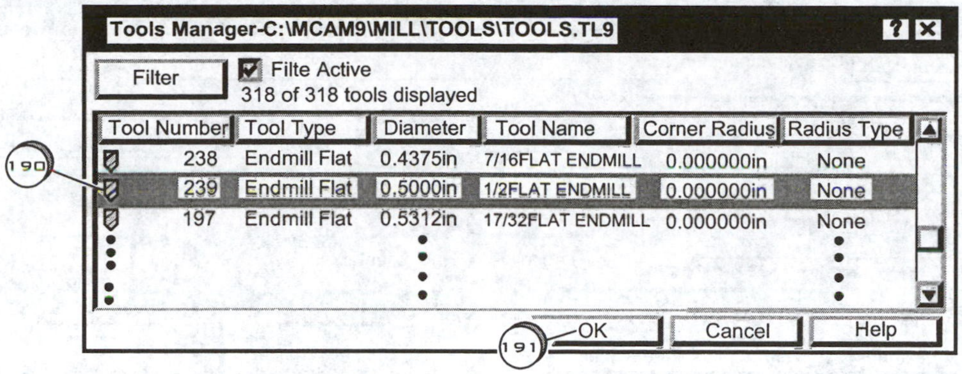

➢ Click (190) on the 1/2 FLAT ENDMILL tool

➢ Click (191) the [OK] button

Mastercam will place these tools into the currently active Tool and Drill Parameters dialog box

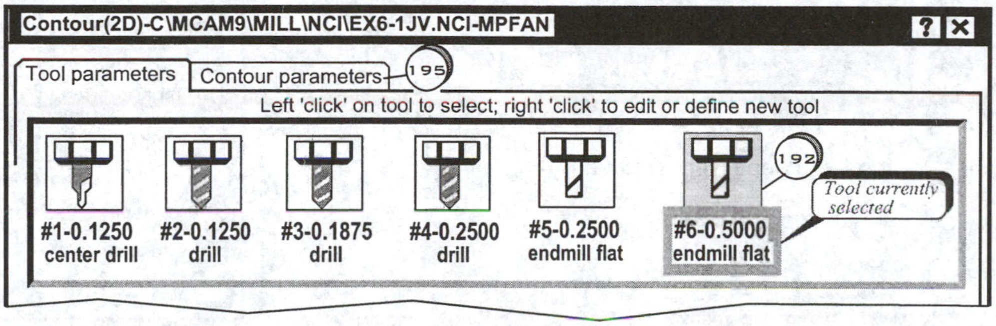

➢ ***Right*** Click (192) on the #6-0.5000 endmill flat tool

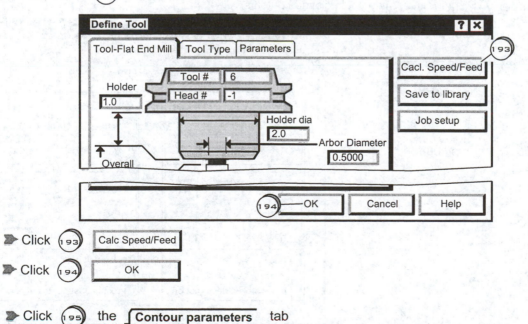

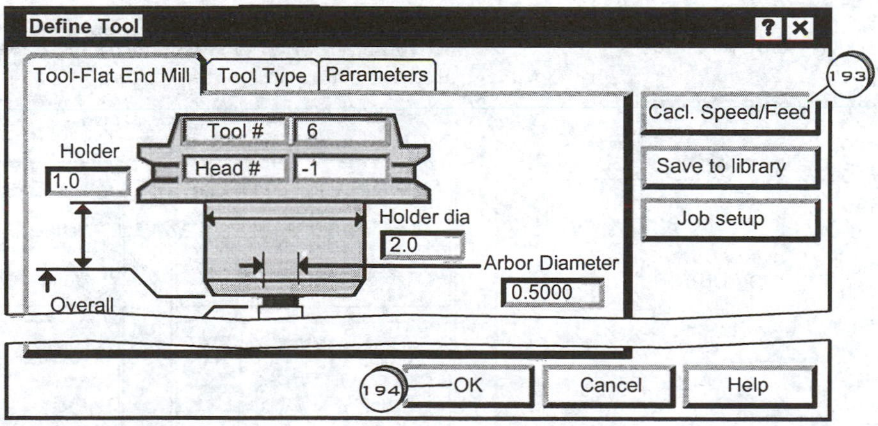

➢ Click (193) [Calc Speed/Feed]

➢ Click (194) [OK]

➢ Click (195) the **Contour parameters** tab

◆ ENTER THE REQUIRED CONTOUR PARAMETERS

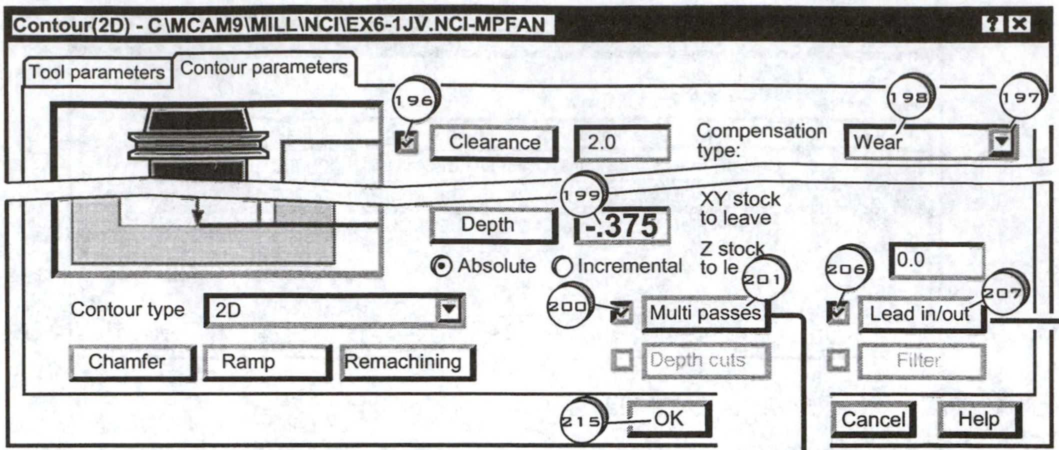

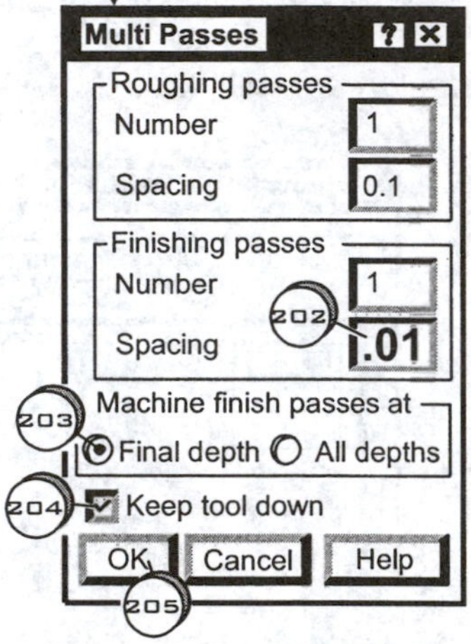

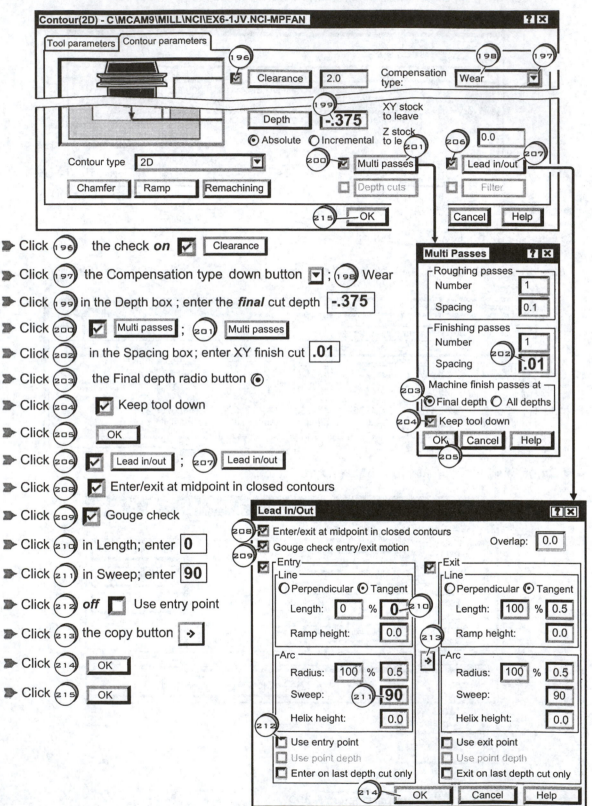

- Click (196) the check ***on*** ☑ Clearance
- Click (197) the Compensation type down button ▼ ; (198) Wear
- Click (199) in the Depth box ; enter the ***final*** cut depth -.375
- Click (200) ☑ Multi passes ; (201) Multi passes
- Click (202) in the Spacing box; enter XY finish cut .01
- Click (203) the Final depth radio button ◉
- Click (204) ☑ Keep tool down
- Click (205) OK
- Click (206) ☑ Lead in/out ; (207) Lead in/out
- Click (208) ☑ Enter/exit at midpoint in closed contours
- Click (209) ☑ Gouge check
- Click (210) in Length; enter 0
- Click (211) in Sweep; enter 90
- Click (212) ***off*** ☐ Use entry point
- Click (213) the copy button →
- Click (214) OK
- Click (215) OK

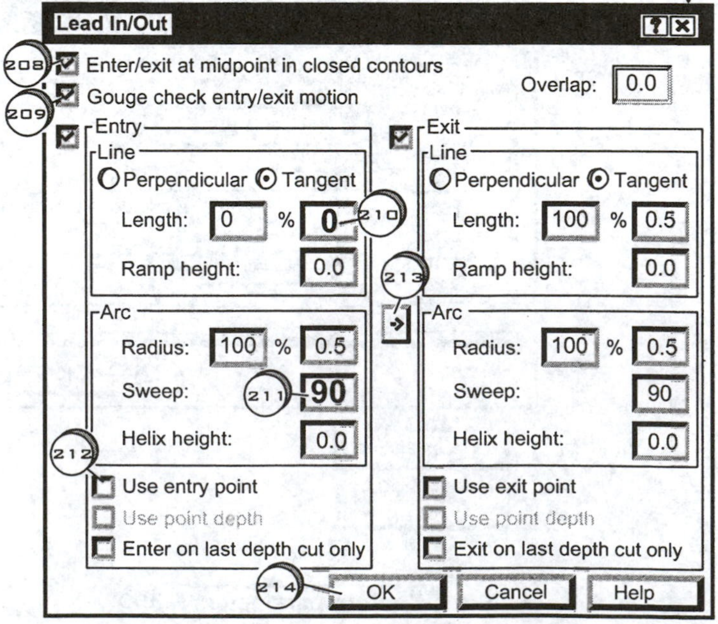

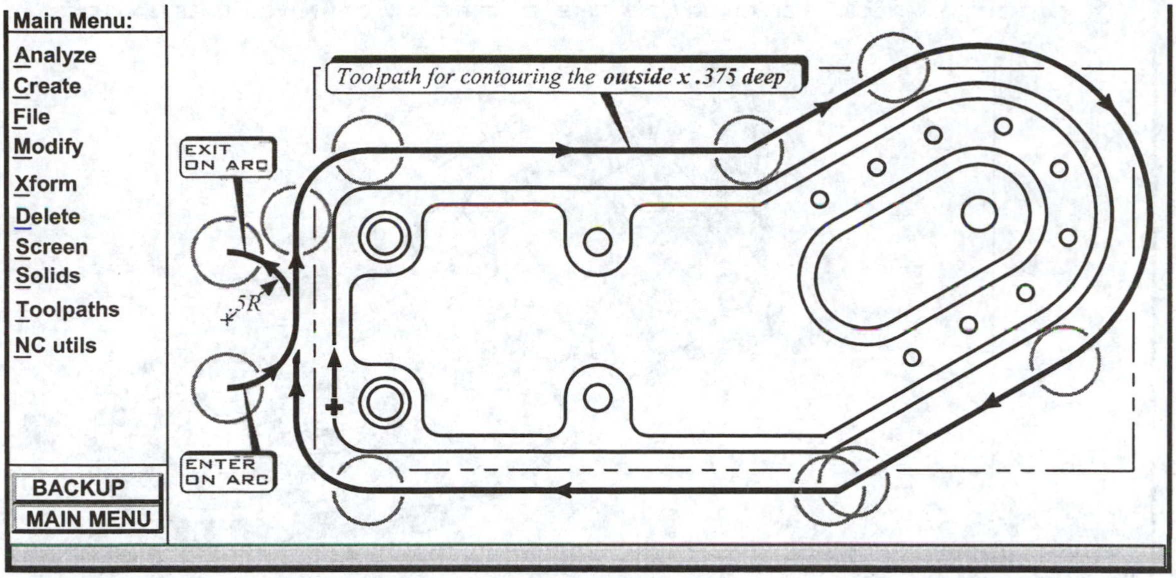

J) VERIFY THE TOOLPATHS FOR ALL THE MACHINING OPERATIONS

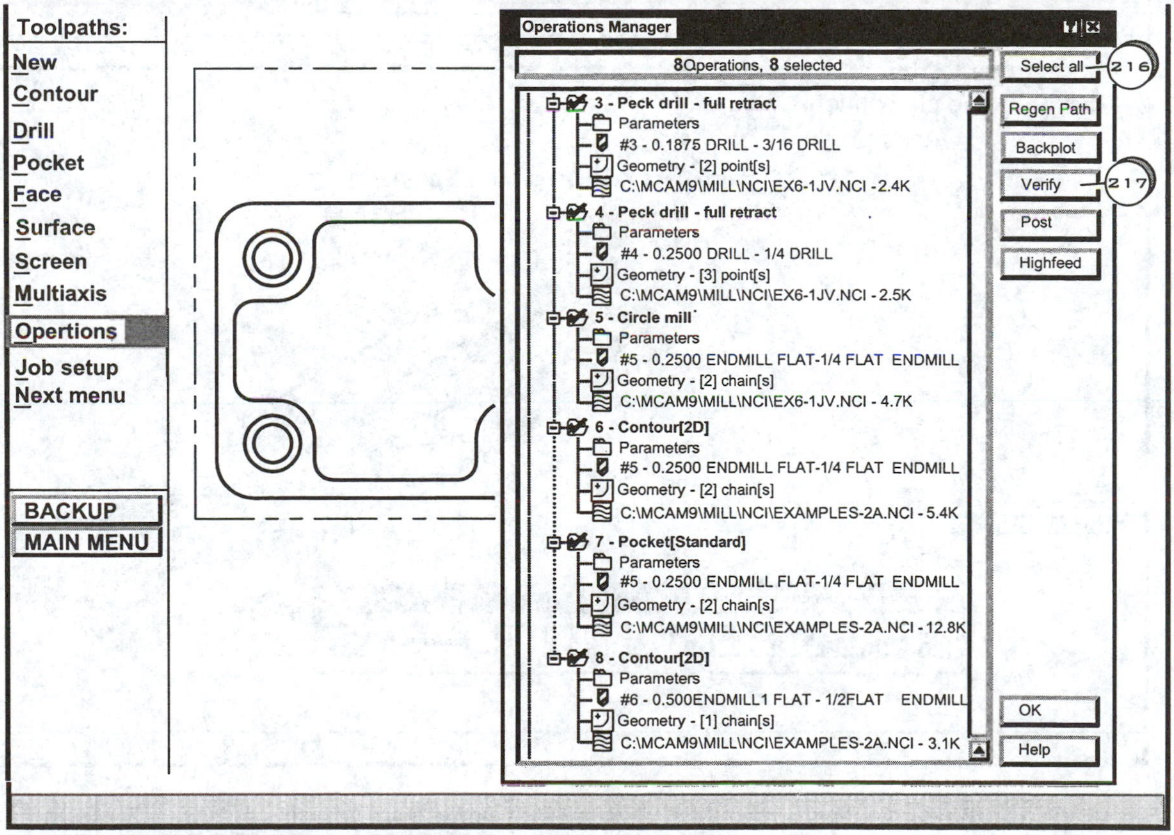

- Click (216) the Select All button
- Click (217) the Verify button

◆ DIRECT *Mastercam* TO ANIMATE THE ENTIRE MACHINING OPERATION

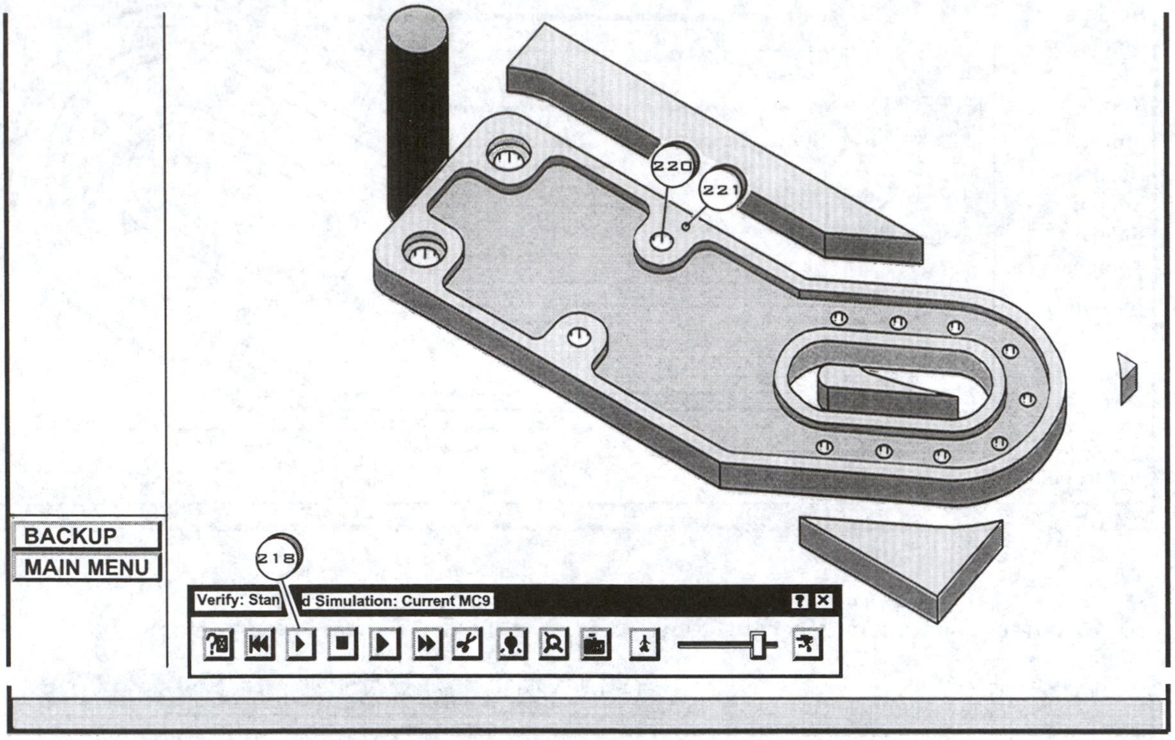

➤ Click (218) the play button ▶

◆ DIRECT *Mastercam* TO SECTION THE PART

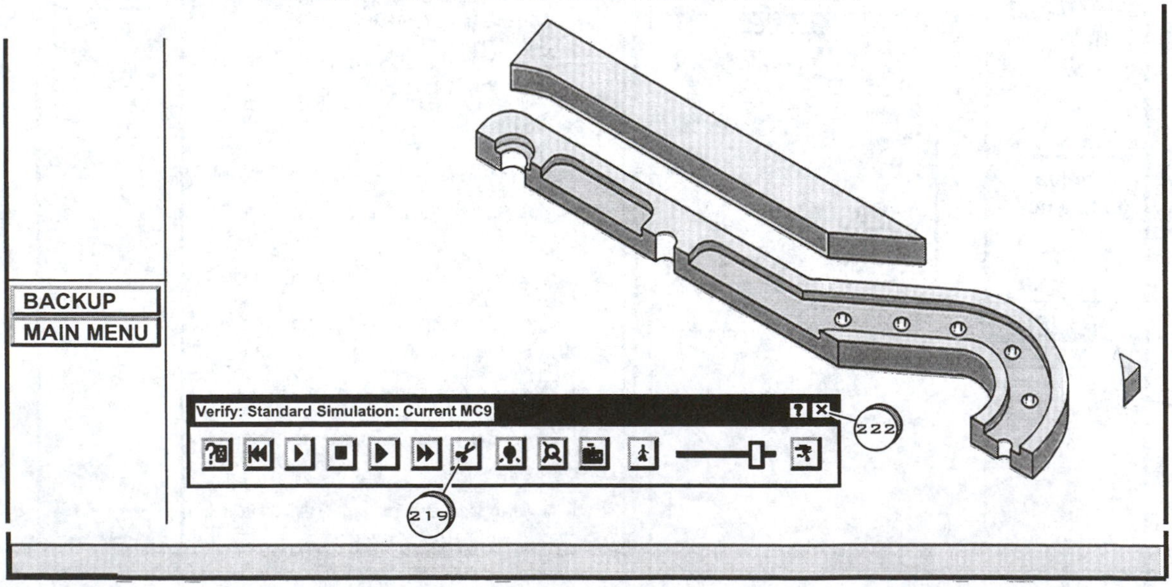

➤ Click (219) the cut section button

Pick point on stock for section reference

➤ Click (220) on the circle

Pick side of stock to keep

➤ Click (221) on the side to keep

➤ Click (222) the exit button ☒ to exit **Verify**

L) Generate the Part Program for All the Machining Operations

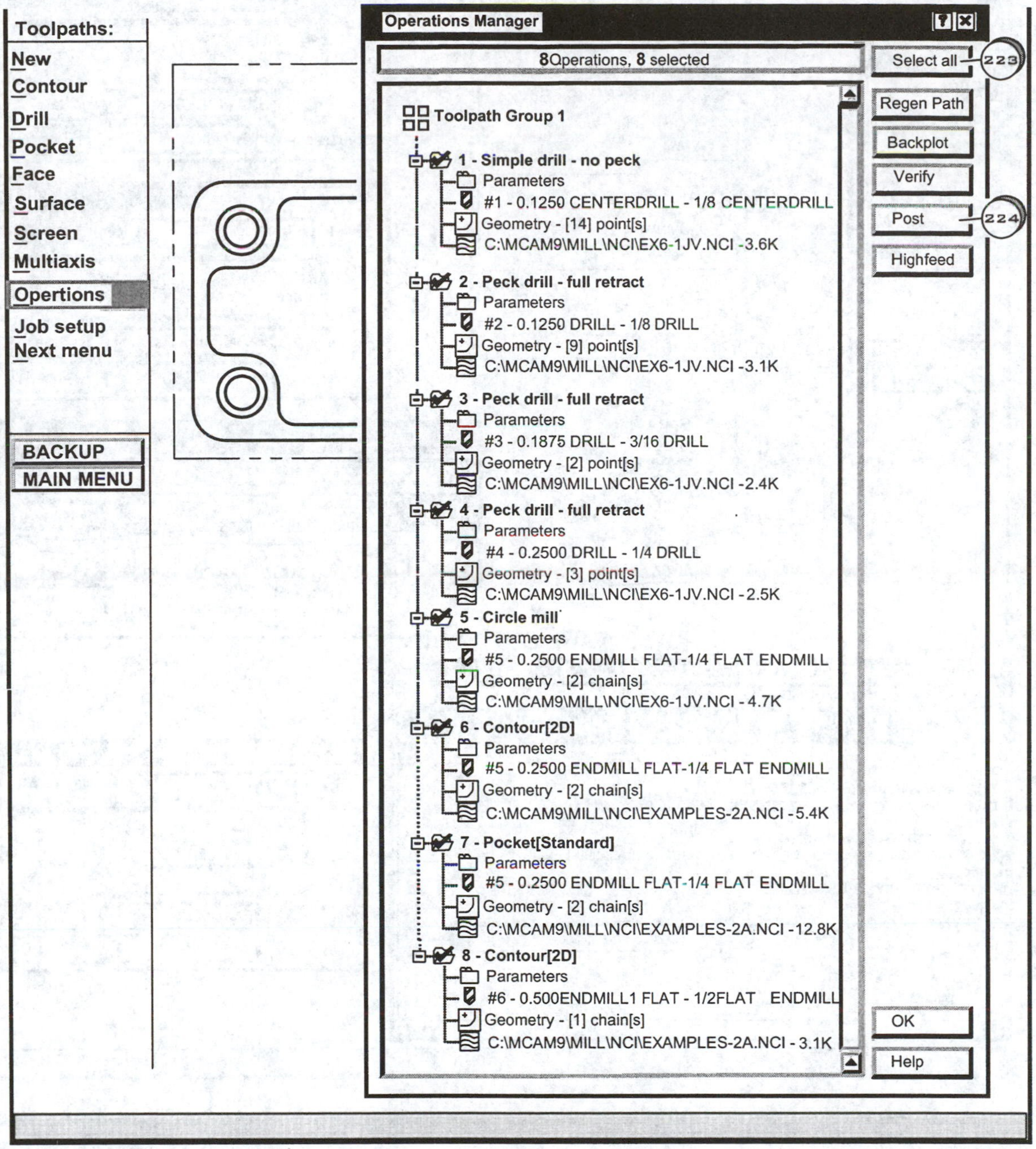

- Click (223) the **Select All** button
- Click (224) the **Post** button

Note: *Mastercam's* sample post file **MPFAN.PST** will be used as the default post for this text

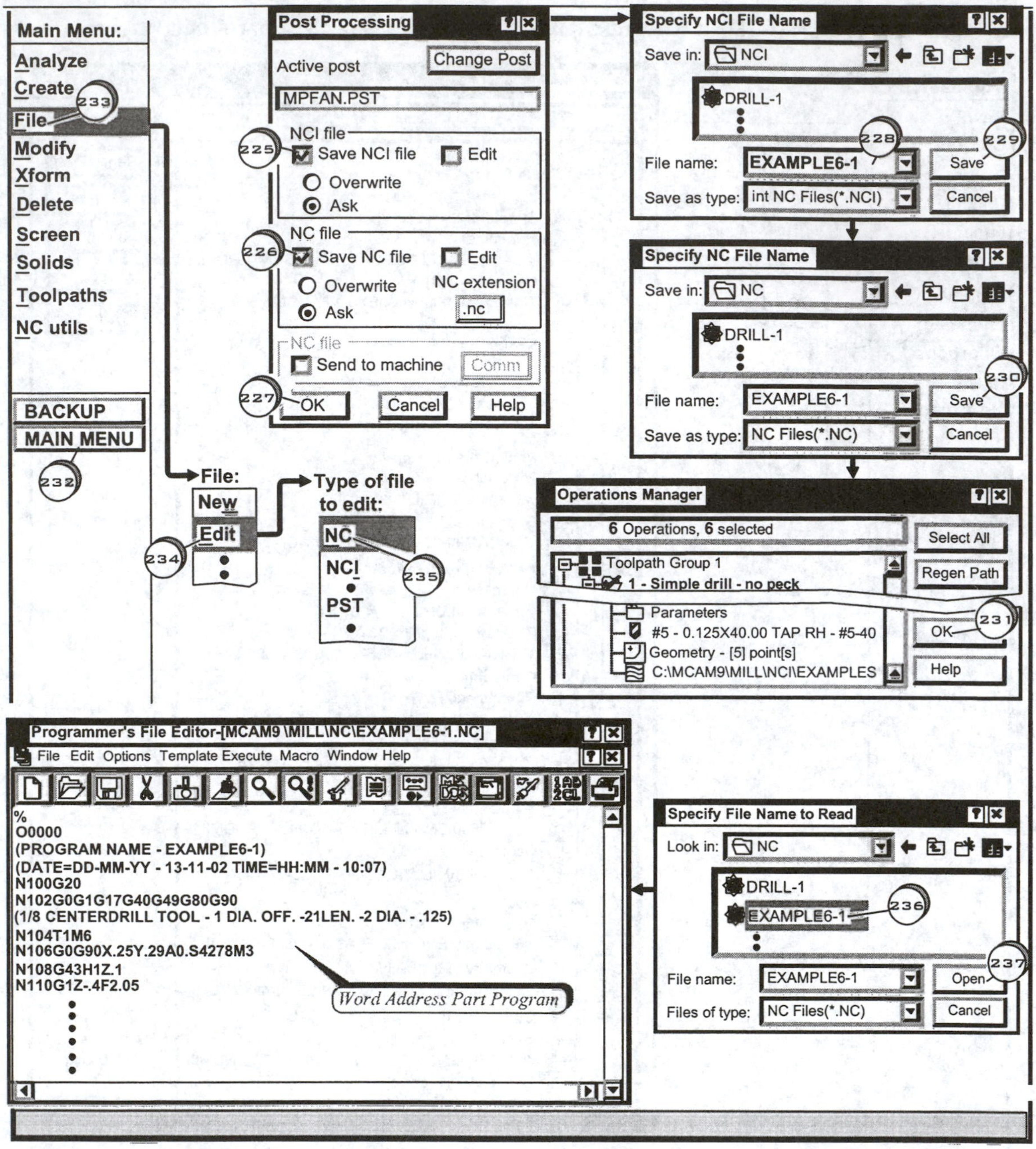

Main Menu:
Analyze
Create
File
Modify
Xform
Delete
Screen
Solids
Toolpaths
NC utils
BACKUP
MAIN MENU
Post Processing
Active post
Change Post
MPFAN.PST
NCI file
Save NCI file
Edit
Overwrite
Ask
NC file
Save NC file
NC extension
.nc
Send to machine
Comm
OK
Cancel
Help
Specify NCI File Name
Save in:
NCI
DRILL-1
File name:
EXAMPLE6-1
Save
Save as type:
int NC Files(*.NCI)
Specify NC File Name
NC
NC Files(*.NC)
Operations Manager
6 Operations, 6 selected
Select All
Regen Path
Toolpath Group 1
1 - Simple drill - no peck
Parameters
#5 - 0.125X40.00 TAP RH - #5-40
Geometry - [5] point[s]
C:\MCAM9\MILL\NCI\EXAMPLES
File:
New
Edit
Type of file to edit:
NC
NCI
PST
Programmer's File Editor-[MCAM9 \MILL\NC\EXAMPLE6-1.NC]
File Edit Options Template Execute Macro Window Help
%
O0000
(PROGRAM NAME - EXAMPLE6-1)
(DATE=DD-MM-YY - 13-11-02 TIME=HH:MM - 10:07)
N100G20
N102G0G1G17G40G49G80G90
(1/8 CENTERDRILL TOOL - 1 DIA. OFF. -21LEN. -2 DIA. - .125)
N104T1M6
N106G0G90X.25Y.29A0.S4278M3
N108G43H1Z.1
N110G1Z-.4F2.05
Word Address Part Program
Specify File Name to Read
Look in:
Open
Files of type:
225
226
227
228
229
230
231
232
233
234
235
236
237
Click 225 check in Save NCI file
Click 226 check in Save NC file
Click 227 OK
Click 228 in File name box ; enter EXAMPLE6-1
Click 229 Save
Click 230 Save
Click 231 OK
Click 232 MAIN MENU
Click 233 File
Click 234 Edit
Click 235 NC
Click 236 EXAMPLE6-1
Click 237 Open

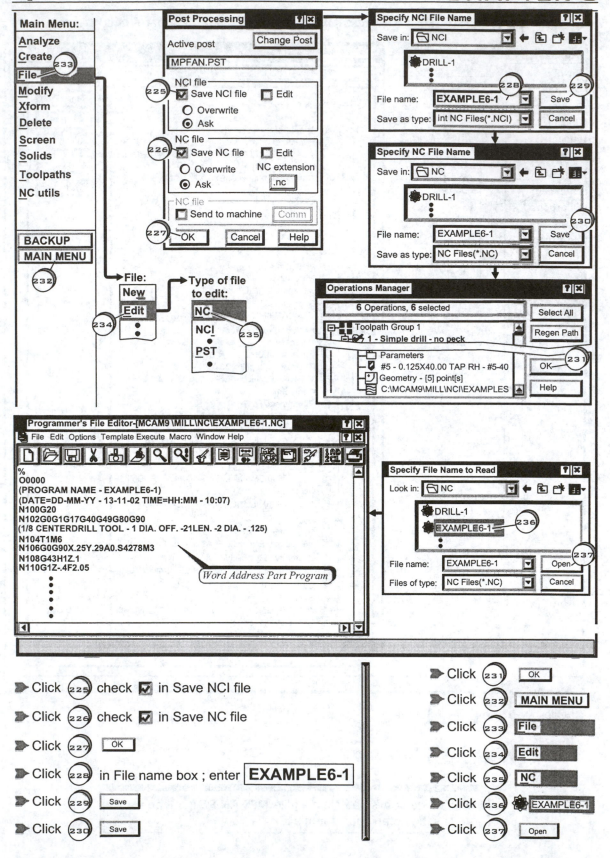

6-2) Get the CAD model file **EX2-1JV** generated in exercise 2-1. Generate a part program for executing the machining listed in PROCESS PLAN 6P-2.

Material: 1030 Steel

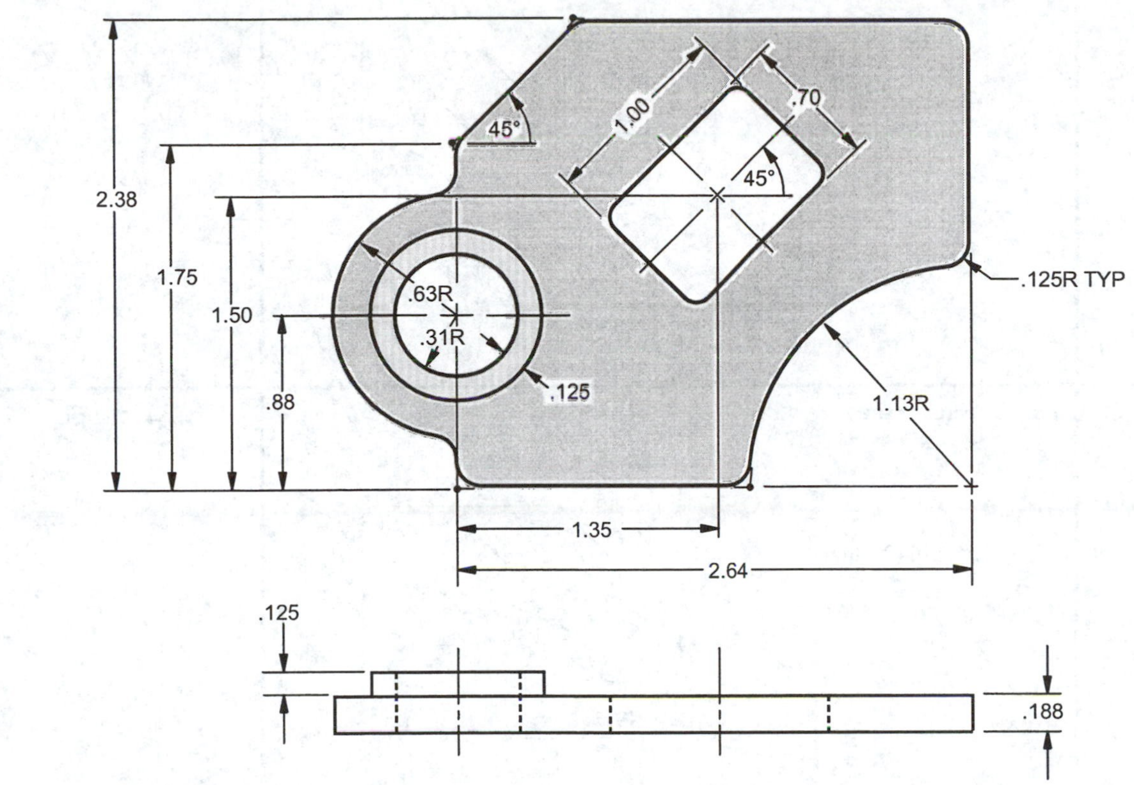

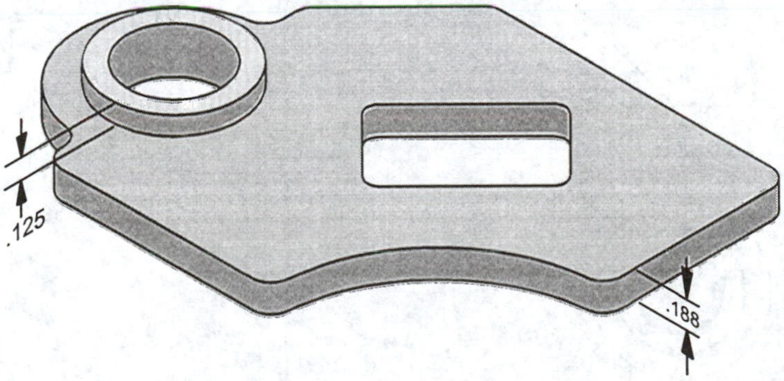

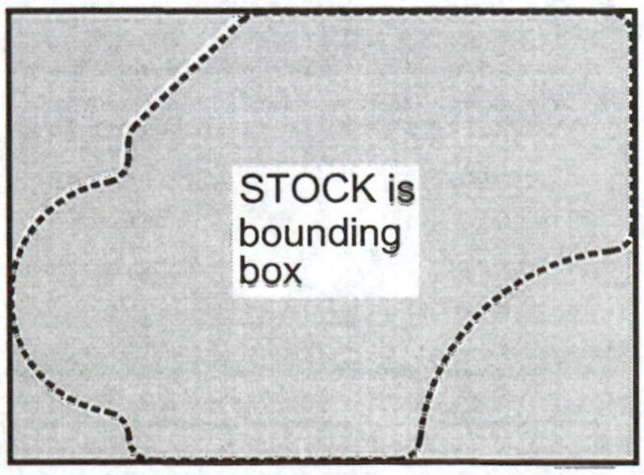

Figure 6-p3

PROCESS PLAN 6P-2

No.	Operation	Tooling
1	CENTER DRILL X .166 DEEP 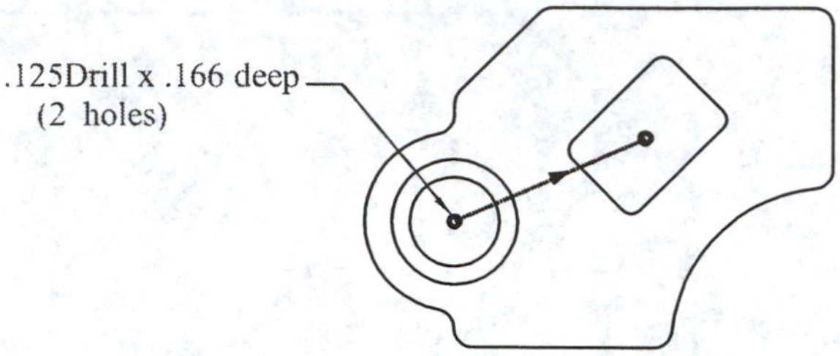	1/8 CENTER DRILL
2	PECK DRILL THRU 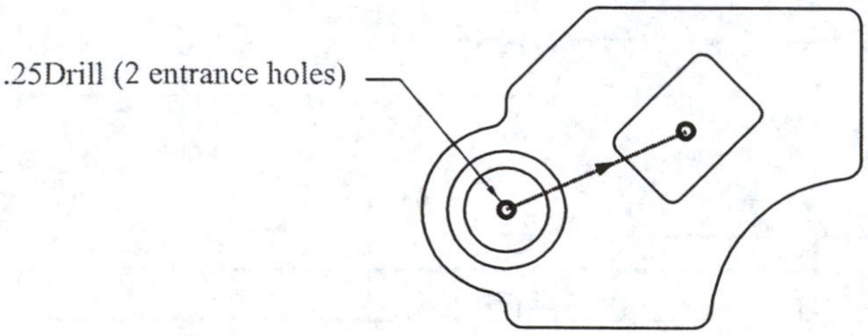	1/4 DRILL
2	CIRCLE MILL .62DIA THRU 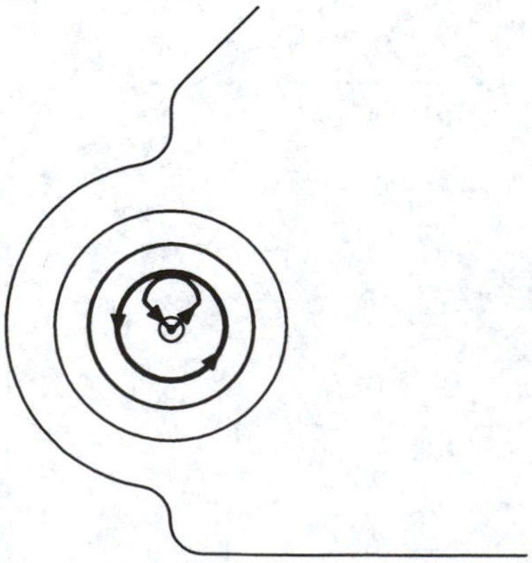	1/4 END MILL

PROCESS PLAN 6P-2(*continued*)

No.	Operation	Tooling
4	ROUGH AND FINISH SLOT X .188 DEEP	1/4 END MILL
5	POCKET X .125 DEEP LEAVE .02 FOR FINISH CUT IN XY AND Z Pocket type: Facing Standard / Facing Facing Overlap percentage: 100 Overlap ammount: 0.75 Cutting method: True Spiral	1/2 END MILL
6	ROUGH OUTSIDE X .188 DEEP LEAVE .02 FOR FINISH CUT IN XY	1/4 END MILL

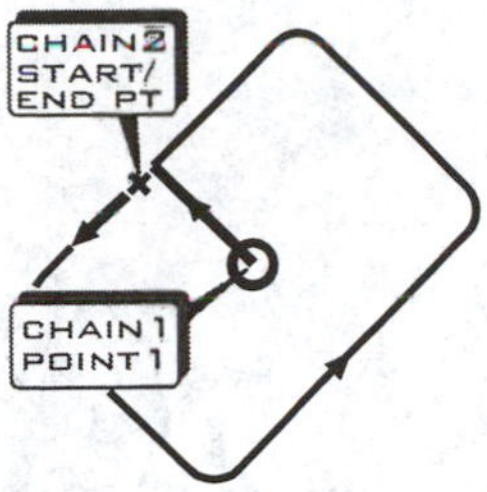

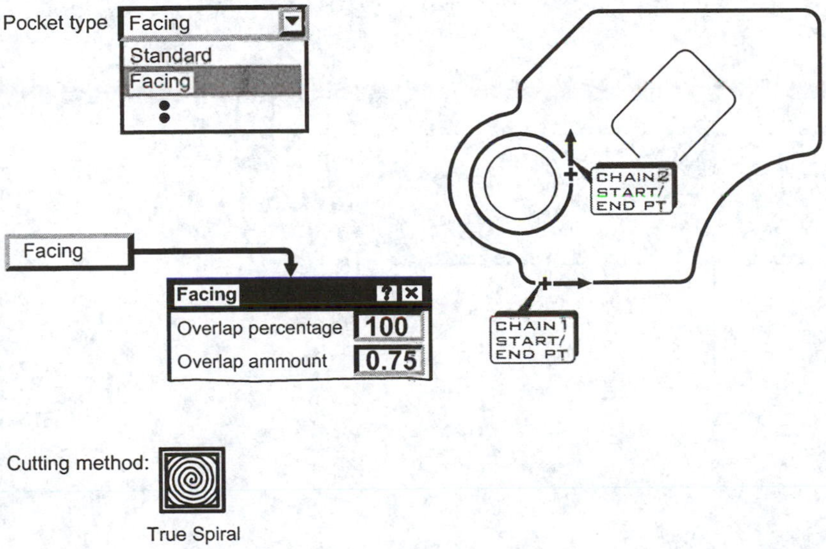

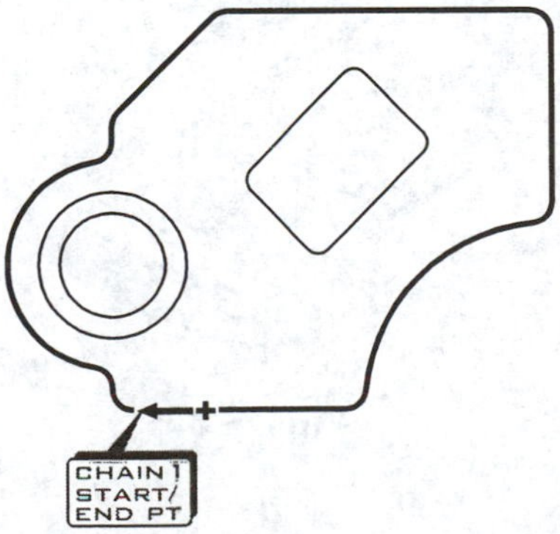

6-3) Get the CAD model file **EX2-7JV** generated in exercise 2-7. Create a part program for executing the machining listed in PROCESS PLAN 6P-3.

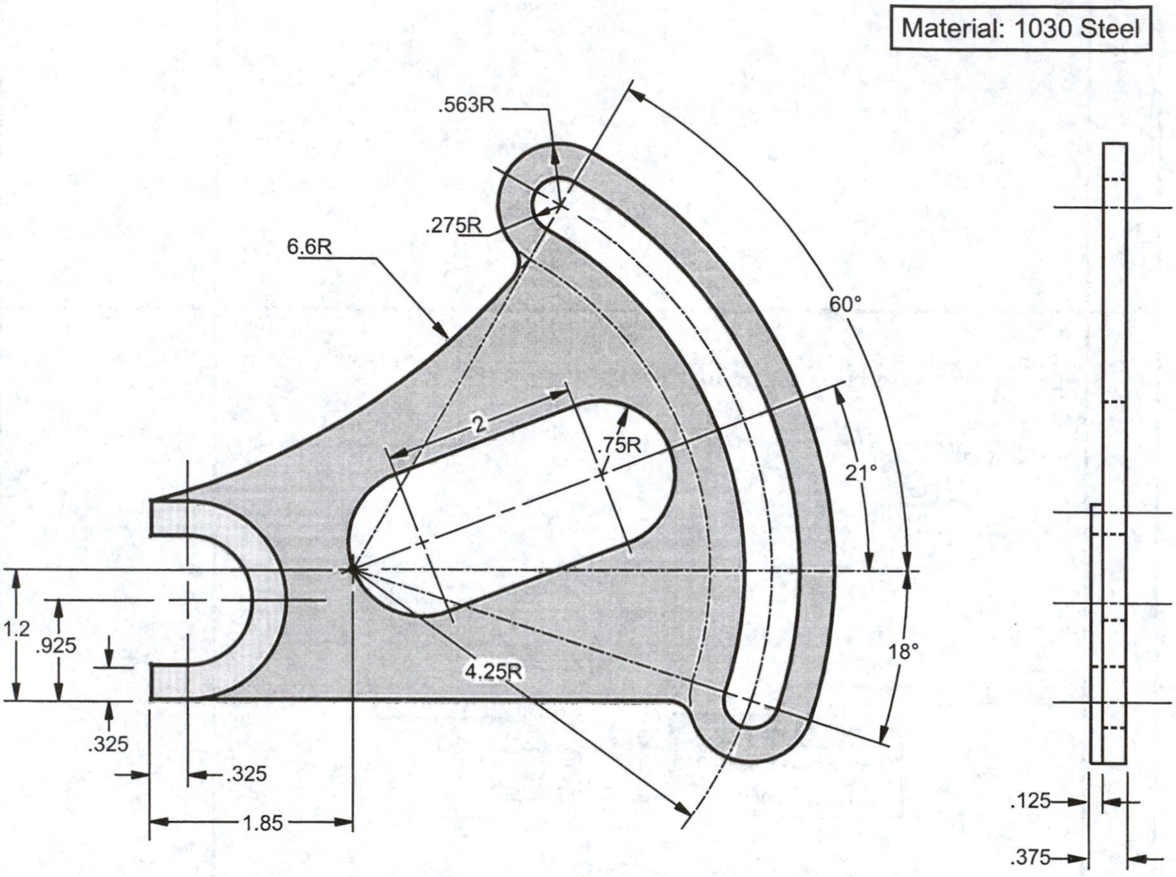

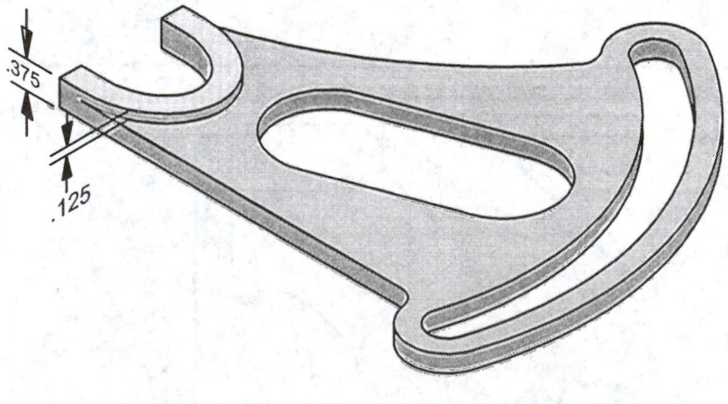

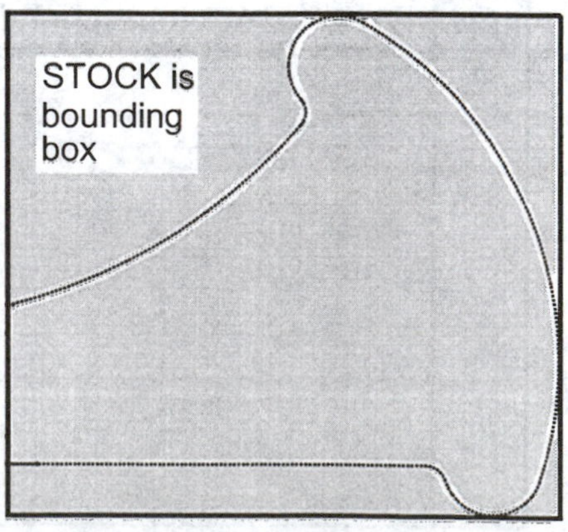

Figure 6-p4

PROCESS PLAN 6P-3

No.	Operation	Tooling
1	CENTER DRILL X .166 DEEP(2 HOLES) 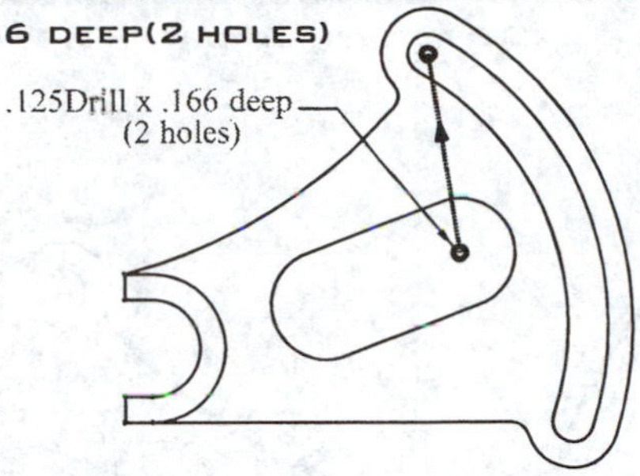	1/8 CENTER DRILL
2	PECK DRILL THRU(2 HOLES) 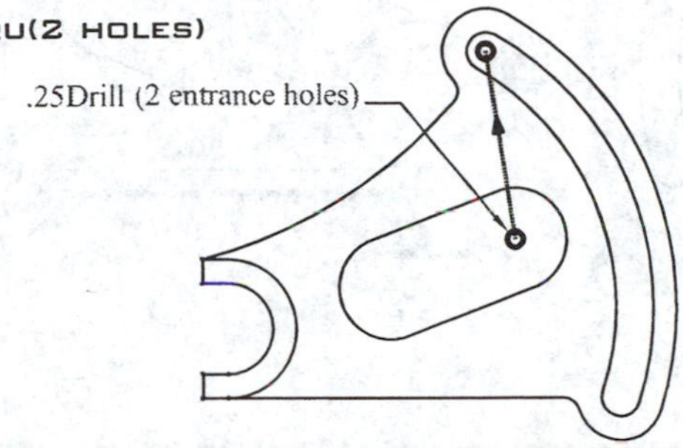	1/4 DRILL
3	ROUGH AND FINISH .75R SLOT X .375 DEEP 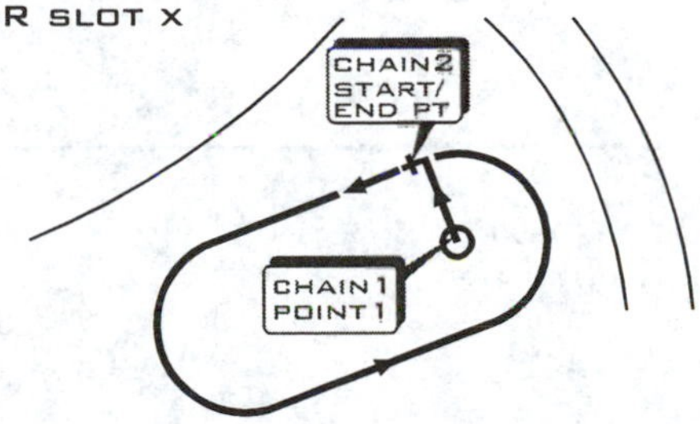	1/4 END MILL
4	ROUGH AND FINISH .275R SLOT X .375 DEEP 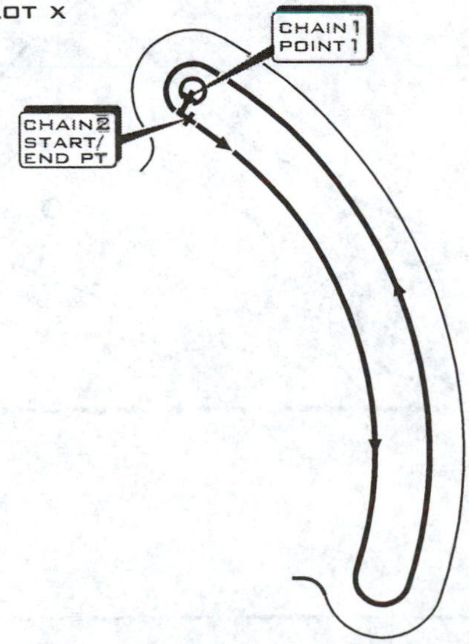	1/4 END MILL

PROCESS PLAN 6P-3(*continued*)

No.	Operation	Tooling
4	POCKET X .125 DEEP. LEAVE .02 FOR FINISH CUT IN XY AND Z 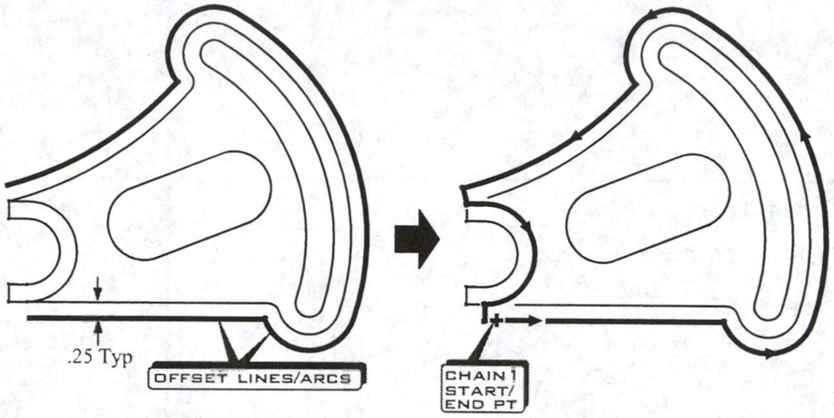 Use **XFORM** ⟶ **OFFSET**	1/4 END MILL
5	ROUGH OUTSIDE X .375 DEEP LEAVE .03 FOR FINISH CUT IN XY 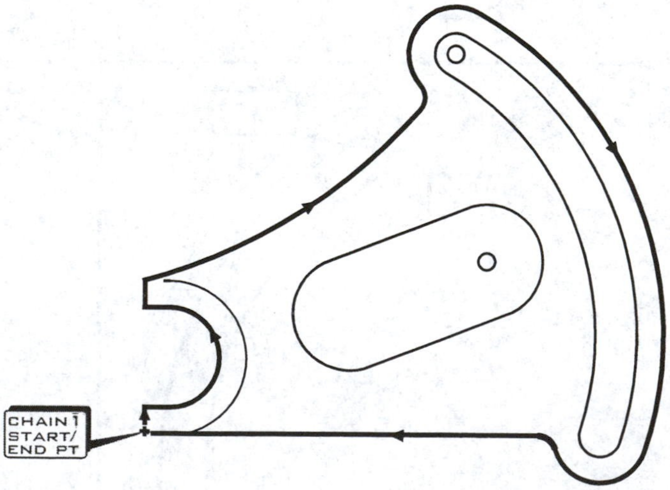	1/4 END MILL

6-4) Get the CAD model file **EX3-2JV** generated in exercise 3-2. Generate a part program for executing the machining listed in PROCESS PLAN 6P-4

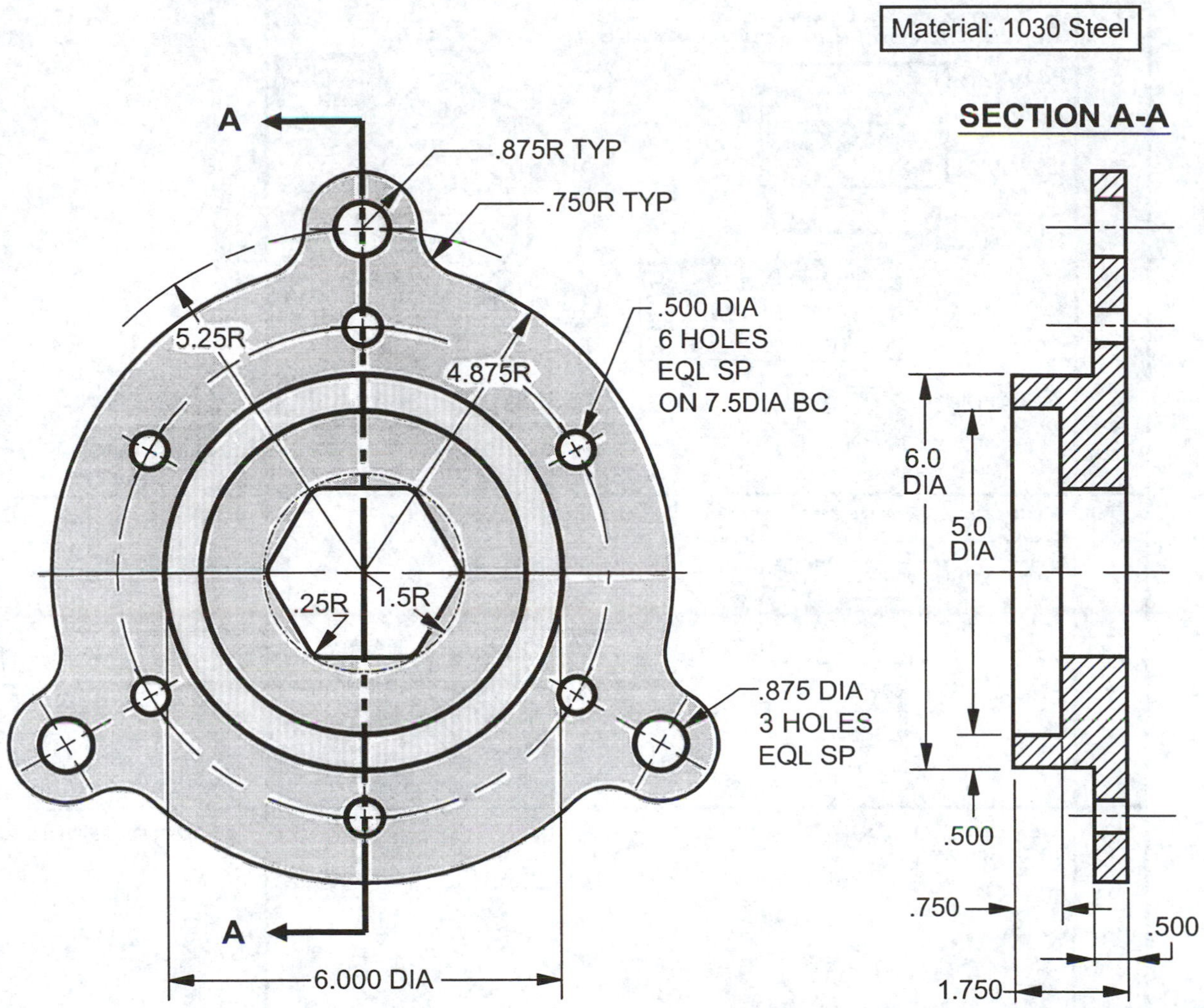

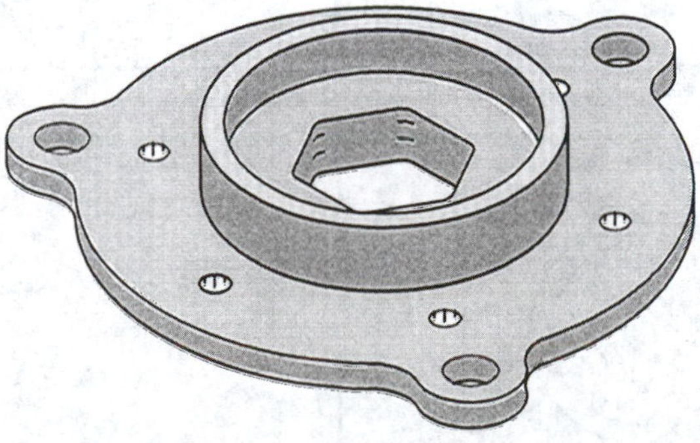

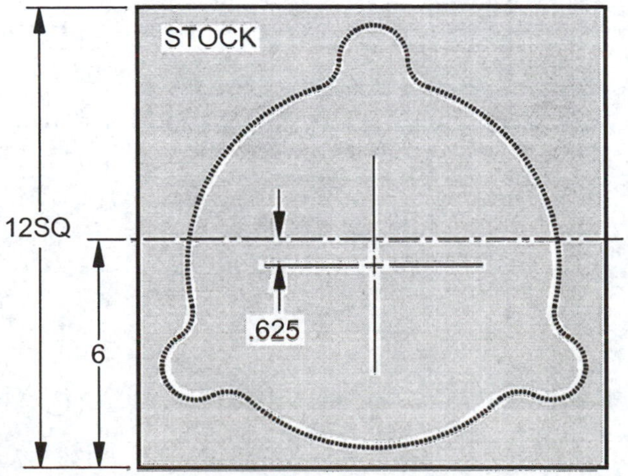

Figure 6-p5

PROCESS PLAN 6P-4

<table>
<tr><th>No.</th><th>Operation</th><th>Tooling</th></tr>
<tr><td>1</td><td>POCKET X 1.25 DEEP
LEAVE .02 FOR FINISH CUT IN XY AND Z
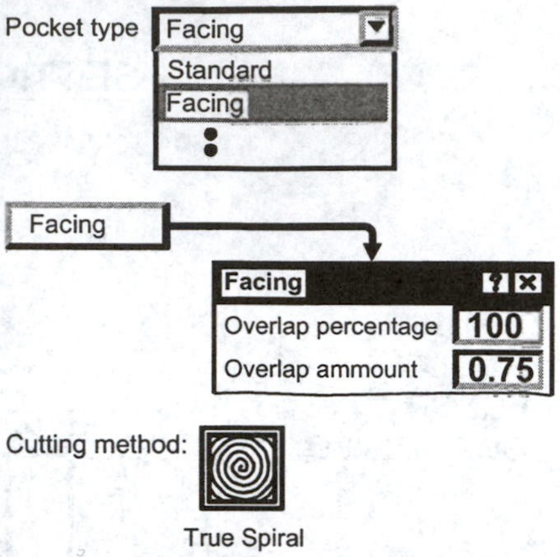
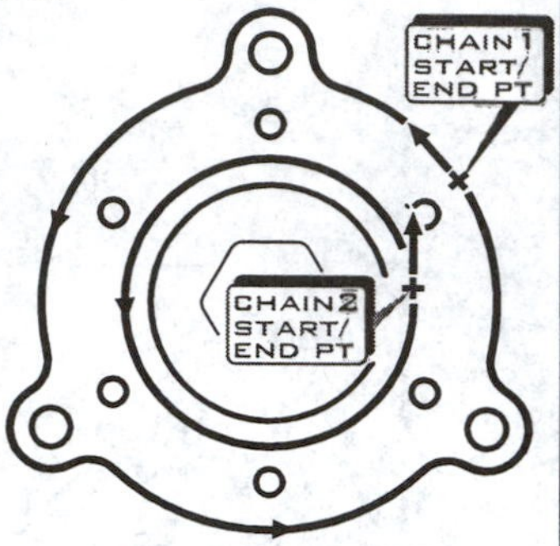</td><td>3/4 END MILL</td></tr>
<tr><td>2</td><td>POCKET X .75 DEEP
LEAVE .02 FOR FINISH CUT IN XY AND Z
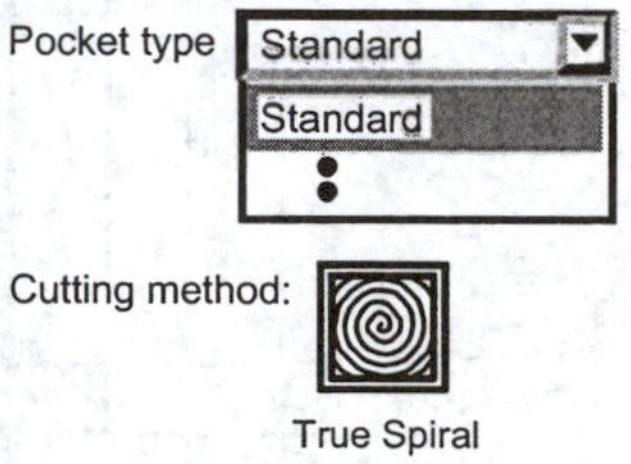
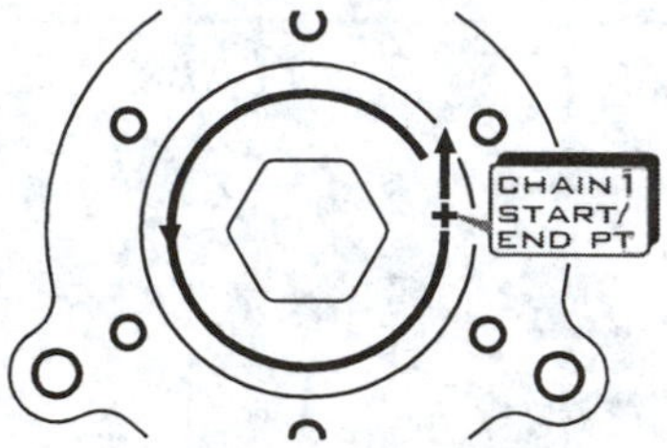</td><td>3/4 END MILL</td></tr>
<tr><td>3</td><td>CENTER DRILL X .25 DEEP
(10 HOLES)
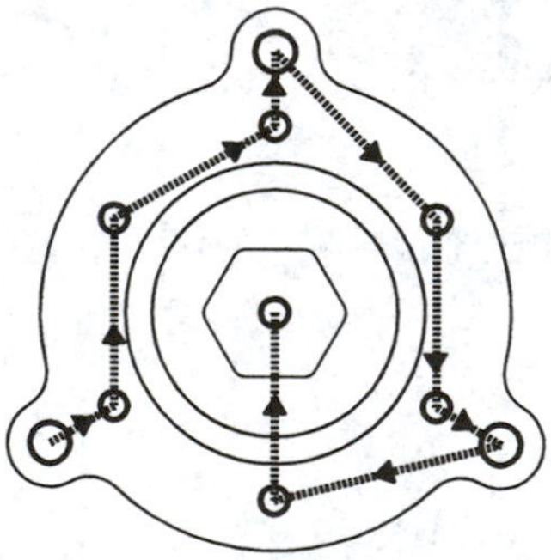</td><td>1/4 CENTER DRILL</td></tr>
<tr><td>4</td><td>PECK DRILL THRU(3 HOLES)
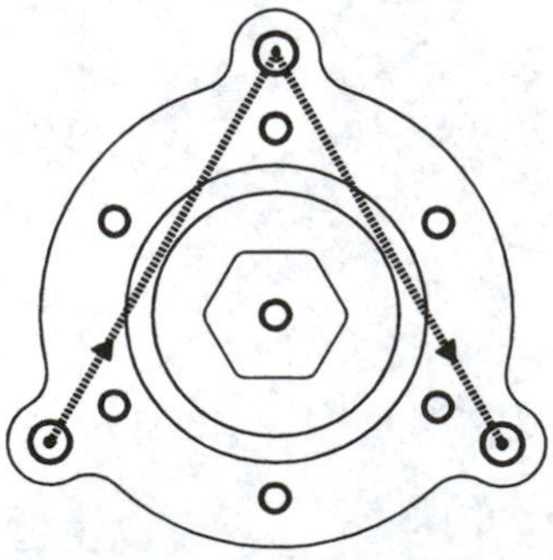</td><td>7/8 DRILL</td></tr>
</table>

PROCESS PLAN 6P-4(*continued*)

No.	Operation	Tooling
5	PECK DRILL THRU (7 HOLES) 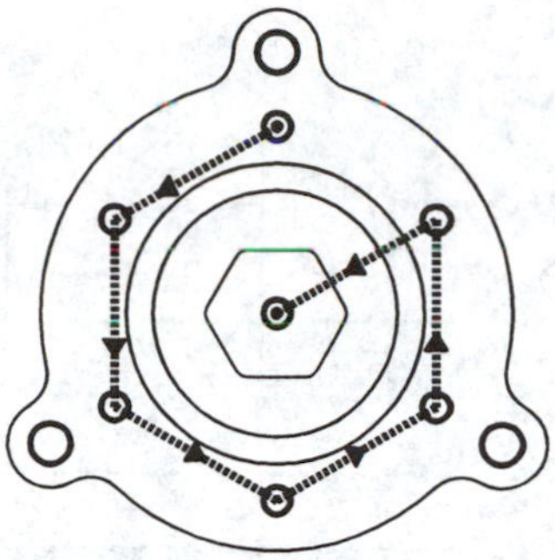	1/2 DRILL
6	ROUGH AND FINISH HEX SLOT X 1DEEP 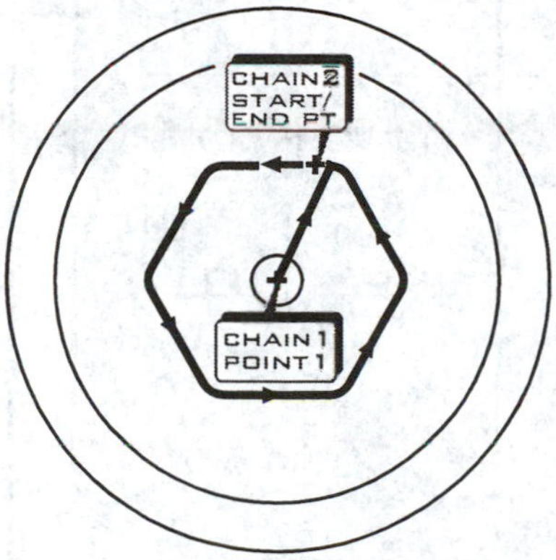	1/4 END MILL
7	ROUGH OUTSIDE X .5 DEEP LEAVE .01 FOR FINISH CUT IN XY 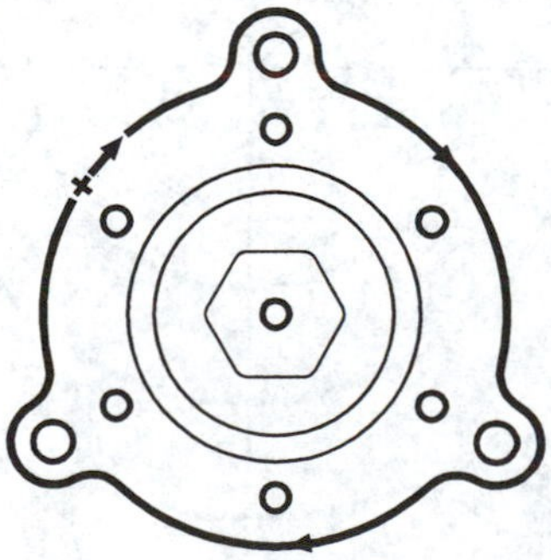	3/4 END MILL

6-5) Get the CAD model file **EX3-3JV** generated in exercise 3-3. Generate a part program for executing the machining listed in PROCESS PLAN 6P-5

Material: 1030 Steel

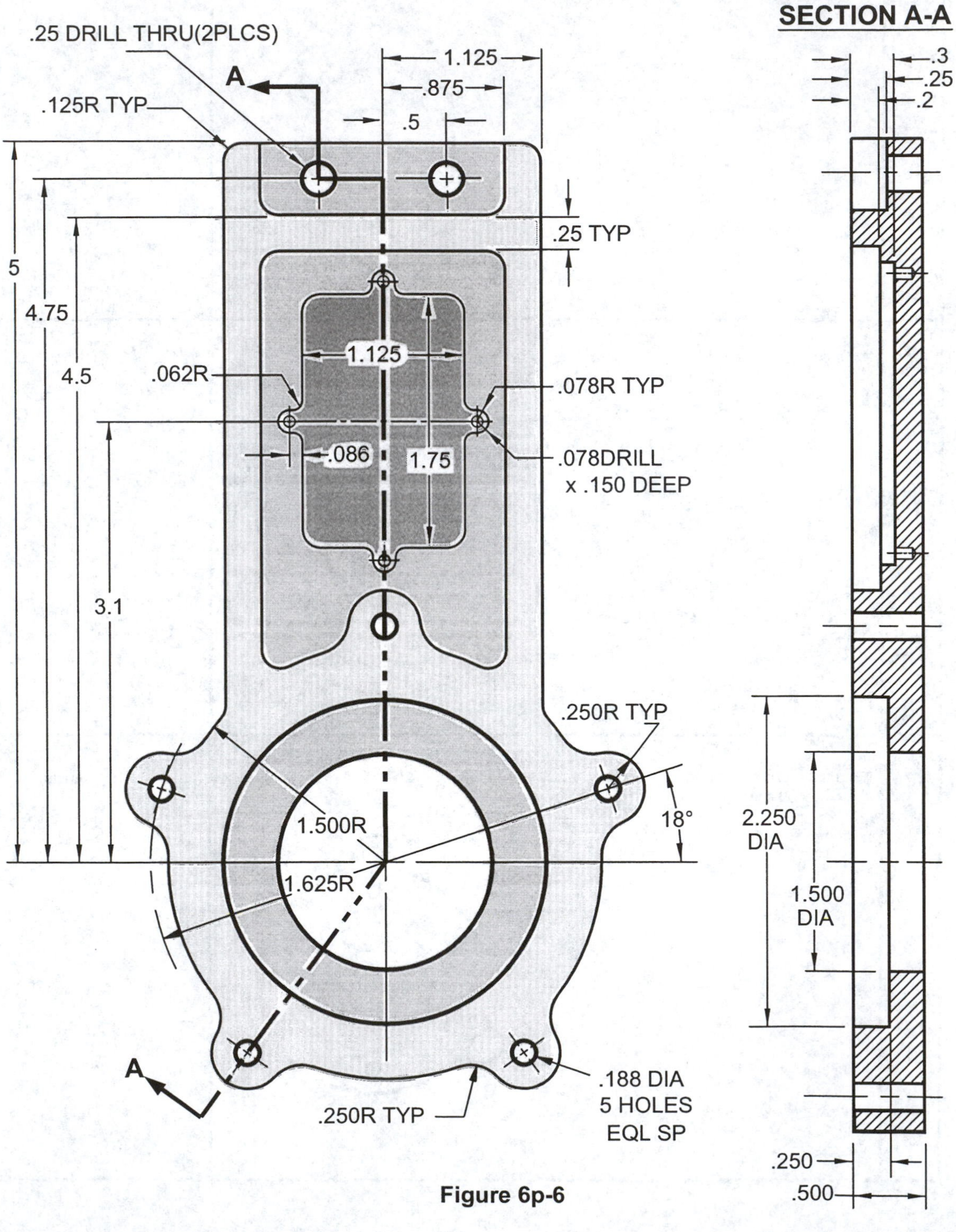

Figure 6p-6

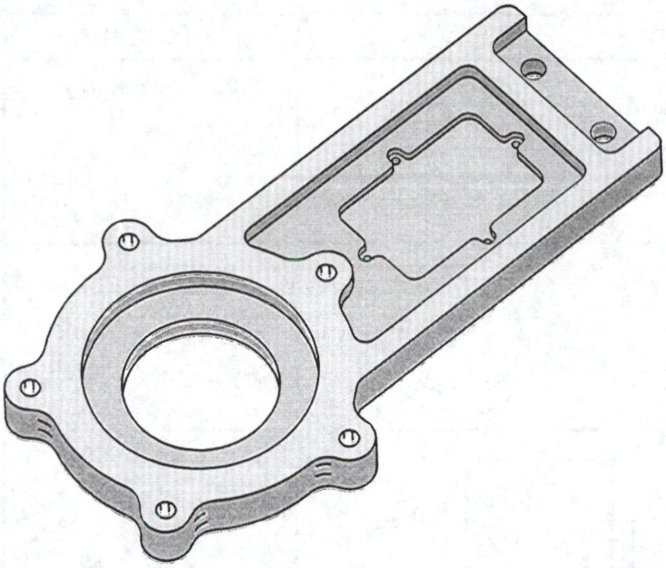

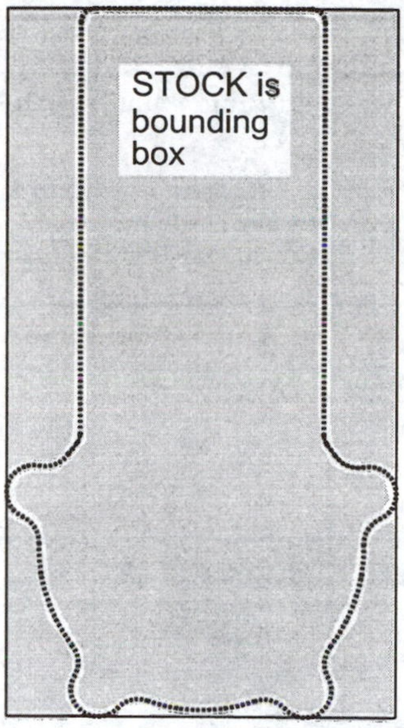

Figure 6p-7

PROCESS PLAN 6P-5

No.	Operation	Tooling
1	CENTER DRILL X .166 DEEP(7 HOLES) 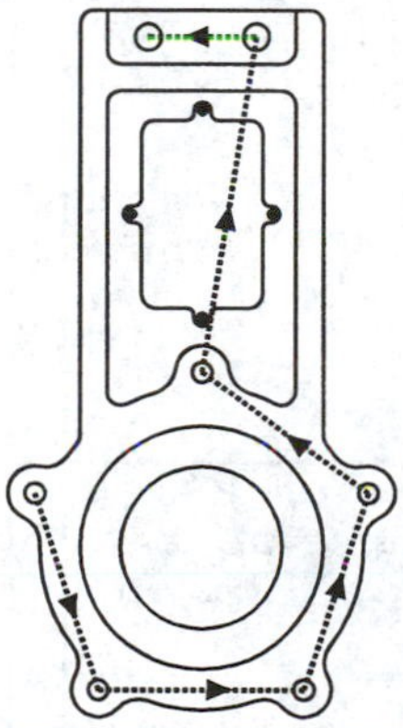	1/8 CENTER DRILL
2	PECK DRILL THRU(5 HOLES) 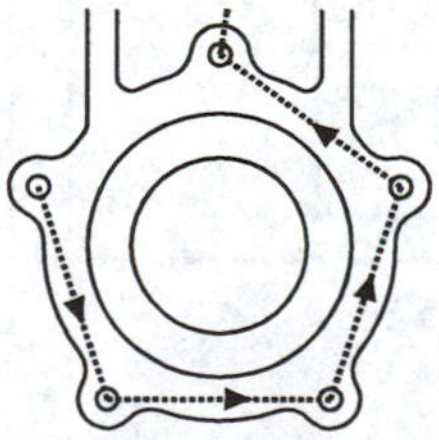	3/16 DRILL
3	PECK DRILL THRU(2 HOLES) 	1/4 DRILL

PROCESS PLAN 6P-5(*continued*)

No.	Operation	Tooling
4	CIRCLE MILL 1.5DIA THRU 	1/2 END MILL
5	CIRCLE MILL 2.25DIA X .25 DEEP 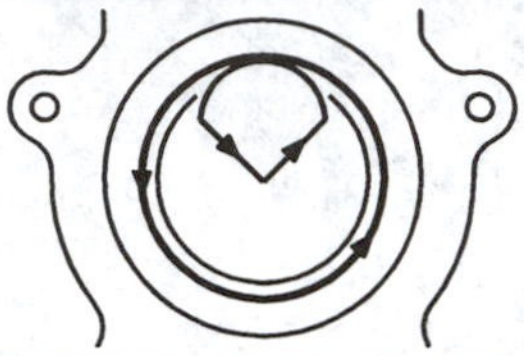	1/2 END MILL
6	POCKET X .2DEEP LEAVE .01 FOR FINISH CUT IN XY AND Z 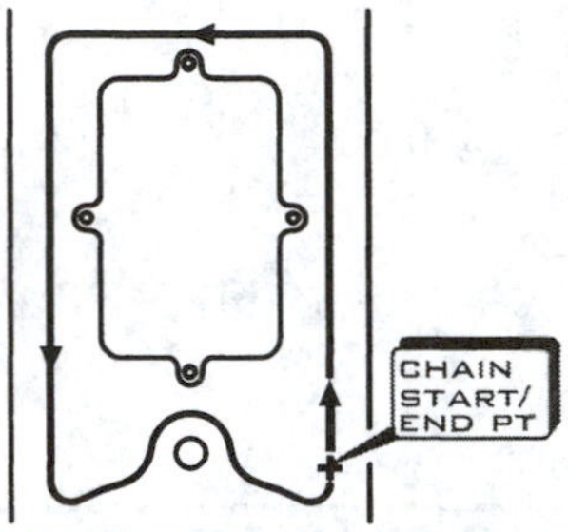	1/4 END MILL
7	POCKET X .3DEEP LEAVE .01 FOR FINISH CUT IN XY AND Z 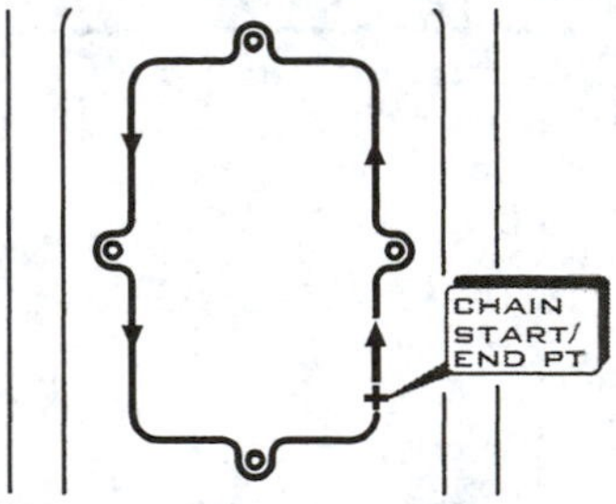	1/4 END MILL
8	POCKET[REMACHINING] X .3DEEP LEAVE .01 FOR FINISH CUT IN XY AND Z 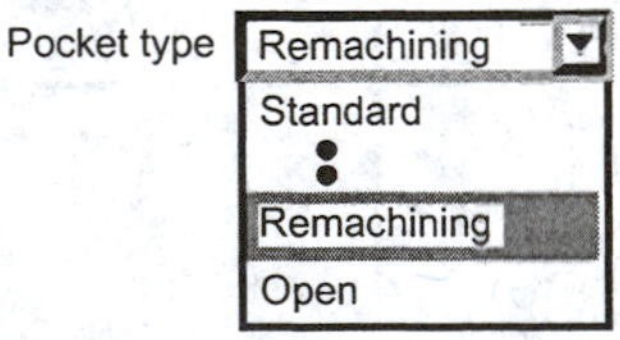 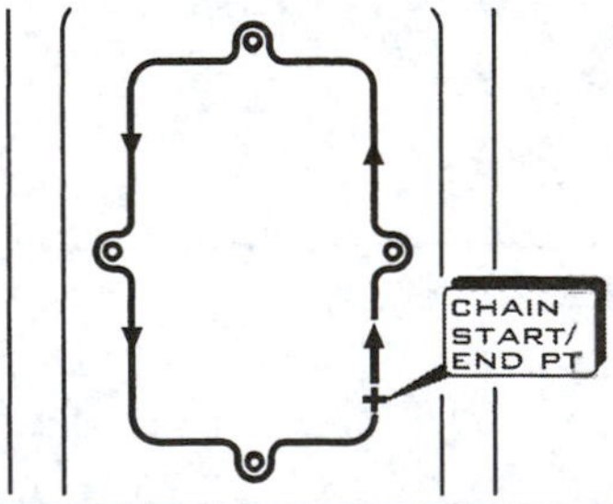	5/32 END MILL
9	PECK DRILL X .450DEEP 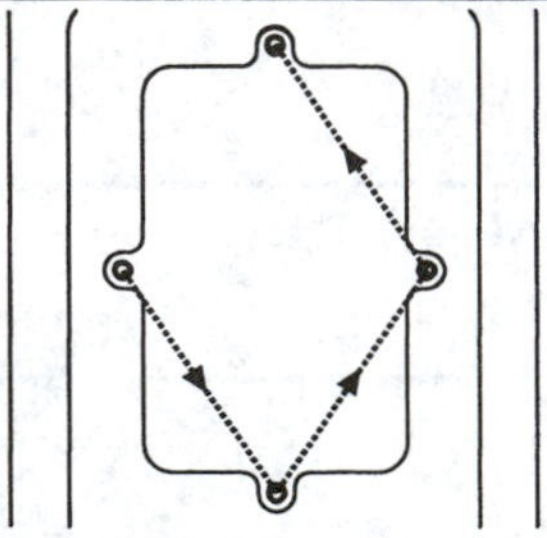	5/64 DRILL

PROCESS PLAN 6P-5(*continued*)

No.	Operation	Tooling
10	POCKET X .25DEEP LEAVE .01 FOR FINISH CUT IN XY AND Z 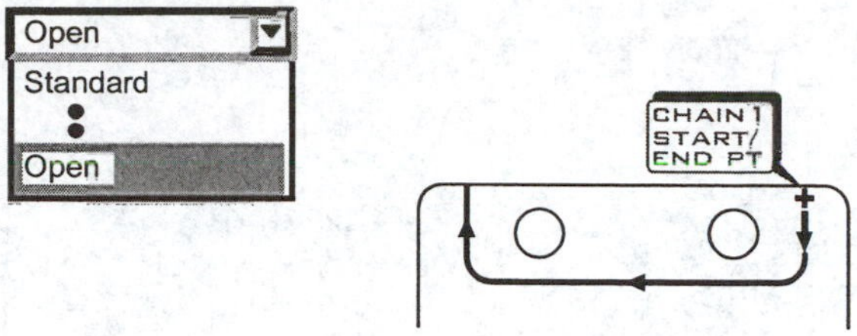	1/4 END MILL
11	ROUGH OUTSIDE X .5 DEEP LEAVE .01 FOR FINISH CUT IN XY 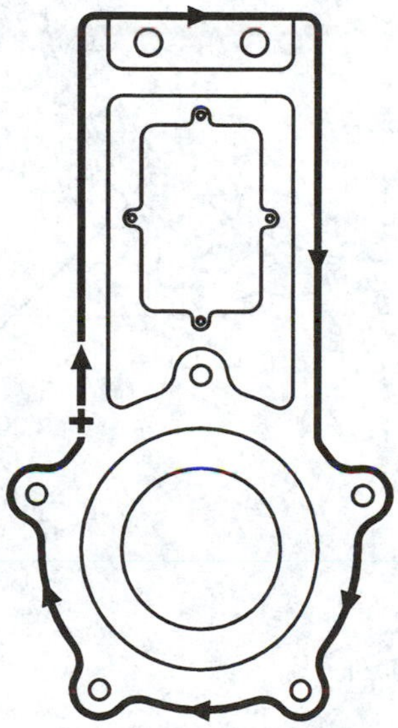	1/2 END MILL

6-6) Get the CAD model file **EX3-4JV** generated in exercise 3-4. Generate a part program for executing the machining listed in PROCESS PLAN 6P-6

Material: 1030 Steel

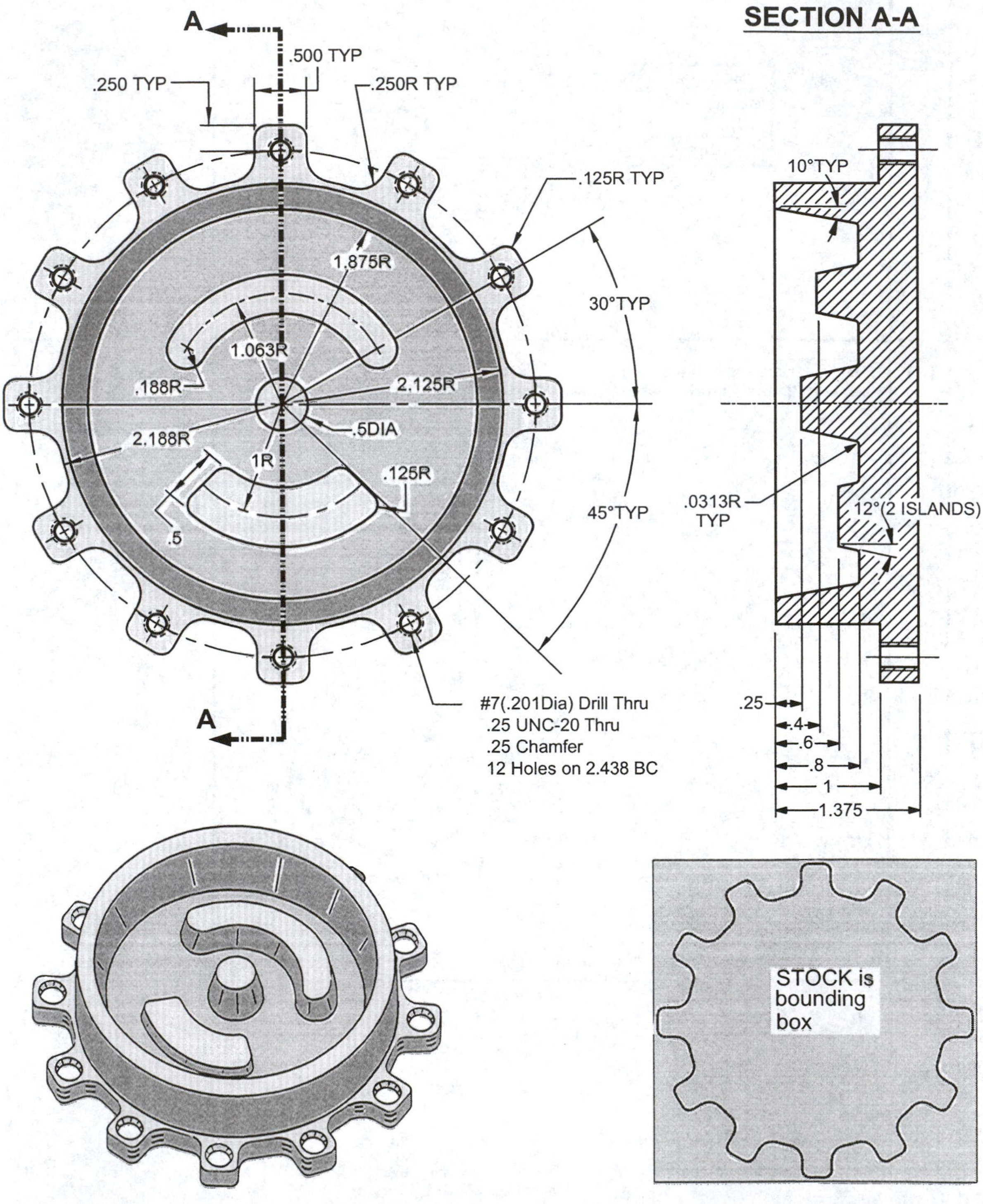

Figure 6-p8

PROCESS PLAN 6P-6(*continued*)

No.	Operation	Tooling
1	POCKET X 1 DEEP LEAVE .02 FOR FINISH CUT IN XY AND Z Pocket type: Facing (Standard, Facing, ⋮) Facing → Facing: Overlap percentage 100; Overlap ammount 0.5 Cutting method:  Constant Overlap Spiral 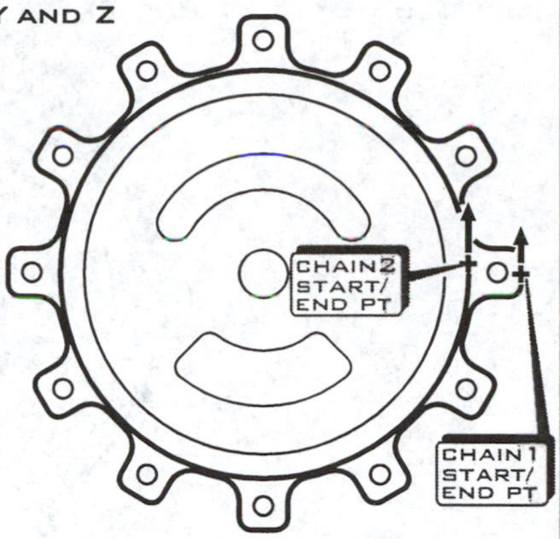	1/2 END MILL
2	POCKET[ISLAND FACING] LEAVE .01 FOR FINISH CUT IN XY AND Z Click ① Area Click ② the location of the point *Mastercam* will ***automatically chain*** all the closed boundaries in the ***area surrounding the point*** 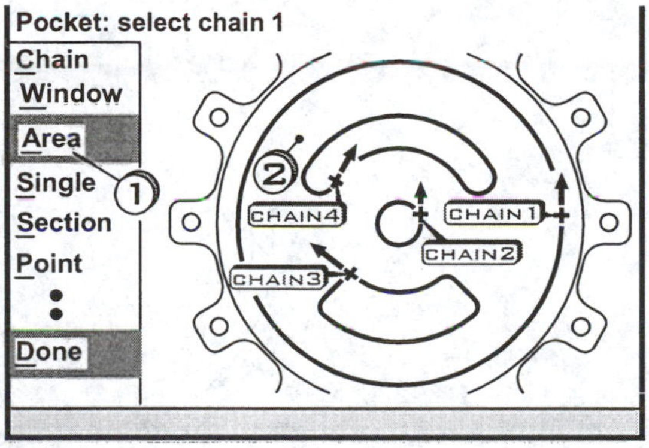 Pocket type: Island facing (Island facing, ⋮) ☑ Depth cuts → Depth cuts: Max rough step: 0.03; # Finish cuts: 1; Finish step: 0.01; Depth cut order: ◉ By pocket ○ By depth; ☑ Keep tool down; ☑ Use island depths (MACHINE TO TOP OF ISLANDS); ☑ Tapered walls (TAPER WALLS); Outer wall taper angle 10; Island taper angle 12; OK / Cancel / Help Cutting method:  Parallel Spiral Clean Corners	1/4 BULL END MILL .0313R

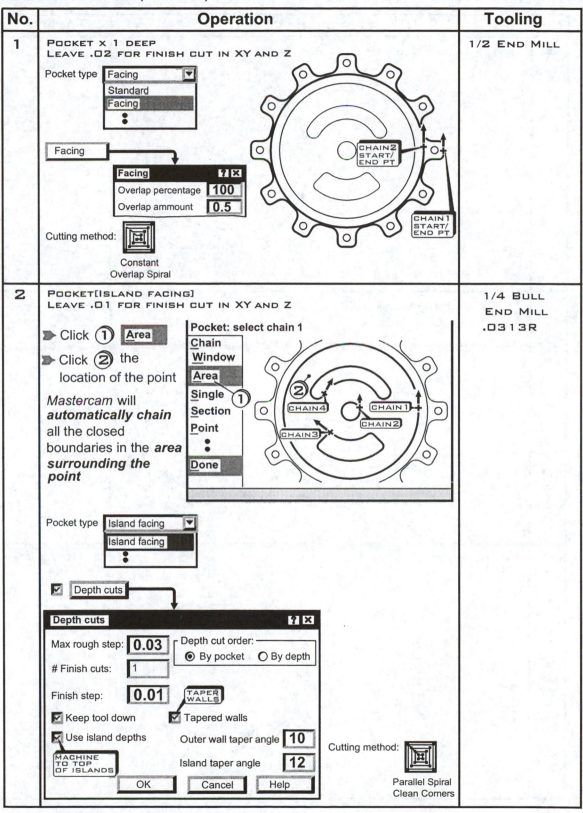

PROCESS PLAN 6P-5(*continued*)

No.	Operation	Tooling
3	CENTER DRILL X .166 DEEP(12 HOLES) 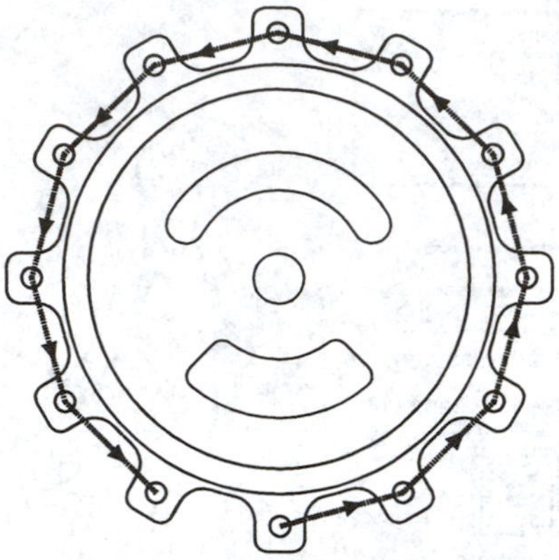	1/8 CTR DRILL
4	PECK DRILL THRU(12 HOLES) 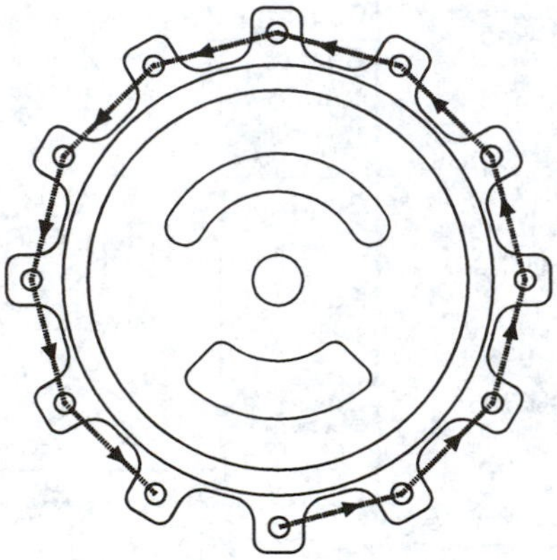	#7(.201) DRILL
5	TAP THRU(12 HOLES) 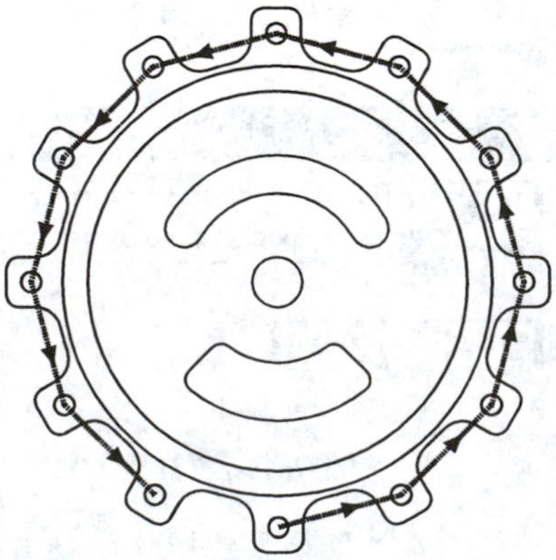	1/4-20 TAPRH

PROCESS PLAN 6P-5(*continued*)

No.	Operation	Tooling
6	CHAMFER X 1.15DEEP(12 HOLES) 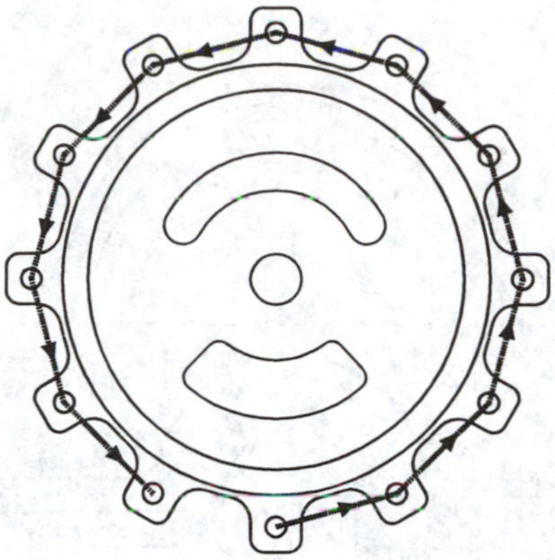	1/2 CHAMFER MILL
7	ROUGH OUTSIDE X 1.375 DEEP LEAVE .01 FOR FINISH CUT IN XY 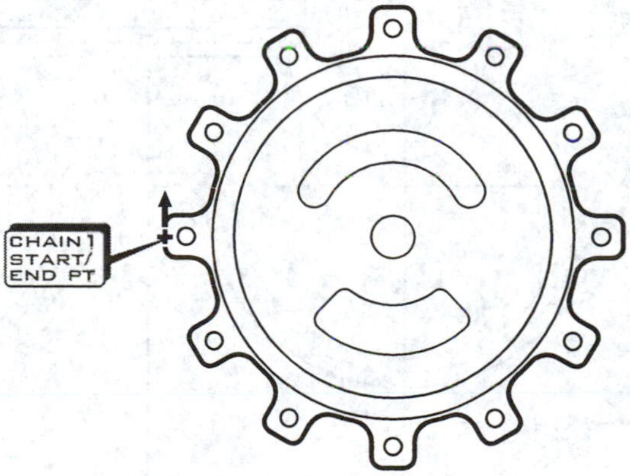	1/2 END MILL

6-7) Get the CAD model file **EX3-5JV** generated in exercise 3-5. Generate a part program for executing the machining listed in PROCESS PLAN 6P-7

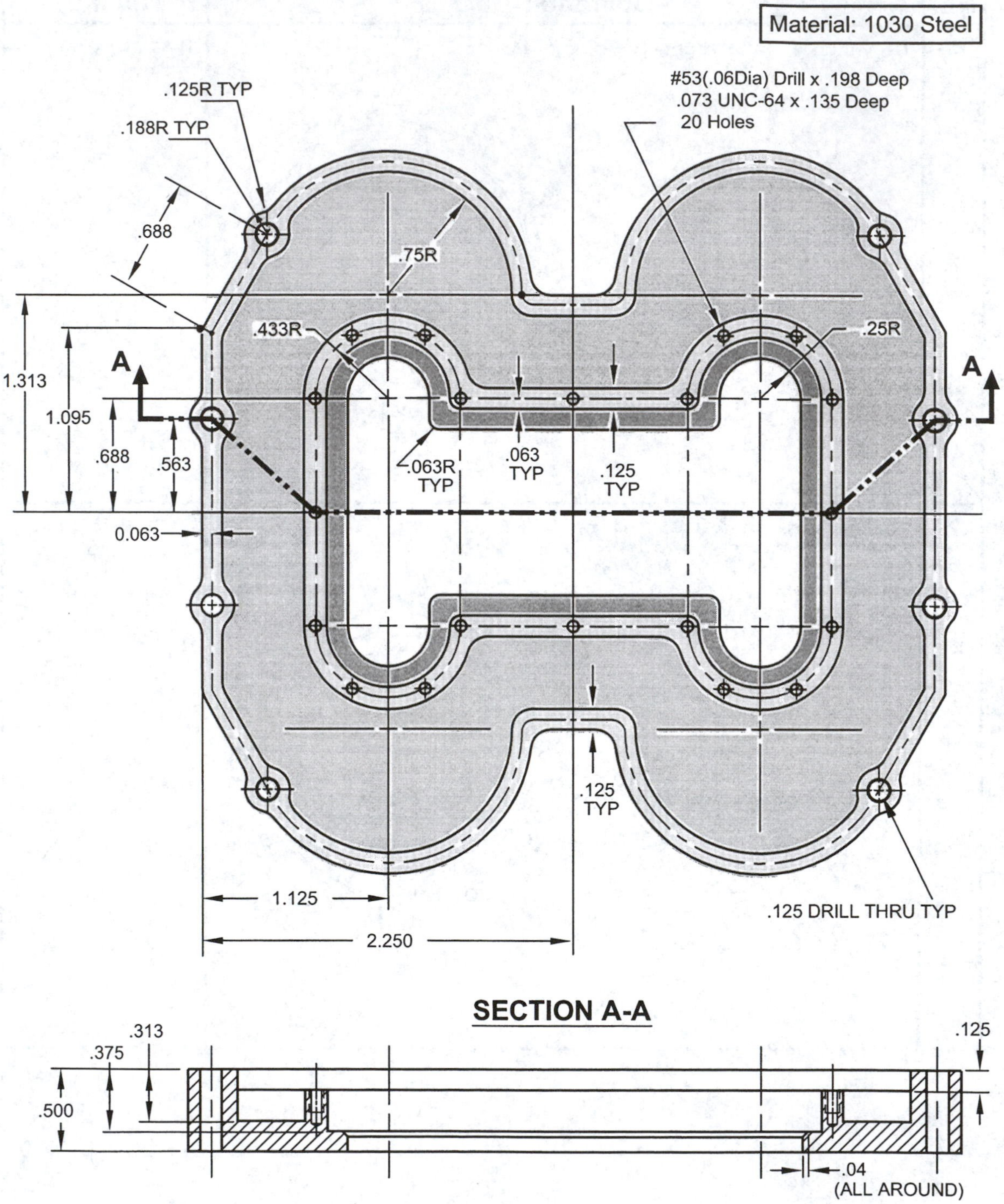

Figure 6-p9

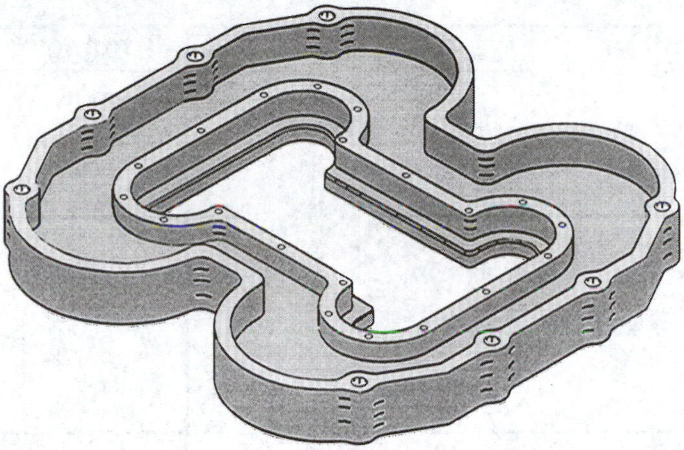

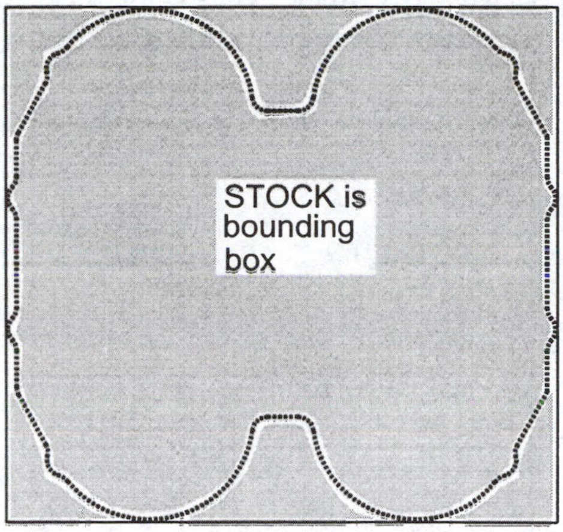

Figure 6-p10

PROCESS PLAN 6P-7

No.	Operation	Tooling
1	PECK DRILL X .198 DEEP(20 HOLES) 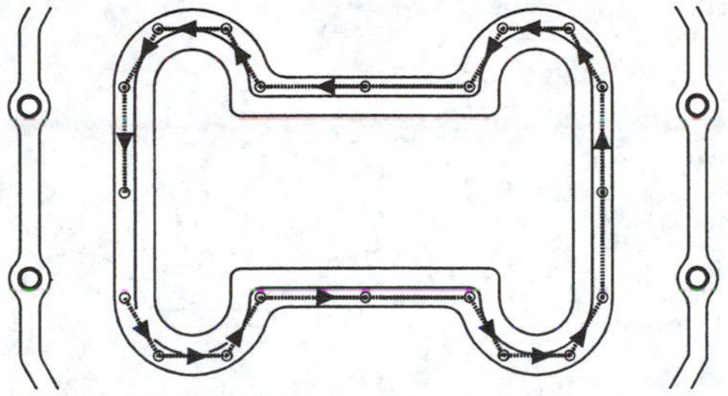	#53(.06 DIA) DRILL
2	TAP X .135 DEEP(20 HOLES) 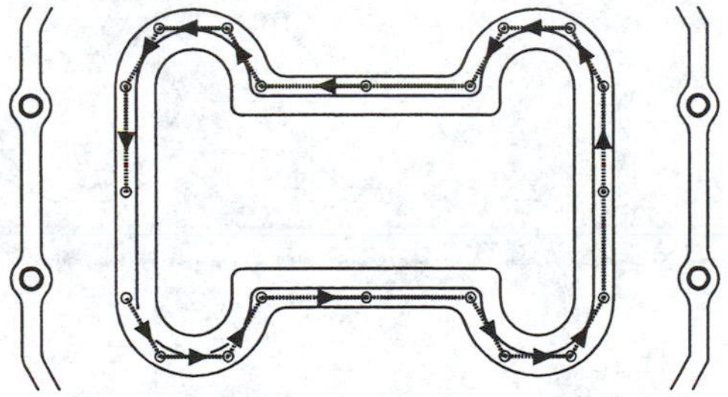	.073-64-TAP
3	CENTER DRILL X .166 DEEP(9 HOLES) 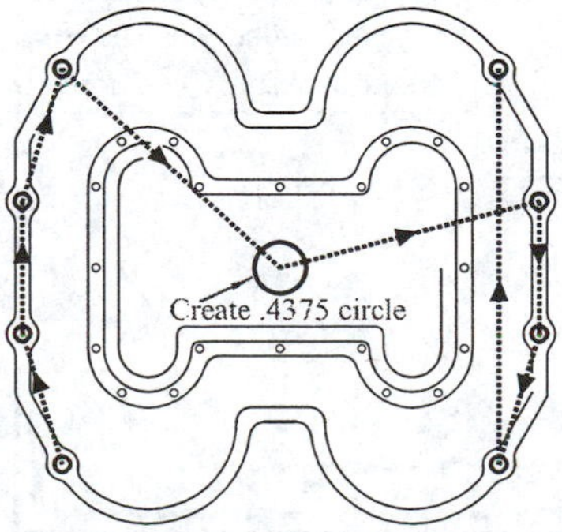	1/8 CTR DRILL

PROCESS PLAN 6P-7(*continued*)

<table>
<tr><th>No.</th><th>Operation</th><th>Tooling</th></tr>
<tr><td>4</td><td>PECK DRILL THRU (8 HOLES)
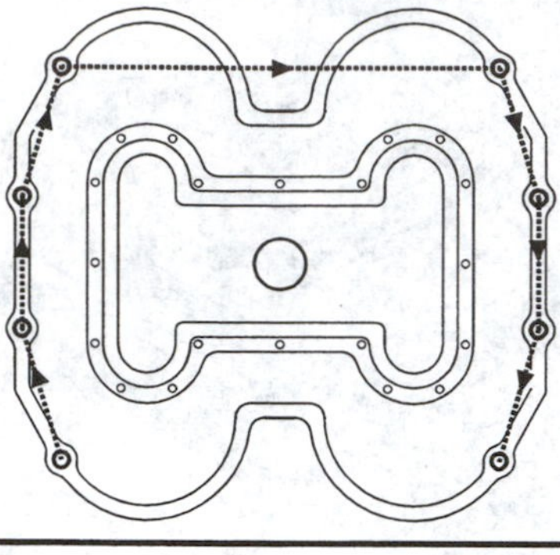</td><td>1/8 DRILL</td></tr>
<tr><td>5</td><td>PECK DRILL THRU
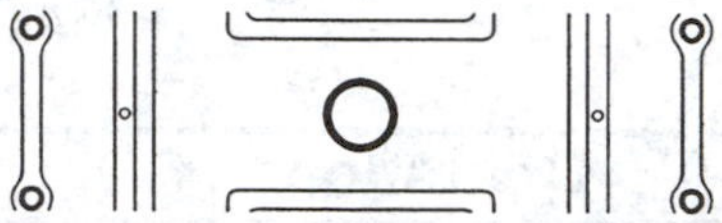</td><td>7/16 DRILL</td></tr>
<tr><td>6</td><td>POCKET[ISLAND FACING]
LEAVE .01 FOR FINISH CUT IN XY AND Z

Click (1) Area
Click (2) the location of the point

Mastercam will automatically chain all the closed boundaries in the area surrounding the point
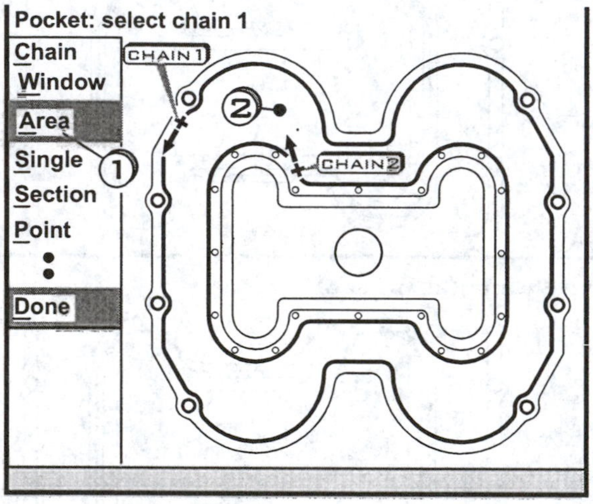
Pocket type: Island facing
Island facing

☑ Depth cuts
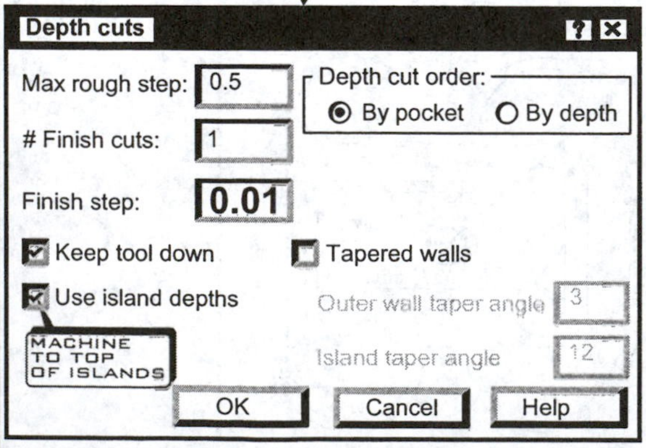
Cutting method:

Parallel Spiral Clean Corners</td><td>1/4 END MILL</td></tr>
</table>

PROCESS PLAN 6P-7(*continued*)

No.	Operation	Tooling
7	ROUGH AND FINISH SLOT X .5 DEEP LEAVE .01 FOR FINISH CUT IN XY 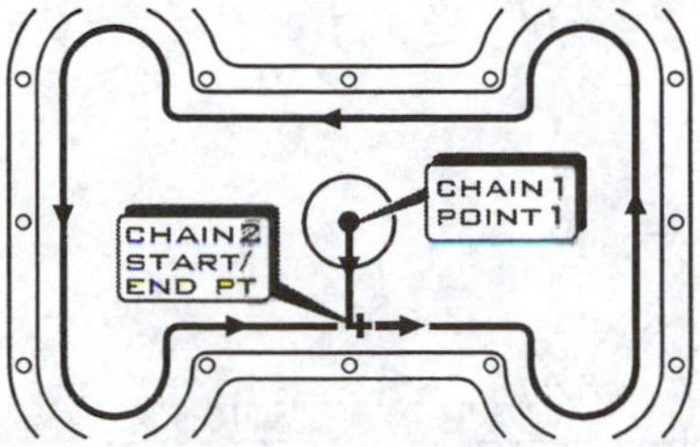	7/16 END MILL
8	ROUGH AND FINISH INSIDE CONTOUR X .375 DEEP LEAVE .01 FOR FINISH CUT IN XY 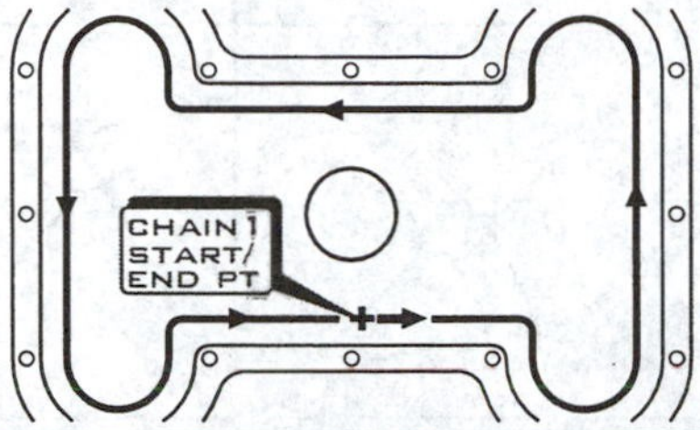	7/16 END MILL
9	CHAMFER INSIDE CONTOUR 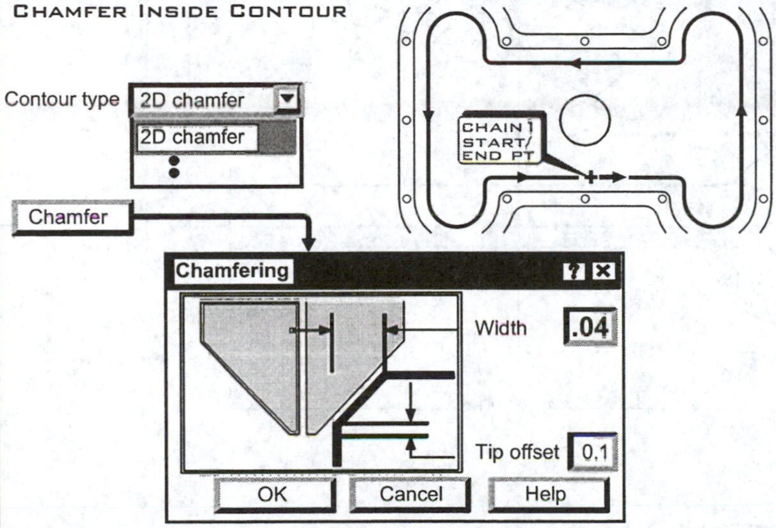	1/2 CHAMFER MILL
10	ROUGH OUTSIDE X .5 DEEP LEAVE .01 FOR FINISH CUT IN XY 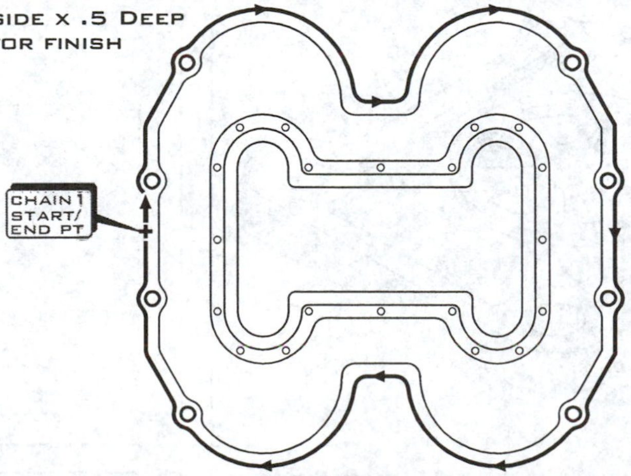	1/4 END MILL

6-8) Get the CAD model file **EX3-6JV** generated in exercise 3-6. Generate a part program for executing the machining listed in PROCESS PLAN 6P-8

Material: 1030 Steel

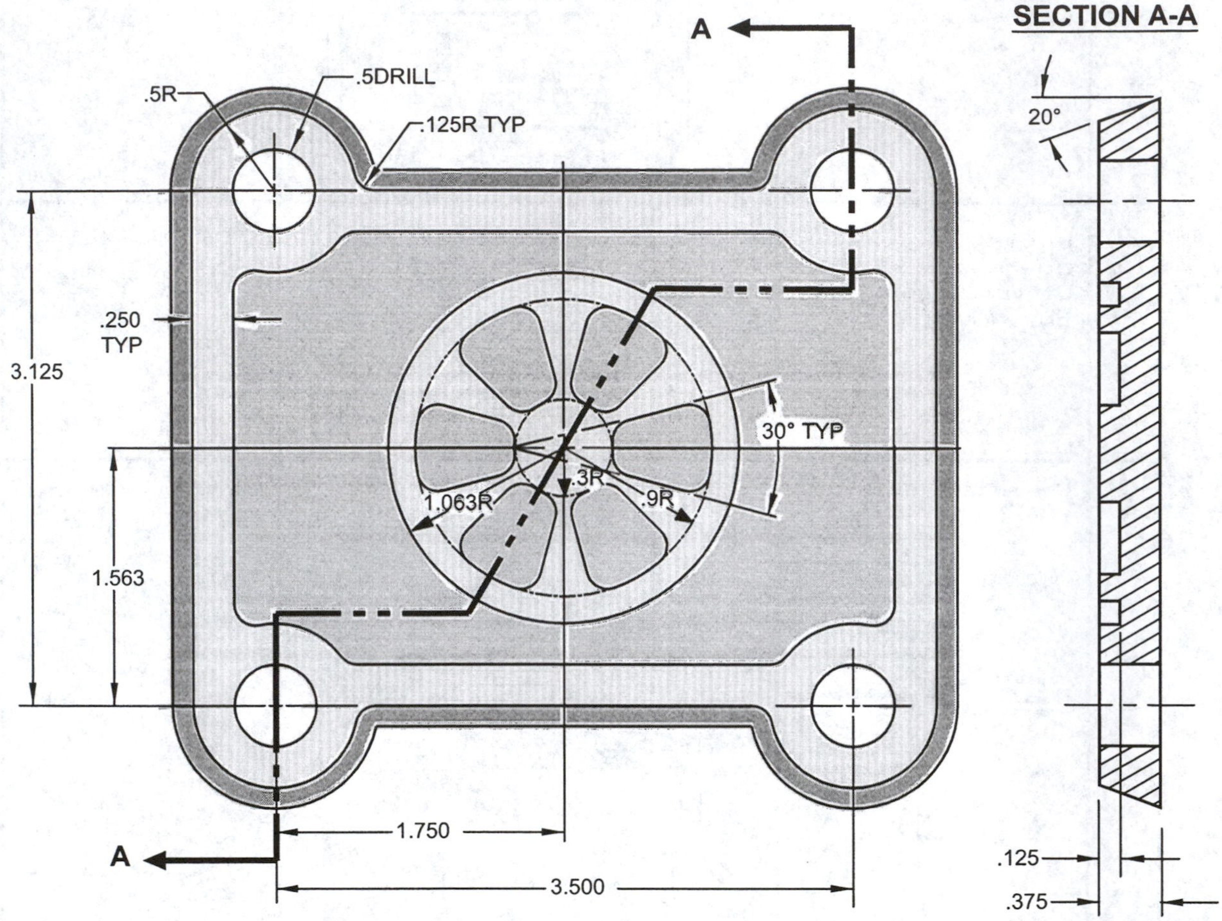

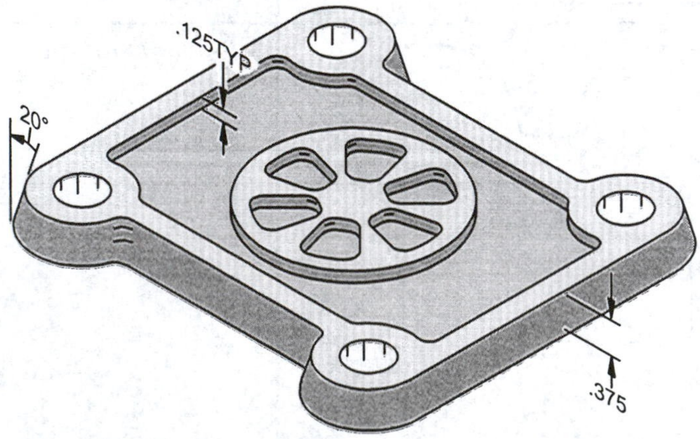

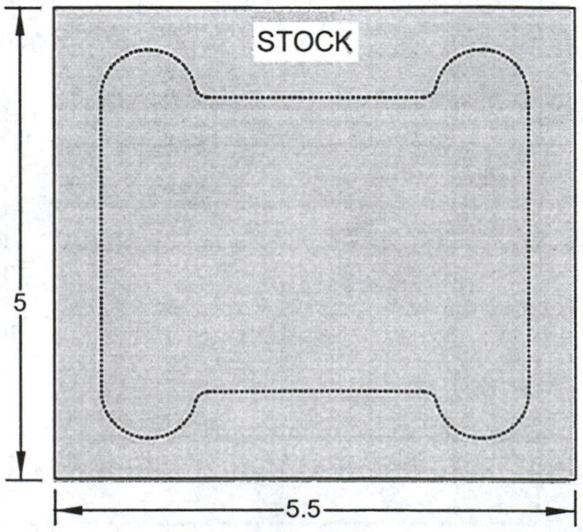

Figure 6-p11

PROCESS PLAN 6P-8

No.	Operation	Tooling
1	CENTER DRILL X .166 DEEP(4 HOLES) 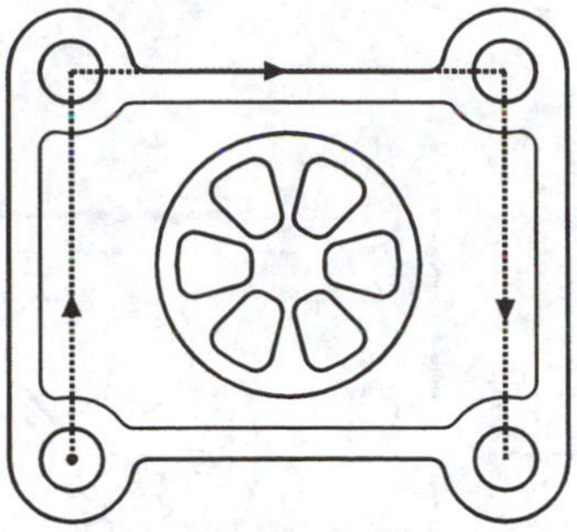	1/4 CTR DRILL
2	PECK DRILL X .375 DEEP(4 HOLES) 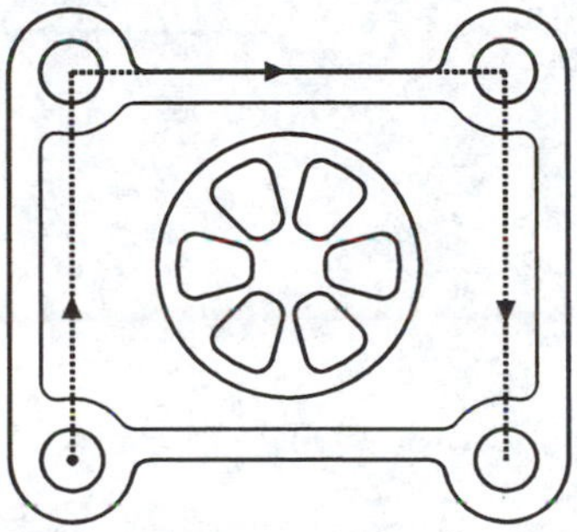	1/2 DRILL
3	POCKET X .125 DEEP LEAVE .01 FOR FINISH CUT IN XY AND Z Click ① Window Click ② ③ the window corners Click ④ the location of the point *Mastercam* will ***automatically chain*** all the closed boundaries ***contained in the window*** 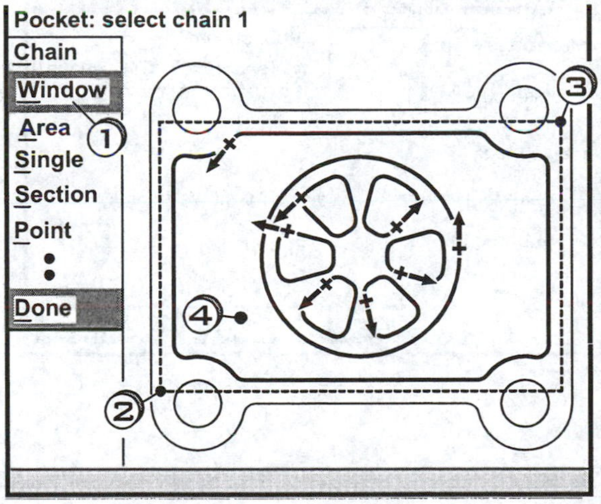 Pocket type 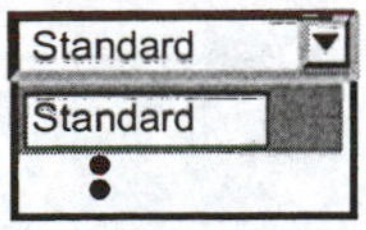 Cutting method:  Constant Overlap Spiral	1/4 END MILL

PROCESS PLAN 6P-8(*continued*)

No.	Operation	Tooling
4	TAPER CUT OUTSIDE 20° X .375 DEEP LEAVE .01 FOR FINISH CUT IN XY 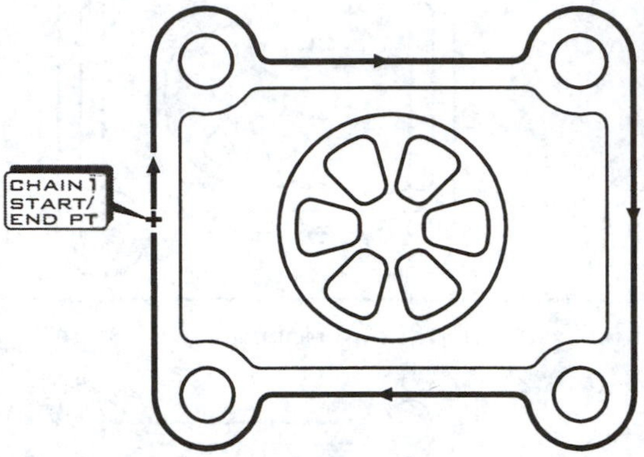 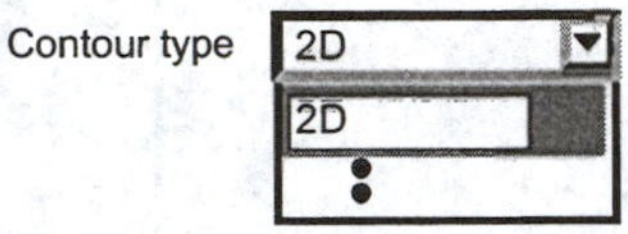 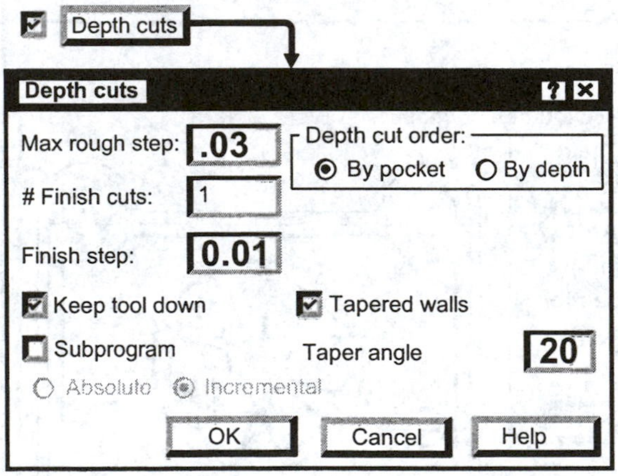	1/4 BALL END MILL

6-9) Get the CAD model file **EX3-8JV** completed in exercise 3-8. Generate a part program for executing the machining listed in PROCESS PLAN 6P-9.

Material: 2024 Aluminum

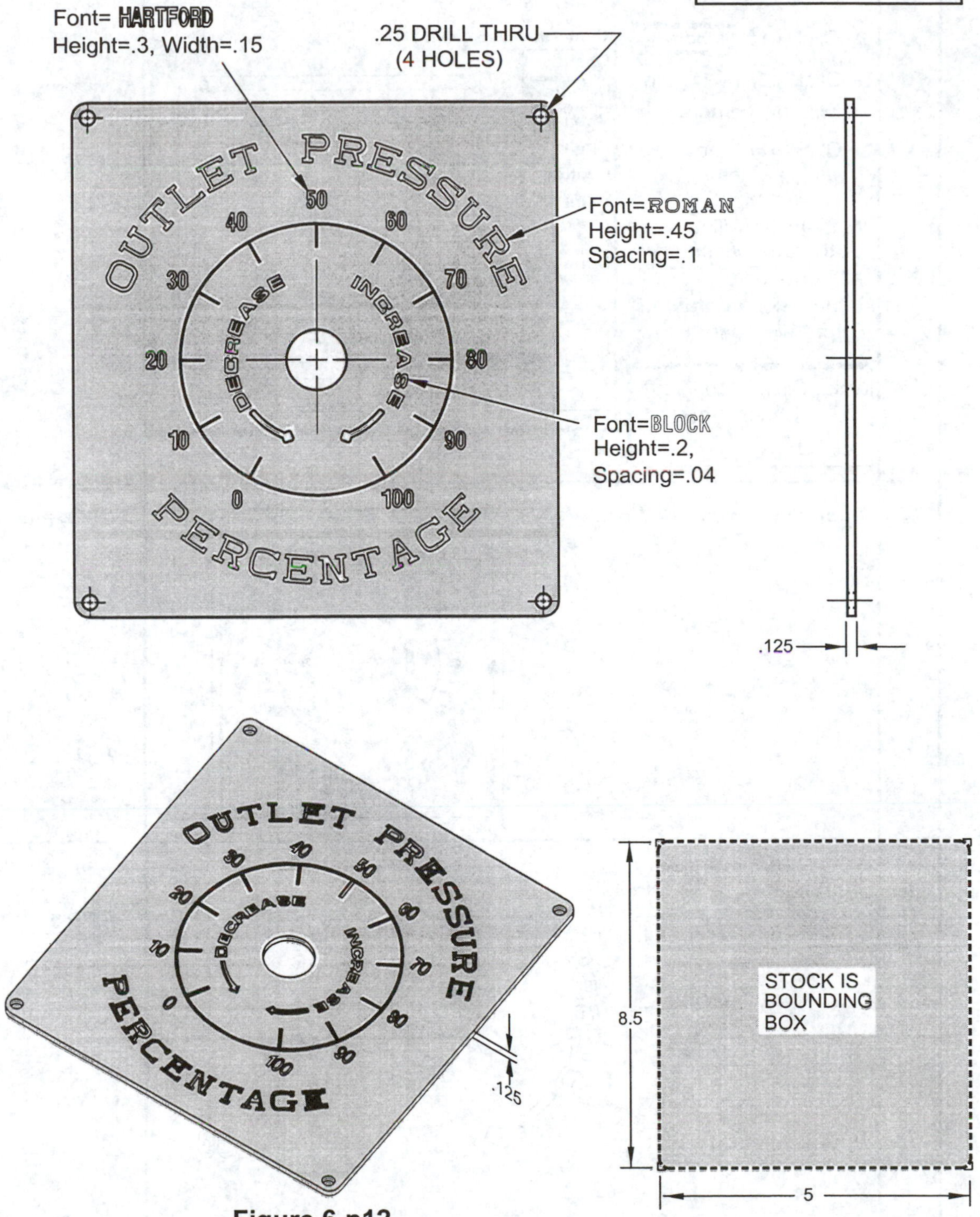

Figure 6-p12

PROCESS PLAN 6P-9

<table>
<tr><th>No.</th><th>Operation</th><th>Tooling</th></tr>
<tr><td>1</td><td>

CONTOUR X .05 DEEP

- Click ① Window
- Click ② ③ the window corners
- Click ④ the location of the search point

Mastercam will ***automatically chain*** all the closed boundaries ***contained in the window***

Note: Click Lead in/out
Check ***off***
☐ Lead in/out

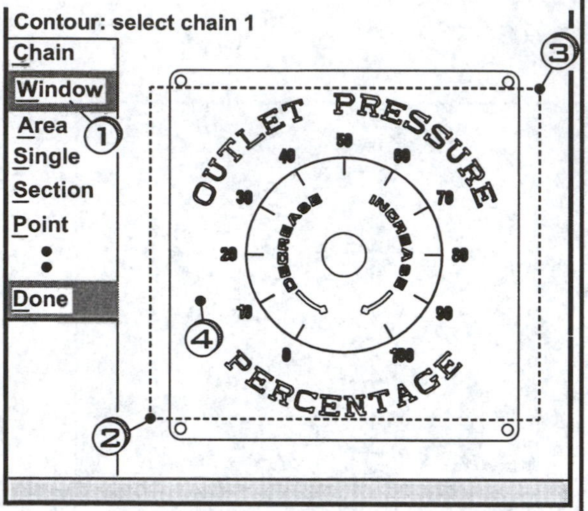

</td><td>1/32 BALL ENDMILL</td></tr>
<tr><td>2</td><td>

PECK DRILL THRU (4 HOLES)

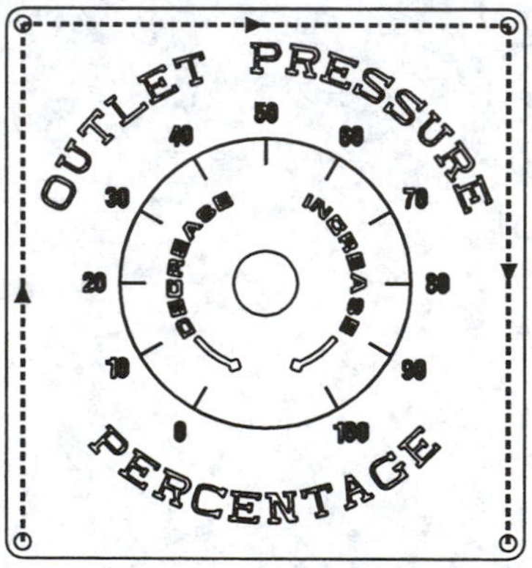

</td><td>1/4 DRILL</td></tr>
<tr><td>3</td><td>

CIRCLE MILL 1 DIA THRU

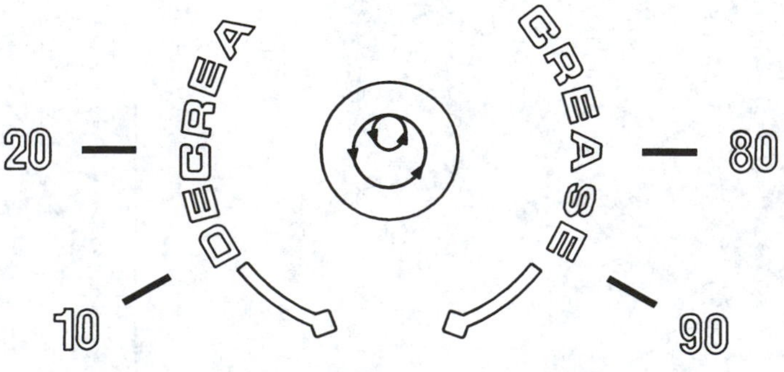

</td><td>1/2 ENDMILL</td></tr>
</table>

PROCESS PLAN 6P-9(*continued*)

No.	Operation	Tooling
1	ROUGH OUTSIDE X .125 DEEP LEAVE .01 FOR FINISH CUT IN XY 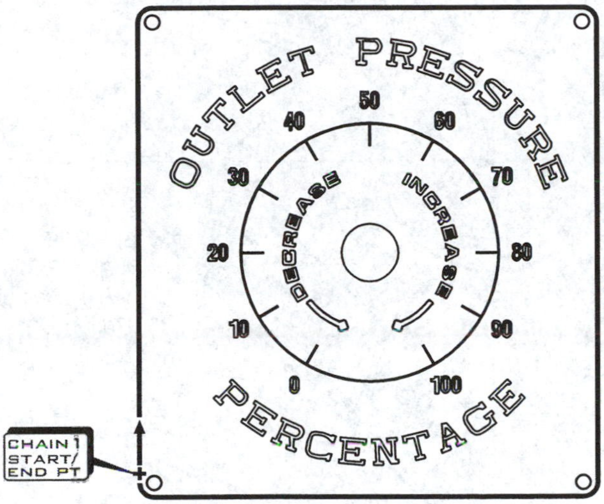	1/2 ENDMILL

6-10) Get the CAD model file **EX2-9JV** created in exercise 2-9. Generate a part program for executing the machining listed in PROCESS PLAN 6P-10.

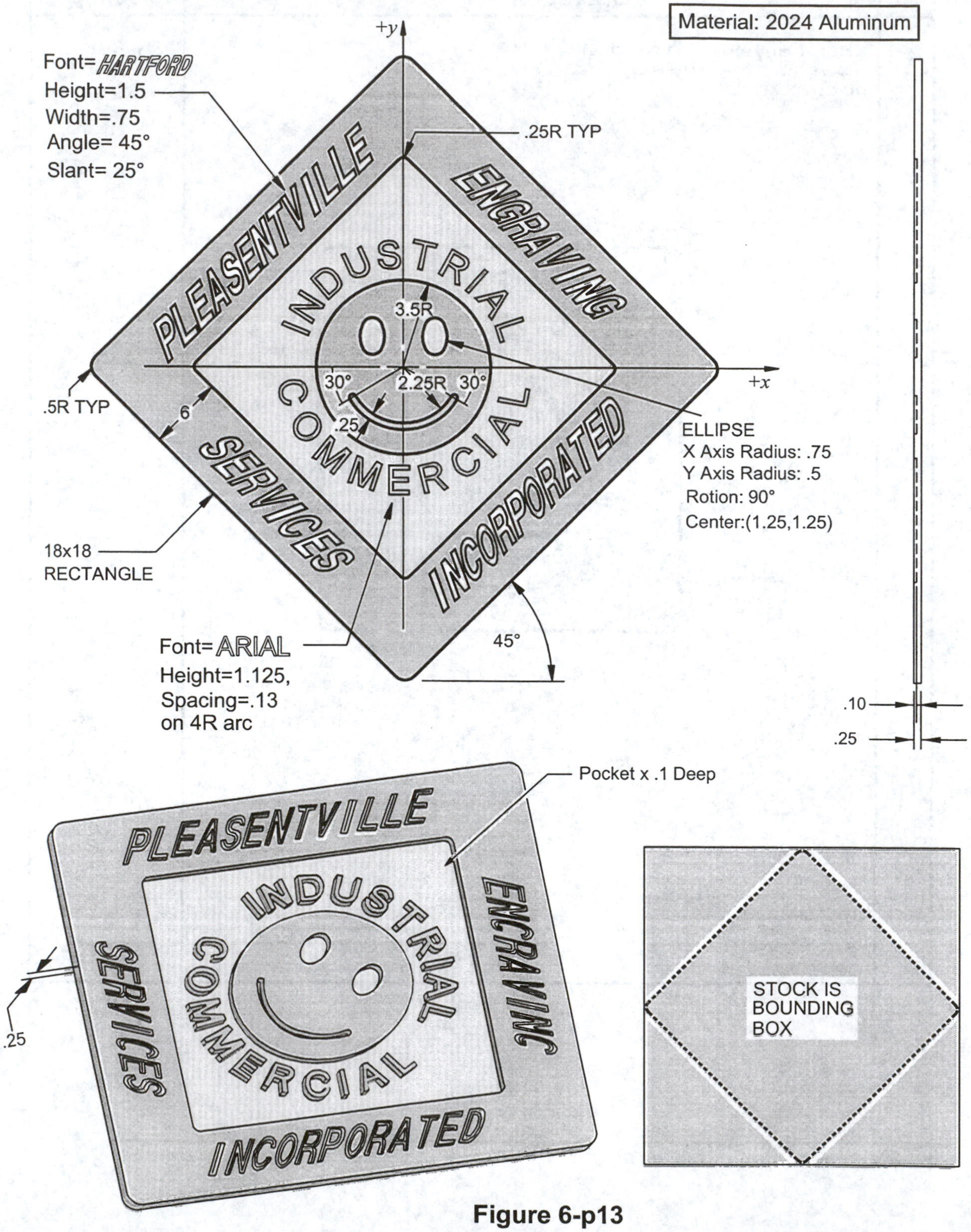

Figure 6-p13

PROCESS PLAN 6P-10

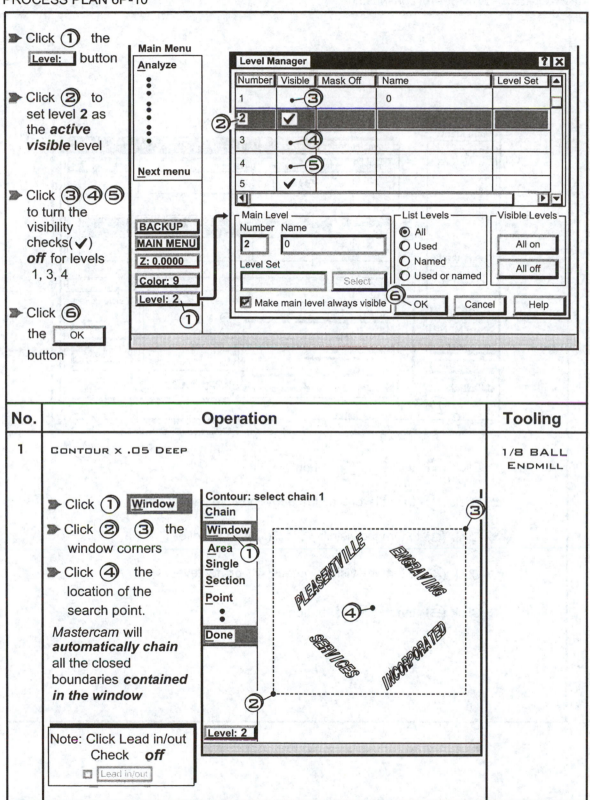

- Click ① the **Level:** button
- Click ② to set level **2** as the ***active visible*** level
- Click ③④⑤ to turn the visibility checks(✓) ***off*** for levels 1, 3, 4
- Click ⑥ the OK button

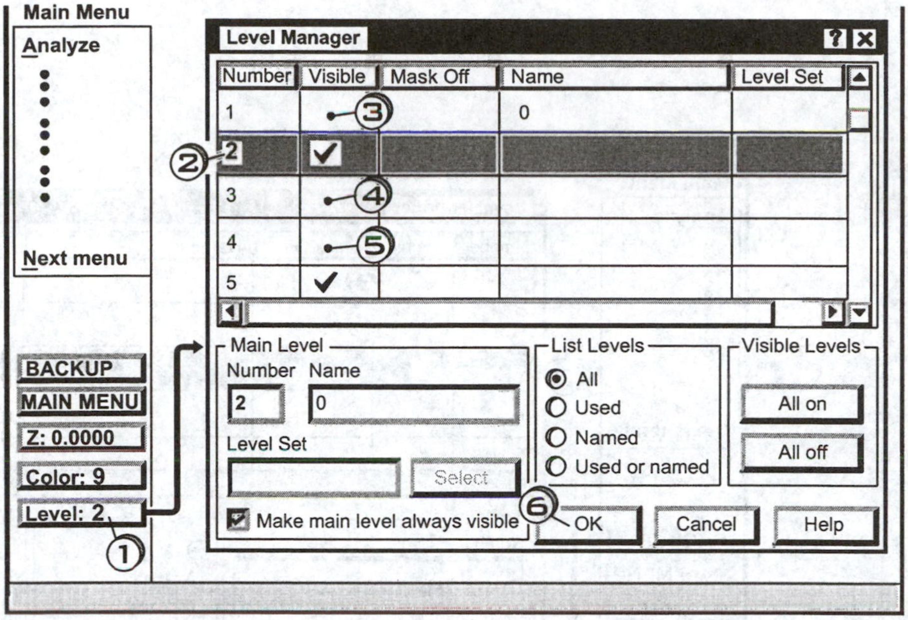

No.	Operation	Tooling
1	CONTOUR X .05 DEEP	1/8 BALL ENDMILL

- Click ① **Window**
- Click ② ③ the window corners
- Click ④ the location of the search point.

Mastercam will ***automatically chain*** all the closed boundaries ***contained in the window***

Note: Click Lead in/out Check ***off***

☐ Lead in/out

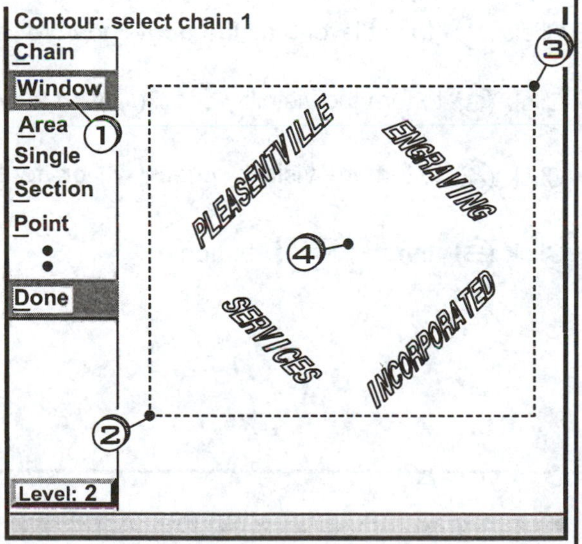

PROCESS PLAN 6P-10(*continued*)

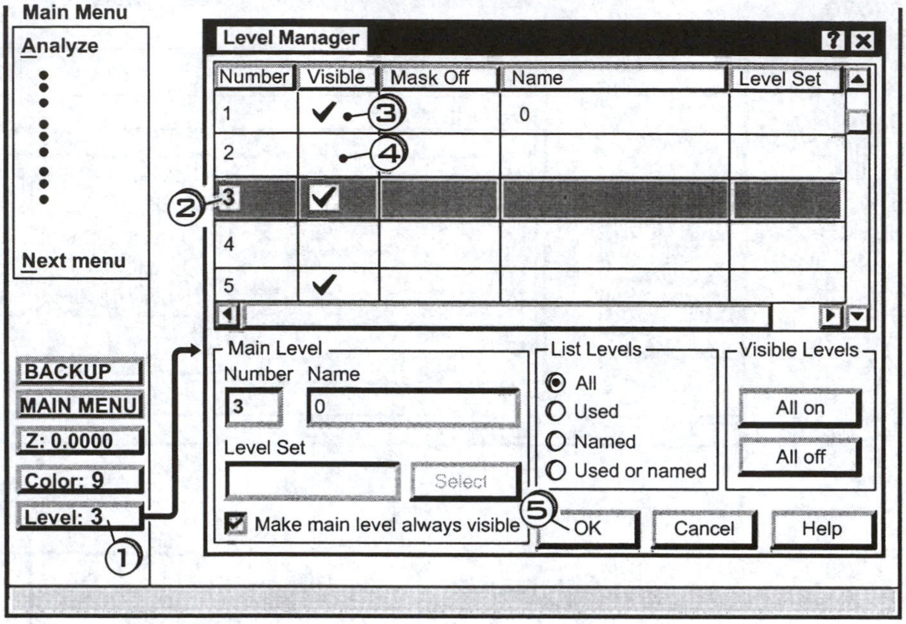

- Click ① the **Level:** button
- Click ② to set level **3** as the ***active visible*** level
- Click ③ to turn the visibility check(✓) ***on*** for level 1
- Click ④ to turn the visibility check(✓) ***off*** for level 2
- Click ⑤ the OK button

PROCESS PLAN 6P-10(*continued*)

No.	Operation	Tooling
2	POCKET X .1 DEEP	1/8 ENDMILL

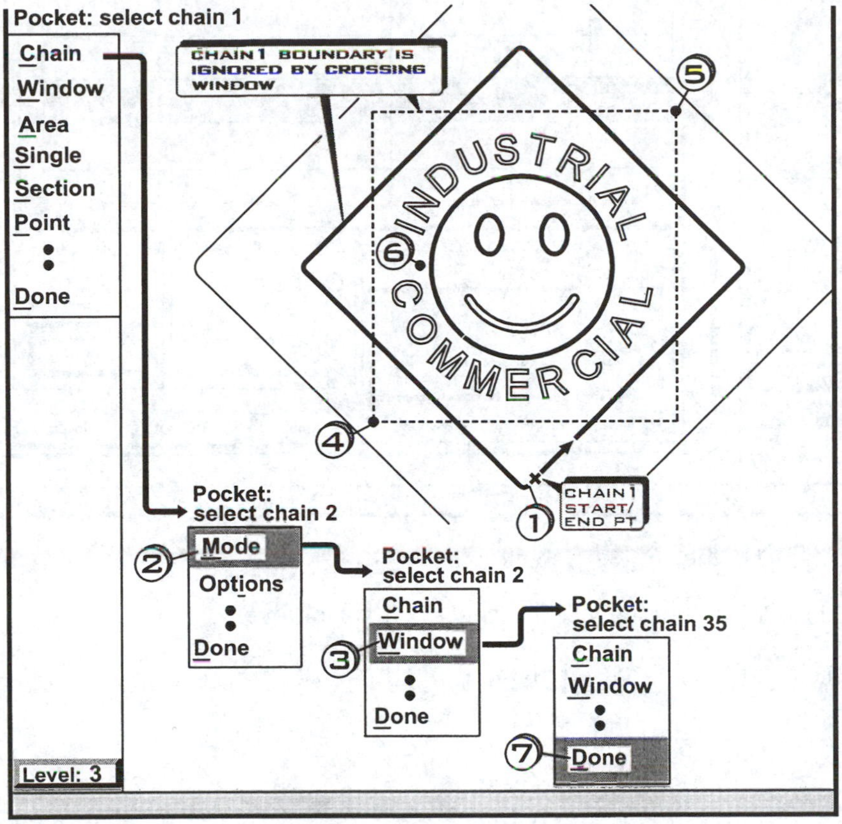

- Click ① the chain start/end point
- Click ② **Mode**
- Click ③ **Window**
- Click ④ ⑤ the window corners.
  the window should *completely enclose* the geometric letters and the face logo. ***The boundary previously chained will not be included since the window crosses its geometry***
- Click ⑥ the location of the search point ***inside the window***
  Mastercam will ***automatically chain all the closed boundaries contained in the window***
- Click ⑦ **Done**

Pocket type: Open (Standard, Open)

Cutting method:

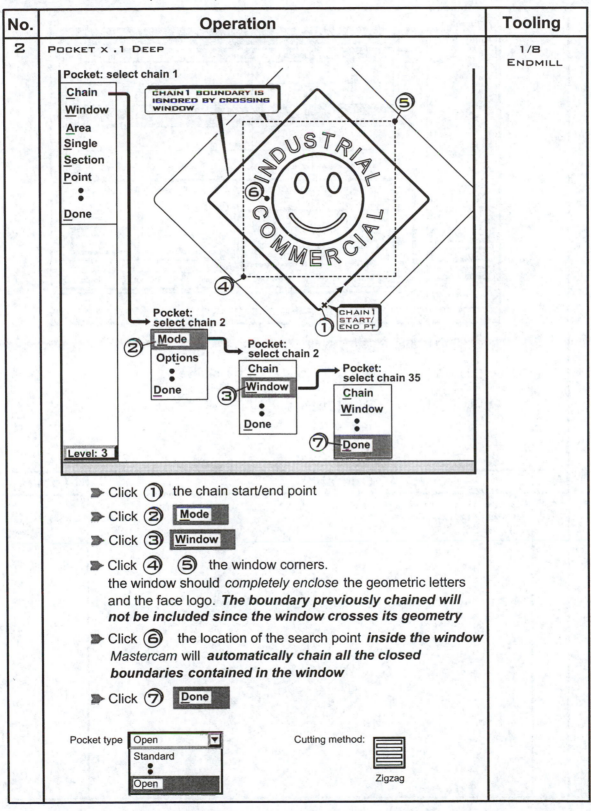

Zigzag

PROCESS PLAN 6P-10(*continued*)

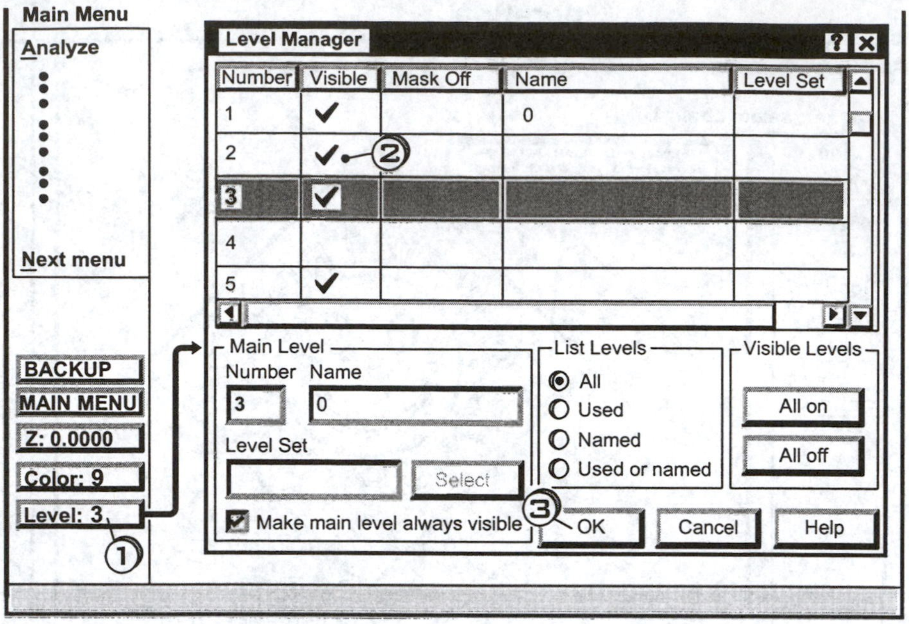

- Click ① the Level: button
- Click ② to turn the visibility check(✓) ***on*** for level 2
- Click ③ the OK button

No.	Operation	Tooling
3	ROUGH OUTSIDE X .25 DEEP LEAVE .01 FOR FINISH CUT IN XY 	1/2 ENDMILL

CHAPTER - 7

EDITING MACHINING OPERATIONS VIA THE OPERATIONS MANAGER

7-1 Chapter Objectives

After completing this chapter you will be able to:

1. Explain what the operations manager is.
2. State the four parts of each operation.
3. Know what effect associativity has on edited machining operations
4. Specify how to use the operations manager to perform the following functions on operations: create new, move, copy and delete.
5. Understand how to use the operations manager to edit existing toolpaths
6. Explain how to enter job setup from the operations manager.

7-2 The Operations Manager

The Operations Manager was introduced in Chapters 5 and 6 as a means for selecting toolpaths to be back plotted or verified. It was also used to edit drill tool paths and to direct *Mastercam* to generate a word address part program for a set of tool paths. In this Chapter we will take a closer look at the Operations Manager and consider its powerful editing features.

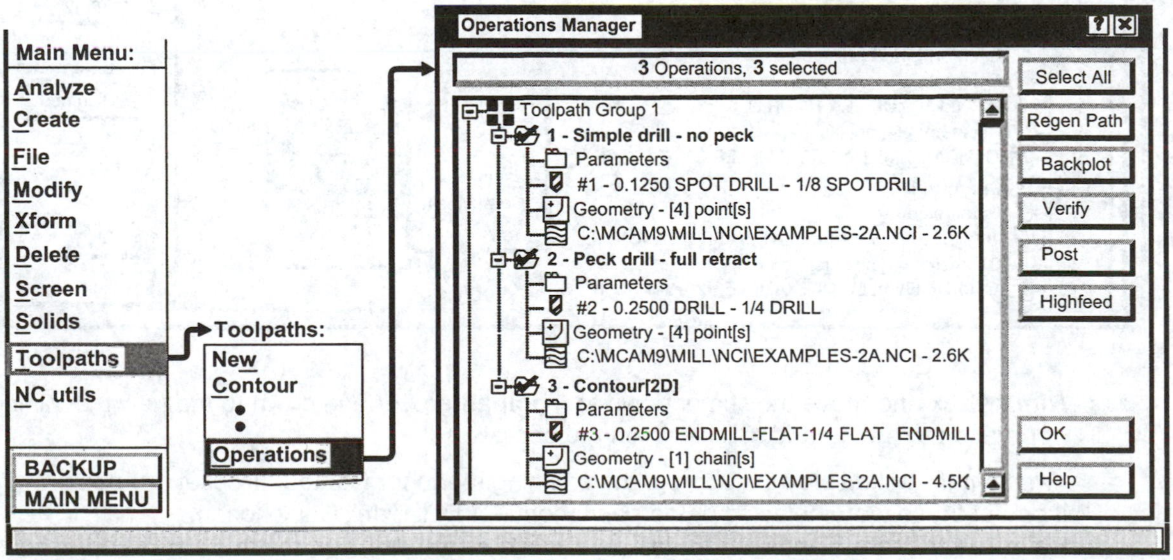

The Operations Manager is the *control center* for *adding new machining operations* or *changing existing operations for* the *current job*. An operation contains all the information *Mastercam* needs to machine a toolpath. Each operation has a name and consists of ***four parts***: *Parameters, Tool definition, Part geometry* and *NCI*

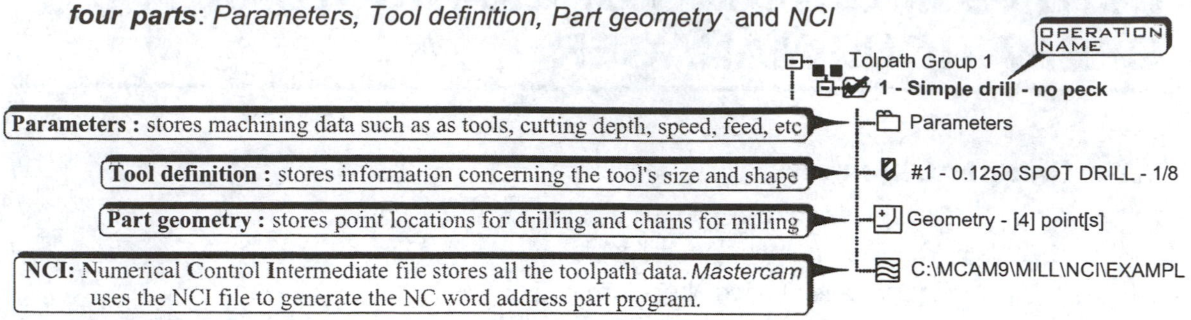

Associativity

The parts Parameters, Tool definition, Part geometry and NCI are all associated or linked together. When the operator enters a different tool and cutting depth in Parameters and adds a drill point in Part geometry *Mastercam* flags these changes. ***Associativity*** enables the operator to direct *Mastercam* to ***automatically regenerate all the other parts of the operation as it incorporates the changes.***

7-3 Creating New Operations

New machining operations can be created within the Operations Manager

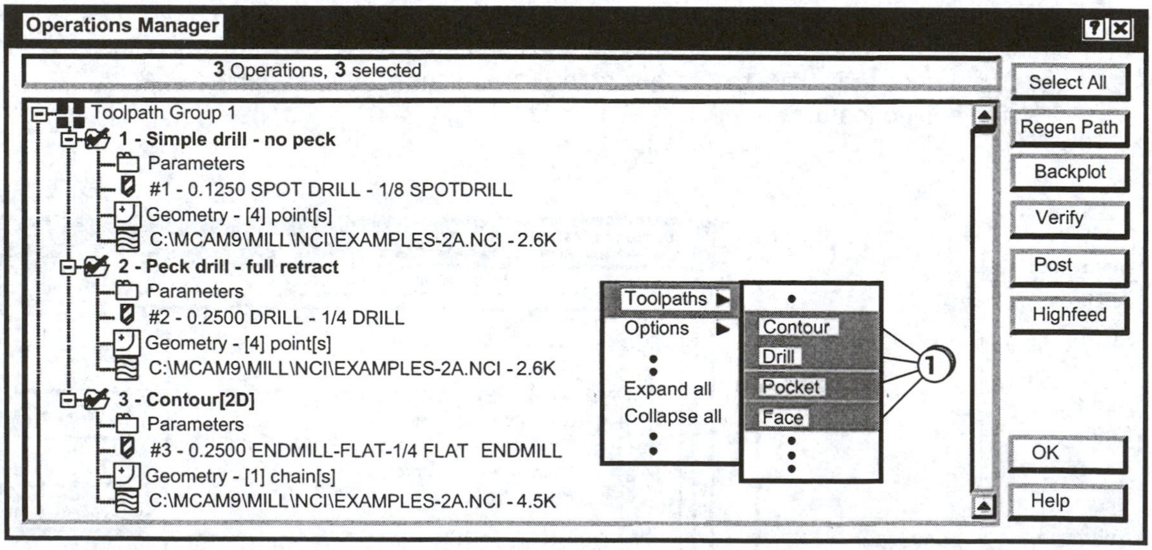

➤ ***Right*** Click and move the cursor down to **Toolpaths** over and down to the new machining operation to be created; Click ① on the operation

Mastercam will display the appropriate chaining menu for Contour, Pocket or Face if one of these were selected or the point menu for Drill if drill was chosen.
Refer to Chapters 5 and 6 for further details in creating drill, contour and pocket operations.

7-4 Moving Existing Operations

The move function can be used to *change the existing machining order.* This is especially useful when trying to rearrange roughing, finishing and chamfering operations to minimize tool changes

METHOD A: PICK / PLACE

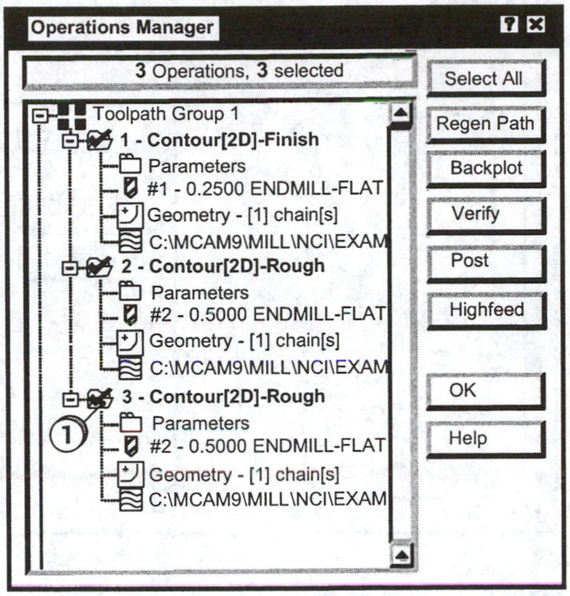

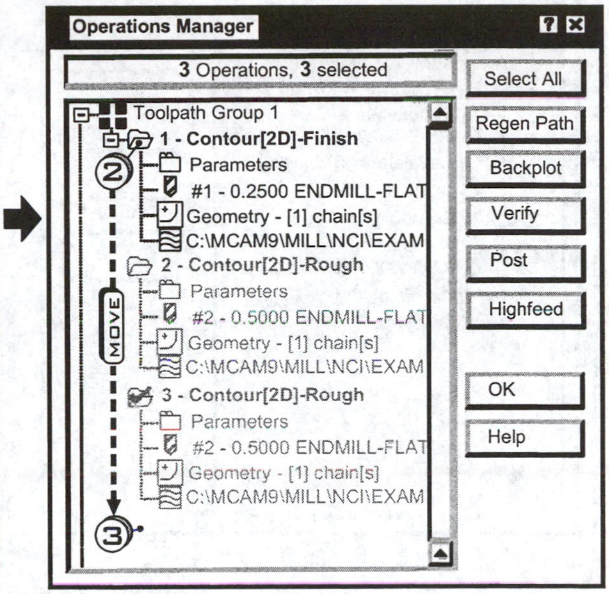

- Click ① on the file folder of an operation *other than* the one to be moved
- Click ② on the file folder of the operation to be moved and ***keeping the Left mouse button depressed*** *move the folder to the new location* ③ *and release.*

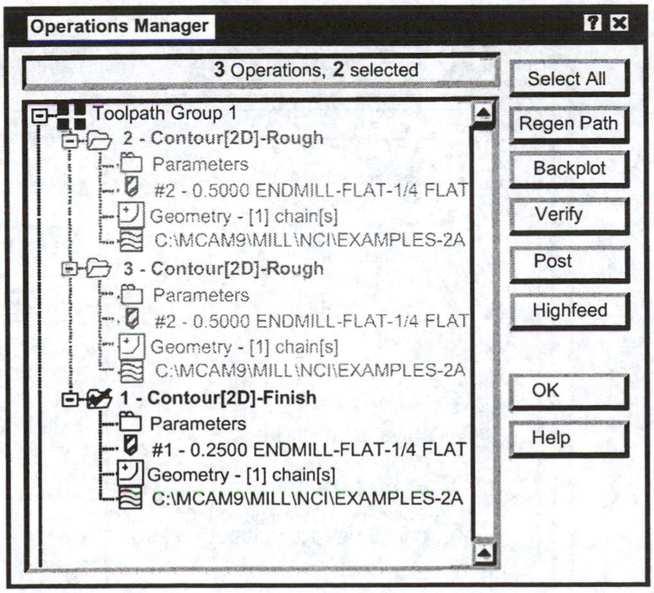

METHOD B: OPTION MENU

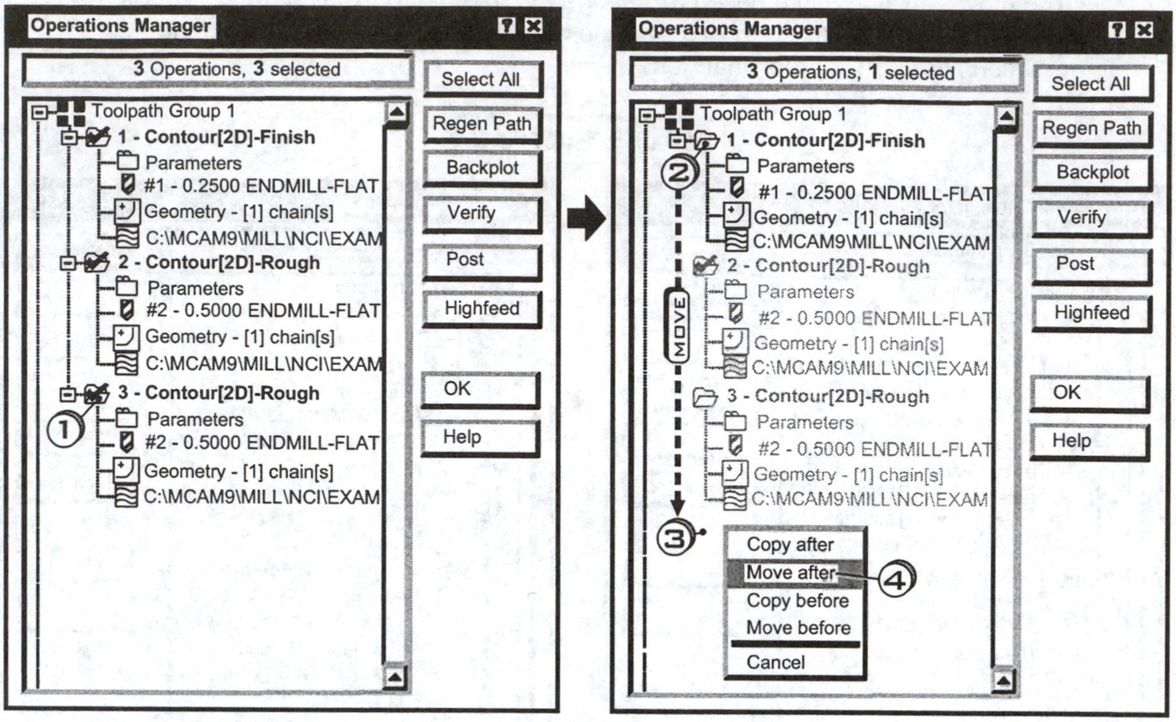

- Click ① on the file folder of an operation *other than* the one to be moved
- ***Right*** Click ② on the file folder of the operation to be moved and ***keeping the Right mouse button depressed*** *move the folder to the new location* ③ *and release.*
- Click ④ the desired move selection from the menu

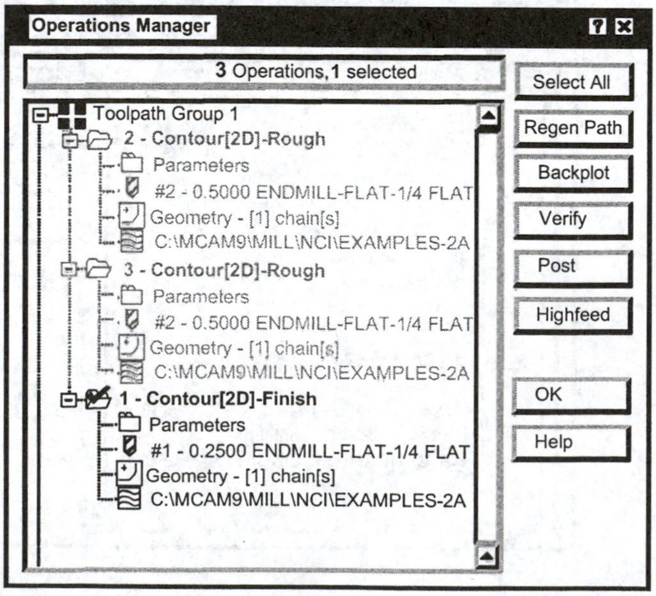

7-5 Copying Existing Operations

Copying is an *efficiency* tool which is especially useful when *different* machining operations need to be performed on the *same* tool path. A drilling example would be a set of holes that first need to be center drilled then drilled and finally tapped. A milling application would involve roughing a contour with endmill-A using speed-A and feed-A and finishing the contour with endmill-B using speed-B and feed-B.

OPTION MENU

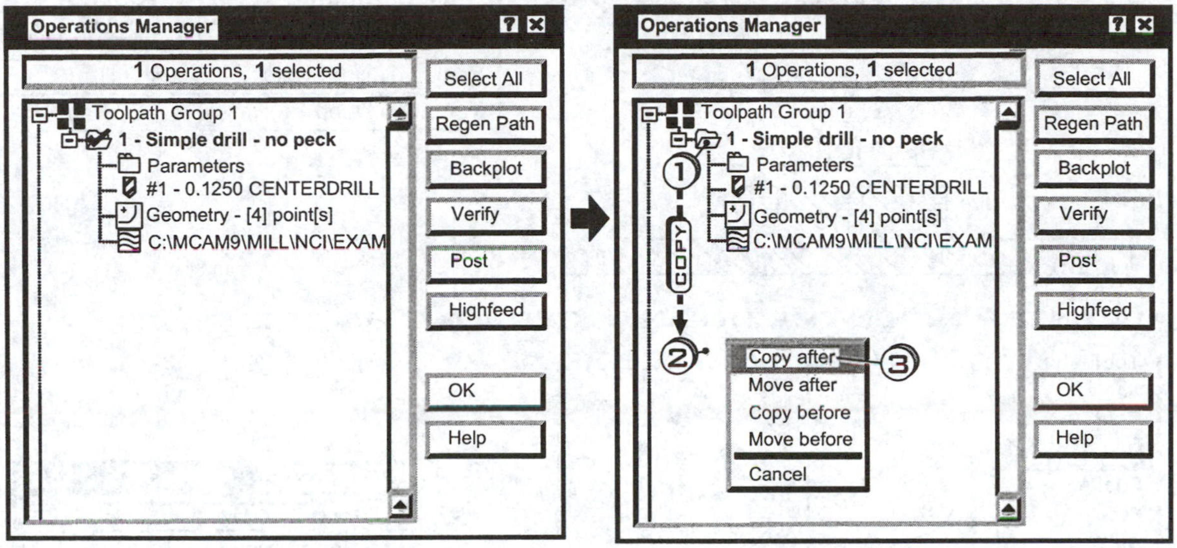

➤ ***Right*** Click ① on the file folder of the operation to be moved and ***keeping the Right mouse button depressed*** *move the folder to the new location* ② *and release*.

➤ Click ③ the desired move selection from the menu

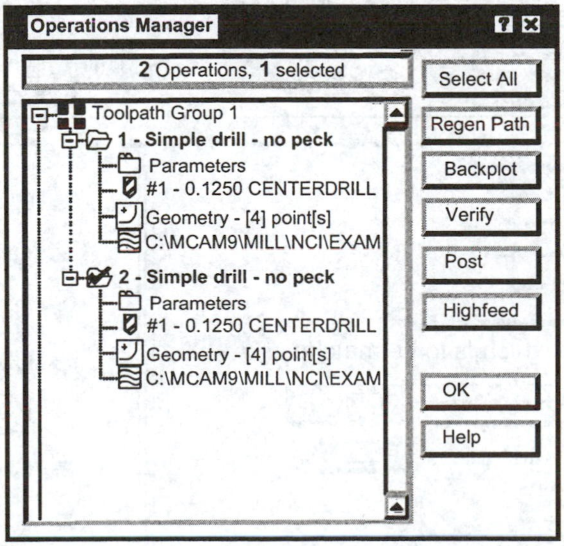

EXAMPLE 7-1

Assume a CAD model the part face shown below has been created in *Mastercam*. Use the Copy function in the Operations Manager as an aid in quickly generating the machining operations needed to produce the drill and tap holes.

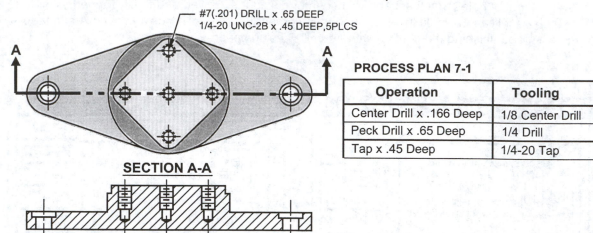

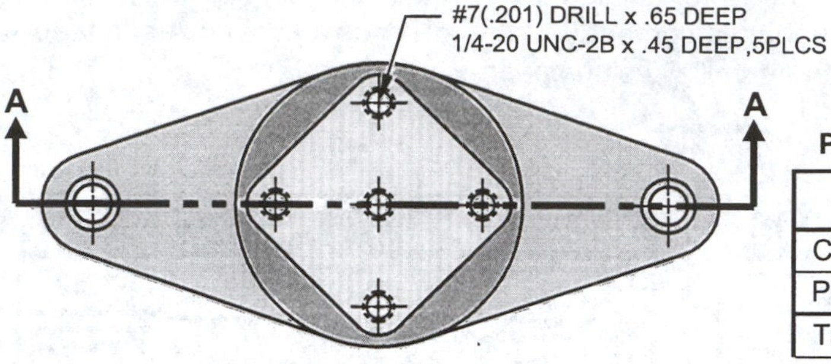

PROCESS PLAN 7-1

Operation	Tooling
Center Drill x .166 Deep	1/8 Center Drill
Peck Drill x .65 Deep	1/4 Drill
Tap x .45 Deep	1/4-20 Tap

SECTION A-A

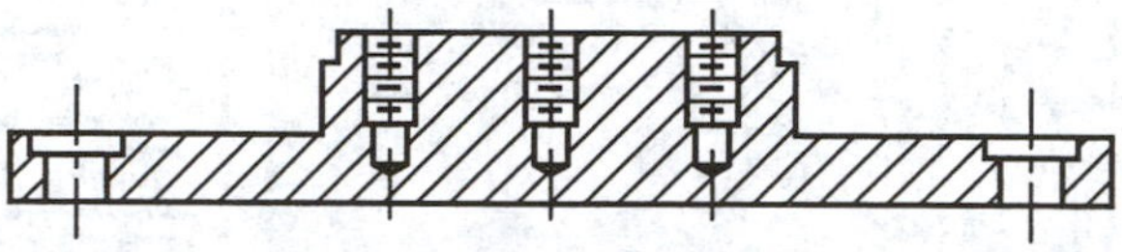

A) CENTER DRILL THE HOLES .125DIA X .166 DEEP

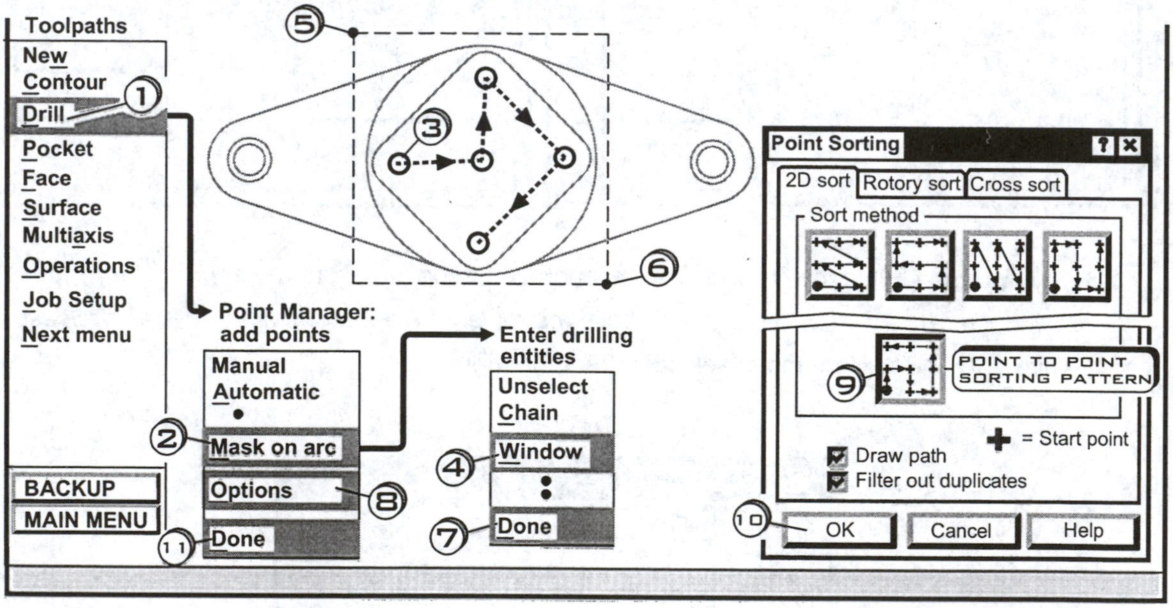

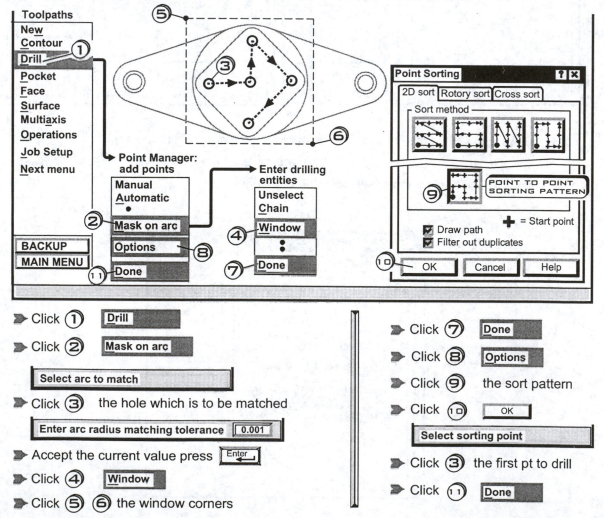

- Click ① **Drill**
- Click ② **Mask on arc**

  Select arc to match
- Click ③ the hole which is to be matched

  Enter arc radius matching tolerance 0.001
- Accept the current value press Enter
- Click ④ **Window**
- Click ⑤ ⑥ the window corners
- Click ⑦ **Done**
- Click ⑧ **Options**
- Click ⑨ the sort pattern
- Click ⑩ OK

  Select sorting point
- Click ③ the first pt to drill
- Click ⑪ **Done**

◆ OBTAIN THE NEEDED .125 DIA CENTER DRILL TOOL

Mastercam will activate and display the Tool and Drill Parameters dialog box after the drill tool path has been created.

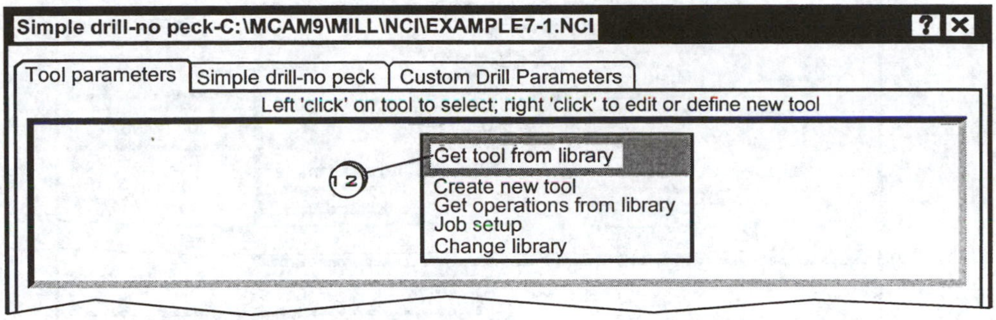

➤ Press the ***right*** mouse button in the ***white area***

➤ Click (12) Get tool from library

Mastercam will display the main Tool Library dialog box

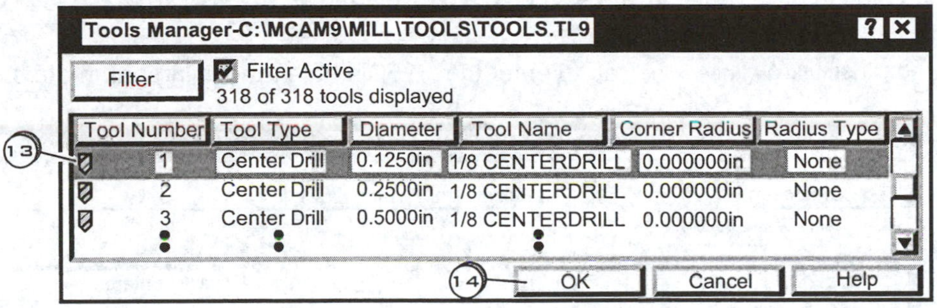

➤ Click (13) on the .125 Center drill

➤ Click (14) OK

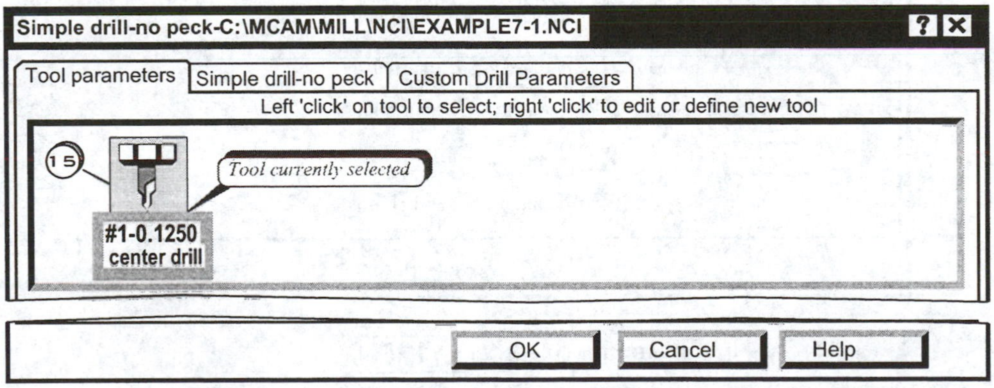

➤ ***Right*** click (15)

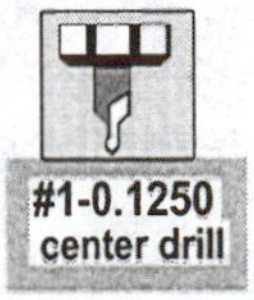

on the #1-0.1250 center drill to select this as the ***active tool*** and direct *Mastercam* to display the Define Tool dialog box

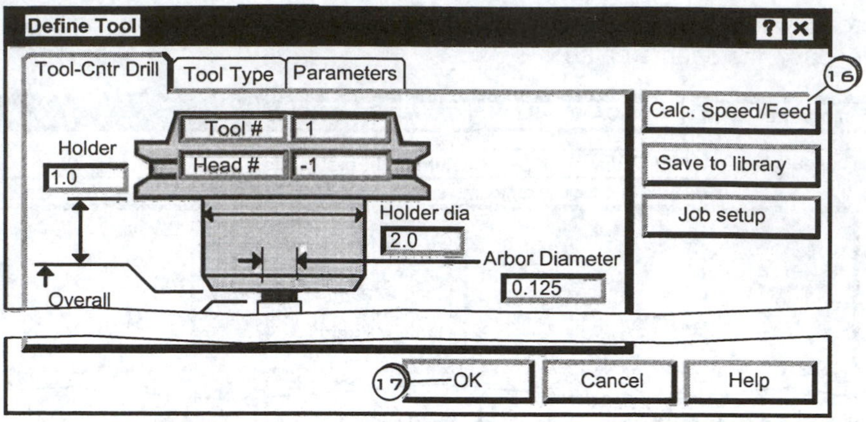

➤ Click (16) [Calc Speed/Feed] ➤ Click (17) [OK]

◆ ENTER THE .125 DIA CENTER DRILL HOLE MACHINING PARAMETERS

➤ Click (18) [Simple drill-no peck] tab to enter the .125Dia hole machining parameters.

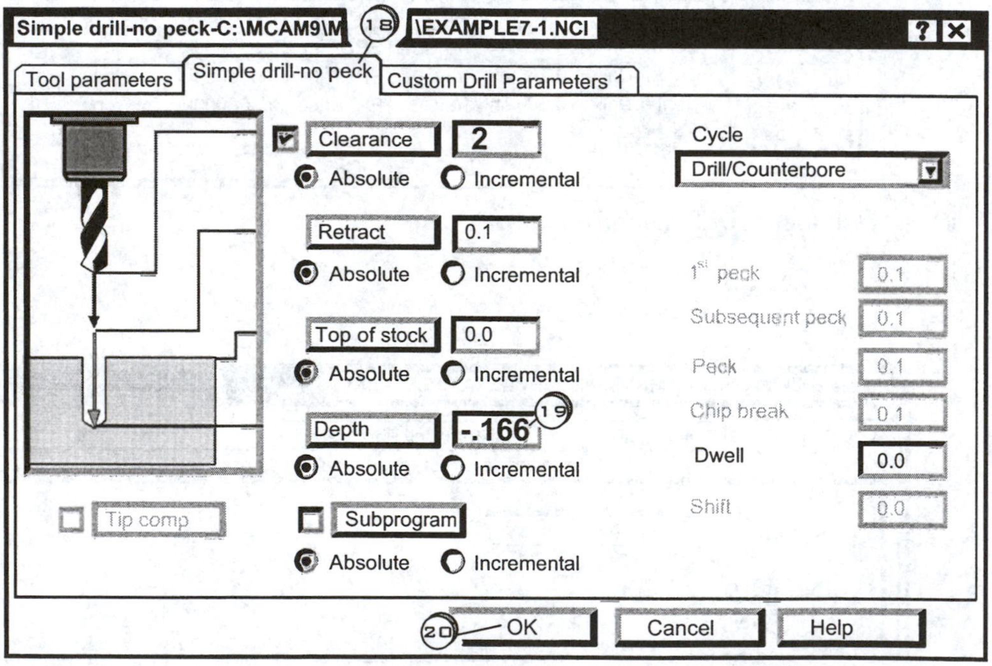

➤ Click (19) in the Depth box; enter [-.125]

➤ Click (20) [OK]

The .125Dia center drilling operation will be added to *Mastercam's* Operations Manager

B) PECK DRILL THE HOLES .201DIA X .65 DEEP

- ENTER THE OPERATIONS MANAGER AND MAKE A COPY OF THE CENTER DRILL OPERATION

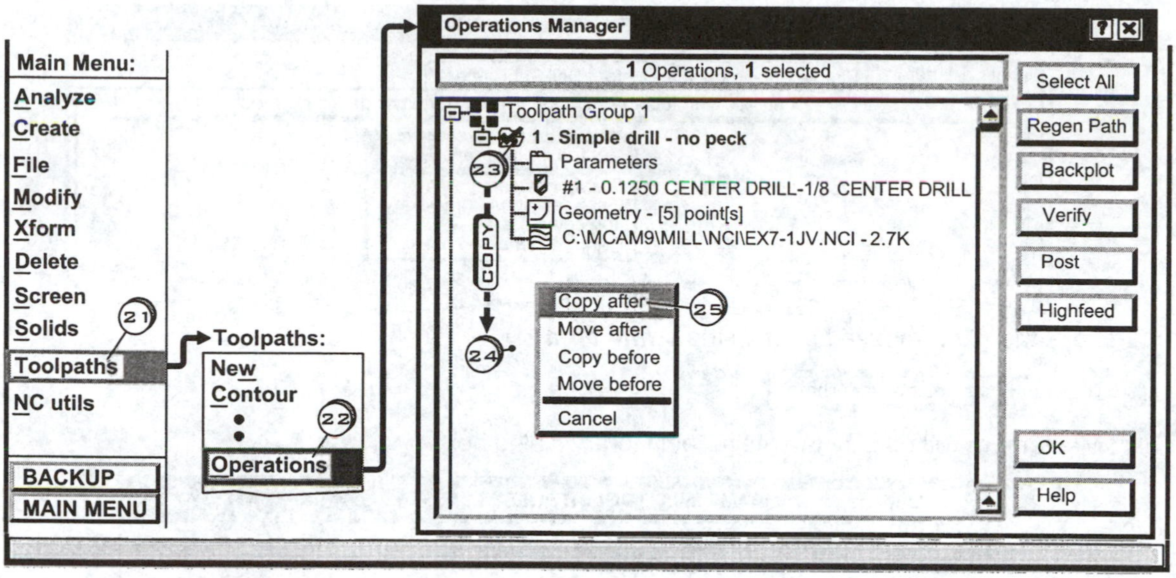

➤ Click (21) Toolpaths

➤ Click (22) Operations

➤ ***Right*** Click (23) on the file folder of the operation to be moved and ***keeping the Right mouse button depressed*** *move the folder to the new location* (24) *and release.*

➤ Click (25) Copy after

- ENTER THE REQUIRED PECK DRILL PARAMETERS INTO THE COPY

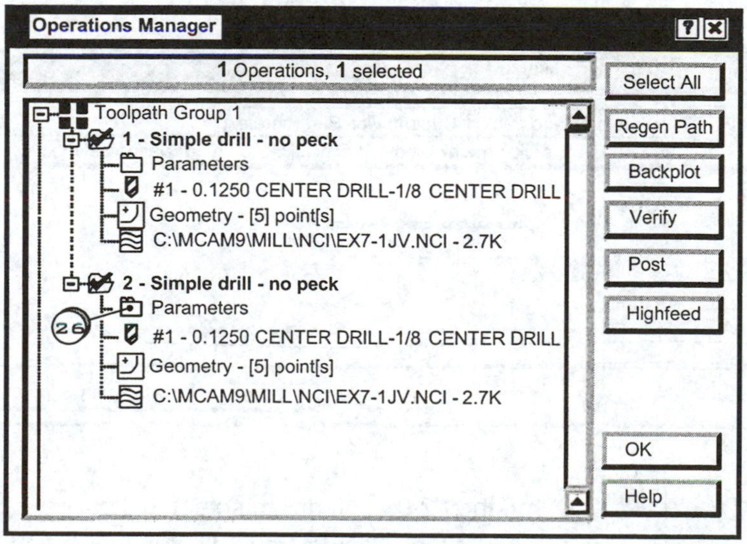

➤ Click (26) on the Parameters icon

◆ OBTAIN THE NEEDED .201 DIA DRILL TOOL

Mastercam will activate and display the Tool and Drill Parameters dialog box after the drill tool path has been created.

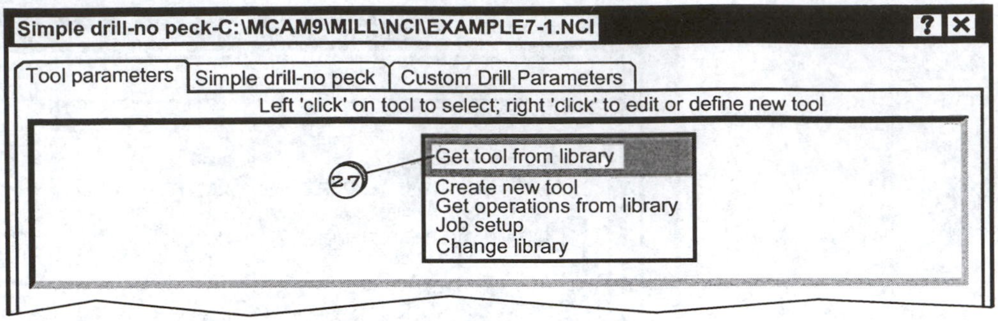

➤ Press the ***right*** mouse button in the ***white area***

➤ Click (27) Get tool from library

Mastercam will display the main Tool Library dialog box

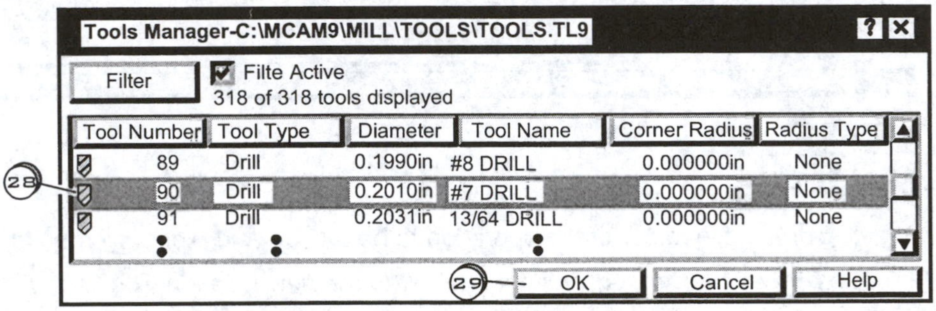

➤ Click (28) on the .201 Dia drill

➤ Click (29) OK

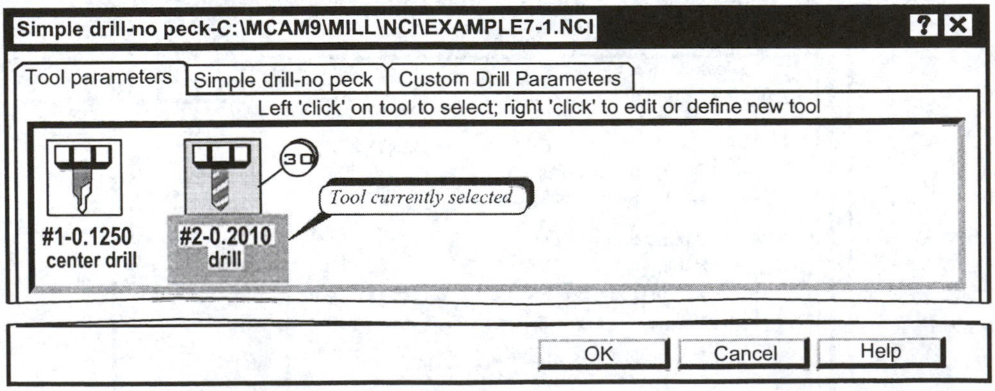

➤ ***Right*** click (30)

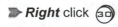

on the #2-0.2010 drill to select this as the ***active tool*** and direct *Mastercam* to display the Define Tool dialog box

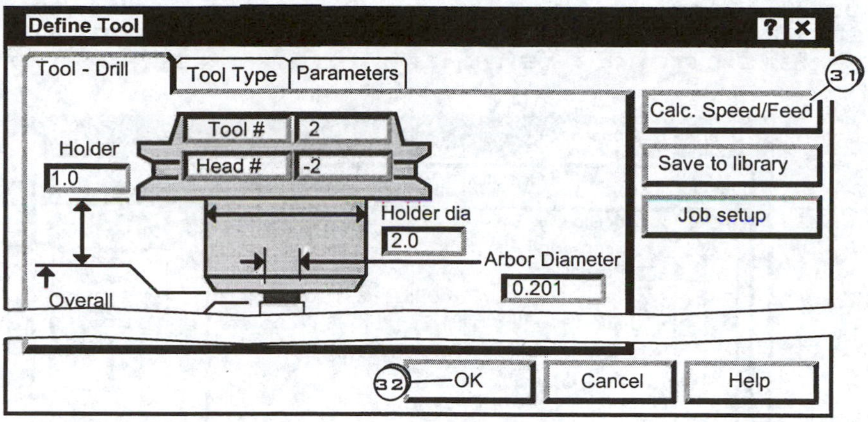

➤ Click 31 Calc Speed/Feed | ➤ Click 32 OK

- **ENTER THE .201DIA PECK DRILL X .65 DEEP MACHINING PARAMETERS**

➤ Click 33 Simple drill-no peck tab to enter the .201Dia peck drill machining parameters.

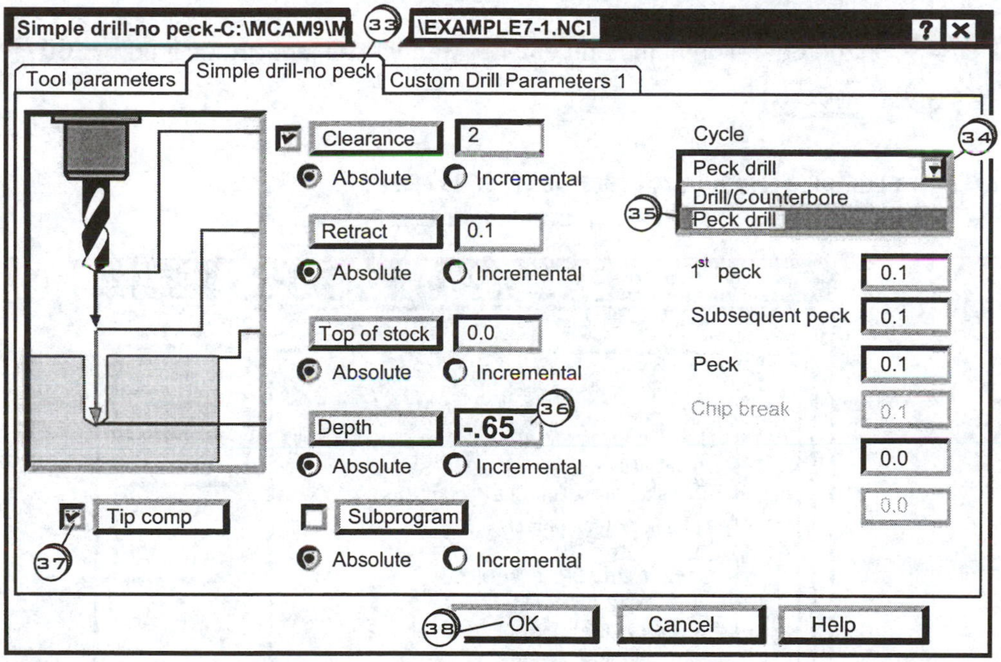

➤ Click 34 the toggle down button

➤ Click 35 Peck drill

➤ Click 36 in the Depth box; enter -.65

➤ Click 37 the tip comp check *on* for drilling *thru* holes

➤ Click 38 OK

The .201Dia peck drilling thru operation will be added to *Mastercam's* Operations Manager

- ADD ANY DESIRED LABELING TO THE COPIED OPERATION REGENERATE THE COPIED OPERATION TO INCORPORATE ANY CHANGES INTO ALL ITS OTHER PARTS VIA ASSOCIATIVITY

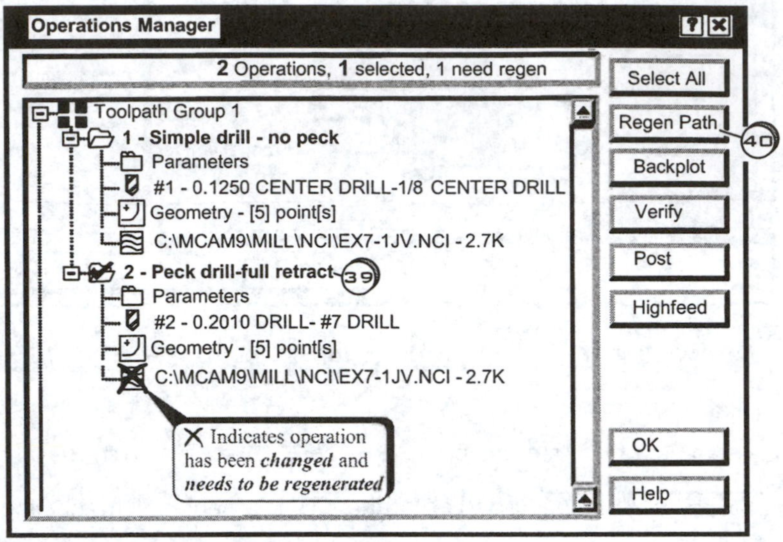

➤ Click (39) on the operation name and enter any additional labeling desired: **x.65 deep**

➤ Click (40) Regen Path

C) TAP THE HOLES 1/4-20 X .45 DEEP

- MAKE A COPY OF THE PECK DRILL OPERATION

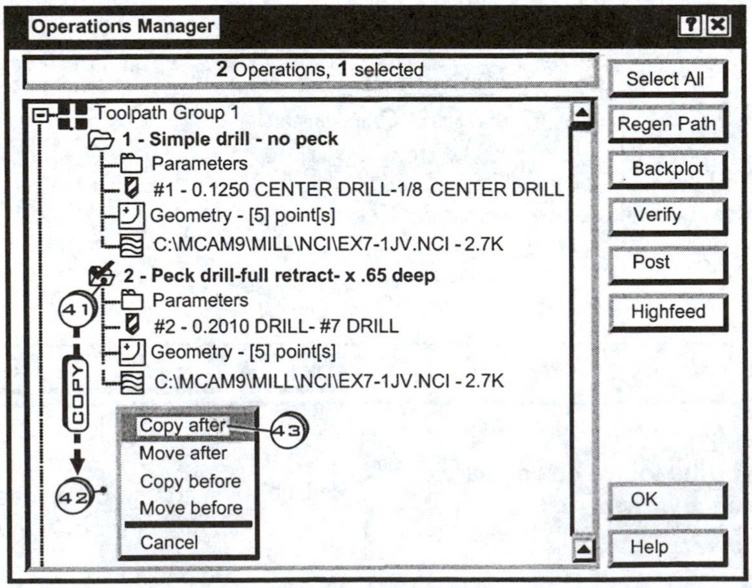

➤ ***Right*** Click (41) on the file folder of the operation to be moved and ***keeping the Right mouse button depressed*** *move the folder to the new location* (42) *and release.*

➤ Click (43) Copy after

- ENTER THE REQUIRED TAPPING PARAMETERS INTO THE COPY

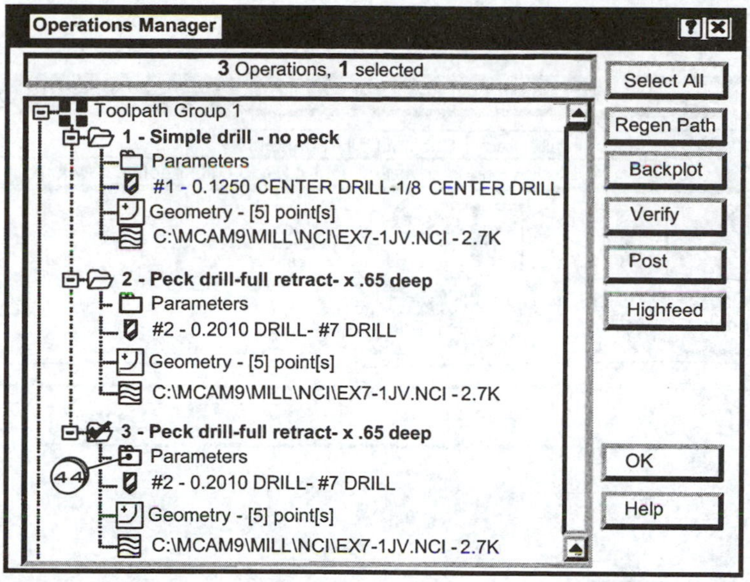

➤ Click (44) on the Parameters icon

- OBTAIN THE NEEDED 1/4-20UNC TAP TOOL

Mastercam will activate and display the Tool and Drill Parameters dialog box

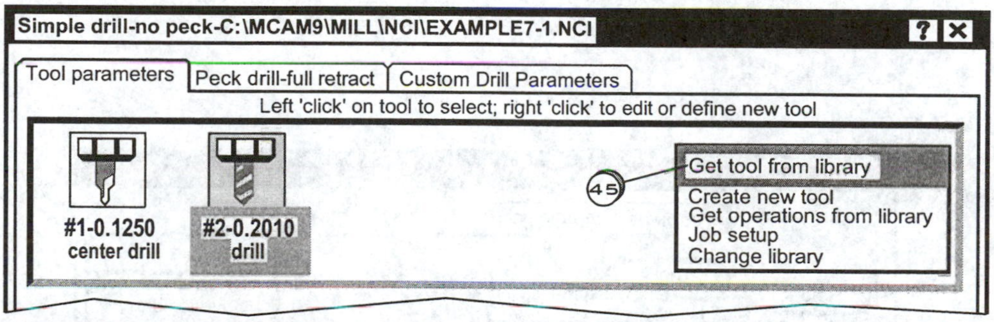

➤ Press the ***right*** mouse button in the ***white area***

➤ Click (45) Get tool from library

Mastercam will display the main Tool Library dialog box

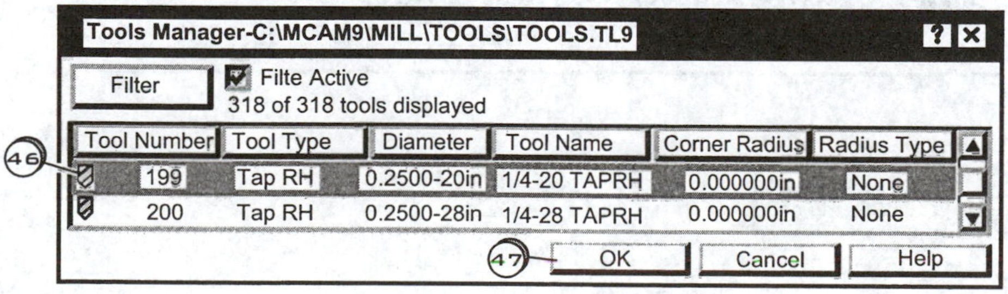

➤ Click (46) the 1/4-20 TAPRH tool

➤ Click (47) OK

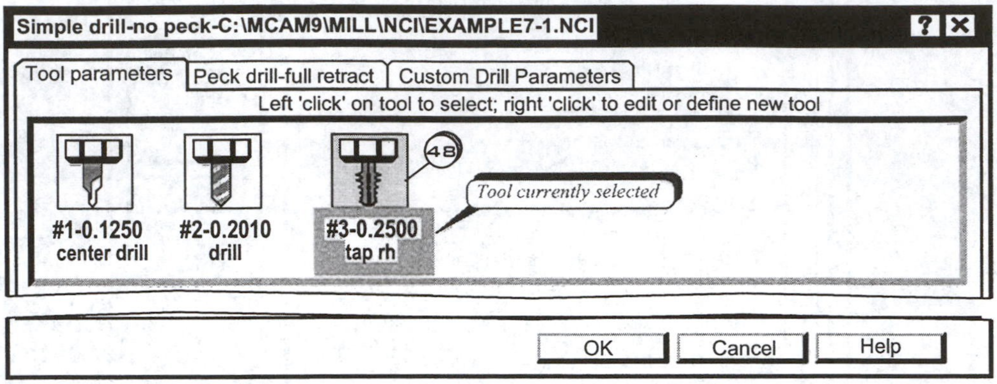

➤ ***Right*** click (48) 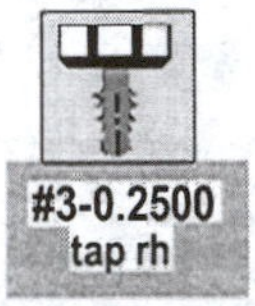 on the #3-0.2500 tap rh to select this as the ***active tool*** and direct *Mastercam* to display the Define Tool dialog box

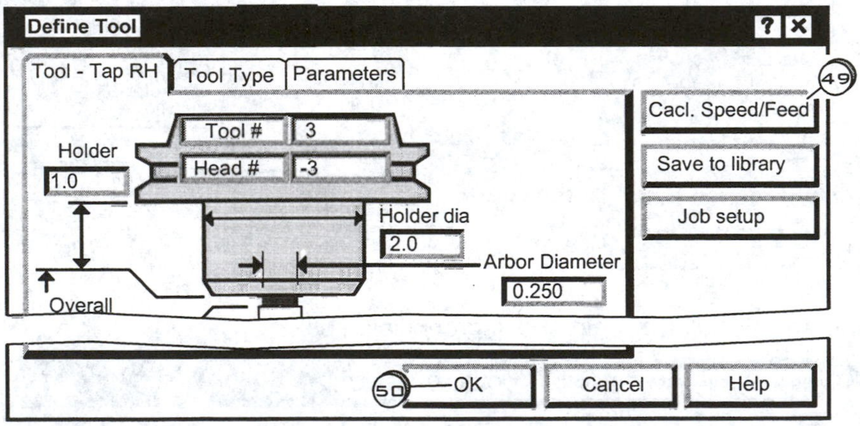

➤ Click (49) 

➤ Click (50) 

- **ENTER THE 1/4-20UNC TAP MACHINING PARAMETERS**

➤ Click (51) [Peck drill-full retract] tab to enter the 1/4-20UNC tap machining parameters.

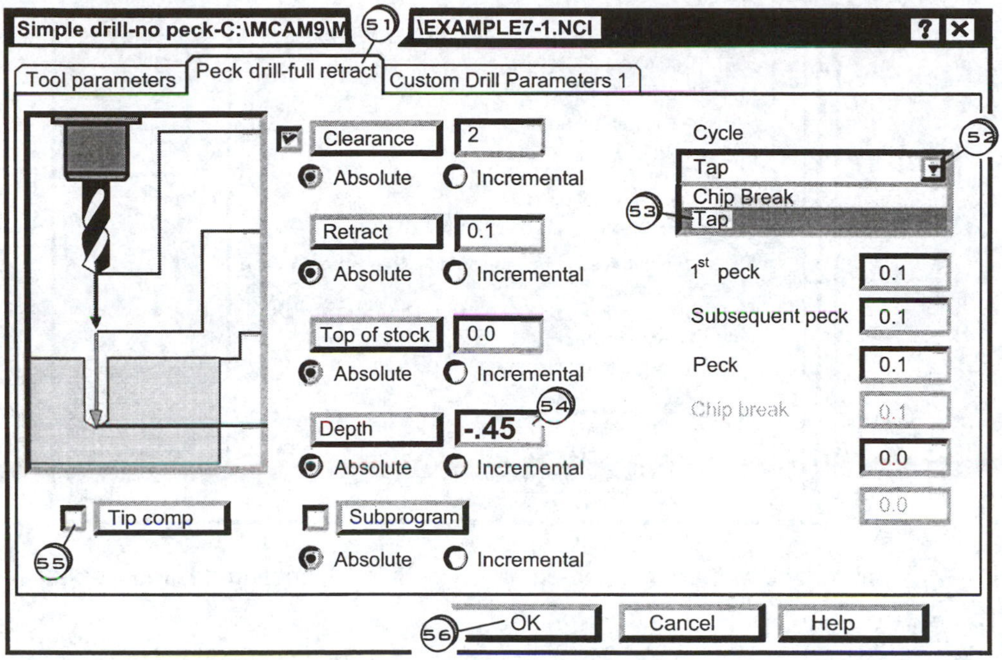

➤ Click (52) [▾] the toggle down button

➤ Click (53) [Tap]

➤ Click (54) in the Depth box; enter [-.45]

➤ Click (55) the tip comp check *off* [☐] when tapping *blind* holes

➤ Click (56) [OK]

The 1/4-20UNC tapping operation will be added to *Mastercam's* Operations Manager

- ADD ANY DESIRED LABELING TO THE COPIED OPERATION REGENERATE THE COPIED OPERATION TO INCORPORATE ANY CHANGES INTO ALL ITS OTHER PARTS VIA ASSOCIATIVITY

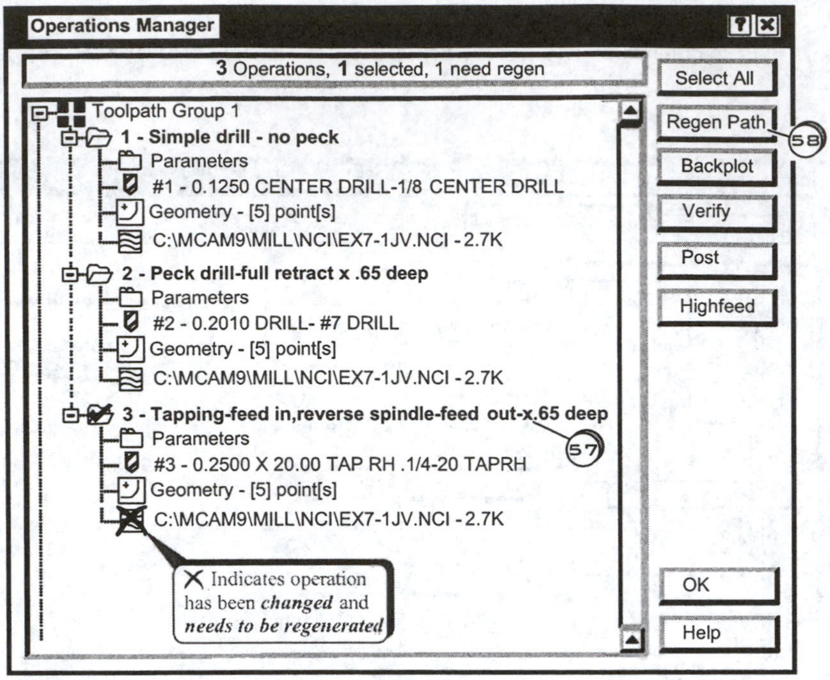

➤ Click (57) on the operation name and enter any additional labeling desired: **x.45 deep**

➤ Click (58) Regen Path

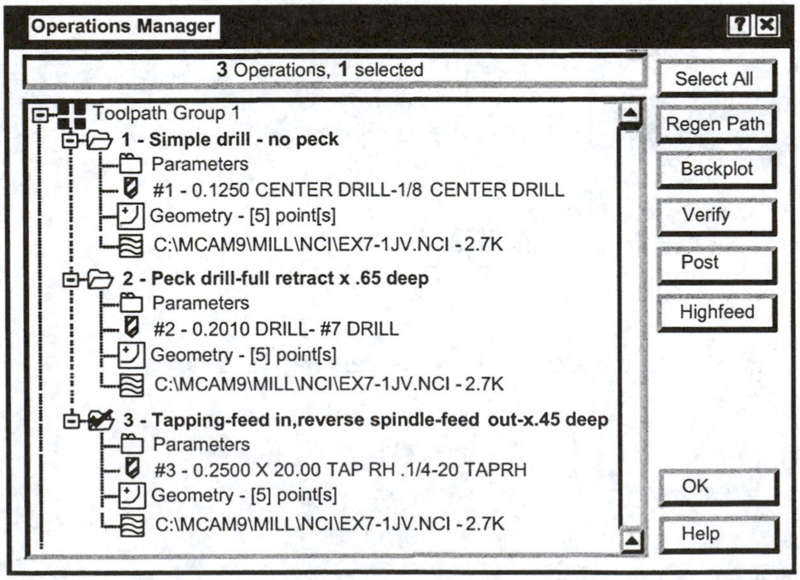

All the machining operations needed to produce the drill and tap holes have now been created in the Operations Manager with the aid of the Copy feature.

7-6 Copying Parts of Existing Operations

Any of the individual parts of one operation: Parameters, Tools or Part geometry can also be copied to another *similar* operation.

Restriction with Parameters: copying can *only* be done between *similar* operation types(contour to contour or pocket to pocket, etc)

Restriction with Part geometry: chains can only be copied between contouring, pocketing or drilling operations.

EXAMPLE 7-2

Copy the Parameters from **1-Contour-[2D]** to **2-Contour-[2D]**

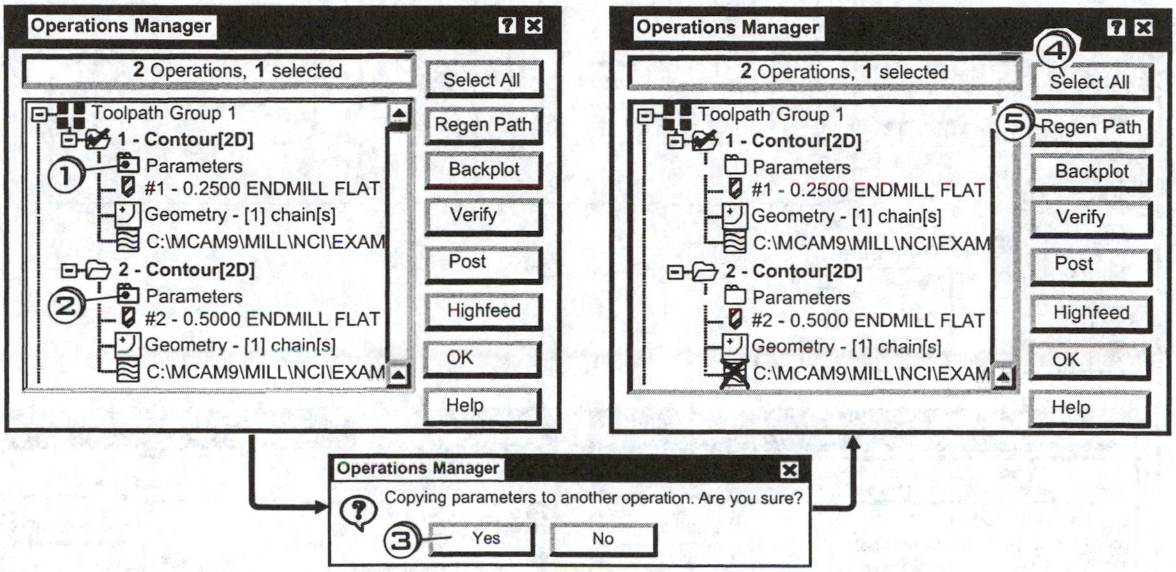

- Click ① on the icon of the Parameter to be copied ***keeping the left mouse button depressed***, move the cursor over the icon of the Parameter receiving the copy ② and ***release***
- Click ③ Yes
- Click ④ Select All
- Click ⑤ Regen Path

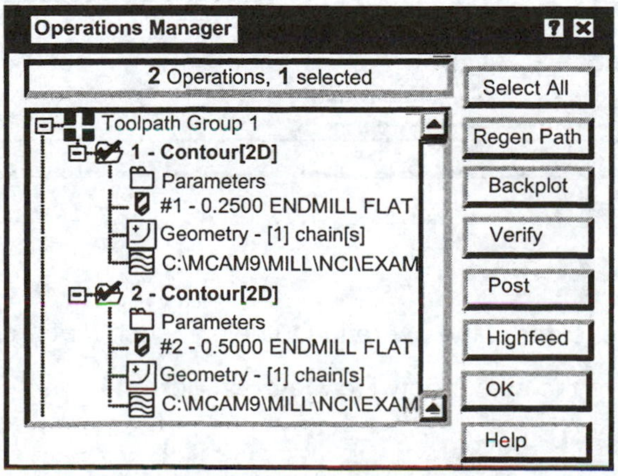

7-7 Deleting Existing Operations

The Operations Manager is used to delete a *selected* operation a specific *set* of operations or *all* the operations for a job.

DELETING A SELECTED OPERATION

EXAMPLE 7-3

Delete the existing contour tool path shown below

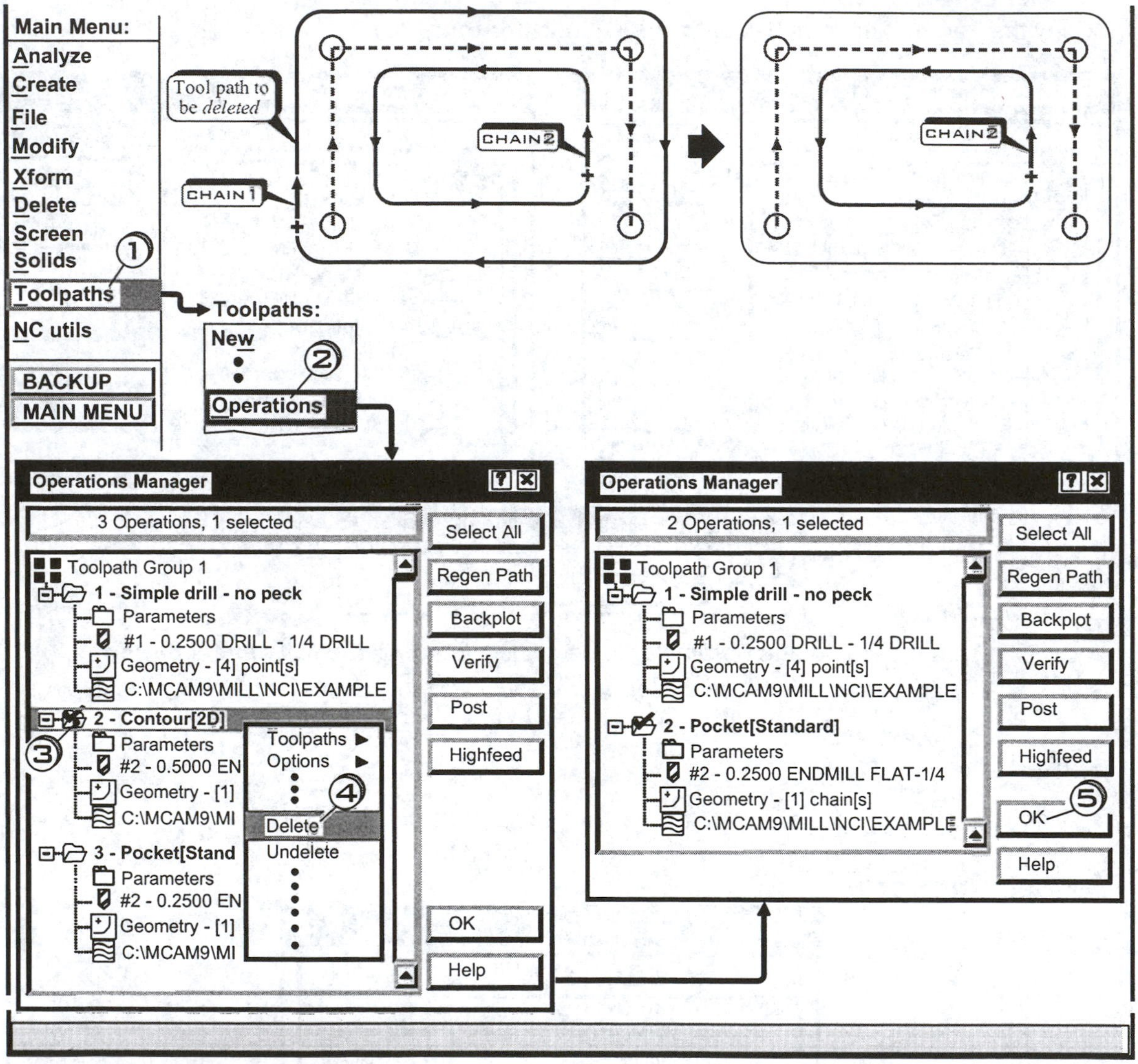

- Click ① **Toolpaths**
- Click ② **Operations**
- Click ③ on the file folder of the operation to delete such that a check appears
- ***Right*** Click and move the cursor down to **Delete**; Click ④
- Click ⑤ the OK button

DELETING A SET OF OPERATIONS

EXAMPLE 7-4

Delete the existing contour and pocket tool paths shown below

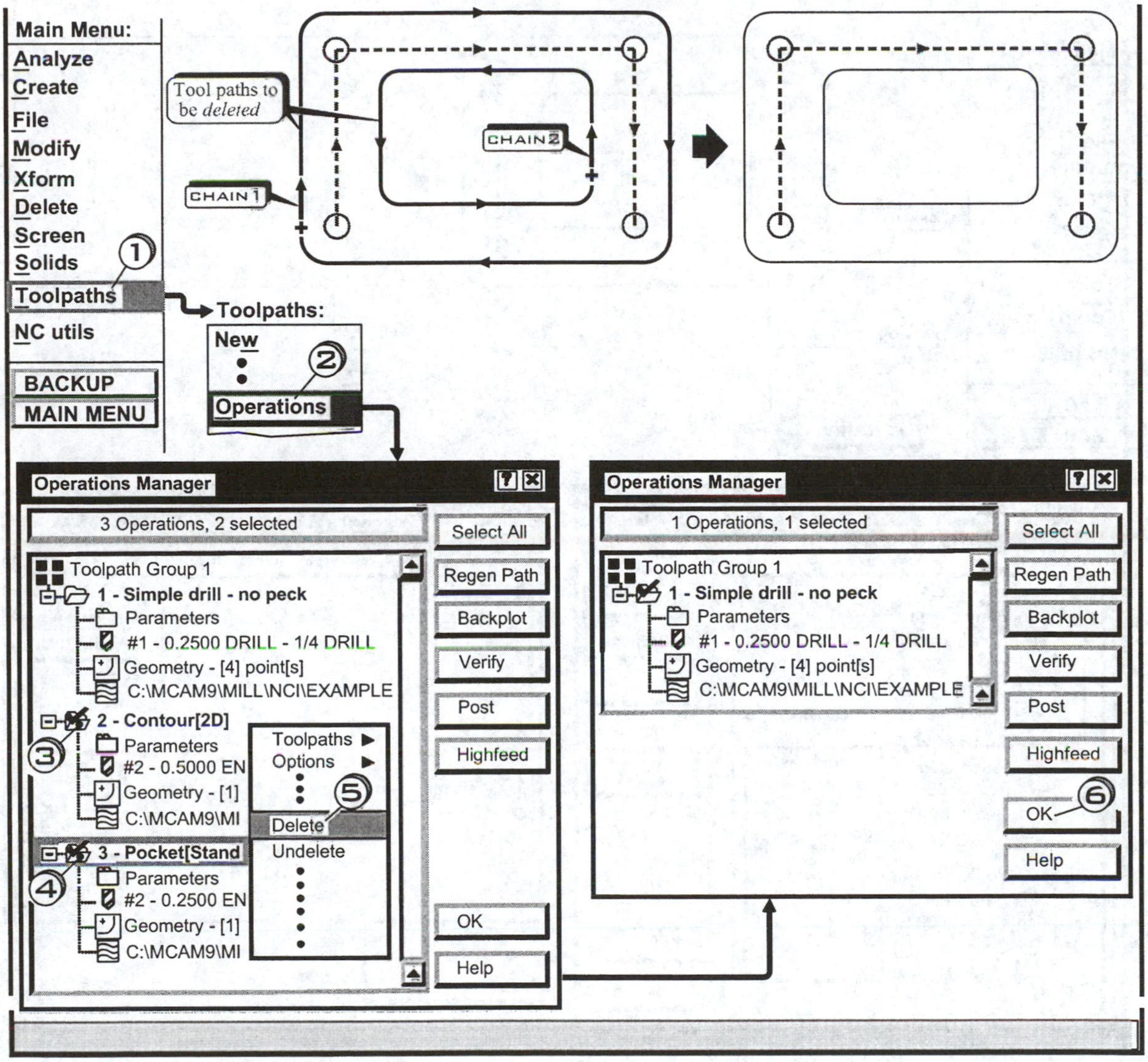

➤ Click ① **Toolpaths**

➤ Click ② **Operations**

➤ Click ③ on the file folder of the operation to delete such that a check appears

➤ Press the Ctrl key and keeping it depressed Click ④ the next operation in the set, etc

➤ ***Right*** Click and move the cursor down to **Delete**; Click ⑤

➤ Click ⑥ the OK button

DELETING ALL OPERATIONS

EXAMPLE 7-5

Delete all the existing tool paths shown below

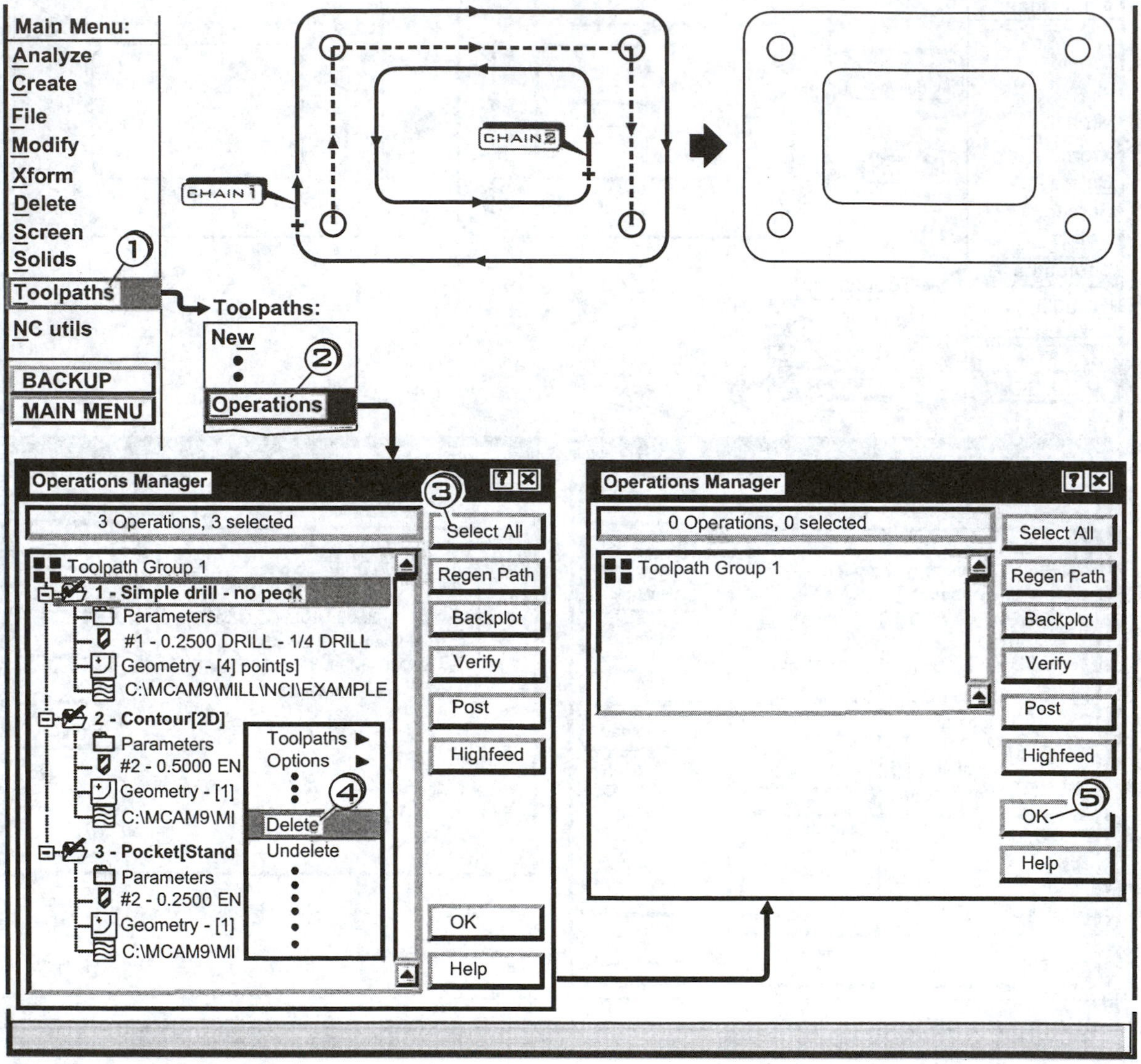

- Click ① **Toolpaths**
- Click ② **Operations**
- Click ③ Select all
- ***Right*** Click and move the cursor down to **Delete**; Click ④
- Click ⑤ the OK button

7-8 Editing Existing Toolpaths

Mastercam provides the operator with a full array of features for changing the parameters of existing tool paths. This allows the operator to quickly respond to design changes that often occur as a part evolves.

All the tool path editing functions are available from the ***Chain Manager*** dialog box. To enter the Chain Manager follow the steps listed below.

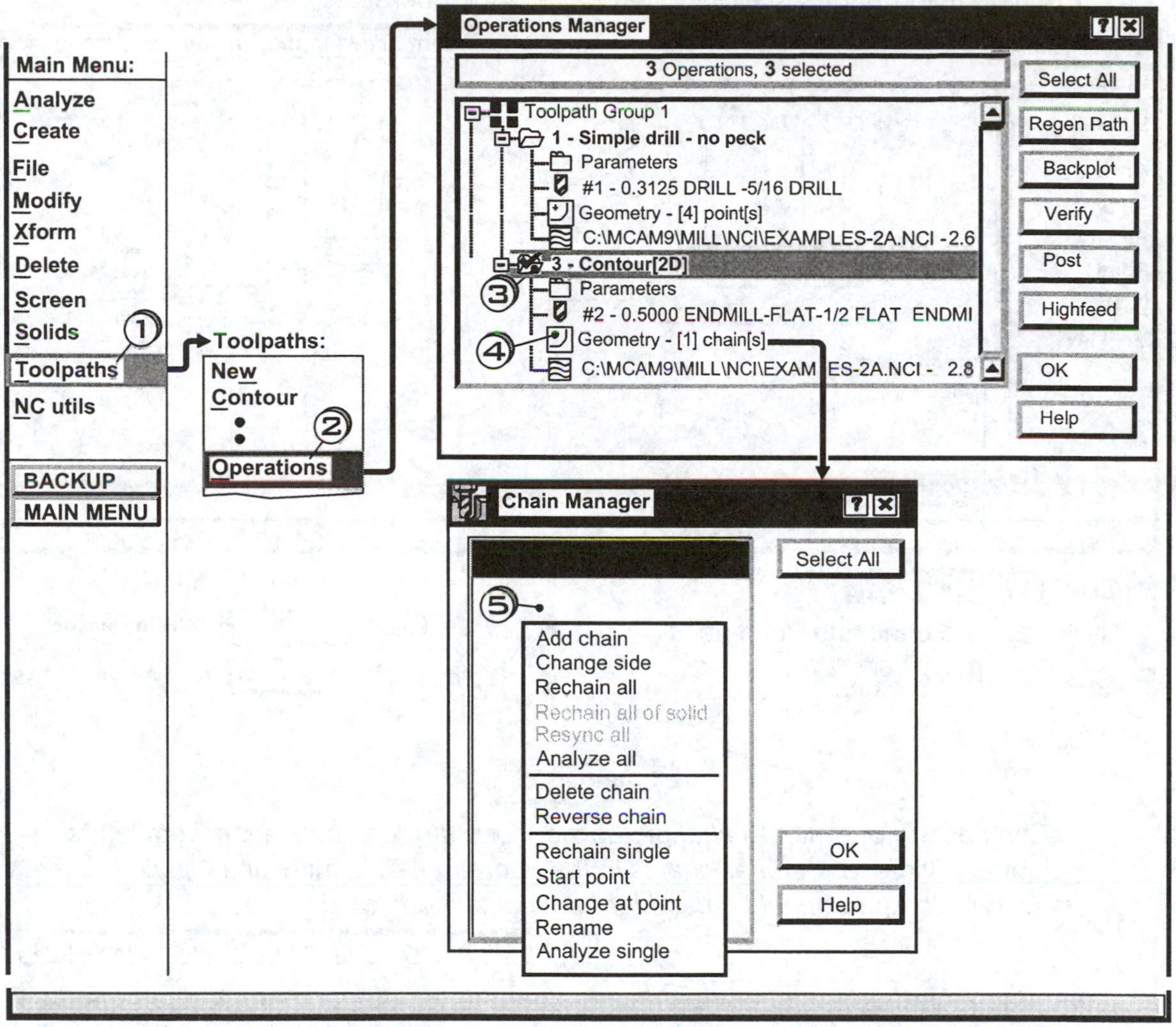

- Click ① **Toolpaths**
- Click ② **Operations**
- Click ③ on the file folder of the operation to be edited such that a check appears
- Click ④ on the Part Geometry icon for the operation to be edited
- ***Right*** Click ⑤ to display the Chain Manager's editing functions

The applications of the editing functions in the pull down menu will now be considered.

Add chain

Enables the operator to *add more chains to the existing operation.* Additional data *need not* be entered into the *tools* and *parameters* dialog boxes. The Regen Path function, however *must be executed* in the *Operations Manager to update* the changes to all the other parts of the operation.

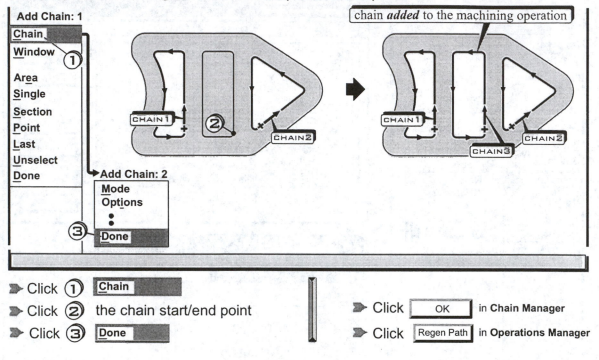

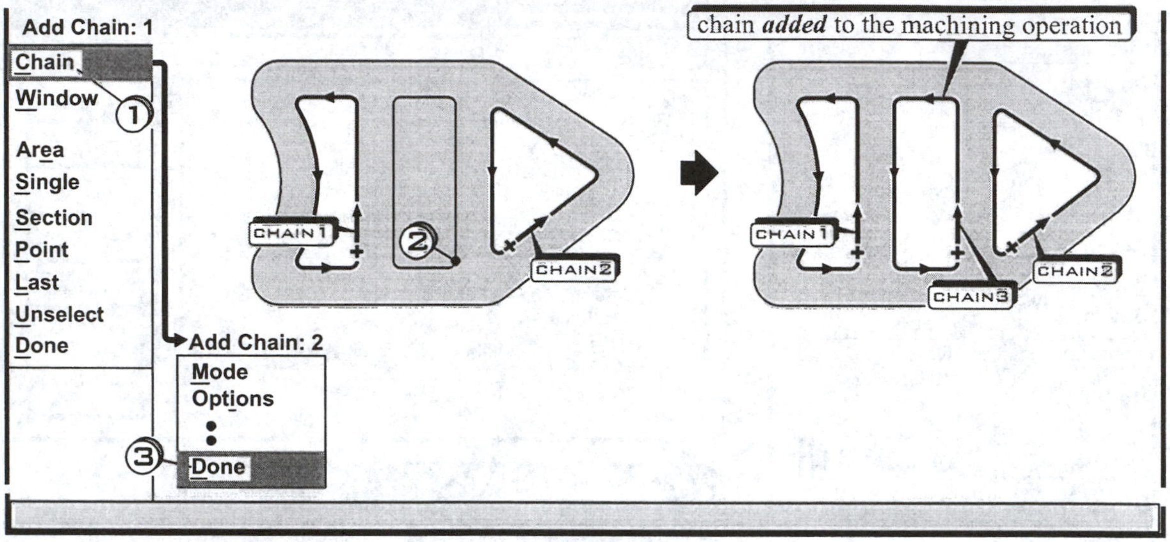

- Click ① Chain
- Click ② the chain start/end point
- Click ③ Done
- Click OK in **Chain Manager**
- Click Regen Path in **Operations Manager**

Change side

Enables the operator to *change the side for cutter compensation* from left to right and vica-versa *for all chains in the operation*. This function is *limited to the 2D module* and *cannot be used for the pocket module.*

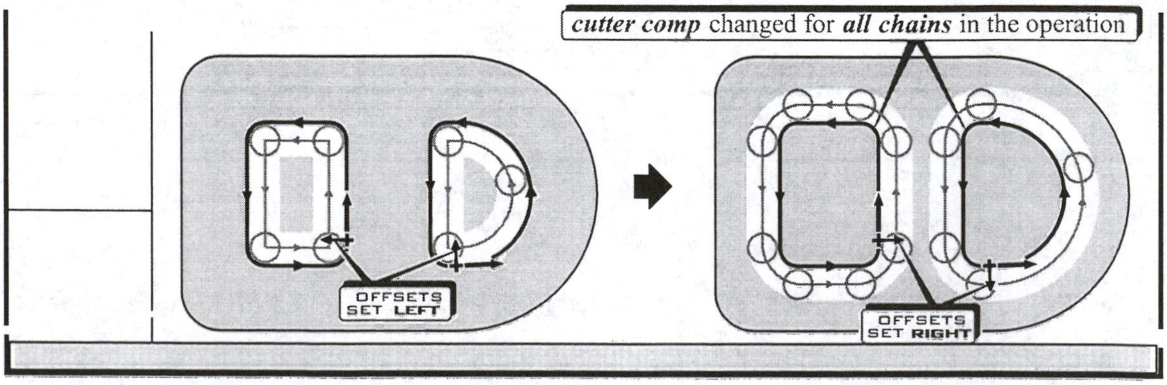

- Click OK in **Chain Manager**
- Click Regen Path in **Operations Manager**

Rechain all

Deletes all the chains in listed in the Chain Manager for an operation and allows the operator to *redefine a new set of chains*.The Regen Path function, however *must be executed* in the *Operations Manager to update* the changes to all the other parts of the operation

Rechain Single

Deletes an existing chain selected in the Chain Manager for an operation and allows the operator to *redefine* it.The Regen Path function, however *must be executed* in the *Operations Manager to update* the changes to all the other parts of the operation

Start Point

Enables the operator to return to the graphics window and *reselect a new start point* for a *chain selected* in the Chain Manager for an operation.The Regen Path function, however *must be executed* in the *Operations Manager to update* the changes to all the other parts of the operation

Delete chain

Deletes a chain selected.The Regen Path function, however *must be executed* in the *Operations Manager to update* the changes to all the other parts of the operation

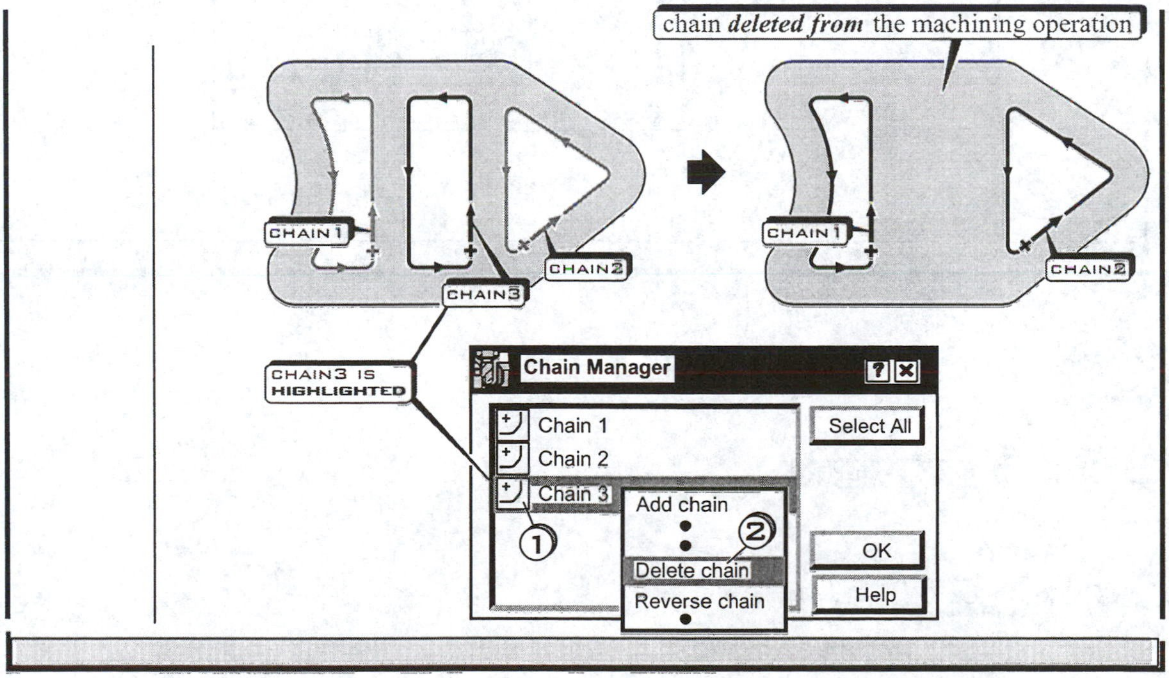

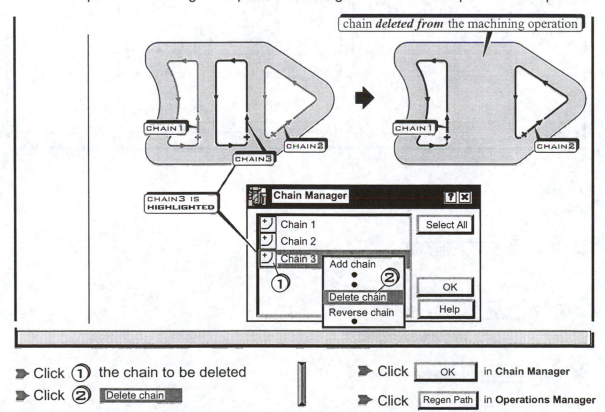

- Click ① the chain to be deleted
- Click ② Delete chain
- Click OK in **Chain Manager**
- Click Regen Path in **Operations Manager**

Reverse chain

Changes the *direction of chaining* for a *chain selected* in the Chain Manager for an operation. Reversing the chaining direction *changes the compensation* from left to right and vica-versa.The [Regen Path] function, however *must be executed* in the *Operations Manager to update* the changes to all the other parts of the operation

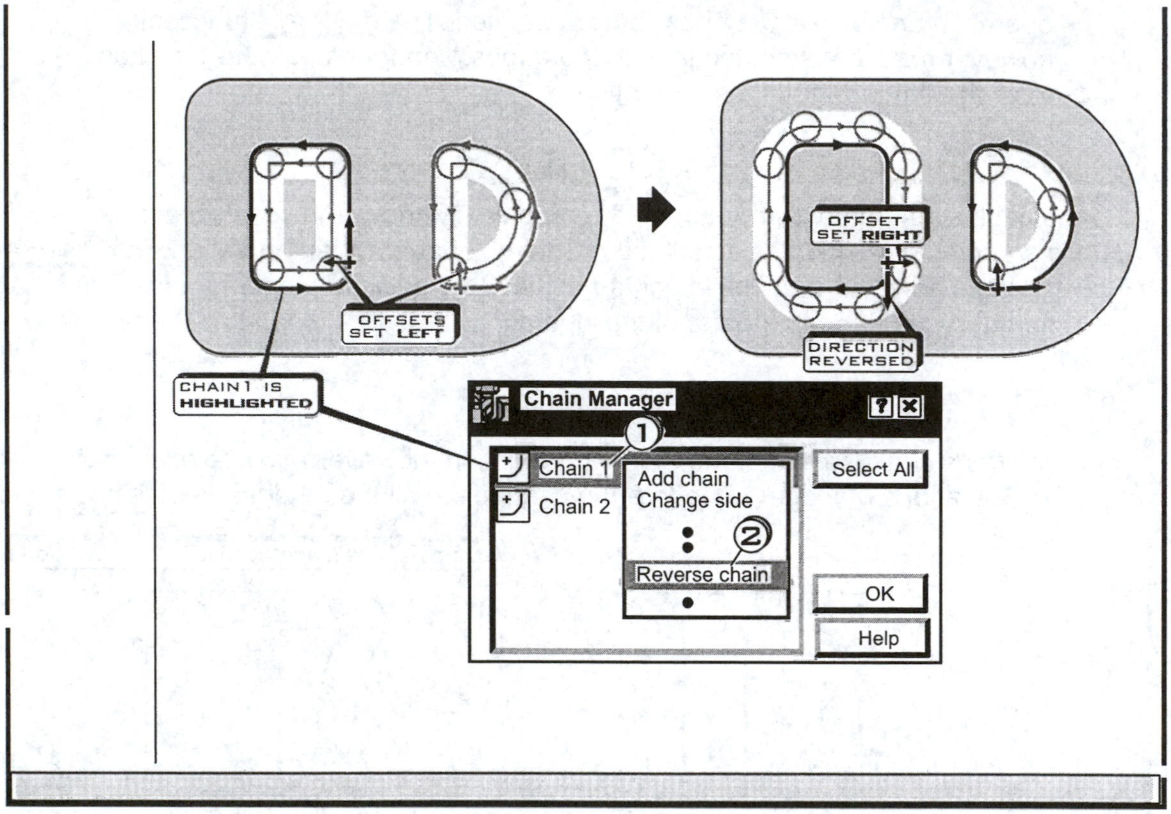

- Click ① the chain to be reversed
- Click ② [Reverse chain]
- Click [OK] in **Chain Manager**
- Click [Regen Path] in **Operations Manager**

Change at point

Enables the operator to *change machining parameters* at a *selected point* on the tool path. This function can *only be applied* to *contour* operations. The Regen Path function, however *must be executed* in the *Operations Manager to update* the changes to all the other parts of the operation

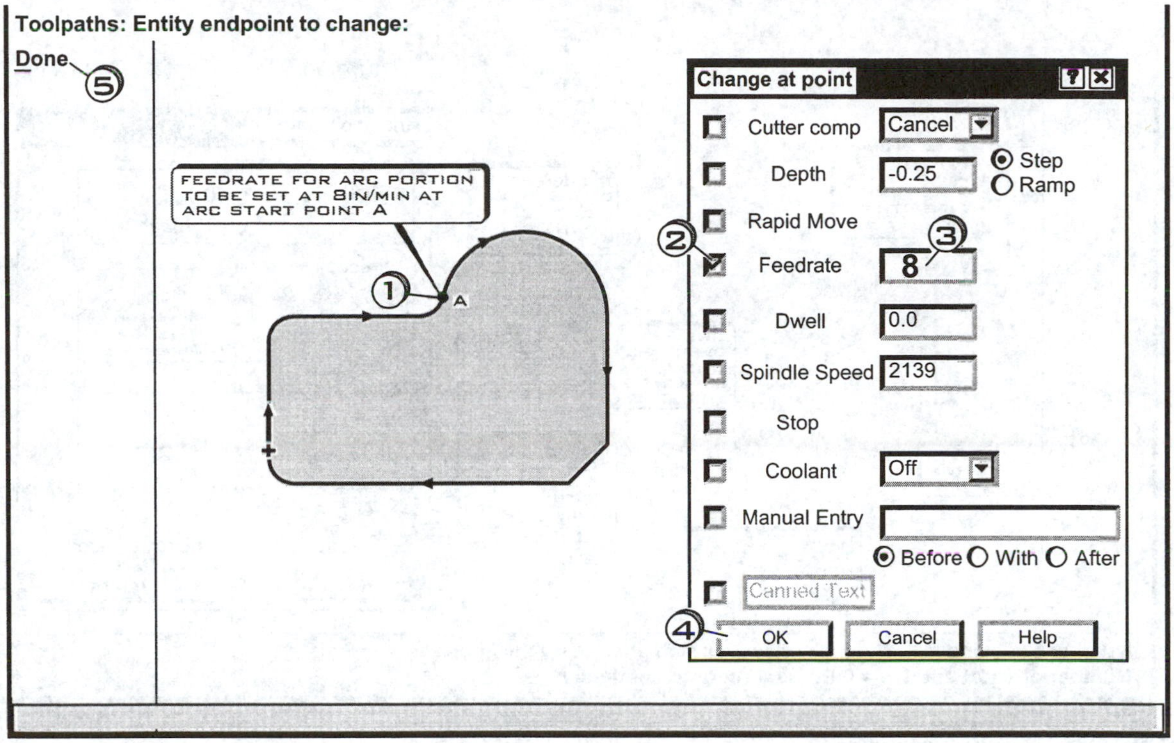

- Click Change at point
- Click ① the point for changing parameters
- Click ② the Feedrate check ☑ ; enter **8**
- Click ④ OK
- Click ⑤ **Done**
- Click OK in **Chain Manager**
- Click Regen Path in **Operations Manager**

Analyze all **Analyze Single**

Enables the operator to direct *Mastercam* to examine *all chains* or a *selected chain* in an operation for *two* types of *problems*: *errors* due to entities that *overlap* and *errors* caused by *reversal* in *chaining direction*. The *maximum length* that *defines a short entity* is also set in the Analyze Chain dialog box The Regen Path function, however *must be executed* in the *Operations Manager to update* the changes to all the other parts of the operation

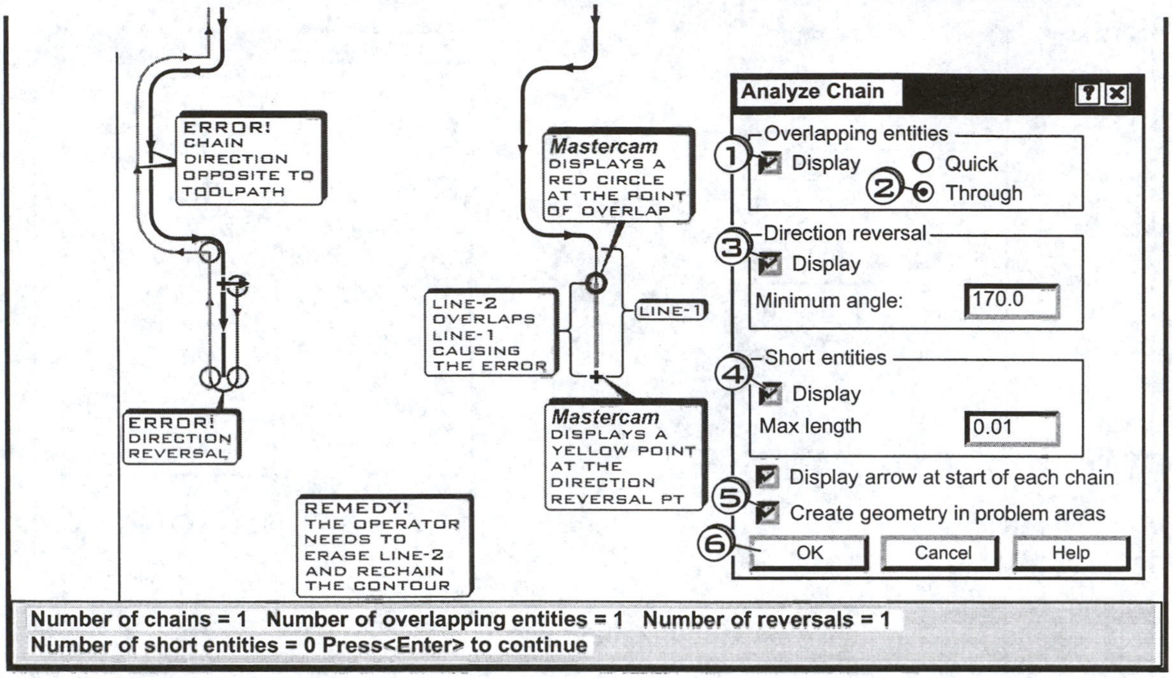

- Click Analyze all
- Click ① the Display check ; Click ② the Through radio button to search for *all* overlapping entities not just adjacent
- Click ③ the Display check to enable Direction reversal checking
- Click ④ the Display check to enable Short entities checking
- Click ⑤ the check to direct *Mastercam* to display: a *red circle* at an *overlap* , a *yellow point* at a direction *reversal* and a *blue circle* on *short entities*
- Press the Enter key
- Click ⑥ OK
- Click OK in **Chain Manager**

Use the **Delete** function and select **Duplicate** and **Entities** to erase line-2, then re-chain.

Rename

Allows the operator to rename a chain.

7-9 Changing the Chaining Order

The order in which the chains are listed in the Chain Manager determines the machining order within an operation. The operator can edit the machining order by changing the order in which the chains are listed.

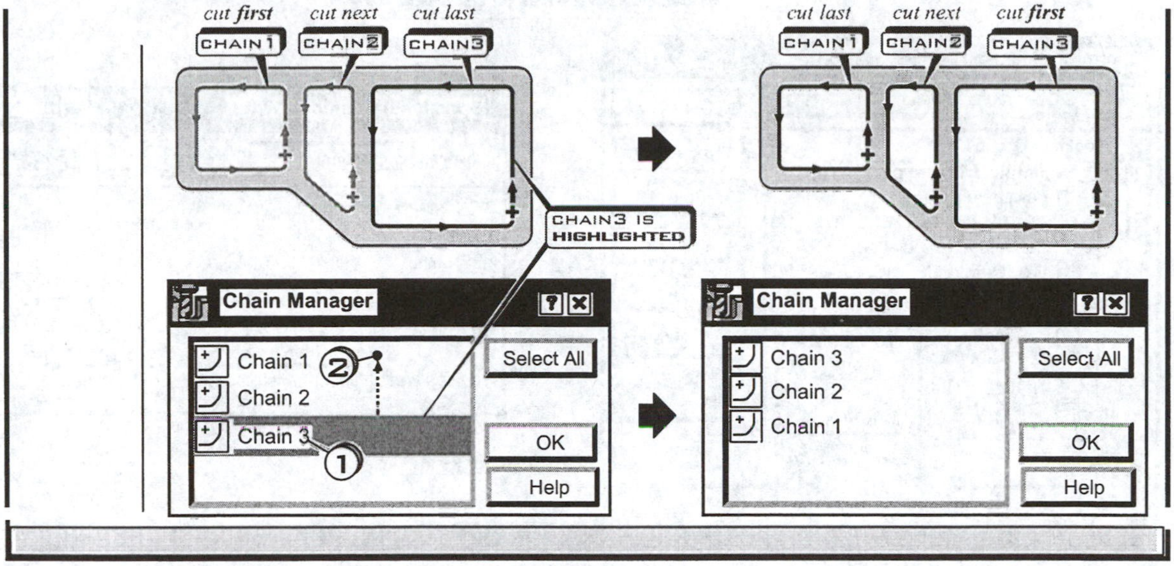

➤ Click ① the chain to be moved; ***keeping*** the ***left*** mouse button ***depressed*** move the chain to its ***new location*** ② and ***release***

➤ Click OK in **Chain Manager**

➤ Click Regen Path in **Operations Manager**

7-10 Expanding and Collapsing the Operations Display Listings

The operations listing for a group can be fully displayed or collapsed as needed.

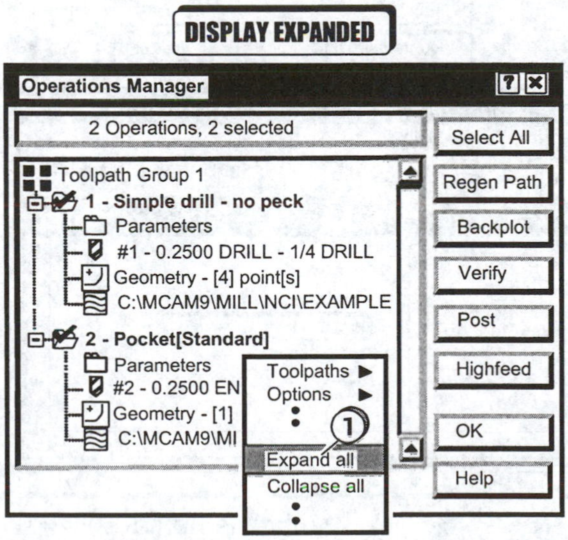

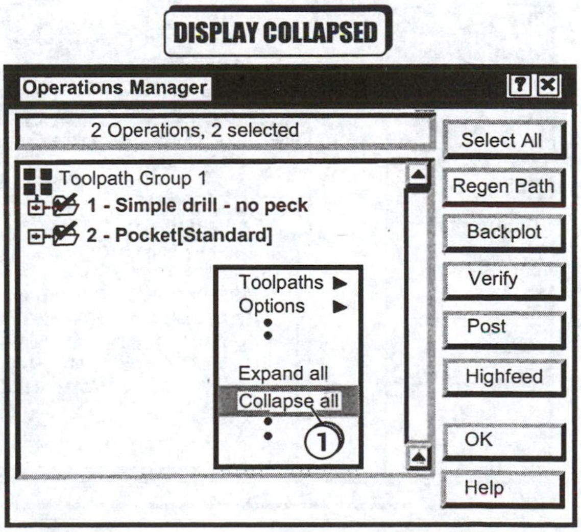

➤ ***Right*** Click and move the cursor down to Expand all ; Click ①

➤ ***Right*** Click and move the cursor down to Collapse all ; Click ①

7-11 Saving and Printing the Operations Manager Listings

The Doc file function is used to direct Mastercam to produce a printable document ASCII file from the information in the Operations Manager list area.

CREATING AND SAVING THE DOC FILE

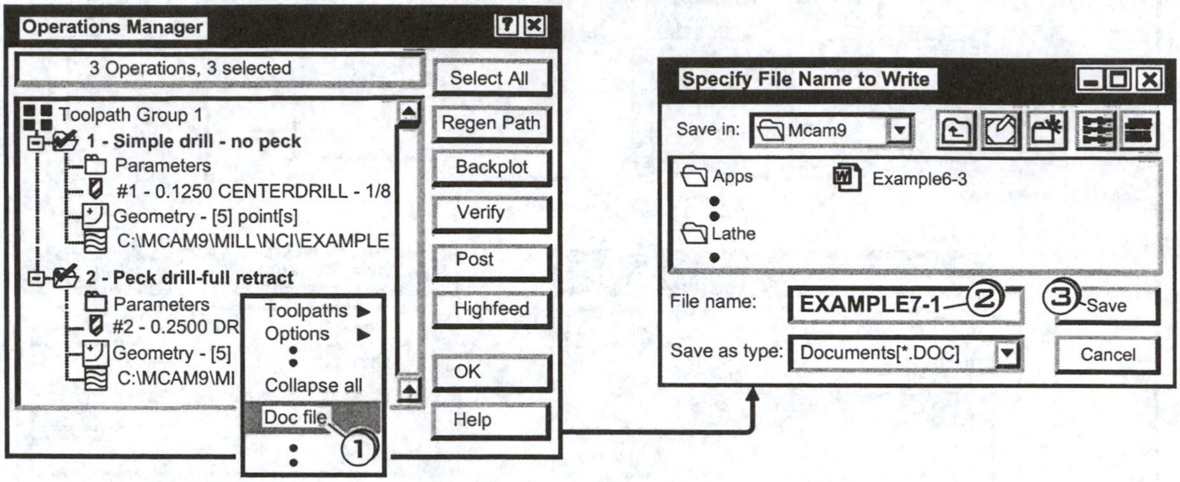

- ***Right*** Click and move the cursor down to Doc file ; Click ①
- Click ② enter the name of the DOC file: **EXAMPLE7-1**
- Click ③ Save

VIEWING AND PRINTING THE DOC FILE

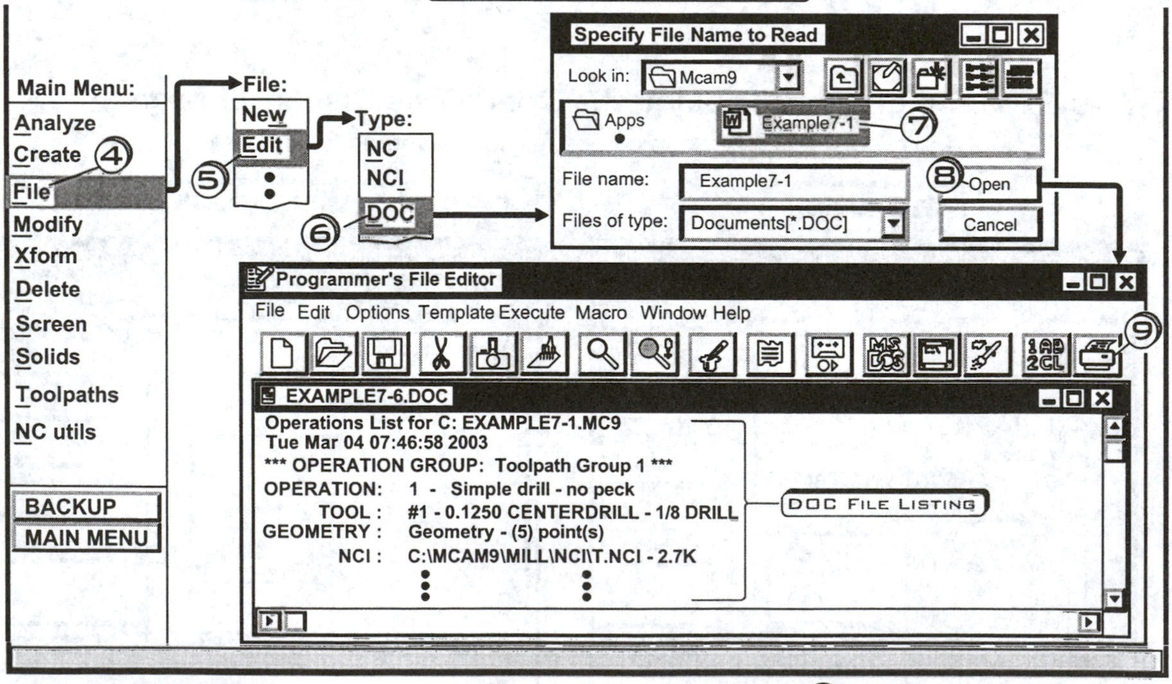

- Click ④ File
- Click ⑤ Edit
- Click ⑥ DOC
- Click ⑦ the file name Example7-1
- Click ⑧ Open
- Click ⑨ the print icon

7-12 Entering Job Setup from the Operations Manager

The Job Setup dialog box can also be entered from the Operations Manager allowing the operator to setup stock, stock material, etc.

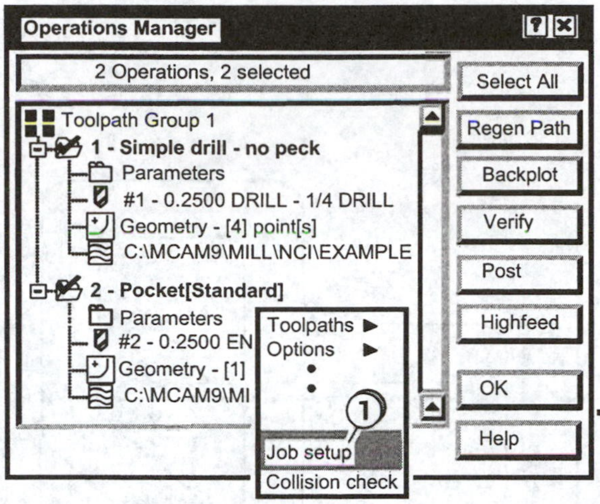

➤ ***Right*** Click and move the cursor down to Job setup ; Click ①

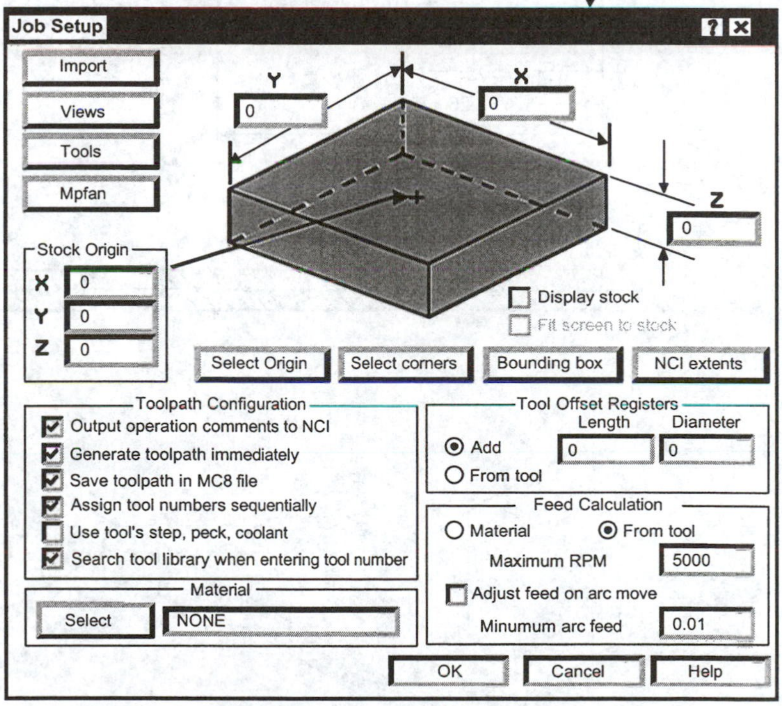

USING THE TOOLBAR MENUS

Toolbar Menu Containing ***Job Setup and Operations*** icons

EXERCISES

7-1) The operations for producing the part shown in Figure 7p-1(left). currently exist in the Operations Manager. Edit the existing operations to produce the re-designed part as shown in Figure 7p-1(right)

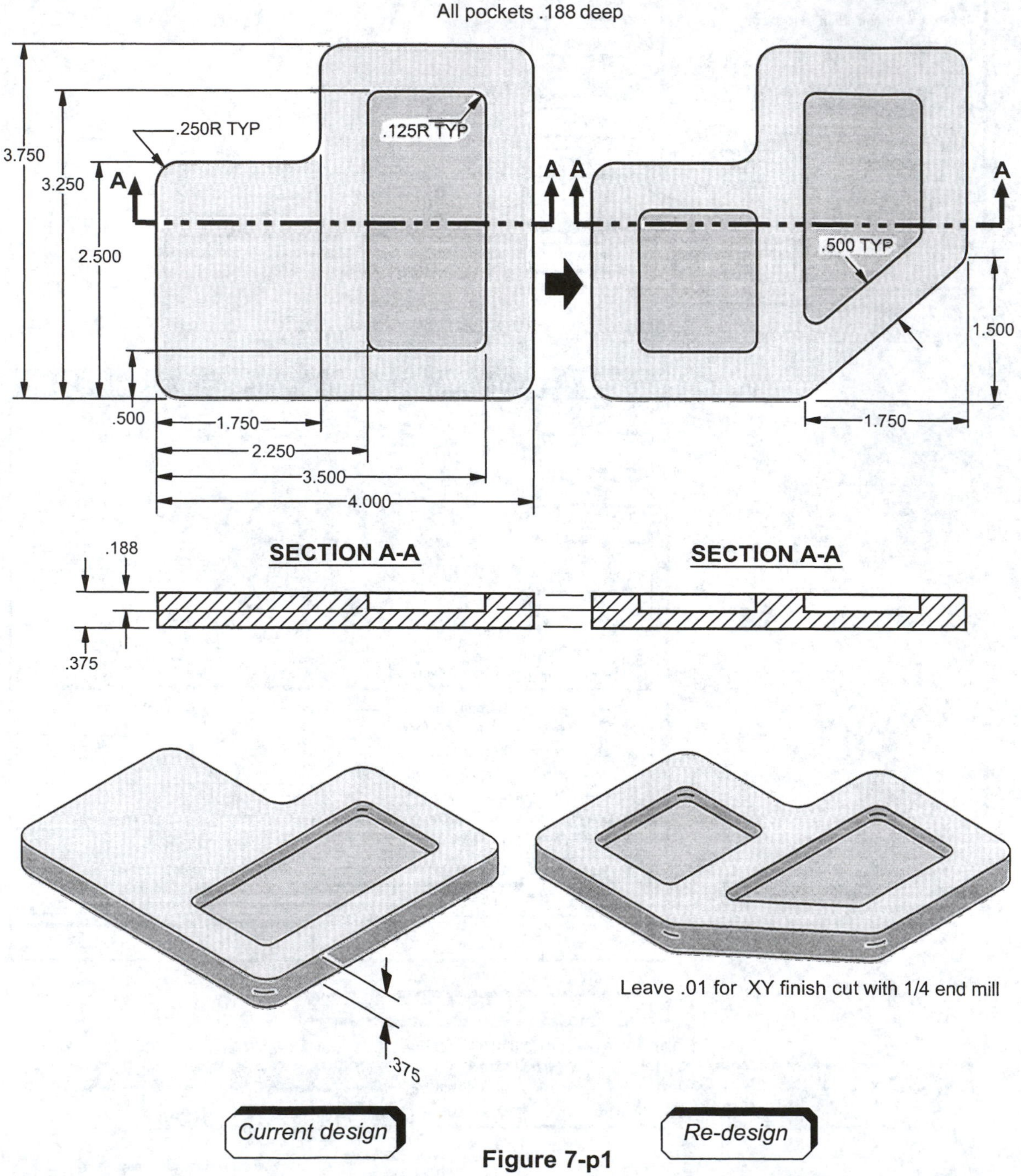

Figure 7-p1

A) Copy the file EX7-1 in the folder CHAPTER7 on CD at the back of this text to your directory C:\MCAM9\MILL\MC9\JVAL-MILL

(your initials)

Open the file from your directory.

Refer to Chapter 1, p1-14 for a discussion on the creation of your own directory.

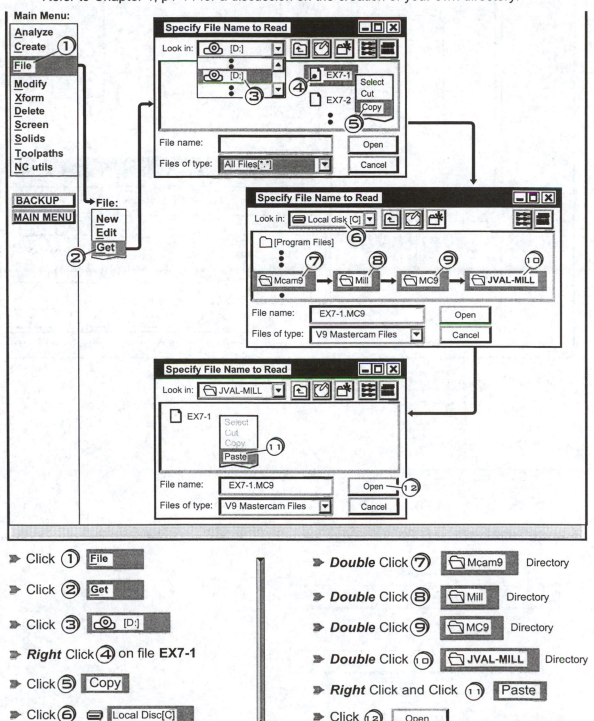

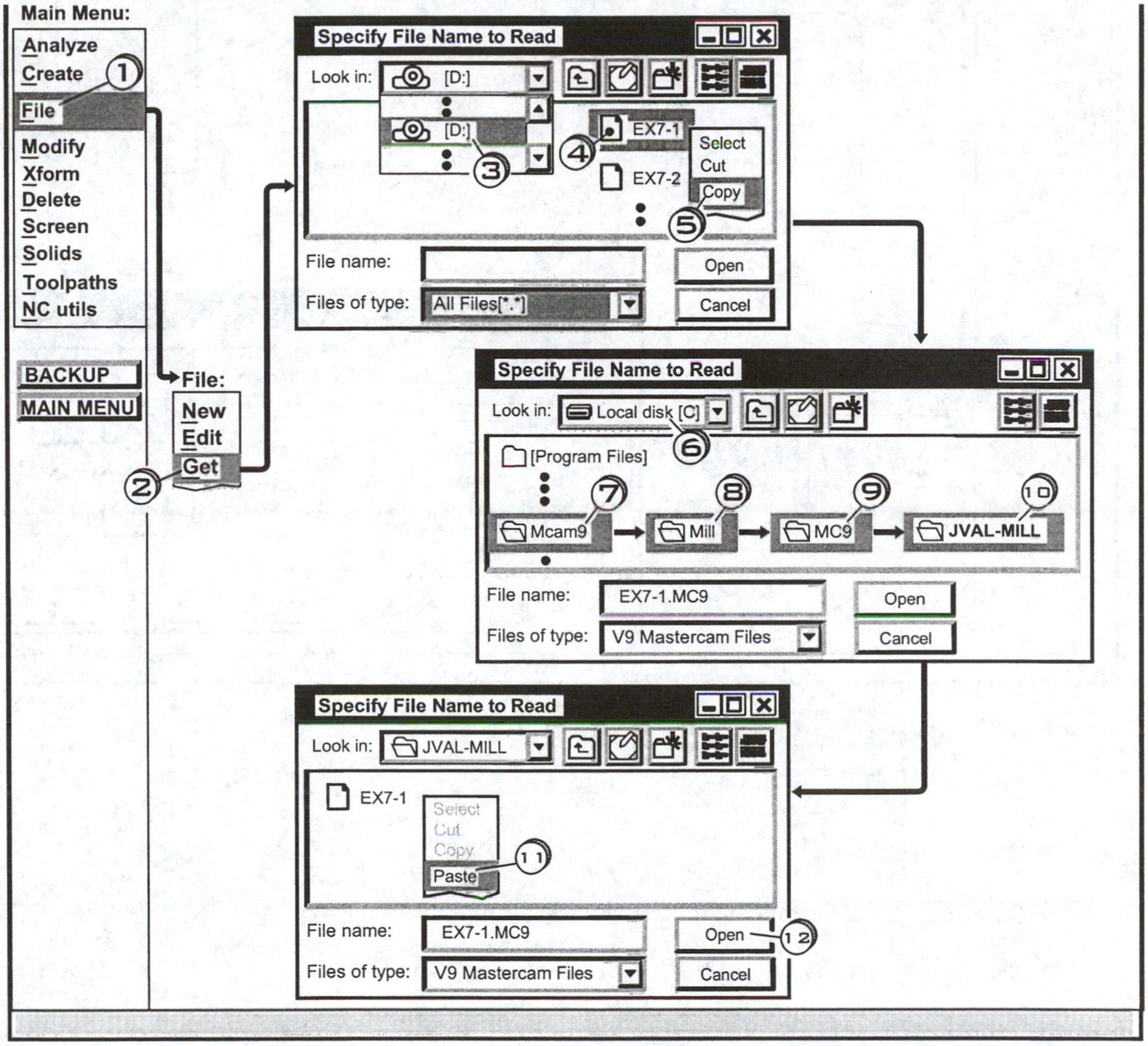

- Click (1) File
- Click (2) Get
- Click (3) [D:]
- ***Right*** Click (4) on file **EX7-1**
- Click (5) Copy
- Click (6) Local Disc[C]
- ***Double*** Click (7) Mcam9 Directory
- ***Double*** Click (8) Mill Directory
- ***Double*** Click (9) MC9 Directory
- ***Double*** Click (10) JVAL-MILL Directory
- ***Right*** Click and Click (11) Paste
- Click (12) Open

B) Edit the Part Geometry

◆ Create the 1.75 x 1.5 Chamfer

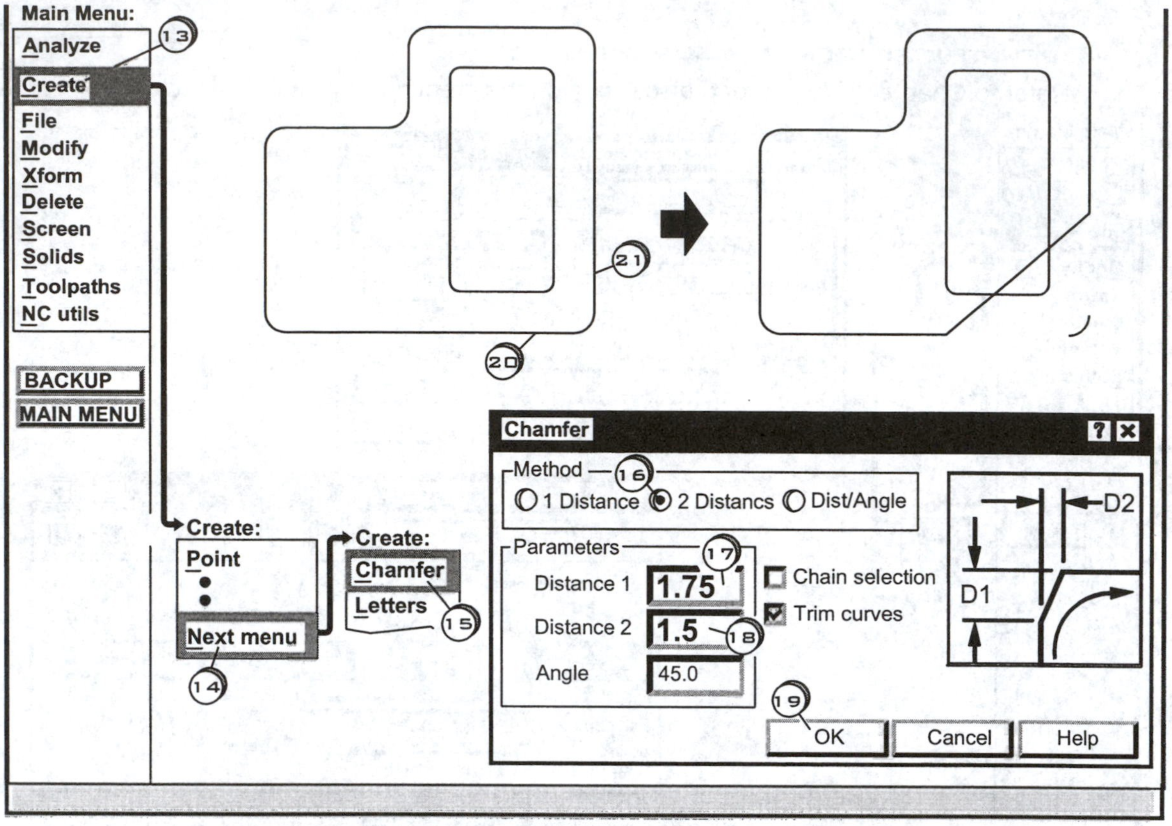

- Click (13) **Create**
- Click (14) **Next menu**
- Click (15) **Chamfer**
- Click (16) the 2 Distances radio button ◉
- Click (17) ; enter Distance 1 **1.75**
- Click (18) ; enter Distance 2 **1.5**
- Click (19) OK

Chamfer: select first line or arc

- Click (20) near the end of the *first* line

Chamfer: select second line

- Click (21) near the end of the *second* line
- Press Esc to cancel the operation

◆ CREATE .5 OFFSET

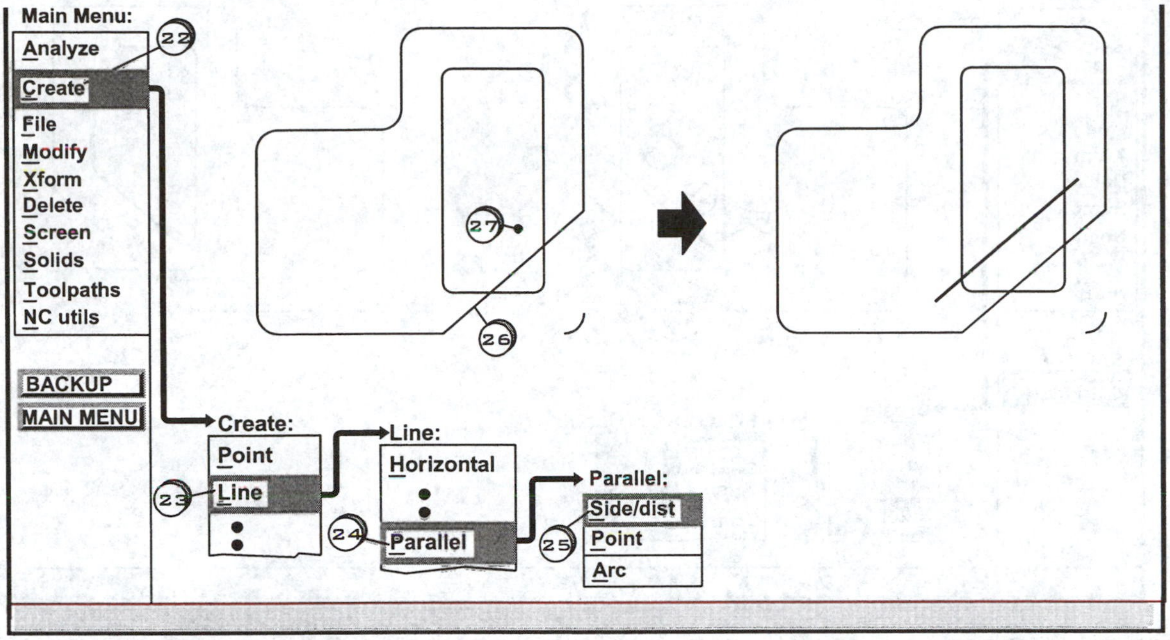

➤ Click (22) **Create**

➤ Click (23) **Line**

➤ Click (24) **Parallel**

➤ Click (25) **Side/dist**

Select line

➤ Click (26) the line entity

Indicate offset direction

➤ Click (27) the side for the parallel line

Parallel line Distance=

➤ Enter the offset distance **.5** ; **Enter**

➤ Press **Esc** to cancel the operation

◆ CREATE .125R AND .25R FILLETS

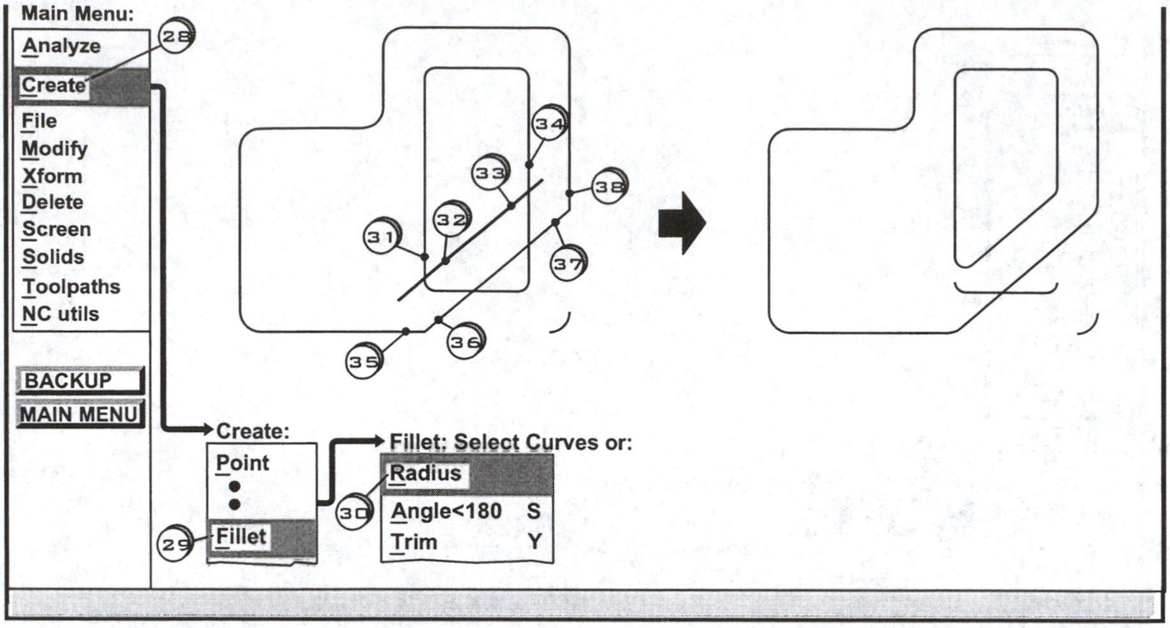

- Click 28 **Create**
- Click 29 **Fillet**
- Click 30 **Radius**

  Enter the fillet radius:
- Enter the fillet's radius **.125** ; **Enter**

  Fillet: select an entity
- Click 31

  Fillet: select another entity
- Click 32

  Fillet: select an entity
- Click 33

  Fillet: select another entity
- Click 34
- Click 30 **Radius**

  Enter the fillet radius:
- Enter the fillet's radius **.25** ; **Enter**

  Fillet: select an entity
- Click 35

  Fillet: select an entity
- Click 36

  Fillet: select an entity
- Click 37

  Fillet: select an entity
- Click 38
- Press **Esc** to cancel the operation

◆ CREATE THE 1.25 X 1.5 RECTANGLE

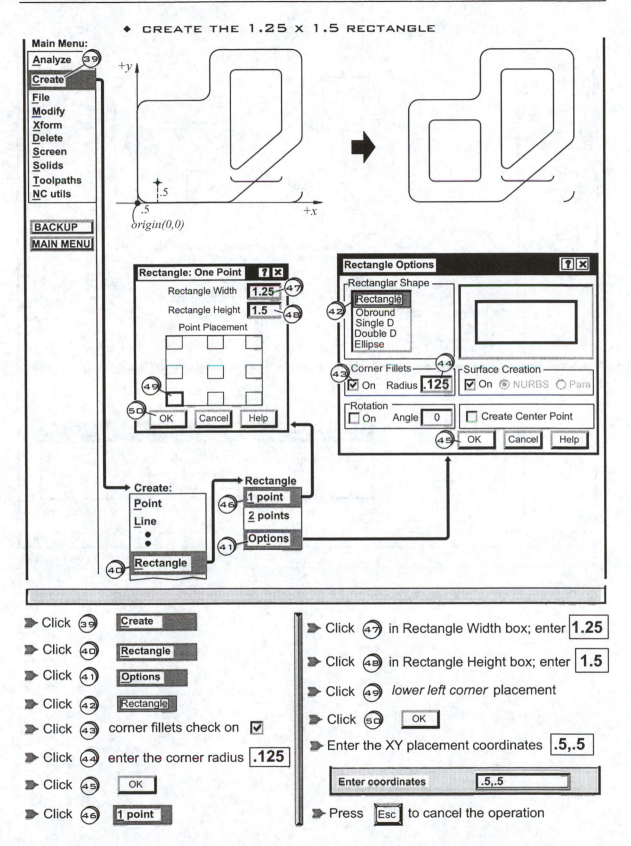

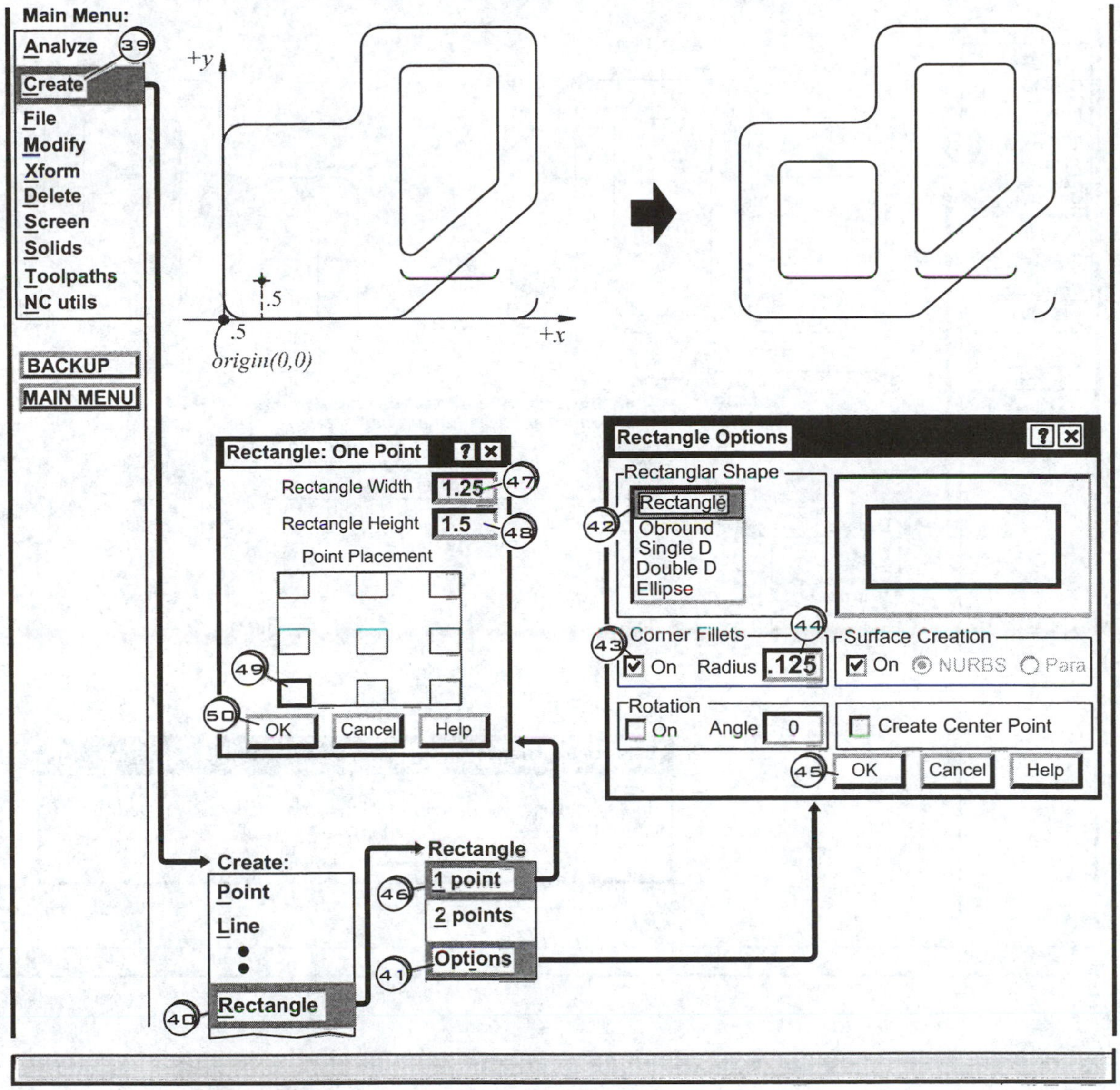

- Click (39) **Create**
- Click (40) **Rectangle**
- Click (41) **Options**
- Click (42) Rectangle
- Click (43) corner fillets check on ☑
- Click (44) enter the corner radius **.125**
- Click (45) OK
- Click (46) **1 point**
- Click (47) in Rectangle Width box; enter **1.25**
- Click (48) in Rectangle Height box; enter **1.5**
- Click (49) *lower left corner* placement
- Click (50) OK
- Enter the XY placement coordinates **.5,.5**

Enter coordinates	.5,.5

- Press Esc to cancel the operation

◆ DELETE ALL UNNECESSARY GEOMETRY FROM THE CAD MODEL

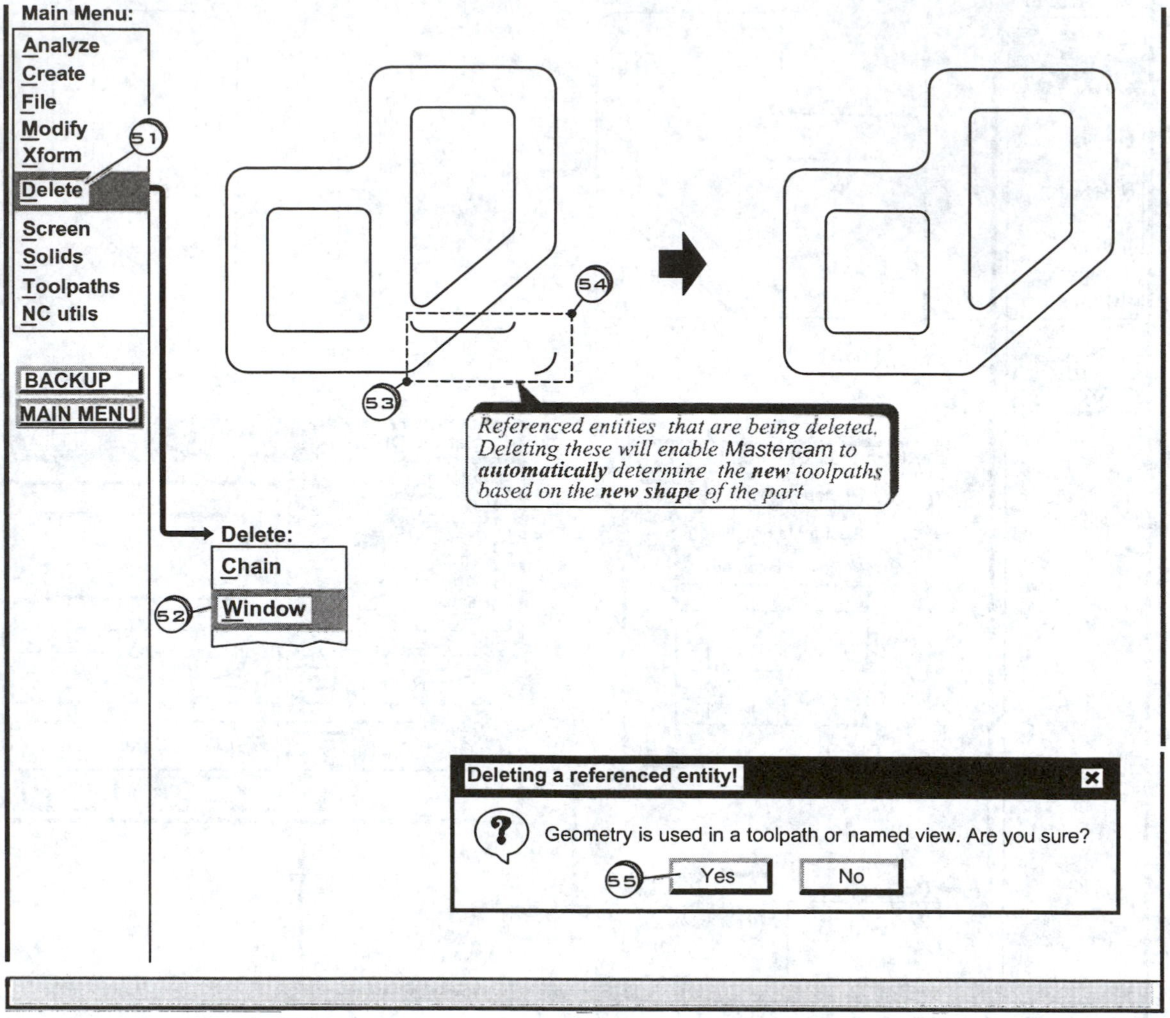
Main Menu:
Analyze
Create
File
Modify
Xform
Delete
Screen
Solids
Toolpaths
NC utils
BACKUP
MAIN MENU
51
52
53
54
55
Referenced entities that are being deleted. Deleting these will enable Mastercam to automatically determine the new toolpaths based on the new shape of the part
Delete:
Chain
Window
Deleting a referenced entity!
Geometry is used in a toolpath or named view. Are you sure?
Yes
No

- Click 51 **Delete**
- Click 52 **Window**
- Click 53 54 the corners of the window
- Click 55 **Yes**

C) Add the Additional 1.25 x 1.5 Pocket Chain

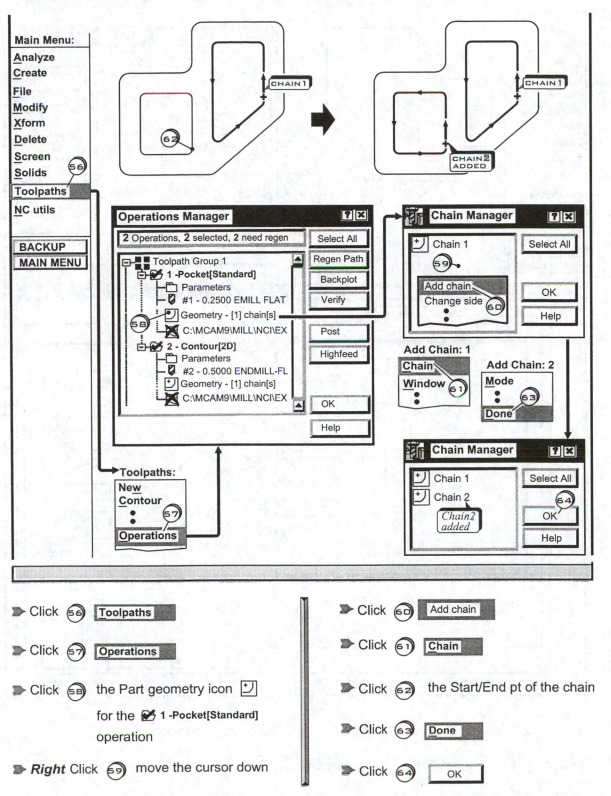

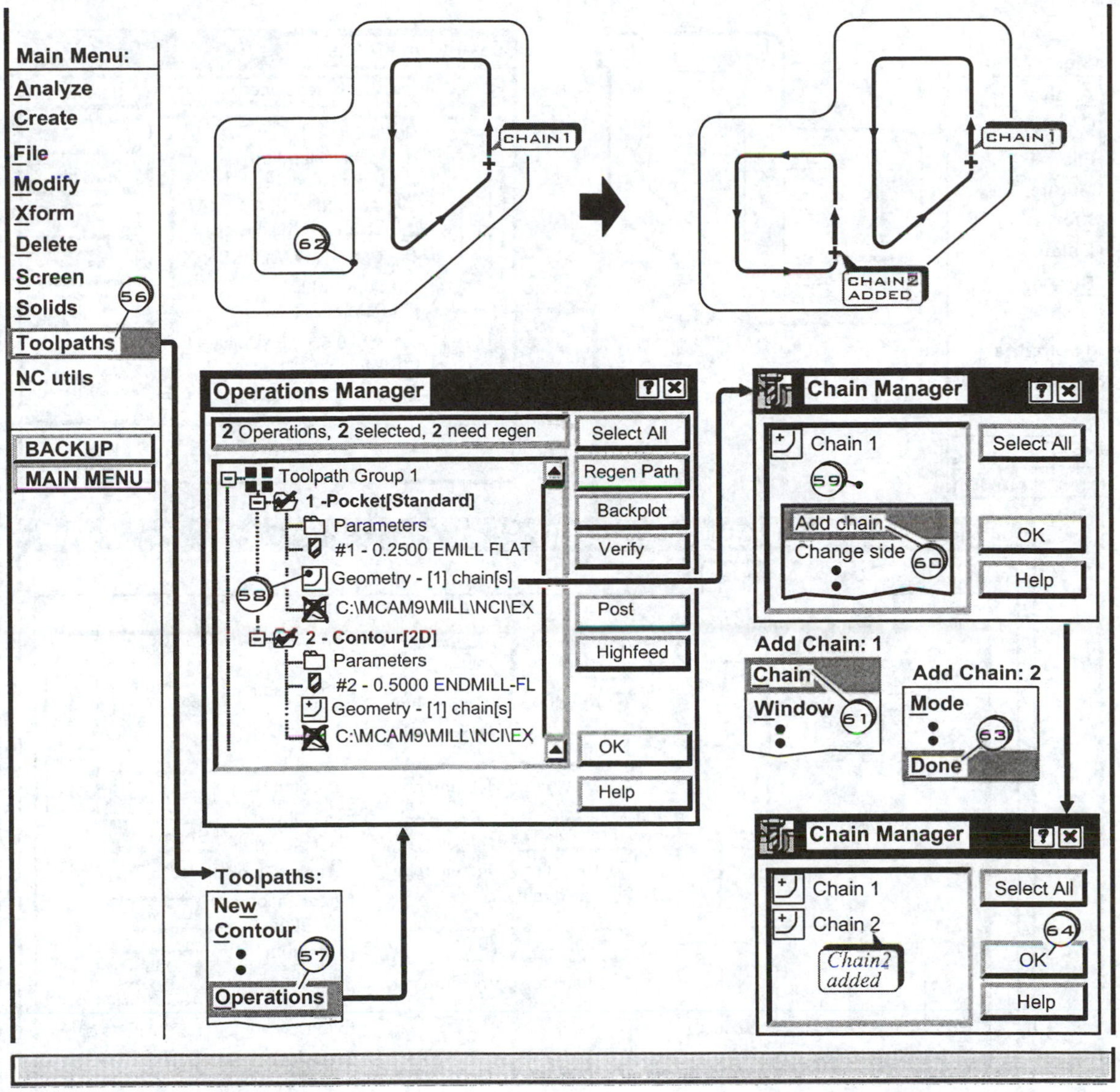

- Click (56) **Toolpaths**
- Click (57) **Operations**
- Click (58) the Part geometry icon for the **1 -Pocket[Standard]** operation
- ***Right*** Click (59) move the cursor down
- Click (60) Add chain
- Click (61) **Chain**
- Click (62) the Start/End pt of the chain
- Click (63) **Done**
- Click (64) OK

D) Edit Contour[2D], Leave .01 for XY Finish Cut

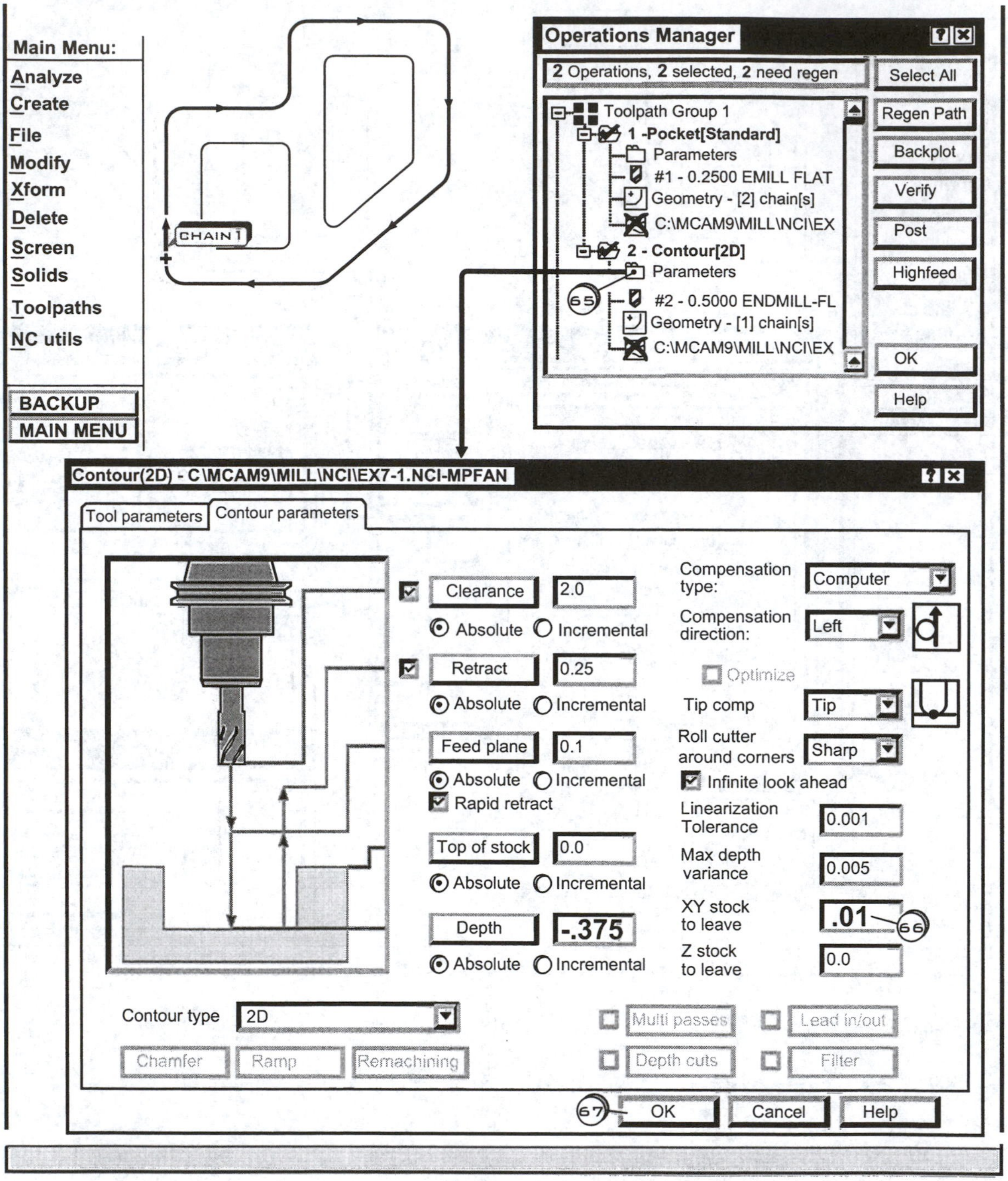

➤ Click (65) the Parameters icon 🗀 for the **2 -Contour[2D]** operation

➤ Click (66) in the XY Stock to leave box; enter **.01**

➤ Click (67) OK

E) FINISH CUT THE OUTSIDE WITH A 1/4 END MILL TOOL

- ENTER THE OPERATIONS MANAGER
- MAKE A COPY OF THE CONTOUR[2D] OPERATION
- EDIT THE PARAMETERS OF THE COPY[REPLACE 1/2ENDMILL TOOL WITH A 1/4 ENDMILL TOOL]

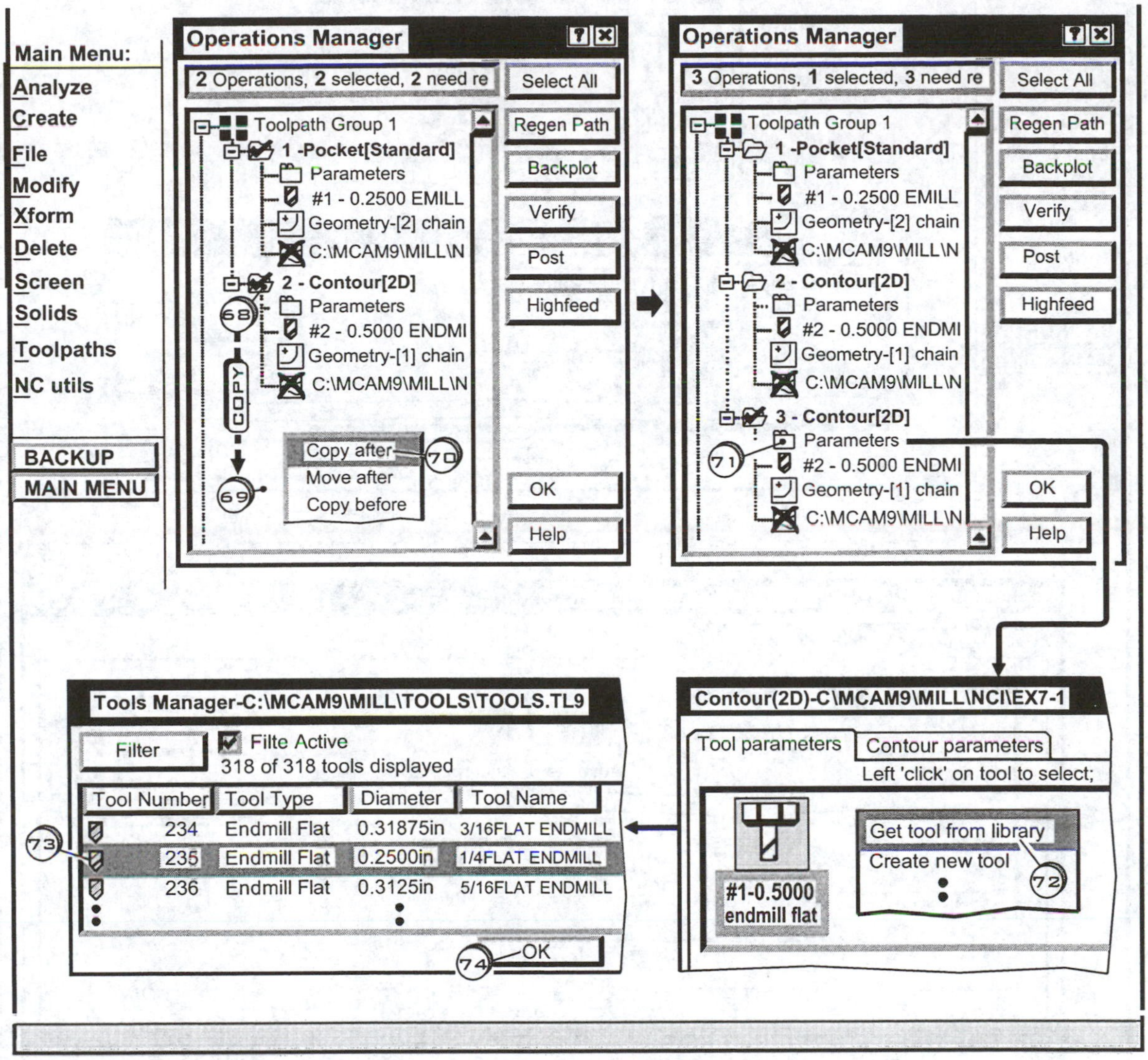

➤ ***Right*** Click (68) on the file folder **2 Contour[2D]** ***keeping the Right mouse button depressed*** *move the folder to the new location* (69) *and release.*

➤ Click (70) Copy after

➤ Click (71) the Parameters icon 🗀 for the **2 -Contour[2D]** operation

➤ Click (72) [Get tool from library]

➤ Click (73) the .25Dia End Mill tool

➤ Click (74) [OK]

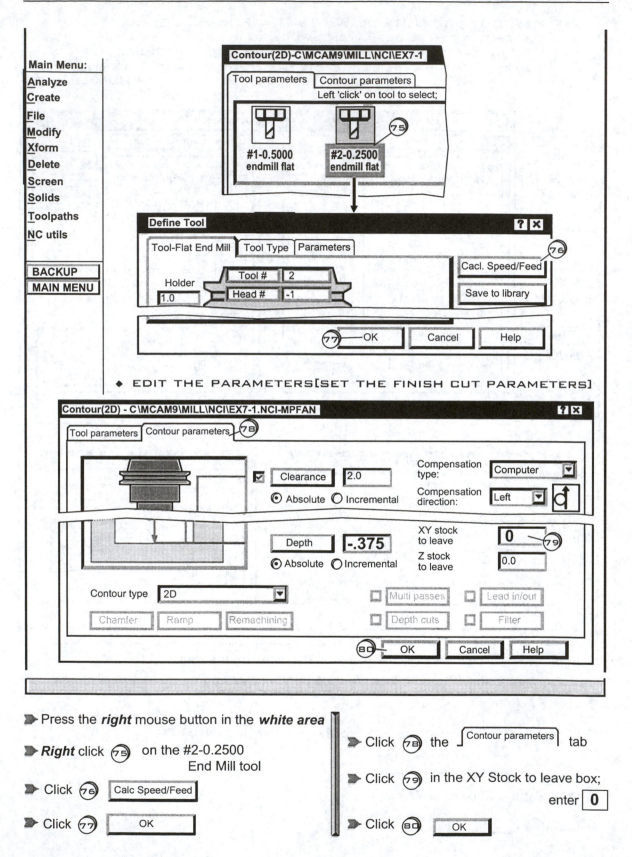

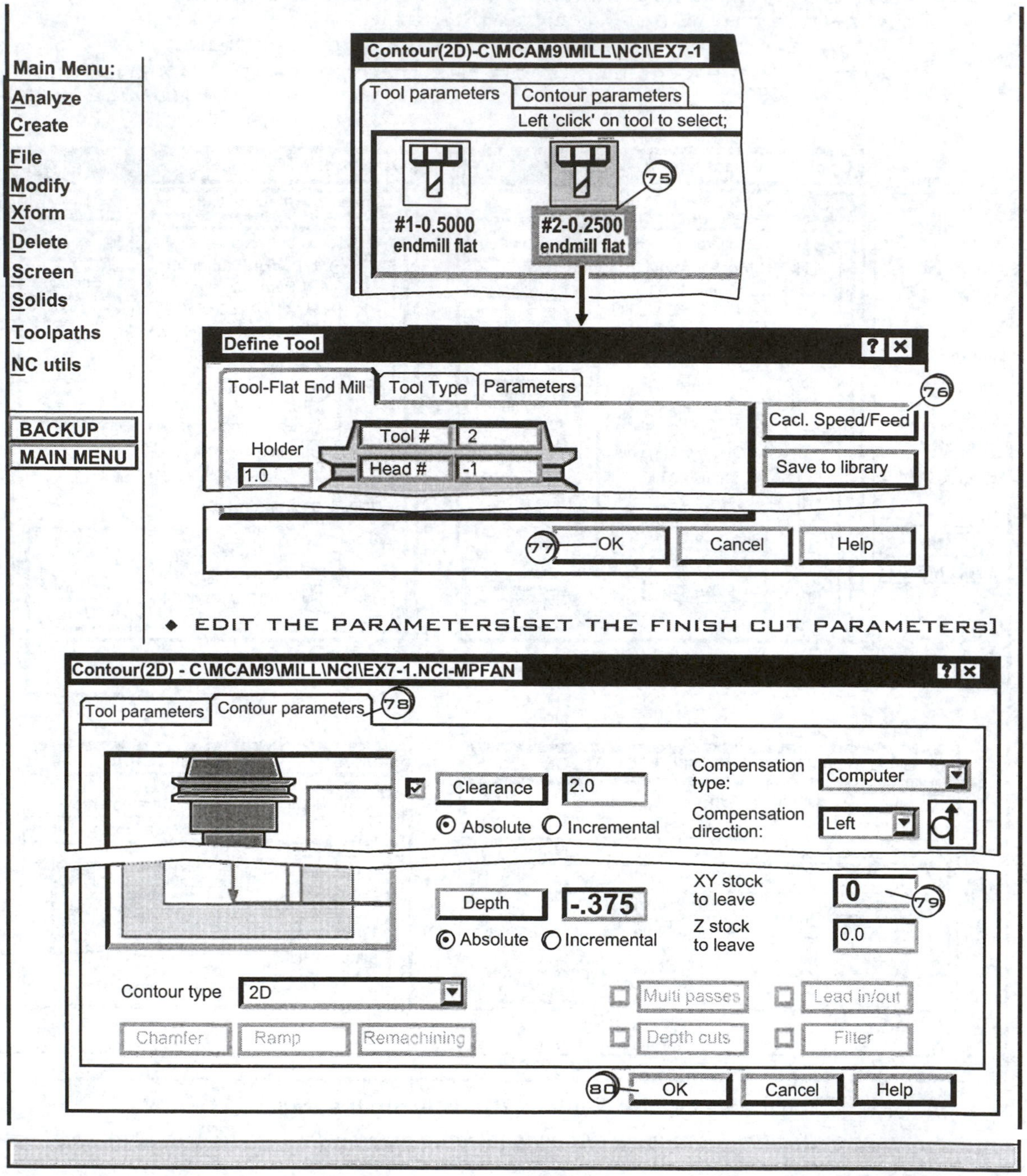

- Press the ***right*** mouse button in the ***white area***
- ***Right*** click (75) on the #2-0.2500 End Mill tool
- Click (76) [Calc Speed/Feed]
- Click (77) [OK]
- Click (78) the [Contour parameters] tab
- Click (79) in the XY Stock to leave box; enter [0]
- Click (80) [OK]

F) Execute the Regen Path Function so that *Mastercam* can automatically update All the Parts of Each Edited Operation Via Associativity

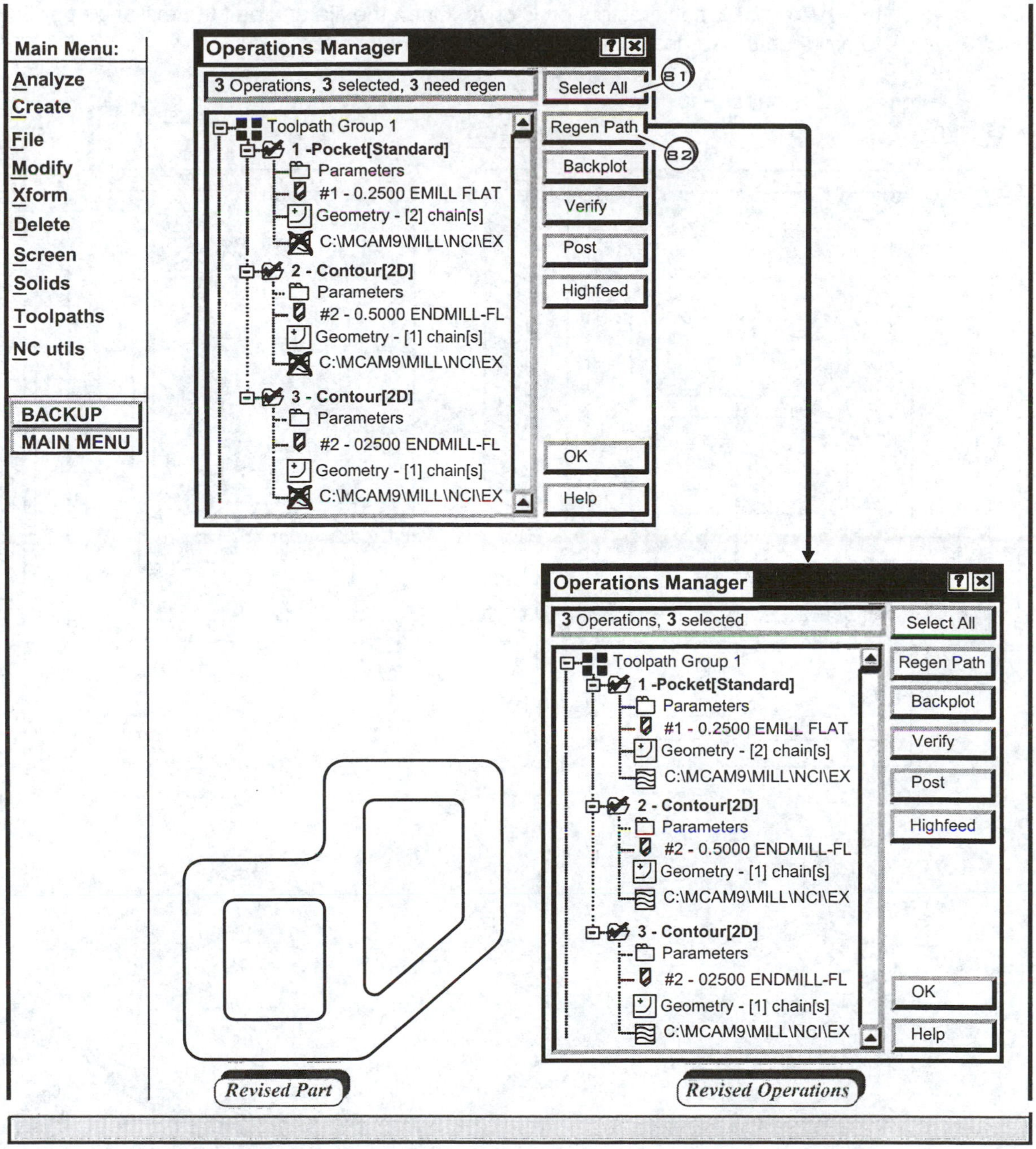

➤ Click B1 Select All

➤ Click B2 Regen Path

7-2) The operations for producing the part shown in Figure 7p-2(left). currently exist in the Operations Manager. Copy the file **EX7-2** in the folder **CHAPTER7** from the CD into the **JVAL-MILL** subdirectory on C drive. Open the file and edit the existing operations to produce the re-designed part as shown in Figure 7p-2(right)

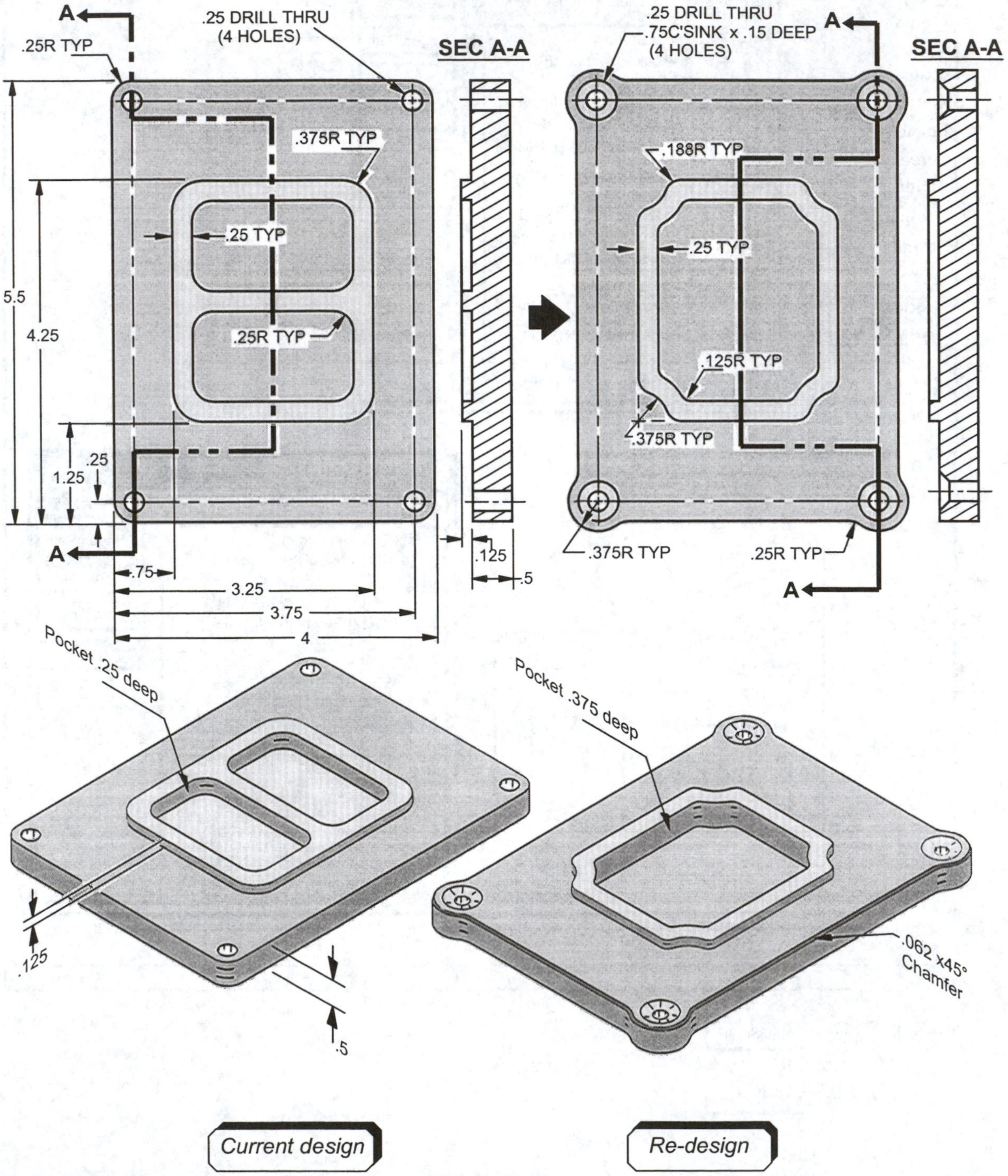

Figure 7-p2

7-3) The operations for producing the part shown in Figure 7p-3(left). currently exist in the Operations Manager. Copy the file **EX7-3** in the folder **CHAPTER7** from the CD into the **JVAL-MILL** subdirectory on C drive. Open the file and edit the existing operations to produce the redesigned part as shown in Figure 7p-3(right)

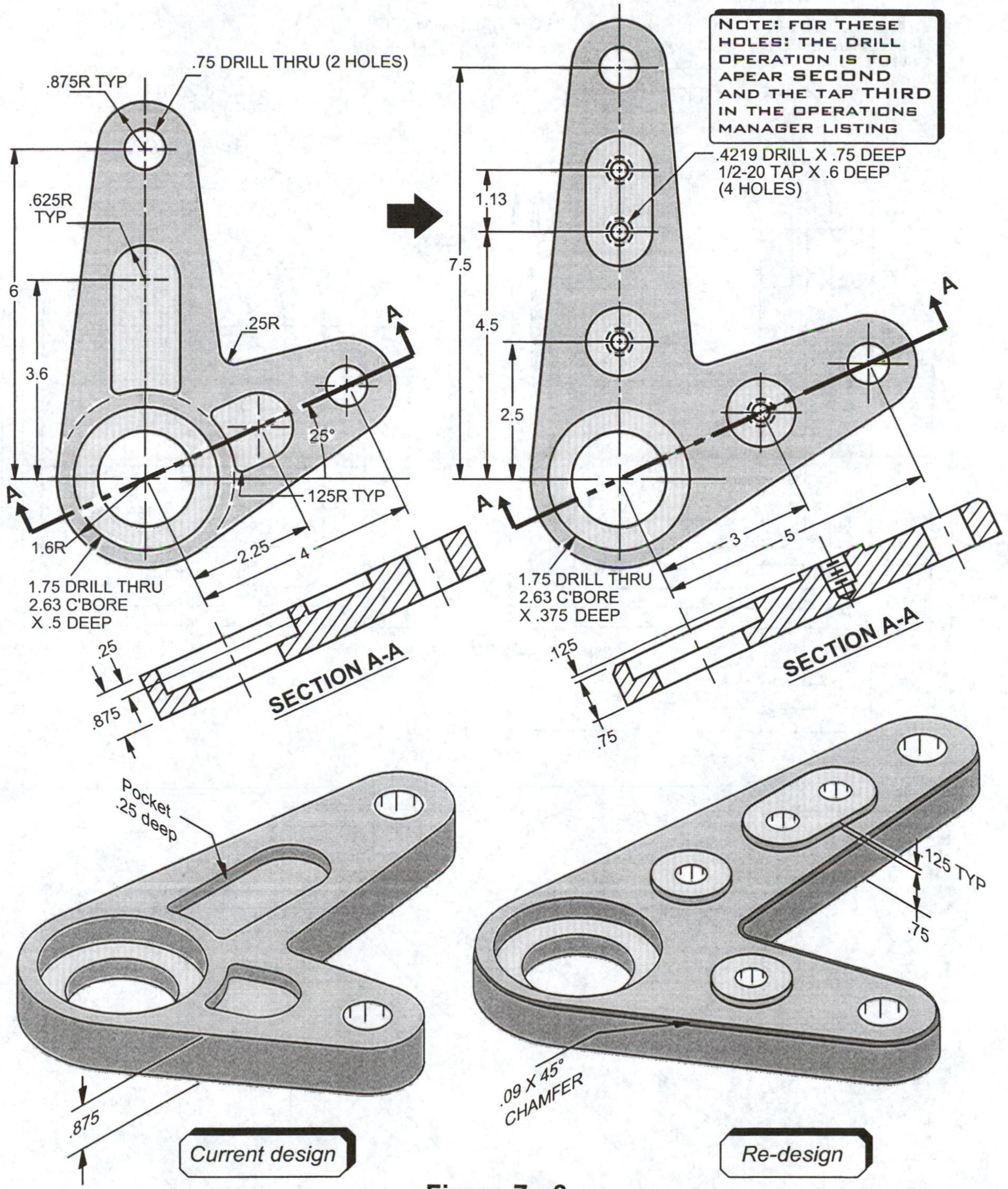

Figure 7-p3

7-4) The operations for producing the part shown in Figure 7p-4(left). currently exist in the Operations Manager. Copy the file **EX7-4** in the folder **CHAPTER7** from the CD into the **JVAL-MILL** subdirectory on C drive. Open the file and edit the existing operations to produce the redesigned part as shown in Figure 7p-4(right)

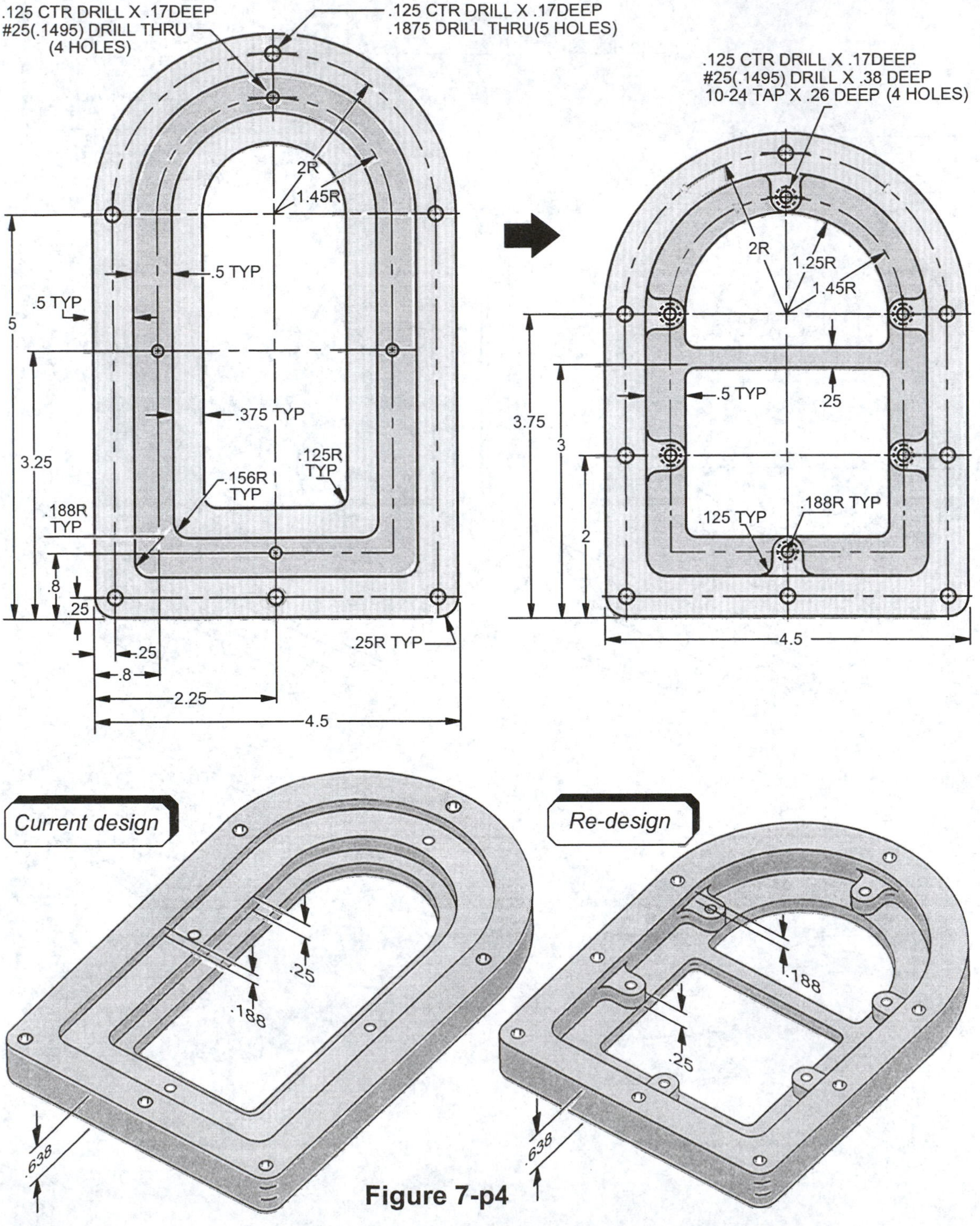

Figure 7-p4

a) Delete the ***inner slot*** geometry

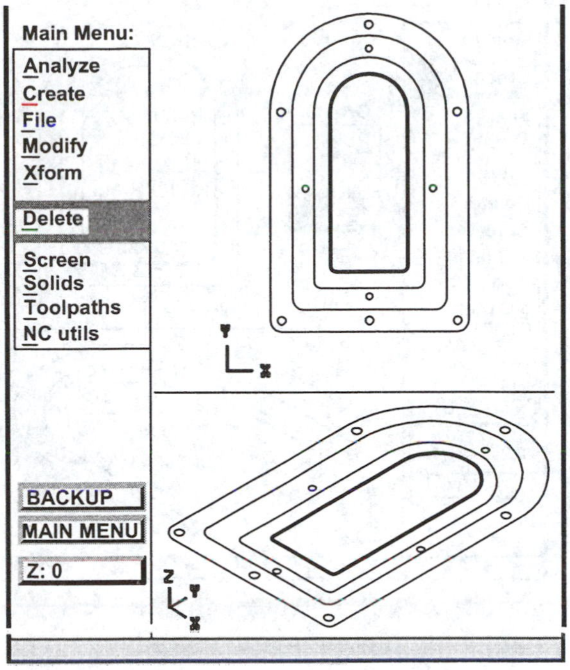

b) Use the ***Xform, Translate, Rectang*** commands to move the windowed geometry down by 1.25

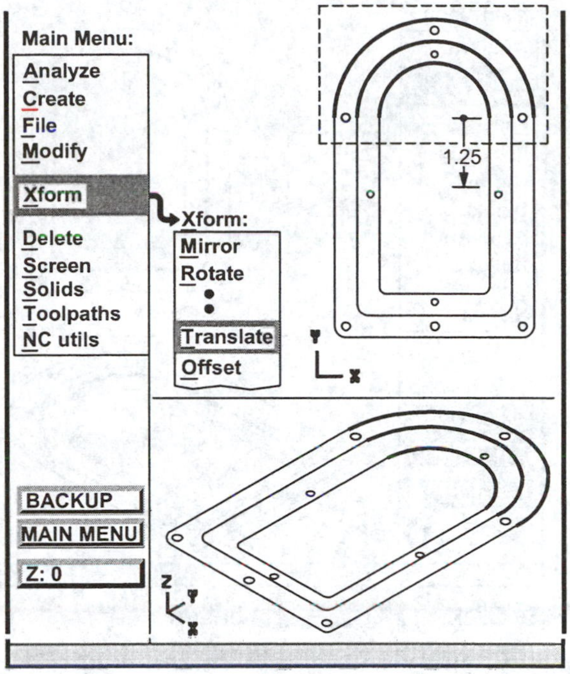

c) Use the ***Modify, Trim, 1 entity*** commands to trim the ***extended line portions***

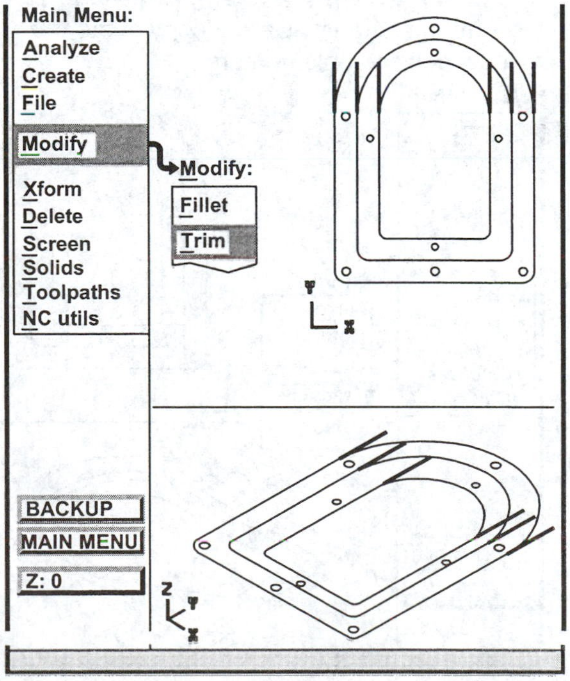

d) Use the ***Xform, Offset, Copy*** commands to create an copy offset .5 from the ***inner contour***

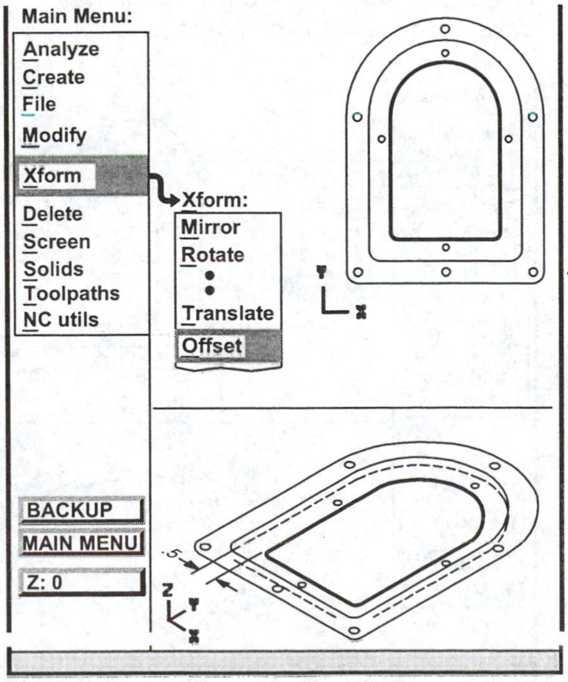

e) Create the .188R corner fillets at a depth of -.188

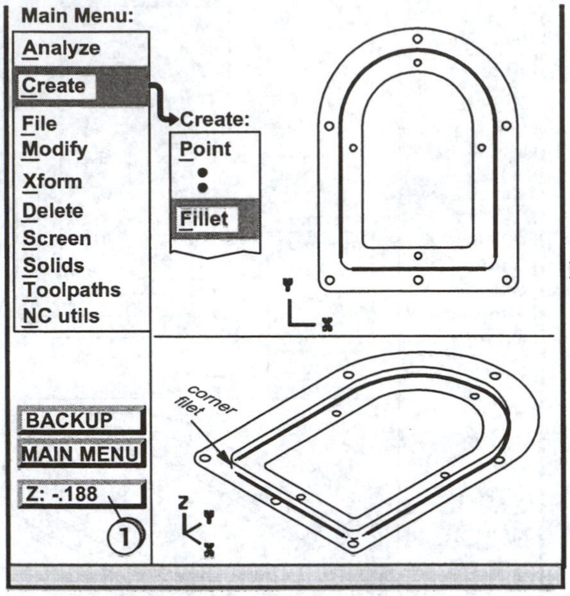

Click (1) [Z:] button

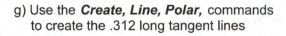

Select point for new construction depth

Enter the new depth for constructions

-.188; Create the .188R fillets

f) Use the ***Create, Arc, Polar, Center*** commands to create the .188R arc at a depth of -.188

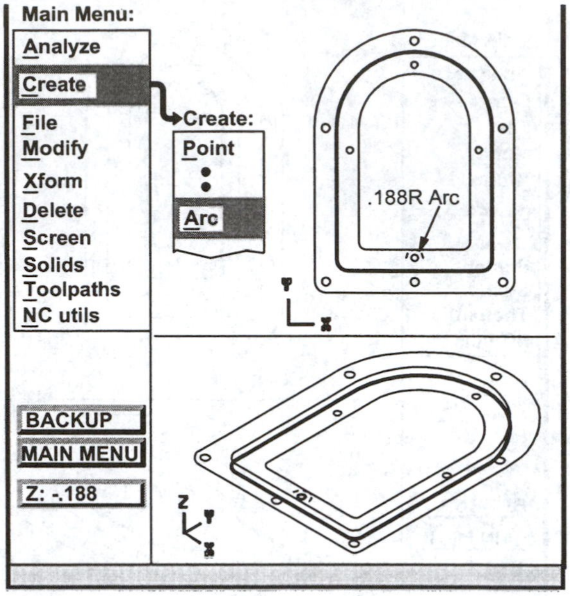

g) Use the ***Create, Line, Polar,*** commands to create the .312 long tangent lines

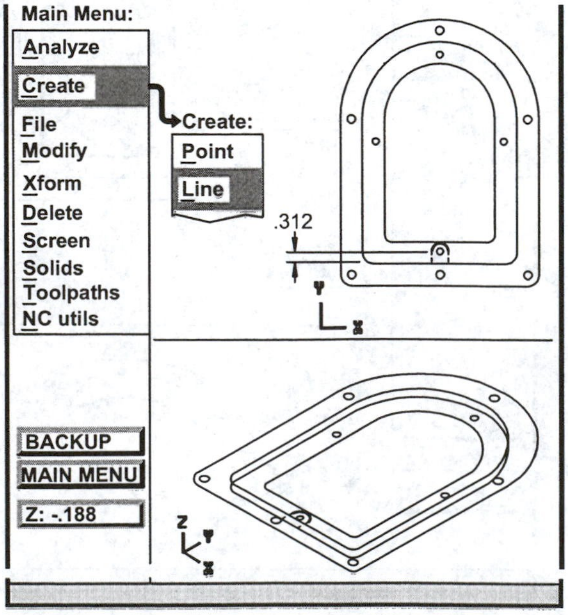

h) Use the ***Xform, Translate, Between pts*** commands to translate a copy of the tab between the hole centers

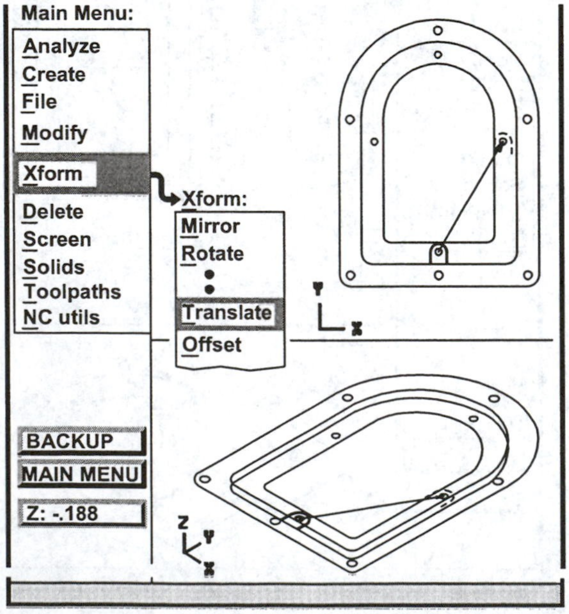

i) Use the ***Xform, Rotate,*** commands to rotate the tab 90°

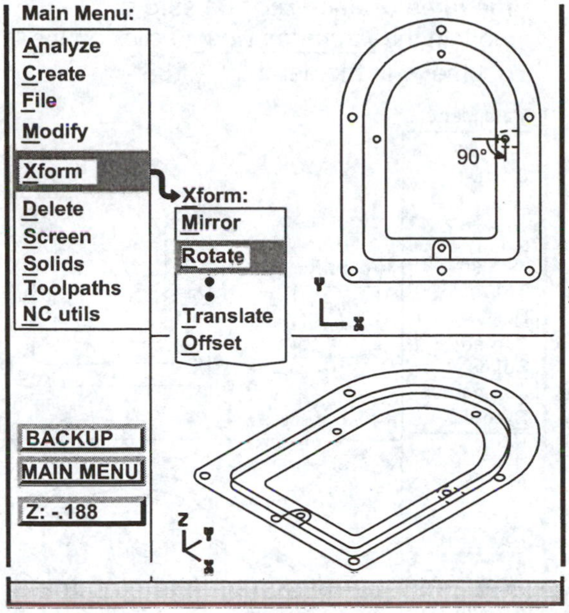

j) Use the ***Xform, Mirror,*** commands commands to create mirror a copy of the tab geometry

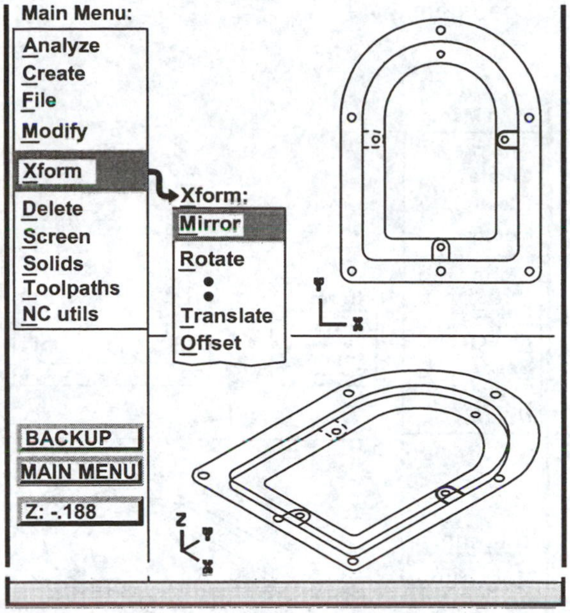

k) Use the ***Xform, Translate, Rectang,*** commands to move the windowed geometry up by .5

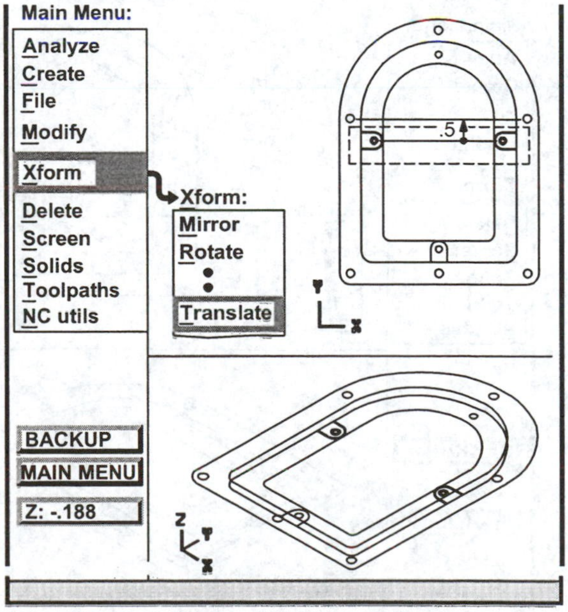

l) Use the ***Xform, Translate, Rectang,*** commands to move a *copy* of the windowed geometry down by 1.75

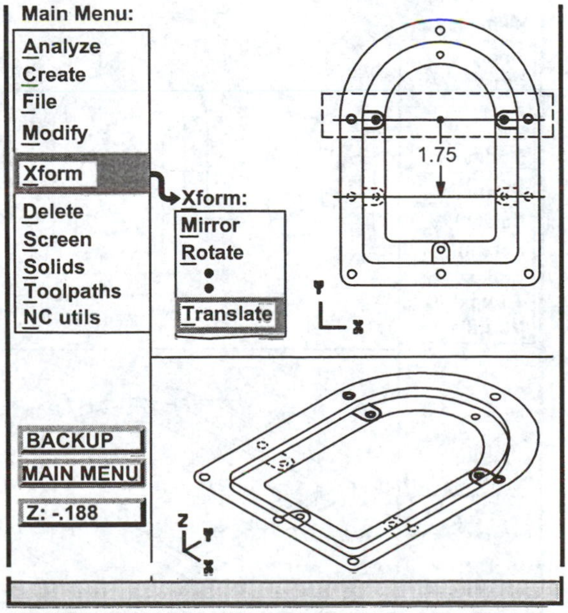

m) Use the ***Xform, Rotate*** commands to rotate by 90° a *copy* of the windowed geometry.

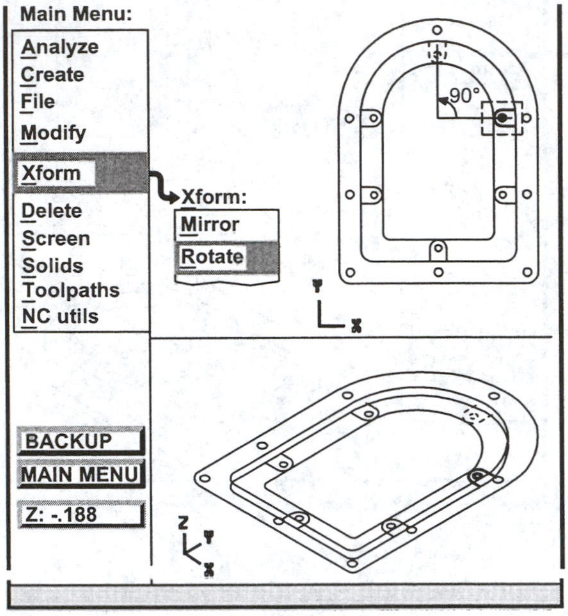

n) Use the ***Modify, Trim, Divide*** commands to remove the ***geometry between the the lines of each tab***. Be sure to work in the <u>*isometric view*</u> to click on the geometry at the depth of -.188.

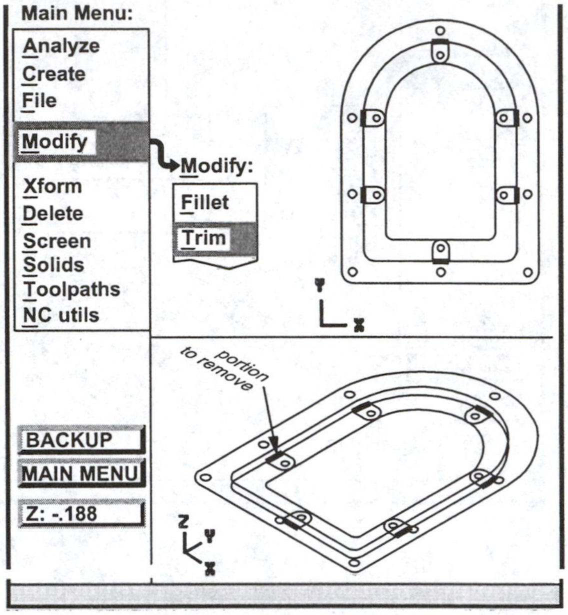

o) Use the ***Create, Fillet*** commands to create the .125R tab fillets at a depth of -.188. Work in the isometric view to click on the depth geometry.

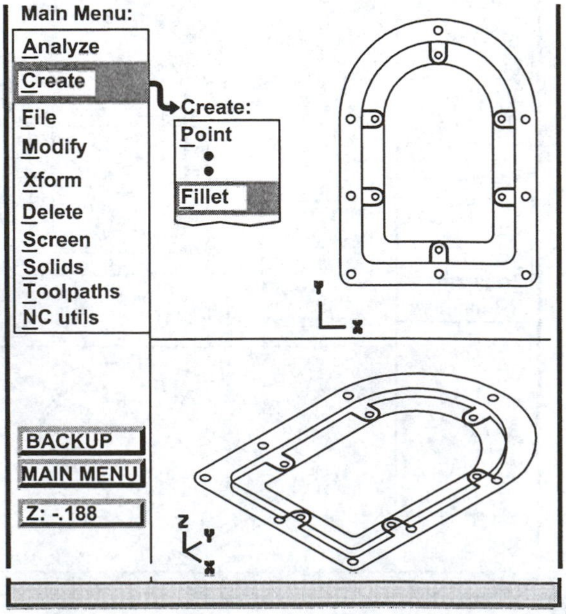

p) Use ***Xform, Offset*** commands to create the top line of the first slot boundary. Use the Isometric view to click the line at -.188 depth

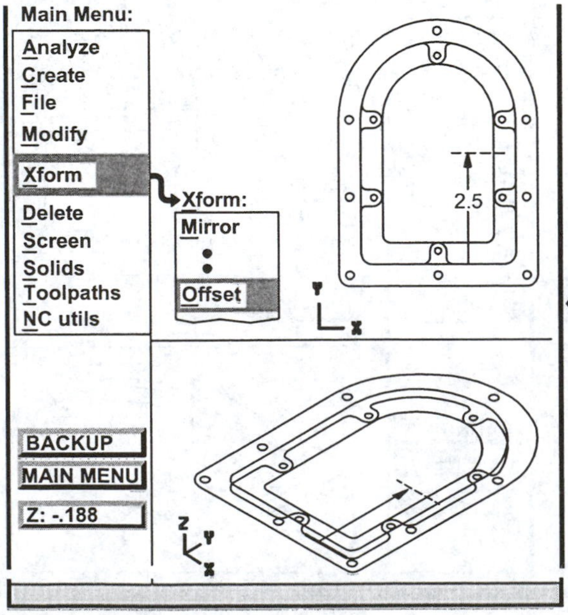

q) Use ***Modify, Trim*** commands to extend the top boundary line

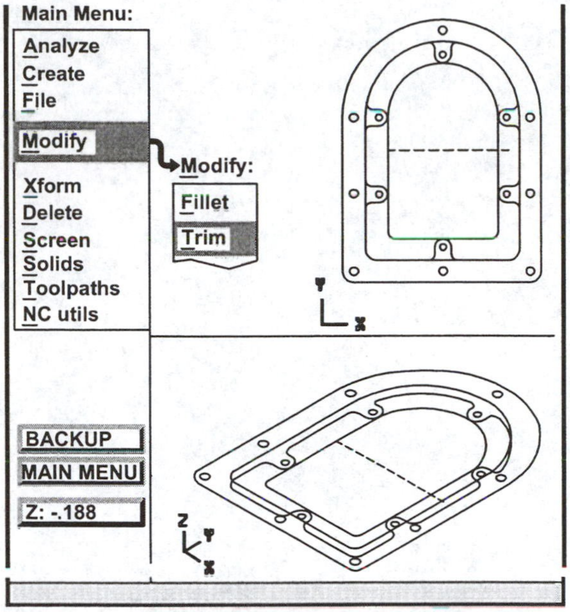

r) Use ***Xform, Offset*** commands to create the bottom line for the second slot boundary

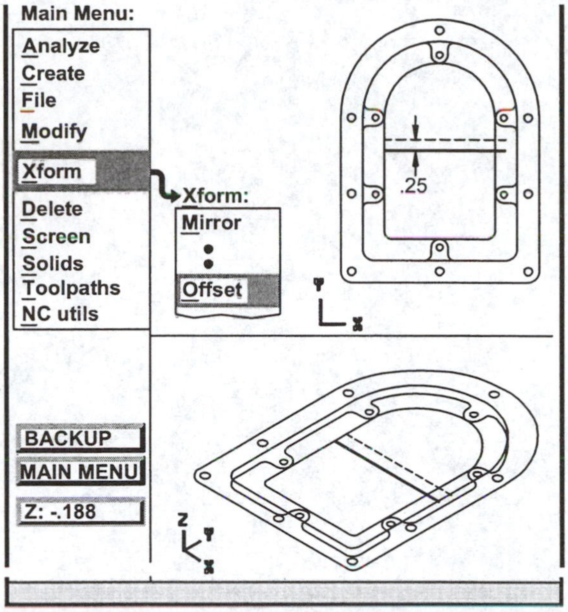

s) Use ***Modify, Trim, Divide*** commands to remove the excess geometry between slots

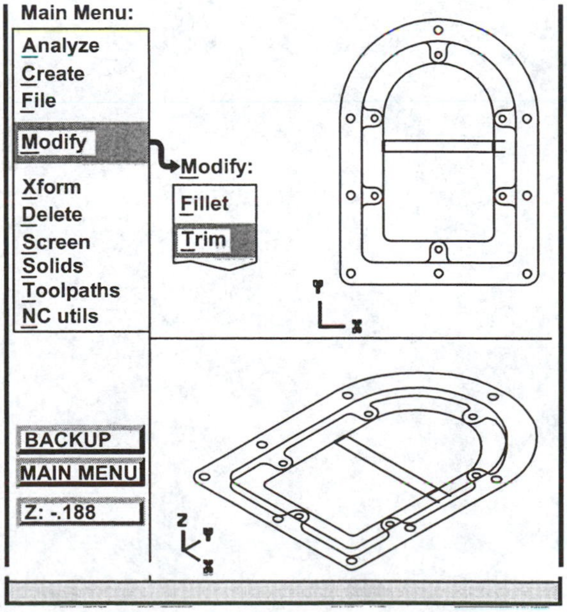

t) Use ***Create, Fillet,*** commands to insert the .125R fillets for the slots and complete the new edited shape

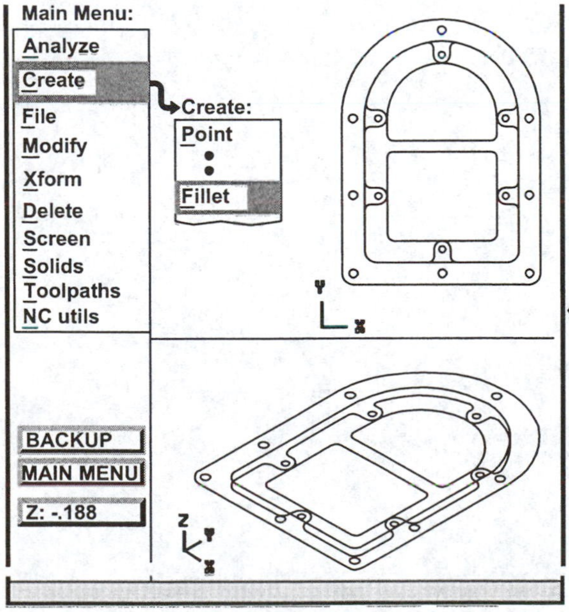

CHAPTER - 8

USING TRANSFORM TO TRANSLATE, ROTATE OR MIRROR EXISTING TOOLPATHS

8-1 Chapter Objectives

After completing this chapter you will be able to:

1. Understand the transform toolpaths function.
2. Use the transform toolpaths function to translate existing toolpaths.
3. Know how to use the transform toolpaths function to rotate toolpaths.
4. Use the transform toolpaths function to mirror existing toolpaths.
5. Explain how to convert a transform toolpath into new geometry and operations.

8-2 The Transform Toolpaths Function

Existing toolpaths can be *translated, rotated or mirrored* via the Transform Toolpaths function. When a toolpath is edited by using Transform all the other information in the operation associated with the toolpath is also updated when [Regen Path] is executed. Thus *associativity is maintained*.

It should be noted that a transform toolpath is ***not an independent toolpath*** but is ***linked to the original*** from which it was copied. If the *original is edited* in any way via the chain manager ***all the transform copies will automatically be edited as well*** .

Many parts contain repetitious features that appear in grid or mirror patterns. Situations also arise where a set of identical parts or left and right hand versions of a part are to be machined in one setup. In these and many other cases Transform toolpath can be used to cut down on CAD geometry creation and dramatically simplify the job of chaining toolpaths.

The Transform Operations Parameters dialog box can be entered as shown below:

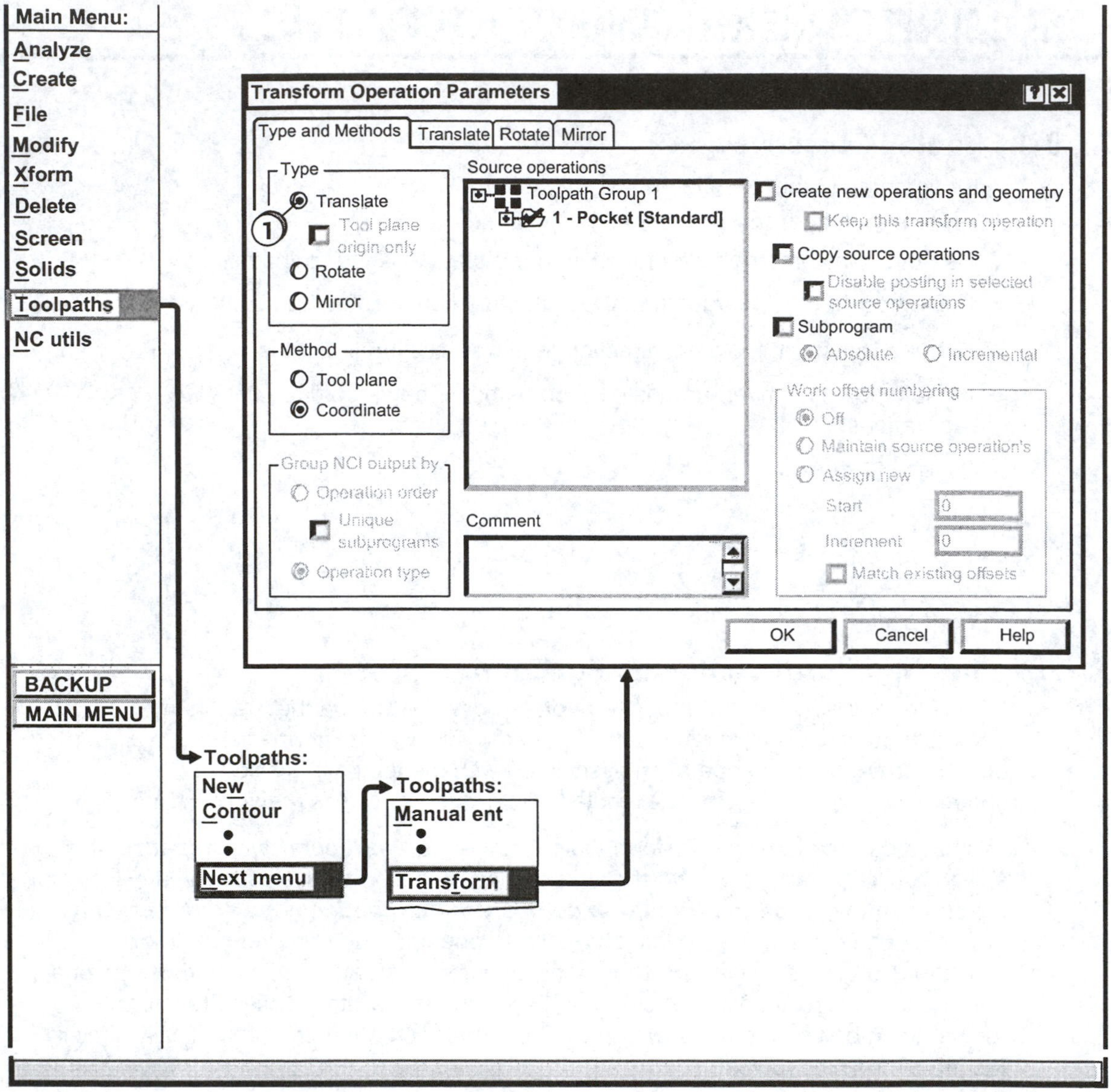

8-3 Translating Existing Toolpaths

The operator clicks the Translate tab in the Transform Operations Parameters dialog box to execute move or copy operations on existing toolpaths.

BETWEEN POINTS

The Operator enters the ***XY location of the existing toolpath*** From pt... and ***the XY location where the copy is to be placed*** To pt... . Mastercam determines the required distance. The integer value entered in Step specifies the number of copies to make each separated by the distance computed..

EXAMPLE 8-1

The Slot toolpath and machining operation currently exists as displayed left.
Direct *Mastercam* to create a copy of the machining operation at the location displayed right.

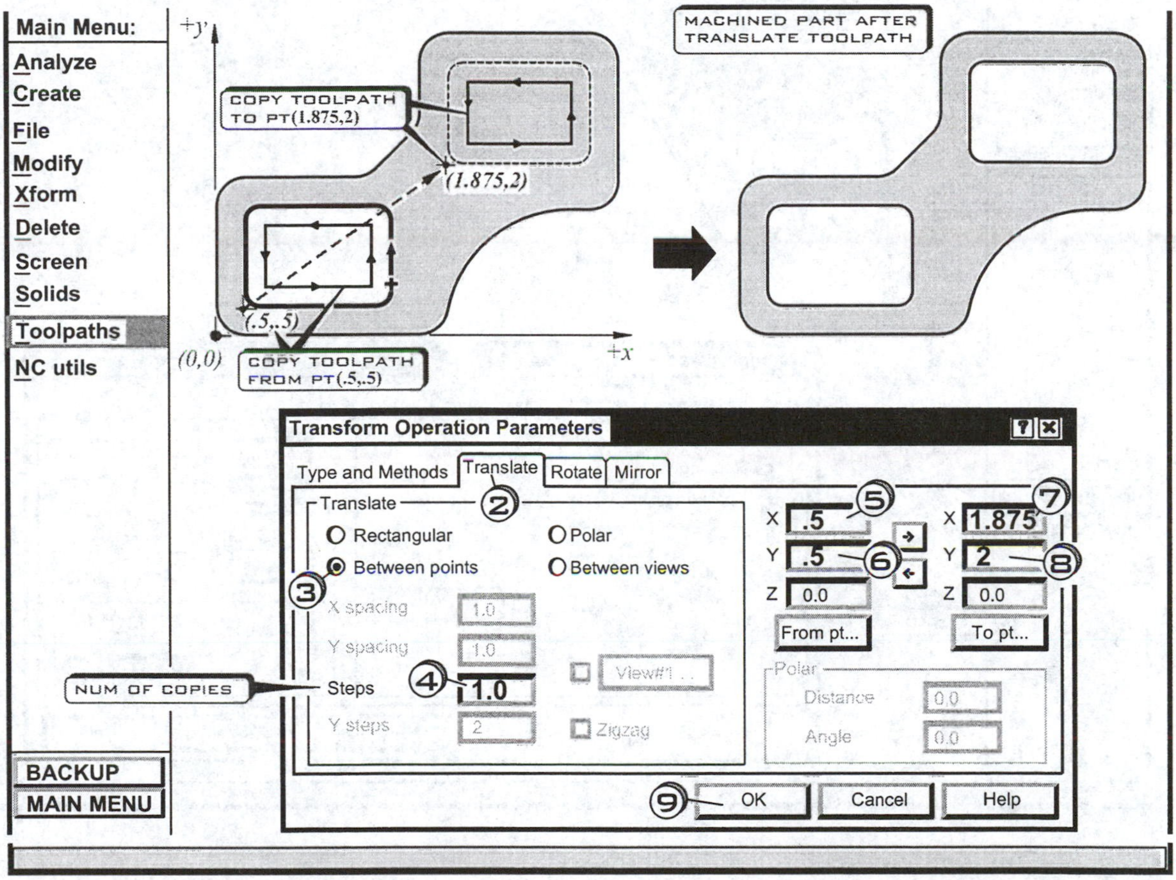

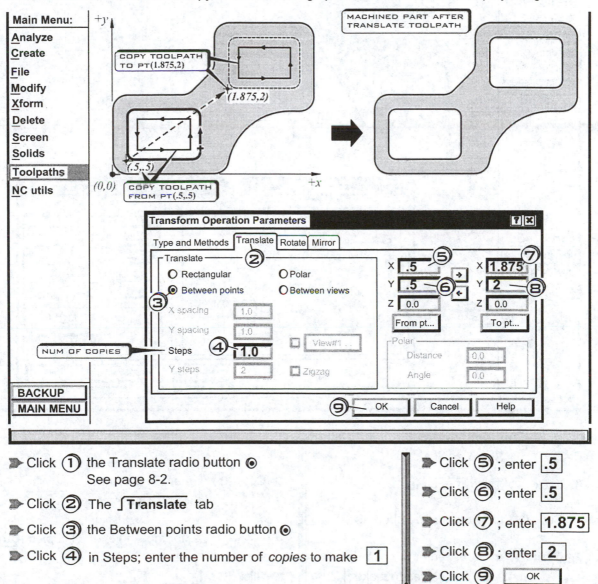

- Click ① the Translate radio button ◉
  See page 8-2.
- Click ② The Translate tab
- Click ③ the Between points radio button ◉
- Click ④ in Steps; enter the number of *copies* to make 1
- Click ⑤ ; enter .5
- Click ⑥ ; enter .5
- Click ⑦ ; enter 1.875
- Click ⑧ ; enter 2
- Click ⑨ OK

POLAR

The Operator enters the ***distance between each copy and the required angle.*** The integer value entered in Step specifies the number of copies to make.

EXAMPLE 8-2

The Pocket toolpath and machining operation currently exists as shown in the display at left. Direct *Mastercam* to create copies of the machining operation at the locations displayed right.

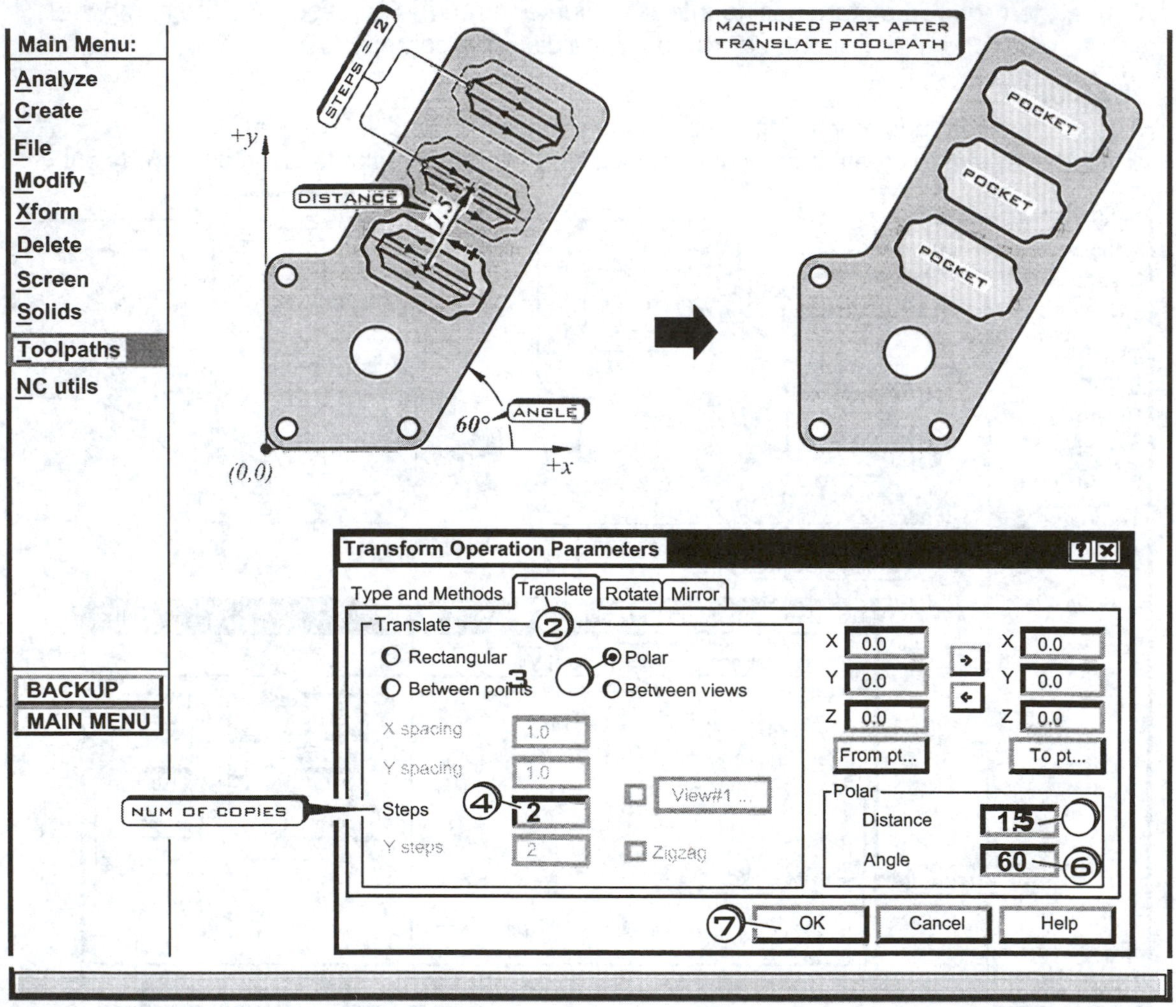

➤ Click (1) the Translate radio button ◉
See page 8-1

➤ Click (2) The **Translate** tab

➤ Click (3) the Polar radio button ◉

➤ Click (4) in Steps; enter the number of *copies* to make **2**

➤ Click (5) ; enter the Distance between each copy **1.5**

➤ Click (6) ; enter the Angle for all copies **60**

➤ Click (9) OK

RECTANGULAR

The Operator enters the ***distance between each copy in the X-direction, distance between each copy in the Y-direction, number of copies to make in the X-direction and number of copies make in the Y-direction.***

EXAMPLE 8-3

The operator has created the slot toolpath and machining operation displayed left.
Direct *Mastercam* to create arrayed copies of the machining operation the locations displayed right.

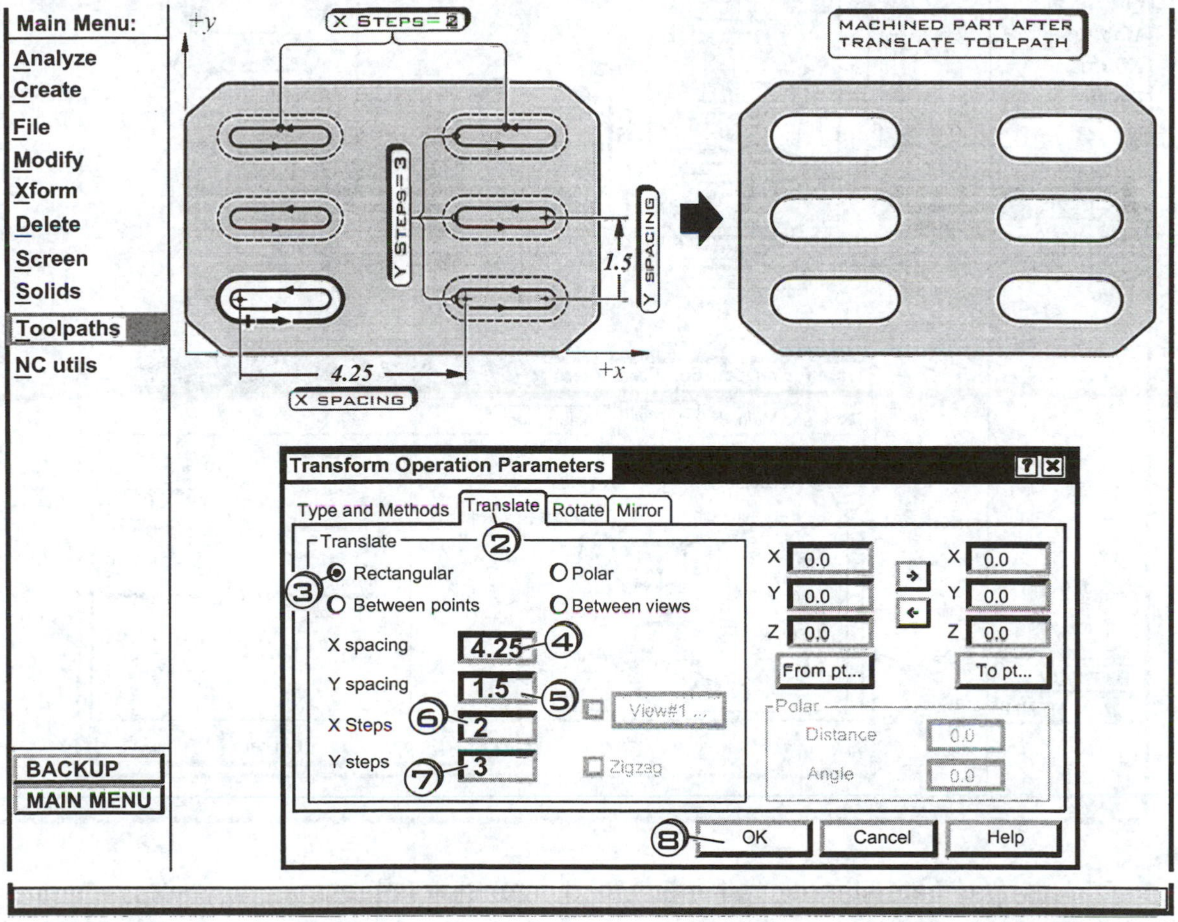

- Click ① the Translate radio button ◉ See page 8-1
- Click ② The **Translate** tab
- Click ③ the Rectangular radio button ◉
- Click ④ in X spacing; enter the distance between each copy in the X -direction **4.25**
- Click ⑤ in Y spacing; enter the distance between each copy in the Y -direction **1.5**
- Click ⑥ the number of X steps **2**
- Click ⑦ the number of Y steps **3**
- Click ⑨ OK

EXAMPLE 8-4

Given the the slot toolpath and machining operation displayed left. Direct *Mastercam* to create arrayed copies of the slot machining operation the locations displayed right.

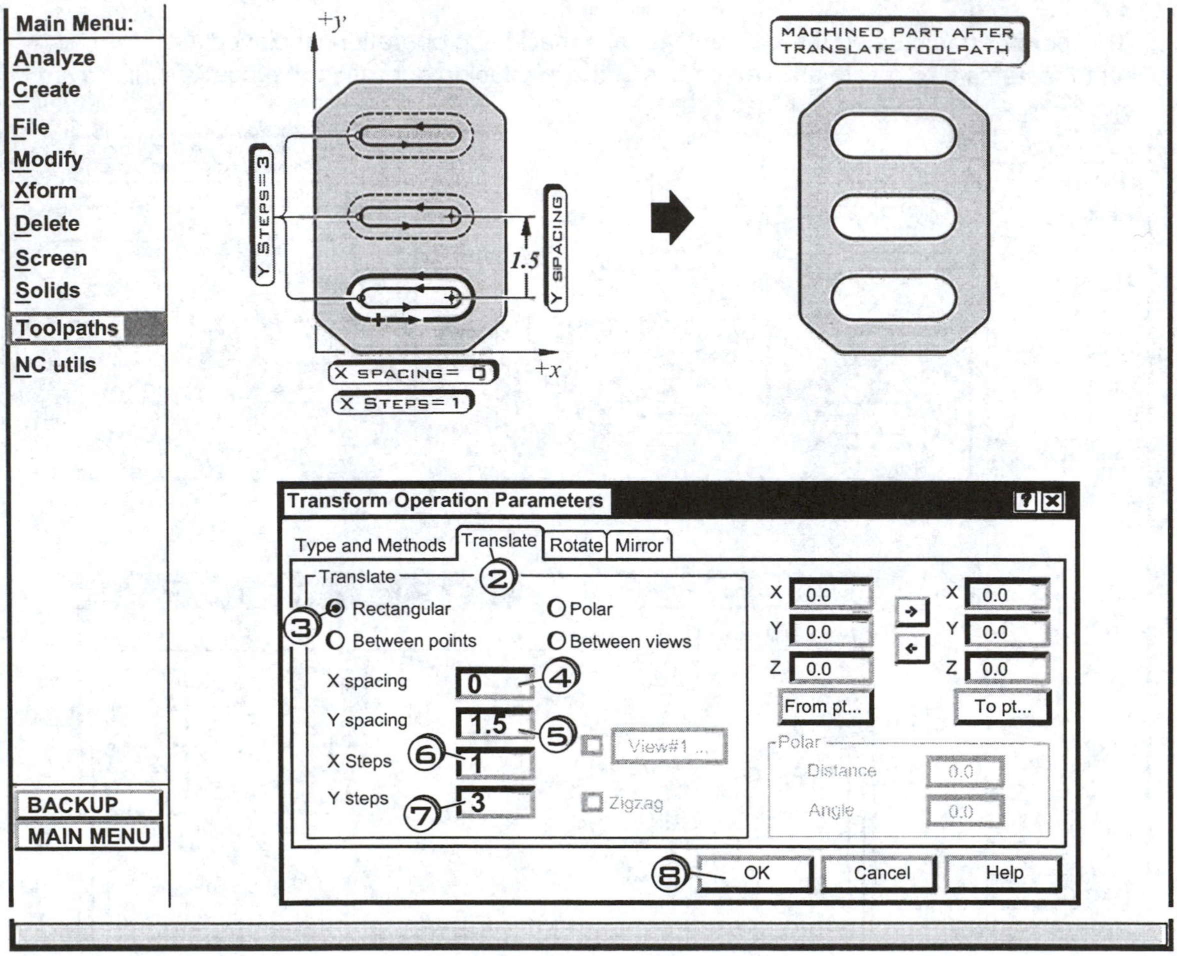

- Click ① the Translate radio button ◉ See page 8-1
- Click ② The ⌈Translate⌉ tab
- Click ③ the Rectangular radio button ◉
- Click ④ in X spacing; enter the distance between each copy in the X -direction **0**
- Click ⑤ in Y spacing; enter the distance between each copy in the Y -direction **1.5**
- Click ⑥ the number of X steps **1**
- Click ⑦ the number of Y steps **3**
- Click ⑨ OK

8-4 Rotating Existing Toolpaths

The rotation function is especially useful for producing parts that have *repeat patterns arranged symmetrically about a point*. In these cases the operator can identify the base pattern create its corresponding toolpath then rotate the toolpath about the required point to produce the complete symmetrical shape.

ROTATE POINT

The Operator enters the ***location of the point about which rotation is to occur,*** the ***Number of steps(copies) to make,*** the ***Start angle for the first copy*** and the ***Rotation angle between each copy.***

EXAMPLE 8-5

The Operator has created the slot toolpath and machining operation as displayed left. Direct *Mastercam* to create rotated copies of the machining operation as displayed right.

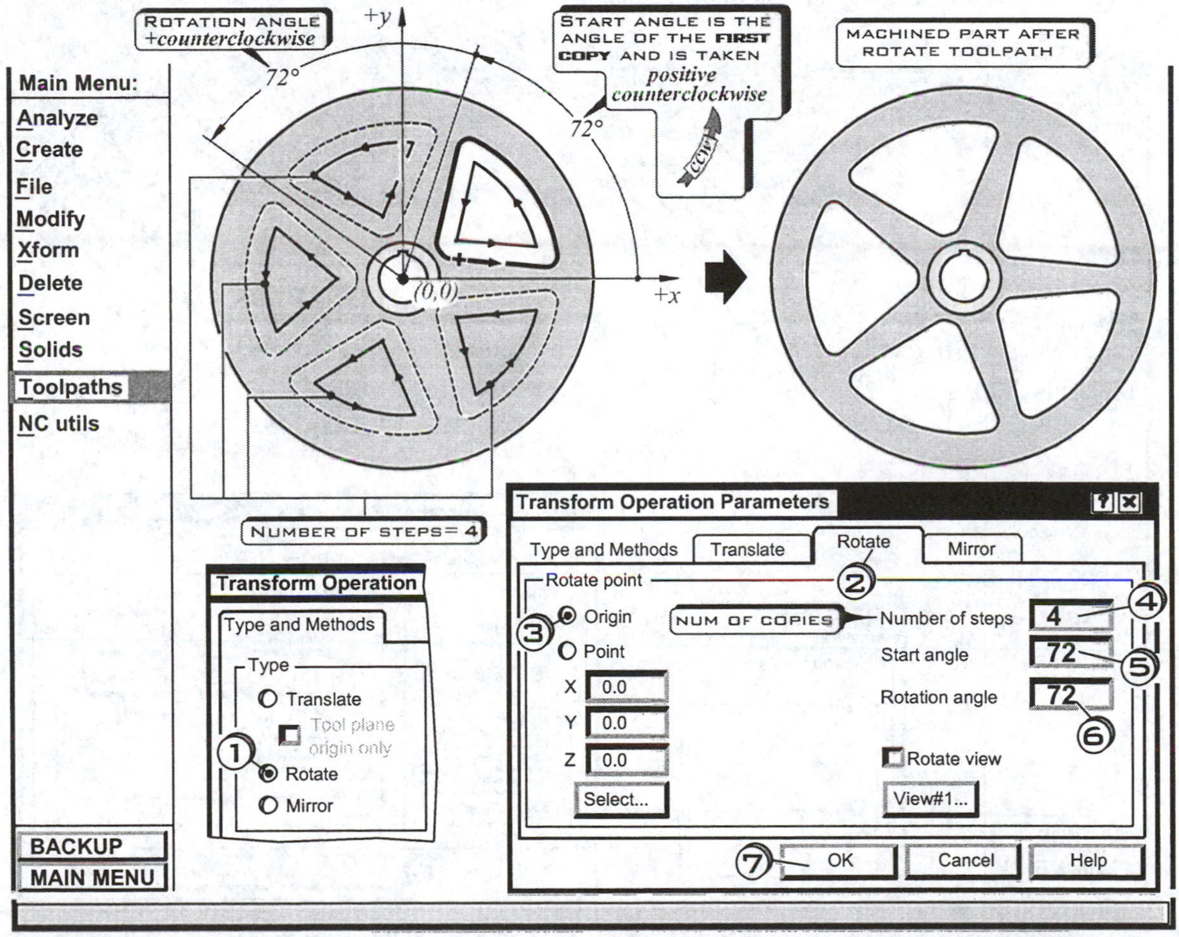

- Click ① the Rotate radio button ◉
- Click ② The Rotate tab
- Click ③ the Origin radio button ◉ to rotate about X0Y0
- Click ④ in Steps; enter the number of *copies* to make **4**
- Click ⑤ in Start angle; enter **72**
- Click ⑥ in Rotation angle; enter **72**
- Click ⑦ OK

8-5 Mirroring Existing Toolpaths

The mirror function is used in to mirror existing toolpaths about a line or axis. It simplifies the job of generating toolpaths for parts that have *repeat patterns arranged symmetrically about an axis or line*. It is also used to quickly generate *toolpaths for left and right hand versions* of the same part in one setup.

ENTITY

The Operator ***clicks the entity about which mirroring is to occur***

EXAMPLE 8-6

The Pocket toolpath currently exists as shown in the left display. Direct *Mastercam* to create a mirror copy of the pocketing operation as displayed right.

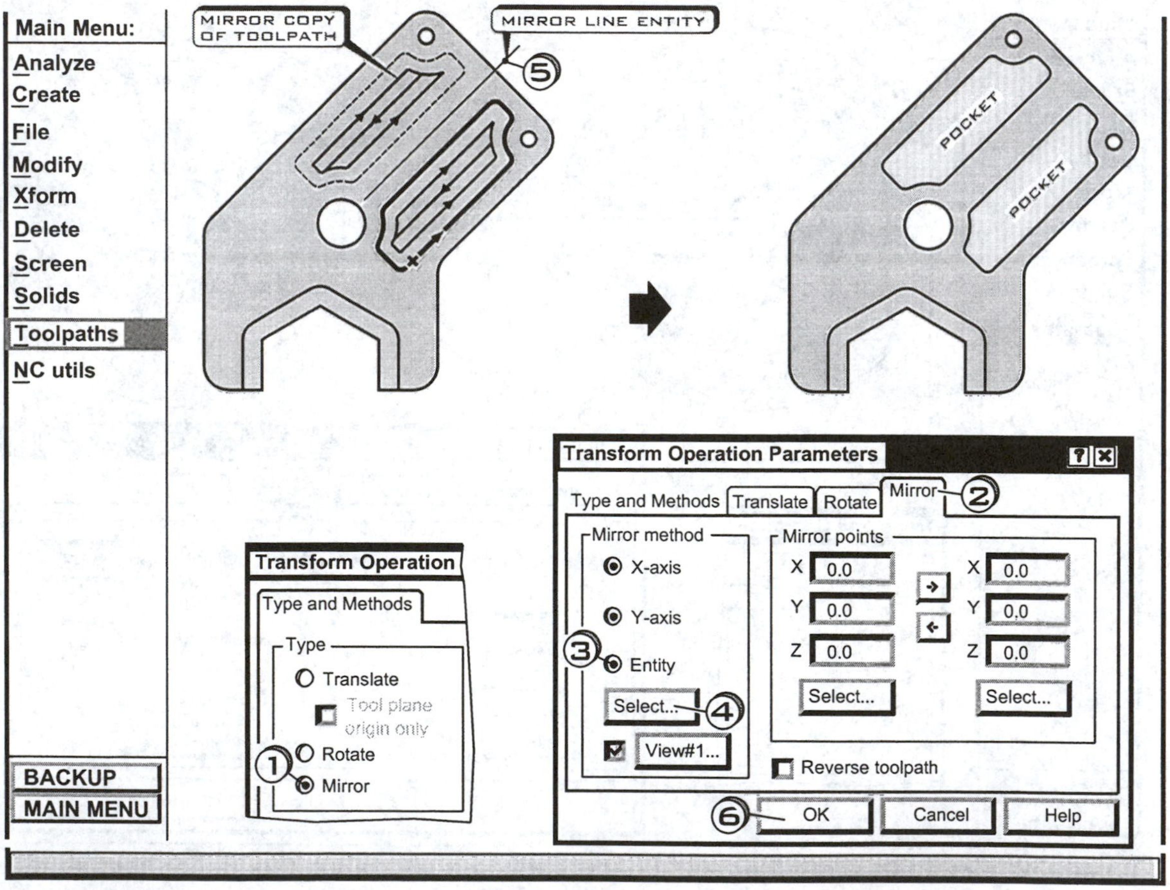

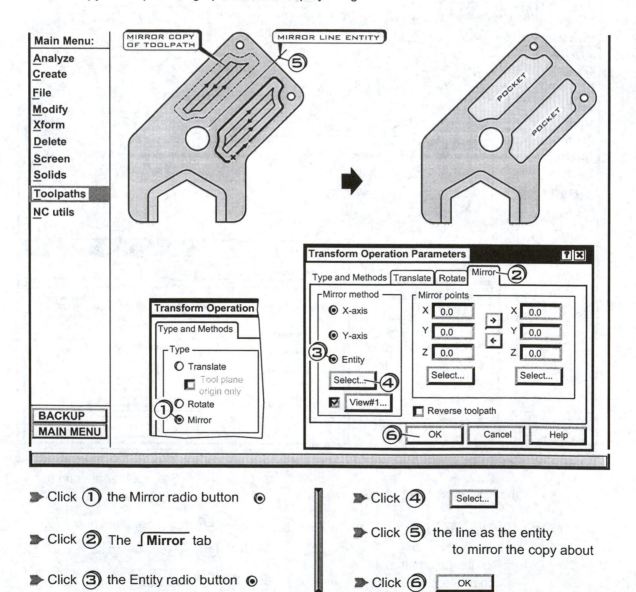

- Click ① the Mirror radio button ◉
- Click ② The Mirror tab
- Click ③ the Entity radio button ◉
- Click ④ Select...
- Click ⑤ the line as the entity to mirror the copy about
- Click ⑥ OK

Note: the mirror function will ***mirror the direction of the toolpath*** and thus should be used with caution.

The existing slot toolpath shown in Figure 8-1 is directed to the ***left*** of upward tool motion(climb milling) but the mirrored toolpath is directed to the ***right*** of upward tool motion(conventional milling).

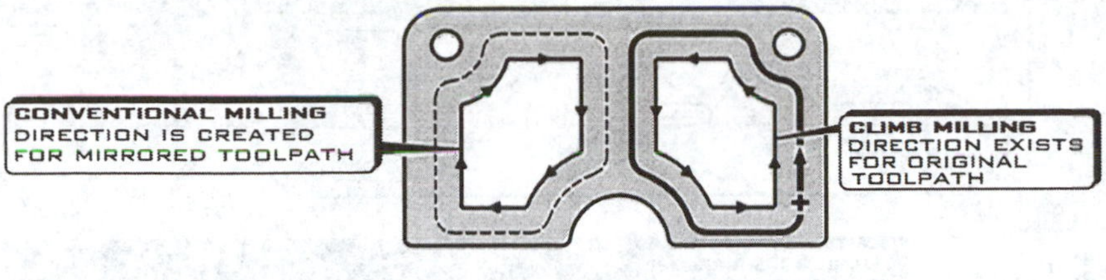

Figure 8-1

In cases like these where *climb milling* is desired for *both slots* it is recommended *not to use the mirror function*. Instead the approach should be to to *mirror the slot geometry* using the **Xform** , **Mirror,** functions. The operator can then chain the second slot in the proper direction for climb milling.

8-6 Converting A Transform into New Geometry and Operations

Transform toolpaths created by the Translate, Rotate or Mirror functions *cannot be edited individually but are tied to the original existing toolpath(s)* from which they were copied. Editing can only be accomplished by *changing the original* existing toolpath(s). When this is done, the changes made to the original(s) will automatically be passed on to their transform copies. This section will consider how to convert a transform toolpath such that it will be listed as a *new and independent operation* in the Operations Manager. The operator can then *edit the operation individually* as needed.

EXAMPLE 8-7

The Original and transform toolpaths have been created in *Mastercam* to produce the part shown in Figure8-2(left) . Edit the existing operations to produce the re-designed part shown in Figure 8-2(right).

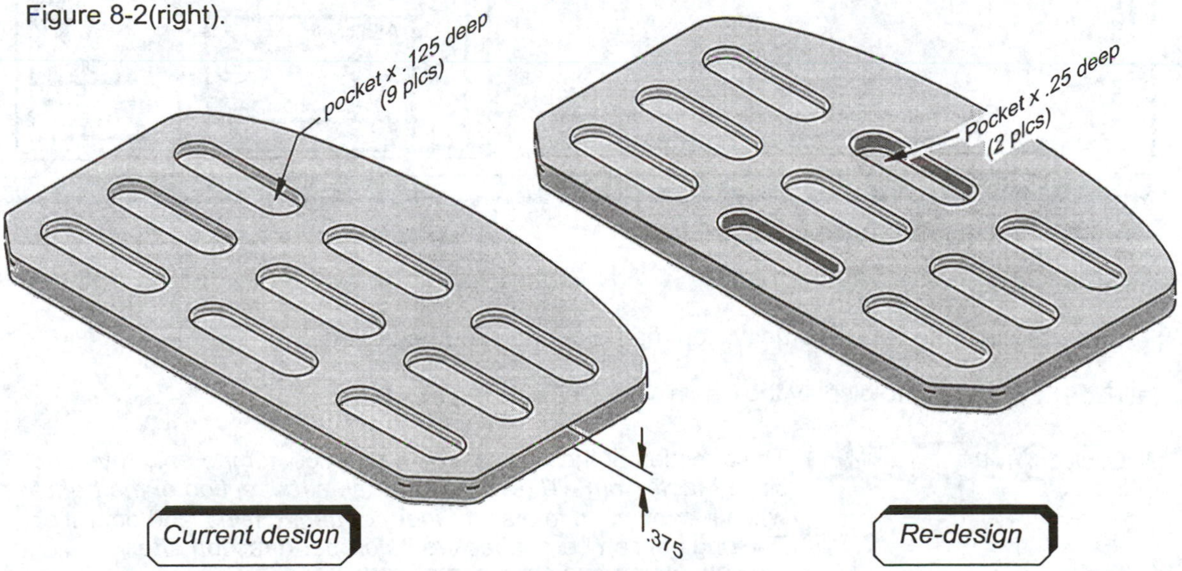

Figure 8-2

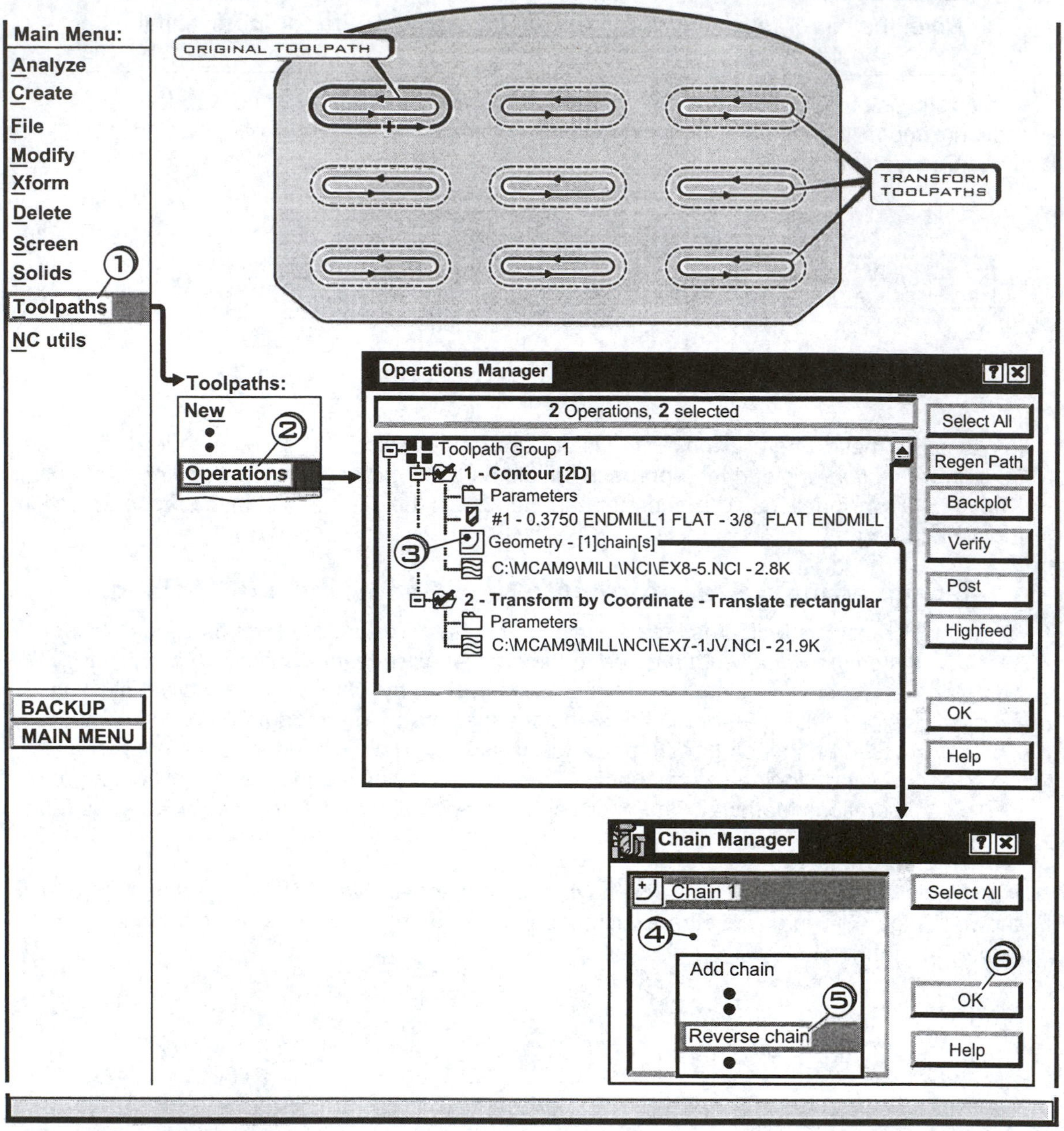

- Click ① **Toolpaths**
- Click ② **Operations**
- Click ③ on the Part Geometry icon for the operation to be edited
- ***Right*** Click ④ to display the Chain Manager's editing functions
- Click ④ Reverse chain — The direction of the original chain must be *temporarily reversed* since *Mastercam will reverse the toolpath direction of the transform* when it *converts it to a set of independent chains* and operations The original can be *set back to its proper direction after* the *conversion has been completed.*

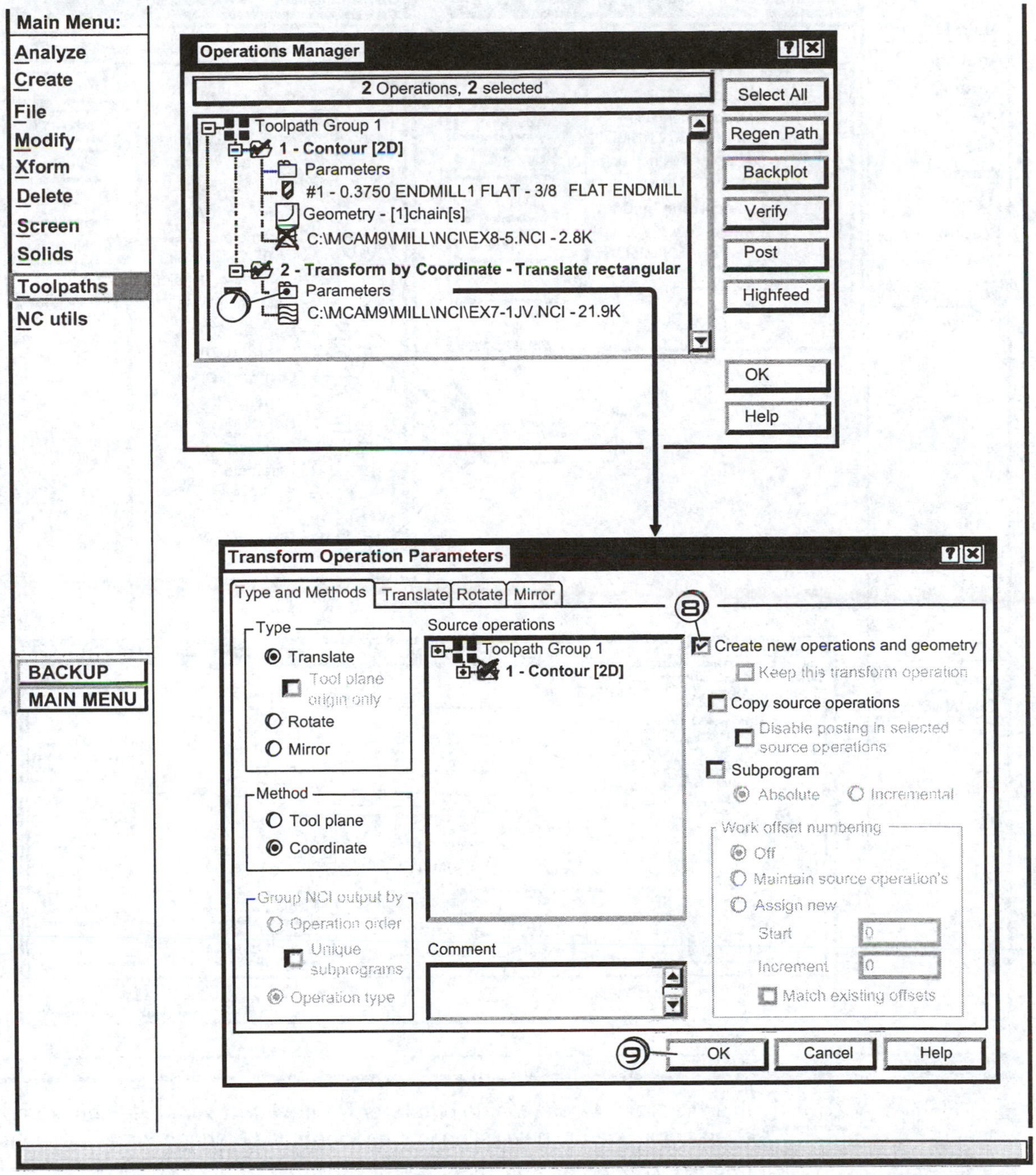

- Click ⑦ on the Parameters icon for the transform toolpaths

- Click ⑧ the check ☑ to enable **Create new operations and geometry**

- Click ⑨ OK

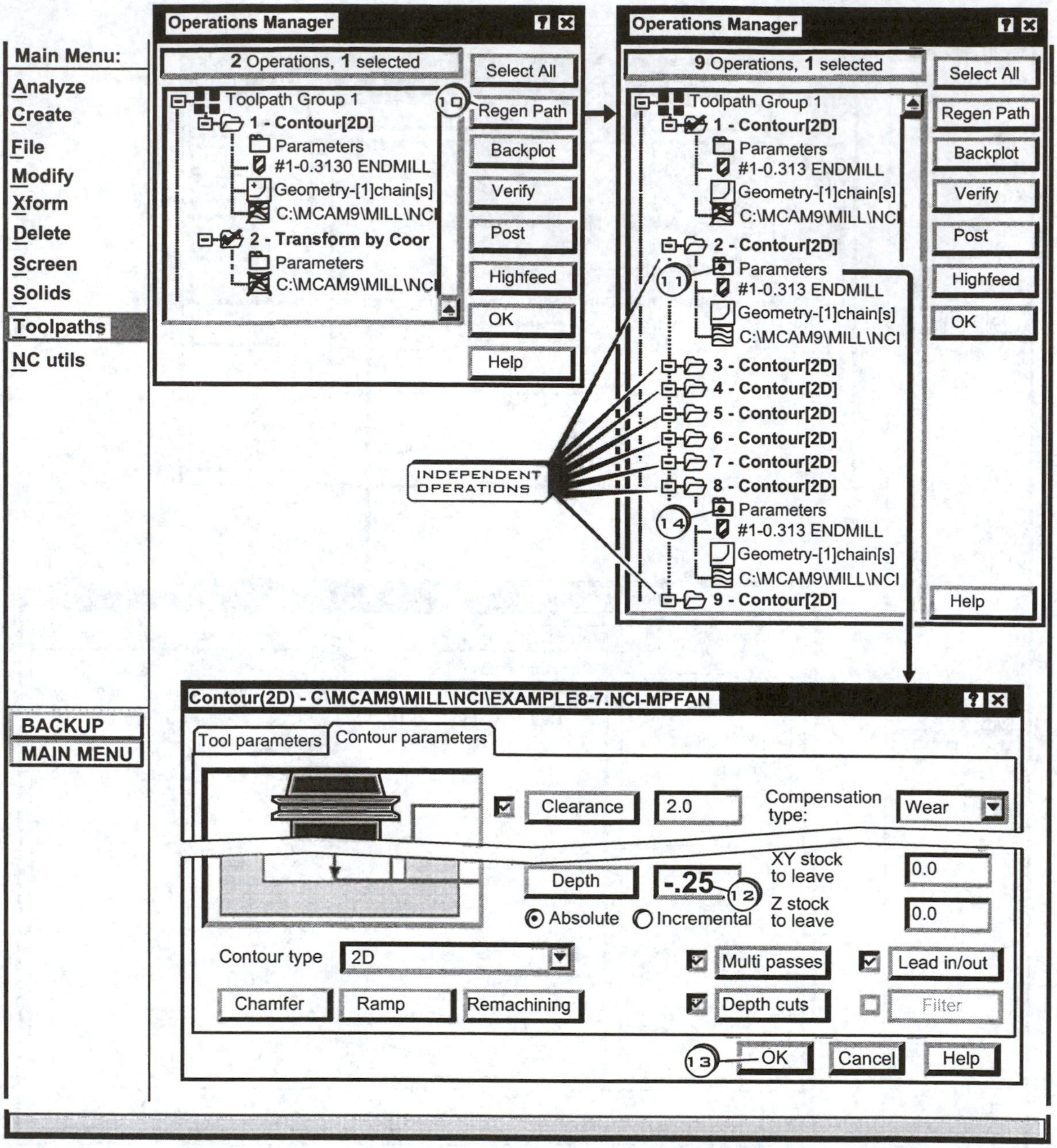

- Click (10) [Regen Path] this will direct *Mastercam* to create ***eight new and separate*** countouring operations.See the new listing in the Operations Manager shown to right
- Click (11) on the Parameters icon for the operation **2- Contour[2D]**
- Click (12) in the Depth box and enter the new depth for the operation **-.25**
- Click (13) [OK]
- Click (14) on the Parameters icon for the operation **8- Contour[2D]**
- Click (12) in the Depth box and enter the new depth for the operation **-.25**
- Click (13) [OK]

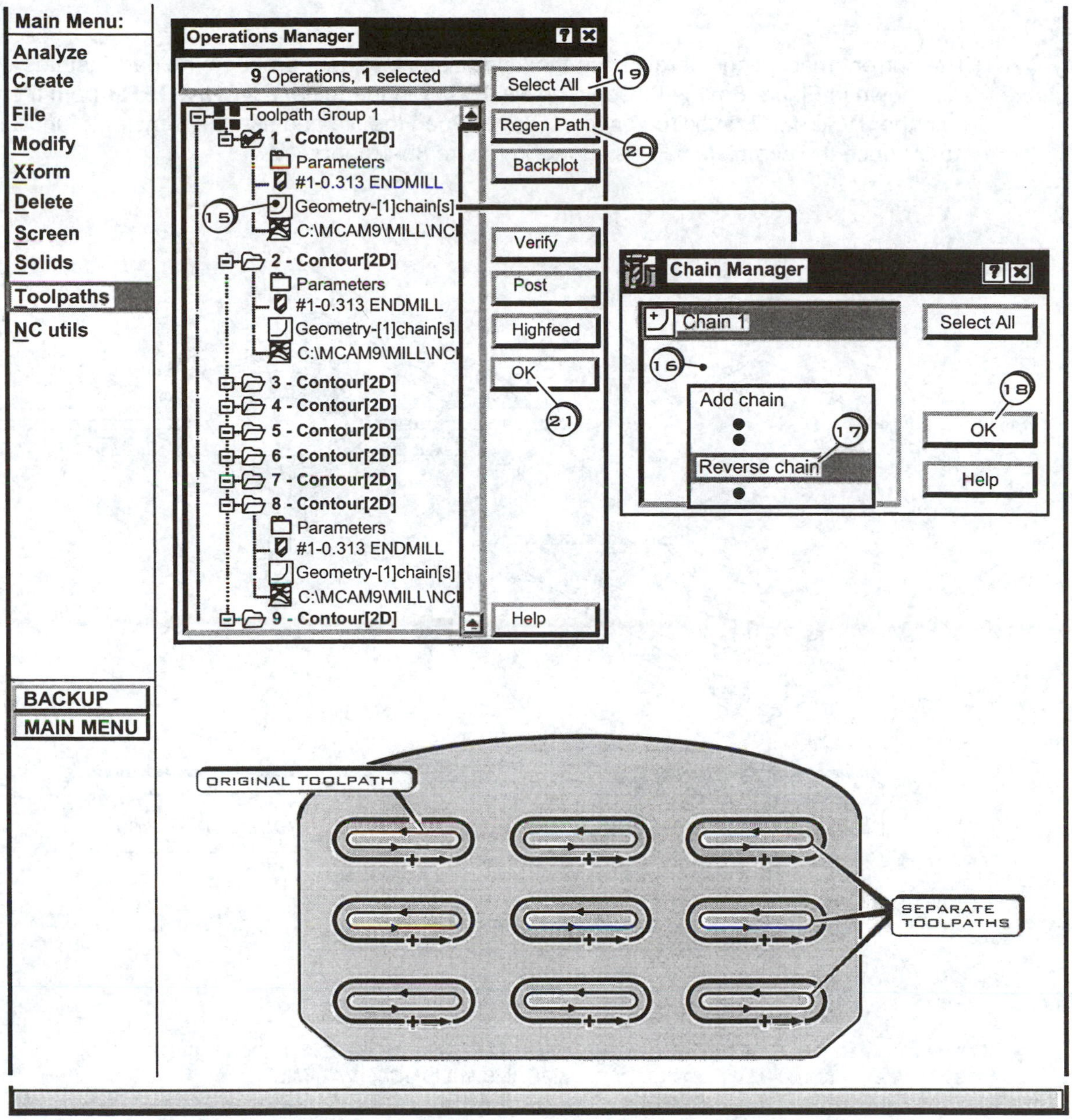

- Click (15) on the Part Geometry icon for the operation to be edited
- ***Right*** Click (16) to display the Chain Manager's editing functions
- Click (17) Reverse chain to reverse the original chain back to the proper direction
- Click (18) OK
- Click (19) Select All
- Click (20) Regen Path
- Click (21) OK

EXERCISES

8-1) The current machining operations in the Operations Manager produce the part features as shown in Figure 8p-1(left). Copy the file **EX8-1** in the folder **CHAPTER8** from the CD into the **JVAL-MILL** subdirectory on C drive. Open the file and use the transform functions to produce the complete part as shown in Figure 8p-1(right)

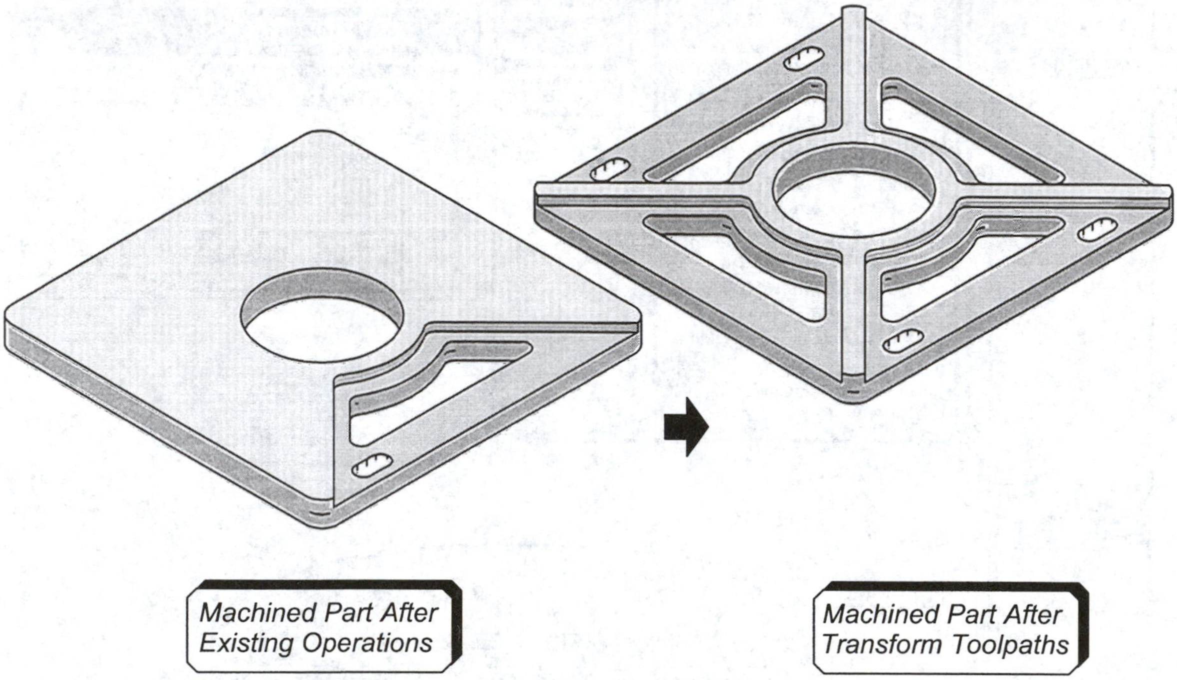

Figure 8-p1

A) COPY THE FILE EX8-1 FROM THE CD AT THE BACK OF THIS TEXT TO YOUR DIRECTORY **C:\MCAM9\MILL\MC9\JVAL-MILL**

your initials

OPEN THE FILE FROM YOUR DIRECTORY.

Refer to Chapter 1 p1-14 for a discussion on the creation of your own directory.

B) Open the Transform Operations Parameters Dialog Box

- ◆ IDENTIFY THE OPERATIONS TO BE ROTATED; SELECT ROTATE

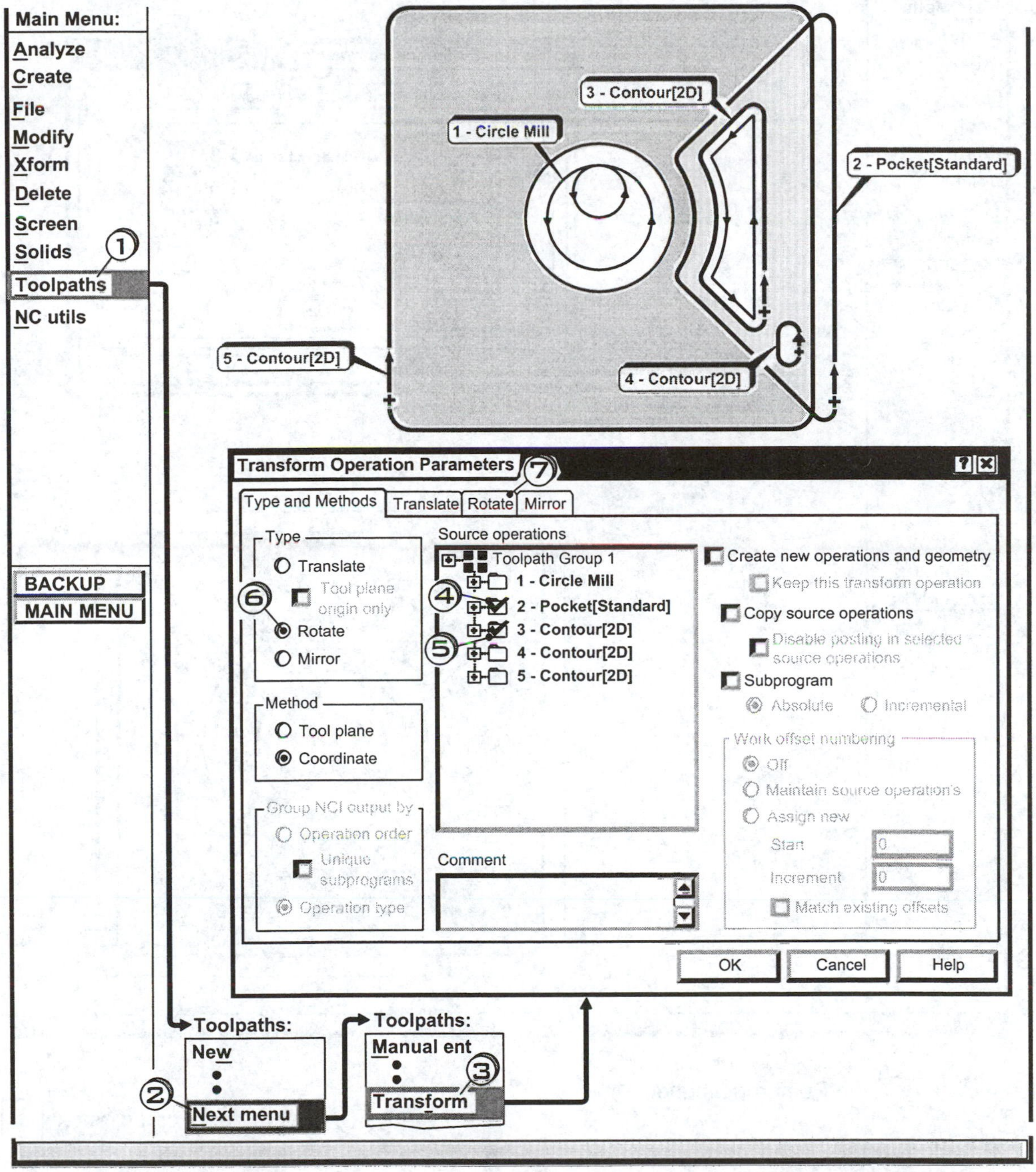

- Click ① Toolpaths
- Click ② Next menu
- Click ③ Transform
- Click ④ check on ✔ for **2-Pocket[Standard]**
- Depress the Ctrl key and ***keeping it depressed***
- Click ⑤ check on ✔ for **3-Contour[2D]**
- Click ⑥ the Rotate radio button ◉
- Click ⑦ the **Rotate** tab

◆ ENTER THE ROTATE TOOLPATH PARAMETERS

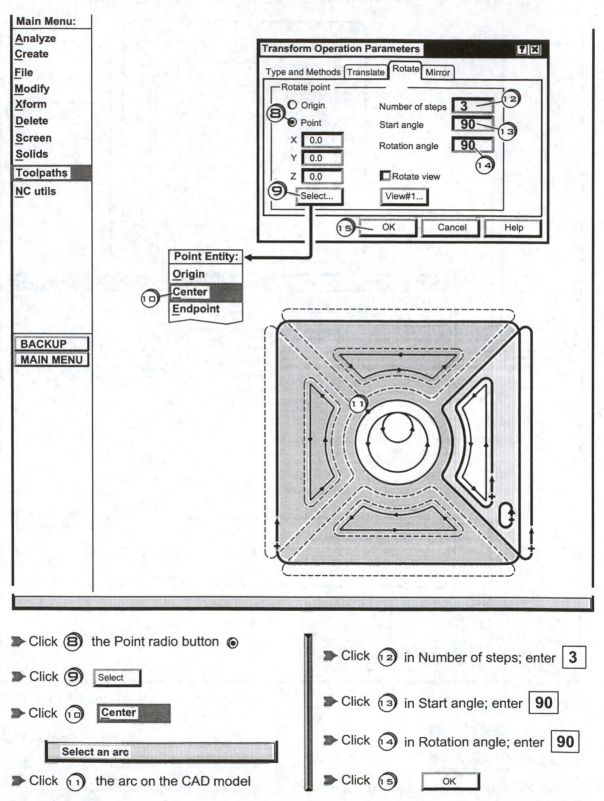

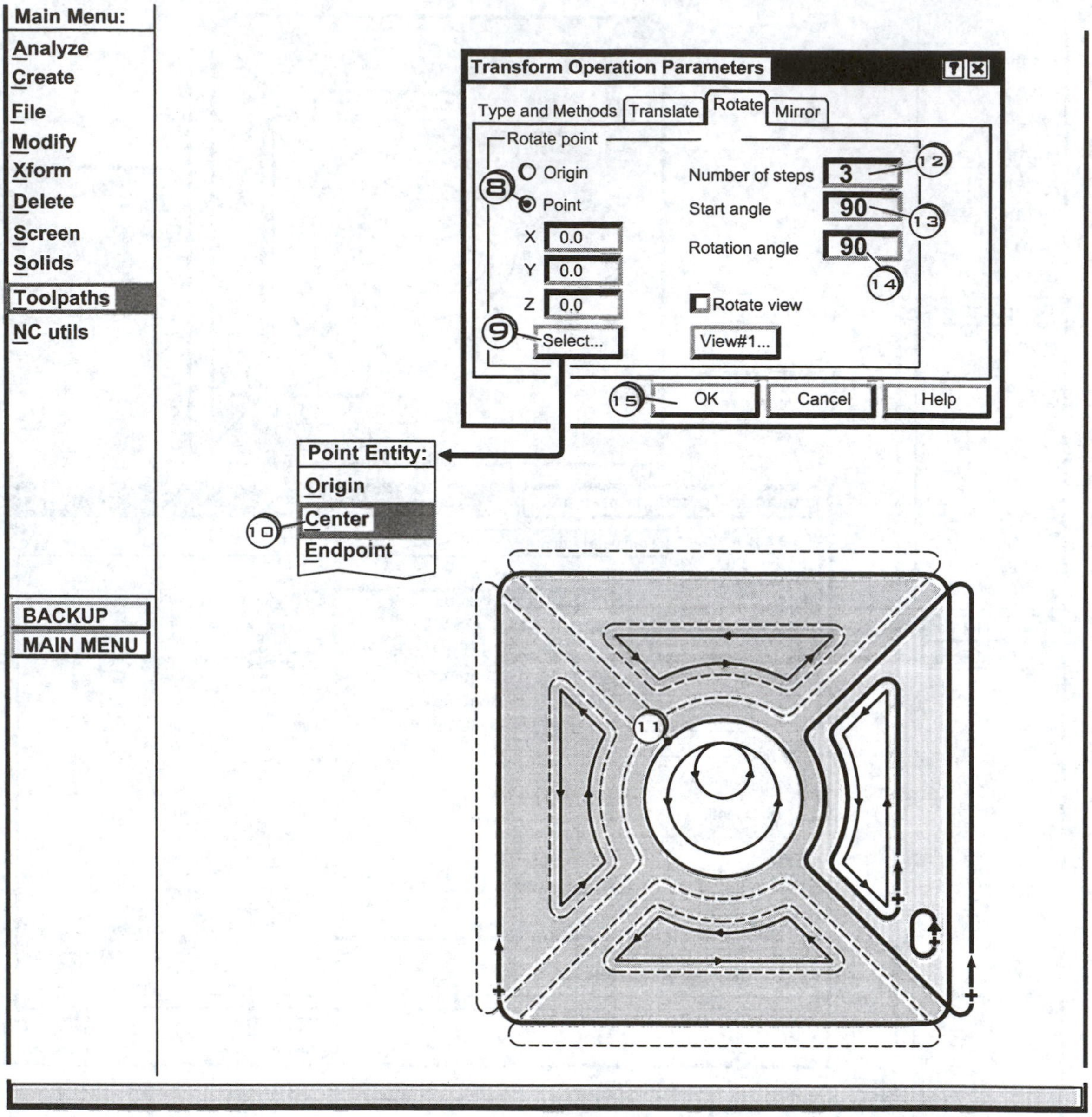

- Click (8) the Point radio button ◉
- Click (9) Select
- Click (10) Center

  Select an arc

- Click (11) the arc on the CAD model
- Click (12) in Number of steps; enter 3
- Click (13) in Start angle; enter 90
- Click (14) in Rotation angle; enter 90
- Click (15) OK

◆ IDENTIFY THE OPERATIONS TO BE TRANSLATED; SELECT TRANSLATE

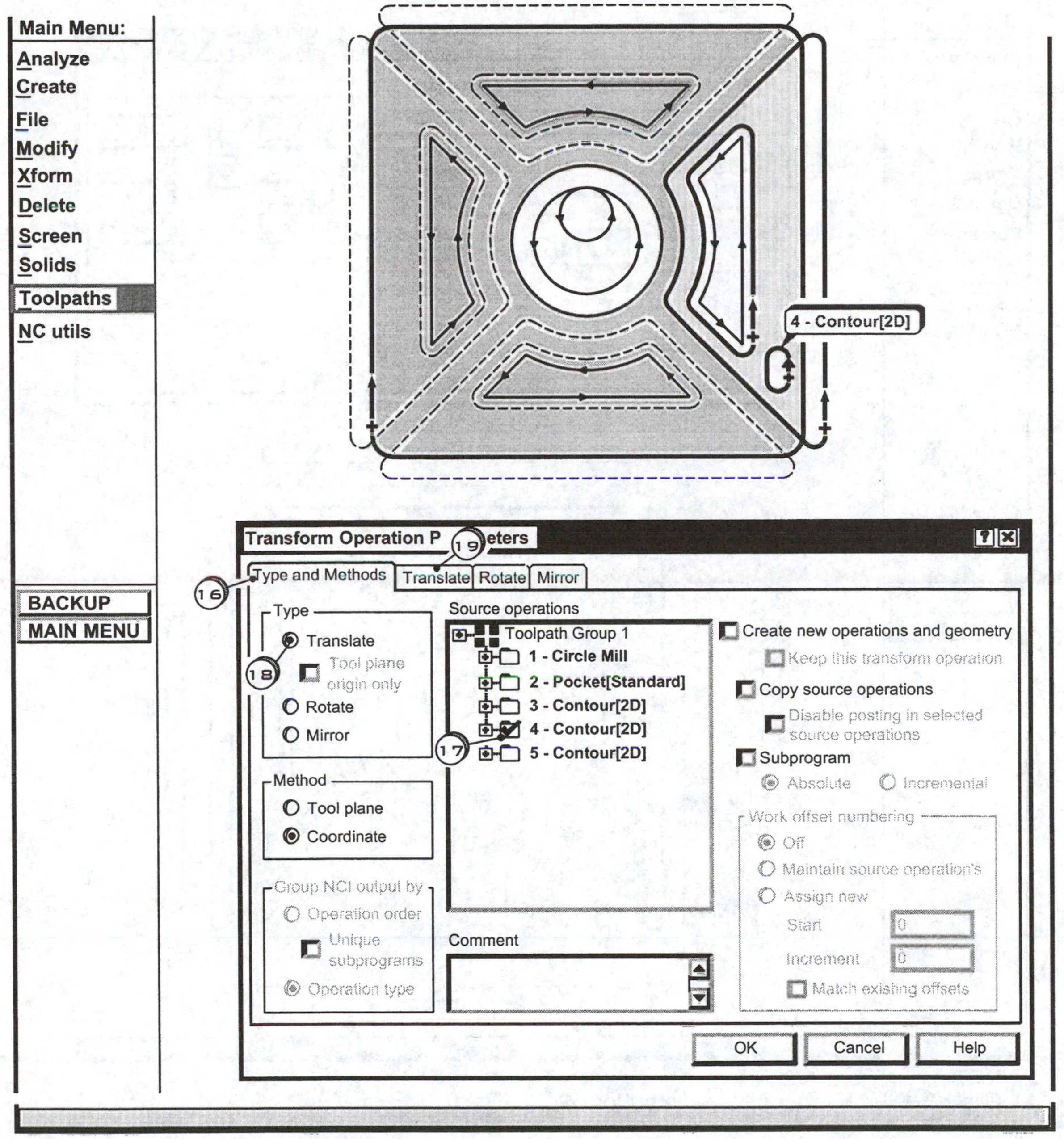

➤ Click (16) the Types and Methods tab

➤ Click (17) check on ✔ for **4-Contour[2D]**

➤ Click (18) the Translate radio button ◉

➤ Click (19) the **Translate** tab

◆ ENTER THE TRANSLATE TOOLPATH PARAMETERS

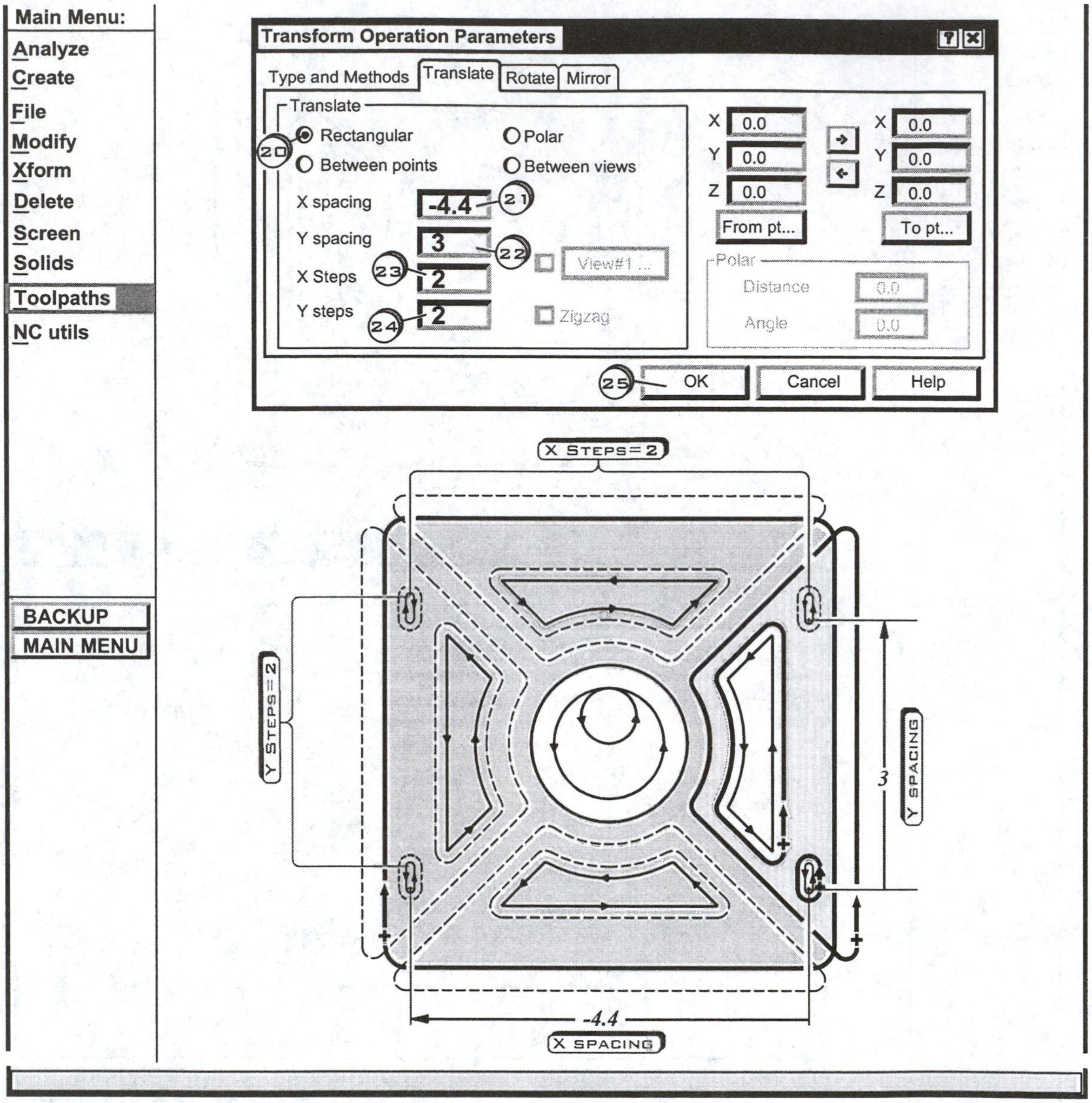

➤ Click (20) the Rectangular radio button ◉

➤ Click (21) in X spacing; enter the distance between each copy in the X -direction **-4.4**

➤ Click (22) in Y spacing; enter the distance between each copy in the Y -direction **3**

➤ Click (23) the number of X steps **2**

➤ Click (24) the number of Y steps **2**

➤ Click (25) OK

8-2) Copy CAD model file **EX8-2** in the folder **CHAPTER8** from the CD into the **JVAL-MILL** subdirectory on C drive. Open the file and generate a part program for executing the machining listed in PROCESS PLAN 8P-1.

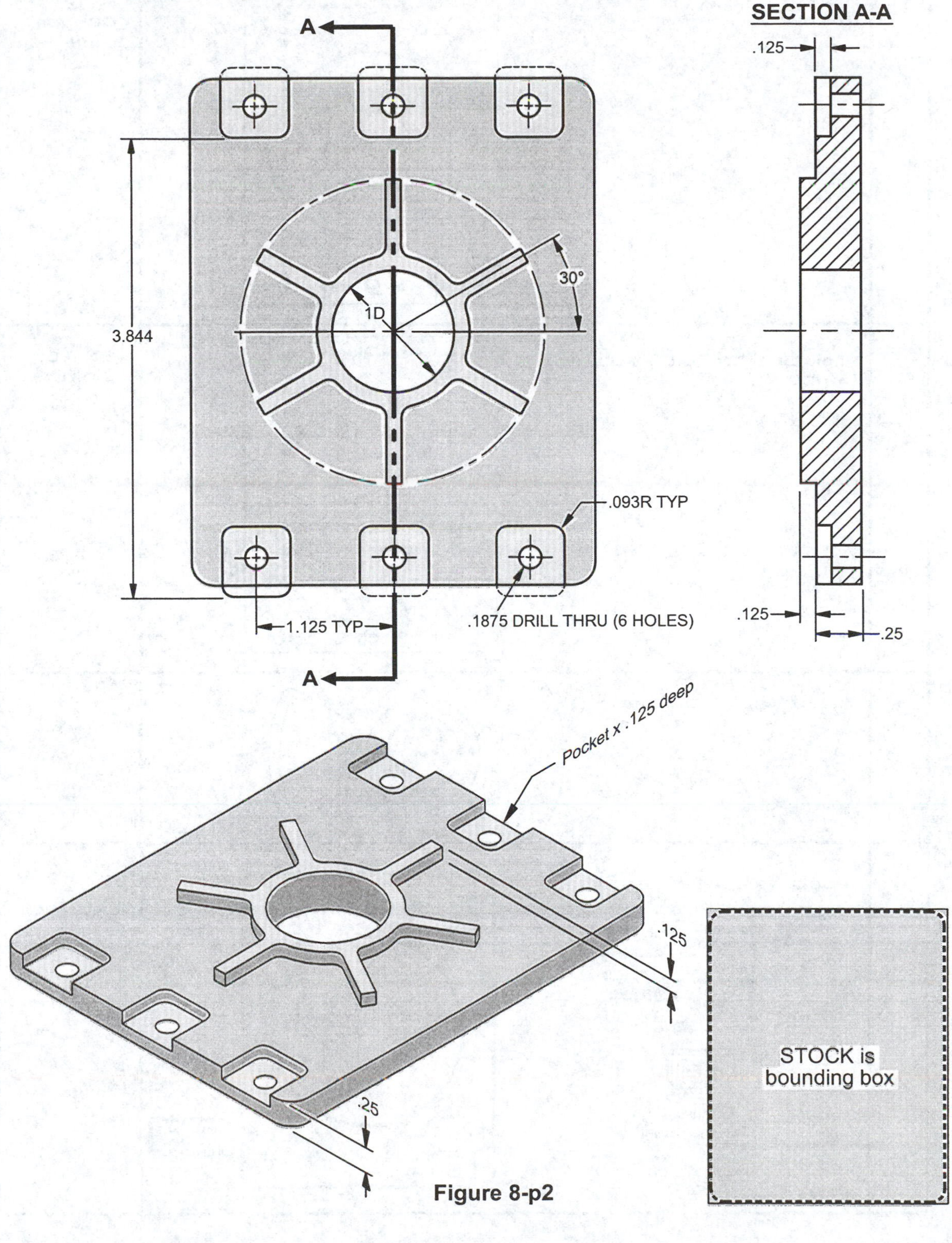

Figure 8-p2

PROCESS PLAN 8P-1

No.	Operation	Tooling
1	CENTER DRILL X .16 DEEP (6 HOLES) 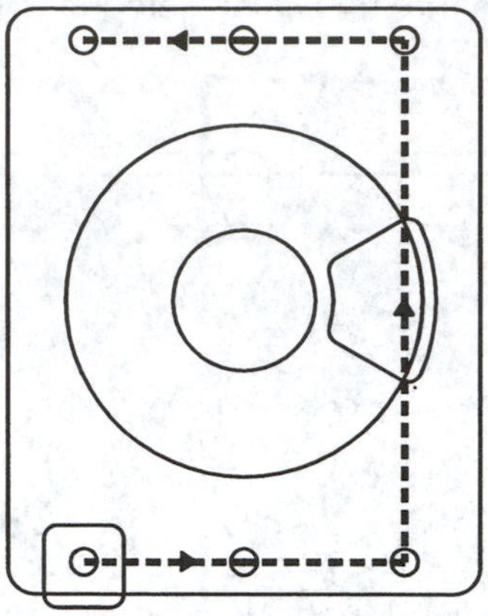	1/8 CENTER DRILL
2	PECK DRILL THRU(6 HOLES) (6 HOLES) 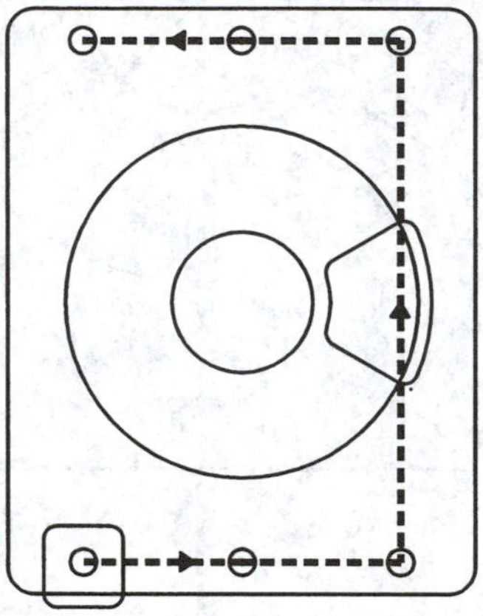	3/16 DRILL
3	CIRCLE MILL 1 DIA THRU 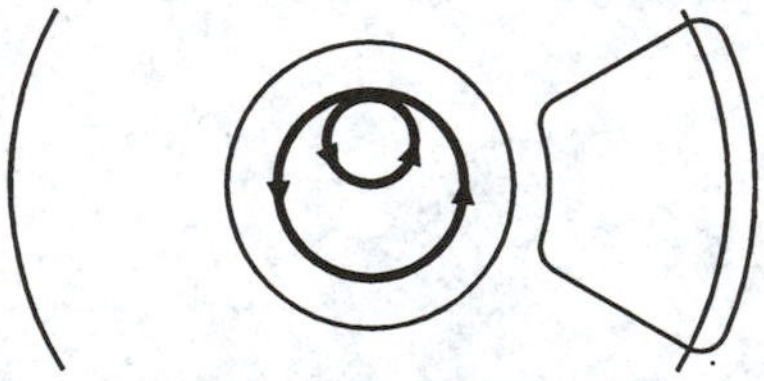	1/2 END MILL
4	POCKET X .125 DEEP 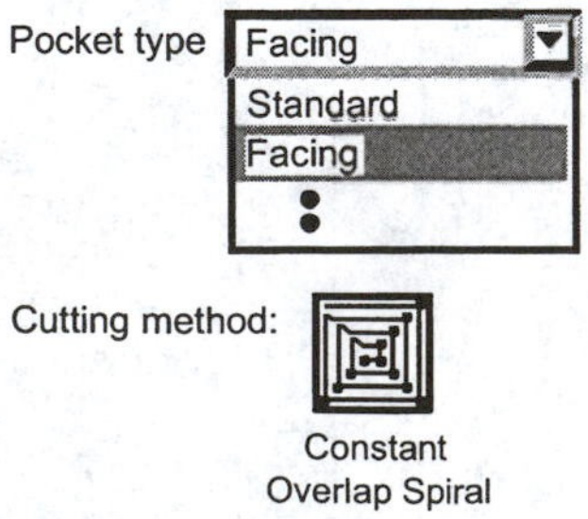 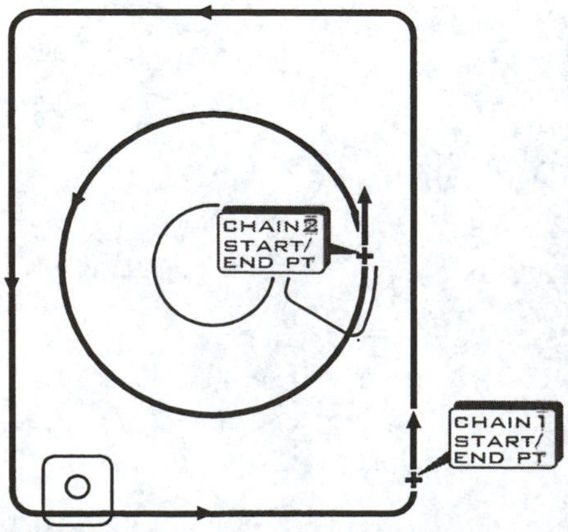	1/2 END MILL

PROCESS PLAN 8P-1(*continued*)

No.	Operation	Tooling
5	POCKET X .125 DEEP Pocket type Standard Cutting method:  Constant Overlap Spiral 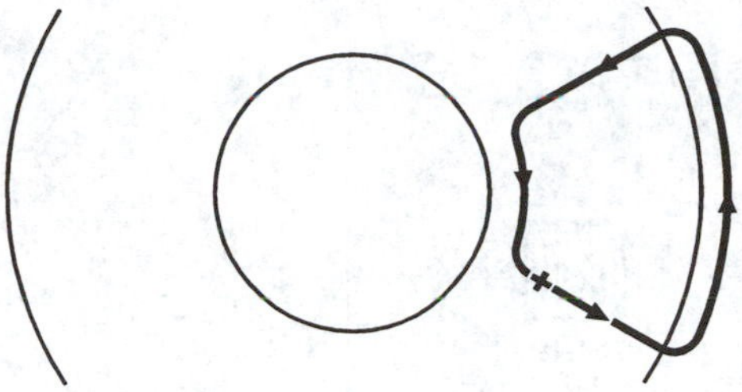	3/16 END MILL
6	TRANSFORM BY COORDINATE- ROTATE ABOUT POINT 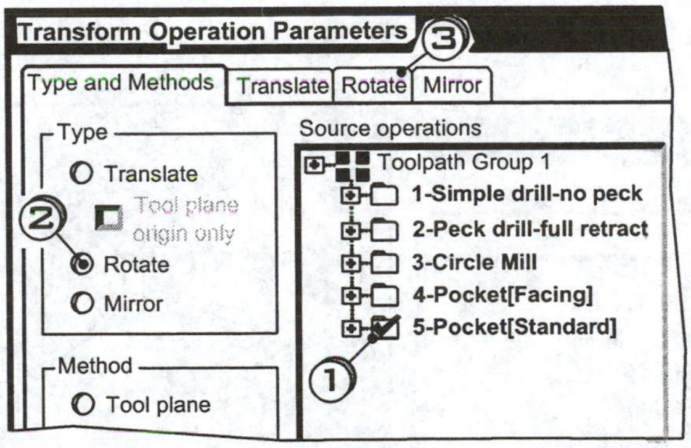 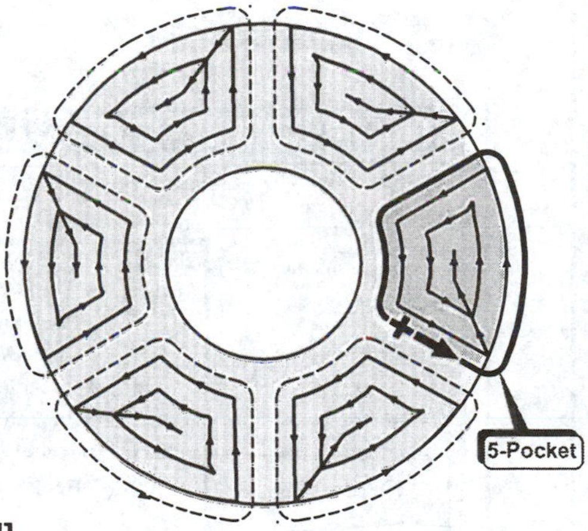 ➤ Click ① check on ✔ for **5-Pocket[Standard]** ➤ Click ② the Rotate radio button ◉ ➤ Click ③ the **Rotate** tab ➤ Enter the Transform toolpaths (rotation) parameters	

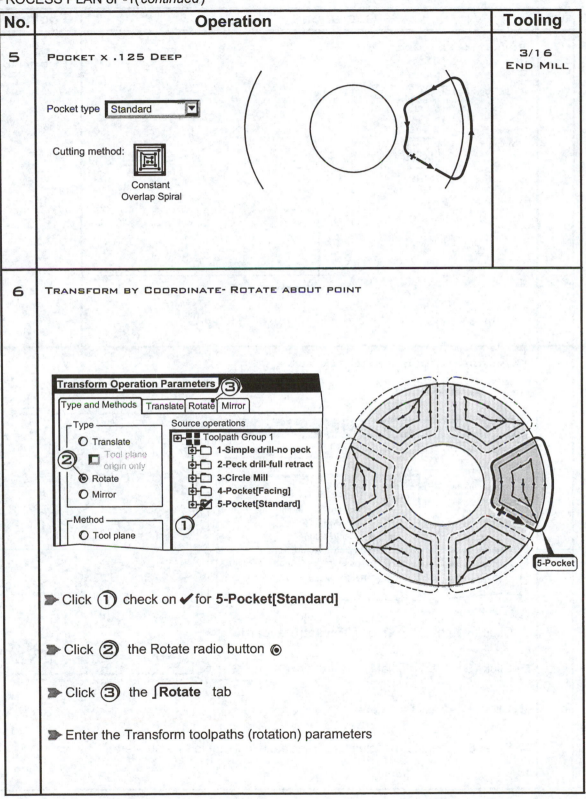

PROCESS PLAN 8P-1(*continued*)

<table>
<tr><th>No.</th><th>Operation</th><th>Tooling</th></tr>
<tr><td>7</td><td>

POCKET X .125 DEEP

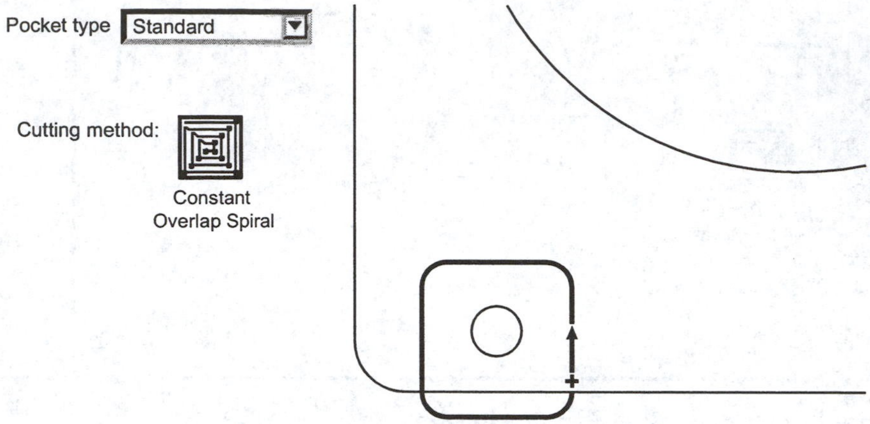

</td><td>3/16 END MILL</td></tr>
<tr><td>8</td><td colspan="2">

TRANSFORM BY COORDINATE- TRANSLATE

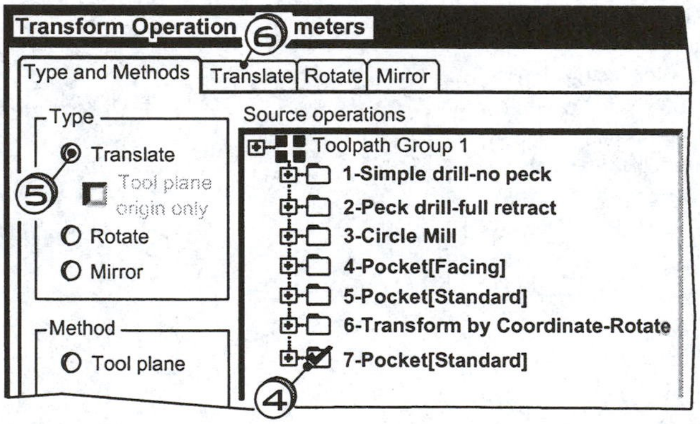

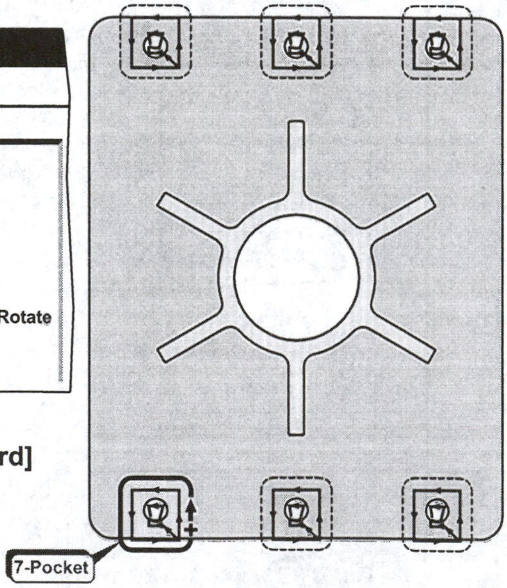

- Click ④ check on ✔ for **7-Pocket[Standard]**
- Click ⑤ the Translate radio button ◉
- Click ⑥ the **Translate** tab
- Enter the Transform toolpaths (Translate, Rectangular) parameters

</td></tr>
</table>

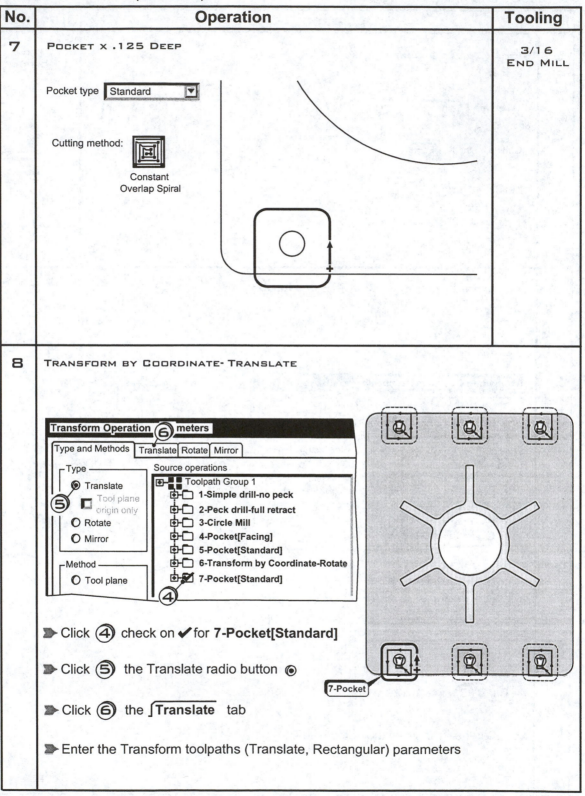

8-3) Copy CAD model file **EX8-3** in the folder **CHAPTER8** from the CD into the **JVAL-MILL** subdirectory on C drive. Open the file and generate a part program for executing the machining listed in PROCESS PLAN 8P-2.

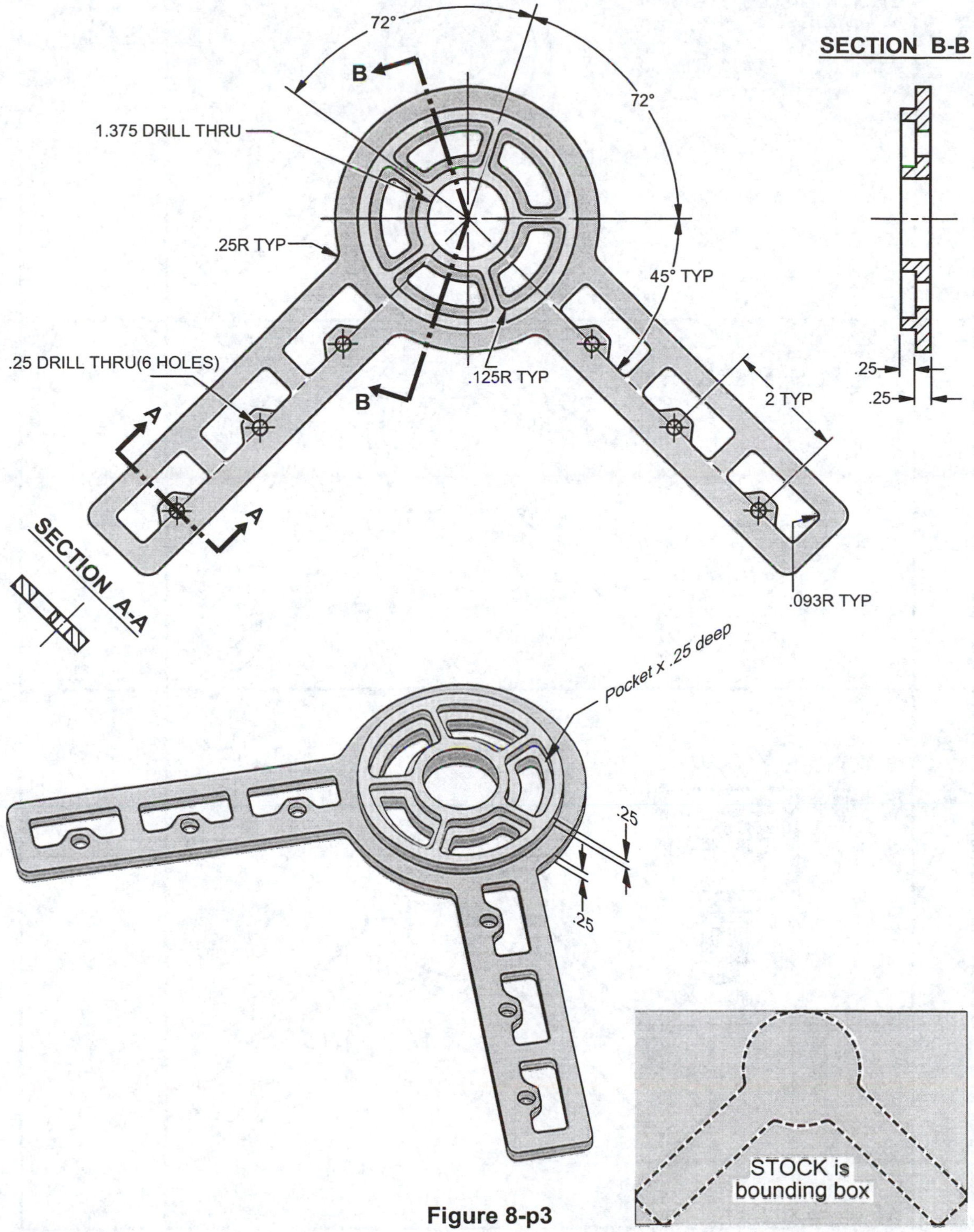

Figure 8-p3

PROCESS PLAN 8P-2

No.	Operation	Tooling
1	CENTER DRILL X .16 DEEP (6 HOLES) 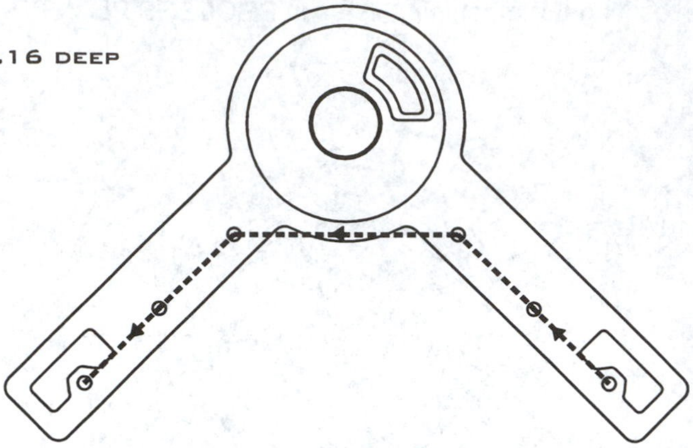	1/8 CENTER DRILL
2	PECK DRILL THRU(6 HOLES) 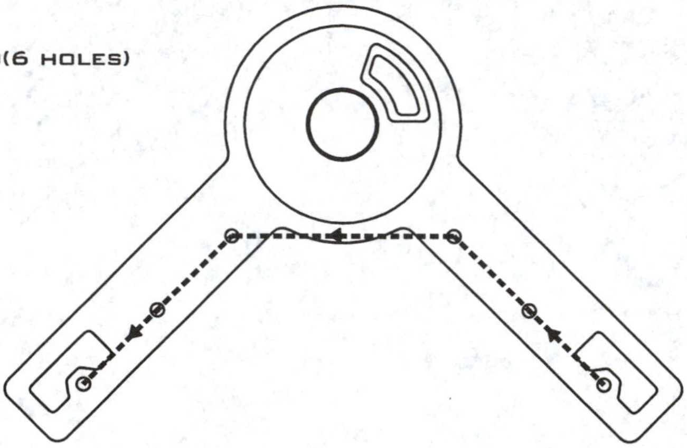	1/4 DRILL
3	POCKET X .25 DEEP 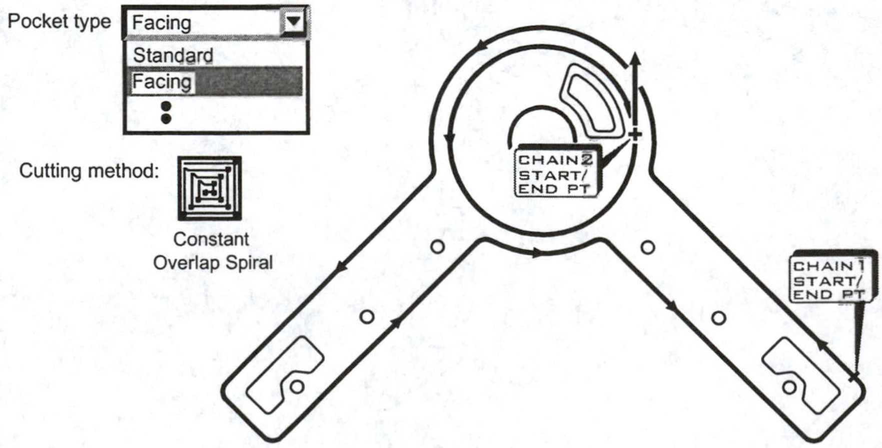	1/2 END MILL

PROCESS PLAN 8P-2(*continued*)

No.	Operation	Tooling
4	CIRCLE MILL 1.375 DIA THRU 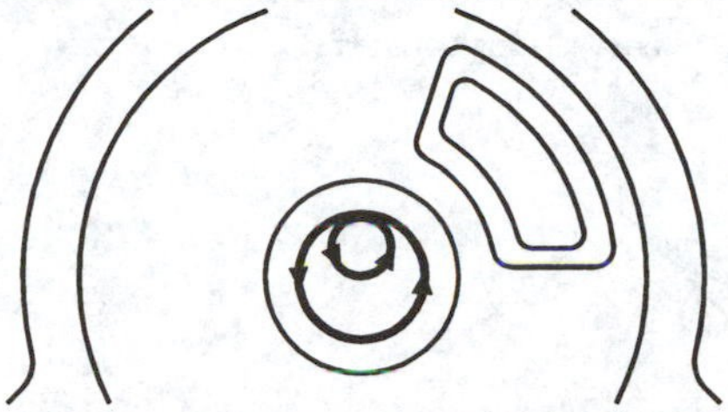	1/2 END MILL
5	POCKET X .25 DEEP Pocket type: Standard Cutting method: Constant Overlap Spiral 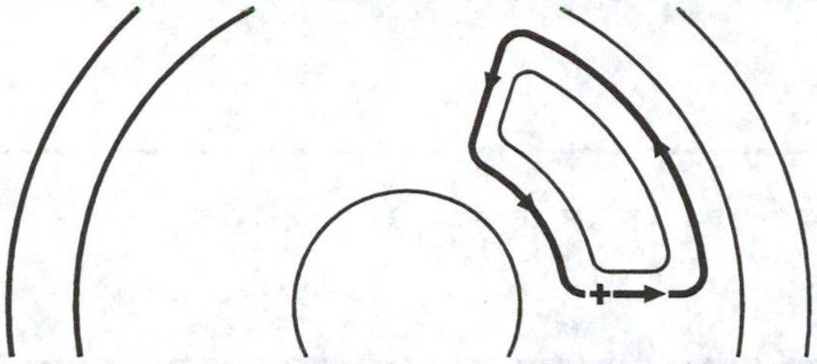	1/4 END MILL
6	ROUGH AND FINISH SLOT X .25 DEEP 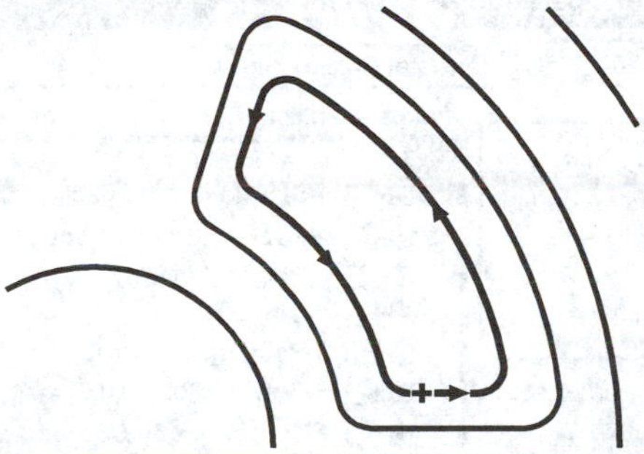	3/16 END MILL
7	TRANSFORM BY COORDINATE- ROTATE ABOUT ORIGIN 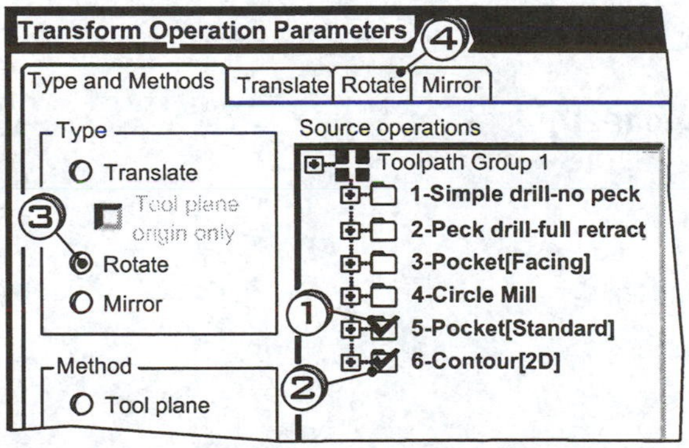 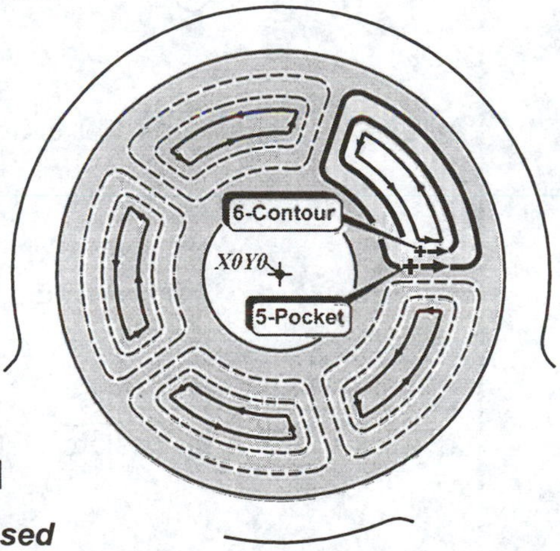 ➤ Click ① check on ✔ for **5-Pocket[Standard]** ➤ Depress the Ctrl key and ***keeping it depressed*** ➤ Click ② check on ✔ for **6-Contour[2D]** ➤ Click ③ the Rotate radio button ◉ ➤ Click ④ the **Rotate** tab ➤ Enter the Transform toolpaths (rotation) parameters	

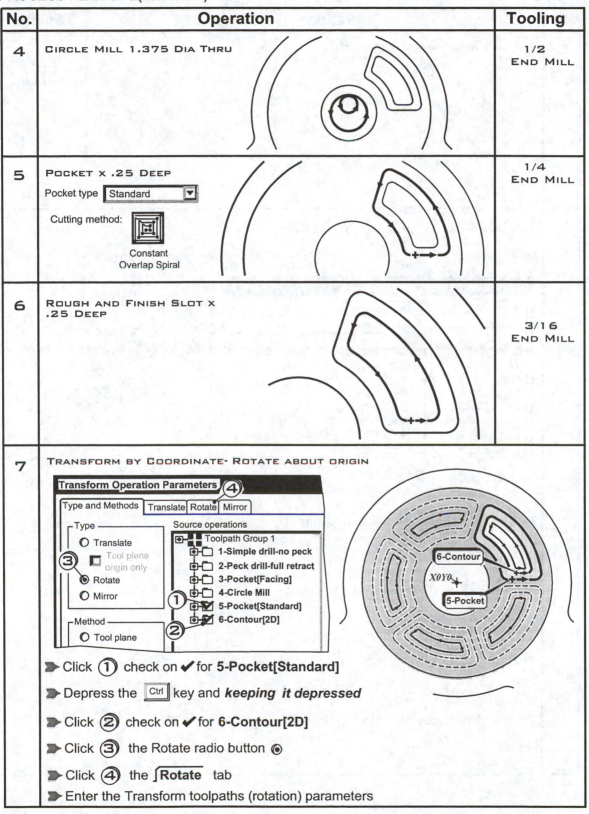

PROCESS PLAN 8P-2(*continued*)

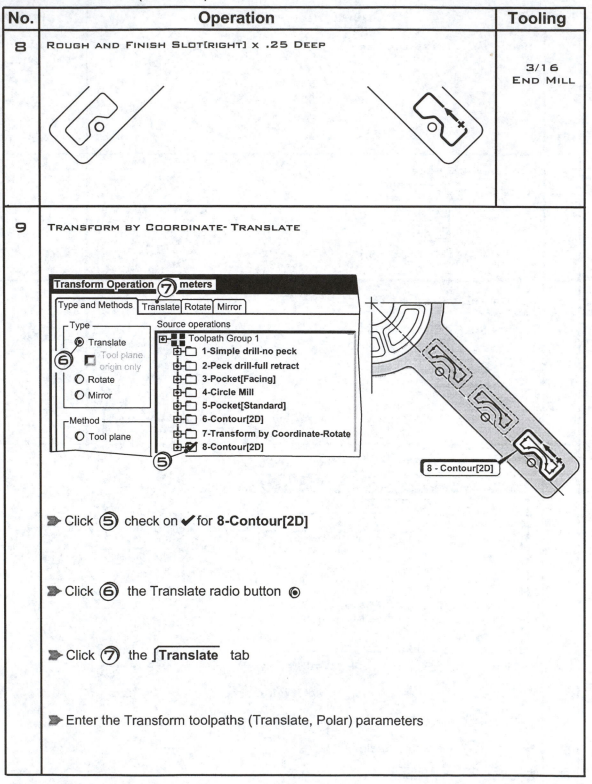

No.	Operation	Tooling
8	ROUGH AND FINISH SLOT[RIGHT] X .25 DEEP 	3/16 END MILL
9	TRANSFORM BY COORDINATE- TRANSLATE 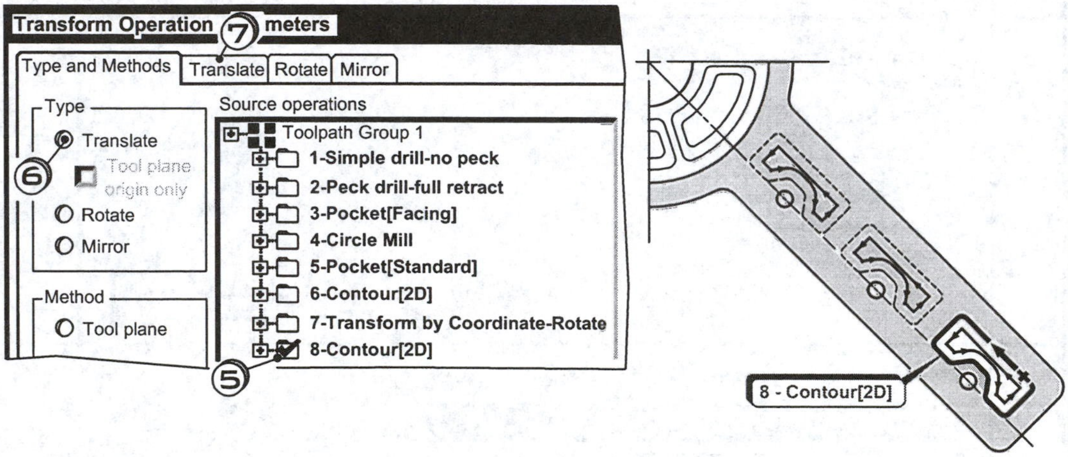 ➤ Click ⑤ check on ✔ for **8-Contour[2D]** ➤ Click ⑥ the Translate radio button ◉ ➤ Click ⑦ the **Translate** tab ➤ Enter the Transform toolpaths (Translate, Polar) parameters	

PROCESS PLAN 8P-2(*continued*)

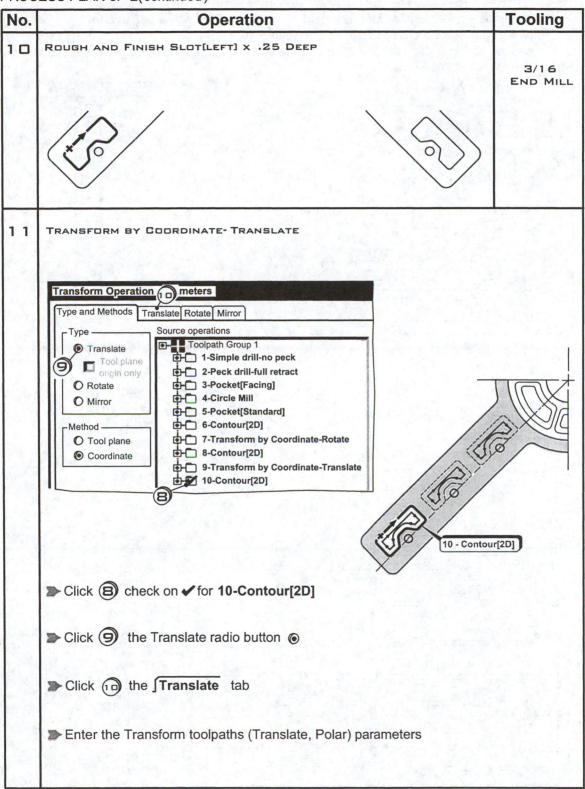

<table>
<tr><th>No.</th><th>Operation</th><th>Tooling</th></tr>
<tr><td>10</td><td>ROUGH AND FINISH SLOT[LEFT] X .25 DEEP
</td><td>3/16
END MILL</td></tr>
<tr><td>11</td><td>TRANSFORM BY COORDINATE- TRANSLATE
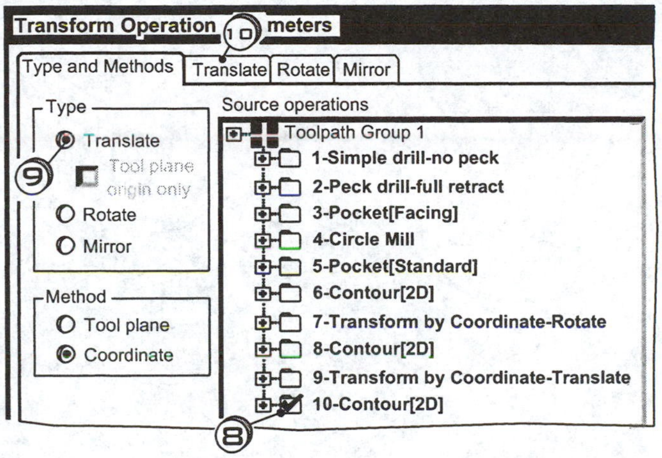

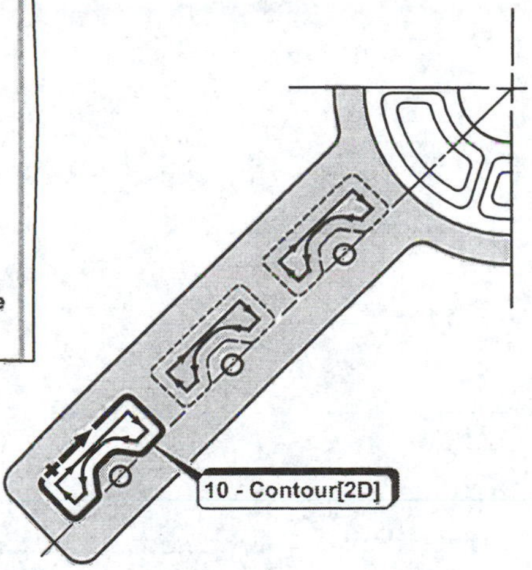

➤ Click (8) check on ✔ for **10-Contour[2D]**
➤ Click (9) the Translate radio button ◉
➤ Click (10) the **Translate** tab
➤ Enter the Transform toolpaths (Translate, Polar) parameters</td><td></td></tr>
</table>

8-4) Open the CAD model file **EX3-7JV** created in exercise 3-7 in chapter 3
Generate a part program for machining four copies of the same part in one table setup at the CNC machine as shown in Figure 8p-4.
The required operations that must be created in the Operations Manager are listed in PROCESS PLAN 8P-3.

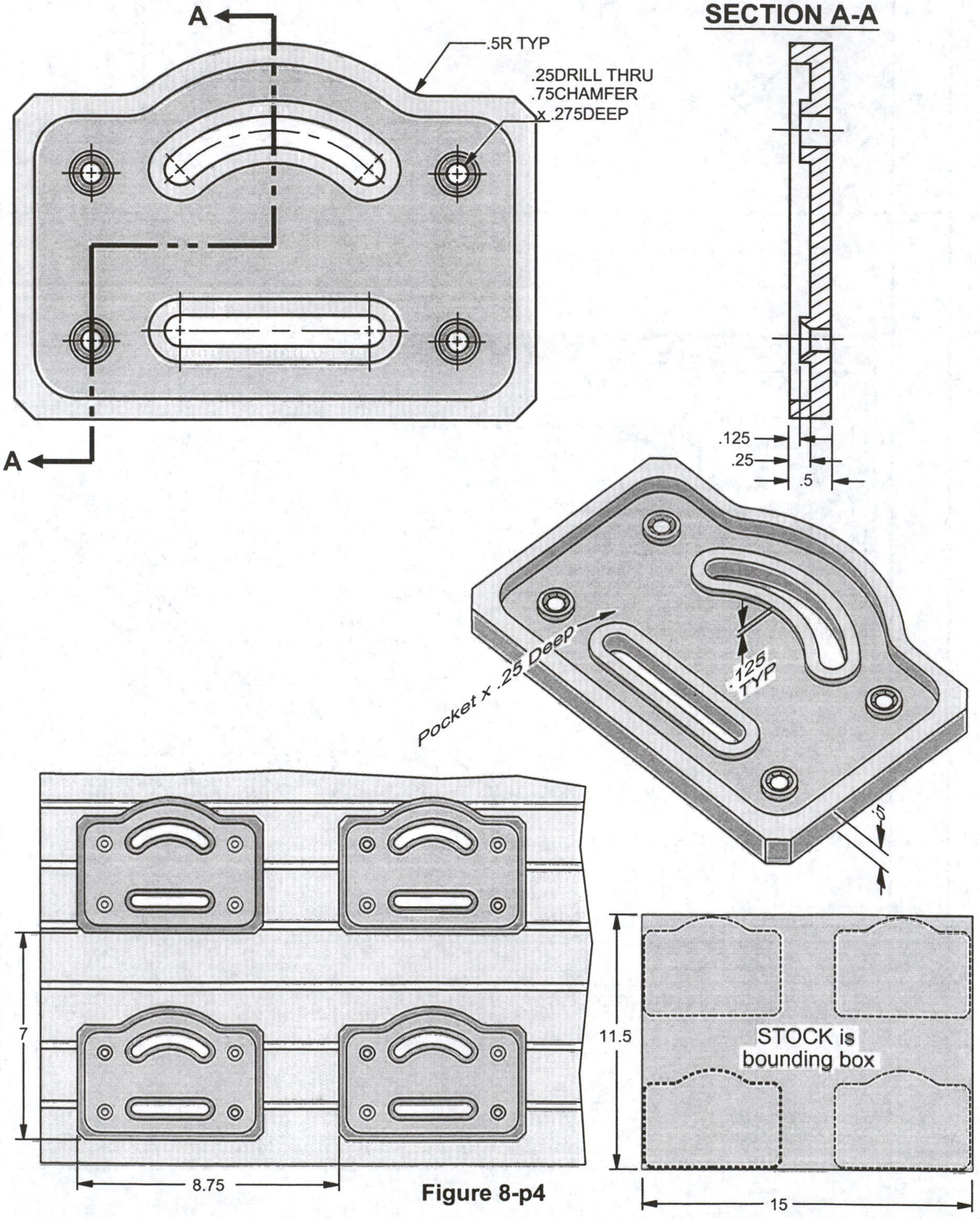

Figure 8-p4

PROCESS PLAN 8P-3

No.	Operation	Tooling
1	POCKET[ISLAND FACING] LEAVE .01 FOR FINISH CUT IN XY AND Z ➤ Click ① Area ➤ Click ② the location of the point *Mastercam* will ***automatically chain*** all the closed boundaries in the ***area surrounding the point*** 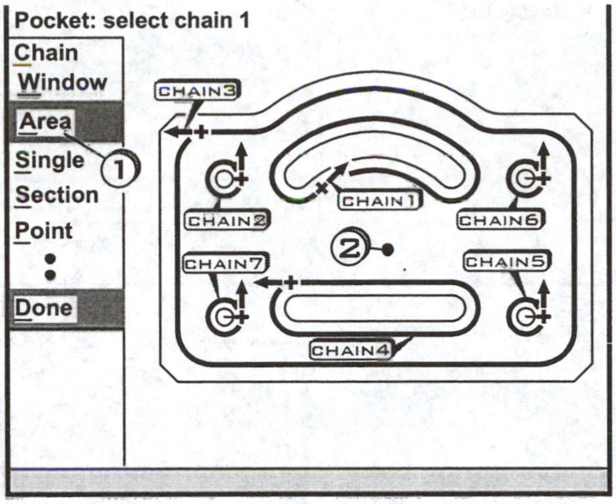 Pocket type 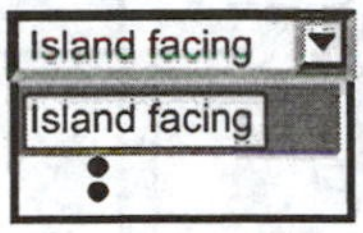 Cutting method:  Parallel Spiral Clean Corners	3/8 END MILL
2	CENTER DRILL X .285 DEEP (4 HOLES) 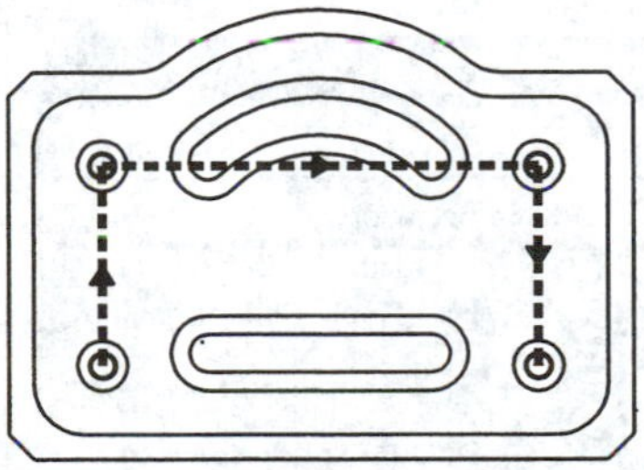	1/8 CENTER DRILL
3	PECK DRILL THRU(4 HOLES) 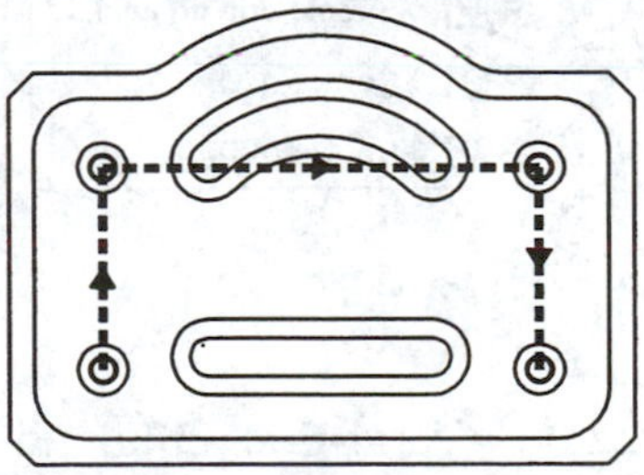	1/4 DRILL
4	CHAMFER X .275 DEEP(4 HOLES) 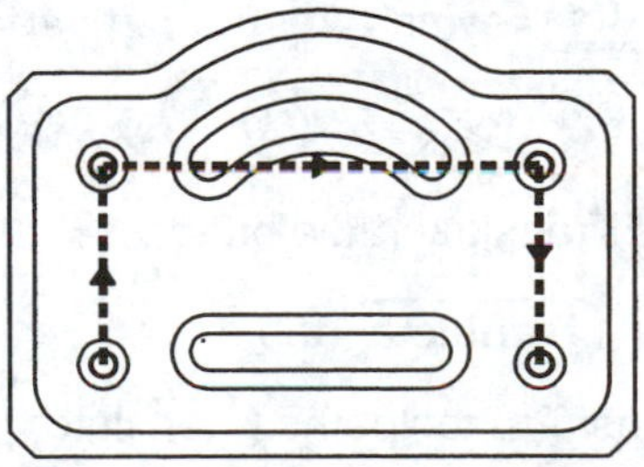	1 CHAMFER MILL

PROCESS PLAN 8P-3(*continued*)

No.	Operation	Tooling
5	ROUGH AND FINISH SLOTS X .5 DEEP 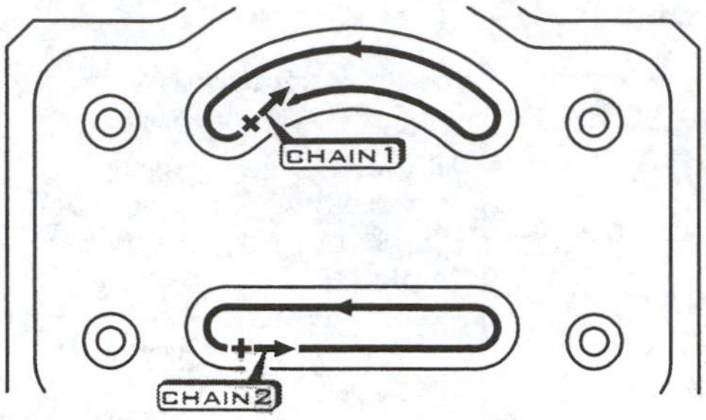 	1/4 END MILL
6	CONTOUR OUTSIDE PROFILE 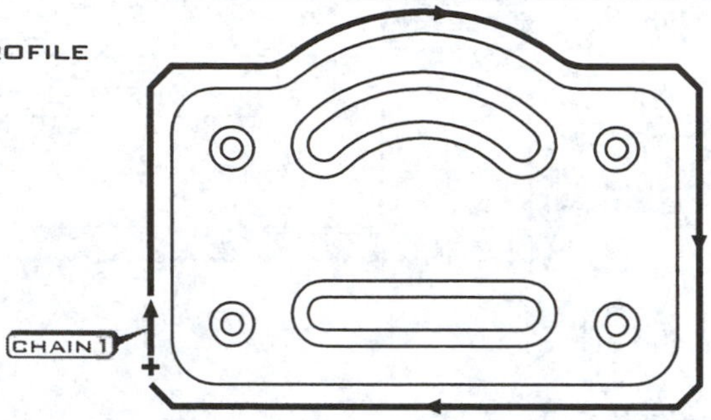 	1/2 END MILL
7	TRANSFORM BY COORDINATE- TRANSLATE 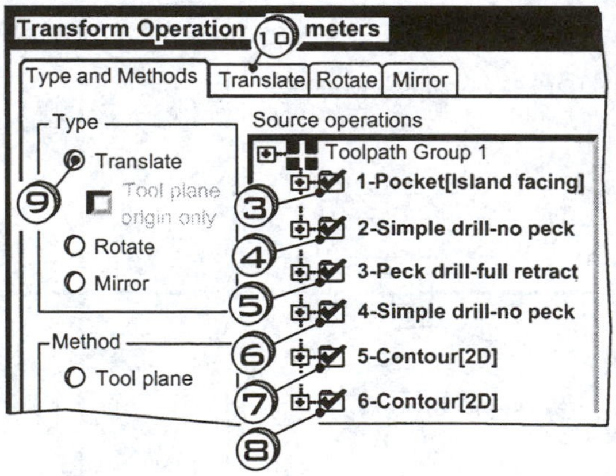 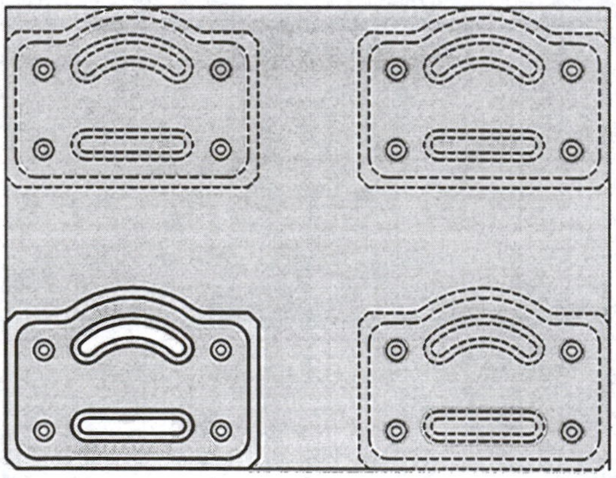 ➤ Click (3) check on ✔ for **1-Pocket[Island facing]** ➤ Depress the Ctrl key and ***keeping it depressed*** ➤ Click check on ✔ for (4) (5) (6) (7) (8) ➤ Click (9) the Translate radio button ◉ ➤ Click (10) the **Translate** tab ➤ Enter the Transform toolpaths (Translate) parameters	

CHAPTER - 9

USING A LIBRARY TO SAVE AND IMPORT MACHINING OPERATIONS

9-1 Chapter Objectives

After completing this chapter you will be able to:

1. Understand the advantages of using *Mastercam*'s operations library
2. Know how to save operations to the operations library
3. Explain how to import operations from the operations library.
4. State how to edit data in the operations library.

9-2 Advantages of Using Mastercam's Operations Library

Mastercam's Operations Library is an important productivity booster. It enables the operator to catalog and store an operation or sequence of operations under a single group name. For example, the group name stored in the library **TAP_1/4-20** could contain the center drilling, drilling and tapping operations required for producing the hole. Later the operator can import the group **TAP_1/4-20** from the library and apply it to any other subsequent part as needed. Producing holes that require center drilling, drilling, tapping, reaming, chamfering and circle milling, etc account for a majority of the machining in many shops. Creating a library of these repetative machining operations will cut down on needless duplication of effort.

9-3 Saving Operations to the Operations Library

EXAMPLE 9-1

Create the operation group names and the operations associated with each group listed in Table 9-1. Save the groups to the operations library for use in other parts.

Table 9-1

Group Name	Operation(s)	Tooling
CBORE-D5/16-E3/8x1/4 CBORE: *c'bore is produced*; D5/16: *first use 5/16 drill thru*; E3/8x1/4: *then use a 3/8 Endmill to machine c'bore to 1/4 deep*	CENTER DRILL X .2 DEEP PECK DRILL THRU CIRCLE MILL X .25 DEEP	1/8 CTR DRILL 5/16 DRILL 3/8 ENDMILL
POC-E3/16x1/8 POC: *pocket is produced*; E3/16x1/8: *use a 3/16 Endmill machine pocket to 1/8 deep*	ROUGH AND FINISH POCKET X .125 DEEP. LEAVE .01 FOR FINISH CUT IN XY AND Z	3/16 ENDMILL

Note: **D** signifies *drill;* **E** *signifies end mill*

A) CREATE THE MINIMUM GEOMETRY NEEDED TO ESTABLISH AND VERIFY THE TOOLPATHS FOR EACH GROUP

- CREATE A .25DIA CIRCLE, AND A 1 X 1 RECTANGLE CENTERED WITHIN A 2 X 2 RECTANGLE

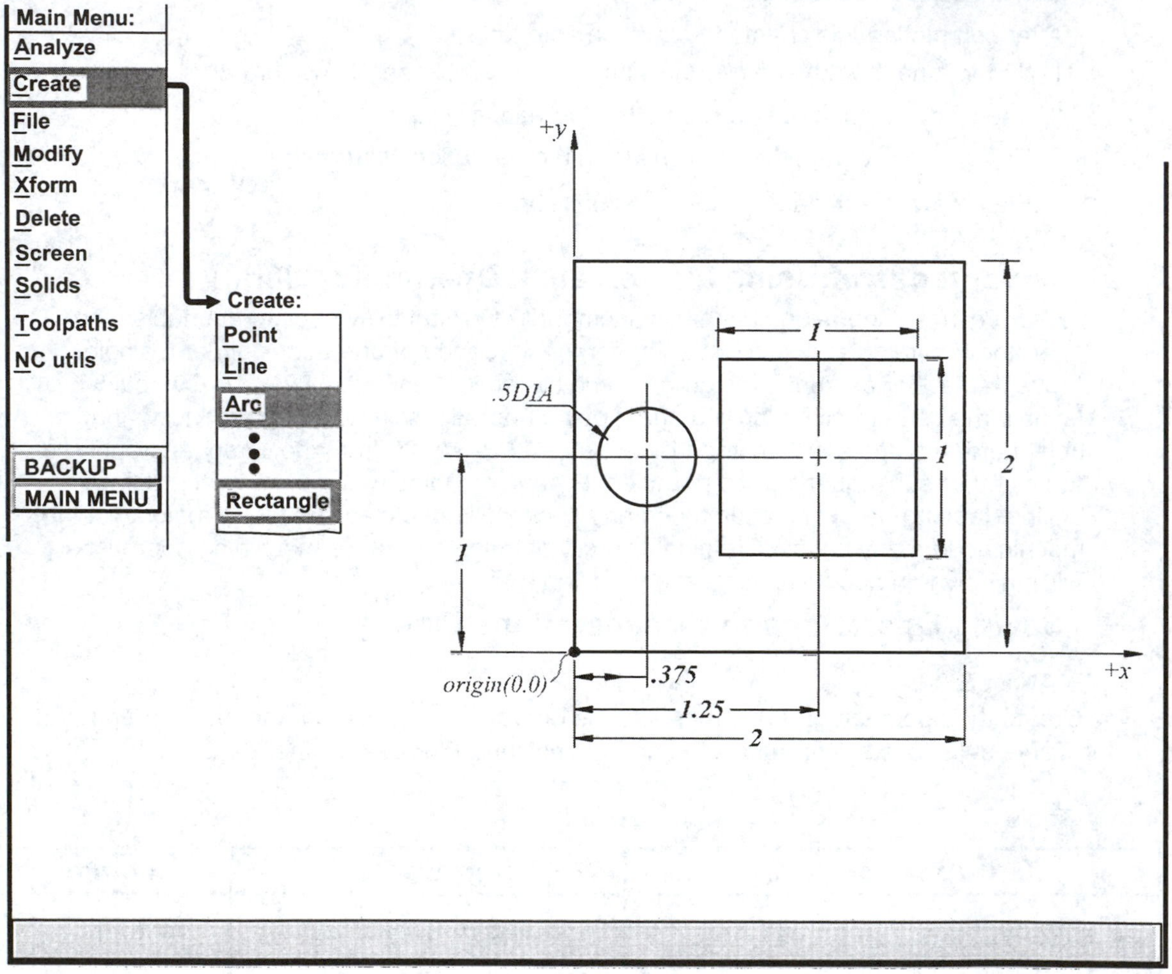

B) Generate operations for Group CBORE-D5/16-E3/8x1/4

- CREATE THE 1/8 CENTER DRILL TOOLPATH

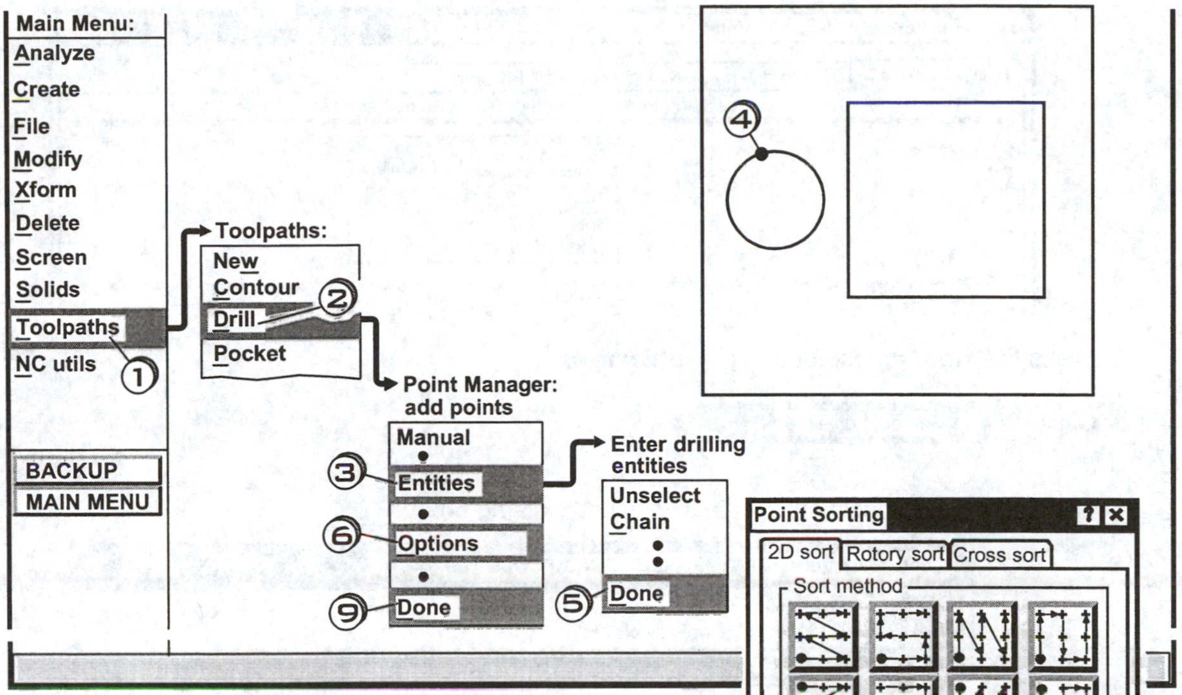

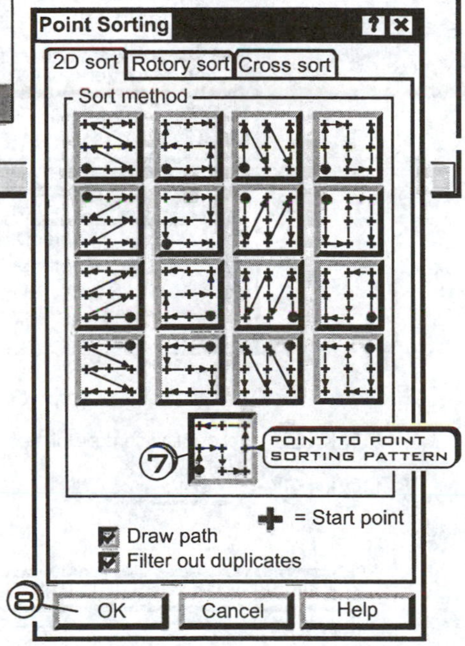

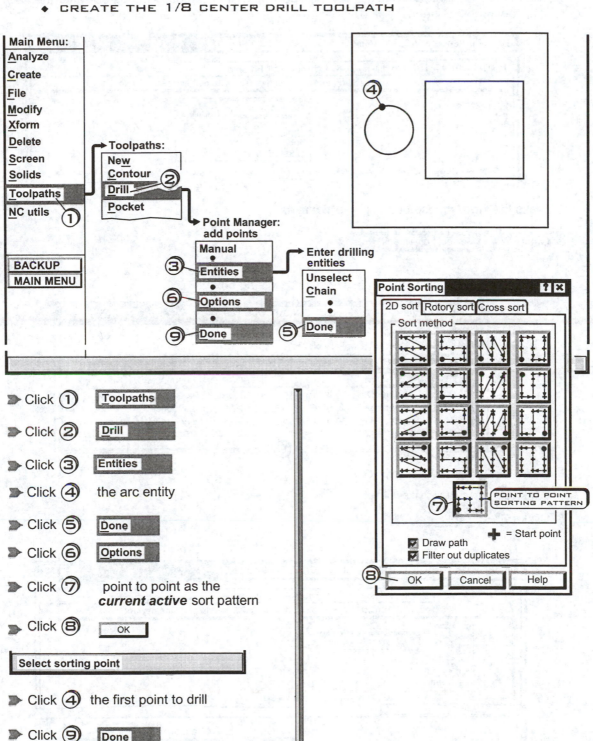

- Click ① **Toolpaths**
- Click ② **Drill**
- Click ③ **Entities**
- Click ④ the arc entity
- Click ⑤ **Done**
- Click ⑥ **Options**
- Click ⑦ point to point as the ***current active*** sort pattern
- Click ⑧ OK

Select sorting point

- Click ④ the first point to drill
- Click ⑨ **Done**

◆ OBTAIN THE NEEDED .125 DIA CENTER DRILL TOOL

Mastercam will activate and display the Tool and Drill Parameters dialog box after the drill tool path has been created.

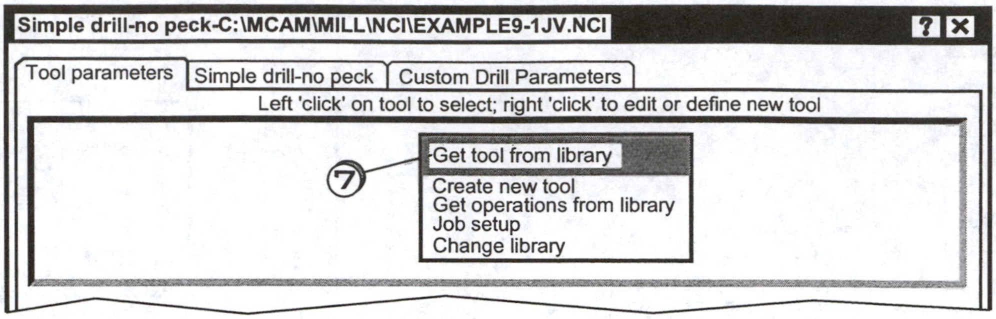

➤ Press the ***right*** mouse button in the ***white area***

➤ Click ⑦ Get tool from library

Mastercam will display the main Tool Library dialog box

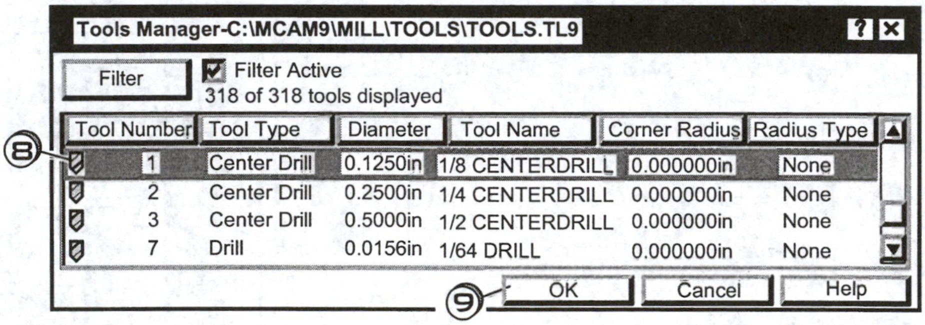

➤ Click ⑧ the .125Dia center Drill tool

➤ Click ⑨ OK

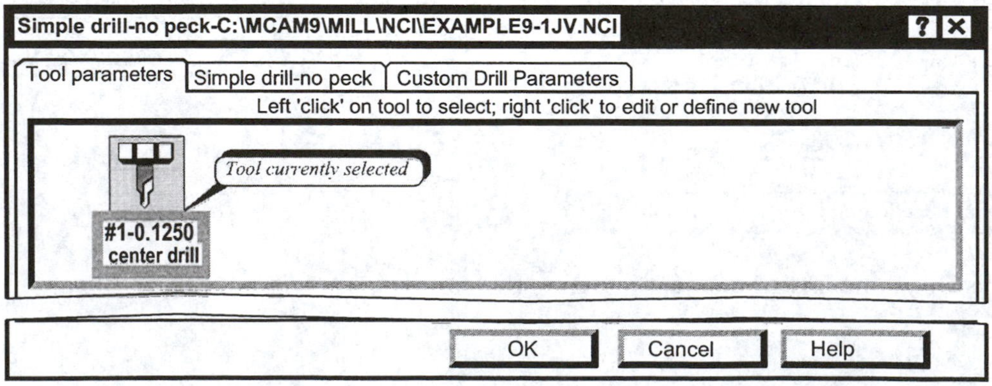

◆ ENTER THE .125 DIA CENTER DRILL HOLE MACHINING PARAMETERS

➤ Click (10) [Simple drill-no peck] tab to enter the .125Dia hole machining parameters.

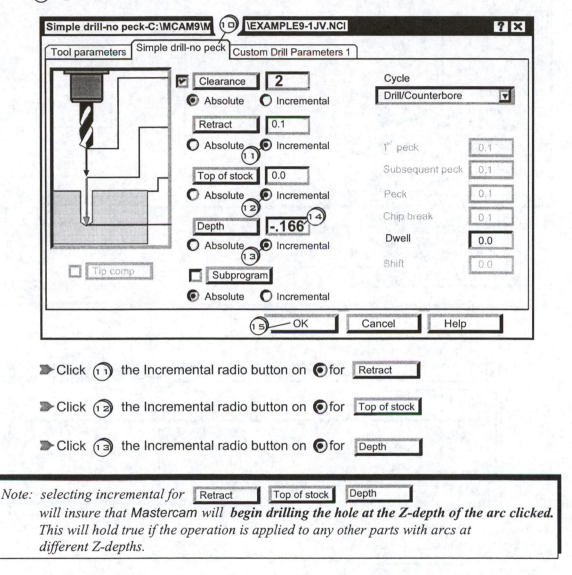

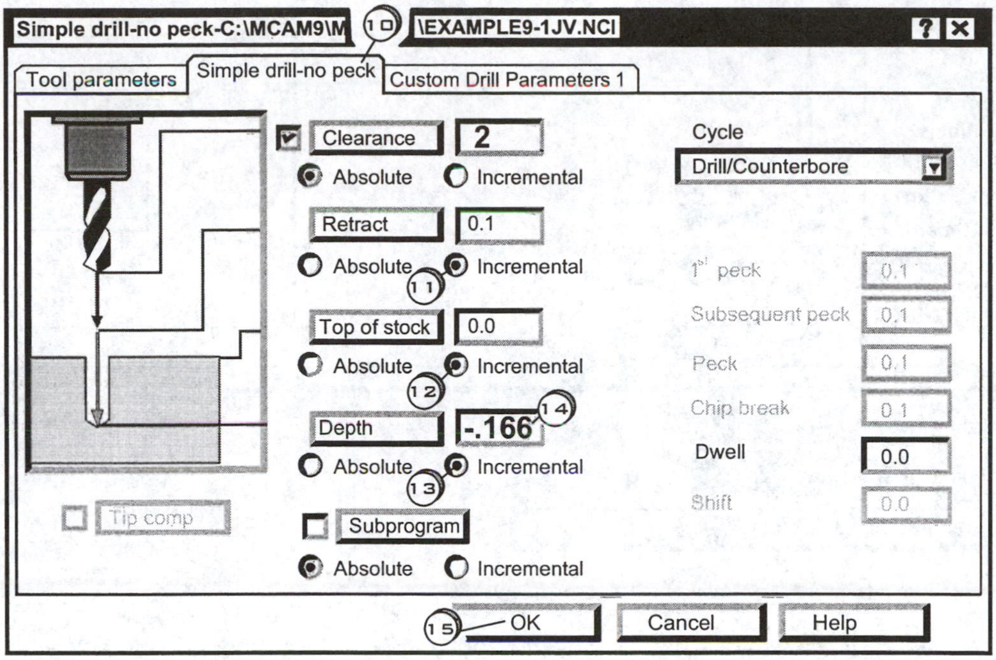

➤ Click (11) the Incremental radio button on ◉ for [Retract]

➤ Click (12) the Incremental radio button on ◉ for [Top of stock]

➤ Click (13) the Incremental radio button on ◉ for [Depth]

> *Note: selecting incremental for* [Retract] [Top of stock] [Depth]
> *will insure that* Mastercam *will* ***begin drilling the hole at the Z-depth of the arc clicked.*** *This will hold true if the operation is applied to any other parts with arcs at different Z-depths.*

➤ Click (14) in the Depth box; enter [-.166]

➤ Click (15) [OK]

◆ CREATE THE .313 PECK DRILL THRU TOOLPATH

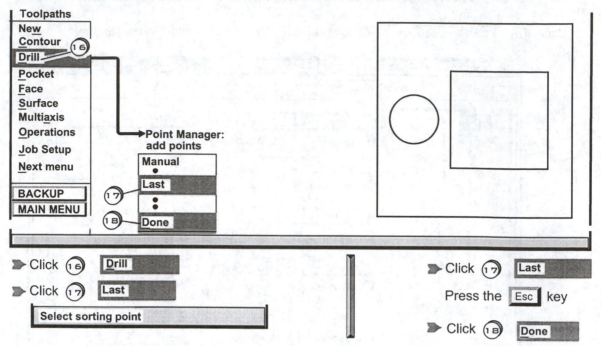

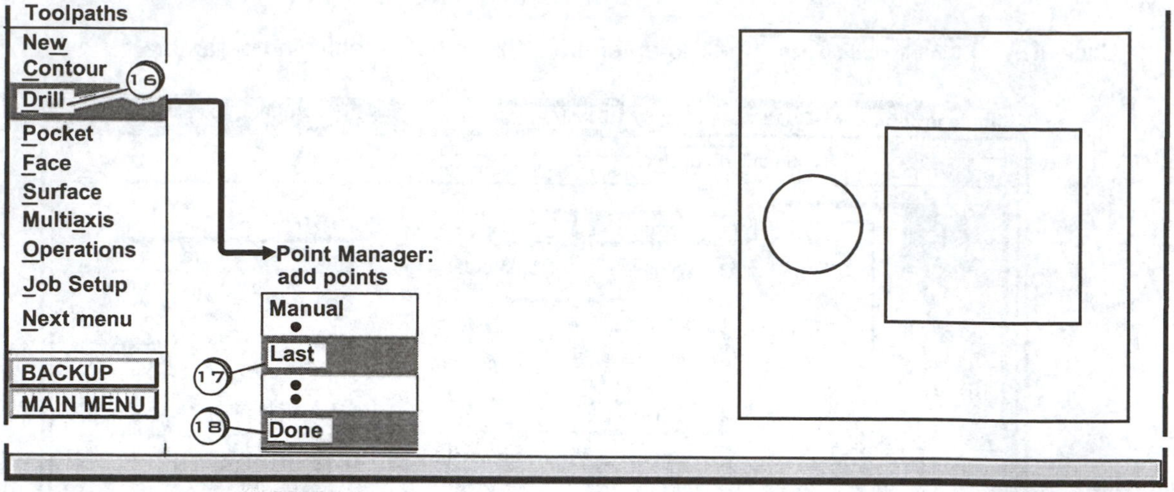

➤ Click (16) Drill

➤ Click (17) Last

Select sorting point

➤ Click (17) Last

Press the Esc key

➤ Click (18) Done

◆ OBTAIN THE NEEDED .201DIA DRILL TOOL

Mastercam will activate and display the Tool and Drill Parameters dialog box after the drill tool path has been created.

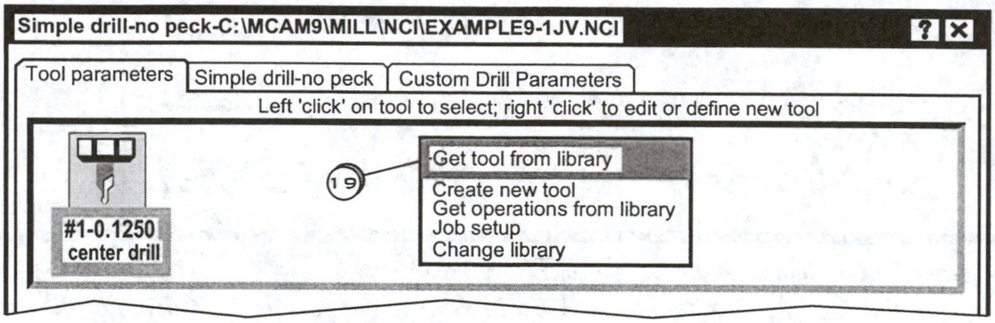

➤ Press the ***right*** mouse button in the ***white area***

➤ Click (19) Get tool from library *Mastercam* will display the main Tool Library dialog box

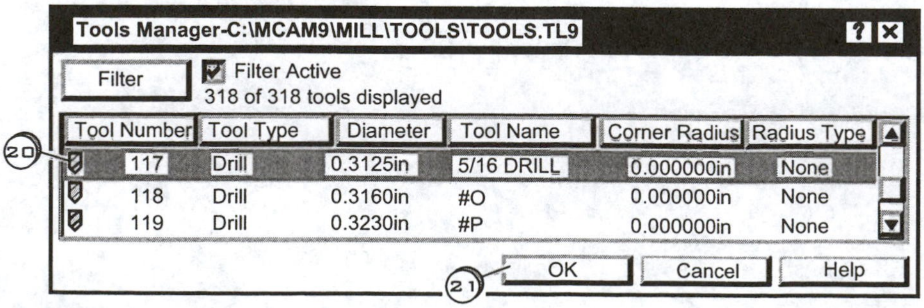

➤ Click (20) the .3125Dia Drill tool

➤ Click (21) OK

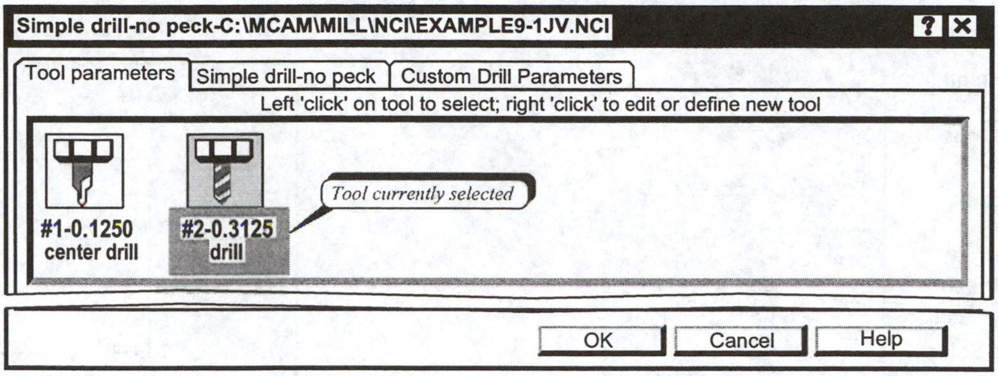

- ENTER THE .313DIA PECK DRILL THRU MACHINING PARAMETERS

➤ Click 22 Peck drill-full retract tab to enter the .313Dia peck drill machining parameters.

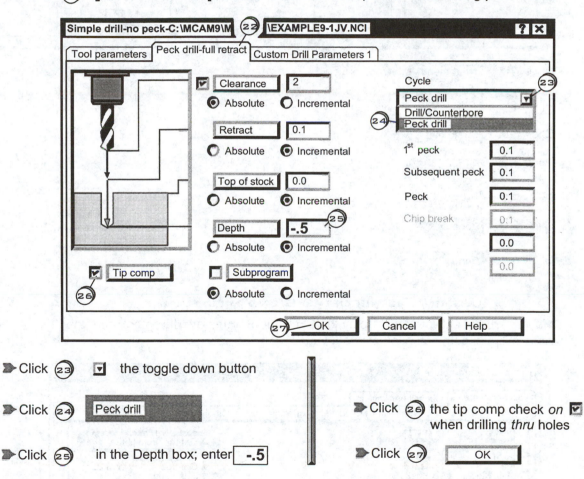

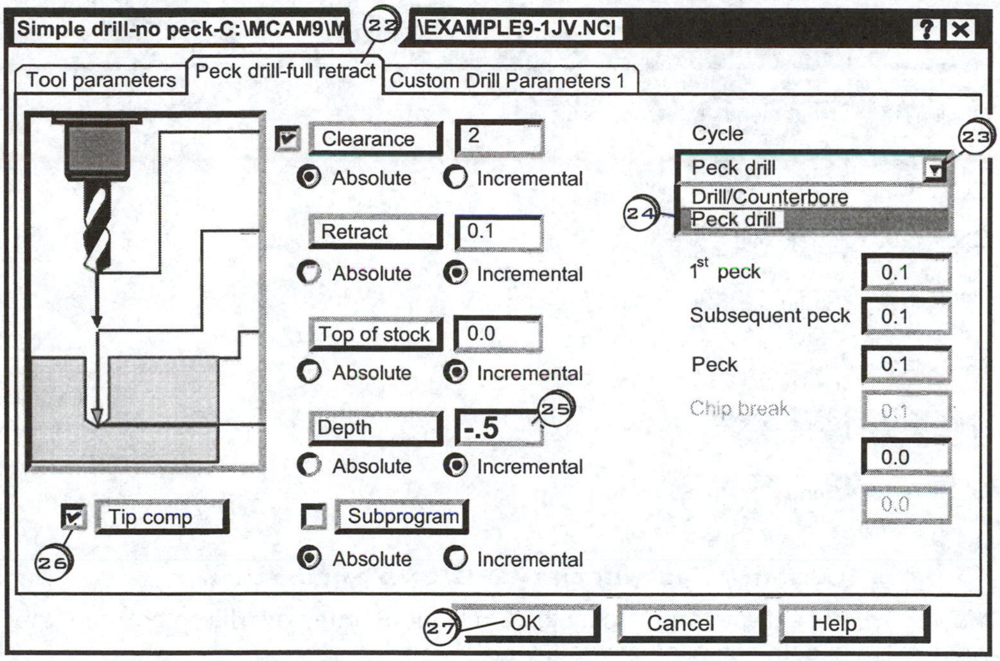

➤ Click 23 the toggle down button

➤ Click 24 Peck drill

➤ Click 25 in the Depth box; enter -.5

➤ Click 26 the tip comp check *on* when drilling *thru* holes

➤ Click 27 OK

The .313Dia peck drilling thru operation will be added to *Mastercam's* Operations Manager

◆ CREATE THE CIRCLE MILL TOOLPATH

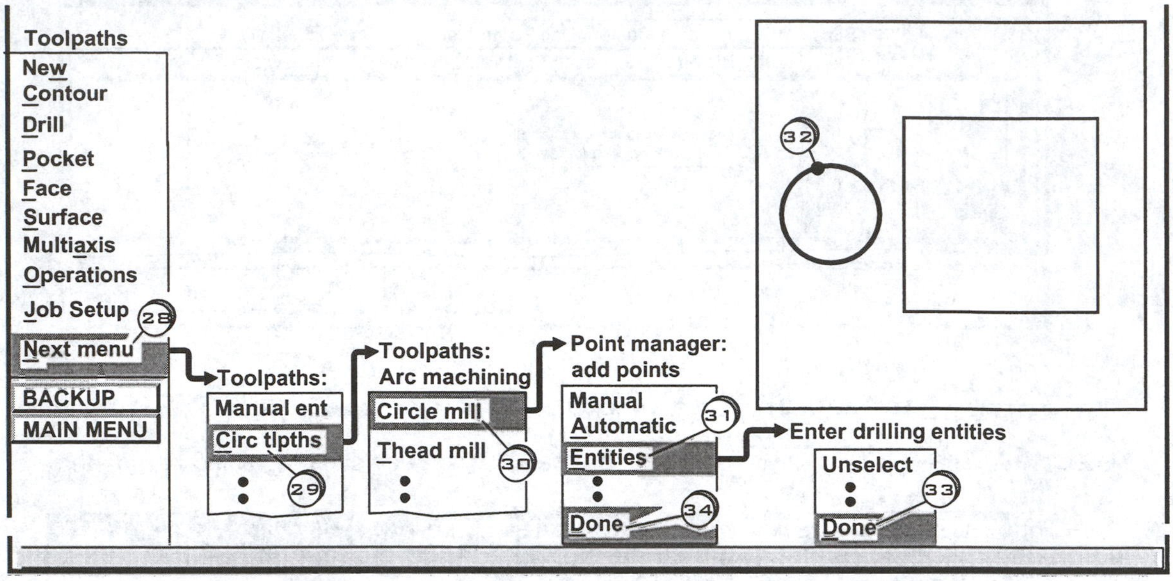

- Click (28) **Next menu**
- Click (29) **Circ tlpths**
- Click (30) **Circle mill**
- Click (31) **Entities**
- Click (32) the arc entity
- Click (33) **Done**
- Click (34) **Done**

◆ OBTAIN THE NEEDED 3/8 END MILL TOOL

Mastercam will activate and display the Circle mill Parameters dialog box after the circle mill tool path has been created.

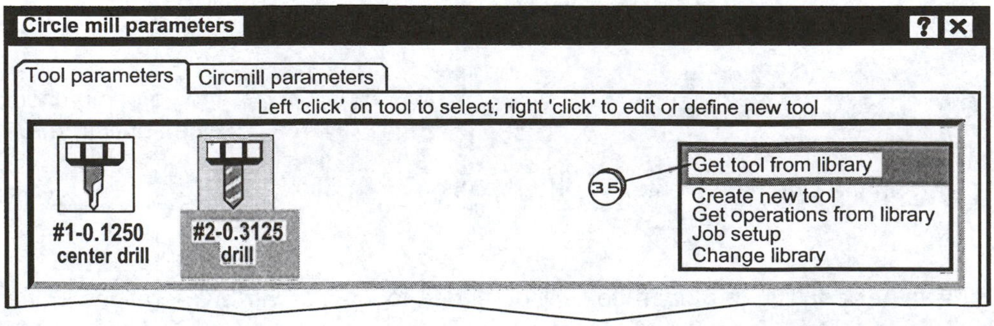

- ***Right*** Click move the mouse cursor *down* and Click (35) Get tool from library

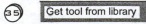

Select the 3/8in Dia end mill tool from the **TOOLS** tool library

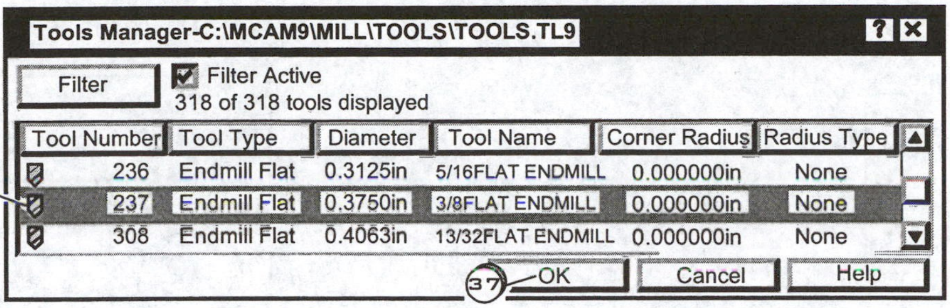

➤ Click (36) on the 3/8 FLAT ENDMILL tool

➤ Click (37) the OK button

Mastercam will place these tools into the currently active Tool and Drill Parameters dialog box

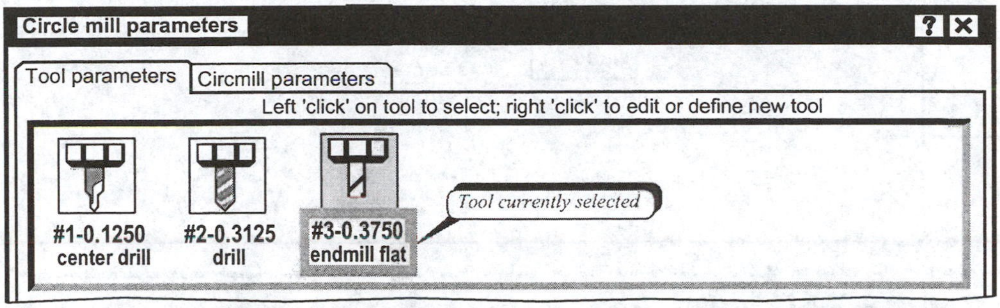

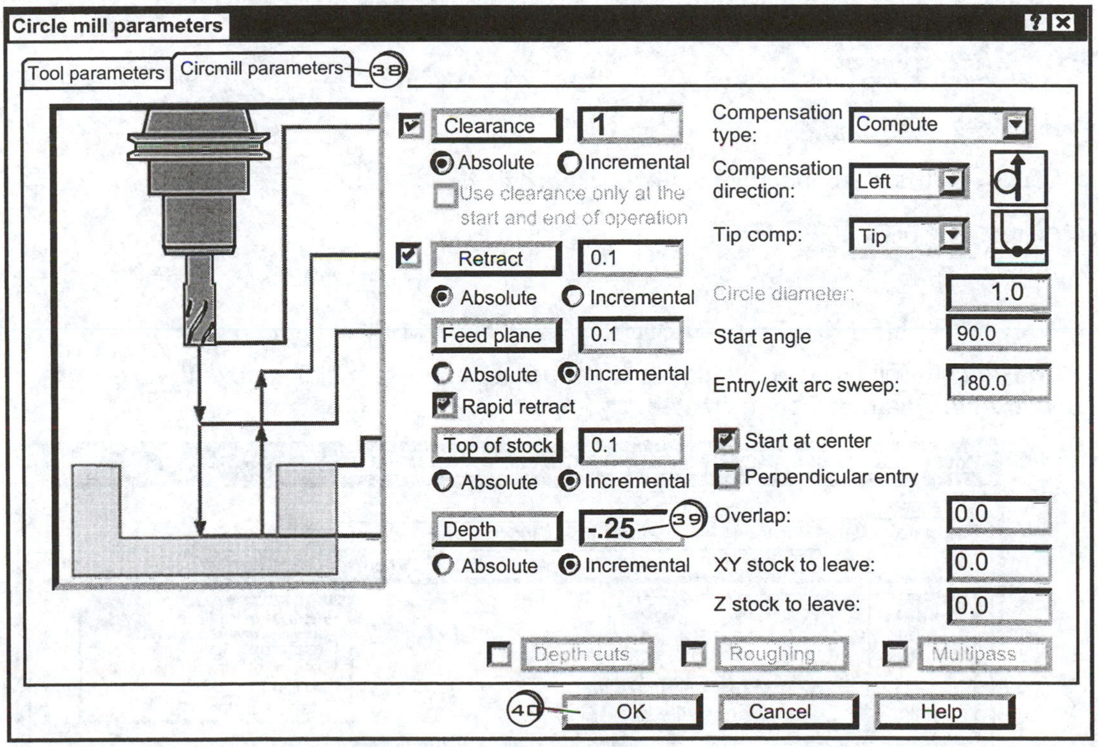

➤ Click (38) the **Circmill parameters** tab

➤ Click (39) in the Depth box and enter **-.25**

➤ Click (40) the OK button

C) Generate operations for Group POC-E3/16x1/8

- Create the pocket toolpath

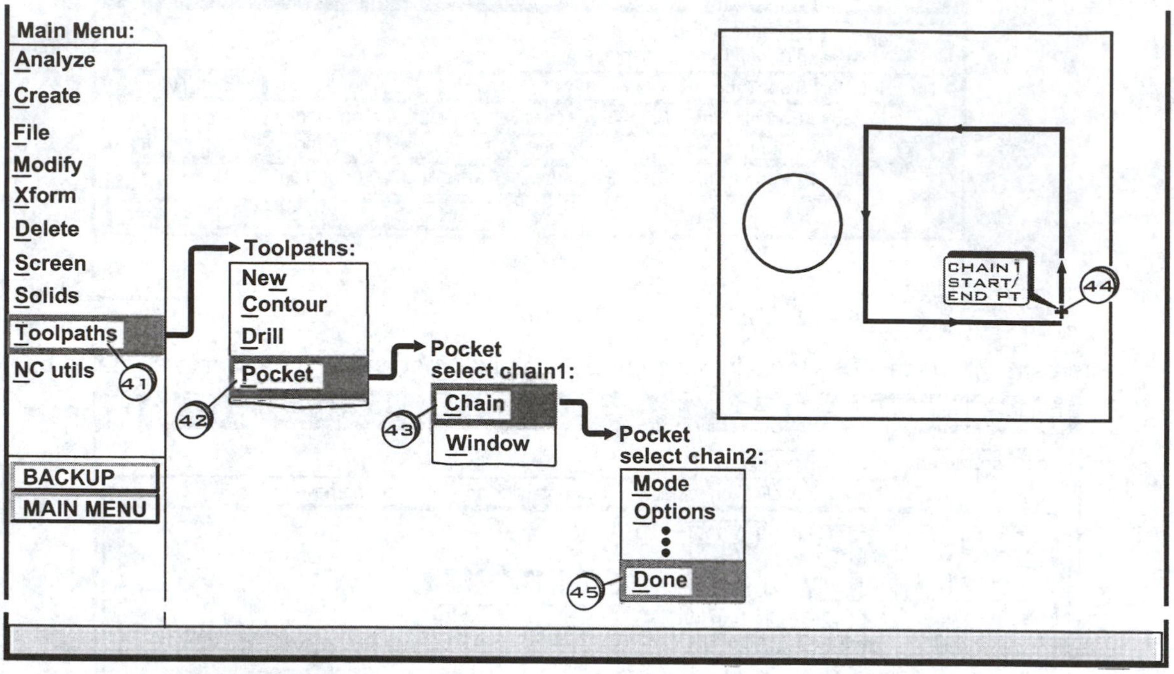

➤ Click (41) **Toolpaths**

➤ Click (42) **Pocket**

➤ Click (43) **Chain**

➤ Click (44) the start/end point of the pocket chain

➤ Click (45) **Done**

- Obtain the needed 1/4 end mill tool

Mastercam will activate and display the Pocket Parameters dialog box after the pocket tool path has been created.

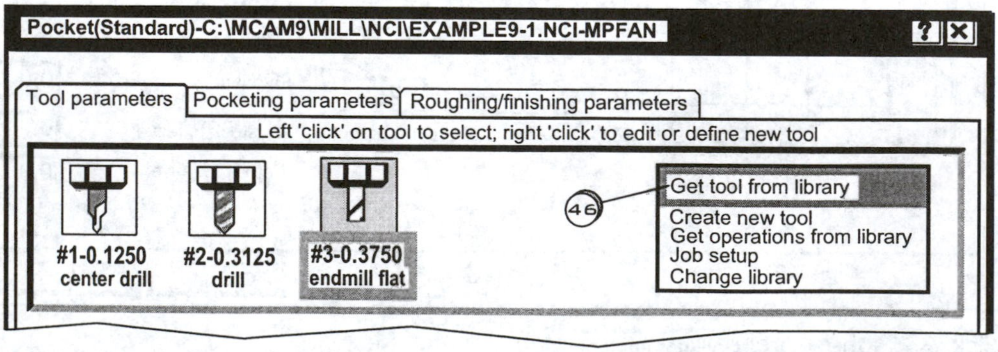

➤ ***Right*** Click move the mouse cursor *down* and Click (46) Get tool from library

Select the 3/16in Dia end mill tool from the **TOOLS** tool library

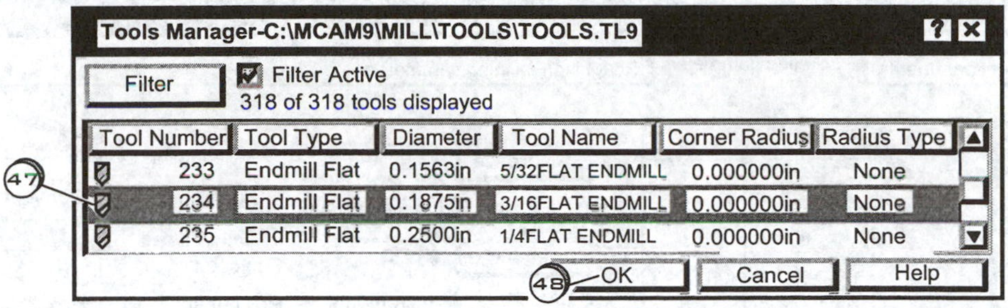

➤ Click (47) on the 3/16FLAT ENDMILL tool

➤ Click (48) the OK button

Mastercam will place these tools into the currently active Pocket Parameters dialog box

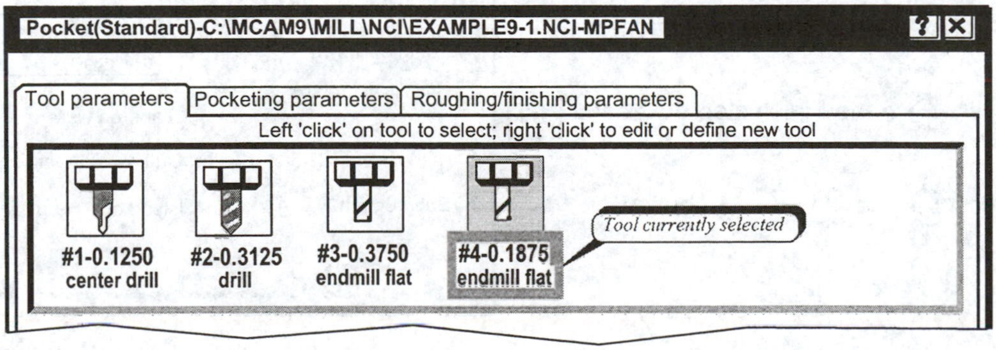

➤ Click (49) the Pocketing parameters tab

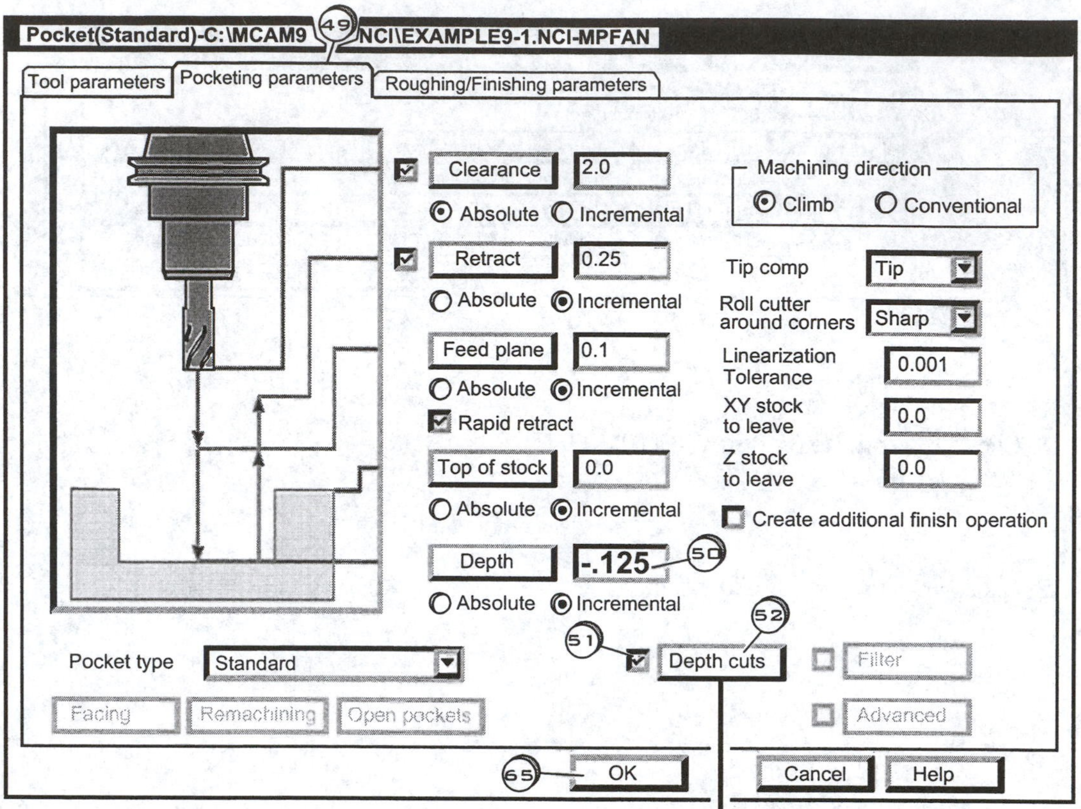

➤ Click (50) in the Depth box; enter -.125

➤ Click (51) activate Depth cuts ☑

➤ Click (52) the Depth cuts button

➤ Click (53) in the Finish step box; enter .01

➤ Click (54) the OK button

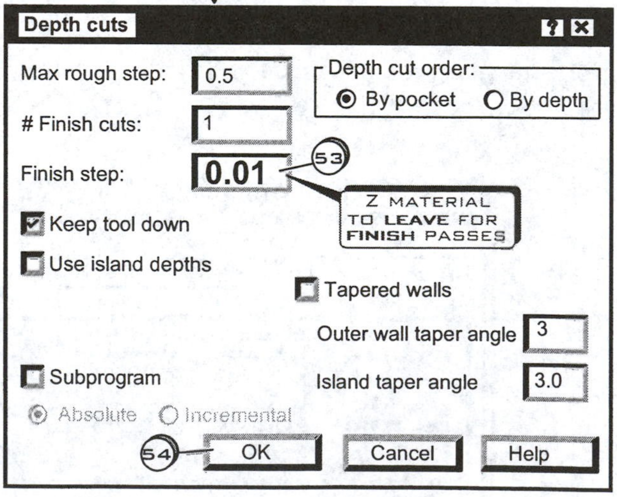

➤ Click (55) the Roughing/finishing parameters tab

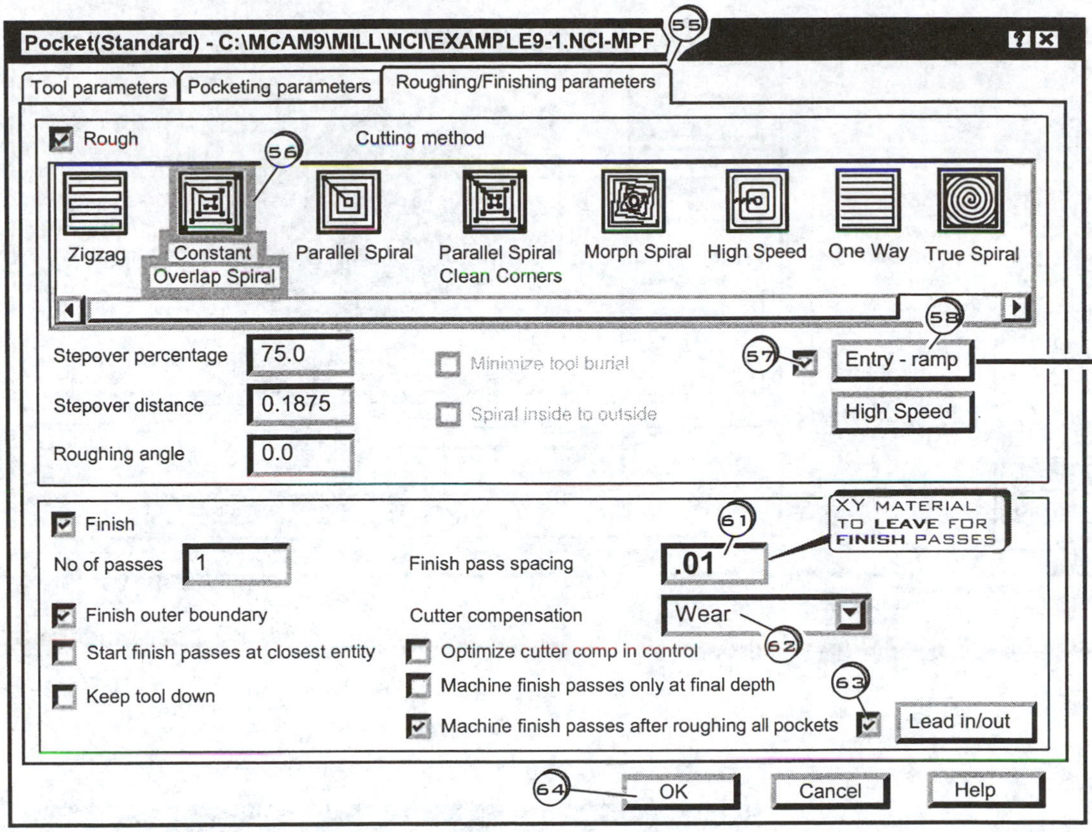

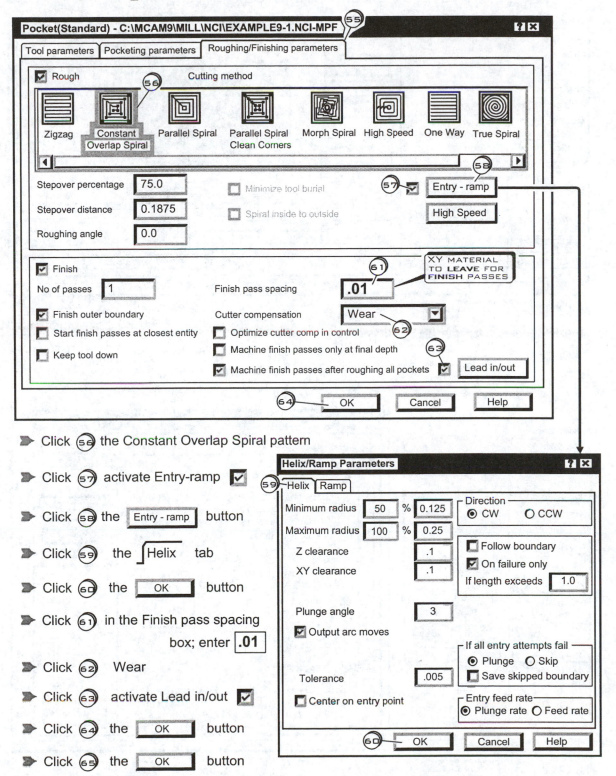

➤ Click (56) the Constant Overlap Spiral pattern

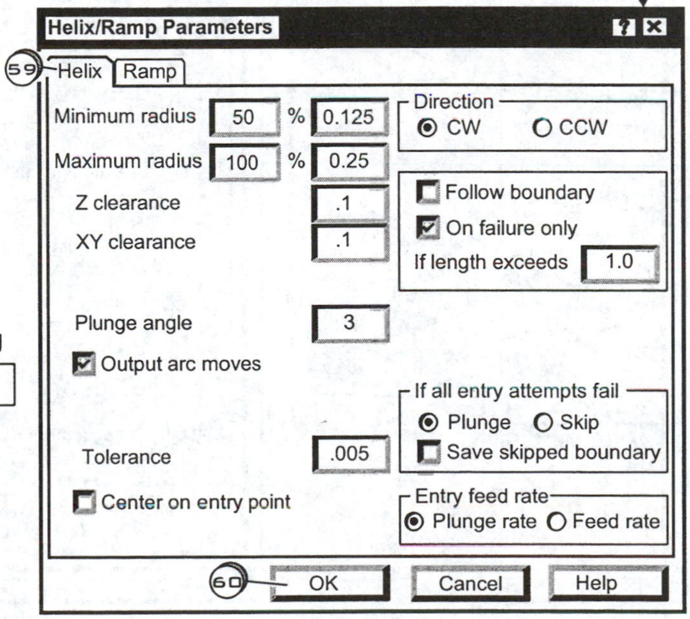

➤ Click (57) activate Entry-ramp ☑

➤ Click (58) the Entry - ramp button

➤ Click (59) the Helix tab

➤ Click (60) the OK button

➤ Click (61) in the Finish pass spacing box; enter .01

➤ Click (62) Wear

➤ Click (63) activate Lead in/out ☑

➤ Click (64) the OK button

➤ Click (65) the OK button

D) Verify the Toolpaths for All the Machining Operations

◆ SELECT SIZE OF THE STOCK

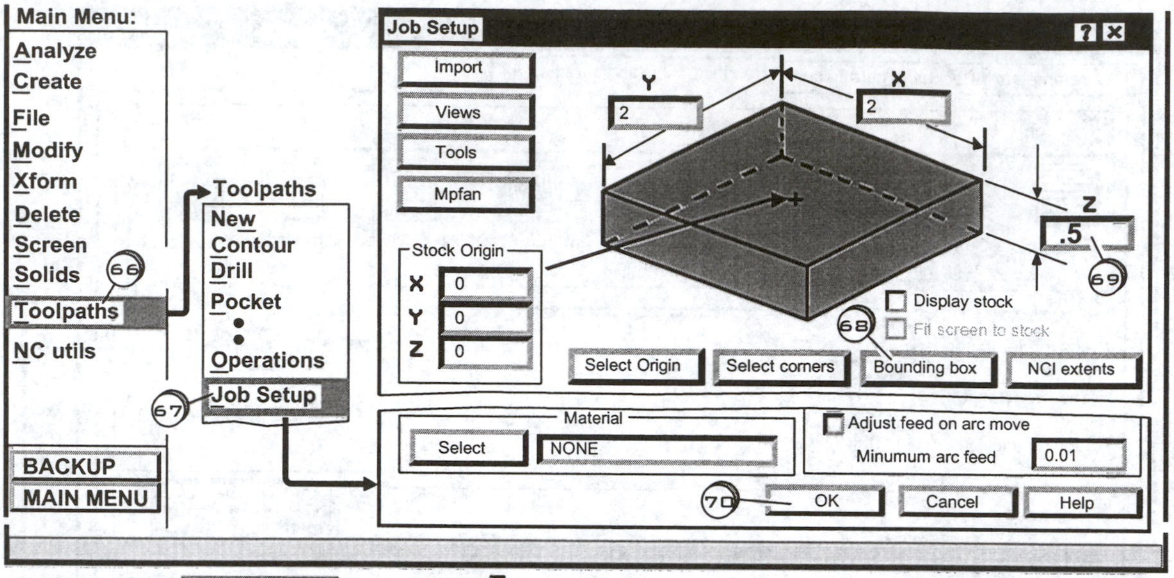

- Click 66 Toolpaths
- Click 67 Job Setup
- Click 68 Bounding box
- Click 69 in the stock Z box ; enter .5
- Click 70 OK

◆ SPECIFY THE OPERATIONS TO BE VERIFIED

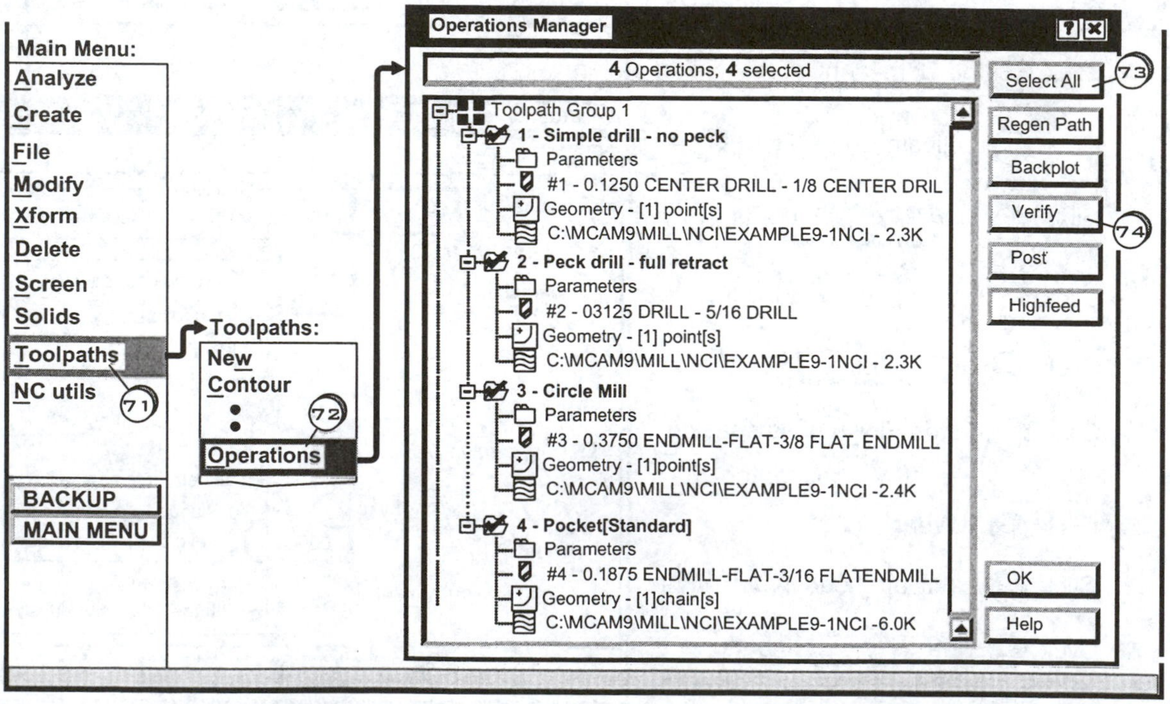

- Click 71 Toolpaths
- Click 72 Operations
- Click 73 Select All
- Click 74 Verify

- DIRECT *Mastercam* TO ANIMATE THE MACHINING OF THE GROUPS CBORE-D5/16-E3/8X1/4 AND POC-E3/16X1/8

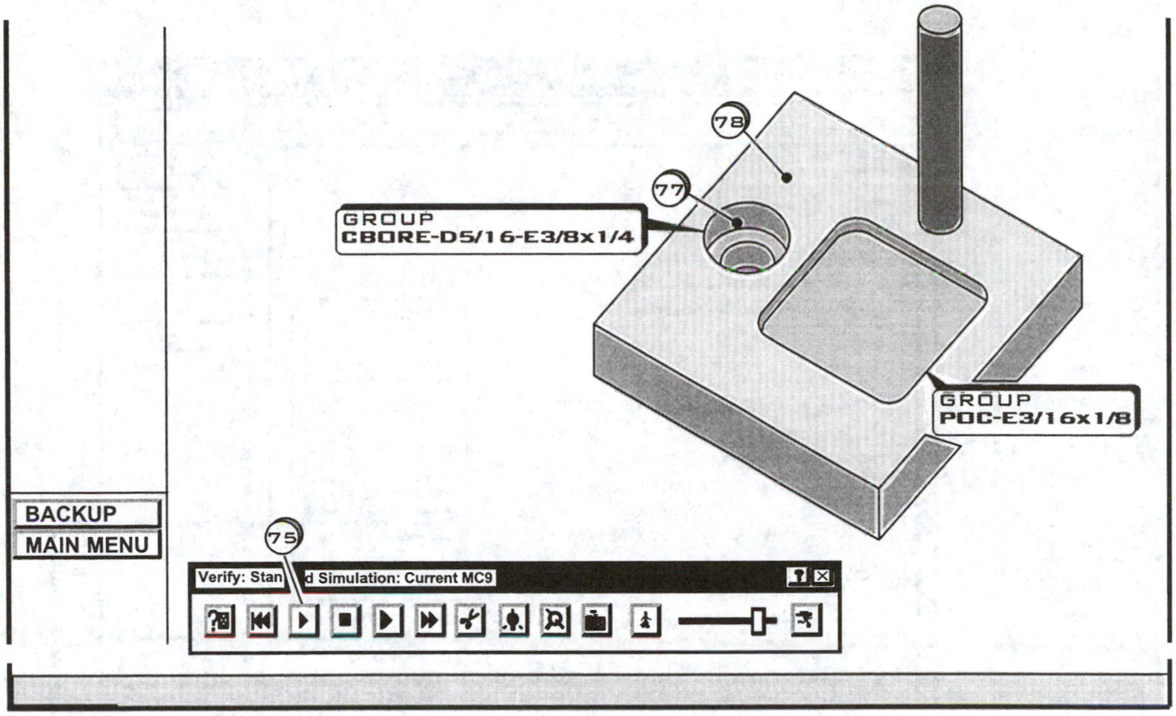

➤ Click (75) the play button

- DIRECT *Mastercam* TO SECTION THE PART TO VISUALLY INSPECT THE CORRECTNESS OF THE MACHINING FOR EACH GROUP

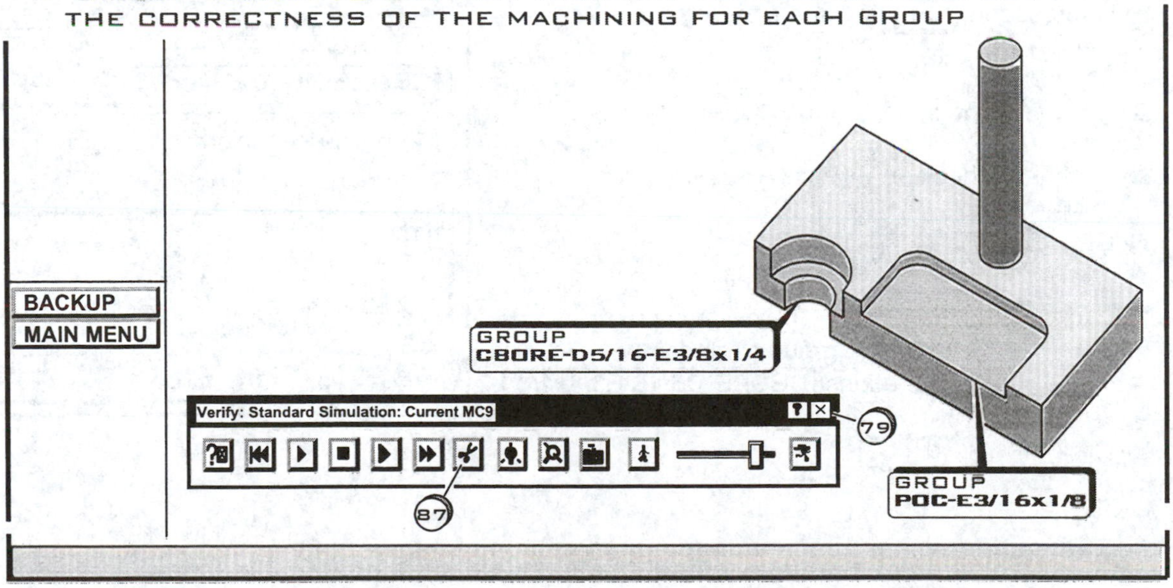

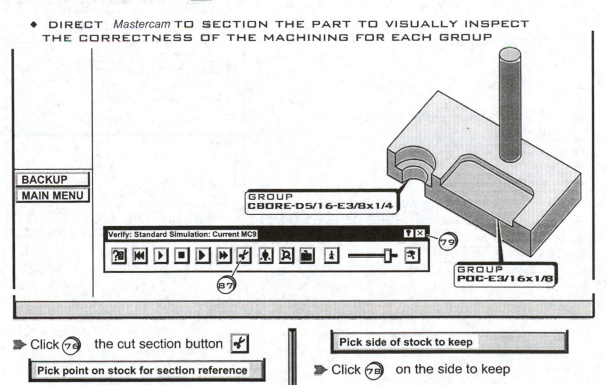

➤ Click (76) the cut section button

Pick point on stock for section reference

➤ Click (77) on the circle

Pick side of stock to keep

➤ Click (78) on the side to keep

➤ Click (79) the exit button ☒ to exit **Verify**

E) Save the operations generated in step B to the operations library under the group name CBORE-D5/16-E3/8x1/4

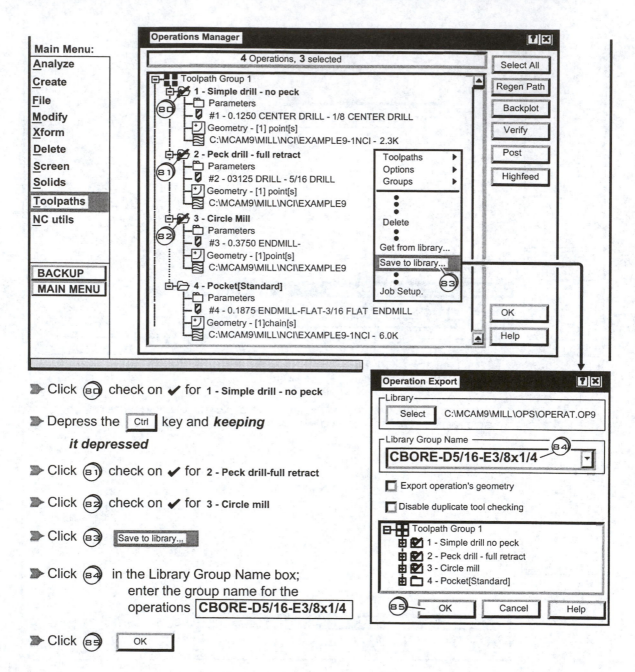

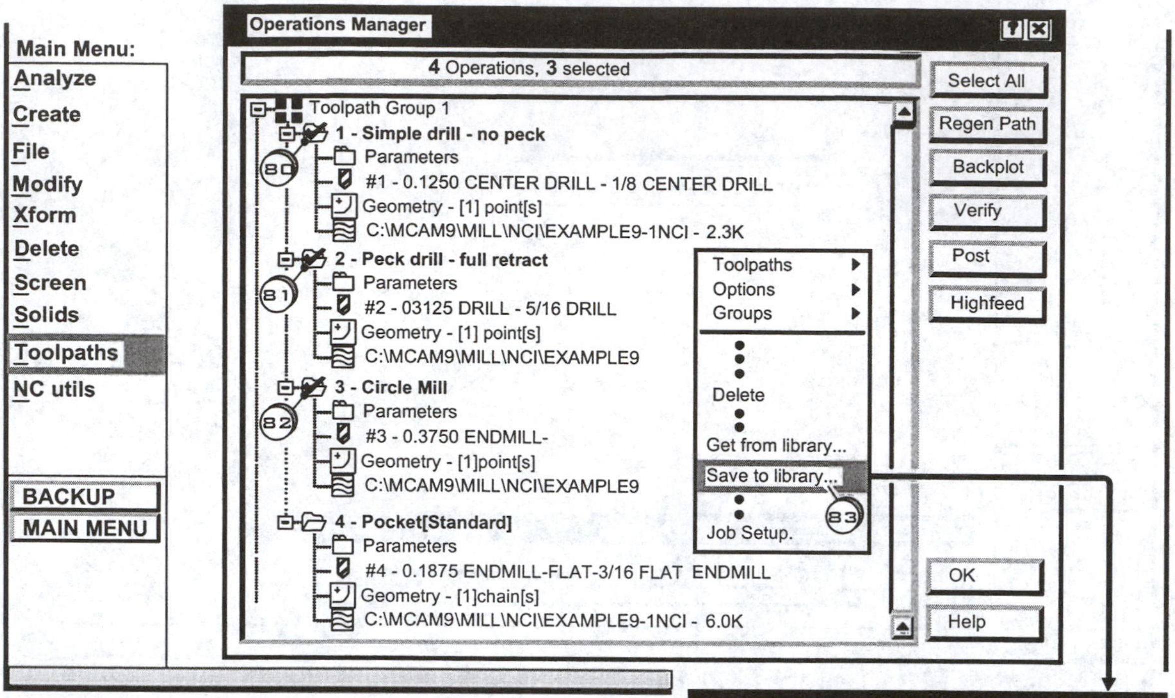

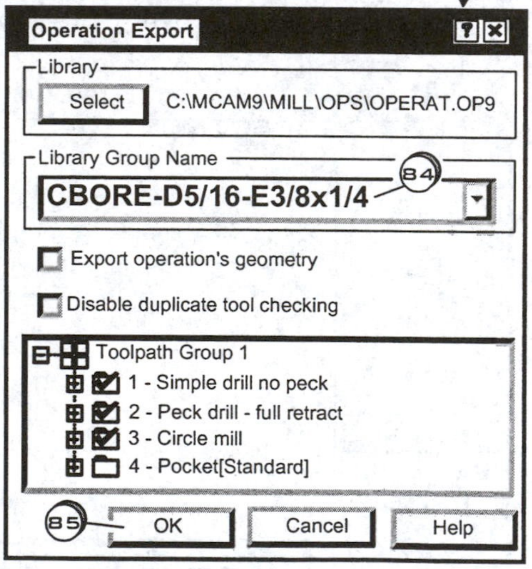

- Click (80) check on ✔ for **1 - Simple drill - no peck**
- Depress the [Ctrl] key and ***keeping it depressed***
- Click (81) check on ✔ for **2 - Peck drill-full retract**
- Click (82) check on ✔ for **3 - Circle mill**
- Click (83) [Save to library...]
- Click (84) in the Library Group Name box; enter the group name for the operations **CBORE-D5/16-E3/8x1/4**
- Click (85) [OK]

F) Save the operations generated in step C to the operations library under the group name POC-E3/16x1/8

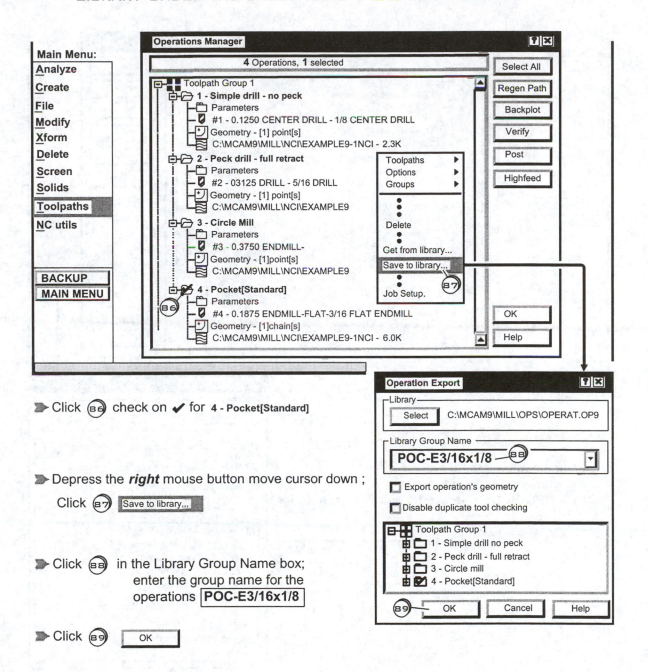

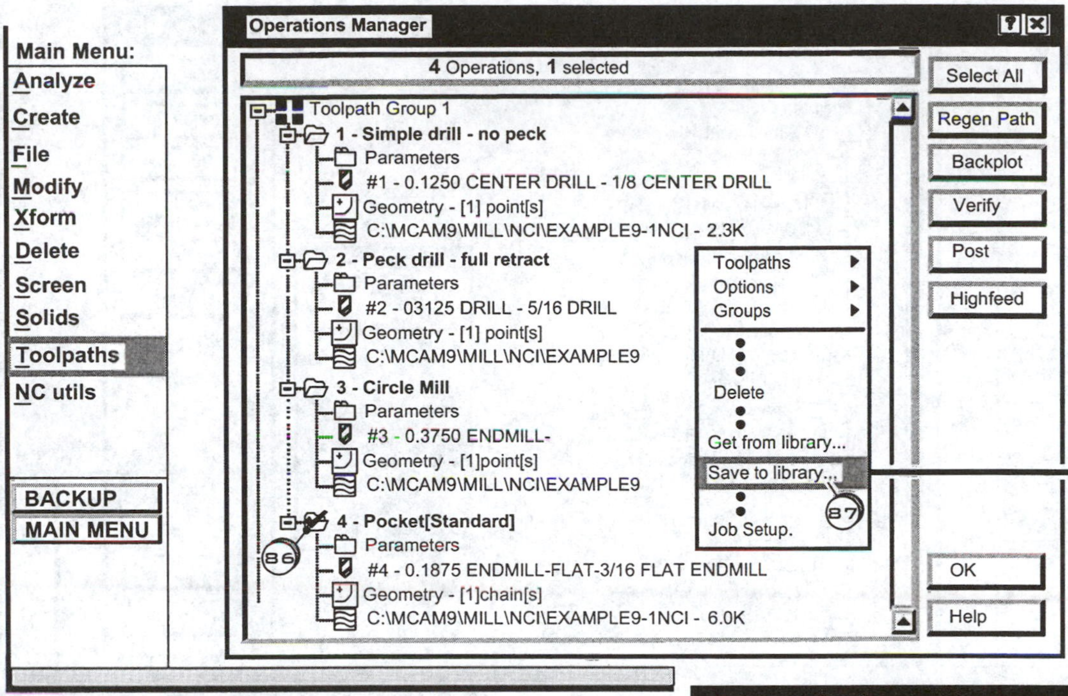

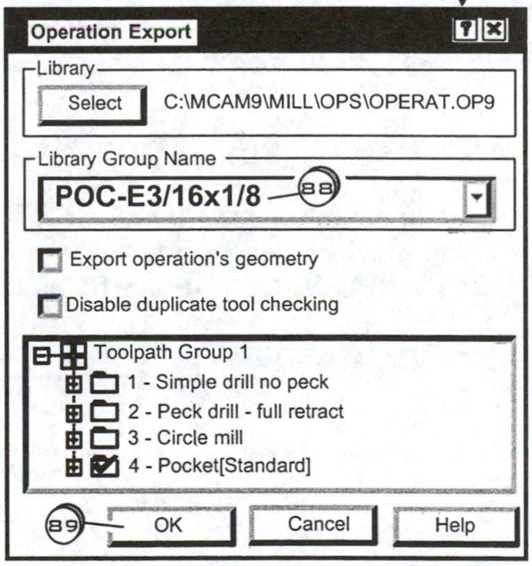

- Click (86) check on ✔ for **4 - Pocket[Standard]**

- Depress the ***right*** mouse button move cursor down ;

  Click (87) Save to library...

- Click (88) in the Library Group Name box; enter the group name for the operations **POC-E3/16x1/8**

- Click (89) OK

G) Verify Operations Library has been Updated to Include the New Groups CBORE-D5/15-E3/8x1/4 and POC-E3/16x1/8

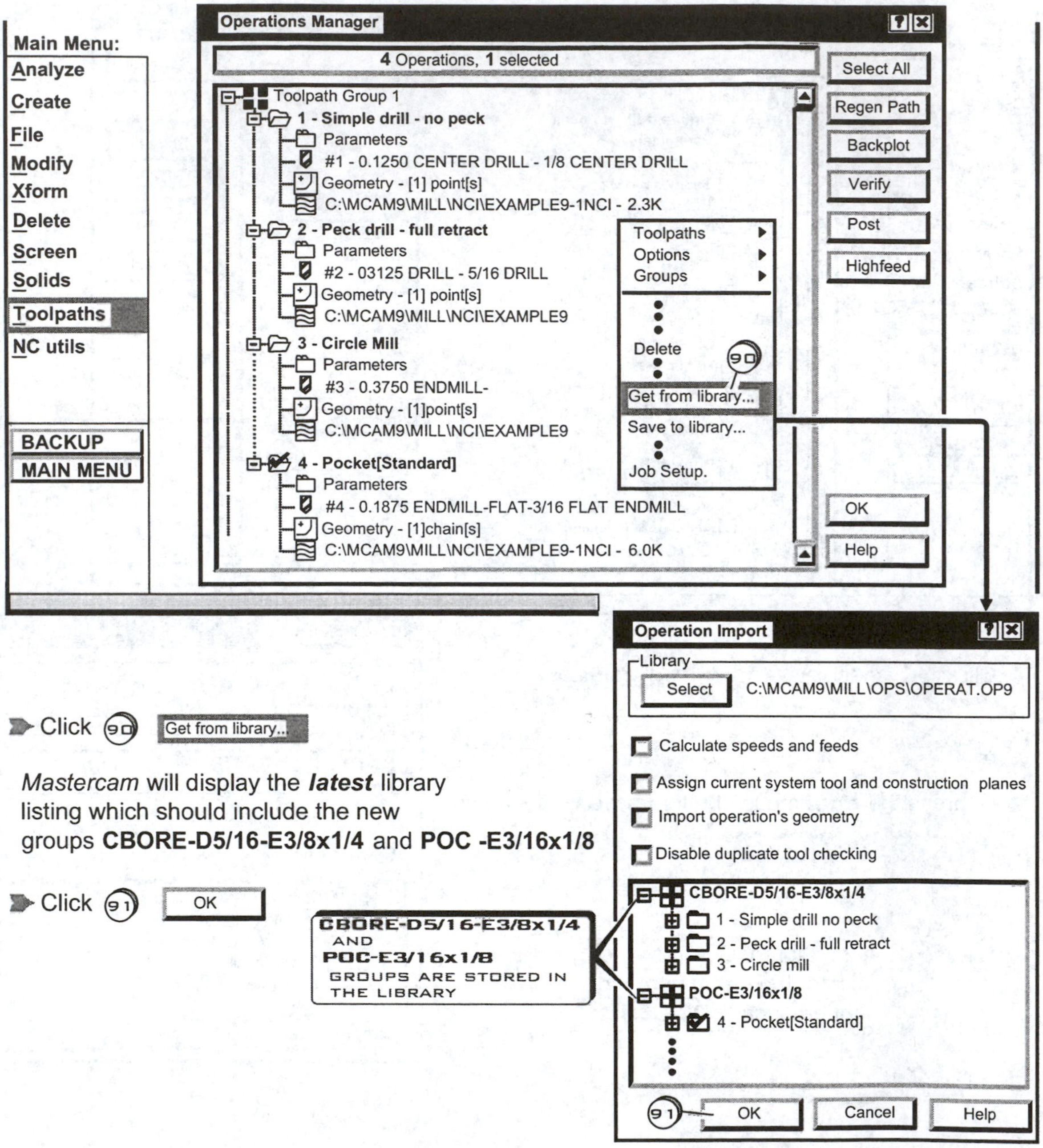

➤ Click (90) Get from library...

Mastercam will display the ***latest*** library listing which should include the new groups **CBORE-D5/16-E3/8x1/4** and **POC -E3/16x1/8**

➤ Click (91) OK

9-4 Importing Operations from the Operations Library

Mastercam provides complete *flexibility* in importing the data stored in the operations library as follows: import of a *single* operation

- import of *any number* of operations clicked
- import of *all* the operations associated with a *group* clicked
- import of *all* the operations associated with *any number* of *groups* clicked.

It was shown in section 9-2 that only a minumim of dummy geometry was needed to create the required operations. In this section it will be demonstrated that the groups and associated operations in the operations library can be quickly applied to a wide array of parts having different physical shapes, thicknesses and made of different materials.

EXAMPLE 9-2

For the part shown in Figure9-1. Use the groups created and saved in the operations library in Example 9-1 as an aid in quickly generating the the required machining operations as outlined in PROCESS PLAN 9-1. Assume the CAD model already exists and is stored as file **EXAMPLE9-2** in the **JVAL-MILL** subdirectory.

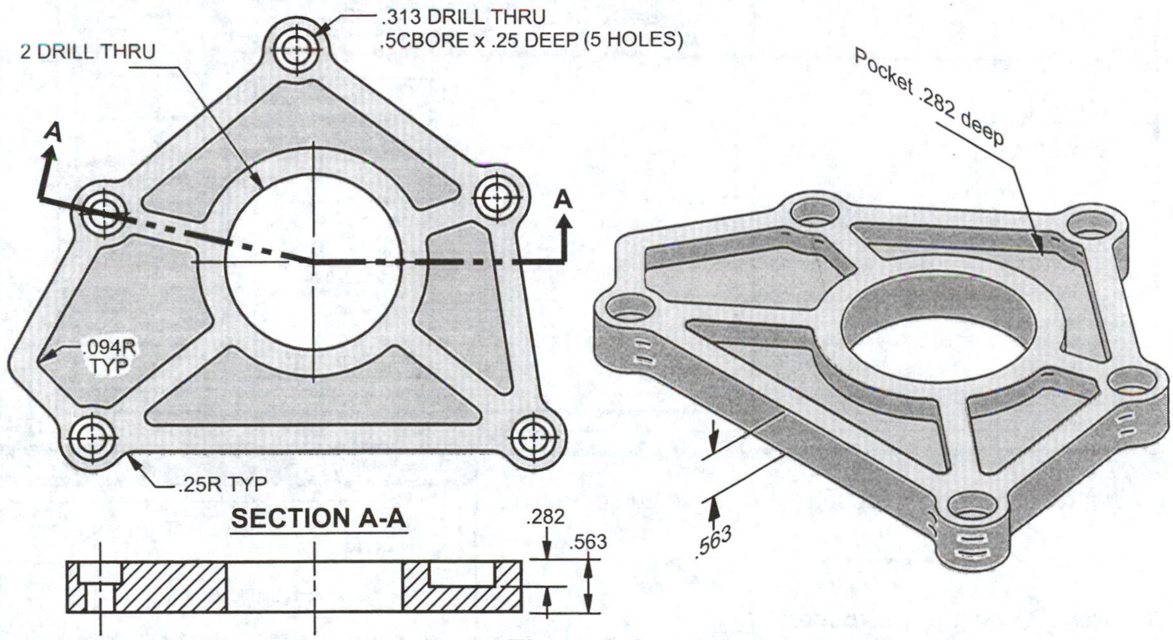

Figure 9-1

PROCESS PLAN 9-1

No.	Operation	Tooling
1	CENTER DRILL X 2 DEEP(ALL HOLES)	1/8 CTR DRILL
2	PECK DRILL THRU(ALL HOLES)	5/16 DRILL
3	CIRCLE MILL X .25 DEEP(ALL HOLES)	3/8 ENDMILL
4	ROUGH AND FINISH POCKET X .282 DEEP LEAVE .01 FOR FINISH CUT IN XY AND Z.	3/16 ENDMILL

CBORE-D5/16-E3/8x1/4 (operations 1–3)

POC-E3/16x1/8 (operation 4)

A) OPEN THE CAD FILE EXAMPLE9-2 CONTAINING THE PART GEOMETRY

B) IMPORT THE GROUP CBORE-D5/16-E3/8x1/4 FROM THE OPERATIONS LIBRARY

- FROM THE OPERATIONS MANAGER SELECT GET FROM LIBRARY

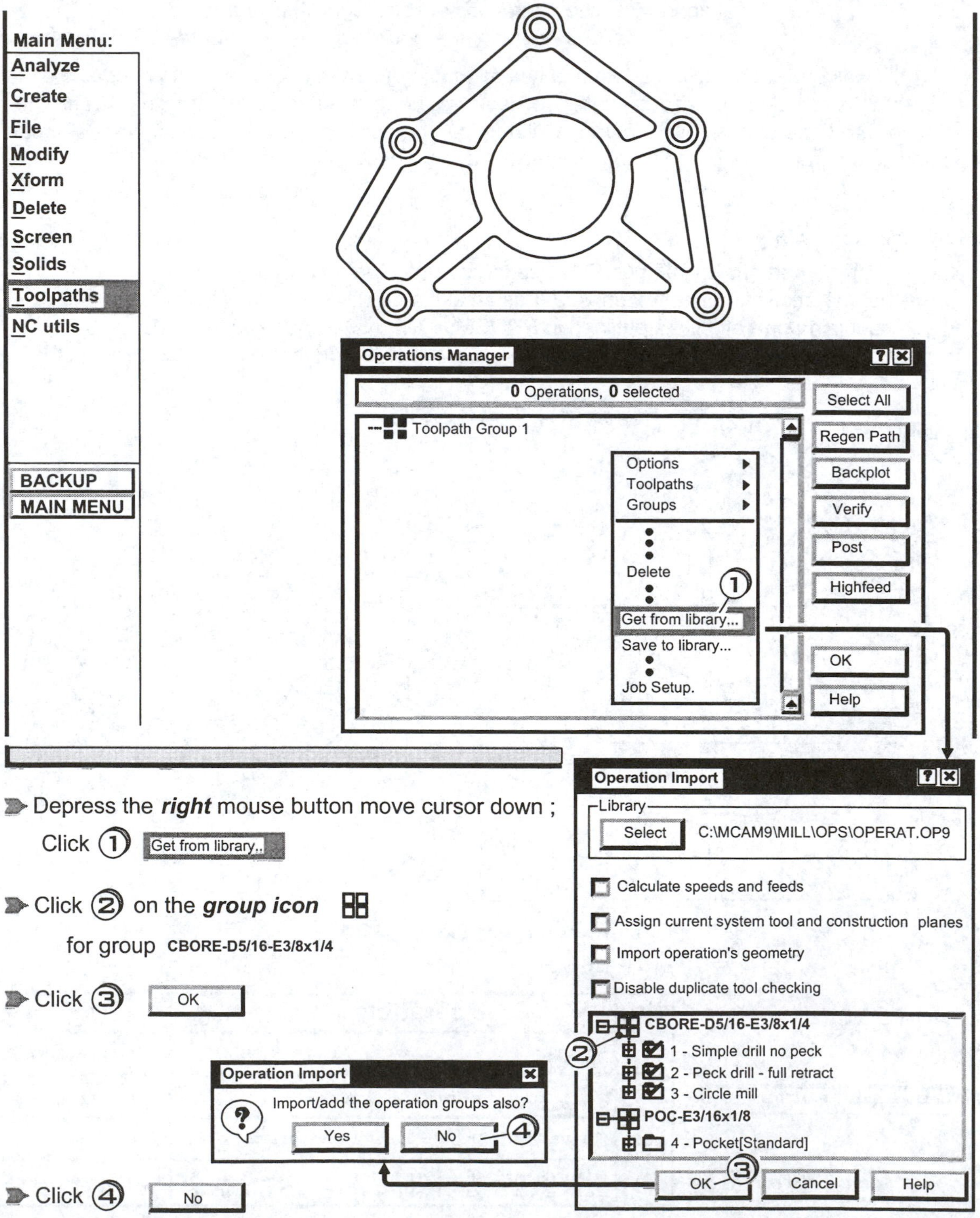

➤ Depress the ***right*** mouse button move cursor down ;

Click ① Get from library..

➤ Click ② on the ***group icon*** for group CBORE-D5/16-E3/8x1/4

➤ Click ③ OK

➤ Click ④ No

C) Add Drill Points to the Drilling Operations

- ADD DRILL POINTS TO OPERATION: 1 - Simple drill- no peck

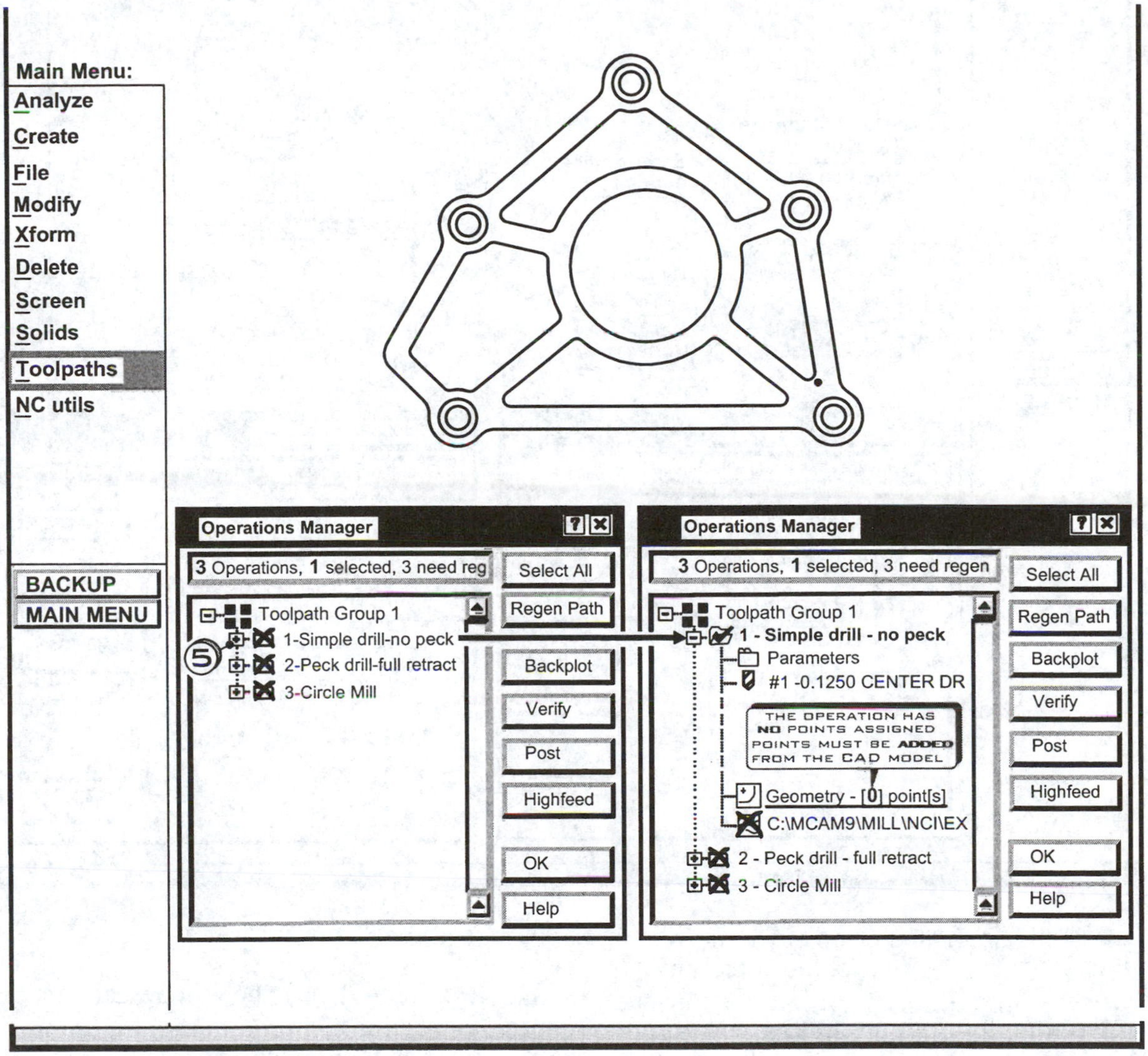

➤ Click ⑤ on the operation expand/contract icon and turn it to expand ⊟

Note: By default, Mastercam will assign ***no drill points*** *and* ***no chains*** *to* ***any*** *operations imported from the Operations library*

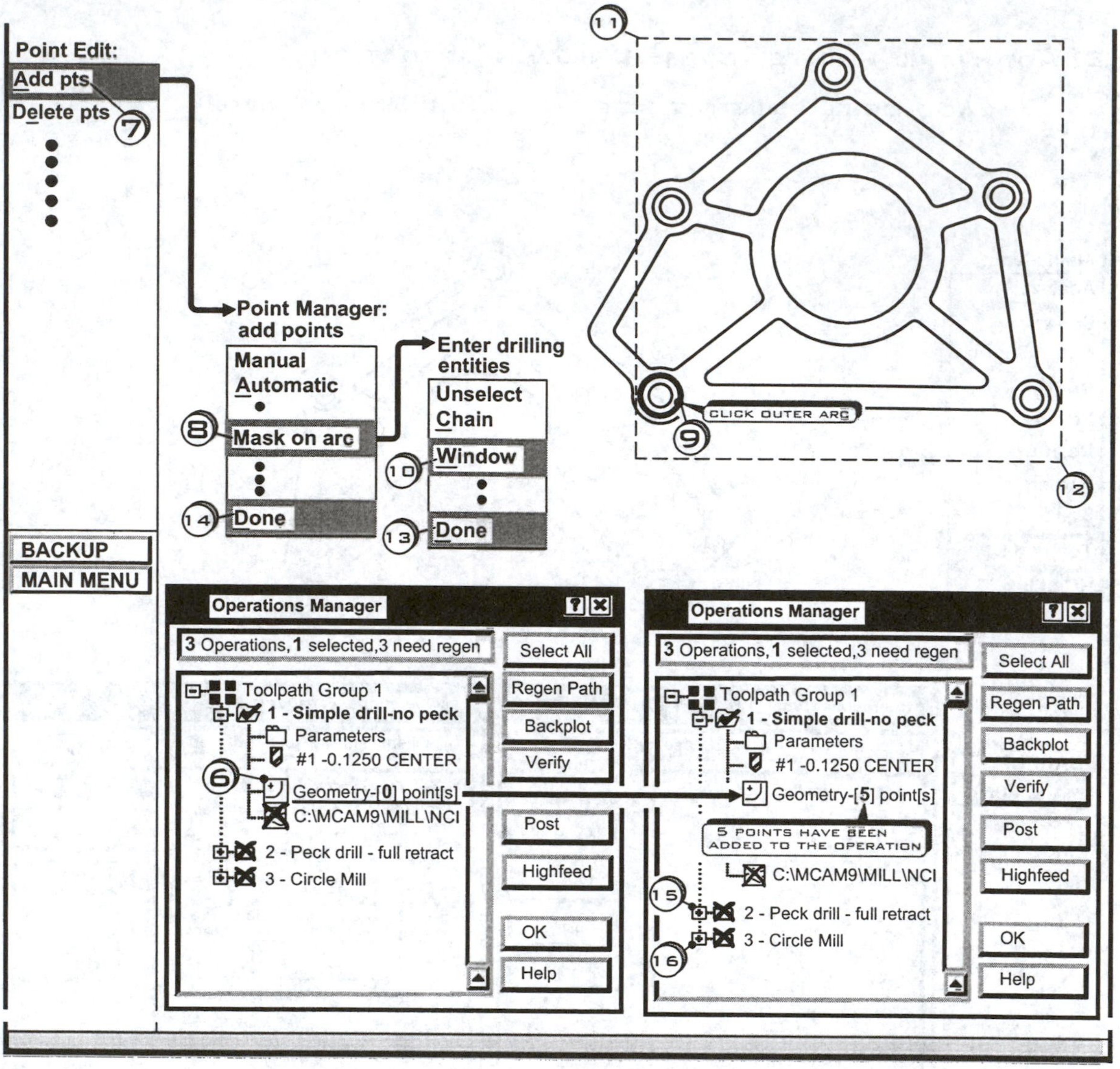

- Click (6) on the Geometry icon for the operation
- Click (7) **Add pts**
- Click (8) **Mask on arc**

Select arc to match

- Click (9) the hole which is to be matched

Enter arc radius matching tolerance **0.001**

- Accept the current value press Enter

- Click (10) **Window**
- Click (11) (12) the window corners
- Click (13) **Done**

Select sorting point

- Click (9) the first point to drill
- Click (14) **Done**
- Click (15) (16) on the operation expand/contract icons and turn them to expand

◆ ADD DRILL POINTS TO THE OPERATIONS: **2 - Peck drill- full retract** AND **3- Circle Mill**

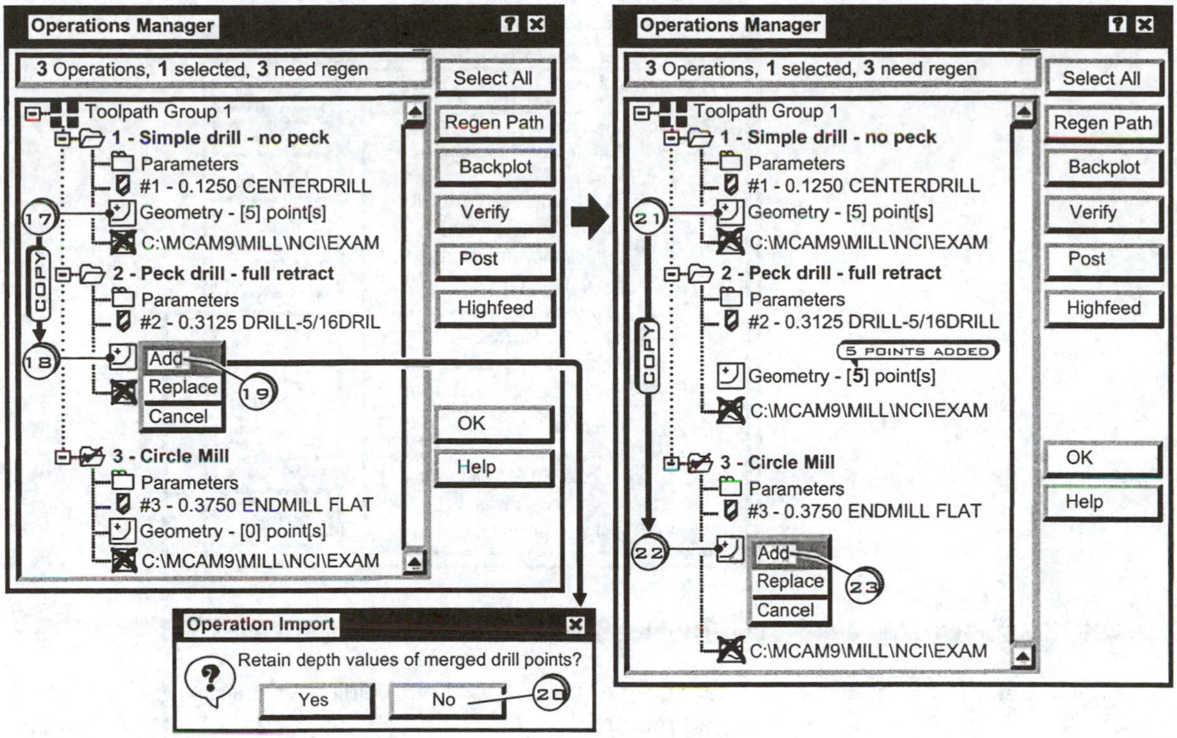

➤ Click (17) on the Part geometry icon for the operation **1 - Simple drill - no peck**
keeping the left mouse button depressed move the icon over the geometry icon for the operation (18) **2 - Peck drill - full retract** and *release*

➤ Click (19) Add

➤ Click (20) No

Mastercam will *add a copy* of the five drill points from opertion 1 to operation 2

➤ Click (21) on the Part geometry icon for the operation **1 - Simple drill - no peck**
keeping the left mouse button depressed move the icon over the geometry icon for the operation (22) **2 - Peck drill - full retract** and *release*

➤ Click (23) Add

Mastercam will also *add a copy* of the five drill points from opertion 1 to operation 3

C) Adjust the drill depth to the material thickness of .563in

- EDIT THE **2-Peck drill-full retract** OPERATION PARAMETERS

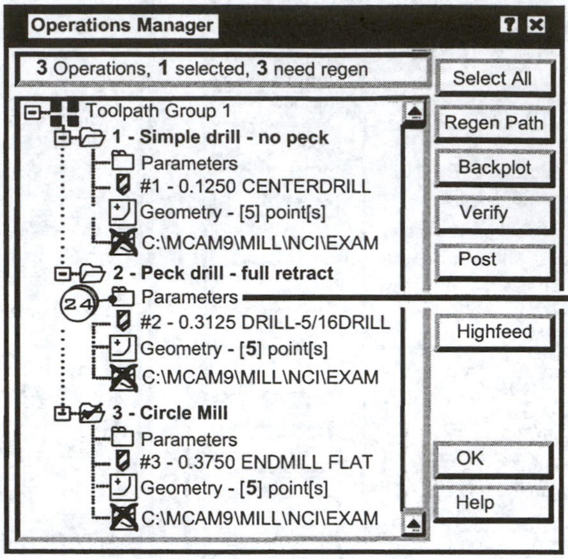

➤ Click (24) the Parameters icon for the **2 - Peck drill-full retract** operation

➤ Click (25) **Peck drill-full retract** tab to enter the .313Dia peck drill machining parameters.

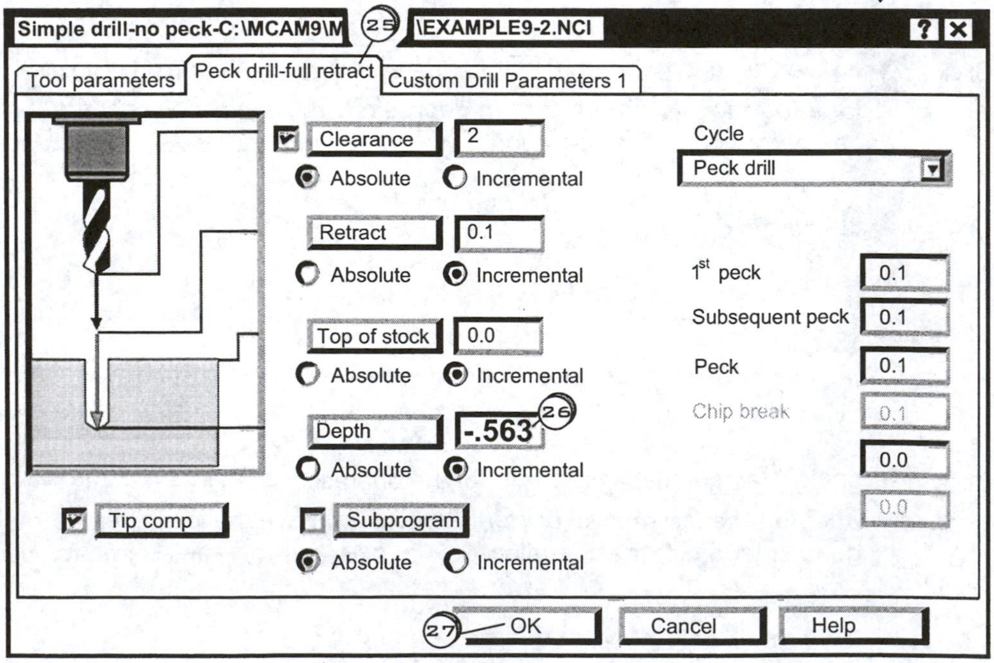

➤ Click (26) in the Depth box; enter **-.563**

➤ Click (27) OK

D) Import the Group POC-E3/16x1/8 from the Operations Library

- From the Operations Manager select Get from Library

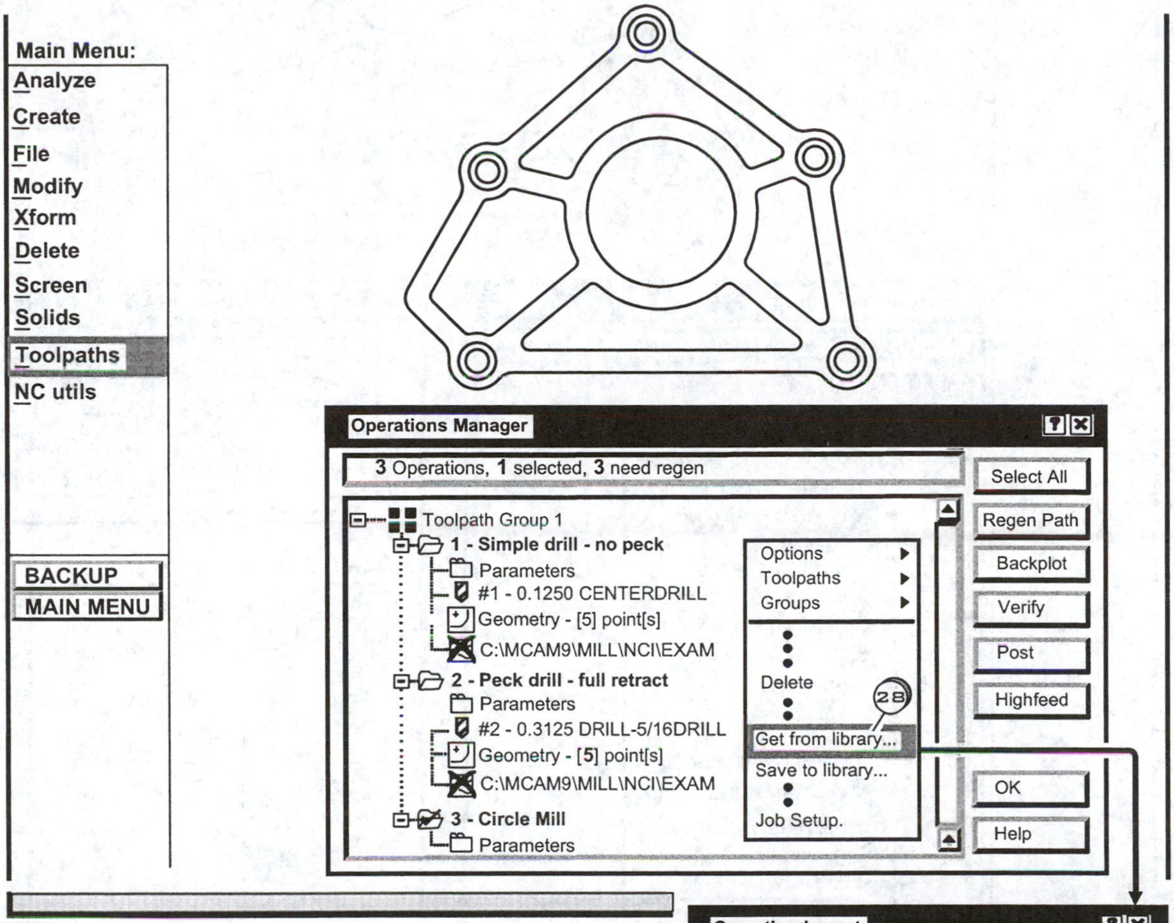

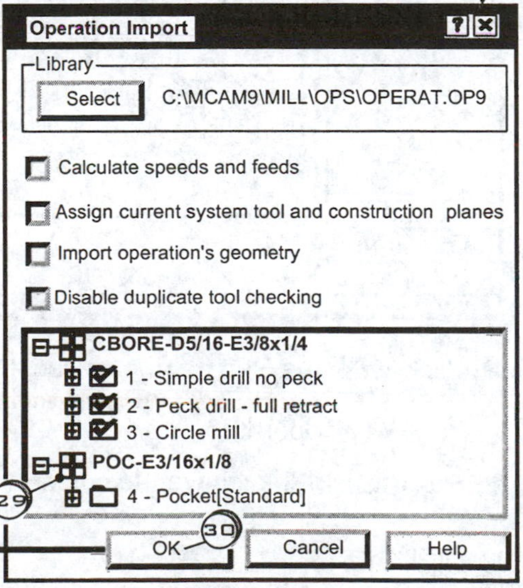

Depress the ***right*** mouse button move cursor down ;

Click (28) Get from library..

Click (29) on the ***group icon***

for group **POC-E3/16x1/8**

Click (30) OK

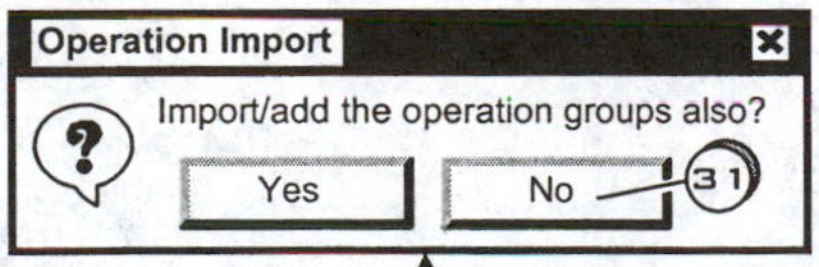

Click (31) No

E) Add Chains to Operation 4- Pocket[Standard]

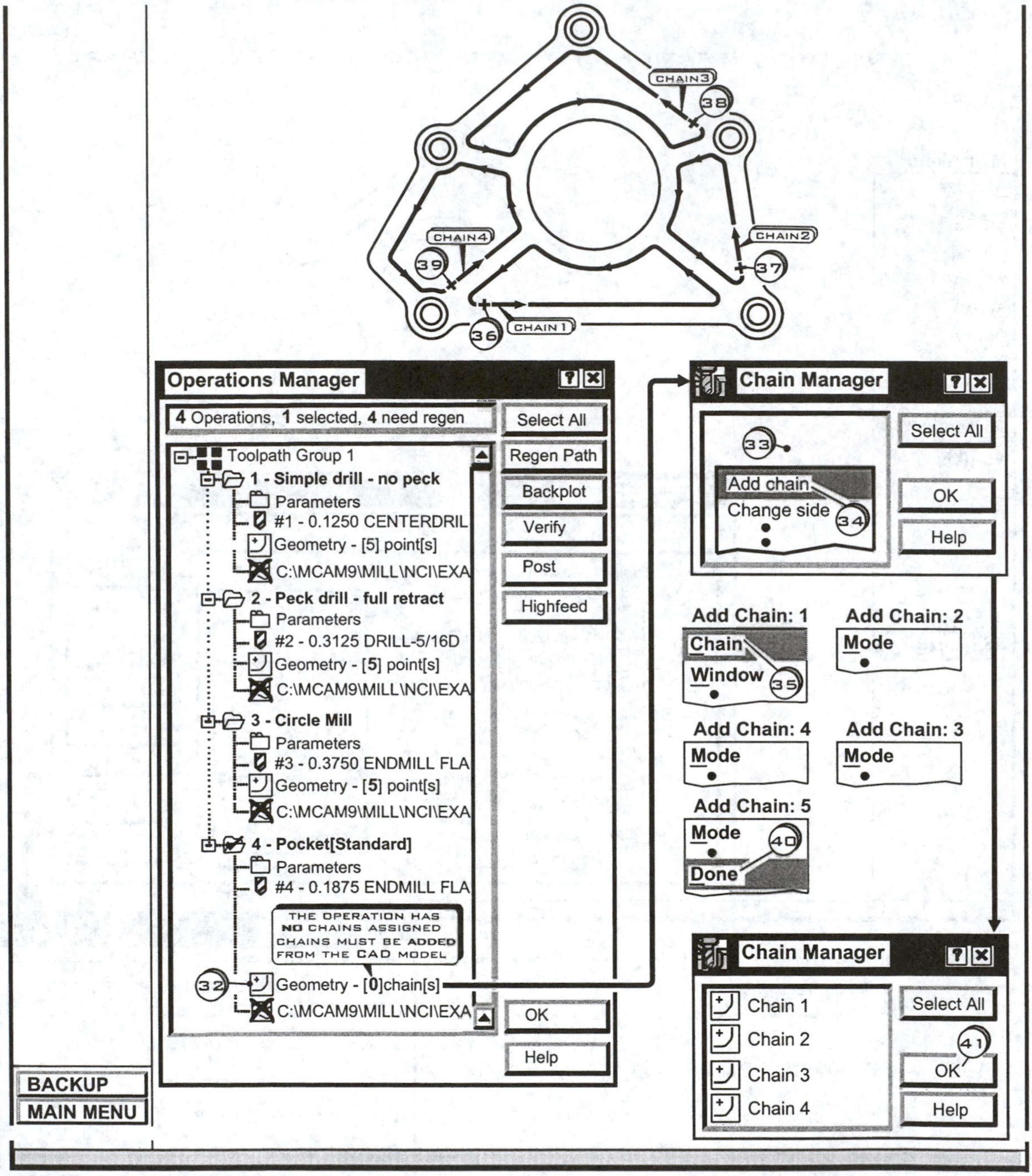

➤ Click (32) the Part geometry icon for the 1 -Pocket[Standard] operation

➤ *Right* Click (33) move the cursor down

➤ Click (34) Add chain

➤ Click (35) **Chain**

➤ Click (36) (37) (38) (39) the Start/End points of the chains

➤ Click (40) **Done**

➤ Click (41) OK

F) ADJUST THE POCKETING DEPTH TO .282IN

◆ EDIT THE **4-Pocket[Standard]** OPERATION PARAMETERS

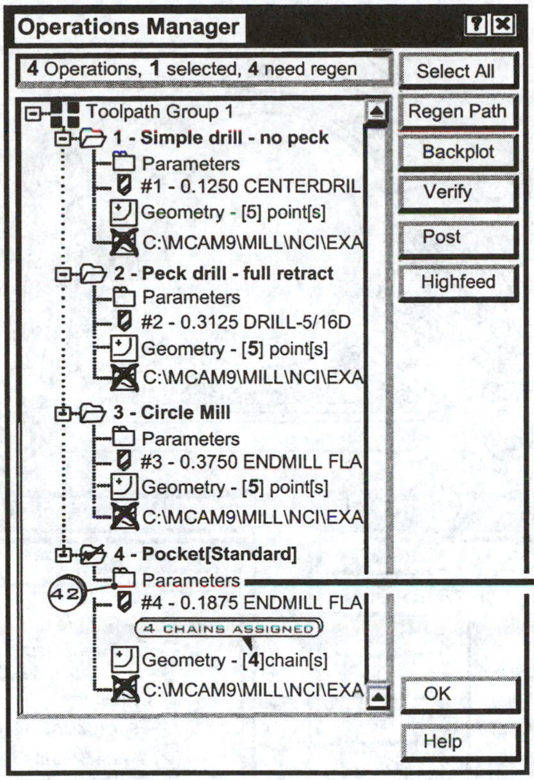

➤ Click (42) the Parameters icon 🗀 for the **2 - Peck drill-full retract** operation

➤ Click (43) the Pocketing parameters tab

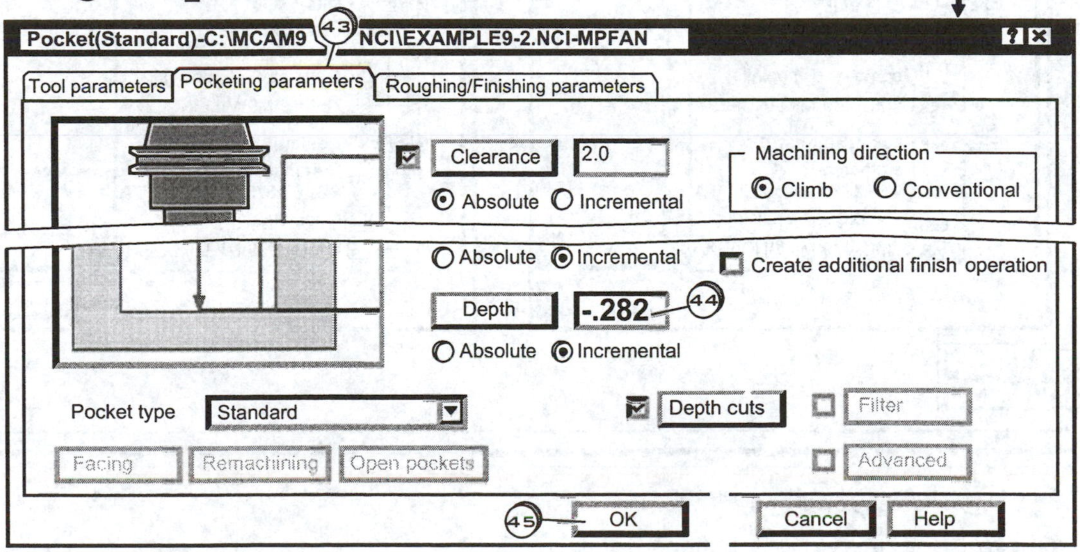

➤ Click (44) in the Depth box; enter **-.282**

➤ Click (45) the OK button

G) Regenerate the toolpaths to reflect the latest changes

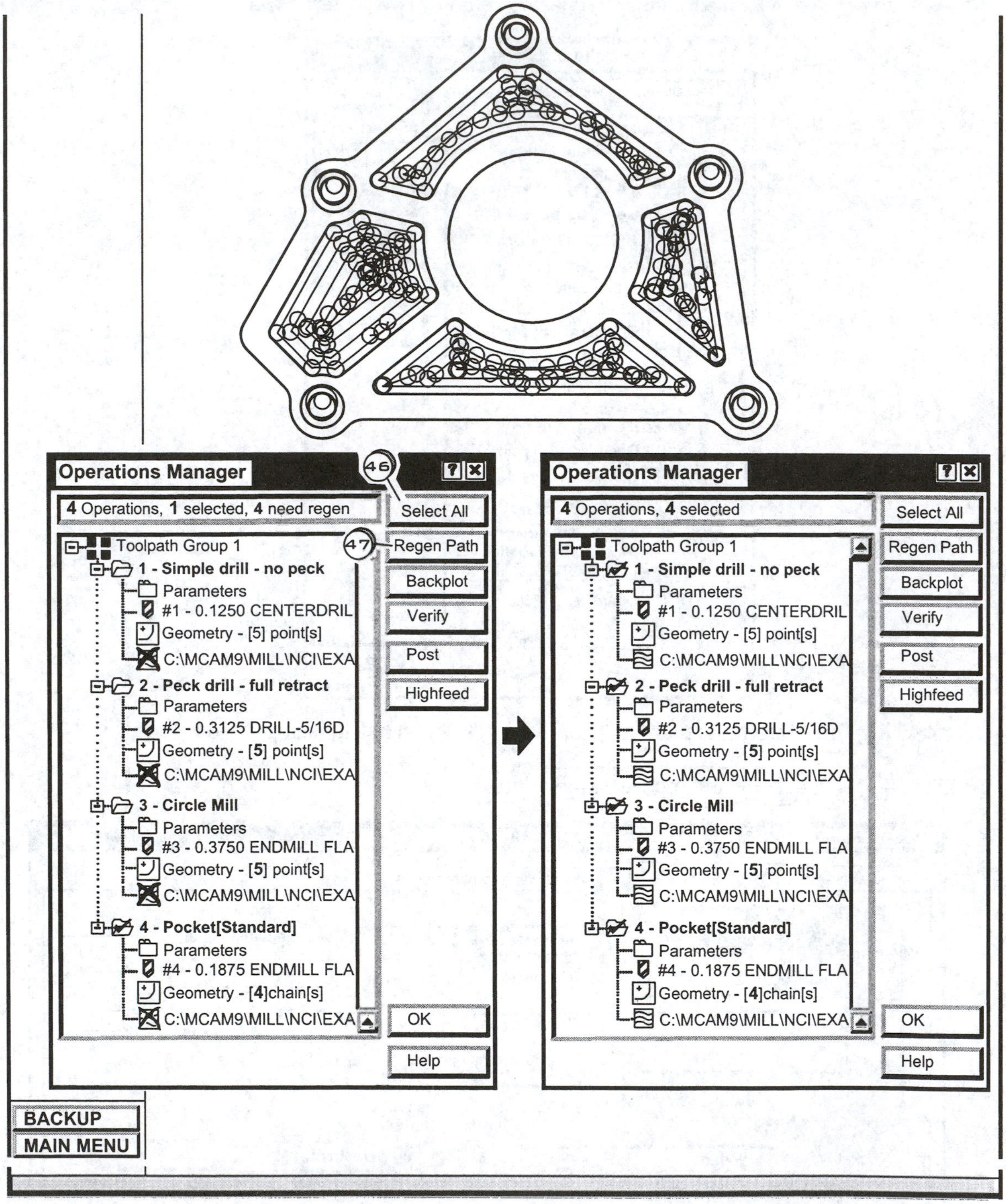

- Click (46) the **Select All** button
- Click (47) the **Regen Path** button

The machining operations as outlined in PROCESS PLAN 9-1 have now been successfully entered in *Mastercam's* Operations Manager.

9-5 Editing Data in the Operations Library

This section will present techniques for editing the data contained in *Mastercam*'s Operations Library. This includes *deleting* unwanted groups and operations as well as *creating new* data by copying and editing existing groups and operations.

ENTERING THE OPERATIONS LIBRARY EDITOR

Main Menu:
Analyze
Create
File
Modify
Xform
Delete
Screen
Solids
Toolpaths
NC utils
BACKUP
MAIN MENU

NC Utilities:
Verify
Backplot
Def. ops

Define Operations:
Defaults
Library
Compress

Specify File Name
Save in: Ops
OPERAT
File name: OPERAT
Save as type: V9 Operations Files(*.OP9)
Save
Cancel

LIBRARY OPERATIONS EDITOR
C:\MCAM9\MILL\OPS\OPRAT.OP9
3 Groups, 6 Operations, 1 Selected, 6 need re
DRILL-D1/4
1 - Simple drill no peck
2 - Peck drill - full retract
BOGUS
3 - Simple drill no peck
4 - Pocket[Standard]
5 - Tapping - feed in, reverse
POC-E1/4x1/8
6 - Pocket[Standard]
Select All
Regen Path
Backplot
Verify
Post
Highfeed
OK
Help

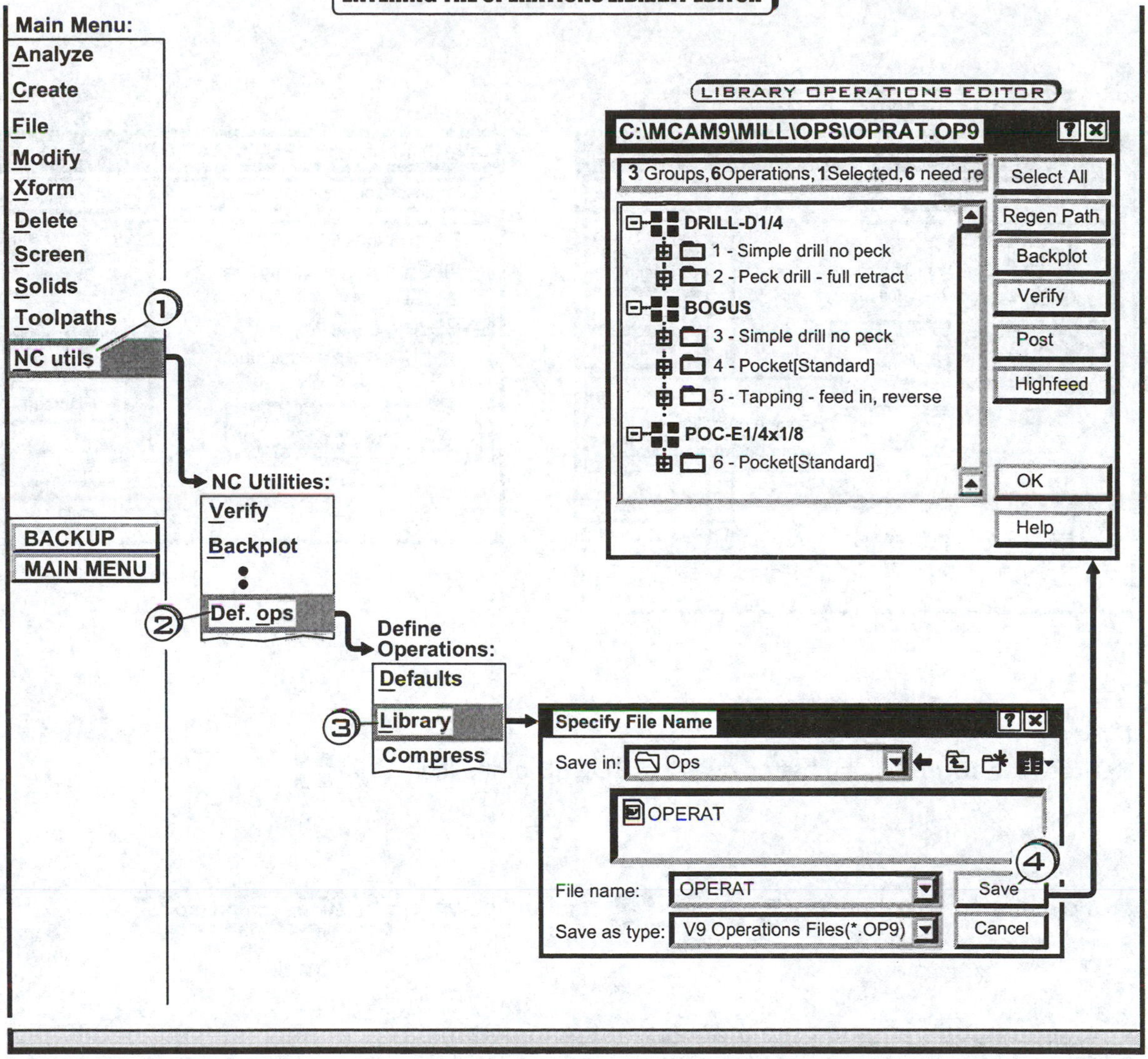

DELETING GROUPS AND OPERATIONS

EXAMPLE 9-3

Use the Library Operations Editor to delete the group **BOGUS** and its associated operatons from *Mastercam*'s Operations Library.

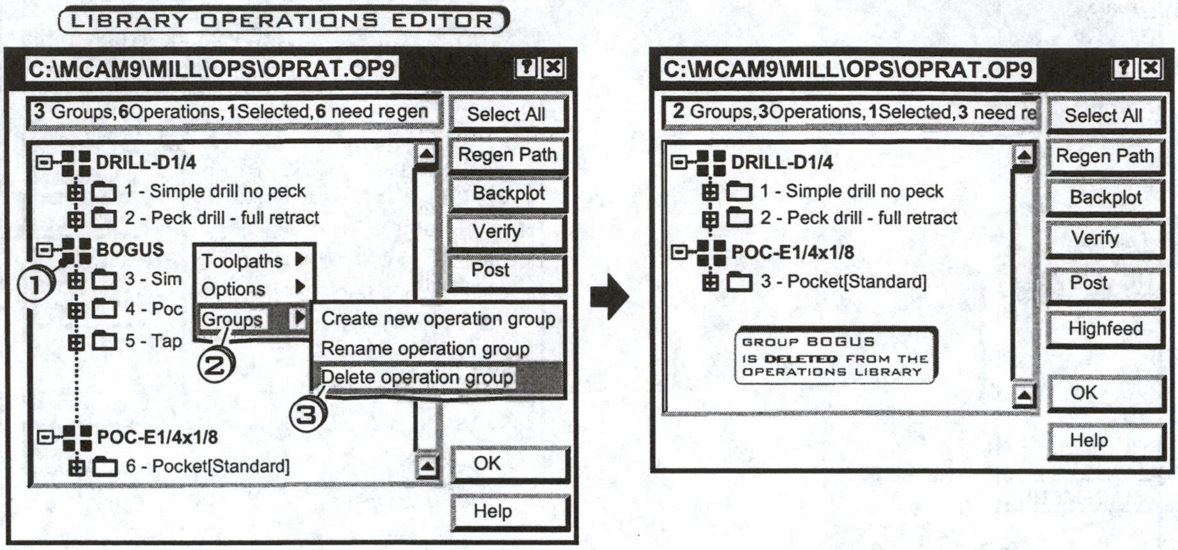

- Press the ***Right*** mouse button over the **BOGUS** group icon ①

- Move the cursor down to ② **Groups** out and down to **Delete operation group**

- Click ③ **Delete Operation group**

Group **BOGUS** and all its operations will then be deleted from the Operations Library

COPYING/EDITING EXISTING GROUPS AND OPERATIONS

EXAMPLE 9-4

Use the Library Operations Editor to create a new group in the library **POC-E3/8x1/8** by copying and editing the existing group **POC-E1/4x1/8** .

A) GENERATE A COPY OF THE GROUP POC-E1/4x1/8

LIBRARY OPERATIONS EDITOR

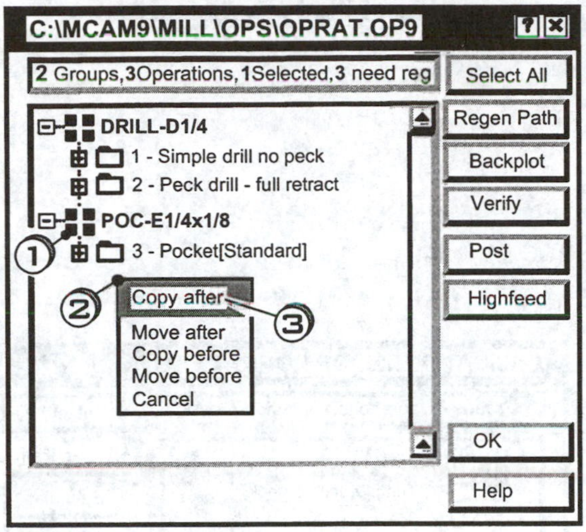

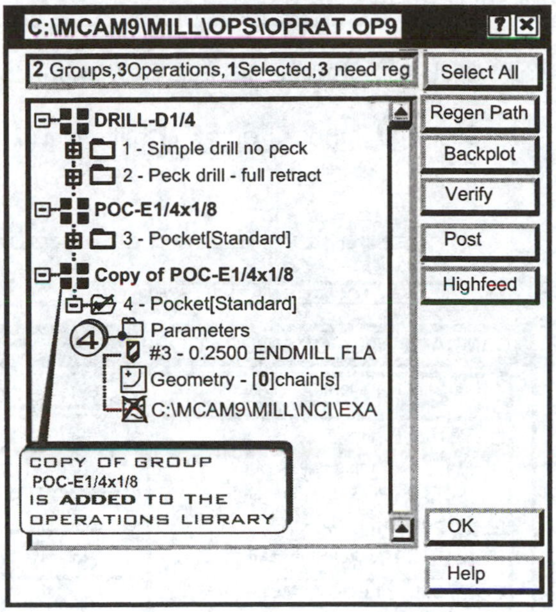

- Press the ***Right*** mouse button over the **POC-E1/4x1/8** group icon ① ***keeping it depressed*** move the cursor down to ② and release
- Click ③ **Copy after**

A copy of group **POC-E1/4/x1/8** and all its operations will then be added to the Operations Library

B) EDIT THE COPY(REPLACE THE 1/4 ENDMILL WITH A 3/8 ENDMILL TOOL)

- Click ④ the Parameters icon for the **2 - Peck drill-full retract** operation
- Click ⑤ the Tool parameters tab

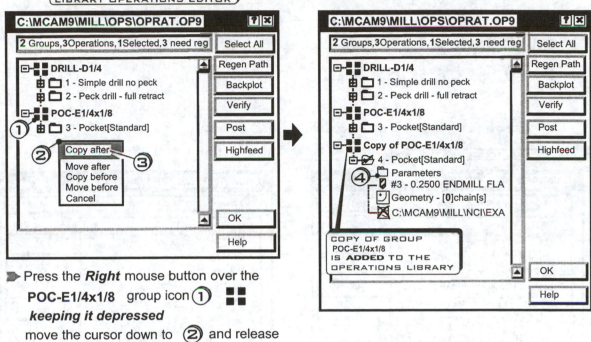

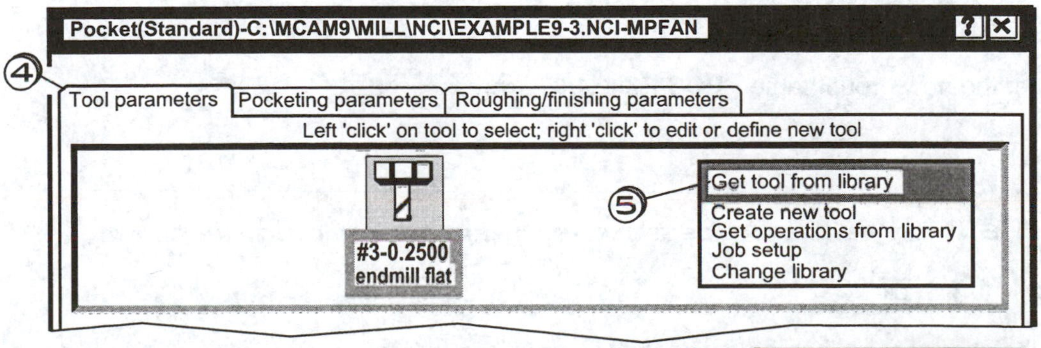

- ***Right*** Click move the mouse cursor *down* and Click ⑤ Get tool from library

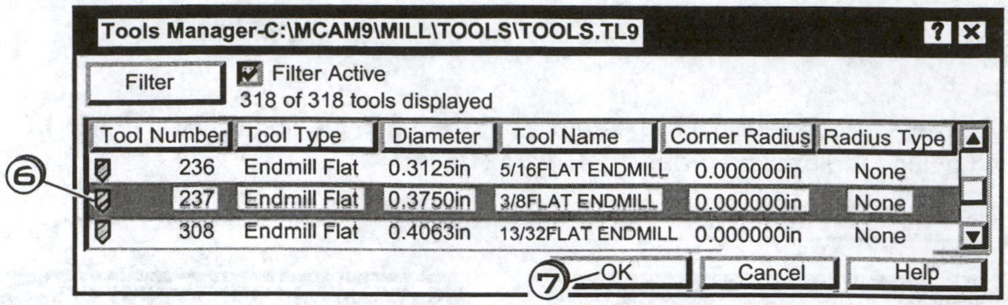

- Click ⑥ on the 3/8 FLAT ENDMILL tool
- Click ⑦ the OK button

C) Rename the edited group to POC-E3/8x1/4

LIBRARY OPERATIONS EDITOR

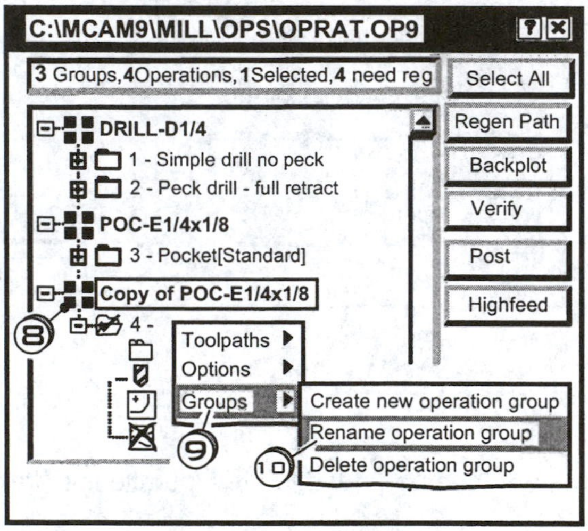

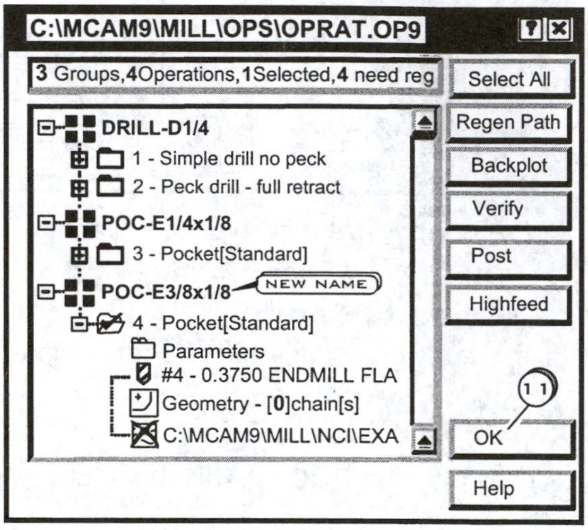

- Press the ***Right*** mouse button over the **Copy of POC-E1/4x1/8** group icon ⑧
- Move the cursor down to ⑨ **Groups** out and down to **Rename operation group**
- Click ⑩ **Rename operation group**
- Enter the new group name **POC-E3/8x1/8** ; press Enter

F) Save the changes to the Operations Library

- Click ⑪ OK

All the changes will then be saved to the Operations Library

EXERCISES

9-1)

PART A:

Create a dummy .25Dia circle and use it to generate a new group **TAP1/4-20BLIND** containing the set of operations listed in Table 9p-1.
Save the new group **TAP1/4-20BLIND** to Mastercam's Operations Library.

Table 9P-1

Group Name	Operation(s)	Tooling
TAP1/4-20BLIND 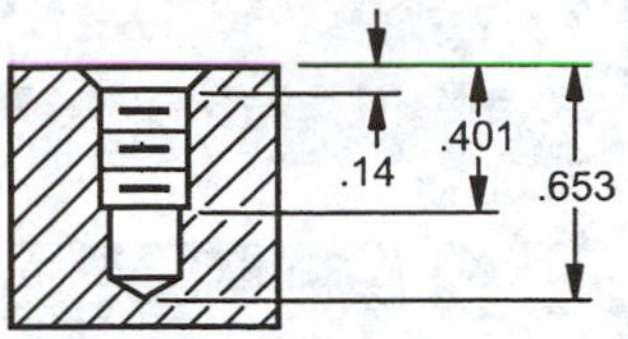	CENTER DRILL x .2 DEEP PECK DRILL x .601DEEP TAP 1/4-20UNC x .401DEEP CHAMFER 1/2 x .14 DEEP	1/8 CTR DRILL #7(.201) DRILL 1/4-20UNC TAP 1/2 CHAMFER MILL

PART B:

The CAD model file **EX9-1** is provided on the CD in the folder **CHAPTER9** on the CD. Copy it into the **JVAL-MILL** subdirectory on C drive. Open it and generate a part program for executing the machining outlined in PROCESS PLAN 9P-1. The part shown in Figure 9p-1 is to be produced.

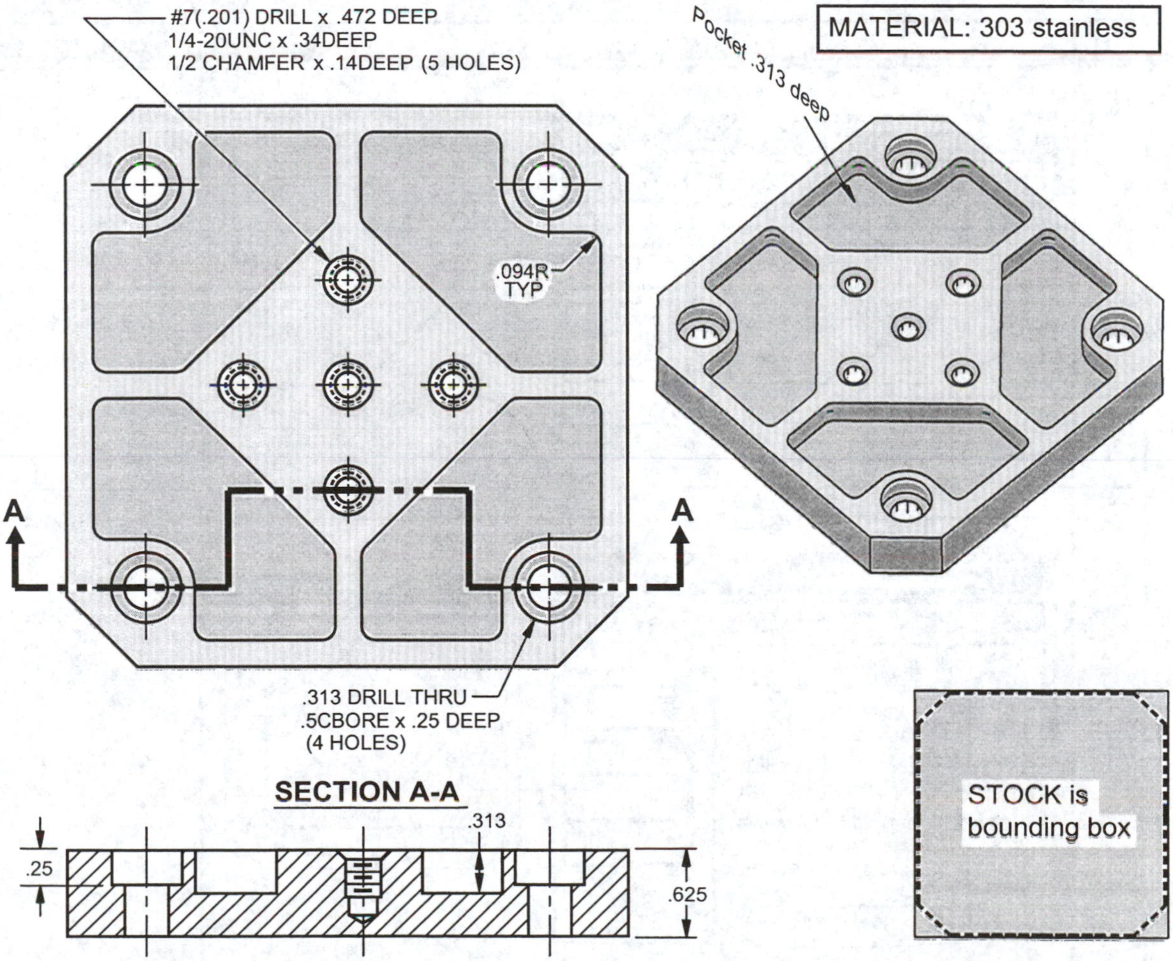

Figure 9-p1

PROCESS PLAN 9P-1

IMPORT THE GROUPS

CBORE-D5/16-E3/8x1/4

POC-E3/16x1/8

TAP1/4-20BLIND

FROM THE OPERATIONS LIBRARY

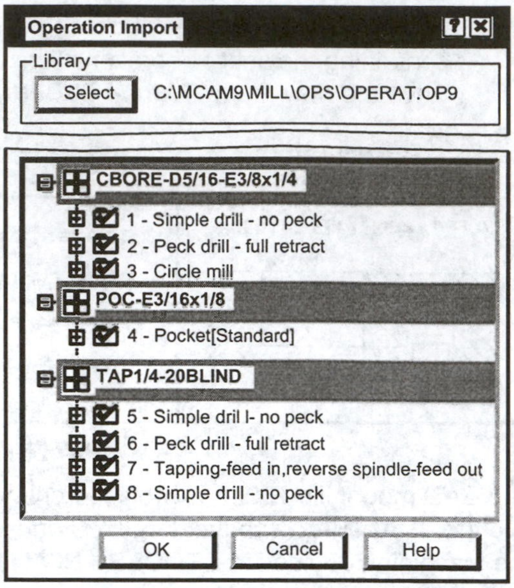

<table>
<tr><th>No.</th><th>Operation</th><th>Tooling</th></tr>
<tr><td>1</td><td>CENTER DRILL X .2 DEEP (4 HOLES)

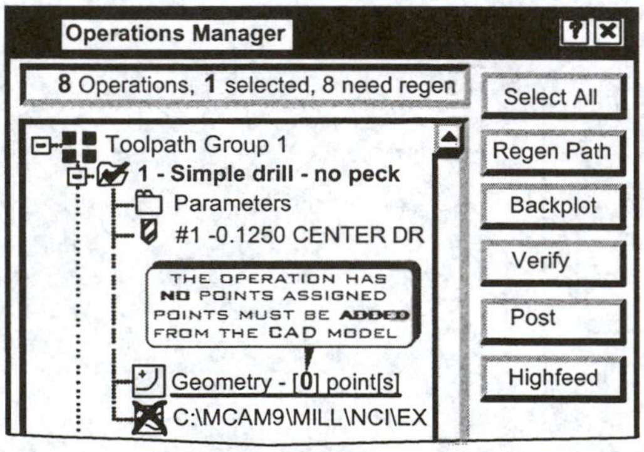

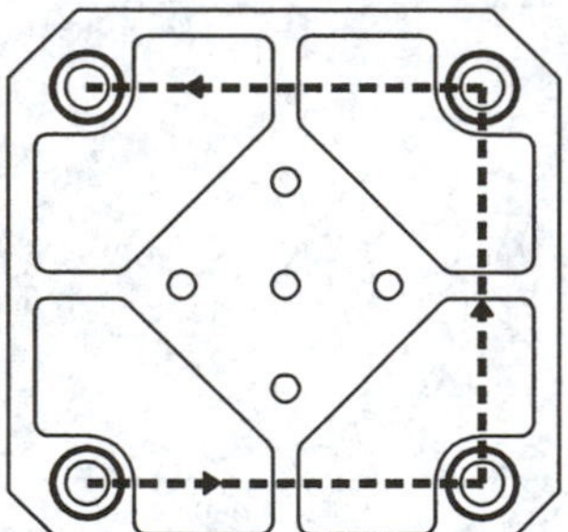</td><td>1/8 CTR DRILL</td></tr>
<tr><td>2</td><td>PECK DRILL THRU (4 HOLES)

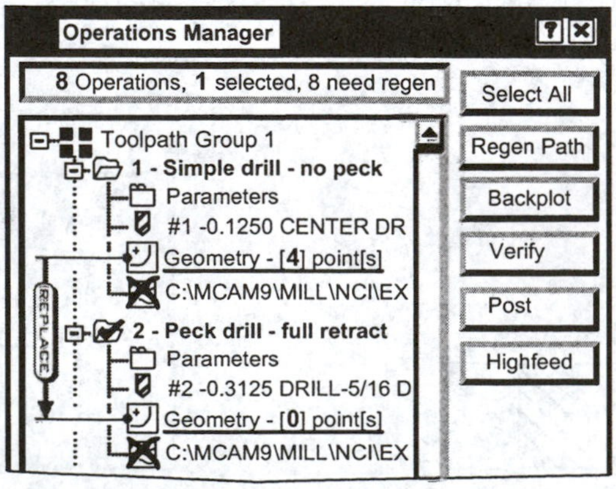

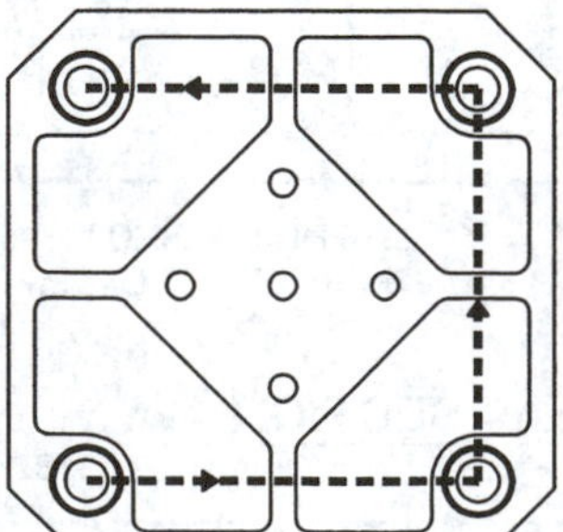</td><td>5/16 DRILL</td></tr>
</table>

PROCESS PLAN 9P-1(*continued*)

No.	Operation	Tooling
3	CIRCLE MILL X .25 DEEP (4 HOLES) 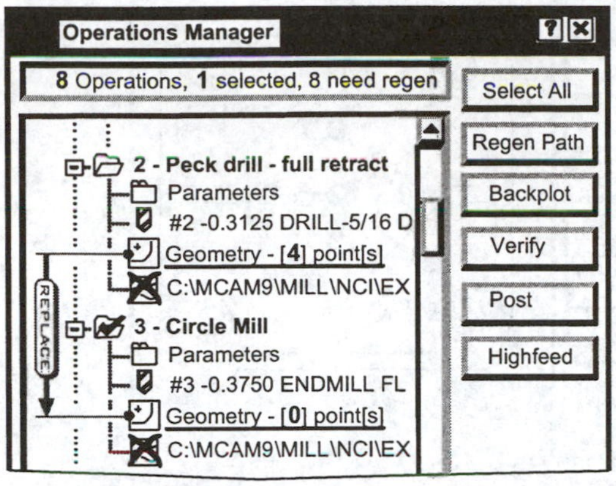 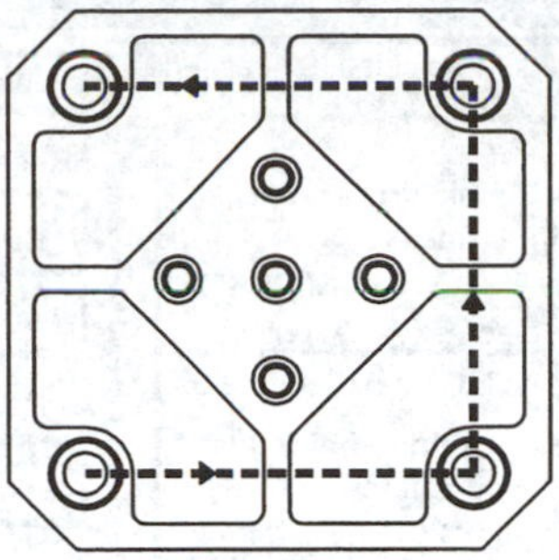	3/8 ENDMILL
4	ROUGH AND FINISH POCKET X .313 DEEP LEAVE .01 FOR FINISH CUT IN X,Y AND Z 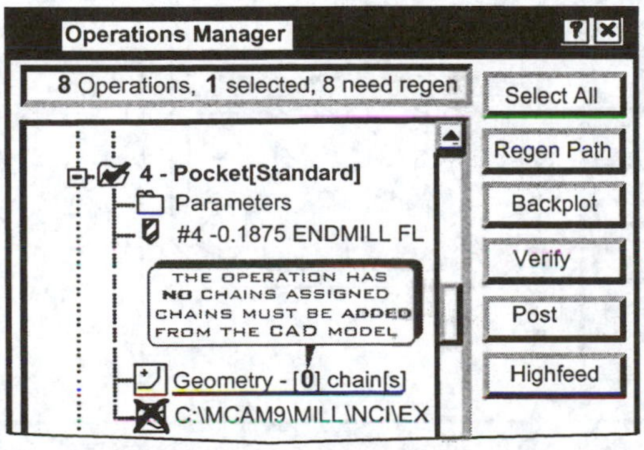 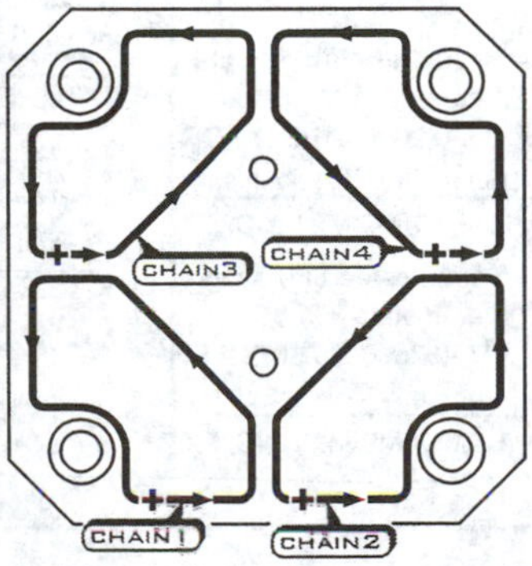	3/16 ENDMILL
5	CENTER DRILL X .2 DEEP (5 HOLES) 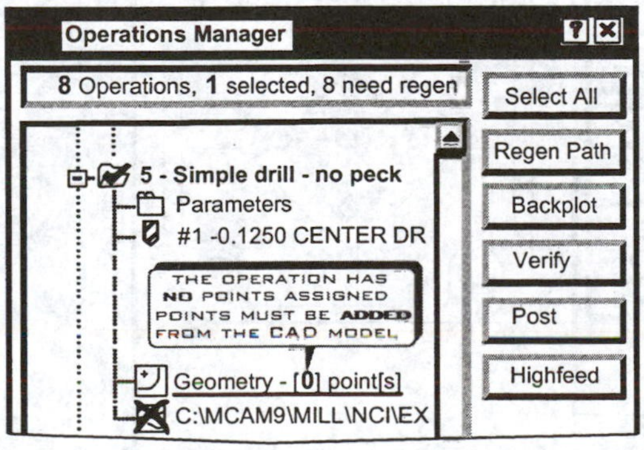 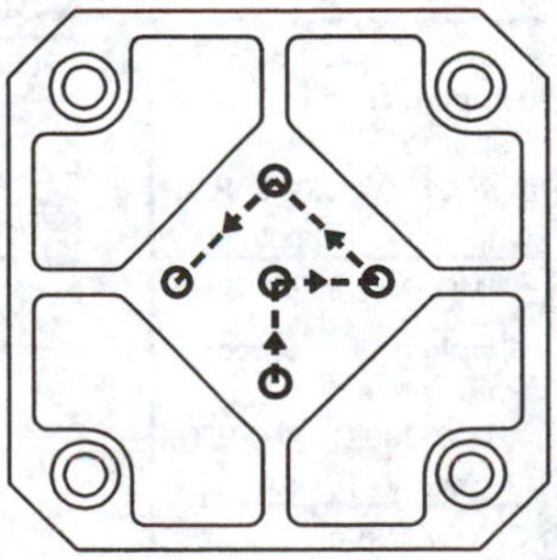	1/8 CTR DRILL

PROCESS PLAN 9P-1(*continued*)

No.	Operation	Tooling
6	PECK DRILL X .472 DEEP (5 HOLES) 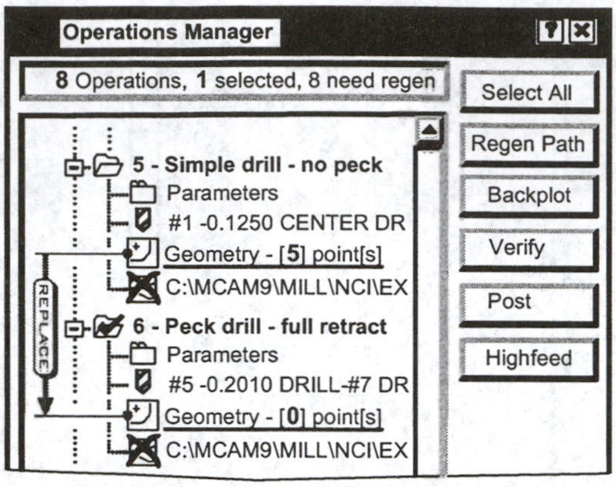 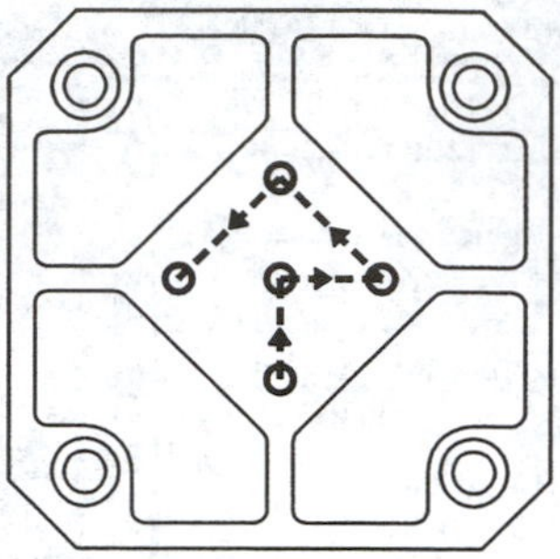	#7(.201)DRILL
7	TAP X .34 DEEP (5 HOLES) 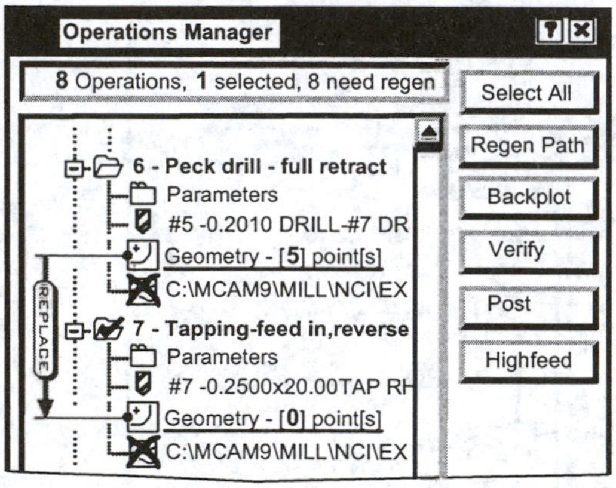 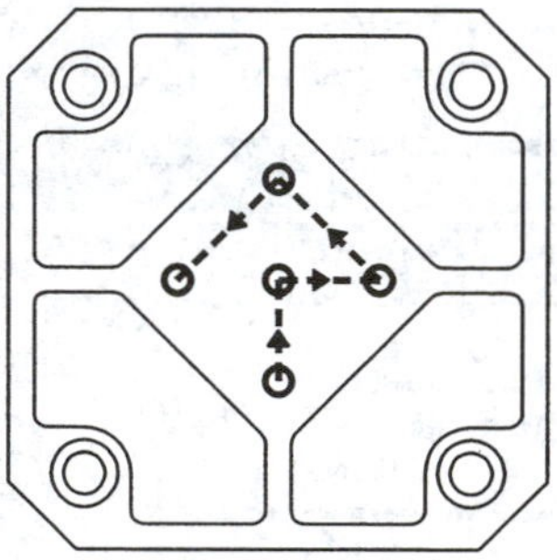	1/4-20 TAPRH
8	CHAMFER X .14 DEEP(5 HOLES) 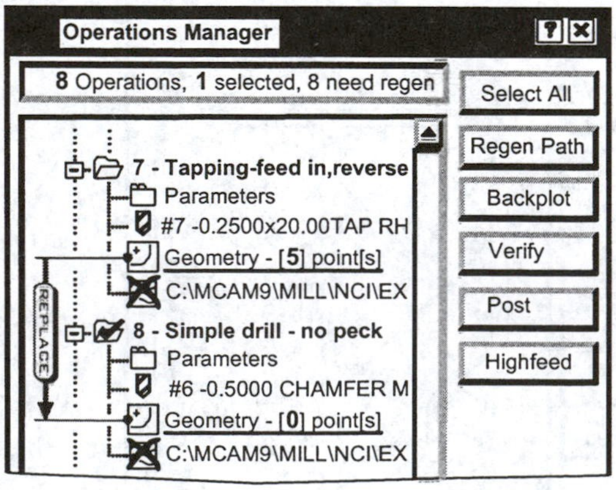 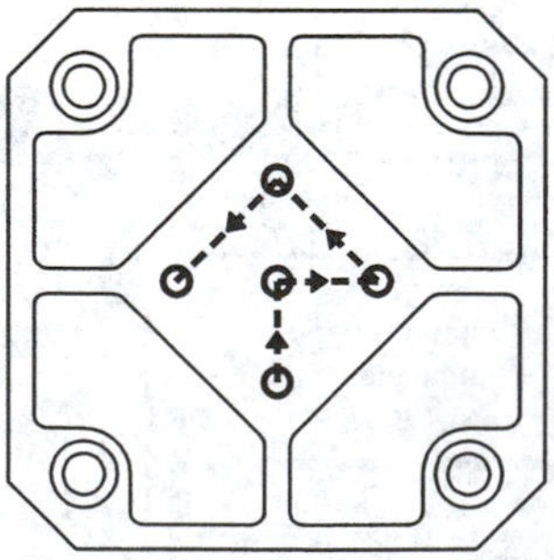	1/2 CHAMFER MILL

9-2) Enter the Operations Library Editor. Generate a new group **TAP#8-32BLIND** by copying and editing the existing group **TAP1/4-20BLIND**. The new group **TAP#8-32BLIND** is to contain the set of operations listed in Table 9p-2. Save all the groups to Mastercam's Operations Library.

PART A:

Table 9P-2

Group Name	Operation(s)	Tooling
TAP#8-32BLIND 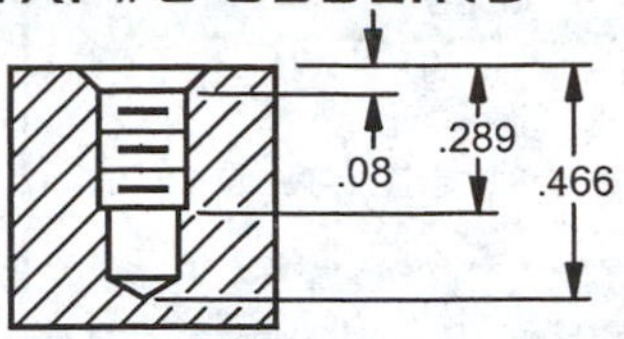 	CENTER DRILL X .2 DEEP PECK DRILL X .414DEEP TAP #8-32UNF X .289DEEP CHAMFER 1/2 X .08 DEEP	1/8 CTR DRILL #29(.136) DRILL #8-32UNF TAP 1/2 CHAMFER MILL

PART B:

The CAD model file **EX9-2** is in the folder **CHAPTER9** on the CD . Copy it into the subdirectory **JVAL-MILL** on C drive. Open it and follow the PROCESS PLAN 9P2 as a guide in generating a part program to produce the part shown in Figure 9p-2

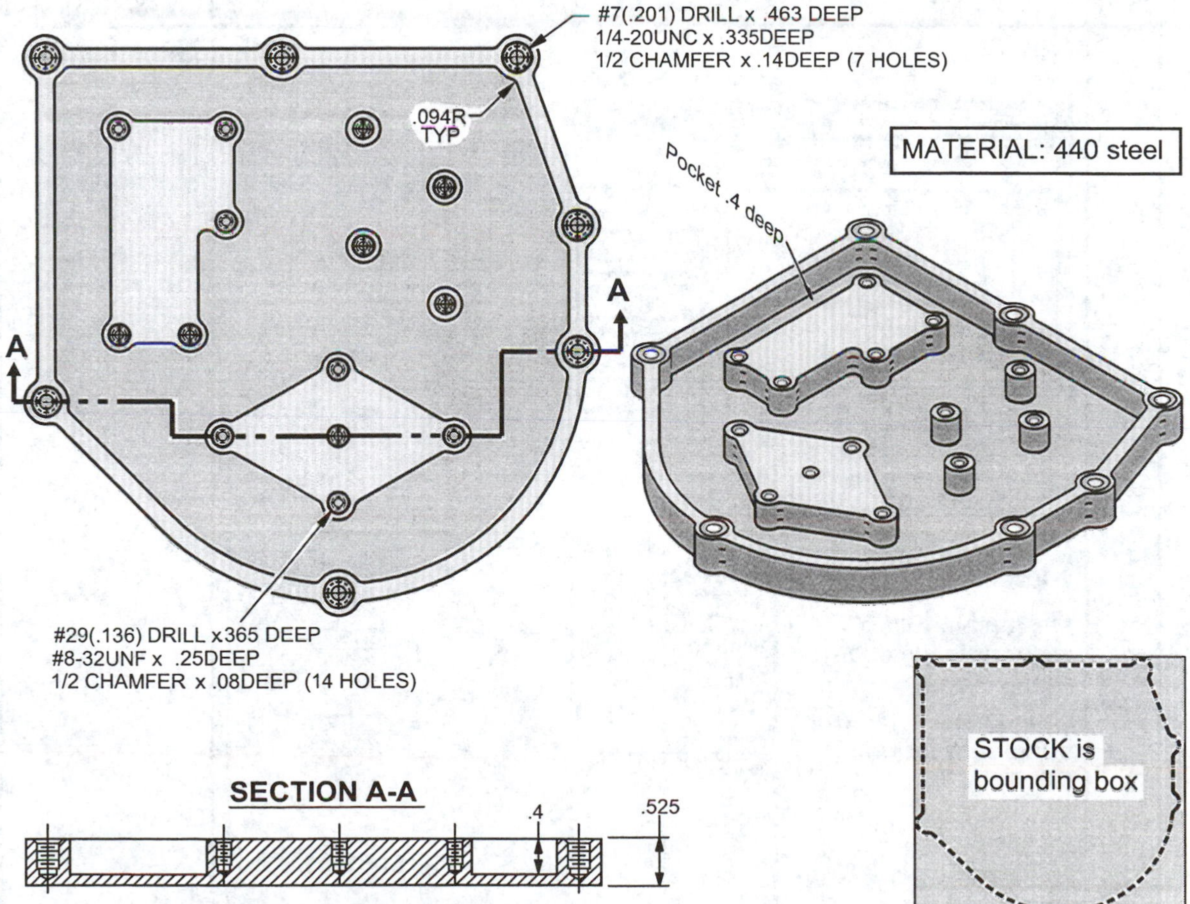

Figure 9-p2

PROCESS PLAN 9P-2

IMPORT THE GROUPS

POC-E3/16x1/8

TAP1/4-20BLIND

TAP#8-32BLIND

FROM THE OPERATIONS LIBRARY

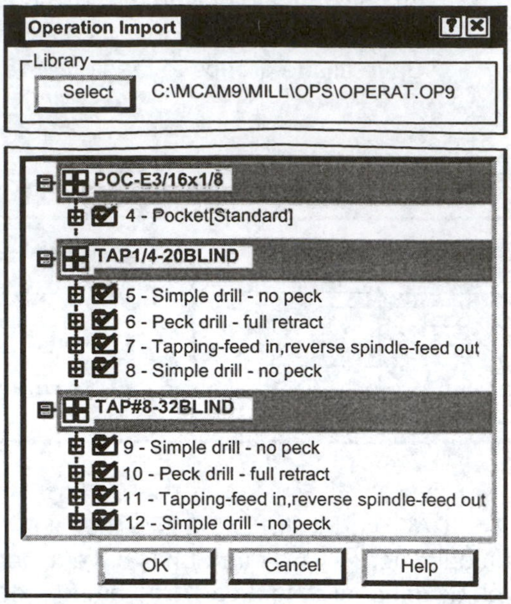

No.	Operation	Tooling
1	ROUGH AND FINISH POCKET X .4 DEEP LEAVE .01 FOR FINISH CUT IN X,Y AND Z	3/16 ENDMILL

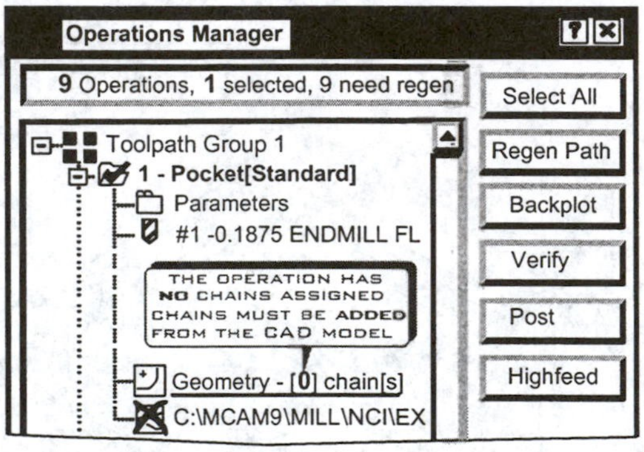

➤ Click ① Area

➤ Click ② the location of the point

Mastercam will ***automatically chain*** all the closed boundaries in the ***area surrounding the point***

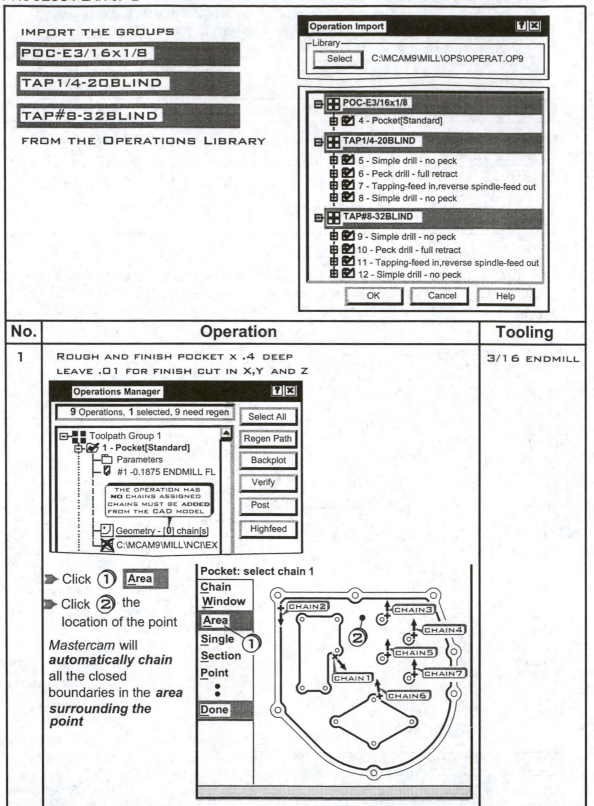

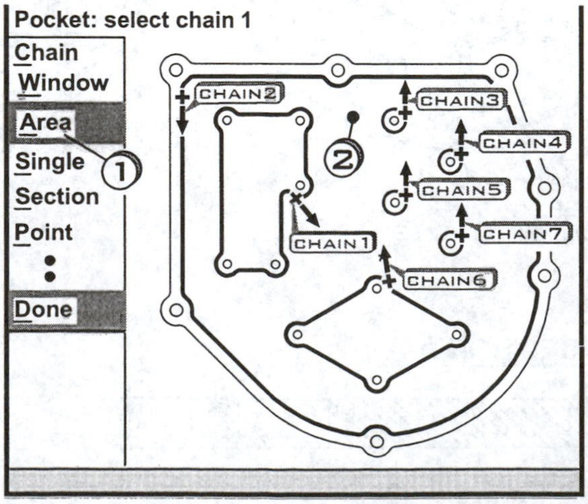

PROCESS PLAN 9P-2(*continued*)

No.	Operation	Tooling
2	CENTER DRILL X .2 DEEP (7 HOLES) 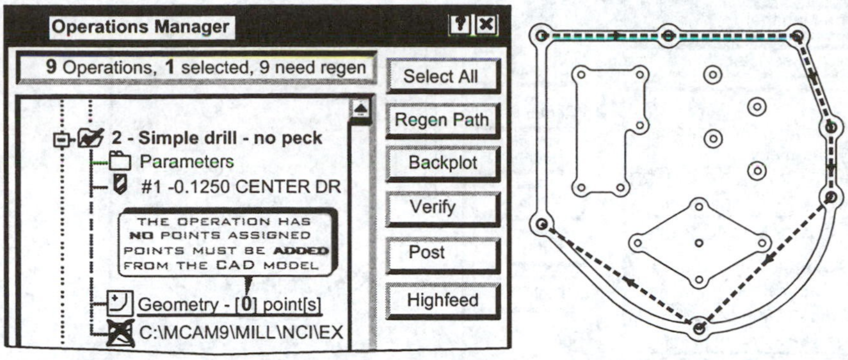 	1/8 CTR DRILL
3	PECK DRILL X .463 DEEP (7 HOLES) 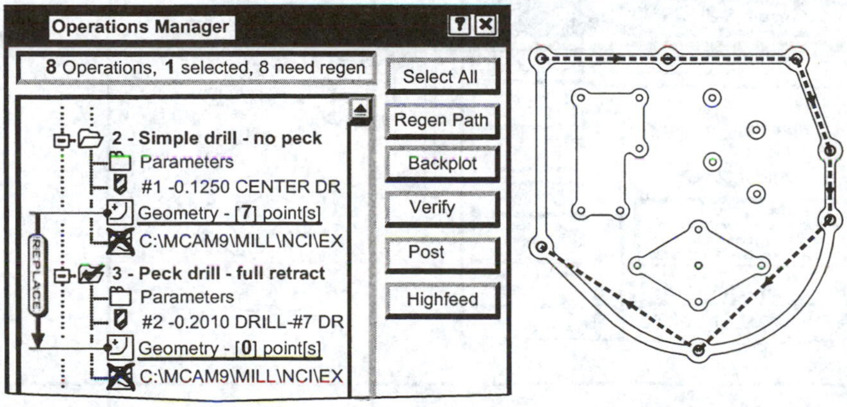 	#7(.201)DRILL
4	TAP X .335 DEEP (7 HOLES) 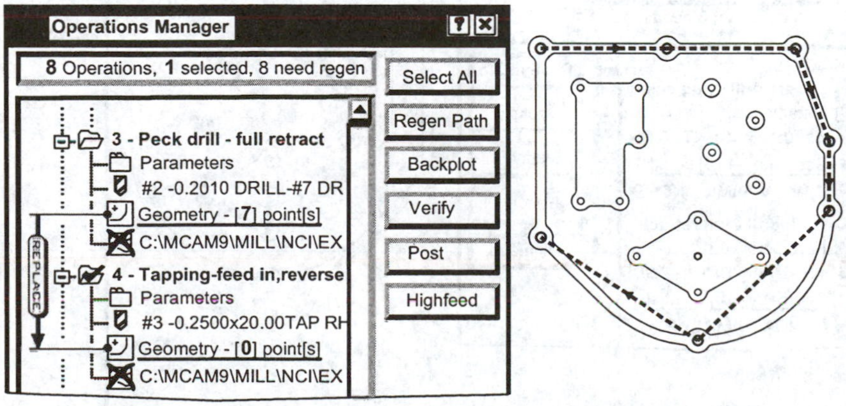 	1/4-20 TAPRH

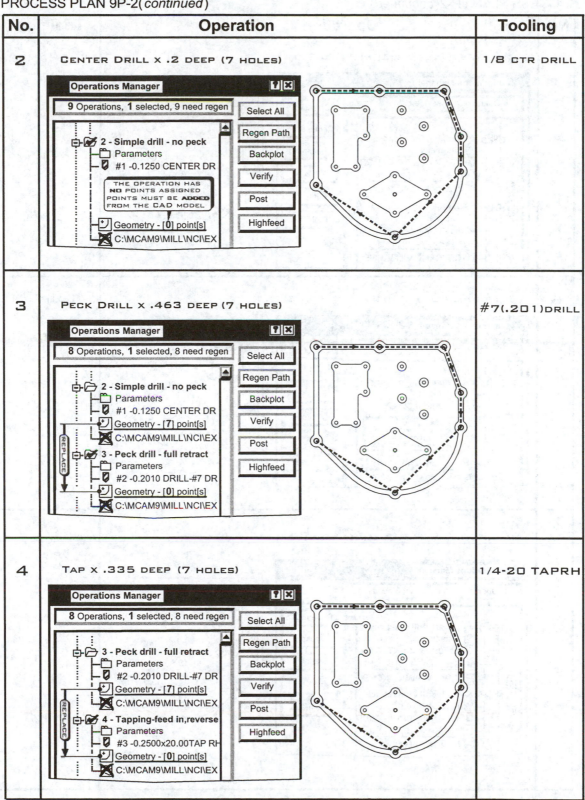

PROCESS PLAN 9P-2(*continued*)

No.	Operation	Tooling
5	CHAMFER X .14 DEEP(7 HOLES) 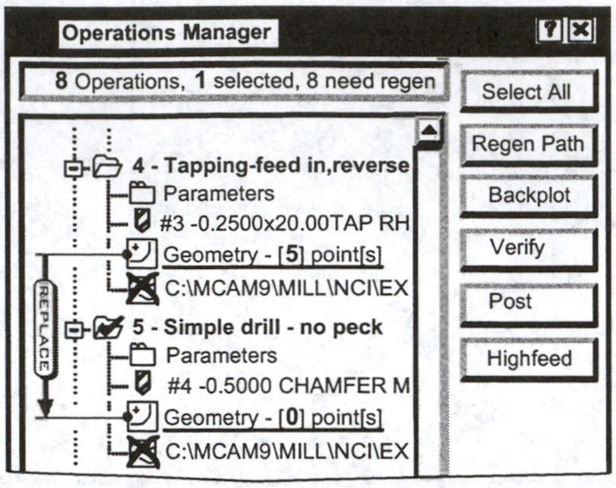 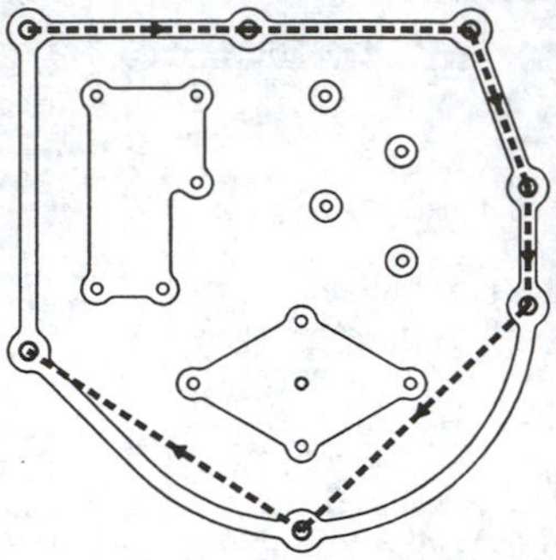	1/2 CHAMFER MILL
6	CENTER DRILL X .2 DEEP (14 HOLES) 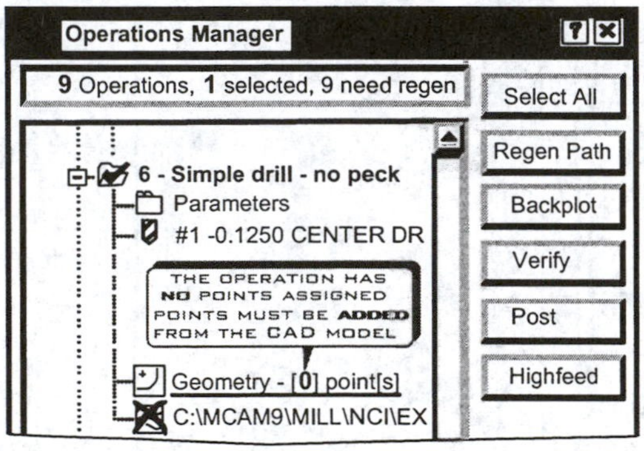 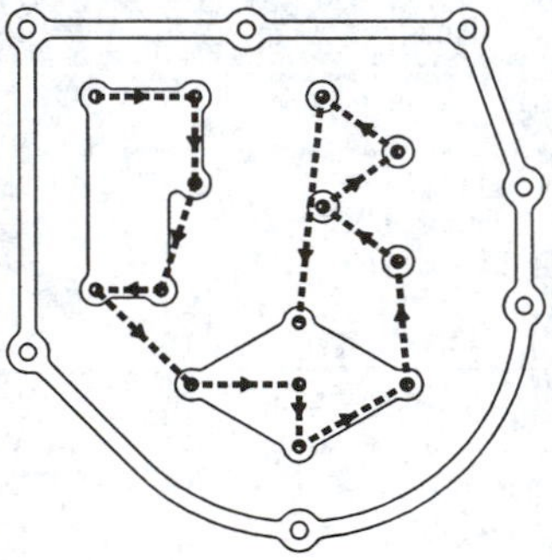	1/8 CTR DRILL
7	PECK DRILL X .365 DEEP (14 HOLES) 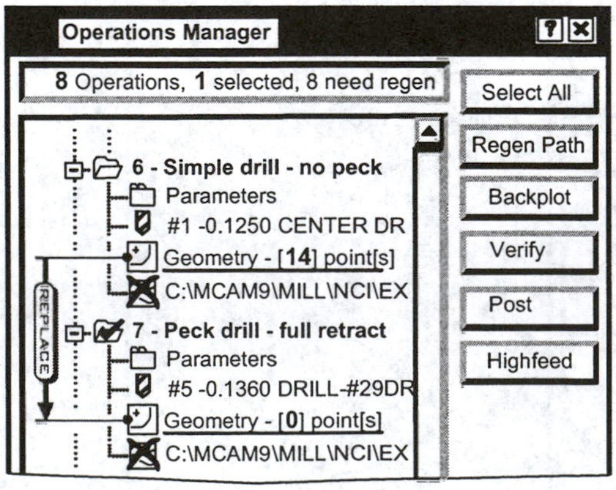 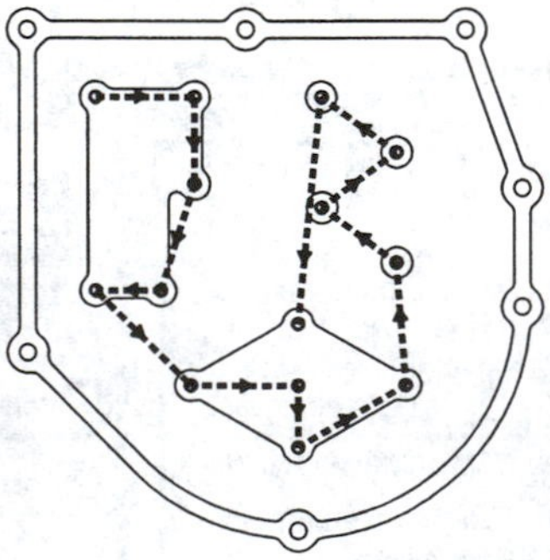	#29(.136) DRILL

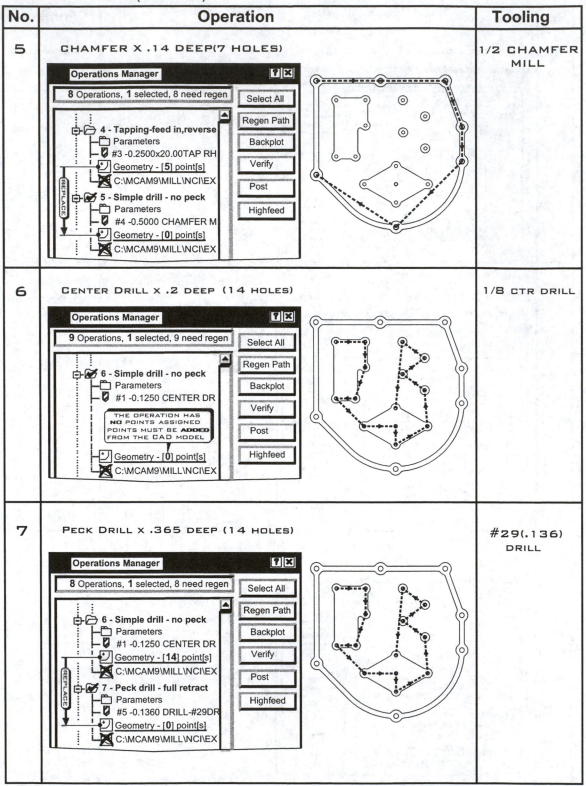

PROCESS PLAN 9P-2(*continued*)

<table>
<tr><th>No.</th><th>Operation</th><th>Tooling</th></tr>
<tr><td>8</td><td>TAP X .25 DEEP(14 HOLES)

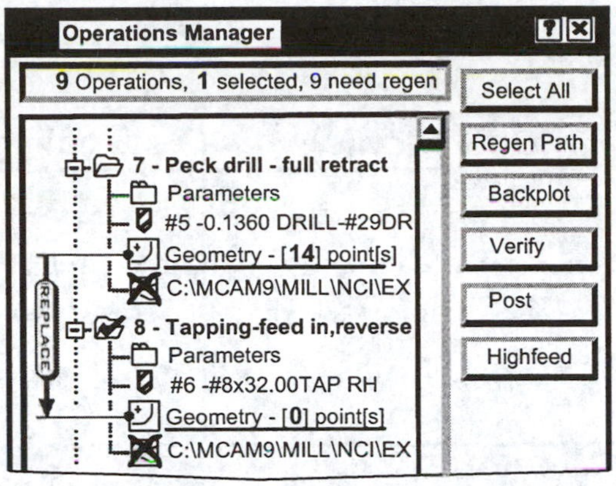

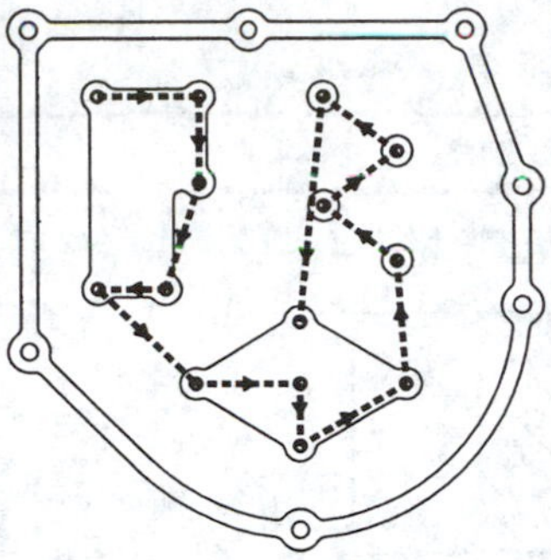</td><td>#8-32 TAPRH</td></tr>
<tr><td>9</td><td>CHAMFER X .08 DEEP (14 HOLES)

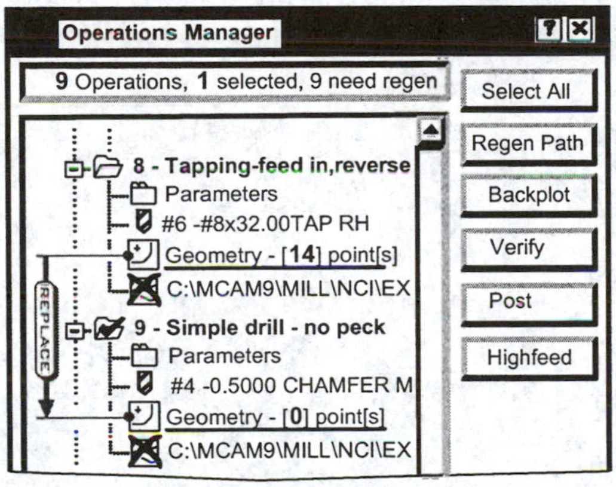

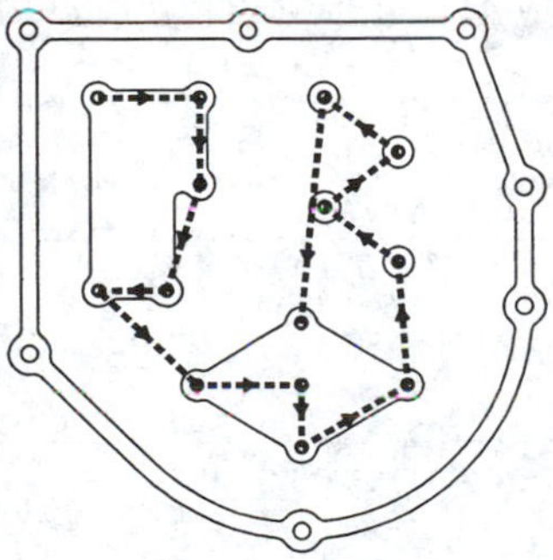</td><td>1/2 CHAMFER MILL</td></tr>
</table>

9-3) Enter the Operations Library Editor. Generate a new group **DRILL-D1/5-CHAMIL1X.719** by copying and editing the existing group **TAP#8-32BLIND**. The new group **DRILL-D1/2-CHAMIL1X.719** is to contain the set of operations listed in Table 9p-3.
Save all the groups to Mastercam's Operations Library.

PART A:

Table 9P-3

Group Name	Operation(s)	Tooling
DRILL-D1/2-CHAMIL1X.719 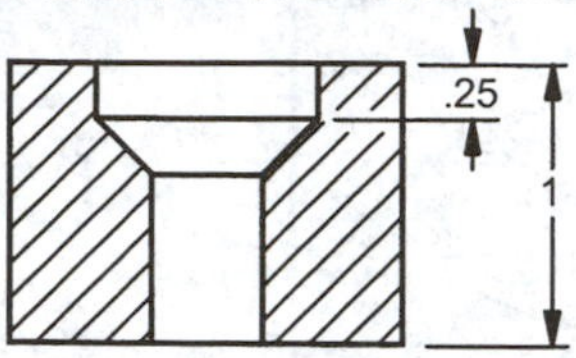	CENTER DRILL X .25 DEEP PECK DRILL THRU CHAMFER 1 X .719 DEEP	1/4 CTR DRILL 1/2 DRILL 1 CHAMFER MILL

PART B:

The CAD model file **EX9-3** is in the folder **CHAPTER9** on the CD . Copy it into the **JVAL-MILL** subdirectory on C drive. Open it and follow the PROCESS PLAN 9P-3 as a guide in generating a part program to produce the part shown in Figure 9p-3

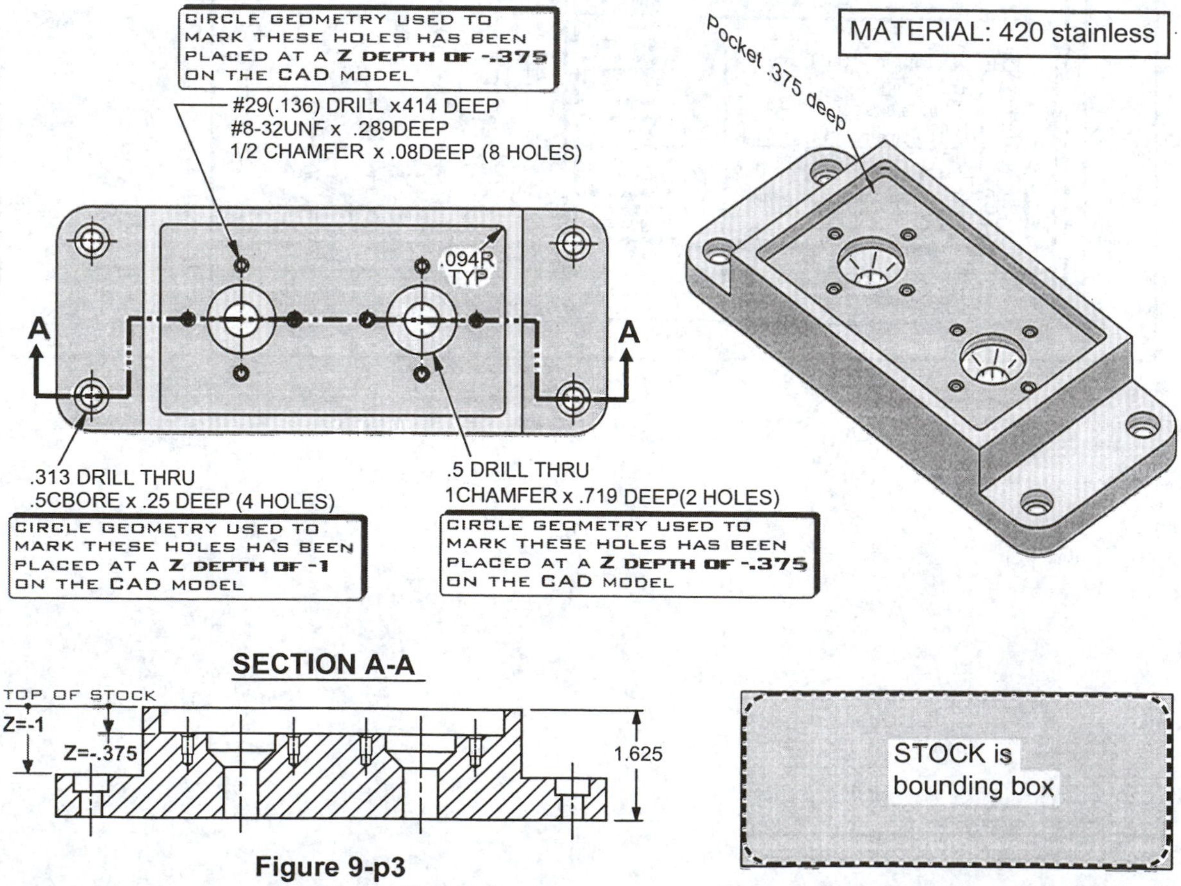

Figure 9-p3

PROCESS PLAN 9P-3

<table>
<tr><th>No.</th><th>Operation</th><th>Tooling</th></tr>
<tr><td>1</td><td>ROUGH AND FINISH POCKET X 1 DEEP
LEAVE .01 FOR FINISH CUT IN X,Y AND Z
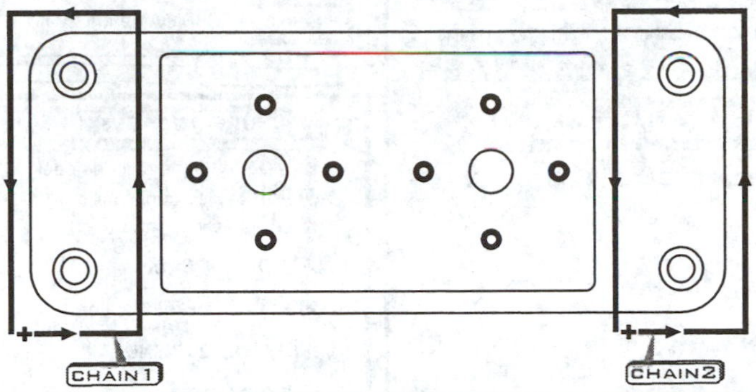
</td><td>1/2 ENDMILL</td></tr>
<tr><td>2</td><td>CONTOUR OUTSIDE PROFILE
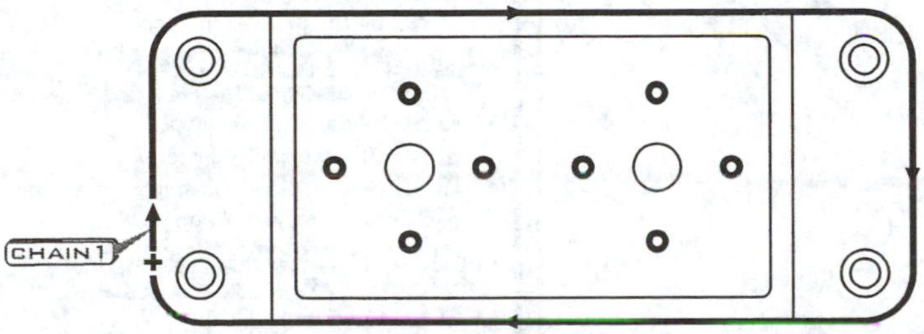
</td><td>1/2 ENDMILL</td></tr>
<tr><td colspan="3">IMPORT THE GROUP
POC-E3/16x1/8
FROM THE OPERATIONS LIBRARY
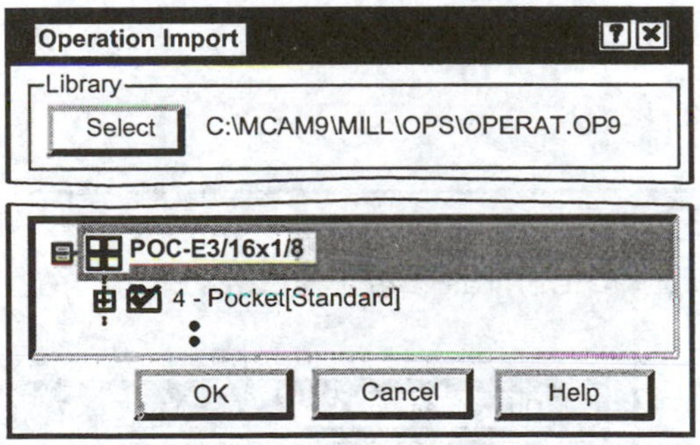
</td></tr>
<tr><td>3</td><td>ROUGH AND FINISH POCKET X .4 DEEP
LEAVE .01 FOR FINISH CUT IN X,Y AND Z
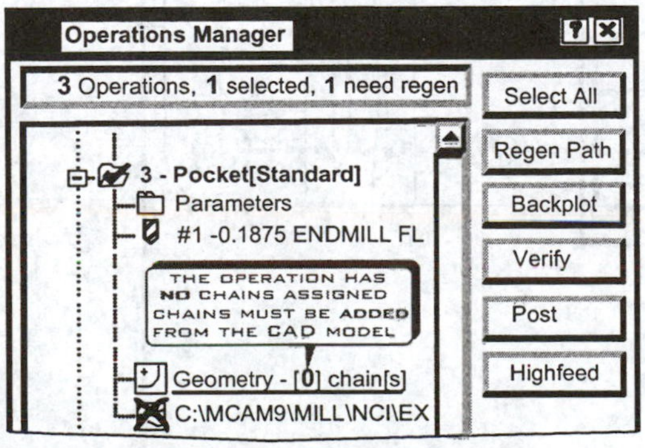

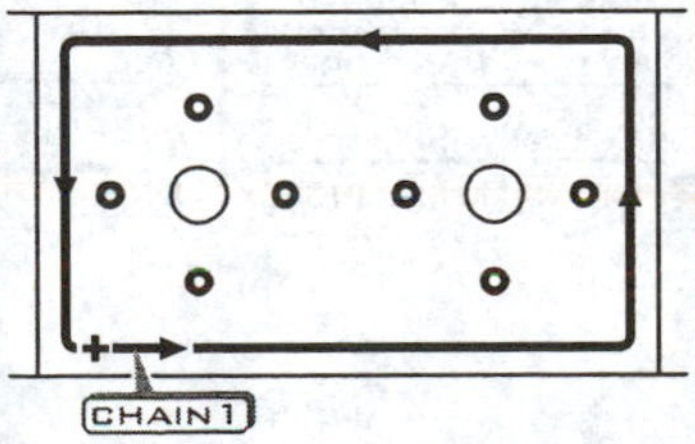
</td><td>3/16 ENDMILL</td></tr>
</table>

PROCESS PLAN 9P-3(*continued*)

IMPORT THE GROUPS

CBORE-D5/16-E3/8X1/4

TAP#8-32BLIND

DRILL-D1/2-CHAMILL1X.719

FROM THE OPERATIONS LIBRARY

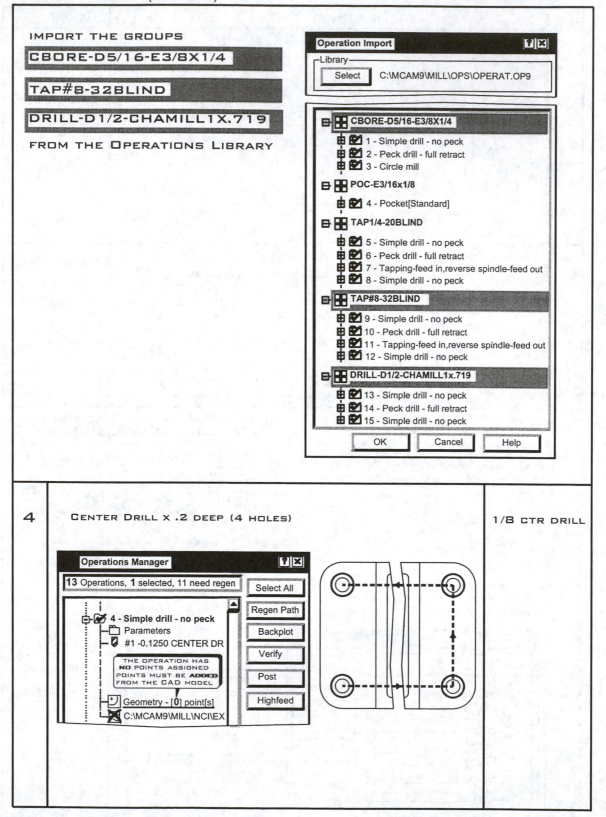

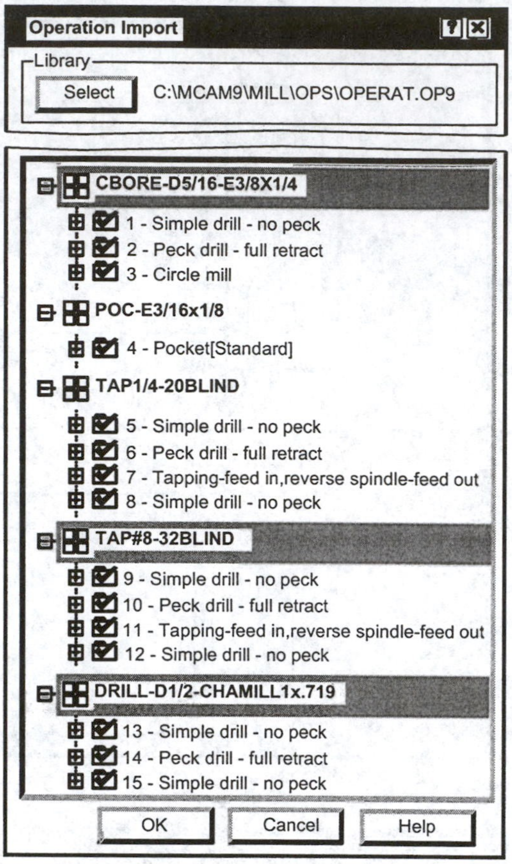

4	CENTER DRILL X .2 DEEP (4 HOLES)	1/8 CTR DRILL

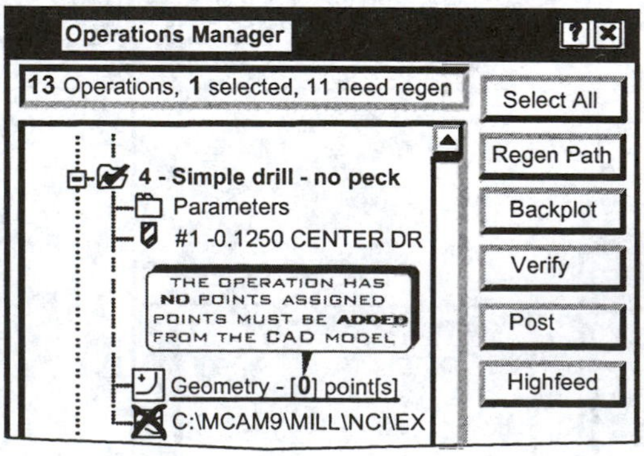

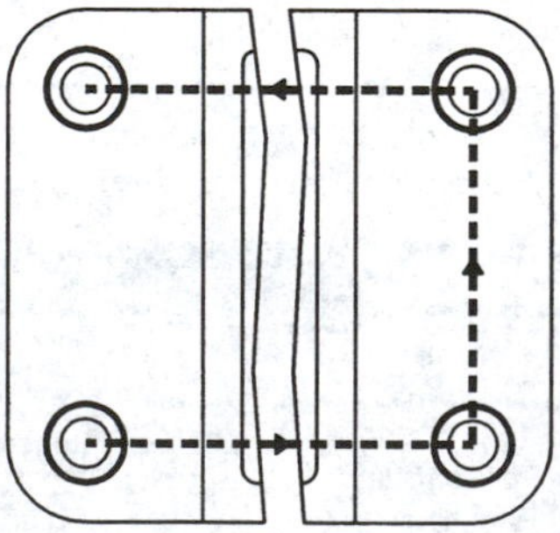

PROCESS PLAN 9P-3(*continued*)

<table>
<tr><th>No.</th><th>Operation</th><th>Tooling</th></tr>
<tr><td>5</td><td>PECK DRILL THRU (4 HOLES)

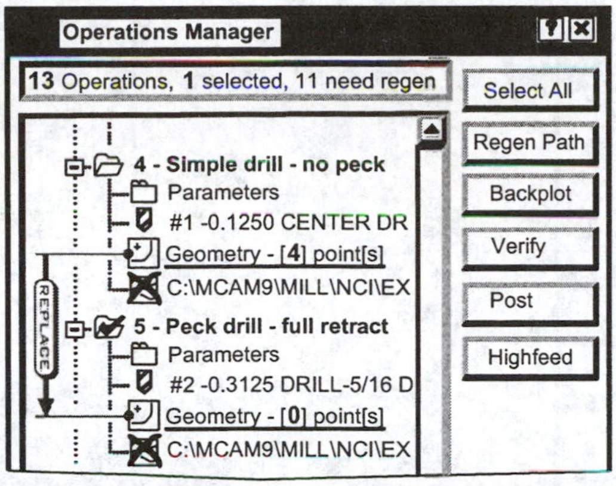

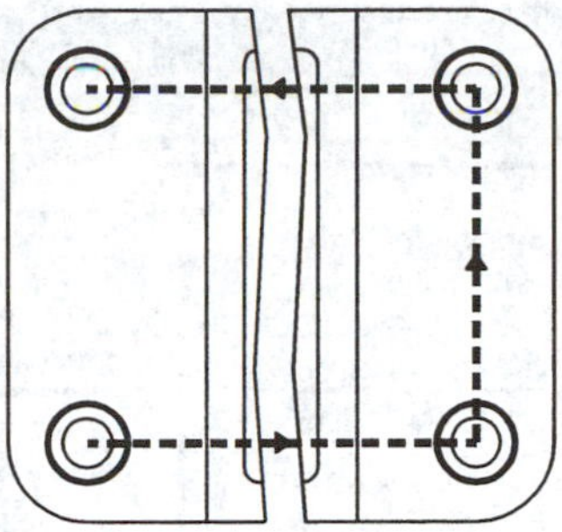</td><td>5/16 DRILL</td></tr>
<tr><td>6</td><td>CIRCLE MILL X .25 DEEP (4 HOLES)

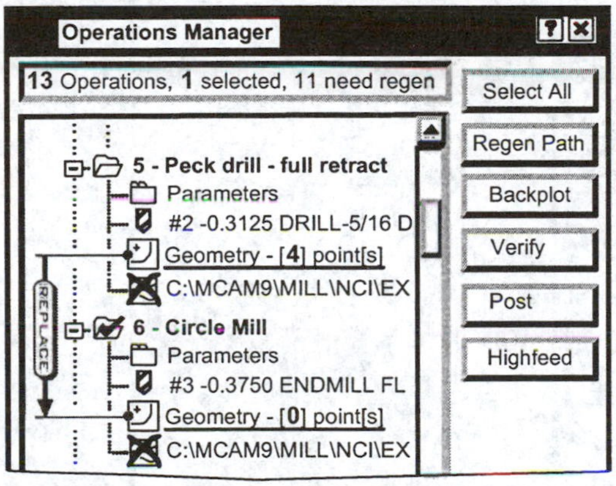

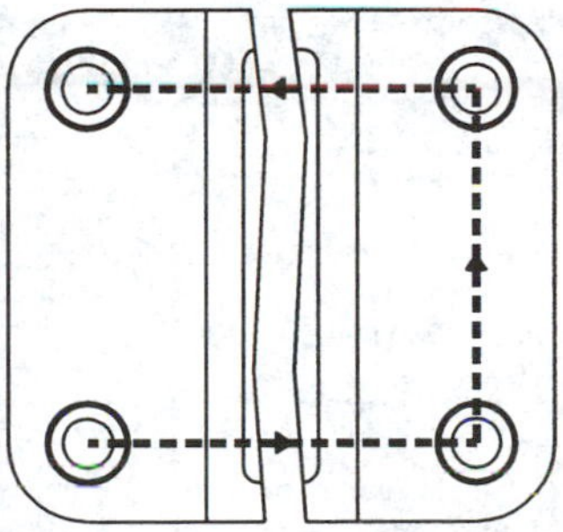</td><td>3/8 ENDMILL</td></tr>
<tr><td>7</td><td>CENTER DRILL X .2 DEEP (8 HOLES)

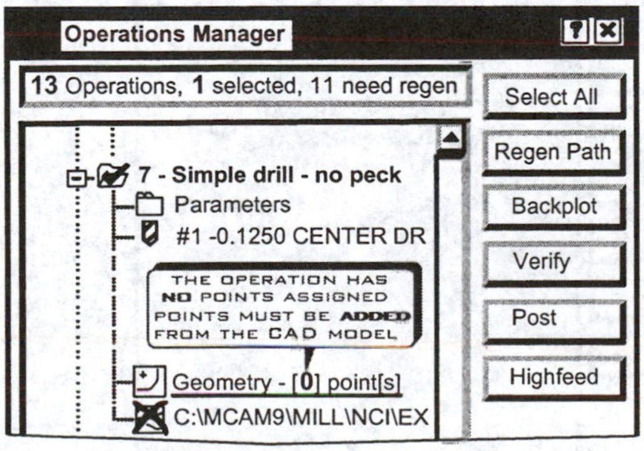

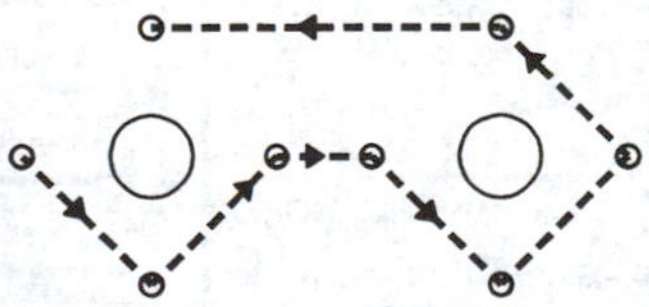</td><td>1/8 CTR DRILL</td></tr>
</table>

PROCESS PLAN 9P-3(*continued*)

<table>
<tr><th>No.</th><th>Operation</th><th>Tooling</th></tr>
<tr><td>8</td><td>PECK DRILL X .365 DEEP (8 HOLES)

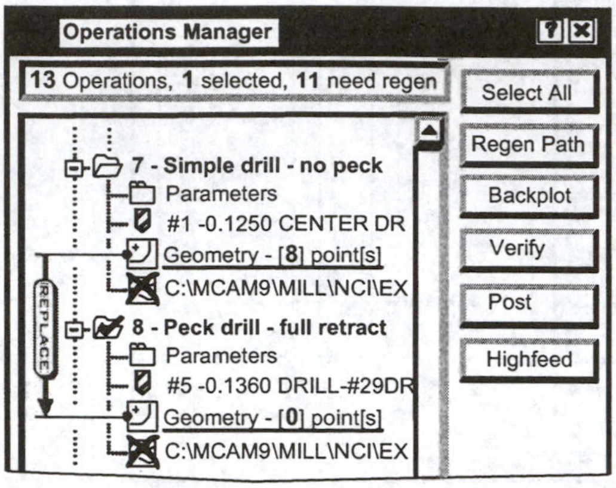

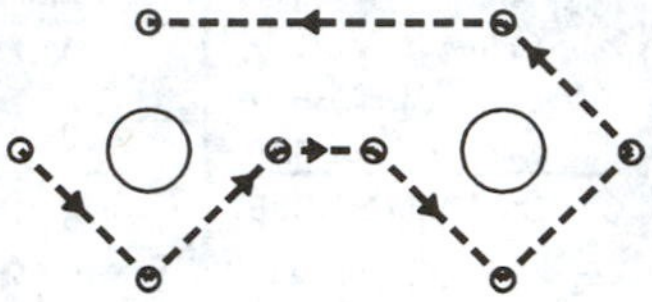</td><td>#29(.136) DRILL</td></tr>
<tr><td>9</td><td>TAP X .25 DEEP (8 HOLES)

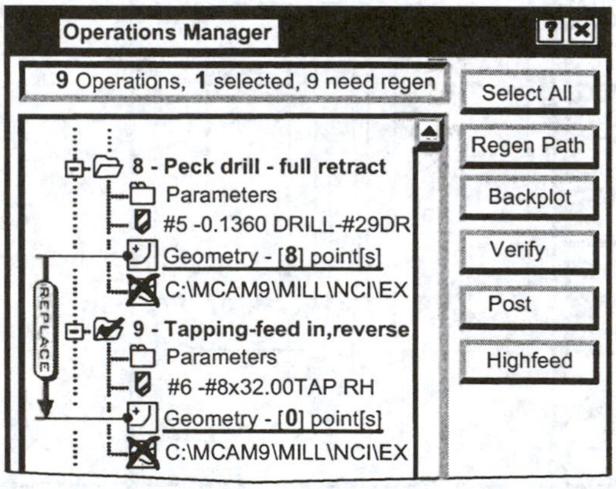

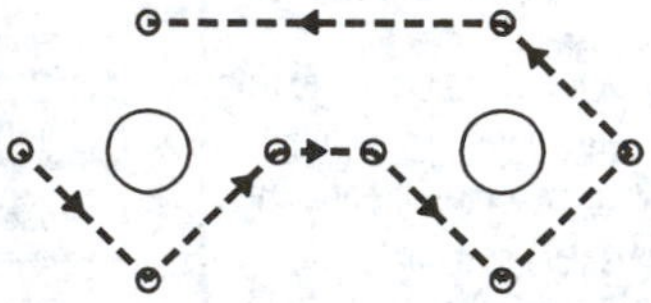</td><td>#8-32 TAPRH</td></tr>
<tr><td>10</td><td>CHAMFER X .08 DEEP (8 HOLES)

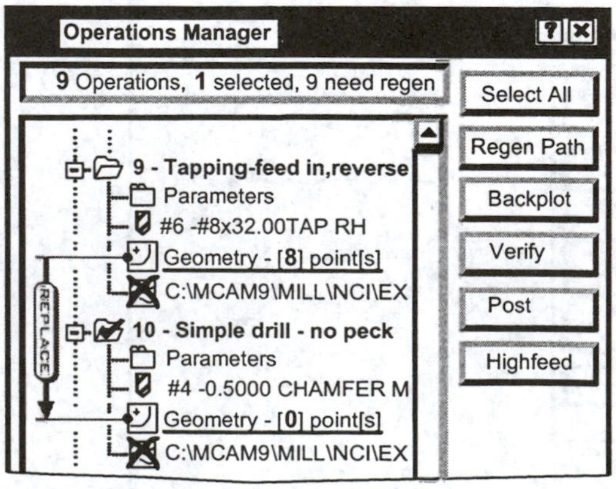

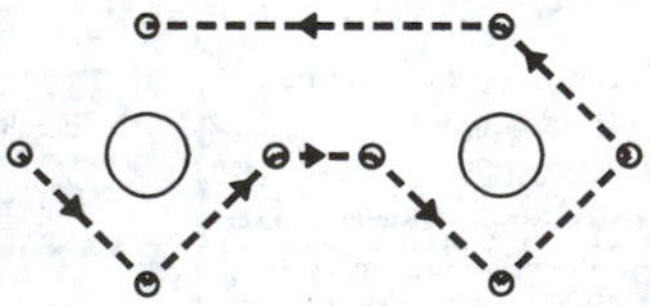</td><td>1/2 CHAMFER MILL</td></tr>
</table>

PROCESS PLAN 9P-3(*continued*)

<table>
<tr><th>No.</th><th>Operation</th><th>Tooling</th></tr>
<tr><td>11</td><td>CENTER DRILL X .25 DEEP (2 HOLES)

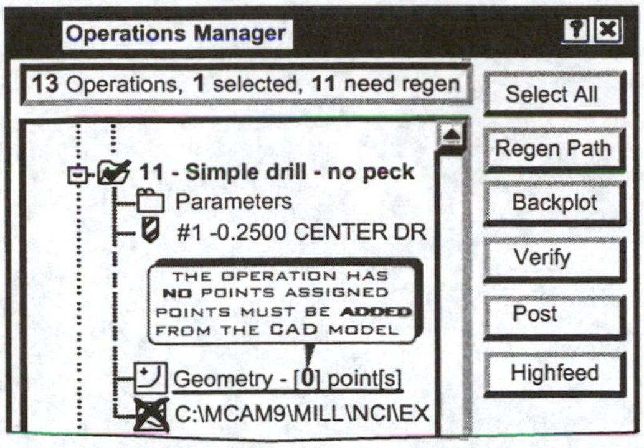

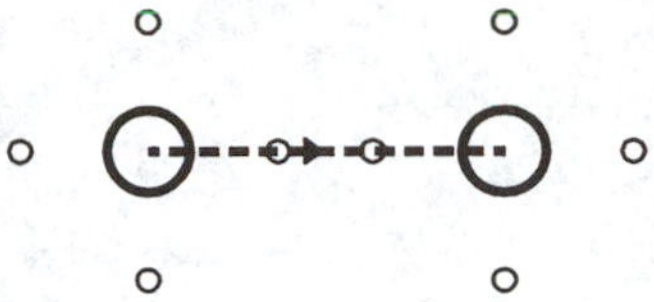</td><td>1/4 CTR DRILL</td></tr>
<tr><td>12</td><td>PECK DRILL THRU (2 HOLES)

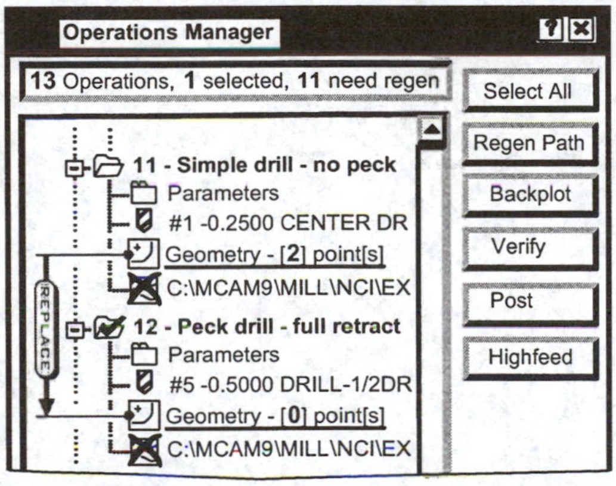

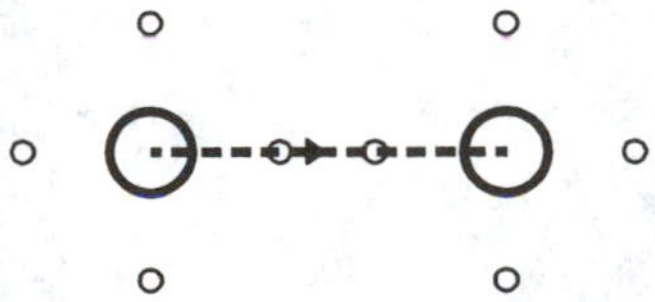</td><td>1/2 DRILL</td></tr>
<tr><td>13</td><td>CHAMFER X .719 DEEP (2 HOLES)

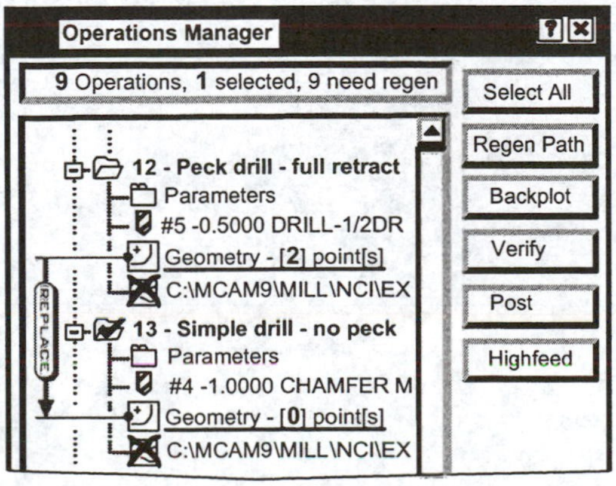

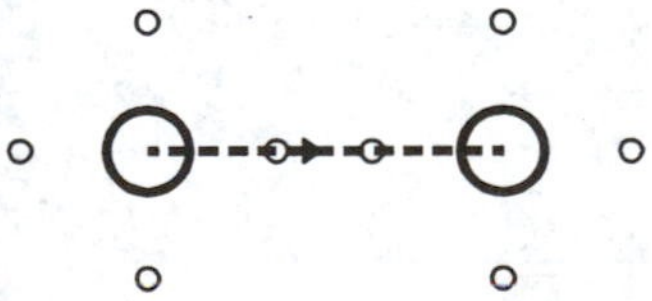</td><td>1 CHAMFER MILL</td></tr>
</table>

Index